KU-782-456

THE NEW ENCYCLOPEDIA OF

BIRDS

THE NEW ENCYCLOPEDIA OF

BIRDS

Edited by Christopher Perrins

OXFORD
UNIVERSITY PRESS

This edition published by:

OXFORD
UNIVERSITY PRESS

Great Clarendon Street, Oxford OX2 6DP

Oxford University Press is a department of the University of Oxford. It furthers the University's objective of excellence in research, scholarship, and education by publishing worldwide in

Oxford New York

Auckland Bangkok Bogotá Buenos Aires Cape Town Chennai Dar es Salaam Delhi Hong Kong Istanbul Karachi Kolkata Kuala Lumpur Madrid Melbourne Mexico City Mumbai Nairobi Paris São Paulo Shanghai Singapore Taipei Tokyo Toronto

Oxford is a registered trade mark of Oxford University Press in the UK and in certain other countries

British Library Cataloguing in Publication Data
Data available

Library of Congress Cataloging in Publication Data
Data available

ISBN 0 19 852506 0

Originated in Hong Kong by AGP Repro Ltd.

Printed in Hong Kong on acid-free paper by Dai Nippon Printing Ltd.

10 9 8 7 6 5 4 3 2 1

Spoonbilled sandpiper
see page 233

AN ANDROMEDA BOOK

Planned and produced by
Andromeda Oxford Limited,
Kimber House,
1 Kimber Road, Abingdon,
Oxfordshire, OX14 1BZ
United Kingdom.

www.andromeda.co.uk

First published 2003

Publishing Director Graham Bateman
Project Manager Peter Lewis
Editors Tony Allan, Rita Demetriou
Art Directors Chris Munday, Martin Anderson
Designers Mark Regardsoe, Frankie Wood
Cartographic Editor Tim Williams
Picture Manager Claire Turner
Picture Researcher Vickie Walters
Production Director Clive Sparling
Proofreader Barbara Fraser
Indexer Ann Barrett

Artwork Panels

Norman Arlott
Trevor Boyer
Ad Cameron
Robert Gillmor
Peter Harrison
Chloë Talbot Kelly
Sean Milne
Denys Ovenden
Laurel Tucker
Ian Willis
Michael Woods

Photos page 1: *Eurasian nuthatch* John Daniels/Ardea London. pages 2-3: *Grey-crowned cranes* Martyn Colbeck/ Oxford Scientific Films. left: *American mourning doves* John Cancalosi/Bruce Coleman Collection.

CONTENTS

Preface 10
Notes on Classification 14

What is a Bird? 18
A Diversity of Nests 32

Ostrich 34
Rheas 38
Emus 41
Cassowaries 44
Kiwis 46
Tinamous .48
Penguins .50
Surviving at the Extremes 58
Loons or Divers 60
Grebes 62
Albatrosses 66
Shearwaters and Petrels 70
Storm Petrels 74
Diving Petrels 76
Pelicans 78
Gannets and Boobies 82
Tropicbirds 88
Cormorants 90
Frigatebirds 94
Darters 96
Herons and Bitterns 98
Storks 106
Up, Down, Flying Around 110
Ibises and Spoonbills 112
Feeding by Touch 118
Hammerhead 120
Shoebill .122
Flamingos 124
Screamers 130
Swans, Geese, and Ducks 132
A Race to Breed 144
New World Vultures 146
Secretarybird 148
Osprey 150
The Fish-hunting Raptor 152
Falcons .154
Shell-shock 160
Hawks, Eagles, and Old World Vultures 162
Nature's Scavengers 174
Pheasants and Quails 176
Constellations of Eyes 182
Grouse 184
Turkeys .188
Guineafowl 190
Megapodes 192
Guans and Curassows 196
Cranes 198
Teaching Cranes how to Migrate . . .202
Leaping and Bowing – The Dance of the Japanese Crane 204
Rails 206
Limpkin .212
Trumpeters 213
Bustards 214
Buttonquails 218
Mesites .219
Kagu 220
Sun Bittern 221
Seriemas 222
Finfoots .223
Plovers and Lapwings 224
Going Out with the Tide 230
Sandpipers and Snipes 232
How Male Ruffs Compete to Mate 238
Phalaropes 240
Avocets and Stilts 242
Jacanas .244
Painted Snipes 246
Oystercatchers 248
Crab Plover 250
Stone Curlews 251
Pratincoles and Coursers 252
Seed Snipes 254
Plains Wanderer 255
Sheathbills 256
Gulls 258
Terns 264
Skuas and Jaegers 270
Skimmers 274
Auks 276
Sandgrouse 284
Pigeons 288
Finding the Way Home 294
Parrots, Lories, and Cockatoos . .296
Lotus-eaters 308
Back from the Brink 310
Cuckoos 312
A Cuckoo in the Nest 316
Hoatzin 318
Turacos 320
Owls 322
Barn and Bay Owls 332
A Face for the Night 334
Nightjars 336
Nightjars and the Lunar Cycle 340
Frogmouths 342
Owlet-nightjars 343
Potoos .344

Opposite: Little bee-eater
see page 374

Finches and Hawaiian Honeycreepers
see pages 613 and 617

Oilbird 345
Swifts 346
Treeswifts 351
Hummingbirds 352
Trogons 362
Mousebirds 364
Kingfishers 366
Motmots 372
Todies 373
Bee-eaters 374
Rollers 380
Cuckoo Roller 381
Hoopoe 382
Wood-hoopoes 383
Hornbills 384
Toucans 390
Honeyguides 396
Barbets 398
Puffbirds 403
Jacamars 406
Woodpeckers 408
Woodland Drumbeats and Dances 414
New Zealand Wrens 416
Pittas 418
Asities 421
Broadbills 422
Tyrant Flycatchers 424
Cotingas 429
Manakins 434
Ovenbirds 438
Woodcreepers 442
Antbirds 446
Gnateaters 450
Tapaculos 451
Australasian Treecreepers 452
Lyrebirds 453
Scrub-birds 456
Bowerbirds 458
Bowerbirds – Avian Artists 462
Fairy Wrens and their Allies 464
Honeyeaters & Australian Chats 466
Australian Warblers 470
Australo-Papuan Robins 472
Logrunners and their Allies 474
Australo-Papuan Babblers 476
Whistlers 477
Vireos 478
Crows 480
Squirreling Crows 485
Calculating Crows and Judicious Jays 486
Birds of Paradise 488
Wood Swallows 492
Magpie-larks and Australian Mudnesters 493
Butcherbirds and their Allies 494
New Zealand Wattlebirds 495
Old World Orioles 496
Cuckoo-shrikes 498
Fantail Flycatchers 501
Drongos 502
Monarch Flycatchers 504
Leafbirds 506
Ioras 507
Shrikes 508
Helmet-shrikes 511
Bush-shrikes 512
Vangas, Wattle-eyes, and Batises 514
Rockfowl and Rockjumpers 515
Palmchat 516
Grey Hypocolius 517
Waxwings and Silky Flycatchers 518
Dippers 520
Thrushes 522
Old World Flycatchers 528
Starlings and Mynas 530
Mockingbirds 534
Nuthatches 536
Holarctic Treecreepers 538
Philippine Rhabdornises 541
Wrens 542
Gnatcatchers 546
Penduline Tits 548
True Tits 550
The Great Tit 554
Long-tailed Tit 556
Swallows 558
Bulbuls 562
White-eyes 566
Old World Warblers 568
Firecrests and Kinglets 573
Babblers and Laughingthrushes 574
Larks 578
The Skylark's Decline 582
Flowerpeckers 584
Sunbirds, Spiderhunters, and Sugarbirds 586
Sparrows and Snowfinches 590
Weavers 592
The Industrious Weaver 596
Wagtails and Pipits 598
Accentors 602
Waxbills and Whydahs 604
Chaffinches 610
Finches 611
Hawaiian Honeycreepers 616
Buntings and New World Sparrows 618
Cardinal Grosbeaks 622
New World Warblers 624
Tanagers and Tanager Finches 628
Icterids 632

Glossary 636
Bibliography 638
Index 640
Picture Credits 656

CONTRIBUTORS

DA David Agro, Toronto, Canada
MA Mark Avery, University of Oxford, UK
AVB Alexander V. Badyaev, Department of Ecology and Evolutionary Biology, Tucson, University of Arizona, USA
AJB Allan J. Baker, Royal Ontario Museum, Toronto, Canada
MJB Michael J. Bamford, Bamford Consulting Ecologists, Kingsley, Western Australia, Australia
JCB Jon C. Barlow, Royal Ontario Museum, Toronto, Canada
KEB Kate E. Barlow, Aberdeen, Scotland
JB Jack Barr, Guelph, Ontario, Canada
GFB George F. Barrowclough, Museum of Natural History, New York, USA
JFB J.F. Bendell, University of Toronto, Ontario, Canada
BCRB Brian C.R. Bertram, Zoological Society of London, UK
MEB Michael E. Birkhead, University of Oxford, UK
TRB Timothy R. Birkhead, University of Sheffield, UK
DB Dieter Blume, Gladenbach, Germany
MdeLB Michael de L. Brooke, Zoology Department, University of Cambridge, UK
DMB Daniel M. Brooks, Houston Museum of Natural Science, Texas, USA
MDB Murray D. Bruce, National Parks and Wildlife Service, Armidale, New South Wales, Australia
DFB Donald F. Bruning, New York Zoological Society, USA
JB Joanna Burger, Rutgers University, Piscataway, New Jersey, USA
KJB Kevin J. Burns, Department of Biology, San Diego State University, California, USA
RWB Robert W. Burton, Great Gransden, Bedfordshire
JC John Carroll, Warnell School of Forest Resources, University of Georgia, Athens, USA
MC Michael Clarke, Department of Zoology, La Trobe University, Bundoora, Australia
PC Peter Clement, English Nature, Peterborough, UK
NJC Nigel J. Collar, Zoology Department, University of Cambridge, UK
PRC P.R. Colston, Natural History Museum, London, UK
LC Lindon Cornwallis, Chipping Norton, Oxfordshire, UK
JAC Jarrad A. Cousin, School of Natural Sciences, Edith Cowan University, Joondalup, Australia
HC Humphrey Crick, British Trust for Ornithology, Thetford, Norfolk, UK
FHJC Frank H.J. Crome, CSIRO, Wildlife and Rangelands Research, Atherton, Queensland, Australia
TMC T.M. Crowe, Department of Zoology, University of Cape Town, South Africa
JPC John P. Croxall, British Antarctic Survey, Cambridge, UK
SJJFD Stephen J.J.F. Davies, Curtin University of Technology, Perth, Australia
WRJD W. R. J. Dean, Knysna, South Africa
AWD A.W. Diamond, Canadian Wildlife Service, Ottawa, Ontario, Canada
VD Veronica Doerr, School of Botany and Zoology, The Australian National University, Canberra, A.C.T., Australia
EKD Euan K. Dunn, University of Oxford, UK
SE Scott Edwards, Department of Zoology, University of Washington, Seattle, Washington, USA
RME R. Michael Erwin, Patuxent Wildlife Research Center, Laurel, Maryland, USA
KE Karl Evans, Lincoln University, Canterbury, New Zealand
MRE Matthew R. Evans, Institute of Biological Sciences, University of Stirling, Scotland
SME Stuart M. Evans, University of Newcastle, UK
CJF Chris J. Feare, Haslemere, Surrey, UK
SF Simon Ferrier, National Parks and Wildlife Service, Armidale, New South Wales, Australia
JWF John W. Fitzpatrick, Field Museum of Natural History, Chicago, Illinois, USA
HAF Hugh A. Ford, Zoology, University of New England, Armidale, New South Wales, Australia
TF Tony Fox, National Environmental Research Institute, Department of Coastal Zone Ecology, Kalø, Rønde, Denmark
CBF Clifford B. Frith, Malanda, Queensland, Australia
DWF Dawn W. Frith, Malanda, Queensland, Australia
CHF C. Hilary Fry, Kentmere, Staveley, Cumbria, UK
RWF Robert W. Furness, University of Glasgow, Scotland
EFJG Ernest F. J. Garcia, Guildford, Surrey, UK
PJG Peter J. Garson, University of Newcastle, UK
AJG Anthony J. Gaston, Ottawa, Ontario, Canada
FBG Frank B. Gill, Academy of Natural Sciences, Philadelphia, USA
LGG Llewellyn G. Grimes, Warwick, UK
JNJ James N. Jolly, Wildlife Service, Department of Internal Affairs, Wellington, New Zealand
MH Michelle Hall, School of Botany and Zoology, Australian National University, Canberra, Australia
KH Keith Hamer, Ecology and Evolution Group, School of Biology, University of Leeds, UK
JH James Hancock, Winchester, Hampshire, UK
JWH John W. Hardy, Florida State Museum, Gainesville, Florida, USA
MPH Michael P. Harris, Institute of Terrestrial Ecology, Banchory, Kincardineshire, Scotland
BJH Ben J. Hatchwell, Department of Animal and Plant Sciences, University of Sheffield, UK
RH Robert Heinsohn, School of Botany and Zoology, The Australian National University, Canberra, A.C.T., Australia
CH Chris Hewson, British Trust for Ornithology, Thetford, Norfolk, UK
GH Graham Hirons, University of Southampton, UK
DH David Holyoak, Tuckingmill, Camborne, Cornwall, UK
JH Jenny Horne, Nanyuki, Kenya
JAH John A. Horsfall, University of Oxford, UK
SH Stephen Howell, Point Reyes Bird Observatory, Stinson Beach, California, USA
RH Robert Hudson, British Trust for Ornithology, Thetford, Norfolk, UK

JH Jane Hughes, Australian School of Environmental Sciences, Griffith University, Nathan, Queensland, Australia

AMH A.M. Hutson, Fauna and Flora Preservation Society, London, UK

AJ Amy Jansen, Johnstone Centre, School of Science and Technology, Charles Sturt University, Wagga Wagga, New South Wales, Australia

ARJ Alan R. Johnson, Station Biologique de la Tour du Valat, Le Sambuc, Arles, France

DJ Darryl Jones, Australian School of Environmental Studies, Griffith University, Nathan, Queensland, Australia

TJ Tony Juniper, Cambridge, UK

JK Janet Kear, Umberleigh, Devon, UK

AK Alan Kemp, Navors, Pretoria, South Africa

RSK Robert S. Kennedy, Maria Mitchell Association, Nantucket, USA

JK Jiro Kikkawa, Department of Zoology and Entomology, University of Queensland, Australia

RWK Richard W. Knapton, Brock University, St. Catherine's, Ontario, Canada

NKK Niels Kaare Krabbe, Zoology Museum, University of Copenhagen, Denmark

JAK James A. Kushlan, US Geological Survey, Smithsonian Environmental Research Center, Edgewater, Maryland, USA

PL Peter Lack, British Trust for Ornithology, Thetford, Norfolk, UK

DRL Derek R. Langslow, Nature Conservancy Council, Huntingdon, UK

NL Norbert Lefranc, Ministry of Environment, France

AL Alan Lill, Biological Sciences and Psychology Department, Monash University, Victoria, Australia

HL Hans Löhrl, Egenhausen, Germany

IJL Irby J. Lovette, Cornell Laboratory of Ornithology, Ithaca, New York, USA

ALu Arne Lundberg, Uppsala University, Sweden

PMcD Paul McDonald, School of Botany and Zoology, School of Life Sciences, Australian National University, Canberra, Australia

KJM Kevin J. McGowan, Cornell Laboratory of Ornithology, Ithaca, New York, USA

PM Philip McGowan, World Pheasant Association/Game Conservancy Trust, Fordingbridge, Hampshire

JM Judith McIntyre, Division of Science and Mathematics, Utica College of Syracuse University, Utica, USA

GLM Gordon L. Maclean, University of Natal, Pietermaritzburg, South Africa

RM Robert Magrath, School of Botany and Zoology, The Australian National University, Canberra, Australia

CM Curtis Marantz, Biology Department, University of Massachusetts, Amherst, Massachusetts, USA

TM Thaís Martins, Institute of Biological Sciences, University of Stirling, Scotland

EM Erik Matthysen, Department of Biology, University of Antwerp, Belgium

DHM Douglas H. Morse, Brown University, Providence, Rhode Island, USA

CJM Christopher J. Mead, British Trust for Ornithology, Hilborough, Norfolk, UK

JBN J. Bryan Nelson, Auchencairn, Castle Douglas, Scotland

IN Ian Newton, Institute for Terrestrial Ecology, Abbot's Ripton, Cambridgeshire, UK

GHO Gordon H. Orians, University of Washington, Seattle, USA

TWP T.W. Parmenter, Natural History Museum, London, UK

SJP Stephen J. Parry, Woolton, Liverpool, UK

CMP Christopher M. Perrins, University of Oxford, UK

HDP H. Douglas Pratt, Museum of Natural Science, Louisiana State University, Baton Rouge, USA

PAP Peter A. Prince, British Antarctic Survey, Cambridge, UK

ROP Richard O. Prum, University of Kansas, Lawrence, USA

NHR Nathan H. Rice, Academy of Natural Sciences, Philadelphia, USA

ASR Andrew S. Richford, London, UK

RSR Robert S. Ridgely, International Bird Conservation, National Audubon Society, New York, USA

MWR M.W. Ridley, University of Oxford, UK

JDR James D. Rising, Department of Zoology, Ramsay Wright Zoological Laboratories, University of Toronto, Canada

RR Robert Robinson, British Trust for Ornithology, Thetford, Norfolk, UK

JRR James J. Roper, Biological Sciences, Federal University of Paraná, Curitiba, Brazil

IR Ian Rowley, Guildford, Western Australia, Australia

ER Eleanor Russell, Guildford, Western Australia, Australia

PS Paul Salaman, Department of Zoology, Natural History Museum, London, UK

EAS Elizabeth Anne Schreiber, Los Angeles County Museum of Natural History, California, USA

RWS Ralph W. Schreiber, Los Angeles County Museum of Natural History, California, USA

KLS Karl-Ludwig Schuchmann, Alexander Koenig Zoological Research Institute and Museum of Zoology, Bonn, Germany

AFS Alexander F. Skutch, San Isidro, Costa Rica

JAS James A. Serpell, University of Cambridge, UK

GHS Greg H. Sherley, Department of Conservation, Wellington, New Zealand

LH Lester Short, Nanyuki, Kenya

DS-C Douglas Siegal-Causey, Museum of Natural History, University of Kansas, Lawrence, USA

GTS G. T. Smith, CSIRO, Wildlife and Rangelands Research, Australia

BKS Barbara K. Snow, Natural History Museum, Sub-department of Ornithology, Tring, Hertfordshire, UK

DWS David W. Snow, Natural History Museum, London, UK

IS Ilse Storch, IUCN/SSC/BirdLife/WPA Grouse Specialist Group, TU Munich, Linderhof Research Station, Ettal, Germany

HT Helen Temple, Department of Zoology, University of Cambridge, UK

GFvT G.F. van Tets, CSIRO, Lyneham, A.C.T., Australia

GT Gareth Thomas, Royal Society for the Protection of Birds, Sandy, Bedfordshire, UK

HST Hazell S. Thompson, BirdLife International, Cambridge, UK

AKT Angela K. Turner, University of Glasgow, Scotland

CAW C. A. Walker, Natural History Museum, London, UK

CW Cliff Waller, Blythburgh, Suffolk, UK

DRW D.R. Wells, University of Malaya, Kuala Lumpur, Malaysia

SLW Sheri L. Williamson, Southeastern Arizona Bird Observatory, Bisbee, Arizona, USA

HW Hans Winkler, Konrad Lorenz Institute for Comparative Behavior Research of the Austrian Academy of Sciences, Vienna, Austria

MW Mark Witmer, Department of Biology, Bryn Mawr College, Bryn Mawr, Pennsylvania, USA

BW Brian Wood, University College, London, UK

RDW Ron D. Wooller, Murdoch University, Western Australia, Australia

JV Juliet Vickery, British Trust for Ornithology, Thetford, Norfolk, UK

PREFACE

birds arouse more interest among the general public than any other animal group, even including mammals. At first sight it is surprising that humankind should respond in this way – we are, after all, mammals. However, most mammals rely on a different mix of sense organs than we do – the senses of smell and hearing are often as important, or more so, than vision; further, few mammals have color vision. Birds, on the other hand, share with man a dependence on color vision and sound as their two main senses and, in keeping with this, they are often brightly colored. Finally, the majority of birds are active during daylight hours.

We can, therefore, perhaps "perceive" more of a bird's world than that of many mammals. Consider, for example, our best-loved pet, the dog. On a walk, it spends a great deal of its time smelling objects: plainly, it is acquiring a large amount of information about its surroundings. Yet, in spite of our long association with dogs, we really do not begin to understand the olfactory messages that it is receiving. Hence it seems easier for us to understand the various activities of birds than of most mammals. However, this may be an illusion. For instance, it is becoming clear that birds "see" ultraviolet as a "color" and hence their perception of the colored world is likely to be substantially different from ours. Likewise, there is some evidence that birds may not hear sounds in the same way that we do.

In this work, we have tried to provide the reader with an encapsulated but up-to-date account of the world's birds. It has been inevitable that we should deal with them by family rather than by individual species, because of the numbers involved, more than twice as many as there are mammals. There is no absolute, agreed number of bird species; some 9,850 are included in this volume. However, a large number of them are virtually unknown; the nests and eggs of many species have not yet been described and, for a number of species, we have inadequate records of their range. The vast majority of these, of course, are species that live in the tropical forests where their ability to fly, their small size, and the density of vegetation mean that they remain virtually unobserved. From the conservation point of view, more seriously, we do not know enough about their status. BirdLife International has listed 1,186 species – or some 12 percent of the total– as threatened (3 Extinct in the Wild, 182 Critically Endangered, 321 Endangered and 680 Vulnerable). A further 727 are Near-threatened. Sadly, some of those classified as Critically Endangered are probably already extinct.

Nevertheless, at the family level, most groups of birds are adequately known for it to be possible to give a comprehensive review of the colorful and varied array of this fascinating assemblage of animals. In particular, there has been during recent years a great increase in the number of detailed studies of the behavior and ecology of many species, including at least one or two from most of the major families, including the tropical ones and these give us at least an indication of how the other related species are likely to live. These studies have not merely filled out what was known before; they have demonstrated the ways in which birds cope with their environment which, we realize as we learn more, is a very complex one. The birds are far more intricately adapted to their ways of life than the casual observer might at first assume; natural selection is a powerful tool.

In particular, recent studies have shown that, in many groups, family life is much more complex than was once thought and there is still much to be learned. Originally, most detailed studies had been conducted in Europe and in North America, where the majority of birds live in simple, monogamous pairs. In contrast, many of the species in the tropics live in "extended families" where six or more birds may share a territory and look after a single nest. The advent of DNA technology has also caused us to revise our thinking. For example, a very large number of birds live as pairs on territories and have been considered to be monogamous. However, DNA "fingerprinting" of their broods reveals that in many species a significant proportion of their young are not fathered by the male territory-holder.

Right *A Bald eagle in Alaska. After serious decline, conservation measures have ensured that this species is thriving again in many parts of North America.*

Blue swallow (listed as Vulnerable) *see page 559*

IUCN CATEGORIES

Ex Extinct, when there is no reasonable doubt that the last individual of a taxon has died.

EW Extinct in the Wild, when a taxon is known only to survive in captivity, or as a naturalized population well outside the past range.

Cr Critically Endangered, when a taxon is facing an extremely high risk of extinction in the wild in the immediate future.

En Endangered, when a taxon faces a very high risk of extinction in the wild in the near future.

Vu Vulnerable, when a taxon faces a high risk of extinction in the wild in the medium-term future.

LR Lower Risk, when a taxon has been evaluated and does not satisfy the criteria for CR, EN or VU.

Note: The Lower Risk (LR) category is further divided into three subcategories: Conservation Dependent (cd) – taxa which are the focus of a continuing taxon-specific or habitat-specific conservation program targeted toward the taxon, the cessation of which would result in the taxon qualifying for one of the threatened categories within a period of five years; Near Threatened (nt) – taxa which do not qualify for Conservation Dependent but which are close to qualifying for VU; and Least Concern (lc) – taxa which do not qualify for the two previous categories.

The book's organization follows a very simple principle. In the overwhelming majority of cases, each article deals with a single family. The sequence of families that is used here is given on page 19, while notes outlining some of the major differences of opinion among ornithological systematists about the way in which birds should be classified will be found on pages 14–15, as well as in the articles themselves. Readers seeking further information on the problematic question of bird classification should also refer to relevant texts listed in the Bibliography.

There is a widespread view held by the "man in the street" that the experts should know how many species of birds (or other animals) there are. This is a long way from being the case. It is not even easy to define a species! Perhaps the commonest way to determine whether an animal is a different species from another one is to base the decision on whether or not the populations to which they belong naturally interbreed. However, in many cases animals that can potentially interbreed in captivity may not meet in the wild – they may, for example, live on different, far-distant continents. Therefore, one has to decide (or test) whether or not they would interbreed if their ranges naturally overlapped. There can be no objective view and, not surprisingly, opinions differ widely. Studies of their biochemistry and anatomy are also important in deciding species relationships. The increasing knowledge of tropical birds is also leading to the discovery of new species. Although a few completely new birds are still being discovered, the large majority of new species are being recognized because detailed studies reveal that wide-ranging forms, formerly thought to be a single species, are in fact best regarded as two (or sometimes more) different species, each occupying only a part of the former extensive range.

As a result, no two books will necessarily list the same families, or identify the same number of species within each family. Since this book is not primarily a work on taxonomy, we have attempted to avoid undue eccentricity. However, the author of each article is often a leading expert on the group concerned and so we have encouraged them to use the results of their own studies.

Each article gives details, where relevant, of physical features, distribution, evolutionary history, classification, breeding, diet and feeding behavior, social dynamics and spatial organization, conservation and relationships with humans.

The textual discussion of each family is supplemented by a Factfile, which provides handy reference to the key features of distribution, habitat, size, diet and so forth. It also contains a scale drawing comparing the size of a representative species with that of a 6ft (1.8m) human or a 12in (30cm) human foot or head. Where there are silhouettes of two birds, they are the largest and smallest representatives of the family. Unless otherwise stated, dimensions given are for both males and females. Where there is a difference in size between sexes, the scale drawings show the larger sex.

In each Factfile, a map summarizes the family's natural global distribution (not introductions to other areas by people). These maps are intended merely as a snapshot of where birds of the family in question might be encountered. Given the necessarily small scale of the maps, they neither distinguish between breeding and wintering grounds for migratory birds, nor show detailed migration routes. However, such information is given in the accompanying Factfile distribution data, and/or is expanded on in the main text.

The reader will find the scientific names of all the species mentioned in text either in the Factfile, or (especially in the case of larger families) in a separate table enumerating selected species, or detailing the family's further division into subfamilies or other groupings. Captions to artwork plates invariably cite the scientific name in parentheses after the common name. Similarly, if a bird not specifically treated in text is shown in a photograph, it is identified by scientific name in the caption.

The fruits of individual contributors' frontline research are encapsulated in a number of "special features." Interspersed throughout the *Encyclopedia*, they span a wide variety of subject areas, including social behavior, breeding biology, and conservation. Similar themes are also developed in smaller "box features" within the text of many entries. Finally, photo stories show certain remarkable species in sequences of stunning wildlife photography.

Most readers have never seen many of the birds mentioned in these pages, but that does not necessarily mean they have to remain remote or unreal. To help overcome our unfamiliarity, superb photographs and color and line artworks bring the animals vividly to life. These illustrations testify to the skill of their creators: they are accurate in minute detail and, what is more, they are dynamic – birds are often shown engaged in an activity or in a posture that enhances or expands upon points made in the text. Moreover, the species have been chosen as representatives of their group.

We live in a changing world where habitats are being destroyed at an alarming rate. Although there is now a great deal of interest in conservation in many countries, in others, especially the developing ones, the need for new resources – particularly for rapidly increasing human populations – make it inevitable that further areas of bird habitat will be lost. Most threatened of the major habitats are the tropical rain forests . Sadly, rain forest birds are among the most specialized – they are poorly equipped to cope with living in impoverished habitats. Some special forests in restricted areas, such as the thin strip of coastal rain forest in Brazil, have already almost completely disappeared, taking with them their endemic birds. Many other rain forest species still exist only because of the vastness of the forests. Unless we can call a halt, man is certain to destroy even the most extensive tracts in time.

To date, we have lost more species from islands than from rain forests. Island species are threatened for a different reason – they only ever exist in comparatively small numbers, since their habitats are restricted in size. Hence such birds are vulnerable to minor, local changes. Doubtless such species will be among those that continue to be lost in greatest numbers in the near future, but over a slightly longer period the wide variety of bird species in the big rain forest blocks must also be considered threatened.

Throughout the Encyclopedia, in treating the conservation status of bird species, we have employed the standard conservation categories defined by the IUCN (World Conservation Union), down to and including Lower Risk (in other words, omitting the categories Data Deficient and Not Evaluated). The precise meaning of each category is outlined in the table opposite.

Today, no single person could hope to put together a complete summary of bird families without the assistance of the large international team of authors we have used here, whose detailed studies and background knowledge far surpass those of any individual. In recognition of the invaluable work done by the authors of the 1985 Encyclopedia, their names and original affiliations have been cited alongside the new contributors who revised and updated their text. In addition, the editorial team at Andromeda Oxford Ltd. has distilled the contributions to consistency, spiced them with the work of the artists and photographers, and skillfully brought all the parts together. If the final product provides a stimulus for readers to discover more about the fascinating world we live in, then all our efforts will have proved worthwhile.

CHRISTOPHER PERRINS

OXFORD, ENGLAND

NOTES ON CLASSIFICATION

A GREAT MANY DIFFERENT CLASSIFICATIONS OF BIRDS HAVE been proposed. As reference to any series of bird books will make clear, there is no simple agreement. However, almost all are based on the so-called Wetmore order that is used in *Checklist of Birds of the World* by J.L. Peters and others (Museum of Comparative Zoology, Cambridge, Massachusetts). This was written over a long period, 1931–1987, and through that time views changed. Even then, no one thought that this was likely to be the final story. In part, this is because birds fossilize so poorly; they are small and have hollow bones. As a result, we have a very incomplete knowledge of their evolution and the relationships between many groups were simply not clear.

Since then, many types of study have shed light on the relationships. Most striking have been those studies involving DNA analyses – primarily, the works of Sibley and Ahlquist (1990) and of Sibley and Monroe (1990 and 1993). Some major changes have been proposed, and these have caused us to rethink much of the bird family tree. In many cases, these new proposals do not affect the classification at the family level, the members of many families are, after all, easily identified to family – penguins, herons, woodpeckers, swallows, wrens, to name but a few. In such cases, all that the new analysis may have done is to make clear the relationships of a few species whose affinities were previously uncertain. However, one of the most important contributions of these studies has been to give us a better understanding of the relationships of the families to one and other.

To some extent, the "jury is still out" on many aspects of these proposals and so it has not been easy to decide how to treat them in this work. As a consequence, we have tended here to be conservative and leave the order of families as it was, while commenting on major changes in the texts. However, there is one major exception to this. The family relationships within the largest order of birds, the Passeriformes (see below), has always been confusing and the new proposals promise to make some sense of these, with new views on many families. These new views also require some re-ordering of the families and we have followed these here.

Equally, it should be stressed that there is no agreement on the number of species in many of the families. Different interpretations and lack of knowledge on many species, especially those of tropical forests, make it impossible to give totally reliable figures. In general, those used here are based on Sibley and Monroe (1990) although we have allowed the specialist author to alter this if he or she felt that this was desirable.

We have also chosen not to follow one other feature of the new classification. In a few cases, large well-marked families have been merged, usually as subfamilies or tribes, within much larger families. For example, the families Passeridae, Motacillidae, Prunellidae, Ploceidae and Estrildidae are sometimes put into a single family, the Passeridae, and likewise the Fringillidae, Carduelidae, Drepanididae, Emberizidae, Parulidae, Thraupidae and Icteridae into a single family, the Fringillidae. This has the effect of putting some very well-marked groups together into very large families indeed (almost 400 and almost 1,000 species respectively). One reason for making some of these mergers are that there are species which do not fall clearly into one or other of the possible families. Nevertheless, since in most cases, these groups are clearly separable and well-known to ornithologists, we have retained them as separate families.

The Order Passeriformes deserves special mention, since it contains some 5,900 out of a total of 9,845 (or, according to Sibley and Monroe, 5,712 of 9,672) birds of the world – almost 60 percent of all living species! The next largest of the other 27 or so orders of birds only contain around 400 species.

The Passeriformes are often referred to as the "passerines" (in contrast to all the other birds, which are termed "non-passerines"). Passerines lack a simple English name, but are sometimes called perching-birds (based on the anatomy of the foot), a term that is hardly helpful when one looks at the many non-passerines that perch! Almost all birds in this order are fairly small, the largest species being the ravens.

This large group of birds is also one of the more recent. It was thought that modern passerines stem from an explosive period of adaptive radiation in the mid-Tertiary; however, recent studies suggest that the forerunners of these birds may have been present in the Late Cretaceous – about 71 million years ago. One current hypothesis is that the passerines originated in Gondwanaland – a southern hemisphere land mass that included the forerunners of South America, Africa, Australia and Antarctica.

The first major division of the passerine stock then took place; into the sub-oscines and the oscines (some 1,151 and 4,561 living species respectively). Again, there are no good English names for these important groups; the oscines, have sometimes been called the songbirds, a name as inappropriate for the suborder as perching-birds is for the order. The Sibley and Monroe classification divides these birds into two suborders: Tyranni and Passeri, equivalent to the sub-oscines and the oscines. The Tyranni contains about 1,151 species in 10 families and is almost exclusively of New World origin, most species being confined to South and Central America.

These sub-oscines separated from the oscines in Gondwana and spread into the lands that were to become South America and Africa. In the case of South America, they remained as the dominant part of the

Cockatoos
see pages 301

Overleaf *King penguins in a colony at Volunteer Point on the Falkland Islands.*

passerine avifauna, but in the Old World only four small families remain, the Asities (Philepittidae), the New Zealand wrens (Acanthisittidae), the Pittas (Pittidae), and the Broadbills (Eurylamidae). Indeed, even these may not be members of the sub-oscine assemblage, since their relationships to other families are uncertain. In view of their distribution (the first two are restricted to Madagascar and New Zealand respectively, the second two are more widely distributed in the Old World tropics), they may well be the remnants of groups that split off from all the other passerines at a very early stage.

In contrast, the oscines or Passeri, although they originated in the Old World have spread to the New World, especially to North America, to a much greater degree than the Tyranni have to the Old. They may have done this either from eastern Asia via the Bering Straits or from Europe via Greenland, the latter sea crossings were much shorter in times past.

One of the most important contributions of the Sibley and Monroe classification may be the recognition that the Passeri is itself divisible into two groups – the Corvida and Passerida. Sibley and Ahlquist thought that these two groups had very different origins, the Corvida evolving in Australasia, the Passerida in either Africa or Europe. Although both groups have since spread into the other's "territory," all this vast array of passerine birds (except for the very small number of Tyranni that seem to have reached the Old World, as described above) can be ascribed to one or other of these groups. Not everyone agrees with this suggestion, but the division of the passerines into three major groups – Tyranni, Corvida, and Passerida – does seem to be helpful.

What is a Bird?

IN 1861, WORKMEN SPLITTING SLATE IN A quarry in Bavaria came across a fossil feather that they passed on to the local museum. After much searching, a fossilized creature was discovered. It was obviously reptilelike, with a long tail; but clearly outlined around it in the fine texture of the slate were the unmistakable imprints of feathers! As a result, this small creature instantly became one of the most famous and important fossils of all time – Archaeopteryx lithographica *(literally, "ancient wing in rock"). Seven further specimens have been found subsequently.*

Above *White terns* (Gygis alba) *on the wing. These elegant birds are superbly well adapted for a life spent traversing broad expanses of sea. From their nesting sites on remote islands, they range widely throughout the tropical and subtropical oceans.*

The slate in which *Archaeopteryx* was found dates from the Jurassic period some 150 million years ago. Despite recent finds of birdlike reptiles, *Archaeopteryx* remains the most important "missing link" between reptiles and birds. *Archaeopteryx* is thought to be a direct descendant of dinosaurs called theropods. Much earlier offshoots of this great lineage of reptiles include the mighty predator *Tyrannosaurus rex*, at first sight not a very likely relative of the birds! However, the line of descent is sufficiently clear to permit the claim that the dinosaurs did not become extinct, since the birds are their living representatives.

Archaeopteryx was about the size of a large pigeon. Apart from its long tail, like a lizard's but feathered, its most obvious differences from modern birds are that it had toothed jaws, front limbs that were modified as wings but still retained claws, and a relatively small breastbone. The feathers – that most obvious characteristic of birds – seem to have been remarkably similar to those of modern birds. The positioning of the tail feathers was odd, because the animal still possessed an "unbirdlike" tail, but the number and arrangement of wing feathers were more or less identical to those in modern birds.

Although warm-blooded animals need more food to survive, they are at a great advantage over the cold-blooded reptiles in that they can be more active in cold conditions – at night, dawn, or dusk, and in temperate climates. It is probable that several groups of dinosaurs developed the ability at least partly to maintain their body temperatures in cold conditions, and it now seems that some of these developed feathers to provide insulation. Likewise, *Archaeopteryx*, and the early mammals of the same period, would have been warm-blooded and would have needed the good insulation that feathers provided. Feathers, therefore, are best regarded as highly complex derivations of reptilian scales, evolved to provide insulation.

In recent years there have been a number of exciting finds of early birdlike reptiles, especially from China. Although they were obviously flightless, some of these (*Sinosauropteryx*, *Protarchaeopteryx robusta*, *Caudipteryx zoui*, and *Confuciusornis*) have feathers or featherlike structures, supporting the view that feathers came before flight and were therefore presumably for insulation. However, all of the feathered reptiles that have been found so far lived later than *Archaeopteryx*, which remains the earliest-known feathered animal. The finds reinforce the idea, however, that feathers can no longer be considered the defining characteristic of birds.

As the ancestors of *Archaeopteryx* clambered about in the trees, using the claws on both pairs of limbs, and started to leap from branch to branch, the gradual extension of the scales at the rear of the forelimb and on the tail probably provided an expanded surface area helping the creature to extend its leaps into glides. Natural selection would have favored the evolution of progressively longer and lighter scales for this purpose. The wing feathers were long enough for *Archaeopteryx* to have been able to glide a reasonable distance, and their shape is such to suggest that the birds were capable of powered flight. The breastbone is badly crushed, however, and some think that it may not have been big enough to provide a base for the large muscles required for prolonged powered flight. In addition, the long tail would have made *Archaeopteryx* somewhat poor at maneuvering in flight, compared with the more compact modern birds. But it undoubtedly flew, if not strongly for long distances.

In this one species lies most of what we know

about the early evolution of birds. Most bird carcasses were probably taken as carrion, or disintegrated before they could become fossils, and because of their relatively small size and thin bones, birds fossilize poorly. It is no coincidence that *Archaeopteryx* and most bird fossils from the later Cretaceous are those of water birds or others that fell into fine mud, for this substance gives the best chance of them being adequately preserved.

Archaeopteryx had a mosaic of bird- and reptile-like features. More fossil birds appear after a gap of some 30 million years in the fossil record, early in the Cretaceous period. They are all obviously similar to modern birds, and include the diverlike *Hesperornis*, whose wing structure shows that its ancestors had been flying birds, although it had already reverted to flightlessness! Hence, within a 30-million-year period, there must have been a considerable evolution of the birds, but we know next to nothing about it.

By the end of the Cretaceous period 65 million years ago, some birds were beginning to show the characteristics of modern families. However, it was some 65–38 million years ago, during the first half of the Tertiary period, that the great radiation of birds took place. From the Eocene (54–38 million years ago) we know fossils of at least 30 modern families. By the end of the Eocene, the birds had truly "arrived."

CLASS: AVES
2 superorders; 28 orders; 172 families; 2,121 genera; 9,845 species

SUPERORDER PALAEOGNATHAE
(Ratites and tinamous)

OSTRICH Order Struthioniformes p34
1 species *Struthio camelus*

RHEAS Order Rheiformes p38
2 species in 2 genera and 1 family

EMU AND CASSOWARIES
Order Casuariiformes p39
4 species in 2 genera and 2 families

KIWIS Order Apterygiformes p46
3 species in 1 genus and 1 family

TINAMOUS Order Tinamiformes p48
46 species in 9 genera and 1 family

SUPERORDER NEOGNATHAE
(All other modern birds)

PENGUINS Order Sphenisciformes p50
17 species in 6 genera and 1 family

LOONS Order Gaviiformes p60
5 species in 1 genus and 1 family

GREBES Order Podicipediformes p62
22 species in 7 genera and 1 family

ALBATROSSES, SHEARWATERS & PETRELS (TUBENOSES)
Order Procellariiformes p66
125 species in 27 genera and 4 families

PELICANS AND RELATIVES
Order Pelecaniformes p78
65 species in 8 genera and 6 families

HERONS, STORKS, FLAMINGOS AND RELATIVES
Order Ciconiiformes p98
117 species in 43 genera and 6 families

WATERFOWL
Order Anseriformes p130
165 species in 52 genera and 2 families

VULTURES, HAWKS, EAGLES, AND FALCONS
Order Falconiformes p146
300 species in 81 genera and 5 families

GAME BIRDS
Order Galliformes p176
285 species in 77 genera and 6 families

CRANES, RAILS AND RELATIVES
Order Gruiformes p198
203 species in 59 genera and 11 families

SHOREBIRDS, GULLS AND RELATIVES
Order Charadriiformes p225
342 species in 87 genera and 18 families

SANDGROUSE
Order Pteroclidiformes p284
16 species in 2 genera and 1 family

PIGEONS Order Columbiformes p286
309 species in 42 genera and 1 family

PARROTS Order Psittaciformes p296
356 species in 80 genera and 1 family

CUCKOOS, HOATZIN, AND TURACOS
Order Cuculiformes p312
164 species in 35 genera and 3 families

OWLS Order Strigiformes p322
205 species in 27 genera and 2 families

NIGHTJARS AND RELATIVES
Order Caprimulgiformes p336
118 species in 20 genera and 5 families

SWIFTS AND HUMMINGBIRDS
Order Apodiformes p346
424 species in 128 genera and 3 families

TROGONS Order Trogoniformes p362
37 species in 7 genera and 1 family

MOUSEBIRDS Order Coliiformes p364
6 species in 2 genera and 1 family

KINGFISHERS, ROLLERS, AND RELATIVES
Order Coraciiformes p366
206 species in 42 genera and 9 families

TOUCANS, WOODPECKERS AND RELATIVES
Order Piciformes p390
403 species in 66 genera and 6 families

PASSERINES Order Passeriformes p416
5,899 species in 1,207 genera and 82 families

Above *With some 5,900 representatives, the passerines or "perching birds" account for well over half of all the world's bird species. The order is cosmopolitan, being absent only from polar regions. Shown here is a Scarlet tanager* (Piranga olivacea)*, a species resident in the Americas.*

Not Too Big and Not Too Small

SIZE CONSTRAINTS

Compared with some other classes of animals, birds are a very uniform group in both structure and size. The mammals, for instance, include horses, lemurs, whales, bats, and tigers, to name but a few diverse examples. Moreover, mammals range from tiny bats and shrews to the great whales, a weight ratio of some 1 to 100 million, whereas flying birds range only from some 2.5g to 15kg (0.09–530oz), a ratio of 1 to a mere 6,000.

The reason for this more limited range of size and form in birds may be found in the requirements for flight. (Flightless birds, evolved from the same ancestors, are freed to some extent from these weight restrictions, only to confront other dangers and, often, extinction.) Flying is, in terms of energy required, an exceedingly expensive method of locomotion, and so it is of paramount importance to do it as economically as possible. Virtually every distinctive characteristic of a bird's anatomy has evolved as an adaptation for flying.

The size of birds has been constrained for different reasons at the small and large ends of the spectrum. Birds need to be able to maintain a constant warm body temperature – between 41° and 43.5°C (105.8–110.3°F) depending on species – in order to function efficiently. However, with decreasing size, the volume (or weight) of a body is reduced in far greater proportion than its surface area. This point is important, because a body loses heat at a rate that can be related to the ratio of surface area to volume. As the surface-to-volume ratio increases (in other words, as objects get smaller), the rate of heat loss increases, so a small bird loses heat relatively faster than a large one. Since the heat that is lost has to be made good by obtaining more food, small birds must eat more, relative to their body size, than large ones. Below a certain size, the time and effort required for energy replacement are so uneconomic that survival is not possible.

It is no coincidence that the smallest birds – for example, the Vervain hummingbird (*Mellisuga minima*) of Jamaica, which weighs only 2.4g (0.085oz) – live in warm parts of the world. Even in the tropics, many hummingbirds go torpid at night to save energy. They have to warm themselves up again at dawn before they can start their day's activities, which include taking up to half their body weight in food.

The upper limit to the size of flying birds also results from problems associated with size and scale. A bird that has linear dimensions that are twice those of another will have a surface area that is four times and a volume (and weight) eight times greater. Hence large birds have a higher weight-to-wing-area ratio than smaller ones: wing loading increases with size. Compared with a small bird, a large one must have relatively bigger wings and/or flight muscles, which in their turn will add further weight.

There is also anatomical evidence that large birds may be more constrained by weight than smaller ones. In the smaller

1

Left and below Reconstructions of prehistoric birds: **1** Archaeopteryx *is the oldest known bird, from the late Jurassic (c.147 million years ago);* **2** Caudipteryx, *dating from some 125 million years ago, was a short-armed, flightless theropod dinosaur with primitive feathers;* **3** Hesperornis, *toothed marine birds of the Upper Cretaceous, some 70–90 m.y.a., were flightless birds that evolved from a flying ancestor;* **4** *and* **5** *From the same period,* Ichthyornis *(left) and* Apatornis *(right) broadly resembled the modern tern;* **6** Odontopteryx *was a long-winged seabird from the Eocene (55–34 m.y.a.), with toothlike serrations on its bill;* **7** *The remains of* Teratornis, *a huge, vulturelike bird of the Pleistocene (1.8 m.y.a.–10,000 y.a.) have been found at La Brea tarpits in Los Angeles – it is thought to have had a wingspan of 7m (23ft);* **8a** *and* **8b** *The in-flight profile of the enormous proto-condor* Argentavis magnificens, *compared with that of the modern Bald eagle;* **9** Diatryma steini *of the Paleocene and Eocene; standing 2.2m (7.2ft) tall, it had an unkeeled sternum and was flightless.*

More recently extinct birds are the giant ratites, the moa and elephant bird. **10** Dinornis giganteus *was the largest of six moa species that inhabited New Zealand; moas finally went extinct in the 19th century;* **11** *The Elephant bird* (Aepyornis maximus) *lived on Madagascar.*

4
5

3
6

2

species, only the largest bones may be hollow (pneumatized), but in larger species more bones may be hollow. For example, the Marabou stork not only has hollow leg bones; most of the toe bones are also hollow.

The actual act of taking off is the most energy-demanding moment of flight; a bird must accelerate rapidly to pass stalling speed. Take-off is not a problem for small birds, which just leap into the air and fly away. A large vulture, however, especially one with a full crop, may have to run along the ground to gain sufficient speed to become airborne; a swan has to do the same on water, and an albatross may have great difficulty in taking off at all, unless it can face into a strong headwind.

The upper weight limit in modern flying birds seems to be of the order of 15kg (33lb). It is perhaps no coincidence that the largest birds of a number of different groups approach this weight. For example, the Great bustard may weigh 15kg (33lb), exceptionally even a little more; the largest swans are about 15kg (33lb), the largest condors about 14kg (31lb), the largest pelicans about 15kg (33lb), and the Wandering albatross about 12kg (26.5lb). However, even for these species such weights are exceptional; most adult individuals are smaller.

This line of argument has one serious flaw: some fossil flying birds were far larger! Until quite recently the largest fossils known were mostly birds of the genus *Teratornis*, which are usually thought of as giant condors, although there is considerable doubt about how they really lived. One of these, *Teratornis incredibilis*, had a wingspan of the order of 5m (16ft) and probably weighed well in excess of 20kg (44lb). A less well-known marine species, *Osteodontornis orri*, had a similar wingspan. Both pale into insignificance against the remains of another species discovered in Argentina, *Argentavis magnificens*, which probably belonged to the same family as *Teratornis*, and may have had a wingspan of some 7–7.6m (23–25ft)!

It has been suggested that these giants specialized in riding the upcurrents off hot, open country, as the vultures in East Africa do today, but this is speculation. For the biologist attempting to

IN SEARCH OF THE LEGENDARY ROC

The tales of Sindbad's voyages in the Arabic stories *A Thousand and One Nights* describe several encounters with a huge bird known as the roc.

Although the roc of the legend can fly, it is thought to be based on a giant, flightless bird that inhabited Madagascar until the mid-17th century – *Aepyornis maximus*, christened the "Elephant bird" by the 13th-century Italian explorer Marco Polo. This huge, robust ratite stood up to 3m (10ft) tall and weighed around 450kg (1,000lb), making it the largest bird that ever lived. Its eggs, fragments of which are regularly found on beaches in the south of the island, were correspondingly massive; some 40cm (16in) long and with a capacity approaching 7.5 liters (2 gallons), they are 15 times larger than ostrich eggs.

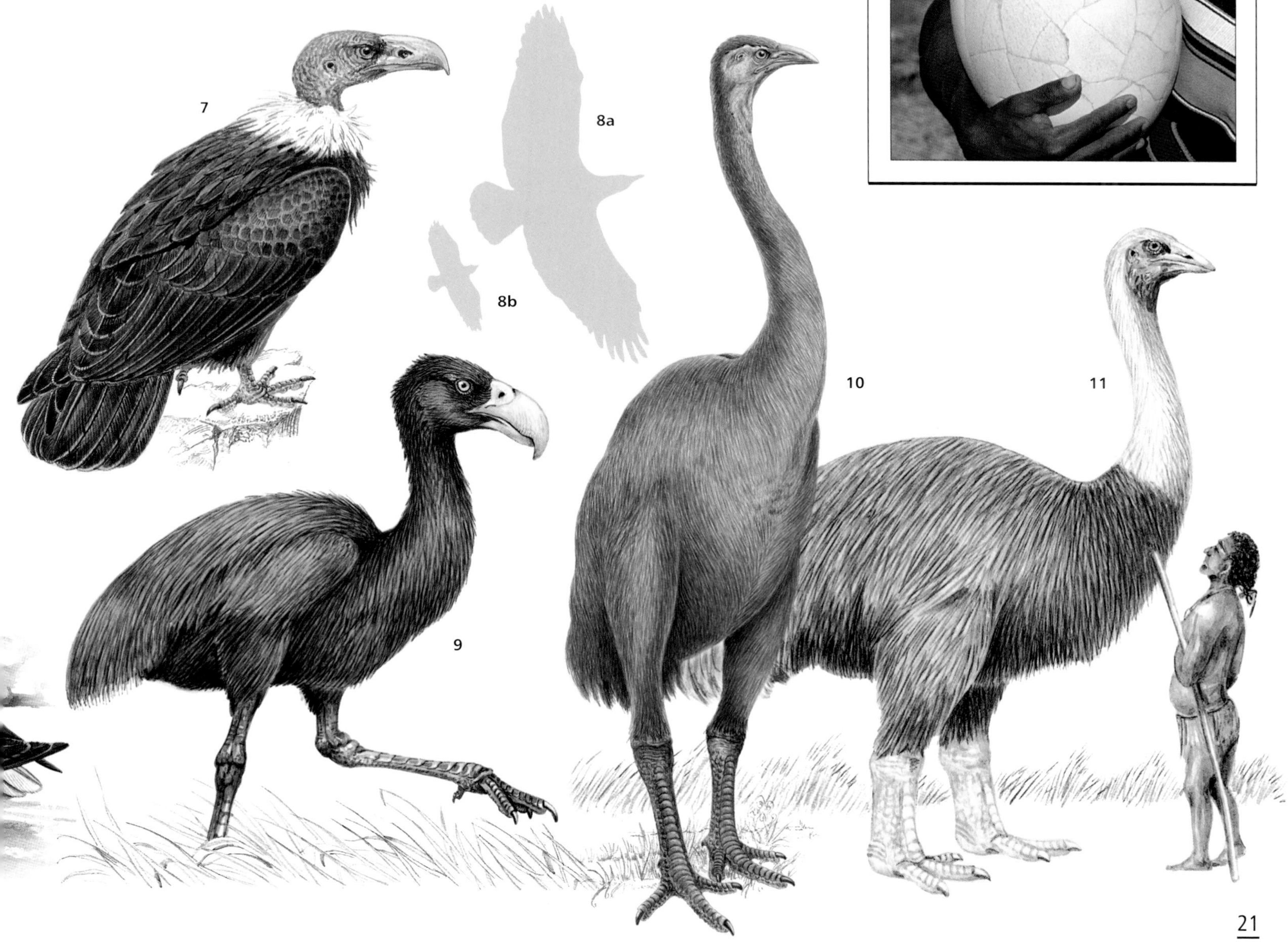

explain how they flew, the giant birds and the gigantic reptilian pterodactyls of an earlier age pose similar problems.

A Body Plan for Flying

MORPHOLOGICAL ADAPTATION

The skeleton and musculature incorporate most of birds' major adaptations for flight (that is, apart from feathers). These adaptations satisfy two main requirements. First, because flight is so expensive in energy, weight has been reduced as far as possible; and second, the need for maneuverability in flight has required the bird to become a very compact unit, with as much weight as possible placed close to the center of gravity.

To begin with, the skull has been greatly lightened. The eyes are large, and their sockets take up a lot of space in the forepart of the skull, virtually meeting in the middle. A major feature is the reduction of the heavy jaws that other vertebrates possess and the complete loss of teeth. The shape and size of the bill varies enormously, enabling different types of bird to obtain and "handle" a very wide range of foods.

At the other end of the skeleton, the tail's bony elements have been greatly shortened and the bones are fused so that all the tail feathers start at more or less the same place. This adaptation has resulted in a structure that is valuable for helping the bird to steer and is much more effective than the long, "floppy" tail of *Archaeopteryx*. The size and form of the tail are varied, matching the flying needs of the bird concerned; in a few species (woodpeckers, woodcreepers, and treecreepers, for example), it is even stiffened so as to act as a prop during climbing.

The main skeleton has been greatly reduced in weight in many places, especially by the evolution of bones that are hollow, including the major limb bones and parts of the skull and pelvis. The light ribs have rearward projections (uncinate processes) that overlap the next rib, giving extra rigidity. In some diving birds such as the guillemots, these are very long, overlapping two ribs and providing extra support to prevent the body cavity being compressed during dives. Many bones have become fused, making a rigid frame without the need for large muscles and ligaments to hold the separate bones together.

The forelimbs, and to a lesser extent the hindlimbs, incorporate some of the greatest changes. The forelimbs have become the wings, and associated areas of the body are adapted to provide the attachment areas for the massive flight muscles. The hand has lost two of its digits, and another is greatly reduced. The main bulk of the wing muscles is at the base of the wing (close to the center of gravity). Although the downbeat results from the direct pull of the muscle, the upbeat (or recovery stroke) requires the "pulley" action of tendons over the shoulder joint. The wing joints are shaped in such a way that there is very

BIRD BODY PLAN

Many aspects of avian form and structure (morphology) and organic functions (physiology) are remarkable, having evolved to facilitate flight – birds' characteristic mode of locomotion. For example, their hollow bones are lighter by far than those of mammals. And, in respiration, a bird replaces almost all the air in its lungs with each breath.

Below *External features of a bird (shown here the sharpbill,* Oxyruncus cristatus*).*

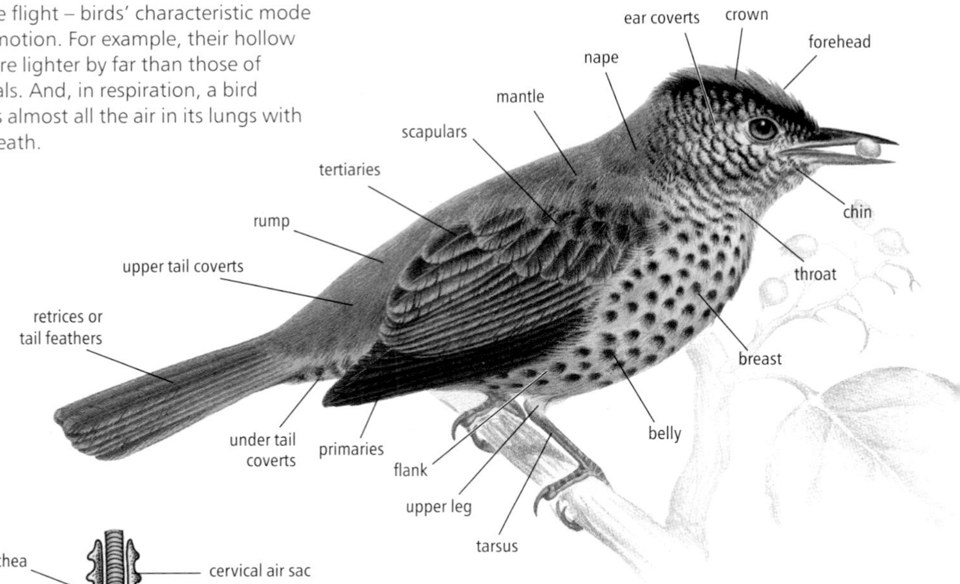

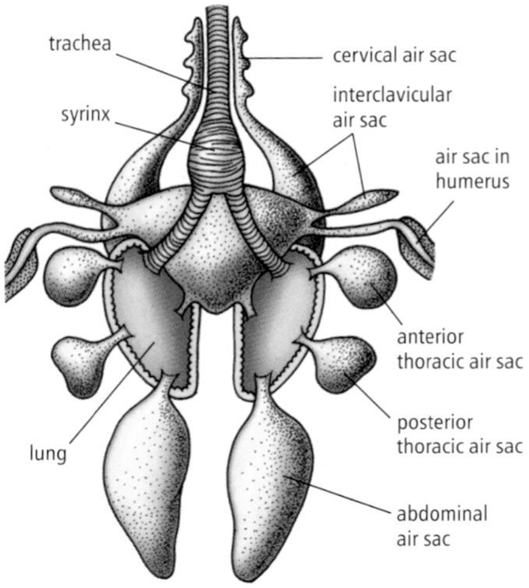

Left *The respiratory system of a bird. Birds' relatively small lungs are supplemented by multiple air sacs, which maximize the diffusion of oxygen into the bloodstream. Most birds have nine sacs – one interclavicular, two cervical, two anterior thoracic, two posterior thoracic, and two abdominal. These thin-walled extensions of the lung act as bellows, pushing a unidirectional flow of air through the lungs. This highly efficient form of respiration allows a constant inflow of fresh air with a very high oxygen content. Greater oxygen content in their cells enables birds to obtain maximum energy from their food – vital when so much energy is expended in flight.*

Right *The digestive system and main organs of the body. After being ingested, food passes into the crop; exclusive to birds, this organ stores food, either passing it on in continuous small amounts to the stomach, or retaining it for later regurgitation to the young. In the stomach, food is first processed in the proventriculus, which secretes digestive juices, and then in the powerful, muscular gizzard, which breaks down food.*

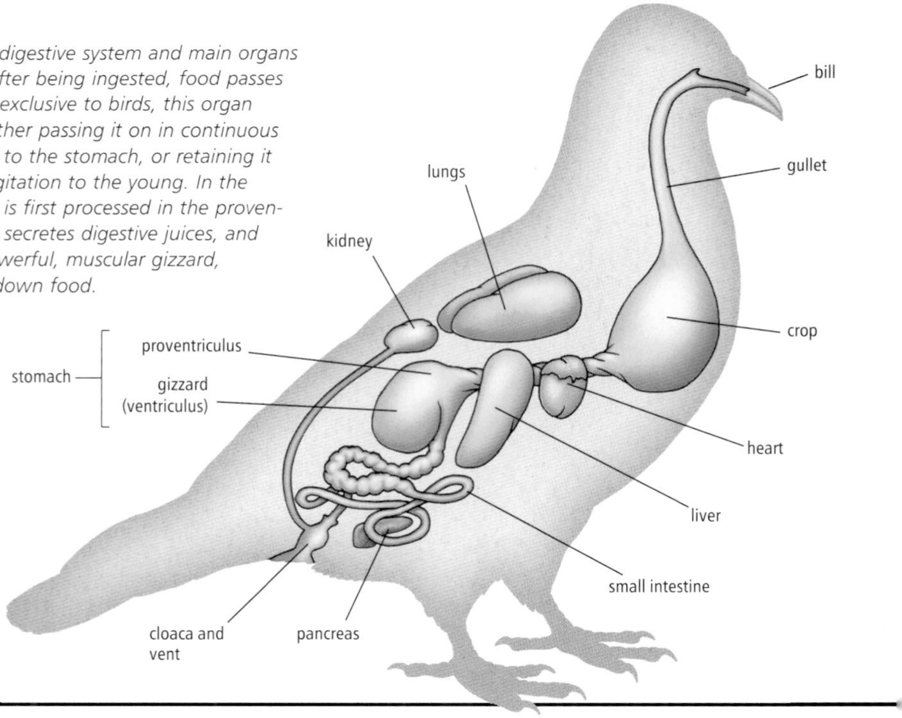

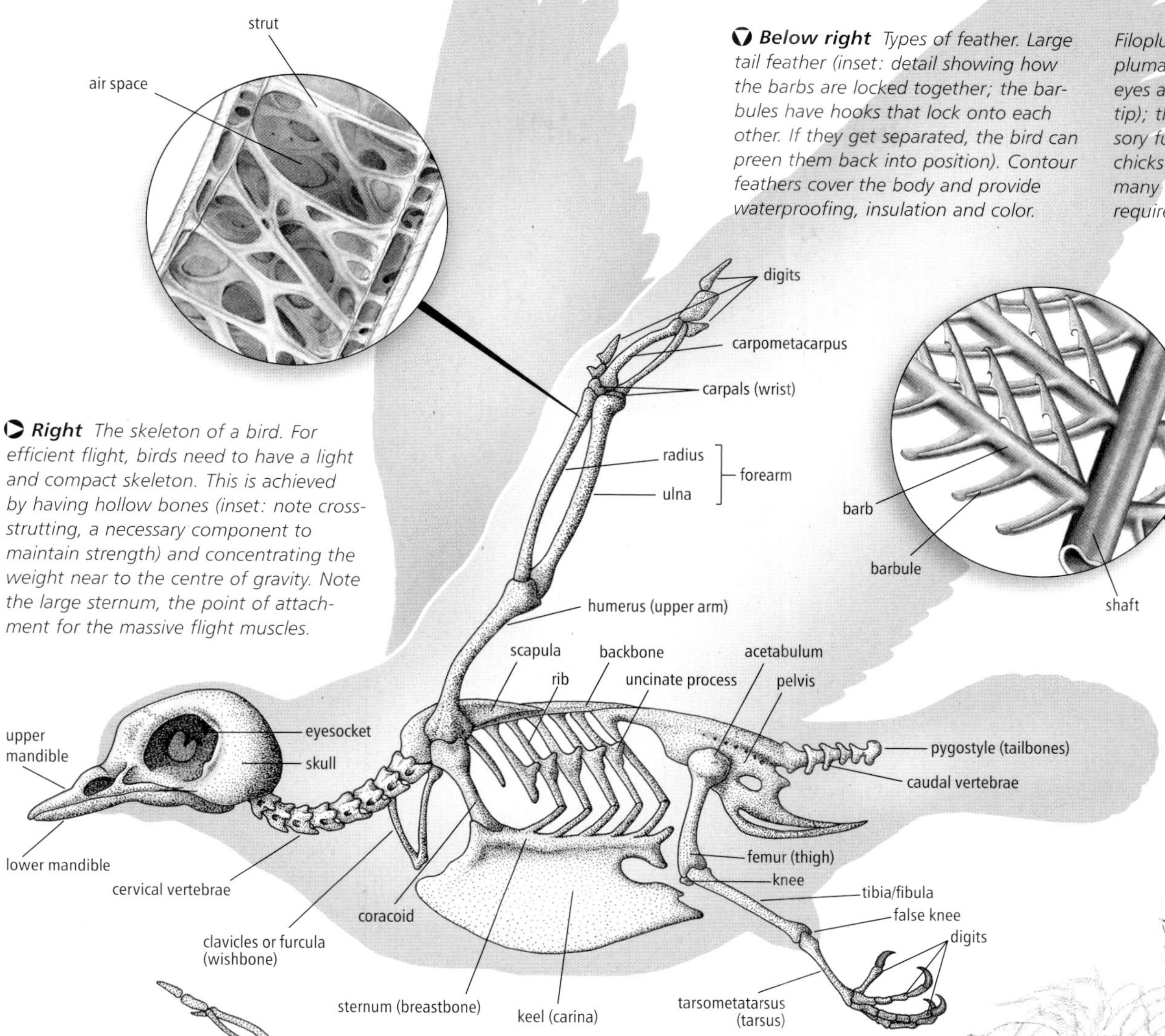

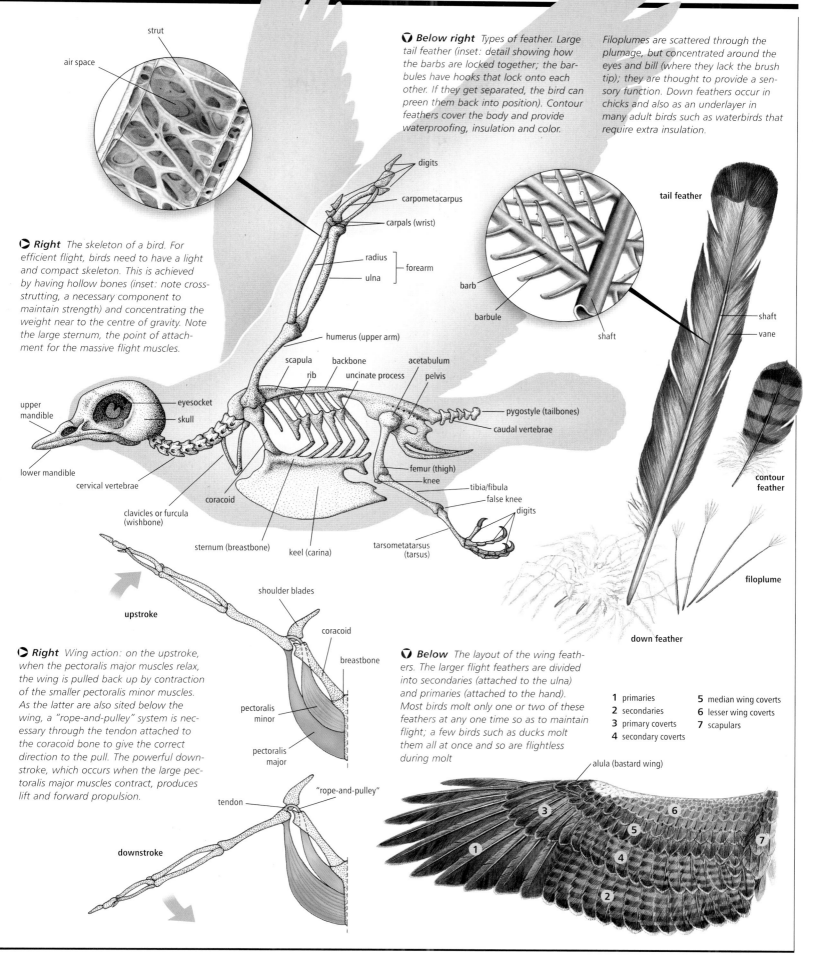

Right *The skeleton of a bird. For efficient flight, birds need to have a light and compact skeleton. This is achieved by having hollow bones (inset: note cross-strutting, a necessary component to maintain strength) and concentrating the weight near to the centre of gravity. Note the large sternum, the point of attachment for the massive flight muscles.*

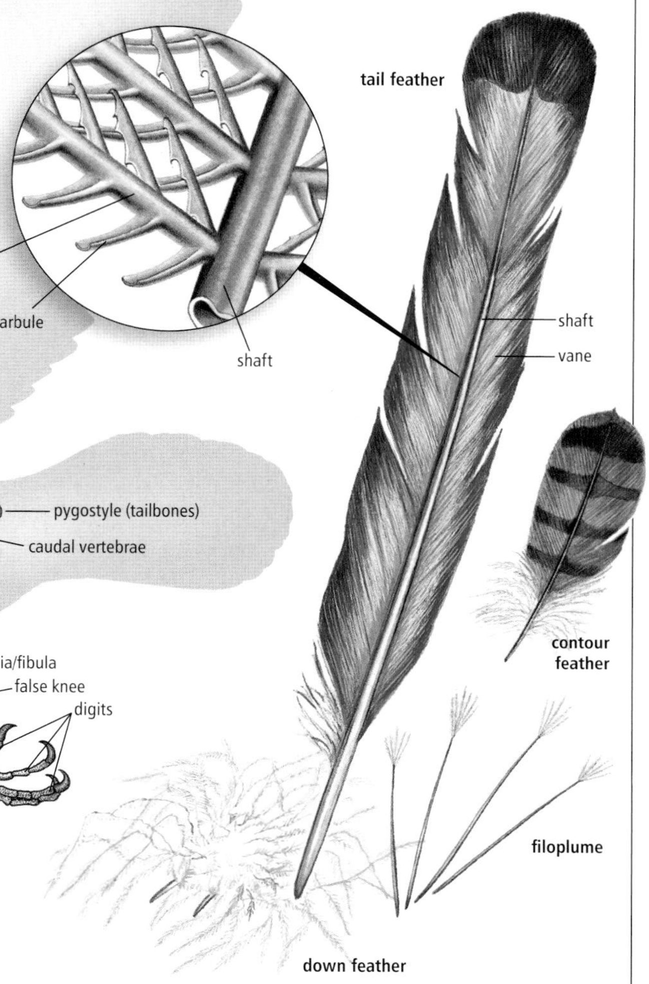

Below right *Types of feather. Large tail feather (inset: detail showing how the barbs are locked together; the barbules have hooks that lock onto each other. If they get separated, the bird can preen them back into position). Contour feathers cover the body and provide waterproofing, insulation and color. Filoplumes are scattered through the plumage, but concentrated around the eyes and bill (where they lack the brush tip); they are thought to provide a sensory function. Down feathers occur in chicks and also as an underlayer in many adult birds such as waterbirds that require extra insulation.*

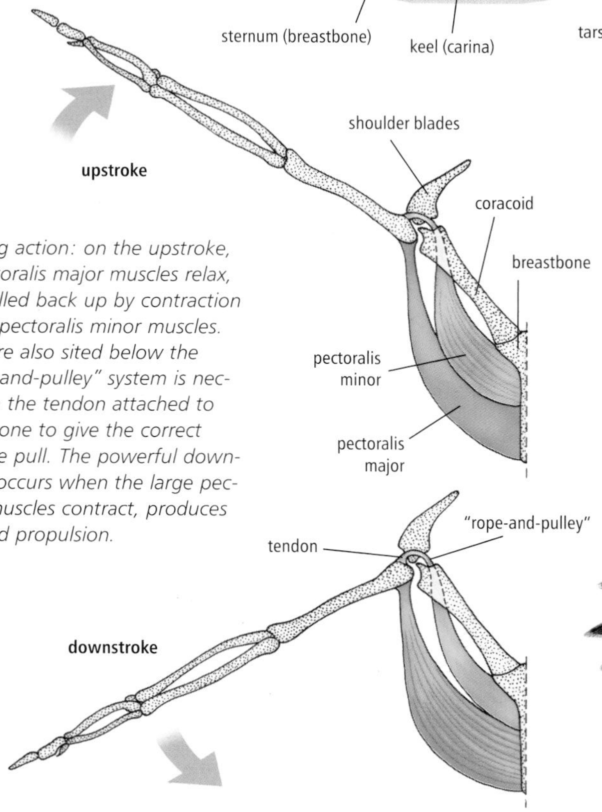

Right *Wing action: on the upstroke, when the pectoralis major muscles relax, the wing is pulled back up by contraction of the smaller pectoralis minor muscles. As the latter are also sited below the wing, a "rope-and-pulley" system is necessary through the tendon attached to the coracoid bone to give the correct direction to the pull. The powerful downstroke, which occurs when the large pectoralis major muscles contract, produces lift and forward propulsion.*

Below *The layout of the wing feathers. The larger flight feathers are divided into secondaries (attached to the ulna) and primaries (attached to the hand). Most birds molt only one or two of these feathers at any one time so as to maintain flight; a few birds such as ducks molt them all at once and so are flightless during molt*

1 primaries
2 secondaries
3 primary coverts
4 secondary coverts
5 median wing coverts
6 lesser wing coverts
7 scapulars

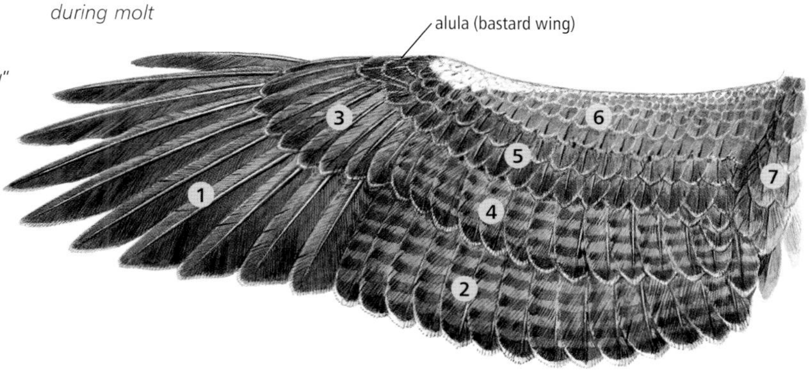

Above *Most birds achieve take-off by first jumping vertically into the air. They then propel themselves forward with their extremely powerful wing and pectoral muscles, generating lift in the process. During flight, the legs are retracted to create an aerodynamically efficient shape and so minimize drag. To slow down, birds then utilize drag, by fanning the tail and lowering the legs. Just before landing, the wings angle and tilt the body into an almost vertical position, acting as brakes.*

Right *In some birds, especially ground-nesting species (here, an American golden plover* Pluvialis dominica*), feather coloration and patterning are cryptic, to enable the bird to camouflage itself and so minimize the risk of being detected by predators.*

little movement possible except in the plane of opening out and closing, so saving the need for muscles and ligaments to prevent "unwanted" movements.

At the base of the upper arm (humerus) there is a broad area for the attachment of the pectorals. These enormous muscles are attached at their other end to the very large, keel-like breastbone (the sternum). When the muscles are contracted to beat the wings downward, they produce a force that would be great enough to crush the bird's body between the sternum and the wing if these were not kept apart on either side by a strong, strutlike bone, the coracoid, supported by the wishbone (the fused clavicles, or furcula) and shoulder blade (the scapula), the ends of which join together and provide a point of articulation for the wing.

Birds are an unusual class of animal in that they use two methods of locomotion: flying (using the forelimbs) and walking and/or swimming (using the hindlimbs). Balance in flight is not a significant problem, since the large flight muscles are situated close to the center of gravity just below the wings. However, partly because of the presence of these muscles, it would be difficult for the legs also to be positioned close to the center of the gravity; in fact the cup-shaped acetabula (hip sockets) in the pelvic girdle, which receive the top ends of the upper legs (femurs), lie some way behind the center of gravity. Balance would therefore be difficult for a walking bird supported directly at this point.

Birds have overcome this problem in a unique way. The femur is inserted at the acetabulum in the normal vertebrate manner, but projects forward along the side of the bird's body and has rather little movement, being bound to the body by muscles. In a sense the lower end of this bone (the knee) acts as a new "hip" joint, to which the lower leg is attached and which is quite well-positioned with respect to the center of gravity. The two sections of leg that are clearly visible are not comparable with ours: the upper section is the equivalent of our lower leg, whereas the lower section or false shin (technically called the tarsometatarsus, or tarsus for short) is formed from parts of the lower leg and sections of the foot bones, and has no human equivalent. This fact explains why birds' legs appear to bend in the opposite way to human ones – the visible joint is not the knee, but is more closely equivalent to our ankle. As with the wing, the leg joints are so shaped that movement in unwanted directions is restricted. Leg movements are controlled – via tendons – by muscles placed near the top of the leg, and so close to the center of gravity.

Warm, Light, and Streamlined

FEATHERS

Even though featherlike structures can be seen in some reptilian fossils, feathers are by far the most characteristic feature of birds, and a major feature in their habits, lifestyles, and distribution. Keratin, the main constituent of feathers, is a proteinaceous substance that is widespread in vertebrates – the hair and fingernails of mammals are made of it, as are the scales of reptiles. The ancestors of *Archaeopteryx* evolved the basic feather for insulation, and this purpose is well served by the evolution in modern birds of feathers that are light and waterproof and trap quantities of air so as to slow down heat loss. The principal body feathers consist of a central quill (or rachis) from which the main side projections – the barbs – spread out on either side. These are locked together by barbules (see Body Plan of Birds).

However, feathers have also evolved to serve a number of other functions important to birds. The feathers along the trailing edge of the wing and on the tail have become greatly enlarged, strengthened, and specially shaped so that they form the

surfaces that provide the lift for flight and for maneuvering. The rest of the visible feathers (the contour feathers) that cover the surface of the body add greatly to the efficiency of flying by streamlining the body as well as providing essential insulation.

Down feathers, found on young birds and also as an insulating underlayer on many full-grown birds, lack the interlocking barbules and are not organized in one plane, and so look more like shaving brushes. Most simple of all are the single shafts of the bristles often found around the eyes or at the base of the bill. In many cases these are thought to have a sensory function.

The wide range of colors in feathers also performs a number of valuable functions. On the one hand feathers may camouflage a bird – such as a nightjar – so well that it is difficult for a predator to spot it. At the other extreme, in the peacock, hummingbirds, quetzal, and other species, feathers provide some of the most dazzling colors found in nature and are important in display.

Feather colors are produced in one of two ways – or a combination of both. Commonest of the pigments in feathers is melanin, which is responsible for the browns and black. Some pigments are very rare, such as the green turacoverdin found only in some turacos. The other type of color is caused by the physical structure of the feather reflecting only a part of the visible wavelengths of (white) light. Such colors include the metallic blue-green of the starling, and most iridescent colors. Feathers reflecting all light wavelengths look white!

Feathers are not just distributed at random but grow in clearly defined tracts. Each feather grows from a papilla – a special ring of cells. As these cells multiply they produce a series of rings of cells that form into a tube. On one side of this tube is a thickened section, the rachis, and on the opposite side a line of weakness. As the feather grows, it breaks along the line of weakness and spreads out. The individual barbs of the feather also "break apart" at lines of weakness.

Feathers are replaced at intervals. They may be molted because they have become worn and need replacing. Almost all birds molt once a year (usually after breeding) for this reason. Some birds put on a thicker covering of feathers for the winter. Feathers are also changed in order to produce a

FLIGHTLESS BIRDS

The birds that diverge most markedly from the "typical bird" structure are flightless; in these, the restrictions on form required for flight have been relaxed. The most obvious characteristic of flightless birds may be the loss, or great reduction, of wings, but in some species there may also be a considerable increase in size. The ostrich averages some 115kg (250lb) and may attain 150kg (330lb). Males share care of the nest and young of several females.

All flightless birds are thought to be descendants of flying ancestors. This fact is fairly obvious in some species, such as flightless rails and coots, which are clearly close relatives of other modern birds that fly. Close examination of other species shows that they have many adaptations for flight, including hollow (pneumatized) bones; in addition, the wing, although sometimes much reduced, is clearly of the form used by flying birds.

A flightless bird is freed from the need to build and carry the very large flight muscles and associated wing structure, and from the energy expenditure of flying. Hence, other things being equal, flying birds would tend to lose out in competition with flightless ones, since the former would need much more food.

The disadvantage of being unable to fly is that the bird is more vulnerable to predators. Flightlessness seems to have evolved where birds were fairly safe from predators, as is often the case on remote islands (instances of this are the kagu of New Caledonia and the kakapo of New Zealand). Yet, if humans, their domestic animals, and introduced predators arrive, such species are highly prone to extinction – the dodo of Mauritius being the quintessential example.

Another group of birds that have tended to lose the power of flight are waterbirds, including the penguins and the extinct Great auk. Very large flightless birds form a third group – the ostrich, cassowaries, emus, and rheas. Most of these are large enough to defend themselves quite well by either strength or great running speed.

different-colored plumage. Many birds put on bright plumage for the breeding season and change to a duller one for winter. It is thought that the birds need the brighter colors in order to display to one other during courtship; duller, often disruptive, camouflage plumage provides better defense from predators, and so birds revert to this at other times of the year.

The ptarmigan's white winter and brown summer plumages help the bird to merge into the background, making it more difficult for predators to see. The drakes of many species of duck remain in bright plumage almost the whole year round, but acquire a camouflaged brown plumage (the so-called "eclipse" plumage) for about four to six weeks during the summer, while they are in full molt and so flightless and vulnerable.

The mature feather is a dead structure. Its replacement requires energy, and while building the new feathers a bird is less well insulated and may be able to fly less well. Some species, such as ducks and most auks, lose the power of flight altogether during the molt. On the other hand, molting allows damaged flight feathers to be replaced (an advantage over, say, bats, which cannot mend a badly damaged wing).

The Advantages of One-Way Breathing

RESPIRATION

In order to be able to fly, birds have to be able to mobilize a large amount of energy quickly. They need a very efficient respiratory system to supply the requisite large amounts of oxygen. The bird lung is indeed efficient, although not more so than that of mammals, at least at sea level. Its particular advantage is its efficiency at altitude. If mice and sparrows respectively are placed in a chamber containing air at the reduced pressure found at the top of Mt. Everest, the mice are soon almost totally exhausted and can barely move around, while the sparrows hop about quite happily – their breathing is not noticeably impaired.

In fact, many birds migrate in a fairly rarefied atmosphere; cranes, ducks, and geese such as the Bar-headed goose often cross high over the Himalayas on their journeys between northern Russia and their winter quarters in India. Although not many have to fly as high as the summit of Everest (8,848m/29,028ft), large birds of prey have been seen from aeroplanes at this height or even higher.

The respiratory system of birds differs from that of mammals in a number of important ways. To begin with, bird lungs are actually smaller than those of mammals of a comparable size. On the other hand, birds possess a large number of air sacs throughout the body spaces, some even penetrating into the hollow bones. No gaseous exchange takes place through the thin, membranous walls of the air sacs themselves, although they may be important in preventing a bird from overheating.

ADAPTATIONS FOR SWIMMING

Of the many different birds that get their food from water, almost all depend on swimming to reach it. Swimming birds include penguins, albatrosses and shearwaters, divers, grebes, pelicans and cormorants, ducks, many members of the rail family, gulls, terns, auks, and the phalaropes. No passerine has become fully aquatic, although the dippers and the Seaside cinclodes get all their food from the water.

Most of these groups have webbed feet **1** with which they swim. Some, however, such as the grebes **2**, coots, finfoots, and phalaropes, have lobed toes rather than webs, while other rails, such as the moorhen (BELOW), simply have a broadened base to their long toes that provides a swimming surface. A few aquatic birds, such as the penguins and the auks, get their main underwater propulsion from their wings, which in penguins (as in the extinct Great auk) have lost the power of flight. One or two flying species, such as the scoters, get some propulsion underwater by use of the folded wing – it is too large to be used fully extended.

For a diving bird, one of the main difficulties is to get below the water's surface. Birds are generally much lighter than water. Most diving birds, by contrast, have relatively high densities and are able to squeeze much of the air (normally important for insulation) out of their feathers as they dive, so helping to reduce their buoyancy. Cormorants have specially wettable feathers that make it easier to lose air – this is why they have to stand with their wings spread to dry off after a fishing session. Grebes are said to increase their specific gravity by swallowing stones.

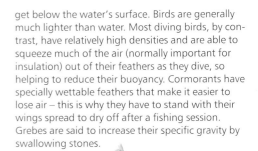

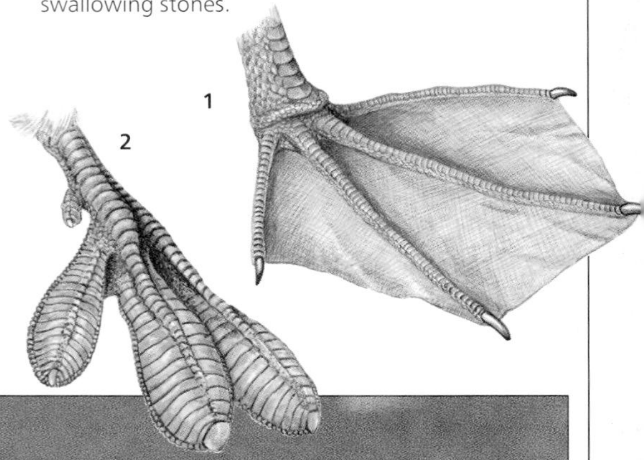

The importance of air sacs to breathing lies in the fact that the inspired air passes first through the posterior air sacs, then into the lungs proper, and finally out of the bird via the anterior air sacs. Thus, air flows in one direction through the lungs, instead of the "ebb-and-flow" system that operates in mammals. Hence almost the entire volume of air in the lung can be replaced with each breath, while humans, for example, only exchange perhaps three-quarters of the volume, even when breathing deeply.

The blood vessels of the avian lung are very efficient in their uptake of oxygen and disposal of carbon dioxide. Since the airflow is always in one direction, the blood vessels can be arranged so that blood continuously flows in the opposite direction to the flow of air. Blood that is just reaching the lungs, and that is low in oxygen, meets air that has flowed some way through the lungs and has had its concentration of oxygen lowered; however, there is still sufficient oxygen there for the blood, with its low concentrations, to be able to take it up. As the blood flows along against the lung wall, it takes up progressively more oxygen, and meets air that contains progressively more oxygen. This system helps to maximize oxygen uptake in a way that is impossible in a mammalian "ebb-and-flow" lung. The same thing happens, in reverse, for the disposal of carbon dioxide. Another advantage of the avian lung over the mammalian one is that birds' air capillaries (alveoli) are very small compared with mammals'. Thus, although a bird's lungs are smaller than a mammal's of similar size, they weigh about the same, due to the greater tissue density, which provides a far larger surface area for gas exchange.

Other Flight Adaptations

Weight-Reduction Strategies

The digestive tract of birds is also adapted for flight. The large, weighty jaws, jaw muscles, and teeth of reptilian ancestry have been lost (although some birds still have remarkably powerful jaws). In birds their function of grinding up food is largely taken over by the muscular portion of their stomach, the gizzard. To get food to the gizzard, some birds may use their bill to tear it into small pieces that are then ingested through the wide gape and swallowed.

Once in the gizzard, food is ground down, often with the help of grit, which certain species – for example grain-eaters such as domestic hens and sparrows – take in for the purpose. Fish- and meat-eating birds such as kingfishers and eagles, and insectivores such as swallows and flycatchers, do not need grit; their food is comparatively soft, so they manage with their strong digestive juices.

Although many birds eat seeds and fruit, few specialize on leaves, as do grouse, or grass (geese and some ducks), at least compared with mammals. Such foods are rather difficult to break down. In the case of many mammals, for instance the cow, the digestion of leaves involves the action of symbiotic bacteria in very large – and so very heavy – stomachs. Since such large guts would be too heavy for flying birds, those that are herbivores have to consume large quantities of material in order to extract their nutritional requirements.

Above *The extraordinary efficiency of birds' lungs means that some species are able to fly at altitudes where the oxygen supply is just one-fourth of that at sea level. Demoiselle cranes* (Anthropoides virgo), *for example, migrate over the Himalayas at a height of over 9,000m (29,500ft).*

Above the gizzard, many birds – particularly seed-eaters – have a rather thin, extensible sidewall to the esophagus – the crop. A bird can cram a large quantity of food into the crop in a very short time and then retire to a place of safety to digest it. Many seed-eating birds, including finches and pigeons, also take quantities of food to their roost in this way, effectively reducing the length of the night's fast. Many species use the crop to carry food to their nestlings.

Birds reduce the amount of water carried in the waste products that are to be excreted, again a weight-saving adaptation. Some birds obtain most of their water from their food. The water is withdrawn from the contents of the hind-gut. Urinary products are highly concentrated and are formed primarily of uric acid. The latter becomes mixed with the feces in the cloaca before excretion (birds have no urinary bladder). Some carnivorous species such as owls do not digest parts of their prey, which are regurgitated in the form of pellets. Some seabirds that have to carry food for their chicks over long distances part-digest the prey to reduce the weight of the load.

The bird's reproductive system also keeps weight to a minimum. For most of the year, sex organs and associated ducts are greatly reduced in size, most markedly in females. As the breeding season nears, they rapidly develop as gametes (sex cells) are produced. All birds lay eggs, and even those – the great majority – that have a clutch of several eggs only lay one per day; some lay only every other day or less frequently. By laying in this way, most bird species are able to produce several relatively large eggs one after the other.

1
2
3
4
5
6
7
8
9
10
11
12
13
14
15
16
17
18
19
20
21
22

While clutch size ranges from one up to an average of 19 eggs in Grey partridges in Finland, the laying bird only carries one fully developed egg in the oviduct at a time (although one or more smaller, developing eggs may be present in the ovary). If the females were to carry all the eggs at the same stage of development as a mammal does her young, then the size of the individual egg would have to be much smaller or the bird would have to have fewer young.

Egg size as a proportion of adult weight varies from about 1.3 percent in the ostrich to 25 percent in the kiwi and some storm petrels. One advantage of having relatively large eggs is that it shortens the development time in the nest and enables the young bird to be able to fly at an early age (12–14 days in many small species); this, of course, tends to reduce the period of threat from predators when the young are helpless in the nest. Most birds lay their eggs early in the day, and so are relieved of the need to carry a fully-developed egg during the morning feeding period when the female needs to be at her most active.

Vision, Hearing, Smell

SENSES

Most animals rely particularly on just one or two of their senses. Most mammals, especially night-active ones, rely chiefly on their powers of smell and hearing. Even mammals for which sight is important mostly lack color vision. In birds, however, the power of vision, including color vision, is almost always the most important sense, with hearing second and smell a very poor third; indeed, in many birds smell may be hardly used at all. In this respect humans are an exception among mammals; our senses rank in importance in the same sequence as those of birds, and like the birds we have good color vision.

This parallel may explain why birds are so popular with people: we rely mainly on the same senses, and also on a pattern of daytime activity, and so can enjoy and appreciate their colors and songs. In contrast, we have relatively little idea of the information that even such familar mammals as the domestic dog or cat obtain from smelling objects, and in this respect have little chance of sharing their world. Visiting a wood, we may see many birds but almost no mammals, even if more mammals are present; they are just less easy for us to perceive, for many are nocturnal, or live underground, or both.

A bird's is a high-speed, aerial life. In such conditions, sight and hearing are clearly more useful than smell. The importance of birds' eyes is reflected in their size. They fill very large portions of the skull; indeed, an eagle's eyes approach those of a human in size, though the eagle itself is far smaller.

The eyes of birds are relatively immobile: the large eye leaves little room in the skull for muscles. However, birds such as owls have very mobile necks that enable them to turn their heads easily, and their actual field of vision is very wide – some birds may be able to see the whole 360°. In fact, a bird such as the woodcock that has eyes placed very high on the sides of its head may be able to see not only all around but also over the top of its head! There is a price to be paid in that, for most birds, the fields of vision of the two eyes barely overlap, so that they have only a small amount of binocular vision. In recompense, however, they can spy movement over the whole of the fields – useful for spotting predators. In contrast, birds with forward-facing eyes, such as the owls, have good binocular vision. Birds also have a large part of this field – perhaps about 20° – sharply in focus at one moment, compared with the 2°–3° across which people can focus sharply.

Left Birds display a huge variation in bill shape, adapted for dealing with different foodstuffs: **1** Brown kiwi (worms, other invertebrates); **2** Anhinga (fish); **3** Toco toucan (fruits); **4** Red crossbill (seeds); **5** Flamecrest (insects and caterpillars); **6** Laughing kookaburra (spiders and small vertebrates, aquatic insects and fish); **7** Avocet (mollusks, crustaceans, small aquatic invertebrates); **8** Hawfinch (hard-shelled seeds); **9** Double-toothed barbet (fruits and hard berries); **10** Greater flameback (arthropods); **11** Giant hornbill (fruit, especially figs); **12** White-tipped sicklebill (nectar from curved-corolla flowers, e.g. Heliconia); **13** Sword-billed hummingbird (nectar from long-corolla passionflowers); **14** Sparrowhawk (small birds); **15** Pennant-winged nightjar (insects); **16** Great crested grebe (fish, crustacea, mollusks); **17** Shoebill (lungfish, frogs, turtles, and snakes); **18** Greater flamingo (algae and diatoms; small aquatic invertebrates); **19** Eurasian spoonbill (small fish and shrimps); **20** Indian skimmer (small fish and crustaceans); **21** Great white pelican (fish, amphibia, small mammals); **22** Yellow-collared lovebird (seeds, nuts, and berries).

Right Birds' external ears are inconspicuous, and usually hidden by the head feathers; on the fleshy neck of the Crested guineafowl (Guttera pucherani), however, the openings are clearly visible. Birds hear well over a narrower frequency range than mammals.

Below A Yellow wagtail (Motacilla flava) calling. Vocalizations are employed by birds for a wide variety of purposes, including attracting a mate, defending territory, and raising the alarm to warn of predators. Some birds that live in caverns even use calls for echolocation, to navigate around their dark habitat.

Above *The courtship display of the Great bustard* (Otis tarda) *is one of the bird world's most flamboyant examples of a male advertising his readiness to mate. The brilliant white wing and breast feathers, which are normally concealed, enhance the bird's striking display.*

Most birds have good color vision, including those species of owls in which the faculty has been tested, although they may see slightly less well at the blue end of the spectrum than we do. The visual acuity of birds of prey and certain other species is perhaps two to three times greater than that of humans, but not more. Some birds, such as the nocturnal owls, have exceptionally good night vision, yet even they probably depend largely on hearing to locate and catch their prey at night. A relatively recent discovery is that many birds can see well in the ultraviolet part of the spectrum, which we cannot. At least some birds thus have tetrachromatic, compared with our trichromatic (and most mammals' dichromatic), vision. Some species – for example, parrots – also have plumages that reflect in the ultraviolet. This means that they are able to distinguish a much wider range of colors than we can, a fact that undoubtedly has important implications for their lives, although these have yet to be worked out.

Hearing is employed in communication between individual birds, and is particularly valuable in wooded areas where it is difficult to keep in visual contact – hence the striking songs and far-carrying calls of many forest birds, such as the Musician wren and the bellbirds. As with vision, birds and people perceive sounds over a range that is roughly similar, although most birds are possibly less good at hearing sounds at the lower frequencies. The hearing of birds also seems to differ from ours in a more important respect: they can distinguish sounds that are very much closer in time than we can. For example, what we hear as a single note may be heard by a bird as up to 10 separate ones. A snatch of "simple" birdsong may therefore convey a lot more information to a bird than appears possible to our ear.

Many birds seem to lack almost entirely the power of smell, although this is not true of certain groups, such as the nocturnal kiwis, which probe for food on the forest floor and have their nostrils close to the tip of the bill. Smell is also known to be used by the vultures of the New World (although not those of the Old World) as they search for carrion on the forest floor. In some other groups, for example some of the petrels, the olfactory lobes of the brain are well developed, indicating that they too are able to use the power of smell.

Taste is by no means strongly developed in all species; in birds as in man, the sense actually involves olfactory information and, as we have seen, birds have poor powers of smell. The tongue of many birds is very horny and would not easily accommodate taste-receptive cells. What taste buds do occur are found toward the back of the mouth, and so birds probably only taste a food item when it is well into the mouth. Nonetheless, birds are capable of distinguishing the four principal taste sensations: salt, sweet, bitter, and sour.

The sense of touch is well developed in the tongues of many birds and also in the bill tips of many species. This trait is especially marked in species such as snipes, godwits, and curlews that probe deep in mud for their prey, and in birds such as avocets, spoonbills, and ibises that "scythe" through water and soft mud with their bill open, ready to snap it shut on contact with a prey item.

Above and below *Among birds, the degree of development of hatchlings varies enormously. Some birds (generally ground-nesters, in immediate danger of predation) hatch chicks that are fully covered with down, have their eyes open, leave the nest within about 24 hours of hatching, and can feed themselves straight away. One species that gives birth to such nidifugous ("nest-fleeing") young is the Common crane – above, a 2-day-old chick is already foraging independently.*

The opposite development strategy is represented by birds such as swallows, whose newly-hatched chicks are born blind and helpless and stay in the nest until they are almost fully grown and able to fly. Such nidicolous ("nest-living") young are characterized by large mouths and digestive systems. These chicks are around 23 days old.

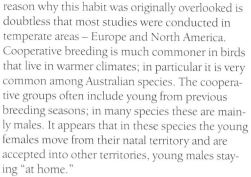

Patterns of Breeding

REPRODUCTIVE BEHAVIOR

The two main senses of birds, vision and hearing, are used strikingly in courtship and breeding. Many birds establish breeding territories by use of song, which functions both to repel would-be intruding males and to attract potential mates. Although not all birds have songs, some of those that do produce sounds that to the human ear are the most beautiful in the natural world. The nightingale and European skylark are among the most famous songsters, but others such as the South American Musician wren are similarly talented, while the songs of some nonpasserine species, such as the kookaburra, are also remarkable. A few species, such as the lyrebirds and the Marsh warbler, show great versatility by being able to mimic the calls and songs of many other species; the cagebirds best at "talking" are mynahs and parrots.

The display of the peacock is renowned for its flamboyance, although in some ways the Argus pheasant can put on an even more dazzling performance, and the birds of paradise are also impressive. All these species are polygamous, the males attracting females to their display grounds (leks), where mating occurs; the females then go off and lay the eggs, incubate them, and rear the young by themselves.

Most birds, however, are monogamous, breeding in pairs – almost all seabirds and birds of prey, for example. For reasons that are still not understood, a very high proportion of the polygamous species are vegetarian fruit- or leafeaters.

It has long been thought that the single pair feeding its young at the nest is the norm for birds. Studies in recent years, however, have shown that in many species it is not at all unusual for there to be several birds attending a single nest. The reason why this habit was originally overlooked is doubtless that most studies were conducted in temperate areas – Europe and North America. Cooperative breeding is much commoner in birds that live in warmer climates; in particular it is very common among Australian species. The cooperative groups often include young from previous breeding seasons; in many species these are mainly males. It appears that in these species the young females move from their natal territory and are accepted into other territories, young males staying "at home."

In temperate areas, birds are faced with wildly fluctuating food supplies – often great abundance in summer and great shortages in winter. Numbers are cut by starvation in winter, so most of those that survive can find a place to breed in spring. By contrast, bird populations in places that are less seasonal tend to be closer, year-round, to the limit of numbers that the habitat can support. With fewer vacant territories becoming available, it seems to pay a young bird to remain longer within its parents' territory, even if it is always on the lookout for a vacant territory nearby. Indeed, it may inherit the territory when its parents die.

Another factor apparently affected by seasonal variation is clutch size. Birds tend to have clutches that are larger at high than at low latitudes. Again, this fact is thought to be a reflection of the greater abundance of food available in summer in temperate areas.

Bird eggs take a considerable time to develop – anything from 12 to 60 days from laying to hatching. During this period, both the incubating parent and the eggs are very vulnerable to predation. Some birds protect the eggs and young by breeding in colonies. Others hide the nest, in foliage or in a hole. Species that do not build a nest often lay eggs with camouflage markings.

The simplest nest is a "scrape" made in the ground – no added materials are used. Species that lay their eggs in scrapes include divers as well as many game birds and waders. King and Emperor penguins do not even keep their eggs on the ground, but carry them around on the tops of their feet to insulate them from the intense cold. Some birds, among them most petrels and auks, lay their eggs on the bare rock or the soil of a burrow. Many others – including hornbills, some pigeons, many birds of prey, and owls – lay their eggs in holes in trees or cliffs without constructing a nest, although some, like the woodpeckers, bee-eaters, and sand martins, excavate holes in trees or banks.

Other birds build highly complex nests, the most outstanding of which are the intricately woven nests of some weavers and New World blackbirds, which may have entrance tubes 1.5m (5ft) long. Other nest-building skills include those of the tailorbirds, which "sew" leaves together, and some swifts and swallows, which build saliva-and-mud nests attached to rock faces. CMP

STORY

A Diversity of Nests

❶ *Almost all birds build nests, from elaborate structures to mere shallow scrapes in the ground. The only exceptions are certain penguins, which incubate eggs on top of their feet, and brood parasites such as the cowbirds and some cuckoos. Some of the largest nests are those of storks (here, White storks* Ciconia ciconia*), which may be added to year on year.*

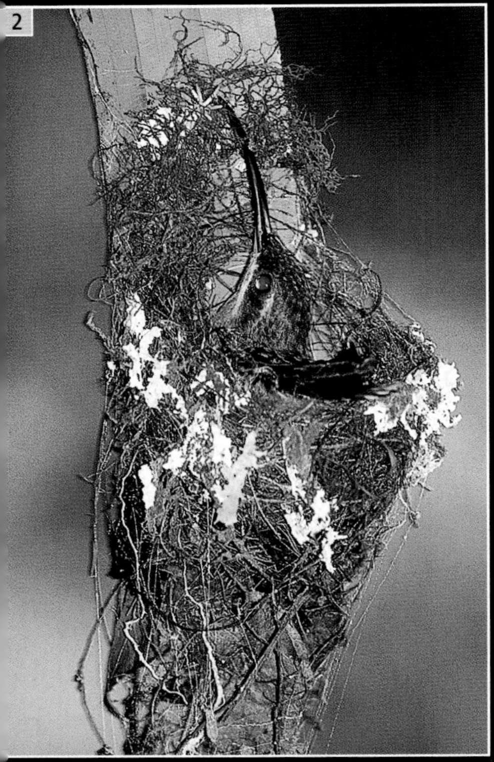

❷ *The tiny nests of hermit hummingbirds (*Phaethornis *spp.) are cone-shaped constructions of vegetation, reinforced with spiders' webs. To prevent the delicate nest from tipping when the bird lands in it, hermits weave small lumps of mud into the base of the cone, which counterbalance the bird's weight.*

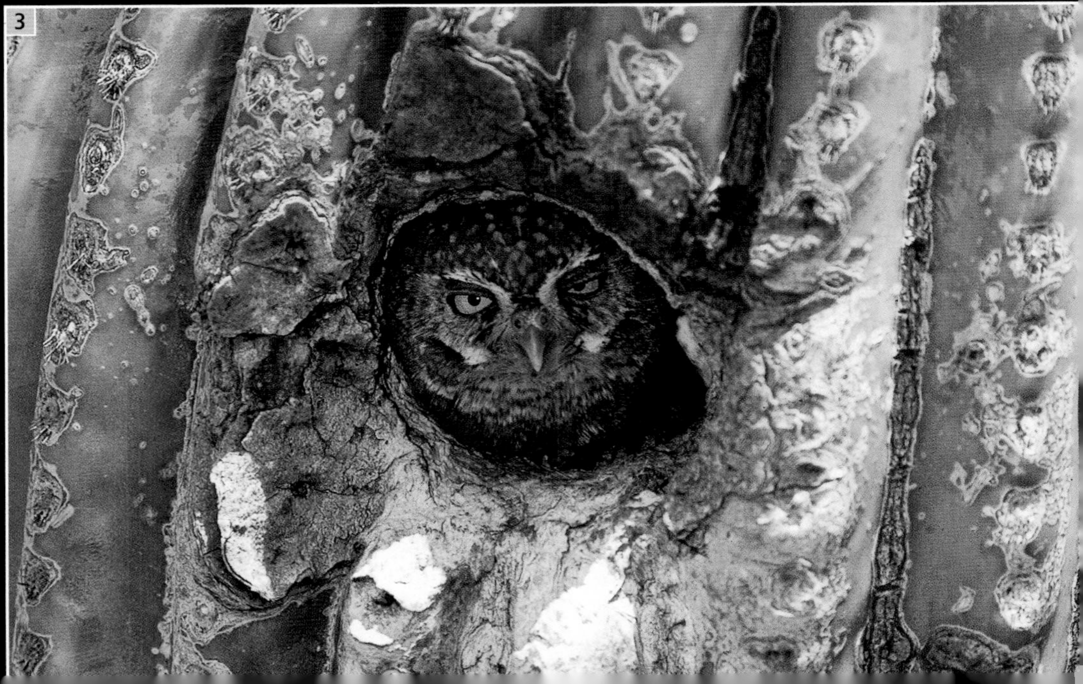

❸ *While many hole-nesting birds excavate their own cavities, others exploit the building skills of another species. In the US Southwest and Mexico, the Elf owl (*Micrathene whitneyi*) uses holes made by Gila woodpeckers and Northern flickers in saguaro cacti. High above the ground, the owls are safe from predatory mammals.*

4 *The White-bellied swiftlet* (Collocalia esculenta affinis) *attaches its nest to sheer rockfaces inside caves. This bird's main building materials are moss and mud; other species use saliva, and their nests are the source of "bird's nest soup."*

5 *One of the most spectacular nests is that of the Crested oropendola* (Psarocolius decumanus). *The female weaves palm fibers and grasses into a pendulous structure some 1–2m (3–6ft) in length, hanging high up in the rainforest canopy.*

6 *Widespread throughout the western USA, Cliff swallows* (Hirundo pyrrhonata *nest communally in bowl-shaped mud nests. These birds have diversified from their "natural" sites on cliffs, and now often nest in barns or under bridges.*

7 *As testament to the great resourcefulness and adaptability of birds, this Spotted flycatcher* (Muscicapa striata) *has raised its young under the protection of a forestry worker's helmet.*

Ostrich

CONTRARY TO POPULAR BELIEF, OSTRICHES have never been observed to bury their heads in the sand. Indeed, when threatened, these enormous, flightless birds invariably rely on precisely the opposite strategy, using their long legs to flee from approaching danger. The ostrich has the distinction of being the world's largest bird.

Ostriches are widely distributed in the flat, open, low-rainfall areas of Africa, in four clearly recognizable subspecies. The North African ostrich, which has a pink neck, inhabits the southern Sahara; the blue-necked Somali ostrich occupies the Horn of Africa; adjacent to this subspecies is the pink-necked Masai ostrich, which lives in East Africa; while south of the Zambezi River is the South African ostrich, also blue-necked. The Arabian ostrich (*Struthio camelus syriacus*) has been extinct since the mid-20th century.

Tall and Flightless

FORM AND FUNCTION

The ostrich's feathers are soft and without barbs. The jet-black plumage of the male, contrasting with his long, white, outer "flight" feathers (primaries), makes him highly conspicuous at long distances by day. The brownish or gray-brown color of females and juveniles renders them well-

camouflaged (newly hatched chicks are fawn with dark brown spots and have a concealing, hedgehog-like cape of bristly down on the back). The neck is long and extremely mobile, the head small, the gape of the unspecialized beak wide, the eyes very large. Vision is acute. The thighs are bare, the legs are long and powerful, and there are only two toes on each foot. The bird can kick forward powerfully, and can run at about 50km/h (30mph). Ostriches are tireless walkers.

Thanks to their large stride, long neck, and precise peck, ostriches are extremely efficient selective gatherers of the sparsely dispersed high-quality food items in their habitat. Eating a wide variety of nutritious shoots, leaves, flowers, fruits, and seeds, their diet resembles that of browsing ungulates rather than birds. The yield from many pecks is amassed in the gullet and then passes slowly down the neck as a large ball (bolus), about 200ml (7fl oz) in capacity, that stretches the neck skin as it goes down. The gizzard can hold at least 1,300g (45oz) of material, of which up to 45 percent may be sand or stones that help grind up hard substances. Ostriches customarily gather in small flocks to feed. At such times, they are highly vulnerable to attack, and will often look up to scan the landscape for predators – most commonly lions, but occasionally leopards and cheetahs.

Left *Evolution has equipped the ostrich with a highly effective early-warning system, in the form of a long, flexible neck and large eyes with keen vision. These enable the bird to spot the approach of impending danger in good time to take evasive action.*

Looking After Others' Eggs

BREEDING BIOLOGY

Breeding seasons vary with locality, but in East Africa ostriches mainly nest in the dry season. A male makes a number of shallow scrapes in his territory, an area varying from 2 to 20sq km (0.8–8sq mi), depending on the fertility of the environment. A female (the "major" hen), who forms a loose pair bond with the male but has a home range of her own covering up to 26sq km (10sq mi), selects one of these scrapes and produces up to a dozen eggs, laying on alternate days. As many as six or more other females ("minor" hens) also lay in the nest, but play no further role there. These minor hens may also lay in other nests in the area. The major hen and the cock share equally, for increasing periods, in first guarding and later incubating the clutch, the female by day, the male by night. Unguarded nests are conspicuous from above and are vulnerable to predation by Egyptian vultures (*Neophron percnopterus*), which drop stones on the huge eggs to break the shells, which are 2mm (0.08in) thick. Even guarded nests are at risk from hyenas and possibly jackals. The rate of attrition is high; fewer than 10 percent of the nests started survive the laying period – roughly three weeks long – and the six-week incubation period.

Ostrich chicks are well-developed (precocial). They are accompanied by both male and female, which try to protect them from the many raptors and ground predators that threaten them. Chicks

FACTFILE

OSTRICH

Struthio camelus

Order: Struthioniformes

Family: Struthionidae

4 subspecies: **Masai ostrich** (*S. c. massaicus*), **North African ostrich** (*S. c. camelus*), **Somali ostrich** (*S. c. molybdophanes*), **South African ostrich** (*S. c. australis*).

DISTRIBUTION Africa (also until recently in Arabia).

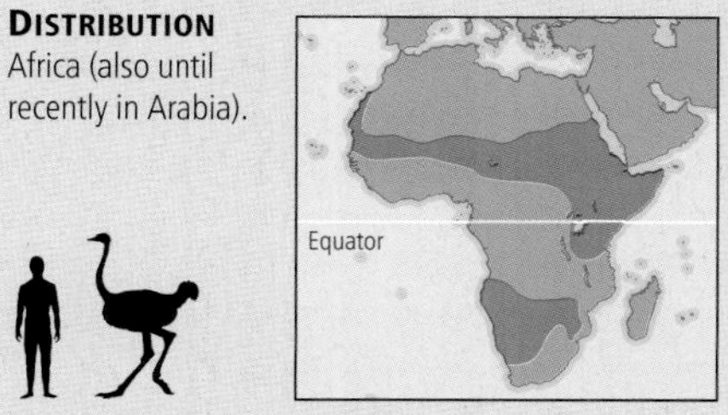

HABITAT Semi-desert and savanna.

SIZE Height about 2.5m (8ft); **weight** about 115kg (250lb). Males slightly larger than females.

PLUMAGE Males black with white primary feathers on wings and white tail, but buff in one subspecies; female gray-brown. Neck and thighs bare, with skin in males blue or pink, according to subspecies; in females, pale pinkish gray.

VOICE Loud hiss and booming roar.

NEST A shallow scrape in the ground.

EGGS 10–40, shiny, creamy white; weight 1.1–1.9kg (2.4–4.2lb). Incubation period 42 days.

DIET Grasses, seeds, fruits, leaves, flowers.

CONSERVATION STATUS African ostriches are not currently under threat. The Arabian ostrich (*S. c. syriacus*) was hunted to extinction, being last recorded in 1966.

Below *A flock of ostriches sprints across the almost lunar landscape of the Etosha Pan, a huge salt flat in Namibia. For ostriches, the ability to sustain a fast running speed is key to survival in a continent that abounds with agile predators.*

◐ **Left** *Three ostriches – two females and a male – stand guard over a nest in the Okavango Delta in Botswana, southern Africa. A shared, round-the-clock vigil is essential in ensuring that the eggs, lying in their vulnerable, shallow scrape, are not snatched by predators. The incubation period lasts for over 40 days.*

◐ **Below** *With his distinctive black-and-white plumage, a male Masai ostrich chases after two displaying females during the breeding season. The females' lowered necks and wings signal their receptivity to the male's advances.*

from several different nests usually combine into large groups, escorted by one or two adults. Only about 15 percent of chicks survive to be 1 year old, when they attain their full height. Females can breed at 2 years of age. Males start to acquire adult plumage when aged 2, and can breed at 3 or 4. They can probably live to be over 40.

Males defend their territories in the breeding season by patrolling, by displaying to and chasing out intruders, and by booming. Their call is surprisingly loud and deep, and is accompanied by inflation of the brightly colored neck, by repeated flicking of the wings, and by postures with both wings raised. Breeding males display to females by squatting and waving their huge, spread wings alternately, the so-called "kantle" display. Females solicit by lowering the head and both wings and quivering the latter. Groups of birds are usually small and not cohesive. Adult ostriches spend much of their time alone.

There are few bird species in which some individuals willingly look after the eggs of others, because natural selection often acts against such apparently altruistic behavior. The large size of ostriches and the vulnerability of their nests to predation are probably the factors selecting for such behavior in this species. An ostrich egg may be the largest of all birds' eggs in absolute terms, but it is nonetheless the smallest in relation to the size of the bird. As a result an ostrich can cover a great many – either more than a female can lay or more than it is worthwhile for her to lay, given the delays and risks involved. The skewed sex ratio among breeding adults, with about 1.4 females per male, and the high rate of nest destruction by predators both mean that there are many hens without their own nests to lay in. It obviously benefits them to lay somewhere. The major hen benefits from the presence of extra eggs in her nest, because her own are protected by a dilution effect against small-scale predators (in other words, her own eggs, probably a dozen among about 20, are less likely to be damaged).

If, as frequently happens, more eggs are laid in a brooding female's nest than she can cover, she rolls away the surplus at the start of incubation into an outer ring outside the nest, where they are not incubated and are doomed. As she is able to discriminate among the many eggs in the nest, she ensures that the eggs she rolls out are not hers. It is an astonishing feat of recognition, since ostrich eggs do not vary much in appearance.

Ratites – Flightless Birds

Seven families of flightless birds, five living and two extinct, are often grouped together as the ratites. They lack a large keel on the sternum (hence the word "ratite," from the Latin *rata*, meaning a raft, as distinct from the carinates, which have a keeled sternum and derive their name from *carina*, meaning a keel). The ratites also share other characters in the palate, the tongue, the pelvis, and the bill that differentiate them from other birds; in addition, adults of all species lack a preen gland.

Fossil evidence suggests that each family may have evolved flightlessness independently. In the Paleocene and Eocene eras, 65–34 million years ago, small, flying, ratitelike birds lived in the northern hemisphere. Whereas it was previously thought that the distribution of ratites in the southern hemisphere indicated that they evolved on Gondwanaland, it now seems possible that they evolved in the north and flew to the great southern continent, where they grew in size and became flightless. With the evolution of flightlessness, large flight muscles and their attachment areas became unnecessary – hence the loss of the keel.

The evolution of large size had a number of other consequences besides flightlessness. Big birds require large quantities of food if they are to remain primarily herbivores (as all but the kiwis have). Without the need to fly, weight is no longer such a limiting factor, and the birds have evolved large intestines that can handle the bulk of vegetable matter that their body size requires. Their legs lengthened to facilitate movement and to allow them to escape quickly from predators. Their necks also lengthened, to enable them to feed more easily on the ground. A longer neck also had the effect of counterbalancing the weight of their enlarged gut, so that the birds' center of gravity remained above the legs, keeping bipedal locomotion efficient.

Above *Prominent wings, a long neck and legs, and great flexibility enable the ostrich to have a great range of highly developed displays. Several different display postures are shown here:* ***1*** *a hen attacks with her wings fully spread;* ***2*** *a hen solicits during courtship;* ***3*** *a hen feigns injury; and* ***4*** *a cock struts about in a threatening posture.*

A Future in Farming?

CONSERVATION AND ENVIRONMENT

Ostrich feathers have long been used for adornment. In ancient Egypt, the ostrich feather, being symmetrical, was the symbol of justice (ostrich brains were also considered a delicacy). Pieces of eggshell are still used in Africa in necklaces and waistbands, and in some places whole eggshells are believed to have magical properties that protect houses and churches against lightning. More mundanely, the Khoikhoi people of southwestern Africa used empty eggshells as water containers.

High levels of predator density or of human activity make nests unlikely to survive. Delinquent hunting drove the once-abundant Arabian ostrich to extinction. Ostrich populations today are decreasing, as human intrusion into their habitat increases, but the species is not yet severely threatened. Many ostriches also survive on farms, where they are kept to harvest their feathers and meat, while their soft skin is well suited to making fine leather. SJJFD/BCRB

Right *Ostrich herding in South Africa. Ostriches have been commercially farmed for over a century. They were originally kept for their feathers, which were once in demand in the fashion industry, but now their meat and skin are the prime commodities.*

Rheas

RHEAS ARE LARGE, FLIGHTLESS BIRDS THAT are often referred to as the South American ostrich. Anatomically and taxonomically, however, rheas are quite distinct from the ostrich, and are more closely related to tinamous; the superficial similarity to the ostrich is a result of convergent evolution, both having adapted to life on open plains.

Rheas may stand up to 1.5m (5ft) tall, but most weigh no more than 40kg (88lb); in contrast, ostriches may reach over 2.5m (8ft) in height and weigh some 115kg (250lb). Apart from size, the most obvious difference between the two occurs in the feet: ostriches have only two enlarged toes, while rheas possess three.

Greater and Lesser

FORM AND FUNCTION

Charles Darwin was the first to recognize and describe the difference between the Greater and the Lesser rheas; one evening in the 1830s, as H.M.S. *Beagle* was making its way south along the Patagonian coast, the great biologist noted a difference in bone structure while eating rhea leg. The Greater or Common rhea was in fact once a widespread inhabitant of the grassland regions from central and coastal Brazil down to the pampas of Argentina, while the Lesser rhea is found on the semi-desert grass and scrublands of Patagonia and on the high-altitude grasslands of the Andes, from Argentina and Chile north through Bolivia into Peru. Each species has three distinguishable subspecies; in the Greater rhea, the smallest, from Brazil, weighs only 20kg (44lb), while the large Argentinian subspecies may weigh up to 50kg (110lb).

The Greater rhea tends to congregate in flocks of 10–100 birds for the winter months and to divide into smaller flocks of 2–7 birds for the breeding season, living at overall densities of 5–19 birds per sq km (13–50/sq mi). The adult rhea is largely vegetarian, feeding on a great variety of plants. It takes

Above *Two Greater rhea foraging in scrubland in the Pantanal, Brazil. Abundant vegetation on the pampas of South America allows both species of rhea to remain sedentary, unlike some of their ratite relatives in Africa and Australia.*

Below *The head of a Greater rhea, displaying classic ratite features – large eyes with excellent peripheral vision to spot danger and a wide, flat beak suitable for grazing. The rhea also has extremely acute hearing.*

some grass, but strongly prefers broad-leaved plants, and frequently eats even obnoxious weeds, such as thistles. Nowadays it feeds largely on forbs, especially lucerne (alfalfa), sorghum, rye, and introduced pasture grasses. Adults take very little animal food. The Lesser rhea, which is generally found in a drier or harsher environment, will consume almost anything green, but also prefers broad-leaved plants. It also eats a few insects and small animals, such as reptiles, when it has the opportunity.

The Child-rearing Male

BREEDING BIOLOGY

Prior to the breeding season (spring and summer), the male Greater rhea displays and fights with other males for possession of groups of females (harems). The breeding system is complex, with four classes of adult males: non-reproductive males; males that incubate only; males that copulate and incubate; and males that copulate only. At the start of the breeding season males construct a nest on the ground and attempt to entice a group of females to it by display and calling. Females move around in groups and are mated by several males during this phase, finally laying in or near one nest. The male on the nest gathers the eggs into his nest by rolling them with his bill. By this process he builds up a clutch of 10–70 eggs (means of 18–26 have been recorded at various localities). As the eggs accumulate, he becomes increasingly aggressive to approaching rheas and less and less willing to gather distant eggs into his nest. Meanwhile, having laid in one nest for several days, the females move on and may lay in other nests, but take no other part in the nesting cycle.

The male incubates for 36–37 days before the chicks hatch. They remain in the nest for a few hours and then leave with the male. In some cases a subordinate male is involved in the breeding group. He and the dominant male establish the nest, but once the clutch is complete the subordinate male incubates it while the dominant male makes another nest, attracts females to lay in it,

FACTFILE

RHEAS

Order: Rheiformes

Family: Rheidae

2 species in 2 genera

Distribution
S America, from Amazon to Patagonia.

GREATER RHEA *Rhea americana*
Greater, Grey or Common rhea
S America, from Amazon S to N Patagonia. Grasslands (both wet and dry). **Size: Height** 1.45m (4.8ft); **weight** typically 25kg (55lb), up to a maximum of about 50kg (110lb); females slightly smaller. **Plumage:** Gray, with white under wings and on rump; in breeding season males have a black bib and collar at the base of the neck. **Voice:** Female voiceless; male has booming call. **Incubation:** 35–40 days. **Longevity:** Less than 20 years in wild; up to 40 in captivity. **Conservation status:** Lower Risk/Near Threatened.

LESSER RHEA *Pterocnemia pennata*
Lesser or Darwin's rhea
S America, in Patagonia and Andes. Scrublands. **Size: Height** 90cm (35in); **weight** 10kg (22lb); females slightly smaller. **Plumage:** Brown, with white flecking throughout. **Voice:** Female voiceless; male has booming call. **Incubation:** 35–40 days. **Longevity:** Less than 20 years in wild; up to 40 in captivity. **Conservation status:** Lower Risk/Near Threatened.

1

2

3

Above *The two species of rhea both exhibit reversed sexual roles in rearing young.* ***1*** *Each male attempts to gather a harem of 2–12 females with his elaborate courtship displays. The male approaches females with his wings spread. Eventually the females begin to follow the dominant courting male. He then drives off all other males and displays vigorously. Once mating occurs, the male proceeds to build a nest consisting of a scrape in the ground surrounded by a rim of twigs and vegetation.*

2 *The females now proceed to lay eggs in the nest, each female laying an egg every second day for a week to ten days. After the second or third day the male remains at the nest and starts incubating the eggs. The females return at midday each day to lay their eggs beside the male who carefully rolls each new egg into the nest with his bill. The females move on from one male to another throughout the three-month breeding season. Males regularly incubate 10–70 eggs.*

Each male incubates the eggs and hatches the chicks alone. The chicks synchronize hatching and most hatch within a 36-hour period. ***3*** *The male must now lead the chicks to food, while protecting and brooding them.*

and incubates it himself. Subordinate males have few matings compared with dominant males; they incubate equally successfully, but are less successful at raising broods. Dominant males with a subordinate are also more successful at raising broods than males without subordinates. Each year many males do not breed at all, and in a given year only 4–6 percent breed successfully, although the proportion of females that do so is higher, at about 30 percent. The pattern of breeding in the Lesser rhea is not well studied, but is probably similar.

Rhea chicks feed largely on insects for their first few days, but gradually follow the example of their fathers and begin feeding on vegetation. Male rheas defend their chicks from all intruders, including other rheas; males with chicks are often seen driving females away. Male rheas with chicks have been known to attack small planes, and regularly charge gauchos on horseback; the threat of a rhea charge, which could cause a horse to shy and bolt, is one reason why most gauchos are accompanied by dogs. By the end of the summer, males, chicks, and females gather into large flocks for the winter. In spring, when males become solitary and females form small groups, the yearlings usually remain as a flock, which lasts until they are nearly 2 years old and ready to breed.

Few Enemies but Humans

CONSERVATION AND ENVIRONMENT

Whenever rheas have come into contact with humans, they have usually been persecuted. They have been hunted for years, and are one of the main animals caught with the gauchos' famous bolas. Rhea feathers are used to make feather dusters throughout South America. Rheas and their eggs are regularly eaten by local people, and the birds are also killed for use as dog food.

Rheas can coexist with cattle and sheep ranching, even though many ranchers accuse the birds of competing with cattle or sheep. In actual fact competition is probably minimal because rheas eat a lot of unwanted plant species, as well as some insects. Once land is put to arable use, however, the rheas are almost invariably eliminated, because they will eat almost any agricultural crop. They now thrive only in remote areas away from man, or where they are protected.

The decline of rhea populations has prompted the Convention for International Trade in Endangered Species (CITES) to list both of the rhea species and all subspecies as either Endangered or Near Threatened, thereby requiring permits for their export or import.

Rheas have few natural enemies other than human beings. The Greater rhea on the pampas of Argentina and in the savanna and riverine grasslands has a virtually limitless food supply. The large predatory cats like the jaguar and mountain lion do not regularly frequent these areas, and smaller predators cannot readily kill an adult bird. However, rhea chicks are vulnerable to a number of predators, including an array of mammals as well as birds of prey such as the caracara. While they are protected by the male, small chicks are safe, but are easily taken by predators if they get separated, as often happens after a sudden thunderstorm. Even the small, kestrel-sized chimango can be a successful predator when chicks have become separated from their parent. SJJFD/DFB

Below A protective male Lesser rhea with his brood in Torres del Paine National Park, Chile. Large groups such as this keep in touch through contact whistles. Lost chicks are often adopted by other males.

Emu

SO FAMILIAR AND WIDESPREAD IS THE EMU in Australia that the bird has been adopted as a quintessential symbol of the country, appearing opposite the kangaroo on its coat of arms. It is one of Australia's largest herbivores and occurs almost everywhere, though it is no longer present in Tasmania.

Having resided there for many millions of years, the nomadic emu is superbly well adapted to life in the harsh Australian outback. Mass migrations (see box) form a vital part of its survival strategy.

Tall and Fast

FORM AND FUNCTION

Emus are large, shaggy birds; their loose double feathers, in which the aftershaft (a secondary feather that branches from the base of the main feather) is the same length as the main feather, hang limply from their bodies. After molting the birds are dark, but as sunlight fades the pigments (melanins) that give the feathers their brown color, they become paler. Emu chicks are striped longitudinally with black, brown and cream, so they blend easily into long grass and dense shrubbery.

Emus' necks and legs are long but their wings are tiny, reduced to less than 20cm (8in). As adults, the birds develop an aperture between the windpipe and the air sac in their necks. This aperture allows them to use the air sac as a resonating chamber, which enhances the carrying quality of their booming call.

Searching Out Succulence

DIET

The emu prefers and seeks a very nutritious diet. It takes the parts of plants in which nutrients are concentrated: seeds, fruits, flowers, young shoots. It will also eat insects and small vertebrates when they are easily available, but in the wild it will not eat dry grass or mature leaves even if they are all that is available. It ingests

FACTFILE

EMU

Dromaius novaehollandiae

Order: Casuariiformes

Family: Dromaiidae

DISTRIBUTION Australia

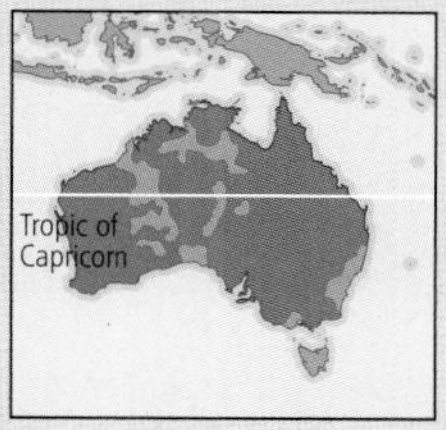

HABITAT In all areas except rain forest and cleared land; rare in deserts and the extreme north.

SIZE Height 1.75m (5.7ft); **weight** 50kg (110lb). Females weigh about 5kg (11lb) more than males.

PLUMAGE Dark after molting, fading to brown.

VOICE Grunts and hisses; females make resonant, booming sounds.

NEST A platform or circle of leaves, grass, bark, or sticks, on ground or under a bush or tree.

EGGS 9–20 eggs, incubated by male for 56 days.

DIET Shoots, seeds, flowers, and fruits of plants; some insects and small invertebrates.

CONSERVATION STATUS Emus are not considered threatened, although two island species, the King and Kangaroo Island emus, were hunted to extinction early in the 19th century.

Left *An emu traveling at a trot. Emus' long legs enable them to walk long distances at a steady 7km/h (4mph) or to flee from danger at a brisk 48km/h (30mph). Emus have three toes, in contrast to the ostrich, which has only two.*

Left *A male emu incubating a clutch of eggs. When the eggs are laid they are dark green, but darken to almost black during the eight-week incubation period. As here, the nest is often built below a tree or bush.*

Above *Emu males are renowned for the solicitous attention they give to their chicks, and for the fierce way they defend them against all comers. The chicks show the cryptic coloration and striping that give them excellent camouflage in grassland.*

large pebbles, up to 46g (1.6oz), to help its gizzard grind up its food, and also often eats charcoal.

Their rich diet enables emus to grow fast and reproduce rapidly, but at a price. Because such rich foods are not always available in the same place throughout the year, emus must move to remain in contact with their foods. In the arid Australian interior, the exhaustion of a food supply in one place often means moving hundreds of kilometers to find another source of food.

The emu shows two adaptations to this way of life. First, it lays down large stores of fat when food is abundant, which it is then able to use while searching for food, so that birds normally weighing 45kg (105lb) are still able to keep moving at bodyweights as low as 20kg (44lb). Second, emus are only forced to stay in one place when the male is sitting on eggs. At other times they can move without limitation, admittedly at a slow pace when with small chicks. Since the male does not eat, drink, or defecate during incubation, he is independent of the local food supply for this period.

The Male–Young Bond

BREEDING BIOLOGY

Emus pair in December and January, each couple defending a territory of about 30sq km (12sq mi). The female lays her clutch of 9–20 eggs in April, May, and June. Once the male starts sitting, many females move away, sometimes pairing with other males and laying further clutches. A few stay to defend the male on the nest, using their characteristic loud, booming call. When the chicks hatch after 56 days, the males become very aggressive; they drive the female away, and will attack approaching humans. The male stays with the chicks for 5–7 months, although they lead him rather than the reverse. After that time, the parent–young bond breaks down, and the male may then remate for the next season's nesting.

Rocky Relations

CONSERVATION AND ENVIRONMENT

Until the late 18th century there were several species and subspecies of emu. The dwarf emus of King Island (Bass Strait) and Kangaroo Island (South Australia), as well as the Tasmanian subspecies, were exterminated soon after Europeans settled in Australia. On mainland Australia, however, the emu remains widespread. It lives in eucalypt forest, woodland, mallee, heathland, and desert shrublands and sandplains. In desert areas it is rare, being found there only after heavy rains have induced the growth of an array of herbs and grasses and caused shrubs to fruit heavily. The emu also lives close to Australia's big cities, but is no longer found where native vegetation has been cleared to provide agricultural land.

Whatever the habitat, the emu must have access to fresh water, usually every day. Emus have probably benefited from man's activities in inland Australia, because the establishment of watering points for cattle and sheep has provided permanent water where there was none before. So much of Australia is unoccupied or used as open rangeland that the emu is in no danger of extinction.

One curious episode in emu–human relations occurred in 1932. Prompted by fears of mass incursions onto wheatfields in Western Australia, the government tried to cull emus there by deploying troops with machine guns. The birds were too elusive and resilient, however, and the "Emu War" ended in ignominious failure. SJJFD

NOMADS OF THE AUSTRALIAN MAINLAND

The emu's biology centers on its need to keep in touch with its food. In Australia the seeds, fruits, flowers, insects, and young foliage that the emu eats become available after rainfall. Emus therefore orient their movements toward places where rain has recently fallen.

Nomadic movements have three phases: initiation, orientation, and termination. A movement is initiated when emus meet many other emus each day. In stable conditions emus tend to avoid one another, but when food is short they move far each day in search of it and tend to meet many other emus. These frequent contacts initiate a movement away from that area. The orientation seems to depend on the sight of clouds associated with rain-bearing depressions, but sound cues from thunder and the smell of wet ground may also be involved. In Western Australia summer rain regularly comes from depressions moving west and south from the north coast, whereas winter rain comes from antarctic depressions moving up from the southwest. Emu movements therefore take on a regular pattern in the western half of Australia.

Once in a movement of this kind, the birds are trapped by their own behavior. Because all emus will be moving in the same direction, the tendency to move will be reinforced. Termination can only occur when the birds at the rear dawdle in areas of rich food and drift away from the movement. The number of birds involved may exceed 70,000, a ravaging horde as far as farmers are concerned. To protect the cereal-growing areas of the southwest from invasion from inland, a 1,000km (600mi) fence has been built.

Emu migrations may be a phenomenon generated by humans' own actions. The establishment of large numbers of artificial but permanent watering points in the inland, where cattle and sheep are grazed, has enabled emus to expand into places from which they were previously excluded by lack of water. They can now breed in places where they could not breed before.

The total emu population of Western Australia (between 100,000 and 200,000) is therefore larger than before, so that when food runs low many more emus are ready to move. With the regular alteration of weather systems, from the north in summer and the south in winter, very large numbers are sometimes drawn south in winter. By spring, as the crops ripen, they would reach the farms were it not for the barrier fence. Although most apparent in the west, similar movements occur in eastern Australia, but are not conspicuous unless the moving birds meet a barrier such as the wide Murray River.

Cassowaries

THERE IS SOMETHING OF THE MYSTERY OF THE jungle about the cassowaries. Many people have seen their footprints and their feces or even heard them call, but glimpses of the birds are rare. For the largest land animals in New Guinea, they have kept their secrets well; little is known of their life history.

The three species of cassowaries live in New Guinea. The largest, the Southern cassowary, also inhabits Australia's Cape York Peninsula, Ceram, and Aru Island. The Northern cassowary, only slightly smaller than the Southern, is confined to New Guinea. The moruk or Dwarf cassowary, little more than 1m (40in) tall, lives in New Guinea, New Britain, and Yapen Island.

Casques and Wattles

FORM AND FUNCTION

All three species have sleek, drooping, brown or black plumage. Their wing quills are enlarged, spikelike structures, used in fighting and defense. The three toes on their feet are also effective weapons, the inner toe being armed with a long, sharp claw: the kick of a cassowary has disemboweled many adversaries. In all species the legs and neck are long, and the head is adorned with a casque (higher in the female than the male). The casque is not horny, but composed of trabecular bone or calcified cartilage, covered with a keratinous skin. Its core is fused firmly with the skull. The neck is ornamented with colorful bare skin and small, fleshy flaps (wattles). The sexes are alike. In chicks the plumage is striped chestnut brown, black, and white, changing to a uniform brown for the first year of life. The glossy black adult plumage begins to grow during the second year and is fully developed at 4 years. The casque on the head is often thought to be used by the birds to push through the thick jungle. A captive bird was observed using it to turn over soil as it sought food, so perhaps wild cassowaries also use the casque to turn over the leaf litter, searching for small animals, fallen fruit, and fungi.

Cassowaries are essentially solitary animals, forming pairs in the breeding season but at other times living alone. The male incubates the 4–8 eggs in a nest on the forest floor. He accompanies the chicks for about a year before returning to his solitary life. The bird's most commonly-heard call is a deep "chug chug," but during courtship the male Southern cassowary approaches the female uttering a low "boo-boo-boo" call, circling her and causing his throat to swell and tremble. Cassowaries are pugnacious, and most contacts between wild individuals observed outside the breeding season culminated in fights. It is assumed from their distribution that individuals, at least of the Southern cassowary, are territorial.

Fallen Fruit

DIET

Cassowaries feed mainly on the fruit of forest trees, which they eat whole. The fruits of laurels, myrtles, palms, and *Elaeocarpus* are most important for the Southern cassowary. As the fruit of these trees grows high in the canopy and the cassowaries cannot fly, they are dependent on finding fallen fruit.

Unlike the other two species, which live exclusively in rainforest jungle, the moruk also has sparse populations in several mountainous parts of New Guinea. In these locations it apparently feeds on those shrubs and heathland plants from which it can take fruits directly. All species will also feed on insects, invertebrates, small vertebrates, and some fungi.

Above *A juvenile Southern cassowary; the brownish hue of the plumage changes to glossy black with age, and the keratinous casque grows larger. This individual is at least three years old, the age by which the birds assume their extravagant neck coloration.*

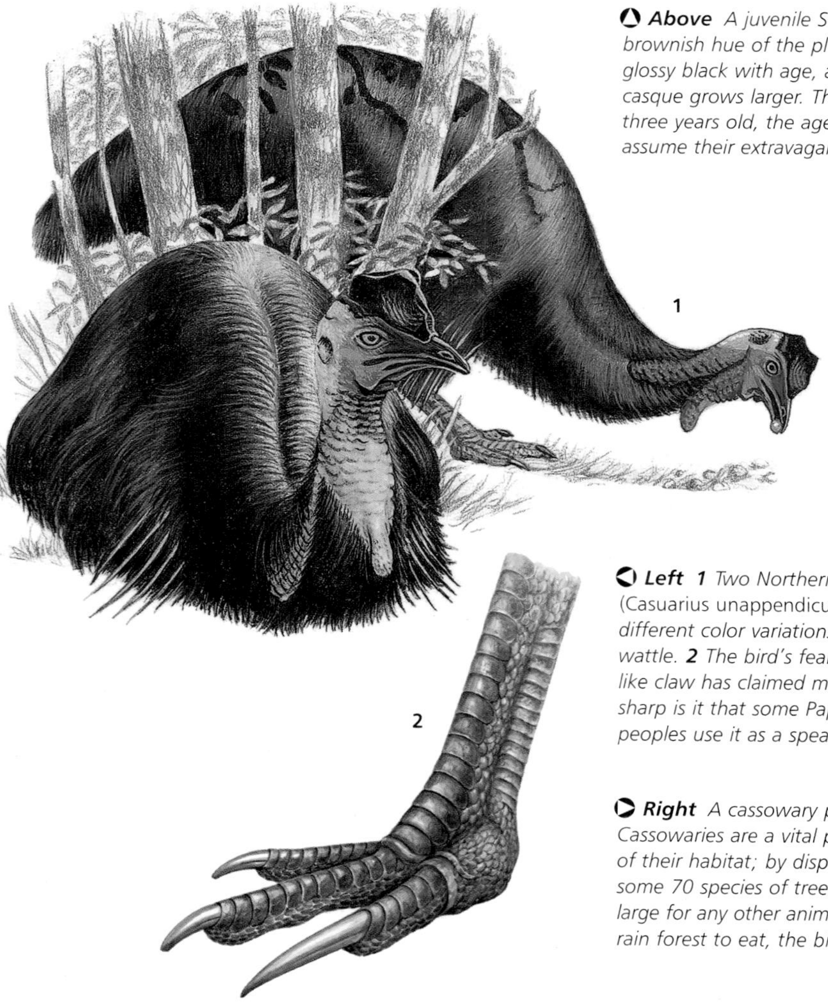

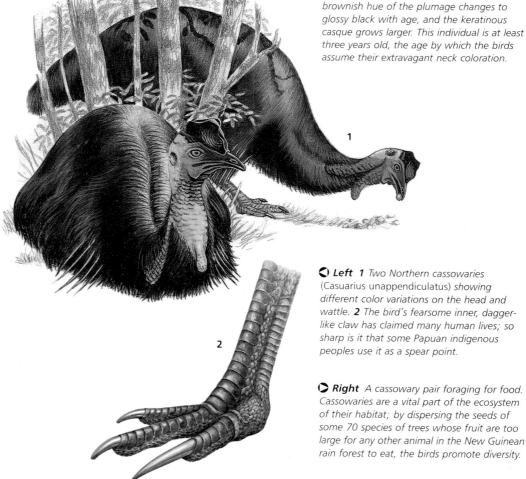

Left ***1*** *Two Northern cassowaries* (Casuarius unappendiculatus) *showing different color variations on the head and wattle.* ***2*** *The bird's fearsome inner, dagger-like claw has claimed many human lives; so sharp is it that some Papuan indigenous peoples use it as a spear point.*

Right *A cassowary pair foraging for food. Cassowaries are a vital part of the ecosystem of their habitat; by dispersing the seeds of some 70 species of trees whose fruit are too large for any other animal in the New Guinean rain forest to eat, the birds promote diversity.*

Sharing the Fate of the Forests

CONSERVATION AND ENVIRONMENT

Cassowaries need a supply of fruit throughout the year, so only large forests with a good diversity of tree species will sustain a population. Studies of the Southern cassowary on Cape York showed that it ate the fruits of 75 species, as well as fungi and land snails. Breeding (in winter) coincided with the period when the maximum amount of fruit was available in the forest. The survival of the cassowaries may therefore depend upon the survival of diverse forests where it is possible for them to obtain food throughout the year. Many forests used for timber production no longer retain their primitive diversity, so in a slow and subtle way they are growing less capable of supporting cassowaries.

Individuals of all three species are kept in captivity by the peoples of New Guinea. The birds' plumes are plucked and used to decorate headdresses, the quills are used as nose ornaments, and the whole animal is finally butchered for a feast.

Cassowaries have been traded throughout Southeast Asia for at least 500 years. When sugar plantations were developed in the late 19th century, cassowary populations were much reduced. The populations of the Southern cassowary on Ceram and of the moruk on New Britain probably arose from the breeding of escaped captives, brought from the main island of New Guinea. SJJFD

FACTFILE

CASSOWARIES

Order: Casuariiformes

Family: Casuariidae

3 species of the genus *Casuarius*: **Southern cassowary** (*C. casuarius*), **Northern cassowary** (*C. unappendiculatus*), **Dwarf cassowary** or **moruk** (*C. bennetti*).

DISTRIBUTION Australia, New Guinea, and adjacent islands.

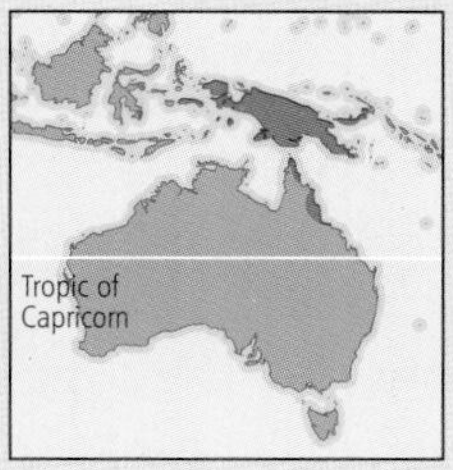

HABITAT Rain forest, swamp forest, montane forest.

SIZE **Height** 1.8m/5.8ft (Southern cassowary); 1.6m/5.3ft (Northern cassowary); 1.1m/3.8ft (Dwarf cassowary); **weight** c.70kg (154lb)

PLUMAGE Chestnut-brown in chicks and young birds; black in adults

VOICE Croaks, squeaks, howls, grunts, and snorts.

NEST A shallow scrape on the forest floor, lined with grass and leaves.

EGGS 4–8 eggs, incubated by the male for 50 days.

DIET Mostly forest fruits; also insects, invertebrates, small vertebrates, and some fungi.

CONSERVATION STATUS Both the Southern and Northern cassowaries are considered Vulnerable; the Dwarf is listed as Lower Risk/Near Threatened.

Kiwis

KIWIS EVOLVED ON NEW ZEALAND AROUND 30 million years ago. Yet this remarkable bird is under great threat from introduced predators, and is declining at a rate of almost 6 percent annually. Expressing this statistic in a more dramatic way, kiwi numbers are halved every decade.*

K iwis have taken flightlessness to extremes. Their tiny, vestigial wings are buried in their feathers, and externally their tails have entirely disappeared. The birds are exceptional in other ways. The hen lays an egg that may be as much as one-quarter of her weight and, unlike most other birds, kiwis use their sense of smell rather than sight to find food.

Hen-sized Ground-dwellers

FORM AND FUNCTION

The kiwi is the size of a domestic hen, but its body is more elongated and has stouter, more powerful legs. Its long, curved bill, which has openings for air passages at the tip, is used for probing the ground for food. There is no breastbone (sternum), to which in other birds the flight muscles are attached; other bones that are hollow in most birds so as to reduce weight for flight are only partly hollow in kiwis. Although their eyes are small for a nocturnal animal, kiwis can see well enough to run at speed through dense undergrowth.

Compared with its flightless relatives, the emu, ostrich, and rheas, the kiwi is a small bird and could probably have evolved only in the absence of mammals. The New Zealand archipelago was formed 80–100 million years ago, before the evolution of efficient land mammals, and by the time that they appeared a sea barrier prevented them from reaching New Zealand, protecting the kiwis and their ancestors from their competition and predation. Other flightless birds (the moas) also evolved in New Zealand, but all are now extinct.

The kiwi is able to detect food by smell, and uses its bill to probe among the forest litter, or for thrusting deep into the soil. It picks up food in the tip of its bill and throws it back to its throat in quick jerks. All species feed mainly on invertebrates that live in the soil and in leaf litter, especially earthworms and beetle larvae, but also take some vegetable material, including fruit.

Resting Out the Days

BREEDING BIOLOGY

Pairs of kiwis maintain territories of 20–100ha (50–250 acres). In these areas they use many dens, shelters, and burrows, in which they rest during the day, coming out at night to feed. The nest is in a burrow or sheltered place beneath dense vegetation, and has little lining. The clutch of one or two eggs represents a great investment by the female. The eggs are highly nutritious and not only sustain the embryos throughout the long incubation period (65–90 days) but also provide the newly hatched chicks with a yolk sac as a temporary food supply. After an egg has been laid, it is thought to be left unattended for several days, but once incubation begins in the Brown and Little spotted kiwis, it is the work of the male; in the Great spotted kiwi, both sexes incubate. There is some doubt whether parents feed the chicks; certainly within a week of their birth the chicks emerge from the nest alone and try to feed themselves.

Kiwis use calls to keep in contact in the dense forest, and also to maintain territories. At closer range the kiwi uses its sense of smell rather than sight, as well as its good hearing, to detect other birds. Strange kiwis are vigorously repelled. Breeding behavior includes loud grunting and snorting as well as wild running and chasing.

Left *Kiwis can sniff out worms and other prey items below ground through the nostrils at the end of their long beaks. So sensitive is their sense of smell that they can detect a few parts per million of scent.*

Below *The Great spotted kiwi lives in rugged terrain in the mountainous regions of South Island. Its preferred habitat has kept it relatively safe from predation and encroachment; some 10,000 pairs remain.*

FACTFILE

KIWIS

Order: Apterygiformes

Family: Apterygidae

3 species of the genus *Apteryx*: **Great spotted kiwi** (*A. haastii*), **Little spotted kiwi** (*A. owenii*), **Brown kiwi** (*A. australis*). The Brown kiwi has 3 subspecies: Stewart Island brown kiwi (*A. a. lawryi*), North Island brown kiwi (*A. a. mantelli*), South Island brown kiwi (*A. a. australis*).

DISTRIBUTION New Zealand

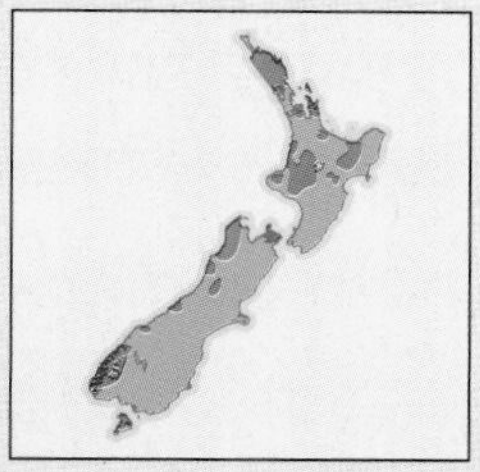

HABITAT Forest and scrub

SIZE Height 35cm/13.8in (Brown and Great spotted kiwis), 25cm/9.8in (Little spotted kiwi); **weight** 2.3kg/5lb (Great spotted kiwi), 2.2kg/4.8lb(Brown kiwi), 1.2kg/2.6lb (Little spotted kiwi). Females larger and up to 20 percent heavier than males in all species.

PLUMAGE Feathers streaked light and dark brown in Brown kiwi, banded light and dark gray in the other two species.

VOICE Loud, repetitive call, high-pitched in all kiwis except the female Brown kiwi, which has a rasping call.

NEST A hollow, unlined or lightly lined with leaves and leaf mold, in a burrow under dense vegetation.

EGGS 1 or 2, white, 300–450g (10.5–16oz). Incubation period 65–85 days.

DIET Invertebrates, including worms, spiders, beetles, insect larvae; plant material, especially seeds and fleshy fruits.

CONSERVATION STATUS All 3 species are listed by the IUCN as Vulnerable; the North Island subspecies of the Brown kiwi is Endangered.

Above A North Island brown kiwi drinking. This subspecies is found in remote regions, especially in the mixed area of primeval forest and farmland known as Northland. Its survival, along with that of other kiwis, is fostered by Operation Nest Egg, an initiative of the New Zealand Kiwi Recovery Program, in which eggs are taken from nests and hatched in incubators.

An Emblem in Danger

CONSERVATION AND ENVIRONMENT

The kiwi has always had great significance for New Zealanders. It provided the Maori with a source of food and of feathers for highly valued ceremonial cloaks. Today it has been adopted as the unofficial national emblem. Yet all kiwi species have suffered greatly since European settlement of the islands began in the mid-19th century. Large tracts of land were turned over to agriculture, and Europeans also brought with them mammalian predators such as cats and stoats. Dogs, which arrived with the earlier Maori colonizers, also attack kiwis.

One of the principal threats now confronting the Brown kiwi is land clearance over much of its range. The population on Stewart Island is doing well, but in the North Island only two substantial populations remain. The South Island population is fragmented, and its status is uncertain. Dedicated catching teams attempt to remove the kiwis before land is cleared, but cannot cope with the large areas involved. Fortunately this kiwi has adapted to some human-modified habitats, and also survives in the few large forest reserves within its range.

The Great spotted kiwi has only two isolated populations on the northwest of South Island, and these suffer from the gin traps that are set to snare introduced possums. The Little spotted kiwi is also endangered, and would be all but extinct but for the foresight shown in the transfer of this species to Kapiti Island, a 2,000ha (4,900-acre) island in Cook Strait. It is believed that the island's population, now numbering over 1,000 individuals, was established from as few as five liberated birds. The success of the liberation is all the more remarkable because at the time the habitat on the island was in a poor state.

Although Kapiti Island is a reserve, the Little spotted kiwi remains in a critical situation, with only one substantial population on one island. Attempts are being made to breed the species in captivity, both to learn more about its breeding biology and to make birds available for liberation on other islands. On Kapiti Island itself the species' habitat requirements, foods, and breeding were investigated, with an eye to suitable locations being selected for establishing further populations. This work led to the transfer of Little spotted kiwis to Hen, Long, and Red Mercury Islands in the 1980s and to the island of Tiritiri Matangi in the 1990s. SJJFD/JNJ

Tinamous

Restricted to the forests, woodlands, savannas, and grasslands of Central and South America, tinamous are secretive birds. The superficial resemblance of certain species of this ground-living group to gamebirds, especially the guineafowl, is no indication of their true relationship.

For some time they have been recognized as belonging to the superorder Palaeognathae (which includes the ratites), an association supported by recent analysis of their egg-white proteins, DNA, and tongue structure. The analysis of fossil evidence has recently shown that the Palaeognathae probably arose in the northern hemisphere, where Eocene fossils of flying forebears have been found. These flying ancestors would have been capable of distributing the group throughout the world, and their existence eliminates the need to suppose that ratites and tinamous walked to their present southern hemisphere ranges. The earliest fossil tinamous come from the Pleistocene (1 million years ago) of Buenos Aires and Minas Gerais, and they are now widely distributed from Mexico to southern Chile.

Clumsy Movers

FORM AND FUNCTION

To the casual observer tinamous outwardly recall the guineafowl in their proportions and carriage, but differ in that the bill is slender, elongate, and slightly decurved. Further, the rear of the body appears arched owing to the great development of the rump feathers, which normally obscure the short tail. The legs are thick and powerful, and possess three forward-pointing toes and one rear toe. The latter is reduced and elevated, or even absent in some species. Although the legs are perfectly adapted for running, the birds soon become tired when chased and often stumble. Their flying ability also leaves much to be desired, for though they have well-developed flight muscles, their flight is clumsy and they often collide with obstacles, sometimes resulting in injury or death. It has been suggested that this apparent weakness is associated with a proportionately small heart and narrow blood vessels.

Because of their limited running and flying capabilities, tinamous rely on their protective coloration to avoid detection. They remain motionless with head extended, attempting to blend in with the vegetation, and will creep away from danger using all available cover. Species living in open areas are known to hide in holes in the ground. At all times they only break cover, or take to the wing, at the last possible moment. The tinamous are not the easiest of birds to see in their natural environment, their presence being indicated only by their persistent, flutelike, whistling calls, which can be uttered by day or night.

Detailed information is available about the diet of some nothuras and the Red-winged tinamou. The crop and stomach contents that have been analyzed indica te that they feed mainly upon seeds, fruits, and other vegetable matter such as roots and buds. Insects and other small animals may also be taken, especially by *Nothoprocta* species. The Red-winged tinamou takes mainly animal food in the summer and mostly vegetable food in the winter. Tinamous use their bill to find food, digging in the ground with it and using it to break up friable termite mounds; unlike game birds, they do not scratch for food with their feet.

Sex Roles in Reverse

BREEDING BIOLOGY

As in ratites, the sexual roles of the tinamous are often reversed; the males take care of the eggs and young, while the female displays the most aggression. In most species courtship involves follow-feeding and the rump-up display. In follow-feeding, the male leads the female(s) through his territory, making feeding movements as he goes. The rump-up display exhibits the colorful plumage around the bird's tail, which is usually hidden by the long rump feathers. The male builds the nest and attracts females to it. Interestingly he will scrape out a nest with his feet in the manner of ratite males, but never uses them to dig for food. Once the eggs are laid the females take no further part in the nesting cycle.

Polygamy is common. One or more females may lay eggs in a single nest, and occasionally a hen will deposit her eggs in different nests tended by different males. Most breeding tinamous maintain territories, varying in size from 0.1 to 30ha (0.25–75 acres), correlating with the size of the

Below *Representative tinamou species:* ***1*** *Red-winged tinamou* (Rhynchotus rufescens) *– when an enemy appears, tinamous stand motionless with their heads held high.* ***2*** *Great tinamou* (Tinamus major) *foraging. Hunting and the loss of forests threaten the future of this species.* ***3*** *Small-billed tinamou* (Crypturellus parvirostris) *chasing after an insect meal.* ***4*** *Little tinamou* (Crypturellus soui).

◐ ***Left** A male Highland tinamou* (Nothocercus bonapartei). *Tinamous are ground nesters; males usually build the nests and apparently always take on the job of incubating.*

species and the productivity of the environment.

The species that live in tropical forests nest in many months of the year, while others may have their egg-laying season governed by rainfall and other climatic factors. All are ground nesters and either lay their eggs directly on the ground between roots in a shallow scrape or else construct a nest of grass and sticks. Clutches are said to consist of 1–12 eggs, although the latter figure may be the result of two females using one nest. The eggs are relatively large and are renowned for their hard, porcelainlike gloss and vivid, clear coloring, with blue, green, violet, and brown hues predominating. During the nest period males often become so tame that they can be picked up off the nest. In the Tataupa tinamou, the cock will pretend to be lame in order to distract an intruder away from the nest, while other species may cover their eggs with leaves, as in *Tinamus* species, or with feathers, as in *Nothoprocta*.

Chicks are downy and buff in color, with darker stripes and mottles. They are well developed, and are capable of running soon after hatching. The chicks grow quickly and, in most species, are able to fly to some degree before they are half grown. Some species can breed seven weeks after birth.

Shy, Cryptic, and Sought for Food

CONSERVATION AND ENVIRONMENT

Throughout their range, tinamous are sought after as food because their meat, although strangely translucent in appearance, is very tender and full of flavor. A number of species are localized in distribution, so that concerns have been expressed for their survival. Deforestation is also a major concern for some species. Others have adapted rapidly to agricultural development, while a few have become agricultural pests. The variety of species and forms in South and Central America is very great; at least 155 subspecies have been described, some of them based on only a few specimens. This group of shy, cryptic birds presents a great challenge for ornithological study.
SJJFD/CAW

FACTFILE

TINAMOUS

Order: Tinamiformes

Family: Tinamidae

46 species in 9 genera. Species include: **Great tinamou** (*Tinamus major*), **Magdalena tinamou** (*Crypturellus saltuarius*), **Tataupa tinamou** (*C. tataupa*), **Red-winged tinamou** (*Rhynchotus rufescens*), **Spotted nothura** (*Nothura maculata*).

DISTRIBUTION S Mexico to southern S America.

HABITAT Forest, woodland, savanna, grasslands up to 4,000m (about 13,100ft).

SIZE Length 14–53cm (6–21in); weight 43–2,300g (1.5–81oz). Females are slightly larger than males.

PLUMAGE Cryptic, mainly browns and grays; usually barred, streaked, or spotted; 2 species have prominent crests. Females are usually brighter, lighter, or more barred than males, sometimes with different leg coloration.

VOICE Loud, mellow, flutelike whistles and trills.

NEST All species are ground-nesters, either building nests of grass and sticks or laying eggs between tree roots in a shallow scrape.

EGGS Usually 1–12; glossy green, blue, yellow, purplish-brown, or nearly black; weight 20–100g (0.7–3.5oz). Incubation period 19–20 days.

DIET Mainly seeds; fruits and insects also recorded.

CONSERVATION STATUS 2 species – the Magdalena tinamou (*Crypturellus saltuarius*) from Colombia, and Kalinowski's tinamou (*Nothoprocta kalinowskii*) from Peru – are listed as Critically Endangered, and both may in fact be extinct. In addition, 5 other species are listed as Vulnerable.

3

4

Penguins

The penguins are a group of flightless seabirds occurring in the southern hemisphere, most numerously in the Antarctic and sub-Antarctic. They form a distinctive group, highly specialized to their marine environment and to the extreme and often harsh regions they inhabit. They are also very attractive birds, thanks to their smart appearance and energetic behavioral displays.

Penguins were first described during the voyages of exploration undertaken by the Portuguese seafarers Vasco da Gama (1497–98) and Ferdinand Magellan (1519–22), which discovered Jackass and Magellanic penguins, respectively. However, most species remained unknown until the 18th century, when the southern ocean was explored in search of the continent of Antarctica.

Diving in Penguins

Penguins are probably better adapted for life at sea than any other group of birds. Decades of research have yielded a wealth of information on their swimming and diving abilities.

Although they were once credited with swimming speeds of up to 60km/h (37mph), all accurate measurements of normal swimming for Emperor, Adélie, and Jackass penguins give speeds of 5–10km/h (3–6mph), although in short bursts, and particularly when "porpoising" (a swimming motion in which penguins briefly leave the water), faster speeds may well be achieved. The purpose of porpoising is uncertain; it might serve to confuse underwater predators, achieve faster speeds by reducing drag when traveling in air, or allow breathing without hindering rapid movement.

Diving behavior has been studied in a number of penguin species by using pressure-sensitive depth recorders. Diving duration varies widely among species: in Gentoo, Chinstrap, and Macaroni penguins, dives usually last 0.5–1.5 minutes, seldom exceeding 2 minutes; in Jackass and Emperor penguins, however, the mean dive time is 2.5 minutes, with dives lasting up to 5 minutes in Jackass and over 18 minutes in Emperor penguins. Dive depths also vary greatly both within and between species, ranging in Emperor penguins, for example, from 45m (150ft) to a staggering 530m (1,740ft). In a study of King penguin dives, half were deeper than 50m (165ft), and two greater than 240m (790ft). In one study of Chinstrap penguin dives, in contrast, 90 percent were shallower than 45m (150ft) and 40 percent less than 10m (33ft), although the birds can dive to depths of 100m (330ft). Similarly, in Gentoo penguins, 85 percent of dives were less than 20m (66ft). Adélie penguin dives range from 20–170m (66–560ft).

How do such relatively small animals manage frequent deep diving? Diving patterns are influenced by three main factors: firstly submergence time, dictated by how long the breath can be held; secondly, the effect of increased pressure at depth; and thirdly, temperature regulation and the ability to minimize heat loss. Breath-hold time is determined by the body's oxygen store. Penguins the size of an Adélie use about 100cc of oxygen per minute at rest, and at that rate their store would be exhausted in a 2.5-minute dive. However, muscles can function on a greatly reduced supply of blood and oxygen during dives; the heart beat is reduced from 80–100 beats per minute when resting to 20 beats per minute, allowing the same oxygen store in practice to sustain a dive of 5–6 minutes.

Human divers who breathe compressed air are vulnerable to inert gas narcosis and the "bends," but in breath-hold diving the inert gas supply is very limited and the risks are much less. Penguins on short, shallow dives should have no problems, but how Emperor and King penguins cope with their long dives to substantial depths is unknown.

Penguins seem well-adapted for reducing heat loss during dives. However, most of their feather insulation is probably due to trapped air, and much of this is expelled under pressure during diving in the form of a trail of bubbles behind the bird, so penguins may need to remain very active in cold water in order to keep warm.

Little is known about how, or how often, penguins catch prey underwater. Species that feed mainly on krill and similar crustaceans have to catch very small prey, usually less than 3cm (1.2in) long. It has been calculated that Chinstrap penguins, which average 191 dives per day during foraging trips when rearing chicks, need to catch 16 krill per dive, or one

Above *In a welter of bubbles and diving bodies, King penguins plunge after prey. This species is renowned for its deep-diving abilities.*

every 6 seconds. In contrast, King penguins only need to catch a single squid or fish on 10 percent of the 865 dives they typically make on each trip.

The nature of the prey may also influence dive duration. Penguins eating relatively slow-moving, swarming crustaceans may be able to catch a lot even on short dives. Birds that pursue fast-moving squid and fish, on the other hand, may need to submerge for much longer. However, they do not need to succeed often to satisfy their food needs. This fact may explain why the dives of Jackass penguins, which feed mainly on surface-shoaling fish, are far longer than those of Gentoo penguins, which also make shallow dives but feed on krill.

FACTFILE

PENGUINS

Order: Sphenisciformes

Family: Spheniscidae

17 species in 6 genera: **Emperor penguin** (*Aptenodytes forsteri*), **King penguin** (*A. patagonicus*), **Royal penguin** (*Eudyptes schlegeli*), **Snares crested penguin** (*E. robustus*), **Erect-crested penguin** (*E. sclateri*), **Fiordland penguin** (*E. pachyrhynchus*), **Macaroni penguin** (*E. chrysolophus*), **Rockhopper penguin** (*E. chrysocome*), **Little penguin** (*Eudyptula minor*), **Yellow-eyed penguin** (*Megadyptes antipodes*), **Adélie penguin** (*Pygoscelis adeliae*), **Chinstrap penguin** (*P. antarctica*), **Gentoo penguin** (*P. papua*), **Galápagos penguin** (*Spheniscus mendiculus*), **Humboldt penguin** (*S. humboldti*), **Jackass penguin** (*S. demersus*), **Magellanic penguin** (*S. magellanicus*).

DISTRIBUTION Antarctica, New Zealand, S Australia, S Africa, S America N to Peru and Galápagos Islands.

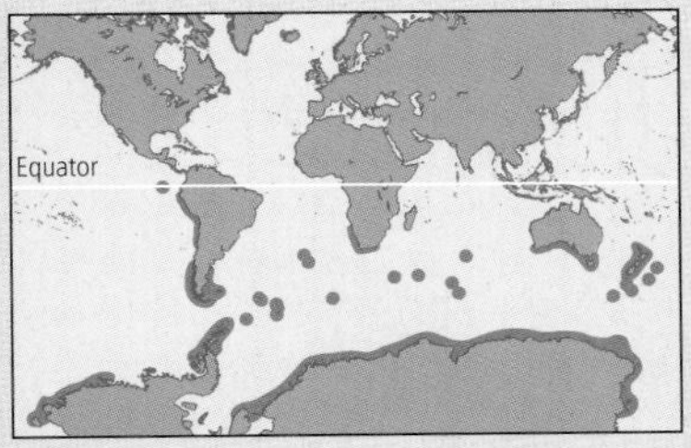

HABITAT Marine only, using terrestrial habitats including ice, rock, islands, and coasts for breeding.

SIZE **Height** ranges from 30cm (12in) in the Little penguin to 80–100cm (32–39in) in the Emperor penguin; **weight** from 1–1.5kg (2.3–3.3lb) to 15–40kg (33–88lb) in the same 2 species.

PLUMAGE Most are dark blue-black or blue-gray above, and whitish below.

VOICE Loud, harsh trumpeting or braying calls.

NEST None in the 2 largest species, the King and Emperor penguins, which incubate standing up; all others employ some sort of nest, made of whatever materials are available.

EGGS 1 or 2, according to genus; whitish or whitish-green. Incubation period 33–64 days according to species.

DIET Crustaceans, fish, squid.

CONSERVATION STATUS 3 species – the Erect-crested, Yellow-eyed, and Galápagos penguins – are currently listed as Endangered; 7 more are Vulnerable.

Birds Built for Swimming

FORM AND FUNCTION

All penguin species are remarkably similar in structure and plumage, but quite variable in size and weight. Their plumage is chiefly blue-gray or blue-black above and mainly white below; species-distinguishing marks, such as crests, crowns, face and neck stripes, and breast bands, are chiefly on the head and upper breast, and so are visible while the birds are swimming on the surface. Males are usually slightly larger than females, those of crested penguins notably so, but outwardly the sexes appear very similar. Chicks are mostly gray or brown all over, or have one of these colors along the back with white on the sides and undersurface. Juvenile plumage is usually very similar to that of adults, only differing in minor ways, for example the distinctiveness of the ornamentation.

Penguin form and structure is highly adapted for marine life; the birds have streamlined bodies and powerful flippers that help them swim and dive. They are densely covered by three layers of short feathers. Their wings are reduced to strong, narrow, stiff flippers, with which they achieve rapid propulsion through water. The feet and shanks (tarsi) are short, while the legs are set well back on the body and are used in the water, with the tail, as rudders. On land, penguins frequently rest on their heels, with their stout tail feathers forming a prop. The short legs induce a waddling gait when on land, but some species move rapidly on ice by tobogganing on their bellies. Despite their inefficient-looking gait, however, some species may travel long distances on foot between their breeding colonies and the open sea. They have comparatively solid bones and, for the most part, weigh only a little less than water, thus reducing the energy required to dive. Bills are generally short and stout, and have a powerful grip. Emperor and King penguins possess long, slightly downcurved bills, possibly adapted for capturing fast-swimming fish and squid at great depths.

As well as having to swim efficiently, penguins must also keep warm in cool, often near-freezing waters. For insulation they have, in addition to a dense, very waterproof feather coat, a well-defined fat layer and a highly developed "heat-exchange" system of blood vessels in the flippers and legs, ensuring that venous blood returning from exposed extremities is warmed up by the outgoing arterial blood, thus reducing heat loss from the body core. Tropical penguins tend to overheat easily, so they have relatively large flippers and areas of bare facial skin that allow them to lose excess heat. They also live in burrows so as to reduce direct exposure to sunlight.

All penguins have an impressive capacity for storing substantial fat reserves, especially before the molting period, when birds spend all their time ashore and cannot feed. Some species – the Emperor, King, Adélie, Chinstrap, and the crested penguins – also undertake long fasts during the courtship, incubation, and brooding periods of the breeding season. During the fasts, which may last for up to 115 days for brooding male Emperor penguins and 35 days for Adélie and crested penguins, individual birds may lose nearly half their initial body weight. By contrast, male and female Gentoo, Yellow-eyed, Little, and Jackass penguins usually swap incubation or brooding duties every 1–2 days and therefore do not undertake long fasts during the main part of the breeding season. In almost all species, however, the parents fatten quickly at the end of the breeding season once the chicks have been reared, in preparation for the 2–6 week molting period, during which time their fat reserves are used twice as fast as in incubation. In Jackass and Galápagos penguins the molt period is less defined, occurring at any time between breeding attempts. Immature birds usually complete the molt before breeding birds start; and at least in crested penguins, the timing of this molt becomes later with age until the first breeding attempt. The differences between penguin species in breeding and molting behavior are driven at least in part by differences in their habitats, and particularly by the short breeding season available to the more southerly species that inhabit colder environments.

Below Representative species of penguins. **1** An adult Yellow-eyed penguin with two chicks. **2** A pair of Rockhopper penguins brooding a chick. **3** Little blue penguins (Eudyptula minor) coming ashore. **4** Two incubating King penguins, one arranging its egg on its feet. **5** A pair of Adélie penguins greeting each other. **6** Adélie penguins tobogganing, **7** leaping out of the sea, **8** porpoising. **9** A Jackass penguin standing and **10** coming ashore.

Chilling Out

DISTRIBUTION PATTERNS

Penguins breed in habitats ranging from sub-Antarctic grasslands and Antarctic sea ice to the bare lava shores of equatorial islands, sandy subtropical beaches, and cool temperate forests. They are, however, basically adapted to cool conditions, and in tropical areas mainly occur where cold-water currents exist, for example within the influence of the Humboldt Current along the western coast of South America and of the Benguela and Agulhas currents around South Africa. Most species occur between the latitudes of 45° and 60° South, with the highest species diversity in the New Zealand area. The greatest numbers of penguins live around the coasts of Antarctica and on the sub-Antarctic islands.

During the breeding season, some species, for example Gentoo penguins, may only travel 10–20km (6–12 miles) from their colonies, whereas others, such as the Emperor penguin, may cover distances of over 1,000km (620 miles) on each foraging trip. Winter distributions and movements are less well known, but tropical and warm temperate species do not migrate. In contrast, some species may travel hundreds of kilometers from their breeding sites during the winter.

Underwater Predators

DIET

The main prey of penguins are crustaceans, fish, and squid, which they chase, catch, and swallow underwater. Different species have different food preferences, which reduces competition among sympatric species; for example, Adélie and Chinstrap penguins favor different sizes of krill. Fish form an important part of the diet of such inshore feeders as the Jackass, Little, and Gentoo penguins, and also of the deeper-diving King and Emperor penguins. The Antarctic silverfish (*Pleuragramma antarcticum*) features in the diet of many penguin species. Squid may predominate in the food of some populations of King penguin and are frequently taken by Emperor and Rockhopper penguins and some *Spheniscus* species. In some sub-Antarctic locations, such as the Falkland Islands, species with a greater foraging range (e.g. Magellanic and Rockhopper penguins) are endangered by conflict with commercial fishing.

Antarctic krill are the principal prey of Adélie, Chinstrap, Gentoo, and Macaroni penguins, and other crustaceans are important to Yellow-eyed and Rockhopper penguins (and probably also to the other crested penguins). In common with many other minute oceanic animals (zooplankton), krill tend to be absent from surface water during the day, which is the time when penguins rearing chicks on this food are mainly at sea. However, some species remain at sea to feed during the night, traveling to and from their colony during the day; alternatively, they use their diving ability to seek prey at depth in the daytime.

Above *Royal penguins occur only on Macquarie Island, in the Southern Ocean midway between Australia and Antarctica. Here, some 850,000 breeding pairs gather in large colonies. They leave the island in May, to spend all winter at sea.*

Life in the Colonies

BREEDING BIOLOGY

Most penguins are highly social, both on land and at sea, and often breed in vast colonies, only defending small areas around their nests. Courtship and mate recognition behavior are most complex in the highly colonial Adélie, Chinstrap, Gentoo, and crested penguins, least so in the species that breed in dense vegetation, such as the Yellow-eyed penguin. Despite living in burrows, Jackass penguins, which usually breed in dense colonies, have fairly elaborate visual and vocal displays; those of Little penguins, whose burrows are more dispersed, are more restrained. The social behavior of these penguins is largely nest-oriented; in contrast Emperor penguins, which have no nest site, show only behavior oriented to their partners and offspring. The great variation in the sequence and patterning of their trumpeting calls provides all the information needed for individuals to recognize one another, even in colonies that may contain many thousands of penguins. A King penguin arriving back at the colony goes near to its nest site, calls, and listens for a response. King and Emperor penguins are the only species in which the two sexes can easily be distinguished by their calls.

The complex behavioral displays of many penguin species are commonly seen at the onset of the breeding season, during courtship. Most penguins normally mate with their partners of previous years: in a colony of Yellow-eyed penguins 61 percent of pairings lasted 2–6 years, 12 percent 7–13 years, and the overall "divorce" rate was 14 percent per annum; in Little penguins one pairing lasted 11 years and the divorce rate was 18 percent per annum. In a major Adélie penguin study, however, no pairing lasted six years and the annual divorce rate was over 50 percent.

Macaroni penguins breed first when at least 5 years old; Emperor, King, Chinstrap, and Adélie penguins are at least 3 (females) or 4 (males), and Little, Yellow-eyed, Gentoo, and Jackass penguins at least 2. Species can, however, be very variable in age at first breeding. For example, in Adélie penguins, very few 1-year-olds visit the colony; many 2-year-olds come for a few days around chick-hatching time, but most birds visit first as 3- and 4-year-olds. Up to about 7 years of age, Adélie penguins arrive progressively earlier each season, make more visits, and stay longer. Some females first breed at 3 years of age, males at 4, but most females and males wait another year or two, and some males do not breed until aged 8.

The timing of breeding is mainly affected by environmental conditions. Antarctic and most sub-Antarctic and cold-temperate penguins breed in spring and summer. Breeding is highly synchronized within and between colonies. In Jackass and Galápagos penguins there are usually two main peaks of breeding, but laying occurs in all months of the year. This is also true of most populations of Little penguins, and in South Australia some pairs are even able to raise broods successfully twice a year. Emperor penguins are unique in their breeding cycle, laying their eggs in the fall and raising their chicks on the ice through the dark Antarctic winter, when temperatures may fall to –40 °C

Left Gentoo penguins with their chicks in December (the southern hemisphere summer) at a colony on the Falkland Islands. Gentoo penguins and the more northerly of the crested penguins have longer breeding seasons and more variation in the timing of their breeding than other species.

Below So well-established is this Rockhopper penguin rookery that toenail grooves made by generations of birds mark the paths that ascend from the shore.

(–40°F). King penguin chicks also overwinter at the breeding colony, but are rarely fed during this period and grow mainly during the previous and subsequent summers. In most species, males come ashore first at the start of the breeding season, establishing territories in which they are soon joined by their old partners or by new mates that they attract to the nest site.

Only Emperor and King penguins lay a single egg; other species normally lay two. In the Yellow-eyed penguin and probably more generally, age affects fertility, so that hatching success in a study colony was 32 percent, 92 percent, and 77 percent of eggs incubated by birds aged 2, 6, and 14–19 years respectively. In species that lay two eggs, hatching is usually staggered, with the first, slightly larger egg hatching first, and this sequence can promote brood reduction, usually by favoring the first hatched chick. (Brood reduction is a widespread adaptation ensuring that, when food is scarce, the smaller chick dies quickly and does not prejudice the survival of its sibling.) Yet, in crested penguins, the first egg of the clutch is far smaller than the second and only one chick is ever reared; only in the Fiordland penguin do both eggs normally survive to hatch. Although several hypotheses have been suggested to explain this unusual phenomenon, none is wholly satisfactory.

There are two distinct periods during chick-rearing in most penguins. First is the brood period lasting 2–3 weeks (6 weeks in Emperor and King penguins), when one parent stays and guards the small chick while the other goes in search of food. This is followed by the crèche period, when chicks are larger and more mobile and form crèches while both parents are away foraging. Crèches may be large aggregations, as in Adélie, Gentoo, Emperor, and King penguins, or small ones involving only a few chicks from adjacent nests, as is the case with Chinstrap, Jackass, and the crested penguins.

Inshore-feeding species such as Gentoo penguins feed their chicks each day. Adélie, Chinstrap, and crested penguin parents, however, are often away at sea for more than a day at a time, so the chick may receive fewer meals. King and Emperor penguin chicks are fed large meals at infrequent intervals, seldom more often than once every three or four days. The Little penguin is unusual in feeding its chicks well after nightfall; as the smallest species it presumably has the shallowest diving capacity, and may be more dependent on feeding around dusk, when a greater proportion of its prey is near the surface.

Chicks grow rapidly, particularly in Antarctic species. Meal sizes increase as the chicks get older, and in larger species meals delivered to older chicks may exceed 1kg (2.2lb) in weight. However, the capacity of even quite young chicks of small penguins is astounding – they can easily accommodate 500g (over 1lb) of food. For much of their early growth, they are little more than pear-shaped food sacks, with large feet and a small head.

Once the molt is complete, chicks usually start going to sea. In crested penguin species, there is a rapid and complete exodus from the colony (almost all leave within one week), and almost certainly no further parental care takes place. In Gentoo penguins, free-swimming chicks return periodically to shore, where they obtain food from their parents for at least a further 2–3 weeks. Some such parental care may also occur in other species, but it is unlikely that chicks are ever fed by their parents at sea.

Once the chicks have fledged, they spend little time at the colony until they return for their first breeding attempt. Juvenile survival in penguins is relatively low, particularly during the first year after chicks fledge and again at the onset of breeding; for example, only 51 percent of Adélie penguin chicks usually survive their first year. This low juvenile survival is balanced by higher adult survival rates, estimated at around 91–95 percent for Emperor and King penguins, a level similar to that of other large seabirds. Survival of smaller penguins is lower, being 70–80 percent for Adélie, 86 percent for Macaroni and Little, and 87 percent for Yellow-eyed penguins.

Below As dusk falls, two Little penguin chicks wait for their parents to return to the nest with food. Not only does this species maximize its chances of catching krill by fishing after dark, it may also avoid day-active land predators by only coming ashore at night.

The Antarctic Refuge

CONSERVATION AND ENVIRONMENT

Although flightless, adult penguins have few natural predators on land because they usually choose isolated breeding sites and their beaks and flippers are effective weapons (they are, however, vulnerable to larger introduced mammals). Eggs and chicks are taken by skuas and other predatory birds. At sea, Killer whales, Leopard seals, and other seals and sharks catch penguins, but the extent of this predation is at most of local significance. In fact, populations of several species of Antarctic penguins (especially Chinstraps) have increased appreciably, following the massive reduction in baleen whales (and hence the great increase in available krill). Some King penguin populations have also grown, recovering at least partially from the time when they were killed for the oil extracted from their blubber. In the past many penguin populations were reduced (and colonies eliminated) by egg collecting for human consumption; in most cases this is not a real problem today. The true Antarctic species have not been exploited, and populations have stayed stable or even increased in recent years. Away from the Antarctic, habitat loss, competition with fisheries and introduced species, and climate change represent the main threats to penguin survival.

Three penguin species are currently Endangered. The Galápagos penguin breeds only on two islands in the Galápagos Archipelago. Once mainly threatened by feral dogs, it was severely depleted in the 1980s and 1990s by El Niño. The numbers of the Yellow-eyed penguin have fluctuated wildly, chiefly because of changing land use and other human disturbance in the coastal dune systems of

New Zealand where it breeds. Populations on offlying islands, where protection is more feasible, have suffered less, and even on the mainland conservation measures have managed to halt habitat decline. The Erect-crested penguin, which breeds in the same area as the Yellow-eyed penguin but has a small breeding range consisting of just two main locations, has declined by 50 percent in 45 years.

Seven other penguin species are considered Vulnerable. For example, both the Jackass and the Humboldt penguins occur in highly productive oceanic systems where an upwelling of nutrients supports large fishing industries. Populations of these species have decreased alarmingly, initially due to egg removal and guano collection and subsequently because of competition for food; the fishing industries off the west coasts of South America and South Africa both depend, as do penguins, on anchovies and pilchards. Only a compromise between commercial fishing and conservation will ensure these species' survival.

All penguins are highly vulnerable to oil pollution, especially the Jackass penguin, many of whose colonies lie near tanker routes around the Cape of Good Hope. Oiled Jackass penguins have been cleaned and returned to their colonies to breed, but this is a time-consuming and expensive operation. Similarly, thousands of Magellanic penguins have died from oil pollution in the Straits of Magellan. In addition, commercial fisheries, whether existing ones for anchovies or developing ones for krill, pose an increasing threat. KEB/JPC

Below *The sheer remoteness and wildness of penguin habitats have ensured that many species have escaped the threats posed by human encroachment.*

SPECIAL FEATURE

SURVIVING AT THE EXTREMES

Breeding strategies in large penguins

WHEN THEY ARE BREEDING, EMPEROR PENGUINS endure the coldest conditions encountered by any bird: the frozen wastes of the Antarctic sea-ice, where the average temperature is –20°C (–4°F) and the mean windspeed is 25km/h (16mph), sometimes reaching 75km/h (47mph). Emperor penguin breeding colonies form mainly on stable, fast sea-ice around the coast of Antarctica in the austral autumn (March–April); the birds may have walked over 100km (62 miles) on ice to reach the sites. After courtship, females each lay one large egg in the month of May; this is incubated by the male for 64 days, during which time the female returns to sea. When the egg hatches, both parents feed the chick for a period of 150 days, from late winter through spring, so the young are ready for independence in the summer before the sea-ice returns.

This breeding arrangement prompts two questions. First, why do Emperor penguins raise their young at the worst time of year? Second, how do the penguins survive the winter conditions?

The answer to the first question seems to be that if Emperor penguins were to breed in the Antarctic summer (a short season of only four months) they would not complete their long breeding cycle before the onset of winter. Even when the chicks fledge, in late spring, they have only attained 60 percent of adult weight – the lowest proportion for any penguin – and juvenile mortality is high. The adults, however, are able to breed annually.

The means whereby Emperor penguins survive in harsh conditions are several remarkable physiological and behavioral adaptations, all arising from the need to minimize heat loss and the expenditure of energy. Their body size and shape give a relatively low surface-to-volume ratio, and the flippers and bill are 25 percent smaller as a proportion of body size than in any other penguin species. Heat loss is further reduced by extreme proliferation of the blood-vessel heat-exchange system, which is twice as extensive as that of any other penguin. Blood vessels flowing to the feet and flippers lie very close to the veins returning blood to the penguin's body core, thus warming the returning blood and allowing the outgoing blood to cool, resulting in minimal heat loss. Emperors also recover heat in the nasal passages through exchange between inhaled cold air and warm air to be exhaled (around 80 percent of the warmth is retained), and by the excellent insulation provided by the very long, multi-layered, high-density feathers that completely cover their legs.

Because open water lies far away across the ice shelf in winter, feeding is extremely difficult. As a result, changeovers at the nest are infrequent and long fasts, of up to 115 days in males and 64 days in females, are essential. The penguins' large size allows them to store the substantial fat reserves needed for these periods of dearth.

Nevertheless, the crucial adaptation is huddling. By reducing activity to a minimum and gathering in large groups of as many as 5,000 birds squeezed together at 10 per sq m (about 11 per sq yd), adults and chicks achieve a 25–50 percent reduction in individual heat loss. The huddle as a

Above *Four ice-encrusted Emperor penguins huddle together for warmth. Once many birds have packed close together against the biting cold, the temperature inside a mass huddle can reach 35°C. This vital huddling instinct makes the Emperor the only penguin species that is not territorial.*

Left *Around 195,000 pairs of Emperor penguins breed in 35 colonies along the Antarctic ice-shelf. As well as enduring frequent blizzards in these inhospitable surroundings, the birds have to contend constantly with katabatic winds that blow down off the Antarctic plateau and further intensify the wind-chill factor.*

Right *Staying on the parent's feet, where it is protected by a warm skin fold, is crucial to a chick's survival; those that fall onto the ice die of exposure within a few minutes.*

whole moves very slowly downwind, and there is regular movement also within the group; the birds positioned to windward move along the flanks and then into the center until they are once again exposed at the rear, so that no birds are continually exposed on the aggregation's edge. Such mobility is feasible only because Emperor penguins have developed the ability to move with their egg balanced on their feet, where it (and, subsequently, the young chick) is covered and insulated with a pouchlike fold of abdominal skin. Another key social adaptation of Emperor penguins is that they have suppressed nearly all aggressive behavior.

King penguins, the other large species, have evolved a very different solution to the problem of breeding in the short summers. They take over a year for a successful breeding attempt and rarely breed successfully more than two years out of three. They have two main laying seasons, in November–December and February–March, during which a single large egg is laid. Both parents incubate the egg and guard the young chick once it hatches (after around 54 days), alternating shifts every few days. Most of the time, any given colony contains adults, eggs, and chicks, at many stages of molt, incubation, and growth respectively.

From the eggs laid in November–December, chicks are reared to 80 percent of adult weight by April and are fed sporadically (undergoing fasts of two months or so, with an overall chick weight-loss of about 40 percent) through the winter. Regular feeding resumes in September and continues until December, when the chicks take their leave. The adults then have to molt, and cannot lay again until February–March of the following year. Chicks from these late eggs are far smaller when the winter comes, and they do not fledge until the following January–February; in practice, many die. KEB/JPC

Loons or Divers

So highly specialized for swimming are the loons that are unable to walk properly on land. With an average of 45 seconds for each immersion – although they can stay underwater for much longer if necessary – they are among the world's most accomplished divers. These skilled aquatic predators hunt by sight, using their feet for propulsion, although they will also occasionally use their wings when turning.

"Diver" – the British name for this group – is self-explanatory, but the American "loon" is thought to derive from the old Norse *lømr*, meaning lame. The term describes the birds' shuffling movement on land; their legs are placed so far to the rear that walking is difficult on a flat surface, although they can readily run uphill or across water.

Underwater Predators

FORM AND FUNCTION

The Red-throated loon is the smallest and slimmest member of the family, and is characterized by a slightly upturned tip to its lower mandible and a rusty-red throat patch when in its breeding plumage. At breeding time, its back is plain gray, but is finely spotted with white in the winter. Unlike other members of the family, which require a long run across the water to become airborne, this species can take off easily from small bodies of water, permitting it to nest in tiny pools, some less than 10m (33ft) across.

The bird's vocalizations are unlike those of other loons and are generally of low frequency. Several resemble those of waterfowl, as suggested by the names given to two of the sounds: "quack" and "kark." Population declines have been reported for both eastern and western populations of this species over the last decade.

In its handsome breeding plumage, the Arctic loon has a glossy black throat patch that shimmers with a green sheen. A row of white feathers runs down the back of the neck; the head feathers are soft gray; and there are 10 or 11 prominent rows of large white spots in distinctive patches along the scapulars. During the winter, the back is a spotless gray, the head gray with white on the underside of neck and belly, and there are extensive white patches along the flanks.

The Pacific loon was formerly considered a subspecies of *Gavia arctica* until the two were discovered breeding sympatrically, first in Siberia and more recently in Alaska. The feathering is nearly similar in the two species, although the Pacific bird's throat sheen is purplish rather than greenish, a difference that is difficult to spot in poor light conditions. In winter, the best way of telling the two species apart is by the minimal white feathering decorating the Pacific loon's flank.

Above *A Red-throated loon taking off. Loons are able to concentrate oxygen in their leg muscles to sustain them during dives as deep as 60m (200ft).*

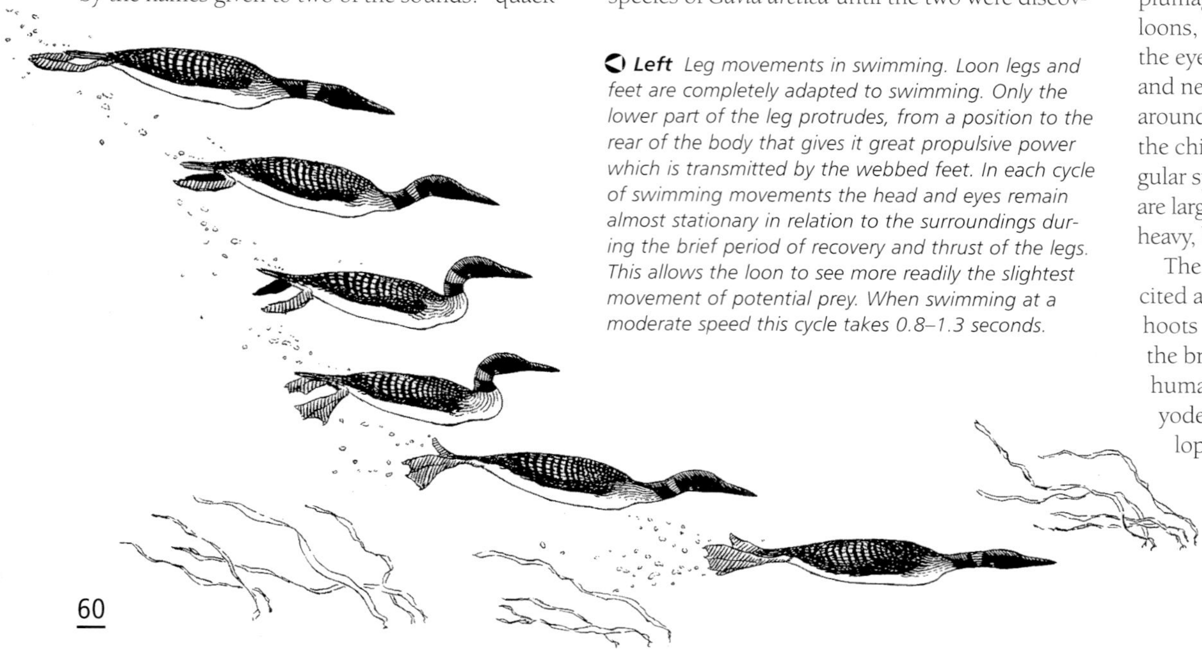

Left *Leg movements in swimming. Loon legs and feet are completely adapted to swimming. Only the lower part of the leg protrudes, from a position to the rear of the body that gives it great propulsive power which is transmitted by the webbed feet. In each cycle of swimming movements the head and eyes remain almost stationary in relation to the surroundings during the brief period of recovery and thrust of the legs. This allows the loon to see more readily the slightest movement of potential prey. When swimming at a moderate speed this cycle takes 0.8–1.3 seconds.*

The Common loon is an altogether larger, stockier bird, with a heavier bill. Its winter plumage is similar to that of Arctic and Pacific loons, but with some white feathering anterior to the eyes. Breeding plumage includes a black head and neck, with one partial row of white feathering around the sides of the neck and a second under the chin; the bird also exhibits bold white rectangular spots along the back in transverse rows that are largest on the scapulars, and a white belly. The heavy, black bill has a slightly downcurved culmen.

The Common loon's vocalizations are often cited as its most distinctive feature. These various hoots and wails (which are all used primarily in the breeding season, and sound mournful to the human ear) famously include a kind of tremolo yodel. This call is given by males to warn interlopers off their territory or drive them away from their young. The Yellow-billed loon is the largest of all the species. It is similar in appearance to the Common loon, although in its breeding plumage there are two or three fewer rows of white rectangular spots over the back, and the individual spots are larger; in winter there is a dark spot behind each eye. The bill is large and ivory-colored with a straight culmen, giving it an upturned appearance. Vocalizations are similar to those of the Common loon, but the tremulous laughs of the alarm calls are slower and of a lower frequency.

Birds of the Northern Hemisphere

DISTRIBUTION PATTERNS

The Red-throated loon ranges across the circumpolar Arctic southward into temperate latitudes. The Arctic loon has two subspecies: *G.a.arctica*, which breeds across western Europe, and *G.a. viridigularis*, in northern Siberia with a small population in Alaska. The closely-related Pacific loon is found primarily in western North America but also in Siberia. The Yellow-billed loon is found across the circumpolar Arctic eastward through Alaska and Canada. Common loons have the widest distribution of all, being found across the northern United States, northward throughout Canada and Alaska, and into Baffin Island, Greenland, and Iceland.

All five species breed on freshwater lakes and winter mainly in coastal waters. While the Red-throated loon may utilize quite small pools, the Arctic and Yellow-billed loons require large bodies of water.

Freshwater Breeding Grounds

BREEDING BIOLOGY

Copulation occurs close to shore. The two chicks of a normal clutch usually hatch within 24 hours, and leave the nest permanently within a day of the younger one hatching. The older chick limits its travels to short swims until both chicks are ready to leave, and for warmth and protection may be brooded under the parental wing or on parents' backs for up to two weeks.

Red-throated loons usually lay 2 eggs that are incubated for about 27 days, although less for the second-laid than for the first. After hatching, competition between the young for food and brooding is severe, and the second-hatched chick frequently perishes. The small ponds where the birds nest seldom contain enough fish to feed the chicks, and adults often fly several kilometers to larger lakes or rivers or to the ocean to obtain food for themselves and their young.

The young of Arctic and Pacific loons first take flight at 50–55 days and leave the nesting territory at 57–64 days. During the winter these two species spend more time in offshore waters and in larger flocks than do other species. For nesting they prefer large lakes with good fish populations, but are able to use smaller ones if necessary; they are thought to take more aquatic invertebrates than other loons.

Large freshwater lakes with quiet bays and good fish populations provide preferred Common loon breeding territories, although the birds may use ponds 4–5ha (10–12 acres) in extent if nothing larger is available. The young are able to secure some fish at 8 weeks, and take flight at 11 weeks, although the adults continue to supplement their diet and provide protection for another 3 months.

Yellow-billed loon nests are located adjacent to the water, preferably on small islands or hummocks resembling the nest sites of other loons; in some parts of their range, they are reported primarily along rivers. The birds breed in lakes of many different sizes, ranging in Alaska from 0.1 to 229ha (0.25–565 acres). The time the chicks take to fledge is unknown. JM/JB

FACTFILE

LOONS OR DIVERS

Order: Gaviiformes

Family: Gaviidae

5 species of the genus *Gavia*: **Arctic loon** or **Black-throated diver** (*G. arctica*), **Common loon** or **Great northern diver** (*G. immer*), **Pacific loon** or **Pacific diver** (*G. pacifica*), **Red-throated loon** or **Red-throated diver** (*G. stellata*), **Yellow-billed loon** or **White-billed diver** (*G. adamsii*).

DISTRIBUTION N hemisphere; N America, Greenland, Iceland, Eurasia.

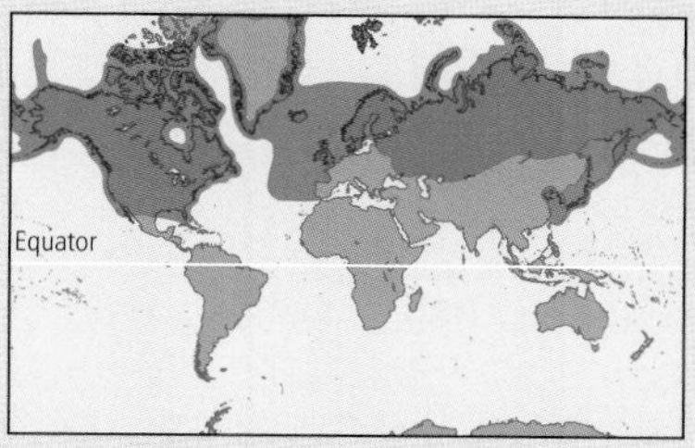

HABITAT The birds breed in freshwater lakes and winter in coastal waters.

SIZE Length 53–91cm (21–36in).

PLUMAGE Breeding plumage variable above, but always with vertical white stripes on neck, the location differing among species; winter plumage grayish above; belly plumage white in adults of all species and in all seasons.

VOICE Highly vocal; duetting between pair members in all species.

NEST Close to shore on solid ground; in mound of damp vegetation when available.

EGGS Usually 2, dark brown to olive, spotted or blotched black or dark brown. Incubation 24–29 days by both pair members.

DIET Mainly fish; also crayfish, shrimps, leeches, frogs, insect larvae.

CONSERVATION STATUS Not threatened.

Left *Representative loon species:* ***1*** *Red-throated loon or diver* (Gavia stellata)*;* ***2*** *Arctic loon or Black-throated diver* (G. arctica)*;* ***3*** *Common loon or Great northern diver* (G. immer).

Grebes

GREBES ARE DUCK- OR COOT-LIKE BIRDS THAT inhabit lakes and marshes. They are almost exclusively aquatic and have representative species on all continents except Antarctica, and are found from sea level to over 4,000m (13,000ft). Fifteen of the 22 species inhabit the Americas.

The grebes are an old group, dating back around 70 million years. Their smooth-edged bills and lobed feet make it unlikely that they diverged from birds with "toothed" bills and webbed feet. Certain characteristics in the neck musculature and the shape of the breastbone suggest that their closest relatives may in fact be the coots (*Fulica*: Rallidae) and the finfoots (Heliornithidae).

Underwater Hunters

FORM AND FUNCTION

Thanks to anatomical adaptations, grebes are well suited to the rigors of aquatic life and underwater hunting. Their feet are at the extreme hind end of the body – the tail is reduced to a downy tuft – and have exceptional flexibility in the ankle and toe joints, allowing them to pivot in all directions and to be used simultaneously as paddles and rudders. The unilateral lobes on the toes further aid maneuverability – a diving grebe can move at about 2m (6.6ft) a second and can turn extremely quickly. When swimming, grebes move their feet parallel to the water surface, and only beat with one foot at a time except when in a hurry. The rear edge of the tarsus is serrate, perhaps an adaptation to cut through submerged vegetation. Grebes can sink low in the water by expelling the insulating air from between their feathers and emptying their air sacs (reservoirs of air); this reduces the energy needed to keep them submerged and allows them to dive silently when hunting and to hide underwater when they are frightened. Also, their flank feathers are modified to absorb water, further lessening buoyancy during diving. Dives typically last 10–40 seconds.

Feet placed so far back make even standing difficult; grebes only stand at the nest and, if obliged to reach nests "stranded" by falling water levels, frequently fall when trying to walk. They need a long take-off run across water to become airborne, but fly quickly with rapid beats and trailing feet; they maneuver poorly in flight. Grebes rarely fly except on migration. Three species and a subspecies are permanently flightless, but many others are so for a large part of the year, partly as a result of the simultaneous molt of the flight feathers (as in wildfowl and cranes) and partly owing to a buildup of fat reserves at one time of year and of leg muscles at another. They may migrate long distances, usually flying at night, when they sometimes mistake wet roads for rivers, land on them, and become stranded.

Sharing Out the Spoils

DIET

Grebes are carnivorous, eating mainly insects and fish but also some mollusks and crustaceans; they take the latter from off and around aquatic water plants or, more rarely, from the bottom. In murky water, grebes spot their prey from below. The larger species chase fish, and the Western and Clark's grebes of North America spear (rather than grab) fish with their dagger-shaped bills. Although fish may form a large part of the food mass ingested in

Below *Representative species of grebes:* **1** *The Little grebe* (Tachybaptus ruficollis) *has no ornamental plumage.* **2** *The Great crested grebe* (Podiceps cristatus) *performs a "discovery" or "cat" ceremony as part of its courtship display.* **3** *Slavonian grebe* (P. auritus) *with its summer coat.* **4** *Pied-billed grebe chick* (Podilymbus podiceps) *riding on its parent.* **5** *Pair of Western grebes* (Aechmophorus occidentalis).

Left *A Slavonian grebe settles on its clutch of eggs. Eggs are laid at intervals of 1–2 days and incubation starts from the first or second egg. Both parents incubate equally and when disturbed the sitting grebe usually covers the clutch with nest material. Although conspicuous when newly laid, the eggs are quickly camouflaged by mud and dirt.*

Below *A Red-necked grebe carrying chicks on its back. Young grebes rely on their parents for food as well as warmth and protection; they ride on their parents' backs for some weeks (except during dives), but after about 10 days the parents may split up, taking one or two chicks each. The young can fly at 6–9 weeks, but in double-brooded species they often stay longer and help their parents feed younger siblings.*

5

many grebes, the quantities of invertebrate items in their stomachs bear witness that they spend much more time catching invertebrates than fish, and some grebes do not eat fish at all. The thick bill of the Pied-billed grebe may be an adaptation to feeding on crabs and crayfish.

Grebe species form "guilds" of aquatic carnivores, vying with each other to exploit foods more efficiently. In Eurasia, for example, the Great crested grebe occurs mostly on open water, eating fish that it usually catches within 8m (26ft) of the surface, whereas Little grebes occur on small ponds covered with floating water-plants, which they are small enough to dive among. Intermediate-sized species, such as Red-necked and Slavonian grebes, may be restricted to lake habitats where they do not compete with larger species. For example, the Slavonian is the only grebe that breeds in Iceland, and there it eats many fish as well as insects, but in Alaska it restricts its

FACTFILE

GREBES

Order: Podicipediformes

Family: Podicipedidae

22 recent species in 7 genera. Species include: **Alaotra grebe** (*Tachybaptus rufolavatus*), **Little grebe** (*T. ruficollis*), **Madagascar grebe** (*T. pelzelnii*), **Least grebe** (*T. dominicus*), **Pied-billed grebe** (*Podilymbus podiceps*), **Hoary-headed grebe** (*Poliocephalus poliocephalus*), **New Zealand grebe** (*P. rufipectus*), **Slavonian** or **Horned grebe** (*Podiceps auritus*), **Red-necked grebe** (*P. grisegena*), **Great crested grebe** (*P. cristatus*), **Black-necked grebe** (*P. nigricollis*), **Silvery grebe** (*P. occipitalis*), **Junín flightless grebe** (*P. taczanowskii*), **Hooded grebe** (*P. gallardoi*), **Western grebe** (*Aechmophorus occidentalis*), **Clark's grebe** (*A. clarkii*).

DISTRIBUTION N and S America, Eurasia, Africa, Australasia.

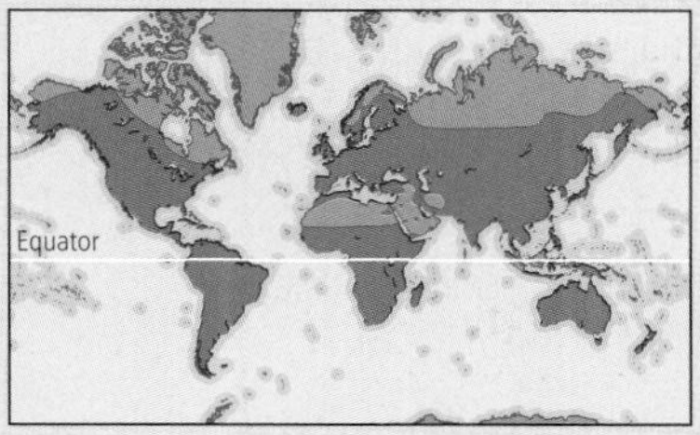

HABITAT Freshwater lakes, marshes, and brackish bays; many species use coastal waters in winter.

SIZE Length ranges from 20cm (8in) in the Least grebe to 78cm (31in) in the Western grebe; weight from 112g (4oz) to 1.8kg (4lb) in the same two species. Females are slightly smaller than males.

PLUMAGE Upperparts mostly drab gray or brown, shading to white on the underbelly; during the breeding season the head, throat, and neck are often brightly colored. Some species have bright, colored tufts and crests on the head, used in courtship. After growing feathers on most of their body, the young maintain a downy, striped head and neck pattern for several months, presumably to avoid aggression from adults.

VOICE A variety of whistling and barking calls; species living in dense vegetation are often more vocal and have replaced part of the ritualized courtship sequences with remarkably well-synchronized duets.

NEST A floating platform made of rotting vegetation and attached to waterweeds.

EGGS Usually 2–4, at high latitudes 3–8, at first light blue but soon becoming white or cream. The eggs are covered by a fine layer of calcium phosphate, allowing them to "breathe" when wet. Incubation period 22–23 days, but because of asynchronous hatching a nest may hold eggs for up to 35 days.

DIET Aquatic insects, crustacea, mollusks, and fish, less commonly tadpoles and annelid worms.

CONSERVATION STATUS 2 species – the Colombian (*Podiceps andinus*) and Atitlán grebes (*Podilymbus gigas*) – are now listed as Extinct, having died out in the 1970s and 1980s respectively; the Alaotra grebe of Madagascar, officially still Critically Endangered, has not been seen since the 1980s. The Junín grebe, which is restricted to a single lake in the highlands of Peru, is also Critically Endangered, and the Madagascar and New Zealand grebes are both Vulnerable.

diet largely to insects and fish fry due to competition from the Red-necked grebe for larger prey; similarly, in eastern Siberia and Alaska a long-billed race of the Red-necked grebe has evolved that takes larger fish than its European counterpart, which has to compete with Great crested grebes for food. Another case of character displacement is that of the Silvery grebe, which has a different bill shape in Peru's Lake Junín, where the similar Junín flightless grebe occurs, than elsewhere in its large range.

Body feathers are numerous and in the Western grebe may amount to 20,000 or more. They are frequently preened and impregnated with oil from the tufted uropygial gland. The feathers are shed throughout the year. Most grebes have a peculiar habit of eating feathers and at the same time drinking large quantities of water. The feathers may fill half the stomach to form a feltlike lining, which is regurgitated at intervals together with indigestible parts of the food. The feather-eating habit may be an adaptation against

__Right__ Complex displays occur when grebes are courting, though the composition of sequences differs from species to species. In the Great crested grebe mating includes: __1__ the discovery display; __2__ the head-shaking display; __3__ preening; __4__ the weed ceremony.

__Below__ Western grebes performing their courtship display known as the "rushing" ceremony. In this display both the male and female rise out of the water, lunge forward and run rapidly side-by-side with their necks arched and bills pointed upward.

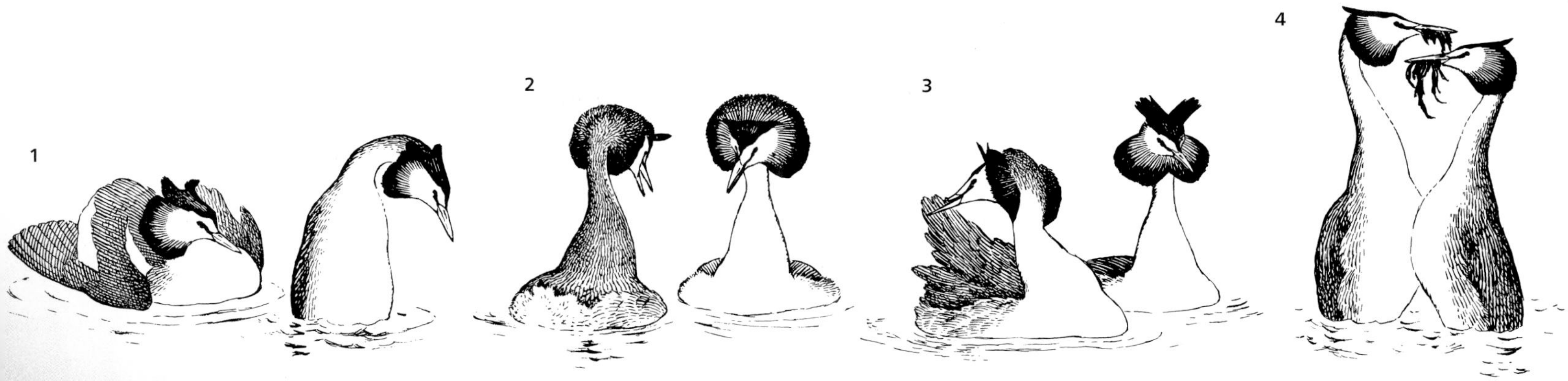

parasites that are abundant in grebes owing to their varied diet. Why the New Zealand and the Hoary-headed grebes do not eat feathers remains a mystery.

The Ceremonies of Courtship

SOCIAL BEHAVIOR

The courtship behavior of grebes is very striking, involving complex sequence of elaborate, ritualized postures and, particularly in *Podiceps* species, much use of the erectile feather ruffs and tufts on the head. Sequences may include running on the water side-by-side, simultaneous diving and surfacing with waterweeds in the bill, and a breast-to-breast standing up while the head is turned from side to side. Sir Julian Huxley's study of Great crested grebe courtship (1914) was a seminal paper in animal behavior. Detailed studies of the courtship of the recently discovered Hooded grebe (1974) have confirmed its close relationship to Black-necked, Silvery, and Junín flightless grebes. The crucial role that these displays serve in pair formation has been emphasized by recent findings on the Western and Clark's grebes: these two species are so similar in morphology, choice of habitat, and distribution that they have been considered color variants (morphs) of one species. However, only birds of the same color morph will pair together, and this segregation is achieved by each morph using a distinctly different "advertising call" to initiate courtship. An interesting feature of grebe displays is that males and females may reverse their normal roles, even to the extent of reverse mountings.

Most grebes are aggressively territorial. Some nest in colonies, but still keep their distance in the feeding areas. A few, however, such as the Hoary-headed grebe, are decidedly social at all times, and Junín flightless grebes hunt in groups, moving in a line and diving synchronously. The Hooded grebe is also very peaceful for a grebe. In winter or during staging, some grebes gather in large numbers. Nearly a million Black-necked grebes, the most numerous of all grebes, gather on a single lake in California in autumn.

The timing of breeding may be flexible – grebes at tropical latitudes seem adapted to exploit opportunistically a good food supply rather than being tied to a specific season. In Africa Little grebes and in Australia Hoary-headed grebes may appear and start breeding within a few days of unpredictable rainstorms producing temporary flood-ponds.

The Hooded grebe is unique in that out of the two eggs it lays, it only takes a single chick away from the nest, abandoning the second egg altogether if the first hatches successfully.

Grebes spend a lot of time sunbathing, often by lying sideways in the water to expose the white belly. The black skin under their feathers quickly absorbs the heat.

Eliminating the Grebe Trade

CONSERVATION AND ENVIRONMENT

Formerly the larger grebes were extensively hunted for "grebe fur," the white belly skin, which was used for women's shoulder capes and mufflers. This led to the near extirpation of these grebes in Western Europe, but they were then imported from other parts of the world. The grebe trade became a major issue for the emerging societies for bird preservation, and thanks to protection the large grebes have become common once again. Grebes are considered ill-tasting, so they are rarely hunted for food.

Presently seven species of grebe are listed in the Red Data Book. A combination of factors are threatening these, including wetland drainage and pesticide pollution. The flightless Atitlán grebe and apparently the Colombian grebe are already extinct, and the Alaotra grebe of Madagascar appears to be beyond salvation. The changing water levels caused by a hydroelectric plant, as well as pollution from mining, threatens to destroy the habitat of the Junín flightless grebe, which is confined to a single lake in the highlands of Peru. The New Zealand grebe is now restricted to the North Island, where the total population numbers about 1,700–1,800 birds. The Madagascar grebe is also giving cause for concern, but its status is not yet critical. Its numbers have declined steadily for half a century owing to reduction of habitat and the introduction of exotic herbivorous fish, which change the habitat sufficiently for a competing grebe (Little grebe) to establish. NKK

Albatrosses

S*UPERSTITIOUS SAILORS ONCE REGARDED the albatross as the repository of the souls of drowned comrades, and believed that to kill one was to court terrible misfortune. Samuel Taylor Coleridge's famous poem* The Rime of the Ancient Mariner *recounts how disaster befalls a ship after an albatross is shot. Even so, many 19th-century sailors were happy to catch and eat these birds to vary their monotonous diet on long voyages, and to turn their feet into tobacco pouches and their wing bones into pipes. The name "albatross" comes from the Portuguese* alcatraz, *used originally for any large seabird and apparently derived from* al-cadous, *the Arabic term for pelican.*

Albatrosses are distinguished from other families of their order (Procellariiformes) by the position of their tubular external nostrils, which lie at each side of the base of the bill rather than being fused on its top. They can be split into four genera: the "great" albatrosses, comprising six *Diomedea* species, which have wingspans averaging 3m (about 10ft); nine smaller species in the genus *Thalassarche,* which are often called "mollymauks" (from the Dutch *mollemok*, a name originally given to the fulmar); four North or tropical Pacific species in the genus *Phoebastria*; and the all-dark Sooty and Light-mantled sooty albatrosses of the genus *Phoebetria*, which have relatively long wings. Recent molecular analysis has increased the number of recognized species from 14 to 21.

Left *The Northern Buller's albatross* (Diomedea bulleri platei) *breeds on the Chatham Islands, New Zealand and is highly migratory, ranging across the southern Pacific from its breeding sites to Chile and Peru. It is then thought to use the Humboldt Current to travel north before returning home.*

Champion Gliders

FORM AND FUNCTION

Albatrosses are renowned for their effortless flight, as they follow ships for hours with barely a wing-beat. One adaptation to reduce the muscular energy expended during gliding is a special sheet of tendon that locks the extended wing in position. Another is the sheer length of the wing, in which the forearm bones are much longer in relation to the hand than in other procellariiforms. To it are attached 25–34 secondaries, as compared to just 10–12 in storm petrels. As a result, the wing is a very efficient aerofoil, the high aspect ratio (ratio of wing length to front-to-rear width) permitting fast forward gliding with a low sinking rate. Such adaptations for rapid, long-distance flight allow the albatrosses to traverse vast tracts of ocean from their breeding bases on oceanic islands.

Mostly Southern

DISTRIBUTION PATTERNS

Albatrosses are typically associated with the belt of windswept ocean lying between the Antarctic and the southern extremities of America, Africa, and Australasia. The greatest number of individuals and species occur between 45° and 70° south, but they also breed in temperate waters of the southern hemisphere, and a few species have spread into the North Pacific. The Waved albatross of the Galápagos Islands and Isla de la Plata off Ecuador breeds on the equator, where the climate is influenced by the cool Humboldt Current. The Steller's or Short-tailed albatross (based on islands off Japan and Taiwan), the Black-footed albatross of the northwest Pacific, and the Laysan albatross from the Hawaiian archipelago all breed in the North Pacific. No albatrosses breed today in the North Atlantic, although this was the case until the Pleistocene 1.8 million–10,000 years ago, and some of the albatrosses that have wandered into

Right *Black-browed albatross foraging trips can take several days and extend hundreds of kilometers. They glide fairly low, taking food from the sea surface or just below, occasionally plunging from heights of up to 9m (about 30ft).*

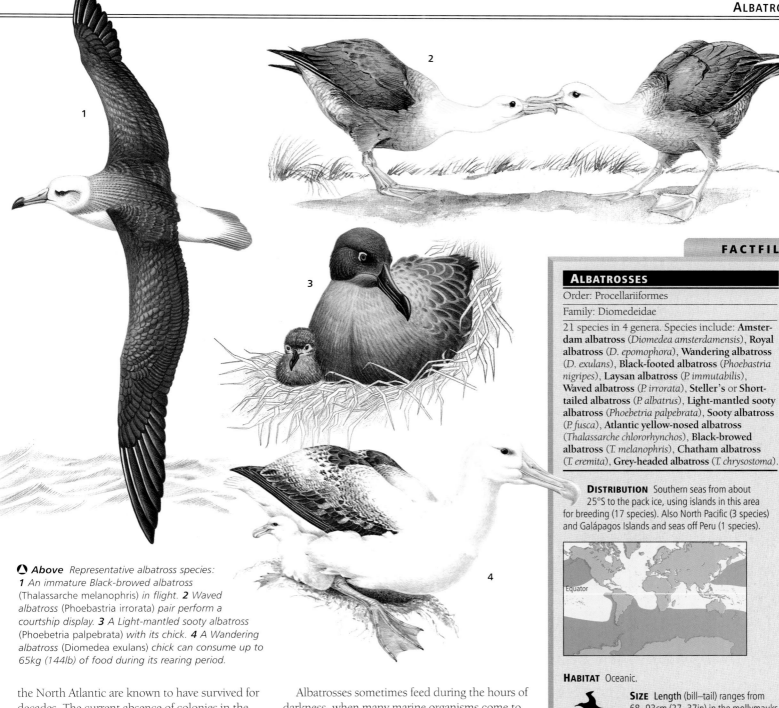

Above *Representative albatross species:* ***1*** *An immature Black-browed albatross* (Thalassarche melanophris) *in flight.* ***2*** *Waved albatross* (Phoebastria irrorata) *pair perform a courtship display.* ***3*** *A Light-mantled sooty albatross* (Phoebetria palpebrata) *with its chick.* ***4*** *A Wandering albatross* (Diomedea exulans) *chick can consume up to 65kg (144lb) of food during its rearing period.*

the North Atlantic are known to have survived for decades. The current absence of colonies in the North Atlantic may simply be because, by chance, post-Pleistocene colonization has not happened.

What the Ocean Provides

DIET

From their habit of following ships, albatrosses are best known as scavengers of offal thrown overboard. They have broad diets, but detailed analysis of stomach contents shows that fish, squid, and crustaceans make up the overwhelming majority of their prey items. These are mostly seized at the surface, although the birds will occasionally plunge in the manner of gannets, and can reach surprising depths, for example 6m (20ft) in the case of the Grey-headed albatross and as much as 12m (almost 40ft) for the Light-mantled sooty albatross.

Albatrosses sometimes feed during the hours of darkness, when many marine organisms come to the surface. Detailed information on the proportion of day and night feeding has been acquired by persuading birds to swallow a sensor that lodges in the stomach and records the sudden drop in temperature that occurs when the albatross swallows a fish snatched from the chilly Southern Ocean waters. The proportions of prey types differ between species, and these profoundly affect the breeding biology of the species.

Single Partners, Single Chicks

BREEDING BIOLOGY

Albatrosses are long-lived, with an average lifespan of 30 years, but they are slow breeders. Although they are physiologically capable of breeding at 3 or 4 years, in practice they do not usually start for several years after that time, and some may not

FACTFILE

ALBATROSSES

Order: Procellariiformes

Family: Diomedeidae

21 species in 4 genera. Species include: **Amsterdam albatross** (*Diomedea amsterdamensis*), **Royal albatross** (*D. epomophora*), **Wandering albatross** (*D. exulans*), **Black-footed albatross** (*Phoebastria nigripes*), **Laysan albatross** (*P. immutabilis*), **Waved albatross** (*P. irrorata*), **Steller's** or **Short-tailed albatross** (*P. albatrus*), **Light-mantled sooty albatross** (*Phoebetria palpebrata*), **Sooty albatross** (*P. fusca*), **Atlantic yellow-nosed albatross** (*Thalassarche chlororhynchos*), **Black-browed albatross** (*T. melanophris*), **Chatham albatross** (*T. eremita*), **Grey-headed albatross** (*T. chrysostoma*).

DISTRIBUTION Southern seas from about 25°S to the pack ice, using islands in this area for breeding (17 species). Also North Pacific (3 species) and Galápagos Islands and seas off Peru (1 species).

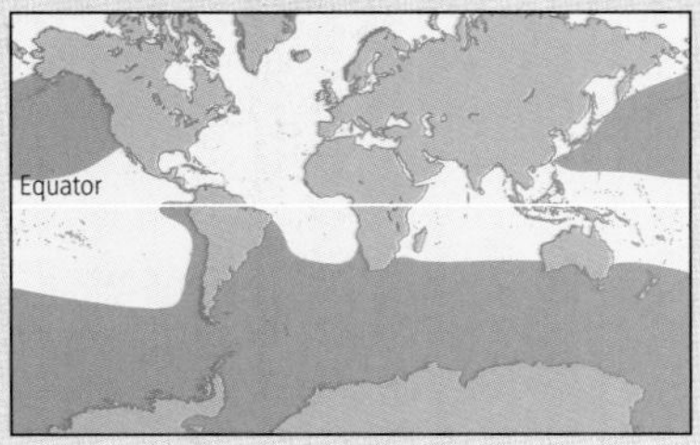

HABITAT Oceanic.

SIZE Length (bill–tail) ranges from 68–93cm (27–37in) in the mollymauks to 110–135cm (43–53in) in the great albatrosses; **wingspan** from 178–256cm (70–101in) to 250–350cm (98–138in) in the same species.

PLUMAGE White with dark wingtips; white with dark brow, back, upperwing, and tail.

NEST Typically a concave mound of mud or soil crudely lined with feathers and grasses, although the tropical albatrosses make scanty nests and the Waved albatross none at all.

EGGS Single white. Incubation period 65–79 days.

DIET Squid, fish, crustacea, fisheries waste.

CONSERVATION STATUS 2 species – the Amsterdam and Chatham albatrosses – are currently listed as Critically Endangered; 2 more are Endangered, and 13 are Vulnerable.

breed until they are 15 years old. When they first mature, birds appear on the breeding grounds for only a short while towards the end of the breeding season, but in subsequent years they spend growing amounts of time ashore courting prospective mates. When a pair has been established, they usually remain together until one partner dies. "Divorce" occurs only after several breeding failures, and it can be costly, since it may be followed by a few years without breeding until a new pair is formed. Indeed a single divorce reduces the lifetime reproductive success of a Wandering albatross by as much as 10–20 percent.

Most albatrosses nest in colonies that sometimes number thousands of pairs with close-packed nests, although the two *Phoebetria* species nest more or less alone on cliff ledges. In several species the nest is a pile of soil and vegetation which may be so large that the adults can have difficulty in climbing onto it. The tropical albatrosses make a scanty nest, however, and the Waved albatross has no nest at all, shuffling about instead with its egg on its feet. Males arrive at the colonies first at the beginning of the breeding season, and mating occurs when the females come to join them.

Incubation is by both parents in alternate shifts of several days' extent over a period ranging from about 65 days in smaller species to 79 in the Royal albatross. Parents first brood and later guard the newly-hatched chick. After the guard phase has finished at about 20 days, the adults visit land only to feed the chick at more or less regular intervals. Black-footed albatross chicks frequently wander up to 30m (about 100ft) from the nest and seek shade during the day, but they rush back whenever food arrives. Adults remain ashore long enough to identify their chicks and provide a meal of undigested marine animals and lipid-rich oil derived from the digestion of prey.

In some species, the parents alternate short trips of 1–3 days with longer excursions of 5 days or more to distant feeding grounds during the

Above *During the breeding season, albatrosses, such as this pair of Wandering albatrosses, perform joint courtship rituals, which involve a strange ballet-like dance, accompanied by bowing, scraping, bill-snapping, and nasal calls.*

Below *Riding the wind. Albatrosses are adapted more for gliding than for flapping, and have developed a gliding practice that makes best use of wind conditions in the southern seas. They take advantage of the phenomenon whereby wind near the sea surface moves at a slower speed than higher, thanks to drag on the sea surface. An albatross glides downwind from a height of about 15m (about 50ft), losing height. Just before it hits the water it turns into the wind and is blown back up to its original height by increasing wind speed.*

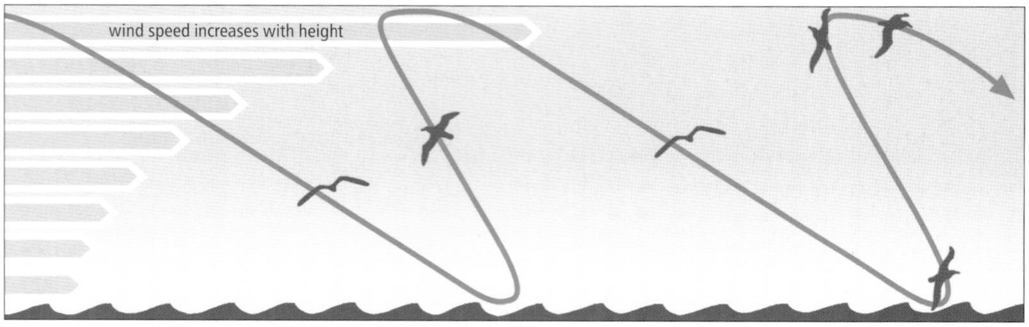
wind speed increases with height

Above *The male Grey-headed albatross incubates the egg for the first 15 days. Thereafter, female and male take turns to sit during the 70-day incubation period. If a chick is successfully raised in one season, the parents will not breed in the next.*

Left *At 8 weeks old, this Laysan albatross chick lives off the squid, fish, crustacea, and stomach oil that are fed through regurgitation by its parents. The rich food and stomach oil is filled with fatty acids and nutrients that can sustain a chick for a few days while the parents are away at sea foraging.*

chick-rearing period. Wandering albatrosses exhibit a further refinement, in that males tend to search for food farther south and therefore in colder waters than females, exposing themselves to stormier conditions. It may be no accident, then, that the males of this species have a higher wing loading (weight:wing area ratio) than the females.

Fledging takes from 120 days in Black-browed and Yellow-nosed albatrosses to 278 days in the Wandering albatross. The extremely long nesting period of the latter (356 days including incubation) means that it can attempt to breed only in alternate years since, after breeding, it must have a period of molt. In fact at least nine species are known to breed biennially, including all the "great" albatrosses, the two sooty albatrosses, and the Grey-headed albatross.

While it once was thought that albatrosses wandered the oceans more or less aimlessly during their off-duty year, modern sensors attached to Wandering albatrosses have shown this not to be the case. Rather, it seems that individual birds head for a particular region of the ocean, and spend most of their time there.

The Threat from Long-line Fishing

CONSERVATION AND ENVIRONMENT

Breeding colonies of albatrosses have long been protected by their isolation on islands with no natural predators, but discovery by seafarers has in the past led to losses through egg-collecting and the killing of adults, followed by massive depredations for the birds' feathers, which were used in clothing and bedding. Steller's albatrosses were almost wiped out by feather-collectors; hundreds of thousands of birds were killed, and breeding ceased altogether in the late 1940s and early 1950s. The species survived because immature birds, wandering the seas in relative safety away from the colonies, returned in due course and "rescued" the species. Since the resumption of breeding in 1954, the size of the main colony on the Japanese islet of Torishima has slowly inched upwards, and now numbers about 200 pairs. The Laysan albatross became a conservation problem when Midway Atoll in the north-central Pacific Ocean was turned into a US military airbase. The birds nested around the runways and installations, and there were many deaths from collisions with aerial wires and aircraft.

Albatrosses face more insidious threats at sea. Contamination by oil spills and chemical pollution occurs, but a more urgent threat arises from fishing operations. Although gill nets are now banned on the high seas, the technique known as long-lining, used to catch bottom fish including Patagonian toothfish and midwater species such as tuna, is widely used. A single tuna long-line may be up to 100km (62 miles) in length. As the line is set, baited hooks stream out from the stern of the fishing vessel. The lure can prove irresistible to an albatross, which snatches the bait, gets hooked, and is then dragged underwater by the line, to be retrieved by the fishermen with the catch some hours later. Up to 44,000 albatrosses per year may be killed this way, a fact reflected in the decline of some southern ocean populations.

Practical measures can help reduce this toll significantly – for example, setting the line at night – and international bodies are striving to persuade countries and their fishing fleets to adopt albatross-friendly methods. As southern seas are increasingly exploited by the world's fishing fleets, however, a new threat is coming to the fore, with the possibility that direct competition for krill, squid, and other marine species will affect albatrosses, along with other animals. MdeLB/RWB/PAP

Shearwaters and Petrels

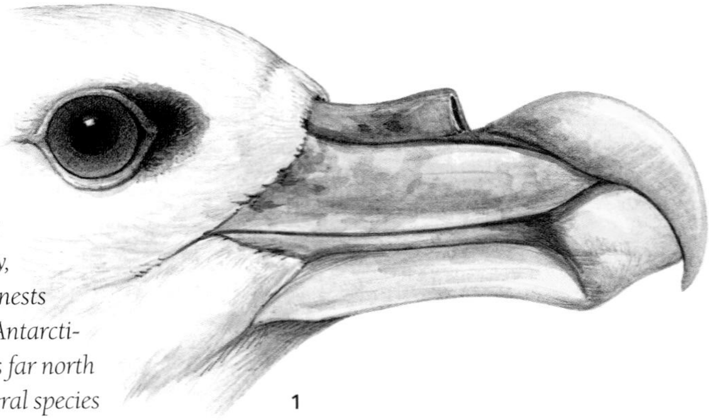

PROCELLARIIDAE HAVE ONE OF THE widest distributions of any bird family, ranging from the Snow petrel, which nests up to 440km (about 280 miles) inland in Antarctica, to the Northern fulmar, which breeds as far north as there is land in the Arctic. Although several species are localized and very rare, others are abundant; many undertake extensive migrations.

Overall, the Procellariidae are an extremely successful family. Some eat plankton, others dead whales, but the bulk catch small fish and squid at the sea surface or by underwater pursuit. Although there is a great variation between species in plumage and habits, the family divides neatly into four groups: fulmars, prions, gadfly petrels, and true shearwaters.

Birds of the Open Ocean

FORM AND FUNCTION

The fulmars are a cold-water group, only venturing into the subtropics along cold-water currents. There are six species in the southern hemisphere, which is where the group probably evolved, given that the single northern species, the Northern fulmar, is closely related to the Southern. Most are medium-sized, but the two sibling giant petrel species, with wingspans of 2m (6ft), are as large as some albatrosses. Indeed, it has been speculated that the "albatross" shot in Coleridge's *Rime of the Ancient Mariner* was, in fact, a giant petrel.

Fulmars' bills are large (enormously so in giant petrels) and broad. The birds probably once fed mainly on plankton, but some species now eat waste from fishing and, until recently, from whaling fleets; the exploitation of these new food resources has led to spectacular increases in numbers. Fulmars are fairly active on land, and giant petrels can walk with upright shanks (tarsi); other

Right *Representative species of shearwaters and petrels:* ***1*** *The bill of the Northern fulmar* (Fulmarus glacialis), *showing the extension of the nostrils that gives the order its familiar name, "tubenoses";* ***2*** *Cahow or Bermuda petrel* (Pterodroma cahow); ***3*** *Greater shearwater* (Puffinus gravis); ***4*** *Broad-billed prion* (Pachyptila vittata), *on land and (background) filter-feeding;* ***5*** *Black-capped petrel* (Pterodroma hasitata); ***6*** *Southern giant petrel* (Macronectes giganteus) *feeding on a dead seal.*

Above *Northern fulmars have a uniformly gray plumage in their dark phase, as opposed to the light phase where their wings are gray, head is white, and tail is white, rimmed in gray.*

groups in this family mostly shuffle with their shanks flat on the ground. In flight they alternate flapping and gliding.

Prions (including the Blue petrel) are another southern group that breed mainly on sub-Antarctic islands, but move into slightly warmer waters at other times of year. These small birds, typically about 26cm (10in) long, all look very similar: blue-gray above and white below, with a dark "W" across the wings. All eat small plankton that they filter out with plates (lamellae) on the bill, but the bill dimensions vary greatly, suggesting subtle differences in diet. Some species pick fish from the surface, while those with broader beaks hydroplane their way through the surface water. A study of the Broad-billed prion concluded that its bill could contain 4.1cu cm (0.25cu in) of water passing across 785sq mm of lamellae.

Prions congregate in areas of high plankton density, and vast flocks typically wheel low over the sea. They were once known as whale birds because they are frequently found in the presence of cetaceans.

Averaging 26–46cm (10–18in)in length, the gadfly petrels include some species barely larger than prions, although others are bigger. Most species are black (or gray) and white above and white below, with white faces; a few are all dark. Identifying birds can be complicated by the fact that some species have various color phases. The short, stout bill, equipped with a powerful hook and sharp cutting edge, is used for gripping and cutting up small squid and fish. The birds occur predominantly in the southern and tropical oceans, where some species such as the cahow are restricted to single islands while others roam far and wide. They are strong fliers and typically arc high above the sea. Their movements are imperfectly known, but some Pacific species migrate across the equator from one hemisphere to the other.

The true shearwaters and petrels are small to large birds, ranging in length from 27 to 55cm (10–22in); most are dark above and black or white underneath. Many species have a black cap, and only one species has white on the head. Widespread and very mobile, they pose considerable taxonomic problems. For example there has been debate about whether the similar, but not identical, black-and-white shearwaters breeding in the northeast Atlantic, Mediterranean, Hawaii, east Pacific, and New Zealand should be considered subspecies of the Manx shearwater; the current tendency is to class them as separate species.

FACTFILE

SHEARWATERS AND PETRELS

Order: Procellariiformes

Family: Procellariidae

79 species in 14 genera. Species include: **Audubon's shearwater** (*Puffinus lherminieri*), **Greater shearwater** (*P. gravis*), **Heinroth's shearwater** (*P. heinrothi*), **Manx shearwater** (*P. puffinus*), **Short-tailed shearwater** (*P. tenuirostris*), **Sooty shearwater** (*P. griseus*), **Beck's petrel** (*Pseudobulweria becki*), **MacGillivray's petrel** (*P. macgillivrayi*), **Blue petrel** (*Halobaena caerulea*), **cahow** or **Bermuda petrel** (*Pterodroma cahow*), **Magenta petrel** (*P. magentae*), **Cory's shearwater** (*Calonectris diomedea*), **Northern fulmar** (*Fulmarus glacialis*), **Southern fulmar** (*F. glacialoides*), **Snow petrel** (*Pagodroma nivea*), **Southern giant petrel** (*Macronectes giganteus*).

DISTRIBUTION All oceans

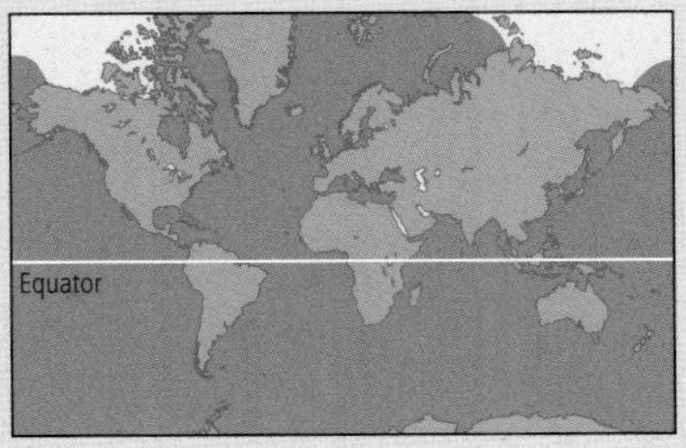

HABITAT Oceanic

SIZE Length 26–87cm (10–34in); wingspan maximum 2m (about 6ft); weight 130g–4kg (5oz–8.8lb).

PLUMAGE Most species black, brown, or gray, with white; a few are all light or dark.

VOICE Unmusical; often nocturnal cacophony at colonies; sexes often differ in call.

NEST All species breed in colonies. Most shearwaters, prions, and gadfly petrels nest in burrows or holes; fulmars incubate in the open, or on scrapes on cliff ledges.

EGGS 1; white. Incubation period 43–60 days.

DIET Fish, squid, crustacea, offal.

CONSERVATION STATUS 10 species are currently Critically Endangered, including the Chatham petrel (*Pterodroma axillaris*) and Townsend's shearwater (*Puffinus auricularis*). A further 6 species are Endangered, and 20 Vulnerable.

Some shearwaters are notable long-distance migrants. Short-tailed and Sooty shearwaters breed in Australia and New Zealand and then spend the northern summer in the North Atlantic (Sooty only) or the North Pacific (both species). Other species move in the opposite direction; Cory's and Manx shearwaters breed in the North Atlantic and spend the northern winter in the South Atlantic.

Shearwater bills are proportionately longer and thinner with smaller hooks than those of other comparable groups, but they nonetheless eat mainly fish and squid. Prey is caught either by the bird plummeting onto it or by swimming after it underwater.

Left *Snow petrels nesting on Queen Maud Land, Antarctica. Breeding colonies of this species are sited both near the sea and further inland; the birds nest in pebble-lined scrapes in rock crevices protected by overhanging ledges.*

Right *When it feeds at sea, the diet of this Hall's or Northern giant petrel* (Macronectes halli) *includes fish and squid. On land, it takes penguins and mammal carrion. Not only is its bill powerful and tough but is also equipped with a sharp hook for holding prey.*

Colonial Couples

BREEDING BIOLOGY

All the species in the family are colonial to a greater or lesser extent. Sometimes this is due to limited suitable habitat – for example, Antarctic-nesting Southern giant petrels are forced to nest on the few patches of stones kept snow-free by the wind – but it is usually a matter of choice, as birds enter and try to nest in seemingly overcrowded areas, even with apparently suitable unused habitat nearby. Nest densities can then exceed 1 per sq m (1.2/sq yd). Colony sites are as diverse as the species themselves, but safety from predators is a prerequisite.

Of the fulmars, only the Snow petrel nests under cover; other species make a scrape on a cliff ledge or incubate in the open. Birds discourage intruders by spitting or regurgitating foul-smelling oil – hence the old name of "stinker" for the giant petrels. Colonies tend to be small and nests dispersed. All the prions nest underground among boulders or in burrows they dig themselves; colonies may be very large. Except for a number of surface-living species on tropical islands, gadfly petrels and shearwaters nest in burrows or under rocks. Some line the chamber with vegetation; others make a mere token of a nest. Typically colonies are large and found on islands, less commonly among forests or high on mainland mountains. Whereas open-nesting species come and go from their nests by day, most burrow-nesters are nocturnal at the colonies, so as to escape predators. While sight is probably the major sense used to guide birds to their burrows at night, there are persistent hints that smell may also play a part.

Breeding is remarkably uniform throughout the family. Birds return to the colonies at least a few weeks prior to nesting and reclaim the nest sites used the previous season. Pairs usually remain together from one season to the next, and probably meet again at the nest sites. Adult survival is extremely high (at least 90 percent per annum) and pairs persist for many years; when "divorces" do occur, they usually follow unsuccessful breeding. In the weeks prior to laying there are noisy aerial displays, and pairs spend days together at the nest. Many species have a "honeymoon" period, when the female leaves the colony for about two weeks to feed so as to lay down reserves for egg-laying. In some species the male is also away at this time, preparing himself for taking the first long incubation stint. In others, however, he returns periodically to check the nest site.

In most species breeding is annual and synchronized. The most extreme case is the Short-tailed shearwater, whose colonies span 11° of latitude; all of its eggs are laid during a 12-day period, with the peak always occurring between 24 and 26 November. In tropical species that frequent the colonies throughout the year, eggs may be laid in all months. In a few species individuals breed at less than annual intervals, but in most pairs breed annually. Even more infrequently, birds at adjacent colonies may breed annually but out of phase.

All species lay a single, very large white egg, which varies from 6 percent (in the giant petrels) to 20 percent (in prions) of the female's weight. Tropical species lay proportionately larger eggs than temperate or polar species, probably because food is often short and the chick needs bigger food reserves to carry it over any shortage. Both sexes have a large, central brood patch (an area denuded of feathers and rich in blood vessels for transferring heat from parent to egg) and incubate in turn for spells of 1–20 days. Often the male takes the first and longest stint, presumably to let the female go back to the sea to recover from the egg-laying. Lost eggs are very rarely replaced. The eggs are tolerant of chilling, especially those safe in the uniform temperature of a burrow. However, if the parents do have periods of absence from incubation, then the incubation period will be longer – by up to 25 percent.

The incubation period is long, but its range (43–60 days) is less than might be expected given the great variety of egg sizes (from 25 to 237g, or almost 1 to 8.4oz). The chick is brooded for the first few days, but may then be left in the burrow while both adults forage. It is fed on a soup of partly digested fish, crustacea, and squid, and on stomach oil. Growth is rapid until the young may be much heavier than the adult. Burrow-living chicks are often deserted and complete their development on stored fat.

Harvesting the Muttonbird

CONSERVATION AND ENVIRONMENT

The young, eggs, and adults of many species were once considered delicacies and were eaten in large numbers, and their fat was also used extensively. Nowadays human predation has declined but not stopped; for instance, Greater shearwaters are still killed at colonies on Tristan da Cunha and in wintering grounds in the North Atlantic. Some shearwaters are known as muttonbirds for their tasty flesh, and the young of two species, the Short-tailed and Sooty shearwaters, are still harvested commercially for their meat, which is sold as "Tasmanian squab." However, a strict quota is now imposed on the number of birds that can be taken, and as the populations of both species exceed 20 million, the harvest has no deleterious effect on total numbers.

In contrast, several gadfly petrels are seriously threatened by habitat destruction and introduced predators. The cahow or Bermuda petrel is a typical example. Occurring only in Bermuda, it once lived inland, where it was first hunted for food and later killed by pigs, cats, and rats. A few pairs survived on offshore rocks, but their breeding success was very low because of competition with tropicbirds for nest-sites. Management has now addressed the problem, and the species' numbers are beginning to climb.

On the Chatham Islands in the southern Pacific Ocean, the population of the Magenta petrel depends on the protection of the New Zealand Department of Conservation and the output of just 12 known burrows. Possibly rarer still are three species (Heinroth's shearwater, Beck's petrel, and MacGillivray's petrel), whose nesting burrows have never been found. MdeLB/MPH

Storm Petrels

THE COMMON NAME OF THE HYDROBATIDAE has an interesting derivation. The epithet "storm" is thought to come from their habit of sheltering from storms in the lee of ships, while "petrel" may be a corruption of "St. Peter" – alluding to the fact that several species appear to walk on water, an illusion created by their hovering just above the surface while feeding and "patting" the water with their feet.

Storm petrels are the smallest and most delicate of all seabirds. They eat mainly planktonic crustacea, squid, tiny fish, and oily scraps snatched from the water while in flight (they rarely settle on the water). They are not worried by the presence of people, and many species feed in ships' wakes; a few haunt fishing boats for scraps.

The Sailors' Companion

FORM AND FUNCTION

Storm petrels are immediately recognizable by their tiny size, fairly small but strongly hooked beak, pronounced tubular nostrils (which are fused together), and steep forehead. Despite the absence of color, many are striking birds, with almost black plumage and a white rump; others are beautiful shades of brown and gray.

The family is clearly divisible into two groups that, judging by molecular data, have a long independent evolutionary history; they presumably evolved in different hemispheres, but now overlap in the tropics. The northern group, typified by Leach's storm petrel, is made up of 14 species in three genera. The birds are black and white, and most have pointed wings and relatively short legs. They feed by swooping down to pick food from the water's surface, rather like terns. The southerners, of which the White-faced storm petrel is representative, consist of seven species in five genera; they have more variable plumage (including species with several color forms), along with rounded wings and long legs that are often held down, as the birds bounce or walk on the sea-surface feeding. The shorter, more rounded wings of the southern group could well have evolved to cope with the fierce winds of the southern ocean.

Absent from brackish water and the Arctic, storm petrels can otherwise be found throughout the world's oceans. They are most abundant in the cold waters around Antarctica and in areas of marine upwelling, like the Humboldt Current off Peru. Some species occur only in areas of the central tropical oceans where there is apparently little food. A few species have very restricted distributions, but others are wide-ranging and undertake long migrations. Wilson's storm petrel, which breeds around Antarctica, spends its non-breeding time throughout the Indian and central Pacific oceans, and in the Atlantic north to Greenland. The opposite pattern is shown by Leach's storm petrels, which breed in the North Pacific and North Atlantic waters and spend the northern winter in equatorial seas.

Colonies that Come Alive by Night

BREEDING BIOLOGY

All species are more or less colonial and highly vulnerable to predators such as rats. Most, therefore, breed on isolated islands lacking ground

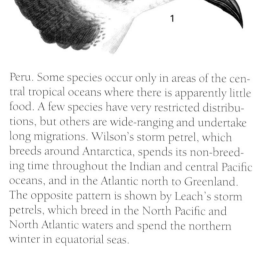

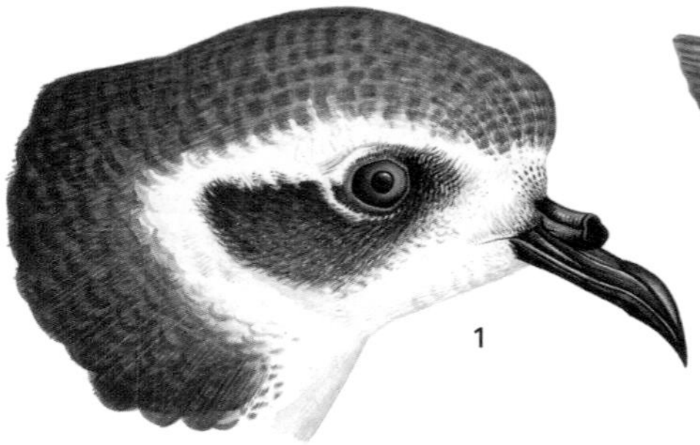

Above *Representative species of storm petrels:* ***1*** *White-faced storm petrel* (Pelagodroma marina). ***2*** *Wilson's storm petrel* (Oceanites oceanicus). ***3*** *Grey-backed storm petrel* (Garrodia nereis).

Right *One of the world's rarest and least known storm petrels, Hornby's storm petrel can be found foraging out at sea off western South America.*

Above *Although rarely seen because they only visit their burrows at night, the presence of a Leach's storm petrel can be detected by the unmistakable musky smell surrounding the nest, which is produced by an oily orange liquid emitted by the bird when disturbed.*

predators. However, Markham's storm petrels breed in burrows in saltpeter deposits in the arid Atacama desert of Peru (and probably also in Chile). Hornby's storm petrels also breed in this region, but their colonies remain undiscovered; they are probably the most common seabird whose nesting sites remain totally unknown. Perhaps they breed well inland towards the Andes, or on coastal slopes. Except for the Wedge-rumped storm petrel, all species visit the colonies at night or, in high latitudes, when light density is at its lowest.

Pairs defend a burrow among rocks, tree roots, or (rarely) under a bush, and remain together from one season to the next. In temperate regions breeding is fairly well synchronized and seasonal, while in the tropics it often appears prolonged, with birds present throughout the year. However, the Madeiran storm petrel has two breeding seasons a year in the Galápagos Islands, where two quite separate populations both breed annually, but roughly six months apart. Remarkably, a similar situation prevails in the Azores: there too, two different populations, showing some body size and genetic differences, breed six months apart. Elsewhere Madeiran storm petrels have a conventional single breeding season.

A single white egg is laid on the ground, where it is incubated for 1–6 days by each bird in turn for a total of 40–50 days. If lost, the egg is only very rarely replaced in the same breeding season, probably because it is so large (up to 25 percent of the female's weight) that it would take too long to produce another.

The chick is fed a mixture of partly-digested food and stomach oil. It fledges alone and at night, after 59–73 days. Although fed less frequently near to fledging, it is not, as has often been claimed, deserted by the parent birds; in fact, the adults may visit the burrow even after it has left.

Audible Decoys for Breeding Birds

CONSERVATION AND ENVIRONMENT

In some species, many immature birds visit other colonies before they first breed (usually at 4–5 years). Such birds can even be attracted to places other than occupied colonies by playing tape recordings of the purring calls associated with these. The potential conservation value of this technique has been demonstrated by the National Audubon Society of the USA, whose members persuaded Leach's storm petrels to nest in artificial burrows on an island where they had never previously bred. More remarkably, Swinhoe's storm petrels have been attracted to North Atlantic coasts, raising the possibility that there are undiscovered Atlantic colonies of this North Pacific species.

MdeLB/MPH

FACTFILE

STORM PETRELS

Order: Procellariiformes

Family: Hydrobatidae

21 species in 8 genera. Species include: **Hornby's** or **Ringed storm petrel** (*Oceanodroma hornbyi*), **Leach's storm petrel** (*O. leucorhoa*), **Madeiran storm petrel** (*O. castro*), **Markham's storm petrel** (*O. markhami*), **Wedge-rumped storm petrel** (*O. tethys*), **Least storm petrel** (*Halocyptena microsoma*), **White-faced storm petrel** (*Pelagodroma marina*), **Polynesian storm petrel** (*Nesofregetta fuliginosa*), **Wilson's storm petrel** (*Oceanites oceanicus*).

DISTRIBUTION All oceans except Arctic seas

HABITAT Oceanic

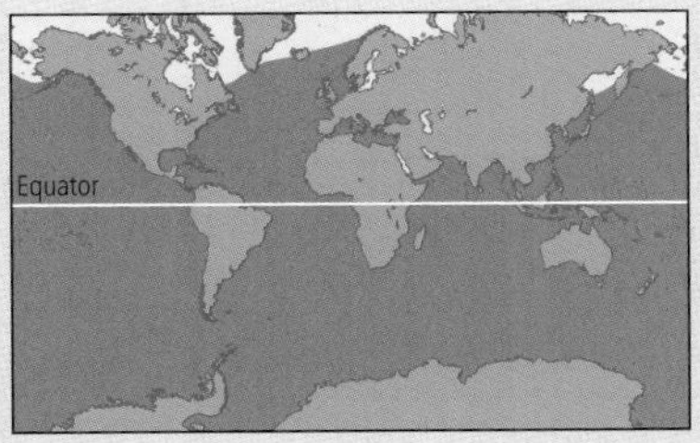

SIZE Length 14–26cm (5.5–10in); wingspan 32–56cm (12.6–22in); weight 25–68g (0.9–2.4oz).

PLUMAGE Chiefly dark brown, black or gray, and white.

VOICE Some species purr and chuckle at breeding colonies.

NEST A burrow among rocks, tree roots, or under a bush.

EGGS 1; white, sometimes with speckles. Incubation period 40–50 days.

DIET Small fish, squid, plankton, fish scraps.

CONSERVATION STATUS The Guadalupe storm petrel (*Oceanodroma macrodactyla*) is listed as Critically Endangered although in fact it may already be extinct, exterminated by feral cats on the island off Mexico's Pacific coast where it bred. In addition, the Polynesian storm petrel is Vulnerable.

Diving Petrels

ONE OF THE AMAZING SIGHTS OF THE SOUTHERN seas is that of a small flock of diving petrels flying through a steep wave, plunging in one side and coming out the other, or else erupting unchecked from the ocean. This behavior is possible thanks to the adaptation of the birds' wings for swimming underwater.

Diving petrels are in fact a southern equivalent of the auks of northern seas; in particular they resemble the Little auk, sharing its whirring, "bumblebee" flight and small wings, adapted for swimming underwater. The structure of the wing bones is remarkably similar in both groups, while another resemblance lies in the fact that the Peruvian diving petrel, like several auk species, becomes flightless during wing molt. The petrel's stomach commonly remains full of food at that time, however, indicating that it can continue to forage efficiently even when "grounded."

Petrels Built to Plunge

FORM AND FUNCTION

The four species of diving petrels are very similar in size and plumage, which makes identification difficult unless the bird is in the hand and the variable amounts of gray or white on the upper side can be seen. Only the Magellan diving petrel, with its white collar, has a distinctive plumage. It also has a distinguishable juvenile, which lacks the white fringes on feathers on the back common to the other species. Also distinctive but only visible from close to are the so-called paraseptal processes that partition the nostrils and are thought to prevent the entry of seawater.

Diving petrels chase their prey by swimming underwater. They feed mainly within the upper 10m (33ft) of the ocean, although on occasion they can dive as deep as 60m (200ft). Their diet consists of small marine organisms. The most detailed analysis of diet has been conducted on Common and Georgian diving petrels at South Georgia and the Kerguelen Islands. In both locations, the two species were found to eat mostly crustacea; but while Georgian diving petrels fed their chicks predominantly on euphausiid krill, Common diving petrels, feeding closer inshore but at greater depths, usually delivered copepods and amphipods.

Southern Strongholds

DISTRIBUTION PATTERNS

Colonies of diving petrels can be extensive; there are an estimated two million breeding pairs of Georgian diving petrels on South Georgia and in excess of one million on the Crozets and Kerguelen Islands in the southern Indian Ocean. These three sites are also the strongholds of the Common diving petrel, the most widespread of the four species. For both species, the most significant current threat is that of predation by rats introduced to the islands.

Only four breeding sites – two in Peru and two in Chile – are known for the Peruvian diving petrel, which nests under boulders and is now classified as Endangered. Many former sites were destroyed by commercial extraction of guano (excrement), in which the bird used to excavate its burrows.

Although common in the fjord region of southern Chile and on the coast of southern Argentina, the Magellanic diving petrel is a very poorly known species that appears to nest in peaty soil.

Left Georgian diving petrels feed primarily on marine crustaceans, together with some small fish and cephalopods. They breed in late summer in the scoria (cooled lava) on volcanic islands. Here, parents feed their chicks with meals that are almost 20 percent of adult body mass.

Right Even though it is still a chick, this Common diving petrel is already displaying its distinctive cobalt-blue legs. They are preyed upon by other birds such as skuas and Kelp gulls.

Above *Representative diving petrel species: **1** Common diving petrels* (Pelecanoides urinatrix) *will nest in tunnels or burrows with entrances just 5–8cm (2–3in) in diameter; **2** A group of Georgian diving petrels* (P. georgicus) *flying into waves.*

Staying Close to Home

BREEDING BIOLOGY

Breeding is confined to cool waters north of 60° south and up the western coast of South America, where the sea temperature is influenced by the cool Humboldt Current. Unlike many other petrels, diving petrels do not appear to travel far, even outside the breeding season; they are usually seen near the breeding area, in waters over the continental shelf.

The breeding sequence has been described in detail for Georgian and Common diving petrels on South Georgia. The birds nest in burrows that are re-excavated each season, flying in at night, presumably to avoid attracting the attention of skuas to the colonies. The Georgian species tunnels into bare, stony soil above the vegetation level, whereas Common diving petrels nest at lower levels in peaty, often waterlogged soil beneath stands of tussock grass. The birds call at night; the Georgian diving petrel makes a series of harsh squeaks, while the Common diving petrel utters a two-syllable phrase rendered as "kuaka," the Maori name for the species.

While calling may be essential for attracting a mate, it can have the less desirable side-effect of attracting predators. One way of making it harder for a call to be located is to introduce noise, in the technical sense of a scrambled multi-frequency signal. It is therefore both logical and fascinating that the calls uttered by Georgian diving petrels on the ground, where they are vulnerable to attack, include just such a scrambled element. The noise disappears when the calls are given in flight, where the birds are not at risk of being attacked by skuas.

Between 7 and 31 December the Georgian diving petrel lays a single egg, which it hatches in late January; the Common hatches on average 29 days earlier. Incubation spells last one to three days. After hatching, Common diving petrel chicks are brooded for at least 11 days, but Georgian chicks for only six. Most nights the chicks are fed by one or both parents. The feed contains very little stomach oil, unlike in other Procellariiformes families, and is little digested, possibly because it is delivered so quickly after capture; it is in fact carried back not in the parent's stomach but in the throat or gular pouch. MdeLB/RWB/PAP

FACTFILE

DIVING PETRELS

Order: Procellariiformes

Family: Pelecanoididae

4 species of the genus *Pelecanoides*: **Common diving petrel** (*P. urinatrix*), **Georgian diving petrel** (*P. georgicus*), **Magellan diving petrel** (*P. magellani*), **Peruvian diving petrel** (*P. garnotii*).

DISTRIBUTION Subantarctic, S America as far N as Peru, S Australia, New Zealand.

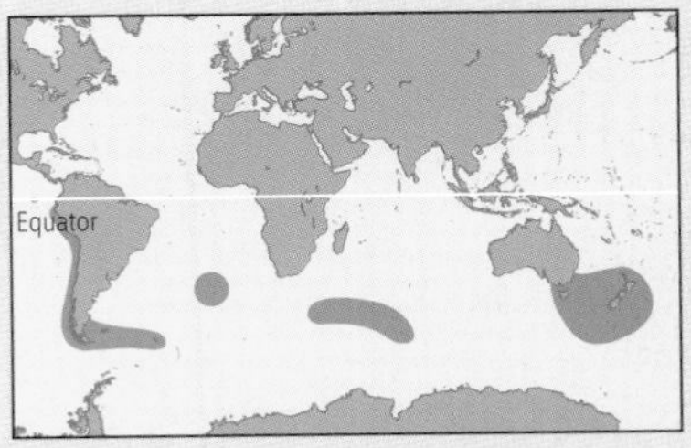

HABITAT Marine, breeding on islands and coasts.

SIZE Length from bill to tail 18–25cm (7–10in); wingspan 30–38cm (11.8–15in); weight 105–146g (3.4–4.7oz).

PLUMAGE Black above, white below.

VOICE Squeaks and croaks, heard at night on the nesting grounds.

NEST Burrows dug in soft soil.

EGGS 1; white. Incubation period 45–53 days.

DIET Small marine organisms, particularly crustacea.

CONSERVATION STATUS The Peruvian diving petrel, which now breeds on four small islands, is Endangered.

Pelicans

WITH THEIR CAPACIOUS THROAT POUCH and comical air, pelicans are all too easy to caricature. In truth, they are highly efficient front-loading feeders, superb fliers, and complex socialites. Their name comes from the Greek pelekon, *derived from* pelekys, *meaning "ax" (from the shape or supposed chopping action of the bill). Yet a more accurate association would be with a shovel, which is exactly how the birds sometimes use their bills.*

Pelicans have often been the subject of folklore. An Indian fable recounts that a pelican once treated her young so roughly that she killed them, only to resuscitate them remorsefully with blood drawn from her own breast. The erroneous yet appealing idea that pelicans nourish their young with their own blood recurs in other cultures, probably inspired by the fact that the birds rest their bill on their breast, while the Dalmatian pelican has a blood-red pouch early in breeding. Because of its supposedly self-sacrificing nature, the pelican became associated with piety in Christian cultures, and features prominently in ecclesiastical heraldry and carvings.

The Heaviest Flying Seabirds

FORM AND FUNCTION

Pelicans divide into two groups, based on plumage color and nesting habits. One, consisting of the Australian, Dalmatian, Great white, and American white species, is made up of birds that are basically all-white and that nest on the ground; the other comprises the Pink-backed, Spot-billed, and Brown pelicans, species that are predominantly gray or brown and that nest in trees. At about 15kg (33lb), the Great white and Dalmatian pelicans are the world's heaviest flying seabirds; in contrast, the Brown pelican may weigh little more than 2kg (4.4lb). Size and weight are essential in those pelicans that wield an enormous pouch, capable of holding 13kg (28lb) of water.

Pelicans have 20–24 short tail feathers, giving them a squarish tail, but their wings are long and broad, with a large number (30–35) of secondary feathers. The birds are superb soarers, able to keep their wings horizontal for gliding flight because of a deep layer of special fibers in the breast muscles. This adaptation enables pelicans, using thermal updrafts, to make daily foraging trips of more than 150km (100 miles), thus greatly enlarging their potential feeding area. Their short, strong legs and webbed feet facilitate swimming. Pelican plumage is waterproof and kept so by secretions of the

Above *After fishing early in the morning, Great white pelicans spend the rest of the day on islands or sandbars, resting, preening, and bathing. This species is abundant in Africa, with some 75,000 breeding pairs, and sometimes congregates in huge colonies numbering tens of thousands of birds. This group is in Amboseli National Park, Kenya.*

Above left *The Australian pelican has the distinction of the longest bill of any bird in the world. It is also characterized by a pink throat pouch and prominent rings around its eyes; this latter feature is echoed both in its scientific name and its alternative common name, the Spectacled pelican.*

FACTFILE

PELICANS

Order: Pelecaniformes

Family: Pelecanidae

7 species of the genus *Pelecanus*: **American white pelican** (*P. erythrorhynchos*), **Australian pelican** (*P. conspicillatus*), **Brown pelican** (*P. occidentalis*), **Dalmatian pelican** (*P. crispus*), **Great** or **European white pelican** (*P. onocrotalus*), **Pink-backed pelican** (*P. rufescens*), **Spot-billed** or **Grey pelican** (*P. philippensis*).

DISTRIBUTION E Europe, Africa, India, Sri Lanka, SE Asia, Australia, N America, northern S America.

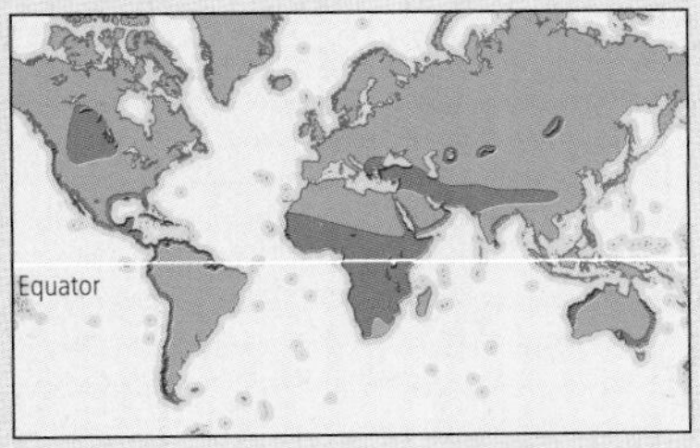

HABITAT On or near coasts and inland waters.

SIZE Length 1.3–1.7m (4.2–5.6ft); wingspan 2–2.8m (6.6–9.2ft); **weight** 2.5–15kg (5.5–33lb). Males slightly larger than females.

PLUMAGE Gray, brown, or white, with black primary and flight feathers; washes of pink or orange on the body; garish facial and bill colours and (in some) adornments; Brown pelican is brown-gray-black.

VOICE Hisses and grunts; nestlings are very noisy.

NEST Tree-nesting species build large structures of dry sticks as much as 30m (100ft) above ground; ground-nesters use shallow depressions, sometimes lined with sticks, leaves, or reeds.

EGGS 1–4; chalky white. Incubation 1 month.

DIET Fish, amphibia, small mammals, young of water birds.

CONSERVATION STATUS The Spot-billed pelican is Vulnerable. The Dalmatian pelican, which was similarly ranked, is now listed as Lower Risk/Conservation Dependent, as its numbers are now rising thanks to conservation measures, notably at Lake Mikri Prespa in Greece.

preen gland, which the bird rubs with the back of its head before transfering the oil to its plumage.

Young pelicans are downy, and either white, gray, or dark brown in coloration. A complete plumage of true feathers is acquired in two months. Adult plumage usually takes at least two years to develop.

Birds of Warm Climates

DISTRIBUTION PATTERNS

Although mainly an Old World family, pelicans occur on all continents except Antarctica, ranging from 65°N to 40°S and across the world. They are absent only from the polar regions and from inland South America. Fossils show that they once enjoyed an even wider distribution. Although largely avoiding overlap, the Great white and Dalmatian pelicans may breed in mixed colonies, and the Great white and Pink-backed overlap in parts of Africa.

Pelicans are primarily birds of warm climates, although some Great whites and American whites return to breeding areas while snow and ice still prevail. They prefer to nest in isolated areas, relatively safe from predatory mammals and humans. They breed, according to species, on the ground or amid low vegetation, on islands in lakes and marshes, on arid sea rocks (in the case of the Peruvian race of Brown pelicans), or in trees (Brown, Pink-backed, and Spotted pelicans).

Outside the breeding season pelicans typically disperse or migrate, although the Great white and American white are resident in some areas but migratory in others. Pre-migratory assemblages can be enormous. Once on the move, the birds may climb high and cross deserts or even mountains. Some species move along coasts in large numbers, but the American white remains almost entirely inland. Occasionally migrating pelicans will descend to barren lakes and, if detained there by bad weather, may die en masse. Mass deaths also ensue if the pelicans persist in their migration despite their food reserves becoming exhausted by adverse weather. No pelicans undertake long sea crossings.

Pelicans combine an attachment to traditional breeding areas – say, a lake, a mangrove swamp, or an island – with a readiness to shift the precise

Below *Diving for fish. All pelicans catch fish in their throat pouch but only the Brown pelican dives into water to take its prey. Each dive seems to be made to capture a particular fish.* ***1*** *On sighting its prey, the pelican enters a dive* ***2*** *and* ***3****, pulling back its wings to form the wings and back into the shape of a triangle.* ***4*** *As the bill enters the water, legs and wings are thrust back, increasing speed.*

location of the colony. Some species, such as the Great white and Australian, breed opportunistically in areas with seasonally unpredictable regimes, although they are usually 3 or 4 years old before breeding for the first time. In the wild, the birds are not particularly long-lived; the oldest recorded individual was an American white of 26 years. However, captives have lived for more than 60 years, and have bred successfully in zoos.

Scooping up Fish

DIET

Pelicans are versatile in their use of feeding habitat and in their hunting methods. They fish, according to species, in anything from inland ditches to open sea; the offshore (pelagic) zone is the only habitat open to some, but not all, species. Feeding behavior includes co-operative hunting, piracy from other water birds, and sea- or land-based scavenging. Except for the Brown pelican, the birds feed largely while swimming, up-ending or partly submerging in search of prey. The ground-nesters often feed co-operatively, swimming in rows and driving fish into the shallows before scooping them up. Brown pelicans plunge-dive, although in a spectacularly ungainly manner that has been likened to throwing a bundle of washing into the water.

Pelicans have to allow the water scooped up with their prey to drain away before they can raise their heads and swallow the fish, at which point they are vulnerable to being robbed of their prey by other seabirds. They take a wide variety of fish, from small fry to large species, and also eat offal, eggs, young birds, amphibia, and crustacea. Most of their prey is commercially unimportant.

Rituals of Courtship

BREEDING BIOLOGY

Pelicans possess a number of highly ritualized courtship behaviors, even though several species are not faithful to a mate, or even to one particular site, in successive breeding attempts. The birds seldom fight, although they may jab or grapple. Defending pelicans point or lunge at one another, or close the bill with a resounding snap.

Pair formation in ground-nesting pelicans is difficult to follow; it involves communal interactions, chases and processions on land (where they strut) or on water, or in the air. By undescribed means a female attracts a number of males that follow her while interacting with each other, using several ritualized postures including pointing, gaping, and thrusting. Amazingly, the entire, complex procedure can be completed in a day.

In the tree-nesting species, there is less group interaction; instead, males tend to advertise for a female from a particular perch. Some of the displays used in pair formation continue as pair-bonding behavior in the early stages of breeding, but later on pair interactions are at best perfunctory. Copulation begins 3–10 days before egg-laying, sometimes within hours of partners coming together; it ceases abruptly after the last egg has been laid.

Ground-nesters may build little or no nest. Those that do so carry nest material in the pouch, which may bulge like a garbage bag. The arboreal pelicans carry material crosswise in the bill.

Eggs are incubated on top of or beneath the birds' webbed feet, and partners share incubation in stints varying from three hours to three days. There may be a conspicuous display at change-over, particularly in the early stages.

Some species (Dalmatian, American white, Brown, and Spot-billed) can rear more than one chick per brood, although most do not. Thus, even though all lay two or more eggs, most raise only one chick. Optimal brood size may be achieved by immediate siblicide or by competition for food, leading to the death of the weakest. In Pink-backed and Spot-billed pelicans, the second-born may survive for many weeks before finally starving to death.

The young grow quickly on frequent, large feeds, achieving flight after 10–12 weeks. Some species bring water to their chicks. Prior to feeding their older young, ground-nesters may treat them extremely roughly, seizing them by the head and dragging them around. Strange young are repelled. After, or even before, feeding, the young may behave as though in a fit, eventually collapsing and lying comatose; this strange behavior still awaits convincing interpretation.

The mobile young of ground-nesting species gather in pods up to 100 strong, and parents find and feed only their own chicks, apparently recognizing them by sight. At 6–8 weeks, they wander more freely, moving to the water's edge and occasionally swimming, especially if they are threatened; they may even practice communal fishing before they can fly. In contrast, young Brown pelicans do not return to the nest once they have flown. In general, young pelicans are fed infrequently or not at all after they have started to fly, although some linger with the adults, roosting and loafing. Thus, despite the claims of some authors, parental care cannot accurately be described as prolonged.

Excluding disturbance and desertion, pelican hatching success can be as high as 95 percent. The percentage of hatchlings that eventually fly is, however, usually below 50 percent. The most significant feature of productivity in pelicans is its inconsistency.

Left *As it enters the water, the pelican positions the mandibles of its beak above and below the fish while the throat pouch expands, trapping the fish. The bird puts its body and head above the water, enabling water to drain from the pouch. The bill is then lifted from the water and the fish swallowed. Since the water in the pouch may weigh more than the bird itself, it has to drain the pouch before it can move, an operation that can take almost a minute.*

Above *Three weeks after hatching, pelican chicks are able to walk. For 8–10 weeks thereafter they live in groups or "pods," though each chick is still fed by its own parents on regurgitated fish. Chicks grow so fast on this rich diet that they become heavier than adults. Once fattened up, they are left to fend for themselves.*

Below *Around the world, drainage for agriculture and human settlement is threatening to reduce or destroy the wetlands on which pelicans depend, together with a great diversity of other animals. Here, Pink-backed pelicans alight on a river in the vast inland delta system of the Okavango, Botswana.*

Running Risks

CONSERVATION AND ENVIRONMENT

While starvation, particularly among newly fledged and immature birds, is the main cause of death, other factors include predation by large mammals, epizootic disease, the effects of severe weather, particularly on birds already weakened by a heavy parasite load, accidents (typically involving power lines in some areas), and, importantly, large-scale massacre by humans, including the incineration of live young by fishermen in parts of America.

The most endangered species is the Spot-billed pelican, with a world population estimated in 1997 at 11,500 birds inhabiting southeastern India, Sri Lanka, and Cambodia, with possibly a few in Myanmar. The birds were once extremely abundant in Asia, where they were counted in millions. The Dalmatian pelican, which breeds exclusively in the Palearctic, was also recently considered endangered, although its population is now rising again in response to conservation measures, numbering an estimated 15,000–20,000 individuals. The states of the former USSR now contain a majority of the world population, notably Kazakhstan.

The most numerous species today is probably the Brown pelican, which includes the many thousand birds of the Peruvian race. The Pink-backed is also widespread in parts of Africa, although the exact size of the population is not known. JBN/EAS/RWS

Gannets and Boobies

GANNETS AND BOOBIES – THE SULIDS – ARE large seabirds, breeding from the Arctic through the tropics to sub-Antarctica. They are notable for their dramatic plunge-diving, gaudy colors, and teeming colonies. They are also uncommonly bold in defense of their nests, and as a consequence have been massively slaughtered.

All sulids breed in colonies, ranging from a million or more Peruvian boobies densely packed at 3–4 pairs to the square meter (3 per sq yd) to Abbott's booby, whose nests are well separated. Colony density varies fairly consistently with species, but within species the dense nesters (the gannets and Peruvian boobies) are less variable than the more widely dispersed ones. Peruvian boobies are one of the "big three" guano birds, the others being the Guanay cormorant and the Peruvian pelican.

Streamlined Plunge-divers

FORM AND FUNCTION

Sulids have a streamlined, torpedo-shaped, ventrally flattened body, a longish, graduated tail, and narrow, somewhat angled wings. The flight muscles are comparatively small, and wing-loading is high. To fly far and fast (necessary foraging adaptations), the body needs low drag. All sulids have high-aspect-ratio wings, but flight and diving ability varies with size, weight, wing proportions, and tail length. The light, long-tailed male bluefoot, which has a short upper arm, can dive with exceptional maneuverability into shallow water, while the heavy, shorter-tailed, long-armed gannet can plunge deeply into turbulent seas. The impact of the dive is cushioned by inflatable air-sacs between the skin and muscles.

Sulid bills differ in shape. Each species has its characteristic cross-sectional profile, which reflects differences in gripping power and functions such as the speed of bill-tip movement. These, in turn, are feeding adaptations. The cutting edges of the bill are saw-edged, and the upper mandible, which curves down at the tip, is movable upward, making it easier to accommodate large prey. There are no open external nostrils, since these would be incompatible with plunge-diving. Finally, sulids possess binocular vision, an important faculty for three-dimensional perception. The legs are stout, and the feet webbed between all four toes. Garish web colors are flaunted in display.

All sulids maintain waterproofing with the waxy secretion of the preen gland, a substance that also inhibits skin parasites. They molt their

Right *1 A Cape gannet* (Morus capensis) *in flight. This species breeds in the hot climes of South Africa and the coast of Namibia.* ***2*** *Atlantic gannets* (M. bassanus) *plunge-diving.* ***3*** *As a female Peruvian booby* (Sula variegata) *prospects for mates, a male "advertises" to her. Courtship usually begins in this way.*

Right *A pair of Masked boobies. This, the heaviest booby species, sometimes has trouble taking off. Colonies are thus usually found near cliffs or other areas where constant upward drafts of air make take-off easier. Some authors argue that the Galápagos race of this species (some populations of which have a distinct orange bill) should be assigned specific status, but this is not the practice in the case of the Brown and (especially) the Red-footed booby, in which populations differ much more markedly than in the Masked.*

flight feathers in stages, so there are always some new, some old, and some part-grown. The tail feathers molt irregularly. Molt occurs when the birds are under least stress, and is discontinued at taxing stages in their annual cycle.

Across the Oceans

DISTRIBUTION PATTERNS

Sulids breed across all three major oceans. One species or another breeds at all points between 67°N and 46°S. Atlantic gannets in Newfoundland may incubate on icy ledges in snow blizzards, while Masked boobies on the equator endure severe heat stress. Only the Masked, Brown, and Red-footed boobies, however, are widely distributed. Abbott's booby is the only sulid restricted to a single small locality (Christmas Island in the Indian Ocean).

Apart from Abbott's, the best documented populations are those of the three gannet species, none of which exceeds three quarters of a million individuals. There are 42 colonies of Atlantic gannets, mostly in Scotland, six African gannetries, and 37 individual locations for the Australasian species. African and Australasian gannets have interbred.

The most abundant sulids are the Peruvian boobies, of which there are millions except when populations have been decimated by periodic famines, and the widely-distributed Red-footed booby. The least numerous is the Abbott's booby, with only about 2,500 pairs in the world.

Gannets and boobies nest in a variety of species-typical habitats, sometimes on coastal headlands but usually on small islands, which provide protection from land predators while being surrounded by potential feeding areas. Precipitous cliffs, slopes, flat ground, low bushes, and climax (stable) forest trees are all used. Where two or more booby species overlap, they almost invariably use different habitats.

Communal Feeders

DIET

All sulids, unlike cormorants or pelicans, are exclusively marine. All except Abbott's forage in inshore waters, and some, especially the gannets, feed up to several hundred kilometers from the colony, even when breeding.

Communal feeding is a sulid characteristic (although not invariable), and fishing flocks may involve many hundreds of birds (especially Peruvian boobies and Atlantic gannets) plunging after prey in a veritable avian hailstorm. The initial dive may take the bird a meter or two beneath the surface, after which greater depth can be achieved by swimming, using the wings and feet. Red-footed boobies especially catch flying fish on the wing. Boobies also take advantage of fish driven to the surface by underwater predators such as tuna and dolphins. Some boobies work surflines, and the bluefoot often dives from the surface like a shag. Gannets commonly scavenge behind fishing-boats, voraciously gobbling up any discarded fish.

FACTFILE

GANNETS AND BOOBIES

Order: Pelecaniformes

Family: Sulidae

9 species in 3 genera: **Atlantic gannet** (*Morus [Bassanus] bassanus*), **Australasian gannet** (*M. [B.] serrator*), **African** or **Cape gannet** (*M. [B.] capensis*), **Blue-footed booby** (*Sula nebouxii*), **Brown booby** (*S. leucogaster*), **Masked, Blue-faced**, or **White booby** (*S. dactylatra*), **Peruvian booby** (*S. variegata*), **Red-footed booby** (*S. sula*), **Abbott's booby** (*Papasula abbotti*). The gannets are sometimes grouped into a single super-species, divisible into three allospecies.

DISTRIBUTION N Atlantic, South Africa, Australasia (gannets); pantropical oceans (boobies).

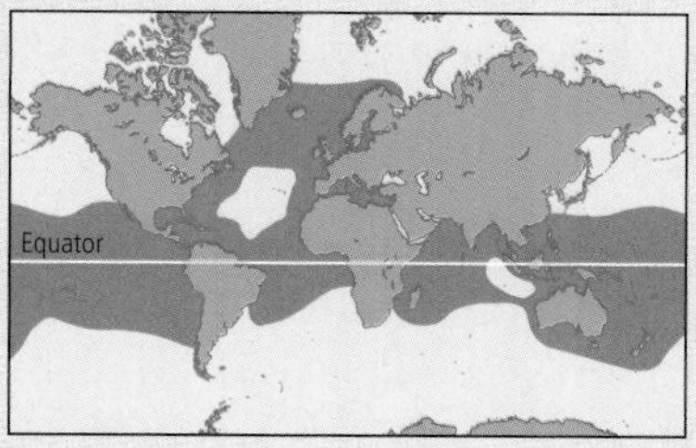

HABITAT Breed mainly on islands and isolated rocks.

SIZE Length 60–85cm (23.6–33.5in); wingspan 1.41–1.74m (4.6–5.7ft); weight 0.8–3.6kg (2–8lb). Females larger than males, or sexes similar.

PLUMAGE Adults of all species have white underparts (except for some races of the Red-footed booby), with variable amount of black or brown above; most have brightly colored bills, faces, and feet.

VOICE Raucous or sonorous single or polysyllabic grunts or shouts; thin whistles.

NEST Colonial, in structures that range from rudimentary to solidly built.

EGGS 1–4, depending on species; plain, whitish, blue-tinged, green, or pink, with a limey coating that becomes stained. Incubation 42–55 days.

DIET Fish, squid, offal.

CONSERVATION STATUS Abbott's booby, which breeds only on Christmas Island, is Critically Endangered, due to land clearance for phosphate mining and the introduction of the yellow crazy ant, which forages in the forest canopy and preys on nestlings. The Cape gannet is Vulnerable.

Conflict in the Colonies

BREEDING BIOLOGY

Before establishing a breeding site, the male reconnoiters the area from the air. He then selects a nest site, which he defends by fighting and display. In the three gannets, especially the Atlantic, territorial fights can be ferocious. In boobies the contests are briefer and less common. Indeed, Abbott's boobies rarely even grapple; the dangers attending a fall to the jungle floor, from which they are unable to rise, are too great.

The core boobies (all but Abbott's and the Red-footed booby) show a basically similar territorial display, although each species has evolved its own variant. In all sulids the display appears to derive from "redirected aggression," that is, aggressive behavior (such as biting the ground or a branch) that has become stylized and now takes the form of a modified, "polished" version of the original. Dispersed ground-nesters, such as the Blue-footed booby, parade around their territory and display at the boundaries. In gannets the nest site itself is the only territory, and the birds display there.

Female sulids prospect aerially and on foot for mates before selecting the male and site that suit them. Male boobies advertise to the females by means of a special call and a conspicuous display that is similar in all species except Abbott's. In the Blue-footed booby especially, the display involves a bizarre spreading and rotating of the wings, so

__Above__ Brood reduction in the Masked booby. The older and stronger chick can be seen attacking and forcing its sibling out of the nest where it will die. This sibling "murder" on average increases the probability of rearing at least one "fit" offspring.

__Left__ Atlantic gannets in flight over Bass Rock, Firth of Forth, Scotland. Gannets often forage and migrate far offshore, alternating steady, soaring flight with short glides. When fishing, they plummet into the sea from a height of 15m (50ft) or more.

that their upper surfaces face forward toward the female. Gannets, in contrast, have an inconspicuous display that does not share the same origins as that of the boobies.

Partners form and maintain a conspicuous bond by mutual behavior, which in the gannets and Abbott's booby takes the form of a spectacular face-to-face display. Mutual preening also helps cement the bond, as does copulation, which, especially in gannets, is frequent and prolonged and entails considerable tactile stimulation.

Sulid nests may be substantial structures, as on cliffs, trees, or potentially wet and muddy ground, or merely symbolic scatterings of material. In the Masked and Blue-footed boobies, for example, the male may bring hundreds of small fragments which have no architectural value but they are important as part of bonding.

Above *Sulid displays. All species have highly developed territorial and pair displays. Each species has evolved variants on common forms, involving exaggerated postures and movements of head, wings, feet and tail, and in some species aerial displays. Display sequences may be long and protracted. Here a male Brown booby makes a territorial display* ***1****, male and female Atlantic gannets greet each other* ***2****, a pair of Cape gannets copulate* ***3****.*

Left *An Atlantic gannet colony in Canada. Gannets are relatively silent birds except during the breeding season, when the roar of a whole colony is almost deafening.*

Eggs are small compared with those of most seabirds, varying in size between 3.3 percent of female weight in the Atlantic gannet up to 8 percent in Abbott's booby. Heavier eggs occur in species whose hatchlings may face dangerously long periods without food. The three gannets and the Red-footed and Abbott's boobies invariably lay a single egg. Masked and Brown boobies lay one or two, the Blue-footed booby two or three, and Peruvian boobies two to four. All sulids can replace eggs lost in the first half of the incubation period, but only Blue-footed and Peruvian boobies can rear more than a single chick to fledging. Eggs are incubated beneath the birds' webbed feet, which become vascularized and hot; sulids lack a brood patch. Both sexes share incubation, the stints varying with species and region. The eggs take 42–55 days to hatch, depending on species; those of Abbott's boobies take the longest.

All sulid chicks hatch essentially naked. Skin color varies with species and region, probably as a result of dietary differences. The down is white (with, in the case of Abbott's booby alone, a cape of black scapular feathers). Masked and Brown boobies often hatch two chicks, some five days apart, but the first-hatched invariably evicts and kills its younger sibling. If the first chick dies due to food shortage, however, the second can often survive, which gives the two-egg clutch its survival value. Attempts to persuade first-hatched Masked booby chicks to accept their sibling by removing the former until the latter caught up proved unsuccessful, and aggression persisted.

Left *The Red-footed booby is unusual among sulids in nesting in bushes and trees. This has had some minor consequences in its behavior; its displays, for example, are slightly less elaborate than those of ground-nesting species. Neither the cause nor the function of the tomato-red feet has been explained.*

Below *Blue-footed booby parents with their two chicks. The eggs are laid on bare ground surrounded by a ring of guano produced during incubation. This ring often marks an imaginary line that defines the nest site, and if a chick oversteps this area then it will not be treated as offspring.*

Blue-footed boobies sometimes do rear two chicks, but if food becomes scarce the first-hatched takes all the food and the other dies. Only the Peruvian booby, which benefits from an exceptionally rich food source of anchovies taken from the Humboldt Current, regularly rears two or three chicks.

The growth rate of sulid chicks varies enormously with species, depending on the abundance of the available food supply. Where food is plentiful and dependable, the young grow comparatively rapidly, fledging in fewer than 100 days in gannets. In contrast, Abbott's boobies take six months.

Hatching eggs and newly-hatched chicks are transferred to the upper surfaces of the parents' webs, otherwise they would be crushed or suffocated. Hatchlings take food directly from the adult's gape, which can be difficult, and adults may turn their head upside down to drop food into the trough of the upper mandible. They are brooded continuously for at least two weeks, after which booby chicks may be left unguarded, a dangerous strategy that is nonetheless essential if the food situation demands foraging by both adults simultaneously. The chick's begging behavior always involves pestering the adult's bill, but the violence of the begging is much greater in ground-nesting species than in Abbott's boobies, for which falling to the ground from the forest canopy is an ever-present danger.

The point at which fledglings become independent from parents is hugely significant. In boobies, post-fledging dependence ranges from one to six months, the latter in Abbott's and some populations of Red-footed boobies. During this period, juveniles learn to fish proficiently before dispersing widely for a long prebreeding period. In the three gannets, however, the fledglings go straight to sea, unaccompanied; there is no post-fledging support or desertion period. This behavior greatly affects immediate post-fledging survival rates, which is relatively low in gannets.

Young birds exhibit clear directional migration only in the three gannet species. The other sulids disperse to differing extents. Adults do not accompany their offspring. After a variable period of nomadism, almost all young of all but Abbott's booby return to their natal colony to breed. Once settled in a breeding colony, however, sulids very rarely breed anywhere else. Some, principally gannets and Abbott's booby, remain faithful to a single site and mate; others may change either or both.

Life expectancy is affected by human activity (pollution, killing, habitat destruction). Some sulids can live for 40 years or more, but average life expectancy is less than half that figure.

Declining Populations

CONSERVATION AND ENVIRONMENT

Abbott's booby is now subject to special protection measures (around 75 percent of Christmas Island, where it breeds, is a National Nature Reserve, and reafforestation is underway). However, plans by the Australian government to build an immigrant detention center next to the largest Abbott's booby colony are causing concern.

Many sulid populations are declining. Despite conservation laws, killing of boobies and egg-collecting still persist in many pantropical regions. Peruvian boobies suffer from periodic natural disasters, usually associated with the El Niño climate events, exacerbated by overfishing of the Humboldt Current. Some African gannets, notably the Cape gannet, are also threatened by overexploitation of their prey and pollution of their waters. JBN

Above *Brown booby nesting colonies are protected in Australia. The Great Barrier Reef World Heritage Area is an important place for breeding boobies, as there is an abundance of suitable islands for nesting in the northern and southern quarters of the Area.*

Tropicbirds

Except when breeding, tropicbirds scorn land entirely, being distributed sparsely over vast areas of ocean. The adults' two long, central tail feathers are highly distinctive; their supposed resemblance to marlinspikes earned the birds their alternative name of "bosun birds" among sailors.

Tropicbirds inhabit regions of warm, clear, and salty water at temperatures of 24–30°C (75–86°F), often occurring in remote ocean reaches bereft of other seabirds. They are characteristic of the tradewind areas of the Central and West Pacific and the southern Indian Ocean.

Built for Flying

FORM AND FUNCTION

Medium-sized, robust but streamlined, tropicbirds have greatly elongated, flexible streamers in the center of the wedge-shaped tail. These wear particularly quickly and are constantly replaced, unrelated to the other flight feathers. The sharply pointed wings span 1m (3.3ft) or more. Although the wingspan of the White-tailed species is nearly 80 percent of the Red-tailed, the former weighs only about half as much.

The pelvis, legs, and webbed feet are small and weak, and tropicbirds are virtually helpless on the ground. In contrast, the pectoral girdle is strong (the high-aspect-ratio wings imposing high wing-loading), and, since tropicbirds depend on rapid and continuous wingbeats, the flight muscles are large and the breast bone deeply keeled. The bill, bright red or yellow in color, is stout, decurved, and sharply pointed, with a finely serrated cutting edge and slitlike nostrils. The eyes are large and dark. Tropicbirds have no brood patch; warmth reaches the egg through the belly feathers. Juveniles are barred with black on a white background. Tropicbird plumage is fully waterproof, and the birds can remain at sea indefinitely.

Often found far from their breeding locations, although nearer during egg-laying and incubation, mostly solitary or in small groups, tropicbirds eschew large, mixed feeding flocks. They dive from a few meters to about 50m (165ft) up with half-closed wings, sometimes spiraling before emerging with prey, held crosswise. A single item may weigh almost one-fifth of the bird's weight. As with other pelagic feeders, food for the chick is carried in the crop or lower in the alimentary tract, where it may become coated with mucus. The birds swoop and plunge for food, catching mainly squid and flying fish up to 20cm (8in) long.

Above A newly hatched White-tailed tropicbird chick, clearly showing its dense grey down. This down can be between 1–3cm thick. After hatching chicks are left alone at the nest while their parents are away foraging. In breeding colonies, these small chicks are vulnerable to attack by adults of the same or related species in search of nest sites.

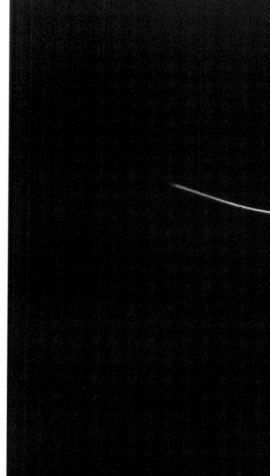

Right The Red-billed tropicbird is found in the tropical western Pacific, the central Atlantic, the Red Sea, and the Persian Gulf. It is the only species in the family Phaethontidae which carries the juvenile black barring plumage on its back into adulthood. When on the ground tropicbirds can only shuffle as their short, set-back legs cannot support their weight on land.

Birds of the Far Ocean

DISTRIBUTION PATTERNS

Red-billed and White-tailed tropicbirds occur in all three major oceans; the Red-tailed is absent from the Atlantic. The White-tailed is the most numerous species, especially in the Caribbean, and is widely distributed. Probably the least populous, numbering perhaps fewer than 10,000 pairs, is the Red-billed, which may have its biggest population in the Galápagos Islands. The Red-tailed, too, is most numerous in the Pacific, although recently it has declined greatly in numbers.

Tropicbirds nest on oceanic islands, in holes in cliffs or on ledges, under boulders on slopes, sometimes beneath vegetation, and even, in the case of the Christmas Island White-tailed population, in holes or crotches in inland jungle trees. Cliff cavities are preferred, for shade and easy take-off; heat-stressed birds may desert the nest.

Colonies are neither large nor dense, but there may be considerable, sometimes bloody, competition for nest-sites, both within and between species. Many pairs nest in semi-isolation. A feature of some colonies is the presence of adult-plumaged nonbreeders, which may interfere with breeders, even to the extent of killing chicks.

No tropicbird is truly migratory, although there may be some significant transequatorial movement. They become pelagic, solitary wanderers over tropical and subtropical seas, although there may be some birds around colonies at all times.

Courtships in the Air

BREEDING BIOLOGY

Tropicbirds first breed between the ages of 2 and 5, generally at 3–4; a breeding cycle takes 21–27 weeks. In experienced pairs, which tend to remain faithful to one another, one partner, usually the male, returns to a previously used site and awaits his mate. Reoccupation of a site may be episodic and prolonged before breeding begins.

Tropicbirds lack ritualized territorial display, but often vigorously contest sites, stabbing, slashing, and interlocking bills, and twisting each other into contorted positions using the wings as levers. There is no ritualized appeasement behavior.

Tropicbird courtship, which functions primarily to form a pair bond rather than to reinforce an existing one, is mainly aerial. It is complex, involving up to 20 birds flying in large gyrations as much as 100m (330ft) up in the sky, with ritualized wingbeats, stiff-winged glides, and, in the case of Red-billed tropicbirds, upward and backward fluttering combined with calling. Such displays appear to be primarily sexual in motivation, serving to advertise the male. Once formed, tropicbird pairs seem notably deficient in structured interactions at the site.

The nest is a mere scrape, being constructed mainly by the male, who loosens the substrate with his bill and kicks it backward with sharp-clawed feet. The single egg is around 8 percent of female weight. Incubation is shared. At up to 13 days, incubation stints are long enough for protracted foraging by the mate.

Hatchlings are covered in long, dense down and are more advanced than those of any other pelecaniform (if, indeed, tropicbirds properly belong in this order). They reach 110 percent or more of adult weight, although they lose some weight before fledging at 70–85 days. Unusually for pelecaniforms, the young do not reach into their parents' throats for food, but instead take the adult's bill into their own gape. Although new chicks are closely brooded, they are left largely unattended by day 4 or 5, releasing both adults for foraging.

When ready to fledge, young tropicbirds, which have generally gone unfed for about a week, have usually had no practice flights. They are not accompanied to sea by the adults, and must achieve the transition to independence without any assistance.

Although not considered threatened, there are some areas in which tropicbirds are still persecuted for their beautiful tail plumes.
JBN/MPH

FACTFILE

TROPICBIRDS

Order: Pelecaniformes

Family: Phaethontidae

3 species of the genus *Phaethon*: **Red-billed tropicbird** (*P. aethereus*), **Red-tailed tropicbird** (*P. rubricauda*), **White-tailed tropicbird** (*P. lepturus*).

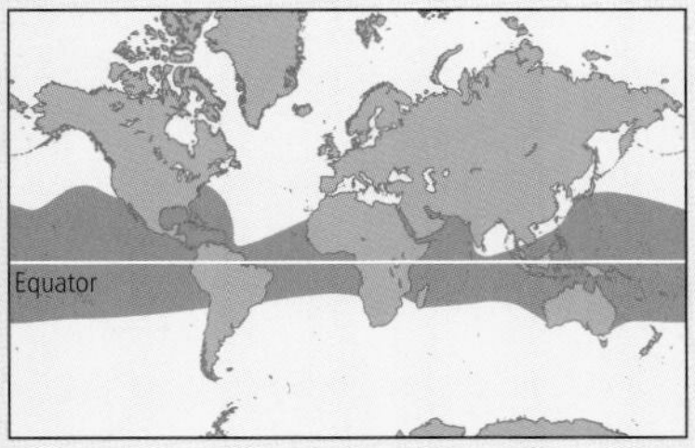

DISTRIBUTION Tropical and subtropical regions

HABITAT Oceans

SIZE Length 80–110cm (31–43in), including tail streamers; **wingspan** 90–110cm (35–43in).

PLUMAGE Silvery white, with black markings on the back and wings; some adults tinged rosy or gold. Sexes similar, but the two greatly elongated inner tail feathers are longer in the male, and absent in juveniles.

VOICE Shrill screams

NEST On bare ground, or in holes in cliffs or trees.

EGGS 1; blotched red-brown. Incubation 40–46 days.

DIET Fish, squid.

CONSERVATION STATUS Not threatened.

Cormorants

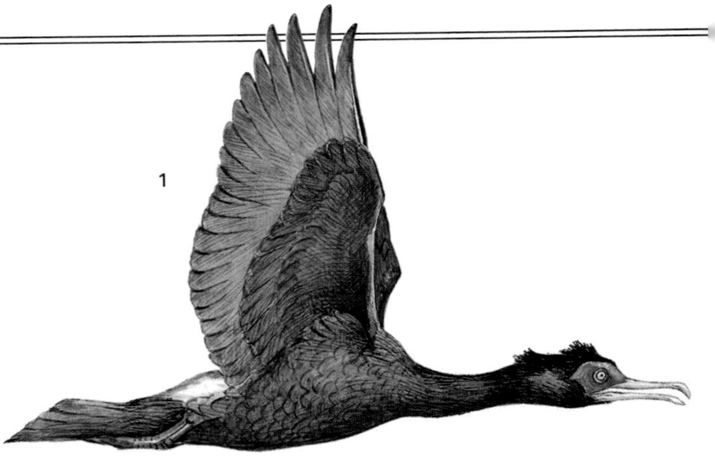

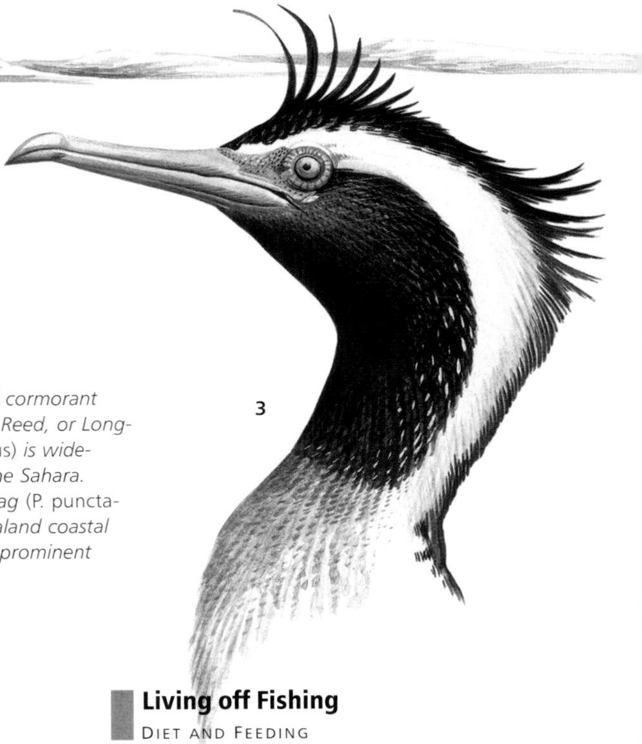

Above *1 The Red-faced cormorant* (Phalacrocorax urile). *2 The Reed, or Long-tailed cormorant* (P. africanus) *is widespread in Africa, south of the Sahara.* *3 The head of a Spotted shag* (P. punctatus). *Found only in New Zealand coastal waters, adults develop two prominent crests during courtship.*

DERIVING THEIR NAME FROM THE LATIN TERM corvus marinus *("sea raven"), cormorants are highly adapted for underwater hunting. Their bodies are streamlined and somewhat flattened beneath, the neck is long and supple, the wings broad, long, and blunt, and the legs powerful and set far back. Using their lean bodies, they thrust through the water and along the seabed to flush out prey. They range in size from the Pygmy to the goose-sized Great cormorant; the heaviest is the flightless Galápagos cormorant.*

Twenty-nine of the 39 or so species are entirely (or almost entirely) marine; four are freshwater residents, and six are at home in inshore waters, estuaries, or inland. They nest on shores, spray-swept rocks, islands, cliffs, trees, bushes, and in reed beds. They can forage in manmade ditches in the tropics and in the icy stormy seas of the Antarctic. Only the open seas are barred to the birds.

Coastal Divers

FORM AND FUNCTION

Cormorants have large webbed feet that produce powerful stern propulsion but are also flexible enough to allow some species to nest and roost in trees. The bill is typically long and sharply hooked; the edges of the mandibles are not toothed. The sides of the mouth and throat are hugely distensible, allowing cormorants to swallow large prey. The shape of the eye lens is highly modifiable, for improved vision underwater.

Like pelicans, cormorants have broad breast bones without the deep keels found in birds with powerful flight muscles. In the Double-crested cormorant, for instance, these muscles weigh only about half as much, as a percentage of body weight, as in birds with strong flapping flight. The birds typically fly low over the water with the neck extended, employing continuous wingbeats. Cormorant bones are much denser than, for example, those of pelicans. Together with a notable absence of body fat, this feature helps to reduce buoyancy, and so facilitates underwater pursuit of prey. The birds swim low in the water or partly submerged. Underwater, they steer with their webs and tail, keeping their wings close to their sides

Although most cormorants are black, often with an iridescent sheen, many species display striking nuptial adornments of crests and filoplumes, together with garishly colored facial skin, gapes, and orbital rings (the ring around the eye). Juveniles are drab brown and white. Like some boobies, certain cormorants exhibit genuine color morphs within the same population.

Inshore or Inland

DISTRIBUTION PATTERNS

Cormorants are widespread along most coastlines and on many inland waters. They breed from well north of the Bering Strait south into the Antarctic and around the globe. Marine species are confined to inshore waters, and most oceanic islands lack them; they are also absent from northern regions of central Asia and from arid continental wastelands. The Australasian region supports the greatest number of species, but the greatest concentrations occur in food-rich inshore waters such as the Humboldt, Benguela, and California Current areas and cool, temperate areas of the sub-Arctic and Antarctic.

The least specialized species, such as the Great cormorant, have the widest distribution. By contrast, several of the Antarctic shags are confined to specific small islands and their populations are extremely small (in one extreme case, just a few hundreds). The only flightless cormorant, found on the Galápagos island of Fernandina, numbers just 700–800 pairs, whereas in contrast there are many millions of guanays in Peru. The world population of the common and widespread Great cormorant is impossible to compute.

Living off Fishing

DIET AND FEEDING

All cormorants pursue prey underwater, using only foot propulsion. They have a large volume of blood in proportion to body weight, and the stored oxygen enables some species to remain submerged for up to four minutes. The ratio of dive duration to time spent on the surface is complex. Diving speed is about 0.69–1.01m/sec (2.3–3.3ft/sec). No cormorant forages far out at sea, but flights from the breeding colony to feeding locations may stretch to scores of kilometers. Often two or three cormorant species overlap, but species-typical differences in feeding habits involving depth, bottom feeding, and other factors tend to reduce competition, although there is considerable overlap in the prey species taken. Precisely what type of prey cormorants take has been determined by studying their mucus-coated pellets, which contain the hard parts of prey and other regurgitates. Although they mainly subsist on a wide variety of fish, cormorants also take amphibia, crustacea, cephalopods, and other invertebrates. Communal and probably cooperative feeding occurs; the extreme example of the former is the vast feeding flocks of guanays off Peru.

Calculations of daily energy expenditure suggest that wild Great cormorants need to eat more than 500g (18oz) of food per day. Cormorants are such efficient hunters that the daily requirements can be met in two fairly short fishing periods. Because their plumage is not waterproof, the periods are generally short, in most cases lasting little more than 30 minutes.

Colonies on the Cliffs

BREEDING BIOLOGY

All cormorants are colonial, although, with a few notable exceptions, the breeding colonies are small. Colonies are of two kinds – large, dense aggregations in superproductive upwelling areas which, because they bring nutrients from deeper layers of the sea, are rich in plankton and therefore in fish, or, more typically, small groups of tens or hundreds. Colonies persist from year to year, especially on small islands, although some species may from time to time shift sites several kilometers. An individual colony of guanays may hold millions of birds, packed so densely that they look like the pile on a carpet. They return from feeding in dense ropes and skeins many kilometers long, and fill the air with a deafening gabble.

In aseasonal colonies breeding may be almost continuous, as in the Galápagos cormorant, whereas in high latitudes it is confined to a short summer period. Some cormorants breed opportunistically, regardless of season. Breeding cycles are very much shorter than those of boobies and,

Below *Great or White-necked cormorant underwater in Lake Tanganyika, Tanzania, Africa. Usually cormorants feed at depths of less then 7m (25ft), but some marine species have been caught in nets 20–30m (70–100ft) deep.*

Below right *A Double-crested cormorant, the most widespread cormorant in North America, eating fish. Cormorants rarely eat their catch under water, but bring it to the surface and toss it about in their bills to get it the right way round before swallowing it whole.*

FACTFILE

CORMORANTS

Order: Pelecaniformes

Family: Phalacrocoracidae

Up to 39 species of the genus *Phalacrocorax*. Species include: **Brandt's cormorant** (*P. penicillatus*), **Cape cormorant** (*P. capensis*), **Double-crested cormorant** (*P. auritus*), **Galápagos flightless cormorant** (*P. harrisi*), **Great cormorant** (*P. carbo*), **guanay** (*P. bougainvillii*), **Olivaceous cormorant** (*P. olivaceus*), **Pelagic cormorant** (*P. pelagicus*), **Pygmy cormorant** (*P. pygmaeus*), **Red-faced cormorant** (*P. urile*), **shag** (*P. aristotelis*), **Spotted shag** (*P. punctatus*), and the **blue-eyed shags** of the sub-Antarctic (several species, including some endemics).

DISTRIBUTION Worldwide; few at high latitudes.

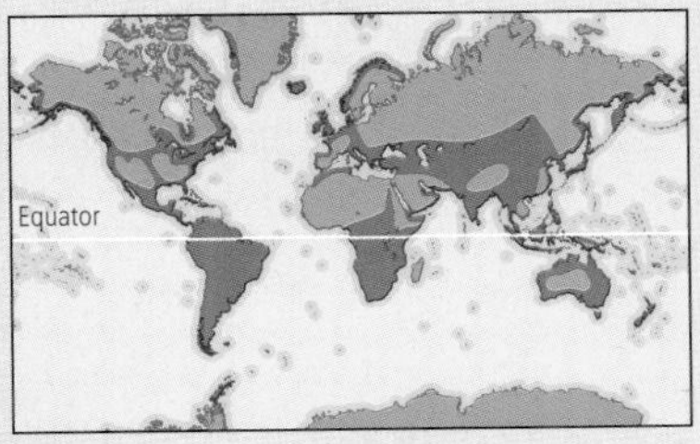

HABITAT Inland waters and marine shorelines.

SIZE Length 45–101cm (17.7–39.8in); wingspan 80–160cm (31–63in); weight 0.9–4.9kg (2–11lb).

PLUMAGE Generally drab black, brown, or blackish with a green sheen; some species have white breasts.

VOICE Grunts, croaks, and whistles, but mostly quiet.

NEST In colonies; the female builds the nest with material supplied by the male.

EGGS 1–6; chalky blue, elongate ovoid. Incubation 28–32 days.

DIET Small fish and marine invertebrates.

CONSERVATION STATUS The Galápagos cormorant and the Chatham Island shag (*P. onslowi*) are listed as Endangered, and 8 other species are Vulnerable. Pallas's cormorant (*P. perspicillatus*), which was restricted to Bering Island, became extinct in the mid-19th century.

especially, frigatebirds. Chicks take only 5–9 weeks to fledge, and the post-fledging feeding period typically lasts for about a month, although it is much longer in a few species. A representative cormorant breeding cycle occupies 19–20 weeks.

In seasonal climates, cormorants breed successfully no more than once a year. However the Galápagos cormorant can breed more often by abandoning the fledgling to the male's care while the female embarks on a new cycle with another partner. Bank cormorants and guanays sometimes manage two broods a year.

Cormorants tend to start breeding when comparatively young. This behavior, along with large broods, rapid growth, and a relatively short life, is part of a cluster of adaptations related to the birds' inshore feeding habits that contrasts with an

Left *Great cormorants of the continental race* (Phalacrocorax carbo sinensis) *wintering at Lake Kerkini, Greece. Along with the Caspian population of the Pygmy cormorant, these are the family's only true migrants.*

opposite cluster in far-foraging, pelagic seabirds.

Even where a colony endures from year to year, cormorants (unlike gannets) are not faithful to a particular nest site, except, occasionally, in the case of species such as the Antarctic shag, in which a durable pedestal remains. Competition for sites may involve brief bill-grappling or seizing the opponent by the neck or a wing, but fights are usually superficial, escalating out of threat behavior (which involves staring, pointing, gaping, and thrusting). Males growl or bark raucously, while females hiss and click. Nest material is seized and quivered in a form of redirected aggression, and the substrate is attacked in lieu of an opponent.

Usually, cormorants change mates for successive breeding attempts. Pairs form in response to a complex advertising display from the male, usually from a potential nest-site. There are strong generic similarities in the display, although each of its various components is modified according to species. Special plumage features, such as head plumes and white thigh patches, develop before the onset of courtship and serve to emphasize the display; for example, wing-flicking may alternately cover and expose the white thigh patch. Another component, in which the head is thrown right back, in some species until the crown rests on the base of the tail, is common to most cormorants. Further displays accompany changeover on the eggs. Before egg laying, there may be copulations involving partners other than the mate. DNA analysis has revealed that, in the European shag, up to 18 percent of chicks result from a female mating with a bird other than her mate.

Males select the nest-site. In all cormorants the nest itself has an important function in raising the eggs and chicks above stony or muddy ground, or else as a platform on ledges or in trees.

During incubation, which is shared, the birds' webs are inserted beneath the eggs, which then receive warmth through the abdominal feathers. Incubation stints rarely last more than six hours. The chicks hatch naked and helpless, but are soon covered in gray or blackish down, which clears from the true feathers after 40–50 days. The young squeal for food with a closed bill but beg silently for water with an open bill. They feed by groping into the parent's throat, a difficult feat for newly-hatched young; many die in their first two or three days of life. Some cormorant chicks suffer from the heat and regulate their temperature by fluttering throat skin, panting, and exposing their webs so as to radiate heat. Many species leave the nest before they can fly, forming crèches in the manner of pelicans. The shortest period before flight is about 5 weeks, in the Crowned cormorant.

Below *Kerguelen shags* (Phalacrocorax atriceps verrucosus) *are endemic to Kerguelen Island in the Southern Indian Ocean. They nest in colonies and build their nests on the ground close to one another.*

Above *A Great cormorant enjoying the sun and drying its wings. After fishing expeditions cormorants emerge from the water thoroughly soaked because of their loose-fitting feathers. Before flying again they stand in the sun to dry them off.*

Breeding success, as judged by the ratio of chicks fledged to eggs laid, varies from 20–60 percent, and cormorants rear on average between 0.3 and 2.5 young per successful breeding attempt. Few cormorants live for more than 10–15 years. Natural mortality figures are hard to come by, because in many species they are vitiated by human persecution and loss of rings, but available figures suggest a prebreeding mortality of 60–80 percent, mostly in the first year.

Rivals and Allies

Conservation and Environment

Many species of cormorant are hated by fishermen because they take prey that humans value, such as salmon smolts. They also prey on fish reared by fish-farmers, especially catfish in the Mississippi area of the United States. Consequently, cormorants, especially the Great and Double-crested cormorants, are subject to often intense persecution by humans. Even rare Antarctic endemics are sometimes killed. However, most cormorants are not endangered.

Since at least the 5th century in Japan and the early 17th century in Europe, cormorants have been used for sport and commercial fishing. This kind of fishing has been described as an ancient form of aquatic falconry. Teams of birds, tethered to a line and with neck collars that prevent them from swallowing the fish, are released, hauled in, and their catch removed. Eventually the birds are permitted to eat some of the fish; if not, it is said, they soon go on strike. JBN/DS-C

Frigatebirds

In the days of sailing ships, frigates were swift vessels, often used for commerce-raiding. By analogy, frigatebirds (also called "men-of-war") are maneuverable brigands of the air. Spectacularly combining highly-developed soaring and gliding skills with matchless speed and agility, they pirate the catches and nest materials of other seabirds or swoop down to snatch prey from the water's surface.

Another especially noteworthy feature of frigatebirds is the male's prominent red throat-patch, which is inflated like a balloon during displays to attract a prospective mate. Throughout the tropics, frigatebirds nest in colonies on remote islands.

The Most Aerial Seabirds

FORM AND FUNCTION

The frigatebird's outline is among the most instantly recognizable of all seabirds'. When soaring, the angular, pointed wings and long, scissor-shaped tail are quite unmistakable. At shorter range the long bill, slightly saddle-shaped and sharply hooked at the tip, is equally distinctive.

Frigatebirds' plumage is not waterproof, and their legs and feet are so tiny (and the latter incompletely webbed) that they rarely swim and have difficulty rising from the water. Yet, despite this they roam far out to sea when feeding and, outside breeding, may cross thousands of kilometers of ocean.

Four of the five species breed in the Atlantic, the Ascension frigatebird only at Ascension Island. The only species not found in the Atlantic, Andrew's frigatebird, is confined to Christmas Island in the eastern Indian Ocean. The two most widespread species, the Great and Lesser frigatebird, share about half their breeding stations with one another.

PIRACY BY FRIGATEBIRDS

Weak legs, tiny feet, and non-waterproof plumage restrict frigatebirds to feeding at, or above, the surface of the sea. Their feeding technique is a spectacular swoop to just above the surface, the head snapping down and back for the bill to pluck a flying fish or squid from the jaws of a pursuing tuna. They use the same method to pick up floating nest material, and can snatch a twig from glass-calm water without making a ripple.

It takes years for a frigatebird to perfect the technique sufficiently to be able to support itself and a chick, so piracy – robbing boobies, terns, and other birds in flight after compelling them to regurgitate their catch, often by upending them – can be a valuable supplementary feeding method. Adult males also chase birds – often of their own species – to rob them of nest material. Piracy is so conspicuous around some colonies that frigates have the reputation of acquiring much of their food that way, although in fact it is mostly used by individual specialists. As a feeding technique piracy is especially valuable because the depth at which the birds can feed for themselves is so limited; by robbing species that dive below the surface, they effectively increase the depth of water they can exploit. AWD

Showing Off to the Females

BREEDING BIOLOGY

The males choose possible nest sites, usually in trees – up to 30m (100ft) high in Andrew's frigate – but where there are none, as on Ascension Island, they will use low bushes or even bare ground. Advertising males display in groups of up to 30; when prospecting females fly overhead, the males spread and vibrate their wings, throw back their heads, and call or clack their bills above the distended scarlet throat-pouch. If unsuccessful in attracting a female, a male may move to display in

Right *A Magnificent frigatebird and her chick on the nest. The hatchling is born naked, but by 4–5 weeks develops a cape of black feathers. White down, scrubby at first but growing longer, then begins to cover the rest of its body.*

Left *A male frigatebird's extraordinary crimson throat-pouch, fully inflated. During their pre-pairing display, these striking adornments fill the trees of the breeding colony like exotic blooms.*

another group. There is consequently little advantage in defending a display site, and advertising males are notably unaggressive to their peers.

When a female lands by a potential mate, the two snake their head and neck across each other, occasionally vocalizing and nibbling at the other's feathers. The males' communal courtship display is noisy, vivid, and spectacular, but the subsequent repertoire of displays between paired birds is low-key and desultory, probably reflecting the comparatively weak pair bond. Male Great and Lesser frigatebirds, even on the same island, show clear differences in the group size of displaying males and in the strength of the tendency to stick with one group until successful.

Males (except in the Ascension frigatebird) collect twigs for the nest, often snapping them off in flight. The female builds the nest while guarding it from theft by other males. The birds have difficulty in procuring twigs, however, and the nest is often flimsy in the extreme. Some eventually disintegrate beneath the chicks, which must then survive precariously on a small perch for many weeks.

The single white egg weighs up to 14 percent of the female's weight, and is incubated by both birds for 6–8 weeks in shifts of up to 12 days at a time. The birds have no brood patch. A puzzling feature of frigatebird colonies is that eggs are sometimes thrown out and chicks killed, apparently by unmated males. This behavior possibly releases the bereaved female back into the pool available to males, although it has yet to be shown that such females remate.

The frigatebird chick grows as slowly as any seabird, spending 5–6 months in the nest and remaining dependent on its parents for food for several months more – sometimes over a year. Frigates probably first breed at about 7 years, and their reproductive output is so low that adults must live at least 25 years on average.

A successful breeding attempt takes more than 12 months, so adults that rear young one year cannot breed the next; thus, like some of the larger albatrosses, most frigatebirds probably breed successfully in alternate years or even less often, since they need a recuperative period. Birds cannot return to the same nest-site each year, reckoning to meet their mate there, because in the interim another pair will have taken it over; the pair-bond is accordingly weak. Breeding success varies with locality, but can be very low, perhaps the lowest of any seabird, due in some areas to difficulties in obtaining enough food. Many newly independent juveniles starve even after their long period of parental subsidy.

One species, the Magnificent frigatebird, shows slight differences from the family pattern. It feeds closer inshore than the others and apparently has a more reliable food supply, for not only are incubation shifts shorter than in any other frigate (averaging one day or less) but the male deserts the colony when the chick is about 100 days old, leaving the female to raise the chick alone. He probably goes away to molt before returning for another breeding attempt with a different female while his previous mate is still feeding their offspring. Such a division of labor between the sexes would be unique, but remains unproven with marked birds.

Several colonies are still significantly persecuted. Because of the frigatebird's long lifespan, a large drop in a local population can go unnoticed for years. There is, however, now cause for concern over the scale of decline in some areas. JBN/AWD

Below *Unable to plunge-dive because of the lack of waterproofing on their feathers, frigatebirds have evolved to become extremely adept at taking prey from the surface. Here, a Magnificent frigatebird fishes in the shallows.*

FACTFILE

FRIGATEBIRDS

Order: Pelecaniformes

Family: Fregatidae

5 species of the genus *Fregata*: **Ascension Island frigatebird** (*F. aquila*), **Andrew's** or **Christmas Island frigatebird** (*F. andrewsi*), **Great frigatebird** (*F. minor*), **Lesser frigatebird** (*F. ariel*), **Magnificent frigatebird** (*F. magnificens*).

DISTRIBUTION Pantropical

HABITAT Oceans

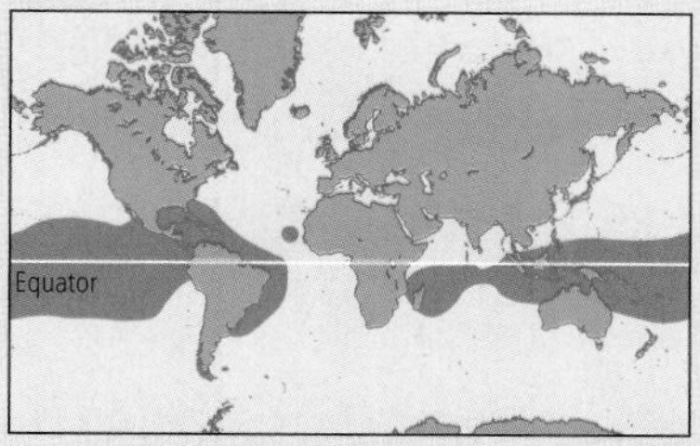

SIZE Length 79–104cm (31–41in); wingspan 1.76–2.3m (5.8–7.5ft); weight 0.75–1.6kg (26–57oz). Females 25–30 percent heavier than males.

PLUMAGE Males all black (Lesser frigatebird with white axillary spurs); females black in Ascension frigatebird, other species blackish brown above and on head and belly, with white breast and (in some species) axillary spurs.

VOICE Varied rattling, whistling, hoarse yakking, and bill-rattling; males give resonant drumming, warbling, or whistling sounds.

NEST Built by the female with twigs supplied by the male; set in trees where available, otherwise on the bare ground.

EGGS 1; white, smooth. Incubation 44–55 days.

DIET Mainly flying fish and squid, caught on the wing.

CONSERVATION STATUS The Christmas Island frigatebird is considered Critically Endangered because of its limited distribution and the risks associated with the introduction of a predatory ant species to its island habitat. The Ascension frigatebird is Vulnerable.

Darters

Left *1 The New World anhinga* (Anhinga anhinga) *nearly always perches in trees and rarely walks.* **2** *When swimming, the African darter* (Anhinga melanogaster rufa) *propels itself with its feet and tail, with its wings often partially spread.*

FACTFILE

DARTERS

Order: Pelecaniformes

Family: Anhingidae

2 species of the genus *Anhinga*: the New World **anhinga** (*Anhinga anhinga*) with 2 subspecies, South American (*A. a. anhinga*) and American (*A. a. leucogaster*); and the Old World **darter** (*Anhinga melanogaster*), with 3 subspecies: African (*A. m. rufa*), Asian (*A. m. melanogaster*), and Australian (*A. m. novaehollandiae*). Some authors treat the subspecies as species.

DISTRIBUTION America, Africa, Asia, Australasia.

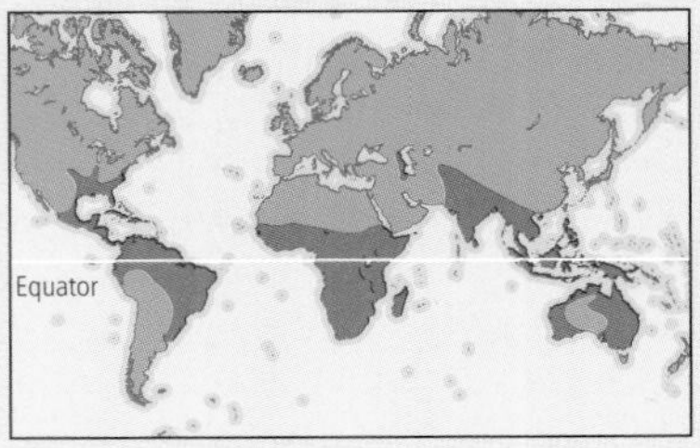

HABITAT Fresh waters (lakes, lagoons, rivers).

SIZE Length 76–98cm (30–39in); wingspan 120–127cm (47–50in); weight 0.9–2.6kg (2–5.7lb).

PLUMAGE Mixture of black, brown, gray, silver, and white; males darker than females.

VOICE Ratchet-like clicking, rattling, hissing (threat); loud or soft caw.

NEST A cup of dry twiglets laid on a base of mainly green twigs, with a lining of leaves; usually built in the fork of a tree branch over water, but sometimes in bushes or reeds.

EGGS 2–6 (average 4); elliptical, pale green, with white, chalky outer layer. Incubation 26–30 days.

DIET Mainly fish.

CONSERVATION STATUS The Asian darter is listed as Lower Risk/Near Threatened.

SLENDER AND LARGE, DARTERS LOOK RATHER like a cross between herons and cormorants. The slim head, sinuous neck, and wings are heron-like, while the plumage patterns, feather structure, rump, and feet more closely resemble those of the cormorant.

Darters are entirely freshwater birds that occur in quiet waters in tropical, subtropical, and warm temperate parts of America, Africa, Madagascar, Asia, and Australasia. They only incidentally come into conflict with humans, although some get caught in nets set for fish. They readily colonize artificial lakes, and increased in numbers in New Guinea following the introduction of the African tilapia fish to the southern lowlands. In Southeast Asia, people use darters for fishing, as they do, in a more organized way, with cormorants.

Spear-fishing Snakebirds

FORM AND FUNCTION

The long, thin, straight bill has finely-toothed cutting edges, and is often used to spear fish. It does not have a terminal hook (unlike the shags and cormorants), but a hinge allows the upper mandible to move upward, thereby increasing the size of the gape. The head is small and flattened, and the neck thin and G-shaped. To reduce buoyancy, the body plumage is permeable to water, as it is also in shags and cormorants. The legs are short and stout with long toes, all four are webbed, as in most other pelecaniforms.

Plumage patterns vary between species, sexes, ages, and individuals; all are dark above, with dark tails. Long, slender scapular and mantle covert feathers have silver, white, gray, or light brown streaks, more marked in adults than in immature birds. Males are blacker than females, and immatures paler than adults, mainly on the head, neck, and underparts. As in ducks and rails, all the flight feathers of the wings are shed at the same time, and darters are then temporarily flightless. The long tail has corrugated outer feathers, which are shed gradually. Although their short legs are set well back, darters are agile among branches, which they grasp with prehensile feet to climb adroitly.

Darters are adapted for soaring on thermal updrafts. Because of their long tails, they resemble flying crosses when gliding. Long-distance flights are usually made by spiraling up on a thermal updraft and then gliding down to the next one. Sometimes they climb trees and bushes before take-off. When they take off from water, both feet kick together, as in other pelecaniforms.

Darters are adapted for moving slowly underwater, submerging gradually and often swimming with only their head and neck exposed, like a swimming snake (hence their popular name of "snakebird"). They stalk their prey underwater, often moving very slowly with the neck coiled and the wings and tail spread. Between the seventh and ninth neck vertebrae there is a special hinge mechanism that enables the birds to dart their neck forward to snap up insects at the water surface or to stab fish, with the bill slightly open. A stabbed fish is usually shaken loose, flipped into the air, and swallowed head first. The same darting and stabbing mechanism is used in defense, to stab at the eyes of predators and unwary human molesters.

Rearing Chicks by the Waterside

BREEDING BIOLOGY

Darters are territorial and may defend not only small resting and nesting sites but substantial feeding, resting, and nesting areas, mainly by threat and display. They do, however, mix with other water birds, such as herons, egrets, ibises, and spoonbills, at roosts and nesting colonies.

Breeding starts with a male selecting a nest-site and claiming a territory around it, which may include a whole tree. He decorates the site with a few fresh, green, leafy twigs, and displays to attract a mate and ward off other darters. There are strong similarities between darter and cormorant displays, which both include wing-waving in which the partly-closed wings are alternately raised, and twig-grasping, in which a nearby twig or stick is grabbed and shaken vigorously; indeed, darters differ more from cormorants physically than in their behavior. After pair-formation, the female builds the nest, often using an old one as a base, with material brought mainly by the male. Over the birds' range as a whole, breeding may occur at any time of year, and the entire cycle takes only three months.

The first egg is laid 2–3 days after the start of a nest, and subsequent eggs at intervals of 1–3 days. Incubation (averaging 28 days) starts with the first egg. The eggs hatch at intervals of 1–4 days. The parents take turns in guarding the nest, from the start of building until the chicks are at least 1 week old. As long as the chicks are at the nest they are guarded, even during the night. Relief takes place at least three times a day, usually at dawn, noon, and dusk. At the nest-site the parents greet each other with several recognition displays before departure and arrival. Some of these displays are similar to those of other pelecaniforms.

Small chicks are fed six to nine times a day on partly digested fish, which flows down the inside of the upper bill. Larger chicks take food directly from the parent's throat, about twice a day at first but only once a day by about 5 weeks of age.

The darter's body-care behavior – preening, shaking, stretching, scratching – resembles that of cormorants. As in their case, it includes spread-eagling the wings to dry, for darters are even less well waterproofed than cormorants. JBN/GFvT

Right A female anhinga lifts her head to swallow a Walking catfish (Clarias batrachus) *whole. Anhingas feed mainly on fish but will also eat a wide variety of other prey, including water insects, crayfish, frogs, and salamanders.*

Below *A pair of anhingas on their nest. The female, on the left, is relieving the male of his nest-guarding duties. Both parents take turns in guarding the nest, using their webbed feet to warm the eggs and their body shade to keep them cool.*

Herons and Bitterns

Members of the heron family can, in the main, be described as long-billed, long-necked, long-legged, long-toed birds, well suited for wading in water searching for prey. A few have secondarily become more terrestrial or arboreal, and have the characteristic features somewhat reduced.

Herons typically wade in water or walk along the shore or on land looking for prey, or else they may stand quietly waiting for it to become apparent. Many species also use more active techniques. The neck and bill are particularly suited for capturing mobile prey. After a rapid thrust of the head and neck, aided by a hinge formed by elongated neck vertebrae, the sharp bill is used like a pair of forceps for grabbing, or like a rapier for impaling, the victim.

Fishers on Stilts

FORM AND FUNCTION

All herons are highly specialized predators on live prey, usually fish and aquatic crustaceans, but also insects, amphibians, reptiles, mammals, and even birds. Larger herons, such as the Goliath or Great blue, catch exceptionally large fish. Smaller species concentrate on small fish and invertebrates. Some are dietary specialists, such as the Yellow-crowned night heron, which concentrates on crabs and crayfish, the Cattle egret on orthopterans (crickets and grasshoppers), and the Reddish egret on shoaling fish.

Large herons have big, strong bills for capturing large prey. Herons eating fast-moving fish have thin, elongated bills, a feature that is most marked in the Agami heron. Terrestrial herons have shorter, thicker bills and necks. The Boat-billed heron has a large, spoonlike bill. Long-necked herons curl their neck and head back in flight, creating a silhouette that is diagnostic of the family. Other herons are more compact. Wings are broad, and their beat typically slow and deep. Herons are capable of flying very long distances.

The bills, legs, irises, and facial skin of many species vary in color according to season. Nonbreeding colors – typically horn, yellow, green, or brown – turn red, orange, or blue in courtship, and sometimes even within seconds during aggressive encounters. Some herons develop exceptional plumes on the head, neck, breast, or back that assume their most luxurious color, length, and texture as courtship commences. Most extravagant of all are the back plumes of egrets such as the Great, Little, and Chinese, which were the subject of market hunting in the late 19th and early 20th century. Head plumes are found throughout the family.

Herons spend much of the nonfeeding portion of the day at roost, where they rest and preen. Special feathers, called powder down, provide a supply of absorbent dust that is rubbed into plumage with the aid of the bill and a comblike toenail. Feather maintenance is apparently quite important to these birds, which spend much of their active day in water.

The typical herons and egrets (subfamily Ardeinae) exhibit a range of size, behavior, and coloration. They range from the huge Goliath heron to the small *Butorides* species, and from herons that feed passively by standing about to others that actively pursue prey. Some are black, others white or multi-hued, but the most usual pattern is darker above, lighter below, with a cryptic neck. The entire range of communality occurs within the Ardeinae.

Largest of the "giant" birds is the massive Goliath heron, although an even larger species, now extinct, existed in Europe within historic times. The best-known of all herons are the three "large" species: the Grey heron of the Old World, its North American equivalent the Great blue, and the South American Cocoi heron. All three have blue, gray, or blackish heads and bodies, with some white and heavily-marked white necks. The Grey and Great blue herons both have reached distant places, breeding from the far north into the tropics; each has developed a pale coastal race, in coastal West Africa and the Caribbean respectively.

The medium-sized egrets and herons are often characterized by distinctive breeding plumes. They include the most adaptable herons, and also some of the most threatened. The Little egret, currently recognized as comprising several quite distinctive subspecies from Europe, Asia, and Africa, has recently invaded the New World via the Lesser Antilles. Conversely, the Slaty, Chinese, and Reddish egrets all have highly restricted distributions, due to narrow habitat requirements.

The night herons are now known to be typical herons. These stocky birds, with relatively short, thick bills and legs that are also short, at least in comparison with other herons, are nocturnal feeders with large eyes that are useful under diverse light conditions. Juveniles have cryptic plumage. The Black-crowned and Rufous night herons are the most cosmopolitan species, together occupying a nearly worldwide range. The best-known of the night herons is the Black-crowned, a gregarious bird often seen in cities. *Gorsachius* night herons are found in wet forests, which have suffered widespread destruction; their ranks include one of the most endangered of all species, the White-eared night-heron of China.

Bitterns are solitary birds that feed mainly in the daytime, stalking their prey with great stealth. Most have brown to yellow plumage, often very

Above *Frozen in the reeds, a Eurasian bittern demonstrates its brilliantly effective camouflage. When the neck is held upright its markings blend perfectly with the reeds.*

Left *Snowy egrets* (Egretta thula) *are easily recognized not only by their pure white plumage but also by their habit of sprinting rapidly through shallow water when hunting. It is thought these two features attract other birds, which then join in the feeding.*

Above *Upon catching its prey – here a frog – the Goliath heron* (Ardea goliath) *will take it to a safe place and stab it to death with its bill before swallowing it headfirst. After eating, the heron will drink and/or clean its bill, preen, and rest.*

FACTFILE

HERONS AND BITTERNS

Order: Ciconiiformes

Family: Ardeidae

62 species in 17 genera

DISTRIBUTION Worldwide except high latitudes.

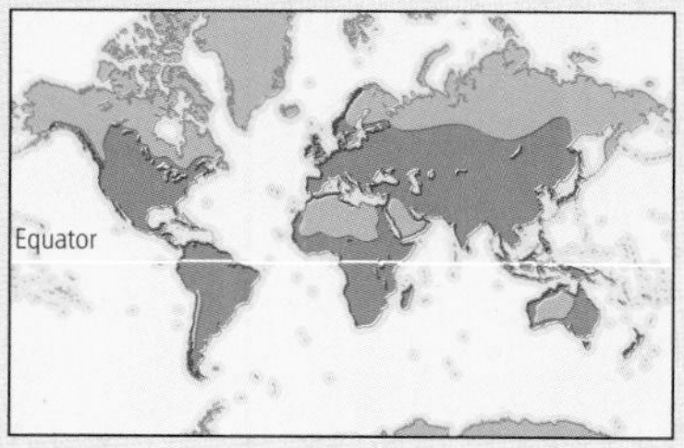

HABITAT Wetlands, marshes, and shallow waters.

SIZE Length ranges from 27cm (11in) in the Least bittern to 150cm (60in) in the Goliath heron; **weight** from 100g (3.5oz) to 4.5kg (10lb) in the same 2 species.

DIET Fish, crustaceans, amphibians; some insects, reptiles, mammals, birds.

CONSERVATION STATUS 3 species, including the White-bellied heron, are Endangered, and 5 more are Vulnerable. The New Zealand little bittern and 3 *Nycticorax* species of night-herons are listed as Extinct.

See Heron and Bittern Subfamilies ▷

heavily streaked to camouflage them in their reedbed habitats. The largest are very stocky, but the Least bittern is the smallest of all herons. When disturbed, they freeze with the bill pointed skyward, in a position called the "bittern posture" that is also used less frequently by many other herons.

The bittern focuses its eyes beneath its upturned bill for close observation, sometimes swaying like a reed in the wind. The larger species can handle very big fish, while all take small fish, frogs, and insects. The Eurasian bittern is famed for its booming breeding call, which can be heard at distances of up to 5km (3 miles).

Tiger herons owe their name to their striped plumage. They tend to live solitarily in dense tropical swamp forests, usually along rivers. Some occur in lowlands, others are mountain-dwellers. Few nests have been recorded, and their solitary habits and camouflage have kept much of their biology a mystery. The voice is not well described, but includes booming roars.

The Agami is an unusual heron, with an exceptionally long bill and neck. This species has only recently come to be understood, through molecular study, as being evolutionarily distinctive as well. It is a bird of stream edges in tropical swamps, where it perches by the waterside waiting to lunge forward, using the full extent of its neck and head.

The Boat-billed heron, with its curious, slipper-shaped bill, is the most atypical of all herons. Once thought to be closely related to the Black-crowned night heron, it has now been shown by molecular study to be distinctive, the similarities probably being due to convergence because of similar nocturnal habits. It feeds in shallow water like a normal heron, but can also use its bill to scoop prey from water and mud.

Group Master Feeders

DIET

The heron family's physical characteristics – long legs, necks, and bills – predispose them for wading in shallow water, using the full reach of their neck and bill to strike at elusive fish and aquatic invertebrates. Some species have secondarily become adapted to more terrestrial feeding on insects, and all herons will take whatever they can

Right *The distinctive Boat-billed heron is actually more strictly nocturnal than the so-called night herons. It is rarely seen during the day away from its mangrove-thicket roosts, where it preens and awaits dusk.*

catch handily. Some individuals specialize – for example, some Black-crowned night herons regularly eat small chicks of other birds in their nesting colony. Nonetheless, the fundamental adaptive suite of the family reflects their ability to catch fish and invertebrates in shallow water.

Herons are mobile birds. Most individuals move significant distances, not only seasonally but also from day to day and even hour to hour. They choose feeding sites, often with other birds, staying for as long as is profitable and then moving on. The white plumage of many species, such as Little egrets, has been shown to attract other birds to a feeding site. These species typically form the core of mixed-species aggregations. It appears that feeding in aggregations on highly concentrated prey offers commensal advantages – the more birds (up to a limit), the more vulnerable the prey.

Herons also feed commensally by following animals. The most specialized commensal forager is the Cattle egret, which follows Cape buffalos in Africa and similarly follows cows in its adopted lands. Like other herons, it will also pursue other kinds of disturbance, including tractors and fire.

Left Capable of carrying twigs and small branches up to 30cm (1 foot) long, the male Great blue heron passes this nesting material on to the female, who then constructs the nest.

Right In an act known as "mantling," a Black heron makes a cowl with its wings while hunting. The shadow may deceive fishes into believing they are fleeing into cover, or enable the heron to see them better.

Below As well as gular-fluttering, herons (here, a Grey heron) can dissipate heat by adopting a sunning posture. This involves standing upright with the wings held out at the side, shieldlike, but not fully spread.

Feeding in groups does, however, have a disadvantage, in addition to prey depletion. In aggregations, herons can steal prey from one another, with large herons being dominant over smaller ones. Other herons are more independent, and fiercely defend their patch of ground, over which they try to hold exclusive feeding rights. Even when feeding together in an aggregation, herons defend the individual space around them, and may try to take over the feeding sites of others.

Herons as a whole have many feeding techniques. All species can feed by standing in one place or by walking slowly; some do so characteristically, especially the very large herons that stand in deep water and the small ones that perch from branches overhanging the water. Others use more active behaviors when the situation allows. They may walk quickly or run after a fleeing prey; hop into the air and fly after an observed prey; hover in flight and dip their bill in the water; or even swim about, catching prey on the surface. Certain species can disturb prey into moving by stirring or scraping the bottom with their feet; several have contrasting yellow feet, apparently for this purpose. Others fly, dragging their feet in the water. The birds also use their wings to frighten prey, flicking them open and shut or else holding them open as they run. They can also attract prey by vibrating their bill in the water.

Various heron species use a succession of behaviors in order to optimize the feeding opportunity of the moment. The Reddish egret uses walking, running, hopping, and flapping wings in order to chase small schooling fish. The Black heron exhibits the most striking behavior of all, running to a chosen spot, forming a full canopy over its head with its wings, then stirring the bottom mud with its feet and stabbing at fish attracted to the canopy or disturbed by the activity.

The use of tools by herons is similarly exceptional. On three continents, Green and Striated herons have been observed using bait and lures. They place the lure (food, feathers, or sticks) in the water and then catch fish attracted to it very much in the manner of an angler presenting a dry fly. At the other extreme, the giant herons will stand for minutes to hours in one place until a suitable fish presents itself. They only need to eat a few per day to meet their energy requirements, and so have perfected a quiet, unobtrusive feeding style. Bitterns behave similarly, and may take many minutes of excruciatingly slow leg movement to take a step.

The Little egret and similar species tend to have the most varied repertoire. They may stand, walk, hop, fly, or use their feet or wings to catch prey, and may feed alone, in aggregations, commensally, or as pirates, as the opportunity arises. Such resourcefulness in first finding and then exploiting ephemerally available patches of prey in effective ways is a key component of the success of the heron family.

Serially Monogamous

BREEDING BIOLOGY

Most herons are gregarious, nesting in colonies with other species, resting in communal roosts, and feeding in aggregations. Bitterns and tiger herons are comparatively solitary, however, and even some typical herons, such as the Capped and Whistling herons, tend to be found in pairs or family groups.

Breeding is often timed to coincide with peaks in food abundance. Herons begin breeding by selecting a display site, where colonial herons gather in groups, using distinctive body postures involving stretching, snapping the bill, site defense, mock preening, flying about, and calling. Solitary herons display using far-carrying calls. Males choose the display site, which usually then becomes the nest site; females choose males by entering their display grounds and withstanding attempts to drive them away. Most species are serially monogamous, although promiscuous mating is common among colonial species. The Eurasian bittern is polygamous, one male mating with up to five females during a single breeding season.

Courtship and pair maintenance ceremonies continue after pairs have formed. Herons build nests of twigs or reeds, often lined with finer material. Depending on species, nests may be massive platforms or token scrapes. Males usually gather the material and give it to the female, who does the building.

A typical heron clutch is 3–5 eggs, with fewer in tiger herons and more in bitterns. Eggs of typical herons are pale blue and unmarked; bittern and tiger heron eggs are white to brown. Both parents guard the nest, incubate the eggs, and feed the young, except in large bitterns, in which the female alone tends the nest. The length of time spent on incubation depends on the size of the heron, being longer in larger species. Incubation begins before the last egg is laid, so the young hatch on different days, giving the oldest an advantage in competing for food, which may help optimize breeding success. It is rare for all the young to survive. Although helpless at hatching, they develop rapidly, especially in the feet and legs, and can scramble out of the nest within days to weeks. Parents feed chicks by regurgitating

Above *Representative species of herons and bitterns:* ***1*** *Great blue heron* (Ardea herodias) with fish. ***2*** *Least bittern* (Ixobrychus exilis) *in reeds.* ***3*** *Black-crowned night heron* (Nycticorax nycticorax). ***4*** *South American bittern* (Botaurus pinnatus). ***5*** *Bare-throated tiger heron* (Tigrisoma mexicanum); ***6*** *Cattle egret* (Bubulcus ibis), chasing a potential insect meal.

4

semi-digested food into the nest or alternatively directly into their mouths.

Most typical herons nest colonially, usually in sites protected from predators. Sometimes they form huge colonies, together with storks, ibises, spoonbills, and other water birds. Some, however – particularly large herons, but also specialized species such as the Whistling heron – nest solitarily. In all but these species, the young disperse after nesting. Adult birds nesting far north or south tend not to remain long in their summer breeding areas before moving toward the tropics.

Bitterns lay their eggs in a nest of reeds at intervals of several days, so the young can be markedly different in age. The chicks leave the nest well before fledging to clamber about among the reeds.

Threatened but Resilient

CONSERVATION AND ENVIRONMENT

Herons are a resilient group. Few bird families have suffered such widespread depredation. In past centuries, hunting by invading humans led to the extinction of island populations. More recently, whole populations were decimated at their breeding colonies to provide plumes for the adornment of ladies' hats. The Royal Society for the Protection of Birds in Britain, the Audubon Society in the USA, and to some extent the entire modern conservation movement owe their existence to the outcry caused by this devastation. Over the ensuing century, most species hardest hit by this trade have survived to reclaim or even expand their ranges. One exception is the Chinese egret, which was long plagued by continued hunting and is now threatened by habitat loss. Herons are still eaten in some areas; in addition, their eggs are sometimes harvested, and adults are killed as unwanted predators.

Habitat loss confronts many species today. Forests and wetlands are under threat throughout the world. The White-eared night heron and the Japanese night heron teeter on the brink of extinction because of relentless habitat destruction. The Chinese egret has been driven from colony sites by coastal development. The White-bellied heron is a rare south Asian species endangered by loss of its wetland and lowland forest habitat.

By and large, however, most species are secure, and many are prospering. Small herons can live almost next door to humans in villages, towns, and even cities. Feeding at dusk and dawn and hiding in deep foliage, the small bitterns, the Green heron, and the confiding pond herons have adapted easily to populated areas. Night herons are a common sight in parks in developed areas. In some parts of the world, almost every zoo and park contains its resident, free-roaming population of herons, which steal the food from the troughs of captive animals. Egrets that nest in rural areas feed alongside cattle, behind tractors, or by the roadside.

Some species are expanding their ranges, reflecting both the dispersal abilities of the group and also its versatility in accommodating to human-made opportunities. The Cattle egret is a good example, having in the past century colonized every continent except Antarctica. As cattle ranching and irrigated pasture expanded, often to the detriment of forests, the bird prospered, spreading to South America and northward into North America as well as in Asia and Australia. And just as the Cattle egret exploited the spread of livestock farming, so other species are now adapting their behavior to take advantage of fish farms, rice fields, reservoir construction, wetland management projects, and similar developments.

5

6

Fish farms and hatcheries in fact provide ideal heron feeding sites. In Britain the Grey heron quickly became an expert at robbing unprotected fish farms, as have many herons around the world. Such opportunism brought the birds into conflict with their human providers; in the late 1970s, no fewer than 4,600 Grey herons were shot annually in England and Wales, reducing the total population at one point to only 5,400 pairs. In fact the problem was quickly resolved by cooperation between farmers and environmentalists, who recommended protecting ponds with cords and chains, and Grey heron numbers have since increased in Britain to record levels. Clearly the interaction of herons and aquaculture must be similarly managed worldwide to assure the continued health of these adaptable birds. JAK/JH

Left *At home near grazing cattle, the aptly named Cattle egret is an opportunistic feeder, which follows large herbivores around, feeding mainly on the insects disturbed by them.*

Right *Both male and female Great egrets care for and feed their young. The chicks stimulate regurgitation by grabbing and tugging their parents' bills. The fledging period is usually around 42 days in this species.*

Heron and Bittern Subfamilies

Typical herons and egrets
Subfamily Ardeinae

42 species in 9 genera. Worldwide, except high altitudes. Usually near water, but some are more terrestrial. Mostly tropical. Temperate-nesting birds migrate toward the tropics; tropical birds migrate with the wet and dry seasons.
Species include: **Grey heron** (*Ardea cinerea*), **Great blue heron** (*A. herodias*), **Cocoi heron** (*A. cocoi*), **Madagascar heron** (*A. humbloti*), **Great egret** (*A. alba*), **White-bellied heron** (*A. insignis*), **Cattle egret** (*Bubulcus ibis*), **Green heron** (*Butorides virescens*), **Striated heron** (*B. striatus*), **Squacco heron** (*Ardeola ralloides*), **Indian pond heron** (*A. grayii*), **Malagasy pond heron** (*A. idea*), **Black heron** (*Egretta ardesiaca*), **Little egret** (*E. garzetta*), **Chinese egret** (*E. eulophotes*), **Slaty egret** (*E.vinaceigula*), **Reddish egret** (*E. rufescens*), **Whistling heron** (*Syrigma sibilatrix*), **Capped heron** (*Pilherodius pileatus*), **Yellow-crowned night heron** (*Nyctinassa violacea*), **Black-crowned night heron** (*Nycticorax nycticorax*), **Rufous night heron** (*N. caledonicus*), **Japanese night heron** (*Gorsachius goisagi*), **White-eared night heron** (*G. magnificus*).
SIZE: Length: 39–150cm (15–58in); weight: 200g–4.5kg (7oz–10lb). Males tend to be larger than females, but seldom obviously so.
PLUMAGE: Generally white, gray, black, and brown, including birds that are entirely white or dark and others that are multi-shaded. Plumage dimorphism occurs. Wide-ranging species vary geographically, particularly island populations. Some juveniles differ from adults, being more cryptically patterned.
VOICE: Flight and disturbance calls are usually harsh, loud croaks, although some, such as that of the Whistling heron, are higher pitched. Calls are used in courtship, and herons can be quite boisterous when together in colonies, roosts, and on communal feeding grounds.
NEST: Generally of twigs placed in trees, bushes, or reeds. Many species nest colonially.
EGGS: 2–7, usually pale blue. Incubation period 18–30 days; nestling period 35–50 days.
DIET: Mainly fish, but also amphibians, small mammals, birds, and insects, the latter especially in terrestrial species such as the Cattle egret.

Bitterns
Subfamily Botaurinae

13 species in 3 genera. Worldwide; large species range farthest N of any heron. Typically found in reed beds, although 1 species occurs in dense, stream-edge tropical forest. Species include: **American bittern** (*Botaurus lentiginosus*), **Australasian bittern** (*B. poiciloptilus*), **Eurasian or Great bittern** (*B. stellaris*), **South American bittern** (*B. pinnatus*), **Least bittern** (*Ixobrychus exilis*), **Little bittern** (*I. minutus*), **Zigzag heron** (*Zebrilus undulatus*).
SIZE: Length: 27–85cm (11–34in); weight: 100–1,900g (3.5–67oz).
PLUMAGE: Highly cryptic mixture of cream, yellow, chestnut, brown, or black; males are larger and more contrastingly colored.
VOICE: Large bitterns have a booming territorial call; smaller bitterns have barking calls.
EGGS: Usually 3–5, but up to 10 in large bitterns; white to pale brown. Incubation 14–55 days, depending on size; nestling 28–55 days.
DIET: Fish, amphibians, small mammals, insects.

Tiger herons
Subfamily Tigrisomatinae

5 species in 3 genera. Disjunctly distributed in New Guinea, W Africa, C and S America. Tropical wet forests. Species include: **Fasciated tiger heron** (*Tigrisoma fasciatum*), **White-crested tiger heron** (*Tigriornis leucolophus*).
SIZE: 60–80cm (24–32in); males larger than females.
PLUMAGE: Brown barred and striped concealment patterns; juveniles patterned more strongly.
VOICE: A booming call.
NEST: Usually in trees.
EGGS: Usually 1, sometimes 2; whitish, blotched red.
DIET: Fish, amphibians.

Agami heron
Subfamily Agamiinae

1 species: *Agamia agami*. S and C America, along tropical forest streams.
SIZE: Length 60–70cm (23–27in); weight: 475–535g (16–19oz).
PLUMAGE: Bright chestnut and blue-green.
VOICE: Alarm call is a low-pitched rattling growl.
EGGS: 2–4, pale blue, unmarked.
DIET: Primarily fish.

Boat-billed heron
Subfamily Cochleariinae

1 species: *Cochlearius cochlearius*. S and C America.
SIZE: Length: 45–51cm (17–20in); weight: 503–770g (17–27oz).
PLUMAGE: Contrasting black and white; juveniles are browner; sexes look alike, but males larger.
VOICE: Characteristic calls are complex, raucous, laughing chants.
EGGS: Usually 3, sometimes 4; pale blue to green, sometimes with cinnamon spotting. Incubation 25–27 days.
DIET: Primarily fish and shrimps.

Storks

For many diverse human cultures, storks are among the most enduringly symbolic birds. The White stork has long been a symbol of pilgrimage and continuity in European and Islamic lands. It nests happily close to people in villages and makes long migrations, but, showing great fidelity, returns to its nest-site on schedule in the spring.

Such reliability has always appealed to humans, and perhaps gave rise to the folk tale (which originated in Germany and Austria but has since spread worldwide) that storks deliver babies. Storks are a reassuring presence wherever they occur.

Birds of Good Omen

FORM AND FUNCTION

The typical storks are large wading birds having long legs, long bills, a stately upright stance, and a striding gait. They are birds of wetlands and water margins, as well as fields and savannas. They prefer warm continental climates and tend to avoid cool and damp regions. As a result, they are widespread in the tropics and subtropics. A few species nest in temperate regions, but they also range into the tropics. The greatest numbers of stork species are found in tropical Africa and tropical Asia.

White and Black storks are particularly widespread, nesting in Europe, East Asia, North Africa, and southern Africa. Both species spend most of the year in Africa or India. Typical storks also include the adjutants and the giant storks, such as the Jabiru, Black-necked, Saddle-bill, and Marabou, the latter having a wingspan of nearly 2.9m (9.5ft). Except for the Jabiru and Maguari, the typical storks are Old World species.

Storks have long, broad wings and are strong fliers. They fly with their necks outstretched, except for the adjutants, which retract the head. Storks can engage in remarkable aerobatics, such as diving, plummeting from the sky, and flipping over in flight. The Black stork, having relatively narrow wings, relies on flapping flight to a degree that is unusual in the family. Most storks alternate flapping flight with soaring on warm air-currents (thermals).

The bills of typical storks are long and heavy. Most are straight, although the massive bill of the Jabiru is slightly up-curved They are used to take a wide variety of aquatic and terrestrial prey, which the birds obtain by walking slowly or standing. A typical stork will walk slowly across fields with its

Above *Despite being widespread throughout most of its range, the Black stork tends to avoid areas of human activity. It prefers forest, where it feeds on fishes and aquatic invertebrates in marshy clearings or stream edges.*

Below *Painted storks, indigenous to South and Southeast Asia, are renowned for their large colonies. For example, in 2002, no fewer than 5,000 birds gathered to nest, under the active protection of villagers, at the hamlet of Veerapura in southern India.*

neck extended and head down, looking for prey. The huge bills of the giant storks allow them to take large prey. The Black-necked stork sometimes hunts by running back and forth, jumping and flashing its wings. The Jabiru, the largest New World stork, feeds by touch, wading slowly and periodically inserting its open bill into the water. The huge bills of the adjutant storks are used for cutting, prying, and tearing, as well as being effective against competitors at the feeding site.

The wood stork tribe includes species having downcurved, ibis-like bills and also the related open-bill storks, which, as their name implies, have a visible gap between their mandibles. Except for the American wood stork, these are Old World species; they are birds of the tropical swamps, where seasonal rainfall leads to water-level fluctuations and large aquatic snails occur in abundance. The four wood storks feed by touch, wading slowly with their partially opened bills inserted in shallow water, the bill snapping rapidly shut when triggered by the feel of a fish. They are particularly effective at catching fish when they are concentrated, generally by falling water levels.

The bill of the open-bill storks is used for dealing with mollusks, especially large water snails. The bill-tip is inserted into the opening of the shell, cutting the snail's muscle, which permits extraction of the body. An open-bill may ride on a swimming hippopotamus to capture the snails it stirs up.

Wood storks, adjutants, and the Jabiru lack feathers on their heads. The sexes look similar, but males are noticeably larger than females. Dark irises distinguish male Black-necked and Saddle-bill storks from the yellow-eyed females. Air sacs lie under the neck skin, and the Marabou and the Greater adjutant have long, bare, pendent throat sacs. Juvenile plumage is dull, reaching full development over the first year. Nestlings of the otherwise white Maguari stork are black, probably serving as camouflage. The African open-billed stork is black, whereas the closely related Asian open-billed stork is white. The colors of the bill, together with the bare skin of the head and legs, are characteristic for each species, and the color of these hallmarks intensifies during courtship. For example, the breeding Maguari stork has a striking blue-gray bill, becoming maroon near its red face. The Jabiru has a pink neckband that changes to deep scarlet whenever it is excited.

Temperate breeding storks, and some in the tropics, undertake seasonal population movements, although movements of tropical species tend to be less long-distance migrations than population shifts in response to rainfall patterns affecting feeding conditions. In contrast, the migrations of European storks have been known since biblical times. The birds' reliance on thermals, which tend to be found over land, restricts routes, as they can only manage short flights over water. In consequence, European Whites use two migratory routes southward to Africa, one down the Iberian Peninsula, the other across the Middle East through Egypt, both of which avoid a long sea crossing over the Mediterranean.

Hunters and Scavengers

DIET AND FEEDING

Most storks feed alone or in small groups, but will also form large flocks when food is abundant, a behavior that is particularly common in the wood storks. The White-bellied stork also often hunts in large flocks, especially near grass fires and locust swarms.

By soaring, storks can forage at long distances from their colonies and roosts. The White stork, wood storks, and adjutants are particularly adept at reaching high altitudes, and then gliding toward distant feeding sites. This behavior helps birds to locate food concentrations, where many birds may forage together. In East Africa, as many as seven stork species may feed in the same location.

The diet of most storks is varied. The White stork's includes aquatic vertebrates, insects, and earthworms, but on its African wintering ground the species is known as the Grasshopper bird, from its habit of following locust swarms. White storks also opportunistically follow mowing machines. Wood storks and open-bills are specialized storks with a much narrower range of prey.

The Greater and Lesser adjutant storks and the Marabou are largely scavengers and carrion-eaters. They are well known for their attendance at carcasses, along with vultures and hyenas; the Greater adjutant, which was formerly common in Indian cities, consumed refuse that included human corpses. Although not adept at tearing flesh, their size and large bills enable them to steal morsels of meat from

FACTFILE

STORKS

Order: Ciconiiformes

Family: Ciconiidae

19 species in 6 genera

Distribution Worldwide, in temperate and tropical areas

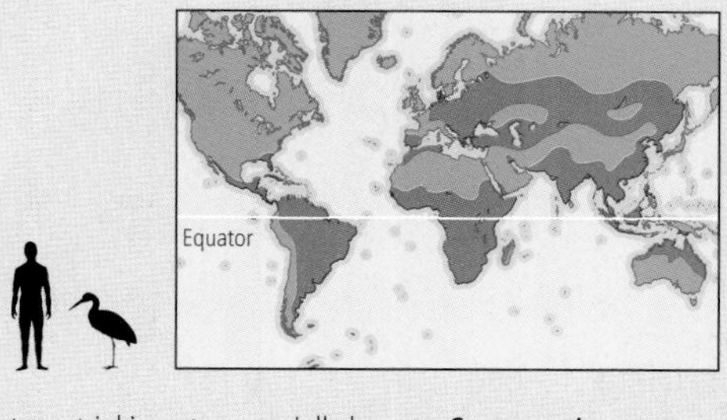

TYPICAL STORKS Tribe Ciconiini

13 species in 4 genera: **Black stork** (*Ciconia nigra*), **White-bellied stork** (*C. abdimii*), **Woolly-necked stork** (*C. episcopus*), **Storm's stork** (*C. stormi*), **Maguari stork** (*C. maguari*), **White stork** (*C. ciconia*), **Oriental white stork** (*C. boyciana*), **Black-necked stork** (*Ephippiorhynchus asiaticus*), **Saddle-bill stork** (*E. senegalensis*), **Jabiru stork** (*Jabiru mycteria*), **Lesser adjutant stork** (*Leptoptilos javanicus*), **Greater adjutant stork** (*L. dubius*), **Marabou stork** (*L. crumeniferus*). Tropical and temperate zones of the Old World, with 1 tropical American species. **Size:** Length: 75–150cm (30–60in); weight: 0.9–7.4kg (33–259oz); males in most species are larger than females. **Plumage:** Chiefly white, gray, and black. **Voice:** Generally silent, although many have a variety of calls when nesting including, according to species, mooing, whistling, and bill-clattering. **Nest:** Large structures of sticks and other materials, including turf and green twigs; those of the European White stork may be reused and become several meters in depth. **Eggs:** Usually 1–4 eggs, up to a maximum of 7; chalky white. Incubation 29–38 days; nestling period 55–115 days. **Diet:** A wide variety of fish, aquatic insects, and other invertebrates; for some species, terrestrial insects, especially locusts. **Conservation status:** Storm's stork, the Oriental white, and the Greater adjutant are Endangered; the Lesser adjutant is Vulnerable.

WOOD STORKS Tribe Mycteriini

6 species in 2 genera: **American wood stork** (*Mycteria americana*), **Milky stork** (*M. cinereus*), **Yellow-billed stork** (*M. ibis*), **Painted stork** (*M. leucocephalus*), **African open-bill stork** (*Anastomus lamelligerus*), **Asian open-bill stork** (*A. oscitans*). Tropical zones of the Old World, with 1 species in tropical America. **Size:** Length: 80–105cm (32–41in); weight: 1–3.4kg (40–118oz); males are larger than females. **Plumage:** 5 species are white (2 having pink tones), and 1 species is black. **Voice:** Generally silent except in colonies, where they give honking or hissing calls. **Nest:** Large structures of sticks. **Eggs:** Clutch size 1–5; chalky white. Incubation 25–32 days; nestling period 35–65 days. **Diet:** The wood storks (*Mycteria*) are specialized fish eaters, while the open-bill storks (*Anastomus*) specialize on snails. **Conservation status:** The Milky stork is Vulnerable.

Above *Representative species of storks:*
1 *African open-bill stork* (Anastomus lamelligerus)*;*
2 *American wood stork* (Mycteria americana)*;*
3 *Marabou stork* (Leptoptilos crumeniferus)*;* ***4*** *White stork* (Ciconia ciconia).

nearby vultures. Marabous frequent predator kills, domestic stockyards, plowed fields, and rubbish dumps, as well as haunting drying pools that contain the natural prey necessary to raise young. The birds are also attracted from great distances to grass fires, where they march along the fire front. The size range of their prey varies greatly; they will stand at termite mounds eating swarming insects, yet also take quite large prey, killing young crocodiles, young and adult flamingos, and small mammals.

Solitary or Colonial

BREEDING BIOLOGY

The nesting cycle of all storks is strongly seasonal, apparently determined by food supplies. Only the White and Black storks regularly leave the tropics to nest, doing so during the temperate spring and summer. The American wood stork nests during the dry season, when prey are concentrated in drying pools and are easily captured by touch, as does the Marabou, as carrion also becomes available. Other species of wood storks nest during the wet season, also the time of maximum prey abundance. The White-bellied stork is considered a "rain bringer" in Ethiopia, in that it nests during the first heavy rains, which produce a flush of its insect food.

Wood storks, open-billed storks, adjutants, and the White-bellied stork all nest colonially, together and with other species of waterbirds. Other species, such as the White and Maguari stork, are either loosely colonial or solitary. The giant storks tend to nest alone. American wood stork colonies may exceed tens of thousands of nests, whereas many European villages have only a single White stork family nesting there.

Most storks nest in trees, but they may also use cliffs or nest on the ground. Non-colonial tropical species, such as Saddle-bill storks, may remain paired year-round, and White storks often pair up with the same mate again because both birds are attracted to the nest site of the previous year. Nests are situated near sites providing suitable food supplies: drying pools for American wood storks; carrion-producing rangeland for marabou; agricultural fields for White storks.

The nest site, selected by the male, is defended against all intruders. The male gives advertisement displays, and the attracted female responds with appeasement behavior (see Up, Down, Flying Around). Advertising displays differ between species, but typically consist of up-down movements, calls, and bill clattering. In one extreme display, a stork bends its neck backward until its head touches its back; in some species, this posture opens a resonance chamber in the throat that amplifies the sound of the snapping of the bird's two mandibles. Even newly-hatched young behave in this way.

Both parents incubate and feed the relatively helpless young by regurgitating food onto the floor of the nest. Storks may also regurgitate water over their eggs and young, presumably in order to cool them. Nesting success is determined by the availability of prey items and weather conditions; indeed, wood storks only fledge young when high densities of food remain available throughout the entire nesting season. The nesting success of White storks is poor in years or locations having very high rainfall.

Above *Jabiru storks build conspicuously large nests, which the birds add to every year when they return to nest. This species occurs in suitable habitat (savanna, coastal lagoons) from Mexico to northern Argentina.*

Below *The bills of storks are perfectly shaped for trawling through shallow water for fishes and amphibian prey. Here, a Woolly-necked stork captures a plump frog, a substantial meal.*

Declining Numbers

CONSERVATION AND ENVIRONMENT

Some stork populations have undergone massive declines. Even the White stork, long regarded as a harbinger of good fortune and many children, has suffered. Between 1900 and 1958, western European populations decreased by 80 percent, and by 92 percent between 1900 and 1973. Storks no longer nest in Sweden or Switzerland, and occur only in small numbers in other countries.

The reasons for the decrease are not certain, but those that have been proposed include cooler and wetter summers, loss of nest-sites, pesticide poisoning, hunting on the winter grounds, and changing agricultural practices. The last hypothesis is lent credence by the fact that, although stork numbers in Europe previously increased following deforestation, they have more recently suffered a steady decline, as more and more foraging sites have fallen victim to the demands of modern agriculture. Hunting pressure in wintering areas in Africa may also be a contributory cause of their decline.

The Greater adjutant population has been critically reduced throughout its range. The Milky stork, confined to the mangrove forests of Southeast Asia, is in jeopardy because of habitat destruction. Some species, such as the Black-necked stork, are now rare over wide ranges, while others, such as the Asian open-bill stork, are still numerous, but only locally. Although it remains abundant elsewhere throughout its range, the American wood stork has decreased in southern Florida as a result of ecological changes in the vast Everglades marsh that have prevented the birds from finding enough food to raise their young. In general, protection of wetlands and other feeding sites is essential for stork conservation. JAK

SPECIAL FEATURE

UP, DOWN, FLYING AROUND

Courtship in storks

COURTSHIP IS CRITICAL TO ANY BIRD, BUT THE great size of storks makes their displays all the more noticeable. The ritualized behaviors and gestures used by the various species are as repeatable and consistent as their bill shapes and plumage patterns.

The advertisement display of the White stork has been known for centuries, and was illustrated in 13th-century manuscripts. It remains familiar to the people of the villages in which they nest to this day. The display is commonly known as the "up-down," and some version of it is found in most stork species.

The up-down is not only the most typical courtship behavior in storks, but in some species is also the most remarkable. It is a greeting issued when one member of a pair returns to the nest. In giving the display, a stork first raises and then lowers its head in a characteristic, stylized manner. Although the behavior is present in all storks, the exact pattern differs among species. The head movement is usually accompanied by a vocalization, which in the typical storks takes the form of bill clattering, the amount of which is species-specific. The White stork, for example, has a loud, resonant clatter that may last ten seconds or more, while the Black stork clatters only infrequently. Such a difference suggests that within the group the two species are not closely related. The White stork also differs from the other typical storks in that its up-down is not accompanied by whistling.

The up-down behavior is simplest in open-bill and wood storks, in which it consists mainly of raising the head and gaping the bill, and then emitting hissing screams as the head and bill are lowered. Even so, the display differs among the four wood-stork (*Mycteria*) species. The American wood stork does not snap its bill during the display; the Yellow-billed stork gives a single or double snap; the Painted stork gives double or triple snaps; while the Milky stork gives multiple snaps.

Differences in the details of the up-down display demonstrate important distinctions between closely-related storks that would not otherwise be obvious. The up-down of the adjutants includes moving the bill to vertical, accompanied by mooing and squealing. The Marabou and Greater adjutant are similar-looking birds, which do not overlap in range. Their displays differ, however, in that Marabous first throw their head upward and squeal with the bill near vertical, before pointing it downward and clattering loudly, while the Greater adjutant clatters while the bill is pointed upward. The difference in this important pair-bonding activity suggests that the birds would not interbreed should their ranges overlap, and so are best considered as separate species.

Right *A pair of Saddle-bill storks in South Africa's Kruger National Park show off the courting display known as "flap-dash," in which the birds run through shallow water flapping their wings.*

Left *Storks have a wide range of aggressive and courtship displays that vary between species and genera.* ***1*** *The last stage of the "clattering threat" in the Yellow-billed stork.* ***2*** *"Display preening" in the Painted stork; in this courting pair, the male in front is preening behind the wing.* ***3*** *A marabou showing the "anxiety stretch" in response to disturbance by people on the ground under the nest.* ***4*** *A male Yellow-billed stork giving an "up-down" display as his newly acquired mate approaches the nest-site.* ***5*** *A courting male Asian open-bill stork performing the "advertising sway" at a potential nest-site.* ***6*** *"Head-shaking crouch" of a male White-bellied stork as a potential mate approaches.* ***7*** *"Full back," a position in the "up-down" display of White storks.*

Black-necked and Saddle-bill storks display infrequently because of their long-lasting pair bond. The up-down of the Black-necked stork is a spectacular greeting that includes rapid fluttering of fully extended wings and clattering of bills, but the head is not raised. These two species and the Jabiru share a distinctive display given on the foraging grounds, the "flap-dash," in which a bird dashes wildly through the water while vigorously flapping its wings.

The typical storks are distinguished from the wood storks in that they alone exhibit a head-shaking crouch, in which the male crouches on its nest and shakes its head from side to side as if saying "No." That is in fact probably the message of the display, since it is given as a warning when another bird approaches the nest.

The wood storks share three unique displays: flying around, in which a male that has just accepted a female leaves its nest and flies in a circle around the site before returning; gaping, in which a bird holds its parted mandibles open; and display preening, a mock behavior in which the male pretends to comb the feathers on its wing with its bill. The displays of the open-bill storks strongly resemble those of wood storks, especially their simple up-down and also their copulation clattering, in which a male clatters his mandibles during copulation while knocking them against the bill of the female. As a result of such resemblances, the wood storks and open-bill storks are thought to be more closely related to each other than they are to other storks. Open-bills also have a unique display, the advertising sway, in which a displaying male bends its head down between its legs and repeatedly shifts its weight from one foot to the other.

Comparative behavioral observations have discovered both similarities and differences among species that not only suggest systematic relationships but also reveal much about the underlying biology of the birds. Even so, much remains to be discovered about the nuances of bird behavior, its evolutionary roots, its real-life importance, and its geographic variability. JAK

Ibises and Spoonbills

Above A Royal spoonbill on its nest. This species nests in colonies ranging from just a few pairs to large numbers, sometimes alongside other waterbirds such as ibises or egrets.

Above A White spoonbill pair courting, with raised crests. Their relationship is temporary and will last for the duration of the breeding season. Despite this bond, paired males sometimes copulate with other females, even those that are not breeding!

IBISES AND SPOONBILLS ARE LONG-LEGGED wading birds of ancient lineage and a long historical pedigree; the Sacred ibis was venerated in ancient Egypt, where it was identified with the scribe-god Thoth. The two subfamilies are fundamentally similar, but differ in their distinctive bills. Ibis bills are relatively long, thin, and down-curved; spoonbills' are long, broad, and flattened.

Both ibises and spoonbills feed mostly by feel rather than by sight (see Feeding by Touch). Ibises use their long bill for probing in shallow water, soft mud, holes, under plants, around rocks, in grass and pastures, and even on hard ground. Aquatic ibises tend to have longer, thinner bills than do terrestrial species. Spoonbills feed typically by swinging their open bill from side to side in the water. The wide bill enhances the birds' ability to encounter fish.

Long Legs and Long Bills

FORM AND FUNCTION

The ibises include an array of species that are similar in being medium-sized, long-legged birds with long, down-curved bills. Relationships among the species are not clearly understood, nor in some cases are species limits. Several species occur in the New World. In North America, the White ibis was historically the most abundant wading bird where it occurred. A very similar South American species, the Scarlet ibis, is a bright-red variant of the American white; significant interbreeding in northern Venezuela suggests strongly that they are conspecific. The three *Plegadis* species also occur in the New World. Of these, the Glossy ibis is the most cosmopolitan of ibises, and is probably a relatively recent arrival in North America, while the Puna ibis is a species of the high Andes. Several other species, such as the Buff-necked ibis, are South American endemics.

More species occur in the Old World. The *Bostrychia* ibises are found in Africa and its islands. The Sacred ibis group of several related species occurs through Africa, Asia, and Australia. The *Geronticus* species (the Bald and Waldrapp ibises) are cliff-nesting birds of semi-arid mountains.

Several Asian ibises are relictual. Although the Black ibis remains locally common in India, its population in Southeast Asia (sometimes considered a separate species, the White-shouldered ibis *Pseudibis davisoni*) is, or is nearly, extinct. In the same area, the status of the Giant ibis is unclear, and it too may be functionally extinct. The Oriental crested ibis, formerly of Japan, China, and Korea, occurs today in only a very limited range.

The spoonbills are long-legged, long-billed wading birds with a distinctively flattened bill. One species in the New World, the Roseate, is pink, the rest are white. A group of three species related to the Eurasian spoonbill is found patchily distributed in Europe, Africa, Asia, and Australia. The Eurasian spoonbill itself is the most northern nesting spoonbill, continuing to breed in small numbers in northern Europe. The group also includes the most

FACTFILE

IBISES AND SPOONBILLS

Order: Ciconiiformes

Family: Threskiornithidae

33 species in 14 genera

DISTRIBUTION Worldwide, in temperate and tropical areas.

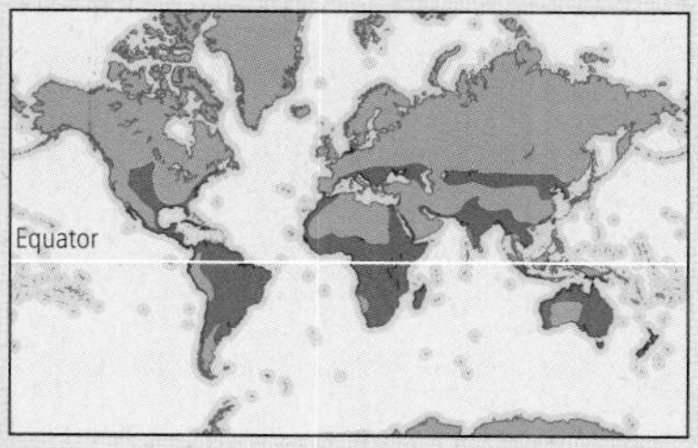

Habitat Usually near water, but some are terrestrial.

Size Length 46–110cm (17–40in); weight 420g–2.1kg (15–73oz). Males are larger than females.

IBISES Subfamily Threskiornithinae
Worldwide, in temperate and tropical zones; most diverse in the tropics. **Size:** Length 46–110cm (17–43in); weight 420g–1.53kg (15–53oz). Males are larger than females. **Plumage:** Diverse, including white, black, brown, gray, buff, and scarlet; darker birds often have glossy bronze or greenish sheens, with no notable sexual dimorphism. **Voice:** Low grunting noises when feeding and noisy during interactions within colony. Distinctive flight and display calls: the American white ibis honks, the Wattled ibis has a deep, raucous call. **Nest:** Ranges from thin platform to substantial structure of sticks, usually lined with finer materials. **Eggs:** 2–3, up to 7; blue to dull white, often with dark spotting. Incubation 20–29 days; nestling period 35– 55 days. **Diet:** Most ibises eat an array of invertebrates. Aquatic insect larvae, crayfish, crabs are common in aquatic diets; grasshoppers, worms, beetles, in terrestrial diets. Fish are consumed when available. **Conservation status:** The Giant, Waldrapp, White-shouldered, and Dwarf olive ibises are Critically Endangered. The Oriental crested ibis is Endangered and the Bald ibis Vulnerable.

SPOONBILLS Subfamily Plataleinae
Worldwide, primarily in tropics, but also in temperate zones. **Size:** Length 60–100cm (23–38in); weight 1.3–2.1kg (45–73oz). Males are larger than females. **Plumage:** White or pink; juveniles of white species are darker, those of Roseate spoonbills lighter, than adults. **Voice:** Spoonbills are relatively quiet birds, with soft flight calls and subdued grunting, hissing, and soft bill-clapping noises at nest. **Nest:** Substantial structure of sticks. **Eggs:** Usually 3–4, up to 7; dull white to buff, sometimes with darker spots. Incubation 20–32 days; nestling period 45–50 days. **Diet:** Small fish, shrimps and other crustaceans, aquatic insects. **Conservation status:** The Black-faced spoonbill is Endangered.

See Ibis and Spoonbill Subfamilies ▷

Left *With its bald, red, vulture-like head, the Waldrapp ibis is a highly distinctive member of the ibis and spoonbill family. With only 65 breeding pairs left in the wild (in Morocco), there are plans to reintroduce the species to central Europe, Spain and Italy, from where it has been absent since the 17th century.*

endangered spoonbill, the Black-faced of East Asia. The African spoonbill is found through much of that continent, and the Yellow-billed spoonbill occurs in Australia.

Plumage and skin color are important characteristics among ibises and spoonbills. Most are basically white, black, or brown. Some species, such as the Sacred ibises and the Yellow-billed spoonbill, develop breeding plumes along the back, and some also on their breast. The otherwise white Oriental crested ibis takes on an adventitious gray feather color during breeding. Some ibises and spoonbills are crested, others have featherless heads, necks, sides, or underwings. Sacred ibises lack feathers over the entire head and neck, and the exposed skin usually shows diagnostic skin colors; for example, the face and an expanded throat pouch of the American white ibis turn bright red during courtship, and the head of the Indian black ibis is covered with bright red bumps. Juveniles are generally duller than adults, markedly so in the American white ibis where the juvenile has a brown back, and are also more feathered. Part to all of the face of spoonbills (and the entire head of the Roseate spoonbill) is featherless and distinctively colored black, yellow, or green

Habitat use is diverse across the family, but ibises and spoonbills most typically occur in open, damp habitats. Spoonbills and aquatic ibises prefer open marshes, ponds, and shallow coastal situations. Terrestrial ibises, such as the Bald and Waldrapp, favor open grasslands, pastures, and semi-arid environments. A few, such as the Olive, Spot-breasted, and Madagascar crested ibises, are forest birds. Ibises such as the Sacred, Buff-necked, and Bald, are attracted to fires, the latter species nesting during the fire season. Most species are typical of lowland wetlands, forests, and coastal habitats. Some species and races, such as the Puna ibis and populations of the Buff-necked and Olive, are found at high altitudes in mountainous regions.

Feeding in Flocks

DIET AND FEEDING

Using their tactile foraging techniques, ibises catch slow-moving or bottom-dwelling prey, while spoonbills catch fish and crustaceans. For example, the American white ibis primarily eats crayfish and fiddler crabs, whereas the Roseate spoonbill eats small fish, shrimps, snails and aquatic insects. In wetlands, both ibises and spoonbills capture a variety of insects, frogs, crustaceans, and fish. Terrestrial species, on the other hand, catch insects, worms, and other invertebrates. Ibises scavenge; the Sacred ibis often feeds on carrion and on broken waterbird and crocodile eggs. Research has found that the Glossy ibis eats considerable amounts of plant material in Cuba, particularly rice. This is a remarkable discovery in what had been considered a predatory family, in which modest consumption of plant material was thought to be limited to terrestrial species.

Most ibises and spoonbills are social. They form flocks that fly from place to place and feed in small to very large aggregations. Many species fly in compact flocks or in long, undulating lines, alternating flapping and gliding flight. Birds returning to roost together can number in the thousands. Most forage socially, and even the more solitary species, such as the Green and Black ibises, tend to forage in pairs and small groups. Social species form aggregations at suitable foraging sites, often with other wading birds, which together can number in the thousands. In such situations, they tolerate other birds in close proximity and often move in unison, probably benefiting from the disturbance caused by their fellow diners. Other wading birds follow ibises, feeding in their wake.

Ibises tend to eat small prey that are swallowed quickly, avoiding loss to piratical attacks. Ibises are diurnal, although spoonbills more often feed crepuscularly and at night. Along the coast, the feeding schedule is determined by tides. Communal roosts, which are located near feeding grounds and may be shared with herons, storks, and cormorants, have been known in some species to number in the tens of thousands of birds. Specific roost sites may be temporary, lasting only as long as nearby food supplies; alternatively, they may persist for years. More solitary species roost alone or with just a few other birds.

Stretching and Bowing

BREEDING BIOLOGY

Most species nest colonially, but some, such as the forest-nesting Olive and Spot-breasted ibises, nest alone. The Hadada, even though highly social when not breeding, nests solitarily. Isolated places, such as islands or trees surrounded by open ground, are often chosen for nesting, as ground predators are less likely to gain access.

4

Ibis and Spoonbill Subfamilies

Ibises
Subfamily Threskiornithinae

27 species in 12 genera: White ibis (*Eudocimus albus*), Scarlet ibis (*E. ruber*), Whispering ibis (*Phimosus infuscatus*), Glossy ibis (*Plegadis falcinellus*), Puna ibis (*P. ridgwayi*), White-faced ibis (*P. chihi*), Sharp-tailed ibis (*Cercibis oxycerca*), Buff-necked ibis (*Theristicus caudatus*), Black-faced ibis (*T. melanopis*), Plumbeous ibis (*T. caerulescens*), Andean ibis (*T. branickii*), Green ibis (*Mesembrinibis cayennensis*), Hadada ibis (*Bostrychia hagedash*), Wattled ibis (*B. carunculata*), Spot-breasted ibis (*B. rara*), Olive ibis (*B. olivacea*), Madagascar crested ibis (*Lophotibis cristata*), Sacred ibis (*Threskiornis aethiopicus*), Australian ibis (*T. molucca*), Black-headed ibis (*T. melanocephalus*), Straw-necked ibis (*T. spinicollis*), Waldrapp ibis or Northern bald ibis (*Geronticus eremita*), Bald ibis (*G. calvus*), Black ibis (*Pseudibis papillosa*), White-shouldered ibis (*P. davisoni*), Giant ibis (*P. gigantea*), Oriental crested ibis (*Nipponia nippon*).

Spoonbills
Subfamily Plataleinae

6 species in 2 genera: African spoonbill (*Platalea alba*), Eurasian spoonbill (*P. leucorodia*), Black-faced spoonbill (*P. minor*), Yellow-billed spoonbill (*P. flavipes*), Royal spoonbill (*P. regia*), Roseate spoonbill (*Ajaia ajaja*).

6

7

Below *Representative ibis and spoonbill species.* ***1*** *Sacred ibis* (Threskiornis aethiopicus)*;* ***2*** *Glossy ibises* (Plegadis falcinellus) *nest in colonies, often along with herons;* ***3*** *The Hadada ibis* (Bostrychia hagedash) *has a distinctive flight call – a loud raucous "haa-daa-daa" – from which its name is derived;* ***4*** *Scarlet ibis* (Eudocimus ruber)*;* ***5*** *The White-faced ibis* (Plegadis chihi) *breeds from the western USA to Argentina;* ***6*** *Eurasian or White spoonbill* (Platalea leucorodia)*;* ***7*** *Roseate spoonbill* (Ajaia ajaja)*, showing the broad, flattened bill characteristic of spoonbills.*

5

Most species place their nests in bushes or trees, but considerable variability occurs. The American White ibis nests in trees, on bushes, in reeds, or on the ground, while Buff-necked ibises nest in pairs in palm trees, in marshes, on mountain cliffs, or on the ground. Several terrestrial species, including the Bald, Waldrapp, and Wattled ibises, nest on cliffs. Hadadas sometimes nest on telegraph poles. Black ibises take over old raptor nests.

In pair formation, display accessories, such as the red throat pouch of the American white ibis, the head color of the Roseate spoonbill, and the black plumes of the Sacred ibis, are at their annual peak. In the few species studied well, the male chooses a potential nest-site from which to advertise, using stretching and bowing displays and vigorous defence of the display site. Females attempt to land near the male, assuming a submissive posture, and he at first repulses them. When he accepts a female, the pair engages in mutual bowing, billing, and preening. Solitary species use loud vocalizations to advertise themselves and to maintain contact.

The male usually gathers nest materials, which he presents to the female, who builds the nest. In the Hadada, stick-passing and mutual nest-building are particularly ritualized. Both sexes defend the nest site from takeover and stick pilfering. Copulation takes place at the nest, and in some colonial species promiscuous copulations among neighbors are frequent. In most species, eggs are laid at 1–3-day intervals, and are incubated before all have been laid, leading to staggered hatching. Both sexes incubate the eggs, and subsequently feed the young regurgitated food, placed in the hatchlings' mouths. Later nestlings insert their bill down the parent's gullet.

The young develop rapidly, especially in their legs and feet, and are soon able to leave the nest.

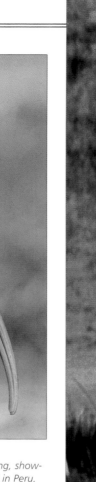

Above *A Puna ibis suns itself after bathing, showing off its glossy green wing feathers. Found in Peru, Bolivia, and Chile, this species frequents large marshes and damp pasturelands where it forages on muddy flats, along creeks, and in short grass.*

Right *Sacred ibises in the Ngorongoro crater, Tanzania, East Africa. Occurring in groups ranging from a few birds to flocks of hundreds, the Sacred ibis was worshiped by the ancient Egyptians as the god Thoth. Today, farmers have good reason to be grateful to it, since it helps control crop pest numbers.*

Below *A pair of Black-headed ibises* (Threskiornis melanocephalus) *with young. With fewer than 100 birds in East Asia, and between 10,000 and 25,000 in Southeast Asia and South Asia, these birds are vulnerable to hunting and collection of eggs and nestlings.*

In colonial species, they roost in groups. Fledging success depends on the availability of food, and nesting failure at any stage is not uncommon if supplies give out. The nesting cycle usually lasts two to three months, with re-nesting sometimes occurring after a failure. In most species, the parents desert the chicks at fledging, but in some, such the Waldrapp ibis, they remain in family groups for some time.

As a result of the dependence on food supplies, local conditions determine breeding seasons. Sacred ibises have quite different breeding schedules in various parts of Africa, coinciding with local rainfall patterns. Even in a single area the nesting schedule may vary from year to year, and not all species in an area nest at the same time. In Venezuela, the Green ibis nests in the wet season, but the Buff-necked ibis nests in the dry, presumably because of different choices of prey. Tropical forest ibises, such as the Spot-breasted, appear to nest through much of the year.

Tropical species, especially solitaries, tend to be sedentary, occupying an area year-round. Those nesting in temperate areas undertake seasonal migrations toward the tropics. In subtropical and semi-arid regions heavily influenced by seasonal rainfall patterns, many species tend to be nomadic; the Australian white ibis is a good example, moving around from place to place following rainfall and water patterns, and nesting when and where water conditions become suitable. The juveniles of most species disperse after nesting.

Migrations of ibises have figured prominently in human cultures. Along the Euphrates, the Waldrapp ibis's return in spring was celebrated by a festival, as a fertility symbol and guide to Mecca. The Waldrapp ibis nested not only in the Middle East but also as far north as the Alps, where it was noted in 16th-century natural history writing. Similarly, the seasonal occurrence of Sacred ibises along the Nile was associated with the annual flood that was crucial to farming.

On the Brink

CONSERVATION AND ENVIRONMENT

Given the importance of the Waldrapp and Sacred ibises to ancient cultures, it is particularly poignant that the two species no longer occur in the areas of their greatest historical significance. Their disappearance is due principally to hunting and habitat destruction, the two main forces affecting ibises worldwide. In the wild, the Waldrapp ibis is now confined to a small area of North Africa. Hunting, harassment, and habitat change led to its decline in Turkey, despite intervention through captive breeding; captive Waldrapp ibises now outnumber those in the wild. The Sacred Ibis has been absent from Egypt since the first half of the 19th century. Similarly, the Oriental crested ibis, widely distributed in Japan and China until the early 20th century and in Korea until World War II, is now among the most endangered ibises, numbering only in the dozens in the wild. Loss of suitable habitat – pine forests surrounded by swampland – probably contributed to its demise.

The spread of scrubby vegetation has affected the habitat of the Bald ibis in southern Africa, and the Giant ibis, populations of the Black ibis, and the Black-faced spoonbill are all nearing extinction. Subfossils from Hawaii and Jamaica show a repeated evolution of flightlessness among ibises on islands, followed by human-induced extinctions. More recently, less distinctive island races have died out or are on the brink of extinction due to a combination of killing and habitat loss. JAK

SPECIAL FEATURE

FEEDING BY TOUCH

Tactile foraging techniques in ibises and spoonbills

Left *In the marshland or mangrove swamps where they habitually forage, Roseate spoonbills detect prey through their sense of touch as they sweep their spatula-shaped bills from side to side through shallow water and silt. During this sweeping action, the bill is held almost vertical and partly open.*

Right *Elegant, slender, and curved, the bill of the Scarlet ibis is an ideal tool for probing in mudflats or around the prop roots of mangroves for crabs and mollusks. The birds' diet accounts for their vivid coloration, through synthesis of a pigment (carotene) present in the bodies of the prey items.*

IBISES AND SPOONBILLS FEED ALMOST ENTIRELY by touch. Ibises probe with their long bills, whereas spoonbills wave their flattened bills back and forth in the water. Although tactile foraging occurs in other species, such as shorebirds and storks, it is nonetheless an evolutionary and ecological characteristic of fundamental importance to this particular family. Clearly the radiation of the family into two groups based on tactile foraging modes was a fundamental turning point in the birds' evolutionary history.

Both groups can, of course, also use sight when appropriate. The birds' line of vision tends to point to the tip of the bill. They use vision to decide where to forage, including where to place the bill. Terrestrial ibises will pick up items they see, although even in these cases they will often probe a little first to locate it. When it comes to actually catching food, however, it is the bill that feels out the prey item.

The bills are equipped with tactile sensors that allow the birds to respond quickly when a potential prey item is encountered. The bill snap captures the prey in a forceps grasp. Ridges on the bill hold the prey. A forward lunge of the head sends the prey item toward the head, either to be caught again closer to the mouth or else to pass straight down the gullet. Large items may be bitten several times before they are swallowed. Some species tear prey apart, stabbing and biting with their bill.

In shallow water, ibises insert the tip of the bill into the bottom or within submerged vegetation. Aquatic ibises tend to have relatively long bills to allow deep penetration. The birds also probe in soft mud, either under the water or exposed above it, in dirt or grass, or in fact in almost any sort of substrate that will allow penetration.

Terrestrial ibises poke, peck, or probe along the ground or in plant cover. They have relatively short bills, which are more efficient on dry land than the longer bills of the aquatic species. Even so, all ibises are perfectly capable of pecking and probing on the ground as well as in water. Various species, including some basically aquatic ones, can commonly be seen in pastures and on lawns, probing into the topsoil.

As a result of the birds' nonvisual approach to feeding, the ibis diet is broad, basically consisting of whatever can be encountered and captured in the places they choose to feed. Typical prey species are slow-moving; in water they tend to live close to the bottom, while on land they may be burrowers, or else found in grass or topsoil. Consequently, insects, crustaceans, snails, and other invertebrates tend to predominate. Fish are readily consumed, but they tend to be caught most effectively when at high densities.

Spoonbills use touch as they swing their open bills in the water. The bills' exceptionally broad surface area makes contact with a potential prey item more likely, and also aids in its subsequent capture. Spoonbills tend to swing their bill through the open water or along the surface of mud or sediment. They are more likely than ibises to encounter small fish, and they also take crabs, prawns, and other demersal (bottom-dwelling) species. So the tactile technique of spoonbills takes more fish than does the probing technique used by ibises.

While probing and head swinging respectively dominate the feeding repertoire of ibises and spoonbills, it is notable that ibises also sometimes swing their bills back and forth, while spoonbills occasionally probe. The birds' ability to use both techniques suggests that both tactile behaviors were developed early in the evolutionary history of the family.

At hatching, the chicks have thick but rather normal-looking bills, which change form rapidly during the growing period. Tactile feeding is not a straightforward technique, and it appears that young birds have to learn to become proficient at it. Practice begins while the birds are still in their natal colony. Maturation is delayed in these species for one to several years, possibly in response to the bird's need to learn how and where to feed effectively. Once developed, tactile feeding techniques can be deployed in many different situations and habitats, from shallow water to semiarid mountains, as evidenced by the diversity of species and the worldwide distribution of the family.

JAK

Hammerhead

SMALL, CHOCOLATE BROWN, AND RELATIVELY short-legged, the hammerhead – also called the hamerkop or Hammerhead stork – gets its name from the combination of a relatively large bill and a conspicuous, rear-pointing crest. It is distantly related to herons, storks, and flamingos, and is probably the sole extant representative of a lineage derived from a common ancestral group.

The hammerhead is locally common to abundant through its African range, and even relatively common in Madagascar, where many other aquatic birds have decreased. It is protected to some degree by local custom, and has benefited from irrigation schemes.

Small and Sedentary

FORM AND FUNCTION

The hammerhead is a sedentary bird, occupying a well-defined territory, although some pairs will move to normally dry areas when seasonal rains fill dry holes and ditches. It is found in a variety of natural and human-made habitats, including savanna, open woodlots, forests, small ponds, estuaries, riverbanks, and many types of irrigated situations. Wherever water is artificially impounded within its range, in the form of dams or canals, the hammerhead will quickly arrive.

Hammerheads feed alone or in pairs by wading in shallow water, standing or walking slowly while visually searching for prey. They may rake their feet against the substrate or flash their wings to startle prey. They also sometimes take prey, especially tadpoles, from the water while flying. Although hammerheads eat a variety of small fish, shrimps, and insects, their principal food is frogs and tadpoles of the genus *Xenopus*, which also form a major part of the nestlings' diet.

The large, flat bill is slightly hooked and is unlike that of any other species in the Ciconiiformes. The neck and legs are relatively short for a wading bird, and the toes are partly webbed, for reasons that are not clear. The bird feeds diurnally, often roosting at midday. It has a short tail and relatively huge, broad, rounded wings that enable it to glide and soar easily. When soaring, the head is stretched out, but when flapping it is tucked back. The bird will usually fly only a short distance when disturbed.

Right *Although active during the day, hammerheads become more active at dusk and are semi-nocturnal in their habits. Despite their distant relationship with storks and herons, the hammerhead's behavior is markedly different from that of any other bird.*

Right *Despite local myths of other bird species helping the hammerhead bring up nesting materials, each nest is the work of a single pair. They are decorated on the outside with brightly-colored objects such as crockery, bones, cloth, and other unusual materials. The huge nest attracts many other animals. Smaller mammals such as genets sometimes take up residence, and small birds such as weaver birds, mynas and pigeons will attach their nests to the main nest.*

Below *During the false mounting display either sex may assume the top position and they may often swap roles a number of times. This behavior is not only confined to mated pairs, and may take place on a tree, on the ground or even on top of a nest.*

Busy Builders

BREEDING BIOLOGY

The hammerhead breeds nearly year-round over much of its range. The nest is built typically in the fork of a tree over water, but if trees are not available, it may build on a wall, bank, or cliff, or sometimes even on the ground. The birds nest solitarily, although more than one nest may be built in an area. The nest can take six weeks to construct, and may exceed 1.5m (5ft) in diameter. It is a domed structure, with a hollow chamber connecting to the outside by a small opening at the bottom, which leads to a tunnel up to 60cm (2ft) long. The opening is plastered with mud. Nests are typically abandoned after a few months, when the bird builds a new one within its territory. Needless to say, the abandoned nests are often used by other birds and by reptiles.

Hammerheads are sometimes to be seen, usually near a nest, participating in group ceremonies that can involve as many as ten birds calling loudly while running round each other in circles. Crests are raised, wings fluttered, and a chorus of cries continues for several minutes. A very distinctive ceremony, called "false mounting," involves one bird mounting another without actually copulating with it.

True mating usually occurs at a completed nest site, often on top of the nest, using displays similar to those seen during the larger gatherings of birds. When the eggs have been laid, both birds incubate them. The young are hatched covered in gray down, but they quickly develop feathers, with the head and crest complete within 17 days and the body plumage within a month. Both adults feed the young, but, unusually for wading birds, may leave them unattended for long periods, presumably because the thick nest walls protect them from predators. When the young are fully fledged, they remain near the nest for another month, roosting in it at night.
JAK/JH

FACTFILE

HAMMERHEAD

Scopus umbretta

Order: Ciconiiformes

Family: Scopidae

DISTRIBUTION Sub-Saharan Africa and Madagascar

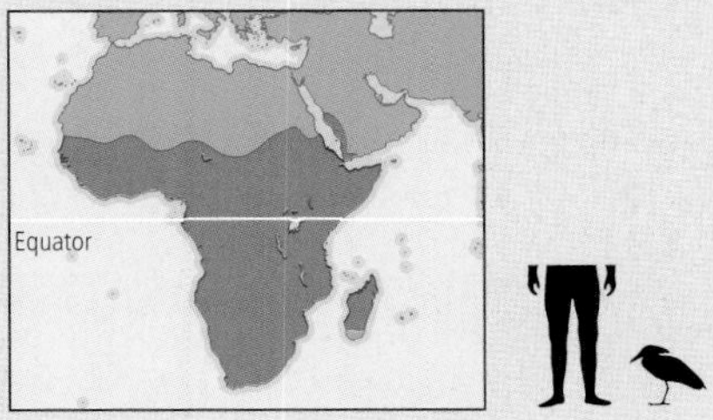

HABITAT Wetlands

SIZE Length: 50cm (20in); **weight**: 470g (17oz).

PLUMAGE Entirely dark brown, with a purple sheen on the back.

VOICE Quiet when alone; in groups, a wide range of vocalizations consisting of crackles and a shrill flight call.

NEST: An enormous dome of sticks, weeds, and grass.

EGGS 3–7; white, but soon stained. Incubation 28–30 days; nestling period 44–50 days.

DIET Amphibians (especially frogs), but also small fish and invertebrates.

CONSERVATION STATUS Not threatened.

Shoebill

ALTERNATIVELY CALLED THE WHALE-HEADED stork, the shoebill is one of the most distinctive of the large wading birds. Its dominating feature is its huge bill, which bears an uncanny resemblance to a Dutch wooden clog. Indeed, early Arab traders in Africa, who encountered the bird in the marshlands of the Upper Nile, dubbed it abu markub*, or "father of the shoe." Its scientific name, which translates as "King Whalehead," is no less descriptive.*

The shoebill is a solitary bird of the vast African marshes, favoring swamps of floating papyrus ("sudd"), lake-edge marshes, and reed beds. Its taxonomic affinities have been much debated. It appears to be related to both the Ciconiiformes and Pelicaniformes, and probably represents an early offshoot of the ancestors of the two groups.

A Bill for Catching Fish

FORM AND FUNCTION

The shoebill's eponymous bill is unique among all birds, being nearly 20cm (8in) long and extraordinarily deep. It is sharply edged and has a terminal hook, unlike the pointed or blunt tips of other ciconiiform birds. This is a strong bill, designed to withstand the pressures of capturing large fish.

Shoebills feed alone, standing on vegetation rather than in the water. Even if members of a pair feed in the same area, they separate themselves. The bird stands quietly, holding its bill

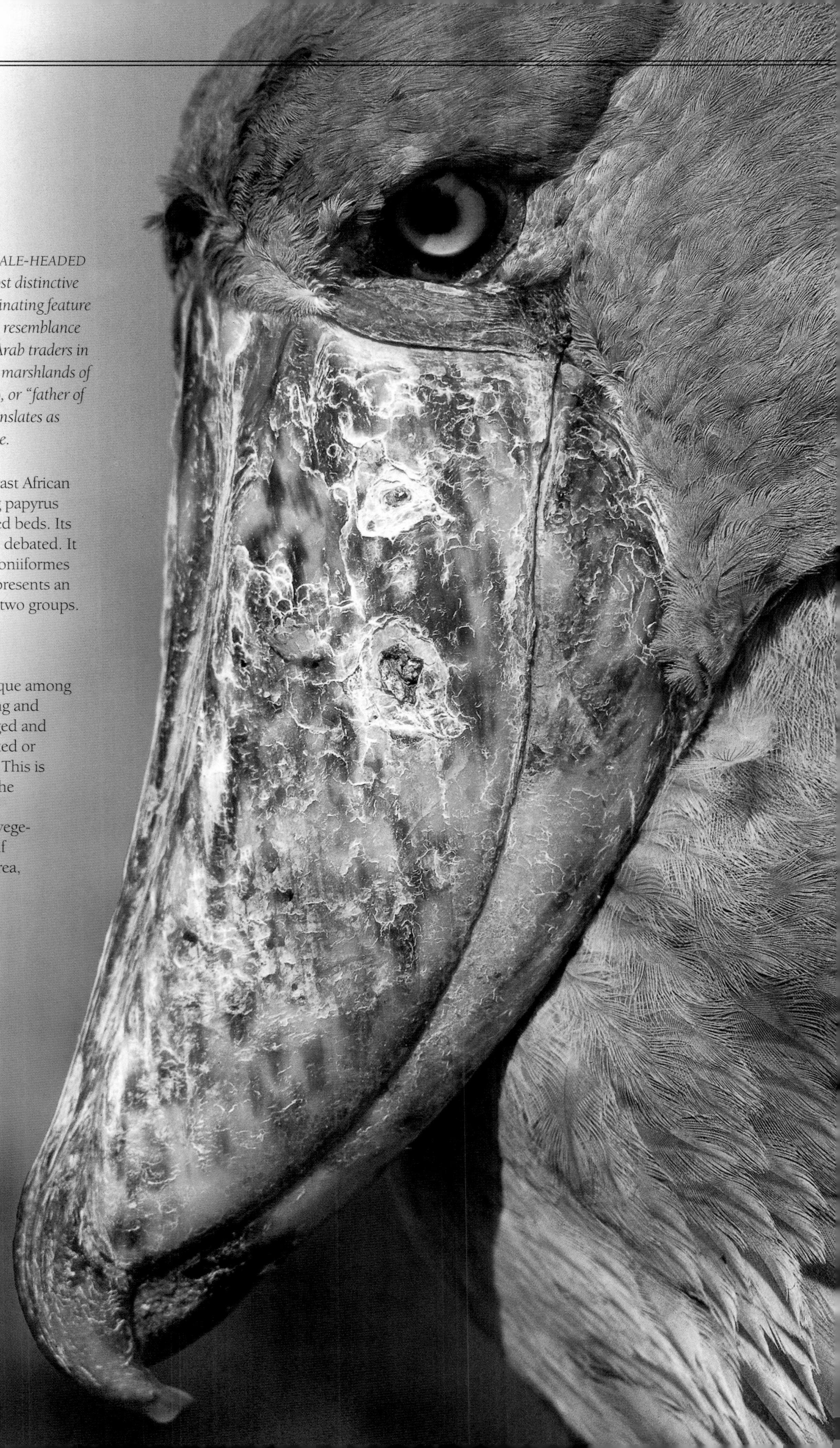

Right *The flaking bill of a shoebill or Whale-headed stork. The huge bill is adapted to probing muddy waters for fish such as lungfish and gars. It will also opportunistically take frogs, young turtles and crocodiles, and other small animals. Additionally, shoebills have been reported to feed on floating carrion.*

downward and peering over it in search of prey. Waiting motionless, or moving very slowly with great care, are apparently vital elements of the bird's tactics, and it may remain frozen for up to 30 minutes or more. Its most characteristic prey items are lungfish (*Protopterus* spp.), which come to the surface of poorly oxygenated water to breathe, and in doing so become vulnerable. The bird strikes bill-first, forward and down, launching its entire body toward the prey. From this collapsed position, it rights itself by jerking back, using its wings if necessary; if it has been successful, the fish will be firmly clamped in its bill. Recognizing the importance of stealth and surprise, the bird moves to another fishing place after each strike.

The shoebill flies well once aloft, and can soar to great altitudes in thermals. It flies with its neck curved over its back, like a heron, in order to support the weight of the head.

Dry-Season Breeders

BREEDING BIOLOGY

The shoebill breeds at different times over its range, but generally in the dry season, when water levels are falling and prey are perhaps easier to come by in the shrinking swamps. As ground nesters, their nesting areas lie well within the swamp and are well defended. Even so, crocodiles may sometimes kill nestlings. Courtship displays are little-known, but appear to include soaring and bill clattering. Given that in most populations the birds are sedentary, it may be that pairs re-form. Hatching is asynchronous, as is usually the case in wading birds. The adult may sometimes use its great bill to carry water to the nest to cool the chicks.

This unusual species is widespread but localized within its range. The remote, impenetrable nature of its habitat means that there is no reliable estimate of its numbers. Habitat protection is the main conservation issue, given the bird's dependence on vast wetlands that fluctuate annually. One important refuge, where there is an estimated 1,500 individual shoebills, is the Bangweulu Swamp in Zambia, one of the largest and most important of the wetlands in Africa. Another stronghold is in the vast Sudd of Sudan, but this swamp, along with others, is now threatened by water diversion and drainage. JAK/JH

FACTFILE

SHOEBILL

Balaeniceps rex

Order: Ciconiiformes

Family: Balaenicipitidae

DISTRIBUTION N C and E Africa

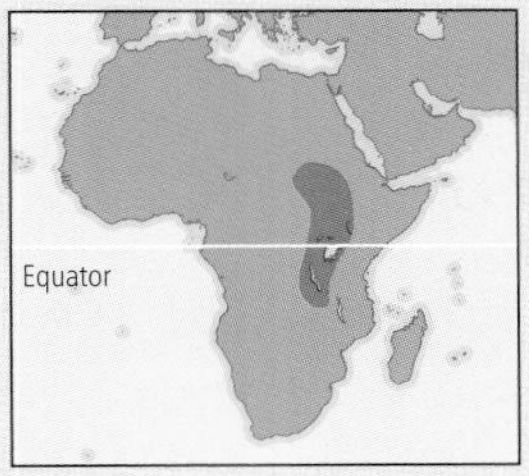

HABITAT Wetlands

SIZE Length: 110–140cm (42–54in); **weight:** 4.4–6.7kg (9.7–14.8lb). Males are larger than females, with longer bills.

PLUMAGE Blue-gray, with dark green sheen on the back. Juveniles are darker.

VOICE Normally quiet; the little-heard flight call is a guttural croak. At the nest, a soft whining or mooing sound; the bird also claps its bill.

NEST A large mound of aquatic vegetation, usually floating in deep water well within a reed swamp or placed on an isolated island. Material is added continually to counteract subsidence and decay.

EGGS 2–3; dull chalky blue to white. Incubation 30 days; nestling period 95–105 days.

DIET Typically fish, especially large ones, but also frogs, turtles, aquatic snakes, and lizards.

CONSERVATION STATUS Lower Risk/Near Threatened.

Left *Resting its heavy bill on its chest, the shoebill can stand motionless for hours, moving only occasionally to alter the position of its head. Yet once potential prey is spotted, it leaps into action. Shoebills lead solitary lives; only during the breeding season are they found in pairs .*

Flamingos

WHETHER FLAMINGOS ARE THOUGHT TO be bizarre or beautiful depends on their numbers: individually they are rather grotesque, but two million pink birds massed around the edge of Lake Nakuru in Kenya's Rift Valley make a breathtaking spectacle.

Flamingos are an ancient group, and fossil evidence dates back to at least the Miocene epoch (about 10 million years ago). Their classification is still very controversial. They have been considered a suborder of the storks (Ciconiiformes), and their egg-white proteins resemble those of herons (Ardeidae). On the basis of behavior and feather lice, they have seemed most like the waterfowl (Anseriformes), but recently affinity with the waders (Charadriiformes) has been stressed because of supposed similarities between flamingos and the Australian banded stilt. Because flamingos are so different from other birds they have also been considered an order of their own, the Phoenicopteriformes. Andean and Puna flamingos differ from the other species in having no hind toe, or "hallux," but these and the Lesser flamingo have a more specialized feeding apparatus than the larger *Phoenicopterus* flamingos.

Oddly, the founder of systematic biology, Linnaeus, described as the typical flamingo of the family Phoenicopteridae the Caribbean flamingo, and not the European Greater, with which one might have thought he would have been more familiar. Early travelers to the West Indies must have provided the specimens that he described.

Long-legged Waders

FORM AND FUNCTION

Flamingos have large bodies, with small heads, long necks, and long legs for wading. The pink and crimson plumage of the adult birds, with black secondary and primary wing feathers, makes them conspicuous and unmistakable. The Caribbean flamingo is the brightest (as are the Caribbean ibis and spoonbill in their respective families). There is a single molt of the feathers of wing and body per breeding cycle. Legs, bills, and faces are brightly colored red, pink, orange, or yellow. The rather small feet are webbed, and the birds can use these webs to swim and to stir up debris from the mud by trampling. Males are larger than females, markedly so in some species, but this discrepancy in size is the only obvious difference between the sexes.

At hatching, the chick has gray-white down, a straight, pink bill, and swollen pink legs, both of which turn black within a week. Juveniles in their first plumage are gray, with brown and pink markings, and their legs and bills are black. In fledged birds the bills are turned down in the middle, with the upper jaw small and lidlike and the lower one large and troughlike; both are fringed, and lined with filtering, comblike structures called lamellae. The tongue is thick and spiny.

Pockets in the Tropics

DISTRIBUTION PATTERNS

Fossils suggest that flamingos once ranged through Europe, North America, and Australia, as well as in the areas where they are found today, but the group now occurs only in widely separated pockets. Many of these are in the tropics, but the Greater flamingo also ranges widely over the southern Palearctic, from the Mediterranean northeast to Kazakhstan. On both sides of the Atlantic flamingos occur on coastal wetlands as well as at inland sites, including some lakes at high altitudes.

Below *Representative flamingo species:* ***1*** *James' flamingo* (Phoenicoparrus jamesi) *feeding;* ***2*** *Lesser flamingos* (Phoeniconaias minor) *can pump water through their bills 20 times per second in order to filter food;* ***3*** *Chilean flamingo* (Phoenicopterus chilensis) *stretching;* ***4*** *The Greater flamingo* (Phoenicopterus ruber) *is the most widespread species.*

FACTFILE

FLAMINGOS

Order: Ciconiiformes

Family: Phoenicopteridae

5 species in 3 genera: **James'** or **Puna flamingo** (*Phoenicoparrus jamesi*), **Andean flamingo** (*P. andinus*), **Lesser flamingo** (*Phoeniconaias minor*), **Chilean flamingo** (*Phoenicopterus chilensis*), **Greater flamingo** (*P. ruber*). The Greater flamingo has 2 subspecies: the **Greater flamingo** (*P. r. roseus*) and the **Caribbean flamingo** (*P. r. ruber*).

DISTRIBUTION Around the world, in a wide range of tropical and warm temperate sites, some at high altitude.

HABITAT Shallow salt or soda lagoons and lakes.

SIZE Length: 80–145cm (31–57in); weight: 1.9–3kg (4.2–6.6lb); females smaller.

PLUMAGE Pink; both sexes similar.

VOICE Loud and gooselike.

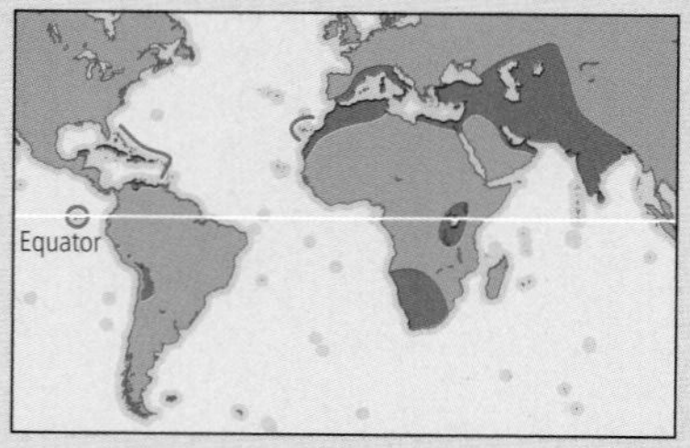

NEST Mud-built mounds

EGGS Usually single; white; weight about 100g (3.5oz). Incubation period 28–30 days; nestling period 75 days.

DIET Algae and diatoms; small aquatic invertebrates, particularly crustaceans, mollusks, and insect larvae.

CONSERVATION STATUS The Andean flamingo is Vulnerable; the Chilean, Lesser, and James's are all classified as Lower Risk/Near Threatened.

2
3
4

Flamingos of all species tend to be erratic in their movements, the intensity of which vary either in relation to the season, water levels, and food availability, or else according to the age of the birds. In some parts of their range, flamingos migrate to and from breeding grounds that are frozen in winter, as is the case, for example, with Greater flamingos in Kazakhstan and Russia. Recoveries and resightings of banded birds reveal sometimes largescale movements over thousands of kilometers, from European or Asiatic breeding sites to African wetlands, which they use outside the breeding season. Some birds originating from the same colony disperse widely, while others have been shown (in France) to be sedentary over many years, reflecting the complexity of movement patterns. Many long-distance flights are undertaken at night.

Suction Filter-feeders

DIET AND FEEDING

Flamingos feed both by day and night. The smaller species are unusual in feeding on the microscopic blue-green algae and diatoms that live in alkaline salt and soda lakes. In contrast, the diet of the larger species consists mostly of small invertebrates, larvae, and crustaceans.

The birds' feeding method is characteristic and peculiar. The bill is held upside-down in the water, with the tongue acting as a piston, so that water and mud are sucked in along the whole gape and expelled three or four times a second past the filtering lamellae. This filter-feeding has been likened to the way Baleen whales feed. In the small Lesser, Puna, and Andean flamingos, which have deep-keeled bills, very fine particles are retained while coarse particles are kept out by stiff excluder lamellae. The "shallow-keeled" Caribbean, Greater, and Chilean flamingos are larger and feed mainly on invertebrates such as brine flies (*Ephydra*), shrimps (*Artemia*), and mollusks (*Cerithium*), which they obtain from the bottom mud, normally by wading in shallow water, more rarely while swimming, and sometimes by upending like ducks. The brilliant red color of flamingo plumage derives from the rich sources of carotenoid pigments (similar to the pigments of carrots) in the algae that the birds consume either directly or secondarily. Blue-green algae are also an extremely rich source of protein, and dense blooms of the planktonic algae *Spirulina platensis* are associated with the gathering of huge flocks of Lesser flamingos in East Africa; success in breeding probably depends upon such blooms.

Above *Lesser flamingos are primarily an African species but there are also significant populations in Pakistan and India.*

Below *At around two weeks old Greater flamingo chicks join other chicks to form a large crèche under the watchful eye of the adults. Flamingos spend most of their day feeding and preening, around 15–30 percent of the time being spent on preening alone.*

Crèches 30,000 Strong

BREEDING BIOLOGY

Flamingos can be very long-lived: several captive birds have reached 60 years of age, and 40-year-old birds are probably not unusual in the wild. Although in captivity pairs may stay together over a period of years, pair bonds in Greater flamingos are not strong and pairs change from one season to the next, or even between successive breeding attempts in the same year, as observed in the Camargue (southern France).

Flamingos of all species tend to be rather erratic in breeding, and whether they nest or not depends mainly upon rainfall and the effect this has on the food supply of the adult birds. The nest mound is made of mud and may be 30cm (12in) high, thus giving protection from flooding and from the often intense heat at ground level. Nests are built by male and female alike using a simple technique of drawing mud toward the feet, or body, with the bill. This activity gradually degrades the nesting sites. In the Camargue, where the breeding island is fully occupied each year, late breeders either attempt to nest in depressions between the mounds, or wait until the mound of an earlier breeder is freed before they themselves can nest. In the Kiaone Islands in Mauritania (West Africa), the Greater flamingo nests on rocky outcrops in the Atlantic Ocean, the eggs being laid on the bare, stony ground.

The single large, chalky egg is incubated by both sexes in turn, attentive periods lasting from 2 to 4 days. Adults incubate with the legs folded under the body, just as other birds do, and not dangling from the nest, as legend formerly suggested.

After hatching, the chick remains in the nest for some days; it is fed by its parents on a secretion from the glands of the upper digestive tract or crop (see Flamingo Milk). Caribbean flamingo chicks can feed themselves at 4–6 weeks of age but, in at least the Greater and Lesser flamingos, parental feeding continues until fledging, by which time the bill of the chick is hooked as in the adult and the youngster is capable of independent feeding.

After they leave the nest, the young move into large crèches, which in the Lesser flamingo may contain as many as 30,000 birds; parents apparently find their own chick in the group and feed it alone, recognizing it by its calls. These feedings take place mainly in the evening and at night, and when the chicks are well-grown, they may last for half an hour or more.

In some of the larger salt depressions where Greater and Lesser flamingos nest in Africa, the groups of chicks may trek over distances of as much as 80km (50 miles) on the hard salt-crust in search of water as their birthplace dries out in the heat of the sun. These natural nesting sites cannot be used in years of drought or when they are flooded – perhaps only once or twice in ten years. In such circumstances the flamingos must race

FLAMINGO MILK

Two groups of birds, flamingos and pigeons, feed their young on "milk." Compared with pigeon milk, the secretion from the crop of the flamingo has somewhat less protein (8–9 percent versus 13.3–18.6 percent) and more fat (15 percent versus 6.9–12.7 percent). There is almost no carbohydrate in either case. About 1 percent of flamingo milk is made up of red blood cells, whose origin is unknown. Thus, bird milk is similar in nutritional value to that of mammals.

As with mammalian milk, secretion is controlled by a hormone called prolactin. In birds, the hormone causes a proliferation of the cells of the crop gland in males and females, so that both sexes feed their offspring, whereas in mammals "nursing" is only ever a female task. From studies of birds in captivity it has been found that a few birds that are not parents produce milk, and even 7-week-old chicks can act as foster-feeders for smaller, orphaned birds. The persistent begging calls of the youngster seem to stimulate hormone secretion. Milk is not produced until the crop is cleared of food, so that food items themselves are never regurgitated.

The crop milk of the flamingo contains initially large amounts of canthaxanthin (the pigment that colors the adults' feathers), which gives the milk a bright red color. After a while this fades and the secretion becomes a pale straw color. Canthaxanthin is stored in the young bird's liver and not in the down nor in the juvenile plumage, which is gray.

Parental feeding of this specialized kind seems to be an adaptation to ensure that the young obtain enough food, especially protein. The high alkalinity of the water in which the adults feed, their unusual feeding habits and bill structure, and the fact that they may nest some distance from their food source all probably encouraged the evolution of crop milk manufacture. The system is successful only because the clutch size of the flamingo is so small (usually a single egg) – even smaller than in most pigeons.

Left *An Andean flamingo chick on its nest mound. Unlike most waders and waterfowl, after hatching the flamingo chick will remain in its nest for 5–8 days. Its straight pink bill and swollen pink legs both turn black within a week of hatching.*

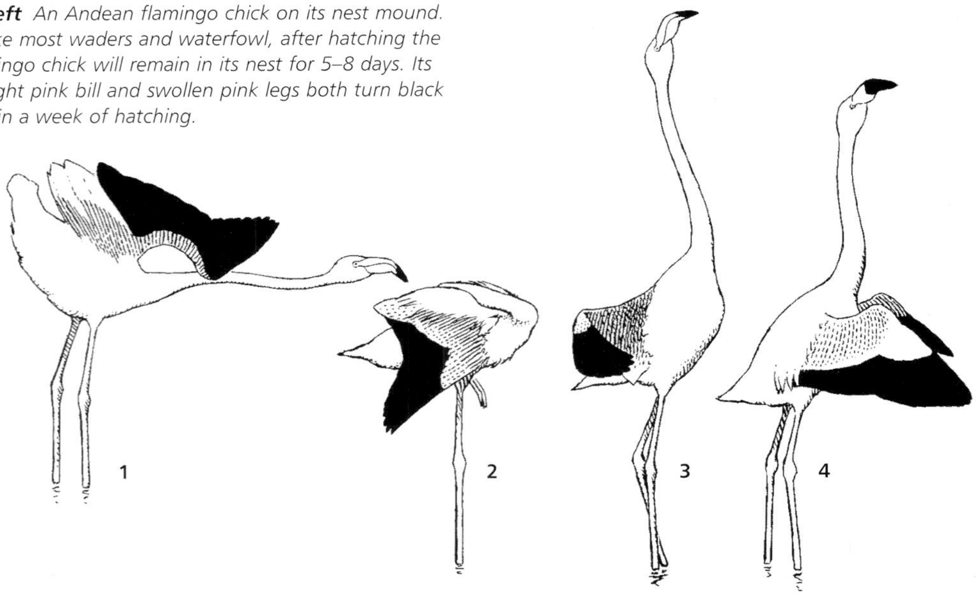

against time to raise their chicks before the depression dries again. The water in such sites can be highly saline, and the less fortunate chicks may be hampered by anklets of soda that form around their legs. Sometimes many thousands of fledglings starve to death as a result, and massive rescue operations have been organized to reduce the losses.

In Europe, studies in Spain have shown that when the breeding site at Fuente de Piedra (Malaga) dries out in summer, the parents of thousands of chicks forage in the marshes at the mouth of the River Guadalquivir, 150km (95 miles) away. Both parents feed the chick, so that most birds undertake such flights only once every two days, but some birds fly back to the feeding grounds the same night. Similarly long flights are undertaken by Caribbean flamingos breeding on Bonaire in the Netherlands Antilles, some birds flying to the coastal wetlands of Venezuela to feed.

The young bird is fledged in about 11 weeks, and gradually loses its gray juvenile color over two or three years; it will not, however, display and breed until it is fully pink. Studies in the Camargue have shown that Greater flamingos never breed before the age of 3 years, and most birds do not do so until they are 6–7 years old. Flamingos did not nest successfully in zoos until the importance of good adult feather color was realized, and an effort made to increase the carotenoids in the diet. Carrots, peppers, dried shrimps, and other such items were tried initially; today synthetic canthaxanthin is added to the food, and the large *Phoenicopterus* flamingos are breeding more regularly in captivity.

The birds are highly gregarious at all stages of their life-history, and the displaying and nesting colonies are noisy places. Breeding in small numbers is almost unknown, although an exception is provided by the Caribbean flamingo, whose isolated population on the Galápagos islands occasionally nests in groups of only 3–5 pairs. Group display seems to bring all the birds of the colony to the same readiness to mate, and thus ensures rapid and synchronized egg-laying in a potentially unstable breeding habitat.

Above *Flamingo displays are similar to the preening and stretching movements that the birds adopt in everyday activity. The displays are only more stiffly performed, more contagious among members of the group, and given in predictable sequences. (Mating displays between male and female are, by contrast, almost nonexistent and quite inconspicuous.) "Head flagging" followed by "wing saluting," in which a bird spreads its wings to the sides and folds them again, is common. The general impression is of a flash of black in a pink field.* ***1*** *In the "inverted wing salute," given here by a Greater flamingo, the bird bends forward and the wings are flashed partly open and held above the back.* ***2*** *A "twist-preen" maneuver sometimes follows the wing salute; here, a Caribbean flamingo twists its neck back, flashing a wing forward to expose the black primaries, and appears to preen behind the wing with its bill. Wing salutes differ between species, for example in* ***3*** *the Greater flamingo and* ***4*** *the Chilean flamingo.*

Below *When taking off, flamingos, like this Greater flamingo, run several steps and begin flapping their wings before lifting into the air.*

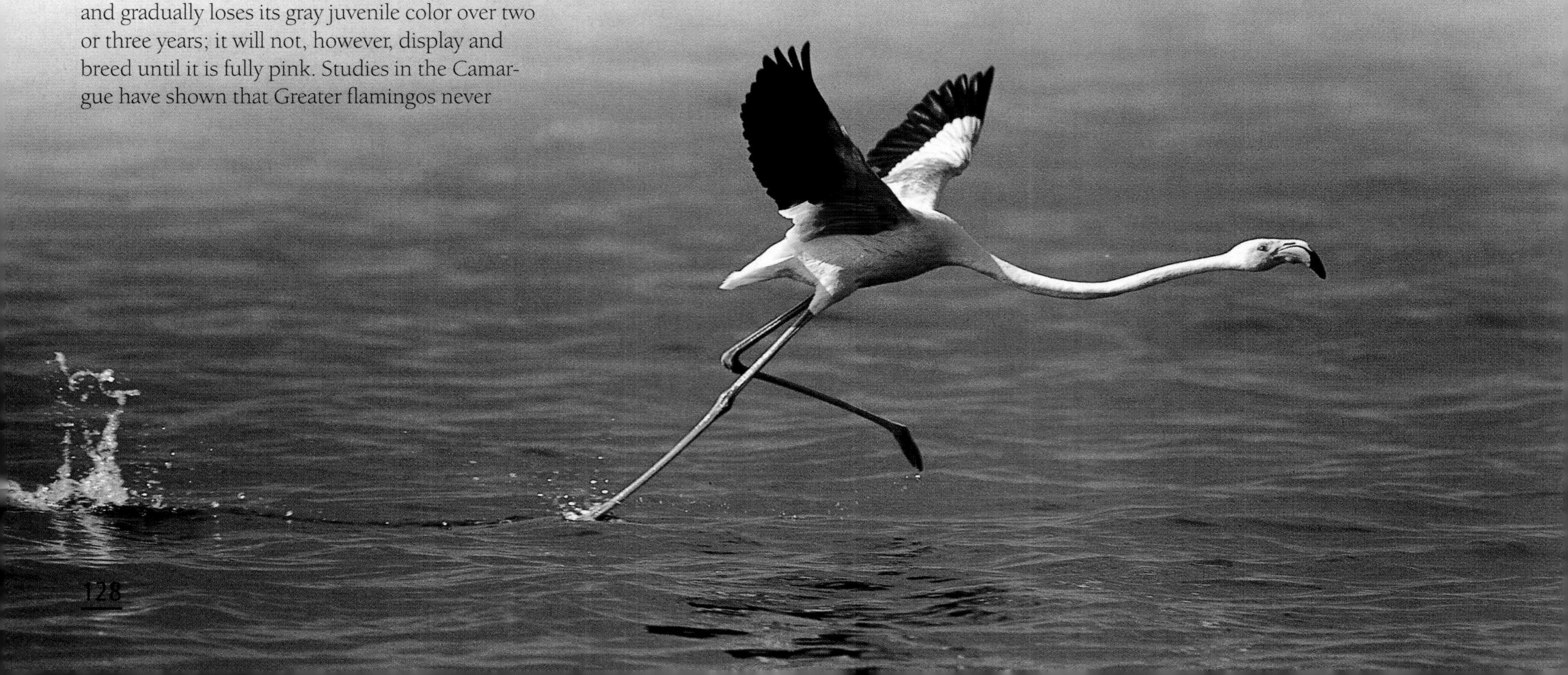

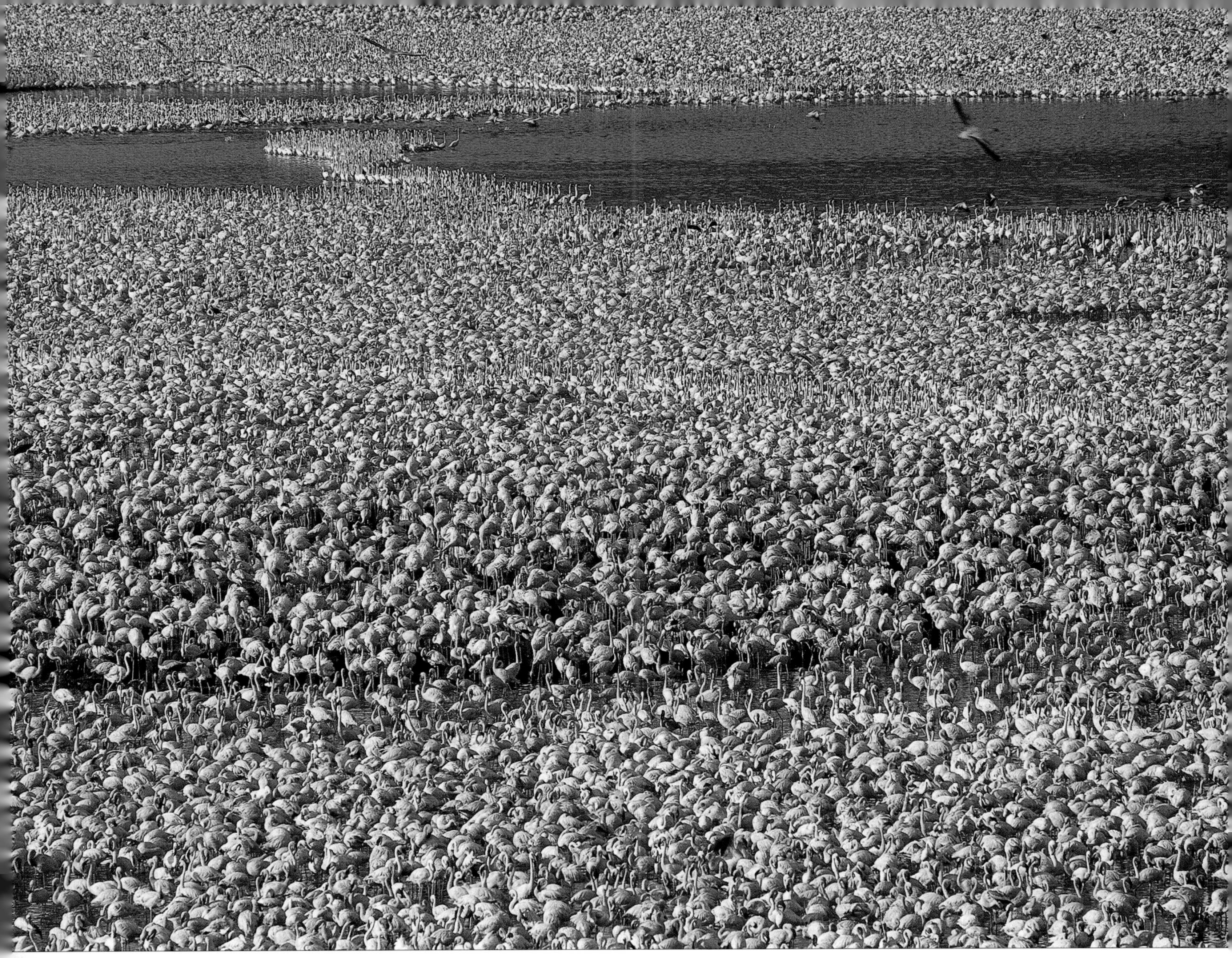

Above *A colony of Lesser flamingos at Lake Bogoria in the Rift Valley, Kenya. Millions of birds gather in the soda lakes throughout the region to extract the plentiful plankton from the alkaline, mineral-rich waters.*

Inhospitable Terrain

CONSERVATION AND ENVIRONMENT

The color of flamingo feathers fades gradually in sunlight, which may have been one reason why large numbers were not taken in the past for the plumage trade. The birds' tongues, however, were once pickled as rare delicacies, and flamingo fat is still considered a cure for tuberculosis by some Andean miners.

The development of salt and soda extraction works has been a particular threat in various parts of the world, but in the Camargue and on Bonaire, for instance, artificial lagoons created by the salt industry have been accepted by the birds with considerable success. At the former location, an island was created in 1970 in the salt pans near Salin de Giraud in response to the lack of suitable natural sites. Flamingos were attracted to the island by the presence of artificial mounds, molded using buckets of mud. The birds have bred on the island since 1974, and it is still the only place in France where the Greater flamingo breeds.

Flamingos have few natural predators, since they tend to live in inhospitable places where the water is so alkaline that the lagoons are barren of vegetation and surrounded by almost desert-like wastes. Some of the colonies established in the Mediterranean are more vulnerable to disturbance, and occasionally suffer egg and chick losses when foxes or stray dogs can reach them. The main predator in this region, however, is the Yellow-legged gull, an invasive species that in the Camargue alone takes hundreds or thousands of eggs and small chicks each year; yet in spite of this loss the birds continue to prosper. In the more northerly parts of their range, Greater flamingos sometimes have to face up to hard winters, and during spells of severely cold weather they can suffer heavy losses; over 3,000 birds died from starvation in southern France in January 1985, when temperatures remained below freezing for two weeks.

All flamingos are vulnerable to habitat change and exploitation. The Puna flamingo is rare, but the crimson Caribbean flamingo is perhaps more threatened because it breeds in only four main colonies around the Gulf of Mexico (in Yucatán, Inagua, Cuba, and Bonaire), and is in great demand by zoos. Many birds die after capture; they tend to travel badly, and are susceptible to stress upon arrival. So far, the three smaller species have bred seldom or not at all in captivity, and birds are still taken from the wild for the zoo trade in regrettably large numbers.

Population estimates have been made for all species of flamingos and the following figures indicate very approximately how many birds there are in the world: Lesser 5 million individuals; Greater and Chilean, 500,000; Caribbean 100,000; Andean and James', 50,000. ARJ/JK

Screamers

Left *All screamers (the Northern species is shown here) have two spurs protruding from each wing that are used in aggressive encounters with rival birds. The huge feet are adapted for stalking in the shallows of the marshy lagoons where the birds live.*

Right *Two Southern screamers stand to attention at the water's edge in the Mato Grosso region of southern Brazil. The birds not only forage in the water for marsh plants but also roost there at night, in nests built of sticks and vegetation close to the shore.*

SCREAMERS ARE WATERFOWL OF THE SOUTH American wetlands that are gooselike in overall shape although with a heavier body, not unlike that of a small turkey. They take their name from their loud, far-carrying voices, used to establish and maintain breeding territories.

There are three screamer species, none very well studied in the wild. Superficially, they bear little resemblance to the other anseriforms, but there are anatomical similarities that link them to the Magpie goose. They look heavy-bodied but are, in fact, relatively light because of a unique feature; beneath the loose skin lies a network of small air sacs, so that the birds crackle when handled. They can produce rumbles by vibrating these sacs, and the sounds seem to function as close-range threat. Flight feathers are molted gradually, so that, like the Magpie goose but unlike other ducks, geese, and swans, screamers do not pass through an annual flightless phase.

FACTFILE

SCREAMERS

Order: Anseriformes

Family: Anhimidae

3 species in 2 genera: **Northern** or **Black-necked screamer** (*Chauna chavaria*), **Southern** or **Crested screamer** (*C. torquata*), **Horned screamer** (*Anhima cornuta*).

DISTRIBUTION South America

HABITAT Open, wet grasslands and shallow, marshy lagoons.

SIZE Length 70–95cm (28–37in); weight 2.5–4.5kg (5.5–10lb).

PLUMAGE Black, gray, brown; both sexes similar.

VOICE Loud screams

NEST Large pile of vegetation, often in shallow water.

EGGS 2–7; whitish, tinged buff or pale green. Incubation 40–45 days; young leave nest early, self-feeding and fed from parents' bill; fledging in 60–75 days.

DIET Vegetation

CONSERVATION STATUS The Northern screamer is listed as Lower Risk/Near Threatened.

Noisy Neighbors

FORM AND FUNCTION

The screamers have long, thick legs, large feet, and long toes with only slight webbing at their base. Their bill is quite different from that of any other waterfowl species, being downcurved and more like that of a turkey. The strongest evidence of a link with the other anseriforms is the presence of lamellae (not particularly well developed) on the inside of the upper jaw of the two *Chauna* species, and in the lower jaw of *Anhima*. But there is also much other skeletal evidence of the link.

The most frequent feeding behavior involves grazing while walking or wading across shallow water within their territory, but the birds are also light enough to walk on top of floating mats of vegetation. They take roots, leaves, stems, and other green parts of succulent plants. Small amounts of insects are probably eaten, particularly by the chicks. The Southern screamer may also feed on crop plants in agricultural areas.

The birds' screaming voices are used to establish a territory up to 240sq m (300sq yd) in size before breeding, and to maintain it while nesting and rearing. Two sharp spurs protrude from the wrists of each wing, and these are used to threaten and attack intruders; broken spurs have been found buried in screamer breast muscle, so fighting must sometimes occur.

The horn of the Horned screamer is a slender, cartilaginous appendage, 15cm (6in) long, protruding forward from the forehead. Its function is uncertain; it is not even known whether it is molted once a year with the rest of the plumage. It has been suggested that the horn, the length of which has been linked with age, may indicate the quality of a potential mate.

Below *The horn that gives the Horned screamer its name is a cartilaginous projection up to 15cm (6in) long. Its exact function is unknown, but its delicate structure suggests it serves only for display.*

Male and female look alike, seem to pair for life, and have a long breeding season, the timing of which is influenced mainly by temperature and rainfall. Breeding tends to be concentrated around September to November, the southern springtime, except in the Northern screamer, which breeds year-round. Courtship display is unstudied, but much of it seems to involve territorial declaration, soaring flights, and calling to neighboring pairs. Pairs of birds often duet their call notes, and they also preen one another's heads and neck feathers.

At Home in South American Marshes

DISTRIBUTION PATTERNS

The screamers are all endemic to South America and are mainly sedentary, although the Southern screamer may congregate in large flocks during winter and move in response to weather conditions and food availability. The Northern screamer has the most restricted range, being confined to the lowlands of Colombia and Venezuela. The Southern screamer is found east of the Andes from northwest Bolivia and central Brazil south to central Argentina. The Horned screamer is distributed in the eastern lowlands from northern Colombia to eastern Bolivia and south-central Brazil.

All three species are marshland birds, but are found also on open savannas, on the banks of ponds, and in slow-moving streams. They fly strongly, soaring to considerable heights, perhaps as part of courtship displays, and can perch easily in trees and on low bushes. When flying high they can look rather like birds of prey.

Nests in the Water

BREEDING BIOLOGY

Nests are placed in shallow water within a few meters of the shore, and are built of sticks plus softer vegetable stems and leaves. Both sexes build; material is not carried to the nest site but passed back in the bill, so that the structure is composed of vegetation within easy reach. Copulation occurs on land, with the male holding the neck feathers of the female while he balances on her back. Clutch size is normally 3–5 eggs, with a range of 2–7, and egg weights range from an average of 155g (5.5oz) in the Horned to 184g (6.5oz) in the Northern screamer. Incubation, which is shared between male and female, lasts approximately 45 days, and the changeover is accompanied by mutual calling and preening.

On hatching, the chicks are covered in dense yellow down and follow their parents from the nest. They are brooded at night for only a few days and have their feeding supplemented with items placed in their open gapes. Wing spurs are present as tiny thorns at hatching. The adults swim reluctantly but their chicks more frequently, especially when accompanying their wading parents. During bathing, the feathers of the adults become thoroughly soaked, and the chicks are sometimes oiled by their parents, presumably to help with waterproofing. Growth is slow; youngsters fly only after about 75 days. They often remain with their parents as a family for a year or more.

No specific conservation measures have been taken to protect the three screamers. The main threats confronting them seem to be habitat destruction and degradation combined, in some places, with hunting. The Northern screamer is designated as Near Threatened by the IUCN, mainly because it is estimated to number only 5,000–10,000 individuals. The population of the Southern screamer is thought to range from 100,000 to 1 million individuals and to have a stable status. The number of Horned screamers is estimated crudely at something less than 100,000, and is thought to be declining. JK

Swans, Geese, and Ducks

For thousands of years, the waterfowl – swans, geese, and ducks – have provided humankind with eggs, meat, and feathers. People have hunted them, pursued them for sport, and domesticated them. Culturally, they have influenced art, music, dance, song, language, poetry, and prose. Their wetland habitat, on the other hand, has often had a bad press, being portrayed as dismal and dangerous and the source of such human ills as malaria. More recently, however, the need to conserve both waterfowl and the wetlands on which they depend has become increasingly obvious.

The closest relatives of the waterfowl and screamers are either the flamingos, the gamebirds, or the storks, ibises, and spoonbills, but the fossil record is not sufficient to reveal their precise relationships and origin. Between members of the waterfowl family itself relationships remain close despite the division into tribes, as demonstrated by the very large number of over 400 hybrids recorded between species; the majority of these have occurred in captivity, but some also in the wild. A number of intertribal hybrids are also known.

The Fifteen Tribes

Form and Function

Nearly all waterfowl nest on or beside fresh water, although several species such as brants, steamer ducks, and scoters live for much of the time in estuaries and on shallow seas. They are essentially aquatic birds, and so have a broad body flattened underneath, a medium to long neck, and shortish legs with webbed feet. The diving habit has evolved more than once within the group, and in diving birds the shape is particularly streamlined. The bill is usually rather broad and flattened, with a horny "nail" toward the tip that ends in some species with a slight hook. The sides of the mandibles have comblike lamellae, which in some species strain food particles from the water, while the tongue is rather thick with short, spiny teeth along its edges, used for grasping and manipulating food items.

Most ducks have their legs set well back on the body – some very far back – and so their progress on land can be slow and awkward. However, the many species that dive and feed underwater, such as pochards, scaup, seaducks, and stifftails, are able to move quickly and to maneuver well. A very few ducks, such as the Red-breasted merganser, also use their wings underwater, but generally the wings are held tight to the body when diving. Geese and sheldgeese are generally more terrestrial, especially when feeding – their longer legs are placed more centrally beneath the body, so they stand upright and can walk with ease.

***Above** A Cape Barren goose* (Cereopsis novaehollandiae) *broods its young. This Australian species, a taxonomic oddity assigned its own tribe (Cereopsini), is largely confined to islands in the Bass Strait, between the mainland and Tasmania. The birds are grazers, feeding off the local tussock and pasture grasses.*

The waterfowl are divided here into fifteen tribes (plus the enigmatic Musk duck, which cannot for the moment be fitted into any tribe), based partly on internal features, such as the skeleton, external features, such as the type and pattern of plumage both of adults and of the downy (less than three-week-old) young, on behavior, and on recent analyses of DNA.

The **Magpie goose** of Australasia is the only member of the tribe Anseranatini. It shares features with the screamers, particularly an overall long-legged, long-necked shape and reduced webbing between the toes. The bill is more typically waterfowl-shaped, although deep and broad at the base, and it rises into a steep forehead ending in an enlarged dome to the head. The flight is steady and direct on broad, slow-flapping wings. The plumage of the Magpie goose is black and white, as it is in many waterfowl, but the wing feathers are not molted all at once, as in nearly all other species, and so the birds do not become flightless during their wing molt. In contrast to most other waterfowl, Magpie goose parents feed their young, passing food items from bill to bill. The usual social unit is one male and two related females, all of which build the nest, incubate, and care for and feed the young. Elsewhere in the waterfowl family polygamy is rare.

Recent research on the Australian **Musk duck** has shown that it is not related to the stifftails with which is was at one time grouped. Any similarity is probably the result of convergence, largely of the hind limb, imposed by diving as a primary means of foraging. As in the Magpie goose, Musk duck young are fed by a parent. Male and female are very different in size (more so than in any other duck), and there is no pair bond; instead, the birds have a lek-type breeding system, by which 20 or more males display dramatically together in an arena and fertilize any female drawn within the circle. The males are pugnacious and belligerent toward one another, often chasing other males underwater.

The **whistling ducks** of the tribe Dendrocygnini have a wide distribution throughout the tropics and subtropics. Most are confined to rather small ranges, which do not overlap. The Fulvous whistling duck, however, has a most extraordinary distribution, occurring in the Americas, East Africa and Madagascar, and southern Asia; yet across all this enormous and discontinuous range there is no detectable variation of form.

Whistling ducks pair for life, and both parents care for the brood. They are mostly quite small, long-legged birds, with an upright stance. They get their name from their high-pitched whistling calls; an alternative name is "tree ducks," from their fairly general habit of perching on branches. They have broad wings and are maneuverable rather than fast fliers. As well as walking well on land, they swim and occasionally dive.

The plumage is the same in both sexes, often highly patterned, particularly with brown, gray, and fawn. The flank feathers of the Fulvous duck and three other species are enlarged in showy ornaments, and the downy young have a distinctive plumage quite unlike that of any other waterfowl except the Coscoroba swan; the unique feature is a pale yellow or white line running right round the head under the eyes; the head is dark-capped. After copulation whistling duck pairs indulge in a mutual display, both sexes performing similar actions by raising the wing on the side nearest the mate.

The **White-backed duck** (tribe Thalassornithini) of Africa is clearly related to the whistling ducks, in that male and female look alike and share a long-term pair bond; both build the nest and incubate the eggs, and both care for the ducklings.

FACTFILE

SWANS, GEESE, AND DUCKS

Order: Anseriformes

Family: Anatidae

162 species in 50 genera

DISTRIBUTION Worldwide except Antarctica

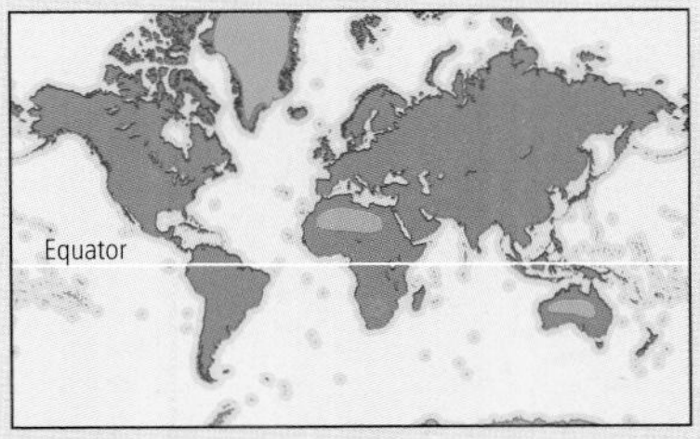

HABITAT Chiefly freshwater and coastal wetlands

SIZE Length 30–150cm (12–59in); **weight** 250g (8.8oz) to 15kg (33lb) or more.

PLUMAGE Variable, majority show some white, combinations of black and white frequent, few all white or all black; grays, browns, and chestnut also common; green or purple gloss often on head and on wing-patch. In some genera, males bright, females and juveniles dull camouflaged brown; sexes similar in others.

VOICE Some genera very vocal (usually quack, cackle, whistle, or hiss), others mostly silent or with soft calls connected with display.

NEST Majority in vegetation on ground, occasionally on rocky ledge or in tree crown; about a third in cavities, either existing tree holes or burrows.

EGGS 4–14; white, creamy, pale green, blue, brown, unmarked. Incubation 21–44 days; young leave nest early; self-feeding except for Magpie goose and Musk duck; fledging period 28–110 days.

DIET Wide variety of animals and plants, including fish, mollusks, crustaceans, insects and their larvae, and aquatic and terrestrial vegetation – leaves, stems, roots, and seeds.

CONSERVATION STATUS 5 species, including the Crested shelduck, the Madagascar pochard, and the Brazilian merganser, are Critically Endangered. A further 7 species are Endangered, and 14 are Vulnerable. In addition, 5 species are listed as extinct, including the Auckland Island merganser, which was last collected in New Zealand in 1902.

See Swan, Goose, and Duck Tribes ▷

◐ ***Left** A male Wood duck rears from the water to shake its wings dry. Once hunted almost to extinction in the USA, the species has been legally protected since 1918. Today the North American population is reckoned to number well over 1 million.*

However, White-backed ducks are highly adapted for diving, and this fact has altered their body shape entirely, so that, like the Musk duck, they resemble the stifftails in form (although without the stiffened tail feathers).

The **Cape Barren goose** (tribe Cereopsini) of Australia is a grazer, like the true geese, and it again exhibits a long-term pair bond. Unlike the geese but like the swans, the male helps build the nest, although he does not incubate. DNA analysis suggests that the Cape Barren goose belongs to an ancient tribe, and is perhaps most closely related to the Coscoroba swan.

The **"true" geese** of the tribe Anserini conduct a mutual display before copulation, and also a triumph ceremony after a rival male or pair has been driven off successfully. The 15 species are confined to the northern hemisphere, their place in the southern hemisphere being taken by the sheldgeese. The geese are variable in color (although male and female are alike), with some species black and white but others predominantly gray or brown. They are able to walk and run well, their moderately long legs being centrally placed beneath their bodies. Their necks are medium to long. Male and female typically pair for life, although only the female builds the nest and incubates, while the male stands guard.

The **swans** (tribe Cygnini) have either all-white plumage, or some combination of black and white, varying from white with black outer wings to black with white outer wings. The sexes are similar in all species and pairs tend to stay with their young for at least a year. Four of the seven swan species are confined (apart from introductions) to the northern hemisphere, and three occur only south of the Equator, of which the Coscoroba swan is probably the most primitive. The swans include the largest of all waterfowl, over 2m (6.6ft) in wingspan, and weighing over 15kg (33lb). They are very long-necked but comparatively short in the leg, and are not very mobile on land. Both swans and geese are strong fliers, and several species undertake regular migrations to the north to breed, covering thousands of kilometers.

The **Freckled duck**, the only representative of the tribe Stictonettini, is rather ducklike in overall shape, and has comparatively short legs. Males and females are similar in plumage, being mottled gray-brown all over, although the male acquires a reddish bill during the breeding season. There are sufficient structural and behavioral similarities to place this Australian tribe next to the geese and swans. The male builds the nest, but thereafter takes no further part in family duties.

The **stifftails** (tribe Oxyurini) are mostly found in the southern hemisphere, with just two species occurring north of the Equator. They are small, dumpy diving birds, with tails of short, stiff feathers, used for steering underwater. Their legs are set very far back on their bodies, and movement on land is limited. Males are mainly dark chestnut or brown, often with black or black and white heads, while many species have a bright blue bill in the courting season. The females are generally dull brown. Flight on their short, stubby wings is rapid and direct, after a long take-off run. The displays by the males are relatively elaborate. Whereas a majority of waterfowl molt their wing and tail feathers once a year, becoming flightless, most, if not all, stifftails do so twice annually.

The **Blue, Torrent, and steamer ducks** may be only temporarily lumped together within the same tribe, the Merganettini; further research is needed. The steamer ducks are confined to southern South America, where they lead a predominantly

◀ ***Left*** *Representative species of swans and geese:* ***1*** *Mute swan* (Cygnus olor), *showing the black knob at the base of the bill (larger in males) typical of the species.* ***2*** *Whooper swan* (C. cygnus) *engaged in triumph display.* ***3*** *Red-breasted goose* (Branta ruficollis) *adopting an aggressive threat posture.* ***4*** *Pink-footed goose* (Anser brachyrhyncus), *a northern species that breeds on Iceland, Greenland, and Svalbard.* ***5*** *Graylag goose* (Anser anser) *retrieving an egg by rolling it with the underside of its bill.* ***6*** *Magpie goose* (Anseranas semipalmata), *an Australasian species with a distinctive dome on top of its head.* ***7*** *Bar-headed goose* (Anser indicus), *flying up to join a migrating flock.* ***8*** *Emperor goose* (A. canagicus) *on its nest in the tundra.* ***9*** *Canada goose* (Branta canadensis) *in characteristic resting posture.* ***10*** *Black-necked swan* (Cygnus melanocorypha) *carrying cygnets on its back.* ***11*** *Hawaiian goose* (Branta sandvicensis) *male seeing off a rival.*

aquatic life, often around the coasts. Three species have such short wings that they are completely flightless, while the fourth flies infrequently and weakly. All are very sturdily built, with short necks and legs and powerful bills. Steamer ducks get their name from their habit of using their legs and wings to thresh the water like a paddle-steamer's wheel in order to move more quickly when escaping from danger. Their plumage is predominantly gray, and the sexes are alike.

Blue and Torrent ducks live in white-water rapids in the southern hemisphere, taking aquatic invertebrates, such as caddisflies, that in the north would be eaten by salmonid fish. The birds are highly territorial, like the steamers, and pairs stay together throughout the year, both helping to rear the young. The Torrent duck, of which there are a number of well-marked geographic races, is adapted for living in the fast-flowing streams of the Andes in South America; it has a very streamlined shape, sharp claws for gripping slippery boulders, and a long, stiff tail used for steering in the fast-rushing water. The Blue duck occupies a similar habitat in New Zealand.

The **Spur-winged goose and the comb ducks**, joined together in the tribe Plectropterini of Africa and South America, are rather gooselike birds that are alike in plumage, with the males usually considerably larger than the females. Black and white plumages, the former often glossed with green, predominate. The pair bond seems rather weak, and the males are only rarely seen with the brood.

The **shelducks and sheldgeese** of the tribe Tadornini include one genus of shelducks comprising seven species (one probably extinct), and four genera and eight species of sheldgeese. Shelducks occur worldwide except for Central and North America, but sheldgeese are confined to the southern hemisphere, with the exceptions of the

Below *A herd of Whooper swans congregates on Lake Kushiro, Japan. The lake, located on the northern island of Hokkaido, lies in an area of hot springs that keep some of its inshore waters open for the birds even when the rest of the surface is frozen.*

Blue-winged and Egyptian geese of Africa. They are medium-sized birds, the sheldgeese upright in stance, and all feed on land as well as in the water. Plumage is very variable, although a majority of species have an iridescent green speculum (a patch on the outer half of the secondaries) and white wing coverts. Other common colors include some white, black, chestnut, and gray. The sexes are similar in some species but completely contrasting in others, although all tend to exhibit long pair bonds with dual parental care. Displays show some similarities with those of the geese. Some, if not all, sheldgeese molt their wing feathers sequentially like the Magpie goose; furthermore, the wing molt may occur only every other year.

The **Pink-eared and Salvadori's ducks** that together make up the tribe Malacorhynchini are also doubtfully related; further research is needed. The small Australian Pink-eared duck has, in both sexes, a zebralike black-and-white plumage with a tiny pink patch on the sides of the head. The massive bill is beautifully adapted, with fringing lamellae, for filtering tiny organisms from the surface of the water. Salvadori's duck of New Guinea lives in a variety of riverine habitats, but mainly in fast-moving waters. This species is territorial; males and females look alike, with striped black-and-white plumage and bright yellow bills.

The **surface-feeding or dabbling ducks** (Anatini) are the largest tribe of waterfowl and contains many highly successful and adaptable species. The great majority are small, short-legged, aquatic birds, feeding on the surface of the water or up-ending to dabble just beneath it. Dabbling ducks occur throughout the world, including on remote oceanic islands, from where a number of non-migrating species including the Laysan duck and

Below *Representative species of surface-feeding ducks:* **1** *Egyptian goose* (Alopochen aegyptiacus) *– actually something of a misnomer, as the bird in fact belongs to the shelduck tribe (Tadornini).* **2** *Male Mandarin duck* (Aix galericulata), *a handsome East Asian species widely introduced into captivity in the West.* **3** *White-faced whistling duck* ((Dendrocygna viduata), *one of the tropical Dendrocygnini tribe.* **4** *Ruddy duck* (Oxyura jamaicensis), *a North American species that breeds from British Columbia and Quebec down to the Mexican border.* **5** *Marbled teal* (Marmaronetta angustirostris), *now listed by the IUCN as Vulnerable as a result of widespread destruction of its wetlands habitat.* **6** *Male Ruddy shelduck* (Tadorna ferruginea), *a southern European species with a distinctive orange-brown body and pale head.* **7** *American wigeon* (Anas americana), *sometimes known as the "baldpate" for its characteristic white crown.* **8** *Mallard* (A. platyrhynchos) *drake exhibiting ritualized preening.* **9** *Shelduck* (Tadorna tadorna), *flying with characteristic arched wings and slow wingbeat.* **10** *Northern pintail* (Anas acuta), *showing the elongated central tail feathers that give the species its name.* **11** *Wood duck* (Aix sponsa), *nesting in a treehole.* **12** *Laysan duck* (Anas laysanensis), *a species confined to the Hawaiian island of Laysan, where the population numbered just 375 birds in the year 2000.*

> ### LEAD POISONING
>
> First recognized as a hazard for waterfowl more than a century ago, lead poisoning has been reported from over 20 countries. The problem has caused increasing concern, particularly in Europe and North America, where it was estimated at one time that between 1.4 and 2.6 million ducks were dying annually after ingesting lead pellets from shotgun cartridges, which they take in mistake for grit or seeds. Intensive shooting beside small flight ponds and over marshland produces a build-up of pellets in the mud and soil. These stay virtually intact and, unless they sink through the subsoil, are available to birds for many years. As a result, the chances of a feeding bird finding a pellet at such sites can be quite high.
>
> In certain circumstances, the provision of plenty of grit can alleviate the problem, but a single pellet that is taken into the gizzard, ground down by grit, and dissolved in stomach acids so that the lead enters the bloodstream, can be sufficient to cause the death of a duck. In consequence, lead was banned for use in shooting waterfowl in Denmark, Finland, the Netherlands, Norway, and the United States in the early 1990s; England followed at the end of the decade, and Spain in 2001. Lead substitutes made of alloys or stainless steel appear to be proving satisfactory.
>
> A second source of lead that has caused waterfowl deaths is lead shot used to weight the lines of anglers. Discarded or lost lengths of line may be eaten by birds such as Mute swans as they pluck underwater vegetation. In addition, some anglers have become accustomed to throwing away the end-line with its weights attached after each day's fishing. Two or three lead weights (or a single large ledger weight) are quite enough to poison a bird the size of a Mute swan, and in some river systems of Britain this species was virtually wiped out by lead poisoning. After much publicity and discussion, most lead weights were banned in England after 1987, and the Mute swan population subsequently recovered sharply.

Auckland Island teal have been described. Many other dabbling ducks such as the Northern pintail, Green-winged teal, and Eurasian wigeon are highly migratory and fly long distances; the smaller species are also very maneuverable in the air, taking off almost vertically from the water. A few can perch on branches and other structures, an ability rare or absent in nearly all other waterfowl. About a third are hole-nesters, and the young of these have sharp claws on their feet and rather stiff tails, both used in climbing out of the holes soon after hatching. Many occur mainly in tropical and subtropical latitudes, extending into the temperate zone in a few cases including the Mandarin ducks. In smaller species such as the pygmy geese, the sexes are dissimilar and the birds show a wide variation in coloring, from dull brown to bright chestnut, green, and white.

In many species the male is brightly colored, generally with an iridescent speculum in the wing, while the female and juveniles are a cryptic brown. However, males lose their bright colors after breeding, and they also are brown while flightless during the wing molt. Brown, green, chestnut, white, and pale blue all occur frequently. Displays involve complex movements by the male that show off his plumage to best advantage. Females are much less demonstrative, and mutual display movements are absent or rare. The female usually cares for the ducklings alone, although the drake of a few tropical species does help guard his offspring against predation.

The **pochards and scaup** of the tribe Aythyini have a worldwide distribution. They are principally freshwater species, although some winter on the coasts, and all obtain their food by diving. Having short legs set well back on their generally plump bodies, they only rarely venture onto land. Becoming airborne usually involves a take-off run over the surface of the water, accompanied by rapid wing beats. Although males have different coloration from females, they are not especially brightly plumed; gray, brown, and black are the commonest colors. They lack the speculum of the shelducks and other members of the family, but often have a whitish wing stripe or bar. Iridescence is frequent only on the dark head and chest. Their courtship displays are relatively simple. Males desert the females at the start of incubation, and court another female the following season.

The **seaducks** (tribe Mergini) are here taken to include the eiders, which are sometimes separated into their own tribe, the Somaterini. Most are principally saltwater species, diving for their animal food, although several species breed beside fresh water. Most are restricted to the northern hemisphere, but the rare Brazilian merganser occurs in South America, and the Auckland Islands merganser (now extinct) once occurred on the island group of that name, south of New Zealand. Many are rather bulky and heavily built,

requiring a long take-off for their rather labored flight. The fish-eating sawbills, however, are much more agile and fast-moving.

Almost all the seaducks are sexually dimorphic, with the males mainly black or black and white, although often with iridescent green or blue heads, or with pastel coloring of green and blue. Displays by the males are often quite elaborate and vary greatly between species. The females are less demonstrative, as in the dabbling ducks. Down, plucked by female ducks from their breast in order to insulate their eggs while they are off the nest, has reached its highest level in the northerly-breeding eiders. The down is a very valuable commodity for Icelandic farmers, who protect colonies of incubating females in order to collect it for cleaning and sale after hatching has occurred.

Everywhere but Antarctica

DISTRIBUTION PATTERNS

Waterfowl are found on every continent except Antarctica, and on all large and many small islands. A few species – the Hawaiian goose and the Brazilian merganser, for instance – are numbered only in the hundreds, and a few island species, such as the Campbell Island teal, with less than one hundred individuals, are likewise rare. Others, such as the Northern pintail, are extremely numerous, perhaps occurring in millions, and are distributed very widely.

Distribution patterns can either result from great adaptability to a considerable range of habitats and food, as in the case of the Northern mallard, or else it may be linked to long-distance migration, as in the White-fronted goose. Alternatively, they can be the consequence of some chance colonization of an oceanic island, followed by the loss by the birds of any migratory instinct and a gradual evolution to fit the conditions found there. Such has been the case in the Laysan duck in the Pacific, the Kerguelen pintail in the Indian Ocean, and the flightless Falkland steamer duck in the south Atlantic.

Grazers, Dabblers, and Divers

DIET AND FEEDING

Grazing on land and plucking vegetation from the water are the commonest feeding methods of geese, swans, and sheldgeese. Some species also dig for plant roots in soft mud, and many geese and sheldgeese, and some swans, have adapted to feeding on agricultural land, at first by grazing grass and growing crops but more recently by picking up waste grain, beans, sugar beet, carrots, maize, and potato tubers. Some of the dabbling ducks, too, graze on land, while most take seeds as well as small insects from the surface of the water. Rice has become a very important item of diet for some species in recent years.

None of the vegetarian waterfowl have bacteria in the gut to help them digest cellulose. The nutrients they obtain from plant leaves and stems are restricted to the cell juices, obtained by breaking down the cell walls in their gizzards with the aid of small particles of grit, which they ingest for the purpose. These species have to feed for long periods, often the majority of daylight hours in temperate winter latitudes, in order to obtain sufficient food. The plant remains pass through the gut in recognizable form, and it is usually possible to identify the species of plant the birds have eaten from their droppings.

While the grazers are known to select the most nutritious parts of the plant, such as the growing tips of grass leaves, dabbling ducks appear to be altogether less selective, taking in relatively large quantities of surface water through the tips of their bills and pumping it out of the sides through

Swan, Goose, and Duck Tribes

Magpie goose
Tribe Anseranatini

1 species: *Anseranas semipalmata*

Whistling ducks
Tribe Dendrocygnini

8 species in 1 genus. Species include: **Fulvous whistling duck** (*Dendrocygna bicolor*).

White-backed duck
Tribe Thalassornithini

1 species: *Thalassornis leuconotus*.

Cape Barren goose
Tribe Cereopsini

1 species: *Cereopsis novaehollandiae*.

True geese
Tribe Anserini

15 species in 2 genera. Species include: **Barnacle goose** (*Branta leucopsis*), **Brent goose** or **brant** (*B. bernicla*), **Hawaiian goose** (*B. sandvicensis*), **Snow goose** (*Anser caerulescens*), **White-fronted goose** (*A. albifrons*).

Swans
Tribe Cygnini

7 species in 2 genera: **Bewick's** or **Whistling swan** (*Cygnus columbianus*), **Black swan** (*C. atratus*), **Black-necked swan** (*C. melanocorypha*), **Mute swan** (*C. olor*), **Trumpeter swan** (*C. buccinator*), **Tundra swan** (*C. columbianus*), **Whooper swan** (*C. cygnus*), **Coscoroba swan** (*Coscoroba coscoroba*).

Freckled duck
Tribe Stictonettini

1 species: *Stictonetta naevosa*.

Stifftails
Tribe Oxyurini

7 species in 3 genera. Species include: **Black-headed duck** (*Heteronetta atricapilla*), **Andean duck** (*Oxyura ferruginea*), **Ruddy duck** (*O. jamaicensis*), **White-headed duck** (*O. leucocephala*).

Blue, Torrent, and steamer ducks
Tribe Merganettini

6 species in 3 genera. Species include: **Blue duck** (*Hymenolaimus malacorhynchus*), **Falkland steamer duck** (*Tachyeres brachypterus*), **Flightless steamer duck** (*T. pteneres*), **Torrent duck** (*Merganetta armata*).

Spur-winged goose and comb ducks
Tribe Plectropterini

3 species in 2 genera. Species include: **African comb** or **Knob-billed goose** (*Sarkidiornis melanotos*), **Spur-winged goose** (*Plectropterus gambensis*).

Shelducks and sheldgeese
Tribe Tadornini

15 species in 5 genera. Species include: **Blue-winged goose** (*Cyanochen cyanopterus*), **Egyptian goose** (*Alopochen aegyptiacus*), **Ruddy-headed goose** (*Chloephaga rubidiceps*), **Common shelduck** (*Tadorna tadorna*), **Paradise shelduck** (*T. variegata*), **South African** or **Cape shelduck** (*T. cana*).

Pink-eared and Salvadori's ducks
Tribe Malacorhynchini

2 species in 2 genera: **Pink-eared duck** (*Malacorhynchus membranaceus*), **Salvadori's duck** (*Salvadorina waigiuensis*).

Surface feeding or dabbling ducks
Tribe Anatini

57 species in 11 genera. Species include: **Muscovy duck** (*Cairina moschata*), **Cotton teal** (*Nettapus coromandelianus*), **Auckland Island teal** (*Anas aucklandica*), **Campbell Island teal** (*A. nesiotis*), **Cinnamon teal** (*A. cyanoptera*), **Eurasian wigeon** (*A. penelope*), **Green-winged teal** (*A. carolinensis*), **Kerguelen pintail** (*A. eatoni*), **Laysan duck** (*A. laysanensis*), **Northern mallard** (*A. platyrhynchos*), **Northern pintail** (*A. acuta*), **Speckled teal** (*A. flavirostris*), **Bronze-winged duck** (*Speculanas specularis*), **Hartlaub's duck** (*Pteronetta hartlaubii*), **Mandarin duck** (*Aix galericulata*), **Wood duck** (*A. sponsa*).

Pochards and scaup
Tribe Aythyini

17 species in 4 genera. Species include: **Pink-headed duck** (*Rhodonessa caryophyllacea*), **canvasback** (*Aythya valisineria*), **Tufted duck** (*A. fuligula*), **South American pochard** (*Netta erythrophthalma*).

Seaducks
Tribe Mergini

20 species in 10 genera. Species include: **Auckland Island merganser** (*Mergus australis*), **Brazilian merganser** (*M. octosetaceus*), **Common merganser** (*M. merganser*), **Red-breasted merganser** (*M. serrator*), **Scaly-sided merganser** (*M. squamatus*), **Hooded merganser** (*Lophodytes cucullatus*), **smew** (*Mergellus albellus*), **Common eider** (*Somateria mollissima*), **Common scoter** (*Melanitta nigra*), **Labrador duck** (*Camptorhynchus labradorius*), **Long-tailed duck** or **Old squaw** (*Clangula hyemalis*), **Common goldeneye** (*Bucephala clangula*), **bufflehead** (*B. albeola*).

Musk duck *Biziura lobata*
The Musk duck, which used to be assigned to the stifftails of the tribe Oxyurini, has recently been shown not to be related to them, and has yet to be reassigned.

the comb-like lamellae with the aid of their tongues. The trapped particles, seeds, and insects are then swallowed at intervals.

Diving ducks feed on underwater plants and invertebrates of all kinds, generally in fairly shallow water often only a few meters deep. The seaducks and sawbills may dive to considerable depths, the latter pursuing and catching fish and large, free-swimming invertebrates, while the former prize mollusks from rocks, or catch crabs and other crustaceans in shallow water or soft mud. Their bills are large and powerful, well adapted for grasping the shells of such prey, which are crushed after ingestion in their gizzards. Some of the shelducks also feed in a marine environment, sifting tiny mollusks and crustaceans from the mud of estuaries, while the steamer ducks, like the eiders, use their strong bills to prize mussels from tideline rocks. The stifftails also feed almost exclusively underwater. They swim along the

Above *Representative species of diving ducks and seaducks:* ***1*** *Common eider* (Somateria mollissima). ***2*** *Red-breasted merganser* (Mergus serrator). ***3*** *King eider* (Somateria spectabilis), *a northern species.* ***4*** *Common goldeneye* (Bucephala clangula) *male, with characteristic white spot in front of eye.* ***5*** *Male and female smew* (Mergellus albellus), *the male in front.* ***6*** *Tufted duck* (Aythya fuligula), *diving for fish with wings held close to body.* ***7*** *Common merganser* (Mergus merganser), *showing the sawtooth lining of its long beak, an adaptation for catching fish.*

bottom, sifting through the silt with their lower mandible just entering it, catching animal prey and in particular chironomid (midge) larvae.

Above *A Falkland steamer duck* (Tachyeres brachypterus) *provides shelter for her downy nestlings. This is one of three steamer-duck species to have lost the use of wings for flight.*

Pairs and Flocks

BREEDING BIOLOGY

The great majority of species breed annually, but some from the near tropics may breed only every other year, or are adapted to wait for favorable conditions before attempting to breed, for example during the irregular rains experienced in parts of inland Australia. In some years, species that breed in the High Arctic may, over wide areas, fail to reproduce at all (see A Race to Breed).

The waterfowl show a wide range of variation in the maintenance of the pair bond and in parental care. Swans and geese mostly pair for life; while only the female incubates the eggs, both parents look after the young, often throughout the first winter of their lives, the family only splitting at the start of the following breeding season. This period may well include a long migration to winter quarters, the young birds thus being shown the route and resting areas by example, a pattern that often leads to highly traditional use being made of certain haunts over many years. Thus White-fronted geese have wintered at Slimbridge on England's Severn Estuary since at least the 18th century.

Many ducks only select a mate for a single breeding season, although pairing may take place during the previous fall, as in the case of many dabbling ducks, through to the spring, as in many pochards and seaducks. The pair bond usually breaks as soon as the female has begun incubation. She generally rears the young until they can fly before deserting them, although some female pochards may abandon their young when they are only half grown. "Crèche care" of several broods by a few females takes place in some of the seaducks, and in some shelducks.

The pair bond and parental care in stifftails are both very short-lived, and most young fend for themselves almost from the day of hatching. Such is the case with the Black-headed duck, which never makes a nest of its own, laying entirely in the nests of others, and whose offspring receive no parental care at all.

The great majority of waterfowl are gregarious, some of them highly so. Swans, geese, and sheldgeese are almost always found in flocks away from the breeding grounds, and some species even breed in semi-colonies. Among these species flock size is very variable, ranging from a few tens to 100,000 or more.

Flocks of swans and geese are made up of family units, pairs, and immature non-breeding birds, while wintering flocks of dabbling ducks and pochards frequently consist principally of males or of females, with considerable geographical separation between the sexes. In several species that inhabit northern hemisphere, especially pochards and scaup, females winter further south than males, which leave the breeding grounds first.

The lifespan of many waterfowl is greatly affected by human hunting activities. If they are not hunted, geese and swans are quite long-lived, with ages of 20 years not uncommon, but heavy shooting pressure can reduce life expectancy to no more than 10 years at most. The majority of dabbling ducks, pochards, and stifftails mature at the age of 1 and have a life expectancy thereafter of 4–10 years, this being greatly affected by whether

Right The Baikal teal (Anas formosa) *was once one of the most numerous ducks in Asia, but is now listed as Vulnerable by the IUCN. Hunting played a major part in its decline, although the loss of the wetland habitat on which it depends to drainage schemes has also been an important factor. In recent years the bird has been legally protected over most of its range, and there are now some tentative signs that the population may be starting to recover.*

or not they are prey species for human hunters. Most swans do not mature until they are 3–4 years old, and geese, sheldgeese, and shelducks until they are 2–3. Similarly, the eiders and larger sawbills are commonly not mature until they reach at least 2 years of age.

Saving the Wetlands

CONSERVATION AND ENVIRONMENT

In the question of waterfowl conservation, the emphasis has moved in recent years from the protection of individual species to an increased recognition of the vital importance of the birds' wetland habitat. A few species, notably the Hawaiian goose, have been the subject of successful captive-rearing programs, aimed at boosting small or declining wild populations.

While conservation measures are probably required for most species of waterfowl, there are circumstances in which they can be regarded as pests, particularly of agriculture, and to a lesser extent of fisheries. While their pest status can be important locally, it is not a major problem worldwide. The extinction in 1875 of the Labrador duck and in 1902 of the Auckland Islands merganser, and the almost certain extinction (through hunting and wetland drainage) of the Pink-headed duck of India and Nepal during the mid-1990s, are reminders that there is much to be done if more species are not to disappear.

The signing of the Ramsar Convention on Wetlands in 1971 was an important advance in international waterfowl conservation. All of the parties to the convention undertook to designate and safeguard at least one wetland region within their borders. Among other criteria set by the convention, the wetland has to regularly support at least 20,000 waterbirds, or alternatively at least 1 percent of the population of any one species. By 2002, 130 countries were contracting parties to the convention, and had listed 1,040 Ramsar sites covering a total area of 917,000sq km (350,000 sq miles). JK/MAO

CAVITY NESTING

Hole nesting appears to have risen independently at least three times in waterfowl evolution. About a third of all duck species prefer to nest in cavities, and this behavior has had a huge influence on their range and life history, meaning for instance that they cannot normally have a distribution that goes further north than the tree line. Being unable to excavate holes for themselves, they have to rely on other factors for their construction; thus, in order to conserve goldeneyes (RIGHT), smew, buffleheads, Hooded mergansers, American wood ducks, and Mandarins, for instance, woods and forests must not only be conserved, but they must also contain a healthy population of woodpeckers that will make the holes that the ducks will use later. The woodpeckers in turn need a supply of insects that feed on or in dead and dying timber, so the woods must be mature, not overly managed, and not disease-free.

The Common goldeneye was not known as a breeder in the British Isles until 1970, when a pair nested in Scotland. Their absence until then was, it is suggested, due to the absence of the Black woodpecker (*Dryocopus martius*). It took 15 years before a supply of nestboxes specially erected for goldeneyes was accepted, perhaps by a female that had hatched in a box on the European continent. The small British breeding population is still largely dependent on nestboxes rather than natural holes.

Some Speckled teals nest in ground burrows of woodpeckers or, in Argentina, choose abandoned chambers of Monk parakeets (*Myiopsitta monachus*), which provide excellent security from predators. In Australia there are no woodpeckers; nevertheless, many of the ducks are cavity nesters, and so moisture, fungus, ants, and termites must be relied upon to provide the hole. Most shelducks use mammal burrows in the ground, except for the Paradise shelduck of New Zealand, which until recently lived where there were no mammals; the female uses rock crevices instead. The Common shelduck must have been far less numerous in Britain until rabbits were introduced by the Normans. The South African or Cape shelduck prefers burrows made by aardvarks, which themselves need active conservation as they are becoming extremely rare. The female Ruddy-headed goose likes to make her nest in penguin burrows, as does the Falkland flightless steamer duck. The Torrent duck may use the abandoned burrow of a kingfisher. So the biodiversity of the whole habitat needs preserving in order to maintain a breeding population of many waterfowl and, if a shortage of safe nesting sites is

limiting productivity, then artificial ones in the form of nestboxes will need to imitate a range of features.

An additional benefit of nestbox schemes is that the construction and erection of boxes can involve the public and help to demonstrate practical conservation techniques. The wild Orinoco goose and Scaly-sided merganser are among those waterfowl that have benefited recently and taken successfully to artificial sites. JK

SPECIAL FEATURE

A RACE TO BREED

Food availability and clutch size in arctic-breeding geese

THE POWERS OF FLIGHT HAVE FREED LONG-distance migratory birds to move great distances over potentially uninhabitable terrain. In doing so, they can exploit seasonally rich food supplies from specific habitats in distant parts of the globe, which may only be available for short periods of time. Nowhere is this more apparent that among the 50 or so populations of herbivorous geese that head south from their arctic breeding grounds each year after the brief summer to seek snowfree latitudes in milder climes in North America, Europe, and Asia.

One cost of such aerial mobility is the need to retain a simple gut structure. Sheep do not fly for good reason! A rumen and a complex digestive system are efficient for digesting poor-quality fibrous herbage, but would be too heavy to carry thousands of kilometers on wings. Instead, herbivorous geese have evolved a short, simple gut, through which they maintain rapid rates of food passage. They have become expert in seeking and selecting high-quality green plant food, selecting, for example, only the very tips of leaves of those species that are most efficient to consume.

Geese therefore compensate for their inefficient digestion by maintaining high ingestion rates and selecting a high-quality diet. In the northern spring, the young green shoots of grasses and sedges are rich in highly digestible soluble protein and carbohydrate, but low in the structural fiber that inhibits nutrient uptake in the gut. Before the leaves grow long, strengthened by investment in cell-wall material, they constitute high-quality goose food. By starting northward from their winter quarters, the geese can follow a successively belated spring toward their ultimate breeding grounds, riding on the fresh flush of young growth revealed by the retreating snows. Indeed, in some populations, such as that of the White-fronted goose (*Anser albifrons*) in west Greenland, geese may continue to follow local patterns of thaw along an altitudinal gradient, following a perpetual spring uphill through the entire summer as solar radiation successively releases young, growing vegetation from snow patches higher and higher into the mountains.

The race to arrive at the breeding grounds as early as possible in the best possible condition can be a tough one. Studies of most arctic-nesting geese show that the earliest birds to lay eggs have the largest broods and the highest hatching success and also raise most young, so to be first carries a considerable premium. But the race is not a simple dash for the post. Weather conditions in the arctic can vary enormously; geese can arrive in one year to encounter deep snow, yet at the same time in another season they may find mild conditions and an ample food supply.

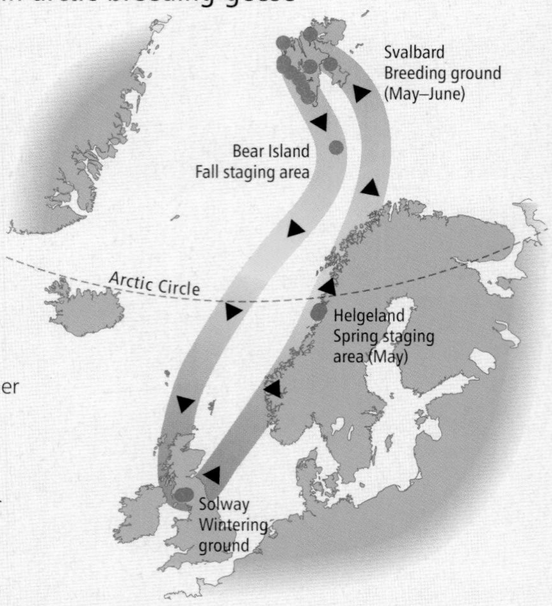

Above *Migration route of Barnacle geese between their wintering grounds on the Scottish–English border and summer breeding grounds in the Svalbard islands (including Spitsbergen) high in the Arctic (to 80°N). On their way north the geese stop off in Norway to feed up on succulent new-growth grasses, laying down food reserves that are vital for successful breeding. On the return journey, Bear Island is an important staging area. A second, entirely separate population migrates to Greenland.*

Migrant geese lay down fat and protein stores in their bodies to sustain and fuel their long flights, during which they have no opportunity to feed. Studies have increasingly shown that these stores are also used by the female to produce her clutch and sustain her and her attendant mate on arrival. Stable isotope studies have used distinct chemical signatures in the food and tissues of Snow geese to demonstrate that female geese use these body stores, supplemented by food obtained at or near the ultimate breeding grounds, to produce the eggs and help sustain the female through incubation. Even geese breeding in the High Arctic such as brant and Greater snow geese, which were thought originally to nest very soon after arrival on the nesting grounds, have now been shown to spend one to three weeks after arrival feeding to supplement the body stores of the female for investment in her clutch. This is the period of rapid follicular development in geese, when the female can hedge her bets to adjust the amount of resources she invests in her eggs, given her internal stores of nutrients and the availability of food on the breeding areas. Geese arriving to deep snow cover and a slow thaw may thus abandon the race and not attempt to breed at all, the female simply resorbing the follicles.

Above and right *In the fall, a flock of Snow geese* (Anser caerulescens) *migrate south to their wintering grounds in the Gulf of Mexico. Each summer, the birds breed in colonies in arctic regions of North America or on Siberia's eastern tip. Clutch sizes vary from 4–8 eggs, laid in a down-lined nest (right).*

In late springs, arriving birds can wait for conditions to improve, and perhaps adjust first-egg dates and clutch size to the nutrients at their disposal. Things may still go wrong: at some North American Lesser snow goose colonies, females apparently gain little supplementary feeding from the breeding areas before the first eggs are laid, and greater reliance is placed on internal reserves. If these stores are insufficient and are invested in too large a clutch, a female may not be able to recoup the necessary nutrients during her short recess breaks from incubation and desert – some birds have even been observed to starve to death on the nest. In a good spring, however, a successful female will arrive in good condition and lay a

large and early clutch to breast the finishing line ahead of her competitors.

Even if a bird survives the attentions of Arctic foxes, skuas, ravens, and a host of other potential predators of the eggs, the race is not fully won with the hatching of the goslings. Although the 24 hours of daylight produce rich plant growth and long, uninterrupted periods of feeding, losses to bad weather or predation before fledging can still be high. Unpredictable weather and competition from conspecifics may mean that individual goslings fail to attain threshold condition for their first flight south to the winter quarters. Studies have shown that lightweight Barnacle goslings (*Branta leucopsis*) from geese nesting in Svalbard are less likely to be seen on the wintering areas on the Solway Firth in Scotland following autumn passage than heavier individuals, which presumably have accumulated sufficient energy to sustain the flight back. For goose parents, the race is not finally run until they are back on the wintering grounds with a complete brood. TF

New World Vultures

1a

NEW AND OLD WORLD VULTURES ARE SIMILAR in appearance, both having the hooked bill, bare head and neck, and large, broad wings that suit them to the scavenging way of life. Yet they are no longer considered to be closely related, being most easily separated by their nostrils, which are perforated in the New World species. Furthermore, fossil evidence indicates that there used to be New World vultures living in the Old World and vice versa, so that even the geographical separation of the two groups is deceptive and only relatively recent.

The condors are the most dramatic of the New World vultures, being among the largest of all flying birds with wingspans of up to 3m (10ft). While these massive birds inhabit mainly open and mountainous habitats with strong air currents, some smaller New World vultures also extend onto open flats and into forest.

A Link to the Storks?

EVOLUTION AND SYSTEMATICS

The fossil record of New World vultures dates back 34 million years to the early Oligocene period, including specimens up to 20 million years old from the European Miocene. Some of the extinct species were even larger than condors; a teratorn from the late Miocene (8–5 million years ago), the fossil remains of which were discovered in Argentina in 1980, was *Argentavis magnificens*. This huge bird is estimated to have weighed up to five times as much as the heaviest living condor and to have had a wingspan of 7.5–8m (24.6–26ft). Old World vultures were formerly present in the New World, perhaps until as recently as 10,000 years ago.

Some taxonomists consider the existing New World vultures to be most closely related to the storks of the order Ciconiiformes, some of which also feed on carrion, despite differences in bill shape and the fact that the cathartids do not build nests. Even so, they share with some storks the practice of carrion-feeding, along with various anatomical similarities, such as a reduced or non-functional hind toe, a naked face and neck, and the lack of a syrinx, which renders them virtually voiceless. They are also similar in aspects of breeding behavior, such as inflating or flushing facial skin during display and dribbling liquids to chicks; and, like storks, they cool themselves down by trickling droppings down their legs for evaporation. However, their distinctness from the other birds of prey may be less important than it seems to be, since secretarybirds also show some storklike behavior, and recent DNA evidence indicates that all raptors may have waterbirds among their more distant relatives.

3

Far left *A Turkey vulture keeps watch from a tree-stump on Mexico's Pacific coast. This species is the most widespread of the New World vultures. An acute sense of smell helps them swiftly locate recently-dead animals.*

Left *A 45-day-old Turkey vulture chick tests its wings. The young are born with down-covered heads, only becoming bald when they assume adult plumage. They may remain in the nest for 12 weeks or more.*

Soaring in Search of Carrion

FORM AND FUNCTION

The head and upper neck of New World vultures are usually bare and brightly colored, while in the larger species there is a ruff of fluffy or lanceolate feathers around the base of the neck. The male Andean condor sports a high comb, the California condor a colorful and inflatable throat, and the King Vulture a multicolored head for use in display. The toes are long and suited to pinning down food, but the claws are only slightly curved and the feet are not well adapted for grasping prey. The large, broad wings and stiff tail open into an extensive surface that enables the birds to soar for long periods on rising air and to search for carrion with minimal effort.

The smaller *Cathartes* vultures are unusual among birds in using smell as well as sight to locate their food. These Turkey and yellow-headed vultures are known to be especially adept at smelling out carrion, even when it is hidden in a cave or under vegetation. Other species, like the King and Black vultures, have minimal powers of smell even though they spend much time in forests, and must rely mainly on the activity of other scavengers to help them in locating carcasses. When searching for food, species that can smell carrion tend to soar at lower altitudes than those that must search large areas by sight alone.

Besides carrion, most New World vultures add eggs, fruit, and some other vegetable matter to their diet, or congregate at rubbish dumps and abattoirs for the scraps they provide. A number of species, especially the sociable Black vulture, also gather at night at communal roost sites, where they may gather information about good feeding sites for the following day by noting how full or empty the crops of their neighbors may be. Such roosts are especially common in the nonbreeding season, and may be a source of guidance and experience for inexperienced juveniles.

Above *Representative species of New World vultures:* **1a** *Andean condor* (Vultur gryphus) *in flight and* **1b** *in close-up, showing the fleshy wattle on the forehead.* **2** *The endangered California condor* (Gymnogyps californianus). **3** *King vulture* (Sarcoramphus papa), *a resident of dense tropical forests.* **4** *Black vulture* (Coragyps atratus), *often incorrectly described as a "buzzard" when seen soaring in American skies.*

Condors in Peril

CONSERVATION AND ENVIRONMENT

The Andean condor is still widespread in South America, but the California condor had declined to just 21 birds by 1983, and became extinct in the wild after the final bird was taken to join 26 others in captivity in 1987. Breeding in cages was so successful, however, that it was possible to release two immatures back into the wild in 1992, and by 1998 there were 147 California condors in the wild, including 47 living free at sites in California and Arizona within previous parts of their range. The birds continue to suffer from human interference in the form of power lines, poisoning, and shooting, but are also the objects of one of the most intensive conservation efforts ever mounted.

Similar negative factors are beginning to affect the Andean condor, once again with serious implications, since condors have a very low reproductive rate: they produce at most only one chick every second year, which in its turn will require six or more years before it attempts to breed. Earlier experiences of captive breeding and release with Andean condors, as well as the subsequent success with the California condor, bode well for the security of both species, although only extensive areas of protected habitat will ensure their ultimate safety in the wild.

AK/IN

FACTFILE

NEW WORLD VULTURES

Order: Falconiformes (or Ciconiiformes)

Family: Cathartidae

7 species in 5 genera: **Andean condor** (*Vultur gryphus*), **California condor** (*Gymnogyps californianus*), **Greater yellow-headed vulture** (*Cathartes melambrotus*), **Lesser yellow-headed vulture** (*C. burrovianus*), **Turkey vulture** (*C. aura*), **Black vulture** (*Coragyps atratus*), **King vulture** (*Sarcoramphus papa*).

DISTRIBUTION S Canada to tip of S America.

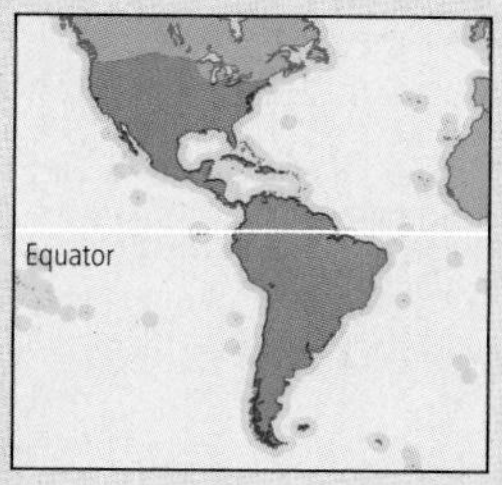

HABITAT Wide range of mainly open habitats, from high Andean steppe to deserts, but including woodland and forest.

SIZE Length 56–134cm (22–53in); weight 850g–15kg (1.9–33lb).

PLUMAGE Sooty brown with paler patches on the underside of the wings, except for the King vulture, which is creamy-white with black flight feathers. Head and neck bare and brightly colored in most species. Sexes similar, juveniles generally dull and brown.

VOICE Silent, except for occasional hissing sounds.

NEST Solitarily in natural cavity on ground, under bushes, in cave or stump, or, rarely, in higher tree cavity.

EGGS 1 or 2, off-white elongated ovals, spotted in Turkey vulture. Incubation 40–60 days; nestling period 70–180 days; nestlings with thick down attended by both parents, fed by regurgitation, initially directly from the bill.

DIET Carrion; also eggs, fruit, and other vegetable matter.

CONSERVATION STATUS The Californian condor is Critically Endangered, while the Andean condor is listed as Lower Risk/Near Threatened.

Secretarybird

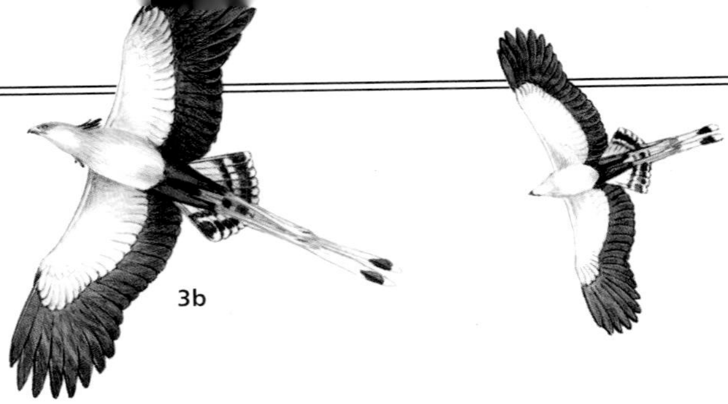

RESEMBLING A CROSS BETWEEN A STORK *and a raptor, the secretarybird is an oddity that is placed in its own family. Described as a "pedestrian eagle," it hunts on foot, pursuing insects, snakes, and small rodents across the African savanna.*

The bird stands up to 1.2m (3.5ft) high on long, pink, storklike legs, and has a wingspan of 2.1m (6.2ft). The long, black-tipped plumes at the back of its head resemble the pen quills of a 19th-century clerk – hence the name.

Africa's Pedestrian Eagle

FORM AND FUNCTION

Although most of its time is spent on the ground, the secretarybird can fly well and often soars, like a stork, with neck and legs extended. It even has a spectacular aerial display, resembling that of some eagles, which involves an undulating, pendulum-like flight in which the bird swings upward, tips gently forward into a steep dive, follows this with another upward swing, and then repeats the pattern, all interspersed with deep, croaking calls. Another behavior, which may serve as a territorial defense or as play between two fledglings, involves one bird chasing another on foot with raised wings, along with bouts of jumping and kicking at one another.

In flight, the two long central feathers of the graduated tail project well beyond the legs. Juvenile secretarybirds are similar to adults, except that their plumage has a brown wash, the white underwing and undertail coverts are finely barred with gray, and the bare skin of the face is a paler orange. The eyes are brown, but turn first gray and then yellow in some adults.

The unique leg morphology and chromosomal karyotype of the secretarybird has been recognized by placing the species in its own family, but the raptorial bill and display flights have always allied it with the birds of prey. Proposed relationships with such birds as New World seriemas and Old World cranes or bustards are probably based on convergence on a common morphology for a terrestrial lifestyle. Although only a single species survives, confined to the open savanna habitats of sub-Saharan Africa, fossil remains of at least two secretarybird-like species with shorter legs are known from Oligocene and Miocene deposits in France from at least 20 million years ago.

Hunting on Two Legs

DIET AND FEEDING

The secretarybird feeds by walking along the ground in search of small animals, most of which are kicked into submission with its stubby toes and stout feet. Small items such as insects may be simply picked up in the bill; large ones, such as rodents, hares, and snakes, may be torn apart, but most are just swallowed whole. The bird hunts while striding rapidly across the veld, normally at a rate of about 120 paces a minute, which equates to a speed of about 3 km/hr (1.9mph); alternatively, it may adopt a slow shuffle, with the crest raised like a spiky halo.

The main items in the diet depend on what is locally available; locusts and rodents, beetles and lizards, termites and snakes are all taken in different areas. Undigested remains of fur, feathers, and skeletons are regurgitated as large pellets below roosts and nest sites.

Right *Confined to sub-Saharan Africa, the secretarybird* ***1*** *is unique among falconiform birds of prey in hunting its prey on foot; it seizes insects, reptiles, and small mammals while stalking across the savanna. Sometimes it takes snakes* ***2****, which it immobilizes with its claws, occasionally killing large ones by dropping them onto rocks. Despite its terrestrial hunting preferences, the bird often takes to the air* ***3a****, flying in a storklike manner with legs and neck extended* ***3b****.*

Above *A secretarybird rests in a treetop nest with her chick. As in most raptors, the young are born relatively helpless, feeding on regurgitated morsels brought by both parents. They take two months or more to leave the nest, and even then stay close by for some weeks, learning vital hunting techniques.*

FACTFILE

SECRETARYBIRD

Sagittarius serpentarius

Order: Falconiformes

Family: Sagittariidae

DISTRIBUTION Sub-Saharan Africa.

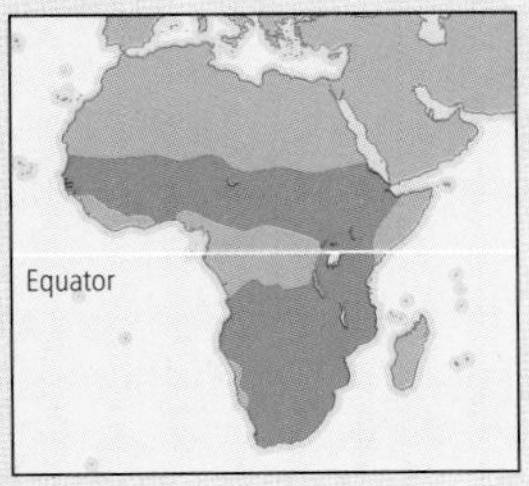

HABITAT Savanna, steppe, and other open habitats.

SIZE Length 125–150cm (49–59in); weight 3.4–4.2kg (7.5–9.4lb).

PLUMAGE Pale gray above, with black wing quills, rump, and thighs; white below. Hooked bill and cere pale gray; face with extensive bare orange skin; nape with loose crest of long feathers. Sexes identical.

VOICE A variety of full-throated, croaking calls.

NEST On a flat-topped tree; often low down on a small, thorny acacia species.

EGGS 1–3, greenish-white. Incubation 42-46 days; nestling 65–106 days.

DIET Locusts, grasshoppers, termites, beetles, rodents, lizards, snakes, birds' eggs and nestlings.

CONSERVATION STATUS Not threatened.

Fecund and Widespread

BREEDING BIOLOGY

Secretarybirds are territorial and monogamous. The territory of a pair may cover 20–200sq km (7.7–77sq mi), depending on regional productivity. In more productive areas a pair may be resident and sedentary, but in areas with fluctuating conditions they are often only nomadic visitors during times of plenty. At night they roost on top of a low tree or on an old nest platform, usually in the core of their territory.

Nests are built on flat-topped trees, often low down on a small thorny acacia species. One to three greenish-white eggs are laid, and the incubation and nestling periods last 42–46 and 65–106 days respectively. A typical brood consists of three chicks, but it is rare for all of them to survive, since food shortages during breeding often lead to the starvation of the youngest sibling. Since the secretarybird cannot carry prey in its feet, it brings food to the nest in its crop and regurgitates it for the young. Less often, it carries large items in its bill for delivery. At first, animal food is torn up and fed to the small chicks or regurgitated as a partly-digested liquid, but within a few weeks of hatching the nestlings are able to gulp down regurgitated food almost before it has landed on the nest floor.

Both adults take part in nest-building, and alternately incubate and provision the nestlings. Each bird hunts for itself while the other is in attendance at the nest, and each will drink water, when available, and bring it back to the nest in its crop. It then dribbles the liquid into the bill of each chick, following a special solicitation behavior by the chick that involves vibrating its bill against that of the parent.

Breeding usually takes place in times of summer rainfall; when good conditions persist, second and even third broods may be reared, with intervals of only a few weeks in between. Such fecundity, added to the variable clutch size and the rate at which the chicks develop, makes the secretarybird one of the most productive large birds of the African savanna. It is still widespread across its range, including in a number of larger conservation areas, and, while it has disappeared from some areas of high human density, it has extended its range into other areas through bush clearance and planting of pastures. AK/IN

Osprey

ONE OF THE MOST DISTINCTIVE LARGE BIRDS of prey, the osprey or Fish hawk is such a fishing specialist that it is usually separated from all other diurnal raptors and classified in its own family.

An osprey in pursuit of fish is one of the spectacular sights of the avian world. From a height of 15–60m (50–200ft), it plunges with its feet foremost and wings half-closed. More often than not, the bird submerges completely before emerging from the water with its prey, which in exceptional circumstances can weigh up to 1.2kg (2.6lb).

Right **1** *The osprey is a familiar sight on many of the world's coasts and rivers.* **2** *This fish-eating specialist relies on its keen eyesight to locate potential targets.* **3** *The rough surface and curved talons of the osprey's feet are adaptations for snatching slippery fish from beneath the water's surface.*

Designed for Diving

FORM AND FUNCTION

The osprey has long legs and very large feet, the soles of which are covered in horny spines and end in long, sharp, recurved claws. These talons are ideal for absorbing the first shock when the bird dives into water, as well as for grasping slippery fish beneath the surface. Furthermore, the outer toe is large and can be swung back, as in the owls, to expand the grip, which is so strong that there are even reports of ospreys being dragged under by large fish. The head is narrow, without the brow ridges common to most eagles, to offer less resistance to the water, and the nostrils are equipped with valves that close during immersion, preventing water from entering. Once a fish is eaten, the long gut typical of a piscivore ensures a thorough digestion. The osprey's wings also have a distinctive shape, being long, narrow, and gull-like, ensuring efficient flight during the bird's prolonged aerial searches for prey.

All Around the World

DISTRIBUTION PATTERNS

Ospreys breed throughout the world, except in the large southern land-masses of South America and sub-Saharan Africa. They are highly migratory and withdraw completely from northern boreal and temperate regions in winter. However, resident ospreys do breed in Australia and on the surrounding islands, from New Caledonia north to Sulawesi and Java.

The European birds migrate to Africa, leaving a small resident population in the Mediterranean and around the Red Sea; North American birds make for Central and South America, leaving a resident population in Florida and the Caribbean. The overall wintering range of ospreys coincides with that of other large fish-eating raptors, such as fish-eagles, except in much of Australasia, which may explain their rather anomalous breeding distribution in the southern hemisphere. In some areas ospreys are primarily coastal, while in others they also occur along lakes and rivers, and such variation may also relate to competition from other fish-eating raptors and owls.

FACTFILE

OSPREY

Pandion haliaetus

Order: Falconiformes

Family: Pandionidae

DISTRIBUTION Worldwide

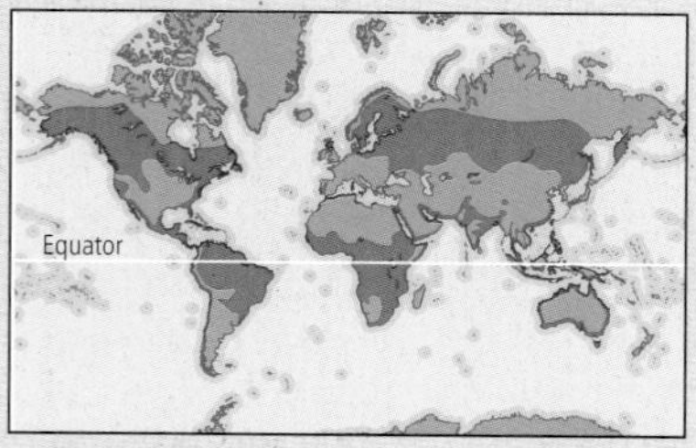

HABITAT Primarily coastal; also lakes and rivers.

SIZE Length up to 62cm (24.5in); weight 1.2–1.9kg (2.6–4.2lb).

PLUMAGE Dark brown above and white below and on the head, with a broad black mask across the yellow eyes.

VOICE A rapid, high-pitched piping call.

NEST A huge pile of sticks, usually on top of a tree, utility pole, pylon, or rocky pinnacle.

EGGS 2–4; red, brown, and chocolate blotches on a white ground. Incubation 35–43 days, nestling period about 50 days.

DIET Fish

CONSERVATION STATUS Not threatened globally, although certain local populations are thought to be at risk.

Plunging Hunters

DIET AND FEEDING

The osprey hunts by flapping and gliding well above the water surface, hovering briefly if it needs to check on prey, and sometimes then plunging in after fish. Its unusual markings, dark above and white beneath, makes it least obvious when viewed from below, a camouflage similar to that used on warplanes. Just before it reaches the water, the talons are brought forward to grab the prey, taking it either from the surface or, more often, after a complete submersion. After a successful catch the osprey rises to the surface with wings outspread, flaps heavily into the air, shakes water from its plumage, and then orientates the fish so that it is carried headfirst so as to offer least air resistance to a suitable feeding perch or the bird's nest. In many parts of the osprey's range, the larger resident fish-eagles will then take every opportunity to rob it of its prey.

Ospreys catch a wide variety of sizable fish, depending on what species are available locally, usually around 150–300g (5–11oz) in weight. They can normally catch all the fish they require in 2–3 hours of hunting, given their 60–70 percent successful strike rate, and they may commute 10km (6 miles) or more between nesting, roosting, and feeding areas.

Nests in High Places

BREEDING BIOLOGY

Before breeding, the male osprey has a spectacular aerial display that involves carrying a fish or nest material. The bird calls, flaps heavily up to an altitude of about 300m (1,000ft) or more, hovers there for a moment with tail fanned and legs dangling to display the fish, and then dives down on closed wings. This performance may be repeated for several minutes at a time and may serve to attract or retain a female, since it often begins and ends at the nest, which is where most courtship and mating take place.

Ospreys nest on the top of a tree or rock pinnacle, but on islands lacking mammalian predators they also nest on the ground. Each pair may nest far apart and defend a large territory, but in some areas the nests are close together in loose colonies of up to 300 pairs. Each nest is built to form a huge pile of sticks, and is normally used over a period of many years.

Two to four eggs are laid, and the incubation and nestling periods span 35–43 and about 50 days respectively. The eggs hatch in the sequence in which they were laid, from 1–7 days apart, and the male then delivers food to the chicks, as he does to the incubating female until she is able to leave the nest and join him in provisioning the young. The young birds do not breed until their third or fourth year. In extreme cases ospreys may live for 15–20 years, but only about 30 percent of fledglings can expect to survive to breeding age.

Back from the Dead

CONSERVATION AND ENVIRONMENT

In parts of the range, particularly in eastern North America, osprey numbers were greatly reduced in the years around 1960 by DDT contamination. Following restrictions on DDT use the species has recovered, and ospreys are now actively conserved by measures that include the erection of nest platforms in suitable areas. In many other parts of the world the birds have also taken to utility poles and pylons as nest sites.

Because it eats fish, including such commercial species as trout and salmon, the osprey has been much persecuted, especially in Europe. It was exterminated completely in the United Kingdom, but, after an absence of 50 years, began to nest again in Scotland around 1955. Since then, under careful and continuing protection, the species has grown in numbers and re-established itself. AK/IN

Below A young osprey descends to the nest, fanning out its tail and lowering the trailing edges of its wings to act as air brakes. To prevent stalling, the bird raises its alulas – tufts of small feathers on the leading edge of the wing that make the flow of air over the wing's surface more laminar, so maintaining lift.

PHOTO STORY

Osprey – The Fish-Hunting Raptor

1

❶ During mid-morning and late afternoon, the Osprey makes several patrols over the large areas of water in its territory. Once its prey – usually fish swimming within 1m (3ft) of the surface – is spotted the Osprey hovers, waiting for the right moment. As the bird strikes the water, its wings are held high and its talons are thrust forward.

❷❸ The Osprey hunts by skimming over water and snatching fish. An alternative strategy is to plunge-dive from a great height, submerging for a few seconds and emerging with the prey clamped firmly in its talons. As it does so, it shakes excess water from its feathers; the Osprey's plumage is fine and dense – especially on its feet – which keeps the bird from getting soaked. Sometimes the Osprey's highly adapted grip is locked so tight that a strong fish will drag the bird under water and drown it.

❹ After catching its prey, the Osprey carries it away using both feet (except in the case of small fishes) and adjusts it in its claws so it is carried headfirst, which helps reduce air resistance.

2

4

5 *Four out of five of the Osprey's hunting forays are successful. The bird takes its prey up to a feeding perch, holding it secure with one foot and tearing it apart with its beak. In the breeding season, the fish is taken back to the nest.*

5

3

Falcons

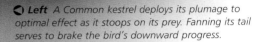

Left *A Common kestrel deploys its plumage to optimal effect as it stoops on its prey. Fanning its tail serves to brake the bird's downward progress.*

Below *Caracaras are not especially noted for their flying skills, spending more time on the ground than most raptors. Even so, they have impressive wingspans. The Striated caracara* (Phalcoboenus australis), *seen here over the Falkland Islands, is one of the lesser-known species.*

For most people, the familiar image of a falcon is likely to be that of a Common or American kestrel hovering at the side of a road, or a Peregrine falcon being used in falconry. Each is a true falcon of the genus Falco, *a long-winged resident of open spaces. The Falconidae, to which the birds belong, is the second-largest grouping of diurnal birds of prey, quite distinct from the hawks and eagles grouped in the Accipitridae. The family is usually separated into two subfamilies, the Falconinae (true falcons and falconets) and the Polyborinae (caracaras and forest-falcons).*

Falconidae species differ most obviously from the Accipitridae in not constructing nests, in the molt sequence of the primary wing feathers (starting in their case with the fourth-outermost feather), and also in details of bodily structure that include a reinforcement of the thorax, a short neck, and a special syrinx. Both families have fossil remains from the Eocene at least 35 million years ago, but each has developed along quite different lines, and falcons have in many respects converged more to resemble owls than their closer relatives, the hawks, buzzards, and eagles. The Falconidae probably originated on the continents of the southern hemisphere, with South America, separated from the north until only 3 million years ago, alone preserving something of their original diversity; up to ten genera are still found there, seven of them endemic.

Caracaras and Forest-falcons

SUBFAMILY POLYBORINAE

The smaller of the two Falconidae subfamilies, the Polyborinae, is largely confined to South America, and more specifically to the neotropical regions of the continent. It includes buzzard-sized caracaras that behave like crows or vultures, hawklike forest-falcons that hunt other birds and reptiles in the canopy, and the unique Laughing falcon, which feeds mainly on arboreal snakes within its more open forest and woodland habitat.

The various caracaras, of which the Crested is the best-known species, are (with the exception of vultures) the least raptorial of the birds of prey, and they may well also be the most primitive. They are large, long-legged birds that inhabit forest, savanna, or semi-desert, where they eat mainly insects, other small animals, some fruit or buds, and any carrion they can find. They are common in several areas and appear rather sluggish, spending much time perched or walking about on the ground in the manner of chickens, although they can run or fly swiftly when necessary. They use their strong feet to turn over heavy objects in search of food, often in association with New World vultures (Cathartidae), but they are cantankerous birds and sometimes force other scavengers to disgorge their carrion. Some species are mainly solitary in their feeding, while others gather regularly in flocks at insect swarms, rubbish tips, or newly-plowed fields. Two forest species are more specialized: the Black caracara adds ticks that it plucks from tapirs to its diet, while the Red-throated caracara lives in territorial groups, feeds mainly on wasp and bee larvae, and provides sentinels for mixed flocks of other forest birds.

Forest-falcons are long-legged birds with a long tail for steering them through the dense forest where they live, and a harrierlike ruff on the face that indicates good powers of hearing. They use their short, broad wings to fly in rapid pursuit of lizards and small birds, and often follow swarms of ants for the insects they disturb and the small animals that these in turn attract. The striking Laughing falcon, with a white crown and black mask, placed alone in its own genus (*Herpetotheres*), may represent a woodland link between caracaras and forest-falcons, nesting like the latter in tree holes.

True Falcons and Falconets

SUBFAMILY FALCONINAE

The Falconinae occur worldwide, but are most speciose in Africa and its islands, especially in the form of kestrels. As well as the true falcons of the genus *Falco*, the subfamily includes the diminutive falconets and pygmy falcons grouped in the genera *Spiziapteryx*, *Polihierax*, and *Microhierax*, among them the sparrow-sized Black-thighed falconet, which is the world's smallest raptor. These miniature falcons occur in the southern continents of Asia, Africa, and South America, each of which used to be united in the supercontinent of Gondwanaland.

The true falcons do not build nests. They probably originated from the smaller falconets, which breed inside the abandoned nests of other birds and lay white eggs, uniquely within the Falconidae but typically for hole-nesters. The Spot-winged falconet of South America breeds inside the bulky nests of ovenbirds (Furnariidae) or the Monk parakeet (*Myiopsitta monachus*), and is so caracaralike in aspects of its anatomy that it is sometimes included among the Polyborinae. The African Pygmy falcon, which nests within the bulky nests of the White-headed buffalo-weaver (*Dinemellia dinemelli*) or the Sociable weaver (*Philetairus socius*), shows differences in coloration between sexes that presage those found in several kestrel species. In Asia, the White-rumped falcon, the other pygmy species, nests in old holes excavated in trees by woodpeckers or barbets, as do the five species of *Microhierax* falconets.

For food, the pygmy-falcons generally specialize on insects, lizards, and small birds, which may either be snatched from the ground or else from the trunks and branches of saplings. The tiny Asian falconets, such as the Philippine falconet, chase large flying insects above the forest canopy or else snatch small birds or lizards with the aid of their strong feet. Asian

FACTFILE

FALCONS

Order: Falconiformes

Family: Falconidae

63 species in 10 genera, separated into 2 subfamilies, the Falconinae and the Polyborinae.

DISTRIBUTION Worldwide, except Antarctica and African rain forests; most diverse in S America, but most true falcons (genus *Falco*) in Africa.

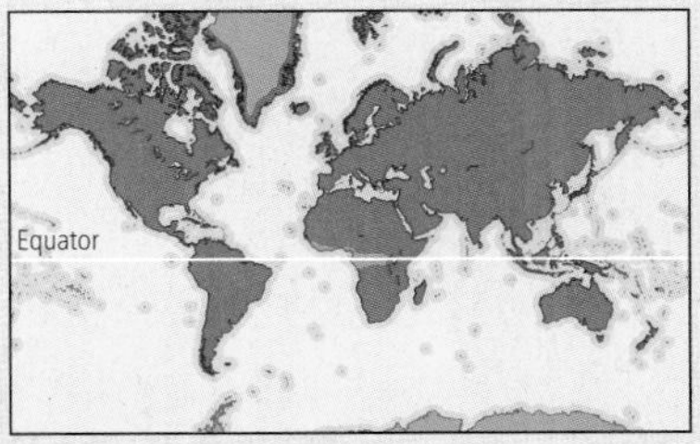

HABITAT Evergreen rain forest to arid desert.

SIZE Length 14–65cm (5.5–25in); weight 28g–2.1kg (1oz–4.6lb).

PLUMAGE Mainly gray, brown, rufous, black and white, underparts usually paler, often marked in patterns of bars or streaks; some species with distinct color morphs. Sexes usually similar, but adults of a few species separable; female often larger. Immature usually different to adult, often with streaked breast, otherwise most like female. Nostril with bony peg in center (except forest-falcons) and upper bill with notch (except caracaras). Cere and bare facial skin bright yellow or orange, but often blue in juveniles. Legs yellow, less often gray, or black in *Microhierax* falconets. Iris usually dark brown, but in a few species pale yellow or brown.

VOICE Various harsh chittering, chattering, cackling, and whining notes. Generally noisy, especially when nesting or living in forests.

NEST Most build no nest, laying their eggs in a tree hole, cliff recess, or old nest of another bird; only caracaras form an untidy platform of sticks and debris.

EGGS 1–7; pale brown, attractively speckled and smeared with brick-red (white in hole-nesting falconets). Incubation 28–35 days, nestling period 28–55 days, depending mainly on body size. Nestlings have two successive coats of thick down, and are usually attended by female while male delivers food to the nest.

DIET Mainly large arthropods and small vertebrates caught alive on the ground and foliage or else in flight. Some caracaras eat vegetable matter and much carrion.

CONSERVATION STATUS The Guadalupe caracara (*Polyborus lutosus*) is now listed as extinct; 4 other species are considered Vulnerable.

See Falcon genera ▷

falconets are characterized by both sexes having a pied plumage, with black feet and cere. Most are found in small groups, in which the members often perch close together, even sharing one another's prey. In at least one species, the Collared falconet, all members also deliver food to a single nest, a form of cooperative breeding that, in the raptors, is apparently only shared with the Red-faced caracara. All the pygmy falcons and falconets pump their tails up and down in excitement and display, as do several smaller kestrels, including the American kestrel.

The true falcons apparently underwent a worldwide radiation within the last few million years, and so are relatively uniform in behavior, calls, and design, including the dark mustache stripe they all share. The smallest of them are the kestrels, most of which (10 species) occur in Africa and its islands. The Seychelles, Mauritius, and Madagascar kestrels are barely larger than falconets, and share their tail-pumping behavior.

Most kestrels belong to one of two color types. The American, Common, Greater, Fox, Spotted, and Australian kestrels are mainly rufous, as are the species of the Indian Ocean islands; in contrast, the Gray and Dickinson's kestrels of Africa and the Banded kestrel of Madagascar are mainly gray. The Greater, Fox, and Banded kestrels have pale yellow eyes when adult, as does the Spot-winged falconet; all other species of Falconidae have brown eyes. Kestrels normally feed on insects and small vertebrates, and most (but not all) are known for their ability to hover for long periods as they search for prey.

In some kestrels the sexes are identical in color, while in others there are distinct differences. Such is also the case with the Western and Eastern red-footed falcons, two other species of small migrant falcons that are similar to kestrels in size and the ability to hover. The red-footed falcons breed in Europe and Asia respectively, but migrate to the open savannas and grasslands of southern Africa for the remainder of each year. The eastern species makes the longest migration of any raptor, traveling at least 30,000km (18,600 miles) from the Amur region of China to southern Africa. On the outward journey it crosses the ocean from India to East Africa, but it flies across Arabia and north of the Himalayas on its return.

In their African winter quarters, these falcons, joined by thousands of Lesser kestrels from Europe, return year after year to the same roosts, the largest of which may hold up to 100,000 birds. The roosts are often in stands of large exotic trees associated with villages or other human habitation, from which the birds spread out and exploit local flushes of such food as termites, crickets, solifugids (sun spiders), and locusts.

While male red-footed falcons look like Gray kestrels, the females, with white or rufous breasts streaked in black, resemble hobbies, another group of small Old World falcons found in Europe, Africa, Asia, and Australasia. Hobbies are long-winged, fast-flying falcons that capture most of their prey on the wing, mainly in the form of insects such as dragonflies, beetles, and termite alates, although small birds, including swifts, form their main diet when breeding. Two other hobby-like species are the Sooty falcon, inhabiting North African deserts, and the larger Eleonora's falcon, which breeds on islands in the Mediterranean Sea and off the African coast. Both these species breed in late summer, so that they have young in the nest at a time when vulnerable young birds are migrating from Europe to Africa. Both, like the European hobby, are themselves migrants, although they migrate almost exclusively to Madagascar, while the European hobby goes to the African mainland.

The larger true falcons are stocky birds of exceptionally swift flight, with long, pointed wings powered by bulky breast muscles and relatively short tails. They habitually kill their prey, principally birds, in full flight, either striking them dead with a blow from the foot or else dragging them to the ground in their talons and executing them with the notched, powerful bill. Best known is the Peregrine falcon, capable of feats of speed and precision flying unequaled by any other bird; its maximum speed, achieved in a bulletlike dive from great height, reaches 180km/h (112mph), or

Above *Representative falcon species (not to scale):* **1** *Laughing falcon* (Herpetotheres cachinnans), *a South American bird with a striking face mask.* **2** *Barred forest falcon* (Micrastur ruficollis) *eating a Black-spotted barbet* (Capito niger). **3** *Female Pygmy falcon* (Polihierax semitorquatus), *an African species.* **4** *Gyrfalcon* (Falco rusticolus), *the largest falcon.* **5** *Red-footed falcon* (F. vespertinus). **6** *Male American kestrel* (F. sparverius), *formerly known as the Sparrow hawk.* **7a** *Common kestrel* (F. tinnunculus) *male perched, and* **7b** *female in hovering flight.* **8** *Peregrine falcon* (F. peregrinus). **9** *Mauritius kestrel* (F. punctatus), *restricted to the island of that name.* **10** *Aplomado falcon* (F. femoralis), *from Central and South America, recently reintroduced to the southern USA.* **11** *Crested caracara* (Polyborus plancus), *the national bird of Mexico.*

maybe more. Whatever the exact figure, a peregrine in full stoop is a most impressive sight, and the bird has traditionally been the first choice of kings, sheikhs, and commoners too as a hunting bird for the ancient art of falconry.

The peregrine is the most cosmopolitan of all the birds of prey, and its distribution overlaps that of many other falcon species. In South America peregrines coexist with the large Orange-breasted, the medium-sized Aplomado, and the small Bat falcons, all of which are black and orange in color; in Africa they share their range with the rufous-breasted Barbary and Taita falcons. Peregrines also overlap with a group of large, pale-colored falcons, known as "desert falcons," since they inhabit open habitats and hunt as much on the ground as in the air. These include the gyrfalcon, the largest of all Falconidae species. This bird, which is adapted to the High Arctic tundra of North America and Eurasia, feeds primarily on ptarmigans and ground squirrels, and occurs in white, gray, and brown color phases that each predominate in different areas. Other desert falcons include the Prairie falcon of North America, the Lanner falcon of Africa, the Saker falcon of the Middle East and the central Asian steppes, and the Laggar falcon of India.

Australia also has two large species in its vast, arid interior, the bird-eating Gray and Black falcons, but their exact relationships remain obscure. The same could be said of Australia's odd, ubiquitous Brown falcon and the New Zealand falcon, which both have relatively long legs and short wings and hunt a wide range of small animal prey as much on the ground as in the air. These Antipodean species highlight the anomalies and problems of understanding falcon relationships. Another pair with uncertain affinities are the dove-sized merlin of the temperate plains of North America and Eurasia and the Red-necked falcon of the deserts and palm savannas of Africa and India, both of which are specialist raptors of small birds.

Re-using Old Nests

BREEDING BIOLOGY

The breeding habits of true falcons are fairly uniform. They start with aerial and perched displays, often centered on the nest-site. The male enters with wings aloft, showing off the underwing colors, and then bows and calls once he has landed. The nests are most often placed in a scrape on a cliff ledge, in an old stick nest appropriated from some other bird such as a crow or a stork, or in a tree cavity, although buildings, nest boxes, utility pylons, and other structures are also now used. Most species breed as widely- spaced territorial pairs, but a few also nest in small colonies, notably Lesser kestrels on old buildings, both species of red-footed falcons in abandoned rook (*Corvus frugilegus*) colonies, and Eleanora's falcon on the cliffs of offshore islands. Successive pairs of

Falcon Genera

Caracaras and Forest-falcons
Subfamily Polyborinae
17 species in 6 genera, ranging from the southern United States (Crested caracara) through the Central American isthmus to the southern tip of South America.

Genus *Polyborus*

2 species: **Crested caracara** (*Polyborus plancus*), **Guadalupe caracara** (*P. lutosus*).

Genus *Daptrius*

2 species: **Black caracara** (*Daptrius ater*), **Red-throated caracara** (*D. americanus*).

Genus *Phalcoboenus*

4 species: **Striated caracara** (*Phalcoboenus australis*), **Carunculated caracara** (*P. carunculatus*), **Mountain caracara** (*P. megalopterus*), **White-throated caracara** (*P. albogularis*).

Genus *Milvago*

2 species: **Yellow-headed caracara** (*Milvago chimachima*), **Chimango caracara** (*M. chimango*).

Laughing falcon
Genus *Herpetotheres*

1 species: *Herpetotheres cachinnans*.

Forest-falcons
Genus *Micrastur*

6 species: **Barred forest-falcon** (*Micrastur ruficollis*), **Slaty-backed forest-falcon** (*M. mirandollei*), **Plumbeous forest-falcon** (*M. plumbeus*), **Collared forest-falcon** (*M. semitorquatus*), **Lined forest-falcon** (*M. gilvicollis*), **Buckley's forest-falcon** (*M. buckleyi*).

True Falcons and Falconets
Subfamily Falconinae
46 species in 4 genera, found worldwide.

True falcons
Genus *Falco*

38 species: **Common kestrel** (*Falco tinnunculus*), **American kestrel** (*F. sparverius*), **Peregrine falcon** (*F. peregrinus*), **Seychelles kestrel** (*F. araea*), **Mauritius kestrel** (*F. punctatus*), **Madagascar kestrel** (*F. newtoni*), **Greater kestrel** (*F. rupicoloides*), **Fox kestrel** (*F. alopex*), **Spotted kestrel** (*F. moluccensis*), **Australian kestrel** (*F. cenchroides*), **Grey kestrel** (*F. ardosiaceus*), **Dickinson's kestrel** (*F. dickinsoni*), **Banded kestrel** (*F. zoniventris*), **Western red-footed falcon** (*F. vespertinus*), **Eastern red-footed falcon** (*F. amurensis*), **Lesser kestrel** (*F. naumanni*), **Sooty falcon** (*F. concolor*), **Eleonora's falcon** (*F. eleonorae*), **European hobby** (*F. subbuteo*), **African hobby** (*F. cuvierii*), **Oriental hobby** (*F. severus*), **Australian hobby** (*F. longipennis*), **Orange-breasted falcon** (*F. deiroleucus*), **Aplomado falcon** (*F. femoralis*), **Bat falcon** (*F. rufigularis*), **Barbary falcon** (*F. pelegrinoides*), **Taita falcon** (*F. fasciinucha*), **gyrfalcon** (*F. rusticolus*), **Prairie falcon** (*F. mexicanus*), **Lanner falcon** (*F. biarmicus*), **Saker falcon** (*F. cherrug*), **Laggar falcon** (*F. jugger*), **Grey falcon** (*F. hypoleucus*), **Black falcon** (*F. subniger*), **Brown falcon** (*F. berigora*), **New Zealand falcon** (*F. novaeseelandiae*), **merlin** (*F. columbarius*), **Red-necked falcon** (*F. chicquera*).

Spot-winged falconet
Genus *Spiziapteryx*

1 species: *Spiziapteryx circumcinctus*.

Genus *Polihierax*

2 species: **Pygmy falcon** (*Polihierax semitorquatus*), **White-rumped falcon** (*P. insignis*).

Genus *Microhierax*

5 species: **Collared falconet** (*Microhierax caerulescens*), **Black-thighed falconet** (*M. fringillarius*), **White-fronted falconet** (*M. latifrons*), **Philippine falconet** (*M. erythrogenys*), **Pied falconet** (*M. melanoleucus*).

Above *In Argentina, two Chimango caracara (*Milvago chimachimo*) chicks call for food. Like the Striated caracara, this species from the southern half of South America occasionally breeds in colonies.*

Left *Peregrine falcons are among the world's most widespread birds, and in many parts of their range have taken to cities. Urban Peregrines like this Denver resident nest on high buildings and subsist mainly on pigeons.*

peregrines and gyrfalcons have been known to use particular nest-sites for periods of over 100 years. Among 49 British peregrine cliffs known to falconers between the 16th and 19th centuries, at least 42 were still in use up to the outbreak of the Second World War.

The handsome, rufous eggs of true falcons are laid directly on the nest floor; held up to the light, the shells appear buff-colored, rather than the pale green of those of Accipitridae species. Incubation is mainly undertaken by the female, while the male hunts for and supplies food, although in smaller species the male may remain on the nest while the female goes off feeding. Any surplus food that the female does not require immediately may be cached, and consumed later in the day.

The eggs are laid at 2–3 day intervals, but incubation only begins towards the end of the clutch, so that the chicks of a brood are of similar age and size. The female later joins the male in bringing food to the chicks. In species that take relatively large and agile prey, especially other birds, the pair may hunt cooperatively, the smaller and more maneuverable male combining with the larger, more powerful female in attacks on quarry; both share the kill when the hunt is successful. Juveniles are usually only supported in the parental territory for 1–2 months after fledging, after which they may wander considerable distances before molting into adult plumage at the end of their first year and then settling down to breed.

Of the Polyborinae, caracaras – uniquely among falcons – build their own nests, which take the form of shallow constructions of small sticks; the Striated caracara occasionally breeds in loose colonies. As for the forest-falcons, their breeding habits have proved so difficult to study that the first nest of any of the six species, that of a Collared forest-falcon, was only discovered in 1978. It was located in a natural hole in a tree.

A Need for Innovative Management
CONSERVATION AND ENVIRONMENT

Peregrine falcons disappeared from large areas of Europe and North America during the 1960s and 1970s, following declines caused by the poisoning of adults, the thinning of egg shells, and an enhanced mortality of embryos through contamination with agricultural chemicals such as DDT and dieldrin. Once these substances were banned, populations recovered, or were reestablished through the use of captive-bred stock (a process known as "hacking"), with many of the new pairs reoccupying nest sites historically used by falcons in the past.

Other falcon species around the world have also suffered declines, and habitat destruction and biocides remain a threat. Even so, the negative effects are being reduced wherever possible. The story of the Mauritius kestrel is exemplary in this respect. The forest habitat to which this bird, with its relatively short, rounded wings and long tail, was originally adapted had been severely reduced, and the total population in the wild had fallen by 1974 to only two pairs. Thanks to intensive captive breeding and innovative management, however, the species has now recovered to at least 50 breeding pairs and a total population in the hundreds. To achieve this end, it had to be conditioned to live in secondary forest and scrub, where the green geckos and small birds on which it feeds, including several exotic species, fortunately remained abundant.

In the Caribbean, an island population of the Crested caracara, often elevated to full species status as the Guadaloupe caracara, was less fortunate. It remains the only modern raptor to have become extinct, sometime around the year 1900. AK/IN

SPECIAL FEATURE

SHELL-SHOCK

The effects of pesticides on birds of prey

Right *At the top of the food chain, birds of prey like these Peregrine falcons ingest pesticides accumulated in the fat of their prey. The risk comes not so much from direct consumption, however, as from the disruption of breeding; DDT in particular thins eggshells, so they often break during incubation (inset).*

TOXIC CHEMICALS THAT KILL LIVING ORGANISMS (biocides), particularly fungal, plant, and animal pests of humans (pesticides), have been widely used since their development in the 1950s. Of these, the cheap and efficient organochlorines have the most harmful effects on wildlife populations, especially predatory birds. Besides being acutely toxic, these chemicals have three further dangerous properties. First, they are stable, and so persist in the environment for many years. Second, they are soluble and accumulate in the body fat of animals, and so pass from prey to predator. Their concentrations therefore increase in successive steps up a food chain, so that raptors, at the top, are especially likely to accumulate large amounts. Third, even small doses may disrupt breeding and reduce the number of young produced.

All bird species that have been studied have been found to be susceptible to organochlorines, but the most marked population declines have occurred in bird-feeding raptors, most notably in the Peregrine falcon in the eastern USA and western Europe but also in the accipiters of these areas, the Sharp-shinned and Cooper's hawks and the Eurasian sparrowhawk respectively. Some fish-eating species also declined, including the osprey and Bald eagle in North America and the White-tailed eagle in northern Europe.

DDT, the most widely-used organochlorine, is in fact not particularly toxic to birds when consumed directly, but its principal breakdown product, DDE, dramatically alters the structure of eggshells, thereby affecting the survival of embryos. Affected shells become thinner and often break during incubation, or they may interfere with the gaseous exchange between the embryo and the air. Both effects, in combination with direct poisoning of the embryos, may lower hatching success to a point where the number of young produced does not offset adult mortality, and the population consequently declines or even becomes extinct.

Different groups of birds vary in their sensitivity to DDE. Birds of prey are particularly vulnerable, partly because they accumulate more DDE than other birds, but also because a given level of DDE produces a greater degree of shell-thinning. The eggshells of herons and pelicans also appear relatively sensitive to DDE, but game-birds and songbirds, frequently prey species, are relatively secure.

Derivatives of more toxic organochlorines, such as HEODs from aldrin and dieldrin, kill adult birds directly and produce an immediate population decline. These chemicals protect cereal crop seeds against attacks from insects, but some seeds are eaten by small birds that later fall prey to raptors. HEODs were held responsible for the elimination of the Peregrine falcon and Eurasian sparrowhawk from large parts of Britain in the late 1960s. Death from organochlorine poisoning is often delayed, since chemicals stored in body fat are only released to affect other, more sensitive tissues when the fat is used to provide energy. Affected birds die during periods of food shortage, such as in winter or on migration, from pesticides that may have been accumulated previously and far away.

Due to their persistence, organochlorines often become dispersed over wide areas, in the bodies of migrant animals or in currents of air and water, and so may affect populations remote from their areas of usage. Hence, even birds of prey that breed in areas where pesticides are not used will not necessarily escape contamination. In winter they may migrate to areas where pesticides are used; for example, Peregrine falcons that nest in the High Arctic migrate south to Latin America, and may become contaminated by DDT in their winter quarters.

In areas where organochlorine use has been curtailed, most bird species have recovered their numbers and range, as with Peregrine falcons in the USA or the endemic Mauritius kestrels. However, DDT residues in the soil are so persistent that they may remain a problem for decades to come, particularly since no bird species has developed any resistance to them, unlike many of their insect targets. In Europe and North America, the use of organochlorines reached a peak in the 1960s, but was banned in subsequent years. However, their use in developing countries remains high, and the raptors there, such as Lanner falcons in South Africa, African fish-eagles, or vultures in India, are showing signs of contamination or decline. Alternative pesticides with less severe environmental effects are now available, but most are more expensive than the original organochlorines.

In the meantime, many other industrial pollutants have emerged that also affect wildlife. Organophosphates, used to kill seed-eating birds in Africa, often kill raptors incidentally; polychlorinated biphenyls, or PCBs, are widespread in developed nations; and heavy metals, such as lead, mercury, and cadmium, kill raptors when prey carry high levels in their blood. Careful use of chemicals that can enter the environment remains essential to the birds' longterm survival. AK/IN

Hawks, Eagles, and Old World Vultures

Diurnal in their habits, the Accipitridae represents by far the largest group of birds of prey. The large number of species and their wide range of body size – from the shrike-sized Pearl kite or Tiny sparrowhawk to the swan-sized Steller's sea-eagle or Lappet-faced vulture – means that this family presents a wide diversity of form and feeding habits.

The birds of prey are noted for their spectacular aerial displays. Some consist of little more than soaring and calling in territorial advertisement, as with the Crested serpent-eagle or Crowned hawk-eagle. This behavior may develop into mock attacks in which one bird dives at another, or even real attacks in which the opponents touch, sometimes even engaging feet to spin down in spectacular cartwheels. Courtship displays often involve repeated undulations, usually by the male, which flaps upward and then dives down on closed wings. In some species such as the Lesser fishing-eagle, the wingbeats are slower and deeper than normal; in others like the African baza, the displays involve a shallow fluttering; while in several, including the Verreaux's eagle, the downward swoop is swift and dramatic, and may even end in a complete loop-the-loop.

Males normally feed the females in courtship, usually passing across the food at a perch. The harriers, however, have a spectacular, aerial food pass in which the flying male drops food for the female to catch in midair.

Death on Wings

FORM AND FUNCTION

The three largest groups within the Accipitridae each exemplify those birds of prey that are most familiar to the general public: hawks, buzzards, and eagles. The **hawks** comprise the best-known group of raptors (58 species, 6 genera), most of which are known as accipiters (after the name of the largest genus); they include the Northern

Above *An aerial food pass. In Western marsh-harriers, the male, when bringing food for the young, does not return to the nest. Instead, the female flies up to him, turns upside down and catches the prey dropped by the male.*

Right *Crested bazas, a tropical species that inhabits New Guinea, the Solomon Islands, and parts of Australia, occasionally prey on small reptiles and frogs (as here), but generally prefer to eat insects, including caterpillars and grasshoppers.*

Below *The unmistakable profile of the Bateleur, with its jet-black plumage, striking red face, and yellow bill. This coloration, along with its short tail and long wings, makes it easy to identify in flight. Some days, the Bateleur may travel as far as 300km (186 miles).*

goshawk and Eurasian sparrowhawk. These are small to medium-sized hawks with short, rounded wings and long tails, adapted for twisting and turning through woodland or forest in dashing pursuit of the small birds, reptiles, and mammals that form the main diet for many species. Most species are rather secretive in their habits and not easy to observe, but some of the goshawks of Africa, such as the Pale chanting-goshawk, occupy open savanna and steppe, where they perch in the open and catch various small animals on the ground; they can also catch guineafowl.

The **buzzards** or **buteonine hawks** are an equally large and even more varied group (57 species, 13 genera) that feed mainly on small mammals and some birds. The true buzzards are particularly widespread; they include the Common buzzard of Europe, the Red-tailed hawk of North America, the Roadside hawk of South America, and the Augur buzzard of Africa. Buteonines are most diverse in the New World. They range in size from the very large and powerful Harpy eagle of South American forests, which eats mainly monkeys and sloths, to the medium-sized, fish-eating Black-collared buzzard or the small white *Leucopternis* hawks from the same forests, which eat mainly insects and small reptiles. Elsewhere in the world buteonines are represented by such diverse species as the rare and striking Great Philippine eagle, with its spiky head plumes and massive bill, the delicate Grasshopper buzzard of Africa, and the long-tailed New Guinea eagle of mountain forests. The latter may comprise part of a special Australasian diversity of kite- and buzzardlike birds, such as the Black-breasted buzzard and Square-tailed kite, although the exact details require further study.

The **true** or **aquiline eagles** (33 species, 9 genera) are distinguished from other eagles by having feathered legs. The largest and best-known species belong in the genus *Aquila*, including the Golden eagle of the northern hemisphere and the Wedge-tailed eagle of Australia, and most take some carrion along with their general diet of mammals and a few birds. All of these "booted" eagles feed mainly on live prey, and many species, known as hawk-eagles, are dashing predators of birds that they capture in flight among the canopy of forests or woodland. A few species are specialists, such as the Black eagle that soars on broad wings above the Asian forests in search of birds' nests, or Verreaux's eagle, which hunts hyraxes among the rocky outcrops of Africa. Many of these eagles lay two eggs, but the first chick to hatch usually attacks and kills its younger sibling, exhibiting the appositely Biblically-named "Cain and Abel syndrome." Siblicide is also known in some other birds of prey, occurring either instinctively or through competition for food, but its origins and advantages remain obscure.

The **kites** and **honey-buzzards** (29 species, 15 genera) show some extreme forms of specialization. The European honey-buzzard digs out wasp grubs with its straight claws and has a feathered face to protect it from stings. The Bat hawk captures bats on the wing at dusk and swallows

FACTFILE

HAWKS AND EAGLES

Order: Falconiformes

Family: Accipitridae

234 species in 62 genera. Most species can be allocated to distinct groups within the family, but the position of a few species remains unresolved.

DISTRIBUTION Worldwide except Antarctica, including many oceanic islands; most diverse in the tropics.

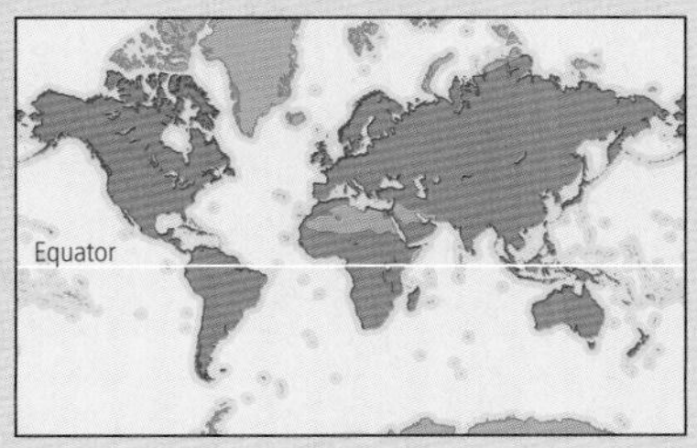

HABITAT From rain forest to desert and arctic tundra; most abundant in forest edge, woodland, and savanna.

SIZE Length 20–150cm (8–59in); **weight** 75g–12.5kg (2.6oz–28lb); female usually larger than male.

PLUMAGE Mainly gray, brown, black, and white, often marked in bold patches or as delicate patterns of bars or stripes. Immature usually colored quite differently from adult, with additional intermediate plumage in larger species. Adult sexes often differ only slightly, but strikingly in a few species. Hooked bill and long, curved talons usually black, but legs, feet, and fleshy cere at base of bill usually yellow, sometimes even orange, red, green, or blue. Iris usually dark brown, but in some species cream, yellow, or red; rarely gray or blue.

VOICE Various whistles, mews, croaks, barks, and yelps, often rather high-pitched. Most species have a limited repertoire and are generally quiet except before breeding, but some forest species are noisy, with daily calling in flight or at perch. Vultures generally silent, except when hissing and fighting at food.

NEST A platform of twigs and branches, often lined with green foliage, and usually built in a tree fork or on a cliff ledge.

EGGS 1–7, depending mainly on body size and latitude. Coloration white or pale green, often with brown and purple marks. Incubation 28–60 days, nestling period 24–148 days, post-fledging dependence a few weeks to 2 years, related mainly to body size. Nestling with 2 successive coats of thick down, usually attended by female while male hunts and delivers prey to the nest. Food usually carried in feet, less often in bill or regurgitated from crop.

DIET All species prefer fresh meat and most kill live animals, but many will take carrion, which is the main diet of vultures. Most feed on as wide a range of animals as they can capture, from earthworms to vertebrates, but some specialize on snails, wasps, bats, fish, birds, mice, or even oil-palm fruits.

CONSERVATION STATUS 8 Accipitridae species are currently listed as Critically Endangered, including the Cuban kite (*Chondrohierax wilsoni*), White-rumped vulture, Long-billed vulture, and Ridgway's hawk. In addition, 4 species are Endangered, and 22 Vulnerable.

See Hawks, Eagles, and Vultures ▷

Above *Representative species of larger accipitrids:* ***1*** *Swainson's hawk* (Buteo swainsoni) *migrates from Alaska to Argentina, flying down the Central American isthmus to avoid a long sea crossing;* ***2*** *Long-legged buzzard* (B. rufinus)*;* ***3*** *The Bald eagle* (Haliaeetus leucocephalus) *is a highly efficient fish-hunter, but will often pirate its food from ospreys in areas where the two species coincide;* ***4*** *Spanish Imperial eagle* (Aquila adalberti)*;* ***5*** *Ornate hawk-eagle* (Spizaetus ornatus)*;* ***6*** *Verreaux's eagle* (Aquila verreauxii)*;* ***7a*** *A vulture's foot – vultures can walk and run well on the ground but do not have the powerful gripping talons found on* ***7b*** *the foot of an eagle;* ***8*** *Asian white-backed vultures* (Gyps bengalensis) *inhabit Indian arable areas;* ***9*** *A Cinereous vulture* (Aegypius monachus) *carries water in its bill to its young;* ***10*** *Bearded vulture or lammergeier* (Gypaetus barbatus)*;* ***11*** *The Palm-nut vulture* (Gypohierax angolensis).

them whole through its exceptionally wide maw. The Black-winged kite hovers kestrel-like in search of rodents, which it stuns and captures with its stout legs and feet. The Snail and Slender-billed kites winkle aquatic snails from their shells with the long tips of their hooked bills. Meanwhile the Black and Brahminy kites have become the most common and versatile of raptors as they scavenge around the towns and villages of Africa, India, and Asia.

The **Old World vultures** (15 species, 9 genera) specialize on eating carrion, although they do this in a number of ways (see Nature's Scavengers). Most are very large birds with a bare or down-covered head and neck, for "wallowing in putridity" as Darwin put it, and broad wings for soaring in search of carcasses. Some have heavy bills for tearing flesh or rending skin and sinew, or else fine bills for extracting morsels from the crevices of a skeleton, but others are specialists. The Egyptian vulture, one of the few tool-using birds, breaks eggs by dashing them to the ground or dropping stones on them. The Bearded vulture or lammergeier drops bones onto rocks to shatter them and gain access to the marrow with its spoon-shaped tongue. The Palm-nut vulture feeds more on the fruit of the African oil palm than on carrion.

The large **fish-** and **fishing-eagles** (10 species, 2 genera) also take much carrion, along with their principal diet of fish and waterbirds, as do their close but smaller and omnivorous relatives, the fork-tailed kites. The most famous of the fish-eagles is the Bald eagle, the national emblem of the United States; once in decline, its numbers have now recovered. Sadly, the same cannot be said for the Grey-headed fishing-eagle, which is declining in its habitat along the Asian rivers, while the Madagascar fish-eagle may now be the world's rarest raptor.

Snake- and **serpent-eagles** (16 species, 5 genera) are large raptors adapted to killing snakes with their short toes and heavily scaled legs. These birds are immediately recognizable by their large, owl-like heads and yellow eyes. They are vulture-like in laying only one egg in a clutch, but unlike vultures they kill live prey, which they carry dangling from the mouth to regurgitate at the nest, as in the case of the Short-toed snake-eagle. Most are species of the forest or dense woodland, including the long-tailed Congo serpent-eagle and the rare Madagascar serpent-eagle, thought extinct for over 50 years until its rediscovery in 1988.

The colorful bateleur, although obviously related to the snake-eagles, has unusual, bow-shaped wings and a very short tail, which enable it to glide low and effortlessly above the African savanna in search of small carcasses and live prey.

11

9

7a

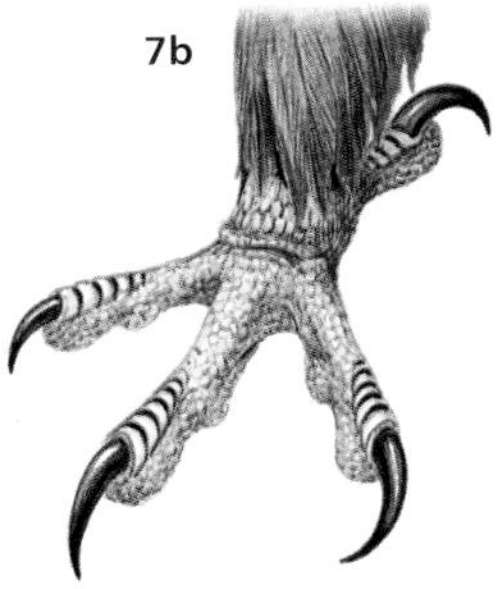
7b

8

10

African and Madagascar harrier-hawks, probably even more distantly related to snake-eagles, have long, gangly legs with unique, double-jointed "knee" joints that enable them to extract small creatures from holes in trees and rockfaces. Their small, naked face also enables them to pick prey from crevices or foliage as they glide slowly and methodically over the woodland vegetation.

The Crane hawk of South America has the same design and habits as the harrier-hawks but is an excellent example of convergence through similarity of biology, since it is probably related to a quite different group, the **harriers** (16 species, 3 genera). These are a uniform group of slim, medium-sized hawks with long tails and broad wings that harry slow and low above grassland (for example, Montagu's harrier) and marshes (the Western marsh-harrier). They feed mainly on small mammals and birds, together with some reptiles and insects, and they have owl-like faces and large ears that are sensitive to the sounds of prey in thick vegetation. Most harriers nest on the ground amid tall herbs or in reeds above water, but the Spotted harrier of Australia, exceptionally, nests in trees, often far from water.

Most raptors in tropical regions are sedentary and live on permanent territories. In more temperate areas, where the climate is more seasonal and unpredictable, most species perform some sort of migration, involving movements of varying distance between breeding and nonbreeding habitats. The longest journeys, of about 20,000km (12,500 miles) each year, are made by raptors that fly regularly between eastern Europe and southern Africa (for instance the Common or Steppe buzzard) or between the extremes of North and South America (Swainson's hawk).

Couples and Colonies

BREEDING BIOLOGY

Most raptors have only one mate in a year; some keep the same mate for several years, and a few large eagles are even said to "pair for life," although this remains unproven. Polygyny, in which one male may breed with more than one female at a time, is frequent among harriers, and polyandry, in which one female may mate with more than one male, is frequent in the Harris's hawk, a desert species of central America. Both mating systems have also been recorded occasionally in other species.

All members of the Accipitridae build their own nests of sticks and stems, often lined with fresh foliage and usually placed on trees, cliffs, or sometimes on the ground or in reeds. Different species sometimes interchange nest sites, while falcons (Falconidae) and owls (Strigidae and Tytonidae), which do not construct their own nests, often take over abandoned Accipitridae nests.

In most raptors there is a marked division of labor during the breeding cycle. The male hunts for food, while the female stays near the nest and is responsible for incubating the eggs and looking after the young. This pattern persists until the young are about half-grown, after which the female may also begin to leave the nest area to assist the male in hunting. Young raptors, in contrast to most other nest-reared birds, hatch covered in down, with their eyes open; they are sufficiently coordinated to reach out for and swallow their own food.

In the majority of raptors, the female initially tears up small pieces of meat that the young pick from her bill, but before fledging they learn to tear up prey for themselves. In vultures and some kites, the partners share parental duties more equally, take turns at the nest, and each parent seeks its own food, some of which it later regurgitates to the young.

Raptors vary in the way that they space themselves within their habitat, depending largely on how their food is distributed. Three main systems exist. In the first, pairs are spaced out in individual home ranges, and each pair defends the vicinity of the nest plus a variable amount of surrounding terrain. This is the pattern in about three-quarters of the raptor genera, including some of the largest, such as *Accipiter*, *Buteo*, and *Aquila*. The hunting or home ranges may be exclusive or overlapping,

Larger than the Male

One of the most interesting features of raptors is the marked difference in size between the sexes, where the female is larger than the male – up to twice as heavy in some species. This sexual dimorphism is evidently connected with the raptorial lifestyle, because the same pattern occurs in other predatory birds, such as owls and skuas. However, it also occurs in a few nonraptorial species such as jacanas (Jacanidae), painted snipes (Rostratulidae), and buttonquails (Turnicidae).

In raptors in general, the difference in size and the separation of roles for each sex increases with the speed and agility of the prey. At one extreme, vultures that feed entirely on immobile carcasses show no consistent difference in the size and role of each sex. In raptors that eat slow-moving prey such as snails, the female is only slightly bigger than the male, and both sexes share several aspects of breeding behavior. Those that feed on insects and reptiles show a somewhat greater size difference, those that feed on mammals and fish rather more, while those that feed on other birds exhibit the largest difference of all. Within these categories, an even greater size difference is shown by those species that take the largest prey in relation to their own size, so that in those bird-feeding accipiters and hawk-eagles that kill prey heavier than themselves the sexes have widely different roles during breeding. In such species, the size difference between the sexes is so marked that it even results in males and females catching quite different sizes and species of prey.

Despite this link with diet, it is not known why the female rather than the male is the larger sex, even though the differences are connected with the role each gender takes during breeding. The main difference, however, is likely to lie in the extent to which each sex competes for essential resources. In raptors, females compete for territories to which they can attract mates and in which they rear their young. Natural selection will drive the body size of the female towards a maximum commensurate with her own best chance of survival and reproduction. The female will select a male with the best match in body size, be it large, small, or similar, to suit the particular ecological niche of that species. In the end, this evolutionary reasoning makes the most sense of the variety of differences in size between the sexes of raptors, nonraptorial species with larger females, and even the majority of bird species that have larger males.

but throughout suitable habitat the nests of different pairs tend to be spaced fairly regularly apart, at distances ranging from less than 200m (650ft) in some small raptors to more than 30km (18.5 miles) in some large ones. Individuals of most species that space themselves in this way usually hunt and roost alone, feed on live vertebrate prey, and show considerable stability in numbers and distribution from year to year.

In the second system, some pairs group closely together and nest in restricted "neighborhoods," from which they range out to forage in the surrounding area. This system is evident, for example, in Black, Red, Black-shouldered, and Letter-winged kites, Grasshopper buzzards, Western marsh-harriers, and Northern and Montagu's harriers. The different pairs may hunt in different directions or at different times from one another, or several may hunt the same area independently, shifting from time to time between different areas. The breeding groups usually contain 10–20 pairs, with nests spaced 70–200m (230–650ft) apart, but larger groups have sometimes been found. In harriers, the tendency to coloniality is sometimes accentuated by polygyny, where each male may have two or more females nesting close together.

Colonial nesting in both harriers and kites is frequently necessary where suitable nesting habitat is patchy, although even where nesting cover is widespread the colonial habit is often still apparent. Such species often exploit sporadically abundant food sources, such as local plagues of grasshoppers or rodents, and this behavior too forces them to be to some extent nomadic, with the effect that local populations may fluctuate substantially from year to year. Kites and harriers often also roost communally outside the breeding season, kites in trees and harriers in reeds or long grass, within which each bird tramples down a platform to stand on. They spread out to hunt over surrounding areas during the day, but at

__Above__ Perched atop a dead tree on heathland, a Golden eagle keeps watch over its domain. Home ranges of Golden eagle pairs vary according to area – in Scotland, for example, they are estimated as being between 4,500 and 7,300ha (11,000–18,000 acres).

__Right__ The Eurasian griffon is one of the few raptors that nests colonially on cliff edges or in caves and roosts in large numbers.

__Left__ Accipitrids use their superb flying skills in courtship displays. __1__ In the "whirling" display, a pair of African fish-eagles grapple with each other's talons, tumbling together in cartwheels or swinging from side-to-side like falling leaves. __2__ Undulating display. These undulations may be shallow or pronounced. In the Hen harrier the bird dives with wings partly closed, then regains height by flapping. __3__ "Pot hooks." In this extreme form of undulating display, shown by Tawny eagles, the bird dives and swoops without wing flapping. __4__ "Pendulum." In this display, shown by Verreaux's eagles, the bird repeats a figure of eight.

Left *Representative species of smaller accipitrids:* **1** *The Long-tailed hawk* (Urotriorchis macrourus) *occurs in the forests of western Africa;* **2** *Western marsh-harrier* (Circus aeruginosus); **3** *Pied harrier* (C. melanoleucus); **4** *Eastern chanting goshawk* (Melierax poliopterus) *calling on top of a termite mound;* **5** *The Eurasian sparrowhawk* (Accipiter nisus) *inhabits the temperate forests of the Old World;* **6** *Black-mantled sparrowhawk* (A. melanochlamys) *with newly captured prey;* **7** *Madagascar harrier-hawk* (Polyboroides radiatus) *hunting for insects;* **8** *Black-chested snake-eagle* (Circaetus pectoralis) *with a kill. Like almost all snake-eagles, this is an African species;* **9** *The Snail kite* (Rostrhamus sociabilis) *lives in American swamps feeding on large snails;* **10** *Swallow-tailed kite* (Elanoides forficatus), *found from the southern United States to South America;* **11** *Red kite* (Milvus milvus) *with rabbit carrion;* **12** *European honey-buzzard* (Pernis apivorus) *digging out a wasps' nest.*

8

9

10

11

12

nightfall tens of individuals gather to roost; at a few places in Africa, several hundred kites or harriers of several species have occasionally been counted. The same roosts may be used yearly, but often by greatly varying numbers of birds.

In the third system, pairs nest in dense colonies and forage gregariously. This system is shown by small snail- or insect-eating kites of the genera *Rostrhamus*, *Elanoides*, *Gampsonyx*, and *Ictinia*, but also by the large griffon vultures in the genus *Gyps*. In these species, the pairs typically nest closer together – often less than 20m (66ft) apart – and in large aggregations. Colonies usually contain at least 20 or 30 pairs, but those of Snail kites sometimes exceed 100 pairs and those of some griffon vultures over 250 pairs. Such species also feed communally, in scattered flocks or, in the case of the vultures, spread out in the air but crowded together around carcasses. The feeding flocks are not stable in size and composition, but change continually as individuals join or leave. The food sources of such species are notably concentrated but unpredictable in space and time, so that food may be plentiful at one place on one day and at a quite different place on the next. Such species roost communally at all times, and may gather in even larger numbers when not breeding.

Most raptors choose a special place for their nests, whatever their dispersion. Such a place may be a cliff, an isolated tree, a grove of trees, or a patch of forest or ground cover, depending on the species. Many such places are occupied over long periods. Particular cliffs have been used for at least a century by successive pairs of Golden or White-tailed eagles. Certain eagle nests, added to year after year, reach an enormous size; one historic Bald eagle nest in America spanned 8sq m (86sq ft) and contained "two wagonloads" of material, while a Verreaux's eagle's nest in South Africa grew to be 4m (13ft) tall. Even patches of ground cover have been used by Hen harriers for several decades. Colonial raptors also tend to nest in the same places year after year, and in southern Africa many cliffs, still in use by Cape griffons today, have names that indicate their use by the same species in previous centuries. As in other colonial birds, each pair defends only a small area around its nest, so that, given enough ledges, many pairs crowd onto the same cliff but leave other, apparently suitable cliffs vacant. In general, of course, sites on rock are likely to be more permanent than those in trees, and those in trees more permanent than those in herbaceous cover.

Nowhere to Nest

BREEDING DENSITY

Body size appears to bear a great influence on breeding trends. The larger the species the later the age at which breeding begins and the longer each successful attempt takes, which means fewer young are produced with each attempt. Furthermore, because of their special nest requirements, raptors are among the few groups of birds whose breeding numbers and success are often limited by the availability of nest sites. For example, cliff dwellers may have their breeding density limited by the number of suitable ledges, and their success by the accessibility of the ledges to predators.

Raptors in open landscapes may be limited by a shortage of trees, especially in prairies, steppes, and grasslands that offer abundant food but few trees. Even in woodlands, nest sites may be fewer than they appear at first sight. In several hundred square kilometers of mature forest in Finland, less than one in a thousand trees was judged by a biologist to be suitable for nests of White-tailed eagles, while in younger forests the suitable, open-crowned trees were even scarcer or even nonexistent. On the positive side, this fact also means that the provision of artificial nest sites, such as platforms on trees, buildings, quarries, or pylons, can be used to increase the breeding density of birds of prey.

Where nesting sites exceed demand, raptor numbers seem to be limited by food supply. Species with varied diets tend to have fairly stable food supplies, and their breeding populations also remain stable in a given area, fluctuating by no more than 10–15 percent of the average over a period of years. In areas where they are free of adverse human influence, birds like the Golden or Martial eagles provide extreme examples of longterm stability in numbers, even though they may vary greatly in breeding density from one area to another, depending on local food supplies.

In contrast, raptors that have diets restricted to prey that fluctuates seasonally will have breeding densities that vary from year to year, more or less in parallel with their prey. Prime examples are the Northern harrier and Rough-legged buzzard, both of which feed on rodents such as lemmings, or the Northern goshawks of boreal forests, which feed on hares and grouse. Rodent numbers peak at intervals of 3–4 years, as do those of their predators, while hares, grouse, and their predators have cycles of 7–10 years. The case of the Northern goshawk is particularly instructive, because the bird's breeding populations remain stable in areas where the supply of prey, such as rabbits, is stable, but fluctuates where supplies are unstable.

In general, small species of raptors feed on smaller, more abundant prey and occur at higher densities than larger raptors that feed on larger, sparser prey. A small African goshawk may hunt over an area of 1–2sq km (0.4–0.8sq mi), a Common buzzard over an area of 1–5sq km (0.4–2sq mi), but a large eagle will cover a much greater area. The Martial eagle, the largest in Africa, which has a diet of small antelopes, monitor lizards, and gamebirds, may occur at one pair per 125–300sq km (48–116sq mi), with 30–40km (18.5–25mi) between nest sites, making it one of the world's most thinly distributed birds. Large fish-eating raptors are an exception, where they form high densities around concentrated stocks, and large colonial raptors, such as griffon vultures, may appear to occur in high numbers at their colonies, although their overall densities are in fact extremely low when their extensive feeding areas are taken into account.

Above *Once hatched, Red-tailed hawk chicks are brooded for a week by their mother, after which both parents have to hunt to supply the voracious young birds with adequate food. After the chicks have begun to fly, they stay close to their parents for a further 6–7 weeks until they learn to forage for themselves.*

Below *Black-shouldered kites can be seen either singly or in pairs, sitting on trees or posts or hovering in search of prey. They are generally silent but do produce various weak, whistling calls. The double-whistled "plee wit, plee wit" is associated with alarm and food.*

Top Predators at Risk

CONSERVATION AND ENVIRONMENT

Since predatory birds feed on other animals, they must occur at lower densities, regardless of their size, than the birds and other animals that constitute their prey. Their position at the top of the local food chain also brings several negative effects, usually as a result of human activity. Firstly, raptors will be most immediately and widely affected by any deterioration in their habitats, such as conversion of unfarmed land to agriculture or the destruction of forests. Secondly, birds of prey will come into conflict with people when they choose gamebirds, poultry, or livestock as their prey, and this competition often leads to their direct persecution by shooting, trapping, or poisoning. Thirdly and most insidiously, raptors are prone to contamination and incidental poisoning – as was the Bald eagle of North America – through the accumulation of toxic chemicals, such as mercury, DDT, PCBs, or dieldrin, taken along with their prey and usually originating as agricultural pesticides or industrial effluents.

It is no wonder, then, that, internationally, some 25 percent (58 out of 234 species) of Accipitridae are listed on the IUCN Red List as near-threatened or threatened, eight of them critically, including the Great Philippine eagle, the Madagascar fish-eagle, the Madagascar serpent-eagle, and the White collared kite. Locally, the situation is even worse, with populations of many other species of raptors in decline or even extinct, although, at this stage, there are still reasonable numbers left of the species as a whole.

Below *In the 1600s the Bald eagle was a familiar sight in North America, with perhaps as many as 500,000 birds patrolling the continent's coastlines. Yet by the 1960s, this imposing national symbol of the USA was believed to be near extinction, due to persecution by humans and poisoning by pesticides, especially DDT. However, by 1995, after two decades of legal protection and a pesticide-free diet (DDT was banned in 1972), its numbers had substantially recovered. With some 100,000 birds now thriving in Canada and Alaska alone, the Bald eagle appears to be out of imminent danger.*

Hawks, Eagles, and Old World Vultures

WITHIN THE ACCIPITRIDAE, IT IS POSSIBLE to assign most species to one of several distinct groups, even though a few species cannot be easily placed and the relationships between groups and their constituent species are still being resolved. Relationships are further blurred by the inconsistent use of common names, such as hawk, kite, buzzard, or eagle, for members of different groups, and the use of names in combinations, such as hawk-eagle, buzzard-eagle, or harrier-hawk, that also do not always comply.

Hawks

58 species in 6 genera. Species include: **New Britain sparrowhawk** (*Accipiter brachyurus*), **Nicobar sparrowhawk** (*A. butleri*), **Northern goshawk** (*A. gentilis*), **Gundlach's hawk** (*A. gundlachii*), **White-bellied goshawk** (*A. haplochrous*), **Imitator sparrowhawk** (*A. imitator*), **Slaty-mantled sparrowhawk** (*A. luteoschistaceus*), **Eurasian sparrowhawk** (*A. nisus*), **Tiny sparrowhawk** (*A. superciliosus*), **African goshawk** (*A. tachiro*), **Chestnut-shouldered goshawk** (*Erythrotriorchis buergersi*), **Red goshawk** (*E. radiatus*), **Lizard buzzard** (*Kaupifalco monogrammicus*), **Doria's goshawk** (*Megatriorchis doriae*), **Eastern chanting-goshawk** (*Melierax poliopterus*), **Pale chanting-goshawk** (*M. canorus*), **Long-tailed hawk** (*Urotriorchis macrourus*).

Buzzards

57 species in 13 genera: Species include: **Grey hawk** (*Asturina plagiata*), **Black-collared hawk** (*Busarellus nigricollis*), **Grasshopper buzzard** (*Butastur rufipennis*), **Augur buzzard** (*Buteo augur*), **Common** or **Steppe buzzard** (*B. buteo*), **Galápagos hawk** (*B. galapagoensis*), **Red-tailed hawk** (*B. jamaicensis*), **Rough-legged buzzard** (*B. lagopus*), **Roadside hawk** (*B. magnirostris*), **Ferruginous hawk** (*B. regalis*), **Ridgway's hawk** (*B. ridgwayi*), **Hawaiian hawk** (*B. solitarius*), **Swainson's hawk** (*B. swainsoni*), **Great black-hawk** (*Buteogallus urubitinga*), **Black-chested buzzard-eagle** (*Geranoaetus melanoleucus*), **Crowned eagle** (*Harphyaliaetus coronatus*), **Harpy eagle** (*Harpia harpyja*), **New Guinea eagle** (*Harpyopsis novaeguineae*), **White-necked hawk** (*Leucopternis lacernulata*), **Grey-backed hawk** (*L. occidentalis*), **Plumbeous hawk** (*L. plumbea*), **Mantled hawk** (*L. polionota*), **Crested eagle** (*Morphnus guianensis*), **Harris's hawk** (*Parabuteo unicinctus*), **Great Philippine eagle** (*Pithecophaga jefferyi*).

True or Aquiline eagles

33 species in 9 genera. Species include: **Spanish Imperial eagle** (*Aquila adalberti*), **Wedge-tailed eagle** (*A. audax*), **Golden eagle** (*A. chrysaetos*), **Greater spotted eagle** (*A. clanga*), **Gurney's eagle** (*A. gurneyi*), **Imperial eagle** (*A. heliaca*), **Tawny eagle** (*A. rapax*), **Verreaux's eagle** (*A. verreauxii*), **Booted eagle** (*Hieraaetus pennatus*), **Black eagle** (*Ictinaetus malayensis*), **Long-crested eagle** (*Lophaetus occipitalis*), **Black-and-chestnut eagle** (*Oroaetus isidori*), **Martial eagle** (*Polemaetus bellicosus*), **Javan hawk-eagle** (*Spizaetus bartelsi*), **Philippine hawk-eagle** (*S. philippinensis*), **Wallace's hawk-eagle** (*S. nanus*); **Black-and-white hawk-eagle** (*Spizastur melanoleucus*), **Crowned hawk-eagle** (*Stephanoaetus coronatus*).

Kites and Honey-buzzards

29 species in 15 genera: Species include: **African baza** (*Aviceda cuculoides*), **Crested baza** (*A. subcristata*), **Hook-billed kite** (*Chondrohierax uncinatus*), **Swallow-tailed kite** (*Elanoides forficatus*), **Black-winged kite** (*Elanus caeruleus*), **Black-shouldered kite** (*E. axillaris*), **Letter-winged kite** (*E. scriptus*), **Pearl kite** (*Gampsonyx swainsonii*), **Brahminy kite** (*Haliastur indus*), **Black-breasted buzzard** (*Hamirostra melanosternon*), **Mississippi kite** (*Ictinia mississippiensis*), **Plumbeous kite** (*I. plumbea*), **White-collared kite** (*Leptodon forbesi*), **Square-tailed kite** (*Lophoictinia isura*), **Black kite** (*Milvus migrans*), **Red kite** (*M. milvus*), **Slender-billed kite** (*Rostrhamus hamatus*), **Snail kite** (*R. sociabilis*), **Black honey-buzzard** (*Henicopernis infuscatus*), **Long-tailed honey-buzzard** (*H. longicauda*), **European honey-buzzard** (*Pernis apivorus*), **Bat hawk** (*Macheiramphus alcinus*).

Old World vultures

15 species in 9 genera: Species are: **Cinereous vulture** (*Aegypius monachus*), **Bearded vulture** or **lammergeier** (*Gypaetus barbatus*), **Palm-nut vulture** (*Gypohierax angolensis*), **White-backed vulture** (*Gyps africanus*), **White-rumped vulture** (*G. bengalensis*), **Cape griffon** (*G. coprotheres*), **Eurasian griffon** (*G. fulvus*), **Himalayan vulture** (*G. himalayensis*), **Long-billed vulture** (*G. indicus*), **Rüppell's griffon** (*G. rueppellii*), **Hooded vulture** (*Necrosyrtes monachus*), **Egyptian vulture** (*Neophron percnopterus*), **Red-headed vulture** (*Sarcogyps calvus*), **Lappet-faced vulture** (*Torgos tracheliotus*), **White-headed vulture** (*Trigonoceps occipitalis*).

Fish eagles

10 species in 2 genera. Species are: **White-tailed eagle** (*Haliaeetus albicilla*), **Bald eagle** (*H. leucocephalus*), **White-bellied fish-eagle** (*H. leucogaster*), **Pallas's sea-eagle** (*H. leucoryphus*), **Steller's sea-eagle** (*H. pelagicus*), **Sanford's fish-eagle** (*H. sanfordi*), **African fish-eagle** (*H. vocifer*), **Madagascar fish-eagle** (*H. vociferoides*), **Lesser fishing-eagle** (*Ichthyophaga humilis*), **Grey-headed fishing-eagle** (*I. icthyaetus*).

Snake eagles

16 species in 5 genera. Species include: **Southern banded snake-eagle** (*Circaetus fasciolatus*), **Short-toed snake-eagle** (*C. gallicus*), **Congo serpent-eagle** (*Dryotriorchis spectabilis*), **Madagascar serpent-eagle** (*Eutriorchis astur*), **Crested serpent-eagle** (*Spilornis cheela*), **Mountain serpent-eagle** (*S. kinabaluensis*), **Nicobar serpent-eagle** (*S. minimus*), **Andaman serpent-eagle** (*S. elgini*), **Bateleur** (*Terathopius ecaudatus*).

Harriers and Harrier-hawks

16 species in 3 genera (including Crane hawk). Species include: **Northern harrier** (*Circus cyaneus*), **Madagascar harrier** (*C. macrosceles*), **Pallid harrier** (*C. macrourus*), **Réunion harrier** (*C. maillardi*), **Black harrier** (*C. maurus*), **Montagu's harrier** (*C. pygargus*), **Western marsh-harrier** (*C. aeruginosus*), **Spotted harrier** (*C. assimilis*), **African harrier-hawk** (*Polyboroides typus*), **Madagascar harrier-hawk** (*P. radiatus*), **Crane hawk** (*Geranospiza caerulescens*).

Below *The Steller's sea-eagle of northeastern Asia is one of the largest eagle species in the world, weighing 6–9kg (14–20lb). Its wingspan, which can measure up to 2.5m (8ft), helps the bird to lift heavy prey items like large salmon and waterfowl.*

Habitat destruction has already accounted for bigger reductions in raptor and other wildlife populations than any other factor. The continuing growth in human population, development, and affluence is still the most serious threat in the long term. Irrespective of any other adverse influence, habitat sets the ultimate limit on the size and distribution of any wild population. Species that live in special or restricted habitats are most vulnerable, since both the total area of habitat and the maximum population size that can be supported are limited – a fact reflected in the large number of forest, marsh, and island-dwelling species that are considered globally threatened.

Since the carrying capacity for raptors in an area is sometimes set by the availability of nest-sites, shortages can on occasion be rectified by adding sites artificially, as mentioned above. Raising the carrying capacity of an area through increasing the food supply is much more difficult, because it usually entails changing the patterns of land-use to promote an increase in prey. Often the best that can be achieved is to preserve existing areas of good habitat, or at least to prevent their further degradation. In North America, Africa, Asia, and Australia, large national parks provide some excellent raptor habitat, capable of maintaining large populations. In more heavily-peopled countries, most areas that are left and could be preserved in this way are too small to support many birds, especially for larger species that require huge areas to sustain them. However, in every area there is a growing need for people to reduce their impact on the extensive areas of land that they occupy alongside wildlife.

Direct human persecution is less serious now than in the past, at least in the countries of the northern hemisphere, where bounty schemes have given way to protective legislation. Such legislation varies widely across the world, and has met with varying success in different countries. It is generally most effective in developed countries, such as those in the European Union, the United States, Canada, Japan, and Australia. However, attitudes towards legislation have ranged from respect to scorn, especially in undeveloped countries, and there often remains a considerable gap between the efficacy of legislation and its subsequent enforcement, since bird protection is so difficult to monitor.

As for the threat from chemical pollution, the only longterm solution is to reduce the use of biocides so that their concentration in the environment falls. In many northern and developed countries, this result has been achieved by substituting chemicals that are less toxic or persistent than the offending ones, even when they are more expensive to use. However, the cheaper and more dangerous chemicals are still manufactured and remain widely used in undeveloped countries, where, apart from the threat to local species, they are also a problem for migrating raptors.

PREYING ON LIVESTOCK

The predatory feeding habits of raptors inevitably lead some species to capture and kill livestock, such as the game animals pursued by hunters or farmers' poultry and other small stock. This rivalry causes a major conflict between raptors and humans, and is the main reason why raptors have often been heavily persecuted. In most cases, the impact of raptors on game or domestic animals is negligible, although in a few cases it can be severe.

One serious problem in parts of Europe is hunting by Northern goshawks (BELOW) of intensively reared pheasants. The young pheasants are hatched in incubators and, when about six weeks old, are put out in open pens, from which they are gradually released into woodland. Goshawks often come to concentrate in these pheasant-rearing areas and, despite regular trapping programs, may take a substantial proportion of the stock. These pheasants would otherwise be available to hunters, who may have paid a considerable amount for their rearing.

Another problem concerns eagles and sheep. Wherever these birds live alongside sheep or goats, they not only scavenge on any dead animals but also occasionally kill some live lambs or kids. This is true of Golden eagles in Europe and North America, White-tailed eagles in Norway and Greenland, Wedge-tailed eagles in Australia, and Verreaux's and Martial eagles in southern Africa. In most areas their impact on the lamb population is negligible, but in some localities it can be serious and often leads to widespread persecution of the eagles. In sheep-ranching areas of western Texas and southeast New Mexico, for example, up to 2,000 Golden eagles were shot down annually from airplanes over a period of 20 years until the practice was banned in 1962. Methods have now been developed to distinguish lambs killed by eagles from those dying of other causes and then eaten by eagles as carrion. This practice allows innocent eagles to be distinguished from the few that become serious lamb-killers, and the latter can then be removed, ideally by capture and subsequent translocation.

In many countries birds of prey are now protected by law, although some exceptions are made to deal with locally troublesome individuals and species. It remains difficult to alter traditionally negative attitudes towards birds of prey, and, since the law is hard to enforce, illicit killing is still common, even in some developed countries.

A number of different measures have been taken to counter the short-term effects of biocides until environmental levels begin to fall. Several species have been propagated in captivity for release to the wild; current projects include the reintroduction of Eurasian griffons in France, Bearded vultures in the Alps of Switzerland – where after 16 years of the scheme 70 birds now fly – Bald eagles in New York State, Great Philippine eagles in the Philippines, and Egyptian vultures in South Africa. Other reintroduction schemes have entailed transplanting young birds from one region to another, as in the current program to re-establish the White-tailed eagle in Scotland. In England and Wales, young birds of the Red kite have been taken from other countries like Spain where the species is more common and then reintroduced to well-wooded places.

Wherever a species has been eliminated by human activities from otherwise suitable habitat, reintroduction is the only option. Populations of many large raptors have been so fragmented that there is now little hope that such species will recolonize isolated patches of habitat naturally – at least not within the foreseeable future. However, reintroductions are difficult and expensive to perform, so that the most cost-effective form of management for any bird of prey is to protect as much good habitat as possible and to reduce any other negative factors to a minimum. AK/IN

SPECIAL FEATURE

NATURE'S SCAVENGERS

Old World vultures

Below *A common sight in the African savannas is a large squabble of White-backed vultures scrambling for their piece of carrion. In parts of East Africa they can sometimes be seen in equal numbers among a group of the slightly larger Rüppell's griffons.*

VULTURES VARY CONSIDERABLY IN THEIR NESTING and foraging habits. The most strongly gregarious are the large griffons: the Griffon vulture of Eurasia, Rüppell's griffon of northern Africa, and the Cape vulture of southern Africa, which nest on cliffs in big colonies numbering up to 100 pairs or more, with some nests only a few meters apart. One of the largest concentrations known is around the Gol escarpment in East Africa. It contains more than 1,000 pairs of Rüppell's griffons, distributed in several colonies, and supported largely by the big-game populations of the Serengeti Plain. Such birds feed entirely from large carcasses and, being dependent on migrant animals, they often have to fly great distances for food, taking more than one day over each trip. They have been followed up to 150km (93 miles) from the colony.

The food searching of Griffon vultures is extremely efficient. Following the Charge of the Light Brigade in the Crimean War (1854), so many birds gathered on the battlefield that shooting squads were posted to protect the injured. The birds' ability to find isolated carcasses and to gather quickly in large numbers in areas where they had apparently been scarce have caused some people to suspect an extreme sensitivity to smell, or even telepathic ability: some native Africans think that vultures dream the locations of food. In fact griffons rely on vision – unlike some species of New World vultures that use smell to help locate carcasses – and most find food indirectly by watching the activities of neighboring birds in the air.

If an Old World vulture spots a carcass, it begins to circle lower. Its neighbors notice this behavior, and fly towards the scene. These birds are in turn noticed by their neighbors, so that, within minutes, vultures are converging from all points of the compass. If trees are available, the birds sit for a while before descending, but once the first few individuals are down, there is a rush for a place at the carcass. A small animal, such as an antelope, can be stripped to the bone in 20 minutes. The birds themselves squabble and fight while feeding, and the more dominant individuals can cram so much food in their crops that they have difficulty in taking wing again.

With their efficient food-searching, griffons are extremely effective scavengers. Their only drawbacks are that they cannot operate at night, nor can they compete with large mammalian carnivores, which can easily drive them from a carcass. Furthermore, their gregarious habits make them extremely vulnerable when carcasses have been deliberately poisoned.

Other, similar species, including the African and Asian white-backed vultures, also feed entirely on carcasses, gathering in large numbers around them. They depend more on resident and less on migrant game, however, and so travel less far. They nest in trees, generally in smaller, more scattered colonies than the cliff-nesting griffons; occasionally they do so as individual pairs. They also weigh less than the big griffons, and so can take wing earlier in the day.

Another group of Old World vultures, including the European black and Lappet-faced vultures, behave in some respects like eagles. Individual pairs nest far apart and hold large ranges around their nests. They feed partly from large carcasses, but also take smaller items, including living prey; they do not fly long distances to forage, so it is rare to find more than one or two pairs at the same corpse.

Although several species of vulture may assemble at the same carcass, they do not all feed in the same way, or take the same tissues. In southern

Right *Lappet-faced vultures are the largest vultures in Africa, frequenting the drier parts from the north-west Sahara and Eritrea and south to the Cape Province. It relies almost entirely on carrion but has been known to take adult and young flamingos.*

Europe, the Griffon vulture eats mainly the softer meat, the large Black vulture more often tears meat and skin off bones, the small Egyptian vulture pecks off tiny scraps of meat remaining on the bones, while the Bearded vulture takes the bones themselves. Moreover, only the Griffon depends entirely on large carcasses; the rest take other foods as well.

The lammergeier, or Bearded vulture, and Egyptian vulture are two species that have had to adapt to a scantier diet because they are smaller and less dominant than their larger vulture relatives. The birds have also developed specialized scavenging skills. Lammergeiers, which inhabit the mountains of Eurasia and Africa, live on a diet that consists of between 70–90 percent bone. They sometimes include hard-shelled turtles in their diet too. In order to get at the bone marrow – or the turtle flesh – and swallow the large bones more easily, they fly high and break the bones/shells by dropping them on ossuaries (breaking grounds). Lammergeiers' tongues are also specially adapted with grooves for removing marrow from the bone. Egyptian vultures, which can be found in Southern Europe, the Middle East, India, and most of Africa, are noted for their specialized egg-eating skills. They are one of the few birds able to use stones as tools, which they use to strike holes in abandoned ostrich eggs. They then use their beaks to enlarge the hole and penetrate the membrane. AK/IN

Pheasants and Quails

FACTFILE

Pheasants and Quails

Order: Galliformes

Family: Phasianidae

187 species in 47 genera

DISTRIBUTION N America, northern S America, Eurasia. Africa, Australia. Introduced to New Zealand (after extinction of native species), Hawaii, and other islands. Some species introduced in Europe and N America.

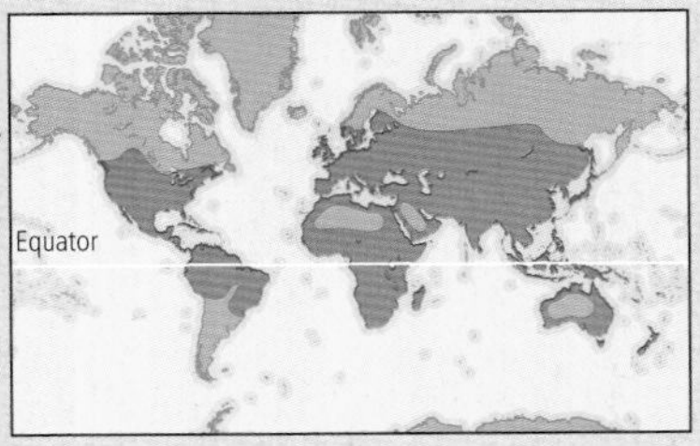

HABITAT A wide variety of open and forested terrestrial habitats, including tropical and temperate forests and grasslands, as well as scrub, desert, and cultivated land.

SIZE Length 12–122cm (4.7in–4ft), excluding display trains; weight 20g–6kg (0.6oz–13.2lb).

PLUMAGE Commonly brown, gray, and heavily marked, but males often boldly patterned with blue, black, red, yellow, white, or iridescent colors. Sexual dimorphism varies from almost none to extreme, with males 30 percent larger than females and equipped with elaborate display structures and spurs.

VOICE Brief, but loud whistles, wails, and raucous crows. Sociable species call often, solitary ones only at dawn or dusk in breeding season or when alarmed.

NEST Chiefly simple ground scrapes, lined, if at all, with grass. Tragopans may nest in trees.

EGGS Usually 2–14, whitish to dark olive, sometimes with markings; weight 4.8–112g (0.15–4oz). Incubation 16–28 days; period in nest no more than a few hours or days.

DIET Varied, chiefly seeds and shoots; also invertebrates, roots, and fallen fruit. Chicks are mostly insectivorous.

CONSERVATION STATUS 3 species – the Djibouti francolin (*Francolinus ochropectus*), Gorgeted wood-quail (*Odontophorus strophium*), and Himalayan quail (*Ophrysia superciliosa*) – are considered Critically Endangered; 8 others are Endangered, and 36 Vulnerable.

See Pheasant and Quail Groups ▷

Ground-dwelling pheasants and quails make up the largest family in the order Galliformes, which also includes the grouse and turkeys. In human terms, the Phasianidae are of unparalleled importance, for the family contains the ancestor of the common chicken, domesticated at least 5,000 years ago; chickens now outnumber humans by almost four to one, for there are allegedly 24 billion on the planet. Yet for all the family's usefulness, many Phasianidae species are now at risk.

Pheasants and quails are often striking in their appearance. The Indian peacock, which is revered in many parts of South Asia by Hindus, is appreciated throughout the world for its beauty. Several species, including the Ring-necked pheasant and Northern bobwhite, are the focus of multimillion-dollar rural industries.

Right *Blyth's tragopan* (Tragopan blythii), *from the Himalayas, is one of five Asian mountain pheasants grouped in the genus of that name. The birds are noted for their plumage, their herbivory, and their habit – unique in the family – of nesting in trees.*

Some species are prized for their singing ability (for example, quails in Pakistan), while others have symbolized bravery (in imperial China, Brown eared-pheasants' tail feathers were worn by generals going into battle). On Java, the train feathers of the Green peafowl were used in traditional costumes. In complete contrast, the family also contains some of the world's least-known species, most notably the Himalayan quail, of which only about 10 specimens have ever been collected, and those well over a century ago. The Manipur bush-quail has also not been recorded in recent times.

Right *Resident in the deserts of the US Southwest, Gambel's quail* (Callipepla gambelii) *has a distinctive, curved black head plume. Males, such as the bird shown here, give persistent, resonant calls.*

Rotund Birds with Rounded Wings

FORM AND FUNCTION

Nearly all pheasants and quails are heavy, rotund birds with short legs and rounded wings. From the tiny Blue quail to the stately Blue peafowl, they are strong runners and rarely fly except to escape danger, when they burst from cover in an explosion of rapid wingbeats, typically relying on glycogen-burning sprint muscles. Some species that inhabit dense forest, such as the Malaysian peacock pheasant, prefer to walk stealthily away through the undergrowth, or will run if suddenly alarmed. Most species cannot remain airborne for long and are therefore sedentary, staying within a few kilometers of their birthplace. However, some of the *Coturnix* quails do undertake long-distance journeys, being migratory or nomadic. The Japanese quail, for example, undertakes regular annual migrations, while some populations of the Harlequin quail in Africa are believed to be nomadic, possibly in response to seasonal rainfall patterns, which may also influence the movements of the Rain quail in South Asia.

The New World quails are most typically small, plump birds, boldly marked with black, white, buff, and gray; some carry firm, forward-pointing crests or "topknots." Perhaps the best known species is the Bobwhite quail, which is a major quarry species in the USA.

Old World quails are found throughout the grasslands of Africa, Asia, and Australia, and although they are few in number, they are widely distributed. Six of the species are often paired in so-called "superspecies," reflecting the very close relationships between the birds concerned: respectively, these are the Japanese and Common quails, the Harlequin and Rain quails, and the Blue and Blue-breasted quails.

The partridges are a diverse collection of stocky, medium-sized game birds found in a range of habitats throughout the Old World. They include the giant snowcocks, which may weigh 3kg (6.6lb) and inhabit the alpine tundras of mountainous Central Asia. In Southeast Asia there are many poorly-known species that inhabit tropical rain forests, including the splendid Crested wood-partridge. Partridges are most commonly found, however, in open habitats such as semi-deserts, grassland, or scrub. Many species adapt well to extensive cultivation, notably the Gray or Hungarian partridge and the chukar, which has become common in farmland throughout much of Europe and has been introduced to North America. In Europe, modern agricultural techniques, and in particular the widespread use of pesticides and herbicides, have, however, caused a steady decline in numbers in recent years. Africa has only two genera of partridges, the bantamlike Stone partridge and the diverse francolins, of which there are 41 species, most of them confined to the continent. These partridgelike birds are sturdy, live in a variety of habitats, and tend to be rather noisy. A new species, the Udzungwa forest partridge, was discovered in the mountains of Tanzania as recently as 1992.

The term "pheasant" is usually reserved for the larger and often more colorful members of the family. Of these 48 species in 16 genera, all but one are confined to Asia. The exception is the extraordinary and beautiful Congo peafowl, the late discovery of which, by W. L. Chapin in 1936, created an ornithological sensation. Pheasants are forest birds; some live in the rain forests of Southeast Asia, others at various altitudes on the mountains of Central Asia. Despite colorful male plumage and loud, raucous calls, most are shy and rarely seen. Extreme examples of this trait are the Golden and Lady Amherst's pheasants, the ruffed pheasants of western China. Males of both species are astonishingly gaudy, the Golden in red, yellow, and orange, Lady Amherst's in white, green, red, and black. For a long time European naturalists dismissed pictures of these birds as figments of the imagination of Chinese artists, so fantastic did they seem.

There is an interesting variation in social organization within the family. Most of the smaller quails and partridges are highly gregarious but monogamous. Some of the larger pheasants, such as peafowl, are also gregarious, but many are solitary, especially those that inhabit dense forest. These species are usually polygynous (one male mating with several females) or promiscuous, forming no pair bonds.

Widespread and Varied

DISTRIBUTION PATTERNS

The pheasants, partridges, and quails are found in most terrestrial habitats throughout large parts of the Americas, Africa, Europe, Asia, and Australia. From the dense rain forests of Southeast Asia to the arid deserts of Arabia or the high peaks of the Himalayas, almost every habitat has its characteristic species. They are absent only from Antarctica, some oceanic islands, the southern half of South America (where their niche is occupied by tinamous), and the tundras and forests of the far north (where they are replaced by grouse).

In all the areas where they are found, there are species that inhabit both open and closed habitats. Broadly speaking, the open-country species have fairly general habitat requirements and are consequently fairly widely distributed. For example, some of the African francolins occur across vast swathes of the continent; the Red-necked francolin has perhaps the largest distribution of all, at more than 4 million sq km (1.5 million square miles). Species of forest and wooded habitats, in contrast, tend to have more specific habitat requirements, along with much more limited distributions. There are exceptions to this rule, of course, such as the Kalij pheasant, which occurs along the Himalayas from Pakistan eastward into Myanmar.

Raking Through the Leaf Litter

DIET

The tragopans, along with the high-altitude snowcocks, are the only phasianid species that are thought to be almost exclusively vegetarian. The Himalayan snowcock, which lives at altitudes up to 4,570m (15,000ft) in the summer, feeds by raking through the soils on snow-free slopes and alpine meadows, where it finds a variety of roots and tubers, seeds, grasses, shoots, and other plant parts. Most other species eat a wide variety of food items, with several feeding strategies evident. Some species, such as the Cheer pheasant and Brown eared-pheasant have heavy bills (as do the snowcocks) and dig over large areas of ground to find plant parts below the surface. Some New World quails, such as the Ocellated quail, use their long legs to dig through the leaf litter. Other species, including the junglefowls, scratch the surface of the soil or leaf litter so as to disturb invertebrates underneath, or else uncover seeds or other plant food. Yet other species, such as the Black wood-partridge and some other tropical forest partridges, pick food items up off the leaf litter. Species inhabiting very dry areas, such as the New World quails

10

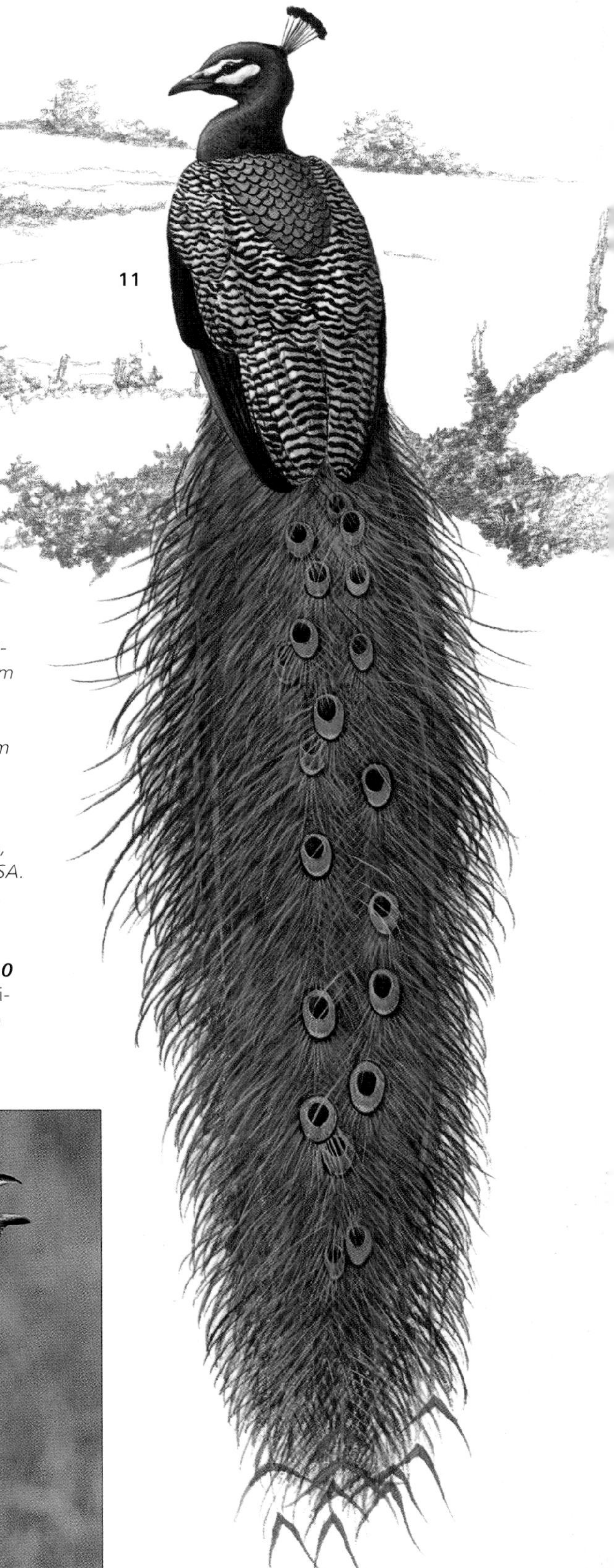

11

in the western USA, also pick food off the ground rather that scratch to uncover food. There remains a vast amount to find out about the diets of many species because they live in habitats, often in remote areas, that make them difficult to observe.

Couples and Coveys

BREEDING BIOLOGY

Among partridges and quails, the basic social unit is the covey, one family party perhaps with a few other birds attached. In species that occupy open habitats such as snowcocks, chukars, and Bobwhite quails, coveys often fuse to form larger flocks. At the other extreme, in forest-dwelling partridges such as the Black wood-partridge of Malaysia or some of the francolins, adults live singly or in pairs throughout the year. Pair formation usually takes place before the

9

8

Left and above *Representative species of pheasants and quails:* ***1*** *Red junglefowl* (Gallus gallus) *from southeast Asia, the wild ancestor of the common domestic fowl.* ***2*** *Lady Amherst's pheasant* (Chrysolophus amherstiae), *an attractive species from the mountains of southern Asia.* ***3*** *Chukar* (Alectoris chukar), *introduced to the western USA from the Mediterranean region.* ***4*** *Common quail* (Coturnix coturnix). ***5*** *Northern bobwhite* (Colinus virginianus), *a common gamebird in the southern and eastern USA.* ***6*** *Grey partridge* (Perdix perdix). ***7*** *Golden pheasant, originally from mountainous areas of China* (Chrysolophus pictus). ***8*** *Red-legged partridge* (Alectoris rufa). ***9*** *Mountain quail* (Oreortyx picta). ***10*** *Common or Ring-necked pheasant* (Phasianus colchicus), *in flight.* ***11*** *Male Indian or Blue peafowl* (Pavo cristatus) *– the familiar peacock, introduced around the world from its original home in India.*

Left *A Yellow-necked spurfowl* (Francolinus leucoscepus) *calling. Clearly visible is the vivid throat patch that gives the bird its name. Francolins are smallish, partridgelike birds of southern Asia and Africa. With 41 species, they comprise the largest genus within the family.*

covey breaks up, although males often join another covey to seek a mate. Recent experiments with Japanese quail have demonstrated that this behavior probably serves to avoid inbreeding, although the quails were found to prefer their first cousins to more distant relatives when choosing a mate.

Among the larger, polygynous pheasants, courtship involves long and spectacular rituals. An extraordinary but rarely seen sight is the display of the male Satyr tragopan from India and Nepal, which lowers a fleshy, electric-blue lappet from its throat and inflates two slender blue horns on its crown. In the Himalayan monal, the iridescent males display in flight over the high cliffs and forests, calling wildly – a breathtaking sight. Perhaps the most exciting of all displays is the dance of the Great argus in the forests of Malaysia. Adult males have huge, broad, secondary wing feathers, each adorned with a series of circular, golden decorations shaded to appear three-dimensional. An adult male prepares a special dance floor on the top of a hill in the middle of the forest. From this site he plucks leaves and stems and blows away leaf litter by clapping his enormous wings. Early each morning he gives loud, wailing cries to attract females. If a female arrives, he begins to dance about her, and at the climax of his dance throws up his wings into two enormous, semicircular fans, revealing hundreds of "eyes." In the gap between his wings, his real eyes can be seen staring at the female.

In the two species of argus pheasants, the display may end with mating, after which the female leaves to rear the brood unaided. In junglefowl and Ring-necked pheasants, however, the male forms bonds with a number of females and guards them as his "harem" until the eggs are laid. This mating system is almost unknown in other birds (although it is common in mammals).

With the exception of tragopans, all pheasants and quails nest on the ground, forming a single

Above Among the Phasianidae, the pheasants are particularly noted for their courting displays. Male Golden (and Lady Amherst's) pheasants can suddenly expand cloaks of feathers normally drawn up on the side of the head to create an impressive ruff effect.

Below A Red-legged partridge stands guard over her brood of chicks. The name "partridge" is applied to various medium-sized, stout-bodied species across the Old World, whose basic social unit is the covey, or extended family group.

Pheasant and Quail Groups

The family Phasianidae has traditionally been divided into four groups: New World quails, Old World quails, partridges, and pheasants. Recent research into the evolutionary history of the Galliformes as a whole is, however, increasingly suggesting that this arrangement may need revision. The most widely agreed change is that the New World quails are very different from the rest of the Phasianidae and should in fact be considered as a separate family, the Odontophoridae. More recently it has been argued that the distinction between the pheasants (subfamily Phasianinae) and the Old World quails and partridges (subfamily Perdicinae) may be artificial, since some partridges are in fact pheasants, and some pheasants partridges!

New World Quails

32 species in 9 genera , from extreme NE Argentina to S Canada. Species include: Northern bobwhite (*Colinus virginianus*), Ocellated quail (*Cyrtonyx ocellatus*).

Old World Quails

14 species in 3 genera, from Africa, Asia, Australia (Common quail migrates to Europe). Species include: Blue quail (*Coturnix adansonii*), Blue-breasted quail (*C. chinensis*), Harlequin quail (*C. delegorguei*), Rain quail (*C. coromandelica*), Japanese quail (*C. japonica*), Common quail (*C. coturnix*), Manipur bush-quail (*Perdicula manipurensis*), Himalayan quail (*Ophrysia superciliosa*). The New Zealand quail (*C. novaezealandiae*) is extinct.

Partridges

92 species in 18 genera, from Africa, Europe, Asia, Australia. Species and genera include: Snow partridge (*Lerwa lerwa*), Black wood-partridge (*Melanoperdix nigra*), chukar (*Alectoris chukar*), Red-legged partridge (*A. rufa*), francolins (*Francolinus*, 41 spp.), Red-necked francolin (*F. afer*), Grey partridge (*Perdix perdix*), Madagascar partridge (*Margaroperdix madagascarensis*), Stone partridge (*Ptilopachus petrosus*), Crested wood-partridge (*Rollulus roulroul*), Sichuan hill-partridge (*Arborophila rufipectus*), snowcocks (*Tetraogallus*, 7 spp.), Himalayan snowcock (*Tetraogallus tibetanus*), Udzungwa forest partridge (*Xenoperdix udzungwensis*).

Pheasants

48 species in 16 genera from Asia; 1 species in Africa. Species include: Satyr tragopan (*Tragopan melanocephalus*), Kalij pheasant (*Lophura leucomelanos*), Red junglefowl (*Gallus gallus*), Himalayan monal (*Lophophorus impejanus*), Cheer pheasant (*Catreus wallichii*), Brown eared-pheasant (*Crossoptilon mantchuricum*), Ring-necked pheasant (*Phasianus colchicus*), Golden pheasant (*Chrysolophus pictus*), Lady Amherst's pheasant (*C. amherstiae*), Great argus (*Argusianus argus*), Crested argus (*Rheinardia ocellata*), Malaysian peacock-pheasant (*Polyplectron malacense*), Indian or Blue peafowl (*Pavo cristatus*), Green peafowl (*P. muticus*), Congo peafowl (*Afropavo congensis*).

Above and right *Some of the most striking pheasant displays can be seen in east Asia. Bulwer's pheasant* (Lophura bulweri), *the Bornean species seen above, has electric-blue wattles and horns that give the head a profile resembling that of a claw hammer. This bird is now classified as Vulnerable by the IUCN. Even more spectacular is the decoration of Temminck's tragopan* (Tragopan temminckii), *which lowers a brilliantly ornamented lappet from its throat, patterned in shades of blue and contrasting strokes of red.*

scrape, usually in dense, herbaceous vegetation. Clutch size varies from two eggs in argus pheasants to nearly 20 in the Gray partridge (the largest clutch size of any bird). Predators often take a heavy toll of the eggs, and female Ring-necked pheasants may make two or more nesting attempts each season. The female Red-legged partridge lays two clutches, one for the male to incubate, the second for herself. Apart from this species, males take little or no part in incubation. In captivity, female Golden pheasants have been found to incubate continuously without food, water, or even moving, for 22 days. In one case a bird sat so still that a spider built its web across her back. Whether this happens in the wild has yet to be established.

The young are well-developed. They feed themselves from birth, leave the nest within a few hours, and can fly as young as one week old. Young blue quails can and do breed when only two months old. Because they are so prolific, pheasants and quails can sustain heavy predation losses, and humans have learned to exploit this fact by managing them for hunting. Many species are hunted, notably the Ring-necked pheasant.

Prey to Humans

CONSERVATION AND ENVIRONMENT

Phasianids are good to eat. This fact, combined with the widespread disturbance and degradation that most terrestrial habitats are suffering, means that their populations are declining more quickly than most other bird families'. The most recent IUCN assessment suggests that, while 11 percent of all bird species are at risk of extinction, a staggering 25 percent of the nearly 200 Phasianidae species are considered threatened. At least one species, the New Zealand quail, is now extinct

In general, the species that live in forest are suffering most. The Himalayas, China, Southeast Asia, and the Neotropics have all been widely deforested, albeit at different times and at different rates. The result is that the area of suitable forest is now much reduced, while the remaining habitat patches are very accessible to humans.

A considerable amount of work is, however, taking place to ensure that the birds do have a future. For many species, especially the pheasants, distribution and baseline ecological studies over the last 20 years have provided a fairly good understanding of the pressures confronting Phasianidae species and an outline of the steps needed to overcome them. Protected areas seem likely to be a crucial component, for parks such as the Great Himalayan National Park in northwest India and Taman Negara in peninsular Malaysia hold large numbers of species. However, these alone will not suffice, and appropriate management strategies in the wider countryside will also be needed. The work currently underway in Sichuan in southwest China could serve as a model; surveys for the endangered Sichuan hill-partridge have evolved into a substantial program designed to manage the forest throughout the southern part of the province, to the benefit of a wide variety of endemic bird species. PM/MWR

CONSTELLATIONS OF EYES

The dance of the peacock and its purpose

THE DISPLAY OF THE BLUE (OR INDIAN) PEACOCK IS a famous symbol of extravagant beauty. The bird spreads behind his proud, blue neck an enormous fan of about 200 iridescent feathers, many adorned with a glowing "eye." Since ancient times, the peacock has had a close connection with man, and has been a graceful sight around many an Indian temple and European garden. Nevertheless, until quite recently few of the details of the peacock's courtship dance were known, and the purpose of that splendid fan was even less understood. (It is not, incidentally, a tail, but consists of enlarged tail coverts.)

Peafowl live for most of the year in small groups or family parties. In the breeding season, however, the cocks become solitary and pugnacious. Each adult male returns to a place he occupied in previous years and establishes his territorial rights, threatening intruders and calling loudly to advertise his presence. Territories are small, from 0.05 to 0.5ha (0.1–1.2 acres), and center on clearings in forest or scrub. These territories tend to be clumped together, so that males are aware of each other's close proximity. Occasionally a junior male will challenge a senior neighbor, and a long and violent battle ensues. The combatants circle each other nervously, looking for an opening, then suddenly spring up to slash out with their claws and spurs. Evenly balanced fights can last for a whole day or more, and are as keenly watched by other peacocks as any boxing contest! Serious injury is rare, and the winner is usually the bird with most stamina, who drives his opponent away.

Within his territory the peacock has from one to four special display sites, where the famous dance takes place. These spots are carefully chosen; a typical one is an "alcove" no more than 3m (10ft) across, enclosed by bushes, trees, or walls. In one English park, a male uses the stage of an open-air theater!

The cock waits near one of these sites until he sees one or more females approach. He then goes to the site and, turning modestly away from the females, spreads his great fan with a long, loud shake to bring each "eye" into place. He then begins to move his wings rhythmically up and down. As the females get nearer, he is careful to keep the unpatterned back of the fan toward them. Peahens have a reputation for being indifferent to the cocks' splendid shows, and it appears that at this stage they are drawn to the site more by chance than purpose.

As soon as a female enters the alcove, a transformation comes over the male. He backs toward her with rapidly fanning wings; she avoids him by stepping into the center of the display site. This is apparently what the cock has been waiting for. He swivels suddenly so as to face her, ceases the movement of his wings, and presses the fan forward, almost engulfing the hen. Simultaneously, spasms of rapid shivering course through the fan, causing the whole structure to rustle with a loud, silvery sound. The female normally responds by standing still for a few moments, and the male then turns away and resumes his wing-fanning. Sometimes she then runs quickly round to the front of the male, and, when he shivers the fan, runs excitedly behind him again. This behavior may be repeated several times.

Charles Darwin recognized that the peacock's fan presents an evolutionary conundrum. Why should the females find this ornament attractive, when it is such an unnecessary encumbrance? An ingenious solution to the problem was proposed by the biologist Ronald Fisher, who suggested that females choose the most ornamented males so that their sons might inherit their fathers' attractions; in other words, they follow fashion. Any female with a different taste would risk having unattractive sons, condemned by other females to evolutionary oblivion. So peahens use colorful rump feathers as a criterion of attractiveness in peacocks, and select the most adorned. Another theory is that the extravagance of the male's train increases with age and that the most ornamented birds are the oldest, who have demonstrated their ability to survive. Therefore, the theory goes, they must be superior males.

How, in practice, do females choose a mate? The answer lies in an extraordinary feature of the peacock's train. The females walk among displaying males, revisiting some, and in most cases ultimately mate with the male exhibiting the most eye-spots. If a group of females assess the males, all will mate with the same one. As the number of eye-spots increases with age, the females are not only choosing the most extravagantly decorated male, but also one of the most experienced survivors. PM/MWR

Right *The display of the peacock, in which the bird courts a mate by spreading a fan of some 200 brilliantly decorated tail coverts, is one of the most spectacular in the entire animal kingdom.*

Below *1 During courtship, the peacock, after backing toward the female, turns suddenly **2**, exposing to her a shivering constellation of eyes. Sometimes, just as the male swivels toward the female, he suddenly launches himself forward with a strangled hoot and attempts to catch hold of her. **3** The female usually dashes out of the way. However, on occasion **4** she hesitates or crouches, and mating ensues.*

Below *As a prelude to mating, peacocks establish small territories that they advertise with harsh, strident cries. The calls serve both to warn away other cocks and to attract potential partners.*

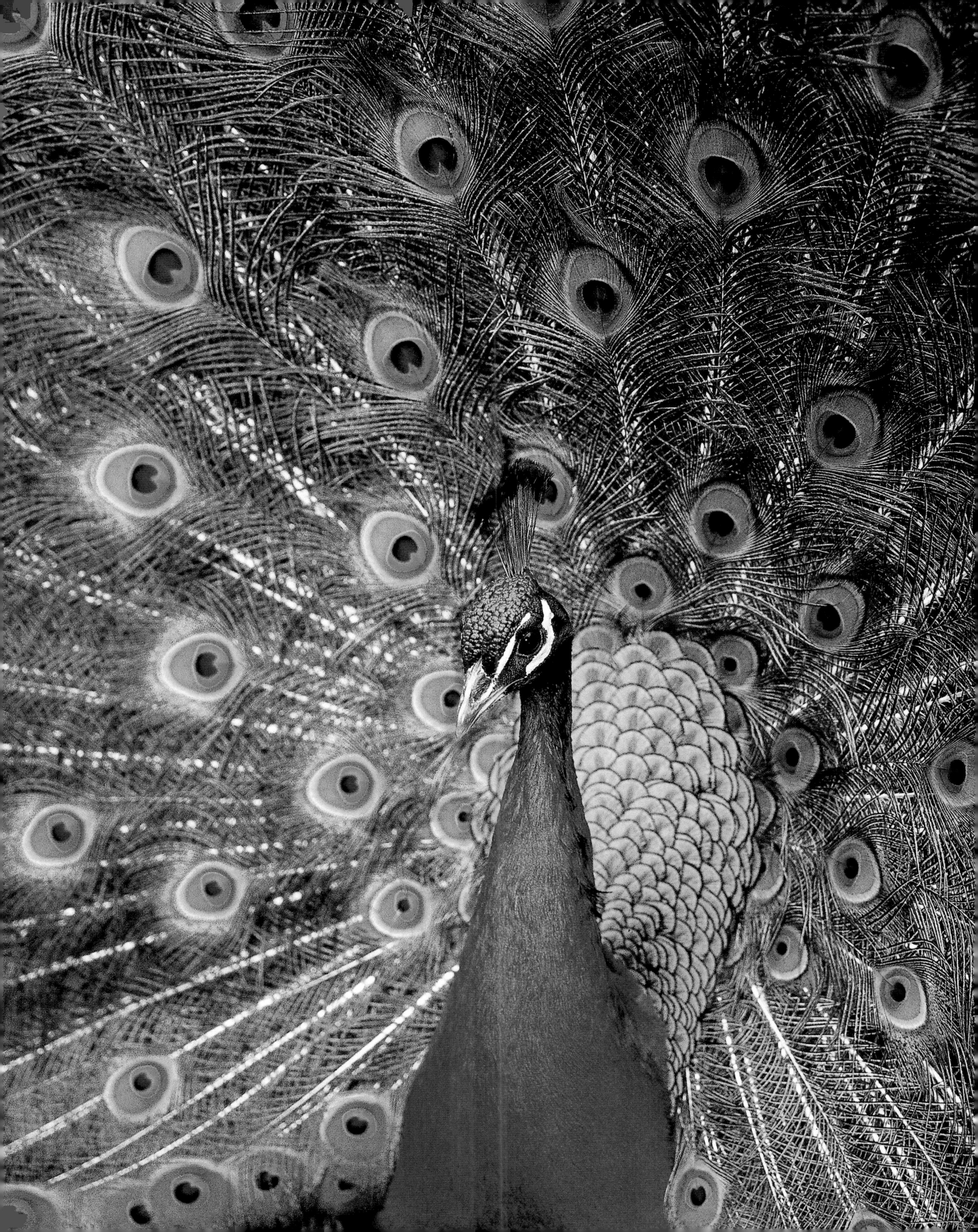

Grouse

GROUSE ARE CHARACTERISTIC OF THE NORTH. Their evolutionary origin is assumed to have been in northern latitudes, and much of their present distribution is in boreal forest and arctic tundra, in which grouse make up a major part of the ecosystems' vertebrate biomass. Their abundance and large size make them an important food source for carnivores such as lynx, martens, foxes, and raptors.

People too have long valued grouse for their meat. Millions of the birds are killed each year for food or sport. In many northern cultures, grouse-hunting plays a significant role in the subsistence of local communities, and grouse have become symbols of regional folklore. Tail feathers of the black grouse adorn the Scotsman's bonnet as well as the hats of Alpine villagers. Lekking black grouse are mimicked in traditional folk dances in the Alps, while some Native American peoples imitate the display of prairie chickens.

Coping with Seasonal Change

FORM AND FUNCTION

In shape, grouse are typical of the order Galliformes, resembling chickens and partridges. In size they range from pigeon to goose; the smallest is the White-tailed ptarmigan with a weight of about 300g (10oz), the largest the male capercaillie, weighing up to 6.5kg (14lb). Grouse are distinguished from other galliforms by their feathered feet and nostrils and the lack of spurs; in addition, their toes are feathered or have small scales along the sides during the winter, helping them to walk on or burrow into snow.

A number of morphological, physiological, and behavioral adaptations to climates with cold and snowy winters allow the grouse to live in environments of enormous seasonal change. Grouse roost in snow-burrows as an insulation against the cold; they feed on low-energy but abundant winter foods; they have large crops and gizzards that permit the holding of enormous quantities of food; they take up grit to ease the mechanical break-up of food; and they have long intestines with well-developed ceca that enable them to digest cellulose with the help of symbiotic microbes.

In some species, especially those with polygamous mating systems, the sexes look very different. The males have a more conspicuous plumage and may be up to twice the size of the smaller, cryptic, camouflage-colored females. Male grouse show bright yellow to red combs above their eyes, and some species have colored, unfeathered patches of skin on their necks (apteria) that can be inflated during courtship. In the monogamous species the differences are less pronounced, and the sexes look almost alike. Seasonal plumages are obvious only in the ptarmigans, which are white in winter; an exception is the Red grouse, the British subspecies of the Willow ptarmigan, which lacks a white winter plumage.

The natal plumage of grouse chicks is a dense, yellow-brown coat of down. The first juvenile plumage develops soon after hatching, and the rapid emergence of the wing feathers allows the chicks to fly short distances in their second week of life. Grouse live mostly on the ground; they break cover in a flurry of wings and fly in a long glide.

In Northern Climes

DISTRIBUTION PATTERNS

Grouse occur throughout the temperate, boreal, and arctic biogeographical zones of the northern hemisphere, and utilize a wide range of natural habitats of the northern Palearctic. Generally, each species is adapted to one or a few vegetation types, although some may use a variety of habitats. Their adaptations match different successional stages as well as different altitudinal and latitudinal zones: there are grouse specialized to alpine and arctic tundra (genus *Lagopus*), to the open grasslands of the North American prairies (*Centrocercus*, *Tympanuchus*), and to the various types and stages of forest from young regeneration to dense deciduous and open, old conifer forests (*Tetrao*, *Bonasa*, *Falcipennis*, *Dendragapus*). In many areas, several grouse species are sympatric, which is to say that they share the same, or at least use overlapping, habitats. Hybrids between some sympatric species are common, but they are usually infertile.

Just as their habitats are widely distributed, so many species of grouse inhabit extended areas within them. As a family, the grouse cover a range above 55° latitude between northern Greenland (the Rock ptarmigan) and the Gulf of Mexico (the Greater prairie chicken). The Willow ptarmigan has the largest range of all, extending in the subarctic and subalpine tundra of both Eurasia and North America from 76°N to between 47° and 62°N. The Rock ptarmigan has the widest latitudinal distribution; in the arctic it occurs as far north as 83°N in northern Greenland, while its southernmost alpine habitats are at 49°N in the Rocky Mountains and at 38°N in the Pamir mountains of Tajikistan. Most forest grouse similarly have extended ranges covering major parts of the boreal and temperate forest of Eurasia and North America. The original ranges of the prairie grouse were more limited, reflecting the natural extension of grasslands in North America.

Several species of grouse with restricted distributions have probably evolved through geographic separation. The Caucasian black grouse in the

Left above *The three ptarmigans grouped in the* Lagopus *genus are smallish, northerly grouse that have different plumages in summer and winter. This White-tailed ptarmigan is in the process of replacing its all-white winter coat with mottled summer feathers.*

FACTFILE

GROUSE

Order: Galliformes

Family: Tetraonidae

18 species in 7 genera: **Black-billed capercaillie** (*Tetrao parvirostris*), **Black grouse** (*T. tetrix*), **capercaillie** (*T. urogallus*), **Caucasian black grouse** (*T. mlokosiewiczi*), **Blue grouse** (*Dendragapus obscurus*), **Spruce grouse** (*D. canadensis*), **Chinese grouse** (*Bonasa sewerzowi*), **Hazel grouse** (*B. bonasia*), **Ruffed grouse** (*B. umbellus*), **Greater prairie chicken** (*Tympanuchus cupido*), **Lesser prairie chicken** (*T. pallidicinctus*), **Sharp-tailed grouse** (*T. phasianellus*), **Gunnison sage grouse** (*Centrocercus minimus*), **Sage grouse** (*C. urophasianus*), **Rock ptarmigan** (*Lagopus mutus*), **White-tailed ptarmigan** (*L. leucurus*), **Willow ptarmigan** (*L. lagopus*), **Siberian grouse** (*Falcipennis falcipennis*).

DISTRIBUTION N America, N Asia, Europe.

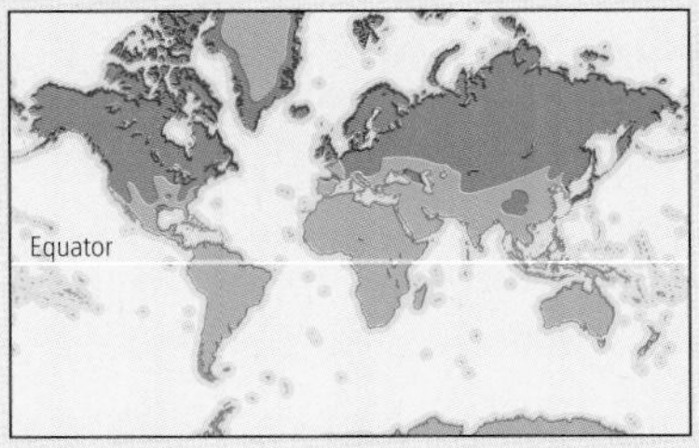

HABITAT Forest, prairie, tundra, heathlands.

SIZE Length 31–95cm (12–37in); weight 0.3–6.5kg (10.5oz–14lb); in some species sexes very different in coloration and size (the male capercaillie is twice the weight of the female).

PLUMAGE Males largely black or brown with white markings; combs red to yellow. Females brown and flecked with black and white. Ptarmigan white in winter. Wings short, rounded; tail of various shapes, often large.

VOICE A variety of hoots, hisses, cackles, clucks, clicks, and whistles; scraping and wing-beating also employed.

NEST A simple depression in the ground.

EGGS Usually 5–12, whitish to light brown and darkly blotched; weight 19–55g (0.7–1.9oz). Incubation 21–28 days, by the female.

DIET Adults eat leaves, needles, buds, twigs, flowers, fruits, and seeds; chicks largely eat invertebrates.

CONSERVATION STATUS The Gunnison sage grouse is Endangered.

Above *A Willow ptarmigan forages in long grass in its mottled summer plumage. Willow ptarmigans are among the most northerly of all grouse species, with a circumpolar distribution that includes northern Canada and Alaska as well as Eurasia.*

Below *Seeking to attract a mate, a male Black grouse displays by spreading its tail, raising the red wattles above its eyes, and half-opening its wings. Found across northern Eurasia, the species is famous for its leks – communal courting display grounds.*

Caucasus mountains between the Black and the Caspian Sea is separated by several hundred kilometers from the Black grouse; the Chinese grouse in central China is isolated by 1,000km (620 miles) from the range of its sister species, the Hazel grouse; the Siberian grouse in the Russian far east is separated from its closest relative, the Spruce grouse, by the Bering Sea. The Gunnison sage grouse in southern Utah and Colorado has only recently been recognized as a full species, and may have originated through geographic separation from the common Sage grouse.

Most grouse are resident and stay on their breeding range throughout the year. All species, however, show some movements between summer and winter habitats, and these may range in extent from localized habitat and altitudinal shifts to long-distance migrations. In some forest grouse such as the Spruce grouse and the capercaillie, many birds show restricted, undirectional movements of about 1–15 km (0.6–10 miles) between individual summer and winter ranges. In other species, movements are directional and related to seasonal variation in species-specific habitat and food availability. In the arctic, many ptarmigan populations are locally migrant. In mountainous areas, ptarmigan may move locally between high-elevation summer ranges and lower elevations in winter, whereas Blue grouse tend to breed at lower and winter at higher elevations. Migratory movements of a few to 100km (62 miles) occur in some prairie grouse populations. Seasonal migrations are most pronounced in the High Arctic, the region with the greatest seasonal change, where Rock and Willow ptarmigans may winter hundreds of kilometers south of their breeding ranges.

Making the Most of Poor Fare

DIET

The diet of the grouse is characterized by pronounced seasonal change. Grouse cope with the general shortage of food during the cold and snowy winters that are typical of most of their range by making the best use of what is available. Most rely on poor but abundant and accessible winter food sources. This strategy is best illustrated by the capercaillie and the Spruce, Siberian, and Blue grouse, which survive the winter almost exclusively on the needles of one or two conifer species. Although readily available throughout the birds' boreal and montane forest habitats, needles provide little energy, and the birds have to eat them in large quantities. Moreover, needles contain oils and resins that are distasteful or poisonous to other animals.

Other species feed on various winter food sources. Typical items are buds, twigs, and catkins of deciduous trees such as willow, alder, and birch, and, when accessible, parts of ericaceous (heath-dwelling) shrubs, mosses, grasses, and herbs. For some populations of prairie grouse, acorns are a common winter food, and in cultivated areas they also take grains such as soybeans and corn. For some species, access to winter food requires localized habitat shifts or even short or long seasonal movements.

During the snow-free seasons, all grouse are largely herbivorous and feed opportunistically but selectively on a variety of plants. With snow-melt in spring, the diet diversifies according to the accessibility of leaves, buds, flowers, and fruits in the ground and shrub layer. Common summer food plants are ericaceous shrubs and various herbs and grasses, but also trees such as willow, birch, and alder. In all species, chicks rely primarily on invertebrates, slowly shifting to a diet of vegetation as they grow older. Adult grouse occasionally eat animal matter, but quantities rarely exceed a few percent of the digested biomass.

The Wonderful Displays

BREEDING BIOLOGY

The mating systems of grouse range from monogamous pair bonds to polygamous leks with clusters of a few to several dozen males on traditional display grounds. The five prairie grouse species in the genera *Centrocercus* and *Tympanuchus* and forest-edge species (*Tetrao tetrix* and *T. mlokosiewiczi*) form leks with small male territories around 0.01ha (0.02 acres) in size that are used only for display. Among the forest grouse, two species – *Tetrao urogallus* and *T. parvirostris* (known by some systematists as *T. urogalloides*) – form leks with larger, permanent territories, up to 10–100ha (25–250 acres) in the case of *T. urogallus*. Of the rest, two species are largely monogamous (*Bonasa bonasia* and *B. sewerzowi*), and four (*B. umbellus*, *Falcipennis*, and the two *Dendragapus* species) are considered intermediate, with more dispersed male territories. The three tundra species (*Lagopus*) are essentially monogamous.

Below Representative species of grouse. **1** *Male Sage grouse* (Centrocercus urophasianus), *displaying at communal courting site (lek) with females in background.* **2** *Greater prairie chicken* (Tympanuchus cupido), *once found across the USA from the Atlantic coast to Wyoming, but now much reduced in range.* **3** *Sharp-tailed grouse* (T. phasianellus), *displaying the pointed tail from which it takes its name.* **4** *Spruce grouse* (Dendragapus canadensis), *which in winter survives largely on conifer needles.* **5** *Rock ptarmigan* (Lagopus mutus) *in its dark summer plumage.* **6** *Willow ptarmigan* (L. lagopus) *molting from its winter coat of white to its summer plumage of rusty red with white wings and belly.* **7a** *Male capercaillie* (Tetrao urogallus) *calling on lek, with* **7b** *drab-colored female adopting a mating invitation posture.*

In spring around the time of snow-melt, usually at dawn and at dusk, males contend for mates by a range of vocalizations such as hoots, hisses, cackles, clucks, clicks, and whistles, accompanied by displays of the neck, tail, and wing plumage and of colored air sacs in the neck, along with wing fluttering, drumming, display flights, tail rattling, and occasional fighting. Mating, courtship display, and most advertising take place on the ground. Females may attend several males before choosing a mate. In lekking species, most females mate with the same dominant male.

Grouse nest solitarily on the ground, and only the female incubates. The nest is a simple shallow depression, sparsely lined with vegetation from the immediate surroundings and often in good cover. Grouse lay one clutch per year, but replacement clutches may occur after egg loss. Females start laying within a week of copulation and produce one egg every 1–2 days. Grouse eggs resemble the eggs of domestic chickens in shape, and are yellowish-white with a few brown spots. Clutch sizes vary between 5 and 12 eggs; incubation takes 21–28 days and begins with the laying of the last or penultimate egg.

In the Willow ptarmigan, both sexes accompany and protect the brood; in all other species this is the exclusive task of the female. Chicks leave the nest shortly after hatching. During their first weeks, grouse chicks depend on high-energy food, and invertebrates comprise the major part of their diet. Broods stay together with the female until autumn. By then, young birds of most species have almost reached adult weights; in the capercaillies, however, males reach full weight only in their second year. All grouse are sexually mature as yearlings, although they may not breed at that time.

Outside the breeding season, the sociability of grouse is variable. In general, the more open the habitat, the more gregarious are the grouse. Forest grouse tend to be solitary, but do not strictly avoid each other and may form flocks in autumn and winter. The prairie grouse tend to be more social, while the tundra grouse may form winter flocks of over 100 birds.

Grouse produce large clutches of eggs, and thus have an enormous reproductive potential. However, the survival of chicks is highly variable between years. In years of poor weather conditions or great predation pressure, most chicks may die; in other years, many recruit into the next year's breeding population. Thus, grouse populations can fluctuate greatly from year to year in relation to random environmental factors.

A Case for Management

CONSERVATION AND ENVIRONMENT

Partly as a result of the birds' extended distribution and often remote habitats, the conservation status of grouse is less critical than that of other galliform taxa; only one of the 18 species, the Gunnison sage grouse, is globally threatened. However, the Chinese and the Siberian grouse are listed as near threatened, and the Caucasian black grouse can hardly be considered safe. At least two subspecies, Attwater's prairie chicken (*Tympanuchus cupido attwateri*) and the Cantabrian capercaillie (*Tetrao urogallus cantabricus*), also qualify as globally threatened according to the IUCN criteria. In addition, many populations of grouse are red-listed at national and regional levels.

Population declines and range contractions were reported for many species in the 20th century, and even before that time the prairie grouse had already lost much of their original ranges to farming and urban growth. Declines have been most severe in densely populated regions; for example, grouse have disappeared from large parts of central Europe and eastern North America. The loss and fragmentation of habitats due to human land use continue to pose major threats. In the forest grouse, declines are pronounced and widespread, as habitats are clear-felled. The tundra grouse, in contrast, still occupy most of their original ranges, being protected by their often remote habitats.

Healthy populations are most likely to occur in extensive landscapes with natural or semi-natural vegetation. Habitat management – common for game birds such as Ruffed grouse in North America and Red grouse in Britain – may help preserve and even increase populations. Because of their attractiveness to people, grouse are also suitable to serve as flagship species to promote ecosystem and biodiversity conservation measures. IS/JFB

Turkeys

BEFORE EUROPEANS CAME TO NORTH AMERICA the Wild turkey was widespread, and was hunted by native peoples. Equally favored as game by early settlers, it was proposed (unsuccessfully) as the US national bird by Benjamin Franklin. By the late 19th century, however, uncontrolled hunting and forest clearance brought the species close to extinction.

Thanks to conservation and reintroduction programs, the Wild turkey experienced a renaissance, and now has a much larger range, and is far more plentiful, than before European settlement. States across the USA maintain healthy population numbers through selective hunting seasons; the Wild turkey has regained its popularity as a game bird, but is now well protected.

Large Birds with Strong Legs

FORM AND FUNCTION

Turkeys are large birds with strong legs, which in the male have spurs. The two members of the family differ in plumage, especially the tail, and the spurs on the males' legs. They generally walk or run, but can fly strongly for short distances. Both species have similar plumages, but the much smaller Ocellated turkey lacks the "chest tuft" of bristles found on Wild turkey males and some females. Both species have naked heads (red and blue in the Wild and blue with red and yellow spots in the Ocellated turkey), bearing wattles and other ornaments used in displays. The spurs are longer and more slender in the Ocellated turkey. The characteristic eyespots on the more rounded tail of the Ocellated turkey give the species its common name.

The Domestic Bird's Travels

DISTRIBUTION PATTERNS

Once thought to require large, undisturbed forest tracks, the Wild turkey is now considered to be rather generalist in its habitats and foods. The bird seems to do best in mixed forest and agricultural areas; however, some subspecies have adapted to the arid climates of the southwestern USA and Mexico.

Hernán Cortes and his conquistadors introduced turkeys to Europe on their return from the New World in 1519, and within a few decades domestic turkeys were well established in several countries. Today's domestic turkey may have originated from either a southwestern US or a Mexican race of the Wild turkey. Conflicting accounts of its origin can probably be explained by the species having been domesticated a number of times by Native Americans in different locations.

Seeds, Nuts, and Berries

DIET

A wide variety of food items has been recorded in the diet of both turkey species. The bulk of the diet is made up of seeds and berries. Acorns are known to be an important part of the diet of the Wild turkey in parts of the United States, and the bird has a large, muscular gizzard to cope with such food items. In oak forests, characteristic, V-shaped scratchings in the leaf litter provide telltale signs of the presence of Wild turkeys. Foods vary greatly by season and location, with larger amounts of hard mast, such as acorns, being consumed in autumn and winter and more green plant material and invertebrates being consumed in spring and summer. A variety of other animals including spiders, snails, lizards, snakes, and salamanders are also consumed.

Left *The colorful Ocellated turkey, with its striking head ornaments, is native to southern Mexico and Central America. The Maya of the Yucatán peninsula kept this species in pens, fattening up the birds for celebratory meals.*

The Contest to Mate

BREEDING BIOLOGY

Wild turkeys are polygynous, one male mating with several females. Females are thought to start breeding at 1 year old, whereas males, although sexually mature, are often inhibited from breeding at that age due to competition from older, more experienced birds. The male birds go through an elaborate display to acquire mates. Spreading their tail fans, drooping and rattling their main flight feathers, and swelling their head ornaments, they strut up and down on traditional "strutting grounds," gobbling as they do so.

After mating has taken place, the females go off by themselves and build the nest. The nests are usually not far from the strutting grounds and are no more than leaf-lined scrapes in the ground.

FACTFILE

TURKEYS

Order: Galliformes

Family: Meleagrididae

2 species in the genus *Meleagris*: **Wild turkey** (*M. gallopavo*), 6 subspecies recognized; **Ocellated turkey** (*M. ocellata*), no subspecies.

DISTRIBUTION Wild turkey widespread in E USA, S Canada to Great Plains and S to Mexico (introduced into W USA, Hawaii, Australia, New Zealand, Germany). Ocellated turkey Yucatán through Guatemala only.

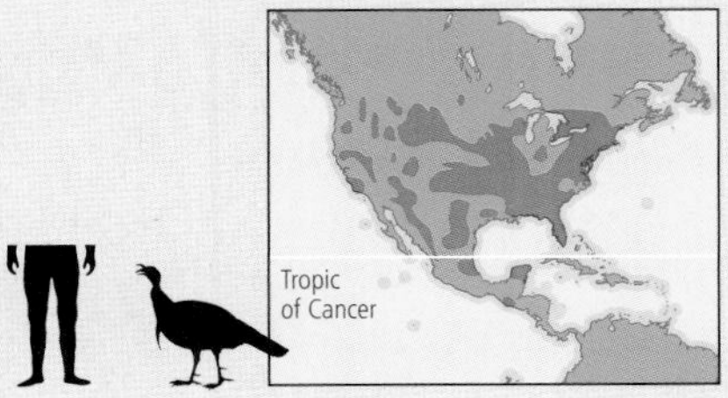

HABITAT Wide-ranging (woodland and mixed open forest preferred), extending from temperate to tropical climates.

SIZE Length for both species 90–120cm (3–4ft): weight 3–9kg (6.5–20lb), to 18kg (40lb) in some domesticated forms; males may be twice the weight of females.

PLUMAGE Generally dark, with brilliant, metallic reflections of bronze and green, especially in males; major secondary wing coverts iridescent copper-colored (Ocellated turkey); head and neck naked. Breast feathers of males tipped in black, whereas females are buff tipped (Wild turkey).

VOICE A variety of gobbles and clucks.

NEST Well concealed, on the ground; built by female.

EGGS 8–15, cream-colored, speckled with brown. Incubation 28 days; young leave nest usually after 1 night.

DIET Mainly seeds, nuts, acorns, fruits, tubers, and agricultural crops such as maize, but also invertebrates and small vertebrates.

CONSERVATION STATUS The Ocellated turkey is listed as Lower Risk/Near Threatened.

Above *Fanning out his imposing tail feathers to attract females to his harem, a male Wild turkey struts around his woodland habitat. Such displays take place during the spring.*

Although clutch sizes range from 8 to 15 eggs, one nest may have 20–30 eggs, as more than one female will often lay in the same nest. The female alone incubates the eggs, and if she leaves the nest, even for a short period, she will make sure that they are covered.

The precocial (well-developed) young are cared for by the female for their first two weeks, and in the evening they are brooded on the ground. Once the young can fly they roost in trees, and are often still brooded by the hen. Chicks grow very quickly on a diet consisting mainly of insects.

The brood flock remains together until the young are around 6 months old, when the males separate off to form all-male flocks. The males in such a sibling group are inseparable – even a solitary male will not try to join them. The juvenile sibling groups usually form flocks of their own, as the older males can normally chase off the younger birds from the larger group. This is a tough time for the young male, as he has to do a lot of fighting, both to determine his dominance among his siblings and to help determine his group's status within the flock. Fights can be very vicious, involving the use of wings and spurs. Contests can last for up to two hours, and fights to the death have been recorded. Once dominance has been established within a sibling group, however, it is rarely challenged. Between groups, fights are usually won by the larger unit and, again, once dominance has been established there appears to be a fairly stable society.

Females, too, establish rank, but it does not appear to be anything like as overt as among the males. In general, older females are dominant to younger birds, and those females from sibling groups that are accustomed to winning contests also seem to win individual contests.

Towards the start of the breeding season the large male flocks break up, but the sibling groups stay together. At the strutting grounds, males of the dominant groups obtain most of the matings.

Very little is known of the social behavior of the Ocellated turkey. However, it is thought to be gregarious all the year round, and flies more readily than it runs when disturbed. JC/MEB

Guineafowl

GUINEAFOWL ARE SEDENTARY BIRDS ENDEMIC to Africa. The Helmeted guineafowl, which in its native habitat is easily recognized as a noisy bird of savannas and semi-desert areas, is also familiar on farms and dinner tables around the world. The other five species are much less well-known, especially the three that rarely venture out of dense rain forest.

The most obvious feature of all species is a virtually naked, richly pigmented head that is usually adorned with wattles and/or a distinctive casque or crest. The Crested and Helmeted guineafowl show some geographical variation in the extent, size, and shape of these adornments, and it seems likely that they help to regulate brain temperature in a range of different climates.

Noisy and Gregarious

FORM AND FUNCTION

The White-breasted and Black guineafowl have unspotted or lightly vermiculated black plumage, red heads, rudimentary wattles, spurs on the ankle or tarsus, and more musical alarm notes than other species. They are the least gregarious guineafowl, associating in parties of less than 10, and are more or less confined to primary rain forest. All other species are primarily black in coloration with white spots, have blue or grayish-blue heads (most with well-developed wattles), are unspurred, and cackle raucously. They are also much more gregarious, especially the Helmeted guineafowl.

The Plumed and Crested guineafowl have distinctive feathered crests, of long, straight feathers in the Plumed, and shorter, curly, downy feathers in the Crested species. The characteristic low pitch of their cackle is perhaps due to the fact that their windpipe (trachea) loops through a hollowed-out extension of their "wishbone" (furcula), which may act as a resonant organ. Both species occur in flocks of 10–30. The Plumed occurs mainly, but not exclusively, in primary rain forest, but the Crested guineafowl prefers secondary growth, forest edges, woodland, and gallery forest,

MOCK FIGHTS AND "DATING" IN THE HELMETED GUINEAFOWL

The break-up of Helmeted guineafowl flocks at the onset of the breeding season results from an increase in aggression between males, mainly in the form of ritualized chasing. In such interactions, one male approaches another side-on in a characteristic, hump-backed display posture **1**. In this display (repeated by the males in courtship), the wings are compressed into the body and elevated, to give the impression of a much larger bird. The approach elicits pursuit by the second male (who also assumes the hump-backed display), but the chaser rarely catches up with the initiator even if the latter slows down. Such chasing is a contagious activity, with up to eight males running in single file. Females view these ritualized chases and presumably assess the stamina of potential mates by seeing which of the rivals can keep up the chase for the longest period of time.

At the same time as this increase in chasing, males and females form short-term pairs. This "dating" behavior probably allows females to compare potential mates more rigorously. After 2–3 weeks stable pairs form, usually lasting until the female begins to incubate the eggs. Although males and females are closely similar, the male is easily identified, since he spends most of his time sitting and resting in alert postures **2**, or in aggressive encounters with males who approach his hen. At this stage of the breeding season he is particularly aggressive, and chasing often leads to fighting (in captivity, where there is no escape route, this can end in the death of subordinate males). The female of a stable pair does little more than feed and preen.

Once the hen begins incubating the eggs (she does all the incubation), the male temporarily deserts her, since he can be sure of the paternity of the eggs and of his hen's commitment to hatching them. He then associates with other females, or may even "rape" solitary hens, forcing copulation upon them without preliminary display. However, when the keets (chicks) are about to hatch, he returns to his original mate and helps to rear them, especially during their first two weeks. If the male is absent at this time the brood will almost certainly fail, since the hen cannot both care for the keets and find food for herself to recoup the energy lost during incubation. TMC

1

2

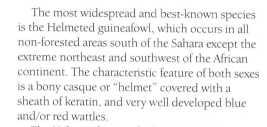

Right *1 The Crested guineafowl, which takes its name from the tuft of feathers on its head, is a bird of scrub forest.* ***2*** *The Helmeted guineafowl prefers open savanna, where dominant males act as scouts for flocks, striking alert postures to watch out for predators.*

Left *Two Vulturine guineafowl forage for seeds and grain in the Samburu Game Reserve, Kenya. These are the largest guineafowl, distinguished above all by their hackle of black, white, and blue feathers. They owe their name to the resemblance of their heads to those of Old World vultures.*

and can even be found in thickets. It will sometimes venture out into the open to feed.

The Vulturine guineafowl is the largest and most striking species, with a distinctive band of chestnut, downy feathers running from ear to ear across the back of its head. It also has red eyes (they are brown in most other species), rudimentary wattles, a well-developed hackle of long, lanceolate feathers striped black, white, and a striking iridescent blue, and several bumps (not spurs) on the tarsus that are conspicuous in courtship displays. It occurs, in groups of 20–30, in the subdesert steppes of northeast Africa, but is reluctant to forage in or even cross large stretches of open ground. It apparently does not need drinking water.

The most widespread and best-known species is the Helmeted guineafowl, which occurs in all non-forested areas south of the Sahara except the extreme northeast and southwest of the African continent. The characteristic feature of both sexes is a bony casque or "helmet" covered with a sheath of keratin, and very well developed blue and/or red wattles.

The Helmeted guineafowl is the family's most gregarious species. In the non-breeding season (the dry season in tropical climes and the winter in temperate areas), gatherings of over 2,000 birds have been observed, but even such large flocks rarely venture more than 2km (1.2mi) from a key resource, which may be a water hole, a roosting tree (or trees), or an important foraging patch.

In the early morning the birds move in single file from their roost to a supply of water. The dominant males that usually lead are probably acting as "scouts," since they spend a much higher proportion of their time than other flock members in alert postures. Later, the flock may advance in line abreast, presumably in a sweep for food. When a potential predator such as a jackal, baboon, or snake approaches a flock containing young birds, the group adopts a third formation, the swarm; in this defensive cluster, the vulnerable young occupy a position in the center or on the far side of the flock, where they are protected by older birds.

All six species are very noisy birds, especially when alarmed. They forage entirely on the ground, mainly for seeds, bulbs, roots, tubers, and fallen grain, although most also eat invertebrates, especially in the breeding season. However they always roost in trees.

The breeding habits of all species except the Helmeted guineafowl are almost unknown. The birds seem to be monogamous, and nest on the ground in rudimentary scrapes that may be thinly lined with grass. Breeding occurs mainly following rain, at which point pairs leave the flock.

FACTFILE

GUINEAFOWL

Order: Galliformes

Family: Numididae

6 species in 4 genera: **Black guineafowl** (*Agelastes niger*), **White-breasted guineafowl** (*A. meleagrides*), **Crested guineafowl** (*Guttera pucherani*), **Plumed guineafowl** (*G. plumifera*), **Vulturine guineafowl** (*Acryllium vulturinum*), **Helmeted guineafowl** (*Numida meleagris*).

DISTRIBUTION Africa S of the Sahara, with an isolated race (probably extinct) of Helmeted guineafowl in NW Morocco.This species has also been introduced to SW Arabia and Madagascar; it is widely domesticated, and there are some feral populations.

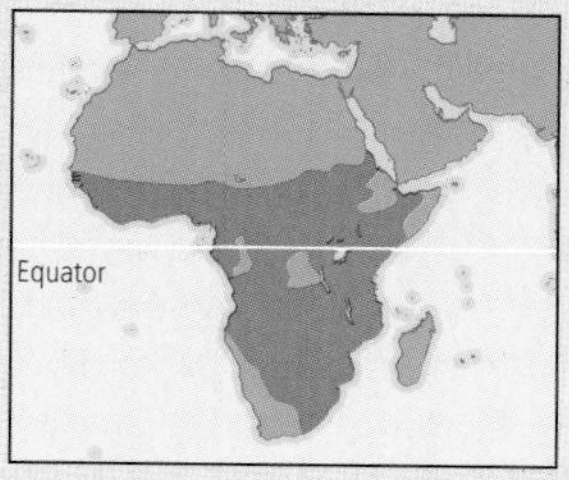

HABITAT Subdesert steppe to primary rain forest.

SIZE Length 39–56cm (15.5–22in); weight 0.7–1.6kg (1.6–3.5lb); males same size or slightly larger than females, although latter may be slightly heavier at times.

PLUMAGE Four species are black, spotted with white; two (*Agelastes*) are unspotted.

VOICE Mostly harsh, loud cackling; a few more musical ringing notes.

NEST Simple scrapes on ground, lined with leaves, grass, or feathers.

EGGS Usually 4–12; white to pale brown, pitted darker; weight 38–40g (1.4oz). Incubation 23–28 days, by the female only; young forage within 1–2 days of hatching, and can fly short distances at 2–3 weeks.

DIET Highly opportunistic; chiefly seeds, bulbs, tubers, roots, and fallen grain in drier times of year, although the birds prefer insects and other invertebrates (often crop pests) when available.

CONSERVATION STATUS The White-breasted guineafowl is listed as Vulnerable.

Declining Numbers

CONSERVATION AND ENVIRONMENT

The Helmeted guineafowl readily ventures into gardens, and thrives on land cultivated with wheat and corn (maize), although in areas of intensive cultivation, such as parts of KwaZulu-Natal, numbers are now declining quite sharply. The isolated Moroccan population is now probably extinct, no birds having been seen since the late 1980s. Its demise was mainly due to hunting and habitat destruction. Otherwise, the only species requiring urgent conservation action is the White-breasted guineafowl, which is severely threatened by the destruction of primary forest across its range in fragmented patches of West Africa. PL/TMC

Megapodes

ALTHOUGH MEGAPODES APPEAR TO BE TYPICAL robust gamebirds of the rainforest floor, their unique incubation techniques set them apart from all other galliforms. Indeed, their exploitation of a variety of naturally occurring sources of environmental heat for the incubation of their eggs is the most astounding aspect of the family.

The megapodes' highly unusual approach to incubation has fascinated naturalists and scientists since the first discovery of the birds in the 16th century. We now know that the methods the birds use have far-reaching influences upon almost every aspect of their reproduction and behavior.

Big-footed Birds

FORM AND FUNCTION

The family Megapodiidae consists of 22 extant species in seven genera, forming the following groups: the brush-turkeys (three species in *Alectura* and *Aepypodius*); the talegallas (three species in *Talegalla*); the megapodes (14 species in *Megapodius*, and also including the single species of *Eulipoa*); and two distinctive species, the malleefowl and the maleo. Recent archaeological studies from sites throughout Oceania suggest that up to 33 additional megapode species have become extinct in the region within the last few thousand years, almost certainly as a result of human activity.

Megapodes are morphologically similar to other galliforms, the group from which they evolved. Their body shape is typically compact, although their legs are long. Their toes and claws, adapted to raking and digging, are also relatively large, giving the family its name: megapode stems from the Greek for "large feet." All species are ground-dwellers, foraging for a wide variety of insects, fruits, and other edible foods on the forest floor. In most species the plumage is generally drab in browns, grays, and blacks, with relatively little in the way of patterning. The notable exception to this generalization is the malleefowl, which has complex patterns of brown, white, and black; this is the only species to live in an arid environment, and the plumage clearly aids in camouflage.

While the plumage is generally dull, most species have virtually featherless areas on their head and necks where the skin is yellow, red, or blue, although the colors are not usually bright. However, one group of megapodes, the brush-turkeys of the genera *Alectura* and *Aepypodius*, possess vividly pigmented heads and necks, with the males of these species bearing a variety of combs

Above *The Orange-footed megapode is the most widely distributed species in the family, with a range that extends over 35° of longitude from central Indonesia to the Trobriand Islands east of New Guinea.*

Right *The Australian brush-turkey is one of the bigger megapodes, at up to 70cm (28in) long. Its brightly colored head contrasts with the generally dull hues of most of the family.*

FACTFILE

MEGAPODES

Order: Galliformes

Family: Megapodiidae

22 species in 7 genera. Species include: **Melanesian megapode** (*Megapodius eremita*), **Micronesian megapode** (*M. laperouse*), **Orange-footed megapode** (*M. reinwardt*), **Polynesian megapode** (*M. pritchardii*), **Australian brush-turkey** (*Alectura lathami*), **maleo** (*Macrocephalon maleo*), **malleefowl** (*Leipoa ocellata*).

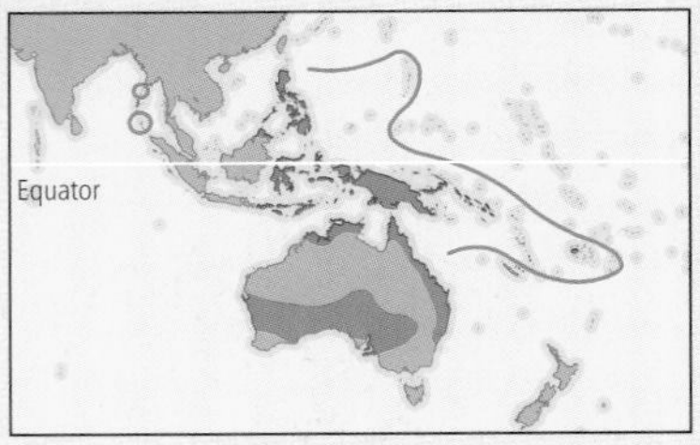

DISTRIBUTION From the Nicobar Islands through the islands of E Indonesia (but not on larger islands or mainland), Marianas and Palau, New Guinea and Australia, and reaching Tonga.

HABITAT Primary and secondary rain forest, monsoonal scrub, dry closed-forests, and (for the malleefowl only) arid low eucalypt woodland.

SIZE Length 27–70cm (10.6–28in); weight 0.29–2.95kg (0.6–6.5lb).

PLUMAGE Generally browns, grays, and black; some species with colored bare skin on necks and heads, and a few with neck sacs and wattles in blue, red, and yellow. Plumage differences between sexes usually slight, although seasonally pronounced in brush-turkeys.

VOICE Unmusical cackles and grunts, some producing booms by inflating neck sacs. Complex duetting in several species.

NEST Eggs laid in incubation mounds, or holes and burrows in sun-warmed soil or volcanic areas.

EGGS White and brownish, in some species pink with chalky covering; laid singly throughout season, between 12 and 30 per female.

DIET Generalist feeders, foraging among litter for insects, fruits, and succulent roots.

CONSERVATION STATUS The Polynesian megapode is now listed as Critically Endangered, and the Micronesian magapode as Endangered; 7 other species are Vulnerable.

and wattles that typically expand and brighten in color during the breeding season. In the case of the Australian brush-turkey, the wattle serves not just as a visual signal but is also inflated to produce a deep, booming call. The brush-turkeys are the only species in which sexual dimorphism is evident; in the others, males and females are usually indistinguishable.

Away from Predators

Distribution Patterns

The present distribution of the megapodes is centered on Australia, New Guinea, and the innumerable islands of Southeast Asia, Micronesia, the Philippines, and the western South Pacific. The easternmost limit of their distribution is the remote island of Niuafo'ou in the kingdom of Tonga, although recently extinct species are also known from American Samoa and Niue. The westernmost megapodes are found on the Nicobar and Andaman Islands in the Bay of Bengal, a considerable distance from any other species. Throughout this large area, it is species of the genus *Megapodius* that are found in the most remote and farflung locations, with one species, the Orange-footed megapode, having by far the largest distribution, stretching from Lombok in Indonesia to Trobriand Island east of New Guinea. (There are numerous authenticated reports of juveniles of this species landing on ships far from land during the night). Undoubtedly, *Megapodius* species are the long-range colonizers of the family, representatives having reached virtually all of the islands of the southwestern Pacific. *Megapodius* species also make up the bulk of the recent extinctions mentioned above.

The major question associated with the distribution of the family is the almost complete absence of any megapodes from larger islands such as Java, Sumatra, and most of Borneo, or from the mainland of Southeast Asia. Having originated in the New Guinea/northern Australia region, megapodes seem to have expanded outward in all directions without being able to colonize the Asian mainland or large islands of the Oriental zoogeographic province. The favored explanation for this surprising fact is the presence of carnivorous predators, especially cats (Felidae) and civets (Viverridae), in these areas. The prolonged presence of megapodes near their incubation mounds makes the birds extremely susceptible to such predation; indeed, there is an almost perfect geographical relationship between the presence of these predators and the absence of megapodes in the region of potential overlap.

Thermal Engineers

Breeding Biology

Every aspect of megapode reproduction is influenced by the species' preferred method of incubation. Three sources of environmental heat are utilized: geothermal activity; solar radiation; and the decomposition of organic matter. In general,

Left *Few chicks have a harsher entry into the world than megapodes like this young Australian brush-turkey emerging from a nest mound. They must dig their way to the surface unaided – a process that can take several days – and then fend for themselves unaided by their parents, whom they never see.*

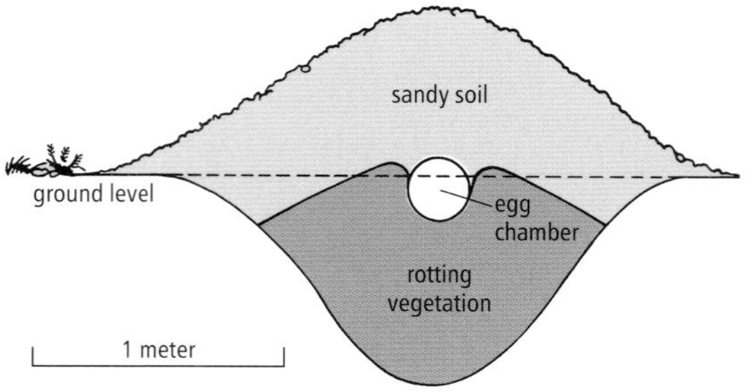

Above *The bizarre nest-mounds of the megapodes are unparalleled among birds. In southern Australia the malleefowl male works on the nest in most months of the year, digging out a hole, then scraping leaf litter into it and covering it with sandy soil: alternatively, an already existing mound may be worked on. Mounds used year after year can reach 5m (over 16ft) in diameter. The female lays eggs in chambers excavated by the male. Rotting vegetation provides heat needed for incubation. Once laid and covered, the eggs themselves are ignored by both parents, as are the chicks, which hatch in 50–90 days.*

Right *Living in the arid lands of central Australia, malleefowl have to work harder on their mounds than any other megapodes. The mounds themselves must be dug deep into the soil to protect the eggs against temperature fluctutations; and throughout incubation the birds have to regularly adjust the thickness of the soil cover to keep them at a constant 34°C (93°F).*

incubation heat from the first two sources, which occurs independently of the birds, is accessed by burrowing into the soil or sand until a suitable laying site – consisting of an appropriate substrate exhibiting the desired temperature – is found.

Species using sun-warmed beach sand may simply excavate a shallow hole, lay an egg, and depart after filling it up again. Those using geothermal heat typically make use of extensive, more or less permanent burrows. Such sites are usually unevenly distributed, and large numbers of birds may congregate at them at certain times of the year to lay their eggs. For example, approximately 53,000 Melanesian megapodes have been estimated to visit a single volcanic site in New Britain. These traditional incubation areas may be located in habitats some distance from the normal range of the species, requiring significant travel for the breeding birds.

In contrast, the species utilizing the heat of decomposing organic matter actually provide their own incubation site through the construction of a mound of moist leaf litter and soil. By the deliberate siting of these mounds to maximize the availability of suitable material and to prevent desiccation, and through daily maintenance, mound-building megapodes are able to control the incubation conditions within.

The species that is most accomplished at mound manipulation is unquestionably the malleefowl. The only megapode to live in an extremely dry environment (in central Australia), it has evolved a sophisticated – and arduous – method of mound construction and maintenance. Careful studies during the 1950s showed that malleefowl are capable of detecting the internal temperature of the mounds to within a few degrees and altering it when necessary. This ability involves the frequent removal and replacement of several tons of sand and soil, needed to prevent the drying out of the damp leaf litter within.

Mound-building is the most common method of incubation among the megapodes, employed by all but three species. The mounds are certainly among the largest constructions of any noncolonial animal. Nonetheless, at least five mound-building species are also known to employ burrow nests in locations where either geothermal areas or suitable beaches are accessible. In New Guinea, some species also appear to lay eggs parasitically in the mounds of other megapode species.

The mating system of most megapodes appears to be monogamous, with paired birds remaining together virtually continuously and sharing in the work of mound construction. Nonetheless, there are numerous reports of polygyny even among these predominantly monogamous species.

In contrast, the brush-turkeys are distinctly non-monogamous. Males defend their mounds from other males, and accept any females that are willing to mate with them and to lay in their incubation sites. In these species, the females appear to be free to choose among the numerous males available, yet rather than mating promiscuously they practice so-called "serial monogamy," mating with one male and laying in his mound before moving on to the next.

Regardless of the type of incubation site – mound or burrow – or the heat source employed, the incubation conditions within a site are typically very similar among all species. In general, the temperature of the substrate surrounding the eggs is between 32° and 35° C (90°–95°F), although there may be considerable daily, seasonal, and individual variation. The incubation conditions are, however, remarkably different from those favored by most other birds, featuring extremely high humidity and low levels of oxygen alongside levels of carbon dioxide that would appear to be critically high.

The physiological adaptations to these conditions that are evident in both the embryos and shells are among the most notable features of the family. For example, megapode eggshells are relatively thin for their size, and the large pores

change shape during the development of the growing chick. These features have been shown to greatly facilitate water loss and gas exchange.

Megapode eggs are relatively large, weighing 75–230g (2.5–8oz), representing 10–22 percent of the female's body weight per egg. They are also extremely rich in yolk, which constitutes 48–69 percent of the whole. The eggs are produced singly and are laid at intervals of several to many days. Because the incubation site is always warm, each embryo begins to develop immediately upon being deposited. Chicks therefore hatch independently, and hatchlings may emerge from the site continuously for several months. Females may lay between 12 and 30 eggs per season.

An additional adaptation to underground incubation is shown at hatching. Unlike most other birds, megapode eggs do not form an air space, normally essential for the first breaths of air made by the fully-developed embryo immediately prior to hatching. In contrast to the typical, slow emergence of other birds' chicks, megapodes hatch explosively, forcing the shell apart with legs, back, and heads. Fluid rapidly drains from the lungs, which then inflate remarkably quickly.

Hatching, however, is only the first challenge facing the young megapode. On emerging from the egg, most find themselves between 30 and 120cm (12–48in) underground, and must therefore dig themselves out. This process, undertaken alone and without assistance, may take anything from a few hours to several days, depending on the depth at which the eggs were buried and the nature of the substrate.

Things do not improve upon reaching the surface. Again, megapode hatchlings receive no aid – or even recognition – from their parents, and are obliged to leave the incubation site quickly. Over the next few weeks or months, the young megapode lives an almost entirely solitary life. It must find food and water, thermoregulate, and evade predators, all without the protection and guidance of adults. Not unexpectedly, the mortality of megapode hatchlings is extremely high.

Eggs at Risk

CONSERVATION AND ENVIRONMENT

All megapode species have had a close relationship with humans, with their eggs providing an important source of food for many indigenous peoples. In many places, the harvesting of eggs has been practiced sustainably for millennia. However, overexploitation has also occurred, leading to the extinction of many species in the past and threatening numerous species today. In addition, the threats of habitat destruction and of predation by introduced animals such as cats, pigs, and foxes are major problems.

Currently the most threatened species is the Critically Endangered Polynesian megapode, which survives only on a tiny volcanic island in the Tonga group. Attempts have been made to translocate populations to two adjoining islands, but it is not yet clear if these have become established. In addition, the Micronesian megapode, from the Mariana Islands, is Endangered, and seven other species are Vulnerable. DJ/FHJC

Guans and Curassows

THIS CENTRAL AND SOUTH AMERICAN FAMILY comprises three groups, the chachalacas, guans, and curassows. They are big-bodied birds with smallish heads, thin necks, short, rounded wings, and long, broad tails. Unusually for game birds, they are chiefly tree-dwellers, although they generally descend to the ground to feed.

FACTFILE

GUANS AND CURASSOWS

Order: Galliformes

Family: Cracidae

50 species in 11 genera. Species include: **Black-fronted piping guan** (*Pipile jacutinga*), **Trinidad piping guan** (*P. pipile*), **Chestnut-bellied guan** (*Penelope ochrogaster*), **Crested guan** (*P. purpurascens*), **White-winged guan** (*P. albipennis*), **Helmeted curassow** (*Pauxi pauxi*), **Horned curassow** (*P. unicornis*), **Highland guan** (*Penelopina nigra*), **Horned guan** (*Oreophasis derbianus*), **Nocturnal curassow** (*Nothocrax urumutum*), **Plain chachalaca** (*Ortalis vetula*), **Great curassow** (*Crax rubra*), **Red-billed curassow** (*C. blumenbachii*), **Sickle-winged guan** (*Chamaepetes goudotii*), **Wattled guan** (*Aburria aburri*).

DISTRIBUTION In the Americas, from extreme S USA (Texas) through N Argentina.

HABITAT Dense tropical forests, low riverside woods and thickets.

SIZE Length 42–92cm (16.5–36in); weight 385g–4.3kg (13.5oz–9.5lb).

PLUMAGE Chiefly plain brown, or black with white patches. Many species crested, some with casques or other bill ornamentation. Wings blunt; tail long and broad.

VOICE A variety of raucous moans and calls, booming notes, and whistles, often repeated.

NEST Usually of twigs and leafy vegetation in trees.

EGGS 2–4; dull white or creamy; weight 62g (2.2oz). Incubation 21–36 days, typically by female.

DIET Chiefly fruits, berries, seeds, leaves, buds, flowers; some also take small animals or insects.

CONSERVATION STATUS 3 species are currently listed as Critically Endangered, 4 as Endangered, and 7 as Vulnerable. In addition, the Alagoas curassow (*Mitu mitu*) from northeastern Brazil, which has not been definitely observed since the late 1980s, is thought to be Extinct in the Wild.

All the cracids are noisy birds – necessarily so, to maintain contact in the dense and often dark forests where they live. The windpipes of some species, notably the guans, are adapted for amplifying calls that are among the loudest and most far-reaching found in any bird species. Curassows utter one or two booming or whistling notes.

Game Birds of the American Tropics

FORM AND FUNCTION

The 11 chachalaca species (genus *Ortalis*) live fairly close to human settlements and are conspicuously gregarious, living in flocks of up to 20 or 30 birds. They are the smallest and dullest of the cracids, being generally plain brown with bare patches on the throat, and there is virtually no sexual dimorphism. They are mostly ground feeders, their plumage providing excellent camouflage, but they take to the trees at the first sign of danger. They prefer low brush woodlands and wooded riverbanks, which has enabled one species, the Plain chachalaca, to survive in the remnant forests of the lower Rio Grande in southern Texas. Whole

Above *A striking feature of many guans is their conspicuous pink wattle, as demonstrated here in a Chestnut-bellied guan. Smaller in size than the curassows, the guans can nonetheless attain weights of 1.9kg (4.2lb) in large species. Most of the 24 species are grouped in the genus* Penelope.

Below *Brown plumage and a reduced head crest mark this Great curassow perching on a tree in Costa Rica as a female. Oddly, the birds take their name from Curaçao, where they do not occur; apparently the first live examples to reach Europe came via the Caribbean island en route from the South American mainland.*

flocks usually call together, especially at dawn or dusk, when a rhythmic, repeated "cha-cha-lac-a" reverberates through the forest.

Guans are larger and more colorful than the chachalacas, with some whitish edges to the body feathers, which range from brown to black, often with a glossy green sheen on the back and wings; many have long crown feathers that form a crest. The outer primaries are rather spinelike in several species, strengthened and curved, and produce a peculiar drumming sound when the wings are vigorously shaken. These feathers are most developed in the Wattled and Sickle-winged guans and the four piping-guan species. The spectacular drumming of their display flight through the treetops is augmented by deep, raucous cackles.

Guans are the most widespread of the Cracidae. The 15 species of the genus *Penelope* are considered to be typical; though tree-dwelling, they too feed on the ground. More specialized and arboreal are the five species in the exclusively South American genera *Pipile* and *Aburria*, which have shorter, less powerful legs and a well-developed wattle on the throat. The two species in the genus *Chamaepetes* are smaller and lack wattles. The two remaining species are both restricted to Central America. The Highland guan is unique in that the female is larger than the male and differs in plumage: the female is brown and the male is black. The Horned guan is the most distinctive, but also shows features of the curassows, to which it is probably closely related. A cylindrical horn, 5cm (2in) long, rises from the center of its crown.

Curassows are the largest and heaviest members of the family and are poor fliers, spending most of their time at ground level. They range in plumage from deep blue to black, invariably with a purple gloss, and the majority have rather curly crests. Their distinguishing features, especially in the genus *Crax*, are the head or facial adornments of wattles and knobs, which vary from yellow to bright crimson and blue in color. The Helmeted and Horned curassows have "horns" on the forehead that are used in elaborate courtship displays. The Razor-billed curassows have more reduced ornamentation, terminating in a sharp, razorlike ridge. The Nocturnal curassow, with its chestnut-colored plumage and red and blue facial skin, is one of the most colorful members of the whole family, yet is entirely nocturnal.

All of the Cracidae are mainly vegetarian, eating principally fruit but also leaves, buds, and flowers; some also take small animals, large insects, or frogs. The chachalacas and curassows, with their long legs, big feet, and strong claws, scratch the litter on the forest floor in chickenlike fashion. Curassows are able to eat nuts and tough seeds by swallowing small stones that aid digestion.

Nests are often built in a tree concealed by cover, and are usually fragile structures that are quite small in relation to the size of the adult bird. The eggs are rather large, and are smooth in some of the guans (genus *Penelope*), but rough and pitted in the chachalacas and most of the curassows. Females care for the young, which can leave the nest after only a few hours and are quick to develop their flight feathers. Indeed, the young of some species are able to fly within a few days.

Easy Targets

CONSERVATION AND ENVIRONMENT

Most cracid species are relentlessly hunted for food and "sport," their tameness and inability to fly far or fast making them easy targets. Over large tracts of land, the rapid destruction of tropical rain forest is also threatening this little-known family of birds.

The White-winged guan was thought to have become extinct in 1870, but was rediscovered in 1977. Estimates of its population do not exceed 200 birds, and most of those live in an area threatened with logging activity. The Red-billed curassow is also close to extinction; at one point it was down to less than 100 individuals, but a captive breeding and reintroduction program has proved successful in increasing the population to some degree. The most precarious of all cracids is the Alagoas curassow, which is extinct in the wild. Approximately 50 individuals exist in captivity, and it is hoped that their progeny may be reintroduced into the wild some day. DB/TWP

Above Representative species of guans and curassows: **1** Great curassow, the largest species; the black coloration and yellow cere (the protuberance at the base of the bill) show it to be male. **2** Horned guan, displaying its unique horn. **3** Trinidad piping guan, one of four piping guan species of the genus* Pipile.

Cranes

Left *Cranes extend their neck and legs in flight to create a streamlined shape. They are strong and high fliers; Common cranes (shown here) migrate over the Himalayas at altitudes of more than 9,000m (29,500ft), the cruising height of jet airliners.*

Right *A Grey-crowned crane (*Balearica regulorum*) displays its magnificent crest. The two* Balearica *crowned crane species are evolutionarily the oldest cranes, and also the only ones to roost in trees.*

Right below *Despite their name, Blue cranes look grayish when seen at a distance in their natural habitat. The species, which is mostly restricted to South Africa, has declined rapidly in numbers since the 1970s, and is now classed as Vulnerable.*

FACTFILE

CRANES

Order: Gruiformes

Family: Gruidae

15 species in 4 genera. Species include: **Blue crane** (*Anthropoides paradisea*), **Black-crowned crane** (*Balearica pavonina*), **Siberian crane** (*Bugeranus leucogeranus*), **Wattled crane** (*B. carunculatus*), **Black-necked crane** (*Grus nigricollis*), **Common crane** (*G. grus*), **Red-crowned crane** (*G. japonensis*), **Sandhill crane** (*G. canadensis*), **Sarus crane** (*G. antigone*), **White-naped crane** (*G. vipio*), **Whooping crane** (*G. americana*).

DISTRIBUTION All continents except S America and Antarctica.

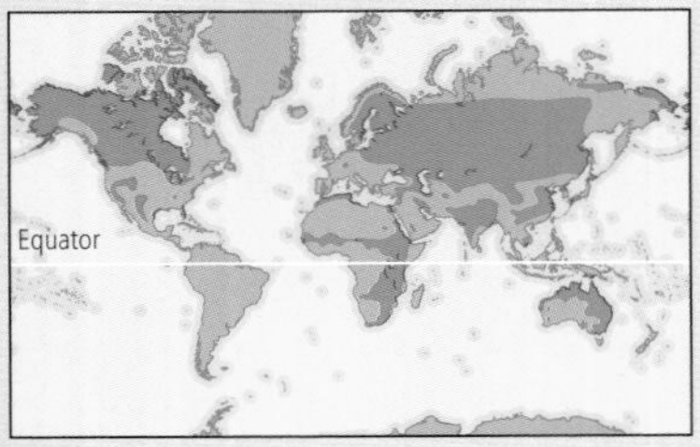

HABITAT Shallow wetlands in breeding season, grasslands and agricultural fields in nonbreeding season.

SIZE Height 0.9–1.8m (3–6ft); wingspan 1.5–2.7m (5–9ft); weight of smallest species 2.7–3.6kg (6–8lb), largest species 9–10.5kg (20–23lb). Males usually larger than females.

PLUMAGE White or various shades of gray, with bright red bare skin or elaborate plumage on head. Long, elaborate secondaries; tail long and overhanging, or ruffled, curled, and raised in display.

VOICE Shrill, carries long distances; in 12 species, sex identifiable from unison call of adult pairs.

NEST A platform in shallow water or short grass.

EGGS 1–3, white or heavily pigmented; weight 120–270g (4–10oz). Incubation period 28–36 days.

DIET Insects, small fish, and other small animals; tubers, seeds, and agricultural gleanings.

CONSERVATION STATUS The Siberian crane is listed as Critically Endangered because of environmental threats to the wetlands on which it depends. The Whooping and Red-crowned cranes are both Endangered, and 6 other species are Vulnerable.

CRANES ARE BIRDS OF SUPERLATIVES. NOT only are they one of the oldest groups of birds, dating back to the Paleocene, some 60 million years ago; they are also long-lived, with captive cranes surviving into their seventies and eighties. Moreover, they are the tallest flying birds; some species stand up to 1.8m (6ft) high.

The cranes' beauty and grace are legendary, and have long been revered by many indigenous peoples. Unfortunately, they are also among the world's most endangered birds; 9 of the 15 species are now at risk, and humankind is entirely responsible for their recent decline.

Long Necks, Long Legs

FORM AND FUNCTION

Cranes have long, powerful, straight beaks, and all species have long necks and legs. They are heavy-set and have loud, shrill calls that carry for several kilometers; indeed, few calls in the bird world are louder. The windpipe of several species is lengthened by coiling within the breastbone: this extended organ amplifies the calls. Cranes fly with their necks extended forward. During flight the legs are usually stretched straight beyond the short, stubby tail, but in cold weather flying cranes fold their legs and tuck their feet under the breast feathers. Although they are predominantly aquatic birds, their feet are not webbed, and they are restricted to the shallows where they breed, search for food, and rest during the night.

Birds of Wide-open Spaces

DISTRIBUTION PATTERNS

Generally cranes are birds of the open marshlands, grasslands, and agricultural fields. Most species usually nest in secluded areas of shallow wetlands, the exceptions being the two species of *Anthropoides*, which often nest in grasslands or semidesert areas.

Only the two *Balearica* species of crowned cranes roost in trees. The crowned cranes are also the "living fossils" of the family. In the remote Eocene (55–34 million years ago), these loose-plumed birds with enormous, gaudy crests flourished in the northern continents for millions of years before the Earth cooled and the cold-adapted

cranes evolved. The Ice Ages restricted the range of the crowned cranes to the savannas of central Africa, where tropical conditions were maintained during the periods when northern continents were covered by ice. Two species of crowned cranes still grace the African grasslands, while 13 of the cold-adapted species now stalk the wetlands of the northern hemisphere and Australia.

Omnivorous Opportunists

DIET

Today's successful crane species are omnivorous, opportunistic feeders that have adapted within the last few thousand years to benefit from agricultural fields. Several *Grus* species, both crowned cranes, and both species of *Anthropoides* have short beaks with which they can effectively grasp insects, pluck ripe seeds from grass stems, and graze on fresh green vegetation in a gooselike manner. In contrast, most of the endangered species have long, powerful mandibles, used to dig for plant roots and tubers in muddy soils or for grasping aquatic animals such as small fish, amphibians, and crustaceans. Aquatic feeders include the larger cranes (Wattled, Sarus, Brolga, Whooping, Siberian, Red-crowned, and White-naped).

Partners Working in Unison

BREEDING BIOLOGY

In the wild, most cranes do not start to breed until they are 3–5 years old. More successful species such as the Sandhill and Common cranes often rear two chicks per breeding attempt. In contrast, the rarer ones, including the Whooping and Siberian cranes, usually rear only one chick. In captivity too, they are often more difficult to breed than the successful species.

Cranes are monogamous, and with the onset of spring or the rainy season mated pairs retreat to secluded grasslands or wetlands, where they establish and vigorously defend a breeding territory that may include several thousand hectares, depending on species or topography.

Mated pairs emit a loud "unison call" duet, during which the male and female calls are distinct yet synchronized. In most species the male emits a series of long, low calls, for each of which the female produces several short, high-pitched calls. In most species this display identifies the sex of each of the birds. The unison call assists in the development of the pair bond and in the defense of the breeding territory. After a stable relationship has been established between two cranes, however, the unison call functions primarily as a threat. At dawn the crane pairs announce their territory with a unison call; neighboring pairs respond with more unison calls that carry for kilometers across the wetlands and grasslands.

The reproductive status of two members of a stable pair are synchronized by their hormonal cycles, which in turn are influenced by the weather, by the length of daylight, and by elaborate displays such as the unison calls and nuptial dances (see Leaping and Bowing – The Dance of the Japanese Crane). Cranes begin copulating several weeks before eggs are laid. For fertility to be assured, a female crane must be inseminated 2–6 days before an egg is laid.

At a secluded spot within the wetland breeding territory, the pair build a platform nest. Crowned cranes often lay a three-egg clutch, while other cranes lay two eggs, with the exception of the Wattled crane, which more often lays a single egg.

Male and female cranes share incubation duties. The female usually incubates during the night, while the male takes his turn during the day. The bird that is not on the nest usually feeds at a considerable distance from it, sometimes in a "neutral area" in company with other cranes. Incubation lasts 28–36 days, depending on the species and the parents' attentiveness at the nest. Crowned cranes do not initiate incubation until

Above *Sandhill cranes at twilight in the Bosque del Apache National Wildlife Refuge, near Socorro in New Mexico. Thousands of the birds winter there each year in wetlands created from the floodplain of the Rio Grande river.*

the clutch is complete, so that their eggs hatch simultaneously. Other cranes begin incubation as soon as the first egg is laid, and chicks generally hatch two days apart.

Crane chicks are well-developed (precocial) when they hatch, and follow their parents around the shallows until they develop flight feathers at 2–4 months of age. The larger tropical species, such as the Wattled and Sarus cranes, have a longer pre-fledging period than species such as the Siberian crane, in which the short arctic summer limits the period when food is available for the fast-growing chicks. Although most eggs will usually hatch, many chicks die, and many of the species that are now considered endangered only rear a single chick per breeding effort. Once fledged, the chicks remain with their parents until the onset of the next breeding season. Migratory cranes learn the migration route by accompanying their parents south to traditional wintering grounds, in some species over distances of thousands of kilometers.

Nurture as well as nature is well demonstrated in cranes. Although they perform their complex visual and vocal displays instinctively, learning determines the context in which the display is performed; for example, young cranes reared by humans prefer to associate with humans rather than with cranes, and will solicit or threaten people. Young cranes are taught by their parents where and on what to feed.

A Perilous Existence

CONSERVATION AND ENVIRONMENT

The aquatic feeders are among the most threatened cranes, with loss and degradation of wetlands a key factor in the birds' decline. In addition, their size and their plumage render them conspicuous and easy prey for hunters and egg collectors.

In North America the Whooping crane – which has the distinction of being the rarest species – has recovered from about 20 birds in the early 1940s to about 400 (wild and captive) today. The Red-crowned crane is the next rarest, at about 1,800 birds in the wild. Siberian cranes number 2,500–3,000 individuals, White-naped and Black-necked each perhaps 5,000, Wattled 8,000, and Hooded 11,000. Fortunately, cranes are appealing birds, and recent efforts in many Asian countries have resulted in the protection of wetlands critical to their survival. But crane-hunting continues in countries such as Afghanistan and the Northwest Frontier province of Pakistan, where 2,000–4,000 live birds are captured each year, and the pressures on the wetlands increase as human numbers soar.

Asia has five endangered species, each numbering in the low thousands. Cranes have special symbolic value in many Asian cultures, and conservationists are optimistic that each of these

Above *Wattled cranes feed in shallow water at the Chobe National Park in Botswana. These large birds, up to 175cm/70in tall, live mainly on tubers and rhizomes of aquatic vegetation, although they also sometimes take small fish, amphibians, and crustaceans.*

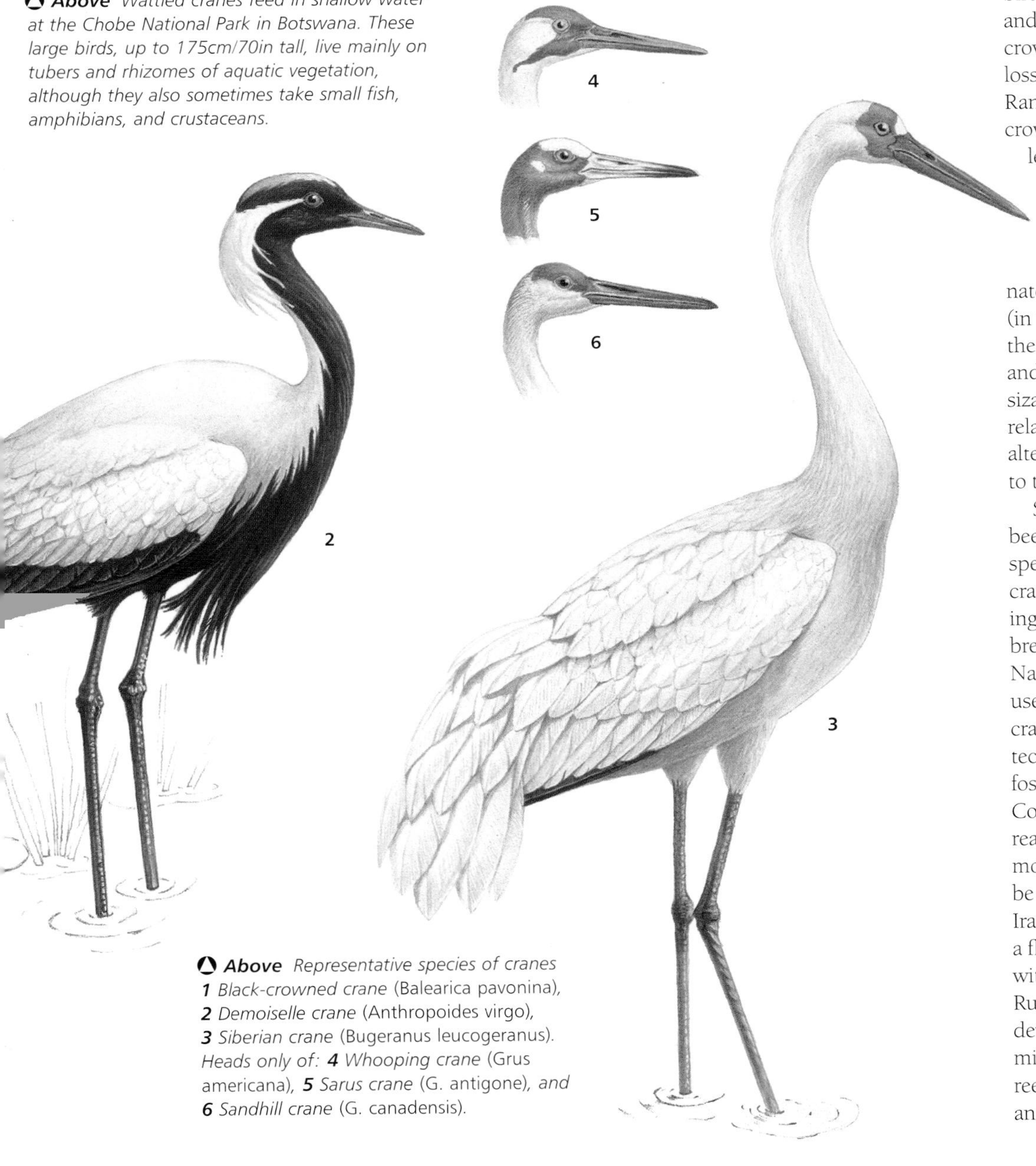

Above *Representative species of cranes* ***1*** *Black-crowned crane* (Balearica pavonina), ***2*** *Demoiselle crane* (Anthropoides virgo), ***3*** *Siberian crane* (Bugeranus leucogeranus). *Heads only of:* ***4*** *Whooping crane* (Grus americana), ***5*** *Sarus crane* (G. antigone), *and* ***6*** *Sandhill crane* (G. canadensis).

species will avoid extinction, although in low numbers, with the possible exception of the highly wetland-dependent Siberian crane. Siberian cranes are adapted exclusively to excavating fleshy roots and tubers of aquatic plants from shallow water, and unlike other species are not able to forage in fields and grasslands during migration and on their wintering grounds. Saving wide expanses of shallow wetlands is a major challenge in China, where the vast majority of the Siberian cranes reside from October through April.

Although the four species of cranes endemic to Africa are locally abundant, there has been a widespread decrease in their numbers in recent decades due to human interference. Aside from perhaps 100 in Namibia, the world's 18,000 Blue cranes are found exclusively in South Africa, where conversion of grasslands to tree plantations, the subdivision of large farms, and poisoning are all major threats to what is the country's national bird, as well as to the local populations of Wattled and Grey crowned cranes. In West Africa Black crowned cranes are similarly threatened by habitat loss, hunting, and trapping for the bird trade. Ranging from East Africa to South Africa, the Grey crowned crane is still relatively secure on a global level. Overgrazing of wetlands and general disturbance by humans are, however, causing a decline in numbers in many regions.

Like the Siberian cranes of Asia, the Wattled cranes of Africa are wetland-dependent. Fortunately, several enormous wetlands in central Africa (in particular the Okavango Delta of Botswana, the Kafue Flats and Bangwelu Swamps of Zambia, and the Zambezi delta in Mozambique) support sizable populations. Water development projects related to the damming of rivers, and the resultant alteration of wetland plains, are the greatest threat to the Wattled cranes.

Several intensive conservation progams have been established to preserve threatened crane species. Parallel to the program for Whooping cranes in North America (see Saving the Whooping Crane), the Russians have set up a captive breeding center for Siberian cranes at the Oka Nature Reserve. Captive-produced birds are being used in trials to boost the numbers of Siberian cranes that migrate to Iran and India. A variety of techniques have been tested, including cross-fostering Siberian crane eggs into the nests of wild Common cranes, and the release of costume-reared cranes with both wild Siberian and Common cranes. Released birds, however, have yet to be located on the known wintering grounds in Iran and India. One bird released in autumn with a flock of Common cranes, however, was reported with two other Siberian cranes on migration in Russia the following spring. The Russians are now developing plans to train Siberian cranes to follow microlight aircraft, with a view to eventually reestablishing the populations that migrate to Iran and India.

GWA

SPECIAL FEATURE

Teaching Cranes how to Migrate

Whooping crane conservation

ALTHOUGH MANY CRANES ARE ENDANGERED, they respond well to protection and management. In an effort to ensure their survival, captive flocks of the endangered species are now being soundly and cooperatively managed at zoos and specialized crane research centers. At the Patuxent Wildlife Research Center in Maryland in the United States, a captive flock of Whooping cranes has been established by collecting one egg from each wild crane's nest containing two eggs. Now captive-produced Whooping cranes are being used to start new wild populations in Florida and Wisconsin. Additional captive populations of Whooping cranes have been established at the International Crane Foundation in Wisconsin, at the Calgary Zoo in Canada, and at several zoos in the United States.

Between 1976 and 1988 captive-produced Whooping crane eggs, and eggs collected from the wild cranes, were substituted into the nests of Sandhill cranes in Idaho. The foster-parents reared many Whooping cranes, and 77 birds migrated south with the Sandhills to wintering grounds in New Mexico. Although Whooping cranes are more aquatic than Sandhills, the rare cranes learned to forage in the more upland niche of the Sandhills. In addition, they learned the migration route of their foster-parents and repeated it each spring and autumn. Unfortunately, though, not a single conspecific pairing occurred. One male Whooping crane paired with a female Sandhill crane, and they produced one hybrid chick before the male disappeared.

Left *The only self-sustaining wild population of Whooping cranes breeds in Canada and winters in Texas. To safeguard these Endangered birds, and to establish a new, non-migratory flock, the Canadian and US governments established the International Whooping Crane Recovery Team.*

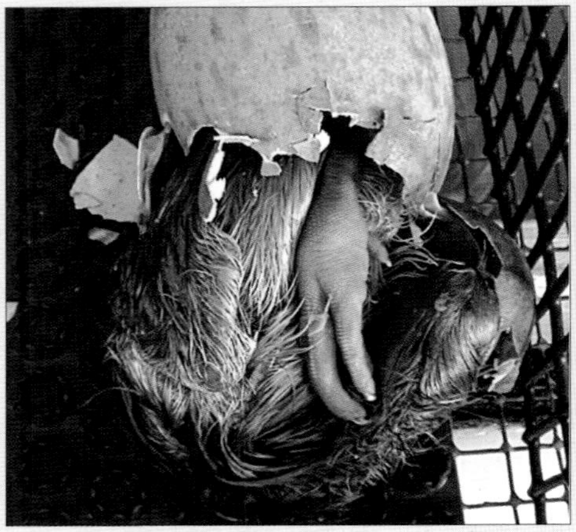

Above *When incubating the eggs in the captive-rearing facility, great care is taken to ensure successful imprinting. The eggs are turned three times daily to simulate what happens in the wild, and during this process a recording of an aircraft engine is played. The first thing that the chick sees once it hatches is a dummy crane head.*

Below *Imprinting plays a major part in persuading young cranes to follow aircraft. Disguised handlers using puppet heads act as surrogate parents, and ensure that the aircraft can always be seen and heard. Taped natural brood calls are also used.*

Above *Flying in formation behind their "parent" – an ultralight aircraft – Whooping cranes embark on their southward fall migration. Speakers mounted on the aircraft broadcast a contact call to the birds.*

Based on studies in cross-fostering of captive cranes at the International Crane Foundation and throughout the experiment in Idaho, it became apparent that chicks sexually imprint on the foster-parent species. Consequently, an ethologist, Dr. Robert Horwich, developed a remarkable method called "isolation rearing," under which captive-reared cranes never see or hear humans. They are exposed to mounted brood models, taped crane calls, hand puppets that resemble the head and neck of their species, adult conspecifics in adjacent enclosures, and crane-costumed humans. The crane costume consists of a suitably colored cloak to hide the human form. A face screen allows the human to observe the cranes, but prevents the cranes from seeing human features. Horwich's initial research on Sandhills indicates that isolation-reared cranes readily join wild cranes, migrate with them, and return to the natal area in spring.

In an effort to establish a non-migratory population of Whooping cranes similar to a population that was extirpated from Louisiana in the 1940s, a release program has been under way in Florida since 1993. More than 200 Whooping cranes have been parent- and isolation-reared at the various captive breeding centers and released into south-central Florida. Many pairs formed, and in the year 2000 one pair hatched two chicks and reared one juvenile until it was almost fledged, when the young bird was unfortunately killed by a bobcat. Although bobcat predation has been the major mortality factor in Florida, the flock now numbers around 97 birds. It is hoped that when drought conditions in Florida end, many pairs of Whooping cranes will be able to breed, and establish a self-sustaining, non-migratory population.

Two techniques have been developed to teach captive-reared cranes a new migration corridor. One technique involves releasing captive-produced cranes with a flock of wild cranes prior to their migration. The released birds then learn the migration route of the wild cranes. The second technique involves teaching isolation-reared captive cranes to follow an ultralight aircraft. By conditioning the cranes shortly after they hatch, fledging cranes can be trained to follow the aircraft, and the young cranes can then be led in a series of short flights to reach the desired wintering area. Weaned from the costumed keeper, crane-costumed pilots, and the ultralight aircraft, the birds remain at the wintering site until the following spring. Then, without assistance from humans, they migrate back to the area where they were reared. Several migrations have been made with groups of Sandhill cranes, and one migration was made with a mixed flock of Whooping cranes and Sandhills. In the latest migration, which ended in November 2002, pilots guided 16 Whooping crane colts from Wisconsin to the central west coast of Florida, a journey that lasted 49 days and covered some 1,940km (1,200 miles). GWA

STORY

Leaping and Bowing – The Dance of the Japanese Crane

1 *Cranes pair for life, the strong bond between birds being maintained and strengthened by courtship "dances." These exuberant displays have long captured the imagination of native peoples. Ceremonial dances based on the nuptial display of cranes are known from many parts of the world, including the Great Plains of the USA, Siberia, China, and Australia.*

The Red-crowned crane of East Asia symbolizes happiness, good fortune, and marital harmony. In Japan, the species was once widespread, but declined as wetlands were drained. The Kushiro marsh, a protected area on the northernmost island, Hokkaido, has been key to its survival. There, the birds' numbers swell over winter as they gather to breed in early spring. Their displays begin with an invitation to dance, as one bird leaps high with outstretched wings in front of its mate. The indigenous Ainu of Hokkaido, who have their own crane dance, call the bird sarurun kamui *– "God of the Marsh."*

1

2

2 3 *The dance proceeds with the partners alternately stretching and bowing and strutting around one another with their wings partially extended. Displaying behavior is contagious, and once one pair has initiated a dance, the entire flock soon takes up the cue. The breeding season lasts from March to July. At the Kushiro International Wetland Center, conservation measures to foster breeding success in this species have been so effective that, from a low point of just 25 cranes in 1952, there are now some 600 birds.*

❹ *Running and short flights are also integral parts of the Japanese cranes' dance ritual. Weighing up to 10.5kg (23lb), this species is the heaviest of the cranes, and requires a run of some 9m (30ft) to get airborne. Unlike Red-crowned cranes on the East Asian mainland, the Hokkaido population is non-migratory; from season to season, birds are estimated to travel no further than 150km (93 miles).*

❺ *An established pair performing a "unison call" duet, a territoral threat posture. Lifting their heads to uncoil their long windpipes, the cranes emit a call so loud that it can be heard from several kilometers away. The stentorian trumpeting of the Red-crowned crane – whose Japanese name is* tancho *– has given rise to the figurative phrase* tancho no hitokoe *("the voice of the crane"), which denotes an authoritative tone.*

Rails

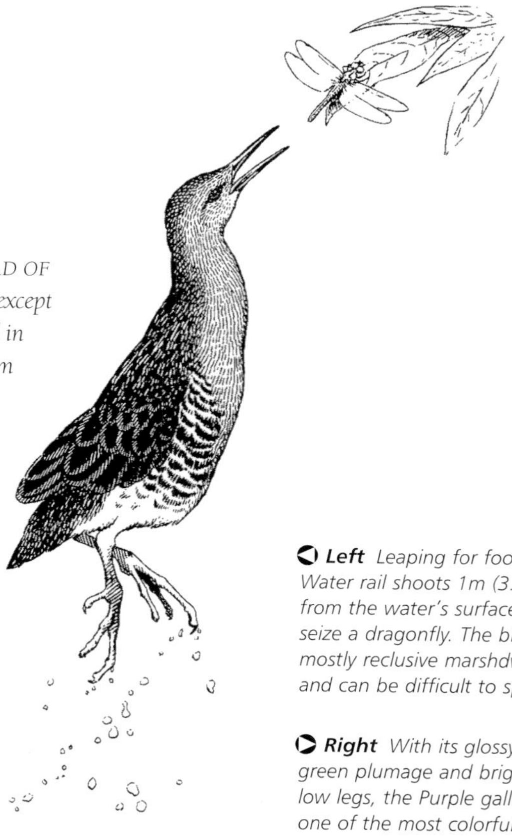

RAILS ARE AMONG THE MOST WIDESPREAD OF bird families. They occur on all continents except Antarctica and on most island groups, and in almost all habitats below the snowline apart from deserts. Yet their love of dense vegetation means that very little is known about most species. They are so secretive that new species are still being found; however, their vulnerability to introduced predators means a large number are close to becoming, or have already become, extinct.

Broadly speaking, the Rallidae can be divided into three (nontaxonomic) groups, too closely allied to be separated into subfamilies or tribes. The rails proper are thin, long-billed birds, adapted for running through marsh grass and underbrush; the crakes and gallinules are similarly conformed, but have shorter, more conically-shaped bills. The third group consists of the largely aquatic coots, which spend much of their time on open water.

◐ ***Left*** *Leaping for food, a Water rail shoots 1m (3.3ft) up from the water's surface to seize a dragonfly. The birds are mostly reclusive marshdwellers, and can be difficult to spot.*

◑ ***Right*** *With its glossy blue-green plumage and bright yellow legs, the Purple gallinule is one of the most colorful of the Rallidae species.*

Marshland Opportunists

FORM AND FUNCTION

Rails are a large group of birds, typically with short, laterally compressed bodies and short, rounded wings. Most rails are weak fliers and some species have entirely lost the ability to fly; paradoxically, however, many undertake long migrations (see To Fly, or Not to Fly...). There are few fossils, although the family probably evolved between 50 and 85 million years ago. Originally birds of the damp African forest, their relatively unspecialized body plan has allowed them to spread throughout the world and into a wide range of habitats. Most are opportunistic feeders, taking whatever food is available, including a wide range of plant material (including some cultivated crops), invertebrates, small amphibians, fish, birds and their eggs, carrion, and a range of artificial foods, from dog food to chocolate. The olfactory system is well developed in most species.

The rails typically have a medium to long bill that is often slightly downcurved. This is a generalist tool that can be used for probing in mud, as in Water and Virginia rails searching for worms, or used more powerfully to smash eggshells, crush grasshoppers, or even kill the occasional frog or duckling. The social organization and behavior of many species remains a mystery, but they seem to be rather territorial, using their loud calls to demarcate territories in the dense vegetation in which they are most typically found.

The crakes and gallinules generally have shorter bills, not long enough to probe into mud. These birds rely more on surface foraging, taking small invertebrates and seeds. Most species are also, to a greater or lesser extent, herbivorous, and some, like the endangered takahe of New Zealand, are almost entirely vegetarian. As they do not depend so critically on soft, marshy ground, they can live in a wide range of habitats. The corncrake was once found in coarse grassland (including farmed land) throughout much of Europe, northern Asia, and North Africa. Seldom seem, this shy bird can easily be detected by its distinctive call, which sounds rather like a knife being scraped over the teeth of a comb. The gallinules and moorhens tend to be associated with aquatic habitats. Most species are thought to be monogamous, but some have quite complicated breeding systems. Both gallinules and coots have frontal shields (see Frontal Shields).

Although all members of the Rallidae can swim, the coots are truly aquatic. Their toes are enlarged with sizable lobes, enabling them to swim and dive well, and they are rarely found far from water. They are even found in the remote, high-altitude lakes of the Andes, where the two largest species, the Horned and Giant coots, make their homes. Not needing to hide in dense vegetation, they tend to be much stouter than other rails. They are omnivores, eating mainly plant material in winter but supplementing this diet with seasonally abundant water insects in spring

FACTFILE

RAILS

Order: Gruiformes

Family: Rallidae

About 133 species in 34 genera, loosely divided into 3 nontaxonomic groups. The long-billed rails include the **Guam rail** (*Rallus owstoni*), **Virginia rail** (*R. limicola*), **Water rail** (*R. aquaticus*), **Invisible rail** (*Habroptila wallacii*), **New Caledonian rail** (*Gallirallus lafresnayanus*), **weka** (*G. australis*), and **Snoring rail** (*Aramidopsis plateni*). Crakes and gallinules include the **Asian watercock** (*Gallicrex cinerea*), **Black crake** (*Amaurornis flavirostris*), **corncrake** (*Crex crex*), **Common moorhen** or **Common gallinule** (*Gallinula chloropus*), **Samoan moorhen** (*G. pacifica*), **Purple gallinule** (*Porphyrula martinica*), **Purple swamphen** or **pukeko** (*Porphyrio porphyrio*), **takahe** (*P. mantelli*), and **White-winged flufftail** (*Sarothrura ayresi*). Coots include the **American coot** (*Fulica americana*), **European coot** (*F. atra*), **Giant coot** (*F. gigantea*), and **Horned coot** (*F. cornuta*).

DISTRIBUTION Europe, Asia, Australasia, N America, S America, and many oceanic islands and archipelagos.

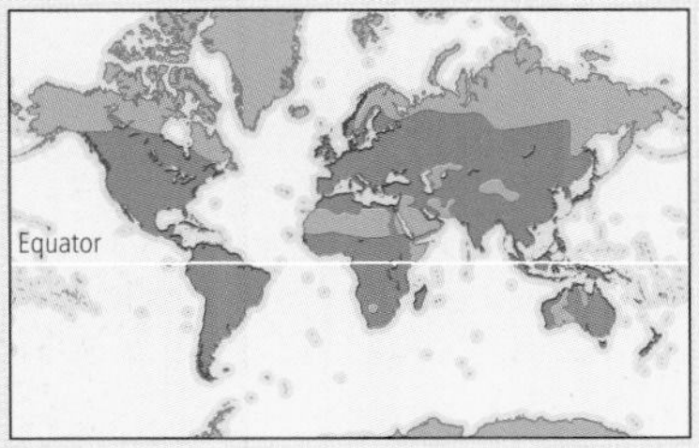

HABITAT Generally damp forest, scrub, meadow, and marshland.

SIZE Length 10–60cm (4–24in); **weight** from 20g/0.75oz (black rail) up to 3.2kg/7.1lb (takahe). Males same size as or 5–10 percent heavier than females.

PLUMAGE Mostly drab brown, gray, and rufous, sometimes with pale spots and flashes; a few species show bright and contrasting colors; differences in color between sexes in some species, but sexes similar in most.

VOICE Whistles, squeaks, and grunts, in combinations from simple to complex. Many sound "unbirdlike."

NEST In wholly aquatic species (coots), conical nest emerges from shallow water on stick or pebble (Horned coot) foundation; others within clumps of grass or reeds, sometimes roofed; a few species in bushes or low trees; always composed wholly of vegetation.

EGGS Usually 2–12, but poorly documented for many species; color off-white to dark tan, often spotted with darker shades of brown, gray, mauve, or black; weight 10–80g (0.4–2.8oz). Incubation 20–30 days.

DIET Medium to large invertebrates, smaller vertebrates, some seeds, fruits, plant shoots etc; a few species are largely herbivorous.

CONSERVATION STATUS 4 species including the Samoan moorhen and the New Caledonian rail, are considered Critically Endangered; a further 11 are Endangered and 17 Vulnerable. In addition, 22 species are listed as Extinct, and the Guam rail is Extinct in the wild.

and summer. Chicks are initially fed almost exclusively on insects, gradually switching to a vegetarian diet as their intestines grow larger. Coots can be quite gregarious, especially during the nonbreeding season. For example, European coots molt all their flight feathers at once and may remain flightless for up to four weeks. During this period they gather on large lakes and seacoasts in flocks that may number thousands of birds, exploiting plentiful food resources and safety in numbers.

All rails have stout, well-muscled legs with three forward-facing toes and one hind toe, which is used as a brace when walking, often with a bobbing head and flicking tail; some species, like the Purple gallinule, can climb trees. The legs, feet, and bill are often brightly colored and become more so as the birds get older; they are also used as weapons during territorial disputes. The sexes generally do not differ in plumage, nor do they have different breeding and nonbreeding plumages, although there are a few exceptions. The downy chicks are almost invariably plain black or dark brown in color, the only exceptions being those of the forest-dwelling Nkulengu rail and the flufftails of Africa; these are thought to be the most primitive species.

As might be expected from their preference for dense vegetation, most rails are very vocal, often calling through the night. They can emit a large range of squeals, trills, grunts, and barks. The Snoring rail from Sulawesi lives up to its name, while the call of the Buff-spotted flufftail is one of the most evocative sounds of the African night: a low-pitched hoot followed by a high-pitched whine that has been variously described as resembling the wail of a banshee or the sound of a chameleon giving birth. Territorial calls tend to be especially repetitive and loud, often making "tack," "kak," or "crek" sounds, and can carry for 2–3km (1.2–1.9 miles) in habitat where birds may not be able to see each other 50m (165ft) apart. The calls often seem to be ventriloqual, and calling birds can be exceedingly hard to pinpoint.

Although many rails call throughout the night, most species are crepuscular, showing greatest activity at dawn and dusk; the coots and gallinules, however, tend to be diurnal. As with other aspects of the birds' biology, roosting behavior has been little studied, but most are thought to roost in dense cover. Some forest-dwelling rails may roost in trees; for example, the Red-necked crake apparently uses communal roosting platforms, while Forbes' forest-rail even constructs a domed roosting nest in which up to seven adults may sleep at any one time. Reports of allopreening (one bird preening another) are common, and some species may sunbathe socially.

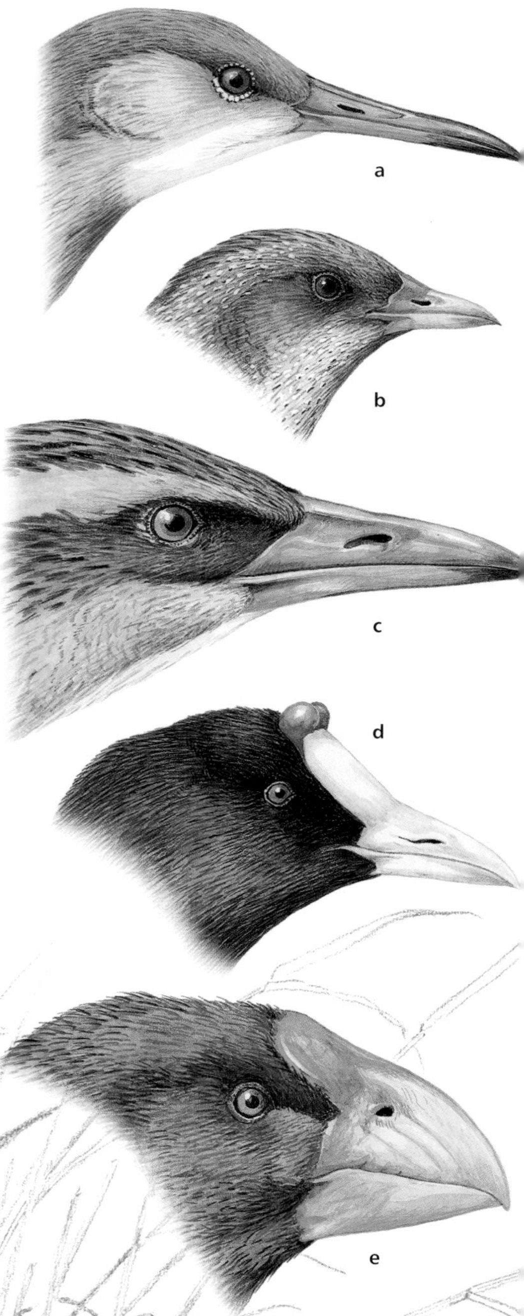

Right *Rail heads, showing variations in bill shape.* ***a*** *Brown-banded rail* (Lewinia mirificus)*: long and slender for probing.* ***b*** *Spotted crake* (Porzana porzana)*, exhibiting the shorter beak typical of the crakes.* ***c*** *Weka* (Gallirallus australis)*, which sometimes takes small mammals.* ***d*** *Crested coot* (Fulica cristata)*, with red-topped frontal shield.* ***e*** *The powerful bill of New Zealand's herbivorous takahe* (Porphyrio mantelli)*, long thought extinct until it was rediscovered in 1948.*

Below *Representative species of rails:* ***1*** *Red-winged wood-rail* (Aramides calopterus)*, a South African resident.* ***2*** *Water rail* (Rallus aquaticus) *removing fragments of shell from hatching young.* ***3*** *Virginia rail* (R. limicola)*, common across the northern USA.* ***4*** *American coot* (Fulica americana)*, keeping watch on chicks.* ***5*** *Common moorhen* (Gallinula chloropus)*, a widespread species equally at home swimming or walking on floating vegetation.* ***6*** *Purple swamp hen* (Porphyrio porphyrio)*, a Eurasian species that migrates from Africa to western Mediterranean wetland regions.*

More or Less Monogamous

BREEDING BIOLOGY

The breeding habitats of all but a handful of rail species are largely unknown. Most are thought to be more or less monogamous, defending a territory during the breeding season; some defend one all year round. Co-operative breeding characterizes certain species. Many species call in duets or chorus, a behavior that is probably related primarily to territorial defense, although in the case of the Black crake it may help members of extended family groups keep in touch with one another. Most rails are sexually mature at 1 year (although they may not actually breed until they are older); however, the White-throated rail of Aldabra – an island group in the Indian Ocean forming part of the Republic of the Seychelles – attains maturity at 9 months, and the Guam rail can breed at only 16 weeks old. The birds' lifespan is largely unknown; some larger species such as coots and the weka can live for up to 15 years, but 5–10 years is probably more representative.

Pair-formation and courtship behavior are also little-known, except in the less secretive species like coots and moorhens. In these, courtship displays are usually simple and often involve displaying boldly-patterned flanks (as in the case of many crakes and rails) or undertail coverts (gallinules). Courtship feeding may be common, and in at least some species aggressive-looking courtship chases often lead to copulation.

Most species nest solitarily, but where habitat is limited loose colonies may form, as in the case of the Purple swamphen. Nests are usually built in thick vegetation, often close to water, and are generally cup-shaped (or, in some forest species, domed). It is well lined, and is constructed from whatever vegetation is available. Some of the more aquatic species may make huge nest platforms. The Horned coot builds a nest of water weed placed on a large, conical mound of stones up to 4m (13ft) across, ending just below the water surface; the whole construction can weigh as much as 1.5 metric tons (1.65 short tons). Generally, only one or two broods are raised each year, although many species can extend their breeding season if conditions prove sufficiently favorable.

The birds usually lay between five and ten eggs, which are incubated for 2–3 weeks, although King rails may lay up 15, and the Chestnut forest-rail lays only one, which it incubates for 37 days. In some species including the Common moorhen, the female may "dump" eggs in the nest of another

TO FLY, OR NOT TO FLY...

Rails are something of a paradox. Although they are for the most part weak fliers, many species undertake long migrations, usually flying at low altitudes by night. Yet flightlessness has also evolved in 32 rail species – about a quarter of the total number.

Some clue to the reasons for this divergence can be gleaned from the fact that all the flightless species occur on islands (although not all island species are flightless), and particularly on those that have no indigenous predators. A further clue comes from the Purple gallinule. This brightly-colored rail normally breeds in northern America and winters in South America, but it is regularly recorded on most of the Atlantic islands, and has even been spotted in the Cape Province of South Africa, some 7,500km (4,650 miles) from its normal range. With their relatively weak flight, the birds are evidently susceptible to being blown off course. When they accidentally land on the continents, they are likely to be outcompeted by local species, or to make a tasty meal for predators, but on islands their lack of specialization increases their chance of survival.

On islands with no natural predators, two factors lead to selection for reduced flight ability and, eventually, for flightlessness. First, flight muscles are expensive, typically accounting for 25 percent of a bird's total weight; they also make great demands on energy. In flightless species, these muscles, and the pelvic girdle to which they are attached, are both greatly reduced. Secondly, in island populations young birds have no need to disperse far from their birthplace; those that do attempt longer flights are likely to find themselves heading for a watery grave.

In fact, rails have a natural predilection for flightlessness; even those that can fly often prefer to run, particularly in dense habitat, and many are also flightless during the molt period, molting most of their flight feathers at one time. So it is perhaps not surprising that flightlessness has evolved many times, and that flighted species like the Purple swamphen and flightless ones such as the takahe may be genetically very closely related, suggesting that such changes can occur rapidly. RR

bird, leaving her to raise them as her own. Incubation may start when the first egg is laid, or when the clutch is complete, depending on species, and incubation is generally performed by both partners, the male apparently taking daytime duties with the female incubating at night.

When the chicks hatch, they quickly become mobile and leave the nest after 2–3 days, but, unusually, they remain dependent on their parents until they fledge, which occurs at 4–8 weeks after hatching. In species in which incubation begins as soon as the first egg is laid, the first chick will hatch earlier than the others, creating differences between siblings of the same brood. Indeed, the first-born may already be capable of foraging for itself while the youngest is still struggling from the egg.

Forced to compete with their larger siblings for food from the parents, the younger chicks often starve; indeed, in some species like the European coot, the parents may actively attack particular chicks, seizing them by the head and shaking them before dropping them back into the water. This behavior is likely to have arisen as a mechanism to best match the size of the brood to an unpredictable food supply.

Most rails are presumed to be monogamous, since the chicks require a period of extended parental care. In around a dozen species, including the Common moorhen of Europe and the African Black crake, extended family groups may form, with young from earlier broods helping to look after the chicks, sometimes assisted by other, adult birds.

In a few species, however, the situation is altogether more complicated, indicating that much about the breeding behavior of rails undoubtedly remains to be discovered. The corncrake, for example, was assumed to be monogamous until it was noticed that other females did not seem to be regarded as territorial intruders. Further research has since revealed that broods in adjoining nests are often associated with several females but only a single male, and that the chicks are brooded by the females alone. Although some pairs are monogamous, serial polygyny is common, with males occupying shifting territories and mating with two or more females, each of whom raises her own clutch of eggs. The African crake and the Yellow rail may have similar mating systems, possibly reflecting particularly productive habitats in which one bird alone can find sufficient food to raise the brood.

The opposite situation, polyandry (in which one female breeds with several males), occurs in some species. The Striped crake breeds in often unpredictable, ephemeral wetland habitat in Africa. Unusually for rails, the sexes in this species have different plumages (males are brownish, females grayish), and the female sets up a breeding territory, mating with two or more males in succession and leaving each one to raise the clutch alone. This situation, which is likely to have evolved to make best use of the short breeding season, is also seen in some waders.

The flightless Tasmanian native hen is also polyandrous, with females forming lifelong bonds with two males, often brothers, and chick-rearing duties shared between all three; the young of previous years sometimes help as well. In this case the behavior is thought to have arisen because of an unequal sex ratio; in most populations there are more males than females.

In Europe and Africa the Purple swamphen is monogamous, but in New Zealand (where the bird is known as the pukeko) a more complicated system has arisen. Here there is a shortage of habitat, and the birds occur in communal groups up to 12 strong, which tend to be fairly stable if the birds are related. Groups consist of from two to seven breeding males, one or two breeding females, and up to seven nonbreeding helpers, usually young from previous years. Within the group mating is promiscuous, with incestuous and homosexual couplings not uncommon. Only the most dominant females breed, and all birds help in raising the brood of up to six young. Groups of nonrelated birds also form, but these tend to be unstable and unsuccessful at producing young. The flexible breeding behavior of the Rallidae partly accounts for the family's widespread distribution.

Above A Common moorhen feeds her chicks at the Wildfowl and Wetlands Trust reserve at Slimbridge in western England. The birds, which are found across much of the world's wetlands except in Australasia, are well-known for their head-bobbing motion as they swim. They are unusual in maintaining several nests for roosting on different nights.

FRONTAL SHIELDS

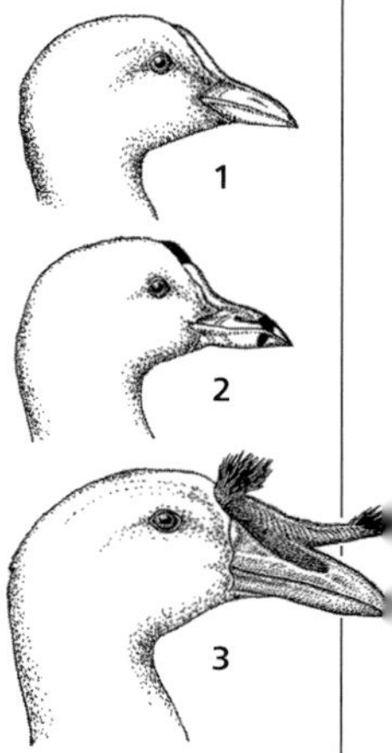

A striking feature of some coots and gallinules is the frontal shield – a fleshy, rearward extension of the upper bill that covers most of the forehead. In the European coot **1** it is a simple white lobe, about the size and texture of the ball of the human thumb, while in the closely related American coot **2** the lobe is overlaid by a smaller, red callus. Easily the most complex ornament is that possessed by the majestic Horned coot of the high Andes **3**, the largest of all coots. In place of a flat shield, this species has a frilled proboscis or horn up to 5cm (2in) long.

Such forehead ornamentation is not peculiar to the Rallidae. The tropical plantain-eaters (Musophagidae) and oropendolas (Icteridae) have very similar shields, while many waterfowl (Anatidae) have knobs or bulbs at the base of the bill.

The puzzle remains as to why these features evolved. One clue is that the size of the shield is related to sex, the male's generally being larger than the female's, and also to the amount of testosterone an individual produces (which is why the male shields are larger). Size also varies seasonally, being largest in the early spring, further reflecting changes in the amount of testosterone present. Not only is testosterone costly to produce, but birds with large shields tend to attract aggression from other birds equally anxious to prove themselves. The shield is, therefore, a good indicator of a bird's quality, showing not only where it stands in the dominance hierarchy within its sex, but also how attractive a mate it would be.

A Rising Tide of Extinction

CONSERVATION AND ENVIRONMENT

In general, rails have little interaction with humans. A few species, such as coots, corncrakes, and some of the larger rails, are hunted for food or sport, and their eggs are generally very tasty. In some Asian countries the watercock is kept for fighting; people are reputed to tie coconut shells around their waists to incubate eggs collected from nests. A few species, such as the Common moorhen and American Purple gallinule, have occasionally been regarded as crop pests. New species continue to be discovered, with one 1997 expedition to the Indonesian island of Karakelong finding two new species.

Current information, however, suggests that around a quarter of all rail species are globally threatened. At least sixteen species have become extinct since the year 1600. Flightless species are particularly vulnerable, with 13 of 18 species currently endangered, and most of those that have become extinct historically, being flightless.

The principal threat to island populations of rails comes from predators such as cats, rats, pigs, and mongooses introduced by the early colonial explorers and, more recently, from habitat destruction and overgrazing. Indiscriminate hunting by the early explorers must also have contributed to the extirpation of some species. Some species living on the continents and larger islands are also threatened, primarily from habitat destruction. The Snoring rail, for example, is threatened by loss of its forest habitat through extensive logging, and the White-winged flufftail through drainage and overgrazing of its preferred wetland habitat.

Two examples may serve to illustrate the threats to rails. In Europe, which over the last century has seen major changes in land-use as farmed land has become increasingly intensively managed, corncrake populations have suffered. The bird, which winters in the African savanna, breeds throughout Europe and central Asia, occurring in open lowland marsh and grass meadows, particularly those managed for hay. The main threats come from loss of habitat as wetlands have been drained, and from agricultural intensification, particularly the conversion of hayfields to silage production, which means more frequent cutting, causing increased destruction of nests and adult mortality.

In addition to protecting remaining marshlands, particularly in Eastern Europe, a number of measures could be introduced to enable corncrakes to survive on cultivated land. These include preserving small uncultivated areas, for example in field corners; delaying mowing; or simply mowing fields from the center out, allowing birds to avoid the harvester. A systematic survey carried out in eastern Europe and Russia in the year 2000 suggests that populations there are around five times higher than previously thought, highlighting how much remains to be discovered. Although trends in these populations are unclear, the gradual eastward spread of "western" agricultural practices suggests reasons for concern for the future.

The Guam rail was once found throughout the Pacific island of that name, where in the 1960s the population probably numbered about 80,000 birds, despite the presence of feral pigs and cats. Numbers rapidly declined, however, along with those of other indigenous species, following the accidental introduction of the Brown tree snake from Australia in 1968; the snake eats both eggs and young birds. By 1981 the population had been reduced to about 2,000 pairs; by 1987 none were left, and the species was declared extinct in the wild. In 1982, however, a captive breeding program was established, which now numbers some 180 breeding birds. In 1998, some of these captive-bred birds were released into a small area protected from snakes. If the snakes can be successfully controlled the future may be relatively bright for this flightless rail. RR/JAH

Left and below *Corncrakes (left) declined rapidly in western Europe as changing farming methods eliminated the tall-grass meadows where they breed; they survive today mainly in eastern Europe. The California clapper rail* (Rallus longirostris obsoletus – below)*, which was once abundant in the wetlands of the San Francisco Bay area, is now the object of a recovery plan intended to guarantee its survival.*

Limpkin

THE LIMPKIN – THE ONLY MEMBER OF THE New World family Aramidae – has anatomical features in common with both the cranes and rails. Biochemical evidence has recently suggested that it may also be closely related to the finfoots.

Limpkins are not globally threatened, but they have disappeared from some regions either as a result of hunting or where marshes have been drained. In the USA the bird is now well protected, with large populations in the Everglades and Lake Okeechobee in Florida.

Left Limpkins use both foot and beak to extract the water snails that make up the bulk of their diet. The birds' feeding places are often littered with piles of empty shells.

New World Marsh-dwellers

FORM AND FUNCTION

The general appearance of the limpkin is not unlike that of a very large rail, with a moderately long, slightly decurved bill, a long neck, rounded wings, a short tail, and long legs. Limpkins walk rather slowly, and their slightly undulating gait, giving the impression of lameness or limping, is supposed to be the basis for their English name. Although they lack webs on the feet, limpkins swim well, and their long toes allow them to walk on floating leaves of water plants. They are strong fliers on broad, rounded wings, with the long, slender neck extended and the feet and legs projecting behind.

The species' range extends from Florida southward through Central America and the West Indies into South America. There the birds reach Argentina but are absent from the arid west coast, the Andes, and the extreme south.

Above A limpkin sits on its nest in the Florida wetlands, where local people sometimes refer to the species as "crying birds" for their eery wails. Mostly nocturnal, they have lost their timidity in the region's nature reserves and are regularly seen by day.

Wading for Snails

DIET

In the swamps, watersides, and marsh areas that are its principal feeding habitats, the limpkin often betrays its presence by the many conspicuous empty shells of large freshwater snails (*Pomacea* spp.) that it leaves in heaps on the muddy banks. To secure this food, it wades in shallow water or walks on floating vegetation, probing with its long, laterally compressed bill, which is slightly downcurved at the tip. When it finds a snail, it carries it to shallower water or the shore and sets it in the mud with the mouth facing upward. With great dexterity the bird then quickly removes the horny operculum that closes the shell mouth, pulls out the body, and swallows it. Limpkins also eat large quantities of freshwater mussels, and at times also may take frogs, lizards, insects, and land snails.

Bulky Nests and Brooding Platforms

BREEDING BIOLOGY

Male limpkins defend territories of several hectares by calling loudly and chasing intruders. A bulky nest of aquatic vegetation or sticks is built, in varied sites ranging from on the ground or in bushes to high in a tree. Clutches are usually of 5–7 eggs. Both parents incubate for a period of about 27 days. The young are well developed at hatching and are soon led to a brooding platform of aquatic vegetation. Both parents feed the young, which are fully grown after seven weeks and disperse after about 16 weeks. DH/PRC

FACTFILE

LIMPKIN

Aramus guarauna

Order: Gruiformes

Family: Aramidae

DISTRIBUTION Florida, W Indies, S Mexico southward to Argentina, chiefly E of Andes.

HABITAT Swamps (wooded or open) or arid brush (as in W Indies).

SIZE Length 56–71cm (22–28in); **weight** 0.9–1.3kg (2–2.9lb).

PLUMAGE Dark olive-brown with bronze to greenish iridescence on upperparts, foreparts broadly streaked with white; sexes alike.

VOICE Noisy; loud wails, screams, and assorted clucks, heard mostly at night.

NEST Shallow, made of rushes or sticks just above the waterline in marshes, or else in bushes or trees.

EGGS Usually 5–7, pale buff, blotched and speckled with light brown; average size 60 x 44mm (2.4 x 1.7in). Incubation about 27 days.

DIET Mainly large snails and freshwater mussels; also some insects, frogs, and lizards.

CONSERVATION STATUS Not threatened.

Trumpeters

Ground-dwelling birds that live in groups in the tropical rain forests of South America, the three trumpeter species take their English name from their distinctive threat call – a loud, long-drawn-out cackle or trumpeting.

The different species are closely related, but they can be distinguished by the color of their innermost flight feathers, inner wing-coverts, and lower backs. These are gray in the Grey-winged trumpeter, white in the Pale-winged trumpeter, and green to black in the Dark-winged trumpeter.

Noisy Residents of the Rain Forest

FORM AND FUNCTION

Trumpeters are about the size of domestic fowl, but are longer legged. Their heads appear small in relation to the body, and the large, dark eyes give the birds a "good-natured" expression. The bill is short, stout, and slightly curved. Soft feathers on the head and the long neck have an almost furlike quality. The wings are rounded, and the very short tail is almost completely hidden by the secondaries. The birds usually walk, but have the ability to run fast; they can easily outpace dogs. They are gregarious, and can sometimes be seen moving in flocks of 100 or more over the forest floor. They are poor fliers, but at dusk their noisy groups flutter rather laboriously some 6–9m (20–30ft) up to roost in forest trees.

The three species are all mainly confined to the Amazon and Orinoco basins, with each one having a discrete range that is generally separated from that of other trumpeter species by a large river. The bulk of their diet is made up of ripe, fallen fruit taken from the ground beneath forest trees, commonly after it has been dislodged by feeding monkeys. A small proportion of the diet consists of insects, including ants and flies, along with other arthropods.

Polygamous Breeding Groups

BREEDING BIOLOGY

Details of the breeding cycle are known only for the Pale-winged trumpeter. Groups of adults of this species share in territory defense, but only the dominant female breeds. Several males in the group mate with her, although the dominant male does so most often. The nest site is a hole in a tree. The average clutch size is three eggs, and both sexes incubate. When the young hatch, they are covered with thick, dark down with a concealing pattern. They soon jump from the nest, after which they accompany the group of adult birds on the forest floor. DH/PRC

FACTFILE

TRUMPETERS

Order: Gruiformes

Family: Psophiidae

3 species of the genus *Psophia*: **Dark-** or **Green-winged trumpeter** (*Psophia viridis*), **Grey-winged** or **Common trumpeter** (*P. crepitans*), **Pale-** or **White-winged trumpeter** (*P. leucoptera*).

DISTRIBUTION S America, mainly in Amazon and Orinoco basins.

HABITAT On ground of tropical rain forests.

SIZE Length 45–52cm (18–20in); weight: 1-1.5kg (2.2-3.3lb).

PLUMAGE Chiefly black, with purple, green, or bronze reflections, especially on lower neck and wing coverts; soft and velvetlike on head and neck; outer webs of tertials and secondaries (white, gray, or brown) form hair-like strands over lower back; sexes similar.

VOICE Loud trumpeting, deep-pitched cries, prolonged cackles.

NEST Hole in tree or crown of palm.

EGGS 2–5, white, rough-shelled; weight about 76g (2.7oz).

DIET Mainly fallen fruit, some insects.

CONSERVATION STATUS Not threatened.

Above and below *The greenish tint of the wing feathers distinguishes the Dark-winged trumpeter **1** from its South American neighbor, the Common or Grey-winged trumpeter **2**. Trumpeters are birds of the forest floor, sometimes kept by indigenous peoples as guards for their noisy call.*

Bustards

Left *Representative species of bustards:* **1** *Kori* (Ardeotis kori), *at up to 18kg (40lb) among the heaviest of flying birds.* **2** *Lesser florican* (Sypheotides indica), *listed as Endangered.* **3** *Great bustard* (Otis tarda) *displaying.* **4** *Red-crested bustard* (Eupodotis ruficrista), *an African species.* **5** *Houbara* (Chlamydotis undulata), *now at risk from overhunting.* **6** *Australian bustard* (Ardeotis australis), *known locally as the Wild turkey.*

BUSTARDS OWE THEIR NAME – BY DERIVATION "slow bird," from their walking gait – to the most northerly and perhaps least typical member of the family, the Great bustard, to which it was first applied. Yet the term is appropriate for the family as a whole, for all bustards are strictly ground-dwellers.

The largest bustards are to the open plains of Africa and Eurasia what cranes are to the world's big marshes: slow-breeding, long-lived birds of ancient lineage, reaching considerable size and weight while retaining the capacity to fly – among birds, the ultimate expression of adaptation to stable habitats. Sadly, like cranes, they are among the first to suffer once those habitats start being exploited and disrupted by humans.

Ground-dwellers that Can Fly

FORM AND FUNCTION

The bustards are a homogeneous family, although there are differences in structure, color, size, and behavior that cloud the relationships between species. All are rather long-necked and long-legged, with robust bodies and short bills, and have lost both the hind toe and the preen gland that most birds possess. These losses, together with the camouflage patterning (often exquisitely delicate) of black on buff, rufous, or brown on their upperparts, are presumably adaptations to the dry, open landscapes they inhabit. Hindclaws are associated with birds that perch on trees or bushes, and oil from preen (uropygial) glands is used by most birds for waterproofing.

The smaller species – the Little, Savile's, Buff-crested, and Red-crested bustards – have relatively short legs and necks. These, and most of the *Eupodotis* bustards, fly with rapid wingbeats; the larger species use slow, deep, powerful beats, but fly deceptively fast. On the ground, bustards are strong but usually slow walkers, characteristically nervous and alert: they move into cover at the first sign of danger, or freeze into a flattened posture and rely on their cryptic plumage for camouflage.

Mostly in Africa

DISTRIBUTION PATTERNS

Africa is the main home of the bustards, and only four species do not breed there: the Australian and Great Indian bustards, and the Lesser and Bengal floricans. Great and Little bustards have only relict populations in North Africa; their patchy distribution extends across the plains of southern Europe into Russia, the Little bustard reaching as far as the northern Kazakh steppes and the Great bustard ranging across Russia to the uplands of Mongolia and northern China. The houbara (which some authorities split into two species) ranges from the eastern Canary Islands through semi-deserts of North Africa, the Middle East, and central Russia to the Gobi desert. The Arabian bustard still occurs in the southern Arabian peninsula and in northwest Africa, but otherwise this and the remaining bustard species are found only in Africa, mostly in the tropics.

Within Africa there are two clear areas where different species have evolved: from the Zambezi southwest to the Cape and from the Nile to the Horn, with six species occurring solely in the former and four in the latter. The kori occurs in both, as do the Black-bellied and White-bellied bustards, although the latter two also have populations across the Saharo-Sahelian savanna belt in

Right *A White-bellied bustard displays the alert posture typical of the Otididae. Spending most of their time on the ground, the birds rely on observation and agility to escape danger as they forage over grasslands in search of worms, insects, and similar prey.*

West Africa, while the White-bellied bustard has a scattered population in Central Africa. The Arabian, Nubian, and Savile's bustards occur in the Saharo-Sahelian zone, the ranges of the first two extending across to the Red Sea coast. Only two species, Denham's and the Black-bellied bustard, are particularly widespread in Africa; the former has become very localized in many areas as a result of human activities.

Feeding on the Move

DIET

Bustards take their food in a slow, meandering walk through an area of grassland or scrub. Their diet is chiefly invertebrates, usually snapped up from the ground or off plants, but also sometimes dug up with the powerful bill. Small vertebrates and large insects such as locusts may also be taken, often after a short pursuit and pounce. All species readily eat vegetable matter, especially plant shoots, certain flowers, and fruit. Some larger species, notably *Ardeotis*, feed on the gum that oozes from acacia trees. Very few bustards appear to seek out fresh water, preferring to get sufficient liquid through their diet.

Concentrations of food may cause a bird to remain in one spot for some time. In Somalia

FACTFILE

BUSTARDS

Order: Gruiformes

Family: Otididae

25 species in 9 genera. Species include: **Arabian bustard** (*Ardeotis arabs*), **Australian bustard** (*A. australis*), **Great Indian bustard** (*A. nigriceps*), **Kori bustard** (*A. kori*), **Bengal florican** (*Houbaropsis bengalensis*), **Black-bellied bustard** (*Eupodotis melanogaster*), **Blue bustard** (*E. caerulescens*), **Little brown bustard** (*E. humilis*), **Red-crested bustard** (*E. ruficrista*), **White-bellied bustard** (*E. senegalensis*), **Denham's bustard** (*Neotis denhami*), **Nubian bustard** (*N. nuba*), **Great bustard** (*Otis tarda*), **houbara** (*Chlamydotis undulata*), **Lesser florican** (*Sypheotides indica*), **Little bustard** (*Tetrax tetrax*).

DISTRIBUTION Africa, S Europe, Middle East and C Asia to Manchuria, Indian subcontinent, Cambodia, Vietnam, Australia, New Guinea.

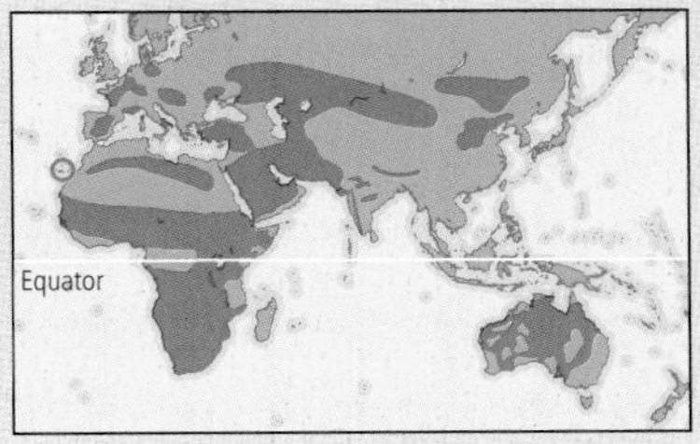

HABITAT Grassland, arid plains, semi-desert, light savanna, thornscrub and acacia woodland.

SIZE Length 40–120cm (16in–4ft); wingspan 1–2.5m (3–8ft); weight 0.45–18kg (1.2–40lb). In some species males are bigger than females.

PLUMAGE Mostly camouflage patterning on uppersides; head and neck with distinctive patterns, combining two or more of gray, chestnut, black, white, and buff. Males of some species more brightly colored than females.

VOICE Larger species generally silent, but many give series of grunts or booming calls during display; smaller ones have distinctive, persistent, usually unmusical wheezing calls during breeding season.

NEST A bare scrape on the ground.

EGGS Mostly 1–2, but up to 6 in some small species; olive, olive-brown, reddish; weight ranges from 41g (1.4oz) in Little bustard to 146g (5.2oz) in Great bustard. Incubation 20–25 days.

DIET Generally omnivorous – shoots, flowers, seeds, berries etc, plus invertebrates (especially beetles, grasshoppers, and crickets), but also small reptiles, amphibians, mammals, and eggs and young of ground-nesting birds.

CONSERVATION STATUS 3 species – the Great Indian bustard and the Bengal and Lesser floricans – are listed as Endangered, while the Great bustard is Vulnerable.

birds have been observed leaping to snatch berries off the higher parts of a bush, and in Zimbabwe Denham's bustard has been seen to wade into water, apparently in quest of young frogs, and to defend a termites' nest at which it was feeding against other birds. Several species gather at bush fires to take fleeing and crippled insects. Bustards have no crop, but their powerful gizzard, their long "blind gut" (cecum), and their habit of taking up quantities of grit assist the digestion of food.

Right *A Black-bellied bustard keeps a wary eye out for intruders. The bird is the largest of the* Eupodotis *species, several of which are sometimes referred to by their Afrikaans name of "korhaan."*

Below *A kori drinks from a pool in a South African national park. Koris are among the few birds – pigeons, sandgrouse, and buttonquails are others – that can actively suck up water, rather than having to scoop it in their beaks and then raise their heads to swallow it.*

Males that Advertise

BREEDING BIOLOGY

No male bustard has been observed to incubate eggs, an emancipation from parental duties that appears to reflect the lack of any pair-bonding in the family. Instead it appears that, in the larger species, males have a lek display in which they advertise themselves to any females in the vicinity. In South Africa, for example, Denham's bustard males seem to display in response to each other and any passing female, and may eventually mate with several females, the males keeping at least 700m (2,300ft) apart. Male Great bustards also operate a similar dispersed lek system, but in this species many males appear not to be territorial, instead moving about, keeping their distance from each other, and displaying at various sites (see Spectacular Courtship Displays). In this and the houbara and Australian bustard, two other species that also do not appear to form pair bonds, the display before copulation is very long and is often impeded by rivals; in these species, territoriality is replaced by simple opportunism and/or a ranking system.

In southern Africa, birds with black underparts in the genus *Eupodotis* occur in quite dense grassland and savanna, give striking aerial displays, and apparently hold group territories within which one pair breeds. The newly-hatched young of all species are precocial and very soon leave the nest, but they are fed bill-to-bill by the mother initially and remain in her company for some months after hatching. Great and Little bustards are notably sociable and sometimes occur in flocks; the remaining species are more solitary, although some are commonly found in small groups.

SPECTACULAR COURTSHIP DISPLAYS

Male bustards are remarkable for their spectacular courtship displays. The Great bustard **1** inflates its neck like a balloon, cocks its tail forward onto its back, and stretches its wings back and down from shoulder to carpal. The primary feathers are kept folded so that their tips are held behind the bird's head, while the secondary feathers are lifted outward to form huge white rosettes on either side of the body. Together with the billowing undertail coverts, this posture transforms a richly colored (if partly camouflaged) animal into one that is nearly all-white. The pose may be held for minutes on end, and the bird is visible over great distances on spring mornings and evenings.

Male Australian **2** and Great Indian bustards cock the tail forward and inflate the neck downward so that it becomes a broad sack that swings around like a punchbag, scraping the ground. The kori simply inflates its neck into a puffy white ball and trails its wings. The oddest of the ground displays is that of the houbara **3**, which raises its white neck-ruff right over its head, fluffs out the body plumage until it resembles a large white puffball, and then begins a high-stepping trot around an irregular path, with very little indication that it can see where it is going.

The small Black-bellied bustards of Africa and India give impressive display-flights or display-leaps above the grass that otherwise conceals them. Both species of florican make vertical leaps of up to 4m (13ft), but the most exciting aerial display is that of the Red-crested bustard, which flies vertically up as much as 30m (100ft) above the ground, somersaults backward with legs uppermost, then drops like a stone, pulling out at the last moment and gliding nonchalantly down to land. On the ground, too, this bird has a beautiful display, fully erecting its crest and employing stiff, clockworklike movements as it approaches a female.

A Pattern of Decline

CONSERVATION AND ENVIRONMENT

Bustards are little known, largely because of the difficulty of studying nervous, well-camouflaged, slow-breeding birds that tend to desert their nests if alarmed. This susceptibility to disturbance is a major cause of their decline, especially in northern parts of the family's distribution, where grasslands are coming under ever more pressure from agriculture. Great bustards can tolerate a degree of disturbance – indeed, they can only have colonized much of Europe thanks to felling of the forests by humans – but the mechanization of agriculture and the reduction of croplands to monocultures by the use of herbicides and fertilizers have been disastrous for them. Farming of the steppes reduced the Russian population from 8,650 birds at the start of the 1970s to an estimated 2,980 by the end of the decade. Similarly, the Little bustard is now nearly extinct almost everywhere in Europe except the Iberian Peninsula, as a result of the disappearance of herb-rich grasslands.

A similar loss of habitat has brought all three Indian species to the brink of extinction. Both floricans are seriously endangered. The Bengal species – which now numbers fewer than 1,000 individuals – survives only in a widely fragmented chain of protected areas along the foothills of the Himalayas, and in two small areas of central Cambodia and southern Vietnam. The Lesser florican appears to be restricted to tiny, scattered patches of grassland (*vidis*) in the far west of India, maintained as reserve grazing but in no other way protected, although the hunting of this species has recently been banned. In the Himalayan foothills the grasslands have largely been converted to tea estates, while in western India they are being converted to pasture; in neither case can the florican concerned survive the change.

Although the Great Indian bustard is perhaps more tolerant of agriculture, its populations have been steadily declining for decades. This species also now numbers fewer than 1,000 birds, but is currently showing signs of recovery in Rajasthan. The decline of its closest relative, the Australian bustard, is commonly attributed to continued, remorseless, indiscriminate shooting from the earliest days of European settlement, but it nonetheless seems certain that conversion of its habitat to farmland is the major factor in its disappearance.

The decline of the houbara is plainly the direct result of hunting. This is the one species of bustard figuring strongly in human culture, as the most prized quarry in the Arab tradition of falconry. In recent years, oil wealth and technology have vastly increased the scale and efficiency of such hunting, and it is feared that in many parts of its range the houbara has been almost completely extirpated. A captive-breeding program has now been established in Saudi Arabia, partly to study the bird's breeding biology and to reintroduce it into areas where it has been drastically reduced through persecution. Unfortunately, bustard hunting by Arabs appears to be growing in parts of Africa, notably the Saharo-Sahelian zone, and certain other species – particularly the Nubian bustard – have suffered alarming declines as a result.

No other purely African bustard is known to be seriously at risk, although the restricted ranges of the Blue and Little brown bustards must be cause for sustained vigilance. Moreover, throughout Africa the pressure to grow more food is unrelenting, and it cannot be long before some of these peace-loving ground-dwellers emerge as new candidates for extinction. PC/NJC

Buttonquails

Above *Sometimes known as "bustard quails," the buttonquails are ground-dwelling birds that inhabit scrub and grassland habitats. The Black-breasted buttonquail seen here is an exclusively Australian species.*

FACTFILE

BUTTONQUAILS

Order: Gruiformes

Family: Turnicidae

16 species in 2 genera, including: **Black-breasted buttonquail** (*Turnix melanogaster*), **Small buttonquail** (*T. sylvatica*), **Sumba buttonquail** (*T. everetti*), **Yellow-legged buttonquail** (*T. tanki*), **Lark buttonquail** or **Quail plover** (*Ortyxelos meiffrenii*).

DISTRIBUTION Africa, S Spain, S Iran to E China and Australia.

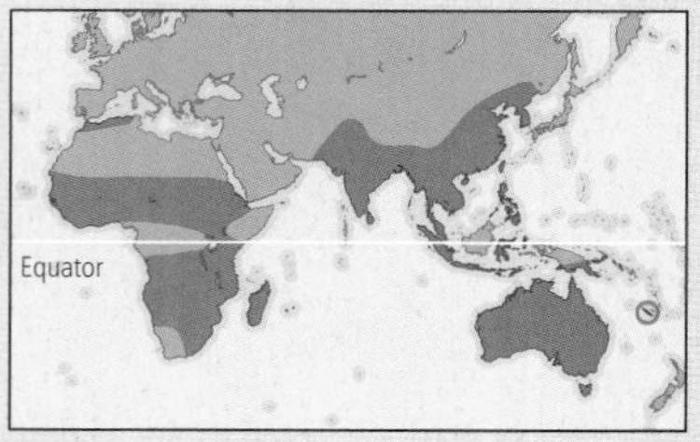

HABITAT Open grassland, thin scrub, crop fields.

SIZE Length 10–23cm (4–9in).

PLUMAGE Buffy-brown with camouflage markings in browns, grays, and cream; tail and wings short; females brighter and sometimes larger than males.

VOICE Females utter a drumming or booming call in the breeding season.

NEST In ground hollows lined with grass.

EGGS 3–7, oval, glossy, richly marked on pale background. Incubation 12–14 days; chicks leave nest almost at once.

DIET Small seeds and insects.

CONSERVATION STATUS The Buff-breasted buttonquail (*Turnix olivii*) is Endangered, and the Black-breasted and Sumba buttonquails are Vulnerable.

BUTTONQUAILS ARE SMALL, QUAIL-LIKE, essentially ground-dwelling birds that inhabit warm, semi-arid regions of the Old World. They owe their alternative name of hemipode ("half-foot") to the fact that the rear toe is missing, providing an easy way of distinguishing them from quails.

The most extraordinary aspect of buttonquail behavior is that the birds display one of the most complete reversals of normal sexual roles known in the avian world. Females are the more brightly colored sex, and where the sexes differ in size, they are the larger. They also defend a territory, pugnaciously driving off other females.

Small and Secretive

FORM AND FUNCTION

Buttonquails run through grassland like small rodents, adopting a typically crouching posture They are secretive and difficult to flush, even then flying only short distances. The one exception is the Lark buttonquail of Africa, which has a strong, larklike flight and flies some distance before alighting. Despite their general reluctance to fly in the face of disturbance, a few species are partially migratory, including the Yellow-legged and the Common buttonquails.

Switched Sexual Roles

BREEDING BIOLOGY

Sexual reversal in the buttonquails extends to their breeding biology. The female initiates courtship, calling with low-pitched, booming cries to attract males and defend her territory; she is able to do so thanks to an enlarged trachea and the presence of an inflatable bulb in the throat that is not present in the silent males.

After the eggs have been laid, it is left to the male to incubate the clutch and tend the chicks. In some species, the female will then set out to court a new partner; she may mate with several males in a single season, producing multiple clutches each with a different father. At 12–14 days, the incubation period is among the shortest for any bird. Yet the young are unusually precocial; they are ready to leave the nest shortly after hatching, following the father who continues to brood them and to supervise their feeding.

In regions lacking any marked seasonal variation in climate, buttonquails will breed opportunistically all year round, and the young can attain sexual maturity at 4–6 months, as in the Old World quails.

Two Australasian species are at risk, one of which – the little-known Buff-breasted buttonquail of northeastern Queensland – is classed as Endangered due to destruction of its grassland habitat. A third species, the Sumba buttonquail from Indonesia, is deemed Vulnerable. Only relict populations of Small buttonquail still exist in Europe, in southern Portugal and Spain. DH/RH

Mesites

THE MESITES ARE A MADAGASCAN *forest-dwelling family. They are thrush-sized birds of the forest floor, whose normal mode of locomotion is walking; they fly only to escape disturbance. With the progressive destruction of Madagascar's forests (only 10 percent of the island remains under tree cover), many species are now threatened, including all three mesites, which are classed as Vulnerable.*

The White-breasted mesite is found in the dry, deciduous forest of northwestern Madagascar, where its total population exceeds 7,000 birds and possibly reaches as high as 38,000. The Brown mesite occurs in damp evergreen forest in the east of the island, where it is present over a wide area but is very unevenly distributed. The Subdesert mesite, with its distinctive decurved bill, inhabits the dry, spiny brush woodland of southwestern Madagascar. This latter species has only a small range, in which it occurs patchily but at high densities.

Somber-colored Ground-dwellers

FORM AND FUNCTION

Mesites have functional wings but rudimentary collarbones, and can only fly weakly for short distances, mainly to ascend to roosting perches. While they build nests above ground, these are invariably sited so that the birds can climb or scramble, rather than fly, up to them.

Breeding takes place from October to December. In contrast to the gender reversal exhibited by the buttonquails, the mating system of the White-breasted and Brown mesites appears to involve "normal" sexual roles, with the male initiating courtship and the female customarily brooding the eggs. The situation may, however, be otherwise in the Subdesert mesite, in which the adult female is more brightly colored than the male. Some parties comprising males and one dominant female have been reported, but in a more recent study females have been observed sharing parental duties, with two birds (paired to one male) laying in one nest. DH/RH

Above *Only found in Madagascar, the mesites are rail-like birds of the forest floor. The Brown mesite **1** lacks the pale, spotted underplumage that gives the White-breasted mesite **2** its common name.*

Left below *Mesites only rarely leave the ground, but if a bird is alarmed, it will fly to a low perch on a spiny shrub and "freeze," as this Subdesert mesite is doing.*

FACTFILE

MESITES

Order: Gruiformes

Family: Mesitornithidae

3 species in 2 genera: **Brown mesite** (*Mesitornis unicolor*), **White-breasted mesite** (*M. variegata*), **Subdesert mesite** (*Monias benschi*).

DISTRIBUTION Madagascar

HABITAT Forest (*Mesitornis*); marginal scrub (*Monias*).

SIZE Length 30–32cm (12–13in).

PLUMAGE Brown to grayish above, paler below and spotted (not Brown mesite); bill straight or (*Monias*) downcurved.

VOICE Various calls, including a nak-nak-nak that has earned the Subdesert mesite the local name of "naka."

NEST A twig platform 1–2m (3–6ft) up in bush.

EGGS 1–3, off-white with brown spots; hatchlings leave nest at once.

DIET Seeds and insects.

CONSERVATION STATUS All 3 species are listed as Vulnerable.

Kagu

The kagu is a strange bird that combines the appearance of a small, pale-colored heron with that of a rail, but also exhibits a peculiar long, erectile crest. It is not closely related to any other birds, although the cranes, rails, and Sun bittern have all been suggested as its closest relative within the order Gruiformes.

The bird is restricted to the Pacific island of New Caledonia, where it is now found only in undisturbed forests. Most kagus are sedentary within their territories, but movements of up to 8km (5 miles) have been recorded.

Flightless Denizens of the Forest Floor

FORM AND FUNCTION

The rounded wings of the kagu are not much reduced in size, but the bird is flightless because of its reduced wing musculature, although it is able to glide when escaping downhill. Its body is stocky, and its stance upright. The birds walk on the forest floor on long, strong, red legs, running to escape or when quarreling over territories. The large eyes are dark red in adult birds, presumably affording good vision in the dark forest interiors that they inhabit. Adults have pale gray to almost white plumage, but with dark bands on the wings that are exposed in displays. Chicks and juveniles have a concealing coloration of brown and gray, with fine dark banding.

Kagus capture their food of millipedes, mollusks, worms, insects, and small lizards on the ground, often among the leaf-litter. Feeding birds watch and listen for prey, which they catch with quick movements of the head; alternatively, they sometimes move leaves aside or dig in the soil with their bill.

Pairs defend their territories by singing duets in the early morning, the repeated calls having been likened to something between a rooster crowing and the barking of a young dog. The breeding season is from June to December. Clutches are of a single egg, laid on a simple nest of dead leaves set on the ground. Incubation lasts from 33 to 37 days and is carried out by both parents in alternating 24-hour shifts. The chick usually leaves the nest after about three days. It is fed by both parents, and is brooded at night until it is about six weeks old. Adults are long-lived, commonly surviving for up to 20 years in captivity.

A Case for Captive Breeding

CONSERVATION AND ENVIRONMENT

The kagu has become restricted to remote forest areas on New Caledonia, where the total population is only about 650 birds. It was formerly hunted and captured for the aviary trade and as a pet. Its range has been reduced by forest destruction, some of it for nickel mines. Today the main threat to the species is from dogs, which kill adult and young birds. Fortunately the kagu can be kept and bred in captivity, and the wild populations are now being supplemented by the reintroduction of captive-bred birds.

DH/RH

FACTFILE

KAGU

Rhynochetos jubatus

Order: Gruiformes

Family: Rhynochetidae

DISTRIBUTION New Caledonia

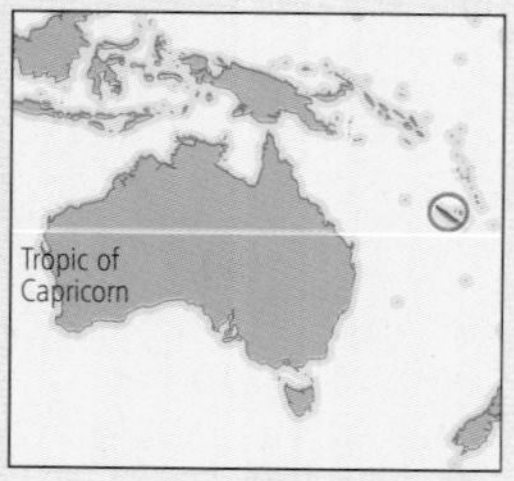

HABITAT Lowland rainforest; drier mountain forest.

SIZE Length 55cm (22in); weight up to 1,100g (39oz).

PLUMAGE Pale gray, but spread wings show striking pattern of dark gray and buff bands; long, erectile crest. Bare parts red; bill slightly downcurved.

VOICE Loud barking at dawn; various quiet hissing and rattling calls.

NEST Leaves on ground.

EGGS Single creamy brown egg with dark brown blotches. Incubation by both parents for 33–37 days; young precocial, leave nest a few days after hatching.

DIET Snails, worms, millipedes, lizards, and insects from forest floor.

CONSERVATION STATUS Endangered.

Above *Restricted to New Caledonia – a small island in the southwestern Pacific – the kagu is a flightless bird notable for its flamboyantly large crest, seen erected in this individual.*

Sun Bittern

THE SUN BITTERN TAKES ITS NAME FROM THE bright patches of golden-buff that are suddenly revealed on its rounded wings in threat displays. It inhabits riversides in tropical and subtropical forests, where it seeks cover within vegetation when disturbed and is easily overlooked.

Sun bitterns are found in lowlands of the Neotropical region from Guatemala to eastern Bolivia and central Brazil; there are three subspecies that differ slightly in size and coloration. Their habitat is in open forest or woodland adjacent to streams, rivers, lake edges, or lagoons. They are not particularly sociable birds, normally being encountered singly or in pairs in spots where the trees provide shade from the full heat of the sun.

A Spectacular Threat Display

FORM AND FUNCTION

The Sun bittern is stout-bodied and rather heron-like, with a long neck and long legs. When feeding or skulking, the bird appears to have somber, camouflaged plumage. However, this impression is belied in flight or display, when the broadly rounded wings are spread to reveal conspicuous patches of golden-buff and chestnut on the primary feathers, and bands of the same color across the tail. Adult males are no more brightly colored than females or even juveniles, and it now seems clear that the spectacular frontal display of the Sun bittern is for threat or defense rather than courtship. In its startling transformation from a reclusive to a large and threatening bird, the patches on the broad wings stand out as big, intimidating "eyes" (supposed also to look like the rising sun, which explains the bird's name).

The birds' diet consists of a wide range of small animals taken from shallow water and its margins, including insects, snails, worms, shrimps and other crustaceans, tadpoles, frogs, and small fish. Sun bitterns often stalk prey in the same manner as a small heron, with a stealthy approach with the head drawn back, followed by a quick lunge of the bill.

Long-lived and Slow to Develop

BREEDING BIOLOGY

The nest of mud, grass, and sticks is a rough cup built on a narrow tree branch. Breeding occurs at the beginning of the rainy season, when mud is available for nest construction. Clutches are of one or two eggs (normally two) that are pinkish-buff with irregular purplish spots. Both parents incubate for a period of about 30 days. Hatchlings have abundant down that is cream-colored to brown with strong black markings. The young fledge after 22–30 days in the nest, but may be dependent on one or both parents for a further two months or more after fledging. Captive birds do not reach sexual maturity until 2 years old, and they may live for over 30 years. DH/RH

FACTFILE

SUN BITTERN

Eurypyga helias

Order: Gruiformes

Family: Eurypygidae

HABITAT Edges of streams and swamps in forests.

SIZE Length 43–48cm (17–19in).

PLUMAGE Sun bitterns show camouflage patterns when at rest, but golden-buff and chestnut marks on wings and tail are conspicuous in display.

VOICE Plaintive, long-drawn whistle.

NEST Made of mud, grass, sticks, stems, and leaves, shaped into a rough cup and set up to 7m (23ft) high in a bush or tree.

DISTRIBUTION Guatemala to Bolivia and Brazil.

EGGS Usually 2, pinkish-buff with dark spots. Both parents incubate (for 30 days) and care for young.

DIET A wide range of small freshwater animals, including mollusks. crustaceans, aquatic insects, frogs, and small fish.

CONSERVATION STATUS Not threatened.

Below *A timid, elegant wading bird of the American tropics, the Sun bittern stalks its prey along stream banks, catching small aquatic creatures with jabs of its beak. It reserves its most striking plumage, the chestnut and black patterns on its wings, for a threat display intended to startle potential predators.*

Seriemas

THE SERIEMAS ARE WARY, DIURNAL BIRDS THAT feed on the ground but roost in trees. The young of both species are often captured for taming, since they make efficient "watchdogs" against approaching predators when kept with domestic fowls.

The two seriemas are placed, on the basis of similarities in musculature and the skull, in the same order (Gruiformes) as the cranes and rails. Superficially, however, they bear a resemblance to the secretarybird (order Falconiformes) of Africa, and are its ecological counterpart on the dry pampas and scrub-savanna of South America.

Tall Birds of the Long Grasses

FORM AND FUNCTION

Seriemas are ground-dwellers, with a typically upright posture. Although their wings are developed, they seldom fly, prefering to run with head lowered when disturbed. Their long neck and legs are well adapted for living in long grass, in much the same way as in the secretarybird and, among much closer relatives, the Old World bustards.

Seriemas live in dry, open habitats of South America east of the Andes and south of the Amazon. The Black-legged seriema has a more restricted range than the better-known Red-legged (or Crested) seriema of the grasslands; it prefers open woodland and scrub areas, and is more arboreal, nesting above ground and roosting higher up.

Most of the birds' diet consists of small animals such as frogs, reptiles, large insects, young birds, and rodents. They also take smaller quantities of vegetable foods, including leaves, seeds, and wild fruits as well as cultivated maize and beans. One notable feature shared with the secretarybird is that their diets include small snakes. Often seriemas feed by walking about close to cattle and horses, presumably taking invertebrates disturbed by the grazing stock.

Nests in the Bushes

BREEDING BIOLOGY

Pairs of seriemas occur widely spaced, so territories are probably defended. Their bulky nests of sticks are built in bushes or low trees. Clutches are of two or three eggs, which are white with a few brownish or purple spots. Incubation is carried out mainly by the female and lasts 24–30 days. Although the newly-hatched young are already well covered with down, they remain in the nest under parental care until they are about 14 days old, after which they jump down to follow their parents on the ground. The young do not attain adult weight or full maturity until they are 4–5 months old. DH/RH

Above Like their African counterpart, the secretarybird, the South American seriemas are ground-feeders renowned for occasionally eating snakes, although in fact these form only a small part of their omnivorous diet.

Left A calling Crested seriema displays the panoply of head feathers from which the species takes its name. The birds' loud, yelping cries often betray their presence before they themselves are seen.

FACTFILE

SERIEMAS

Order: Gruiformes

Family: Cariamidae

2 species in 2 genera: **Black-legged** or **Burmeister's seriema** (*Chunga burmeisteri*) and **Red-legged** or **Crested seriema** (*Cariama cristata*).

HABITAT Dry, open woodland and scrub in W Paraguay, N Argentina (Black-legged seriema); grasslands in Brazil to Uruguay and N Argentina (Red-legged seriema).

SIZE Length 70–85cm/28–34in (Black-legged seriema); 75–90cm/30–36in (Red-legged seriema).

PLUMAGE Black-tipped tail, negligible crest, black bill and legs (Black-legged seriema); white-tipped tail, large frontal crest, red bill and legs (Red-legged seriema).

VOICE Loud, far-carrying yelping noises during courtship.

DISTRIBUTION Southern S America.

NEST In bush or low tree, up to 3m (10ft) above ground level.

EGGS 2 or 3, white to buff with brown markings. Incubation mainly by female (24–30 days); both parents tend young.

DIET Omnivorous.

CONSERVATION STATUS Not threatened.

Finfoots

FINFOOTS ARE A FAMILY OF THREE WIDELY *scattered species; one inhabits Africa, another central and southern America, the third southern Asia. They share characteristics with various families of waterfowl, but the balance of the available evidence suggests that they are most closely related to the rails.*

▲ ***Above*** *On the water, the finfoots can recall grebes or cormorants. The African finfoot* ***1*** *is the largest of the species, almost twice the size of South America's sungrebe. The slightly smaller Masked finfoot* ***2*** *is a reclusive bird of the Southeast Asian tropics.*

All three species of finfoots are shy and retiring, and they are not easily observed in their tropical waterside settings. Even if the rails are their nearest relatives, they resemble grebes when seen on the water, an affinity suggested by their alternative name of sungrebes.

Shy Waterbirds of the Tropics

FORM AND FUNCTION

Finfoots combine the lobed feet of grebes and coots with the long neck of cormorants; the African finfoot also has the cormorant's stiffened tail. When taking flight they patter across the water before becoming airborne, but their normal reaction to disturbance is to run ashore into dense undergrowth, or to swim away with the body submerged and only the head and neck visible. They dive infrequently. Although finfoots are most often seen on the water, they also spend some time – how much is still unclear – on land, in the cover of thick vegetation.

All three finfoot species are essentially birds of the tropics and subtropics. The sungrebe occurs in the American tropics from southern Mexico to northern Argentina; the African finfoot is found over most of tropical Africa south to South Africa; while the Masked finfoot ranges from northeastern India through southeast Asia to Sumatra, where it occurs in coastal and inland wetlands, including mangrove swamps and flooded forest. All species are mainly sedentary, and none of them makes regular migrations.

Little detailed information is available on their diet, but insects picked from the water or its edges are the most frequently reported food, with notes also of mollusks, crustaceans, worms, other small invertebrates, tadpoles, frogs, and, at least occasionally, seeds and other vegetable matter.

Pouches for the Chicks

BREEDING BIOLOGY

Sungrebe parents carry their young in flight as well as on the water. The adult male has special folds of skin beneath each wing, forming cavities into which the chicks fit, muscular control probably helping to hold them firmly clutched against the parent's body. The naked, helpless hatchlings are carried from the nest even before their eyes have opened.

It is not yet known whether this unique adaptation applies also to the two Old World finfoot species, which have young that are downy and better developed when hatched. The African finfoot may well share the habit, for it is known, like the sungrebe, to lay just two eggs; the same is, however, unlikely to be true of the Masked finfoot, which is reputed to lay a clutch of 5–7 eggs. In the sungrebe, both parents share the incubation, although only the male is confirmed as carrying young; in the African finfoot, females have been seen accompanying well-feathered juveniles.

The Masked finfoot is listed as Vulnerable, having suffered a major population decline in recent years. The surviving birds are patchily distributed across at least nine countries, but throughout their range there has been destruction of the wetland habitat on which they depend. Besides deforestation and drainage, they have also been affected by the pressures of increased human traffic on the waterways by which they live. DH/RH

FACTFILE

FINFOOTS

Order: Gruiformes

Family: Heliornithidae

3 species in 3 genera: **African finfoot** (*Podica senegalensis*), **Masked finfoot** (*Heliopais personata*), **sungrebe** (*Heliornis fulica*).

DISTRIBUTION C and S America, Africa, SE Asia.

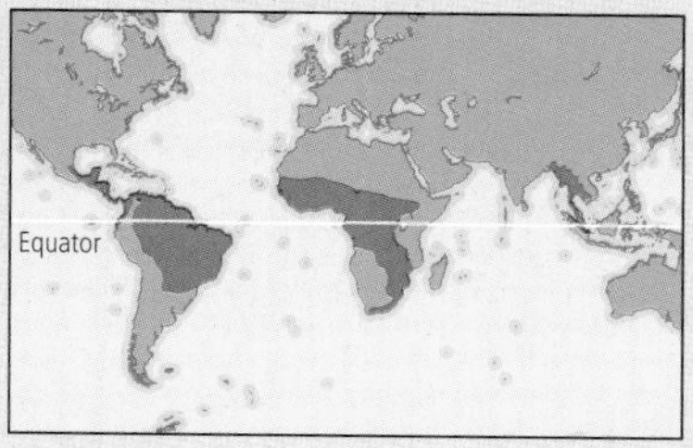

HABITAT Rivers and lakes, mainly in forest.

SIZE Length 35–59cm/14–23in (African finfoot); 49cm/19.5in (Masked finfoot); 30cm/12in (sungrebe).

PLUMAGE White-spotted back, bill and legs orange-red (African finfoot); back plain, face and throat black, yellow bill, green legs (Masked finfoot); olive-brown, with scarlet bill and black and yellow feet (sungrebe).

VOICE Various barking, growling, and bubbling sounds associated with territorial defense and courtship; otherwise, largely silent.

NEST On low branch or flood debris.

EGGS 2–4 (Masked finfoot 5–7); reddish-brown or cream. Incubation perhaps 11 days (sungrebe).

DIET Insects, crustaceans, mollusks, occasionally seeds and frogs.

CONSERVATION STATUS The Masked finfoot is listed as Vulnerable.

Plovers and Lapwings

THE NAME PLOVER COMES FROM THE LATIN pluvia, *meaning "rain," and may have arisen because plovers sometimes appear after a downpour; alternatively, it may derive from the fact that the plumage of some plovers is dappled, as though it has been rained upon! Despite the range in size, the proportions of plovers are conservative, the birds being moderately long-legged but short-billed, with rounded heads and large eyes.*

Plovers are very uniform in shape, with short beaks, the upper mandible of which is slightly swollen, round heads, large eyes, moderately long legs, and neat, rather compact bodies. The different species even forage in the same way, standing and watching, running forward and pecking, then standing and watching again. Where they are not conservative is in their patterns of movement. The family includes species that undertake some of the greatest migrations in the world, breeding in the high latitudes of the northern hemisphere but spending the nonbreeding season close to the equator or in the southern hemisphere. In contrast, many others are sedentary or undertake only local movements.

A Family Divided

FORM AND FUNCTION

The plover family is divided into two distinct subfamilies: the Charadriinae and the Vanellinae. The Charadriinae tend to be small but large-headed, and most species are called plovers or sand-plovers. Exceptions are the Eurasian dotterel and several Australian and South American species to which the name dotterel has also been applied, such as the Black-fronted (Australia) and Tawny-throated (South America). Use of the name dotterel has no taxonomic basis. The wrybill of New Zealand is also a member of the Charadriinae. Up to eight genera are recognized within the Charadriinae.

The Vanellinae tend to be larger but with proportionately smaller heads. Many species are called plovers, such as the Blacksmith plover of southern Africa, but the name "lapwing" is being increasingly used for this subfamily. Lapwings have broad wings that are boldly marked in black and white, and they tend to fly with an exaggerated flap that makes them very conspicuous and is the source of their name, originally used only for the Northern lapwing of Europe. All Vanellinae are usually placed in the genus *Vanellus*, but the Pied plover of South America is sometimes placed in a genus on its own (*Hoploxypterus*), and some authors have suggested that it belongs in the Charadriinae.

Members of the Vanellinae are typically boldly marked in black and white, especially on the wings and tail and often on the head, with juveniles generally duller than adults. Members of the Charadriinae that do not undertake long migratory flights have similar simple but strong plumage patterns as adults. In contrast, species that are strongly migratory have distinct breeding and nonbreeding plumages, the differences between the two being subtle in many cases, but very distinct in many of the Holarctic-breeding Charadriinae. Breeding plumage can be very bright, with black, white, gold, and chestnut featuring prominently in many species, especially on the head and neck. Nonbreeding plumage generally consists of drab grays and browns, resembling juvenile plumage.

Species that have distinct breeding plumage must molt their body feathers twice a year, but replace their flight feathers once during the nonbreeding season. These molts have to be tightly synchronized to fit in with the demands of seasonal migration and breeding. This necessity results in a partial (body) prebreeding molt and a complete (body and flight feathers) postbreeding molt. Species without a distinct breeding plumage may follow the same pattern, or may replace body and flight feathers only once a year. Molts may be poorly synchronized and extended if breeding occurs over a long period or if the timing of breeding varies depending on seasonal conditions. Juveniles have an incomplete molt, replacing juvenile body plumage (consisting of pale-edged dorsal feathers) within a few months, but often retaining at least some flight feathers until their second year of life.

The Vanellinae have a bony knob and, in some species, a distinct spur on the carpal joint at the front of the wing. Most have four toes, although the fourth (hind) toe is vestigial, whereas most Charadriinae have only three toes. The toes generally lack webbing, but are partly webbed in a few species, notably the Semipalmated plover of the Americas. The Vanellinae often have boldly-colored, fleshy wattles and bare facial skin, especially well-developed in the Masked lapwing of Australia, while the Charadriinae have, at most, a colored orbital ring. Several Vanellinae, like the Northern lapwing, have a crest.

Left *Ringed plovers* (Charadrius hiaticula) *nest in Eurasia, the eastern Canadian Arctic, and Greenland. Laid on shingle beaches, the eggs are camouflaged among the pebbles, but still suffer predation from Arctic foxes, crows, dogs, and badgers.*

FACTFILE

PLOVERS AND LAPWINGS

Order: Charadriiformes

Family: Charadriidae

Approximately 65 species in up to 10 genera.

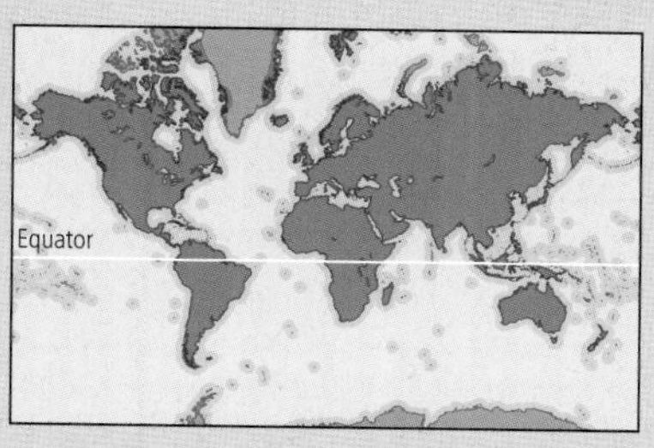

DISTRIBUTION Worldwide except for Antarctic, with nonmigratory species and races concentrated in tropics and in temperate regions of the S hemisphere. Nearly half of Charadriinae breed in high latitudes of the N hemisphere but migrate S during boreal winter, while many species and even races of Charadriinae, and most of Vanellinae, are sedentary or undergo only local movements.

HABITAT Shorelines, grasslands, tundra; some species abundant on agricultural land.

SIZE Length 14–31cm (5.5–12in) in the Charadriinae, 25–41cm (10–16in) in the Vanellinae; **weight** 30–250g (1.2–10oz) in the Charadriinae, 150–400g (5–16oz) in the Vanellinae. Males may be slightly larger than females.

PLUMAGE: Many Charadriinae molt into a distinct and bright breeding plumage, but have a nonbreeding plumage of gray and brown. In contrast, many nonmigratory Charadriinae and most Vanellinae retain simple but bold color patterns throughout the year when adult. Males may be more brightly colored than females.

VOICE Calls of Vanellinae strident and scolding, with repeated harsh syllables. Those of many Charadriinae are plaintive, and melancholy, although also repeated often.

NEST Usually a simple scrape on bare or open ground.

EGGS Typically 4 (range 2–6), pale with small flecks to large, dark blotches. Incubation 18–38 days; precocial young fledge at 21–42 days. Sexual maturity within first year for some species, but not until second year or later for others.

DIET Plovers forage visually, standing, watching, and then chasing prey that includes terrestrial and aquatic invertebrates, occasionally small vertebrates, and sometimes berries.

CONSERVATION STATUS The Javanese lapwing (*Vanellus macropterus*) is currently listed as Critically Endangered, but is probably already extinct, not having been recorded since 1940. 2 other species are Endangered, and 5 are Vulnerable.

See Plover and Lapwing Subfamilies ▷

Out of Gondwana

EVOLUTION

The resident Charadriinae of the southern hemisphere are taxonomically rich at the generic level, with up to eight genera being recognized. Six of these occur in Australasia and five are endemic to that region. There are three genera (two endemic) in South America, but only one in Africa. In contrast, only three genera of Holarctic migrants are recognized, and one of these also contains southern-hemisphere species. This pattern of endemicity and generic richness suggests that the Charadriinae are an ancient group of southern-hemisphere, and probably Gondwanan, origin.

The Holarctic migrants, in comparison, are a recent evolutionary innovation that have probably flourished over the past 20,000 years, since the retreat of the Pleistocene glaciation made extensive but seasonal breeding habitat available in the northern hemisphere. Interestingly, the Holarctic migrant species have secondarily given rise to some of the resident southern-hemisphere species. For example, the Red-capped plover of Australia and the Malaysian plover of Southeast Asia are closely related to the Holarctic-breeding Kentish plover. Similarly, the Palearctic-breeding Little ringed plover has a resident race (*jerdoni*) in Southeast Asia. In a number of Holarctic-breeding species with broad breeding ranges, the southernmost breeding populations may be virtually sedentary.

With endemic races of Charadriinae only in Australasia and South America, and the Vanellinae most species-rich in Africa, it has been postulated that ancestral plovers evolved in Gondwana, with ancestral Vanellinae soon isolated in Africa by continental drift. Africa separated from other southern continents about 125 million years ago. South America and Australia remained united as recently as 45 million years ago, and were home to the ancestral Charadriinae.

Above *The English names of* Pluvialis squatarola *(foreground) reflect seasonal change in its plumage. As the Grey plover, its winter color is mottled silver-gray, while its US name, the Black-bellied plover, alludes to its breeding plumage. In the background is a turnstone.*

Below *Common on farmland and marshes, the Northern lapwing* (Vanellus vanellus) *is known by its tumbling flight, mournful call, and long, wispy crest.*

Birds of the Open Spaces

DISTRIBUTION PATTERNS

Plovers occur on all continents except Antarctica, and different species can be found from the tideline to grasslands above the treeline. They are birds of open spaces, and although they are referred to as waders or shorebirds, few species actually wade when foraging. Most species search for food along the shores of oceans, estuaries, rivers, streams, or lakes, but some of the Charadriinae and many of the Vanellinae are grassland species that have no particular association with wetlands. Several of these species have learned to exploit pasture and croplands. The Inland dotterel (Charadriinae) occurs in semi-arid shrublands and plains of Australia. The Long-toed lapwing (Vanellinae) is exceptional in being convergent with jacanas by foraging on floating vegetation. The Vanellinae generally avoid oceanic coastlines and are species of inland environments, whether wetland or grassland, whereas many of the Charadriinae occur on tidal coasts.

For migratory species, breeding and non-breeding environments may differ. Most migratory Charadriinae breed in the high latitudes of the northern hemisphere, including the Arctic tundra. Nesting may be on beaches, gravelly river and lake margins, or away from the water's edge. In contrast,

the nonbreeding environment is usually tidal coastline shared with sandpipers (Scolopacidae). Like some grassland species, a number of these coastal species have learned to forage in agricultural environments. For example, the Eurasian golden plover, normally a coastal species in the nonbreeding season, will forage with Northern lapwings on fields being prepared for sowing.

The Vanellinae are largely tropical and subtropical in distribution and are best represented in Africa, with 11 of the 24 species confined or largely confined there. There are no Vanellinae in North America, three species in South America, four species in southern and southeastern Asia, and two species in Australasia. In the southern hemisphere, the Southern lapwing (South America) and the Masked and Banded lapwings (Australasia) extend well into temperate regions. In the northern hemisphere, the Northern lapwing, Sociable plover, White-tailed plover, and Grey-headed lapwing occur in high latitudes during the breeding season. All migrate south in the nonbreeding season, although the Northern lapwing remains well to the north in western Europe, where the Gulf Stream causes some climatic amelioration.

Patterns of movement among other Vanellinae are variable. The Brown-chested plover is an intra-African migrant, breeding in West Africa but migrating to central Africa in the region of the Lake Victoria Basin. One population of the Spur-winged plover migrates from northern Africa to Greece to breed, but other African populations of the species are sedentary or undertake local, seasonal movements in response to seasonal patterns of rainfall. Other tropical, subtropical, and southern-hemisphere species are also sedentary or undergo local, seasonal movements. Such movements can be regular or may vary from year to year with climatic conditions.

The Charadriinae are more widespread than the Vanellinae. Eighteen species are Holarctic migrants, breeding in the northern hemisphere but migrating south during the boreal winter. Some of these remain within the northern hemisphere, but others breed in the high Arctic of North America, Europe, and Asia, and migrate to South America, southern Africa, and southern Australasia during the nonbreeding season.

Plover and Lapwing Subfamilies

Plovers
Subfamily Charadriinae

41 species in 9 genera, including: Eurasian dotterel (*Eudromias morinellus*), Red-kneed dotterel (*Erythrogonys cinctus*), Black-fronted dotterel (*Elseyornis melanops*), wrybill (*Anarhynchus frontalis*), Double-banded plover (*Charadrius bicinctus*), Inland dotterel (*C. australis*), Kentish plover (*C. alexandrinus*), Little ringed plover (*C. dubius*), Malaysian plover (*C. peroni*), Red-capped plover (*C. ruficapillus*), Semipalmated plover (*C. semipalmatus*), St. Helena plover (*C. sanctaehelenae*), Tawny-throated dotterel (*Oreopholus ruficollis*), Eurasian golden plover (*Pluvialis apricaria*), Grey plover (*P. squatarola*), Shore plover (*Thinornis novaeseelandiae*).

Lapwings
Subfamily Vanellinae

24 species in 1 genus, including: Banded lapwing (*Vanellus tricolor*), Blacksmith lapwing (*V. armatus*), Brown-chested lapwing (*V. superciliosus*), Masked lapwing (*V. miles*), Northern lapwing (*V. vanellus*), Pied lapwing (*V. cayanus*), Southern lapwing (*V. chilensis*), White-tailed lapwing (*V. leucurus*), Long-toed lapwing (*V. crassirostris*).

Below Representative species of plovers and lapwings: **1** Golden plover (Pluvialis apricaria) *on nest;* **2** Killdeer (Charadrius vociferus) *performing a "broken wing" display to distract predators away from a nest site. Its curious name comes from its cry;* **3** Wattled lapwing (Vanellus senegallus) *in alert posture, showing spurs on the carpal joint of the wing.*

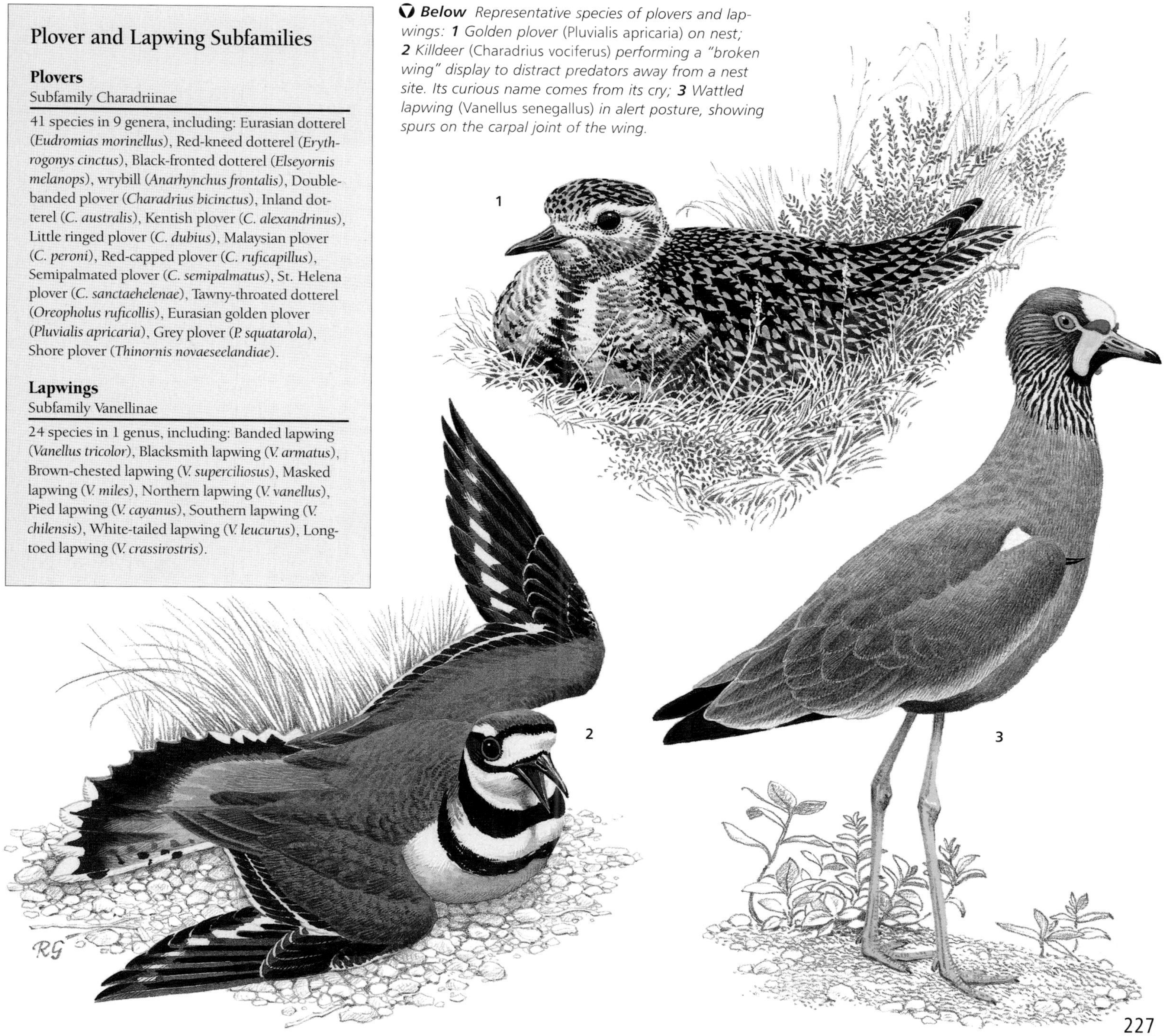

While almost all Charadriinae of the northern hemisphere are migratory, the southern hemisphere supports not only many of the Holarctic migrants but also species that are residents or local migrants. For example, South America has six species that remain within South or Central America and are sedentary or undergo local movements, and another six that are nonbreeding visitors from North America. In contrast, North America has only eight species, all being Holarctic migrants. Two of them stay within North America during the nonbreeding season. Similar patterns occur in Africa and Australasia, Africa having seven resident species and Australasia nine. Several Australasian species have distinct migratory patterns within the region, with the wrybill migrating the length of New Zealand and one population of the Double-banded plover migrating between New Zealand and southeastern Australia.

"Watch-Run-Peck" Foragers

DIET

Plovers forage on open ground, such as the moist mud left by a receding tide. Unusually, the Red-kneed dotterel (Charadriinae) and the White-tailed lapwing (Vanellinae) wade and even swim when foraging, and are the only species that regularly put their head underwater when searching for food. Plovers rely heavily on sight, hence their large eyes, and typically pluck food from on or close to the surface with their short bills. In contrast, the sandpipers forage largely by touch with their sensitive and often highly specialized bills.

Plovers feed largely on small invertebrates and occasionally small vertebrates, but will also take some plant matter. The Inland dotterel feeds on plant material during the day, to obtain moisture, but on invertebrates at night. Most species are essentially opportunistic feeders; food items for a typical Charadriinae species include mollusks, annelid and other worms, crustacea, insects (both aquatic and terrestrial), and spiders. Terrestrial invertebrates are not only taken by terrestrial plovers but are an important part of the diet of many plovers that forage around wetlands. Such invertebrates are trapped on the water's surface and subsequently become washed up along the shore, where plovers forage. Most prey items are eaten whole, but large crabs are carried to dry ground and dismembered before being eaten.

Typical foraging behavior of plovers consists of "watch-run-peck-stop-watch-run-peck" and so on, but some species search and probe for food rather than rely just on sight. Kleptoparasitism by the Grey plover on whimbrels and oystercatchers has been observed. Some species, such as the Double-banded plover, use foot-trembling in shallow water to flush prey. The timing of foraging is

Above *Blacksmith lapwings* (Vanellus armatus) *feeding in the shallows of a lake in Ngorongoro Crater, Tanzania. Lapwings are characterized by a feeding technique that involves running, halting abruptly, and then stooping to snatch food.*

Right *Plover young are cryptically patterned with brown and black speckling, and are looked after and defended by the adult birds until they fledge at 21–42 days. Here, newly hatched Ringed plovers lie exposed but well camouflaged in their shallow scrape nest.*

Above *Displays of breeding plovers and lapwings sometimes include spectacular aerial advertisement of the territory, as in the tumbling flight of the lapwing* **1**. *On the ground, when warning rivals off the mate or territory, the effect of the display may be enhanced by vocalizations and display of plumage, including crest, and colored wattles. Eurasian golden plovers perform a "song duel"* **2**. *Spur-winged lapwings adopt a threat posture* **3** *or run at the opponent* **4**.

small shells and stones. The Long-toed lapwing is unusual in having a comparatively well-built nest that may even rest on floating vegetation. The Red-kneed dotterel may also construct a substantial nest, including using mud to raise one side of the bowl. The Shore plover nests beneath vegetation, and may even use an abandoned petrel burrow. The clutch size is usually 4 for Holarctic-breeding plovers, but among southern-hemisphere Charadriinae clutch sizes of 2 to 3 are the norm. The eggs are oval to pear-shaped and well-camouflaged, and a completed clutch may be up to 70 percent of the female's weight. Incubation, ranging from 18 days in the smallest species to 38 days in the largest, does not begin until the clutch is complete.

Pairs may breed singly or in loose aggregations in suitable habitat, with territory defended against immediately adjacent pairs. In the Double-banded plover, the distance between nests ranges from 4 to 150m (13–500ft). Breeding success varies annually in the Holarctic, which may be linked to population cycles in other arctic fauna; for example, Arctic fox predation on ground-nesting birds is greatest when small mammals like the Arctic lemming are scarce. In the Red-capped plover, 10 percent of eggs yielded fledged young in one study. Average annual mortality of the Double-banded plover is 23–29 percent, while banding studies have demonstrated minimum ages of 8 years for the Red-capped plover and 9 for the Grey plover.

Dangerous Migrations

CONSERVATION AND ENVIRONMENT

Some plover species are amazingly abundant and widespread; there are an estimated 7 million Northern lapwings in Europe alone, while many Holarctic-breeding plovers have ranges that span several continents. In contrast, the population of the Shore plover of New Zealand is put at 140, and the species is confined to the remote 220ha (550 acre) Southeast Island of the Chatham Archipelago. Other plovers with very restricted ranges include the St. Helena plover, confined to the island of St. Helena, 2,800km (1,750 miles) west of Angola in the Atlantic Ocean, and the extinct Javanese lapwing that occurred only on the island of Java in the Indonesian archipelago.

The Javanese lapwing succumbed to hunting and habitat loss. The Shore plover disappeared from the main islands of New Zealand in the late 19th century due to introduced predators (rats, stoats, weasels, and cats), and was threatened on Southeast Island by uncontrolled scientific collecting. These two species were vulnerable because they had naturally small populations and restricted distributions, but similar problems can face more widespread species, while the migratory nature of many plovers can make them especially sensitive. Migration takes birds across international boundaries, and means that they rely on widely dispersed sites where protection levels vary. MJB

dictated by when food is accessible. Especially in tidal environments, this can be at any time of the day or night, although the reliance on sight means that plovers are preferentially diurnal, only foraging at night when necessity demands.

Complex Courtship Rituals

BREEDING BIOLOGY

Plovers are seasonal breeders, with breeding being very synchronous in the Holarctic migrants that exploit a brief seasonal period of productivity, but extended and sometimes opportunistic in response to rainfall in less predictable environments, such as inland Australia. Many species are monogamous and faithful to the same partner, with both parents incubating the eggs and tending the chicks, which are able to feed themselves from hatching. The Eurasian dotterel is unique among plovers in being polyandrous, with the female more brightly colored than the male. She may produce several clutches, each for a different male. Where polyandry and polygamy occur, the single female or male may or may not assist with brood care.

Breeding and courtship behavior includes complex aerial displays, often with butterflylike flights, and ground displays of running, wing-dropping, tail-fanning, and bowing and curtseying. These displays may also be very vocal, and calls can include melodious warbles and trills. There are also distinctive patterns of behavior associated with territorial defense and antipredator strategies. Plovers in general, and particularly the Vanellinae, are aggressive around the nest, calling loudly and swooping on intruders, and the practice of feigning injury to distract predators is widespread.

The nest is usually a simple scrape in bare ground, sometimes poorly lined with grass or

SPECIAL FEATURE

GOING OUT WITH THE TIDE?

The conservation of shorebirds

MANY WADERS ARE INTERNATIONAL MIGRANTS that breed in the high latitudes of the northern hemisphere in summer and fly south for the nonbreeding season. Migration allows these birds to exploit seasonally productive habitats, but it poses problems for their survival. A population may depend, for example, on breeding areas in northern Russia, migration stopover sites in Sweden, a molting area on the Dutch coast, and several sites used during the nonbreeding season in Britain and France. Similarly, birds breeding in the Russian Far East may head south, via stopover sites in South Korea, to spend the nonbreeding season in eastern Australia. Particularly during migration but also in the nonbreeding season, whole populations that may breed at a low density across vast areas of tundra depend upon relatively small coastal and wetland sites elsewhere in the world. The sites that shorebirds use are like stepping stones across the globe, but the sheer concentration of birds makes them vulnerable to human activities. What happens at sites in one country can affect the abundance of a species throughout its range.

Humans impact upon shorebirds through hunting, habitat loss, and disturbance, and these impacts can be severe because, like the shorebirds, human populations are concentrated along coastlines. Reclamation of coastal environments has taken place for thousands of years, but the pressure on coastlines is increasing with demands for deepwater ports, land for industry, aquaculture, and tidal power projects. In east Asia, recent and proposed reclamation projects have already destroyed or are threatening tens of thousands of hectares of tidal mudflat. Even when birds have flight ranges that exceed 5,000km (3,000mi), the loss of major foraging areas used to refuel on their epic journeys may make such migrations impossible.

When all appropriate habitat at a site disappears, then so do the waders. Sometimes, however, only part of the habitat is lost. When 60 percent of intertidal land was lost at Seal Sands in the Tees Estuary of northeastern England, for example, the decline in abundance of different species varied from 0 to 95 percent. This variation was related to changes in duration of tidal exposure and the differing feeding times required by the various species, and to their adaptability to using alternative feeding areas and prey. Following drainage of wet agricultural "polder" land in the Netherlands, Common snipe (*Gallinago gallinago*) appeared to adapt by shifting their main southward-migration molting area from the Netherlands to Britain. They may have displaced birds already using the sites in Britain, however, and the species subsequently declined as a result of further habitat loss.

Habitat loss involves not just coastal environments, as many shorebirds use inland wetlands, pasture, and moors for breeding. In Europe, government policies have encouraged the conversion of wet meadowland and other pasture to intensive agriculture, while the afforestation of moorlands is a threat to upland-breeding shorebirds in some areas, including Britain.

Coastal environments are used by people for commercial fisheries as well as for recreational activities. At migration refueling sites, shorebirds must replenish their reserves before the next leg of their journey, while in nonbreeding areas the birds must replace their flight feathers. In tidal environments, shorebirds have a limited window of time when they can feed, and disturbance at low tide can prevent them from putting the time to use. During high tides they must roost, and when roosting birds are disturbed they waste energy by taking to flight. Shorebirds will abandon roosting and foraging areas if disturbance levels are such that they can no longer effectively utilize them.

Although the breeding grounds of many shorebirds are in parts of the tundra where habitat loss and disturbance are not currently serious issues, very large areas will need protection, because the birds often nest at low densities, and in regions such as Alaska oil-related developments are beginning to take place. Furthermore, many shorebirds do not migrate to remote parts of the northern hemisphere to breed. The Hooded plover (*Thinornis rubricollis*) of Australia breeds on sandy beaches in summer, when the recreational use of such beaches is at its highest. Breeding success is often very low, and the species is declining in abundance.

The hunting of shorebirds is a traditional activity in many countries. In parts of Southeast Asia and Africa, hunting is seen as an economic necessity, the birds being an important part of the diet and a source of cash. During the 1980s, the annual take of all waterbird species in East and Southeast Asia ran to hundreds of thousands. The Guangdong Markets in China were at the time a major source of recovered bird bands, particularly of Great knots (*Calidris tenuirostris*) banded in Australia. Such a level of hunting is almost certainly neither traditional nor sustainable. As a consequence, hunting threatens not just the birds but also, ultimately, the livelihood of the hunters themselves.

Recreational hunting of shorebirds still takes place in the Americas and parts of Europe, with restrictions varying between, and even within, countries. Australia only banned the recreational hunting of Japanese snipe (*Gallinago hardwickii*) in the mid-1980s, when it was demonstrated that nearly 30 percent of the population was being shot each year. In a conservation sense, there is nothing wrong with hunting, as long as it is sustainable. The conservation dilemma lies in determining the level of sustainability.

Paradoxically, the international scale of the problem has its advantages, in that international conventions encourage national governments to conserve sites and control hunting. Many countries are now signatories to the 1971 Ramsar Convention on Wetlands of International Importance especially as Waterfowl Habitat, which requires governments to conserve at least one designated site of international importance. The European Economic Community Directive on Conservation of Birds (1979) requires the designation of areas of habitat, especially wetlands, and recognizes the need both for further research and

for updating its own provisions. Similar developments have also taken place recently in Australia and the east Asian region. In North America, the Treaty on Migratory Birds, signed in 1916 between the USA and Canada and extended in 1936 to Mexico, was a response to excessive commercial hunting; some species such as the Hudsonian godwit (*Limosa haemastica*) have still not recovered from the effects.

Although there is not always consistency between the wording of agreements and what actually happens on the ground, the conservation of shorebirds is receiving more and more attention. They are becoming recognized as an integral part of productive, dynamic, and at times finely balanced ecosystems. In one case, the decline of shorebirds on tidal mudflats led to the accumulation of silt in a nearby harbor. Previously, the silt had been stabilized by seagrasses, but the mudbanks had been denuded by herbivorous shrimps. The numbers of shrimps had earlier been kept in check by the shorebirds.

Opposition to coastal reclamation projects is growing, not just for conservation but also for cultural and economic reasons. Coastal environments are important for fisheries and recreation, and can be shared with shorebirds. Work on the Snowy plover (*Charadrius alexandrinus nivosus*) in the USA has found that by identifying and protecting the small percentage of shoreline that the birds need for roosting, people and plovers can coexist. In the Netherlands, plans to contruct barrages to prevent flooding in the Oosterscheldt Estuary were modified to accommodate a tidal surge barrier, so permitting some of the tidal nature of the estuary to be preserved.

Above *Oystercatchers feed on tidal mudflats, on an estuary occupied by heavy industry. As human use of coastal sites grows, so shorebird habitat dwindles.*

The implications of global climate change for shorebird conservation are only starting to be considered. Rapid change could be devastating in its effects. When sea levels rose or fell in the past, the tidal environments where so many shorebirds feed moved accordingly. Today, however, many coastlines have been secured on their landward side with seawalls and dikes, and in these cases the tidal environment cannot move and will drown. How will this affect shorebirds and the other wildlife that depend on such places? MB/MWP

Sandpipers and Snipes

In the northern winter, the estuaries of Atlantic Europe hold over 2 million waders, most of them sandpipers. The Waddenzee in the Netherlands is a major overwintering and stopover site, but wader numbers here have declined following overharvesting of shellfish stocks. Nearly half the wintering birds are to be found in Britain, with concentrations of 100,000 in larger estuaries and bays, such as the Wash in the east and Morecambe Bay in the west, which are of huge importance as feeding and roosting sites. Equivalent sites in North America include the Fraser River delta in British Columbia and Delaware Bay.

FACTFILE

SANDPIPERS AND SNIPES

Order: Charadriiformes

Family: Scolopacidae

86 species in 20 genera.

DISTRIBUTION Most species breed in N hemisphere, a few in tropics, Africa, and S America. Most migratory.

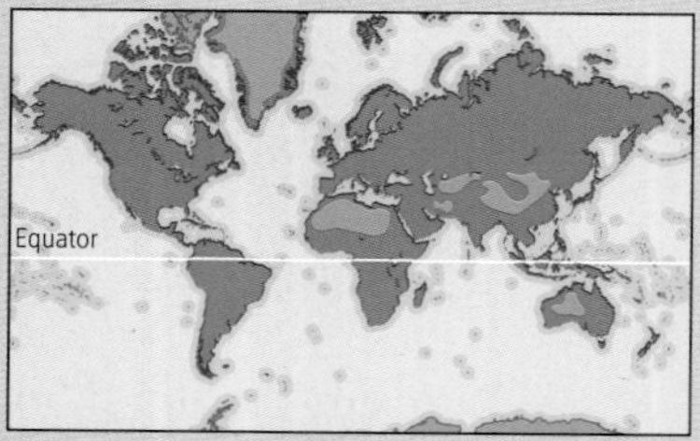

HABITAT The birds breed in wetlands and grasslands, mainly in tundra, boreal, and temperate zones; they winter on coasts, estuaries, and wetlands.

SIZE Length 13–66cm (5–26in); weight 18–1,040g (0.6–37oz).

PLUMAGE Upperparts mottled browns and grays; underparts light; markings cryptic. Many much brighter in breeding plumage.

VOICE Twitterings, rattles, shrill calls, and whistles.

NEST Made in tussocks or on dry ground; exceptionally in trees or holes.

EGGS Typically 4 (range 2–4, rarely more); pear-shaped; buff or greenish backgrounds with variable markings; 5.8–80g (0.2–2.8oz). Incubation 18–30 days; young precocial, fledging at 16–50 days.

DIET Mollusks, crustaceans, aquatic worms and flies, some plant material at times.

CONSERVATION STATUS 2 species – the Eskimo and Long-billed curlews – are currently listed as Critically Endangered. In addition, 3 species are regarded as Endangered and 5 as Vulnerable.

Flocks of dunlin and Red knot, sometimes tens of thousands strong, together with smaller numbers of other species, are marveled at by birdwatchers. At low tides all are widely dispersed over the sands and muds and are busy feeding. As the tide comes in and the feeding grounds become covered, the waders are concentrated into tighter and larger groups, and they are forced into the high salt marshes or arable farmland to roost or rest during the high tide period. Before settling, the large flocks wheel and turn in the sky like billowing plumes of smoke. The roosting flocks break up when the tide recedes, and the birds then stream back to the shore to start another feeding cycle.

Long Bills and Long Wings

FORM AND FUNCTION

The sandpipers and snipes in the family Scolopacidae make up the largest family of waders. They arose in the late Tertiary about 35–40 million years ago, and have evolved into an amazing variety of ecomorphological types, ranging from tiny little stints and the endangered Tuamotu sandpiper to the Spoon-billed sandpiper and the Long-billed curlew (*Numenius americanus*). Their closest relatives are the jacanas, painted snipes, and seedsnipes, together with the Plains wanderer of Australia.

Commensurate with this diversity of forms, there is a wide variety in mating systems and in degrees of parental care among sandpipers and snipes. Many have the normal system of monogamous pair-bonds, but females of certain species can produce up to three clutches of eggs per year for different males to incubate. Sometimes females can lay twice and incubate the second clutch themselves. At the other extreme are the lek-breeding species, in which the males compete for access to females and contribute only sperm to the reproductive effort.

All sandpipers have relatively long wings and a short tail. Their legs and neck are also often long, as in the shanks. All have three fairly long front toes and, except for the sanderling, a short hind toe. There is a wide array of bill shapes and sizes (see Fitting the Bill). The bill is at least the length of the head in all species, but is usually much longer. Plumage patterns are generally cryptic – mottled browns and grays above, with paler underparts, sometimes with streaks and spots. The sexes look alike, but in some species there may be some sexual differences in breeding plumage. All are quick runners, can wade in water, and can swim if necessary.

*Above Representative species of sandpipers and snipes: **1** Common sandpiper* (Tringa hypoleucos)*; **2** Common redshank* (Tringa totanus) *in courtship display; **3** Common snipe* (Gallinago gallinago) *in "drumming" display flight; **4** Short-billed dowitcher* (Limnodromus griseus)*; **5** Red knot* (Calidris canutus)*; **6** Dunlin* (Calidris alpina)*, with its distinctive black belly patch; **7** Bar-tailed godwit* (Limosa lapponica)*; **8** Eurasian curlew* (Numenius arquata) *feeding; **9** Spoon-billed sandpiper* (Eurynorhynchus pygmeus) *in breeding plumage; **10** Sanderling* (Calidris alba) *foraging along the shoreline.*

Great Migrators

DISTRIBUTION PATTERNS

Most species breed in the northern hemisphere, especially in the arctic and subarctic regions. The breeding range of many species is circumpolar, and only a few sandpiper species breed in tropical areas. Most are highly migratory, and the most northerly breeders tend to undertake the longest journeys. For example, Red knots wintering in Tierra del Fuego, at the southern tip of South America, make annual journeys of about 30,000km (18,600 miles) to and from their breeding sites in the Canadian Arctic. This means that a bird that lives to be 13 or more years old probably flies in a lifetime the distance from the Earth to the Moon (385,000km/239,000 miles)!

2
3
4
5
6
7
8
9

Left *A Common sandpiper plucks an insect from a grassy bank. Insect larvae and adults are the staple diet of sandpipers in the breeding season; hatching of young in June and July coincides with the greatest abundance of arthropods.*

Right *Dunlins tend to gather in large flocks; for example, just nine estuaries play host to half of all the dunlins that overwinter in Britain.*

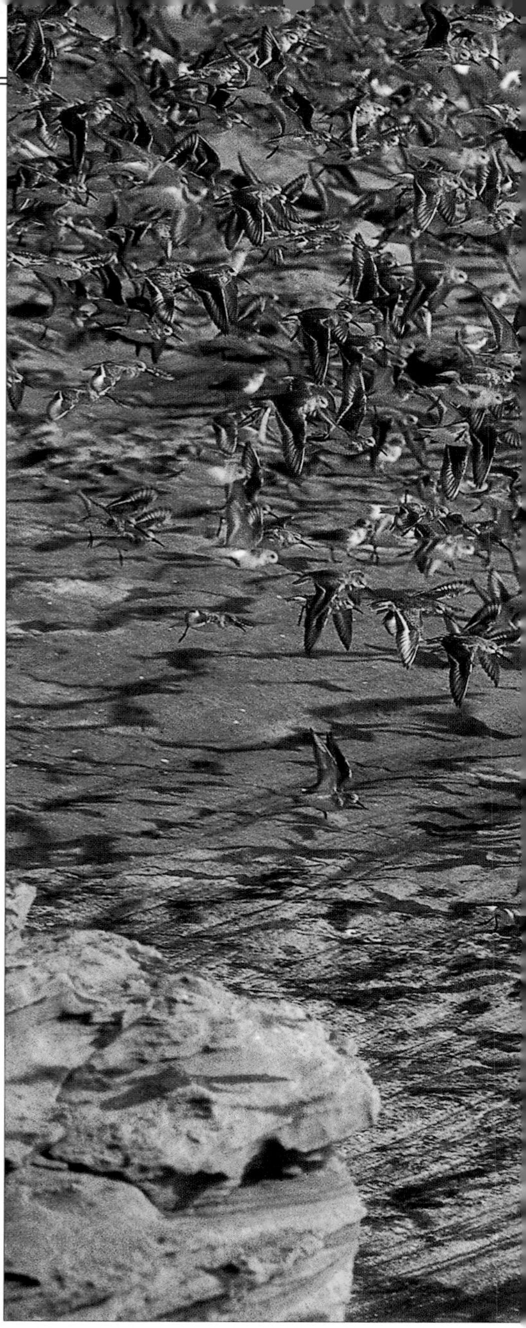

Migrating birds may travel singly or in small groups (as in Common sandpipers), but most species are gregarious and travel in flocks of several hundred. On the wintering grounds some species are highly gregarious and are found in mixed-species flocks numbering tens of thousands. Species that nest in the high Arctic, such as the Red knot, Ruddy turnstone, and sanderling, migrate south over most of the coasts of the world as far as Australia, Chile, and southern Africa. Ruffs breeding in Siberia fly westward to northwest Europe, and then continue on a remarkable journey over the Mediterranean and Sahara deserts. Up to 1 million have been recorded in the Senegal delta in West Africa.

Sandpipers breed in all types of wetlands and grasslands, from coastal salt marshes to mountainous moorland. Temporary pools and areas of tundra freed from winter snows are favored by many species. Some nest on prairies and along rivers. Wintering areas are primarily sand and mudflats in estuaries, although some species use inland freshwaters, pastures, or rocky shores. In 1981 Australian ornithologists discovered that northwestern Australia was one of the world's major overwintering sites for northern-breeding waders. At peak times, about 750,000 birds in 50 species occur there, representing almost 25 percent of all wader species in the world. Scolopacids are the dominant species in the region.

Probing for Worms and Mollusks

DIET

The main foods of most species during the breeding season are two-winged (dipterous) flies, especially crane flies and midges. Sometimes plant foods are taken before the insects emerge. The main foods of the shore feeders in winter are mollusks such as tellins and spire snails, crustaceans such as *Corophium*, and marine worms such as ragworms and lugworms. Snipes and woodcocks mostly take oligochaete worms from damp soils, the snipes from marshy ground, the woodcocks often in damp woodland. Surface foods may be located by sight and picked up, but those beneath the surface are probed for and located by touch. The Bristle-thighed curlew is, rather unusually, fond of eating the eggs of other birds, particularly seabirds. The eyes of feeding waders are so placed in the head as to give the birds a wide field of view; this adaptation is so complete in the woodcock that it has all-round vision.

Males on Display

BREEDING BIOLOGY

Arctic-nesting species either arrive at the breeding grounds already paired or pair within 2–10 days, to take advantage of the short time available for breeding. Temperate-nesting species have a more protracted breeding season, and individuals may spend several weeks at the breeding grounds before nesting. All species have elaborate display flights or song flights, and there are ground displays involving winglifting prior to copulation. Sandpiper calls may be mono- to trisyllabic, and range from the noisy, piping calls made by the redshank to the twittering of the true sandpipers.

Most species nest in tussocks on dry ground or in vegetation that conceals them well. The Black-tailed godwit often fashions the vegetation into a cupola over its nest to give increased cover. The Subantarctic snipe nests in burrows made by other birds. Green sandpipers, Solitary sandpipers,

FITTING THE BILL

Sandpipers' bills show a great variety of forms, which are related to variations in feeding behavior among species. They range from the short, straight bill (1.7cm/0.7in) of the Little stint **1** to the outsized, down-curved bill (20cm/8in in females) of the Long-billed curlew **2**. The short, thin bills of the stints allow them to pick at surface prey, such as crustaceans, which they detect by sight. The Curlew sandpiper's **3** is the largest and most decurved bill of the small true sandpipers and is used to probe for a variety of marine animals, including small mollusks.

The Broad-billed sandpiper has a heavy bill **4** that allows it to feed on relatively large prey, including mollusks. The Spoon-billed sandpiper has a broad and flattened tip to both mandibles **5**; the function of this design is not fully understood. The head and bill are sometimes moved from side to side when the bird feeds (for example on insects and other larvae) – sometimes in quite deep water. The two species of turnstones have short, thickset bills **6** that they use adeptly to turn over stones and seaweed to expose such foods as sandhoppers and crabs.

The curlews probe with their long, decurved bills for such animals as the deeper-burrowing shellfish and marine worms, which are out of reach of most other shore feeders. Godwits, with their long straight bills **7**, also probe, often rapidly, deep into wet substrate for prey such as aquatic worms. The dowitchers **8**, woodcocks **9**, and snipes **10** have long, straight bills that are proportionately longer than those of all other waders except godwits.

In these, and other deep-probing species, the parts of the bill toward the tip are well endowed with Herbst's corpuscles. These touch-sensitive organs are vital in helping locate the buried prey by a pressure mechanism. Another adaptation in some of the long-billed waders is the ability to move the portion of the bill near the tip independently of the rest – a useful trait when manipulating prey, either on the surface or while probing underground.

Sandpiper and Snipe Subfamilies

Sandpipers, Curlews, and Allies

Subfamily Tringinae

61 species in 16 genera, including: Ruddy turnstone (*Arenaria interpres*), Upland sandpiper (*Bartramia longicauda*), Curlew sandpiper (*Calidris ferruginea*), Sanderling (*Calidris alba*), Little stint (*C. minuta*), Pectoral sandpiper (*C. melanotos*), Sharp-tailed sandpiper (*C. acuminata*), Western sandpiper (*C. mauri*), Red knot (*C. canutus*); Dunlin (*C. alpina*), Spoon-billed sandpiper (*Eurynorhynchus pygmeus*), Broad-billed sandpiper (*Limicola falcinellus*), Long-billed dowitcher (*Limnodromus scolopaceus*), Short-billed dowitcher (*L. griseus*), Black-tailed godwit (*Limosa limosa*), Bar-tailed godwit (*L. lapponica*), Bristle-thighed curlew (*Numenius tahitiensis*), Eskimo curlew (*N. borealis*), Eurasian curlew (*N. arquata*), Long-billed curlew (*N. americanus*), ruff (*Philomachus pugnax*), Tuamotu sandpiper (*Prosobonia cancellata*), Green sandpiper (*Tringa ochropus*), Solitary sandpiper (*T. solitaria*), Wood sandpiper (*T. glareola*), Common sandpiper (*T. hypoleucos*), Common redshank (*T. totanus*).

Snipes and Woodcocks

Subfamily Scolopacinae

25 species in 4 genera, including: Subantarctic snipe (*Coenocorypha aucklandica*), Common snipe (*Gallinago gallinago*), Eurasian woodcock (*Scolopax rusticola*).

and Wood sandpipers sometimes lay their eggs in abandoned songbird nests in trees and bushes. Green sandpipers seek out well-wooded areas in which to breed. Most species are highly territorial, at least early in the breeding season, and nesting densities range from 1 pair per sq km (2.6/sq mi) in the Long-billed curlew to 510 pairs per ha (206 per acre) in Sharp-tailed and Western sandpipers.

Eggs are laid at 1–2 day intervals; incubation begins after the last egg is laid, and for most species lasts for 21–24 days. The pear-shaped eggs are relatively large and fit together neatly in the nest bowl. The eggs are proportionally very large in the stints, with a clutch of four eggs representing some 90 percent of the female's body weight. Typically, both males and females incubate, but the division of labor varies between species; in the ruff and Pectoral sandpiper, however, only the females incubate. The female of the Arctic-nesting sanderling lays two clutches, one of which she incubates herself, while the other is incubated by her mate.

The chicks hatch out within 24 hours of each other. They are well camouflaged and are able to fend for themselves. Once dry, they are typically tended by both parents and are led to suitable feeding grounds. However, only the female Pectoral and Curlew sandpipers tend their young. The dowitchers are most unusual, in that only females incubate the eggs and only males look after the young. Female dunlins may leave the males to tend the chicks with the help of non-breeding or failed breeding birds. Male and female Common snipes may divide the newly hatched brood between them. Woodcocks and redshanks are said to carry their young in flight between their thighs, but this behavior has never been seen by field biologists. The fledging period varies from about 16 days in

Well-traveled Waders

Sandpipers and their relatives the plovers are the two biggest families of the waders or shorebirds, the group of waterbirds renowned for undertaking extensive annual migrations. All but two sandpipers and about one-third of plover species undertake long migrations; for many species this means traveling twice a year between breeding grounds in the high latitudes of the northern hemisphere, and nonbreeding areas well into the southern hemisphere. Some species travel 20,000km (12,500 miles) a year or more; some Bar-tailed godwits are thought to fly nonstop from Alaska to New Zealand, the longest single flight of any bird. This sort of migration is associated with the Holarctic regions of the world, because the present layout of the Earth's continents makes this region a worthwhile destination.

Migration is a strategy that enables waders to exploit seasonally productive environments that are inhospitable at other times of the year. The arctic tundra, for example, is tremendously productive during the brief boreal summer. The days are long, so, with abundant sunlight, plant life and invertebrates flourish, creating ideal conditions for waders to breed. These conditions are short-lived, however, so the birds have to be able to travel between their breeding grounds and the less productive nonbreeding areas, where they spend most of the year.

In order to undertake this sort of migration pattern, waders must carry enough fuel to be able to fly continuously for in excess of 36 hours, and they must be able to travel from their departure point to an arrival point where they can find shelter and food. Furthermore, juvenile waders, migrating south from the breeding grounds for the first time, must do so unaccompanied by adult birds.

Prior to migration, small species may almost double in weight as they accumulate the fat to be used as fuel during sustained flight. In addition, some atrophication of unnecessary body mass, such as muscles around the crop, occurs. During long flights, some muscle tissue may be metabolized, resulting in extremely low body weights among newly-arrived birds.

The mechanism for achieving migration is not clearly understood. Migratory birds are capable of orienting themselves through the use of a variety of cues, including the positions of celestial bodies (especially the Sun and stars), the Earth's magnetic field, and ultrasound, such as that produced by waves along a shoreline. Site fidelity between years at both breeding and nonbreeding sites indicates that familiarity with coastlines is also important.

These facts alone cannot, however, explain the ability of inexperienced, unaccompanied birds to successfully reach approximately the same destinations as more experienced ones. Observations at departure points have found that birds only begin to migrate under particular weather conditions, leading to the suggestion that they fly on a specific bearing at times when prevailing winds will carry them roughly to the right destination. In the general area of that destination, familiarity with the coastline probably becomes important, at least for experienced birds. MB

Below *Ruddy turnstones resting at Delaware Bay, on the eastern seaboard of the USA. Around 1 million waders pass through this important stopover site each May, doubling their mass there in just 2–3 weeks to take large energy reserves to their arctic breeding sites.*

Right *The Little stint winters in Africa and the Middle East and breeds in Asia and northern Europe; this individual is nesting in Siberia. In North America, stints and other calidridine sandpipers are commonly known as "peeps."*

the smaller species to 35–50 days in the curlews.

Displaying Common snipes are vociferous (producing "chipping" noises) but are best known for their aerial "drumming" display. They dive at an angle of about 45° with the tail fanned out. The two outer tail feathers have highly asymmetrical vanes, the leading edges comprising very narrow strips. When the bird reaches about 65km/h (40mph) the air passing over these feathers causes them to vibrate and make a resonant drumming sound that can be heard from far away. Most drumming is done by males, although females may drum early in the breeding season. Drumming continues in the rain, but not in windy conditions.

A Dangerous Reliance on Stopover Sites

CONSERVATION AND ENVIRONMENT

Major food-rich estuaries and bays around the world harbor huge flocks of waders, mostly sandpipers, that rely on these key sites for their survival during brief migratory stopovers or in longer overwintering residencies. Such high dependency on a few sites makes these species extremely vulnerable to habitat destruction, pollution, or overharvesting by humans of the invertebrate fauna they rely on.

Hunting has severely depleted several Old and New World species. Upland sandpipers, now scarce in their prairie breeding grounds, were massacred in North America in the 1880s and 1890s and eaten as delicacies. Vast numbers of Eskimo curlews were shot in the 1870s and 1880s, particularly when they journeyed north from the Argentine pampas to the tundra breeding grounds. Today no wintering or breeding sites are known for this near-extinct species, but there have been a few recent sightings of individuals in passage and on former breeding grounds. AB/GJT

PHOTO STORY

How Male Ruffs Compete to Mate

1 Assigned their own genus as Philomachus pugnax, the ruffs are the largest sandpipers and the only ones not to form permanent pair-bonds. Instead, males seek to attract mates on the display grounds known as "leks" by showing off the spectacular collars of plumage from which the species takes its name. Unusually, two separate types of males display there together, as seen here: so called "residents," which have dark-colored ruffs and defend territories, and "satellites" with white ruffs, which are allowed onto residents' territories to help them attract females.

2

2 Male ruffs fight on a lek in Finland. Residents are aggressive in defense of their territories, and will attack any rival male that comes into them uninvited. As a result, many males avoid the leks altogether, forming a third category of "marginals" that take little part in breeding.

3 Satellites wander about the lek, seeking to appease the resident birds' aggressive instincts by adopting a submissive demeanor and also by displaying their own magnificent white plumage, as shown in the "flutter-jumping" display at right. There are fewer satellites than residents, which apparently permit them to intrude on their territories because of the competitive edge that their presence provides in attracting females to their particular patch.

3

5

④ *A resident male displays to a visiting female (at left) that may have been partly attracted to its territory by the satellite bird half-concealed behind its raised collar of feathers. In winter, the male and female birds look quite similar. The males only develop their decorative head tufts and pectoral ruffs, which vary in details of coloration individually from bird to bird, in time for the breeding season each spring.*

⑤ *A resident mates with a visiting female. Among male ruffs, the residents enjoy the vast majority of copulations – almost 90 percent according to one study. Even so, the satellite birds have a greater success rate than marginals, and account for most of the remaining 10 percent. As for the females, they move from territory to territory, and over half have broods fathered by more than one male.*

4

Phalaropes

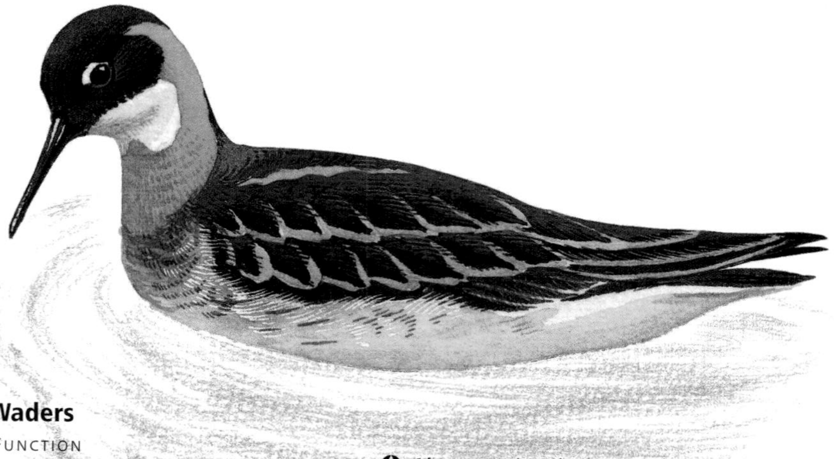

SMALL AND GRACEFUL, THE PHALAROPES ARE the most specialized swimmers among the shorebirds. The three species have lobed, partially webbed toes, laterally flattened tarsi (lower legs) that reduce underwater drag, and plumage like that of a duck on their underparts that provides a layer of trapped air on which they float as lightly as corks. Indeed, phalaropes are so buoyant that they cannot remain waterborne in strong gales.

Outside the breeding season vagrant individuals may appear almost anywhere in the world, as they get blown before strong gales, their weak flying skills unable to bail them out of such bad weather conditions. In fact, in Newfoundland they are known as "gale birds."

FACTFILE

PHALAROPES

Order: Charadriiformes

Family: Phalaropodidae

3 species of the genus *Phalaropus*: **Grey** or **Red phalarope** (*Phalaropus fulicarius*), **Red-necked** or **Northern phalarope** (*P. lobatus*), **Wilson's phalarope** (*P. tricolor*).

DISTRIBUTION Breed in N hemisphere, winter in tropics or S hemisphere.

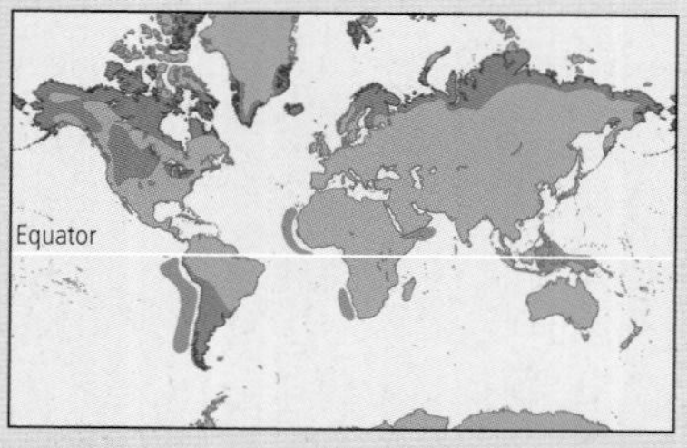

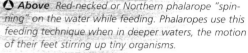

HABITAT Breed beside shallow bodies of water; in winter, 2 species oceanic, 1 found in inland waters.

SIZE Length 16.5–19cm (6.5–7.5in); weight 30–85g (1–3oz); females larger than males.

PLUMAGE Breeding – reds, white, buff, gray, and black; males considerably duller. Nonbreeding – dark above and light underparts.

VOICE Some short calls; noisy at times.

NEST In tussocks near water.

EGGS Usually 4, oval to pear-shaped, olive-buff with irregular black or brown spots and blotches; weight 6–9g (0.2–0.3oz). Incubation 16–24 days; young precocial, fledging at 18–21 days.

DIET Chiefly insects and plankton in oceans.

CONSERVATION STATUS Not threatened.

Delicate Waders

FORM AND FUNCTION

Phalaropes have relatively long necks and beautiful breeding plumages, acquired in late spring but quickly lost. All species have some white parts on the head, plus red and black markings, with the amount of red being variable. In all plumages, the Gray and Red-necked display dark rumps and show a white wing-bar in flight. Wilson's phalarope on the other hand has a white rump and there is no wing-bar. The Grey phalarope is actually the reddest of the three: "grey" describes its winter coat (or, more specifically, the predominant color of its upperparts), but in the United States the bird is commonly known as the "Red phalarope," after its summer plumage.

The female birds are markedly different from the males, both larger (10 percent so in the Northern, 20 percent in the Gray or Red, and 35 percent in Wilson's phalarope) and gaudier in their breeding plumage. The males, who tend the nests, are more cryptically colored. The differences reflect the reversed role of the female in defending a territory, performing displays, and competing for and courting males with whom she will breed.

The bill is straight and needlelike in Northern and Wilson's phalaropes, thicker in the Grey. All are active feeders, scarcely stopping as they peck at their prey. All can feed while "spinning," or swimming in tight circles, a technique that may serve to stir up invertebrate foods, causing them to rise to the water's surface.

All phalaropes forage in shallow water, along the shoreline or among wracks of seaweed. Their fine bills and large eyes help in catching their prey, which is chiefly insects, especially midges and gnats. The range of prey taken is large, however, and includes water snails, water beetles, caddisflies, and large plankton. Phalaropes are highly opportunistic feeders at their breeding grounds – Northern phalaropes will quickly change from swimming and pecking at newly-emerged midges to walking along the shoreline and pecking at emerging caddisflies as they dry off on partly submerged stones. The oceanic plankton taken includes tiny fishes, crustaceans, and small jellyfish. Grey phalaropes sometimes pick parasites off the backs of whales.

Above Red-necked or Northern phalarope "spinning" on the water while feeding. Phalaropes use this feeding technique when in deeper waters, the motion of their feet stirring up tiny organisms.

Reversing Sex Roles

BREEDING BIOLOGY

The genetically divergent Wilson's phalarope breeds inland in central parts of North America. The breeding distribution of the other two species is circumpolar, the Grey phalarope in the high arctic tundra and boreal zone and the Northern in the subarctic tundra. All three species select shallow waters, ponds, and lakes near to marshy and grassy areas. The Northern phalarope favors permanent bodies of fresh water, Wilson's fresh to salt semipermanent waters, and the Grey temporary ponds of the tundra. Wilson's phalarope winters at inland and coastal waters in South America, while the other two species winter in the open oceans.

Adult females are known to live at least five years, and breed first when they are one year old. The breeding season is very short. Generally, females arrive at the breeding grounds in June, ahead of the males, but the sexes sometimes arrive together, apparently paired. Otherwise, pairing takes place within a few days. The females may participate in nest-site searches, and both sexes make nest scrapes; it appears that the female selects the site and the male prepares the final nest, hidden in a dry grass or a sedge tussock near to water. Incubation of the eggs and care of the young are exclusively carried out by the male.

The two birds are seldom far apart for the duration (often brief) of the pair bond, and keep in touch with repeated short calls. Nesting is solitary or loosely colonial. The two arctic-nesting species often nest in or near colonies of Arctic terns, which may help in providing increased defense against predators. Grey phalaropes copulate on land, Northern while swimming, and Wilson's either while standing in water or swimming.

Normally one egg is laid every 24–30 hours. The male begins to incubate sometime between the production of the first and third eggs. After the initial egg-laying, female Northern and Grey phalaropes may mate with other males if there are enough males around. A female starts to lay a clutch of eggs for a new male some 7–10 days after completing her first. She remains at the breeding site, and provides second clutches for the original or second male should the first ones be destroyed.

The incubating male rarely leaves the nest. The small eggs, brief courtships, and polyandrous habit (one female mating with several males) all seem to be adaptations for the short breeding season. The chicks of one clutch hatch more or less at the same time. The hatchlings are well developed and leave the nest when they are 3–6 hours old. They are cared for by the male alone. He broods them frequently by day and night during the first few days, and also in times of bad weather. They swim and feed like ducklings under his supervision, before he abandons them at 14–20 days, at or near to the time of fledging.

After breeding, large concentrations of Wilson's phalaropes occur at saline lakes like Utah's Great Salt Lake, where 600,000 birds have been recorded. There they molt and fatten before migrating to South America. Up to 2 million Northern phalaropes concentrate in the Bay of Fundy in Newfoundland before migrating south. AB/GJT

Left *A flock of Wilson's phalaropes in flight over Mono Lake, California. Unlike the other two species, Wilson's phalarope does not have fully lobed toes and so rarely swims, spending little time at sea. It is confined to the interior of North and South America.*

Below *Displaying its rich chestnut breeding colors this male Grey phalarope is brooding its young chick. After egg-laying, incubation and tending to the newly hatched chicks are roles reserved solely for the male.*

Avocets and Stilts

THE MOST STRIKING PHYSICAL FEATURE OF THE *avocets and stilts is the proportions of their bills and legs. The long, slender bills may either be straight or curved upward. Their legs, too, are long, and in the case of stilts extremely long. Avocets and stilts are thus adapted for feeding in deep water, and the three species of stilts usually feed in slightly deeper water than the avocets.*

Stilts have characteristic black bills and bright pink legs and feet, while avocets have black bills and blue-gray legs and feet. The closely-related ibisbill, which is sometimes classified in its own family, has a distinctive long, crimson, decurved bill and grayish-purple legs.

The Longest-legged Shorebirds

FORM AND FUNCTION

Avocets and stilts fly with quick wingbeats, and their legs trailing behind. The front toes are well webbed in avocets and in the South Australian Banded stilt, but only partly in other species. Variation in the amount of webbing is related to the importance of swimming habits in each species.

THE FALL AND RISE OF AVOCETS IN BRITAIN

Avocets bred regularly in eastern England until the early 19th century, when they were wiped out by wetland reclamation followed by persistent egg collecting and shooting. Subsequently there was no regular breeding for over 100 years, until in 1947 Havergate Island and Minsmere in Suffolk were colonized. During the Second World War, the cattle-grazed fields of the Minsmere level had been deliberately flooded as a defensive measure, and, as a result of damage to a main sluice caused by a stray shell from a nearby firing range, the pastures on Havergate also became inundated.

Breeding numbers on Havergate Island steadily increased to almost 100 pairs over the next 10 years, with an average at 1.5 chicks successfully fledging for each pair of breeding adults. However, this success coincided with a steady increase in the numbers of Black-headed gulls (*Larus ridibundus*) breeding in the same area, to about 6,000 pairs. There was intense competition for nest sites; gulls took over avocet nests within hours of their completion, and also killed newly-hatched avocets as they were being led through the densely packed gulleries. In the eight years after 1957, the fledging success of avocets dropped to about 0.5 per breeding pair, and breeding numbers fell to just 48 pairs.

By 1965, however, constant removal of the gulls' nests and eggs by conservationists caused the larger birds' breeding numbers to drop to 1,000–1,500 pairs – a level at which they have since been maintained. The effect on avocets was marked. Over the next five years, from 1965 to 1969, their breeding population recovered, reaching 118 pairs in 1969, and fledged chicks averaged 1.7 per pair.

Since the 1970s avocets have increased in numbers and expanded their breeding range to the counties of Norfolk, Suffolk, Essex, and Kent. In 1996 the breeding population was estimated at 592 pairs. A similar increase has also occurred in the Netherlands, due to increased food supplies in the Waddenzee and the creation of new breeding sites. Some of these Dutch birds have probably immigrated to Britain.

Havergate and Minsmere are now nature reserves, and the avocet has become the emblem of Britain's Royal Society for the Protection of Birds (RSPB), symbolizing its most successful work in British wildlife conservation to date.

Below *The Black-winged stilt* (Himantopus himantopus) *has an elegant gait, raising its long, pinkish legs high and taking long, slow steps. The stilt's legs are longer in proportion to the remainder of the body than in any other birds except flamingos.*

The ibisbill is found breeding in shingle-bed river valleys between 500 and 4,400m (1,640–14,450ft) above sea level in parts of Asia, while the Andean avocet is restricted to lakes and marshes above 3,600m (11,800ft) in the Andes. The six other species are found in a variety of lowland wetlands that include freshwater marshes, brackish and coastal salt marshes, and coastal and inland salt lakes. The Black-winged stilt is the most widespread, with five subspecies found over six continents. Some populations of the most widespread species migrate over considerable distances; up to 30,000 Eurasian avocets have been counted wintering in the Great Rift Valley of East Africa.

All species take a wide range of aquatic invertebrates and small vertebrates in relatively deep water. Important prey include mollusks, crustaceans including brine shrimps, insect larvae, annelid worms, tadpoles, and small fish. Stilts seize their prey from above or below the water surface with their long, needlelike beaks. Avocets either seize their prey directly in the water or with a scything motion of the bill. In this last method, the curved part of the submerged bill locates the prey by touch as it moves from side to side in the water or soft mud. Ibisbills wade breast-deep in mountain rivers and can probe for prey under stones and boulders.

Above *With its upturned bill, the adult Eurasian avocet* (Recurvirostra avosetta) *is ideally adapted to feeding in shallow water, filtering the upper silt layers for small crustaceans and worms. The chicks have short, straight bills until they are about 10 days old.*

Territorial and Colonial

BREEDING BIOLOGY

Avocets appear to prefer nesting on islands, where predation may be reduced. In all species incubation is undertaken by both sexes. Most birds first breed when they are two years old. The Banded stilt is highly colonial: colonies of up to 27,000 pairs have been recorded, with the nests only about 2m (6.6ft) apart. Stilts usually nest in loose colonies of 20–40 pairs; in American avocets colonies range between 15–20 pairs, and Eurasian avocet colonies range between 20–70 pairs. Ibisbills defend linear territories of about 1km (0.6mi) along rivers. All species seem highly territorial, have fairly aggressive displays, defend nesting territories, and noisily mob intruders and potential predators. The young leave the nest within 24 hours of hatching, and the brood-rearing areas, which may change during the one-month fledging period, may be vigorously defended by the adults.

Growing Numbers

CONSERVATION AND ENVIRONMENT

The habitats of most species, although sometimes specialized, do not appear to be universally threatened. The number of breeding pairs in Europe has increased substantially since the 1940s, to about 30,000 pairs by the end of the century. Two geographical populations exist in Europe, one in western Europe and the western Mediterranean, the other in the Black Sea and eastern Mediterranean. An exception to the generally benign picture is the Black stilt of New Zealand, which, at less than 100 birds, is Critically Endangered, and now hybridizes with the Black-winged stilt. AB/GJT

FACTFILE

AVOCETS AND STILTS

Order: Charadriiformes

Family: Recurvirostridae

8 species in 4 genera: **American avocet** (*Recurvirostra americana*), **Andean avocet** (*R. andina*), **Eurasian** or **Pied avocet** (*R. avosetta*), **Red-necked avocet** (*R. novaehollandiae*), **ibisbill** (*Ibidorhyncha struthersii*), **Black-winged** or **Black-necked stilt** (*Himantopus himantopus*), **Black stilt** (*H. novaezelandiae*), **Banded stilt** (*Cladorhynchus leucocephalus*).

DISTRIBUTION Europe, Asia, Australasia, Africa, N and S America.

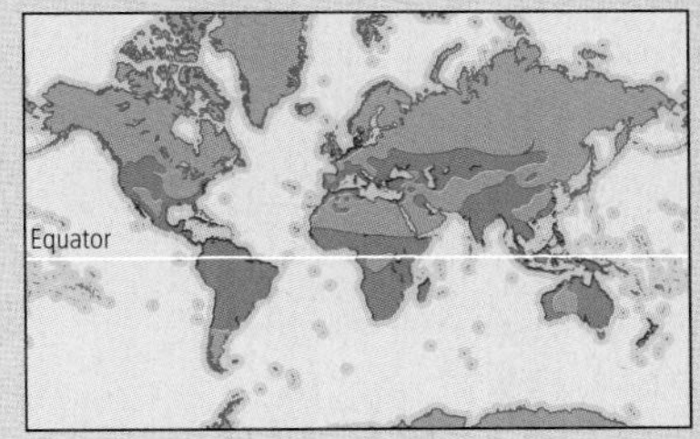

HABITAT Fresh, brackish, and saline waters.

SIZE Length 30–46cm (12–18in); **weight** 140–435g (4.9–15.3oz); females usually slightly smaller.

PLUMAGE Basically brown or black and white on body and wings; both sexes similar.

VOICE Mostly mono- or disyllabic yelping calls.

NEST Scrapes, sometimes lined, on bare ground or short vegetation near water.

EGGS Usually 4 (range 2–5), light background with dark markings; weight 22–44g (0.8–1.6oz). Incubation 22–28 days; young precocial, fledging at 28–35 days.

DIET Wide range of aquatic invertebrates and small vertebrates.

CONSERVATION STATUS The Black stilt is one of the world's most threatened shorebirds; listed as Critically Endangered, its worldwide population once fell as low as 18 breeding pairs.

Below *Stilt mating ceremony. The female solicits the male by adopting, usually in shallow water, the rigid posture* **1** *which she keeps for most of the time. The male responds by dipping his bill in the water* **1**, *shaking it, and preening himself, a procedure repeated several times* **2**. *Then the male mounts the female* **3**. *The male mates* **4** *and dismounts almost in one movement, the two birds crossing bills, and the male's wing across the female's back breaking the descent* **5**. *After mating, in the "leaning" ceremony* **6**, *the two may stand apart and lean toward one another several times.*

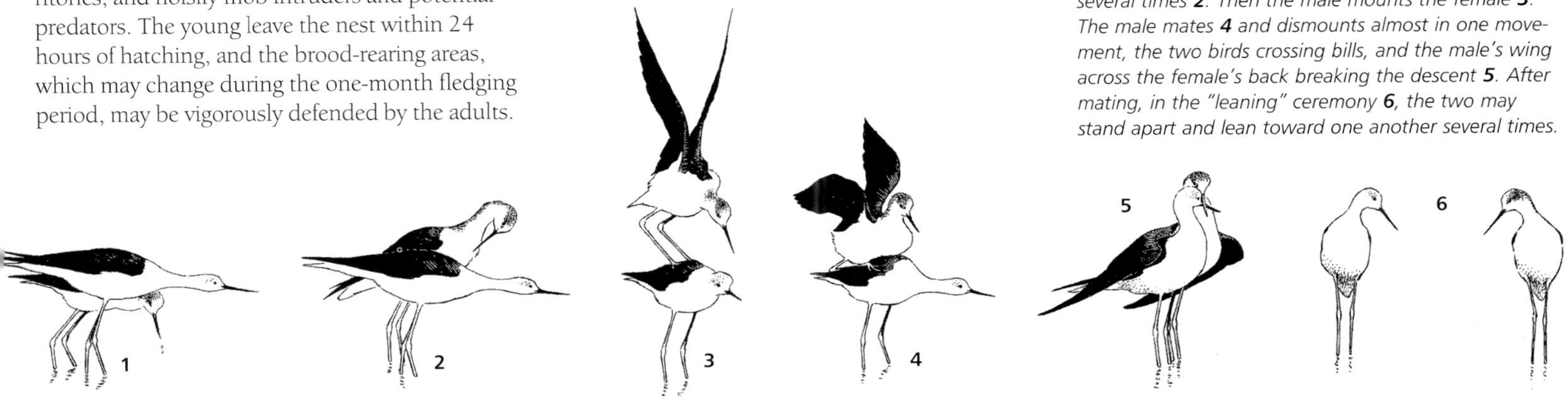

Jacanas

THE MOST DISTINCTIVE FEATURE OF JACANAS is their extremely long toes. These extended digits enable them to walk about easily on floating or just-submerged vegetation, especially water lilies, giving rise to their alternative name of lily-trotters.

Despite their superficial resemblance to the rails (Rallidae), the jacanas are in fact more closely related to the waders and especially the painted snipes (Rostratulidae), with whom they share various osteological, biochemical, and behavioral features, particularly with regard to breeding behavior.

Long-toed Lily-trotters

Form and Function

Jacanas' plumage is predominantly bright chestnut, black, and white in various bold patterns, although young birds are rather duller and the Lesser jacana gives the impression of being a young bird even in adult plumage. The Northern and Wattled jacanas have yellow flight feathers, and the extraordinary Pheasant-tailed jacana has mainly white wings with black tips. Six species have a fleshy shield above the bill; this protuberance is bright blue in the African, pearly gray in the Madagascar, livid red in the Bronze-winged, lobed yellow in the Northern, and lobed red in the Wattled jacana; the Comb-crested jacana has yellow or red lobes developed into a vertical comb. All species have short tails, except for the Pheasant-tailed jacana, which, in its breeding plumage, has long, dark brown, central tail feathers that add an extra 25cm (10in) to its length, otherwise 30cm (12in). All species have a short spur that may be either sharp or blunt on the carpal joint of the wing.

The wings are fairly short, making the birds rather weak fliers, but the Pheasant-tailed jacana nevertheless migrates quite long distances after breeding, and the African jacana may also at times move quite extensively within its range, sometimes being forced to as a result of droughts. In wetter seasons all species will utilize temporary flood waters, retreating again to permanent wetlands when these dry up. The legs and feet dangle when they make short flights, or flights for display purposes, but on longer flights they trail behind the body. When the birds alight they usually hold the wings up vertically for a moment before closing them, which makes them seem to disappear, especially in those species that have distinctive markings on the wings.

Despite their bright colors, the birds can be remarkably inconspicuous, even in the open, and they readily disappear into vegetation or underwater when danger threatens. They are quite noisy at times, with a series of raucous notes that are very variable both within and between species, and indeed between individuals. They also, however, have some soft calls, which are uttered mainly by males accompanying chicks.

Jacanas occur in wet places throughout the tropics and subtropics. They are more aquatic than most Charadriiformes species, occurring in and along the margins of freshwater lakes and slow-moving rivers, in marshes and in rice fields. Their long toes enable them to walk safely on

Below *Using the backs of hippos as stepping stones, an African jacana crosses a waterhole. Clearly in evidence are the remarkably long toes that allow these birds to spread their weight and so nimbly traverse their wetland habitat.*

floating vegetation of all kinds, and sometimes also on submerged vegetation. They can swim well despite their long toes, and young birds may remain submerged with just the tip of their beak above water when danger threatens. Adult African jacanas also exhibit this behavior during the molt period, when they are flightless.

Jacanas are predominantly carnivorous, with aquatic insects, mollusks, and other invertebrates forming most of the diet, although they occasionally also take small fish and seeds of aquatic plants. The birds spend most of their time foraging while stalking nimbly over vegetation, occasionally swimming across a patch of open water or jumping over it with a flick of their wings.

Sexual Role-reversal

BREEDING BIOLOGY

In most species the breeding season is protracted and coincides with the local wet season, when insect food is more abundant. All species except the Lesser jacana show reversed sexual roles, with the males, which are often considerably smaller than the females, doing all the nest-building, incubating, and brooding of the chicks. In several species, however, the female is also often present during chick rearing, apparently serving as a guard.

The females are sometimes polyandrous. In the Pheasant-tailed and Bronze-winged jacanas polyandry seems to be the rule, but in other species it is apparently determined, at least partially, by habitat. For example Northern jacanas breeding in uniform marshland habitats are usually monogamous and each pair holds a large territory, but in areas in Mexico and Puerto Rico that contain scattered ponds, the females have from one to four mates. In the monogamous Lesser jacana both sexes incubate (and have brood patches), and in this species and the Comb-crested jacana both sexes look after the young.

Females lay four eggs (three in the Lesser jacana). Incubation lasts 3–4 weeks, but rearing the chicks can be a prolonged process whose duration is highly variable both within and between broods. If it becomes necessary to move young chicks from a nest, adult birds will often carry them under their wings, with their long legs dangling down.

No jacana species appears to be directly threatened at present, although the numbers of Pheasant-tailed jacanas have declined in parts of China. All jacanas, however, are entirely dependent on wetlands for their survival, and they will obviously suffer if these fragile habitats are drained or polluted. PL/ASR

Right A Comb-crested jacana guarding its intricately patterned eggs in Kakadu National Park in northern Australia. Jacana nests, built by the males, are often quite insubstantial structures, which leave the eggs semi-submerged. If the first nest becomes too dilapidated and flimsy, male birds will sometimes move the eggs to a new site.

FACTFILE

JACANAS

Order: Charadriiformes

Family: Jacanidae

8 species in 6 genera: **African jacana** or **African lily-trotter** (*Actophilornis africana*), **Madagascar jacana** (*A. albinucha*), **Northern jacana** (*Jacana spinosa*), **Wattled jacana** (*J. jacana*), **Bronze-winged jacana** (*Metopidius indicus*), **Comb-crested jacana** (*Irediparra gallinacea*), **Pheasant-tailed jacana** (*Hydrophasianus chirurgus*), **Lesser jacana** (*Microparra capensis*).

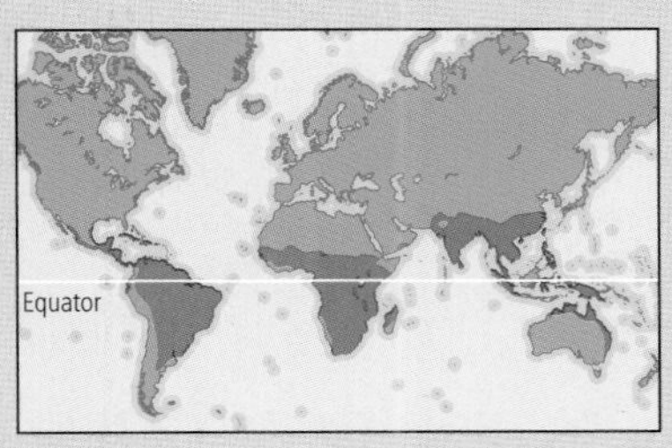

DISTRIBUTION Africa S of the Sahara, India, SE Asia, New Guinea, N and E Australia, C and S America.

HABITAT Marshes; still and slow-moving water covered in floating vegetation.

SIZE Length 15–30cm (6–12in), except Pheasant-tailed jacana (see text); **weight** 40–230g (1.4–8.1oz): female larger and up to 75 percent heavier than male in most species.

PLUMAGE Striking; mostly black and white on head and neck; upperparts various chestnut browns, some darker below and some have yellow or white on flight feathers; no pronounced sexual dimorphism.

VOICE Generally noisy, with a variety of high, staccato, squawking calls.

NEST Simple, made of aquatic leaves, usually on floating vegetation or raised platforms; occasionally partially submerged.

EGGS Normally 3–4, highly polished with dark spots, streaks, and lines. Incubation about 21–26 days; fledging period several weeks at least.

DIET Insects and aquatic invertebrates, occasionally seeds of aquatic plants.

CONSERVATION STATUS Not threatened.

Painted Snipes

THE TWO SPECIES OF PAINTED SNIPE ARE distinctive birds that inhabit tropical wetlands. They are found, respectively, in the Old World from Africa south of the Sahara to Australia (the Greater painted snipe) and in the New World (the South American species).

The exact relationships of Painted snipe have not been resolved. Some authorities, such as Sibley and Monroe (1990), place them near the jacanas and sheathbills, while others suggest a closer relationship to true snipes of the genus *Gallinago*.

Both species are beautifully and intricately colored, semi-iridescent olive-brown and chestnut to almost black on the upperparts and with some striking buff markings on the head and shoulders – a very disruptive pattern which, when combined with their skulking, crepuscular behavior and their habit of freezing when disturbed, makes them very hard to see. The Old World species also has some striking displays, and the eggs and chicks are looked after entirely by the duller-colored male.

Old and New World Species

FORM AND FUNCTION

The plumage of the South American bird is almost black on the head and breast, and finely patterned olive-brown with pale spots on the upperparts and wings. The female of the Old World species has a bright chestnut head and breast and very finely vermiculated olive-brown back and wing coverts, whereas the slightly smaller male is much duller, especially on the head. Both species have white underparts, broad, round wings, a characteristic, rail-like flight with dangling legs, and long bills, downcurved at the tip, with the nostrils set in deep, narrow grooves that extend over half the bill's length. When on the ground both species show a prominent pale "V" on the back.

Birds of the Marshes

DISTRIBUTION PATTERNS

Both species are thought to be largely sedentary and are primarily lowland species, although when conditions become too dry some more or less extensive local movements may occur, some of which may be seasonal migrations over short distances; displacements of up to 900km (560mi) have been recorded from recoveries of ringed birds in India. The South American species visits parts of its range only as a seasonal visitor, while in China Greater painted snipes only use the northern parts of their range in summer for breeding.

Below *The South American painted snipe was once considered a congener of the Greater, but they are now widely regarded as separate. Distinguishing features of the South American species are a more decurved end to the bill and webbed toes; also, in contrast to the Old World species, the sexes are alike.*

Both species have in addition been recorded outside their normal range at times. At all seasons the birds occur mainly in swamps and marshes, normally quite close to thick vegetation, although they also exploit open grassland, pasture, and flooded agricultural land. South American painted snipes have also been recorded regularly on mudflats in estuaries.

Both species are crepuscular, feeding mainly at dawn and dusk. This habit makes them hard to spot, the more so as they are reluctant to fly during the day, preferring instead to remain motionless when disturbed. In their feeding habits, they are typical waders, probing soft soils and mud for seeds, insects, and other invertebrates.

Above *A female Greater painted snipe spreads its wings in a territorial display. This species shows strong sexual dimorphism, the female being by far the more strikingly marked.*

Females that Display

BREEDING BIOLOGY

The South American painted snipe breeds in the austral spring and summer. It is monogamous, and both sexes incubate and look after the young. This species is semi-colonial, breeding in groups of up to 6 nests scattered over an area of 1–2ha (2.5–5 acres), but there is no evidence for polyandry.

The reproductive behavior of the Greater painted snipe is different. In Africa it usually breeds immediately following rain, but its behavior is more variable elsewhere. It is generally polyandrous, with the male doing all the incubation and looking after the chicks single-handedly. In this respect, Painted snipes exhibit the same sexual role-reversal that occurs in the phalaropes and the Plains wanderer. However, in some parts of its range, such as southern Africa, it is monogamous, with the female helping in nest-building and raising the chicks.

At the start of the breeding season, females start to make evening displays. They call from the ground, or from a low display flight rather similar to the "roding" flight of woodcocks (*Scolopax* spp). The display call is a succession of low, hooting notes, which, thanks to the bird's long, convoluted, resonating trachea, are audible from well over 1km (0.6 miles) away. The female also has an esophageal crop that performs no known digestive function but instead serves as an auxiliary resonance chamber. With these displays a female defends a territory, usually around 200m (650ft) in diameter, and attracts male birds. Intruding females are repelled with a display in which the wings are outstretched and turned forward to show the spotted flight feathers, and with the tail fanned and bill pointed down; the same display is used to court males and defend the territories against predators. Later on, when competition for males intensifies, females may actually fight to defend their mates from the advances of rivals.

Females take the initiative in courting, circling the male in the spread-wing posture and giving a melodious "boo" call, reminiscent of the sound produced by blowing across the neck of a bottle. After mating, the male builds a nest into which the female lays her eggs (usually 4) before leaving to court other males. When conditions are good, one female may attract 3–4 males in quick succession, with new clutches laid before the previous one hatches. In very poor conditions, however, a female may attract only one male, in which case she may help with the nesting.

The males are not territorial, and those mated to a single female usually nest quite close together. The chicks can run as soon as they are dry, and are brooded exclusively by the males, which may give the spread-wing threat display to distract or deter predators or human intruders. After breeding, small flocks comprising two or more males and their broods may form before the birds finally disperse. Males first breed when they are 1 year old, but females do not do so until they are aged 2.

Neither species is thought to be threatened, although information is sparse due to the birds' elusive habits. Painted snipes used to be shot in some numbers in parts of India. A recent, troubling trend is human encroachment onto Painted snipe habitat; for example, railroads and property development are threatening wetland sites in Hong Kong. PL/ASR

FACTFILE

PAINTED SNIPES

Order: Charadriiformes

Family: Rostratulidae

2 species in 2 genera: **Greater painted snipe** (*Rostratula benghalensis*), **South American painted snipe** (*Nycticryphes semicollaris*).

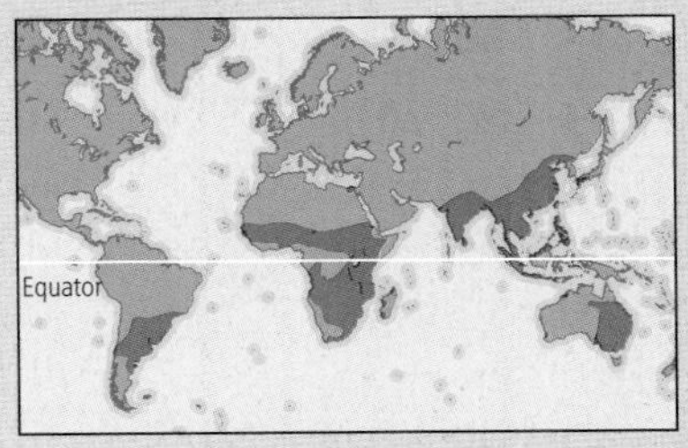

DISTRIBUTION Africa, India, SE Asia, Australia (Greater painted snipe); southern South America (South American painted snipe).

HABITAT Swamps, marshes, rice paddies, edges of rivers and lakes.

SIZE Length 20–28cm (8–11in); weight 65–165g (2.3–5.8oz); females slightly larger than males.

PLUMAGE Old World species olive-brown and chestnut on upperparts and breast, marked with black and buff, New World species almost black on upperparts; both species white/cream below, with buff stripes on head and across shoulders and round, yellow spots on wing feathers; females of the Old World species more brightly colored than males.

VOICE Soft, booming notes, growls, and hisses.

NEST Simple cup of stems and leaves hidden on ground in tall vegetation.

EGGS 2–4, cream to yellow-buff, heavily spotted with black and brown; weight 13g (0.5oz). Incubation 15–21 days; young are nidifugous (leave nest soon after hatching) but dependent for a few days.

DIET Omnivorous: insects, crustaceans, earthworms, snails, and seeds.

CONSERVATION STATUS Not threatened.

Oystercatchers

OYSTERCATCHERS ARE NOISY, CONSPICUOUS coastal birds found throughout the world except in the polar regions and on oceanic islands. Their most distinctive feature is a prominent beak, shaped to crack bivalve shells and chisel oysters, limpets, and mussels off rocks.

The birds are gregarious and flock throughout the year, except while breeding, when pairs take up territories on beaches or rocky platforms, in fields near the coast, or occasionally inland by lakes and rivers. Pairs breeding on small rocky islets often have very restricted territories and nest under overhanging rocks, in caves or under bushes.

Bills for Cracking Shells

FORM AND FUNCTION

The oystercatchers are a morphologically uniform family characterized by long, straight, red bills that may be either blunt or pointed, pied or black plumage, and relatively short legs. All are coastal, although the wholly black species (Sooty, African, Black, and Blackish) generally prefer rocky shorelines to the sandy or muddy ones favored by the others.

The species inhabiting sandy or muddy shores have longer, thinner and more pointed bills for probing for bivalves and worms, whereas birds specializing in hammering mussel shells or chiseling limpets and chitons off rocky surfaces have much blunter and stouter bills. The longest-billed oystercatcher is a relatively unstudied subspecies (*H. o. osculans*) found in Asia. Species that forage on larger food items in rocky habitats tend to have larger body size and relatively bigger feet, presumably reflecting adaptations to life on these hard substrates.

The Mollusks' Nemesis

DIET AND FEEDING

Adult oystercatchers feed mainly on various types of mollusks, typically bivalves buried in soft sediments and limpets, mussels, and oysters attached to hard substrates. Two main methods of opening these prey are used. Should the bird be able to surprise an open shell, either by stalking it on the surface or probing in the sand, the bill is driven in between the shell halves. The adductor muscle that holds the shell halves together is cut with a scissorlike bill action and the flesh is chiseled or shaken out of the defenseless mollusk. If the shell halves are closed, one side may be broken into by hammering with the bill, and the flesh removed. Individual birds tend to specialize on one particular type of prey and develop their own technique for consuming it.

Young oystercatchers may rely on their parents for food for months before learning the basic techniques, and may eventually acquire their parents' own peculiar methods and prey preferences. It may be as long as two years after independence

Above *Watched by two turnstones, a Eurasian oystercatcher cracks open a mussel. Shaped like a double-edged oyster knife, the oystercatcher's bill is ideal for splitting open mollusks and crustaceans.*

Above *A Variable oystercatcher incubating its eggs. Oystercatchers defend their nest in several ways, including attacking the predator and distracting it by false-brooding or injury-feigning where there are no eggs.*

Below *During the piping display of Black oystercatchers, birds run around with their necks arched and bills pointed to the ground, while simultaneously uttering a rapid high-pitched trill.*

FACTFILE

OYSTERCATCHERS

Order: Charadriiformes

Family: Haematopodidae

11 species of the genus *Haematopus*: **African oystercatcher** (*H. moquini*), **American oystercatcher** (*H. palliatus*), **Australian pied oystercatcher** (*H. longirostris*), **Black oystercatcher** (*H. bachmani*), **Blackish oystercatcher** (*H. ater*), **Chatham Island oystercatcher** (*H. chathamensis*), **Eurasian oystercatcher** (*H. ostralegus*), **Magellanic oystercatcher** (*H. leucopodus*), **Sooty oystercatcher** (*H. fuliginosus*), **South Island pied oystercatcher** (*H. finschi*), **Variable oystercatcher** (*H. unicolor*).

DISTRIBUTION Europe, Asia, Africa, Australia, N and S America except in high latitudes.

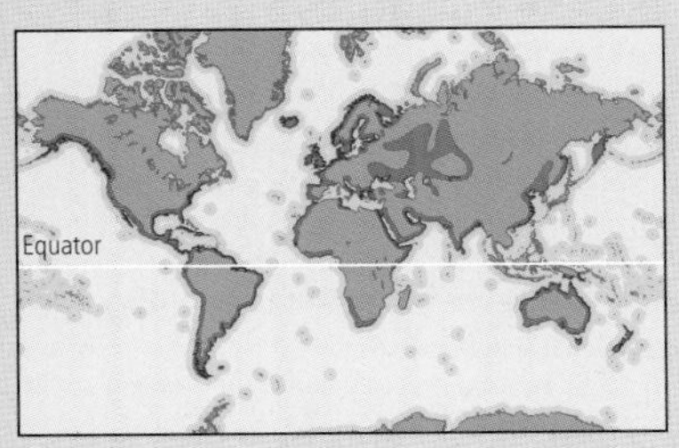

HABITAT All types of coast and fresh and brackish waters and marshlands.

SIZE Length 37–51 cm (14.6–20.1 in); weight 425–900 g (15–31.8 oz).

PLUMAGE Pied or uniform black; sexes alike.

VOICE Wide variety of simple and complex piping calls.

NEST Shallow scrape in sand or shingle, generally unlined.

EGGS 2–3 (range 1–5), brown or gray with black, gray, and brown spots and streaks; generally 40–60 g (1.4–2.6 oz). Incubation 24–35 days; fledging period 28–35 days.

DIET Chiefly bivalve mollusks; also limpets, crabs, worms, echinoderms (starfishes, sea urchins).

CONSERVATION STATUS The Chatham Island oystercatcher is Endangered, while another insular species, the Canary Islands oystercatcher, is now regarded as extinct, not having been sighted since 1981.

before Eurasian oystercatchers are sufficiently skilled to survive on mussels alone, and during this period they feed extensively on ragworms and other invertebrates found on mudflats and fields by the coast. In Virginia, American oystercatchers in their first year of life feed mainly on scraps left in oyster shells opened by adults; the hinge between bill and skull is not sufficiently ossified or strong enough for the birds to open the shells for themselves until the second year of life.

Butterfly Flights and Piping Displays

BREEDING BIOLOGY

Oystercatchers do not breed until they are 3–5 years old, but usually form pairs earlier than this, the non-breeding birds roosting in flocks during the breeding season. Territories are proclaimed and boundaries contested by means of a ritualized "piping display" in which a bird stands at its territory boundary, neck arched and bill pointed downward, and gives a succession of piping notes. The display normally attracts the owners of adjacent territories, and up to 10 pairs may gather and display, either standing or rushing up and down. A version of this display may also be given on the wing, and is perhaps the most familiar sight for a casual intruder in oystercatcher territory. Another typical display is "butterfly flight," in which a calling bird flies over a territory with slow, exaggerated wingbeats resembling those of a giant butterfly.

The modal clutch size varies among species between two and three eggs. The eggs are laid in a scrape and are brooded by both parents. On rare occasions three birds are seen incubating at a nest, indicating communal breeding. Predation rates are often high, and several replacement clutches in a season are common. The chicks can run within a short while of hatching, but due to the oystercatcher's specialized diet they are unable to feed themselves and must follow their parents, who forage for them. The first chick to hatch is fed first, and soon a size hierarchy develops in which the largest chick has first refusal of any food the parents offer.

The threatened Chatham Island oystercatcher is at high risk of extinction due to its small population size and very limited range. Intensive recovery efforts by the Department of Conservation in New Zealand saw the population increase to 58 breeding pairs and a total of 171 birds by the year 2000, which was 68 more than in 1987. The Canary Islands oystercatcher, last collected in 1913, is almost certainly extinct, and its taxonomic status remains unknown. AB/ASR

Crab Plover

THE SOLE REPRESENTATIVE OF ITS FAMILY, THE Crab plover is instantly recognizable by its massive bill and pied plumage. Looking somewhat like a cross between an avocet and an oversized plover, this curious, little-studied bird inhabits the sandy beaches and mudflats of the African and Asian coasts of the Indian Ocean.

The bird is probably a relative of the stone curlews (Burhinidae), but DNA evidence indicates a closer link to the pratincoles (Glareolidae). It is so distinct though that it is placed in a family by itself. When walking, the long neck is often carried hunched between the shoulders, giving a silhouette that is somewhat gull-like but made distinctive by the long wader legs with their short toes and partial webs.

Left *A young Crab plover accompanies its parent, begging for food but at the same time watching and acquiring the skills needed to catch crabs.*

Above *In flight, Crab plovers hold their necks straight out in front with their legs trailing behind. This is in stark contrast to their hunched position on land.*

Gregarious Crab-hunters

DISTRIBUTION AND DIET

Crab plovers are highly gregarious at all seasons and usually feed in flocks of 20 or more, searching the intertidal zone for crabs, which are easily broken by their powerful bills. Crab plovers tend to be nocturnal, as they feed on tropical fiddler crabs that emerge at dusk to feed and find mates.

Outside the breeding season, the birds spread out along the coasts of Africa and India, reaching as far south as Natal and, sporadically, east to Malaysia and north to Turkey. In the wintering areas their gregarious habit persists, large flocks forming to fly to and from the traditional roosting sites. On several occasions, flocks have been observed to gather in order to mob a human hunter who has shot one of their number.

Nests Tunneled in the Sand

BREEDING BIOLOGY

Compared to the large area of occurrence, the species' breeding range is restricted to a small number of offshore islands between Somalia and Iran. Crab plovers breed colonially, and thousands may congregate in an area of just a few thousand square meters (up to about 1.5 acres).

Colonies are confined to areas of sand and dunes. The birds dig a tunnel up to 2m (6.6ft) long, using both bill and feet, and enlarge a chamber at the end in which the single egg is laid. This feature is surprising in a long-legged bird, and unique among shorebirds.

Unusually for a shorebird, the chicks remain in the nest until well after they are fully grown, possibly as a precaution against predation by the crabs that will later be their prey. Food is brought to a chick by both parents. After leaving the nest, the chicks apparently take some time to learn how to successfully tackle an angry crab, and food-begging by young birds is frequently seen in family groups in the wintering areas. PC/ASR

FACTFILE

CRAB PLOVER

Order: Charadriiformes

Family: Dromadidae

1 species: *Dromas ardeola.*

DISTRIBUTION Africa, Madagascar, Middle East, India, Sri Lanka, Andaman Islands.

HABITAT Tropical coastline of sand dunes, mudflats, coral reefs, and estuaries.

SIZE Length 33–36cm (13–14in); weight about 250–325g (8.8–11.5oz).

PLUMAGE White with black back and primary feathers; sexes alike.

VOICE Noisy, with variety of harsh, barking calls and sharp whistles.

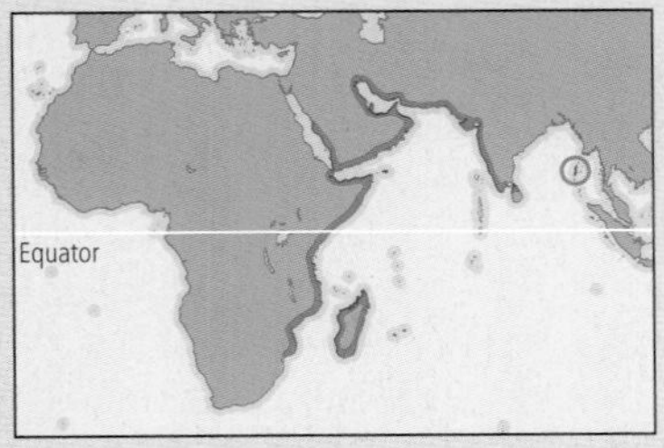

NEST Unlined hollow at end of tunnel in sand.

EGGS 1, occasionally 2, white; weight about 45g (1.6oz). Incubation and fledging periods not known.

DIET Chiefly crabs, also mollusks, worms, and other invertebrates.

CONSERVATION STATUS Not threatened.

Stone Curlews

Named for the fact that they occur on open, stony ground and call like curlews, this family's alternative name of "thick-knees" refers to their knobbly leg joints. Yet their most striking feature is their large, piercing yellow, hawkish eyes, which stare out from under pronounced eyebrows and have evolved for a crepuscular and nocturnal lifestyle.

Recent conservation measures in Britain have increased the Stone curlew population from about 160 pairs to well over 200, with several producing two broods in a year. Many nests on farmland are marked and moved when agricultural operations are in progress: the birds happily return when the eggs are replaced. In general, however, their numbers are down in Europe. They thrive on barren, remote land, and such places are becoming rarer.

Stout Bills

FORM AND FUNCTION

Stone curlews of the genus *Burhinus* are of uniform appearance, with cryptic brown plumage heavily streaked with brown and black above, pale underparts, and pronounced white wing-bars shown in flight. The two *Esacus* species (which some authorities subsume within *Burhinus*) are plainer above and pale below and have prominent black markings around the eye and crown. All have stout to massive (*Esacus*) bills well able to cope with their varied diet. They are terrestrial and run swiftly, flying low only when necessary, with their long legs trailing behind. They usually inhabit dry, open country and semi-desert, although the Water thick-knee and Senegal thick-knee favor lake and river margins, and the Great and Beach stone curlews frequent coasts. Out of the breeding season, and at the end of it, they often gather in small groups.

Shared Duties

BREEDING BIOLOGY

Pairs of most species breed in areas with sparse vegetation allowing a wide field of vision. The European species defends its territory with an aggressive display, in which the bird raises itself until its body is almost vertical, with its tail fanned and pointing downward and its wings held folded away from the body. Pairs collaborate in territorial defense, and the period during which a territory is established may be prolonged and very noisy (especially at night). The courtship posture is quite the opposite, the pair standing together with necks arched and bills pointed downward.

Eggs are laid in a scrape in the ground and incubated by both birds in turn; the off-duty parent may stand guard behind its sitting partner. The chicks leave the nest within 24 hours of hatching and quit the area soon after. The parents feed them for the first few days and they gradually learn to feed themselves in the areas chosen by their parents – sometimes 1km (0.6 miles) or more from the nest. If danger threatens, Stone curlews have occasionally been seen to pick up very small chicks and carry them away. Distraction displays are uncommon but are very strenuous, with the bird jumping and fluttering to the ground, where it rolls around hissing with extended wings. Some colonial nesting may occur among Senegal thick-knees; 21 nests were found on a flat roof in Egypt! CJM/ASR

Below A Stone curlew returns to its nest with a beetle meal. When catching their prey, stone curlews take on a heron-like stalking posture and stab swiftly.

FACTFILE

STONE CURLEWS

Order: Charadriiformes

Family: Burhinidae

9 species in 2 genera: **Double-striped thick-knee** (*Burhinus bistriatus*), **Peruvian thick-knee** (*B. superciliaris*), **Senegal thick-knee** (*B. senegalensis*), **Spotted** or **Cape dikkop** (*B. capensis*), **Stone curlew** (*B. oedicnemus*), **Water thick-knee** or **dikkop** (*B. vermiculatus*), **Bush thick-knee** (*B. grallarius*), **Beach stone curlew** (*Esacus magnirostris*), **Great stone curlew** (*E. recurvirostris*).

DISTRIBUTION Europe, Africa, Asia, Australasia, S America.

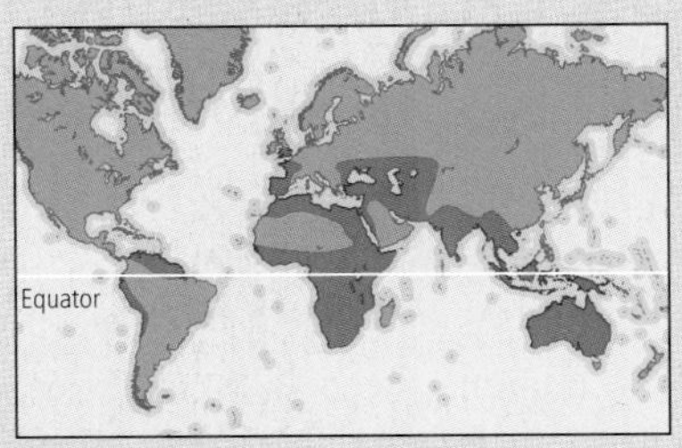

HABITAT Sparsely vegetated open country, near water or in more arid regions, steppe, deserts, mudflats, savanna, and open coastlines and estuaries.

SIZE Length 32–59cm (12.6–23.3in); weight 300g–1.1kg (10oz–2.4lb).

PLUMAGE Pale, sandy browns above, streaked black; pale, streaked breast and whitish underparts; many have white marks on the wings.

VOICE Wailing, bisyllabic whistles given communally at dusk; variety of other hoarse, rasping calls, hissing, and creaking sounds, often at night.

NEST Shallow scrape in soil, shingle, or sand, unlined or lined with debris.

EGGS Generally 2, creamy white, ocher, buff, pale brown, black and purplish-gray spots and blotches; in Stone curlew, average 38g (1.3oz). Incubation 24–27 days, fledging 36–42 days.

DIET Terrestrial invertebrates, small vertebrates (frogs, rodents), crabs and mollusks in coastal species.

CONSERVATION STATUS The Bush thick-knee and the Beach stone curlew, are Lower Risk: Near Threatened.

Pratincoles and Coursers

The pratincoles and coursers are distinct groups within the family Glareolidae, which also contains the rather anomalous Egyptian plover and, possibly, the Australian dotterel. The Egyptian plover is usually considered to be a closer relative of the coursers; as regards the contentious Australian dotterel, some systematists include it with the pratincoles, while others assign it to another family altogether, the true plovers (Charadriidae).

This family of shorebirds have slender bodies and, in the main, elongated wings. Many species inhabit hot and dry parts of the world, and consequently are crepuscular or nocturnal, resting during the day.

Pratincoles

GENERA *GLAREOLA, STILTIA, PELTOHYAS*

Pratincoles are wonderful fliers, their long pointed wings, forked tails, and agile maneuvers in pursuit of winged insects being strongly reminiscent of swifts or marsh terns. However, unlike these groups, they also feed on insects on the ground and are fast runners. Most have short legs, the sole exception being the Long-legged pratincole, which is mainly terrestrial in its habits. Pratincoles congregate on flat, open terrain, both near water (mudflats on estuaries and lakes are especially favored sites) and in more arid areas, where there is an abundant supply of insect food. They are gregarious, forming flocks both inside and outside the breeding season, and nest in small, loose colonies within which each pair maintains its own territory. The chicks stay in or near the nest for 2–3 days, fed by both parents, after which they gradually learn to feed themselves. As in other waders, elaborate "broken wing" displays are commonly given by the parent to lure predators away from their chicks.

Coursers

GENERA *CURSORIUS, RHINOPTILUS, SMUTSORNIS, PLUVIANUS*

With their long legs, short tails, and characteristic upright stance, the coursers are running birds that take flight only when they are forced to do so. Their cryptic, sandy plumage is plainer than that

Above and below *Representative species of pratincoles and coursers:* ***1*** *Egyptian plover* (Pluvianus aegyptius) *using wing-raising display to greet its mate;* ***2*** *Cream-colored courser* (Cursorius cursor) *at the run;* ***3*** *Collared pratincole* (Glareola pratincola) *in a characteristic "broken wing" distraction display.*

of the pratincoles, although many sport conspicuous black-and-white eye stripes. They favor arid and desert areas and are often crepuscular or even nocturnal. When disturbed in their open habitat, coursers will run a little and then stop with neck stretched upright to look at the intruder.

The species in the genus *Cursorius* are gregarious, gathering in small flocks and family parties after breeding. Poorly known, they appear to be monogamous and territorial, both parents tending the young and staying with them until well after fledging. Adults, however, do not give elaborate distraction displays.

The Egyptian plover is a beautiful and striking bird with blue-gray wings, orange belly, and bold black-and-white marking on the head. It prefers the margins of inland lakes and rivers and neighboring fields and grasslands, feeding on insects and invertebrates taken on the ground, particularly at the water's edge. It is extremely tame and often frequents the vicinity of human habitations. The Egyptian plover's alternative common name, Crocodile bird, arises from an account by the Greek historian Herodotus, telling how certain birds on the Nile – which later commentators took to be this species – fed on the scraps left between the teeth of basking crocodiles. Despite being corroborated by anecdotal evidence from two eminent German ornithologists in the 19th and 20th centuries, this alleged behavior has never been properly authenticated!

FACTFILE

PRATINCOLES AND COURSERS

Order: Charadriiformes

Family: Glareolidae

17 species in 7 genera: **Australian** or **Inland dotterel** (*Peltohyas australis* – sometimes placed in Charadriidae), **Black-winged pratincole** (*Glareola nordmanni*), **Grey pratincole** (*G. cinerea*), **Collared pratincole** (*G. pratincola*), **Oriental pratincole** (*G. maldivarum*), **Small pratincole** (*G. lactea*), **White-collared pratincole** (*G. nuchalis*), **Madagascar pratincole** (*G. ocularis*), **Australian pratincole** (*Stiltia isabella*), **Cream-colored courser** (*Cursorius cursor*), **Indian courser** (*C. coromandelicus*), **Temminck's courser** (*C. temminckii*), **Egyptian plover** or **Crocodile bird** (*Pluvianus aegyptius*), **Jerdon's courser** (*Rhinoptilus bitorquatus*), **Three-banded courser** (*R. cinctus*), **Bronze-winged courser** (*R. chalcopterus*), **Two-banded courser** (*Smutsornis africanus*).

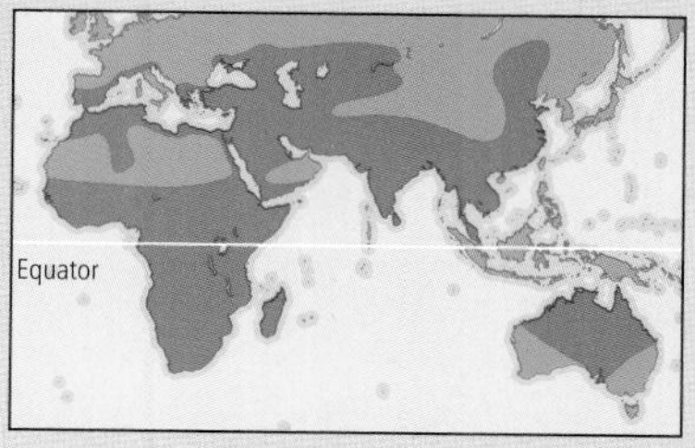

DISTRIBUTION Europe, Asia, Africa, Australasia.

HABITAT Open or scrub country, usually in arid regions.

SIZE Length 18–30cm (7–12in); weight 80–100g (2.8–3.5oz)

PLUMAGE Pratincoles generally brown above with white rump and belly; many have colored throats bordered with black. Coursers generally buffs and sandy browns; many have bold black markings on head and breast; sexes alike.

VOICE High-pitched trilling and melodious, piping notes. Very vocal on the wing; on the ground mostly silent – principal vocalizations are contact or alarm calls.

NEST Unlined or sparsely lined scrape in sand or gravel.

EGGS 2–3 (rarely 1,4, or 5); yellow-brown, cream, or buff, speckled with black, brown, and gray; generally about 15g (0.5oz) where known. Incubation 17–31 days; fledging period 25–35 days where known.

DIET Chiefly insects, occasionally other invertebrates.

CONSERVATION STATUS Jerdon's courser from southern India is regarded as Critically Endangered, as a result of human disturbance of its habitat.

Right Two- or Double-banded courser with its young, seeking the shade of a shrub to escape the heat of the day. Chicks usually acquire the distinctive chest bands at 3 months and full adult plumage at 4.

Below Like other pratincoles, the Australian or Long-legged pratincole lays near water. It makes no nest, laying its clutch of two eggs on bare soil among stones or gravel. The eggs, seen here, look very much like stones in color and texture.

Although Egyptian plovers sometimes flock outside the breeding season, they are basically solitary, nesting in pairs in territories. Members of a pair greet each other on landing with an elaborate wing-raising display in which the exquisite wing markings are shown to advantage. When leaving the nest, the incubating bird buries its eggs in the sand, smoothing over the side so that the nest is completely invisible. Even if the bird is surprised, it still performs this task, although haste often makes for a less than perfect job. Newly hatched chicks frequently receive the same treatment, being completely covered with sand; older nestlings may also be concealed in this way. Should the eggs become too hot, they are cooled with water gathered in the adult's belly feathers. Young chicks are cooled by the same method, and may also drink from the soaked feathers. Distraction displays are not common, but adults may feign a broken wing on occasion. After having been shown food by the adults, the chicks begin to feed themselves from a very early age.

The only threatened species of the family Glareolidae is Jerdon's courser, which was, until relatively recently, feared to be extinct. However, in January 1986, after a lapse of 86 years in reliable recording, a small number of these birds were rediscovered at a site 170 km (105 miles) away from where the species had last been sighted. An inhabitant of thorny scrubland in the south-central Indian state of Andhra Pradesh, Jerdon's courser is Critically Endangered. The dwindling population of this fully nocturnal species is constantly under threat, most recently by a planned irrigation scheme that would have made serious inroads into its habitat. ASR/CJM

Seed Snipes

The squat, short-legged, short-billed seed snipes of South America are, together with the sheathbills, perhaps the least wader-like of the shorebird-related families. The two Thinocorus *species are more like larks or buntings, while the* Attagis *species most closely resemble small partridges. As the common name for the family indicates, they have adapted themselves to a mainly vegetarian diet.*

Different species favor different habitats. The Least seed snipe prefers the arid coastal and dune areas and inland dry plains that occur at lower altitudes. The White-bellied and Grey-breasted seed snipes range further into the higher, dry slopes of the Andes, and the Rufous-bellied species chooses wetter moorland in the high mountain pastures.

FACTFILE

Seed Snipes

Order: Charadriiformes

Family: Thinocoridae

4 species in 2 genera: **Grey-breasted seed snipe** (*Thinocorus orbignyianus*), **Least seed snipe** (*T. rumicivorus*), **Rufous-bellied seed snipe** (*Attagis gayi*), **White-bellied seed snipe** (*A. malouinus*).

Distribution Montane western S America

Habitat Open country, often at high altitudes.

Size Length 16–30cm (6.5–12in); weight 50–400g (2–14oz).

Plumage Upper parts generally various browns, marked with black, gray, and cinnamon; females like males but less strongly marked.

Voice Series of rapid, often trisyllabic, notes, sometimes nocturnal; also short, rasping, and peeping alarms.

Nest Scrape in the ground, lined with any available loose material.

Eggs Usually 4, buff or cream, speckled dark brown and lilac or olive, blotched with black. Incubation about 25 days; young nidifugous; fledging 49–55 days (in Least seed snipe).

Diet Succulent vegetation and, probably, seeds.

Conservation Status Not threatened.

The Seed Snipe Ascending

Form, Function, and Breeding Biology

Seed snipes are all fast-running, ground-dwelling birds, and tend to be gregarious, often gathering in coveys like partridges. Their short, stout bills are an adaptation to their diet, now thought to be mainly the succulent parts of growing plants rather than seeds. These also provide them with all their water. Their plumage is generally cryptic, with brown feathers edged with black and buff above, but the male Grey-breasted and Least seed snipes have a gray forehead and upper breast with black and buff markings on the sides of the head and neck, which are marked like the back in the females.

Unlike other waders, the seed snipes have an operculum covering their nostrils; this thin flap of skin is thought to be an adaptation to protect them from dust in their arid habitat.

Males of the two *Thinocorus* species take up territories and perform a display flight that is reminiscent of a lark's, flying up then gliding down on stiff wings while singing a rapid, staccato song. The ascent of the Least seed snipe is higher than that of the Grey-breasted and the display lasts rather longer. Both species may also sing from a perch, and the Least seed snipe has been observed performing a "butterfly" flight, beating

Right Least seed snipes have a wide distribution in South America, ranging from the Falkland Islands to Ecuador. Southern populations migrate north in winter.

Below In its high Andean habitat, the gamebird-like appearance of the Rufous-bellied seed snipe has earned it the colloquial name perdiz cordillerana, or "partridge of the mountain range."

its wings in a stiff, exaggerated manner. There is much song-flight activity at night.

In all but one species, the buff or cream eggs with darker markings blend with the dry stony soil by the scrape. The dark olive and black eggs of the White-bellied seed snipe are better camouflaged for the mossy soils of its heathland habitat. In the *Thinocorus* species, and probably the others also, the female alone incubates, kicking loose material over the eggs to hide them if she wants to leave the nest or is disturbed. In the latter case she may use a distraction display, fluttering low over the ground or trailing a wing. The young are very precocious and feed themselves soon after hatching. Both parents guard them, and the male, too, may use distraction displays, running to and fro with hunched back and drooping wings to lure intruders away. CJM

Plains Wanderer

SUPERFICIALLY, THE PLAINS WANDERER might appear to be a fairly typical buttonquail (family Turnicidae), with its broadly similar camouflage color patterning and habits. Unlike buttonquails, however, it retains the hind toe; it also lays pear-shaped (rather than oval) eggs and tends to adopt a more upright posture, while the carotid arteries in the neck are paired.

Taxonomists traditionally placed the Plains wanderer in a family of its own, next to the buttonquails within the order Gruiformes, but research since the 1980s suggests that this listing is wrong and that the Plains wanderer really belongs with the plovers, dotterels, and sandpipers in the order Charadriiformes. The first doubts concerning its former classification came over comparisons of skeletons, and they were reinforced when studies of DNA suggested that the Plains wanderer's closest relatives were the Neotropical seed snipes (Thinocoridae) within the Charadriiformes. The anatomy of the bird's liver and thigh muscles and the shape of its eggs, as well as other features, all reinforce this conclusion.

The Plains wanderer was formerly widespread in eastern and southeastern Australia, but it has declined as agriculture has become more intensive. It is also threatened by predation by introduced foxes. It is now scarce or rare, its range is fragmented, and it is found chiefly on unimproved grasslands and on fields left fallow in alternate years; such conditions provide the mix of grass and weed seeds needed by the birds for food. There is no evidence that the Plains wanderer is migratory: the concentration of April–June records in the south merely reflects the quail-shooting season there.

Eggs have been found in all months except March and April, suggesting that there is breeding year-round as the opportunity arises. As in the buttonquails, the brighter-colored female initiates courtship, then leaves the male to incubate the clutch and raise the young; in captivity, however, females sometimes help with incubation. Once the parent–young bond has broken, the birds become separated in ones and twos. They seldom fly, instead crouching for concealment or else running away when alarmed, and they have the habit of standing on tiptoe, neck outstretched, to gain a better view through or over vegetation. DH/RH

Above *The grassland habitat of the Plains wanderer in eastern Australia has been severely depleted, and fewer than 11,000 birds may now remain in the wild.*

FACTFILE

PLAINS WANDERER

Order: Charadriiformes

Family: Pedionomidae

1 species: *Pedionomus torquatus*

DISTRIBUTION SE and C Australia. Greatest density near the borders of New South Wales and Victoria.

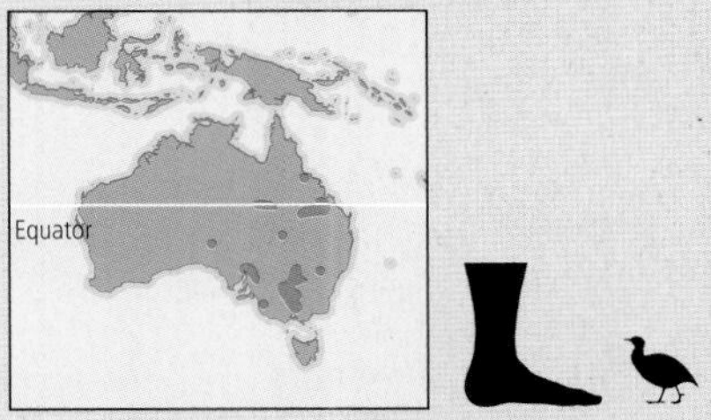

HABITAT Open grassland and stubble fields; avoids scrub areas.

SIZE Length 15–19cm (6–7.5in).

PLUMAGE Cryptic coloration of buff, brown, and black, with collar of black spots on white; females larger and more brightly colored, being distinguished by a black-and-white collar and a chestnut-brown patch on the breast.

VOICE Low, sonorous "whoo-whoo" call.

NEST Grass-lined ground hollow

EGGS 2–5 (usually 4); gray and olive markings on yellowish backgound. Incubation 23 days; well-developed hatchlings leave nest at once.

DIET Small seeds and insects.

CONSERVATION STATUS The decline of the species' favored lowland grasslands has seen numbers fall, and it is consequently classed as Endangered.

Sheathbills

SOME EUROPEAN PEOPLE ONLY BECAME aware of the Antarctic sheathbills for the first time when a lone stowaway bird was discovered on board a British troopship returning from the Falkland Islands in 1982, during the conflict with Argentina. These curious birds look somewhat like a cross between a pigeon and a domestic hen.

The two species of sheathbill are very similar in appearance and habits but are geographically separated. The Black-faced sheathbill is resident on some of the most remote sub-Antarctic islands, one distinct subspecies inhabiting each of the four main island groups of the southern Indian Ocean – Prince Edward and Marion islands, Heard and Macdonald islands, and the Crozet and Kerguelen archipelagos. The slightly larger Snowy sheathbill lives farther to the west, breeding on the Antarctic peninsula, South Georgia, and the South Orkney and South Shetland islands in summer but migrating to the Falklands and both coasts of Argentina during the austral winter. During their migration, the birds will often rest on passing icebergs or ships; some individuals that have hitched a ride on vessels have traveled as far as South Africa, and even Europe!

Neither Shorebird nor Seabird

FORM AND FUNCTION

The Chionididae has the distinction of being the only bird family that is endemic to the Antarctic and sub-Antarctic. Taxonomically, sheathbills are peculiar birds and, according to some systematists, may represent an evolutionary link between the shorebirds and the gulls. They certainly share a number of anatomical features with both groups and have a rudimentary spur on the carpal joint of the wing – a typical plover feature. However, unlike gulls and other true seabirds, their feet are not webbed, an anatomical feature that has obliged them to adopt a scavenging lifestyle. They have a bald patch around the eye and naked carbuncled skin above the pale yellow (Snowy sheathbill) or black (Black-faced species) bill. The bill is broad and strong, with a horny sheath that partly covers the nostrils; the sheath is more prominent

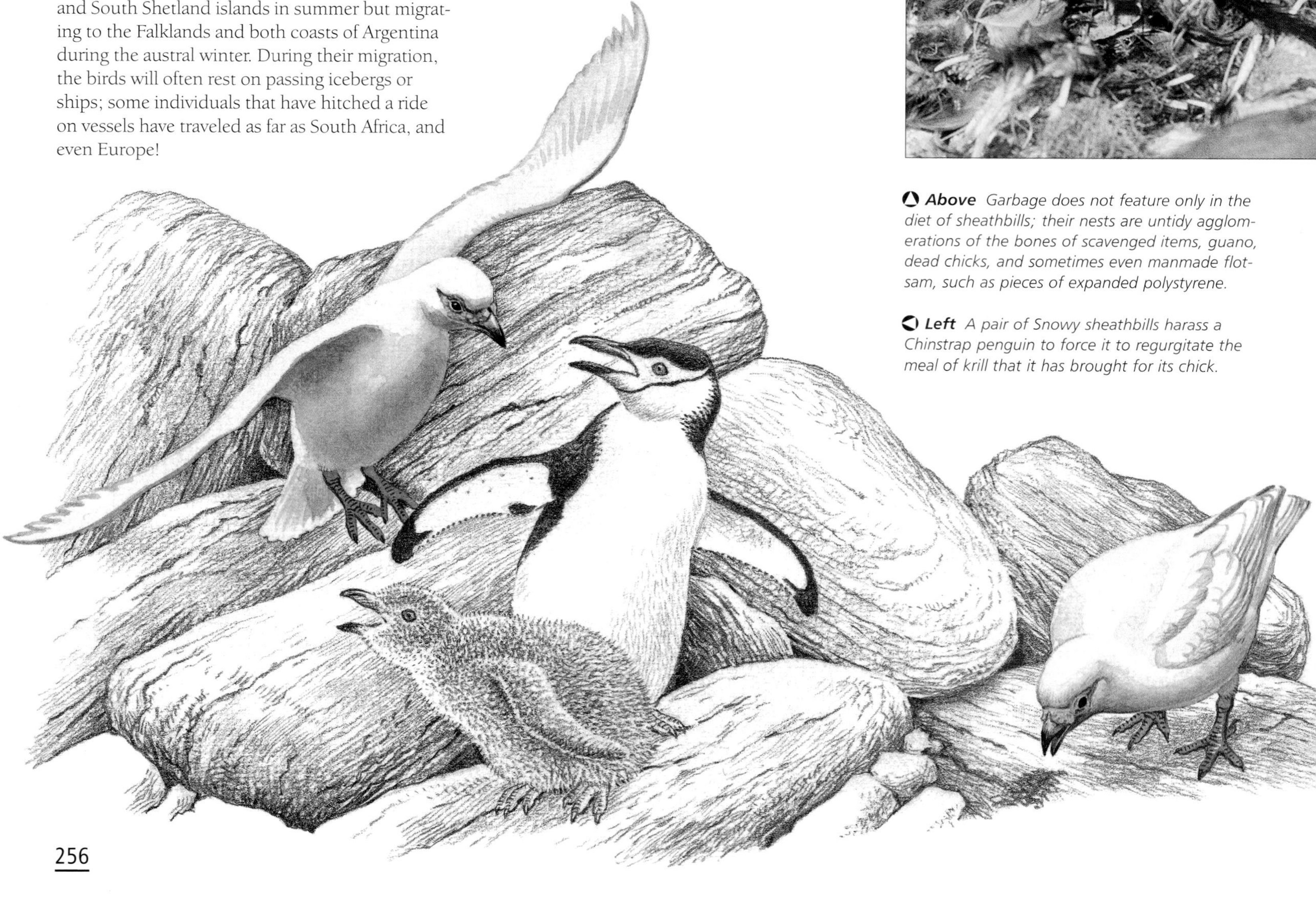

Above *Garbage does not feature only in the diet of sheathbills; their nests are untidy agglomerations of the bones of scavenged items, guano, dead chicks, and sometimes even manmade flotsam, such as pieces of expanded polystyrene.*

Left *A pair of Snowy sheathbills harass a Chinstrap penguin to force it to regurgitate the meal of krill that it has brought for its chick.*

Above *A Crozet Lesser sheathbill* (C. m. crozettensis). *The four lesser sheathbill subspecies have been separated from one another for some 10,000 years, after the southerly retreat of the Antarctic pack ice left the different island groups isolated.*

in the Black-faced species. The plumage is white; under the feathers is a thick layer of down that acts as insulation from their bitterly cold environment. Their flight is pigeonlike, while their gait recalls that of a rail. In their general demeanor, they are somewhat like a crow or gull, being bold and inquisitive, and not at all shy of people. They are generally seen on the ground but will fly if forced and during migration. As befits a migratory bird, the Snowy sheathbill's wings are more elongated than those of the nonmigratory species, for aerodynamic efficiency.

Antarctic Garbage Collectors

DIET

Sheathbills are opportunistic feeders and their diet is extremely varied. In winter they are frequent scavengers at the refuse heaps of whaling stations and Antarctic survey bases, consuming offal and household scraps. Away from the few human sites, they forage among seaweed for intertidal invertebrates and scavenge the tideline. When the seals come to breed, they gorge themselves on stillborn pups and the afterbirths, in much the same way as crows do at lambing. In their own breeding season, they turn for food to the large colonies of breeding penguins. Pairs of sheathbills take up territories containing a number of pairs of penguins, which they defend vigorously from other sheathbills, and harass the penguins in order to steal the eggs, young chicks, or the boluses of krill that the parents regurgitate for their chicks. Both members of the pairs often join forces to surprise the adult penguin, drive off the chick and scoop up the krill as it is regurgitated. This rich food is ideal for the rapidly growing sheathbill chicks, which wait at the nest to be fed. The penguins clearly resent these depredations and may lunge forward to strike with their bills, but they are generally too slow to trouble the bold and agile sheathbills.

After fledging, the young disperse to the shoreline to scavenge for scraps of fish, limpets, and kelp. Because of their messy feeding habits, especially their propensity for eating feces, generations of Antarctic explorers have crudely dubbed them "shitehawks"(an epithet elsewhere applied to the Black kite). Yet, precisely because of the distasteful nature of their diet, sheathbills take great pains to keep themselves clean, spending a considerable proportion of their time in bathing and preening. They have no obvious predators, except for the occasional egg lost to other sheathbills.
CJM/ASR

FACTFILE

SHEATHBILLS

Order: Charadriiformes

Family: Chionididae

2 species of the genus *Chionis*: **Black-faced** or **Lesser sheathbill** (*C. minor*), **Pale-faced, American,** or **Snowy sheathbill** (*C. alba*).

DISTRIBUTION Antarctic and sub-Antarctic islands.

HABITAT Coastal

SIZE Length 34–41 cm (14–16 in); weight 450–780 g (16 – 25 oz).

PLUMAGE Entirely white; sexes alike.

VOICE Harsh, crowlike calls and guttural, rattling croaks.

NEST In crevice in rocks or in former petrel burrow near seabird colonies; may be lined with stones, debris, seaweed, or lichen.

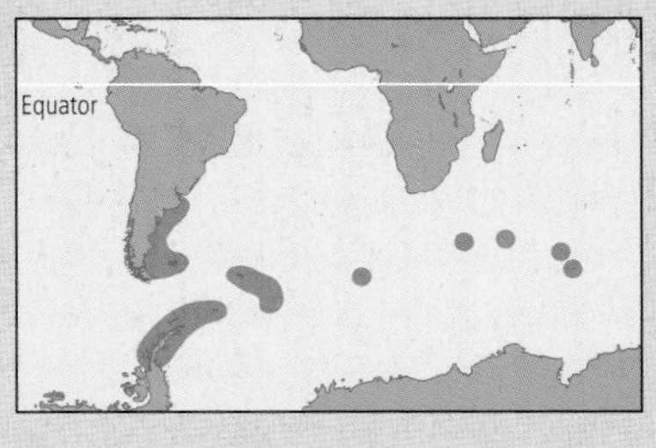

EGGS 2, sometimes 3 or 4, white or grayish with dark brown blotches. Incubation 28–32 days, fledging 50–60 days.

DIET Omnivorous, opportunist feeders and scavengers often at seabird colonies.

CONSERVATION STATUS Neither species is at risk; however, on the islands where the Black-faced sheathbill is endemic , some populations are in decline, due to competition with introduced species such as mice and rats.

Gulls

GULLS ARE THE MOST FAMILIAR SEABIRDS OF northern temperate regions, regularly seen hunting along the shore for intertidal prey, or straying far inland in search of food and breeding sites. Some species, such as Franklin's gull, breed entirely in inland regions. As a group, they are outstanding opportunists in their feeding behavior, which partly distinguishes them from their close but more specialized relatives, the terns, skimmers, and skuas.

In adapting to a wide variety of lifestyles, gulls range enormously in size, from the dainty and diminutive Little gull to the heavily built, piratical Great black-backed gull. The larger species have robust, slightly hooked bills, while the smaller species have slender, forceps-like ones. In most species the underside is white, with gray or darker coloration on the back and upper surfaces of the wings. The generally pale coloring below is thought to make the flying bird less conspicuous to its fish prey in the surface waters of the sea. A minority, notably the aptly named Lava gull and the Sooty gull, are much darker all over.

Elegant Fliers

FORM AND FUNCTION

The bill and legs of gulls are usually yellow or bright red, but there is much variation, both between species and at different times of year within the same species. However stoutly built they may be, gulls are always graceful in flight, switching easily from powerful forward flight to gliding and soaring, their maneuverability serving them well on updrafts near cliff nesting sites. They are equally at home on the water surface, where their webbed feet provide ample propulsion. They are less accomplished divers than terns, but some, such as the kittiwake and Lesser black-backed gull, do make use of aerial plunging to catch fish.

Most gulls fall into one of two major subgroups, distinguished by their summer plumage pattern. The "white-headed" group contains the largest species, including the well-known and highly successful Herring gull of Europe and North America, and its close relative, the more strictly marine Lesser black-backed gull (see "Nature" and "Nurture" Set Species Apart). Further north, various species, including the Glaucous gull, form a circumpolar distribution. The second major group, the "dark-hooded" or "masked" gulls, are mostly of slighter build, and characterized in breeding plumage by a bold chocolate-brown or sooty-black head; in winter

Left *As its common English name indicates, the Ivory gull is characterized by its pure white plumage, a coloration in accord with its high Arctic habitat. This individual is from the Canadian Arctic.*

Above *A flock of kittiwakes in front of the rocky terrain of Shetland. Across a huge geographical range, this shoreline species nests on offshore islands, sea stacks, and other inaccessible stretches of coast.*

this hood is molted, and the birds then have a predominantly white head with a residual dark patch or collar behind the eye. In Europe, the best-known representative is the ubiquitous Black-headed gull. American allies include the Laughing gull, Franklin's gull, and Bonaparte's gull. Less closely related, and sometimes placed in a separate genus *Xema*, is Sabine's gull. Two other gulls, strikingly different in appearance from these, are placed in separate genera: Ross's gull (genus *Rhodostethia*) breeds almost exclusively in northeastern Siberia; it has rosy-pink plumage on the head and underparts, as well as a black collar instead of a complete hood. Also confined to the high Arctic is the Ivory gull (genus *Pagophila*), resplendent in its all-over pure white plumage. The two cliff-nesting species of kittiwakes are placed in the genus *Rissa*. The nocturnal Swallow-tailed gull is the sole representative of the genus *Creagrus*.

Prevalent in the North

DISTRIBUTION PATTERNS

Although the gulls enjoy a worldwide distribution, the largest concentrations occur in the northern hemisphere, where they have succeeded in colonizing the harshest of marine environments. The Ivory gull, for example, breeds in the presence of pack ice and snow where none but the hardiest vegetation survives. Although gulls are well represented in temperate and subtemperate latitudes, they are more sparsely distributed in the tropics; this has been attributed to a relative scarcity of shore food. Although most gulls live on or near the coast throughout much of the year, others live deep in the heart of the continents. The Great black-headed gull and the rare Relict gull thus breed on islands in the inland seas and lakes of the central Asian steppes, many hundreds of kilometers from the nearest ocean.

At the end of the breeding season, when adherence to a colony on land is no longer required, many gulls disperse into offshore waters and some, such as the kittiwake, then lead a truly open-sea (pelagic) existence, British-bred birds journeying as far as the coasts of Canada. While there may be a strong random element in such dispersal, birds often congregate at food-rich, cold-water upwellings on the edge of continental shelves. Compared with the terns and skuas, however, relatively few gull species are true migrants. A notable exception is Sabine's gull, which has a circumpolar breeding distribution but at the end of summer migrates south through the Atlantic and Pacific to winter off Africa and South America; the bulk of the Atlantic migrants cross the Equator and make for the productive Benguela Current off Namibia and South Africa. Gales in the fall sometimes blow Sabine's gulls toward the British coast, where enthusiastic bird-watchers delight in identifying these rare and elegant visitors. Some species regularly migrate overland, like Franklin's gull, which passes in the spring and fall over the Great Plains of North America and the highlands of Mexico.

Outside the breeding season, gulls typically continue to be highly gregarious, often assembling in massive flocks for feeding, roosting, and bathing. The favored roosting sites are extensive open areas that offer a good all-round view for early detection of ground predators. Gulls often loaf on the flat expanses of airfields, where they represent a serious hazard to aircraft on their take-off run or landing approach (for example, gulls from the Jamaica Bay wetland preserve in New York are responsible for numerous "birdstrikes" to airliners entering or leaving the nearby JFK International Airport).

FACTFILE

GULLS

Order: Charadriiformes

Family: Laridae

51 species in 10 genera.

DISTRIBUTION Worldwide, but few in tropics.

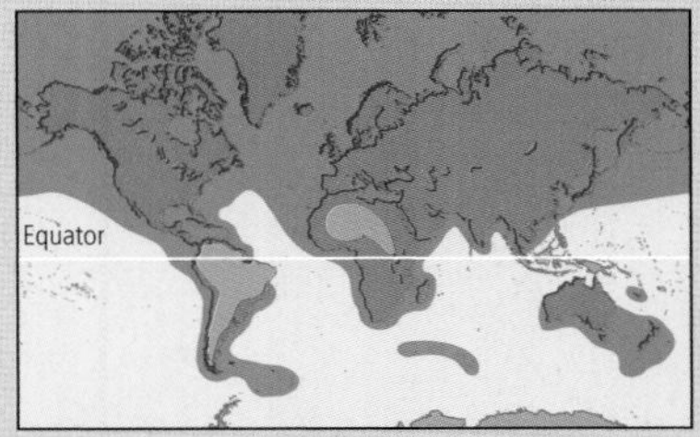

HABITAT Chiefly coastal waters.

SIZE Length 25–78cm (10–30in); weight 90g–2kg (3oz–4.4lb); males somewhat larger than females.

PLUMAGE Chiefly white, gray, and black in adult, streaked or mottled brown when immature. Sexes similar.

VOICE Wide repertoire includes ringing, laughing sounds, yelping, mewing, and whining notes.

NEST Typically a cup of vegetation, seaweed, or other plant matter, sometimes substantial, often on a cliff ledge or on the ground, some on marshes, bushes, or trees.

EGGS Usually 2–3, olive, brownish, or greenish, heavily mottled; weight ranges from 19g (0.7oz) in the Little gull to 117g (4.1oz) in the Great black-backed gull. Incubation 3–5 weeks; young fledge after 3–7 weeks.

DIET Fish, crustaceans, mollusks, worms, and (in smaller species) insects; also vegetable food, refuse, and carrion, some preying on birds and mammals.

CONSERVATION STATUS Six species, including the Red-legged kittiwake (*Rissa brevirostris*) and the Relict gull, are classed as Vulnerable.

See Selected Gull Species ▷

Supreme Opportunists

DIET

All gulls can store substantial quantities of food in their crops, from which they regurgitate when feeding mates or young. Birds usually settle at their night roosts with full crops, and leisurely digestion follows. Indigestible parts of the meal are periodically disgorged in the form of pellets. Analysis of these gives a good idea of the diet.

Gulls have a range of feeding habits unparalleled in almost any other group of birds. In the Arctic, for example, where there is a limited variety of prey, Glaucous and Ivory gulls regularly eat the feces of marine mammals, and also associate with whales to exploit the invertebrates they force to the surface. Swallow-tailed gulls are exceptional among gulls in feeding entirely at night, their large eyes apparently helping them detect and capture fish or squid. In temperate latitudes, the flexibility and ingenuity of foraging habits is just as striking. Herring gulls and their kin smash open shellfish by carrying them to a height and dropping them onto hard surfaces such as roads or the roofs of houses.

Many Herring gulls have capitalized on food waste on garbage dumps, and as a result have increased hugely in numbers. However, as landfill sites are supplanted by waste recycling and incineration, their numbers are falling again. Gulls that breed inland also enjoy a wide variety of natural foodstuffs, the smaller species such as Little and Franklin's gulls dipping tern-like to pluck insects as small as midges from the water or land, or hawking insects in the air. Many species, including the Black-headed gull and Franklin's gull, follow plows for earthworms and other soil creatures. In some places, the Lesser black-backed gull, which is essentially a fish-eating species, includes large numbers of worms in its diet. In the fall, stubble grain on farmland provides a valuable bonus.

The larger gulls often prey on birds and mammals that share their breeding stations. The Glaucous gull is an important predator of Little auks, while the Great black-backed gull plunders a wide variety of seabirds, notably puffins and ducks, and can dispatch a good-sized rabbit. Gulls have also learned. skua-like, to harry other seabirds, forcing them to disgorge their food; terns that share colony space with, for example, Black-headed or Silver gulls regularly suffer from such piracy. Sometimes gulls shadow foraging ducks, cormorants, or pelicans, and rob them as soon as they surface with prey. In the same way, Black-headed and Common gulls are frequently found in fields among flocks of lapwings, which are far superior in the art of locating and extracting earthworms; the gulls are quick to pounce on a successful lapwing to relieve it of its worm. Ring-billed gulls will also pirate from starlings and other birds. In Australia, the Silver gull has begun to forage on plowed fields far inland, repeating a pattern found in other parts of the world where people have developed arable farming on a significant scale.

In most regions, gulls breed once a year during a well-defined season that corresponds with the summer flush of food. In the tropics, where the food supply is less seasonal, more complex patterns may occur. The Swallow-tailed gull, for example, is known to breed in every month of the year; pairs that raise young successfully will mount another breeding attempt 9–10 months later, unsuccessful birds sooner still. In southwest Australia, the Silver gull may breed twice a year, in spring and fall, an unusual pattern.

◐ **Left** *Representative species of gulls:* ***1*** *Juvenile kittiwake* (Rissa tridactyla)*;* ***2*** *Great black-backed gull* (Larus marinus)*, largest of all gulls, in first-winter plumage, scavenging on a dead razorbill;* ***3*** *Ivory gull* (Pagophila eburnea) *in first-winter plumage;* ***4*** *Little gull* (Larus minutus) *juvenile, the world's smallest gull;* ***5*** *Sabine's gull* (Xema sabini) *on nest;* ***6*** *Ross's gull* (Rhodostethia rosea).

Mostly Monogamous

Breeding Biology

Gulls are generally monogamous, and usually pair for as long as both members survive. However, in species such as the kittiwake, divorce is not uncommon, especially among inexperienced birds, and individuals will seek a new mate if the existing pair-bond proves unfruitful. In some species, where sex ratios are unbalanced, female–female pairs develop; females obtain copulations from males and both females lay a full clutch in one nest. As the breeding season approaches, gulls typically assemble in large, dense colonies, frequently reclaiming their nest site of the previous year. Many species breed on cliff ledges or atop coastal islands, while inland species often seek the safety of a marsh. Common gulls may build their nests on the stump or fork of a tree up to 10m (33ft) above the ground, also not uncommonly on stone walls and buildings. In keeping with their growing use of man's domain, Herring gulls are also favoring roof-tops, chimney stacks, and other buildings. The kittiwake, a gull of the open seas for much of the year, also sometimes adopts a window ledge as a substitute for its usual cliff-nesting habitat. The Grey gull breeds in one of the most inhospitable habitats chosen by any bird, the hot, arid deserts of Peru and Chile, while Ivory gulls sometimes nest on stony patches on ice floes.

The density of nesting depends partly on the local food supply, on the size of the gull, and on the degree of cannibalism. In temperate regions, where fish stocks are high in summer, many gull species breed in huge colonies, siting their nests only 1m (3.3ft) or so from one another, and defending a territory little larger than the nest itself. Such gulls are notably successful in ousting smaller competitors from their nesting space, and populations on the increase often completely expel less competitive seabirds from islands. On the Isle of May, in Scotland, where Herring gulls increased from a solitary pair in 1907 to over 14,000 pairs in the space of 60 years, terns were forced to abandon the island as a breeding site. Where the food supply is less plentiful, gulls may nest much more sparsely. In an extreme case, Lava gulls, which number about 300–400 pairs and occur only on the Galapagos Islands, typically nest over 3km (1.9mi) from each other.

In species that nest densely, there is much rivalry as the pairs stake out territories at the start of

"Nature" and "Nurture" Set Species Apart

The Herring gull and the Lesser black-backed gull are two distinct species: they are dissimilar and they do not normally interbreed when they occur together. However, the two do illustrate how closely related species may be induced to interbreed if the recognition barriers that separate them are broken down.

During the ice ages of the Pleistocene epoch, 2 million to 10,000 years ago, the ancestors of today's Herring gull diversified, as a result of being isolated into a number of refuges separated by extensive ice sheets. A yellow-legged form confined to Central Asia later gave rise to the Lesser black-backed gull, while a pink-legged form spread from northwest Asia via North America to Europe, where it came into contact with Lesser black-backed gulls. By this time, however, the Herring gull had diverged in ecology and behavior to such an extent that the two did not readily interbreed.

Today, Lesser black-backed gulls differ from Herring gulls in Britain not only in appearance but also in being more migratory, and, to a lesser extent, in diet. How influential is nature as opposed to nurture in maintaining these differences and keeping the species distinct? In an experiment designed to explore this question, eggs in a mixed colony were interchanged between nests of the two species. On hatching, the resulting Herring gull chicks were raised by Lesser black-backed parents, and vice versa. There was not much change in the migratory tendencies of the cross-fostered gulls. However, their breeding habits were markedly affected. On reaching adulthood, many of the fostered young, especially females, mated up with a member of the "wrong" species, and successfully raised hybrid young. Thus the isolating mechanisms between the two species may fail if the young of each are brought up to "think" that they belong to the other species. EKD

the breeding season; males are the main aggressors, but the females also join in. Gulls command an impressive repertoire of aggressive and appeasement displays and calls during these contests. Although prolonged fights sometimes take place, most of this behavior is ritualized and injuries are avoided. Black-headed gulls, for example, regularly avert their heads ("head-flagging") when squaring up to one another, so hiding the provocative black mask and bill. Rival Herring gull males may symbolically tug and tear at the vegetation along a contested territorial boundary, then each may claim victory by throwing their heads back and "long-calling" vociferously, before resuming hostilities. Such shows of strength also attract females, which typically approach bachelor males tentatively in a submissive, cowed posture. Once accepted, the female is fed by the male as a prelude to egg-laying.

The clutch is typically 2–3 eggs; the tropical Swallow-tailed gull is unique in laying only one. Both sexes share incubation, changing over several times each day until the eggs hatch, usually after about four weeks. The emerging young are mobile as soon as their down dries, but remain in or near the nest for a week or so, where their parents can brood and tend them closely. In most species the chicks remain on the territory until fledging. When small, they jostle for food by pecking at the parent's bill; in some species a brightly colored spot near the bill-tip serves as a target and stimulus for this begging action. Once liberated from the need for brooding, the young may seek refuge in vegetation or other cover around the nest. If they trespass onto neighboring territories they are often fiercely attacked by the owners. Injury and even death may result, especially in larger species.

Above On its cliff-side nest, a kittiwake regurgitates food for its young. Kittiwakes do not scavenge at garbage tips like many other gull species, preferring to take small surface fish and invertebrates.

Below A Herring gull chick hatching. The first weeks of life are a perilous time for Herring and other large gull species, since adults frequently cannibalize other broods to nourish their own young.

In the larger gulls, some adults specialize in killing the young of other broods and feeding them to their own offspring. At one Herring gull colony it was observed that almost a quarter of all hatchlings were cannibalized in this way, and many eggs were also pirated. In species that lay three eggs, the last laid egg is typically the smallest. Gull clutches hatch over a few days, so that the last chick has to compete with larger siblings. This third "runt" chick is therefore prone to succumb if food is short, and more likely to fall victim to adult cannibals. Occasionally, a cannibal Herring gull has difficulty in distinguishing the instinct to nurture its own offspring from the urge to kill and eat the young of other pairs. One such adult ate over 40 chicks while sharing incubation of its own clutch. When its own brood hatched, it continued to bring live chicks to its nest site, but failed to kill them. Over a week, it added eight live, healthy young to its own brood. The problem of raising this extended family ultimately proved insurmountable, but one of the adopted young was successfully raised to fledging.

By the time the young leave the nest at 3–7 weeks (depending on species), they are fully feathered, but in a mottled brown garb quite different from their parents. This dress is lost by degrees, until breeding age is reached. The parents usually continue to feed their offspring for some time after fledging, up until six weeks afterwards in some of the larger gulls.

Like other seabirds, gulls that survive the rigors of juvenile life can, on average, look forward to a relatively long life. Ringing studies show that Black-headed gulls and Herring gulls can live over 30 years. Presumably because breeding is a hazardous venture, requiring considerable experience, gulls generally do not breed until they are several years old – two years in Little and Black-headed gulls, usually five in Herring and Lesser black-backed gulls. When they approach breeding age, some birds may return to the colony where they were born in order to establish a territory, and they sometimes settle remarkably close to their natal nest site. However, other gulls may travel considerable distances to join other colonies, a dispersal that probably helps to mitigate the possible adverse effects of inbreeding.

Boom Time

CONSERVATION AND ENVIRONMENT

Many, but not all, gull species have likely never been as numerous as they are now, given the new food supplies provided by humans. Yet a number of gull species nevertheless are threatened. The Chinese black-headed or Saunders's gull is Vulnerable because there are fewer than 2,500 pairs. Foremost among scarce gulls is the Relict gull, of which no more than 1,500–1,800 pairs are known from Lake Alakul and Lake Barun-Torey deep in the Russian interior. The other gull species classified by the IUCN as Vulnerable at time of writing are the Lava gull, Black-billed gull, and Olrog's gull. A number of other gulls are considered as being at lower risk, but near-threatened; for example, Heermann's gull (*L. heermanni*) of Mexico and Southern California experiences periodic fluctuations in its population, caused by the adverse effects of the El Niño Southern Oscillation and overfishing on fish stocks. JB/EKD

Selected Gull Species

In Arctic latitudes: Glaucous gull (*Larus hyperboreus*), Iceland gull (*L. glaucoides*), Ivory gull (*Pagophila eburnea*), Ross's gull (*Rhodostethia rosea*), Sabine's gull (*Xema sabini*). **In temperate latitudes**: Black-headed gull (*Larus ridibundus*), Bonaparte's gull (*L. philadelphia*), Common or Mew gull (*L. canus*), Franklin's gull (*L. pipixcan*), Great black-backed gull (*L. marinus*), Great black-headed gull (*L. ichthyaetus*), Herring gull (*L. argentatus*), Laughing gull (*L. atricilla*), Black-billed gull (*L. bulleri*), Lesser black-backed gull (*L. fuscus*), Little gull (*L. minutus*), Relict gull (*L. relictus*), Ring-billed gull (*L. delawarensis*), Western gull (*L. occidentalis*), kittiwake or Black-legged kittiwake (*Rissa tridactyla*). **In Mediterranean latitudes**: Audouin's gull (*Larus audounii*), Mediterranean gull (*L. melanocephalus*), Slender-billed gull (*L. genei*). **In comparable, or warmer, climate in the southern hemisphere**: Grey gull (*Larus modestus*), Silver gull (*L. novaehollandiae*), King or Hartlaub's gull (*L. hartlaubii*), Olrog's gull (*L. atlanticus*). **In the tropics**: Grey-headed gull (*L. cirrocephalus*), Lava gull (*L. fuliginosus*), Sooty gull (*L. hemprichii*), White-eyed gull (*L. leucophthalmus*), Saunders's gull (*L. saundersi*), Swallow-tailed gull (*Creagrus furcatus*).

Right *Kelp or Southern black-backed gulls* (Larus dominicanus) *scavenging sheep carcasses on a garbage dump in the Falkland Islands. Extreme adaptability has made gulls one of the most successful bird families.*

Below *Gulls have an extensive range of ritualized displays to signal aggression and appeasement:* **1** *Long call;* **2** *Begging for food;* **3** *Tugging at grass on the territory border;* **4** *"Head flagging";* **5** *Threat.*

Terns

AMONG THE MOST GRACEFUL AND APPEALING *inhabitants of shorelines and marshes, terns have narrower, more elongated bodies than gulls, and proportionately longer wings. Their streamlined body is adapted for plunge-diving so that they can easily catch their staple food of fish.*

Many terns are familiar summer visitors to north temperate coasts, catching the eye with their winnowing flight and spectacular headlong plunging. Some larger terns, like the massive Caspian tern, are closely related to gulls, while a resemblance of other members to skimmers is also apparent.

Black Caps and Pink Blooms

FORM AND FUNCTION

Most terns (24 of 44 species) belong to the "black-capped" group of *Sterna* species. These sea terns (or "sea swallows," from their tail shape and agility in flight) have a slender form, with long, tapering wings and a deeply forked tail. The typical plumage pattern is white, gray and black. Some species, such as the Roseate tern, have a delicate pink bloom to the feathers of the breast at the start of the breeding season. This color fades quickly, and may disappear soon after the birds arrive on the breeding grounds. Juveniles are often mottled brown, especially on the back, and may take 2–3 years to adopt adult form. In the marsh terns (three species of *Chlidonias*) and noddies (three species of *Anous*), the plumage is generally darker or even black. Conspicuously different is the blue Inca tern, with its yellow gape wattles and white mustache. The larger terns, such as the Caspian and Royal terns are less agile and graceful.

Terns' bills – often bright yellow, red, or black – vary in shape from pincer- to dagger-like, depending partly on the size of the prey taken. The flight, though buoyant, is strong, often allowing a sustained hover. Although the feet are webbed, most terns seldom settle on water for long.

Above Representative tern species: 1 Blue-gray noddy (Procelsterna cerulea)*; 2 Lesser noddy* (Anous tenuirostris)*; 3 White tern* (Gygis alba)*; 4 Inca tern* (Larosterna inca)*; 5 Arctic tern* (Sterna paradisaea)*; 6 Juvenile Black tern* (Chlidonias nigra)*; 7 Large-billed tern* (Phaetusa simplex)*; 8 Sooty terns* (Sterna fuscata) *are seldom seen near land except when breeding 9 Adult Caspian tern* (Sterna caspia)*; 10 Caspian tern in first-winter plumage.*

FACTFILE

TERNS

Order: Charadriiformes

Family: Sternidae

44 species in 7 genera. Species include, in temperate latitudes: **Aleutian tern** (*Sterna aleutica*), **Arctic tern** (*S. paradisaea*), **Common tern** (*S. hirundo*), **Gull-billed tern** (*S. nilotica*), **Little** or **Least tern** (*S. albifrons*), **Roseate tern** (*S. dougallii*), **Royal tern** (*S. maxima*), **Sandwich tern** (*S. sandvicensis*), **Black tern** (*Chlidonias nigra*), **Whiskered tern** (*C. hybrida*), **White-winged black tern** (*C. leucoptera*); in the tropics: **Black noddy** (*Anous tenuirostris*), **Brown noddy** (*A. stolidus*), **Inca tern** (*Larosterna inca*), **White tern** (*Gygis alba*), **Bridled tern** (*Sterna anaethetus*), **Cayenne tern** (*S. eurygnatha*), **Damara tern** (*S. balaenarum*), **Peruvian tern** (*S. lorata*), **Sooty tern** (*S. fuscata*).

DISTRIBUTION Worldwide

HABITAT Primarily coastal and offshore waters, some up rivers and in marshes.

SIZE Length 20–56cm (8–22in); weight 50–700g (1.8oz–1.5lb); males somewhat larger than females.

PLUMAGE Typically white below, gray mantle and upper wings, with black crown (crested in some) during breeding.

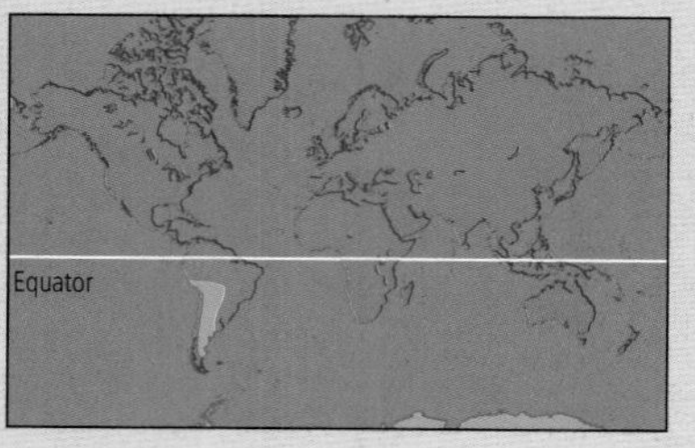

VOICE Most varied repertoire from shrill to hoarse, penetrating calls to soft crooning notes.

NEST Usually a simple scrape, occasionally well lined; some make floating rafts (marsh terns); others in trees and on cliff ledges (White tern and noddies), in holes in cliffs (Inca tern), sometimes under boulders or down burrows.

EGGS 1–3, pale cream to brown or greenish, with darker blotches; most weigh about 20g (0.7oz), but range from 10g/0.4oz in the Little tern to 65g/2.3oz in the Caspian tern. Incubation 18–30 days; fledging mostly at 1–2 months.

DIET Chiefly fish, squid, and crustaceans; in marsh terns insects, amphibians and leeches.

CONSERVATION STATUS The Chinese crested tern (*Sterna bernsteini*), is Critically Endangered. Five other *Sterna* species are Lower Risk: Near Threatened status.

Almost Ubiquitous

DISTRIBUTION PATTERNS

Terns are found worldwide, extending to all but the highest, ice-fast latitudes. Unlike gulls, which concentrate in the northern latitudes, the greatest number of tern species are found in subtropical and tropical regions. Terns breed on all continents, including Antarctica. Some species are truly pelagic during the non-breeding season. The Sooty tern, the most pelagic species, remains at sea from the time it fledges until returning to land to breed for the first time (at 3–7 years). Arctic terns wintering in Antarctica often roam through loose ice along the edge of the ice pack. Despite worldwide distribution, some species, such as Damara terns, have a restricted range.

Habitat preference divides the species broadly into two groups, sea terns and marsh terns. The sea terns generally nest on sandy beaches or

4
5
6
7
8
9
10

Left *After executing a plunge-dive, a Common tern emerges from the sea with its catch. Small fish and shrimps are the staple diet of this species.*

islands, where nests are often mere depressions in the sand. Some sea terns, such as the Common tern, nest in salt marshes, where they construct nests of grasses or make depressions in wrack. Others, such as the Roseate tern and Caspian tern, are among the most cosmopolitan of all birds. While the majority prefer warm tropical and subtropical waters, others favor colder latitudes for breeding, and the sea terns thus range from the Arctic to the Antarctic. By contrast, the marsh terns have adopted a largely inland existence, on freshwater marshes, lakes, and rivers, often deep in the heart of continents. They build floating nests of vegetation, which they anchor to water weeds to prevent movement during floods.

Terns undertake prodigious migrations, many journeying in summer to the food-rich waters of higher latitudes to breed, and resorting to tropical climes for the winter. The Arctic tern undertakes possibly the longest migration of any bird species. Many breed north of the Arctic Circle and move south to the Antarctic for the northern winter, an each-way journey of some 17,500km (11,000 miles) "as the crow flies." By doing so, they exploit the long daylight for prolonged feeding time in both hemispheres. Ringing terns and plotting their movements has done much to unravel the routes taken; many Canadian Arctic terns, for instance, cross the Atlantic on westerlies to the European coast on their way south. While most travel by sea, feeding as they go, overland routes are not uncommon. Many marsh terns, for example, cross the Sahara en route from their breeding grounds to their African winter quarters.

APPRENTICE PLUNGE-DIVERS

Plunge-diving for fish is the hallmark of many terns. A typical plunge-diver in European waters is the Sandwich tern. The bird flies upwind, usually 5–6m (16–20ft) above the surface, and, on spying its prey just below the surface, hovers briefly before plummeting into the water. The prey is seized in the vice-like grip of the bill, just behind the gills, and quickly eaten. When young are being fed, the fish is held crosswise in the bill and carried back to the colony.

As a technique for obtaining food, plunge-diving is remarkably successful. An adult tern often secures a fish in one dive out of three. But it is a difficult technique and many factors can make the task more so. Fishing success is greatly reduced in strong winds, partly because the shoaling prey, often sprats or sand eels, sink deeper to avoid the turbulent wave action. Very calm seas also appear to pose problems, possibly because the fish can see the tern overhead and take evasive action. It is also possible that in calm conditions the fish can sense the tern's splash on entry just soon enough to veer off.

The young Sandwich tern not only has to learn the best places to fish, but also needs to develop skills over time. Its first efforts are often shallow, unrewarded belly flops, but it gains practice by picking up bits of seaweed and other flotsam. Faced with the additional hazards of migrating south, the young bird continues to be fed by its parents for perhaps 3–4 months after fledging. Meanwhile, it gradually learns to dive from greater heights, and gains access to prey at greater depths, down to a maximum of about 1m (3.3ft) below the surface. However, even at 7–9 months old, in their West African winter quarters, some juveniles are still less adept than their parents at catching fish.

This long apprenticeship probably helps to explain why most young Sandwich terns stay in the winter quarters for two years. EKD

Headlong Divers

DIET

The sea terns are primarily fish-eaters, although squid and crustaceans are also relished. The black-capped terns are bold plungers, spotting their prey as they hover into the wind, before diving headlong (see Apprentice Plunge-divers). In general, the bigger the tern, the higher and deeper it dives; the Caspian tern may plunge from 15m (50ft). Unlike gannets, terns do not swim underwater, and prey is seized near the surface. Many species, such as Common and Roseate terns, rely on the behavior of predator fish to force prey to the surface, where it is easy to catch. Noddies typically dip to the surface and may use their feet for pattering, like storm petrels. They often catch flying fish in mid-air. Noddies and some other tropical terns range far offshore to feed, swallowing their prey for later regurgitation to the young.

Left *Arctic terns mass in Namibia while migrating from the Arctic to the Antarctic. They make this journey in order to spend a second summer in the southern hemisphere. The round trip between the two poles is about 35,000km (21,750 miles).*

Though most terns are daytime feeders, some, such as Sooty terns, have been recorded feeding at night. The dainty marsh terns are well adapted for hawking insects or hovering to pluck them off vegetation. They also make shallow plunges for frogs and other aquatic animals. The Gull-billed tern is the most terrestrial of all, and swoops to seize large insects, lizards, and even small rodents from the ground. The rate of feeding visits to the young varies according to the distance the parents must travel to hunt. While a marsh tern may feed its young every few minutes, the Sooty tern, which forages hundreds of kilometers away, may only deliver a meal once a day.

Lifelong Partners

BREEDING BIOLOGY

In common with many other seabirds, most terns are long-lived, if they survive to adulthood. Arctic terns have been shown by ringing to live 33 years or more, and a life span of 20 years is probably not unusual. Breeding may begin as early as two years, but more often at three or four in temperate breeding species (for example, Grey-backed and Bridled terns delay breeding until they are four). In tropical species, it generally occurs later; most Sooty terns, for example, do not reach sexual maturity until they are at least six years old.

In higher latitudes, terns usually have a well-defined breeding season once a year, in Europe and North America, from about May to July. In the tropics, breeding is generally not synchronized to a particular time of year. In a few populations, however, terns breed both at intervals of less than a year and synchronously. On Cousin Island in the Indian Ocean, Bridled terns breed every seven months, while the highly adaptable Sooty tern breeds, depending on location, at intervals varying from six to 12 months; in some cases it seems likely that food is equally abundant all year.

Terns customarily pair for life. Even though the pair bond breaks down outside the breeding season, there is a strong tendency to return to a previously successful breeding site, which enables former mates to rendezvous at the start of each new breeding season. Although most terns are philopatric, returning to the same colony year after year, species that breed in ephemeral nesting habitats shift sites as conditions dictate. Freshwater marshes, riverine sand bars and coastal sandspits may be used for only one or a few years before they become unsuitable. Often the adults return to the colony site, find their mates, and then select a new site.

Most terns breed in bustling colonies, often at high density. They also roost en masse and may join together to mob predators at the colony. Nesting colonies range in size from a few widely scattered pairs (Damara tern) to colonies of a million or more (Sooty tern). Intermediate-sized terns, such as the Common tern, nest in colonies of tens to hundreds of pairs. The large terns nest in colonies on the order of hundreds of pairs.

Some species of terns nest in single-species colonies, but many nest with other species of terns, gulls, skimmers, boobies, alcids, albatrosses, cormorants, and ducks. These mixed-species assemblages may occur because of habitat restrictions, not necessarily by choice. In many cases, the terns nest in monospecific groups within these larger mixed-species colonies. Some species, such as Forster's and Black terns, choose to nest with other species, and select their nest sites only after other species have settled and are nesting.

The colony site is usually on flat, open ground, often on an island or reef, making them inaccessible to ground predators. Noddies, however, crowd on trees, bushes, and cliff ledges, while Inca terns seek crevices in rock. The White tern is celebrated for building no nest, opting to lay its single egg directly onto (most commonly) the branch of a tree. Most ground nesters are scarcely more constructive, merely fashioning a shallow scrape, at the best thinly lined. Noddies and marsh terns, however, build a more substantial platform of vegetation, the latter species anchoring a raft of reeds, say, to submerged plants.

Above *The White or Fairy tern has the curious habit for a tern of nesting in low bushes. It perches its single egg on a bare branch little wider than the egg and incubates it there, with each parent taking turns.*

Below *A large nesting colony of Sooty terns in the Seychelles. These birds are the world's most abundant terns, breeding in their thousands on tropical oceanic islands.*

Terns normally spend only 2–3 weeks in the vicinity of the colony site before settling to establish territories. Courtship is an elaborate ritual, especially in birds seeking a mate for the first time. In many terns, the first stage of pairing is the "high flight," in which the male ascends at speed, as if to demonstrate his prowess, often to several hundred meters, while the female pursues him. At the end of the climb the prospective pair glide and zigzag earthwards. With growing familiarity, the male increasingly courtship-feeds his mate; this has more than just symbolic value – it also helps the female to form eggs and perhaps allows her to gauge the fishing skill of her partner. Ground courtship often occurs near the male's chosen nest site, and involves much elegant strutting and pirouetting with raised tail and drooped wings. This is usually the prelude to copulation. During courtship, both mated and unmated birds engage in fish flights, which begin when a mater male brings a fish to a female. After he lands on the territory, both take off and fly high, often joined by one or two other birds. They glide and circle on bowed wings, uttering unique flight calls.

Nest site selection occurs when both members of the pair agree on a given location. Both sexes defend the nest territory, often only 1sq m (11sq ft) or so in extent, while the "crested" terns, which nest most densely of all, may be within jabbing distance of neighbors. Territory size is inversely related to the size of the tern species – the larger the tern, the smaller the territory size. In some species of terns the female remains on the territory to defend it while the male goes on foraging trips. He returns with fish to courtship-feed the female, and immediately goes back to fishing. This division of labor allows the female to defend the nest, and the male to feed her. While the male is away, unmated males may solicit the female. Mating usually follows courtship feeding.

The normal clutch varies from one egg in tropical species to 2–3 in higher latitudes. Incubation is shared by the sexes, and lasts 3–4 weeks, although it can range as high as 37 days. On hatching, the downy chicks are soon actively exploring their surroundings, but seldom stray far unless disturbed. Then they take refuge in vegetation, or under stones, driftwood and the like. When disturbed, parents may move their chicks a considerable distance. The well-grown young of "crested" terns may seek safety in numbers, forming a mobile crèche. The chicks in a crèche move about together, remain densely packed, and separate only to seek food from their parents. Parents

Left *An adult Arctic tern feeds its chick on the wing. Occasionally the adult will catch more than one fish and always carries its prey crosswise in its bill before taking it to its mate or chicks.*

returning with food recognize their own young in the crèche by voice, and feed only them. Parents usually recognize their chicks by 4 to 7 days of age, at the time when chicks become mobile and might wander from their nests. The chick period lasts from 20 days in the Peruvian tern to 65 in the Bridled tern. For most terns, the chick-rearing period is about 20–30 days. The total breeding season may last only two months for Arctic species, 3–4 months for temperate species, and 3–5 months for tropical species.

After the young fledge, they have much to learn about catching prey for themselves and are fed by their parents for some time, before gradually being weaned off. Young not only have to learn prey types and foraging habitats, but how to plunge-dive, a difficult task. Post-fledging parental care in terns is known to range from seven days, in the Cayenne tern, to over 200, in the Royal tern. It is likely that post-fledging care extends for longer than previously thought because of the difficulties of observing this care in birds that have migrated to wintering grounds.

Under Human Pressure

CONSERVATION AND ENVIRONMENT

Terns suffer heavy and increased predation by mammals introduced by man, including cats, dogs, and rats. Reproductive success is also reduced by predator populations that have increased because of man's activities, including Herring gulls, foxes, and raccoons. When such predators reach offshore breeding islands, ground-nesting terns can be eliminated or their populations reduced in a short period of time.

The isolation sought by terns for breeding purposes is an ever more scarce resource, as people turn increasingly to the coast for leisure, commercial fishing, and other activities. Recently, terns are facing new pressures from personal watercraft that can come closer to nesting colonies than traditional motorboats. As more and more people move to coastal regions, terns nesting on barrier beaches and islands are forced to shift colony sites to less suitable places. At the same time, increasing Herring gull populations have resulted in additional loss of habitat because the gulls arrive earlier, are larger and more able to win territorial encounters, and are predators on the eggs and chicks of the terns. Pressure on land use in South Africa has reduced the Damara tern to a precarious 1,500 pairs, while snaring for food and sport in its West African winter quarters is believed to have contributed to the decline of the European Roseate tern to around 1,000 pairs. Some tern populations were devastated by plume hunters at the end of the 1800s, and populations only recovered slowly. In some parts of the world, people still collect tern eggs for food or as an aphrodisiac. Many populations continue to flourish in more remote regions, and on Christmas Island, the Sooty tern is numbered in millions.

The Chinese crested tern is Critically Endangered; it has, on several occasions, been listed as Extinct, but there are a few recent sight records of wintering birds. Its breeding area is unknown. Several other species of tern are threatened, including the Kerguelen and Damara terns. The status of other species is poorly known (River and Saunders's terns, Black and Grey noddies). Conservation measures include protecting colonies from direct exploitation, creation of suitable nesting space, construction of artificial nesting islands or platforms, removal of predators, and reduction of human disturbance. Protecting foraging sites may become increasingly important as fisheries exploit the prey base of terns. JB/EKD

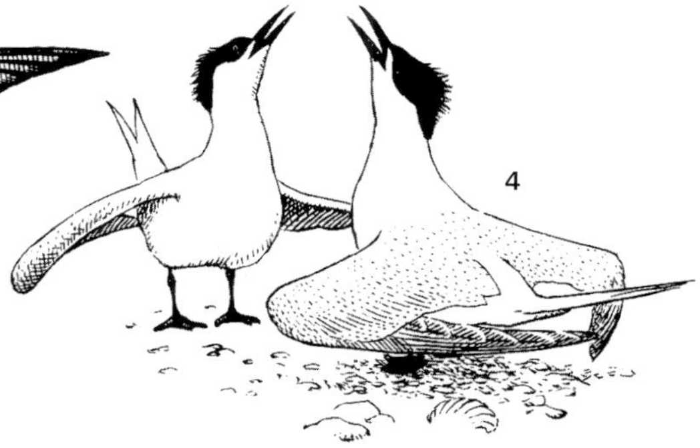

Left *Elaborate courtship of terns.* ***1*** *Female Arctic tern pursues male upward in "high flight."* ***2*** *Male Little tern feeding his mate.* ***3*** *Common terns mating.* ***4*** *Erect or "pole" stance of Sandwich tern pair, seen after copulation and also after high-flight.*

Skuas and Jaegers

DURING THEIR BREEDING SEASON, SKUAS AND jaegers are the pirates and predators of the skies in high latitudes. Skuas frequently harry other seabirds, such as terns and kittiwakes, until they drop their catch or disgorge their last meal, which the rapacious skuas then snatch in mid-air.

In North America and elsewhere the small skuas are known as jaegers, from the German word meaning "hunter." In Shetland, Arctic jaegers were given the local name *skooi*, which may be derived from *skoot*, meaning "excrement," upon which the birds were believed to feed by scaring other seabirds. The local name for Great skuas in Shetland is "bonxie," which may come from the Norse *bunksi*, denoting an untidy heap or an unkempt, dumpy woman!

A Varied Diet

FORM AND FUNCTION

The four species of large skua are generally brown with lighter flecking on the dorsal feathers. The Chilean skua has conspicuous rufous underwing feathers, while the South polar skua has both dark and light phases (dimorphic plumage), with light-phase birds increasing in frequency toward the Pole. All three jaegers display two color phases, although the dark phase is extremely rare in the Long-tailed jaeger. In the Arctic and Pomarine jaegers the proportions of birds of each phase within the population vary geographically. In Shetland less than 25 percent of Arctic jaegers are light; the proportion tends to increase northward, with nearly 100 percent light in Svalbard and arctic Canada. Birds with light plumage appear to have a selective advantage in the north of the species' range but not in the south, where the proportion of dark birds has increased at many colonies since the 1950s. The elongation of the two central tail feathers, characteristic of adult skuas, is prominent in the Arctic jaeger and extreme in the Long-tailed jaeger. The Pomarine jaeger has twisted, club-shaped central tail feathers. All juvenile jaegers are barred below.

The skuas have feet that are gull-like, but with prominent, sharp claws. The bill is hard and strongly hooked at the tip, adapted for tearing flesh. In gulls, as in many other bird families, males are slightly larger than females, but the opposite is true for skuas, as for birds of prey. In both of these groups, the male does most of the hunting and the female remains in the territory to guard the nest or young.

FACTFILE

SKUAS AND JAEGERS

Order: Charadriiformes (suborder: Lari)

Family: Stercorariidae

7 species in 2 genera: *Catharacta* and *Stercorarius*.

Distribution High-latitude regions in both hemispheres. Antarctic, sub-Antarctic, southern S America, Iceland, Faeroes, N Britain, Norway, Svalbard, and northern Russia. Arctic and boreal regions.

Habitat Tundra and coastal heath.

LARGE SKUAS Genus *Catharacta*
Antarctic, sub-Antarctic, southern S America, Iceland, Faeroes, N Britain, Norway, Svalbard, and N Russia. 4 species: **Great skua** (*Catharacta skua*), **South polar skua** (*C. maccormicki*), **Antarctic** or **Brown skua** (*C. antarctica*), **Chilean skua** (*C. chilensis*). **Size**: Length 50–58cm (20– 23in), Weight 1.1–1.9kg (2.4–4.2lb); females slightly larger than males. **Plumage**: Brown, with white wing flashes. **Voice**: Limited range of yelps and screams. **Nest**: Scraped depression on ground. **Eggs**: Normally 2, occasionally 1; olive with brown blotches; weight 70–110g (2.5–3.9oz). Incubation 30 days; nestling period 45–55 days. **Diet**: Diverse, particularly fish, krill, seabird eggs and chicks, adult seabirds.

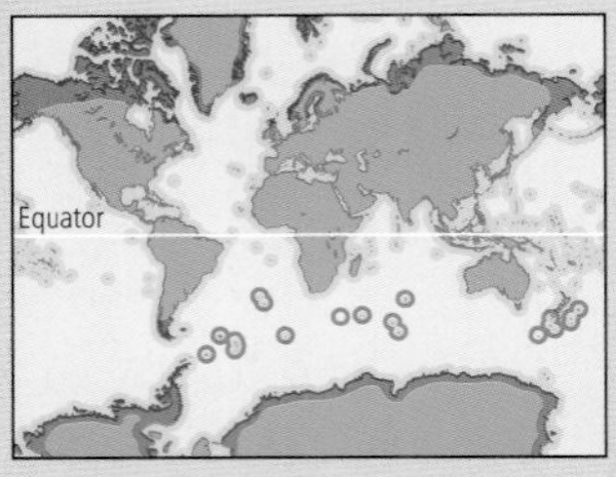

SMALL SKUAS OR JAEGERS Genus *Stercorarius*
Arctic and boreal regions. 3 species: **Parasitic jaeger** or **Arctic skua** (*Stercorarius parasiticus*), **Long-tailed jaeger** or **Long-tailed skua** (*S. longicaudus*), **Pomarine jaeger** or **Pomarine skua** (*S. pomarinus*). **Size**: Weight 250–800g (8.8oz–1.8lb); females slightly larger than males. **Plumage**: All brown or brown above and creamy white below. Elongated central tail feathers in adults. **Voice**: Mewing cries. **Nest**: Scrape on ground. **Eggs**: Normally 2, occasionally 1; olive with brown blotches; weight about 40–70g (1.5–2.5oz). Incubation 23–28 days; nestling period 24–32 days. **Diet**: Small mammals, insects, berries, birds' eggs, fish (often robbed from other seabirds).

Above *A South polar skua on its breeding ground in Antarctica in December. After breeding, Antarctic skuas migrate northward; they have been recorded as far north as the Aleutian Islands and Greenland.*

Left *Representative species of skuas and jaegers:* ***1*** *Great skua* (Catharacta skua) *long-calling;* ***2*** *Pomarine jaeger* (Stercorarius pomarinus) *in its breeding plumage;* ***3*** *Arctic skua* (S. parasiticus) *harassing Atlantic puffins; piracy of prey from smaller birds accounts for much of this species' diet.*

Skuas take many types of food. Pomarine jaegers feed largely on lemmings in summer and on small seabirds in winter, less regularly by fishing or by harassing other birds (kleptoparasitism or piracy). Long-tailed jaegers feed on lemmings, insects, berries, small birds, and eggs in summer and by piracy, chiefly by harrying terns, in winter. In arctic tundra areas, Arctic jaegers feed on insects, berries, small birds and eggs, as well as some rodents; in coastal areas they feed almost exclusively by piracy of terns, kittiwakes, and auks. The large skuas in the southern hemisphere feed by predation and scavenging on penguins, by nocturnal predation of oceanic seabirds (petrels) that return to land at night, and on a wide range of other prey including fish, crustaceans (krill), and rabbits. Great skuas eat mainly fish, sometimes obtained by scavenging or kleptoparasitism. However, they are very adaptable and have been seen to kill prey items many times heavier than themselves, such as Grey herons, Greylag geese, shelducks, and Mountain hares. They are also important predators of seabirds during the breeding season, and they rely mainly on seabirds at some colonies, especially on the island of St. Kilda off western Scotland, where they prey nocturnally on storm-petrels on land. In addition, they can be cannibalistic, and will readily take chicks from neighboring territories, especially in conditions of poor food supply. In Shetland and Orkney, Great skuas take mainly one-year old sand eels, and many aspects of their breeding ecology – from diet and chick growth to adult survival and the number of nonbreeders – are closely related to sand-eel abundance.

Epic Migrants

DISTRIBUTION PATTERNS

Skuas have been seen closer to the South Pole than any other vertebrate except man. In the northern hemisphere, Great skuas have recently expanded their breeding distribution northward and eastward into Norway, Svalbard, and northern Russia, and all of these new colonies include birds originally ringed as chicks at colonies in Scotland. Southward expansion into more temperate regions has, however, been much more limited, probably because skuas have a number of adaptations to cool environments (high basal metabolic rate and body temperature, heavy insulative plumage, and thick scutes over the legs) that prevent them from breeding further south.

Outside the breeding season, skuas migrate over all the world's oceans, the jaegers also traveling in some numbers directly overland. Records of Arctic jaegers in Austria and Switzerland in the autumn are not uncommon. Great skuas, on the other hand, tend to remain some distance offshore, but a few storm-tossed young Great skuas have been picked up, exhausted, in central Europe. One, ringed as a chick in Shetland, was rescued from the central reservation of a motorway in Germany only to be shot a week later as it attacked hens on a farm in the east of Austria. Others have been found in a Swiss town and attacking ducks on a pond in Poland. Southern-hemisphere skuas have a very wide range of migration patterns: Brown skuas at the Chatham Islands are resident throughout the year, while Brown skuas from other colonies and Chilean skuas show limited dispersal or short-distance migrations and South polar skuas are long-distance transequatorial migrants. One South polar skua ringed as a chick at Anvers Island, Antarctica, was shot five months later in Godthabsfjord, Greenland – one of the longest migrations of any bird ever recorded by ringing.

Skuas are closely related to, and presumably evolved from, the gulls, which almost certainly originated in the northern hemisphere. Early in the evolution of the skuas, one form must have colonized the southern hemisphere, where it gave rise to the three very similar large skua species in the Antarctic. One of the latter then relatively recently (possibly as late as the 15th century) colonized the northern hemisphere, where some individuals appear to have hybridized with female Pomarine jaegers before giving rise to Great skuas. This hybridization, along with a comparison of behavior and feather lice, has led some taxonomists to suggest that Pomarine jaegers, which are the largest of the jaegers, should be placed within the same genus as the large skuas (*Catharacta*).

Right A Parasitic jaeger atop its solitary nest in the arctic tundra of Svalbard. In such bleak, open terrain, nests are typically widely separated from one another, though elsewhere within their range, skuas and jaegers can form dense colonies.

Below Nesting on the fringes of penguin colonies, skuas wait tirelessly for the opportunity to snatch unattended chicks or eggs. Here, an Antarctic skua on the Falkland Islands scavenges a penguin egg.

Defense by Dive-bombing

SOCIAL BEHAVIOR

Skuas are long-lived (the annual adult survival rate usually exceeds 90 percent) and socially monogamous, although in New Zealand and on Marion Island Brown skua trios comprising one female and two males occur regularly, a social system not yet found in any other seabird. Skuas normally pair for life, although some pairs (usually less than

10 percent) do divorce each year, most often following one or more years of breeding failure. Some established breeders (usually less than 10 percent) also take an occasional year off from breeding. In some cases this is forced by the loss of a partner or breeding territory, but in other cases nonbreeding may result from birds failing to regain breeding condition after the winter.

Distances between skua nests vary enormously, from 2km (1.2 miles) apart (a common distance on the arctic tundra), down to just 5–10m (16–33ft) within the largest Great skua colonies in Shetland. On Foula, Shetland, well over 100 pairs of Arctic jaegers breed in a colony occupying 1.7sq km (0.7sq mi), approximately the area defended by a single pair breeding on arctic tundra. Part of the explanation for this difference is that skuas nesting in Shetland do not obtain food within their territory, but feed at sea. Skuas defend their nests by dive-bombing intruders, including humans. Jaegers also employ a "broken wing" distraction display to lure predators away from a nesting-site.

In common with most seabirds, laying dates of skuas advance and the number and size of eggs increase with age up to the onset of senescence (c. 14–18 years in Great skuas), after which birds lay progressively later and smaller clutches as they enter old age. Hatching and fledging success also increase with age except when the food supply is very poor (when breeding success is uniformly low) or very good (when success is uniformly high). These changes are probably influenced more by experience than by age per se; for example, in Great skuas, young males are less efficient foragers and young females are more likely to leave chicks unattended, and thus vulnerable to predation by neighboring adults. Skuas have only two brood patches, so they could not effectively incubate more than two eggs. Most pairs with two eggs manage to hatch both, but if food is scarce, the older chick, which hatches 1–3 days before the second, will sometimes attack and kill its smaller sibling. In Shetland, Arctic jaegers begin breeding when 3–6 years old, while Great skuas first breed when 4–11 years old. Presumably this long period of immaturity helps them to learn the many skills that are needed to become an effective hunter and kleptoparasite of other seabirds. KH/RWF

Skimmers

DEFTLY SNAPPING FISH FROM LAKES, RIVERS, and lagoons with their uniquely adapted bills, skimmers are highly appositely named – they do, literally, skim the water surface as they hunt. They crowd by the hundreds into nesting colonies on sand bars, where the contrast between their brilliant bills and legs, and their stark plumage, make them a prized "target" of wildlife photographers and artists.

Even though there are only three species of skimmer in the world, the family is widely distributed. Three subspecies (or races) of the Black skimmer are found in the New World: the North American subspecies inhabits the ocean coasts and the Salton Sea in the western USA. The two South American subspecies are almost exclusively riverine, using coastal areas outside the breeding season. African skimmers are most abundant in East and Central Africa on the larger river systems. Indian skimmers range from Pakistan across India to the Malay Peninsula, mostly in close association with large rivers.

FACTFILE

SKIMMERS

Order: Charadiiformes

Family: Rynchopidae

3 species of the genus *Rynchops*: **African skimmer** (*R. flavirostris*), **Black skimmer** (*R. niger*), **Indian skimmer** (*R. albicollis*).

DISTRIBUTION Tropical Africa, S Asia, C and S America, southeastern N America.

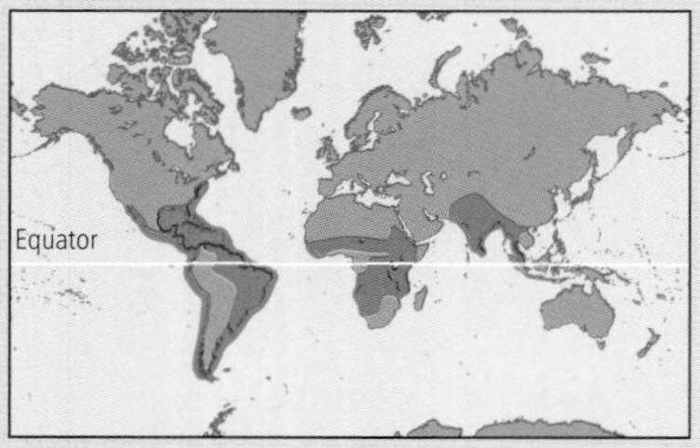

HABITAT Major river systems and ocean coasts.

SIZE Length 35–45cm (14–18in); weight males 400g (14.1oz), females about 300g (10.6oz).

PLUMAGE Chiefly black or dark brown above, white or light gray below; sexes similar.

VOICE Barks (Black skimmer) or shrill, chattering calls.

NEST Simple scrape in sand or on shell bank.

EGGS Usually 3–4 (range 2–5), whitish or beige with dark brown or black blotches and irregular spots; about 4x3cm (1.6x1.2in). Incubation 22–24 days; nestling period 25–30 days.

DIET Primarily small fish, also small invertebrates such as shrimps, prawns, and other small crustaceans.

CONSERVATION STATUS The Indian skimmer is classed as Vulnerable due to habitat degradation.

Scooping up Fish

FORM AND FUNCTION

The Indian and Black skimmers have orange-red bills (but the Black skimmer's is black toward the tip) and vermilion legs and feet in the breeding season (duller at other times), while the African skimmer has yellow legs and feet and a yellow-orange bill. The young of all species are lighter brown above and less white below, and the tail is mottled, unlike the mostly white tail of the adults. The wings are very long, with a span $2\frac{1}{2}$ times the length of the bird.

Yet the most striking feature in all species is the large, scissorlike bill with its flattened "blades," the upper mandible fitting into a notch between the edges of the lower mandible, which is one-quarter to one-third as long again as the upper. The lower mandible was once thought to be very touch-sensitive, but this has been found not to be the case. When feeding, skimmers hold the bill open in such a position that the tip of the lower mandible slices the water. When the lower mandible touches a prey item, usually a small fish, the head flexes downward rapidly, trapping the prey sideways between the "scissors" (hence the popular name "scissorbill"). The musculature of the head and neck is well developed and acts as a shock absorber. Skimmers often feed at dusk and during the night, especially in the non-breeding season. The skimmer eye has a vertical pupil, like that of a cat, which may enhance its light-gathering properties. Skimmers prefer to feed in waters with little surface turbulence, such as lakes, pools, marshes, and river edges. After "cutting a trail" in the water, birds often retrace their course, snapping up prey in their wake. They usually feed alone or in pairs, but on occasion groups of 10–15 may engage in brief bouts of intense feeding in a certain spot. On the coast, feeding increases at low or ebbing tides.

Male Black skimmers are larger than females. Measurements of wingspan, bill, tail, and weight show males ranging from about 10 percent (in the wing and tail) to 25 percent (in weight and bill length) larger than females. Such dimorphism has not been verified for African and Indian skimmers.

Skimmers used to be considered more closely related to terns than gulls. However, unlike gulls, neither skimmers nor terns use their wings during

Right A Black skimmer feeding in its unique way. Fishing rates in this species have been measured under various conditions at one fish per half minute to one per 6 minutes of skimming.

Below In this aerial fight between two Black skimmers, their unique bills can clearly be seen. Skimmers are the only birds whose lower mandible is longer than the upper one.

aggressive encounters. However, further analysis of breeding behavior suggests that the skimmers split from the ancestral stock before the divergence between terns and gulls. Of the three families, only skimmers have a "broken wing" distraction display, although it is not used often.

Frequent Disputes

SOCIAL BEHAVIOR

Skimmers are highly social birds in all seasons of the year. When they reach breeding age (probably at 3–4 years), they gather on open, sandy bars and small islands where courtship begins. Vertical flights and aerial chases by courting birds are common at this time. The breeding colonies, established on these sites after a few weeks, range in size from a few pairs up to 1,000 or more. Skimmers often form mixed colonies with terns, and they benefit from some terns' greater display of aggression in driving off predators.

Skimmers have small nesting territories, with nests spaced 1–4m (3.3–13ft) apart, depending on vegetation and terrain. The degree to which birds nest at the same time (synchrony) can be very high in certain areas in the colony. Aggression is high during the period of territory establishment and egg-laying, and both sexes engage in disputes over space and mates. The males are more aggressive toward other skimmers, while females more frequently interact with other species nesting nearby. Males incubate and brood more than females, at least during the day. Males and females often switch incubation duties, especially in the hottest part of the day. Foot and belly-wetting by adults helps to regulate the temperature of the egg. After the young hatch, the females feed the young more than the males do. Parents continue to feed beyond the nestling period, and the fledged young accompany adults on feeding forays, perhaps learning to fish.

The nesting period at a skimmer colony is sometimes longer than that of most of their gull and tern relatives – along the eastern seaboard of the USA the Black skimmer may nest from May to October, although they usually nest from May through August. After the nesting season, skimmers gather in loose flocks at certain "staging areas." They follow major river systems and coastal routes when migrating to wintering areas. Some populations of skimmers are nonmigratory.

Skimmers are not yet considered threatened. However, damming of rivers in India, Africa, and South America continues to reduce the nesting habitat. This practice, coupled with destruction of tropical forest, also diminishes water quality and productivity which, in turn, affects skimmers' diet. In North America, many coastal habitats have been disturbed, forcing Black skimmers (and other species) to nest on small saltmarsh islands and even roofs of buildings. RME/JB

Auks

RELATED TO GULLS, TERNS, AND SHOREBIRDS, the auks are familiar seabirds that are often considered the northern hemisphere equivalents of the penguins, being well adapted to underwater swimming. The flightless Great auk became extinct in the 19th century, and all extant auk species retain the power of flight.

Auks are among the most abundant seabirds on earth, with many populations numbering in the millions; for example, Common and Thick-billed murres, Least and Crested auklets, and dovekies all have world populations of more than 10 million individuals. They are the dominant seabirds over large areas of arctic and subarctic waters, and they may be significant predators on large zooplankton and small fishes in areas around their breeding colonies. They are compact, robust birds that swim underwater, using their wings as paddles. They include some of the bird world's deepest divers and, for their size, they have exceptional underwater speed and endurance.

A Variety of Plumages and Bills

FORM AND FUNCTION

Auks are small to medium-sized birds with short tails and small wings. In fact the size of their wings is a compromise, large enough for flying but small enough to use in the denser medium of water – hence their rapid wingbeats and whirring flight.

The legs of most auks are positioned toward the rear of the body, which accounts for the upright posture of many species. Others, especially the murrelets, tend to lie on their belly while on the ground and rarely stand. The legs of some species are compressed laterally, an adaptation for swimming. The three toes are connected by webs. In species such as the Atlantic puffin and Black guillemot the legs and feet are bright orange or red, while in murrelets and Cassin's auklet they are blue. The mouth is also brightly colored in some species: red in the Black guillemot, yellow in the razorbill.

The shape of the bill varies markedly, partly reflecting differences in diet and feeding methods. Species that feed mainly on plankton tend to have wider, shorter bills, while those of fish-eaters are longer and more daggerlike. Inside the mouth, the palate and tongue of plankton-feeders are covered with horny tubercles, which presumably assist in manipulating the prey. The Parakeet auklet, which feeds on jellyfish as well as crustacea, has a peculiar, scooplike bill, different from anything seen among other seabirds. In the razorbill and the puffins, the bill is laterally compressed.

The large, colorful bills of the three puffins play an important role in pair formation and courtship. The bill is encased in nine distinct plates that are shed each year during the molt. Outside the breeding season, the puffin's bill is much reduced in size and is less brightly colored. The Rhinoceros auklet, closely related to the puffins, is unique in having a solid, fleshy "horn" projecting about 2cm (0.8in) above the base of the bill. This is not shed in winter and is found even in juvenile birds. Its function is unknown.

Many auks have distinct summer and winter plumages. The Black and Pigeon guillemots are black with a white wing patch in the summer, but during the winter they are mainly white and gray. The Marbled and Kittlitz's murrelet have cryptic brown summer plumage, but outside the breeding season are mainly black above and white below. In the Crested and Whiskered auklets and the Tufted puffin, long head plumes and "whiskers" are lost during the winter, while species that have black throats in summer change them to white in winter. Only the Rhinoceros and Cassin's auklet and Craveri's and Xantus's murrelet look similar throughout the year.

During the molt, which in most species occurs soon after breeding is over, the larger auks are flightless, probably for some 45 days, as the flight

feathers are dropped simultaneously. The wing-loadings (ratio of body weight to wing area) of the Common and Thick-billed murres, razorbill, and other large species are so high that losing feathers one after another at intervals, as in most birds, might jeopardize the ability to fly at all; by losing the feathers simultaneously, the duration of the flightless period is kept as short as possible. The wing-loading of the smallest species (the Least auklet and Whiskered auklet) is lower; they can molt their flight feathers one at a time and still retain the ability to fly. The flightless Great auk also molted in this way.

Above *1 Adult Parakeet auklets* (Cyclorhynchus psittacula) *sit high on cliffs keeping watch near their nesting holes while their mates tend the nest.* *2 Standing erect, the razorbill* (Alca torda) *resembles its larger, extinct relative, the Great auk. 3 A Japanese murrelet* (Synthliboramphus wumizusume) *emerges from its nest cavity. 4 Little auks* (Alle alle) *tend to fly in large flocks, massing like starlings. 5 Atlantic puffins* (Fratercula arctica) *swallow their food underwater unless they are feeding their young, in which case they can return to the nest with up to 30 fish in their bills.* *6 Black guillemots* (Cepphus grylle) *are mostly black with white wing patches, but in the winter they molt to become mostly white and gray.*

FACTFILE

AUKS

Order: Charadriiformes

Suborder: Alcae

Family: Alcidae

23 species in 12 genera. Species: **Ancient murrelet** (*Synthliboramphus antiquus*), **Craveri's murrelet** (*S. craveri*), **Japanese murrelet** (*S. wumizusume*), **Xantus's murrelet** (*S. hypoleucus*), **Atlantic** or **Common puffin** (*Fratercula arctica*), **Horned puffin** (*F. corniculata*), **Tufted puffin** (*F. cirrhata*), **Black guillemot** or **tystie** (*Cepphus grylle*), **Pigeon guillemot** (*C. columba*), **Spectacled** or **Sooty guillemot** (*C. carbo*), **Cassin's auklet** (*Ptychorampus aleuticus*), **Common murre** or **guillemot** (*Uria aalge*), **Thick-billed murre** or **Brunnich's guillemot** (*U. lomvia*), **Crested auklet** (*Aethia cristatella*), **Least auklet** (*A. pusilla*), **Whiskered auklet** (*A. pygmaea*), **Kittlitz's murrelet** (*Brachyramphus brevirostris*), **Long-billed murrelet** (*B. perdix*), **Marbled murrelet** (*B. marmoratus*), **Little auk** or **dovekie** (*Alle alle*), **Parakeet auklet** (*Cyclorhynchus psittacula*), **razorbill** (*Alca torda*), **Rhinoceros auklet** (*Cerorhinca monocerata*).

DISTRIBUTION N Pacific, N Atlantic, and Arctic oceans and coastal regions.

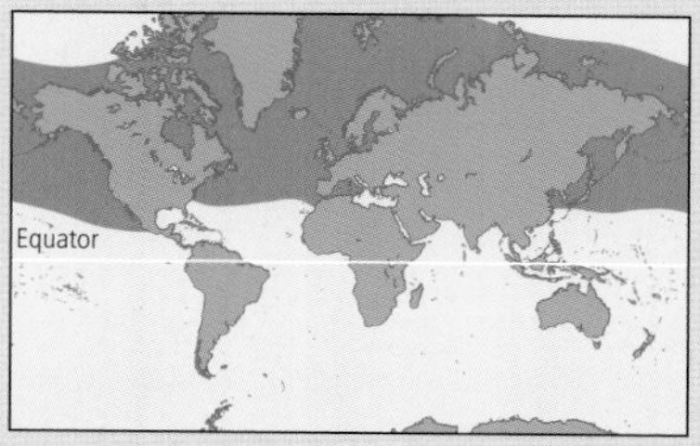

HABITAT Breeding, mainly along coasts on islands and headlands; non-breeding, mainly in coastal and continental shelf waters.

SIZE Length from 16cm (6.5in) in the Least auklet to 43cm (17in) in the Common guillemot; **weight** from 85g (3.1oz) to 1.1kg (2.4lb) in the same two species (extinct Great auk about 6kg/13lb). Sexes very similar in size, but males usually with slightly larger bills.

PLUMAGE Most species dark above, pale below; some have colored (red or yellow) bill or feet.

VOICE A wide range of whistling, growling, and yelping noises: some species almost silent.

NEST Breeds on open flat rock or cliff ledges (no nest), or in a very simple nest in crevices or burrows; Marbled murrelet on large branches of conifers.

EGGS 1 or 2; weight 16–110g (0.7–4oz); pear-shaped and variable in color and markings (razorbill and guillemots) to ovoid and plain (puffins, auklets). Incubation period 29–46 days; time chick spends at nest site very variable, from 2 to 50 days.

DIET Fish or marine invertebrates, caught by diving from the surface.

CONSERVATION STATUS 4 murrelets – the Crested, Marbled, Craveri's, and Xantus's – are listed as Vulnerable by the IUCN. The Great auk (*Pinguinis impennis*) was hunted to extinction in the mid-19th century.

In Northern Parts

DISTRIBUTION PATTERNS

Auks are almost entirely confined to arctic, subarctic, and temperate waters and are marine throughout the year, being most common in waters of the Continental shelf and slope. In the Atlantic they extend south to Portugal, to the Mediterranean in the east and to Massachusetts in the west, and in the Pacific from the subtropical waters of Baja California to the Yellow Sea off China. The dovekie or Little auk is the most northerly breeder, with no breeding sites south of 65°N. Both this species and the Black guillemot breed to latitudes above 80°N.

Six species of five genera occur in the North Atlantic, and 20 of 10 genera in the North Pacific south of the Chukchi Sea (21 of 11 if the Little auk, with a few hundred in the Bering Sea, is included). Two murres, the Little auk, and the Black guillemot are circumpolar in distribution. The large number of species in the Pacific, particularly in the Bering Sea, and the fact that all genera except *Alca* (the razorbill) occur there, suggest that the family might have originated in the Pacific, a theory strengthened by the fact that fossil *Alca* specimens have been found in California.

Seafood Specialists

DIET

Auks obtain all their food from the sea, by diving from the surface and pursuing prey underwater (see Master Divers). They occur both inshore and offshore, but only the puffins are found regularly away from continental shelves. All species feed on fish or plankton (invertebrates and larval fish). Some species, like the auklets of the Pacific and the Atlantic dovekie, feed almost entirely on plankton; others, like the Common murre, eat mainly fish. In the North Atlantic the dovekie is in fact the only exclusive plankton-feeder, whereas in the North Pacific there are at least six such species, among them the Least, Cassin's, and the Parakeet auklet. This difference may be due to a greater total biomass and diversity of plankton in the North Pacific.

In the dovekie and other plankton-feeding auks, food is carried back to the chick in a throat pouch in the form of a "plankton paste." Dovekies feed their young 5–8 times a day, and each meal contains on average 600 items, with a total weight of 3.5g (0.1oz). In contrast, the Common murre feeds its young mainly on fish that school in shallow water, such as capelin, sand eels, or sprats. A single fish weighing 10–15g (0.4–0.5oz) is carried back lengthwise in the bill, with the head held in the mouth; the chick is fed 2–5 times a day. The Black and Pigeon guillemots specialize in bottom-dwelling fish of similar size, such as blennies and sculpins; a single fish is carried crosswise in the bill about nine times a day. Puffins usually carry several fish (up to 60 larval fish have been recorded), held crosswise in the bill.

Left Common murres breed in the open on sea stacks or on narrow, seaward-facing cliff ledges. Despite their seemingly precarious position and exposure to the elements, their densely packed nests afford them protection against gull and crow raiders.

Right A Common murre with its single egg. The egg is pear-shaped, which reduces its chances of rolling off the cliff ledge. As the embryo grows, the egg's center of gravity changes as well as its rolling radius, from 17cm (fresh) to 11cm (fully incubated). The color and patterning of these eggs are more varied than within perhaps any other bird species. The different base colors (turquoise to white) and markings (red, brown, or black) help the parent to recognize its own egg in a crowded colony.

Small Clutches and Long Lives

BREEDING BIOLOGY

Auks are long-lived birds, and there are several records of Common and Thick-billed murres that were ringed as adults and found still breeding 20 years later! Like many other seabirds, auks show a number of features associated with longevity. They lay small clutches of only one or two eggs each year. Young birds may spend two or three years at sea before they start breeding. Few auks breed until they are at least 3 years old. The Atlantic puffin provides a typical example. It remains at sea, away from the breeding area, until its second summer, when it may visit land for at most a few weeks; usually such two-year-olds spend very little time on land. In its third summer the young puffin returns a little earlier and attempts to find a mate and a suitable burrow for breeding. Egg-laying may occur for the first time in the fourth or fifth summer The larger auks generally breed with the same partner, using the same site year after year, whether it is a tiny rock ledge (as in the Thick-billed murre) or an earth burrow (puffins); auklets, however, frequently swap partners from year to year.

Although arctic-breeding auks may return to the colony only a few weeks prior to breeding, Common murres in Britain may be absent from their colonies for little more than two months after they finish breeding: just time to complete their wing molt. Time spent at the colony prior to breeding is spent re-establishing pair-bonds and mating. Copulation takes place at or near the breeding site (in murres and the razorbill, for example) or on the sea (puffins, auklets). Mating is frequent and, in some species, noisy. In Common murres, mating may occur three or four times a day in the two to three weeks prior to egg-laying, and each copulation lasts on average for 20 seconds; in most songbirds, by comparison, the duration is only 1–2 seconds. Monogamy is the rule in all species, the male and female cooperating to rear the chick(s). Despite this, male Common murres will also attempt to mate with any unattended female in the colony.

The arrival of a female can result in a frenzy of activity, with up to 10 males simultaneously trying to mount the newcomer.

The eggs of auks are relatively large, constituting between 10 and 23 percent of the female's body weight. Among burrow-nesting auks, the eggs are mainly white, with a few darker markings. The eggs of Kittlitz's, the Long-billed and Marbled murrelets are cryptically colored olive green with dark blotches. All these species nest alone and in the open; Kittlitz's murrelet on the ground in tundra vegetation or on rocky mountain slopes, and the other two mainly on the moss- and lichen-covered branches of large trees. Marbled murrelet may nest up to 40m (135ft) above ground in the branches of coniferous trees and up to 60km (36 miles) or more inland. The eggs of the Common and Thick-billed murres are among the most striking, and most variable, of any bird species, ranging in color from bright blue or green to white.

Most auk species, in common with most other seabirds, breed a considerable distance from their food supply, and can only provide enough food for a single chick. However, two-egg clutches occur in species that either feed close inshore, such as Black and Pigeon guillemots, or have chicks that are well developed on hatching (precocial). The two eggs are laid several days apart: 2–3 days in the Black guillemot and 7–8 days in

MASTER DIVERS

Auks are exceptionally good underwater swimmers, and can dive to over 200m (650ft). Only the largest penguins plunge deeper.

Recently, depth recorders small enough to be carried by auks have been developed. These devices, which record the depth every few seconds, have shown that murres may make up to 20 consecutive dives; staying at the surface for less than a minute between each one. Such bouts are followed by rest periods of several minutes. Most dives reach a maximum depth of about 60m (200ft). During bouts, some individuals have been recorded carrying out several consecutive dives lasting more than three minutes with only brief pauses at the surface, showing that the birds can recharge their oxygen stores very rapidly. Prey are swallowed underwater to avoid having to return to the surface after every capture.

At present, the Thick-billed murre holds the depth record for auks, with one bird reaching 210m (690ft). The maximum speed of descent is about 2m (6.5ft) a second. Returning to the surface is slightly faster, because murres are less dense than water, so they are assisted by buoyancy during their return journey. Underwater videos show that the wings are not used in the last stage of ascent to the surface: the birds shoot up like rockets, trailing bubbles, without needing to flap.

Overleaf Plunging from an icefloe, Common murres show off their diving prowess. Like penguins, they use their wings for propulsion; their graceful motion makes it seem as if they are flying through the water. They can stay submerged for up to three minutes at a time.

Below Some Atlantic puffins excavate their own breeding burrows by digging with their beaks 90–120cm (3–4ft) down into slopes near the sea. They shovel the soil out with their webbed feet.

the precocial murrelets (see When Young Auks Leave Home).

The male and female parent change places on the nest, at intervals varying from a few hours in the Black and Pigeon guillemots up to 5–6 days in the Ancient murrelet. The duration of these incubation shifts probably reflects the distance that the off-duty birds must travel to find food.

In most species in most years, 50–80 percent of pairs successfully rear young. Failures usually occur because of predation or infertility or, in cliff-nesting species, from eggs rolling off the breeding ledges. The disappearance of important food resources such as schooling fishes may, however, cause total reproductive failures; one population of Atlantic puffins in Norway affected in this way managed to rear chicks only once in a 10-year

1

When Young Auks Leave Home

In most bird families, the newly-hatched young are either nidicolous – naked and helpless, remaining in the nest for days – as in songbirds, or they may be covered with down and have their eyes open, like semi-precocial young gulls, which are fed by their parents, or ducks, whose precocial nestlings feed themselves. Unlike other bird families, however, young auks show a range of developmental patterns.

Most auks, including puffins, auklets, and guillemots, hatch semi-precocial young that are down-covered and open-eyed but dependent on their parents for food; these chicks remain at the nest site for 27–50 days, leaving when almost fully grown. Dovekie chicks are accompanied to the sea by their fathers, but all the other semi-precocial species are independent of their parents after leaving the nest.

At the other extreme are four Pacific species: the Ancient, Japanese, Xantus's, and Craveri's murrelets. They produce a clutch of two eggs, hatching precocial young that leave the nest site after only 2–3 days, accompanied by both their parents. The chicks are well-developed at birth, with feet that are already almost adult sized, and they can run, swim, and dive actively as soon as they leave the nest site. However, they do not begin to feed themselves until deserted by their parents at the age of about 6 weeks.

A third pattern is found only in the murres and the razorbill: in their case, the young birds are fed at the colony until they are 18–23 days old, when they leave while still only one-quarter adult size and flightless. Where the rearing site is on a cliff, the young bird glides down to the sea on wings made up of only the primary covert feathers – the true primaries develop later. Once they have left the colony, they are fed by their father for several weeks until fully grown. The mother remains at the breeding site for a few days and then departs to begin her molt.

The semi-precocial strategy, as seen in puffins and auklets, is typical of the auks' closest relatives, the gulls and terns, and as it is found in several lineages of auks, it is probably the ancestral strategy. The intermediate departure of the young murres and razorbill seem to have evolved in relation to the amount of food the parents can bring to their chicks. These are large species that have relatively small wings, so they can deliver little food to their young at each visit. By leaving the nest when only partially grown, the young can be convoyed to the feeding area, saving the parents the work of commuting.

The precocial murrelets breed in areas where the adults run a high risk of predation by falcons and eagles. By taking their young to sea soon after hatching, the parents avoid the risk of predation while visiting the colony. At the same time, taking the young to the feeding area eliminates the energetic expense of commuting, allowing them to rear two chicks rather than one as in most other auks. This strategy is not without costs, as the females must produce relatively large eggs to provide the hatchling with the dense down and energy reserves that it needs to survive at sea and to travel to the feeding area.

In the precocial species, all activities on land are nocturnal. Consequently, the chicks must find their way to the sea at night. To ensure that chicks can identify their own parents in the dark, the calls of both adults and chicks are individually recognizable.

◐ **Left** *To reduce losses to predation by gulls, some razorbills choose to lay their eggs in rock crevices. In any event, both parents take turns to watch over the chick constantly. Here, on a sheer cliff-face in the Outer Hebrides, Scotland, a razorbill guards its offspring.*

◑ **Right** *Whiskered auklet juveniles are dark gray with traces of three white stripes on the head. When fully grown, these auklets are sparrow-sized with three distinctive, white, ornamental plumes that grow from the face like a long mustache.*

period. Similar failures, of shorter duration, have been observed for Tufted puffins.

The young of most species are fed by their parents at the colony. Among those species that breed on open ledges, such as the murres and the razorbill, only one adult can forage at a time. The other parent must remain to brood the chick and protect it from predators such as gulls.

The process of young auks leaving the colony, conveniently but inaccurately referred to as "fledging," varies between species. Young puffins, auklets, and Black guillemots fledge alone at night, unaccompanied by their parents. They probably fly several hundred meters before alighting on the sea and dispersing away from the colony. By contrast, the young of the two murres and the razorbill are still flightless when they leave, accompanied by their father. For those that live on 300m (1,000ft) high cliffs, fledging can be a spectacular process. For several hours before leaving, the chicks become more and more excited, jumping up and down as they exercise their tiny wings, and uttering shrill, piping calls. Chicks may deliberate for several minutes before launching themselves into the air and fluttering down to the water's surface, closely followed by the father. In some areas where razorbills and murres do not have direct access to the sea, the chick may have to scramble over boulders to reach the water. Father and chick recognize each other's calls, and the adult may find the chick and guide it on its way.

Some colonies may be enormous, containing over a million birds. Only the *Brachyramphus* murrelets breed in solitary pairs. The Black and Pigeon guillemots breed in small, loose colonies. Solitary or loosely colonial species feed inshore on a predictable food supply. In contrast, colonial species typically feed further offshore and exploit patchily distributed, unpredictable prey, such as shoaling fish or plankton. Where prey is unpredictable, it may be advantageous to be part of a colony so that information on the whereabouts of food can be obtained from other colony members. At large colonies the traffic between the colony and feeding areas may be sufficiently dense to enable birds to find their way simply by following the flight line of incoming birds.

In the Great Auk's Shadow

CONSERVATION AND ENVIRONMENT

Four species of auks are listed as vulnerable by the IUCN, of which three (the Crested or Japanese, Craveri's, and Xantus's murrelets) are members of the genus *Synthliboramphus*, the precocial murrelets. All of these species occur at the southern edge of the auks' range in the Pacific, in areas where problems such as introduced mammalian predators, increased numbers of predatory gulls and crows, and high levels of marine pollution are especially acute. The fourth species, the Marbled murrelet, lives in rather remote areas of the North Pacific, but is threatened by logging of its preferred breeding habitat in old-growth forests.

Humans have eaten auks and their eggs for thousands of years. Traditional hunting, using nets and snares on a local scale, had little effect on auk numbers, but the use of firearms and commercial egg-collecting have caused the extinction of many colonies and one entire species. Excessive egg-collecting of Common murre eggs at the Farallon Islands, off California, caused the population to fall from about 400,000 birds in 1850 to just a few hundred by the 1920s; subsequent protection allowed a recovery to over 100,000 birds by the 1990s. The Thick-billed murre suffered a similar fate in Novaya Zemlya, Russia, where both adults and eggs were overexploited. Large numbers of the Great auk once bred on Funk Island, Newfoundland, but in the 17th century this became a regular stopping-off place for sailors and fishermen, who took many adult birds for food and later for feathers: by 1800, the colony was extinct. The last Great auks were killed in 1844 on Eldey, a small island off Iceland.

Auks are particularly vulnerable to oiling, because they spend so much of their time on the sea. Fuel oil released onto the sea, either deliberately when cleaning tanks or as a result of accident, coats the birds' plumage and destroys their waterproofing. In some cases, oil is ingested as the birds try to clean themselves and may prove toxic. During the 20th century, oil pollution became a major cause of deaths for Atlantic auk populations.

Commercial fishing activities have also taken a toll. A salmon gill-net fishery off Greenland, for example, destroyed 500,000–750,000 Thick-billed murres annually between 1968 and 1973. A more widespread phenomenon is competition for fish. Until recently, species such as capelin, sand eel, and sprat were not fished commercially, but now that stocks of larger fish such as cod are depleted, there is competition from humans for these small species on which many auks depend. Despite this threat, however, auks seem to be doing well in the North Atlantic, the most heavily fished marine area in the world. AJG/TRB

◒ **Below** *In Common murres a threat* ***1*** *is sometimes employed to warn off an intruder. Fights, although common, are usually very brief, because they are cut short by appeasement displays. These include:* ***2*** *side-preening;* ***3*** *stretching away or turning away; and* ***4*** *a ritualized walk when passing other birds in the colony.*

Sandgrouse

SANDGROUSE ARE CAMOUFLAGED PERFECTLY to blend in with their their habitats in dry, sandy deserts and scrubland. They are rarely seen, except when they gather in the tens, hundreds, or even thousands to fly in to drink at waterholes in the morning or evening. Although they look somewhat like grouse (hence their name), this resemblance is no more than superficial.

Formerly, there was much debate among taxonomists as to whether sandgrouse were most closely related to pigeons or waders (shorebirds). However, new evidence from molecular analysis strongly supported the view that they share their ancestry with the waders. By contrast, their skeleton resembles very closely that of the doves, to which they are undoubtedly related, but which arose earlier from the same evolutionary line.

Protection Against Heat and Sand

FORM AND FUNCTION

Most sandgrouse species are cryptically spotted, barred, or streaked; they crouch on the ground to avoid detection, but their long, pointed wings also enable them to make a quick getaway in swift, direct flight, rather like that of a plover. Their plumage is dense, the entire body being covered with a thick undercoat of dark down. This is a highly unusual feature, since most other species of bird have distinct lines of feathers on their bodies, separated by areas of bare skin. This down insulates sandgrouse against the temperature extremes that occur in deserts between night and midday, and between winter and summer. Even the base of the bill is feathered to protect the nostrils against sand and dust blown on the wind.

Despite having short, feathered legs, sandgrouse can walk and run well. Their feet are equipped with three broad, stout front toes that spread their weight evenly and enable them to walk efficiently on loose sand. The two genera of sandgrouse differ with respect to the presence or absence of hind toes and to the degree of feathering on their tarsi and toes. In the majority of species (namely, those in the genus *Pterocles*), the hind toe is present but much reduced and above ground level, and just the front of the tarsi are feathered. However, in the two species of *Syrrhaptes*, which are from the steppes and mountains of central Asia, the hind toe is lost and the toes and tarsi are completely feathered.

The Search for Food and Water

DIET

Sandgrouse eat mainly small seeds with a relatively high protein content (in particular those of legumes) and a low water content (less than 10 percent water, as a rule). These they pick up by walking with small steps, pecking frequently with their short bills. The crop of an adult Black-bellied sandgrouse was found to contain about 8,700 indigo plant seeds, while that of a Namaqua sandgrouse chick just a few days old contained 1,400 tiny seeds. Sandgrouse take up grit to help break down the seeds in the gizzard. They may also eat small bulbs, green leaves, berries, and even insects (termites and ants), especially during the breeding season. They feed for most of the daylight hours, resting only in the extreme heat of midday in summer, usually in the shade of a bush. Non-breeding flocks of thousands of birds are known (such as in Namaqua sandgrouse), but these are exceptional other than at waterholes. Flocks usually number 10–100 birds on their feeding grounds.

Sandgrouse need to drink every 2–3 days, possibly every day or even twice a day in hot weather. Large flocks of hundreds or thousands of birds gather daily at set times at waterholes. Most

◐ *Left* *Representative species of sandgrouse:* ***1*** *Pallas' sandgrouse* (Syrrhaptes paradoxus)*;* ***2*** *Painted sandgrouse* (Pterocles indicus).

◑ ***Main picture*** *Namaqua sandgrouse gather at a waterhole in South Africa. Hundreds or thousands of birds congregate daily at set times to drink. When not breeding, flocks mass around the waterhole for half an hour before all approaching the water at the same time.*

species drink in the morning only, but three (Painted, Lichtenstein's, and Double-banded sandgrouse) are exclusively night-time drinkers and form a subgenus Nyctiperdix, characterized also by barred plumage in both sexes, and bold black-and-white frontal patches in the males.

Sandgrouse may fly up to 80km (50 miles) one-way to water, although seldom more than 20–30km (12–20 miles), cruising at a speed of about 70km/h (43mph). The birds call in flight as they travel to waterholes, thereby gathering ever-increasing numbers near the water. They then fly or run to drink quickly, taking about 10 gulps of water, raising the head to swallow between each gulp. This may only take 10–15 seconds, but if both members of a pair are present, the first to finish will wait for the other before flying off. Some species, like Burchell's sandgrouse of the Kalahari,

FACTFILE

SANDGROUSE

Order: Pteroclidiformes

Family: Pteroclididae

16 species in 2 genera. Species: **Black-bellied sandgrouse** (*Pterocles orientalis*), **Black-faced sandgrouse** (*P. decoratus*), **Burchell's** or **Variegated sandgrouse** (*P. burchelli*). **Chestnut-bellied sandgrouse** (*P. exustus*), **Crowned** or **Coroneted sandgrouse** (*P. coronatus*), **Double-banded sandgrouse** (*P. bicinctus*), **Four-banded sandgrouse** (*P. quadricinctus*), **Lichtenstein's sandgrouse** (*P. lichtensteinii*), **Madagascar sandgrouse** (*P. personatus*), **Namaqua sandgrouse** (*P. namaqua*), **Painted sandgrouse** (*P. indicus*), **Pin-tailed sandgrouse** (*P. alchata*), **Spotted sandgrouse** (*P. senegallus*), **Yellow-throated sandgrouse** (*P. gutturalis*), **Pallas's sandgrouse** (*Syrrhaptes paradoxus*), **Tibetan sandgrouse** (*S. tibetanus*).

DISTRIBUTION Africa, S Iberia, and France; Middle East to India and China.

HABITAT Desert, semi-desert, dry grasslands, arid savanna, and bushveld.

SIZE Length 25–48cm (9.8–19in); weight 150–650g (5.3–22.9oz).

PLUMAGE Mainly dull tones of buff, ocher, rufous, olive, brown, black and white. Males usually spotted or barred: most have black, white, or chestnut chest bands. Females usually ocher or buff with black streaking and barring. This sexual dimorphism is marked and invariable. Central tail feathers very long in six species.

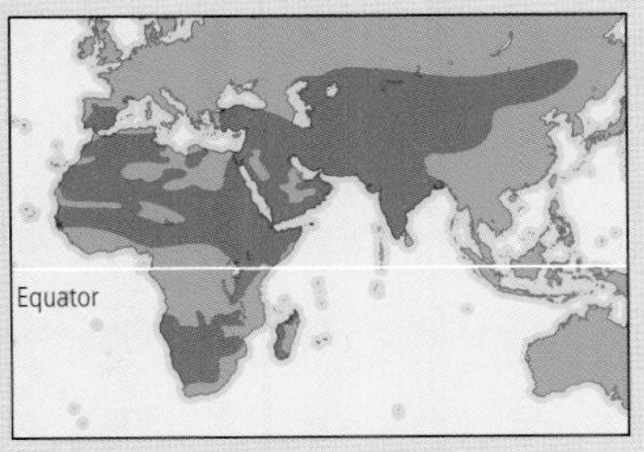

VOICE Mellow, whistled, or chuckling calls in set phrases of 2 or more syllables, usually given in flight, and highly characteristic for each species.

NEST Simple scrape in open or by a bush, stone, or grass tuft; sometimes scantily lined with dry plant fragments or small stones.

EGGS Almost always 3, rarely 2; elongated, and equally rounded at each end; light cream, grayish, greenish or pink, blotched, smeared, and spotted with brown, red-brown, olive-brown, and gray. Incubation 21–31 days; fledging period about 4 weeks.

DIET Almost exclusively small, dry seeds; some other plant material, insects, small mollusks, and grit.

CONSERVATION STATUS Not threatened.

land right at the water or even on its surface, floating like ducks while drinking, and taking off without effort. Sandgrouse will not normally drink water with a salt content higher than about 40 percent of that of seawater, since their kidneys are poorly adapted to excreting high salt concentrations. Furthermore, unlike most shorebirds, they lack a salt gland with which to excrete excess salt.

At high temperatures (above about 37°C/99°F) sandgrouse tend to become inactive, seek shade, and cease feeding, drooping their wings and holding their wrists well away from the body to increase heat loss. In the evenings birds may dustbathe; for example Pin-tailed sandgrouse (which occur in the region just north of the Sahara) are known to sandbathe by turning onto their backs with their feet in the air. And at night, certain species are known to make shallow roosting scrapes, which may be used on successive nights.

Most species are generally resident in one area year-round or are "nomadic," that is they range widely over an area depending on local availability of food and water in their unpredictable, arid habitat. However, a few are genuinely migratory, moving long distances between regular breeding and non-breeding grounds. The most southerly population of the Namaqua sandgrouse migrates up to Namibia and Botswana, while the Zambian and Botswanan race of the Yellow-throated sandgrouse emigrates south and east to Zimbabwe and South Africa; and the Indian population of Black-bellied sandgrouse is also truly migratory. Pallas's sandgrouse of the central Asian steppes does not strictly migrate but can undergo explosive "eruptions," with large numbers moving out of their normal range, probably because of food shortages caused by deep snow. Major invasions into Europe and as far east as Beijing can occur, and breeding was recorded in Britain in 1888–89 following one such event.

Above *A Chestnut-bellied sandgrouse chick nestles in a shallow scrape on the ground. Nests are usually built in the open, but can sometimes be found against a stone, grass tuft, or shrub.*

Below *The male Yellow-throated sandgrouse can be recognized by its black chest band (left); the female lacks a chest band but has black streaking and barring all over except on its throat (right).*

Nesting in the Desert

BREEDING BIOLOGY

All species so far studied are monogamous, but are not very territorial, so they can occur in small colonies. Northern-hemisphere species breed in spring and summer, and southern-hemisphere species mainly in winter, but in the Namib and

Kalahari deserts of southern Africa times may vary, depending at least partly on rainfall and the consequent food supply. Courtship involves head-down, tail-up chasing displays, similar to some threat displays. The female usually incubates by day and the male at night, although this pattern may be somewhat different in the four members of the subgenus *Nyctiperdix*. Studies of the Namaqua sandgrouse show that the parents attempt to keep their eggs cool when temperatures are hot by raising the body to shade the clutch while resting on lowered wings. However, the eggs must be able to tolerate relatively high temperatures because soil temperatures can sometimes rise to 50°C.

The downy chicks leave the nest as soon as the last hatched is dry. They are not fed by either of the parents and begin to feed on small seeds within a few hours of hatching. The pecking movements of the female parent show seeds to them. Although they usually lay three eggs, one chick is generally lost early. The young can fly a little at about four weeks, but are provided with water by the male (see Flying Water Carriers) for at least another month, when they can fly well enough to accompany the parents to the waterhole. Parents fly to water separately so that the chicks are not left unattended until the age of about 3 weeks. The chicks attain sexual maturity at about a year. Depending on conditions, some species, such as Black-bellied sandgrouse in Israel and Spotted sandgrouse in Morocco, may be double-brooded.

Friends and Enemies

CONSERVATION AND ENVIRONMENT

Sandgrouse are vulnerable to predation because of their habit of descending in large flocks to drink at waterholes at predictable times in the morning or at dusk. They are among the favorite prey of raptors, especially the Lanner falcon (*Falco biarmicus*), which hunts mainly at the waterholes, as well as of such mammalian carnivores as foxes, jackals, and mongooses, to which they are particularly vulnerable when nesting. The chicks are also vulnerable to avian predators such as kestrels and crows.

Sandgrouse are no longer in great demand for the pot and for sport, as they once were (attempts to introduce them from India and Pakistan into arid regions of the USA for sporting purposes have failed). Poor agricultural practices, exacerbated by drought, may be increasing the extent of suitable habitat. Where they occur close to human settlements, some species, such as the Pin-tailed sandgrouse, may move onto farmland to feed on wheat, oats, and other cereals, as well as lentils. In Egypt, Spotted and Crowned sandgrouse gather in large flocks to feed on raw grain spilled from trucks trading between ports on the Red Sea and Nile Valley markets. Combined with the provision of watering places fed by boreholes, conditions for most sandgrouse species have undoubtedly been improved by human activities. None are considered to be threatened. HC/GLM

FLYING WATER CARRIERS

Young sandgrouse have a diet of dry seeds and are unable to fly or walk to the nearest isolated waterhole. They must have water, however, and it is the male parent that is uniquely adapted to provide it. From the day they hatch, until at least two months later, young sandgrouse are brought drinking water in the soaked belly feathers of the adult male.

When the chicks are very small, the female flies off first to drink. On her return the male takes his turn while she takes over brooding. Before walking into the water, he rubs his belly in dry sand or soil to remove waterproofing preen oil; then he wades in belly-deep, keeping wings and tail well clear of the water, intermittently rocking his body up and down to work the water deeply into the belly feathers; this may take a few seconds or as much as 20 minutes. The returning male stands erect and the chicks run to drink from the central groove in his belly plumage. Once the chicks have drunk their fill, the male walks away, rubs his belly on a patch of sand to dry the feathers, and the family moves off to feed for the rest of the day.

The male's belly feathers have a unique structure that allows them to hold relatively large amounts of water on their inner surfaces, where evaporation is kept to a minimum. The barbules of the central portion of each belly feather are spirally coiled when dry, and they lie flat on the feather vane, tightly coiled together to give the feather structural cohesion. When wet, the barbules uncoil and stand at right angles to the feather vane, forming a dense bed of hairs about 1mm (0.04in) deep, which holds water like a sponge.

With one exception, all sandgrouse species so far studied employ this water-carrying mechanism. (The Tibetan sandgrouse does not need to, as snow meltwater is always close by in the Central Asian mountains that are its home.) It certainly prevents the male parent from depleting his own internal water supply, as would happen if he were to regurgitate water from his crop in the same way as the doves. Furthermore, although the parents start with three young, usually only one survives to fly, so the demand for water transport is limited. GLM

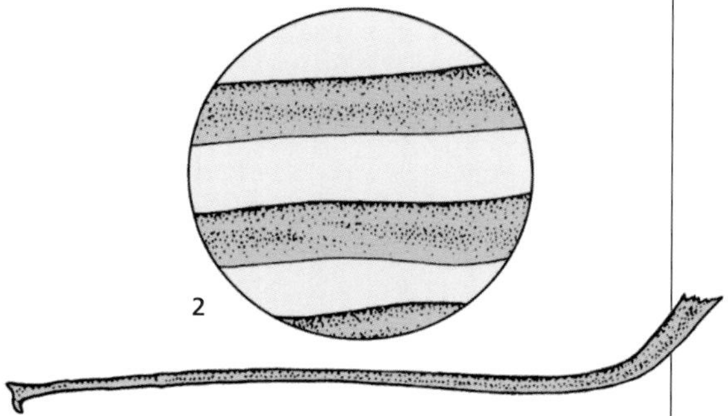

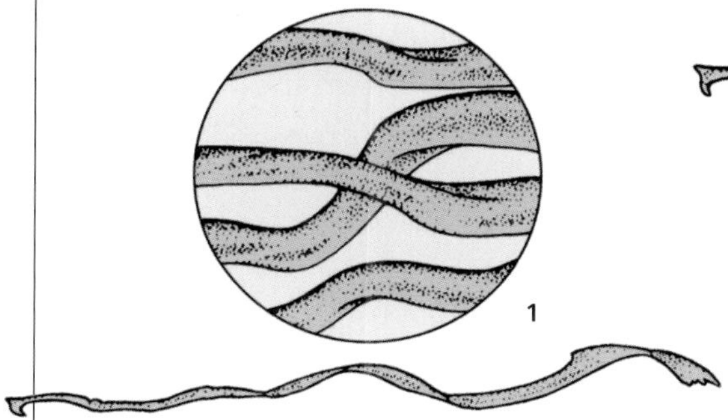

Left *When dry, **1**, the barbules of the belly feathers of the male sandgrouse have a spiral structure; when wet, **2**, the barbules uncoil, allowing the feathers to hold more water.*

Below *Chicks drink from the belly feathers of a male Namaqua sandgrouse on his return from the waterhole. The male can hold some 15–20 milliliters of water in these feathers.*

Pigeons

AMONG THE MOST SUCCESSFUL OF ALL BIRDS, pigeons live almost everywhere that humans do. There are many millions of descendants of the Rock dove inhabiting cities all over the world. In these urban environments, the lack of birds of prey, the pigeons' habit of nesting and roosting on buildings, and the fact that humans habitually feed them allow them to thrive – sometimes in such numbers that they cause fouling problems. In the countryside other species have benefited from the spread of agriculture. Some other Rock dove descendants are used in the sport of pigeon racing, which exploits their well-known homing abilities.

Pigeons are a distinctive family with a worldwide distribution. They occur on all continents except Antarctica and almost everywhere harbors two or three species, although only the Rock dove and Eurasian collared dove occur to any extent north of the Arctic Circle. They are found throughout temperate and tropical regions; however, in most of the central Sahara and Arabian deserts, they are only seen in passage. Pigeons are good dispersers, and have reached most offshore and oceanic islands. They have proliferated on the islands of Southeast Asia and the South Pacific. The islands of the central Atlantic ridge and Hawaii are among the few groups without at least one species.

The family is divided into four subfamilies. First, the typical pigeons are primarily seedeaters and are found throughout the range; second, the fruit pigeons occur in Afrotropical and Oriental regions; third, a group of three crowned pigeons is endemic to New Guinea; and, finally, the Tooth-billed pigeon of Samoa represents a single-species subfamily.

*◆ **Below** Two Spinifex pigeons* (Geophaps plumifera) *mutually preening. Named for the type of grassland that they inhabit, across northern Australia and its interior, Spinifex pigeons often sit still for long periods.*

*◆ **Above** Occurring south of the Sahara, the African green-pigeon* (Treron calva) *prefers to occupy woodlands, especially riverine areas, where it feeds on figs and the fruit of other trees.*

A Muscular Build

FORM AND FUNCTION

Pigeons are rather stocky, heavily-built birds, mostly medium-sized, and with soft plumage that comes out easily. They range in size from small ground doves, which look and behave like sparrows, to the crowned pigeons, which exceed 2kg (2.2lb) in weight. In most species the sexes are alike, although the female is commonly slightly duller. In some species, there is a distinct difference between the sexes. For example, the male Orange dove of Fiji is bright orange, whereas the female is dark green (they both have a similar yellow head), and the male Namaqua dove of Africa and Madagascar has a black mask that is absent in the female.

Wing muscles make up as much as 44 percent of the average pigeon's body weight, and perhaps even more in birds that have been specially bred for their speed and homing abilities (see Homing). These muscles also enable the birds to take off almost vertically. Good "racing pigeons" can achieve mean flight speeds of around 70 km/h (44mph).

Most of the typical pigeons are various shades of gray, brown, or pink. In addition, many have obvious white, black, or iridescent patches on the sides of the neck or on the wings or tail. Some of these patches are emphasized in displays. A few have a small crest, and the Australian bronzewings have long, pointed crests.

The fruit-eating forest pigeons of the Old World tropics are much more colorful. The plumage of the African and Asian *Treron* species is mostly of a soft, rather bright, green hue, often interspersed with areas of yellow and mauve; the Indian Ocean *Alectroenas* species are primarily blue; and the Asian and Pacific fruit pigeons (such as the *Ptilinopus* fruit doves and *Ducula* imperial pigeons) are a wide range of brilliant colors arranged in striking patterns.

Below *Bar-shouldered doves* (Geopelia humeralis) *eat mostly seeds of grasses and cereal crops. This has led to the proliferation of this species in agricultural areas across southern New Guinea and Australia.*

FACTFILE

PIGEONS

Order: Columbiformes

Family: Columbidae

309 species in 42 genera. Species and genera include: **Australian bronzewings** (*Phaps* spp.), **crowned pigeons** (*Goura* spp.), **Eurasian collared dove** (*Streptopelia decaocto*), **European turtle dove** (*S. turtur*), **imperial pigeons** (*Ducula* spp.), **Pink pigeon** (*Columba mayeri*), **Rock dove, Feral pigeon** or **Racing pigeon** (*C. livia*), **Stock dove** (*C. oenas*), **Scaly-breasted ground dove** (*Columbina passerina*), **Tooth-billed pigeon** (*Didunculus strigirostris*), **Namaqua dove** (*Oena capensis*), **Orange dove** (*Ptilinopus victor*), **Purple-crowned pigeon** (*P. superbus*), **Eared dove** (*Zenaida auriculata*), **American mourning dove** (*Z. macroura*).

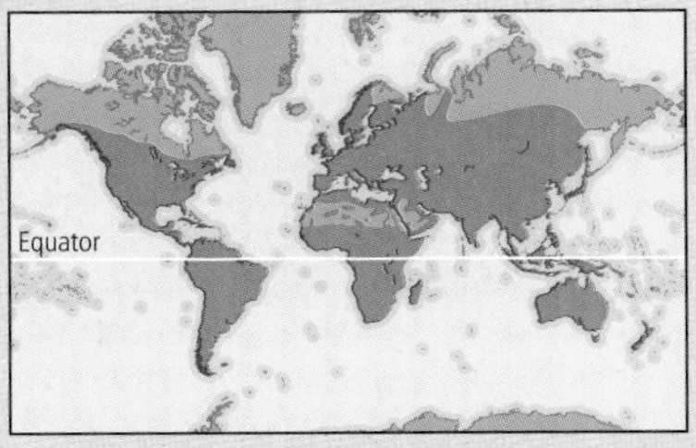

DISTRIBUTION Widespread except Antarctica, the high northern latitudes and not resident in the drier areas of deserts; members of the family have reached many isolated islands.

HABITAT Most in woodland or forest; some in open country or near cliffs. The feral and domesticated forms of the Rock dove are now commonly found in towns and cities all over the world.

SIZE Length 15–82cm (6–33in); weight 30g–2.4kg (1oz–5lb 5oz).

PLUMAGE Majority are soberly colored in grays and browns: some more brightly colored; birds of the large tropical genus *Treron* mainly bright green. Some with crests. Sexes usually similar.

VOICE Wide range of soft calls and coos; song usually simple, consisting of only a few notes.

NEST The large majority build simple twig nests on the branches of trees, a few nest in holes or on the ground.

EGGS 1 or, more often, 2, white; weight 2.5–50g (0.1-1.8oz). Many species have several broods in succession. Incubation period 13–18 days in most species but 28 days in the largest; nestling period not well known for many species, but up to 35 days, possibly longer in some. However, in many species the young leave the nest before they are fully grown and complete their growth later.

DIET Primarily vegetable matter: fresh green leaves, fruit or seeds; several are serious pests of crops as a result. Young fed on crop milk.

CONSERVATION STATUS 61 pigeon species are threatened, of which 13 are Critically Endangered, 14 Endangered, and 34 Vulnerable. The Socorro dove (*Zenaida graysoni*) of Mexico is now Extinct in the Wild. 11 pigeons are listed as Extinct; 8 of these have become extinct since 1600.

The three crowned pigeons are mainly grayish with pink or chestnut on the underparts or on the wing-coverts and with a large white patch on the wings. They are characterized by being much larger than any others and sporting a large, laterally flattened crest.

The Tooth-billed pigeon has glossy, dark green plumage on its head, neck, breast, and mantle, a chestnut color on its back and wings, and a very robust, somewhat raptorlike red and yellow bill.

Some Seasonal Migrants

DISTRIBUTION PATTERNS

Pigeons occur in almost all terrestrial habitats from tropical and temperate forests to steppes and semi-desert thorn-scrub and from sea level to above the snowline in the Himalayas. Most are seedeaters, and therefore need to drink regularly. Consequently, they are rarely very far from water. Pigeons are typically tree-dwelling, but some cliff-dwelling and ground-dwelling species occur too. Many species nest in trees but feed on the ground.

Most species are sedentary but they are strong fliers and a few migrate large distances. Some are extensively nomadic, especially dry-country species such as the Namaqua dove in Africa and several Australian species. Some others are seasonal migrants; for example, the European turtle dove breeds over much of Europe, central Asia, and North Africa but migrates across the Sahara to winter in the Sahel zone just south of the desert. Similarly, in the New World the American mourning dove moves south from its breeding grounds to Mexico for the winter.

Seeds or Fruits

DIET

Typical pigeons primarily eat seeds gathered from the ground, which they grind down in their strong, muscular gizzard, often with the aid of grit. In some seasons, most will also consume green leaves, buds, flowers, and often some fruit. As a result, some species are serious agricultural pests, attacking both ripe grain and newly germinated crops. The fruit pigeons subsist almost exclusively on the pulp of fruit, and in these species the gizzard is adapted to strip off the pulp, leaving the seed to be voided intact. This adaptation makes the birds excellent dispersers of seeds, and there are numerous examples of the co-evolution of fruiting plants and pigeons.

Many species take a limited amount of snails or invertebrates, especially in the breeding season, and feral pigeons in towns will eat almost anything at times.

The seedeaters at least need to drink regularly and, unlike most birds, pigeons drink actively by immersing their bill in water up to their nostrils and sucking without raising their heads. Some species may fly considerable distances to water, where they may gather in large flocks, especially at dawn and dusk.

Paired for Life

SOCIAL BEHAVIOR

Pigeon calls are variously cooing, crooning, or booming sounds, and are, in the main, relatively quiet. Certain species repeat single notes at regular intervals, while others emit a more or less recognizable song. In addition, some have whistles or harsher calls. They do not appear to utter alarm calls as such, although several species clap their wings very loudly during escape flights, a behavior that is also repeated during their display flights.

At the start of the breeding season, the flocks start to break up and the birds pair off. As far as it is known they are all monogamous and pairs seem to stay together throughout a season; in some species the pair bond may endure over several seasons, or even for life.

The songs of male pigeons are usually made up of a simple, rather monotonous, sequence of "coos." Some of these sounds are far-reaching, even though they may comprise nothing more than a repeated, single note. Many species also have display flights, which are used both as a threat and as part of courtship. The latter may be simple, as in the turtle doves (*Streptopelia* spp.), but in others courtship is a very elaborate series of bowings and other movements, often accompanied by cooing.

Arboreal species build a fragile-looking, but actually quite tightly interwoven, nest of twigs,

Below *Representative species of pigeons:* ***1*** *The American mourning dove* (Zenaida macroura) *is the common dove of North America;* ***2*** *The Victoria crowned pigeon* (Goura victoria) *is the largest pigeon, and the male uses its crest in a bowing display during courtship;* ***3*** *European turtle doves* (Streptopelia turtur) *winter south of the Sahara and migrate to Europe in summer;* ***4*** *The Wompoo or Magnificent fruit dove* (Ptilinopus magnificus) *has occasionally been observed feeding in flocks.*

THE DODO – AN EXTINCT COLUMBID

Biologists long surmised that the dodo, the flightless bird of Mauritius that was driven to extinction in the 17th century, was a member of the pigeon family. However, conclusive proof of this affinity had to await the advent of DNA analysis.

Researchers from Oxford University and the Natural History Museum, London, took samples from the only dodo specimen with soft tissue still intact. These were then compared with genetic material from the solitaire, an extinct dodo-like bird from neighboring Rodrigues Island, and from extant bird species, including pigeons and doves. The molecular analysis indicated that the dodo and solitaire evolved from a common pigeon-like ancestor in Southeast Asia. This bird separated from its relatives some 42 million years ago, and thereafter migrated across the Indian Ocean, settling on the first of the volcanic Mascarene islands chain, which was formed around 26 m.y.a. Around the same time, the dodo and solitaire speciated from one another. Over time, as the first-emerging islands in the chain submerged, it is thought that the birds "island-hopped" to Mauritius (8 million years old) and Rodrigues (1.5 million years old).

The dodo's nearest living relative is the Nicobar pigeon (*Caloenas nicobarica*); other relatives include the crowned pigeons of New Guinea and the Tooth-billed pigeon of Samoa.

usually placed loosely on a branch or among twigs of a tree. Others nest on cliffs, sometimes on man-made structures, and a few species nest in open situations on the ground. Certain others, such as the Rock dove, naturally nest in crevices or caves (although the domesticated and feral forms of the Rock dove now customarily nest on buildings). Exceptionally, a few species such as the Stock dove nest in a true hole in a tree or in a burrow. All these usually build some sort of nest but in some hole-nesters in particular it is extremely rudimentary. It is usually the female who builds the nest, but the male brings most of the material. All species lay unmarked, white, or near-white eggs. The majority of species lay two eggs, but the larger species and most of the tropical fruit pigeons generally lay one-egg clutches. Pigeon eggs are exceptionally small in relation to adult body-size when compared with those of other birds and this,

*▲ **Above** A Giant or Notu pigeon* (Ducula goliath) *feeding on Pandanus fruit. This particular species is considered to be Lower Risk/Near Threatened by the IUCN.*

*▶ **Right** The feral form of the Rock dove* (Colomba livia) *originally lived on cliff ledges but now typically nests and roosts on buildings in urban areas.*

combined with the small clutch size, means that the pigeons have the smallest total clutch weights in relation to adult weight (about 9 percent) of all the families of nest-reared land birds.

On the other hand, breeding seasons are often very long and many species have many successive broods, with up to eight in a year in some cases. This fecundity is helped by having short incubation (13–18 days) and fledging periods compared with other birds of similar size (young can often fly by the time they are 2 weeks old), and by having successive clutches overlapped; that is, a new clutch is laid (sometimes in the same nest) while the parents are still tending young from the previous brood. Both sexes share in incubation and the care of young, and both produce pigeon milk (see box), which is rich in energy and nutrients and helps the nestlings to grow very rapidly indeed. Chicks of all open-nesting pigeons studied to date fledge when still not adult size, and when still well below adult weight (usually about 65 percent, but as little as 26 percent in the case of the Purple-crowned pigeon). However, chicks of hole-nesting Stock doves fledge at about adult size and weight.

Some pigeons can breed at an exceptionally early age – five months in the case of Scaly-breasted ground doves – and many species are relatively long-lived, especially when kept in captivity.

Pigeon Milk

Pigeons are very unusual among birds, in that they produce a crop milk with a chemical composition similar to that produced by mammals. Only Greater flamingos (*Phoenicopterus ruber*) and Emperor penguins (*Aptenodytes forsteri*) share this feature.

Pigeon milk is a secretion of the adult crop which forms the complete diet of nestlings for the first few days of life (RIGHT), and in this period it appears to be important that it is not contaminated. Thereafter, they are fed a growing percentage of food items obtained by the parents, but the actual quantity of milk stays fairly constant until the young are well grown.

Crop milk is produced by both sexes in response to secretion of the pituitary hormone prolactin (as in mammals). From about the midpoint of incubation the tissue of part of the crop begins to thicken and blood vessels grow there. Growth of the crop wall can more than triple its weight over the last half of incubation, and by the time the young hatch, reddish folds with a honeycomb texture are visible in it. From there, cells containing the "milk" are successively detached into the crop, and these are then regurgitated to the young. Milk cells are initially sloughed off only when the crop is empty, so ensuring that it is not contaminated by other foods. Later, crop milk production is confined to periods when adults tend the young, but the milk is mixed with other foods.

Crop milk is a thick solution (containing 19–35 percent dry matter) with the consistency and look of cottage cheese. It contains 65–81 percent water, 13–19 percent protein, 7–13 percent fat, 1–2 percent mineral matter, and vitamins A, B, and B_2, but no carbohydrates. The dry matter is mostly protein, and (at least qualitatively) contains almost all the amino acids needed for growth. Crop milk is low in calcium and phosphorus, but high in sodium.

Milk production seems to be an adaptation to ensure that nestlings receive the adequate and predictable supply of energy and nutrients that are required for the high growth-rates that are characteristic of pigeons, particularly as pigeons are more exclusively vegetarian than most birds. Most seedeaters switch their diet to invertebrates to a far greater extent when feeding young. HR/PL

Winners and Losers

CONSERVATION AND ENVIRONMENT

The spread of agriculture has greatly benefited pigeons. Many species were adapted to feed on grains and fruits before these were developed for human use, and in certain areas they have made serious inroads into commercial crops, for example the Eared dove in South America. Pigeons have the ability to fly into an area, quickly fill their crop with food, and return to the safety of woodland to digest their meal. Some species are potential carriers of agricultural diseases and even pose a health hazard to humans, especially feral pigeons in cities.

A huge increase in range was shown by the Eurasian collared dove in the mid-20th century. Originally it bred only in the far southeast of Europe. In the early years of the century, however, it began to spread slowly through the Balkans. From about 1930 it spread very rapidly northwestward through Europe. It first bred in England in 1955, and within 15 years had spread to almost all of the British Isles. By 1974, it had reached Portugal, and is now spreading into western North Africa.

At the other end of the spectrum, some pigeons have very limited distributions, especially certain island species, and many are seriously threatened by habitat loss. This is usually human-induced, but natural disasters such as hurricanes also play a part. Numbers of the Mauritius pink pigeon, for example, have been increasing following the reintroduction into the wild of birds that have been raised in captivity. However such programs only succeed if the original cause of the decline, in this case introduced predators, is first dealt with.

A large population is not necessarily a safeguard. The Passenger pigeon (*Ectopistes migratorius*) of North America was thought to number 3,000 million birds late in the 18th century, making it one of the most abundant species in the world. It nested in huge colonies occupying several square kilometers. Even as late as 1871, one dispersed colony in Wisconsin was estimated at 136 million individuals. As with all pigeons, the birds were good eating, and this species was so easy to shoot that it was hunted commercially even when it was greatly reduced in numbers. It seems that the Passenger pigeon was very dependent on crops of mast, which were very unpredictable over space, and the bird may have relied on large numbers to be able to find the good feeding areas. The combination of habitat destruction and shooting led to its extermination in the wild in about 1900, and the last specimen died in Cincinnati Zoo in 1914. Even today the American mourning dove is still harvested in large numbers in the USA, while in parts of South America the Eared dove is an important protein source, as are many of the domestic varieties of several species. PL

Above *This 18-day old squab is a precious addition to the population of endangered Pink pigeons on Mauritius. Since 1996 over 250 birds have been successfully raised in captivity, taking the species off the Critical list, but it is still under strict management.*

Below *Wood pigeons* (Columba palumbus) *are the most abundant and widespread of Europe's pigeons. Outside the breeding season, they are usually seen in flocks, sometimes of a considerable size.*

SPECIAL FEATURE

FINDING THE WAY HOME

How do pigeons navigate?

MANY BIRDS DISPLAY AN ABILITY TO FIND THEIR way home if blown off course or transported to a distant location. Just how they do so has been investigated in several species, but most studies are based on pigeons – domesticated varieties of Rock dove that are easy to raise and keep in captivity and that adapt well to experimental conditions.

The homing capacity of Rock doves was first exploited by the ancient Egyptians, who used the birds as messengers, a role they preserved until the advent of the telegraph and the wireless. Even when new modes of communication became available, carrier pigeons continued to be used, and many soldiers and airmen in World War II owed their lives to the messages they transported. Nowadays, the principal use of their homing ability is in the sport of pigeon racing.

For a transported animal to get home successfully requires both an ability to steer in the correct direction – in other words, a compass – and an ability to work out which direction to follow – namely, a map. There is now a substantial body of experimental evidence to show how birds (and other animals) navigate when homing or when migrating over longer distances. What is less well understood, however, is how they initially manage to work out where they are in relation to home.

When the birds are close to home, they probably use local landmarks as cues. Slightly farther away – between about 10–20km (6–12mi), the farthest extent of navigation by familiar landmarks, and 80km (50mi), where other navigational cues come into play – there is a zone where they are seemingly far less adept at working out where they are. Homing from such areas is often successful, but it may take some time, and it seems almost as though the birds simply fly around at random until they see landmarks that they recognize.

Most of the evidence for the factors used for navigation has been obtained by placing birds in cages and showing them a restricted range of cues (for example, only parts of the night sky), by artificially changing their positions (say, with mirrors), or by subjecting them to controlled experimental conditions (for example, by artificially moving dawn and dusk by a few hours).

During the day, the Sun is clearly the first choice for determining direction. The birds have an internal clock that allows them to compensate for the Sun's apparent movement across the sky. At night, the Moon is used in a similar way; the birds can also use the constellations, and in the northern hemisphere the area around the pole star is the most important. However, if all these visual cues are obscured, radar tracking has demonstrated that they are still able to hold the right bearing,

Above *Scientists researching pigeons' internal clocks have kept birds under artificial light conditions, in which daybreak appeared to occur either earlier or later than was actually the case. Confused, the birds flew in the wrong direction when released, following routes that would have been right if sunrise had occurred when the artificial conditions suggested.*

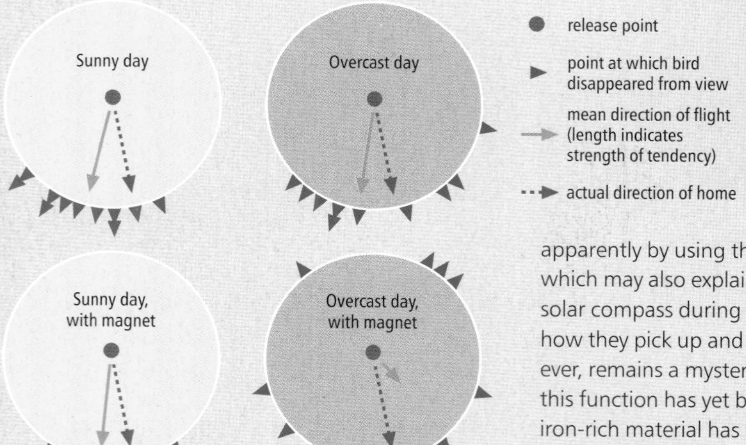

Above *The effect of the Earth's magnetic field on pigeon navigation has also been the subject of study. Researchers set out to disorient birds by attaching bar magnets to their backs. On sunny days, the magnets had relatively little effect, but when the sky was overcast the birds apparently lost all sense of direction. The obvious conclusion was that pigeons respond first and foremost to cues provided by the Sun, but that when such cues are not available, they rely heavily on the Earth's magnetic field.*

apparently by using the Earth's magnetic field, which may also explain how they calibrate the solar compass during daylight navigation. Exactly how they pick up and use magnetic fields, however, remains a mystery. No internal organ with this function has yet been identified, although iron-rich material has been found at many places in the skull and embedded in muscles in the neck.

Keeping to a fixed direction is relatively easy in comparison to working out one's position in relation to home. Pigeons released at a previously unknown location normally circle a few times and then set off in more or less the right direction. In doing so, they appear to be working out where they are before deciding where to fly.

Broadly speaking, three hypotheses have been advanced to explain this ability. One suggests that they use the position of the Sun to work out which

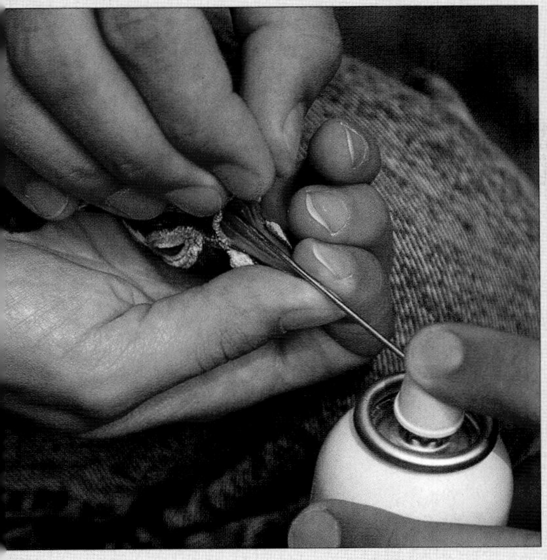

Left *Scientists investigating a Rock dove's navigational abilities spray an anesthetic into its beak to mask olfactory cues that might otherwise guide it.*

Above *Birds released from a truck in eastern England head home to the north-east, a region famed for its love of pigeon racing.*

way to go – a method that would only work if the birds could remember the height of the Sun above the horizon at home as well as having an accurate clock. Experiments on clock-shifted pigeons have shown, however, that only the compass sense of the birds was affected, not their map.

Second, magnetic fields vary around the globe and could possibly be used, although nobody yet knows how. Yet attaching low-power magnets to birds, while disrupting their orientation, again has little effect on the map.

The third suggestion is that the birds can smell their way home. Famously, salmon use the scent of their natal stream to return to breed after years at sea. In experiments, pigeons that had their sense of smell impaired or that were exposed to a strong odor (for example by painting one onto their beaks) performed less well than normal birds. Yet these findings – while suggesting that smell plays a part in navigation – nonetheless fell short of proving that birds can smell home from a long way off.

In sum, a huge number of experiments have yielded a mass of conflicting results, many of which have, moreover, been found to be unrepeatable in different laboratories. The only permissible conclusion is that birds navigate and work out where they are using several different cues, some of which may yet remain to be discovered. PL

Parrots, Lories, and Cockatoos

An infamous incident involving the pet parrot of US President Andrew Jackson testifies to these birds' feisty character and keen intelligence – sometimes so acute as to be embarrassing. At Jackson's funeral in 1845, his Yellow-naped amazon, "Pol," began to yell obscenities (presumably learned from its plain-speaking master), causing such a furore that it had to be ejected from the solemn proceedings!*

FACTFILE

PARROTS

Order: Psittaciformes

Family: Psittacidae

356 species in 80 genera.

DISTRIBUTION S and C America, southern N America, Africa and Madagascar, S and SE Asia, Australasia and Polynesia.

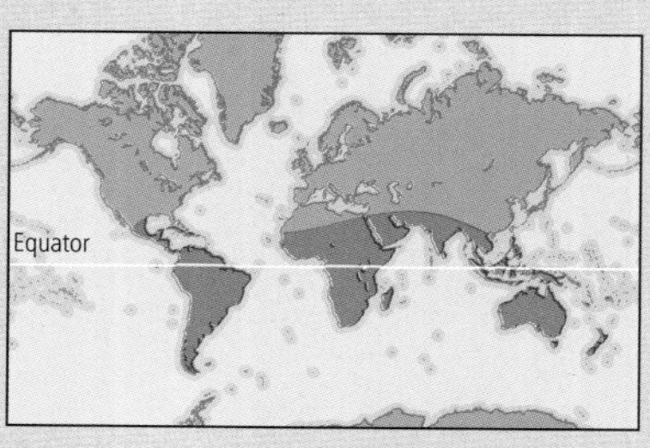

HABITAT Principally lowland tropical and subtropical forest and woodland; occasionally in mountain forest and open grassland.

SIZE Length 9–100cm (3.5–39in).

PLUMAGE Extremely varied: many species brilliantly colored, others predominantly greenish or brownish. Males and females usually similar or identical in appearance and coloration, with some notable exceptions.

VOICE Great variety of typically noisy and unmusical calls. In captivity, some species become skilled vocal mimics.

NEST Usually in holes in trees: rarely in burrows in cliffs, soil, or termitaria. A few species nest communally in large grass or twig nests.

EGGS Generally 1–8, depending on species: invariably white and relatively small: length from 16–54mm (0.6–2.1in). Incubation period 17–35 days; nestling period 21–70 days.

DIET Chiefly vegetable matter: fruit, seeds, buds, nectar, and pollen. Insects occasionally eaten.

CONSERVATION STATUS 91 species are currently at risk, 12 of which are listed as Critically Endangered, while 34 are Endangered and 45 Vulnerable.

See Parrot Subfamilies ▷

Parrots are known not just for mimicry but also for longevity. In captivity, some larger species (such as cockatoos and macaws) live for up to 65 years. Yet, despite parrots' long history of captivity, the budgerigar of Australia is the only truly domesticated species; after the dog and cat, it is probably the most common household pet in the West.

Gaudy and Highly Vocal

FORM AND FUNCTION

The parrots form a distinctive and fairly uniform order of birds comprising the single family Psittacidae. The majority of species are predominantly green, with highlights of bright yellow, red, or blue. Others are mainly white or yellow, and a few are blue. Parrots vary hugely in size, from the tiny pygmy-parrots that weigh in at about 10g (0.4oz) up to the adult male kakapo, which can tip the scales at 3kg (6.6lb), some 300 times larger! They are also very variable in shape, with many graceful and slender and others short and stocky.

Although a few atypical species, such as the Australian Ground parrot, seem to be largely solitary, the vast majority of parrots are sociable and gregarious birds that are usually observed in pairs, family parties, or small flocks. Occasionally, when conditions are appropriate, some of the smaller species aggregate in very large numbers. Observers in Australia, for example, sometimes report flocks of wild budgerigars so vast that they darken the sky. Perhaps reflecting the principle of safety in numbers, many species of parrot also roost communally at night. Communal roosts are often in traditional locations and tend to be used year after year. Favored sites often consist of exceptionally tall or isolated trees where the birds can get a good view of approaching predators. Asian hanging-parrots have the distinction of roosting suspended upside-down like bats. At a distance, it is hard to distinguish a dead tree full of roosting hanging-parrots from a tree with normal foliage.

Parrots are highly vocal birds, with generally harsh and unmelodic voices. Calls include a range of chatters, squeaks, shrieks, clicks, squawks, and screams, many of them loud and discordant. The Crimson rosella of Australia has a pleasant, whistle-like call, and another Australian species, the Red-rumped parrot, produces a melodious, trilled whistle that is the nearest any parrot comes to a song. In some species, pair-mates perform vocal

➲ ***Right*** *Native to Central and South America, the macaws, with their vibrant colors and noisy calls, are most people's image of the quintessential parrot. Here, Scarlet macaws* (Ara macao) *show their vivid plumage.*

Left *Representative species of parrots:* ***1*** *St. Vincent parrot* (Amazona guildingii), *whose decline to extinction has been halted by conservation measures, but which remains Vulnerable;* ***2*** *Red-capped parrot* (Purpureicephalus spurius)*;* ***3*** *Hyacinth macaw* (Anodorhynchus hyacinthinus), *using its zygodactylous claws to grip a Brazil nut;* ***4*** *Rainbow lorikeet* (Trichoglossus haematodus)*;* ***5*** *Fischer's lovebird* (Agapornis fischeri)*;* ***6*** *Blue-crowned hanging-parrot* (Loriculus galgulus)*;* ***7*** *Crimson rosella* (Platycercus elegans)*;* ***8*** *Eclectus parrot hen* (Eclectus roratus)*;* ***9*** *Black-capped lory* (Lorius lory), *showing the adaptation of the tongue for feeding on nectar;* ***10*** *Night parrot* (Geopsittacus occidentalis), *a Critically Endangered species;* ***11*** *Kea* (Nestor notabilis).

duets – rapidly alternating sequences of calls exchanged between partners.

Because of their many unique and specialized features, it is difficult to determine the relationship between parrots and other groups of birds. They are usually, however, classified somewhere between pigeons and cuckoos, although their affinity with either of these orders is, at best, tenuous and uncertain. Despite the recent rise of genetic techniques, it is still not possible to assign the evolutionary pathways that gave rise to the parrots. This suggests that parrots diverged from other lineages at a comparatively early stage in bird evolution and are an ancient group. The oldest parrot fossil, from about 55 million years ago, comes from a bird named *Pulchrapollia gracilis*. Its remains were found in early Eocene London Clay deposits at Walton-on-the-Naze, Essex, England. Another fossil that may have come from a parrot-like bird was found in Cretaceous deposits in the Lance Formation, Wyoming, USA.

The most obvious defining feature of birds included in the Psittacidae (parrot family) is the characteristic parrot bill, which consists of a downward-curving and somewhat hooked upper mandible, which fits neatly over a smaller, upward-curving lower mandible. The upper half of the bill is attached to the skull by a special kind of hinge, and this gives it greater mobility and leverage. The parrot bill is a highly adaptable structure. It can be used to perform delicate tasks such as preening, but at the same time in many species is powerful enough to crush the hardest nuts and seeds. The bill also serves as a third "foot" – a kind of grappling-hook that the birds use in conjunction with their feet when clambering about among the treetops. In the Great-billed parrot of Indonesia, the bill is abnormally large and bright red in color. This conspicuous structure is thought to serve as a form of visual display.

Parrots are renowned for their gaudy plumage, and some of the larger, tropical species such as the South American macaws are undoubtedly among the most brilliantly colored of all birds. Yet despite their bright plumage, most species are surprisingly well camouflaged in tree foliage, where their colors blend in with flowers and dappled light. The large cockatoos of Australia are highly conspicuous, however. They are generally either white, salmon-pink, or black in color, and most of them have prominent, erectile crests on their heads. Males and females in the majority of parrots are either very similar or identical in appearance. But there are some notable exceptions to this rule. For example, male king-parrots have brilliant scarlet plumage, whereas females and juveniles are almost entirely green. In the Eclectus parrot of New Guinea and Australia, males and females are so different in coloration that for many years they were thought to be different species. Males are bright emerald green with scarlet underwings and flanks, while the female is a rich crimson red with a violet-blue belly and lower breast. This species is also unique among parrots in that the female is gaudier and more conspicuous than the male. The feet of parrots are also unusual: the two outer toes point backwards and grip in opposition to the two forward-pointing inner toes. This arrangement (zygodactyly) not only provides parrots with an extremely powerful grasp, but also enables them to use their feet like hands for holding and manipulating objects close to the bill. In terms of manual dexterity, parrots are unsurpassed by any other group of birds. However, this ability is absent in some species that habitually feed on the ground. Like humans, parrots also exhibit both right- and left-handedness (or, in their case, left-footedness). One study found that, from a flock of 56 Brown-throated parakeets, 28 consistently used the right foot to hold food

8

9

10

11

1

while the other 28 used the left. When walking along a perch or on the ground, most parrots are noticeably pigeon-toed, and have a characteristically comical rolling gait.

Parrots are variable in their powers of flight. In general, flight is swift and direct in small species and relatively slow and laborious in the larger ones. There are, however, some notable exceptions. The South American macaws, for instance, are fast fliers despite their size. Species such as the budgerigar and many lories are highly nomadic and are capable of flying considerable distances in search of food. Parrots are generally not long-distance migrants, but the Swift parrot and Blue-winged parrot are exceptions. Both these species, from southeastern Australia, are migratory and every year fly across Bass Strait – a distance of 200km (124 miles) – in order to breed in Tasmania. As its name suggests, the Swift parrot flies with exceptional speed and directness.

Differences in flying ability in parrots are linked to their varying ecological requirements, which in turn are reflected in differences in wing structure. In general, species that fly rapidly have comparatively narrow, tapering wings, while the wings of slow-flying forms are correspondingly broad and blunt. The kakapo of New Zealand (see Back from the Brink) has very short wings, and is the only wholly flightless parrot.

The structure of the tail in parrots is also extremely variable. In macaws and in the Papuan lorikeet, for example, tails are especially long and elegant and may account for almost two-thirds of the bird's total length. Long tails likely serve an important signaling function. At the other extreme, the tail of the Blue-crowned hanging-parrot is so short and blunt that it is almost hidden by the tail coverts. The racket-tailed parrots of Indonesia and the Philippines have distinctive, elongated central tail feathers, consisting of long, bare shafts with flattened, spoon-shaped tips; their function is unknown. The tail feathers of the New Guinea pygmy-parrots also terminate in short, bare shafts. These are stiffened like woodpecker tail feathers, and similarly help to support these tiny birds when they are climbing about and feeding on tree-trunks.

Apart from man, the most important predators of parrots are hawks and falcons, although monkeys and other arboreal mammals also take many eggs and nestlings. When feeding in flocks, parrots are often noisy and quarrelsome and appear oblivious to danger. However, when danger threatens, the flocks fall perfectly silent before exploding suddenly from the treetops with harsh screams. Most predators find this furore disconcerting.

Left and right *Representative species of cockatoo: **1** Salmon-crested cockatoo* (Cacatua moluccensis), *an endangered species; **2** Sulphur-crested cockatoo* (Cacatua galerita)*; **3** Yellow-tailed black cockatoo* (Calyptorhynchus funereus)*; **4** Palm cockatoo* (Probosciger aterrimus)*; **5** Galah or Rose-breasted cockatoo* (Eolophus roseicapillus).

Talking Parrots

Very few species of parrot (with the exception of African greys and galahs) have been observed by scientists mimicking non-parrot sounds in the wild, and it remains a mystery why this ability appears so well developed in captive birds. Research, however, shows that the enormously diverse and variable calls of wild parrots are used in complex communications. Observations of parrots in different areas of forest have revealed how dialects develop in different flocks, thereby underlining the importance of mimicry in shaping the vocal repertoire of separate populations. In the unnatural social conditions experienced by captive parrots, the disposition to imitate their own species is transferred to the adopted human "flock members" and their vocalizations. The talking abilities of captive parrots have long intrigued humans. In ancient Rome, parrot tongues were cut out and fed to people to "cure" speech impediments.

Most people assume that parrots merely mimic sounds at random and are incapable of using speech in appropriate contexts – hence the phrase "to learn something parrot-fashion." However, decades of research in the USA by animal behaviorist Professor Irene Pepperberg, beginning in 1977, suggest that some parrots can be trained to use human language to communicate intelligently with people. After months of careful tuition, Pepperberg first succeeded in training Alex – a young male African grey parrot – to learn verbal labels for 23 different objects or materials such as paper, cork, nut, rock, and water. He also learned five different colors, four different shapes, numbers up to five, and commands such as "Want," "Come here," and "Tickle me." More to the point, Alex was able to combine these vocalizations to identify, request, or refuse over 50 different items – even some that were not included in his original training schedule. During his second year of training, and without any formal instruction, Alex also began using the word "No" when he did not want to be handled – a particularly interesting development, since linguists regard negation as a relatively advanced conceptual achievement.

Several more African greys have since joined Alex on the program. One important application of Pepperberg's research has been to the improvement of teaching methods for autistic children. JAS

At Home in the Tropics

DISTRIBUTION PATTERNS

Parrots have become potent symbols of tropical climates and exotic forests, and it is principally in tropical latitudes and forests, especially in the southern hemisphere, that they are distributed. By far the largest numbers of species are found in South America and Australia and New Guinea, whereas relatively few species occur in Africa and Asia. Although mainly tropical, some parrots penetrate temperate latitudes; the Austral parakeet and Antipodes parakeet, for example, are respectively found in Tierra del Fuego at the southern tip of South America and on Antipodes Island in the South Pacific. The Carolina parakeet of North America was at one time the most northerly representative of the family. However, the species was wiped out by the early 20th century, and this distinction now belongs to the Slaty-headed parakeet of eastern Afghanistan.

Some parrots also brave cold climates at high altitudes within the tropics and subtropics, for example the Papuan lorikeet, the Derbyan parakeet from the Himalayas, the Yellow-fronted parrot from Ethiopia, and the Grey-hooded parakeet from the Andes. The kea of the Southern Alps of New Zealand is perhaps the strangest of all highland

forms. These large, bronze-green birds live among snow-covered mountains between 600 and 2,000m (2,000–6,500ft), and seem to enjoy playing in the snow. Around ski resorts and human habitations they are bold and inquisitive, and have even been known to climb down chimneys in order to steal food.

Several species are also now established outside their natural ranges, for example the Monk parakeet in the eastern USA. The Rose-ringed parakeet has the widest natural range of any parrot. It extends from North Africa to the Far East and, recently, it has been introduced accidentally to parts of Europe, the Middle East, and North America. The parrot with the most restricted natural distribution is probably Stephen's lorikeet, a small species confined to the 35sq km (13.5sq mi) Henderson Island in the South Pacific.

The vast majority of parrots dwell in forest habitats, and these birds tend to be most plentiful in and around lowland tropical forests. A number, such as the budgerigar of Australia and Fischer's lovebird of East Africa, inhabit more open, grassy habitats, but even these species are generally never seen very far from the cover of trees. Two exceptions are the completely terrestrial Ground and Night parrots of Australia. The former inhabits coastal heaths and sand dunes; and the latter, which until very recently was thought to be extinct, is confined to arid desert grassland.

Variegated Vegetarians

DIET

Parrots are primarily vegetarian and consume a wide variety of plant parts, including seeds, fruits, buds, and flowers. Some parrots have quite specialized diets; for example, Fuerte's parrot and its close relatives in the little-known genus *Hapalopsittaca*, living in the high Andes, are believed to feed mainly on mistletoe seeds, the pygmy-parrots of New Guinea on lichens and mosses, the great blue macaws of South America on palm nuts, while the lories and lorikeets of Australasia specialize in feeding on tree pollen and nectar (see Lotus Eaters). Most parrots procure their food in the treetops, where their zygodactylous feet and hooked bills enable them to climb about with consummate skill and agility. However, some of the smaller parakeets, parrotlets, and lovebirds feed extensively on grass seeds on or near the ground.

There are, not surprisingly, exceptions to the vegetarian rule. Cockatoos and rosellas hunt for grubs on trunks and tree limbs – perhaps exploiting a niche filled elsewhere by woodpeckers (which are absent from Australia). In South America, the Blue-throated parakeet and its relatives in the genus *Pyrrhura* are partial to insect grubs too, while Hyacinth macaws, from the same continent, have been seen eating aquatic snails. In addition to their normal diet of insects and low-growing plants, keas have also taken to feeding on garbage dumps and carrion. One community of keas preys on the fledglings of Sooty shearwaters (*Puffinus griseus*), which it excavates from nest holes, and keas have even acquired a widespread, although exaggerated, reputation for killing sheep.

Because of their predation of crop plants, some parrots in some localities are seen as agricultural pests. In certain parts of the world grain and fruit crops suffer extensive damage because of the attentions of feeding parrots, a problem often made worse by the destructive foraging practices of some species. In South Australia, for instance, rosellas devastate cherry orchards, eating both the flower buds and the fruit. When wild populations of parrots reach levels where they are capable of causing serious economic damage, their numbers are usually kept in control by natural mortality due to starvation and diseases such as ornithosis (psittacosis), to which parrots are particularly susceptible. Even so, some parrots have suffered major persecution that has led to local population declines, and in several cases appears to have been a key contributory factor to extinction, the North American Carolina parakeet being perhaps the best example. This species was ruthlessly shot and trapped because of its consumption of crops.

◐ ***Left*** *In common with many other parrots, the Yellow-naped amazon parrot* (Amazona auropalliata) *thrives on seeds, grains, and fresh fruit. This individual is raiding a bunch of developing bananas on a plantation in Honduras, Central America.*

Partners for Life

BREEDING BIOLOGY

The timing and duration of the breeding season in parrots depends very much on their geographic location and on the principal types of food on which they depend. In general, species living outside the tropics, where food availability tends to be seasonal, have more regular and better defined breeding seasons than those in tropical regions. For instance, the Purple-crowned lorikeet of southern Australia breeds from August to December, whereas the Varied lorikeet from Australia's northern tropics will breed at any time of the year.

Most parrots are monogamous, and males and females often pair for life. Pairs remain together constantly, and the bond between them is reinforced by mutual feeding and preening. Courtship details have only been described for a few species. Prior to copulation, the males of most species display to the females with a variety of relatively simple movements and postures, including bowing, hopping, wing-flicking and flapping, tail-wagging, and strutting. Areas of conspicuous plumage are often incorporated in these movements, and, in many species, the brightly-colored irises of the eye are expanded – a phenomenon known appropriately as "eye-blazing." When the female is ready to mate, she adopts a characteristic crouching position and allows the male to mount. The male's attempts at copulation are often interspersed with curious treading movements performed on the female's back. Their function is unknown.

The majority of parrots – large and small – nest in holes in the limbs or trunks of trees, often at a considerable height above the ground. They generally take over the holes of other cavity-nesting species, such as woodpeckers or barbets, or take on a hollow caused by rot or a branch fall. Most

Above *Blue-headed parrots* (Pionus menstruus) *at a clay-lick at Manu, Peru. Clay-licks are patches of eroded earth on cliff faces or river banks; the birds gather there to eat the soil, possibly to obtain nutrients or as a way of neutralizing toxins in their foods.*

Below *Galahs are the most common parrots in Australia. Congregating to forage in large flocks, they can devastate grain crops; controversially, the widespread use of poisons by farmers has been officially sanctioned in an attempt to limit such damage.*

species do not build a nest, although a platform is often created for the eggs from a layer of decayed wood-dust scraped from the inside of the nest hole. The African lovebirds and the hanging-parrots of Asia line their nests with grasses, leaves, and strips of bark. In some species of lovebird, the female carries material to the nest tucked under the feathers of her rump. Some parrot species excavate a hollow in termite colonies; for example, the Golden-shouldered parrot of Australia makes its nest burrow in terrestrial termite mounds, while the Buff-faced pygmy-parrot of New Guinea nests in tree-borne termitaria. The termites perhaps provide some protection from predators.

The Rock parrot nests only under rocks just above the high-tide mark on the coast of southern Australia, while the Burrowing parakeet excavates nesting burrows up to 3m (10ft) long in the cliffs and river banks of Patagonia. The Ground parrot of Australia makes its nest in a shallow depression under a bush or grass tussock. Several species of parrots nest colonially. The Rosy-faced lovebird either constructs its own nesting colonies from grasses and leaves or, more often, invades and takes over the existing colonies of weaver-finches. The most advanced communal nesting behavior among parrots is found in the South American Monk parakeet. This species nests communally in sometimes quite immense structures built from twigs in the tops of trees. Within the main structure, each pair has its own separate nest-chamber.

Two New Zealand species, the kea and the kakapo or Owl parrot, are both polygamous. In the former, males sometimes mate and share parental duties with several different females at the same time. The mating system of the nocturnal kakapo is highly unusual. The males congregate at night in specific areas, known as leks, and advertise their location with loud, booming calls. Females then visit these sites and mate with the male of their choice. As far as is known, male kakapos play no part in parental care.

Above *A female Northern rosella* (Platycercus venustus) *emerges from her nest-hole in the trunk of a eucalyptus tree. Incubating her clutch of eggs alone, she only comes out two or three times a day to forage with the male or to receive food from him.*

Below and right *Mutual preening, as exhibited here by budgerigars (below) and Military macaws* (Ara militaris; *right) is common among monogamous species, strengthening the pair-bond. Studies have shown that many of the feathers that parrots use in courtship displays are fluorescent, to attract potential mates.*

Most parrots attain sexual maturity between their second and fourth years. The clutch varies from one to about eight eggs, with the larger species generally laying fewer than the smaller ones. Incubation begins with the first egg, leading to staggered ages of chicks in the nest. The practical result is that younger chicks die if food is scarce. The Golden parakeet of the Amazon adopts a curious breeding strategy, whereby several females lay their eggs in a single nest.

In all but a few species, the eggs are incubated exclusively by the female. The male, however, keeps her supplied with food during this critical period. The young are blind and helpless when they hatch, and develop rather slowly. In small species such as the budgerigar they leave the nest 3–4 weeks after hatching, but in the much larger Blue-and-yellow macaw the nestling period may be as long as three and a half months. As a rule, both parents play an equal role in feeding the young. Juvenile parrots are generally noticeably smaller than the adults of either sex, and they also tend to have duller plumage colors.

The Struggle for Survival

CONSERVATION AND ENVIRONMENT

Although as a group parrots are comparatively successful, many species have become extinct within the last few centuries and many more are facing serious threats to their existence. Today, the parrot family is the most endangered of all major bird groups. Of 350 species, more than 90 were listed as being at some risk of extinction at the beginning of the 21st century.

Several factors have brought about this tragic state of affairs, the most important being the effects of habitat destruction and degradation, trapping and collecting for the pet market and live bird trade, and the predations of introduced species such as rats and cats. The situation has been made more acute by the fact that many species of parrot – more than one-third in total – are confined to very restricted ranges, in many cases no bigger than a single, small island.

Given these statistics, it is perhaps fitting that the world's rarest bird should have been a parrot. During the 1990s, the wild population of the Spix's macaw, a species confined to northeastern Brazil, stood at a single bird. In 2000 this bird disappeared, leaving the species extinct in the wild and with only a small captive population to offer some hope of reintroduction in the future. The reason for the loss of this species was, firstly, the

long-term degradation of its habitat, and subsequently merciless exploitation for the lucrative rare-bird market. For several years the last remnant population was plundered, first through the removal of chicks from the nest and then through the capture of adults. The parrots were illegally exported from Brazil to international buyers via dealers in Paraguay, who falsified export papers.

The Spix's macaw is not unique in appealing to people. Parrots have been valued as show birds for the cage and aviary and as pets since ancient times. Our earliest written account of a pet parrot is a Greek description of a Plum-headed parakeet, dating from about 400BC. The author, a historian and physician called Ctesias, was clearly captivated by the bird's ability to speak the language of its homeland, India, and also observed that it could be taught to speak Greek. From then on, it seems, exotically-colored talking parrots became sought-after status symbols among the ruling elites of classical Greece and Rome, and, later, medieval Europe. Exploration in the New World in the 15th and 16th centuries brought to light many new parrot species, and these and subsequent discoveries in the East Indies and Australia helped sustain the interest of European collectors. Nowadays, a huge range of species is kept in captivity and, unfortunately, the aviculturist's obsession with novel or exotic forms is undoubtedly accelerating the extinction of some species. Many parrots fetch a small fortune on the black market for rare birds, and some are still directly threatened with extinction as a result.

One of the most mysterious disappearances of a parrot species was that of the Carolina parakeet of the southeastern USA. In the early 19th century, this species was common throughout its range east of the Great Plains, but by 1831 it was already

Parrot Subfamilies

True parrots
Subfamily Psittacinae

265 species in 57 genera. Species include: Rosy-faced lovebird (*Agapornis roseicollis*), Australian king-parrot (*Alisterus scapularis*), Imperial parrot (*Amazona imperialis*), Hispaniolan parrot (*A. ventralis*), Puerto Rican parrot (*A. vittata*), Red-tailed parrot (*A. brasiliensis*), Glaucous macaw (*Anodorhynchus glaucus*), Blue-and-yellow macaw (*Ara ararauna*), Brown-throated parakeet (*Aratinga pertinax*), Grey-hooded parakeet (*Bolborhynchus aymara*), Carolina parakeet (*Conuropsis carolinensis*), Black parrot (*Coracopsis nigra*), Burrowing parakeet (*Cyanoliseus patagonus*), Spix's macaw (*Cyanopsitta spixii*), Antipodes parakeet (*Cyanoramphus unicolor*), Austral parakeet (*Enicognathus ferrugineus*), Eclectus parrot (*Eclectus roratus*), Golden parakeet (*Guaruba guarouba*), Fuerte's parrot (*Hapalopsittaca fuertesi*), Swift parrot (*Lathamus discolor*), Blue-crowned hanging parrot (*Loriculus galgulus*), budgerigar (*Melopsittacus undulatus*), Monk parakeet (*Myiopsitta monachus*), Blue-winged parrot (*Neophema chrysostoma*), Rock parrot (*N. petrophila*), Yellow-eared parrot (*Ognorhynchus icterotis*), Ground parrot (*Pezoporus wallicus*), Crimson rosella (*Platycercus elegans*), Yellow-fronted parrot (*Poicephalus flavifrons*), Golden-shouldered parrot (*Psephotus chrysopterygius*), Red-rumped parrot (*P. haematonotus*), Mauritius parakeet (*Psittacula echo*), Plum-headed parakeet (*P. cyanocephala*), Rose-ringed parakeet (*P. krameri*), Slaty-headed parakeet (*P. himalayana*), Derbyan parakeet (*P. derbiana*), African grey parrot (*Psittacus erithacus*), Blue-throated parakeet (*Pyrrhura cruentata*), Great-billed parrot (*Tanygnathus megalorynchos*).

Lories and Lorikeets
Subfamily Lorinae

54 species in 12 genera. Species include: Black lory (*Chalcopsitta atra*), Papuan lorikeet (*Charmosyna papou*), Pygmy lorikeet (*C. wilhelminae*), Purple-crowned lorikeet (*Glossopsitta porphyrocephala*), Varied lorikeet (*Psitteuteles versicolor*), Scaly-breasted lorikeet (*Trichoglossus chlorolepidotus*), Stephen's lorikeet (*Vini stepheni*).

Keas
Subfamily Nestorinae

4 species of the genus *Nestor*: kea (*Nestor notabilis*), kaka (*N. meridionalis*), Norfolk Island kaka (*N. productus*), bluebonnet (*Northiella haematogaster*).

Owl parrot
Subfamily Strigopinae

1 species: kakapo (*Strigops habroptilus*).

Cockatoos
Subfamily Cacatuinae

20 species in 5 genera. Species include: Sulfur-crested cockatoo (*Cacatua galerita*), Gang-gang cockatoo (*Callocephalon fimbriatum*), galah or Rose-breasted cockatoo (*Eolophus roseicapillus*), Palm cockatoo (*Probosciger aterrimus*).

Cockatiel
Subfamily Nymphicinae

1 species: Cockatiel (*Nymphicus hollandicus*).

Fig- and Pygmy-parrots
Subfamily Micropsittinae

11 species in 3 genera. Species include: Double-eyed fig-parrot (*Cyclopsitta diophthalma*), Buff-faced pygmy-parrot (*Micropsitta pusio*), Large fig-parrot (*Psittaculirostris desmarestii*).

on the decline, and the last known specimen died in Cincinnati Zoo on 21 February 1918. It is not known for certain what finally brought about the extinction of the species, although it was regarded as an agricultural pest and there is little doubt that human persecution played a major part in its initial decline. Its forest habitat was also clear-felled, and large numbers of birds were taken so their plumes could be used for decorative purposes.

Although the bird trade continues to take a tragic toll on many species, the most serious overall threat to parrots today is the continued destruction and degradation of the tropical and subtropical forests where the great majority of them live. The reasons for forest clearance vary, but a combination of logging, agriculture, mining, and urbanization, often backed by damaging economic policies, is leading to the annual loss and degradation of thousands of square kilometers of forest. In southeastern Brazil, for example, forest cover has been so reduced that the Glaucous macaw has already been driven to extinction through the clearance of the groves of Yatay palms (*Butia yatay*) on which it depended for food. Other species, such as the Red-tailed parrot, are seriously threatened. A similar story can be told of the highland forests of the Andes. Several species that were once widespread there are now reduced to small populations hanging on in tiny pockets of forest. The Yellow-eared parrot of Colombia and Ecuador is one such species that is now close to extinction.

Right *The kea's domain is the snowy, mountainous western coast of New Zealand's South Island. After some individuals graduated from scavenging sheep carcasses to attacking live merino sheep on hill farms, this species was killed in great numbers, before being accorded full protected status in 1986.*

Far right *The Great green macaw* (Ara ambigua) *of Central America is threatened by logging of* Dipteryx panamensis, *its main feeding and nesting tree.*

Below *Changes to the ecology of its remote habitat on the Cape York peninsula of northern Australia are imperiling the Golden-shouldered parrot* (Psephotus chrysopterygius). *In particular, the introduction of cattle is reducing the savanna grassland it relies upon.*

Island species of parrot are especially vulnerable to human encroachment. Most have small populations and relatively slow breeding rates and, because they have evolved in isolation, they tend to be more sensitive to destruction of their habitat and less able to cope with introduced competitors, predators, and diseases. Most of the parrots known to have become extinct during the last few centuries were island species. For example, many species of parrots, including several macaws, that were formerly native to the islands of the Caribbean have already been lost, and most of the parrots that remain there are Critically Endangered.

Happily however, conservation programs to save these spectacular birds have met with some success. For example, in 1975 there were only 13 Puerto Rican parrots left in the wild, and it looked as though the species was doomed. However, an emergency conservation program involving strict control of hunting and trapping, artificially increasing the number of suitable nest-sites, and cross-fostering of eggs and nestlings between the Puerto Rican parrot and the closely related and non-endangered Hispaniolan parrot, has enabled the population to gradually increase. By 1996 the wild population was up to 48 birds, with 87 more carefully maintained in captivity.

Another story of an island species struggling back from the brink comes from Mauritius, former home of the dodo. Here, the native Mauritius parakeet, which had already long gone extinct on the neighboring island of Réunion, had been reduced by the late 1980s to fewer than 10 birds, most of which were males. A recovery program involving predator control, captive breeding, and the provision of artificial nest sites and feeding helped the species to recover. There were estimated to be about 120 wild birds in 2000, with numbers still rising. The long-term aim is to establish a wild population of 300 birds by the year 2010. TJ/JAS

SPECIAL FEATURE

LOTUS-EATERS

Nectar-feeding lories and lorikeets

THE LORIES AND LORIKEETS ARE FLAMBOYANT, theatrical birds. The Rainbow lorikeet possesses up to 30 different ritualized gestures, including a variety of stylized hopping, walking, flying, and preening movements, which it incorporates into elaborate "dances." Most of these performances are aggressive and are used to intimidate rivals of the same species, but males also use similar displays to impress females during courtship.

The lories and lorikeets form a distinct subgroup within the parrot family. They occur throughout much of Indonesia, New Guinea, Australia, and the Pacific, and they differ from other parrots in their habit of feeding mainly on pollen and nectar from flowering trees and shrubs.

Typically, lories and lorikeets have sleek, glossy plumage, and the group includes some of the most brilliantly colored of all parrots. In the wild, they are mostly gregarious and their behavior is generally noisy and conspicuous. The most widespread species is the Rainbow lorikeet, which is divided into 22 distinct island races or subspecies distributed throughout eastern Indonesia, New Guinea, northern and eastern Australia, and the western Pacific islands. New Guinea has by far the highest diversity of different species, and is also close to the geographic center of their distribution.

Above *Lorikeets have a large repertoire of displays, which they incorporate into "dances" performed in the face of rivals and, by males, in courtship. Among the gestures employed are* ***1*** *the hiss-up,* ***2*** *strong fluttering,* ***3*** *ritualized scratching,* ***4*** *bobbing, and* ***5*** *bouncing.*

Right *Feeding on a flower in its home in the Australian tropics, a Scaly-breasted lorikeet* (Trichoglossus chlorolepidotus) *displays its specialized tongue, adapted to mop up nectar.*

Above *Brilliantly-colored Rainbow lorikeets cluster at a feeding table. They are the most widely distributed of the lories and lorikeets, small to medium-sized parrots that feed mainly on pollen and nectar.*

In order to cope with their specialized diet, the birds have evolved structural modifications of the bill, tongue, and alimentary canal. The bills of lories are narrower, more elongate, and less powerful than those of other parrots, and the gizzard – the muscular organ used by most other species to pulverize hard or fibrous foods – is relatively thin-walled and weak. Their most striking adaptation is the tongue, which is rather long for a parrot and equipped with a tuft of threadlike papillae at its tip. These papillae are normally enclosed within a protective, cuplike sheath when the bird is at rest or feeding on fairly substantial foods such as fruit or seeds, but they can be expanded like the tentacles of a sea anemone when the tongue is extended to feed on flowers. In this state, the tongue is an effective instrument for mopping up pollen and nectar. Fringe- or brush-tipped tongues are also found in several other families of nectar-feeding birds (the same adaptations are found in some species of nectar-feeding bats).

Few of the tree and shrub species exploited by lories for food have distinct flowering seasons. Individual flowering trees of the same species are often highly dispersed, and pollen and nectar production can vary considerably from year to year, as can the length of the flowering period. The locally abundant but highly erratic nature of this food resource has a number of important consequences for the birds. Most species are highly nomadic, for example, and cover considerable distances in search of food. In the Pacific region, Rainbow lorikeets have been observed flying up to 80km (50 miles) between neighboring islands. Lories also tend to be opportunist breeders – in other words, instead of confining themselves to a fixed breeding season, pairs generally start to breed whenever sufficient pollen and nectar is available. In practice, breeding tends to peak during the rainiest part of the year, since this also corresponds to the period when most trees come into flower. Lories are monogamous and, as far as is known, males and females pair for life. As in many other parrots, enduring pair-bonds have probably evolved in response to ecological factors. The absence of a well-defined breeding season favors a continuous, year-round association between pair-mates so that their reproductive cycles are always synchronized and they can commence breeding whenever conditions are suitable.

Lories are exceptionally pugnacious, and many have evolved unusually elaborate threat displays. This behavior may also be an adaptation to feeding on flowers. When trees come into flower in the tropics, they tend to attract large numbers of birds of different species, all eager to exploit the temporary abundance of pollen and nectar. Within this highly competitive environment, the more aggressive species such as lories seem to be at an advantage. In Australia, lories have also become exceptionally bold and opportunistic in their relationships with people. Scaly-breasted and Rainbow lorikeets inhabit city suburbs and are easily persuaded to visit birdtables. In parts of Queensland, huge flocks of these two species are fed publicly for the entertainment of tourists.

The group includes a number of very rare and endangered species. Stephen's lory, the most easterly representative of the group, is confined to Henderson Island in the Pitcairn Archipelago. Several members of the eastern Polynesian genus *Vini* are currently threatened by habitat destruction, illegal trapping, and the effects of introduced avian malaria.

TJ/JAS

3
4
5

PHOTO STORY

BACK FROM THE BRINK

1 *The flightless kakapo, which thrived in New Zealand before humans arrived, was hunted for its meat and feathers by the Maoris, and then fell victim to introduced species such as Polynesian rats (kiore), cats, and stoats. This is the gruesome result of a cat attack; by the 1960s, the bird was feared extinct.*

2 *Isolated kakapo colonies were, however, discovered in the 1970s. Their survival first entailed moving them to three predator-free sites, Little Barrier in the Hauraki Gulf, Maud Island in the Marlborough Sounds, and Whenua Hou (Codfish Island) off Stewart Island; here, Don Merton of the Kakapo Recovery Program examines a fledgling. Yet a key problem was how to get the aging birds to breed.*

1

2

3 Kakapo breed on Whenua Hou in synchrony with the superabundant fruiting of the Rimu (a native conifer), an event that occurs at 2–5 yearly intervals. Anticipating a huge crop in 2002, the conservationists airlifted all females of breeding age from other islands to this site. Detailed monitoring of nest sites and long, nightly vigils then ensued.

4 Uniquely among parrots, male kakapo gather at a lek to signal their readiness to mate. Using "booming bowls" such as this, which reflect their foghorn-like calls for several kilometers, they display to attract females.

5 Extra food is supplied in hoppers to bring the females to peak breeding condition. Supplementary feeding is carefully regulated to maximize the number of female chicks for future breeding; if, prior to laying, females gain too much weight, they tend to have male chicks.

6 The 2002 breeding program enjoyed great success; all but one of the world population of 21 adult females mated and laid eggs. Twenty-four chicks (15 of them females) hatched and survived, bringing the total number of kakapo to 86. Plans are now afoot to move some of the burgeoning population to a larger island. DVM

Cuckoos

A HARBINGER OF SPRING, AND THE SOURCE of fascinating studies of the interactions between brood parasites and their host species, the European cuckoo is famous, even infamous. So firmly established is its bad reputation for usurping the place of others that it has become implanted in the English language, in the word "cuckold," denoting a husband deceived by a philandering wife. Yet this well-studied bird, which breeds throughout temperate Europe and Asia, and which is more familiar from its call than its looks, is the exception; most cuckoos are poorly known tropical species. Moreover, not all species are parasitic.

The cuckoos are a very diverse family: the sturdy Greater roadrunner of North American deserts bears little resemblance to the delicate Klaas's cuckoo of the African bush. Details of internal anatomy, as well as the possession of feet with two toes pointing forward and two back (zygodactyly), distinguish cuckoos from the superficially similar songbirds and align them with parrots and nightjars. This unusual foot structure enables cuckoos to climb stealthily up slender reed stems or run swiftly over the ground with almost equal poise.

Above *A Channel-billed cuckoo eating figs. The world's largest cuckoo, it lays its eggs in the nests of various species of crow, the Australian magpie* (Gymnorhina tibicen), *and the Pied currawong* (Strepera graculina).

Only Partly Parasitic

FORM AND FUNCTION

The six subfamilies, three in the Old World and three in the New, are varied. In the Old World, the largest subfamily of 54 species contains exclusively parasitic cuckoos. The other two contain, respectively, 28 species of coucal living in Africa, Southeast Asia, and Australia, and 26 species of coua and malkoha. The couas are confined to Madagascar, the malkohas to Southeast Asia. In the New World, there are 18 nonparasitic species, also called cuckoo, but in a different subfamily to their Old World relatives. The three species of communally-nesting ani plus the Guira cuckoo belong in the second subfamily. Finally, there is the 10-strong subfamily of ground cuckoos, of which three species are parasitic.

Many species resemble small hawks, having a distinctly downcurved bill and long tail, and share with hawks the discomfort of being "mobbed" by small songbirds. The reason for these unwelcome attentions is that many cuckoos reproductively parasitize the smaller birds. About 57 species have no other habit of reproduction than to place their eggs in the nests of other species. These include all the species in the subfamily Cuculinae plus the three parasitic American ground-cuckoos. Debate continues about how many times the parasitic habit has arisen in the evolutionary past.

Below *Inhabiting the gallery forest of Madagascar, the Crested coua* (Coua cristata) *is characterized by its electric-blue eye marking and rufous chest patch.*

Right *Even after the young of parasitic cuckoos have left the nest of their hosts, they continue to receive succour. Sometimes, passing birds will respond to the plaintive begging call of the cuckoo by bringing it food. Here, a Red-chested cuckoo* (Cuculus solitarius) *gets a meal from a Cape wagtail* (Motacilla capensis).

Sophisticated Deception

BREEDING BIOLOGY

The European cuckoo female defends a territory within which she keeps a close eye on the comings and goings of the resident songbirds. In fact she is primarily concerned with only one of the resident species, for her eggs are characteristic in color and will closely match the eggs of only one potential host species. When a suitable nest becomes available, usually one in which laying is underway, the cuckoo flies warily down, takes one or more host eggs in her bill, and quickly deposits a single egg in the nest (see A Cuckoo in the Nest). She then departs, all within 10 seconds. The mimicry of the egg color, and the relatively small size of the cuckoo's own egg ensure that the clutch appears untouched when the rightful owner returns. Having successfully completed this delicate operation, the cuckoo eats the stolen egg as reward for its stealth!

The cuckoo egg develops extremely rapidly and, even if some host eggs were already partly incubated when parasitism occurred, is generally the first in the nest to hatch. At this point the European cuckoo nestling displays the remarkable

FACTFILE

CUCKOOS

Order: Cuculiformes

Family: Cuculidae

140 species in 28 genera. Species include: **Black-billed cuckoo** (*Coccyzus erythropthalmus*), **European cuckoo** (*Cuculus canorus*), **Great spotted cuckoo** (*Clamator glandarius*). **Groove-billed ani** (*Crotophaga sulcirostris*). **Klaas's cuckoo** (*Chrysococcyx klaas*), **Asian koel** (*Eudynamys scolopacea*), **Guira cuckoo** (*Guira guira*), **Channel-billed cuckoo** (*Scythrops novaehollandiae*)

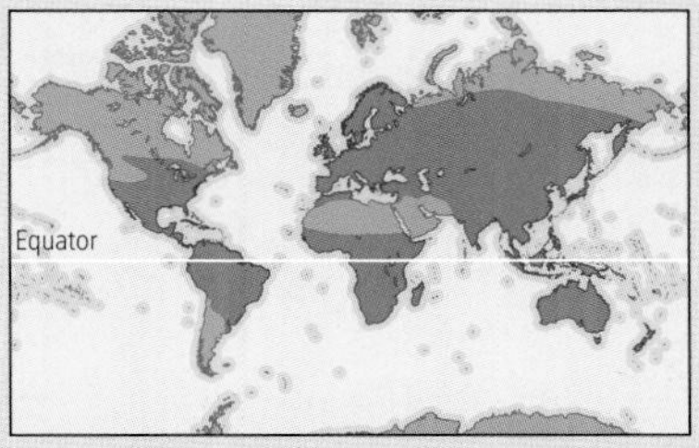

DISTRIBUTION Europe, Africa, Asia, Australasia, N America, S America. Most species sedentary, tropical, or subtropical, though a number of migratory species extend to temperate latitudes.

HABITAT Arid desert to humid forest and even moorlands (European cuckoo), but most species typical of light to heavy scrub and woodland, often with an affinity for watercourses.

SIZE Length 17–65cm (7–26in); weight 30–700g (1–25oz). Sexes usually similar in size, males sometimes slightly larger. A family characteristic is that sizes and weights are unusually variable within each sex.

PLUMAGE Generally subdued grays and browns, underparts often barred and/or streaked, tail sometimes conspicuous with spots or flashes when opened.

VOICE Generally simple flutes, whistles, and hiccups exemplified by the disyllabic note that gives the group its name. Also many harsh notes, especially by fledglings. In at least some species, voice differs between the sexes.

NEST Nonparasitic species construct a platform of sticks in trees, bushes, or on open ground.

EGGS Parasitic species may lay 8–15 per season, although more may be stimulated by unusually high losses; nonparasitic species 2–5; weights from 8–70g (0.3–2.5oz). The eggs of nonparasitic species are very heavy relative to female body-weight while those of some parasitic species are small. Incubation period 11–16 days; nestling period 16–24 days. Egg color variable due to mimicry of host eggs in parasitic species.

DIET Almost completely insectivorous, with most species taking noxious prey (e.g. hairy caterpillars) unavailable to other groups of birds. Larger forms take some smaller vertebrates; one genus (*Eudynamys*) largely vegetarian.

CONSERVATION STATUS 2 species – the Sumatran ground cuckoo (*Carpococcyx viridis*) and the Black-hooded coucal (*Centropus steerii*) – are Critically Endangered. 1 species is Endangered, and 6 Vulnerable.

adaptation of egg or hatchling ejection, which was first described in detail in 1788 by the English physician Edward Jenner (the inventor of vaccination). The cuckoo continues to push out all other objects around it until it is left as the sole occupant of the nest. In so doing, it eliminates any possible competition and thereby ensures that its foster parents concentrate their reproductive efforts on one thing – the raising of the voracious cuckoo! Even if the foster parents are sitting on the nest while this tragedy unfolds, they do not intervene to stop the cuckoo's slaughter of their own young.

This pattern is not invariable among the cuckoos. Many species, for instance the Great spotted cuckoo, the Channel-billed cuckoo, and the Asian koel, do not show ejection behavior. Instead, the young share the nest with offspring of the host species, in these species usually crows. Yet the rapidly growing and more active cuckoo nestling either tramples the crow chicks to death underfoot or, more subtly, monopolizes the food brought to the nest by its foster parents.

Even after hatching, the cuckoo nestling must continue the deception to be able to obtain food from its foster parents. This trickery seems to be achieved by the cuckoo exploiting the signals that usually pass between parent and offspring in the nest. The Great spotted cuckoo produces a passable imitation of the begging calls of nestling magpies, and its wide-open gape is even more vivid than those of the host nestlings with which it shares a nest.

So powerful is the stimulus of the cuckoo's vigorous begging that it can induce other birds to care for it even after it has left the nest; passing small birds that are neither the true nor the foster parents will bring food to the imploring bird!

In an evolutionary sense, it is surprising that cuckoo nestlings succeed in obtaining parental care from foster parents, when selection against this behavior must be very strong indeed. Mimicry of egg color and size may increase the chances of host acceptance, and appear to be very finely tuned to the resident host population. For example, the Brown babbler (*Turdoides plebejus*) is a central African host species that, over most of its range, lays clear blue eggs, but in one part of northern Nigeria lays pink or mauve eggs. Incredibly, its cuckoo parasite has evolved to mimic faithfully these changes in coloration. The accuracy of this kind of local mimicry is presumably dependent upon a high degree of breeding site tenacity in both migratory and nonmigratory cuckoos: a female which tried to breed other than close to where she was hatched might find it more difficult to locate the correct hosts. It seems that this system is maintained because young females not only inherit egg color from their mothers but also tend to parasitize the same host as reared them. Thus, there are female-based lineages of cuckoos that are genetically distinct. However, they are not in the process of becoming new species, since males mate with females from any lineage, and promote gene flow.

An evolutionary "half-way house" between full-scale cuckoo parasitism and the part-time parasitism of some other species (for example, the House sparrow, starling, and moorhen, which occasionally deposit eggs in the nest of conspecifics) is represented by Yellow-billed and Black-billed cuckoos. In most years, these species nest "normally." When food, such as cicadas, is unusually abundant, however, each female also attempts to parasitize the nests of either conspecifics or other species, in addition to raising a brood herself. What ecological factors have caused the evolution of this mixed policy? The interesting features of these cuckoos is that they have very large eggs, relative to body size, and that these large

3

2

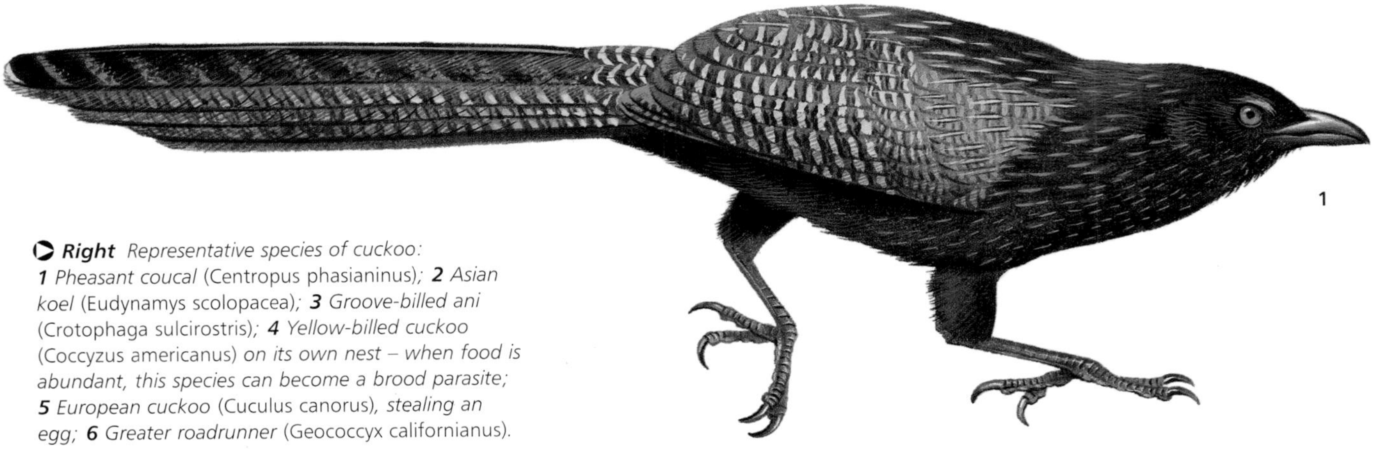
1

Right *Representative species of cuckoo:* **1** *Pheasant coucal* (Centropus phasianinus)*;* **2** *Asian koel* (Eudynamys scolopacea)*;* **3** *Groove-billed ani* (Crotophaga sulcirostris)*;* **4** *Yellow-billed cuckoo* (Coccyzus americanus) *on its own nest – when food is abundant, this species can become a brood parasite;* **5** *European cuckoo* (Cuculus canorus), *stealing an egg;* **6** *Greater roadrunner* (Geococcyx californianus).

eggs have an extraordinarily fast development time, hatching in only 11 days – the shortest incubation period of any bird! Rapidly hatching eggs are, of course, essential for successful parasitism because the host's nest will already contain developing eggs when discovered by the cuckoo. If the cuckoo's eggs hatch much later than those of the host, their chance of success will be very small.

The extraordinary behavior of parasitic cuckoos tends to obscure the fact that about two-thirds of the species appear to be nonparasitic in their breeding habits, mating monogamously and remaining together while the offspring are reared. The Pheasant coucal is one such species, producing one or two grotesque black nestlings which, in common with many cuckoos, excrete a foul liquid when they suspect the presence of a predator. However, our knowledge of most species is extremely poor.

Relatively Safe

CONSERVATION AND ENVIRONMENT

Unlike several other groups of fascinating and understudied birds, many cuckoo species are characteristic of scrub, secondary forest, and other types of disturbed ground which, if anything, increase under human interference. This habitat preference has stood them in good stead, keeping the proportion of threatened species relatively low, at under 10 percent. Most of the threatened species occur in Southeast Asia, have restricted ranges, and are at risk from habitat loss. For example, the forest habitats of the two Critically Endangered species, the Sumatran ground cuckoo and the Black-hooded coucal, respectively on the Indonesian island of Sumatra and the Philippine island of Mindoro, have been subject to extensive deforestation for slash-and-burn agriculture. MdeLB/JAH

THE DESERT SPECIALIST

Even the nonparasitic members of the cuckoo family turn out to be remarkable in other ways, a case in point being the Greater roadrunner. This ground-dwelling bird, which lives in the desert chapparal of Mexico and the US Southwest (with the greatest abundance in Arizona), was once heavily persecuted in the mistaken belief that it was harmful to populations of gamebirds. In fact, it eats an assortment of large invertebrates and is also a voracious predator on small lizards. A good turn of speed enables it to run down prey; it has been recorded running at 24km/h (15mph). Travelers' tall tales have embellished the reputation of the roadrunner: in one popular account the bird delicately wields a cactus leaf, teasing an angry rattlesnake until it strikes and becomes impaled on one of the bird's thorny weapons.

Rather more reliably attested is the fact that this species displays a physiology that is most unusual among birds, being, to some extent, cold-blooded! When air temperatures become very low, during the desert night, most birds need to increase metabolism to maintain their internal body temperature at a constant, high level. This, of course, means burning internal food reserves at a greater rate. The roadrunner adopts a more economical approach – it simply allows its body temperature to fall slightly, turning down the "central heating" with no ill effects, and thereby saving energy. The bird does in fact go into a slight torpor and may not be able to respond as quickly to sudden danger, but for a creature with few predators, this disadvantage is of little consequence.

As the first rays of dawn break the cool of the desert night, the roadrunner has a neat trick for warming up. Areas of skin on its back, located just between the wings, are darkly pigmented and absorb the energy of sunlight, warming the skin and the underlying blood vessels. To accelerate this process, the bird fluffs the feathers covering the patches so that the light penetrates more effectively. This rapid warming mechanism can save the roadrunner up to 50 percent of the energy it would otherwise need to warm up to a working temperature. MdeLB

4

5

6

PHOTO STORY

A Cuckoo in the Nest

1 A female European cuckoo surveys her territory, waiting for an opportune moment to visit the nest of a songbird pair of the species she habitually parasitizes. Reed and Sedge warblers are common victims. The cuckoo steals one of the host's eggs, which she later eats.

2 The cuckoo, which lays about eight eggs each breeding season, deposits a single egg in the host's unattended nest. To lessen the chances of detection, the cuckoo egg is as close as possible in size and color to the resident clutch.

3 The cuckoo egg is usually the first to hatch. About a day after hatching, the tiny cuckoo thrashes about in the nest until a hollow in its back makes contact with another object, either an egg or another small chick. Using surprisingly powerful legs, the cuckoo climbs the inside of the nest until it reaches the rim and ejects its little load into oblivion. It repeats this task until it has completely cleared the nest.

4 A Reed warbler feeds the single young cuckoo at about the same rate as it would have provisioned its own brood of three or four young. The key factor ensuring that the host works hard for the cuckoo is the interloper's persistent begging call, a rapid "si..si...si...," which mimics an entire brood of host young. The cuckoo's precise frequency of calling makes up for the fact that the area of its gape is rather less than the total gape area of four Reed warbler chicks.

5 Sustained by the Reed warblers for about three weeks, the cuckoo grows rapidly and soon dwarfs its unwitting foster parents. In some areas, cuckoo brood parasitism is so successful that it reduces the host population to a point where the cuckoo is obliged to shift to a different host species. MdeLB/IAH

Hoatzin

THE REMARKABLE HOATZIN HAS DEFIED conventional methods of classification since it was first described in 1776. One especially fascinating feature is the presence of two claws on the "elbows" of juvenile birds, enabling them to grip onto branches; most other birds exhibit this character only as embryos, and its retention in the hoatzin is adduced by paleontologists as evidence of the reptilian ancestry of birds.

Systematists agree that the hoatzin should be assigned to its own monotypic family. This family has traditionally been aligned with the Galliformes – an assemblage of "fowl-like" birds (including the turkeys, pheasants, and curassows) whose relationships to each other are not fully understood. With its stout legs, coarse plumage, chestnut primaries, and weak flight, the hoatzin certainly bears some resemblance to a domestic hen. However, recent studies of skeletal features, genetic sequence data from mitochondrial and nuclear DNA, egg-white proteins, and scleral ossicles surrounding the eye all point consistently to a relationship with the Cuculiformes. What remains uncertain is whether, within the Cuculiformes, the hoatzin is more closely related to the cuckoos (Cuculidae) or the turacos (Musophagidae).

Bovine Digestion

FORM AND FUNCTION

Classification of the hoatzin has undoubtedly been complicated by the fact that the species is one of the most refined and specialized herbage eaters among the birds. A key adaptation is the foregut, which constitutes approximately 25 percent of the bird's weight. So large is it that the flight muscles are reduced, and the hoatzin is a correspondingly weak flier.

Green leaves of many plant species, particularly young leaves that are low in tannins, make up approximately 80 percent of the hoatzin's diet, with the remainder of the food derived from flowers and fruit. Food is foraged in the early morning and early evening. Unlike other herbivorous birds, such as grouse, which digest plant material in the hindgut, hoatzins ferment vegetation within the spacious foregut. There, assorted microbes break down cellulose plant cell walls, exactly as they do in ruminants like cows or sheep and in kangaroos. To achieve this breakdown, food remains in the gut for a long time, 18 hours for liquids and 1–2 days for solid plant material. This period is similar to the retention time of sheep and explains why the feces are sloppy, in contrast to the fibrous cylinders that are voided by, for example, grouse. It also explains why hoatzins are commonly seen resting on trees, their sternum supported by a branch. They are allowing time for their digestive enzymes to act.

The singular digestive system is thought to be at the root of the hoatzin's reputation for smelling awful. Despite this, however, the birds' flesh is said to be perfectly palatable, and hoatzins are sometimes hunted. Provided they are unmolested, they can live in the vicinity of humans, for example along the edges of canals. Some individuals have survived in zoos, on a green leaf diet, for more than five years.

Additional Helpers

SOCIAL BEHAVIOR

Hoatzins live in social groups usually of 2–5 birds but sometimes with as many as 8 members. Ringing has ascertained that these groups contain a breeding pair plus their offspring from previous years. The extra birds are mostly male, since the young females disperse. Except for copulation and egg-laying, these extra birds take part in all breeding activities, such as territorial defense, incubation, and feeding the young on a regurgitated liquid vegetable soup. This nourishment is rich in bacteria, thereby inoculating the young's gut with the microbes it will need for fermentation. The young begin to fly when about two months old, and begin feeding independently when some 50–70 days old. Around this time the claws are also lost. MdeLB/JAH

FACTFILE

HOATZIN

Order: Cuculiformes

Family: Opisthocomidae

1 species: *Opisthocomus hoazin*

DISTRIBUTION Northern S America, in the Amazon and Orinoco basins.

HABITAT Rain forest.

SIZE Length 60cm (24in), weight 800g (28oz).

PLUMAGE Dark brown on back, buff below, shading to chestnut on abdomen and sides; facial skin electric blue; head quills chestnut with dark tips; tail with a broad buff tip.

VOICE Various calls, including a clucking courtship call, a mewing feeding call, a wheezing alarm call, and a sharp screech like a guinea fowl.

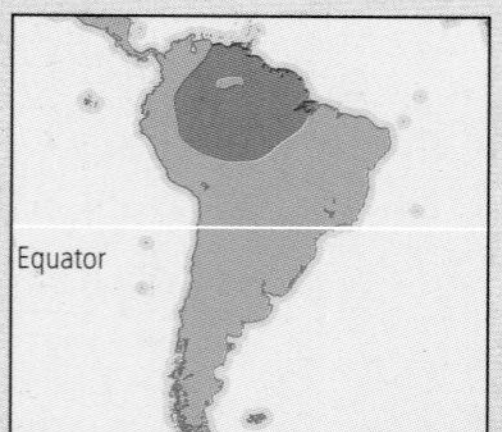

NEST Twigs and sticks, in trees or large bushes, usually over water.

EGGS 2–3 (occasionally 4 or 5), buff, dappled with brown or blue spots; weight 30g (1oz). Incubation period 28 days; chicks start to feed themselves at 10–14 days.

DIET Leaves, flowers, and fruits of marsh plants.

CONSERVATION STATUS Not currently at risk.

THE REMARKABLE HOATZIN CHICK

High above the shallow, muddy waters of a South American river, a newly-hatched hoatzin chick weakly lifts its head above the nest rim while its parents are away foraging. The chick is naked and ugly and, like many other young birds, it appears to be virtually helpless. The latter could hardly be further from the truth. Upon the noisy return of its clumsy parents, the dark-skinned nestling leaves the nest, climbing gingerly among the thin branches, to intercept any tender young leaves which the parent may be carrying back to its hungry brood.

Two unique events are occurring here! The hoatzin is the only tree-living bird that feeds its young on foliage to any great extent. It is also the only one in which the chicks habitually leave the nest soon after hatching. Such a lightly muscled body hardly has the strength to balance on the swaying twigs, but the task is eased by the use of tiny claws emerging from the "elbow" bends of the unfeathered wings.

These appendages are not unique among birds – very young European coots bear single claws to aid their frequent climbs back into nests that may tower high above the water surface, and some species of geese carry sharp spurs on the wing-edge even as adults. Many species of animal in South America show similar adaptations to a precarious life over water: the American monkeys and anteaters possess gripping tails as an added insurance against a hazardous fall. The hoatzin chick, however, has an additional safety net. Even if it should fall into the brown waters several meters below, all is not lost – the leathery bundle simply swims to the nearest branch and begins a slow, deliberate climb back up to parental care! While ascending, it utters a cheeping call to enable the parents and helpers to locate it. The climbing young do not make for the nest. Instead they are cared for on branches within the territory.

Left *With its spiky crest and heavy body, the hoatzin is a distinctive denizen of Amazonia. Its preferred habitat is in swamps and sluggish backwaters.*

Right *Hoatzins live socially throughout the year, but particularly so at breeding time, when a single tree may contain several nests.*

Turacos

THE TURACOS – THE ONLY LARGE FAMILY OF birds found solely in Africa – were once allied with the cuckoos, mostly because both groups have a zygodactylous arrangement of toes. On the other hand, evidence from feather parasites once hinted at an affinity with gamebirds (Galliformes). Yet DNA analysis tends to reaffirm a relationship between turacos and cuckoos, while suggesting they might be placed in two separate orders, and not just two families.

Turacos are so poorly known that for years they were called "plantain-eaters" (the literal translation of "Musophagidae"). However, once it was realized that these medium-sized birds, with long tails and short, rounded wings, rarely or never ate plantains or bananas, the misnomer was supplanted by the name "turaco." In any case, long-term survival in Africa would have been impossible for plantain-eaters, since these fruits were introduced to the continent by people only comparatively recently!

Fruit a Favorite

DISTRIBUTION PATTERNS

The turacos occupy a great variety of wooded habitats, from montane forest to woods to savanna to suburban gardens. Foraging among foliage in parties of up to a dozen birds, they take mainly a wide variety of fruits, including certain berries that are highly poisonous to humans. Up to 80 percent of the seeds ingested are deposited away from the parent tree, indicating that turacos are important seed dispersers. The few available reports suggest that the nestlings are also fed largely on a fruit diet supplemented by the occasional invertebrate, especially snails. This is unusual among young land birds, most of which are fed on a high-protein diet of invertebrates during their growth between hatching and independence.

Unique Pigmentation

FORM AND FUNCTION

One characteristic that the turacos share with no other living birds is the possession of two vivid feather pigments, both copper compounds – a green, turacoverdin, and a red, turacin. Turacoverdin produces the rich green body feathers found in 14 species, and is the only green pigment found in birds. (Most birds' feathers produce their green colors by refracting light with specialized feather structures that produce iridescence.) Turacin colors the crimson wing flashes and head ornaments found in most turacos. The long period of about a year taken by a young turaco to develop full adult coloration may perhaps be related to the difficulty of acquiring the relatively scarce copper for the pigment. It has been estimated that a turaco would need to eat about 20kg (44lb) of fruit to obtain enough copper to color its plumage.

Stealthy Feeding

BREEDING BIOLOGY

The common observation that turacos forage in groups might suggest that these birds, like some others in the tropics, are social breeders, organizing themselves so that individuals other than the parents contribute to the nesting chores of incubation, brooding, and feeding the hungry young. Our knowledge of their breeding habits is, however, so poor that this mode of reproduction has been confirmed in only a single species. Rather, the norm seems to be monogamous pairs breeding in strongly defended territories, with courtship often intense at the start of the rainy season. The close similarity between the sexes is typical of other monogamous bird species—the only sexual difference seems to be one of bill color.

For several weeks before egg-laying, the male regurgitates gifts of fruit pulp for his female. The clutch size is 2–3 in savanna species, normally two in others. Once reproduction proper has begun, both birds contribute equally to incubating, brooding, and feeding the chicks. These nestlings are covered in a fine down, of varying color and thickness, and they advertise their hunger with a large, orange-red gape. Parents respond by regurgitating the mixture of fruit and insects directly into the nestlings' throats. Unlike in some birds, this operation takes place in silence, perhaps because of the high density of predators in the forest habitat.

A remarkable affinity with the peculiar hoatzin emerges once the young have recovered from hatching and increased their strength with a few meals. The silky nestlings are endowed with tiny claws on their wing-joints and can use these,

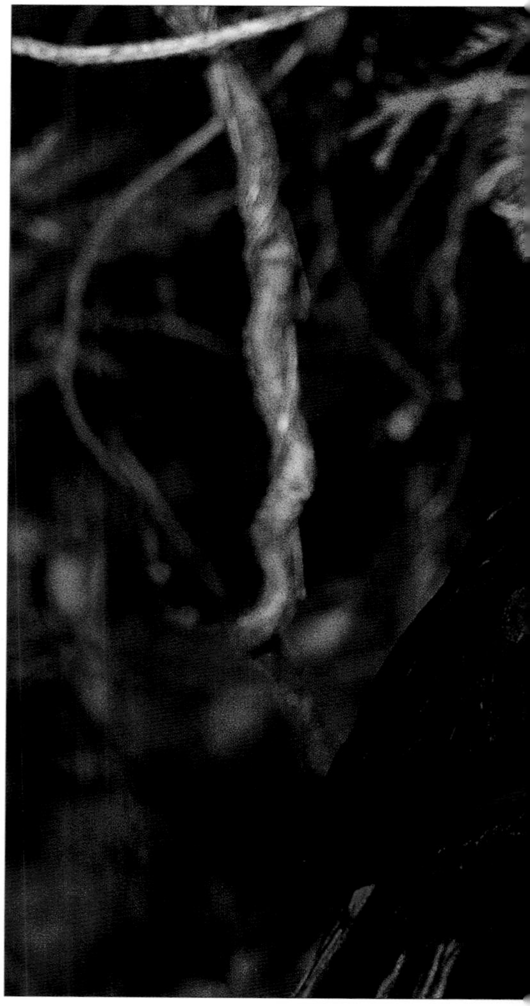

Below The White-bellied go-away bird inhabits East Africa, from Ethiopia and Somalia down through Kenya to Tanzania. These birds' curious common name derives from the sound of their calls.

Below *Hartlaub's turaco* (Tauraco hartlaubi) *is a spectacularly colored species, with its silky green and iridescent purple plumage, a distinctive white comma before the eye, and underwings of a bright crimson. Some of its favored haunts are the highlands of Kenya, such as Mount Kenya and the Aberdares range.*

Above *A denizen of the low-lying, humid forests of West Africa, the Violet turaco specializes in eating fruits, with a predilection for figs. It was one of the species wrongly designated a "plantain-eater."*

FACTFILE

TURACOS

Order: Cuculiformes

Family: Musophagidae

23 species in 5 genera. Species include: **Prince Ruspoli's turaco** (*Tauraco ruspolii*), **Bannerman's turaco** (*T. bannermani*), **Guinea turaco** (*T. macrorhynchus*), **Great blue turaco** (*Corythaeola cristata*), **Violet turaco** (*Musophaga violacea*), **White-bellied go-away bird** (*Corythaixoides leucogaster*), **Western grey plantain-eater** (*Crinifer piscator*).

DISTRIBUTION Central and S Africa.

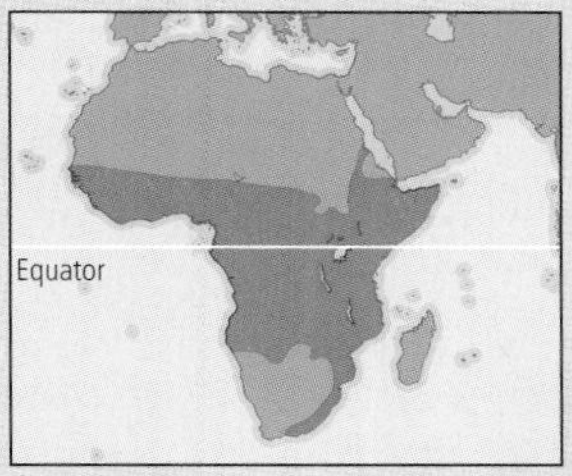

HABITAT Evergreen forest, wooded valleys; more rarely savanna.

SIZE Length 35–75cm (14–30in); weight 230–950g (8–34oz).

PLUMAGE Green bodies with green, blue, or purple wings and tail, or largely blue and purple, or mainly gray.

VOICE One- or two-syllable barks, with some longer wailing notes.

NEST Flat, insubstantial nests of twigs in trees or low bushes.

EGGS 2–3, usually glossy white or pale blue/green: weight 20–45g (0.7–1.6oz). Incubation period about 17–30 days; nestling period 10–12 days.

DIET Fruit, some invertebrates.

CONSERVATION STATUS Bannerman's turaco and Prince Ruspoli's turaco are Endangered.

together with their adaptable foot structure, to leave the nest and sit on the periphery or even on adjoining twigs. In fact the young leave the nest for good when aged about 2–3 weeks, a week or two before they can fly. Independence from parental feeding seems to be gained at about six weeks in many species, although young Great blue turacos are fed for about three months.

Refuge at Kilum-Ijum

CONSERVATION AND ENVIRONMENT

There are two threatened species. Bannerman's turaco is restricted to highland areas of Cameroon, where the population of under 10,000 is confined to less than 500sq km (193sq mi) of montane forest. The largest patch, the Kilum-Ijum forest, is protected by a community-based conservation project. Local people take great pride in "their" turaco. The second threatened species is Prince Ruspoli's turaco, which also has a very restricted distribution in Ethiopia, where it is under pressure from habitat loss. MdeLB/JAH

Owls

BECAUSE THEY ARE NOT AS EASY TO OBSERVE *as birds that are active during the day, owls are not well known to scientists, bird-watchers, or the general public; indeed, it is often only their calls that alert us to their presence. New owl species continue to be discovered, especially in tropical regions.*

Less than 3 percent of all bird species are active at night and over half of these are owls; they are the nocturnal counterparts of the day-hunting hawks and falcons. Although the largest species is 100 times the weight of the smallest, all owls are instantly recognizable as such. This uniformity derives from their unique adaptations for their role as nocturnal predators. Owls occur wherever there are animals on which they can prey. Most are associated with trees, but others are adapted to living in grasslands, deserts, marshes, or even arctic tundra. The diets, biology, and behavior of many tropical owls are unknown, but a significant proportion of all owl species are thought to be primarily night-hunting. Most of the rest can hunt at any time, but do so especially at dusk and dawn.

*▶ **Right** The sight of an Eagle owl landing in a tree is much like the fleeting glimpse that countless small, terrified rodents get of these fearsome nocturnal predators as they swoop out of the night sky for the kill.*

A Characteristic Shape

FORM AND FUNCTION

All owls are easily recognized by their shape: an upright stance, short tail, large head, and dense covering of feathers, giving them a neckless, rounded outline. Equally characteristic are the huge, frontally placed, often orange or yellow eyes, which stare out from saucer-shaped disks of radiating feathers (see A Face for the Night). Daytime-hunting species have smaller eyes and ill-defined facial disks. Many owls have flexible tufts of feathers above the eyes, used in visual communication; these "ear tufts" have no connection with hearing. All owls have powerful, usually feathered legs with sharp, curved talons for gripping prey. The short, hooked beak is curved downward and may be hard to see among the feathers.

Birds active only in darkness do not require striking plumage; owls mostly spend the day roosting in quiet places, often pressed tight against a tree-trunk, and so both sexes are usually patterned in various somber shades of brown to aid concealment. If discovered by small birds, owls are mobbed to advertise their presence and to persuade them to move on.

Owls that live in open habitats are paler than those from woodland: desert forms are often sandy-colored, and the Snowy owl is mainly white to match its arctic surroundings. Some woodland owls have two distinct color-phases – gray in northern coniferous forests, brown in deciduous woods further south. With few exceptions, juveniles look similar to adults. In most owls the female is larger than the male, although the difference is not usually as marked as in some of the day-hunting birds of prey.

The relationships of owls to other birds are poorly known. On the one hand it has been suggested that they are related to raptors such as hawks and falcons, on the other to other nocturnal birds, such as nightjars and their allies. There is little support among avian taxonomists for either position. The relationships among the orders of birds, including owls, remain an enigma.

FACTFILE

OWLS

Order: Strigiformes

Family: Strigidae

189 species in 25 genera.

DISTRIBUTION Almost cosmopolitan, except Antarctica.

HABITAT Chiefly woodlands and forests; some grasslands, deserts, and tundra.

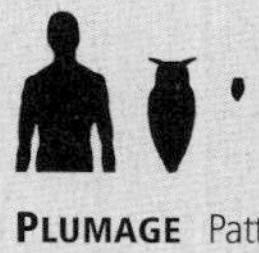

SIZE Length 12–71cm (4.7–28in); weight 40–4,000g (1.4–141oz). Sexual dimorphism slight, but females usually larger.

PLUMAGE Patterned brown or gray; one white, several black and white.

VOICE A wide range of shrieks, hoots, and caterwauls.

NEST Chiefly holes, or abandoned nests of other species; a few on the ground or in burrows.

EGGS 1–14, depending on food supply; usually 2–7, white and rounded; weight 7–80g (0.2–3oz). Incubation period

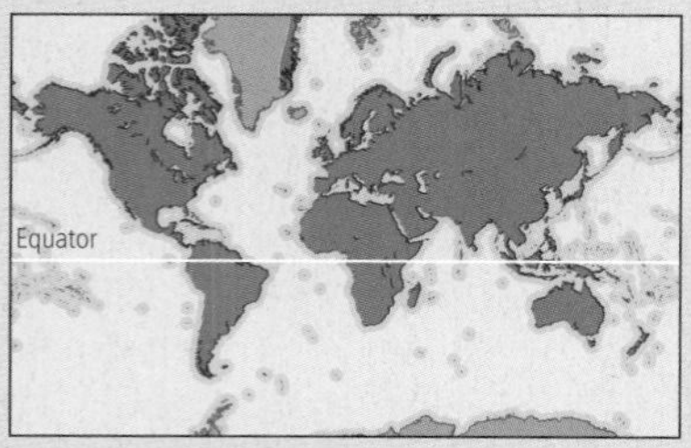

15–35 days; nestling period 24–52 days, but young may leave nest before able to fly (15–35 days after hatching).

DIET Mostly small ground-living rodents, also birds, reptiles, frogs, fish, and crabs (fishing owls); earthworms and large insects (especially small owls).

CONSERVATION STATUS 7 species, including the Christmas Island hawk owl, are Critically Endangered; 6 are Endangered, and 10 Vulnerable. The Laughing owl (*Sceloglaux albifacies*) of New Zealand may now be extinct.

See Selected Owl Species▷

Far and Wide

DISTRIBUTION PATTERNS

Most owls are "typical" owls of the family Strigidae. The largest genus (*Otus*) contains 63 screech and scops owls – a far-flung group, although absent from Australia; all are small to medium-sized, unspecialized, "eared" owls of temperate or tropical woodland or scrub. Most are nocturnal and feed on insects, but the few temperate species switch to rodents in winter. They are among the most abundant owls in many habitats; in North America, screech-owls often occur in urban and suburban areas. The New World and Old World members of this genus are not close relatives.

The 18 eagle owls are the powerful nocturnal equivalents of the large day-hunting eagles and buzzards. They occupy habitats ranging from open country to moderately dense forest in the New World, Africa, and across Eurasia; in the northern tundra they are represented by the large Snowy owl, a related species that hunts by day during the long days of the arctic summer.

The six large fishing owls of Asia (*Ketupa*) and Africa (*Scotopelia*) are the food specialists among the owls – the nocturnal counterparts of the osprey and fish eagles and equivalent to the fish-eating bats of tropical America. They occur in forests along rivers, lakes, and swamps in Asia and Africa; the Brown fish owl occurs from the plains of India to the Malay Peninsula and Pel's fishing owl occurs widely in the river basins of southern Africa. The 31 pygmy owls (*Glaucidium*) are widespread in Eurasia, Africa, and the Americas. Some of the species are quite active during daylight. This group includes the sparrow-sized Least pygmy owl of tropical South American forests, which shares with the Elf owl of the US southwest the distinction of being the smallest owl.

Woodland owls of the genus *Strix* are found in forested areas throughout the world except Australia and remote islands. The extensively studied Tawny owl, whose range extends from Britain across Europe and northwest Africa to the mountains of Myanmar and China, is perhaps the best known of the 19 species in the genus. The Barred owl is common in damp woodlands and swamps of North America, and the Spotted owl is the woodland owl of old-growth forests of the western USA. The seven *Asio* species fall into two ecologically distinct groups: "long-eared owls," found in broad-leaved or coniferous woodland, and "short-eared owls" that frequent marshes, grassland, and other open country.

The genus *Athene* takes its name from the Greek goddess of wisdom, Pallas Athene. One member of this group is the Little owl, which sometimes hunts by day. It is a familiar sight in open habitats from Western Europe and North Africa to China. In the late 1800s it was introduced into Britain and New Zealand. The only New World representative of *Athene* is the Burrowing owl, a long-legged, daytime-hunting, terrestrial species of open, treeless grasslands; it is commonly found in Prairie-dog towns. The genus *Aegolius* is a basically New World family of four geographically separated, small, nocturnal forest owls. The most widespread is the Boreal or Tengmalm's owl. Like several other owls of northern coniferous forests (for example, the Great grey and the Northern hawk owl), its range extends in a belt across the high latitudes of the New and Old Worlds.

Most medium-sized owls of Indonesia and Australasia are hawk owls (*Ninox*). This large genus is geographically restricted, but quite important as it seems to represent one part of the most ancient division within the typical owls. Of this group only the Brown hawk owl of the Asian mainland has a wide distribution; most of the other 17 species are confined to single islands where their ranges do not overlap. In Australia, where most of the major owl genera are absent (including *Bubo*, *Otus*, *Strix*, and *Glaucidium*), three *Ninox* species do exist side by side, including the small Southern boobook owl that also occurs in New Guinea.

Right Representative species of owls: **1** Barking owl (Ninox connivens) with nestlings at the nest-hole. **2** White-faced scops owl (Otus leucotis) listening for prey. **3** Tengmalm's owl (Aegolius funereus) about to catch a vole. **4** Pel's fishing owl (Scotopelia peli) with fish. **5** Spotted wood owl (Strix seloputo) being mobbed by passerines. **6** Malaysian eagle owl (Bubo sumatranus) with dead bird. **7** Spectacled owl (Pulsatrix perspicillata) on the look-out. **8** Elf owl (Micrathene whitneyi).

1
2
3
4
5
IW

Predators Par Excellence

DIET AND PREY

Owls feed on a wide variety of animal prey; what exactly they eat depends mainly on their size and the habitat they occupy. Tawny owls living in woodland feed mainly on mice and voles, but in the towns they feed on birds, especially house sparrows. Small owls are for the most part insectivorous; medium-sized ones feed mainly on small rodents or birds; the largest species take mammals (up to the size of hares or even small deer) and medium-sized birds – including other owls and birds of prey!

Owls catch most of their prey on the ground in the open. Woodland owls have short, rounded wings, and when hunting sit quietly on a low perch watching and listening for small mammals. On hearing a likely noise they rapidly rotate their head until the sound registers equally in both ears; they are then directly facing it. When the source of the sound is precisely located, the owl glides silently down toward it; at the last second it swings its feet forward to hit the prey, often killing it outright. Many owls of open country hunt mainly in flight. They have long wings that enable them slowly to quarter the ground like the day-hunting harriers, with little expenditure of energy. Long-eared owls spend about 20 percent of the night hunting. Once prey is located, it is pounced on from a low height in the manner of perch-hunting owls; about one in five attempts is successful, although this figure varies substantially. Owls are opportunistic hunters and will often try to catch prey any way they can: insects (and sometimes birds) may be chased in flight, birds are grabbed while roosting, several species (for

OWL PELLETS

Owls usually swallow their prey whole, and much of what is indigestible, such as fur, bones, teeth, claws, beaks, or the head capsules and wing-cases of insects, are compressed into a sausage-shaped pellet that is cast back out through the mouth. To ease the pellet's passage, the softer fur or feathers enclose the hard parts. Pellets can be collected, teased apart, and their contents identified, to provide a record of what the owl has been eating. Mammals can usually be identified and counted from their skulls, jaws, or teeth, and birds from their beaks, feet, or certain bones. Even the hard remains of insects and other invertebrates can often be recognized.

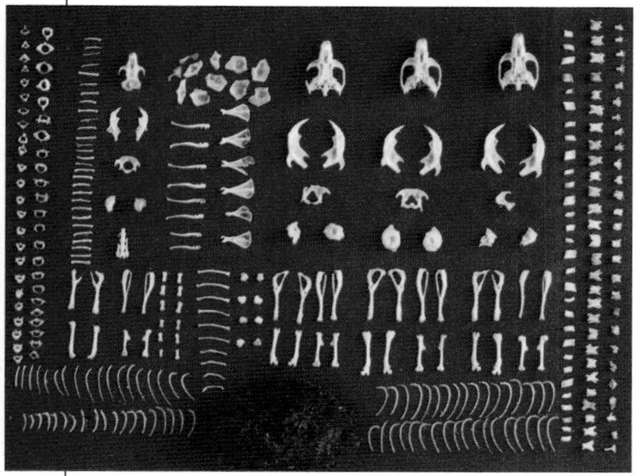

The ease with which pellets can be collected and analyzed depends on the species. Barn owls roost in the same place day after day, and it is an easy job to collect the accumulation of pellets at regular intervals. Other species, like the Tawny owl, deposit their pellets at widely scattered nocturnal roosting or feeding sites, making them difficult to find; they also seem better at digesting bones than other owls, and sometimes decapitate their prey.

Most owls of temperate regions produce one or two pellets per 24 hours, depending on season. In summer, when nights are short, they produce one large pellet at their daytime roosts. In winter they often cast a second, smaller pellet about 7 hours after the first, while hunting at night.

The size and appearance of pellets is often characteristic of a particular species. In general the largest owls produce the largest pellets. Those of the Eagle owl average 77x31x28mm (3x1.2x1.1in), those of the Pygmy owl only 27x11x9mm (1.1x0.4x0.35in). Barn-owl pellets are black, shiny and hard, those of the Tawny owl gray and crumbly.

Although the analysis of remains in pellets has drawbacks, it provides the best clue to the diets of owls, allowing comparisons to be drawn between different seasons and species. Owl pellets have also contributed to historical ecology and paleontology. Generations of owls may occupy the same roost, and the long-term deposition of large numbers of pellets can provide information about the local distribution and abundance of prey based on the frequency of bones in the deposits. In some places, for example, caves have been used by owls as roosting places for centuries, and the build-up of deposits of pellets may contain bones from species of birds and rodents that were formerly present in the area but are so no longer.

In some cases, the owl pellets may even contain skulls and other diagnostic elements of species that have subsequently become extinct. A particularly interesting example of this phenomenon was uncovered in the Hawaiian Islands. There, several extinct species of Hawaiian honeycreepers were described from pellet deposits from an owl that itself has also since died out. GWB/GH

instance, Little and Burrowing owls) bound across the ground in search of invertebrates, and Tawny owls will plunge into water to catch frogs; Barred owls may wait near bird-feeding stations and ambush squirrels that come to take grain and seeds. The specialist fishing owls swoop down to pluck fish from the water surface. Roadsides also provide good hunting areas for owls, and in developed countries many are killed by traffic. Owls that winter in the far north sometimes hunt from perches and can hear rodents moving under the snow; Snowy and Great grey owls can pounce on their prey through up to 30cm (1ft) of soft snow.

Unlike hawks and falcons, owls carry all but the largest prey in their bill. Some owls first remove the head of larger prey and swallow only the body and tail; small prey is swallowed whole, head first. The nutritious parts of the prey are digested, while the indigestible remains, such as bones and fur, are regurgitated as pellets, which provide a good record of what owls eat (see Owl Pellets). Owls have no crop in which to store food, but sometimes they cache prey.

Owls can have a substantial impact on prey populations: in one study, a pair of Tawny owls, consuming a maximum of seven 20g (0.7oz) rodents per day, were found to have removed 18–46 percent of the Bank voles and 28–70 percent of the Wood mice present in their hunting range in each two-month period.

Periodic fluctuations in the numbers of rodents have striking effects on owl populations. In years when prey is scarce, many owls either do not breed or lay reduced numbers of eggs. At such times, some species are found far outside their normal ranges: for example, large numbers of

Left *As snowflakes drift past its perch in a Wyoming forest, a Great grey owl keeps watch for prey on the ground below. With a wingspan of 1.5m (5ft), this is one of the largest owl species.*

Right *A North American Northern hawk owl rests on a branch with its deer-mouse prey. Hawklike in appearance, this day-flying predator relies, again like the hawks, more on sight than on hearing when hunting.*

Snowy owls appear in the USA when lemming or hare populations have crashed in the Arctic. The birds are often very tame when in unfamiliar surroundings: North American bird-banders catch the impressive-looking Great grey owl by casting out a dead mouse attached to a fishing line (without a hook) – a hungry owl will pounce on it and can be reeled in!

Good Timing

SOCIAL BEHAVIOR

Most owls are monogamous. There are a few species in which males are known to have two or more mates on occasion. However, this situation only arises when prey become very abundant, for example, during the peak of rodent cycles, because the male has to provide all the food for both females plus all of the growing nestlings.

Generally, the breeding of owls is timed so that food is most plentiful when the young are learning to hunt for themselves and the adults are undergoing their annual molt (which reduces their hunting efficiency). In northern climates, this can result in breeding happening very early in the year; for instance, Great horned owls lay their eggs as early as February in New York, when the ground is still covered by snow. In that case, the female will be incubating her eggs when the temperature is well below freezing. In tropical regions, breeding is geared to rainfall – small species have young in the nest when the onset of the rainy season produces a flush of insects. Many owls have only a single brood per year, but some open-country species can breed whenever rodents are abundant, and may raise several broods in a year. Most owls can breed in their first year if conditions are suitable. Some species, such as the Spotted owl, do not attempt to breed at all in years during which prey are scarce.

Owls are not great nest-builders: most breed in holes in trees, rocks, or the ground, but some open-country owls line depressions in the ground and Burrowing owls can dig their own underground nest chambers (although they usually take over Prairie-dog lairs). Small species occupy old woodpecker holes. Large owls unable to find appropriate holes, and most woodland owls, readily occupy nest-boxes. Large owls unable to find suitably-sized natural fissures take over abandoned tree-nests of crows or birds of prey.

Incubation tends to be by the female alone, with the smaller male providing all the food from before egg-laying until the young – initially born blind, helpless, and covered in sparse grayish-white down – no longer need brooding or their prey torn up for them. This division of labor allows the female to accumulate fat reserves and remain on the nest even when the male finds hunting difficult, for example in wet weather. In many species, the larger female vigorously defends the young against intruders, including humans (some people have even lost eyes to them). Other species have threat displays, in which the female tries to make herself look larger and even more fearsome. To further reduce the chances of predation, the young of open-nesting owls grow faster than those that are reared in holes and often leave the nest before they are fully feathered.

Fledged young beg loudly for food and are often dependent upon their parents for several months before they disperse. Newly independent owls suffer a high mortality: over half of young Tawny owls die in their first year, many of starvation, but once settled they can expect to live for at least four or five years, and some have survived for

more than 15. Larger species probably live even longer – a captive Eagle owl survived for 68 years.

Most owls are territorial and non-migratory, especially those living in the tropics or woodland. Here, pairs often spend all their lives in strictly defended territories, switching to alternative prey if one kind becomes unavailable; the populations of such species remain stable over long periods. Northern owls and those of open country that feed mainly on rodents have a narrower range of quarry available to them. They usually defend territories in the breeding season, but their populations tend to fluctuate in parallel with those of their prey. A few owls undertake regular north–south migrations, like the Common scops owl, which exploits the summer flush of insects in southern Europe; the Short-eared owl is nomadic, settling wherever prey is temporarily abundant. In mountainous regions, some species may have local vertical migrations – that is, individuals may move from a mountain into a nearby valley during winter storms and periods of deep snow.

Owls that are territorial throughout the year live in pairs but forage alone so as not to interfere with each other's hunting. Non-territorial owls usually live alone outside the breeding season, except some owls of open country that congregate in areas where prey is plentiful; these often roost communally but disperse at dusk to hunt alone. Only the Burrowing owl is sometimes colonial.

Breeding territories tend to be smaller where more prey is available. In Britain, Tawny owls defend 12–20ha (30–50 acres) in open deciduous woodland where small rodents are abundant, but

Left Deserting its breeding grounds in Alaska and northern Canada, a Snowy owl enjoys the sun on a beach in New York State. The birds regularly fly south for the winter, and are fairly often seen on beaches, where they scavenge for dead fish. Other favorite prey include rabbits and waterfowl.

Above A Long-eared owl spreads outs its wings to reveal the full panoply of its body and wing feathers in a threat display intended to make the bird look as large as possible and scare off intruders.

Below A family of Burrowing owls stand watchfully near their burrow. These small owls can dig their own nest-tunnels, but usually prefer to borrow the abandoned lairs of Prairie dogs or Pocket gophers.

over 40ha (100 acres) in more sterile conifer plantations. The huge Eagle owl takes bigger, less common prey and needs a correspondingly larger territory: their nests are usually spaced 4–5km (2.5–3mi) apart.

Different Species, Different Calls

VOCALIZATIONS

To communicate over large distances at night, owls have well-developed vocabularies, and they are much more vocal than day-hunting birds of prey. The familiar territorial hooting of many owls is equivalent to the song of other birds; it serves to warn off rival males and to attract a mate. The hoots of male Eagle owls can be heard 4km (2.5mi) away, and like many other owls the pair frequently answer one another in a duet, probably to maintain the pair bond between them. Day-hunting owls are generally less vocal than nocturnal species, but all owls have calls. Some of their sounds may be quite soft and heard only over short distances; for example, contact calls between members of a pair. In addition, owls can produce loud bill-snapping sounds when they are frightened or angry.

Many owls will answer imitations of their hoots, and territories can be mapped in this way. Local names for owls often reflect their distinctive calls: for example, the names "boobook" and "saw-whet" are, respectively, a phonetic rendering and a description suggesting the noise made when a saw is sharpened.

Because much of the communication between owls is vocal rather than visual, plumage pattern and coloration may be more important for camouflage than for species or individual recognition. Therefore, the plumage of owls tends to be conservative. Closely related species of owls may be very similar in appearance, and their major divergence may occur as different cadences in their hoots or different notes in their songs. In fact, as field ornithologists have become increasingly familiar with tropical species such as the pygmy owls, hawk owls, and screech owls, they have learned that if owls in different regions have different calls, it is frequently the case that they represent different species. Thus, in the period from 1980 to 2000, the number of generally recognized species of owls has grown from 123 to 189, almost entirely as a result of scientists' greater knowledge of owl calls and their importance. Even in the United States, where birds have been intensively studied for over a century, it was only in 1983 that the Eastern and Western screech-owls were recognized as two species, largely on the basis of their divergent songs and calls.

Above *The Saw-whet owl, so called because its cry supposedly resembles the noise of a saw being sharpened, is a small nocturnal species widely distributed across northern and western America.*

Below *The splendid "eyebrows" of the Crested owl* (Lophostrix cristata), *a Middle American forest species, are actually extended versions of the white ear tufts found in many of the Strigidae.*

Loggers Versus Owls

CONSERVATION AND ENVIRONMENT

Populations of owls face a multitude of threats, including the continuing use of pesticides and widespread loss of habitat. In addition, raptorial birds such as hawks and owls have often been thought of as a danger to domestic fowl or as competitors with humans for game. Sometimes this has resulted in persecution; for instance, the Eagle owl has been eliminated from densely populated regions of Europe. However, active persecution was principally a problem of the past. Today an enlightened public, at least in the developed world, generally recognizes that many raptorial birds perform a beneficial function by eating rats, mice, and other granivorous rodents.

Probably the greatest threat to owls is the destruction and fragmentation of their habitats. Unfortunately, the fact that owls are predators means that they exist at reduced densities compared with populations of other birds, such as songbirds, that are lower in the food chain. In order to secure a sufficient supply of prey, many owls require large individual foraging territories. For large owls, these home ranges can stretch across ten or more square kilometers (over 4 square miles). Consequently, when natural habitat is developed for agriculture or other purposes, owls and hawks may be disproportionately affected, and the consequences are particularly severe for those owls with limited ranges or specialized ecological requirements.

A number of owls have very restricted ranges, especially species endemic to tropical islands, such as several of the scops, hawk, and barn owls. For example, the Siau scops owl from the small Indonesian island of Siau, north of Sulawesi, has not been reliably recorded for over a century, and is now listed as Critically Endangered, although it may well be extinct. In addition, several continental species are also known only from tiny areas; the Sokoke scops owl is restricted to very limited areas of coastal forest in East Africa, while the Forest owlet currently is known from only a single river valley in west central India. Any substantial habitat change would pose a grave danger to these species.

Even for species with relatively large ranges, there may be conservation issues as a result of the owls' specialized ecological needs. The Spotted owl of the old-growth forests of the west coast of the United States depends for its food supply on woodrats and flying squirrels; these rodents do not become abundant until a tract of forest is more than 100 years old and has developed a distinct understory of shrubs and ground vegetation. Thus, the continued existence of Spotted owl populations is not compatible with modern forestry practices that involve clear-cut logging of forests at frequent intervals. This dilemma has led to conflict between the goals of environmentalists and the timber industry.

Above *These Sokoke scops owls are two of only about 2,500 endangered survivors. The East African species is threatened by habitat clearance and the illegal felling of the large forest trees in which it nests.*

Selected Owl Species

Species include: Saw-whet owl (*Aegolius acadicus*), Tengmalm's owl (*A. funereus*), Long-eared owl (*Asio otus*), Short-eared owl (*A. flammeus*), Burrowing owl (*Speotyto cunicularia*), Forest owlet (*Athene blewitti*), Little owl (*A. noctua*), Eagle owl (*Bubo bubo*), Great horned owl (*B. virginianus*), Least pygmy owl (*Glaucidium minutissimum*), Pygmy owl (*G. passerinum*), Brown fish owl (*Ketupa zeylonensis*), Elf owl (*Micrathene whitneyi*), Brown hawk owl (*Ninox scutulata*), Christmas Island hawk owl (*N. natalis*), Southern boobook owl (*N. boobook*), Snowy owl (*Nyctea scandiaca*), Eastern screech-owl (*Otus asio*), Common scops owl (*O. scops*), Siau scops owl (*O. siaoensis*), Sokoke scops owl (*O. ireneae*), Western screech-owl (*O. kennicottii*), Pel's fishing owl (*Scotopelia peli*), Barred owl (*Strix varia*), Great grey owl (*S. nebulosa*), Spotted owl (*S. occidentalis*), Tawny owl (*S. aluco*), Northern hawk owl (*Surnia ulula*).

Introduction of exotic species also can threaten owl populations. The Christmas Island hawk owl is known only from a single island in the Indian Ocean. This species is considered Critically Endangered because the population expansion of an introduced ant is changing the ecology of the small island and eliminating most of the owl's prey. The Laughing owl of New Zealand is probably extinct; it declined following the introduction of European stoats and weasels that destroyed its nests and competed with it for food.

The International Union for Conservation of Nature (IUCN) and Birdlife International list the status of 23 species of typical owls as cause for concern. Seven of these species are Critically Endangered, six are Endangered, and ten are Vulnerable. All of them have small populations combined with either a limited range (frequently a single island) or extensive habitat destruction. Twelve of the 23 species occur in Southeast Asia or the surrounding Australasian region. Nine occur in Africa or on adjacent islands, and two in South America. At the present time, there is no globally threatened species of owl in Europe or North America. GFB/GH

Barn and Bay Owls

The barn and bay owls are distinguished from the typical owls (family Strigidae) by several physical characteristics: they have heart-shaped, rather than round, faces; their middle and inner toes are of equal length (the inner is shorter in strigids); their middle claws are serrated; and their wishbones are fused to the breastbone.

Barn owls rely primarily on their extremely acute hearing, rather than their vision, to locate prey. Laboratory experiments in conditions of total darkness have demonstrated conclusively that the owls responded to sounds made by mice as they moved in leaf litter.

Widespread Predators

DISTRIBUTION/CONSERVATION AND ENVIRONMENT

The rodent-hunting Barn owl of open country is one of the most widely distributed of all birds, being found on every inhabited continent and many remote islands. The other 13 barn owls occur in Africa, Southeast Asia, and Australasia, as well as on islands in the Caribbean and the Indian Ocean. They reach their greatest diversity in Australia, which is home to five species. The little-known Common bay owl is found in Asian forests from India to Java and Borneo.

The Itombwe owl is known only from a single specimen collected in the Rift (or Itombwe) Mountains in the far east of the Democratic Republic of Congo in 1951 and a bird that was caught and released in the same region in 1996. Based on the appearance of the museum specimen and photographs of the 1996 bird, it is probable that the Itombwe owl will eventually be found to be related to the barn owls rather than to the Common bay owl. The extremely limited range of this bird and continuing human encroachment on its habitat have led to it being classified as Endangered. The other endangered African barn owl, the Madagascar grass owl, which inhabits the northeast of the island, is under threat from deforestation for slash-and-burn agriculture.

The two other threatened tytonid species, the Sula barn owl and the Minahasa masked owl, are from the Indonesian islands of Taliabu and Sulawesi, respectively. Here, they are imperiled by habitat loss and degradation as a result of extensive lowland rainforest clearance by commercial logging concerns. This threat is set to intensify as most remaining forest is under timber concession.

Above *The Common bay owl* (Phodilus badius) *has a distinctive, angular facial disk with dark, vertical markings. As with all forest owls, it is susceptible to habitat loss from deforestation.*

Right *As its diet consists mainly of agricultural pests such as mice and rats, the Barn owl is probably the most economically beneficial pest control product to farmers the world over.*

Left *Highly nocturnal, the Barn owl hunts more by sound than by sight, enabling it to accurately locate its prey in total darkness. The design of its wings also makes it a deadly silent predator; it will fly as far as three and a half miles in search of food.*

Man's Friend?

SOCIAL BEHAVIOR

The Barn owl's nocturnal lifestyle, ghostly white appearance, and association with ruins or churches (where it likes to nest) have earned it a place in the folklore of many cultures. However, as its English name implies, it is most commonly associated with agriculture. Wherever crops of grain are grown and stored, there are inevitably large populations of rats and mice, which attract Barn owls.

Barn owls will readily use artificial nesting boxes that have been specially provided for them, but are also opportunistic in their choice of nest site. For example, in the United States, they often nest in boxes that have been placed near water for the use of Wood ducks (*Aix sponsa*). In the Netherlands, farmers actively encourage owls by installing special "owl doors" to allow them easy access to their buildings, and by providing food in hard weather, which otherwise causes heavy losses. Unfortunately, the Barn owl's close association with agriculture was almost its undoing in Western

FACTFILE

BARN AND BAY OWLS

Order: Strigiformes

Family: Tytonidae

16 species in 2 genera. Species include: **Barn owl** (*Tyto alba*), **Madagascar grass** or **Madagascar red owl** (*T. soumagnei*), **Manus masked owl** (*T. manusi*), **Minahasa masked owl** (*T. inexspectata*), **Sula barn owl** (*T. nigrobrunnea*), **Common** or **Oriental bay owl** (*Phodilus badius*), **Itombwe** or **African bay owl** (*P. prigoginei*).

DISTRIBUTION Europe except far north; Indian subcontinent, SE Asia, Africa, N America to Canadian border, S America, Australia, many island groups.

HABITAT Open areas, including arid and semi-arid lands, farmland, scattered woodland, forests.

SIZE Length 23–53cm (9–21in); weight 180–1,280g (6.3–45.5oz). Females usually slightly larger than males.

PLUMAGE Orange-buff to blackish brown above, white, rufous or blackish-brown below.

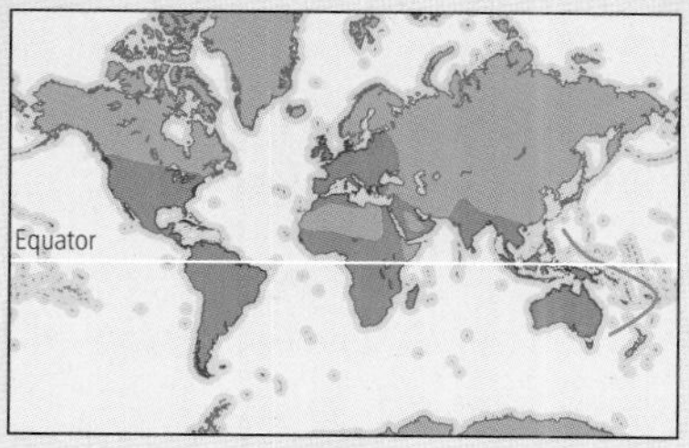

VOICE Shrill hissing, screeching, or whistling; also audibly snaps bill.

NEST In barns and other infrequently used buildings, nest boxes, holes in banks, rocks, or trees, or on the ground.

EGGS Usually 2–9, sometimes up to 11, white, elliptical: most weigh 17–42.5g (0.65–1.5oz). Incubation 27–34 days; nestling period 49–64 days.

DIET Small mammals (up to size of rabbits); birds, fish, frogs, lizards, and large insects.

CONSERVATION STATUS 3 species – the Itombwe owl, the Madagascar grass owl, and the Sula barn owl – are Endangered and the Minahasa masked owl is Vulnerable.

Europe, since they suffered far more from poisoning by pesticides and herbicides than other owls. In Malaysia, Barn owls and chemicals are actually used together to control the plagues of rats that cause severe damage in oil-palm plantations. Formerly rare, Barn owls invaded the plantations after nesting boxes were erected for them to breed in. The owls can raise several families per year and often congregate in flocks of up to 40 birds. Each owl family eats about 1,300 rats per year, which can slow the recovery of rodent populations that have first been reduced by poisoning.

Not every pest-control scheme with Barn owls has been a success. In the 1950s they were introduced to the Seychelles, again to control rats. Unfortunately, the owls found the native birds easier to catch and in 12 years they eliminated the White tern (*Gygis alba*) from two islands. GH/GFB

SPECIAL FEATURE

A FACE FOR THE NIGHT

Why an owl looks like an owl

EVERYONE RECOGNIZES AN OWL; THEY HAVE A distinctive appearance not shared with other kinds of birds. This uniqueness resides in their face: a round disk punctuated by large, forward-facing eyes. In fact, this feature is an adaptation to their nocturnal way of life.

Owls are generalized predators – their specialization lies not in feeding on a particular type of prey but in catching it in darkness. The modifications that enable owls to do so create their characteristic appearance.

Owls have especially acute hearing and vision, and need oversized skulls to accommodate ear openings and eyes far larger than those of other birds; the largest owls, weighing just 4kg (8.8lb), have eyes the same size as those of a full-grown human.

What, then, is the advantage of these large, frontally placed eyes? Large eyes can have large pupils to allow more light to fall on the retina (the light-sensitive layer at the back of the eye). A Tawny owl's eye has 100 times the light-gathering power of a pigeon's and produces a large retinal image to provide the visual acuity necessary to discriminate potential prey. Owls have tubular (rather than spherical) eyes, placed frontally in order to accommodate the huge lens and cornea. The cause of the tubular shape is a ring of bone around the eye called the scleral ring. Unfortunately, tubular eyes have a reduced field of view and are virtually immobile, giving owls a visual field of only 110° compared with a man's 180° and a pigeon's 340°. To overcome this limitation, owls have remarkably flexible necks, enabling them to invert their heads as well as to look directly behind! Frontally-placed eyes can also provide binocular vision, in which both eyes view the same area from different aspects, allowing better judgement of distance.

Although owls see much better at night than birds that are active by day, the popular belief that their eyes are vastly superior to man's in the dark but function poorly in bright light is not correct. The Tawny owl has color vision, sees in daylight as well as a pigeon, and has eyes only some two to three times more sensitive than man's in the dark. In fact owls that live at high latitudes, such as the Snowy owl, need good diurnal vision, as they hunt in permanent daylight for almost half the year.

Owls can hunt successfully at night because their visual sensitivity is allied to exceptional hearing. Owls are especially sensitive to sounds with a high-frequency component, such as the rustling of dry leaves. Experiments in total darkness have shown that some species can even locate and capture small rodents just from the noise they make in moving across the floor of a large cage.

The characteristic facial disks of owls are part of this specialized hearing apparatus. The tightly-packed rows of stiff feathers that make up the rim of the disk reflect high-frequency sounds that are channeled by the mobile facial disks into the ears behind, in the same way that mammals use their large, fleshy external ears.

Owls have an exceptionally broad skull that helps in sound location – a noise from one side will be louder, and perceived fractionally sooner, in the ear nearest to it. Owls can locate sounds in the horizontal direction four times better than a cat, but an owl hunting from above needs to pinpoint sound in the vertical direction as well if it is to fix the prey's position exactly. Barn owls can manage this feat with an accuracy of 1–2° in both the horizontal and vertical directions – an astonishing achievement, as 1° equates to about the width of a little finger at arm's length! Owls achieve this aural acuity by moving their ear flaps to alter the size and shape of the ear openings to make reception different for the two ears. In some highly nocturnal owls (for example, Tengmalm's and Northern saw-whet owls) the ear openings themselves are placed asymmetrically on the skull.

To be able to hear their prey without frightening it off, owls are equipped to fly silently. From head to toe, they are covered with an enormous number of soft, downy feathers, which give them their characteristic rounded outline and make them look much larger than they really are. Long-eared owls have over 10,000 feathers.

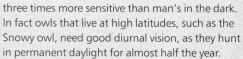

Above *Not only do owls have keen vision, they can also rotate their heads through 270°, as shown by this Ferruginous pygmy owl* (Glaucidium brasilianum).

Right *A full-frontal view of a Short-eared owl emphasizes the disklike facial mask as well as the huge, circular eyes, used to seek out prey.*

Below *The eyes and ears of owls are extremely sensitive.* **1** *The eye differs from a typical mammalian eye* **2** *in the size of the pupil and lens (the owl's are larger) and in having the retina close to and equidistantly spaced from the lens. The retina has a high proportion of rods (which only detect black and white) and relatively few cones (enabling color vision), permitting it to function at very low light intensities. The pecten is a structure thought to provide nutrients to the eyeball.* **3** *At 110°, the owl's field of vision is not large, but provides good stereoscopic coverage over an angle of 70°; the intense tunnel vision that results is an adaptation to allow the owl to judge the distance of prey with lethal precision.*

1

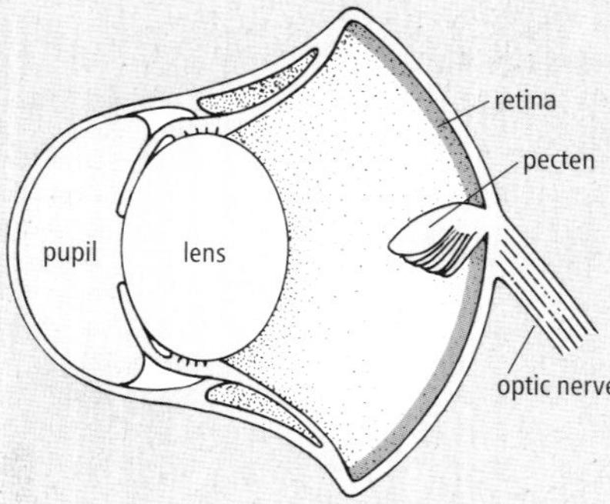

2

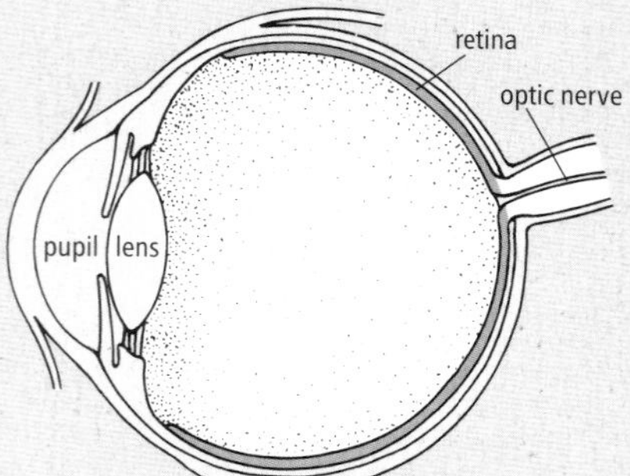

3

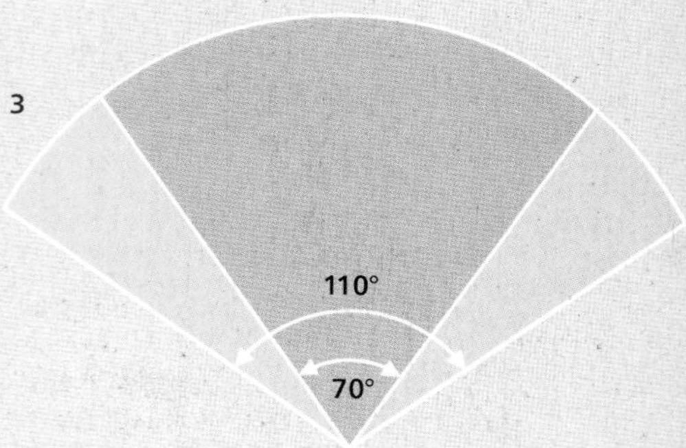

Most owls' wing feathers lack the hard sheen of those of other birds, and have soft fringes with fluted leading edges to ease the airflow over them. These adaptations, coupled with extreme lightness in relation to their wing area, enable owls to hunt stealthily. Fishing owls lack these adaptations, since their prey cannot hear them approaching.

Like hawks, owls have relatively short legs; they are feathered to the toes, probably to aid thermo-regulation, but possibly also to protect against bites from prey. Exceptions are species like the Burrowing owl, which has noticeably longer legs, and the fishing owls, which have bare legs and osprey-like feet with spiny soles to grasp slippery fish. One peculiarity of owls is a reversible outer toe that can point forward or backward to increase the "catching area" of the feet and improve grip. GFB/GH

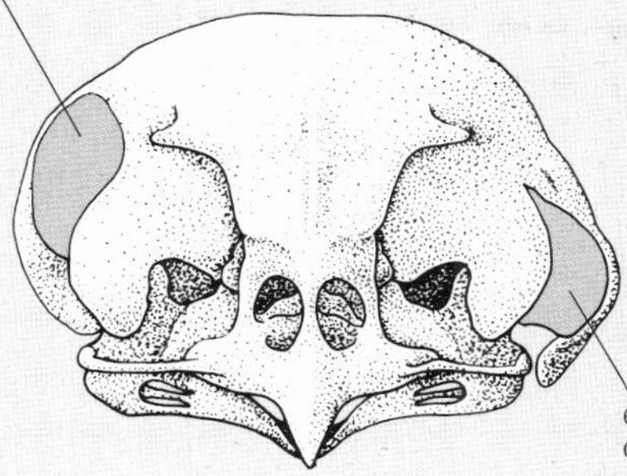

Right *Some owls' ears are set asymmetrically, to aid in locating sounds. The ear openings are large, vertical slits running almost the whole depth of the skull and hidden beneath feathered flaps of skin.*

Nightjars

Best known for their strange songs and calls, nightjars are nocturnal aerial insectivores. In their daytime roosts, their immobility and beautifully camouflaged plumage keep them well hidden. Unsurprisingly, many superstitions have grown up around these elusive birds; for example, the Satanic eared nightjar (Eurostopodus diabolicus) *of northern Sulawesi derives its name from the belief that it plucked out people's eyes!*

The European nightjar's curious alternative English name "goatsucker" (which also exists in other languages – for example, Italian: *succiacapre*) has its origins in the story that the birds sucked milk at night from the teats of goats, a superstition that goes back at least to Aristotle (384–322 BC). The myth may have its roots in the unusually wide gape of nightjars, together with their habit of foraging for insects around livestock, including goats suckling their young.

Flying the Flag

FORM AND FUNCTION

Most nightjars look like big, soft moths, dressed as they are in variegated patterns of brown, buff, cinnamon, and gray. More conspicuous patches of white or black and white are mainly hidden in the folded wings and tail or on the upper throat, and revealed when the male birds display. The mouth is very wide, allowing them to swallow large moths whole. Their wings are long and tapered, the tail typically long and rather broad, but shorter or notched in some species.

Typical nightjars of the subfamily Caprimulginae have strong rictal bristles surrounding the base of the bill, which may function partly as a net to direct prey into the bird's open mouth and partly to protect the eyes from hard prey such as beetles. The nighthawks of the subfamily Chordeilinae lack these bristles.

Adult males of a few tropical and subtropical species have extraordinarily long wing or tail feathers that make their courtship displays very conspicuous. In the African Pennant-winged and Standard-winged nightjars, a single inner secondary is greatly elongated to form a pennant or flag, while several South American species, such as the Long-trained nightjar, have greatly lengthened tail feathers. Evidence is accumulating that these highly ornamented species differ from other nightjars in being polygamous. The specialized inner secondaries of males of the two African species break off or are broken off shortly after the breeding season.

Above *Rictal bristles on a Long-tailed nightjar. The role of these bristles has been much debated. One theory is that they act like whiskers and detect prey, so the bird knows when to snap its bill shut on insects.*

Life in the City

DISTRIBUTION PATTERNS

As largely insectivorous birds, nightjars live mainly in tropical climates. Consequently, the family has relatively few species that migrate to temperate zones for the warm seasons. The Common nighthawk, the species that breeds farthest north in North America, migrates to spend the winter in South America, where it ranges southward to northern Argentina. The European nightjar is the only nightjar that breeds across most of Europe and northern Asia; it too is a long-distance migrant, spending the winter in Africa from the tropics southward to South Africa. Some nightjars within the tropics make shorter migrations to avoid the worst of insect shortages during the dry season.

The nighthawk subfamily Chordeilinae has long been regarded as being restricted to the New World, but it has recently been suggested that the tropical African genus *Veles* (1 species) and the Asian and Australian *Eurostopodus* (7 species) are also nighthawks. The typical nightjar subfamily, the Caprimulginae, has representatives in all the temperate and tropical continents, with no fewer than 55 species in the largest genus, *Caprimulgus*.

Several nightjar species have adapted to city life, nesting on flat roofs of buildings and catching insects around street lights or high over the city. The Common nighthawk has become a familiar urban dweller in North America, nesting on gravel-covered roof-tops, but similar behavior is exhibited by the Band-winged nightjar, which has colonized Rio de Janeiro since about 1955, and the Savanna nightjar, in the Indonesian cities of Jakarta and Surabaya.

FACTFILE

NIGHTJARS

Order: Caprimulgiformes

Family: Caprimulgidae

89 species in 15 genera. Species include: **Band-winged nightjar** (*Caprimulgus longirostris*), **Chuck-will's-widow** (*C. carolinensis*), **Fiery-necked nightjar** (*C. pectoralis*), **European nightjar** (*C. europaeus*), **Itombwe nightjar** (*C. prigoginei*), **Little nightjar** (*C. parvulus*), **Puerto Rican nightjar** (*C. noctitherus*), **Rufous nightjar** (*C. rufus*), **Savanna nightjar** (*C. affinis*), **Scrub nightjar** (*C. anthonyi*), **Sooty nightjar** (*C. saturatus*), **Vaurie's nightjar** (*C. centralasicus*), **Whip-poor-will** (*C. vociferus*), **Common nighthawk** (*Chordeiles minor*), **Short-tailed nighthawk** (*Lurocalis semitorquatus*), **Standard-winged nightjar** (*Macrodipteryx longipennis*), **Pennant-winged nightjar** (*M. vexillarius*), **Long-trained nightjar** (*Macropsalis forcipata*), **Common poorwill** (*Phalaenoptilus nuttallii*), **Jamaican poorwill** (*Siphonorhis americana*), **Brown nightjar** (*Veles binotatus*).

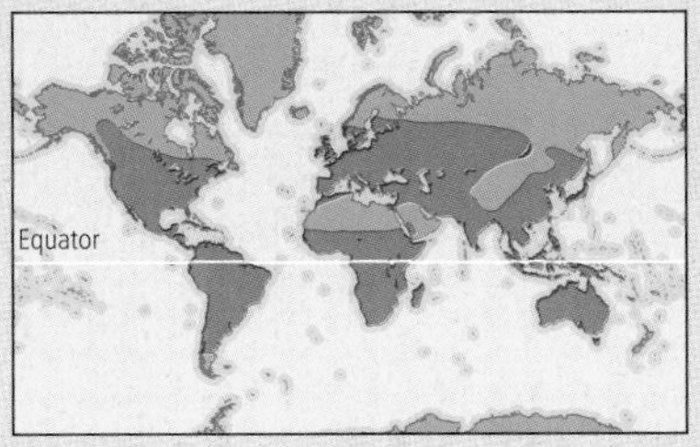

DISTRIBUTION Throughout the tropical and temperate world, except New Zealand, southernmost S America, and most oceanic islands.

HABITAT Mostly forest edge to savanna and desert, a few forest dwellers; crepuscular and nocturnal.

SIZE Length 15–40cm (6–16in); weight 25–120g (0.9–4oz).

PLUMAGE Concealment patterns of browns, grays, and black with patches of white on tail, wings, and head; females often differ from males in having less white on the wings and tail. Some tropical forms with elongate wing or tail feathers.

VOICE Loud, repetitive churring, trilling, or whistled male song; other calls and sounds made with wings.

NEST Usually none; eggs are laid on the bare ground.

EGGS 1–2, white or buff-colored, often with blotches. Incubation period 16–22 days; nestling period 16–30 days.

DIET Insects

CONSERVATION STATUS Its declining, fragmented range has led to the Puerto Rican nightjar being classified as Critically Endangered. The Jamaican poorwill also has Critically Endangered status; however, it has not been sighted since 1860 and may already be extinct. Two further nightjar species are classed as Endangered, and 3 as Vulnerable.

Hawking and Sallying

DIET

Nightjars have two main feeding techniques. Many of them sally out from perches to catch insects in much the same manner as that used by flycatchers. Others hawk insects in sustained flight much like swallows. Some species use both sallying and hawking techniques at different times, but most are more specialized. Nighthawks, with their very long wings and

Left *Common nighthawks frequently roost in suburban settings in North America. This species' name is a misnomer: early European settlers mistook it for a hawk; moreover, it hunts not at night, but at dusk.*

Right *A Rufous-cheeked nightjar* (Caprimulgus rufigena) *sunning itself. After breeding across a wide range in southern and southwestern Africa, these birds fly north to overwinter in Nigeria and Cameroon.*

strong flight, are supremely adapted to hawking insects, whereas others, such as the rather round-winged Fiery-necked nightjar, specialize in sallying. Whichever technique is used, flying insects form the main prey, often with a predominance of beetles and moths, but with many other types taken, including flies, bugs, crickets, mayflies, lacewings, termites, and flying ants.

Suggestions that nightjars use echolocation to catch nocturnal insects have not been confirmed by detailed study. Instead, there is much evidence that they hunt by sight and that this is the reason why on dark nights most of their feeding occurs around dusk and dawn, with a lull in the middle of the night when it is too dark for them to see flying prey. Old notions that nightjars fly around with their bills open like a net to trawl for insects also receive little support from modern studies, which show that individual insects are pursued, although "trawling" may sometimes be employed in dense swarms of tiny mosquitos or termites.

There are several records of the Chuck-will's-widow of the southern United States, the largest of all the nightjars, capturing small birds such as wood warblers, especially when it is on migration. This species has also been recorded catching tree frogs on the ground.

Solicitous Parents

BREEDING BIOLOGY

Breeding seasons are timed to coincide with seasonal peaks of insect abundance, during late spring and summer at temperate latitudes, often around the end of a wet season in the tropics. Some species have only one brood each year, many are double-brooded, and most, if not all, will lay replacement clutches if the first attempt fails. No nests are built, the clutch of one or two eggs being laid on the ground in most species. Exceptionally, nests of a few species are situated above the ground, on a horizontal tree branch in the Short-tailed nighthawk, and perhaps on the midrib of a palm frond in the Brown nightjar.

The newly hatched young have beautifully patterned soft down that helps hide them from predators, but eggs and young are rarely left unattended for long and small young are continuously brooded. The parents feed the young on regurgitated insects. The young birds are active and able to walk within a few hours of hatching; their parents commonly encourage them to walk away from the nest site to a safer place several meters distant. Erroneous accounts of nightjar parents carrying their young in flight are probably based on occasional sightings of a small chick becoming entangled in the body plumage of a parent as it flies away. Many nightjars have elaborate distraction displays to draw potential predators away from their eggs and young, often of the

Above European nightjar camouflaged with its chicks. The destruction of open heathland – this species' favored habitat – for agriculture and development has brought a serious decline in its numbers.

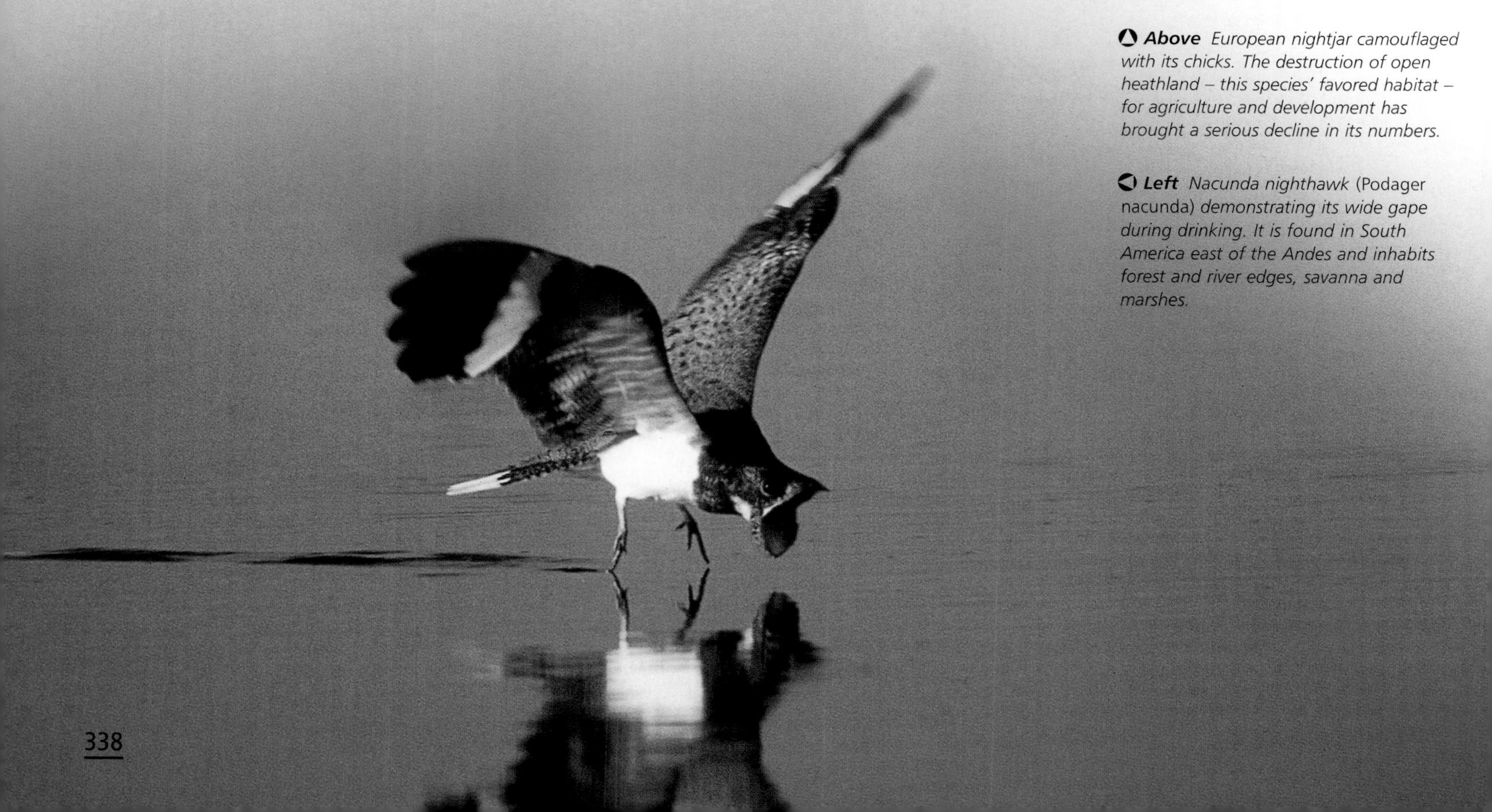

Left Nacunda nighthawk (Podager nacunda) demonstrating its wide gape during drinking. It is found in South America east of the Andes and inhabits forest and river edges, savanna and marshes.

"disablement-lure" type, in which the parent flutters along the ground as if it were injured.

The sexes take different shares in incubation and care of the young in different species, often with a division of labor, as for example in the Fiery-necked nightjar, where the more cryptically-colored female carries out most of the daytime incubation, the male incubates at night, and both sexes feed the young. In the European nightjar the male hardly incubates the eggs, but does take full charge of the young of the first brood if the female lays a second clutch. In contrast, the males take no part in parental duties in the polygamous Pennant-winged and Standard-winged nightjars.

The Common poorwill is the only species of bird known to become torpid for long periods in winter. The Hopi Indians called the poorwill *hoechko* – "the sleeping one." Their folk wisdom about the bird was confirmed by the discovery of a hibernating poorwill in a rock crevice in southern California in 1947. It is now established that the species may become immobile for months on end, with its body temperature at a remarkably low level of around 18°C (64.5°F), allowing very sparing energy use during a season when there is no insect food.

Fragmentary Evidence

CONSERVATION AND ENVIRONMENT

Seven nightjar species are globally threatened. Of these, the Jamaican poorwill may already be extinct, as it has not been spotted since 1860. Several of the other threatened species are very poorly known birds, with Vaurie's nightjar known from a single specimen from western China, the Nechisar nightjar of southern Ethiopia known from a single wing salvaged from a roadkilled bird in 1990, and the Itombwe nightjar known from a single specimen from mountain forest in the eastern Democratic Republic of Congo. The Puerto Rican nightjar was rediscovered in 1961 and is Critically Endangered because of the small world population (712 singing birds in 1989–92) and continuing loss of its forest habitat. DH/JWH

Above *Nightjar courtship. Some nightjars have dramatic courtship displays.* ***1*** *On its breeding grounds the male Common nighthawk dives from a great height, then swoops upward near the female, making a booming sound by the rush of air over the soft inner vanes of certain wing feathers.* ***2*** *The male Standard-winged nightjar slowly circles his mate in fluttering flight, the wings stiff and vibrating. This movement causes an updraft that elevates the extended inner primaries into flapping pennants.*

WHIP-POOR-WILL: NIGHTJAR MUSIC

In many birds with stereotyped vocalizations, song is an inherited trait that may offer clues to relationships between species. Nightjars communicate mostly by voice, and males of a given population all sing basically the same songs.

The whip-poor-will is a widespread species from southern Canada to Central America. The whip-poor-will of the eastern USA (*C. v. vociferus*) has a clear, slightly warbled song **1**, while that of the whip-poor-wills of the southwestern USA and Mexico (*C. v. arizonae*) has a guttural quality **2**, yet both subspecies clearly utter "whip-poor-will."

A little-known species, the Sooty nightjar of mountainous Costa Rica and Panama, has a song that is strikingly similar to that of the western whip-poor-wills, being higher-pitched but containing the same syllables **3**. Based on this close resemblance and its general physical similarity to the whip-poor-will, the Sooty nightjar should perhaps be regarded as a Central American representative of the whip-poor-will. The Puerto Rican nightjar is also regarded as a relative of the whip-poor-will, although its "will-will-will-will" song is rather different **4**.

The Little nightjar **5** is widespread in tropical South America, the birds in Venezuela singing a complex series of clear, resonant "tick-tock" notes **6**. Birds of very similar appearance in eastern Peru were previously regarded as a subspecies of the Little nightjar, but these sing a charmingly guttural song (rather froglike), and are now treated as a separate species, the Scrub nightjar.

Based on physical appearances, the Chuck-will's-widow of the southern USA **7** and the Rufous nightjar of northern South America **8** are each other's closest relative; voice evidence strongly supports their kinship.

The song of the European nightjar and of some African species is very different from that of any American nightjar, consisting of a prolonged churring reminiscent of a loud cricket or a distant engine. Indeed, the English name "nightjar" derives from the "jarring" nature of this species' song. As with many other nightjars, the European nightjar devotes large amounts of energy to singing in the early part of the breeding season, apparently because the song is a key element in advertising and defending the breeding territory and perhaps also in attracting a mate.

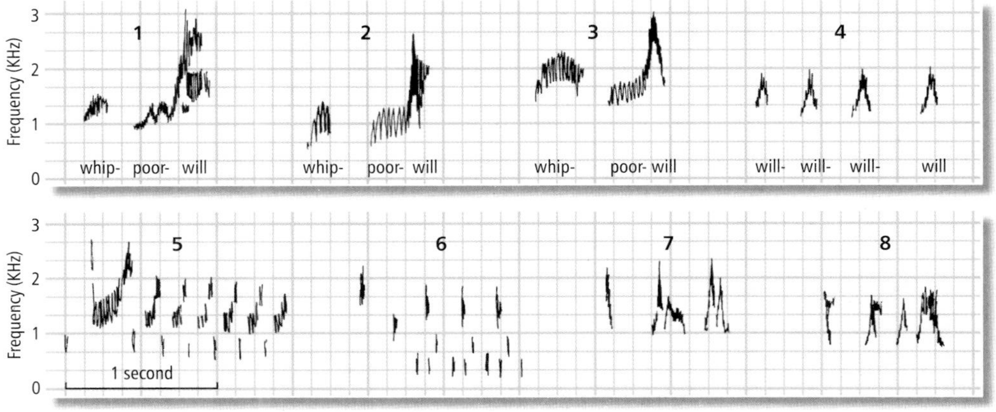

SPECIAL FEATURE

NIGHTJARS AND THE LUNAR CYCLE

Hunting by the light of the moon

ALMOST ALL THE ACTIVITY OF NIGHTJARS OCCURS between dusk and dawn, and so is difficult to study. As well as having very large eyes to improve nocturnal vision, the birds have a special structure – the tapetum lucidum – that consists of microscopic oil droplets that serve to enhance the light-catching power of the retina. The presence of the tapetum accounts for the bright eyeshine seen when nightjars are observed by torchlight.

The detection of flying insect prey by nightjars is now known to depend mainly or entirely on vision. It has been demonstrated that they cannot echolocate and do not have other special adaptations of hearing, or of their other senses. Nor can they utilize infrared or ultraviolet light. Because some light is essential for any vision to be possible, the question arises of how little light a nightjar needs in order to see and catch flying insects.

Certainly, it would appear that, on its own, starlight in a moonless night is insufficient for hunting. Naturalists long ago noticed that much of the singing and calling by nightjars occurs around dusk and again before dawn, with far less activity in the middle of the night. At the time it was not clear, however, whether feeding activity followed a similar pattern, with a lull in the middle of the night, especially since observations in bright moonlight revealed that, under such conditions, feeding often continued long after dusk.

Intensive research of the activity patterns of feeding nightjars had to await the advent of miniature radio transmitters, which could be attached to the birds. Studies using this new technology quickly confirmed that Common nighthawks ("hawking" feeders), Common poorwills ("sallying" feeders), and European nightjars (which both hawk and sally) all cease feeding in the middle of dark nights. Indeed, the nighthawks only feed for quite short periods around dawn and dusk, while the other two species feed mainly before or after the dark part of the night. On moonlit nights, however, the pattern can be very different, with both Common poorwills and European nightjars feeding for much longer.

Studies of the prey of the Common poorwill show that the bird mainly takes large insects measuring 5mm (2in) or more across, and that (as with other nightjars) it detects the insects visually as silhouettes against the sky. It seems that the middle of a dark night offers insufficient light for the bird

Right *A male Standard-winged nightjar* (Macrodipteryx longipennis) *prepares to take a moth as it hunts at night during the breeding season. This African species grows the striking "standards" on its inner secondaries as part of its breeding plumage, and molts them when it travels north to overwinter in the Sahara.*

Left *Eyeshine caused by reflected light from the tapetum lucidum – an adaptation to the retina that improves nocturnal vision – betrays the presence of a European nightjar* (Caprimulgus europaeus), *as it waits to sally forth from a dead branch after insect prey.*

Below *It has been established that the breeding biology of the Fiery-necked nightjar* (Caprimulgus pectoralis) *of southern Africa is related to the lunar cycle. Chicks of this species tend to hatch in synchrony with a waxing moon.*

to see its prey well enough to make sallying flights, despite the fact that it chooses large insects and possesses a tapetum and other visual adaptations.

The time available for nightjars to feed on moonless nights is therefore surprisingly short, probably less than one hour around dusk and a similar period around dawn. Flying insects are much more abundant at night than during the daytime, so even the restricted access to these hordes of insects around dusk and dawn must have been sufficient to allow nightjar feeding habits to evolve. Given their brief feeding opportunities, then, it is no surprise that nightjars commonly cram their stomachs with food in a frenzy of activity at dusk and dawn. On cloudless nights, however, the period around full moon allows far more prolonged opportunities for nocturnal feeding.

In 1985–86, studies of the whip-poor-will in North America and of the Fiery-necked nightjar in Zimbabwe revealed that the precise timing of laying in these species was related to the phases of the moon. The overall timing of the breeding seasons of both species is, of course, linked to the time of greatest insect abundance, caused by seasonal climate change, so that both breed in late spring to early summer. However, the precise day of laying was found to fall much more often in the phase following a full moon than at other phases around the new moon to first quarter.

There are several advantages for nightjars in timing their breeding to coincide with the full moon and the better feeding opportunities associated with it. Thus, the laying female is able to collect more food and so is better nourished when forming eggs. A month later – again coinciding with the full moon – there will be better opportunities to find food for the young chicks. Another month later, the fledged young will have a longer period in which to begin catching their own food. Some confirmation of such advantages has been obtained for the whip-poor-will, since adults have been found to feed their chicks more frequently on moonlit than on dark nights.

In contrast, studies of the European nightjar appeared to show no relationship between the timing of laying and the phases of the moon, probably because of its need to begin breeding quickly in the short northern summer. However, study of second broods begun in June did show some correlation with the lunar cycle, again with most laying occurring around full moon. DH

Frogmouths

One of the most striking features of the frogmouths is their beautifully patterned brown or gray plumage, resembling tree bark. If alarmed during the day, they use this cryptic coloration to good effect, freezing and adopting a "broken branch" posture, with their bill pointed upward. Like the true nightjars, frogmouths are active at night.

The two genera of frogmouths are somewhat similar in structure and appearance, but biochemical studies suggest they are not closely related and may be better placed in separate subfamilies (Batrachostominae for the nine, mainly smaller, Asian species; Podarginae for the three, mainly larger, Australasian species).

Hawking after Insects

FORM, FUNCTION, AND DISTRIBUTION

Frogmouths are nocturnal birds with large heads, and large eyes that are commonly yellow, orange, or red. Walking is usually limited to a few steps, in accordance with the small size of their legs and feet. The base of the strong bill is partly covered by strong rictal bristles, and *Batrachostomus* commonly have other long facial bristles, the function of which may be to protect the eyes from contact with insect prey rather than any sensory role as "whiskers." Several of the species are rather variable in plumage coloration, a few having well-defined color morphs. Males are often duller in color than females, apparently because they need concealing coloration for daytime incubation and brooding, whereas females may be more active in nocturnal territory defense.

Frogmouths hunt mainly by catching prey on short flights from their perches, their rounded wings and tail giving them great maneuverability. Their hunting method has been described as shrikelike or rollerlike, the birds pouncing on prey on the ground or taking it from trunks or foliage. The smaller species are predominantly insectivorous, but the larger *Podargus* species, including the Tawny frogmouth, also take small vertebrates such as frogs and mice; their largest prey is taken and beaten against a stone before it is swallowed.

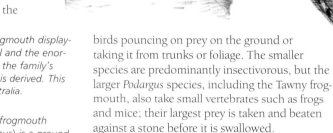

Batrachostomus occur in southern Asia from northern India and Sri Lanka to Vietnam and southward to Java, with the largest numbers of species in Borneo. Two *Podargus* species occur in New Guinea and northeastern Australia. The Tawny frogmouth is more widespread throughout Australia, and it often occurs in much more open woodland and scrub habitats than all other species of frogmouths. All species are non-migratory.

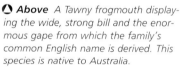

Above *A Tawny frogmouth displaying the wide, strong bill and the enormous gape from which the family's common English name is derived. This species is native to Australia.*

Left *The Philippine frogmouth* (Batrachostomus septimus) *is a ground feeder, preying on beetles, centipedes, caterpillars, and worms. It occurs in both a rufous and a gray form.*

Cup-shaped Nests

BREEDING BIOLOGY

Frogmouths differ from nightjars in building cup-shaped nests in trees or saplings. Those of *Batrachostomus* are tiny and lined with the birds' own down, while those of *Podargus* are bulkier and built of twigs. Clutches in *Batrachostomus* are of one or two white eggs, whereas the Tawny frogmouth sometimes lays as many as four eggs. Daytime incubation and brooding is carried out by the male. Young hatch with white down that is replaced by a darker, grayish down before juvenile plumage is acquired. Both sexes feed the young, on regurgitated food. The young frogmouths remain in the nest until they are able to fly. DH

FACTFILE

FROGMOUTHS

Order: Caprimulgiformes

Family: Podargidae

12 species in 2 genera. Species include: **Gould's frogmouth** (*Batrachostomus stellatus*), **Large frogmouth** (*B. auritus*), **Marbled frogmouth** (*Podargus ocellatus*), **Papuan frogmouth** (*P. papuensis*), **Tawny frogmouth** (*P. strigoides*).

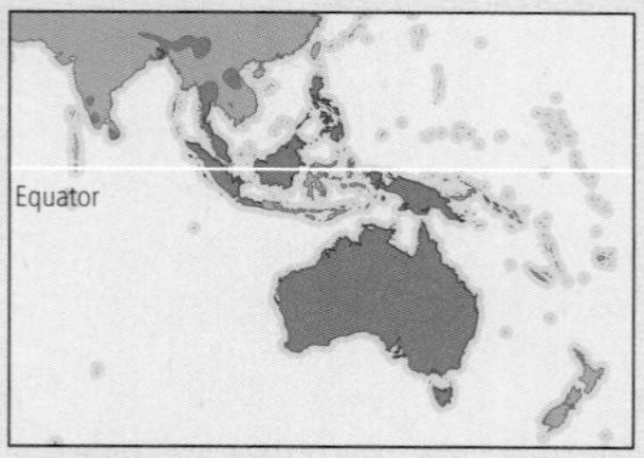

DISTRIBUTION Sri Lanka, SE Asia, Indonesia, Australia.

HABITAT Mostly tropical rain forest.

SIZE Length 19–60cm (7.5–24in).

PLUMAGE Brown to gray; most species have intricate color patterns.

VOICE Usually a low, repetitive booming in *Podargus*; varied short whistles, trills, or mews in *Batrachostomus*.

NEST In trees, a tiny neat cup in *Batrachostomus*, larger structure of twigs in *Podargus*.

EGGS 1–4, according to species, white; size 30x20 to 51x33mm (1.2x0.8 to 2x1.3in). In the Tawny frogmouth, incubation period about 30 days, nestling period about 30 days.

DIET Mainly insects, but *Podargus* is known to take a wide range of small animals, including small vertebrates such as frogs and mice.

CONSERVATION STATUS 5 *Batrachostomus* species are classified as Lower Risk: Near Threatened.

Owlet-nightjars

MOST SPECIES OF OWLET-NIGHTJARS ARE mysterious forest birds that roost in tree holes or among thick vines, so they are rarely seen by ornithologists. Most information is available for the widespread Australian species, whereas the nests and habits of some of the others are still unknown.

Above *Extensive fossil remains found in New Zealand show that this young Mountain owlet-nightjar had a closely allied ancestor in the Pleistocene to Recent, described as* Megaegotheles novaezealandiae.

As their name suggests, owlet-nightjars are small relatives of the nightjars that resemble owls, with large heads and large eyes that face forward. They have a rather small bill with long bristles at the base, rounded wings, a relatively long tail, and slender legs and toes. Coloration of the intricately marked plumage is variable in some species.

Short Sallies

DISTRIBUTION PATTERNS AND DIET

Most owlet-nightjar species are restricted to New Guinea, where some, such as the Barred and Spangled owlet-nightjars, live in lowland forests, while others, including the Large and Mountain owlet-nightjars, are restricted to montane forests. The Australian owlet-nightjar occurs throughout most of Australia and Tasmania, as well as in a small area of southern New Guinea. Unlike the other species, it occurs in open scrub country and among scattered trees, as well as in forests. Outlying island groups are occupied by the Moluccan owlet-nightjar and New Caledonian owlet-nightjar. The New Caledonian owlet-nightjar was known from a single specimen collected in 1880 and feared extinct until a sighting was reported in 1998; it is now classed as Critically Endangered.

Fossil records, like the 38 million year old Eocene deposit found in France, indicate that owlet-nightjars formerly had a much wider range. Fossil occurrences in New Zealand are only a few thousand years old, and are thought to represent a species that was flightless, or at best a weak flier.

Owlet-nightjars hunt mainly, if not entirely, by making short, "sallying" flights from perches to catch insects and other prey in flight, on the ground, or from tree trunks or foliage. Old accounts suggesting they catch prey in prolonged hawking flights are incorrect. The diet of the Australian owlet-nightjar consists of a wide range of insects, especially beetles, ants, grasshoppers, and crickets, along with spiders and millipedes. The only other species for which much information is available is the Mountain owlet-nightjar, which takes mainly beetles, with fewer moths, flies, crickets, and earthworms.

Thick Eggshells

BREEDING BIOLOGY

Information is very sparse except for the Australian owlet-nightjar, which usually nests in tree holes, laying a clutch of 1–5 white eggs on a pad of dead leaves. Incubation, of about 26 days, is mainly by females. The young hatch with white down, which is soon replaced by longer gray down before juvenile feathers grow. Both sexes feed the young, which fledge after 21–29 days.

Barred and Mountain owlet-nightjars are also known to lay unmarked white eggs in tree holes or hollows in dead tree stubs. Two nests of the Mountain owlet-nightjar were found to contain only a single egg, but it is uncertain whether these were full clutches. Eggs of all owlet-nightjars have unusually thick shells, for unknown reasons. DH

FACTFILE

OWLET-NIGHTJARS

Order: Caprimulgiformes

Family: Aegothelidae

9 species of the genus *Aegotheles*: **Archbold's owlet-nightjar** (*A. archboldi*), **Australian owlet-nightjar** (*A. cristatus*), **Barred owlet-nightjar** (*A. bennettii*), **Large owlet-nightjar** (*A. insignis*), **Moluccan owlet-nightjar** (*A. crinifrons*), **Mountain owlet-nightjar** (*A. albertisi*), **New Caledonian owlet-nightjar** (*A. savesi*), **Spangled owlet-nightjar** (*A. tatei*), **Wallace's owlet-nightjar** (*A. wallacii*).

DISTRIBUTION Australia, New Guinea region.

HABITAT Forest in New Guinea, open scrub or wooded country in Australia.

SIZE Length 18–30cm (7–12in); weight 29–85g (1.0–3oz).

PLUMAGE Predominantly brown, rufous, or gray, often intricately marked with cryptic streaks and vermiculations.

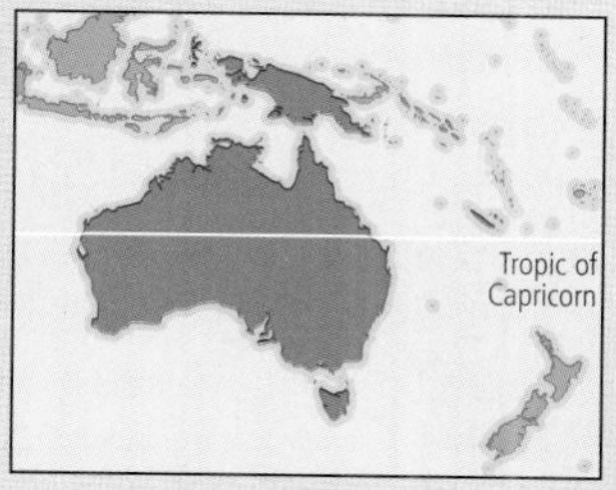

VOICE Shrill whistles, squawks, and churring sounds.

NEST Pad of leaves in tree hole, rarely a hole in a cliff face.

EGGS 1–5, white.

DIET Mostly insects.

CONSERVATION STATUS The New Caledonian owlet nightjar is Critically Endangered; this species is known only from one specimen and one confirmed sighting.

Potoos

POTOOS HAVE THE ABILITY TO "FREEZE" motionless in an upright posture, with sleeked plumage, if they are disturbed. This behavior, along with their superb cryptic coloration, enables them to blend into a tree trunk, and appear like just another snag on a branch they are using as a perch. Although "freezing" birds seem to have their eyes shut, the so-called "magic eye" – two small notches in the closed upper eyelid – allow them to remain watchful.

The range of potoos extends from the subtropics in Mexico southward as far as northern Argentina. The Northern potoo also occurs in the West Indies, on Hispaniola and Jamaica. Most species inhabit lowland rain forests, although the Andean potoo is found only in montane forest. Common and Northern potoos occur not only in forests, but also in plantations and more open savanna or farmland with trees.

FACTFILE

POTOOS

Order: Caprimulgiformes

Family: Nyctibiidae

7 species of the genus *Nyctibius*: **Andean potoo** (*Nyctibius maculosus*), **Common potoo** (*N. griseus*), **Great potoo** (*N. grandis*), **Long-tailed potoo** (*N. aethereus*), **Northern potoo** (*N. jamaicensis*), **Rufous potoo** (*N. bracteatus*), **White-winged potoo** (*N. leucopterus*).

DISTRIBUTION C and S America.

HABITAT Forest.

SIZE Length 21–58cm (8–23in); weight 155–557g (5.5–20oz).

PLUMAGE Concealment coloration of barred and mottled browns or grays; juvenile plumage of several species is paler than in adults.

VOICE Barks, croaks, snarling, or whistles.

NEST No nest is constructed, the egg being laid in crevice of tree branch or stump.

EGGS A single egg, spotted.

DIET Mainly insects.

CONSERVATION STATUS Not threatened.

Left *Potoos have short bills with a projecting "tooth" on the maxillary tomium, but as shown by this Common or Grey potoo, their flesh-colored mouths are large and wide, ideal for trapping flying insects!*

Nocturnal Insect-hunters

FORM AND FUNCTION

Potoos resemble nightjars not simply in their nocturnal habits, but also in certain physical characteristics: a small bill with a large gape, large eyes, and intricately patterned plumage. Where they differ from nightjars is that they lack a comb on the middle claw, and only the small Rufous potoo possesses rictal bristles at the base of the bill. They are small to medium-sized birds with long, rounded wings and a relatively long tail, but small legs and feet.

Large flying insects such as moths, beetles, and crickets form the bulk of the potoos' diet, but there are a few records of the larger species taking small bats and even, on one occasion, a tiny bird. Food is caught on sallying flights from an exposed perch, being taken on the wing or plucked from foliage or tree trunks.

Shared Duties

BREEDING BIOLOGY

Like most nocturnal birds, potoos have loud calls – distinct for each species – that serve to advertise and defend territories. They vary from gruff barks and booming in the Great potoo to a series of rasping notes or whistles in the smaller species.

No nest is built; the clutch of a single egg is laid in a natural crevice in the top of a tree stump or in the bark on a large tree branch. The egg is white with gray, brown, or purplish speckles. In the Common potoo, incubation and brooding are carried out by the male during the day and the female at night; in the Great potoo daytime incubation is also apparently usually by the male. The incubation period is at least 30 days in the Common potoo; the fledging period is 40–60 days in all potoos. The young are fed on regurgitated insect food. Incubating and brooding birds are inconspicuous, thanks to their beautiful camouflage. DH

Below *When alarmed, a potoo stiffens and lengthens its body, with its bill pointing straight up while its dark brown plumage blends into the tree. Here a Great potoo is cleverly camouflaged as a snag.*

Oilbird

*Discovered in 1799 by Alexander von Humboldt in a cave near Caripe, northern Venezuela, the oilbird became famous for its nestlings, which are fed on the oily fruits of palms and other trees and attain weights up to half as much again as those of the adults. The English and scientific names (*Steatornis caripensis *– the "fat bird of Caripe") derive from this peculiarity.*

The oilbird is related to the large, worldwide group of nightjars and allied birds, but is so specialized and individual in its anatomy, behavior, and ecology that it is placed in a family of its own.

Gregarious Cavedwellers

FORM, FUNCTION, AND DISTRIBUTION

Oilbirds have long wings, an ample tail, short legs, a strong, hawklike bill surrounded by long bristles, and large eyes. They are the world's only nocturnal frugivorous birds and use extremely sensitive night vision to navigate. They also have unusually large and sensitive olfactory lobes in the brain, which are thought to help them locate aromatic fruit.

Oilbirds are sedentary in Trinidad, but in some other areas, especially the Ecuadorean and Peruvian Andes, there is evidence that the birds leave their caves for part of the year, perhaps due to seasonal changes in fruit availability, although it is not known where they go. What is certain is that some oilbirds, probably young birds, occasionally wander widely. Stragglers have reached Panama, and the island of Aruba off the coast of Venezuela.

Even during nesting, oilbirds spend all day deep within caves, emerging at nightfall to forage and returning before dawn. They are highly social, and large caves may contain thousands of birds. The essential requirement of large caves, with ledges on which they can roost and nest, restricts them to mountainous areas, especially limestone formations. In Trinidad, however, they also occupy a few large sea caves on the island's north coast.

Outsize Young

BREEDING BIOLOGY

An intruder into an oilbird cave would be immediately struck by the unearthly sound of the birds' snarling screams of alarm, an almost deafening noise in a large colony. Not surprisingly, indigenous peoples have traditionally regarded them as devils or lost spirits: a widespread Spanish term for them is *guácharo* ("the one who cries"), while in Trinidad they are called *diablotin* ("little devil"). The oilbird's other main call, a staccato click, is used for echolocation; by picking up echoes from the cave walls and other surrounding objects, the oilbird is able to avoid obstacles in darkness. However, their echolocation is not as sophisticated as the ultrasonic system employed by bats.

Adult oilbirds occupy their nests continuously, returning to them to roost even when they are not breeding. In Trinidad, where oilbirds have been most thoroughly studied, the breeding season is long, lasting for most of the year. It is just possible for a pair of birds to fit two breeding cycles into a year, but most pairs nest only once annually.

The nests are built of regurgitated fruit stones and grow year by year, taking the form of a truncated cone with a saucer-shaped depression in the top. Eggs are usually laid at intervals of three days or more, an extreme of nine days being recorded. Incubation starts with the first egg, so that after the incubation period, they hatch in the sequence in which they were laid. At hatching, the young are sparsely downy; juvenile plumage appears at about five weeks. During the nestling period, both parents feed the young on fruit. By about the 70th day, the nestling reaches its maximum weight, and for the remaining 30 or so days it loses weight as the plumage finishes growing, until both wing-length and weight reach adult levels and the young bird is able to fly. DH/DWS

Left *In appearance, oilbirds resemble something between a large nightjar and a hawk.*

FACTFILE

OILBIRD

Steatornis caripensis

Order: Caprimulgiformes

Family: Steatornithidae

Sole member of family.

DISTRIBUTION S America from Guyana and Venezuela along the Andes to Bolivia; Trinidad.

HABITAT Forested country with caves.

SIZE **Length** about 45–52cm (18–21in); **weight** 375–430g (13–15oz); nestlings reach up to 50 percent heavier.

PLUMAGE Mainly rich brown with a scattering of white spots which are especially large and conspicuous on wing-coverts and outer secondaries; males slightly more gray and darker than females.

VOICE A variety of harsh screams, squawks, and clucking calls; also series of short, staccato clicks, used for echolocation in darkness.

NEST On ledges in caves, from half-light to pitch darkness.

EGGS 2–4, white, 41 x 22mm (1.7–0.9in), 17–22.5g (0.6–0.8oz). Incubation period 32–35 days; nestling period 88–125, usually 100–115 days.

DIET Exclusively the flesh of fruits from forest trees (the seeds being regurgitated), especially of the palm (Arecaceae), laurel (Lauraceae), and incense (Burseraceae) families.

CONSERVATION STATUS Not threatened.

Swifts

THE COMMON ENGLISH NAME FOR THIS FAMILY of birds captures their most familiar aspect – the rapid, restless movement as they wheel and dart through the air, never alighting on the ground or on vegetation. Equally apt are the generic name Apus, *from the Greek term meaning "without a foot," and the former name of the order, Machrochires, "with large hands" (referring to the outer wing); the family is characterized by very short legs and exceptionally long wings. The seasonal appearance of migratory species to breed during the warmer months has made the swift a quintessential symbol of summer in temperate zones.*

The sight and sound of swifts are familiar even to urban dwellers. Several species, such as the Common swift of Europe, regularly nest in or on buildings within large cities. Such artificial nesting sites are commonplace, but not exclusive. Although hardly any breeding records exist for this common species at "natural" sites in Britain, in the primeval forests of Europe, such as those remnants that survive in Poland (notably the Bialowiecza Forest), swifts' nests have been found in high, broken-off hollow branches and in holes in the trunks of ancient, rotten trees.

Sophisticated Flying Machines

FORM AND FUNCTION

Swifts' wings are made up of ten long primaries and a short block of secondaries. The narrow, elongate, sickle shape of the wings determines the birds' mode of flight, giving them great speed in flapping flight, but more importantly allowing them to conserve a great deal of energy while gliding. The configuration of the wings may also explain the comparatively low flight metabolism exhibited by this family, as well as the reduced breast-to-body-mass ratio, since such a wing shape does not require very strong breast muscles. Being highly aerial in their lifestyle, swifts are not well adapted to the ground. In fact, the ratio of wing to leg length makes it difficult for them to take off from the ground.

FACTFILE

SWIFTS

Order: Apodiformes

Family: Apodidae

92 species in 19 genera. Species and genera include: **Alpine swift** (*Apus melba*), **Common swift** (*A. apus*), **Black swift** (*Cypseloides niger*), **Chimney swift** (*Chaetura pelagica*), **Edible-nest swiftlet** (*Collocalia fuciphaga*), **Palm swift** (*Cypsiurus parvus*), **swallow-tailed swifts** (*Panyptila* spp.), **White-throated spine-tail swift** (*Hirundapus caudacutus*), **White-throated swift** (*Aeronautes saxatilis*).

DISTRIBUTION Worldwide except high latitudes and some islands.

HABITAT Aerial feeders rarely coming to rest.

SIZE Length 10–30cm (4–12in); weight 9–150g (0.3–5oz).

PLUMAGE Most species dull black or brown, many with conspicuous white or pale markings.

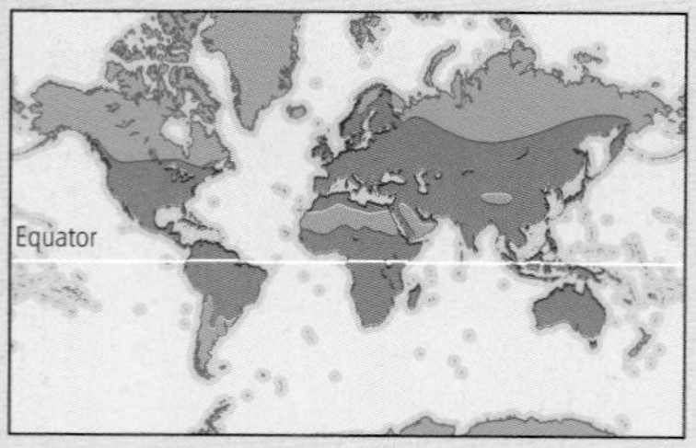

VOICE Shrill, piercing screams.

NEST In rocks, crevices, or caves; a variety of materials cemented with saliva (except for Cypseloides, Streptoprocne and Hirundapus).

EGGS 1–6, white, weight 1–10g (0.04–0.35oz). Incubation period 17–28 days; nestling period 34–72 days.

DIET Flying insects; other airborne arthropods.

CONSERVATION STATUS The Guam swiftlet (*Collocalia bartschi*) is Endangered. 3 further *Collocalia* species are Vulnerable, plus the Dark-rumped swift (*Apus acuticauda*) and Schouteden's swift (*Schoutedenapus schoutendeni*).

Nevertheless, their tiny feet are surprisingly strong, and their sharp claws are well designed for gripping onto vertical surfaces (as anyone who has handled swifts can testify!). Other adaptations include highly sensitive hemoglobin in the bloodstream for optimum delivery of oxygen in conditions of low oxygen pressure, namely at high altitude. This sophisticated flying machine is also equipped with a short, weak beak that opens to a huge gape, making it easier for the swift to catch aerial insects in flight.

Swifts are almost always seen in flight, when they appear to be flying very fast. However, when feeding they must be able to sight their prey and catch it on the wing, and too fast a speed would make this much more difficult. Nevertheless, during displays they can fly very fast and are often able to take advantage of the wind to cover the ground very rapidly even when their airspeed is not exceptional.

It has been proved that the Common swift regularly spends the night on the wing. Birds have been observed rising in the evening long past the time when they would be able to find their way into a nest. They have been watched from airplanes and gliders and regularly tracked by radar. They may well never come to land at all except when breeding; this means that some individuals will complete a nonstop flight of 500,000km (well over 300,000 miles) between fledging late one summer and their first landing at a potential nesting site two summers later!

Most species have rather dull coloration, although several species' plumage is shot through

Right *Most people's experience of swifts is a fleeting glimpse of them as they wheel around the sky. These Common swifts display their beautifully streamlined, aerodynamic shape.*

Left *At least three different swift species habitually nest on vertical cliff-faces behind waterfalls, entering and exiting through the spray. Here, Great Dusky swifts* (Cypseloides senex) *shelter behind a cascade at Iguaçu Falls on the Paraná river, Argentina.*

with a blue, green, or purple iridescent gloss. Individual Common swifts undoubtedly recognize each other by voice – their shrill screams – but apparently not by sight, probably due to their dark nesting sites. Many have forked tails, although the spine-tailed swifts have the shafts of their tail feathers extending beyond the vane to form a row of spines. The stiff tail feathers act as a supporting prop for the birds when they are clinging to vertical surfaces. For instance, Chimney swifts are so named for their habit of breeding and roosting inside tall, industrial chimneys, which must be a recent habit.

All swifts feed exclusively on insects and spiders, taken mainly in the air. Their diets have been studied through analysis of stomach contents, droppings, pellets, and food boluses. All these methods have revealed the most common prey items to be Hymenoptera (bees, wasps and ants), Diptera (true flies), Hemiptera (true bugs), and Coleoptera (beetles).

SURVIVAL IN COLD WEATHER

Birds like the Common swift that feed on flying insects have a major problem in the northern part of their range: when the weather turns cold, wet, and windy, their food becomes scarce, or even non-existent. Outside the breeding season, the birds may make good their escape by flying south to warmer areas but this is impossible while they are nesting.

Studies reveal a range of adaptations to combat this problem. Adult birds can store a large amount of subcutaneous fat that they can use when food is scarce. They can roost at the nesting site for several days on end with a very slow metabolism and a depressed temperature. Swift eggs can withstand chilling much more than those of most species; the parents may even practice a crude form of birth control, for many observers have reported eggs being removed from the nest-cup in times of cold weather.

Much the most interesting adaptation is the ability of the young to withstand long periods of starvation and cold. This is not just crucial in enabling chicks to survive but will also often allow adults to recover their condition in time to re-establish parental care. Young which have not been fed for a week or more at some stage of the nestling period may still fledge successfully if conditions improve and their parents start to feed them again. Indeed, fasts of up to 10 days have been reported, with weight losses of over 50 percent. The young may be so cold and torpid that they seem dead even to the experienced observer. Indeed, the torpor is so profound that the chick dies if it is warmed up too quickly. In such circumstances, the fledging period approaches 60 days instead of the 5 weeks expected in summers with consistently good weather.

Torpidity occurs in several other species of swift and also in nightjars and hummingbirds. In Europe, swallows have also been found in a similar state and it is suspected that this condition may have given rise to theories of hibernation in birds. A record of eight White-throated swifts in a torpid state during January in California was probably a response to poor weather locally and not an example of true hibernation – so far the only bird to approach true hibernation is the poorwill. CJM

Warmer Winter Climes

DISTRIBUTION PATTERNS

Because swifts rely on catching insect prey on the wing, they have to winter in areas where the temperature allows insects to keep flying in sufficient numbers to feed them (see Survival in Cold Weather). Therefore, when the cold season approaches in the temperate parts of their range, most species retreat south. Common swifts from Britain winter in East Africa, Chimney swifts from Canada in the upper Amazon Basin, and White-throated spinetail swifts from China and Japan in Australia. Such long-range migrations are easily within their flying capability; of all groups of land birds, swifts seem most at ease in the air and do, in any case, fly hundreds of kilometers every day while feeding, even when they are not migrating. In experiments, nesting Alpine swifts have successfully returned home from 1,620km (c.1,000 miles) in three days. Likewise, a ringing record from Britain showed that a newly fledged young Common swift had reached Madrid from Oxford in the same time.

Above *Representative species of swift:* **1** *Seychelles swiftlet* (Collocalia elaphra), *a colonial cave-nester. This species has been badly depleted by nest collection, insecticide use, and wetland destruction;* **2** *Common swifts* (Apus apus) *mating on the wing;* **3** *Palm swift* (Cypsiurus parvus); **4** *Alpine swift* (Apus melba) *chasing an insect;* **5** *Indian swiftlet* (Collocalia unicolor).

A Variety of Nests

BREEDING BIOLOGY

The temperate, migratory species that have been studied in detail are long-lived and faithful both to their breeding site and to their mate. They have to undertake their breeding attempt quickly, as there may only be sufficient aerial food for them for 12–14 weeks, even in areas where they commonly breed. For instance, Common swifts arrive in Britain to breed in early May and leave at the end of July. The males generally arrive first and take possession of the breeding site. This is nowadays almost always in a roof and the birds will make a small ring of material, taken in flight and glued down with saliva, at the place where the eggs will be laid. The chicks are brooded by one or other of the parents for the first few days after hatching and food is brought to them as "boluses" – gobs of insects stored in the parent's throat. These can weigh up to 1.7g (0.06oz) and may contain over 1,000 tiny insects and spiders. In fine weather, feeds may be brought in every 30 minutes or so, totaling some 30–40g (1–1.4oz) of food in a single day. In such ideal circumstances, the minimum nestling period is around five weeks but, if the weather is inclement, it may extend to eight weeks.

Colonial breeding can lead to several dozen pairs of Common swifts breeding in a single roof-space or, more often, in adjacent buildings. In this species, the first-year birds seldom return to the breeding grounds and many do not breed until their third or fourth year. The immature birds form large screaming parties that display in mid-summer and often fly up to occupied nesting sites in a very excited state – much to the annoyance of the resident birds.

The other 70 or so species of swift include some swiftlet species (*Collocalia*), whose colonies in vast Asian caves may contain several hundred thousand individuals. These birds include species that make their nests entirely from dried saliva and stick them to the roof and walls of the caverns. They are of economic importance, since the nest is the source of the delicacy known as "Bird's nest soup." Collecting the nests is a very hazardous undertaking, involving precarious ropeways and ladders up to 100m (330ft) long. The nests command a high

Right *Nest types:* **a** *the Fork-tailed palm swift* (Tachornis squamata) *builds a bag-shaped nest hanging from a fan frond of a palm;* **b** *the nest of the Lesser swallow-tailed swift* (Panyptila cayennensis) *is tubular, and is attached to a rock or tree trunk;* **c** *the Edible-nest swiftlet* (Collocalia fuciphaga) *builds cup-shaped nests mainly of saliva;* **d** *Palm swifts glue their nests to the underside of palm leaves.*

Right *Because there is no light in the caves where they nest, many Asian swiftlets (here, a White-rumped swiftlet,* Collocalia spodiopygia) *use echolocation to find their way around.*

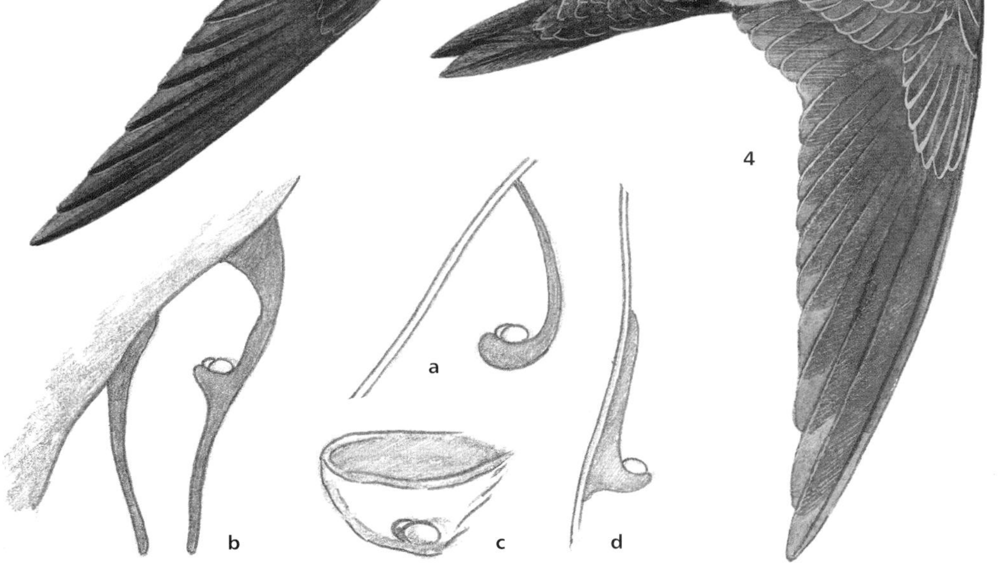

price and the number taken can be considerable – more than 3.5 million nests were exported in one year from Borneo to China. In such vast colonies, the droppings (guano) also accumulate rapidly and are extracted from the cave floor for use as a fertilizer.

During the nest-building phase of the breeding season, even those species that only use saliva as cement to glue together other nesting material have greatly enlarged salivary glands. The glands of the Chimney swift, which glues its nest of twiglets to a vertical wall, undergo a 12-fold enlargement. In common with others that use twigs for nest-building, this species breaks them from trees in flight. Other materials – feathers, seeds, grasses, straw – are gathered while they are being blown about; during World War II, the metal foil strips ("chaff") that were dropped from aircraft to confuse enemy radar were incorporated into swift nests.

The shape of the nest, and the method used to construct it, are often peculiar to individual species. For instance, the Palm swift of the Old World is only found where the Fan palm (*Livistona rotundifolia*) grows. Its nest is made along the vertical channel on the inside of one of the palm's leaves, from feathers and fibers, and has a small lower rim. This is for the bird to perch on, as it incubates while clinging vertically to the nest into which its two eggs are firmly cemented.

In the New World, the palm swifts belong to a different genus (*Tachornis*) and build their nest inside the vegetation hanging from the crown of the palm trees. In this case, the bag-shaped nest is glued to the leaf and the bird enters along the leaf side and lays its eggs in a cup formed inside the lower, outer edge. Very complex nests are also constructed by two other New World species, the swallow-tailed swifts, which may form a tube 70cm (28in) long hanging vertically from a rock face. The birds have a nest at the top of the tube, close to the point of attachment. These nests can be very durable and may be used year after year.

The setting of the Chimney swift's nest, down a vertical chimney, is the "artificial" equivalent of nesting in a hollow tree (which the species also still does). Several other species will fly down to nest in potholes – as far as 70m (230ft) underground – while the Black swift has been described as nesting on cliffs facing rough seas, in sea-caves whose entrances are covered by each successive breaking wave.

These varied nesting sites and structures are an especially interesting illustration of how a group of aerial birds, without the opportunity for collecting much nesting material, has managed to make use of safe nesting sites. Most of them are out of reach of mammalian or reptilian predators; this inaccessibility safeguards not only the eggs and the young but also the vulnerable parents. None of the adult swifts are at all maneuverable when on the ground or perched.

Adapting to Survive

Conservation and Environment

Swifts are currently confronted by a number of different threats. Human destruction of habitats has reduced foraging areas for some species; over-harvesting for the lucrative bird-nest trade is taking its toll on Southeast Asian swiftlet populations; while the increasing use of insecticides and pesticides has severely reduced both the range and number of insect prey in many areas. On the credit side, partially counteracting this depletion of natural habitat, many species have adapted to ready-made nests in artificial environments, to the extent that several now rarely utilize natural sites. However, the near-total conversion of certain species to nesting in buildings – a widespread phenomenon – has brought its own problems, since roofing repairs are seldom undertaken with the birds in mind. It is to be hoped that "swift-friendly" building regulations will be widely adopted in order to forestall further declines in the swift population. TM/CJM

Above *One of the most remarkable aspects of the behavioral ecology of swifts is that they have, in many cases, adapted to human encroachment by nesting in the roofs and eaves of buildings in urban environments.*

Treeswifts

FOR A LONG TIME, TAXONOMISTS WERE unclear whether treeswifts were more akin to swifts or swallows. The name Hemiprocne *(from Greek* hemi progne, *or "half swallow"), given in 1829 to the only genus in this family, helped foster the confusion, yet tells us much about the birds' morphology.*

Right A Crested treeswift (Hemiprocne coronata) *perches on a branch while incubating its eggs.*

Below The Whiskered treeswift *is under increasing threat, as logging reduces its rainforest habitat.*

In many regards, treeswifts are more like swallows than swifts, using a nonreversible hind toe to perch firmly on branches in an upright posture, and displaying a very deep forking of the tail (which in some species can account for 70 percent of the standard tail length, far greater than any true swift). This swallowlike feature aids maneuverability when foraging. Yet their nest construction and wing structure hint at a closer affinity with the true swifts.

Canopy and Forest Edge

FORM AND FUNCTION

The closely related Grey-rumped and Crested treeswifts are very similar, with a forehead crest 2.5–3 cm (1–1.2in) tall present in both sexes, and usually raised when perched. The main differences between the two species are the extent of black in the upperparts and the proportionate length of their tail streamers. Unlike the Grey-rumped treeswift, the Crested treeswift has tail streamers that project well beyond the tips of the folded, scythe-shaped wings.

The Whiskered treeswift and the Moustached treeswift are only slightly crested, but share a bold face pattern, displaying two striking white lines both above and below the eye. The Whiskered treeswift differs from the other three species in its habitat preference and in being substantially smaller. It also uses a unique foraging technique presumably related to its smaller size; as a genuine member of the continuous-forest community, it chases flying insects within the forest canopy. By contrast, all the larger species strongly prefer forest-edge habitat, eschewing continuous canopy just as often as they do open ground. As far as foraging techniques are concerned, they all seem to be tied to their favorite perches and exploit relatively limited, fixed airspace daily and over long periods. However, their feeding behavior recalls that of shrikes or flycatchers, perched birds flying out to catch prey and then returning to the perch.

FACTFILE

TREESWIFTS

Order: Apodiformes

Family: Hemiprocnidae

4 species of the genus *Hemiprocne*: **Grey-rumped treeswift** (*H. longipennis*), **Crested treeswift** (*H. coronata*), **Moustached treeswift** (*H. mystacea*), **Whiskered treeswift** (*H. comata*).

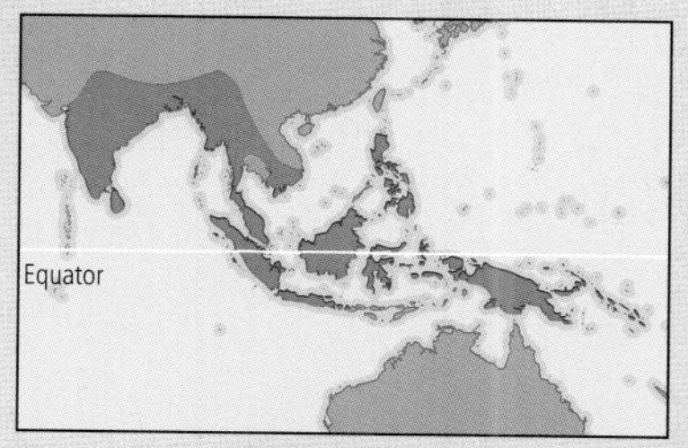

DISTRIBUTION S and SE Asia, from India to the Solomon Islands.

HABITAT Deciduous savanna woodland to evergreen forests.

SIZE Length 17–33cm (6.5–13in); weight 70–120g (2.5–4oz).

PLUMAGE Mainly soft gray or brown, with white stripes on sides of head and crest on front of crown.

VOICE Shrill twittering, especially when returning to roost for the night. Generally louder song than in Apodidae.

NEST Tree bark and feathers glued to a branch.

EGGS 1, white, weight 6–10g (0.2–0.35oz).

DIET Flying insects, including bugs, beetles, and small flies.

CONSERVATION STATUS None of the treeswift species is currently threatened, although habitat loss through forest clearance may make treeswifts vulnerable, especially the Whiskered treeswift.

Delicate Nests

SOCIAL BEHAVIOR

None of the treeswift species migrate in the true sense; on the other hand, only the Whiskered treeswift is strictly sedentary, remaining in the nesting territory all year. Other species are territorial during breeding but rather more gregarious during the winter months, when there are records of large parties of up to 50 birds occupying communal trees.

The overall breeding season of all species is long. Where adequate information is available, they show a peak of activity in early spring or early summer, or experience two peaks during the season. As with typical swifts, both pair members cooperate in nest building. The nest is a tiny structure, with paper-thin walls stuck on the side of a thin branch. The nest is only about 2.5cm (1in) across, just large enough to accommodate a single egg, glued in with saliva for safety. The nest is often sited on a fully exposed, pencil-thin twig at the edge of the canopy, presumably to give uninterrupted flight access from all directions. The thinness of the twig helps give early warning of the approach of climbing nocturnal predators, such as snakes. The nest is so small that, when the bird incubates, it looks as if it is merely perching. In fact, so insubstantial are the nest and its support that the bird will often perch on the branch with the nest between its legs. (For the same reason, shortly after hatching, the nestling also takes to perching on the branch.) Male and female share incubation, but the females do two to three times the work of the males. Food is brought to the nestling in boluses and regurgitated into the chick's mouth.

TM/CJM

Hummingbirds

A UNIQUE GROUP COMPRISING WELL OVER three hundred species, the hummingbirds form one of the largest avian families of the New World. As a family, they are surpassed in species numbers only by the tyrant-flycatchers (Tyrannidae), with over 370 taxa.

Hummingbirds' uniqueness resides in their great agility in flight (in particular their ability to hover), their brilliant iridescent plumage, and their generally tiny size. Through their symbiotic relationship with flowers, they have evolved to fill an ecological niche occupied by no other bird; they feed almost exclusively on high-calorie nectar, and in so doing pollinate the plants on which they specialize.

Although hummingbirds form an undisputed phylogenetic unit, their relationship to other bird groups is still the subject of debate. Traditionally, trochilids have been placed within the Apodiformes alongside the treeswifts (Hemiprocnidae) and the true swifts (Apodidae). This classification has recently been supported by biochemical studies. Swifts and hummingbirds share a unique form of the enzyme malate dehydrogenase. This phylogenetic treatment is corroborated by DNA comparisons of Sibley and Ahlquist (1990). Generally, two subfamilies are distinguished within the Trochilidae: the hermits (Phaethornithinae) and the trochilines (Trochilinae).

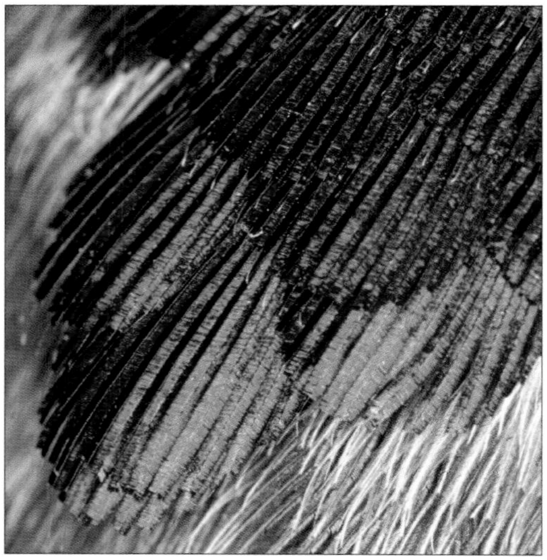

Above *Flattened barbules give an iridescent sheen to the rose-colored gorget of an Anna's hummingbird, a North American species that is resident in California and winters from British Columbia south to Arizona.*

Right *Beating its wings up to 80 times a second, an Allen's hummingbird takes nectar from a red hot poker plant* (Kniphofia *sp.*). *To supply the energy requirements of its hovering flight, a hummingbird may consume up to half its body mass in sugar a day.*

Below *Hummingbird bills are often extraordinarily lengthy, but the birds' tongues are even longer. Here a Jamaican mango* (Anthracothorax mango) *protrudes the tip to suck water from a wild plantain flower.*

Hermits and Non-hermits

FORM AND FUNCTION

The Trochilinae or typical hummingbirds, which with 96 genera account for more than 90 percent of hummingbird species, are an extremely diverse subfamily. As might be expected in such a large assemblage, species are more heterogeneous than in the hermits, a fact that has led several taxonomists to seek further possibilities of splitting at the subfamily level. For instance, one proposal is to place the coquettes and thorntails in a subfamily of their own (Lophornithinae) and to consider the distinctive members of the genera *Androdon* and *Doryfera* as another subfamily (Doryferinae). Whereas the classification of the Lophornithinae is based primarily on the somewhat superficial location and development of the nasal operculum and the primaries, the Doryferinae are markedly different, in their hindneck musculature and song structure, from all other previous subdivisions. Systematists widely agree that the genera *Androdon* and *Doryfera* are not members of the hermit assemblage to which they were formerly assigned, but their advocated subfamily status should be treated with caution in the absence of detailed comparative anatomical and molecular information.

The Trochilinae exhibit a subfamily-specific pattern of the humeral tendon. Their often straight to slightly decurved bills range in length from a few millimeters in the Purple-backed thornbill (*Ramphomicron microrhynchum*) to 12cm (5in) in the Sword-billed hummingbird. Most show marked sexual dimorphism in their plumage color. Iridescent bright colors of many metallic shades of red, orange, green, and blue frequently occur on the head, upperparts, and underparts of males. In some species, jewel-colored adornments like extendable gorgets, crests, and elongated modified tail feathers are striking male characteristics. Females are of duller appearance, often lacking brightly colored feathers. However, in some species plumage dimorphism is minimal or even absent, such as in the members of the genera *Colibri* and *Eulampis*, whose males and females are almost indistinguishable to the human eye. Males of the brightly colored trochilines are generally territorial. Lek behavior is known only for a few species and courtship in many non-hermits features characteristic aerial displays. Nests are predominantly cup-shaped and attached to branches or forked twigs, although pendular and domed constructions are also known. Trochilines live in extremely diverse habitats, ranging from desert borders, mangroves, and tropical rain forests to

the perennial meadows below the snowline on the peaks of the high Andes.

The six genera of hermits (*Ramphodon, Eutoxeres, Glaucis, Threnetes, Phaethornis,* and *Androdon*) differ from the true hummingbirds in their characteristic humeral tendon and their predominant pigmentation: brownish, grayish, and reddish colors. Iridescent colors are sparse and mostly limited to the back feathers. Hermits are generally nonterritorial, occur in the understory of dense scrub in tropical forests, and usually have long bills adapted to flowers with tubular corollas. They have a particular affinity to *Heliconia* spp. plants. In all species studied, males gather at leks, where they attract females. In their noisy displays, in which they make repeated clicking sounds, male hermits employ fanned rectrices (tail feathers) and wide-open bills showing the yellow or red interior base of the lower mandible (gape display). Hermit nests are pendent, often cone-shaped, and are either attached to the inner tip of a long, narrow, pliable leaf or fastened to a vertical root or twig if shelter is provided by a riverbank, cave vegetation, or even a bridge.

Hummingbirds are extremely small. Most species are just 6–12cm (2.5–5in) in size and have a body mass of around 2.5–6.5g (0.08–0.2oz). The Reddish hermit (*Phaethornis ruber*) from Guyana and Brazil and the Bee hummingbird (*Mellisuga helenae*) of Cuba each weigh less than 2g (0.07oz), making them not only the smallest birds, but also the smallest warm-blooded animals in the world. Members of the genera *Eutoxeres*, *Ensifera*, and *Pterophanes*, on the other hand, are heavier than average, at 12–14g (0.4–0.5oz). As its name suggests, the largest of all is the Giant hummingbird (*Patagona gigas*), weighing 19–21g (0.6–0.7oz), a body mass comparable to that of a small swift.

FACTFILE

HUMMINGBIRDS

Order: Apodiformes (suborder: Trochili)

Family: Trochilidae

328 species in 108 genera

DISTRIBUTION Americas from Alaska to Tierra del Fuego; also W Indies, Bahamas, Juan Fernández islands. Many species of extreme northern and southern latitudes migratory.

HABITAT Wherever nectar-producing flowers occur, from sea level to below snowline of the Andes.

SIZE Length 5–22.1cm (2–8.7in); body mass 1.9–21g (0.06–0.7oz).

PLUMAGE Most have glittering green body feathers, often with other brilliant metallic colors on head, back, throat, breast, belly, or rectrices. Males often more iridescent, some with crests and/or elongated tail feathers.

VOICE Territorial and flight calls high-pitched, brief, given by both sexes. Male advertisement song often complex, with short, repeated sequences of guttural and warbling notes. Females of some species sing, but less elaborately.

NEST Phaethornithinae (hermits') nests attached by cobweb beneath leaf or to rock . Trochilinae nests are cup-shaped, often rather small, and attached to branches or forked twigs, although pendular and domed constructions are known.

EGGS 2 (rarely 1) elongated white eggs; size about 10–15 percent of adult female body mass. Incubation period 14–24 days, nestling period 18–41 days.

DIET Mostly nectar (90 percent); small insects and spiders.

CONSERVATION STATUS 9 species are classed as Critically Endangered, 11 as Endangered, and 9 as Vulnerable.

See Hummingbird Subfamilies ▷

Adaptations for Hovering

HUMMINGBIRD PHYSIOLOGY

Trochilids are highly evolved nectarivores that depend almost entirely on the carbohydrate-rich sugar secretions (nectar) of ornithophilous (bird-pollinated) flowering plants. The approximate composition of their diet is 90 percent nectar and 10 percent arthropods and pollen. They reach this liquid food with thin, elongated bills of various shapes, which protect their specialized long, sensitive tongues. Their feeding behavior necessitates a specific locomotor performance, namely hovering flight, which allows them to remain apparently motionless in the air when feeding on exposed flowers. The humming sound produced by the wings during hovering has given the family its English name. However, as a result of this unique foraging behavior, they can no longer walk or climb with their feet, which serve only for perching. While hovering, their pointed, uncambered wings are moved mainly in the horizontal plane, describing with their tips a flat figure-of-eight, in the manner of a variable-pitch rotor on a helicopter. By slightly altering the wing angle, this technique allows them to perform all kinds of controlled forward, backward, and sideways movements in the air, including upside-down maneuvers. When hovering, the wingbeat rate in smaller trochilids like the Amethyst woodstar (*Calliphlox amethystina*) averages 70–80 per second, compared to just 10–15 in the Giant hummingbird. Yet the record wingbeat rate is held by certain North American hummingbirds, such as the Ruby-throated hummingbird (*Archilochus colubris*), which produces over 200 beats per second during its courtship flight.

The hovering flight mode involves specific skeletal and flight-muscle features. The deeply keeled and elongated hummingbird sternum is relatively large compared with those of other flying birds. Eight pairs of ribs, two more than in most birds, help to stabilize the body during flight. The strong coracoids of the pectoral girdle are exceptional in their structure: only hummingbirds and swifts have a shallow cup-and-ball joint where the coracoids are connected to the sternum. Tendons connect the flight muscles with the humerus. The modified humerus bone of hummingbirds moves freely at the shoulder joint, permitting optimal wing movement in all directions, including axial rotations of nearly 180 degrees. It is only the

humerus that moves at this joint. Slow-motion pictures, however, indicate little flexure of the short arm bones.

The two major muscles for trochilid flight are the mitochondria-rich *Musculus pectoralis major*, attached to the sternum, clavicle, and humerus, and the *Musculus supracoracoideus*, located beneath the pectoralis and also attached to the sternum. Both muscles consist exclusively of dark red fibers, providing the energy for powerful flight. The two flight muscles together make up over 30 percent of the body mass of a hummingbird, much more than in other strong fliers such as migratory birds, in which these muscles account for no more than 20 percent of body mass.

Due to their energy-demanding hovering flight, hummingbirds have the highest oxygen requirements of all vertebrates. Their respiratory system, two compact symmetrical lungs for gas exchange and nine thin-walled air sacs acting as bellows for their ventilation, is adapted to utilize high gas volumes. At rest, the breathing rate is 300 times per minute, which may rise under hot conditions or during flight to over 500 per minute, whereas a starling or a pigeon breathes roughly 30 times per minute, and humans about 14–18 times per minute. In trochilids, the tidal volume of each respiratory cycle ranges from 0.14 to 0.19cu cm (0.0085–0.0116cu in), which is twice as great as in mammals of a similar size (shrews).

The daily energy requirement of a 4–5g (0.15oz) hummingbird lies roughly in the range of 30–35 kJ (125–145kcal), which amounts to five times the estimated basal metabolic rate. In order to meet their daily energetic needs, hummingbirds have to consume the nectar of some 1,000–2,000 flowers each day. Estimated daily water intake with the nectar meals amounts to around 160 percent of their body mass. This significant water excess is eliminated in chronic diuresis, causing a salt-balance problem. Hummingbirds have solved this physiological constraint by having kidneys that contain a poorly developed renal core consisting largely of collecting ducts with only a few looped nephrons, or waste-extracting units. Hummingbirds thus do not concentrate their urine like other birds and mammals, but reclaim valuable salts thanks to a reduced osmotic concentration in the region of 15–24 percent of plasma concentration. Although 76–85 percent of the solutes is conserved each day, over 10 percent of body sodium and potassium is lost. These salts are normally replaced from trace amounts in floral nectars. It has been suggested that the evolution of nectar-feeding in trochilids led to their small body size. Therefore, constraints on kidney size to process this relatively large flow of excess water could only be handled by foregoing nephrons capable of producing concentrated urine (mammalian-type nephrons).

Trochilids are found between Alaska in the north and Tierra del Fuego in the south of the American supercontinent, occupying all types of habitat from sea-level to about 4,500 m (15,000ft) where flowering plants are available. Over 50 percent of all species occur in the montane regions and are exposed to a daily temperature stress of more than 15°C (60°F). For physiologists, it was therefore of particular interest to investigate the amount of heat production necessary to offset the difference between ambient and body temperature. If thermal conductance, a measure of physical characteristics influencing heat exchange, is high, then the bird is poorly insulated. One would expect trochilids to be poorly insulated, due to their low body mass (in birds, thermal conductance increases exponentially by a factor of –0.5 with decreasing body mass), and by the relatively small number of feathers covering their bodies. Nevertheless, hummingbirds have very high heat production per gram of body mass. Due to their low mass, their energy requirement for temperature regulation is far less than it is for larger animals.

The main problem for small endotherms is to store sufficient energy to compensate for the expenditure involved in thermoregulation. In general, animals store energy in linear proportion to their body mass with a rate of increase of 1.0, but the rate of energy loss increases linearly by 0.75. The energetic problems experienced by extremely small animals result from the differences between these parameters. Thus, small endotherms like

Left *A Green-crowned brilliant* (Heliodoxa jacula) *feeds on a ginger plant* (Zingiber *sp.*). *Hummingbirds are unique in gathering nectar entirely in flight.*

Right *A Sword-billed hummingbird's wingstroke sequence demonstrates the great flexibility of the hummingbirds' hovering flight. A reduction in the size of the upper arm relative to the wrist and hand bones, along with wing muscles that average 25 percent or more of body mass, give great leverage and mobility.*

hummingbirds are under considerable time pressure: on the one hand, they must meet their daily food requirements, and on the other, accumulate sufficient energy reserves to survive their nocturnal starvation. The following factors are therefore of decisive importance for extremely small hummingbirds: quality and accessibility of food; and mechanisms to reduce energy consumption. In trochilids, these mechanisms take the form of long periods of inactivity between daily meals and torpor. During torpor, gaseous metabolism and body temperature are adjusted to levels of ambient temperature and regulated to remain within the range of 18–20°C (64.4–68°F). In torpor, hummingbirds become lethargic and incapable of reacting in a coordinated way to external stimuli. Energy savings during this immobile state are considerable; up to 60 percent of the total energy accumulated for the nocturnal resting phase. During observations on Nearctic migratory hummingbirds, an irregular occurrence of torpor was noted. Birds reacted with torpor when energy levels fell below lower limits during the night. According to these studies, torpor is an energy-regulated mechanism that is triggered below a threshold value, equivalent to a physiological regulation pattern coming into force in extreme conditions. As reasons for the irregularity of torpor, physiologists cite the risk and energetic costs of the lethargic state. The risks involved would be the danger of predation while immobile, and insufficient residual energy for thermoregulation which could prevent awakening from torpor.

Physiological adaptations for nectar feeding and energy regulation patterns resulting from limiting temporal and environmental conditions are fundamental to understanding the general habits of trochilids. The utilization of energy-rich nectars most likely fostered strong individual competition

Left *Representative species of hummingbird:* ***1*** *Reddish hermit* (Phaethornis ruber). ***2*** *Snowcap* (Microchera albocoronata). ***3*** *Ruby-topaz hummingbird* (Chrysolampis mosquitus). ***4*** *Scale-throated hermit* (Phaethornis eurynome). ***5*** *Red-tailed comet* (Sappho sparganura). ***6*** *Sword-billed hummingbird* (Ensifera ensifera) *feeding.* ***7*** *Bearded helmetcrest* (Oxypogon guerinii). ***8*** *White-tipped sicklebill* (Eutoxeres aquila). ***9*** *Fiery-tailed awlbill* (Avocettula recurvirostris) *on nest.* ***10*** *Ruby-throated hummingbird* (Archilochus colubris). ***11*** *Giant hummingbird* (Patagona gigas).

Hummingbird Subfamilies

Hermits
Subfamily Phaethornithinae

49 species in 4 genera, including: sicklebills (*Eutoxeres*), Hairy hermit (*Glaucis hirsuta*), Green hermit (*Phaethornis guy*), Saw-billed hermit (*Ramphodon naevius*), barbthroats (*Threnetes*).

Typical Hummingbirds
Subfamily Trochilinae

293 species in 104 genera, including: Tooth-billed hummingbird (*Androdon aequatorialis*), lancebills (*Doryfera*), sabrewings (*Campylopterus*), jacobins (*Florisuga*), violet-ears (*Colibri*), mangos (*Anthracothorax*), Crimson topaz (*Topaza pella*), caribs (*Eulampis*), Antillean Crested hummingbird (*Orthorhynchus cristatus*), Violet-headed hummingbird (*Klais guimeti*), Plovercrest (*Stephanoxis lalandi*), coquettes (*Lophornis*), thorntails (*Discosura*), streamertails (*Trochilus*), emeralds (*Chlorostilbon*), Fiery-throated hummingbird (*Panterpe insignis*), Violet-capped hummingbird (*Goldmania violiceps*), Broad-billed hummingbird (*Cynanthus latirostris*), woodnymphs (*Thalurania*), sapphires (*Hylocharis*), goldenthroats (*Polytmus*), Rufous-tailed hummingbird (*Amazilia tzacatl*), Versicolored emerald (*Agyrtria versicolor*), Blue-chested hummingbird (*Polyerata amabilis*), Indigo-capped hummingbird (*Saucerottia cyanifrons*), Snowcap (*Microchera albocoronata*), plumeleteers (*Chalybura*), Blue-throated hummingbird (*Lampornis clemenciae*), brilliants (*Heliodoxa*), coronets (*Boissonneaua*), sunbeams (*Aglaeactis*), Andean hillstar (*Oreotrochilus estella*), incas (*Coeligena*), Sword-billed hummingbird (*Ensifera ensifera*), Giant hummingbird (*Patagona gigas*), sunangels (*Heliangelus*), pufflegs (*Eriocnemis*), Booted racquet-tail (*Ocreatus underwoodii*), trainbearers (*Lesbia*), Red-tailed comet (*Sappho sparganura*), metaltails (*Metallura*), sylphs (*Aglaiocercus*), visorbearers (*Augastes*), Marvellous spatuletail (*Loddigesia mirabilis*), starthroats (*Heliomaster*), Ruby-throated hummingbird (*Archilochus colubris*), Anna's hummingbird (*Calypte anna*), woodstars (*Chaetocercus*), Rufous hummingbird (*Selasphorus rufus*).

3
4
5
6
7
8
10
9

for this food source, favoring the evolution of specific maintenance and survival strategies. As a general consequence, males and females of almost all hummingbird species studied live solitarily, often aggressively defending nectar sources in the form of flowering shrubs and trees against any potential food competitor. The sexes are polygamous and associate only briefly to fertilize the eggs.

Locked in Combat

SOCIAL BEHAVIOR

In general, the males of species with bright, iridescent colors establish feeding territories at flowering bushes that allow them to meet their daily energy requirements. In order to defend their nectar resources they often perch high on nearby exposed branches. These serve as vantage points from which predators can be detected with ease and from which the area can be defended against possible intruders, including females, by vocal warning signals and agonistic flights. Often the territory holder first empties the nectar from peripheral flowers to remove or reduce the feeding reward for newly arriving competitors. Trespassers ignoring the threat calls of the territory owner are robustly attacked in flight, sometimes resulting in actual physical combat. An airborne fighting pair may then be locked in strong claw grips, tumbling to the ground like a falling stone. These fights are rarely harmful to the birds but occasionally small featherless areas can be seen on the upperparts as a result of such aggressive encounters.

Hummingbirds bathe several times a day. Some sit in shallow water and splash like sparrows, while others cling to rocks beside waterfalls gathering moisture and spray from above, vibrating their wings, and ruffling their body feathers. Hermits and many trochilines hover above gently flowing forest streams, then abruptly drop into the water, sometimes almost completely submerging the body. These dives are sometimes repeated over several minutes.

In their generally crowded habitats, hummingbirds are often heard before they are seen. Unmelodious, high-pitched, monosyllabic chirps and whistles are the sounds most commonly heard during initial encounters with these birds. Their calls, often lasting less than half a second, are usually uttered by both males and females between feeding probes or from exposed perches in shrubs or tree-tops, and indicate territorial birds occupying nectar-rich food sources. Chase calls – a series of aggressive rapid chatterings known from several species – are heard from individuals defending a feeding territory. These loud vocal signals, employed by either sex, are species-specific and are an important feature for field identification.

Some hermit species of the genus *Phaethornis* and some trochilines, such as the violet-ears, belong to the most persistent daytime singers of all hummingbirds, producing their vocal repertoire tirelessly from early morning to sundown. They are only silent during the months of body molt, when singing ceases in most of the hummingbirds studied.

Besides the squeaky, high-pitched, persistent advertising songs performed by many lekking males (*Phaethornis*, *Amazilia*, *Polyerata*), a low-volume warbling subsong of fledglings and adults of both sexes is typical for many trochilid species. Very often aerial displays are accompanied by a specific vocal repertoire and by mechanical sounds, such as tail-feather and wing noises in *Selasphorus*, that are not employed at other times.

Three Developmental Stages

BREEDING BIOLOGY

Male hummingbirds mate with several females during a reproductive period. All remaining reproductive responsibilities like nest-building, incubation, and rearing the young are carried out solely by the female. The onset of breeding in hummingbirds is very variable from species to species and from region to region. As a general rule, the peak of reproduction in most trochilids is closely associated with the months of mass-flowering of many ornithophilous plants. In the high Andes of Ecuador, violet-ears, metaltails, trainbearers, and pufflegs begin breeding during the wet season, often around mid-October, and continue to March or sometimes April. At similar altitudes further north or south, reproduction starts about three months earlier or three months later than in the mountain regions near the Equator, the breeding season often lasting only a few weeks. At lower altitudes, seasonality of the reproductive cycle declines and nests of several species may be found throughout the year, with decreasing numbers during peaks of the dry and wet season.

Females of many non-hermits select nest-sites after a nearby rich nectar source has been located. Suitable branches for nesting are first inspected by hovering above the surface and touching down over the spot repeatedly. Nest-sites of hermits, however, are not associated with nearby food sources. Females of this subfamily often cling with

Above *Many hummingbirds exhibit sexual dimorphism in their plumage, with the males sporting the brighter colors to attract partners for mating. Here the red hues of the male Rufous hummingbird (right) contrast with the relatively dull colors of the female (above).*

Left *A violet-ear (Colibri coruscans) emerges from a quick dip in a South American river. Hummingbirds need to bathe regularly to keep their wings clean and in good condition, as well as to cool down on hot days.*

their feet to suitable green leaves of palms or *Heliconia* stands, to which they later attach their cone-shaped nests. This behavior may serve to test the strength of the stratum for the purpose of nest construction.

Nests can be found at all heights, from a few centimeters above the ground to tree-top level (10–30m/33–98ft). Even within a species, nest-sites may vary from low secondary vegetation to canopy level. Although all hummingbird nests are accessible in flight, only a few species build them completely exposed. Nests are commonly placed in locations giving some protection from direct sun and rain by overhanging leaves. For nest-site selection balanced microclimatic conditions like temperature and humidity seem to be major requirements for ensuring the successful development of the embryos. Thus nests are often located near waterfalls, forest streams, or lake shores. The process of constructing a nest lasts around 5–10 days for most species studied. Females repair nests regularly, especially during the incubation period.

The clutch of all hummingbirds consists of two eggs, which are white, non-glossy, and of elliptical oval shape. Only in the Stripe-tailed hummingbird (*Eupherusa eximina*) of Central America have bright pinkish eggs been reported, besides the regular white clutches. This difference in egg coloration is not genetic, but caused by the red oak lichen sometimes used in the lining of the nest. Rain leads to a permanent chemical color reaction with the eggs and the belly feathers of the incubating female, which are regularly stained with the same pinkish hue.

The incubation period for most trochilids lasts for 16–19 days, about 2–5 days longer than in songbirds. Depending on the time of the start of incubation, eggs may hatch at intervals of 48 hours or almost synchronously. Hatchlings of all hummingbirds are altricial, blind, and helpless. During their long nestling period of 23–26 days (or, in high-Andean trochilids, 30–40 days) the following three well-differentiated morphological stages of development can be observed.

In the first, from day one to about day five after hatching, the nestling is nearly naked, except for

*◐ **Left** A Long-tailed hermit* (Phaethornis superciliosus) *roosts in a cone-shaped nest attached to the underside of a palm leaf.*

*◑ **Right** On the Argentine pampa a Glittering-bellied emerald* (Chlorostilbon aureoventris) *feeds her well-feathered chicks. By this late stage of nestling development, the disturbance of the air caused by the mother's beating wings triggers the gape response.*

two dorsal rows of neossoptile down about 5mm (0.2in) long, and the eyes are still closed. During this stage the chicks, generally two in number, are inactive in the nest. When the female arrives with food she lands on the edge of the nest and touches the nestlings behind the eye-bulges with her bill. In response to this stimulus, the young birds gape and are fed by the female, which inserts her fine bill into each nestling's mouth and regurgitates food consisting of nectar and tiny arthropods from her crop into that of the chick. Gaping in nestlings of that age can easily be artificially induced several times by touching the eye-bulges with a matchstick. No begging call can be heard at this stage.

The second stage, from day six to day nine, when the eyes begin to open, is the period of major feather development on the wings, tail, and back. The dorsal down is not shed but remains attached to the contour feathers. Begging calls are still not heard at this time.

The third stage of nestling development runs from day ten until fledging. By the beginning of this period chicks are almost completely feathered and often sit on the edge of the nest facing outward, but they still fail to give begging calls.

*◐ **Left** A Giant hummingbird chick panting on a nest atop a prickly pear in Ecuador. Few hummingbirds build such exposed nests, although here the cactus spines at least provide protection from land predators.*

*◒ **Below** All hummingbirds commonly lay two white eggs, whose tiny dimensions reflect the birds' own. This nest belongs to the Magnificent hummingbird* (Eugenes fulgens)*, actually a medium-sized species.*

The exposed nest-site and low reproductive output probably favored the development of the highly specific gape-response behavior. Loud begging calls of the offspring, as well as uncontrolled begging and gaping movements through non-specific causes like vibations of the nest by wind, are types of behavior that could potentially betray the nest-site to predators. Thus the highly specific stimuli that elicit gaping by unfledged hummingbird chicks are most likely adaptations for reducing predation on exposed nest-sites. Consistent with this observation is the fact that chicks of those hummingbirds, such as sylphs and metaltails, that build domed nests give begging calls soon after hatching, presumably in response to tactile stimuli from the female as she enters the nest.

Slaughtered for Show

CONSERVATION AND ENVIRONMENT

In the second half of the 19th century, millions of hummingbird skins were exported from Central and South America, not only to decorate ladies' hats and clothes but also for the manufacture of feather pictures, ornaments, and artificial flowers. For example, in London in 1888, 12,000 trochilid skins were sold in one month; at one sale held in the city, 37,603 hummingbird skins were auctioned. A single delivery contained 3,000 Ruby topaz skins. Some species were probably hunted to extinction in this slaughter, since a few that were first described from such skins were never subsequently seen in the wild, although the hybridization between species, and even genera, often reported for trochilids might perhaps have been responsible for some of these forms. KLS

During the second and third stages, the female gradually approaches the nest and begins to hover over the feathered young with an increasing wing-beat frequency that is clearly audible. When the dorsal neossoptile down attached to the chicks' contour feathers is visibly agitated by the resulting air movement, the chicks invariably begin to gape; gaping can easily be triggered in chicks of this age by blowing with a straw on their dorsal neossoptiles. The gape-releasing stimulus during this stage of development has changed and no touching of the eye-bulges by the female can be observed. When gaping, and also during feeding, the chicks raise themselves only slightly and remain in a rather stooped position in the nest.

From about15 days old, chicks may sit at the edge of the nest during the day, mostly with their backs turned away from the cup. At feeding time the female hovers just above the chicks, causing movements of their dorsal neossoptile plumage; only after this stimulus will the nestlings gape and be fed. In all trochilids raised in open, cup-shaped nests, begging calls are not heard at this nestling stage, but only after fledging. They normally give loud calls, irrespective of the presence of the female, but more vigorously when she is in view.

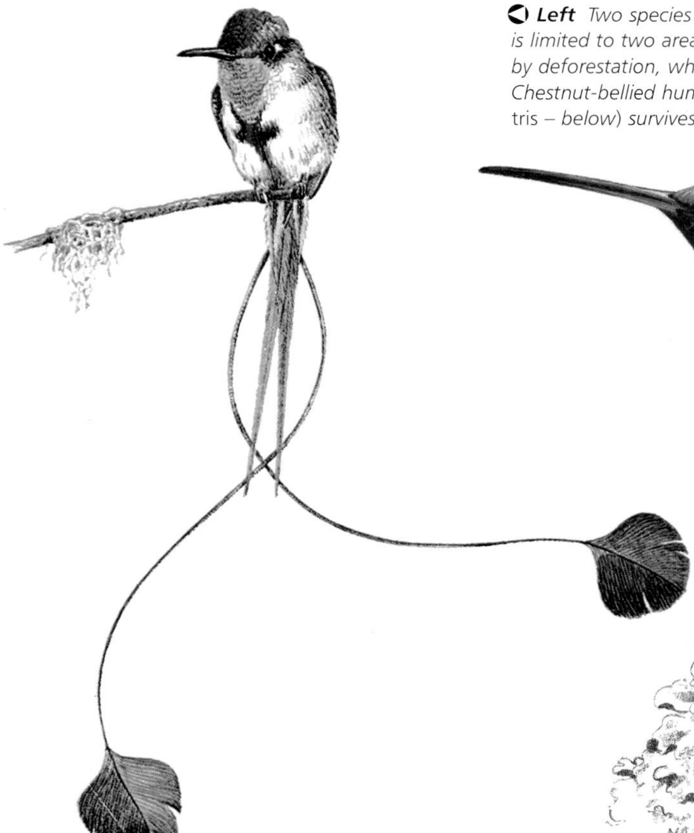

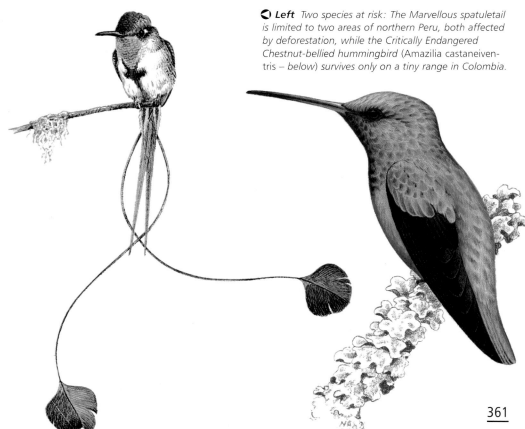

Left *Two species at risk: The Marvellous spatuletail is limited to two areas of northern Peru, both affected by deforestation, while the Critically Endangered Chestnut-bellied hummingbird* (Amazilia castaneiventris – *below) survives only on a tiny range in Colombia.*

Trogons

TROGONS ARE COLORFUL BIRDS OF TROPICAL forests. Although they are superficially parrotlike, with an upright stance, strong bills, and short legs, they have no close relatives among living birds and are placed in an order by themselves. While the arrangement of their toes, with two pointing backward, is similar to that of parrots, a closer look reveals that it is the first and second toes that are turned back – an arrangement unique to the trogons.

The New World boasts the greatest diversity of trogons, with 23 species in three genera distributed from the southwestern United States (southeastern Arizona) south to northern Argentina; a fourth genus consists of two species, the Cuban and Hispaniolan trogons, which are endemic to islands in the Caribbean. The Elegant trogon occurs over a remarkably wide range of elevations and habitats, from the humid tropical forests of Costa Rica to the cool pine-oak woodland of the US Southwest up to 2,500m (8,000ft).

FACTFILE

TROGONS

Order: Trogoniformes

Family: Trogonidae

37 species in 7 genera. Species include: **Bar-tailed trogon** (*Apaloderma vittatum*), **Narina trogon** (*A. narina*), **Eared trogon** (*Euptilotis neoxenus*), **Orange-breasted trogon** (*Harpactes oreskios*), **Red-naped trogon** (*H. kasumba*), **Resplendent quetzal** (*Pharomachrus mocinno*), **Cuban trogon** (*Priotelus temnurus*), **Hispaniolan trogon** (*Temnotrogon roseigaster*), **Elegant** or **Collared trogon** (*Trogon collaris*).

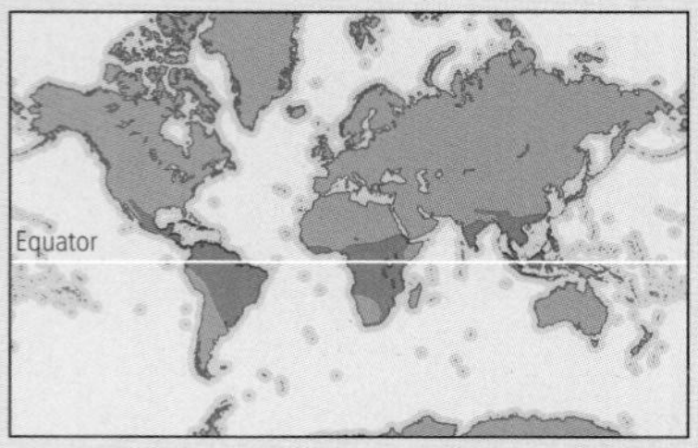

DISTRIBUTION Southern half of Africa, India, and SE Asia; Malaysia, Philippines; SE Arizona, Mexico, C and S America, W Indies.

HABITAT Forest, woodland, and secondary growth, sea level to over 3,000m (9,850ft).

SIZE Length 23–38cm (9–15in).

PLUMAGE Soft and dense, with adult males having the belly and undertail coverts shades of red, orange, or yellow and the head, breast, and upper parts often brilliant iridescent green or blue. Females and juveniles may be similar to adult males or duller.

VOICE A variety of simple calls, including hollow whistles, gruff barks, hoots, churrs, whines, and squeals.

NEST In cavities.

EGGS 2–4, white or buff to greenish blue. Incubation period 17–19 days; nestling period 17–28 days.

DIET Insects, spiders, small frogs, lizards, snails, small fruits.

CONSERVATION STATUS 10 species, including 6 in the genus *Harpactes*, are currently classified as Lower Risk/Near Threatened.

Left *The Elegant trogon has the widest range of any trogon species, stretching from the Costa Rican rain forest to a small area of the southwestern United States, where it is eagerly sought out by local birders.*

In Africa, three species are distributed across the humid tropical zone. The commonest and most widespread is the Narina trogon, found from lowland forest up to 3,300m (almost 11,000ft) in montane forest. The Bar-tailed trogon overlaps in range with the Narina but is confined to highland forests, mainly above 1,600m (5,000ft).

The 11 species of Asian trogons belonging to the genus *Harpactes* range from western India to southeast China, the Southeast Asian mainland, and the islands of Indonesia. The Malay/Indonesian region is by far the richest in trogon species, with eight present in Sumatra and six in Borneo. Several species are widely distributed, among them the Orange-breasted trogon, which lives in evergreen forest from Myanmar to Malaysia, from Thailand and southwest China to Java.

A Rainbow of Colors

FORM AND FUNCTION

Trogons are built for life in the trees. Their stubby legs are virtually useless for walking, while their short, rounded wings and long tail provide for excellent maneuverability in flight. They can even hover for brief periods, a skill useful for plucking fruits or small animals from foliage or branches. The stout, often serrated bill is used to soften hard fruits, kill small prey, and excavate nest cavities in rotting wood or termite nests. Their voices are not melodious by human standards but can carry long distances through dense vegetation.

The 25 species of trogons found in the Americas are richly iridescent on the upperparts in shades of green, bronze, blue, or violet, contrasting with red, pink, orange, or yellow underparts. Females are usually very similar to adult males, differing mainly in the overall intensity of coloration or merely in the markings on the outer tail feathers. However, in females of some species ruddy brown, taupe, or charcoal gray replaces the bright green or blue iridescence of the male. The three African trogons in the genus *Apaloderma* are strikingly similar to their New World counterparts except for small patches of brightly coloured bare skin on the face. The 11 *Harpactes* trogons native to Asia lack the vivid metallic colors of their African and American relatives, but they compensate with patches of bright scarlet, pink, orange, or cinnamon on the head, rump, or underparts. Females are usually duller than males, but both sexes have a bare patch of skin around the eye.

The most widely known member of the family, the Resplendent quetzal of southern Mexico and Central America, is one of the world's most beautiful birds. The male is entirely glittering metallic green above and crimson below, with a narrow, ridged crest formed by bristly, hairlike feathers. The metallic green wing coverts are very long and curved, with pointed tips extending beyond the edge of the folded wing. The uppertail coverts are equally well developed and normally extend just past the tips of the dull black central tail feathers. In the breeding season, the adult male grows a central pair of coverts up to twice the length of the body, forming a graceful curving train that hangs below the bird when perched and ripples behind it in flight. Four other quetzals native to South America are equally colorful but have coverts that barely reach past the tip of the tail.

Swallowing Fruit Whole

DIET AND BREEDING BIOLOGY

Fruit and invertebrates make up the bulk of most trogons' diet. Large-seeded fruits, including relatives of the avocado, are swallowed whole; the stone is regurgitated after the nutritious flesh has been digested. Trogons use their strong bills to pluck prey animals from leaves and branches in much the same way that they harvest fruit, hovering briefly before swooping away to a perch to feed. Prey usually includes medium to large insects such as caterpillars and cicadas, but larger trogons often eat small vertebrates such as lizards and frogs. Animal food is particularly important during the nesting cycle, and breeding is often timed to coincide with the greatest abundance of prey.

Both sexes use calls to attract mates and defend territories. Typical nest sites include natural cavities, abandoned woodpecker holes, or new cavities excavated in rotten tree trunks. A few tropical species excavate nest cavities within an arboreal termite nest; the termites quickly seal the walls between the new chamber and their nest and will reclaim the cavity once the birds have abandoned it. In the Elegant trogon, the male scouts potential nest cavities before inviting the female to inspect them. The pair make low, muttering croaks to each other during this ritual before the female signals acceptance or rejection of the site. The eggs may be buff, whitish, or blue, as in the five species of quetzals and the closely related Eared trogon. Both parents cooperate in incubation and care of the young, which may remain dependent for weeks or months after leaving the nest.

The Aztec Survivor

CONSERVATION AND ENVIRONMENT

Although still common in many areas, trogons are threatened by the widespread destruction of tropical forests for wood products and agriculture. Unlike many other brightly colored tropical birds, they are difficult to maintain in captivity and so are not sought after for the caged-bird trade.

Among the most threatened is the Resplendent quetzal, whose plumes were prized more than gold among the ancient cultures of southern Mexico and northern Central America. The bird's very name derives from Nahuatl, the language of the Aztecs, and may be translated as "feather," "precious," or "beautiful." So valuable were these plumes as a renewable resource that they were harvested from live birds that were captured and then released unharmed. Among the Maya, the killing of a quetzal was an offense punishable by death, but the Spanish conquest brought this protection to a sudden end. For almost 400 years the birds were slaughtered with impunity, reducing thriving populations to scattered remnants in remote mountains. Despite legal protection in modern times, the Resplendent quetzal remains rare in many areas where it was once abundant. The greatest threat to its existence is the relentless destruction of its highland habitat for timber and agriculture, especially coffee plantations. SLW/PRC

Above The greatest glory of the Resplendent quetzal, found from southern Mexico to Panama, is its tail feathers, which can grow 60cm (24in) long.

Below The Red-headed trogon is one of 11 closely-related Asian species grouped in the genus Harpactes. A distinctive feature is the oddly patterned tail.

Mousebirds

SUPERFICIALLY, MOUSEBIRDS LOOK AND behave like mice, hence their name. They are dumpy, medium-sized, brown or gray birds with a long tail and are usually seen in small groups running or clambering under or through thick bushes. Mousebirds are very common and highly sedentary. They have no close living relatives and have been assigned their own order.

Above *Speckled mousebirds* (Colius striatus) *in flight. They move from tree to tree with rapid wingbeats and long glides.*

Mousebirds occur throughout Africa's bushy savanna and woodland, as well as in areas of secondary growth, forest edges, and arid thorn scrub. They are even found in gardens and on cultivated land. The only places they do not inhabit are dense forests, deserts, or mountain peaks. They are often found near watercourses, although they do not appear to need to drink copiously.

Color-coded Climbers

FORM AND FUNCTION

All six species are very similar, being rather loosely feathered in pale brown or gray all over with paler underparts. The two genera are distinguished mainly by various skeletal features. Most have a distinctive colored patch – for example, a chestnut rump on the Red-backed mousebird, a white rump on the White-backed mousebird, a pale blue back to the head on the Blue-naped mousebird, and a white crest on the White-headed mousebird. Almost all have a crest (the exception is the White-backed, with a bare patch around its face) and a long, stiff, strongly graduated tail. Their wings are short and rounded and flights usually comprise a series of rapid, whirring wingbeats followed by a longish glide. When clambering around among twigs, they will customarily hold the two central toes forward, while the other two toes, being extremely mobile, can be pointed forward, backward, or to the side. This enables them to climb rapidly and with great agility through bushes – clinging on underneath twigs rather than perching on top of them – and to use their feet to transfer food to their bill.

Mousebirds have short, stubby, finch-like bills and feed mainly on berries, fruits, and other plant matter, including flowers and leaves of plants known to be poisonous to other vertebrates. They have even been recorded eating the poisonous dogbane *Acokanthera oblongifolia*, an extract of which ("bushman's poison") is used by some indigenous peoples to tip hunting arrows. They occasionally eat animal matter, and have been accused of cannibalizing nestlings. Some species have been seen taking moist or clayey earth, presumably in order to aid their digestion (especially in the afternoons, when they have been eating predominantly leaves). Feeding occurs throughout the day but is interspersed with many rest periods, during which groups may huddle together.

A large amount of time is also spent preening, dustbathing, and sunbathing, perhaps to reduce their ectoparasite loads (which can be high). Sunbathing also helps the birds maintain a constant body temperature.

Sociable Territory Defenders

SOCIAL BEHAVIOR

All species live in family groups generally numbering 3–20 individuals. These groups are maintained year-round, although larger aggregations will sometimes form at fruiting trees. Mousebirds are very sociable, with flock members allopreening regularly, frequently hanging belly-to-belly from a twig or branch; at night they will often roost huddled closely together. At least four species have the capacity to allow their body

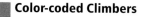

FACTFILE

MOUSEBIRDS

Order: Coliiformes

Family: Coliidae

6 species in 2 genera: **Speckled mousebird** (*Colius striatus*), **Red-backed mousebird** (*C. castanotus*), **White-backed mousebird** (*C. colius*), **White-headed mousebird** (*C. leucocephalus*), **Red-faced mousebird** (*Urocolius indicus*), **Blue-naped mousebird** (*U. macrourus*).

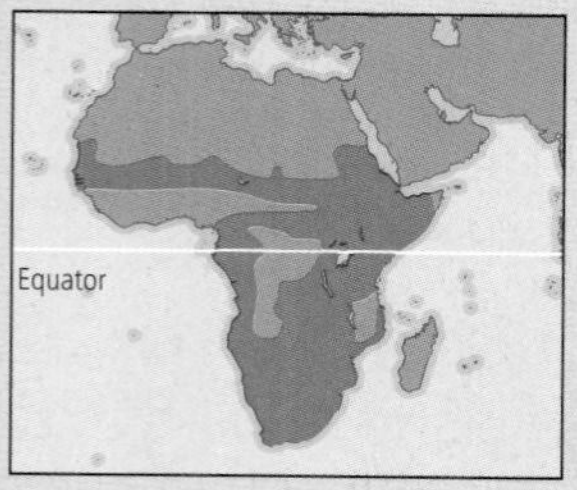

DISTRIBUTION Africa S of the Sahara.

HABITAT Open woodland, bush country, thorn scrub, gardens; avoids dense forest and deserts.

SIZE Length 30–35cm (12–14in), most of which is made up of a tail some 20–25cm (8–10in) long; weight 35–70g (1.3–2.5oz).

PLUMAGE Light brown or gray, lighter below; loose crests; some have bright face or neck marks (white, red, or blue); long tail strongly graduated; sexes similar.

VOICE A single, whistle-like note or a series of more twittering notes.

NEST In an open cup, sometimes bulky and untidy, usually in a thick, thorny bush.

EGGS Usually 2–4, whitish with blackish or brownish streaks; size 20–23 × 15–18mm (0.8–0.9 × 0.6–0.7in).

DIET Leaves, fruits, and berries; will occasionally take nestlings of other species.

CONSERVATION STATUS Not currently at risk.

male, but not exclusively) sitting on a branch or the ground and jumping up and down rhythmically for several minutes; the performance culminates in copulation.

The nests are usually rather untidy, open bowls made of twigs situated a few meters up in a thick or thorny bush. In Namibia the nests of Speckled and Red-faced mousebirds are often sited near that of an aggressive wasp. The eggs (usually 2–4 in a clutch) are the smallest in proportion to the female's weight of any bird except parasitic cuckoos. Eggs are incubated by both members of the pair, sometimes even both together, and occasionally by a helper, for 11–15 days. The young may leave the nest, before they can fly, from 10 days old, but they continue to be fed by all members of the group on regurgitated plant material for 4–6 weeks, and survivors will usually remain with the natal group for a while after fledging. Females are more likely to leave the group than males.

No species is known to be threatened. However, two species – the White-headed mousebird in East Africa and the Red-backed mousebird in western Angola – have very restricted ranges. The Speckled mousebird can be a pest to ornamental shrubs in gardens and to fruit in smallholdings, but the birds rarely range far into extensive plantations. They have also been accidental victims of pesticide campaigns aimed at such persistent nuisance birds as the Red-billed quelea. PL/CMP

Left *Pair of speckled mousebirds roosting belly to belly with their tails pointing straight down and heads up. While resting these birds have a low metabolic rate, which allows time to loaf, preen, sleep, and play.*

Below *A Red-faced mousebird feeds its chick with partly digested food by regurgitation. When only a few days old, the young leave the nest and crawl about on the nearby branches but return to be brooded at night.*

temperature to drop at night, an adaptation that is associated with a reduction of body weight below the norm. This behavior may happen quite regularly, since their specialized diet can result in an energy deficiency.

Vocalizations are a variety of whistles and twittering contact notes and are used especially in defense of territory. Often a whistle will precede a succession of birds leaving one bush in single file and flying across open ground before crashing into an adjacent bush and resuming feeding.

Groups defend a territory year-round and, although most are cooperative breeders with helpers being younger members of both sexes, they seem to be mainly or exclusively monogamous. Breeding occurs mainly in wetter months, with the precise timing closely related to available fruit supplies. Courtship involves the male feeding the female and rubbing bills with her, but may also include a display known as "jumping," which has been noted in four of the six species. This display takes the form of an individual (normally a

Kingfishers

A STAB OF ELECTRIC BLUE CONTRASTED WITH the warm chestnut orange of the underparts – such is the image Europe's Common kingfisher most often presents. And then it is gone, leaving behind the fleeting impression of a living cobalt and azure jewel. If legend is to be believed, the Common kingfisher was once a dull gray bird that acquired its fantastic colors when it left Noah's Ark.

Similar small, brilliant-blue kingfishers live on the other side of the world too, but in addition Australia and New Guinea boast the kookaburras. Large, noisy and much-loved birds of gardens and dry woodlands, kookaburras are quite gregarious in their habits, and live in trees, from which they pounce to the ground for their animal food. New Guinea and adjacent archipelagos have the greatest number of kingfisher species in the greatest diversity of forms, although Africa and south Asia are rich in species too. Other species, nearly all of them colorful and attractive, are to be found in the Americas and on hundreds of islands scattered throughout the Pacific Ocean.

Sit-and-Wait Plunging Predators

FORM AND FUNCTION

Kingfishers are bright-plumaged, monogamous, more or less solitary birds of forests, savannas, and watersides. The great majority of species are tropical, but one or two species from each subfamily have extended as migrant breeders into temperate latitudes. Primitive species are forest-dwelling predators feeding mainly on forest-floor insects. More specialized types plunge into shallow water for small animals, flycatch for airborne insects, forage in leaf-litter for earthworms, prey on birds and reptiles, and deep-dive for fish from a perch or (particularly the Pied kingfisher) from hovering flight.

Like other birds of their order, kingfishers are large-headed, short-necked, stout-bodied, and short-legged, with weak, fleshy feet on which the second and third toes are partly joined. The bill is straight, strong, and long, flattened from top to bottom in insectivorous species and from side to side in fish-eaters. The extraordinary Shovel-billed or Earthworm-eating kingfisher of New Guinea has a short, wide, conical bill. Other forms have the bill sharp-pointed and daggerlike, but in the adult African dwarf kingfisher it is blunt-tipped (sharp in the juvenile). For no obvious reason, several not-closely-related lineages of kingfishers are three-toed, having lost the fourth toe. Plumage and other characters show that three-toed species are very closely allied with some four-toed species in the genera *Ceyx* and *Alcedo*, and that the three-toed kingfishers do not comprise a single natural assemblage, as they were formerly held to do.

Although kingfishers are colorful, the colors are in general muted, with shades of blue and red predominating. Shoulders and rump are usually a

Right *Sometimes mistaken for the Pygmy kingfisher, the Malachite kingfisher* (Alcedo cristata) *bears a lovely cobalt blue crest with black bands, whereas the Pygmy does not have a crest. The Malachite also differs in that it is a waterside, not a woodland, bird.*

Left *Representative kingfisher species:* ***1*** *Blue-breasted kingfisher* (Halcyon malimbica)*;* ***2*** *Belted kingfisher* (Megaceryle alcyon)*;* ***3*** *Amazon kingfisher* (Chloroceryle amazona)*. These three species give a hint of the huge variety of colors within the family Alcedinidae, ranging from blues, greens, and reds to plain black-and-white.*

shining azure blue, and a dark cap and back are commonly separated by a white or pale collar. Juveniles of paradise kingfishers are dusky, differing markedly from the adults, but in other species juveniles are bright in plumage, although duller than adults. There is little geographic variation within a species, and evolutionary color conservatism has led to allied species looking much alike. Notable exceptions are the Variable dwarf kingfisher, whose subspecies on islands from the Philippines to the Solomons vary from red to blue or yellow, Africa's Grey-headed kingfisher, and the much larger Black-capped kingfisher of China. Although the last two differ in appearance, biochemical and biological characteristics, as well as the geographical relationship of their ranges, suggest strongly that they are of immediate descent from a common ancestor.

Kingfishers living on dry land are sit-and-wait predators of small animals on the surface of the ground. Those living by water are true fishers. All have very good eyesight, but the fishers have particular problems to overcome, because of light refraction, which makes prey appear to be nearer to the surface of water than they really are, and because of light reflection from rippling or choppy surfaces. Kingfishers have a limited degree of movement of the eyes in their sockets, and make up for it by moving the whole head, fast and flexibly, in order to track rapidly-moving prey. Sacred kingfishers are thought to be able to spot a small prey animal up to 90m (300ft) away. Belted kingfishers are sensitive to near-ultraviolet light, which may also assist in prey detection.

FACTFILE

KINGFISHERS

Order: Coraciiformes

Family: Alcedinidae

86 species in 14 genera

Species include: **Blue-eared kingfisher** (*Alcedo meninting*), **Common kingfisher** (*A. atthis*), **Green-backed kingfisher** (*A. monachus*), **Moustached kingfisher** (*A. bougainvillei*), **Little kingfisher** (*A. pusilla*), **Scaly-breasted kingfisher** (*A. princeps*), **African dwarf kingfisher** (*Ceyx lecontei*), **Oriental dwarf kingfisher** (*C. erithaca*), **Pied kingfisher** (*C. rudis*), **Sulawesi dwarf kingfisher** (*C. fallax*), **Variable dwarf kingfisher** (*C. lepidus*), **American pygmy kingfisher** (*Chloroceryle aenea*), **Green kingfisher** (*C. americana*), **Green-and-rufous kingfisher** (*C. inda*), **Lilac-cheeked kingfisher** (*Cittura cyanotis*), **Earthworm-eating** or **Shovel-billed kingfisher** (*Clytoceyx rex*), **Laughing kookaburra** (*Dacelo novaeguineae*), **Beach kingfisher** (*Halcyon saurophaga*), **Black-capped kingfisher** (*H. pileata*), **Collared kingfisher** (*H. chloris*), **Grey-headed kingfisher** (*H. leucocephala*), **Mangrove kingfisher** (*H. senegaloides*), **Ruddy kingfisher** (*H. coromanda*), **Woodland kingfisher** (*H. senegalensis*), **Crested kingfisher** (*Megaceryle lugubris*), **Ringed kingfisher** (*M. torquata*), **Giant kingfisher** (*M. maxima*), **Stork-billed kingfisher** (*Pelargopsis capensis*), **Common paradise kingfisher** (*Tanysiptera galatea*), **Sacred kingfisher** (*Todirhamphus sanctus*), **Tuamotu kingfisher** (*T. gambieri*).

DISTRIBUTION Worldwide, except for very high latitudes.

HABITAT Interior of rain forests, woodlands far from water, desert steppe, grassy savannas, streams, lakeshores, mangrove, seashores, gardens, montane forest, oceanic islands.

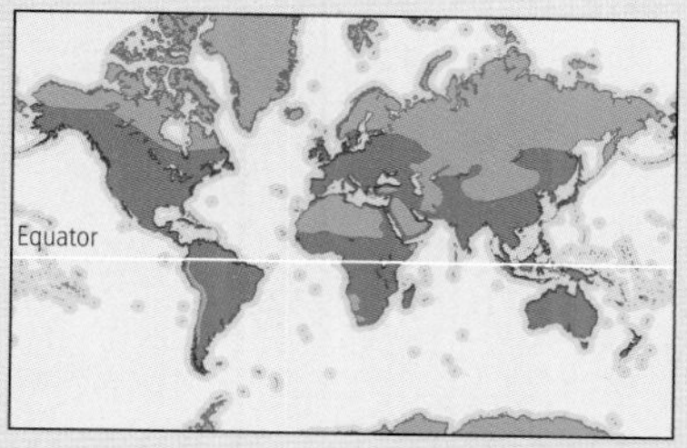

SIZE Length 10–45cm (4–18in), excluding any tail streamers; **weight** 8–500g (0.3–18oz); females in many species slightly larger than males.

PLUMAGE Azure blue above and reddish below; also light and dark blue, green, brown, white, and black; bill and legs vermilion, brown, or black. Males and females similar in most species, marked variations in a few.

VOICE Ringing notes in slowing tempo and falling cadence; single loud, coarse cries; occasional weak, quiet notes.

NEST In earthen holes excavated by the birds, including in termitaria on the ground or in trees, and in tree holes.

EGGS Clutches vary from 2–3 in tropics up to 10 at high latitudes: white; weight 2–12g (0.07–0.4oz). Incubation period 18–22 days; nestling period 20–30 days.

DIET Terrestrial arthropods and small vertebrates, aquatic insects and fish.

CONSERVATION STATUS The Marquesas kingfisher (*Todirhamphus godeffroyi*), is classed as Endangered. A further 11 are listed as Vulnerable, including the Blue-capped kingfisher (*Actenoides hombroni*).

Left When plunging into water, the Common kingfisher's eyes are protected by a membrane, which means that when it catches its prey it relies on touch to decide when to snap its bill shut.

Right The Pied kingfisher is among seven species that are able to hover over their prey. They can dive from 12m (39ft) high; in rough waters they make four times as many dives from hovering as from perches.

All kingfishers have two foveae in each eye – depressions on the retina with very numerous light-detecting cone cells. Fields of vision overlap to the front, giving binocular vision there, and one of the foveae in each eye looks out into the binocular field; the other looks into the monocular field at each side of the head. Experiments show that the hunting kingfisher instantly detects prey as its image crosses the monocular fovea; then, when the head is angled in the usual manner with the bill pointing down at an angle of about 60°, a slight rotation of the head transfers the image to the binocular fovea in one or both eyes, allowing the prey's distance away to be precisely evaluated. Pied kingfishers can catch a fish 2m (6.5ft) below the surface, the bird achieving that depth by diving vertically from a height of 2–3m (6.5–10ft) above the water. In the split second when a Common kingfisher enters the water surface, it turns its wings backward at the shoulder joint, and a cover of translucent skin, the nictitating membrane, passes over the eye from front to back, protecting it. Shooting down like an arrow, the bird uses its wings as brakes and in the same instant seizes its prey between its mandibles, retracts its neck, turns its body, and flies up through the water and then the air, often nearly along the same course as its dive.

The evolutionary history of other groups of kingfisher species is better understood than for most groups of birds. The family almost certainly arose in tropical rain forest, partly in the northern Australasian region (in the case of the insectivorous woodland kingfishers of the subfamily Daceloninae) and partly in adjacent Indonesia, Borneo, and southeast Asia (forest insectivores, evolving into waterside fishers of the subfamily Alcedininae). Both subfamilies extended into Asia and repeatedly invaded Africa, on as many as 12 separate occasions; the Alcedininae invaded the New World to give rise to the Green and Giant kingfishers there (in addition to the Cerylinae).

The several Pacific archipelago species of woodland (*Halcyon* and *Todiramphus*) kingfishers have clearly evolved from the wideranging complex formed by the Collared and Beach kingfishers and the more southerly Sacred kingfisher. Mangrove, Woodland, and Blue-breasted kingfishers are similarly of recent descent from a single ancestor; their habitats keep them apart, although they are acquiring sufficient ecological differences to permit some degree of geographical overlap. Belted, Ringed, Giant, and Crested kingfishers, respectively in North America, tropical America, Africa, and southern Asia, are all very closely allied, and it is thought that the Giant and Crested descended from small populations of the first two that crossed the Atlantic (Belted kingfishers occasionally still arrive in Europe as vagrants).

Species multiplication is also demonstrated by the four green kingfishers of the neotropics. Long ago, their common ancestor there separated into two geographically distinct populations that duly happened to evolve differences of size, enabling them to overlap as distinct species. Later, each of the two species repeated the separating process, and the result today is four species all occupying much the same range, having body-weights close to the proportions 1:2:4:8, with the smallest and second-largest (the American Pygmy kingfisher and the Green-and-rufous kingfisher) being almost alike in appearance, and the largest and second-smallest (Amazon and Green kingfishers) also being remarkably similar.

Mostly in the Tropics

DISTRIBUTION PATTERNS

The end result of all of these ancient and recent historical movements within and between continents and across the oceans is a wealth of diverse kinds in many regions. Temperate zones in the northern tropics do not do so well, and only single species reach as far north as the Gulf of Fin-

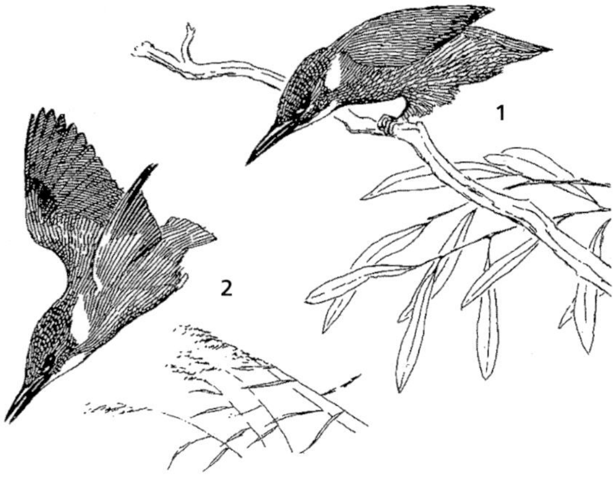

Above The kingfisher's dive. **1** The kingfisher has spotted a fish and tenses for the dive. **2** A 45° plunge with powerful wingbeats takes it to the water. **3** The kingfisher enters the water, having made last-second adjustments to its aim by fanning the tail feathers. **4** With eyes closed, the fish is seized. **5** The kingfisher, its eyes still closed, emerges from the water with the fish. It returns to its perch and swallows the fish head-first after beating it against a branch.

Above *A Red-backed kingfisher (*Todirhamphus pyrrhopygia*) returns to its nest hole with a small amphibian catch. Like other forest kingfishers, it hunts a variety of arthropods and small vertebrates.*

land and the western shores of the Sea of Okhotsk (Common kingfisher) and Alaska and Newfoundland (Belted kingfisher). Central and South America possess five, the huge Ringed kingfisher and the four small or medium-sized green kingfishers. Africa and Madagascar have 18 species. From India to Japan and Cambodia 12 species reside, more passing through on migration. The Philippines have 11 species, six of them occurring nowhere else; Malaysia and Indonesia boast another 11 species; and Sulawesi possesses 11 also, of which five are found there alone.

In New Guinea, the Bismarck Archipelago, and Australia's tropical Cape York Peninsula there are 16 woodland kingfishers and kookaburras and three fishing ones. The rest of Australia has six; and Oceania, from the Solomon Islands and New Zealand to Tahiti and the Tuamotu Archipelago, hosts 11 species, seven of them endemic. To single out the smallest of these areas, Sulawesi is home to a variety of kingfishers representing two of the three main groups or subfamilies (families, according to some researchers). Unique to the forests of that island are the Lilac-cheeked, Green-backed, Scaly-breasted, and Sulawesi dwarf kingfishers. Common paradise, Ruddy, Black-capped, Collared, Sacred, Blue-eared, and Common kingfishers inhabit Sulawesi too, each as part of a much more extensive range.

Not Just Fish

DIET

All fishing kingfishers take a certain amount of invertebrate prey in addition to fish; insects, for example, make up about 21 percent of the Common kingfisher's diet; most are aquatic, but some are caught on dry land. Pied kingfishers, fishing from hovering flight more than from a perch, are in that sense at the peak of the family's evolution; in Africa they live entirely upon fish (but in India take insects and crabs too, and can even "hawk" for flying termites). Not having to rely on a perch means that they can fish far from shore: on Lake Kariba they fish up to 3km (2 miles) offshore at dawn and dusk, catching sardines, deep-sea fish that rise to the surface at those times. In Natal, 80 percent of their fish food consists of *Sarotherodon mossambicus*, mainly in the 1–2g (0.035–0.07oz) weight class, and on Lake Victoria they prey almost exclusively on fish from the genera *Haplochromis* and *Engraulicypris*. When foraging close to the shore in windy conditions, they dive from hovering flight, as ruffled water seems to make fishing from a perch unrewarding; only when the surface is calm do they fish from perches to a greater extent than from hovering. A Pied kingfisher flies low over the water to a desired hunting station, then rises 10m (33ft) above the surface and hovers on rapidly beating wings. With the body held almost vertically and the bill pointing acutely down, the bird keeps station for 5–10 seconds, then dives steeply to penetrate possibly 2m (6.5ft) underwater, occasionally catching more than one fish at a time. The Belted kingfisher of North America behaves in a similar way.

The divide between preying on dry-land animals and aquatic ones is not precise. Sacred kingfishers live in woodland but often hunt from shrubs along ditches and lakeshores. They take a wide range of prey, including insects, spiders, earthworms, mollusks, crustaceans, centipedes, fish, frogs, tadpoles, reptiles, and even small birds or mammals. A study in a single region of the food and foraging relations of South America's five fishing kingfishers showed that fish-eating was proportional to the abundance of fish at the surface near the shoreline, with all types being taken as available; the larger kingfishers perched higher, and presumably dived deeper, and the average prey sizes were in proportion to the birds' different body sizes and bill lengths.

Shared Parenting

BREEDING BIOLOGY

Most kingfishers are monogamous and territorial, a pair defending an area of woodland on a stretch of river against incursion by other birds of the same species. Several species are migratory, both in the temperate zone and within the tropics; others are sedentary. What little is known of their breeding habits suggests that most species breed at the end of their first year, and are quite long-lived. Woodland kingfishers (*Halcyon* species) have a territorial advertising display, singing loudly and repeatedly from a conspicuous treetop perch, spreading the wings widely with the patterned undersides facing forward and rotating the body about the vertical axis. Other species have little by way of any courtship display. Both sexes dig the nest tunnel, and the male takes a minor role in incubation. The eggs hatch at about daily intervals, in the same sequence as they were laid, so nestlings vary considerably in size. They are fed by both parents equally.

Laughing kookaburras in Australia and Pied kingfishers in Africa have a more complex social system. Each has adult helpers at the nest, and in Pied kingfishers these are of two kinds: primary helpers (those helping their own parents) and secondary helpers (those helping an unrelated pair). A pair seldom has more than one primary helper but, particularly in places where the food resources are not so good, they usually have several secondary helpers. "Helping" includes defending territory and feeding the young in the nest and after fledging. This species breeds in loose colonies, the only kingfisher to do so.

At Risk on the Islands

CONSERVATION AND ENVIRONMENT

Kingfishers have not, in general, come into direct conflict with man. As fish-eaters, a few species have sometimes been viewed as pests on fishing streams and persecuted; but usually they are treated with respect and often with admiration. Formerly, great numbers of Common kingfishers were shot or netted to make fishing "flies" from their feathers, and in earlier times (in Britain at least) superstition caused the destruction of many, for a dried kingfisher corpse in the house was supposed to avert thunderstorms and keep out moths! Today, the harmful effects of humans on kingfishers are more accidental than deliberate, in the pollution of fresh waters and the modification of habitats, especially rain forest. Bird-catchers destroy many; at Jatinga in Assam great numbers of migrating Common, Stork-billed, Ruddy, and Oriental dwarf kingfishers are killed (and presumably eaten) when they are attracted to light beacons around the villages at night. In some Mediterranean countries, many kingfishers are killed by netting, shooting, and liming, although they are not target species.

Few populations of kingfisher are at great risk. However, so many species are confined to tropical rain forests, or to small Pacific islands or archipelagos, that their fate depends largely on the preservation of their habitats. The distinctive race of the Tuamotu kingfisher that lived until about 1922 on the island of Mangareva in the central Pacific is almost certainly extinct; the other race, only found on Niau Island in the Tuamotus, numbers only a few hundred birds and is classed as Vulnerable. The very poorly-known Moustached kingfisher is another threatened bird. It is restricted to hill forest on the islands of Bougainville and Guadalcanal in the Solomons; the total population is estimated at less than 1,000 individuals in all, and perhaps as few as 250. CHF

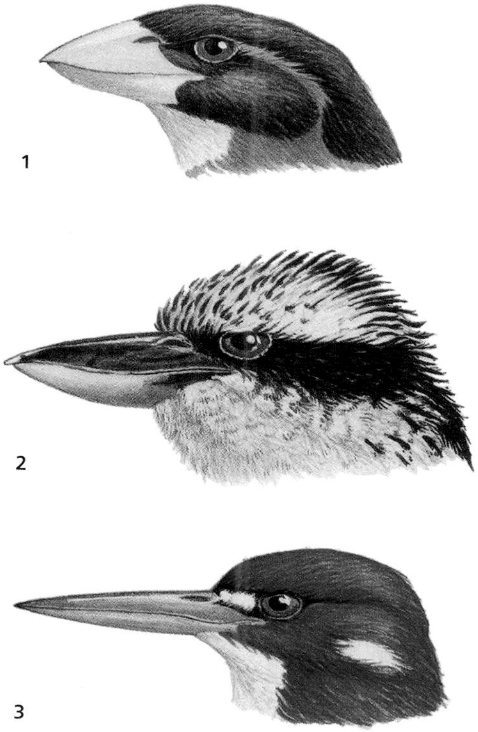

Above *Kingfishers' bills are adapted to different diets:* ***1*** *The Earthworm-eating kingfisher has a short, conical bill;* ***2*** *the Laughing kookaburra's stout bill helps it tackle lizards;* ***3*** *the pointed bill of the Little kingfisher characterizes a fishing species.*

Below *A male Common kingfisher takes his turn at brooding his 5-day-old chicks. Both sexes of kingfisher share in incubating, brooding, and feeding their young. The young hatch naked and blind, but within a week start growing their feathers.*

Motmots

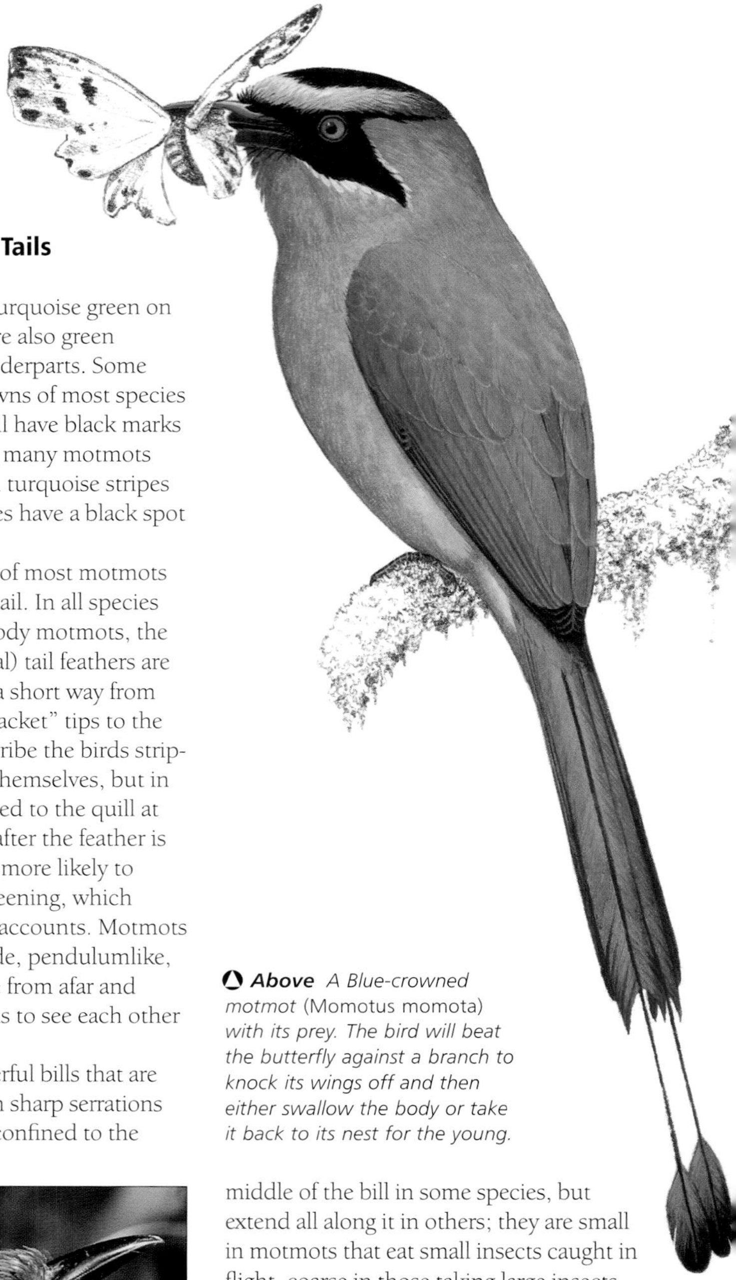

a COLORFUL BIRD SITTING ON A BRANCH IN A South American forest swinging from side to side its long tail, shaped at the end like the flights of a pair of darts, is almost certainly a motmot – a distant relative of the kingfishers. As in kingfishers, only one of the motmots' toes is directed backward.

Motmots are medium-sized, insectivorous birds that are usually found in pairs, well separated from their neighbors. Although they are now confined to the New World tropics, an ancient fossil found in Florida and an even older one in Switzerland show that the family was once far more widespread.

FACTFILE

MOTMOTS

Order: Coraciiformes

Family: Momotidae

10 species in 6 genera. Species include: **Blue-crowned motmot** (*Momotus momota*), **Blue-throated motmot** (*Aspatha gularis*), **Keel-billed motmot** (*Electron carinatum*), **Tody motmot** (*Hylomanes momotula*).

DISTRIBUTION C and S America

HABITAT Forests below the canopy

SIZE Length for most species 28–45cm (11–18in), but for the Tody motmot only 17cm (6.5in); the figures include long tails in all species except the Tody motmot.

PLUMAGE All species bright green above (some with blue in wings and tail), several with green or brown crowns; a mix of browns and greens below depending on species; several have a black spot on the breast, and most have a black mark through the eye. Sexes similar.

VOICE A wide range of hoots and squawklike notes, many far-carrying.

NEST Sometimes in crevices in rocks, but mostly in burrows in banks excavated by the birds themselves.

EGGS 3–4, white, incubated by both sexes.

DIET Mostly insects; sometimes lizards, snakes, frogs, and berries.

CONSERVATION STATUS The Keel-billed motmot is listed as Vulnerable.

Long Beaks and Longer Tails

FORM AND FUNCTION

All species are bright green or turquoise green on the back and tail, and several are also green beneath; others have brown underparts. Some have brown heads, but the crowns of most species are turquoise, blue, or black. All have black marks through or near the eye, and in many motmots this mark is highlighted by thin turquoise stripes above and below. Several species have a black spot on the breast.

The most distinctive feature of most motmots is their long, highly graduated tail. In all species except the Blue-throated and Tody motmots, the vanes of the two longest (central) tail feathers are missing for 3cm (1in) or more a short way from the tip, leaving the bird with "racket" tips to the feathers. Some old reports describe the birds stripping the barbs off the feathers themselves, but in fact the barbs are weakly attached to the quill at this point and wear off shortly after the feather is fully grown; doubtless they are more likely to break loose while the bird is preening, which probably explains the different accounts. Motmots swing their tails from side to side, pendulumlike, which makes the rackets visible from afar and doubtless helps mates and rivals to see each other through the trees.

Motmots have longish, powerful bills that are slightly downcurved, some with sharp serrations along the edges. Serrations are confined to the middle of the bill in some species, but extend all along it in others; they are small in motmots that eat small insects caught in flight, coarse in those taking large insects from the ground, or (in the Tody motmot) entirely absent. The bill of the Keel-billed motmot is strongly keeled. The birds feed mainly on insects and use their powerful bills to crush them; they also sometimes eat lizards, frogs, and berries. They hunt by perching beneath the forest canopy and sallying forth, flycatcherlike, to catch flying insects or to pounce on small animals on the ground.

***Above** A Blue-crowned motmot* (Momotus momota) *with its prey. The bird will beat the butterfly against a branch to knock its wings off and then either swallow the body or take it back to its nest for the young.*

***Above** The elusive Tody motmot of the forests of southern Mexico to northwestern Colombia, is the smallest motmot species.*

Burrowing out Nests

BREEDING BIOLOGY

All motmots nest in holes, usually in a bank or sometimes in a burrow they have dug in the ground. At the end of the burrow they excavate a largish chamber, in which the female lays the eggs. Both sexes incubate the eggs and feed the young. The young do not leave the nest until they are fully able to fly, but their elongated tail feathers have not developed by this time. CHF

Todies

Todies are close relatives of the motmot family of Central and South America, but although the two families share many characteristics of structure and nesting, the todies are much smaller. Todies have a distinctive crimson throat patch which puffs out every time they call out.

Todies are confined to the Caribbean, where each species has a limited distribution. The Cuban tody lives on Cuba and the Isle of Pines, the Jamaican tody on Jamaica; Hispaniola has two species, the Broad-billed tody at low altitudes and the Narrow-billed tody in the mountains. The Broad-billed tody is also found on Gonave, off Hispaniola. In most lowland areas the birds are very common, and they are extremely tame and approachable. But the poorly-known Narrow-billed tody is uncommon, or only locally common, in humid moss-forests at an altitude of 1,000–3,200m (3,300–10,500ft).

FACTFILE

TODIES

Order: Coraciiformes

Family: Todidae

5 species of the genus *Todus*: **Broad-billed tody** (*T. subulatus*), **Cuban tody** (*T. multicolor*), **Jamaican tody** (*T. todus*), **Narrow-billed tody** (*T. angustirostris*), **Puerto Rican tody** (*T. mexicanus*).

DISTRIBUTION Larger islands only of the Caribbean

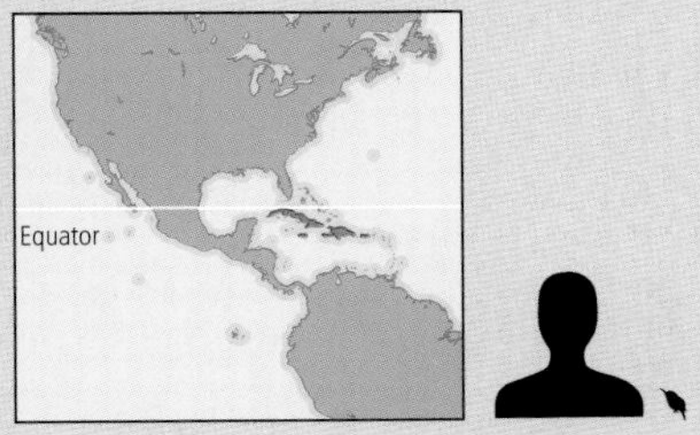

HABITAT Forest and woodland, often along streams.

SIZE Length 10–11.5cm (4–4.5in).

PLUMAGE All species bright iridescent green above, with red throats, white mustaches, and white, yellow, and pink underparts. Sexes similar.

VOICE Simple, unmusical, buzzy notes, guttural "throat-rattling" and "beep" sounds.

NEST A short burrow in a perpendicular earth bank.

EGGS 2–5, white, round.

DIET Mostly insects; occasionally small lizards and other animals.

CONSERVATION STATUS Not at risk

Tiny Insect-catchers

Form and Function

All todies are a brilliant, iridescent green above, with a bright carmine-red throat, a white and bluish-gray stripe behind it, a pale gray eye, and yellow undertail coverts. The color of the breast, belly, and flanks varies from species to species: whitish, pale gray, brownish, yellow, green, or pink, the colors merging. Unlike the motmots, todies have short tails. All have long, straight bills, varying somewhat in width but all flattened from top to bottom, the upper mandible black and the lower mainly red; such flattening is typical of flycatchers and other birds that catch small insects on the wing. In flight, both sexes may make a whirring noise; this is apparently produced by the wings and may be associated with courtship display.

The todies are also like miniature versions of their mainland motmot cousins in their ecology. They live in wooded country, usually in forests and frequently along the edges of streams or rivers. They spend much of the day sitting still, either alone or in pairs, perched on small twigs from which they sally diagonally upward to catch small insects moving underneath large leaves; although mainly insectivorous they may occasionally pounce on tiny lizards or other small animals. Todies are tiny, weighing only 5–7g (0.2oz), and everything about them is small-scale. The Broad-billed tody makes foraging sallies of only 2.2m (7ft) on average, and has never been known to make a single tree-to-tree flight longer than 40m (130ft). Its legs are small and its feet tiny, the third and fourth toes united at their bases. On twigs the birds move little, but may sometimes hop sideways or sidle like miniature parrots. When they snatch an insect from the ground they seldom alight, but if they do visit flat ground they may pursue their prey with a few hops.

During the breeding season todies use their beaks and feet to excavate tiny burrows in a steep bank by a stream or road. They lay their eggs in a chamber at the end. These chambers are defended vigorously and are rarely used more than once. Both parents incubate the eggs and care for the young, although they are both surprisingly inattentive during the incubation period. The young hatch naked, and remain in the nest until they are able to fly. During the nestling period the parents become more attentive, expending a lot of energy in provisioning their chicks with the utmost speed – up to 140 feeds per chick per day! Indeed, this is the highest rate recorded for any species in the bird world. CHF

Above *A somewhat inactive bird with a large, neckless head, the Cuban tody sits on its perch and watches for insect prey, constantly swinging its head and eyes with quick, jerky movements.*

Below *Known locally as Robin Redbreast (not to be confused with the Eurasian robin!), the Jamaican tody is almost silent over the non-breeding season. Its distinctive guttural "throat-rattling" call is given during territorial displays from where the tody is perched.*

Bee-eaters

THOROUGHLY ATTRACTIVE, SLEEK, GRACEFUL, melodious, restrainedly colorful, and sociable, bee-eaters really are rather special birds. Wherever they occur, throughout the Mediterranean, Africa, Madagascar, South and Southeast Asia, and Australia, they are regarded with great affection (except by bee-keepers, since bee-eaters certainly can make serious inroads into beehive populations).

The European bee-eater, better named the Golden bee-eater, is on every young birdwatcher's "must-see" list. In Australia the arrival of their only species, the Rainbow bee-eater or Rainbowbird, is taken to herald the spring. Throughout Africa, villagers take a proprietary pride in any nearby colony of Carmine bee-eaters, while a clamorous nesting colony of 50,000 Rosy bee-eaters is one of the seven wonders of the bird world.

Big Beaks and Long Tails

FORM AND FUNCTION

Bee-eaters are highly colored birds: most are green above and green, buff, or chestnut below, but one is predominantly black, one blue, one pink and gray, and one carmine. All have a black eye mask, most have a black band on the upper breast, and the intervening chin and throat are strikingly bright yellow, red, reddish, blue, or snowy, often with a cheek stripe of contrasting color. Wings are rounded (in forest-dwelling bee-eaters) or long and pointed (in open-country species, particularly those that hunt or migrate over long distances). In most species the wings are green with a broad black trailing edge. The tail is quite long, not much patterned, but often with slightly or greatly elongated central feathers or, in the Swallow-tailed bee-eater, elongated outer feathers. In other respects, all species are physically much alike: large-headed, short-necked birds with a long, slender, downcurved bill, very short legs, and weak feet. When perched, all move the tail backwards and forwards through a small arc – these are balancing movements which have come to have a social function. All sunbathe using a number of postures, the commonest being to sit back-to-the-sun with mantle feathers acutely raised.

Because of differences between the species in overall size, wing shape, tail shape, and the shape of individual feathers in the forehead, chin, and throat, bee-eaters were formerly classified in up to

Above *Representative species of bee-eater: **1** Blue-bearded bee-eater* (Nyctyornis athertoni)*; **2** European bee-eater* (Merops apiaster) *in flight; **3** Purple-bearded bee-eater* (Meropogon forsteni)*; **4** Little bee-eater* (Merops pusillus).

eight genera. Certainly two large Indian and Malaysian species, the Red-bearded and Blue-bearded bee-eaters, are sufficiently distinctive to be separated into their own genus (*Nyctyornis*). They are large, quite heavy, relatively sluggish birds with robust, gray-based bills, with the curved top ridged and grooved; they have long, pendent throat feathers and long, square-ended tails that are yellow below. They are not very vocal, and their calls are somewhat coarse. The Red-bearded bee-eater is a bright grass-green bird with an amazing combination of colors around the head: lilac-pink, scarlet, and a little blue, set off with an orange-yellow eye. In forests on the island of Sulawesi lives the Purple-bearded bee-eater, green, russet, dark maroon, and blue, which is more slimly built but shares the same character of a full throat with long, pendent feathers. It possesses six ribs, as opposed to five in the Red-bearded and Blue-bearded bee-eaters, and so is placed in its own genus (*Meropogon*).

There is a broad measure of agreement nowadays that all other bee-eaters should fall into the single genus *Merops*. Opinions are changing, however, in regard to the number of species that should be recognized, and that is in large part due to the fact that contemporary ideas about the very nature of species and how to define them are changing too. Three bee-eaters look very alike: the Blue-tailed bee-eater that breeds around the western Sahara and east to the western Himalayan foothills, the Olive bee-eater of Madagascar, the east African seaboard, and the dry Angolan coast, and the Blue-tailed bee-eater of southeast Asia. Appearances and voices are very alike, and all three are long-distance migrants. Their breeding ranges are separate, or only just meet, so it is hard to put them to the test of separate species and to determine how they interact with each other.

The Little green bee-eater is a sedentary bird with an enormous range, from Senegal to Vietnam. It is divided into several separate populations, four of them in the greater Arabian peninsula alone, which differ quite strikingly in crown color (green, olive, russet) and throat color (green, yellow, blue). Perhaps the Little green is really several species.

FACTFILE

BEE-EATERS

Order: Coraciiformes

Family: Meropidae

24 species in 3 genera. Species include: **Black bee-eater** (*Merops gularis*), **Blue-cheeked bee-eater** (*M. persicus*), **Blue-tailed bee-eater** (*M. philippinus*), **Carmine bee-eater** (*M. nubicus*), **European bee-eater** (*M. apiaster*), **Little bee-eater** (*M. pusillus*), **Little green bee-eater** (*M. orientalis*), **Olive bee-eater** (*M. superciliosus*), **Rainbow bee-eater** (*M. ornatus*), **Red-throated bee-eater** (*M. bullocki*), **Rosy bee-eater** (*M. malimbicus*), **Swallow-tailed bee-eater** (*M. hirundineus*), **White-fronted bee-eater** (*M. bullockoides*), **White-throated bee-eater** (*M. albicollis*), **Blue-bearded bee-eater** (*Nyctyornis athertoni*), **Red-bearded bee-eater** (*N. amictus*), **Purple-bearded bee-eater** (*Meropogon forsteni*).

DISTRIBUTION Eurasia, Africa, Madagascar, New Guinea, Australia.

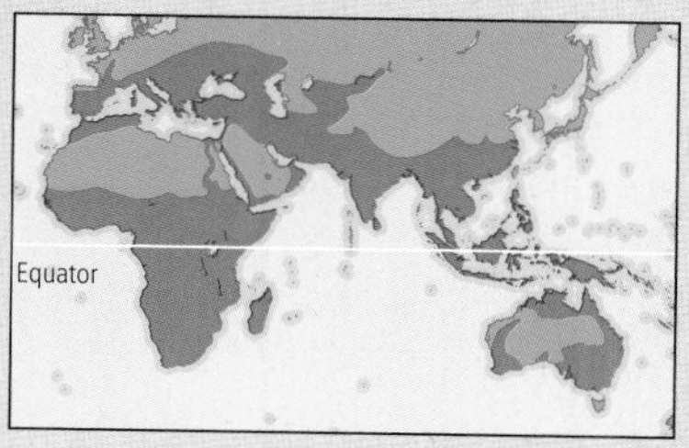

HABITAT Mainly open country: woodland, savanna, steppe; 6 species in rain forest.

SIZE Length 17–35cm (6.5–13.5in), including tail; weight 15–85g (0.5–3oz).

PLUMAGE Mostly green above, buff below; some species black or blue or carmine, with black eye mask, black gorget, and colored throat. Males and females very similar; some males brighter than females, with longer tail streamers.

VOICE Rolled, melodious, liquid syllables; alternatively, hoarse cawing.

NEST Unlined chamber at end of a tunnel 5–7cm (2–3in) in diameter and up to 3m (10ft) long, dug in cliff or flat ground.

EGGS 2–4 in tropics, up to 7 in Eurasia; white; weight 3.5–4.5g (0.12–0.16oz). Incubation period 18–23 days; nestling period 27–32 days.

DIET Airborne insects, mainly wasps and bees.

CONSERVATION STATUS None currently at risk.

Essentially Tropical

DISTRIBUTION PATTERNS

The bee-eater family is essentially tropical, and its more primitive members, the Blue-bearded, Red-bearded, and Purple-bearded bee-eaters, inhabit south and southeast Asian rain forests: this and other clues suggest that bee-eaters arose there, and spread over to the African continent, where they proliferated. Ancestral populations of these birds evolved separately as a result of having been isolated in rain forest between northern and southern tropical savannas.

Northern and southern Carmine bee-eaters are thought to have diverged from a common ancestor only about 13,000 years ago, and the northern tropical Red-throated bee-eater and southern tropical White-fronted bee-eater diverged from their common ancestor about 75,000 years ago. Only two species of bee-eater penetrate the temperate zone of Europe and Asia to any distance. No species has ever invaded the New World, and only the ancestral Rainbow bee-eater has ever entered Australia.

Left *The Black bee-eater is among three forest-dwelling* Merops *species that are non-migratory. The other two species are the Black-headed* (Merops breweri) *and Blue-headed* (M. muelleri) *bee-eaters.*

Right *Breeding populations of the Swallow-tailed bee-eater in southern Africa are believed to have originated from overwintering migrant visitors from Europe.*

Hunters of Honeybees

DIET

The world distribution of the family was remarkably congruent with the world distribution of honeybees, genus *Apis*, before the spread of bee-keeping to the temperate zone and the Americas. That is for the simple reason that by far the commonest prey of most bee-eaters are honeybees. When readily available – near their hives, or around flowering trees and herbs – they are taken in preference to other, equally abundant flying insects. All four species of honeybee are hunted and eaten, including the Giant honeybee (*Apis dorsata*), a dangerous stinger living in vast colonies that Oriental people regard with great respect. Other insects taken by some species of bee-eaters include bumblebees, wasps, hornets, dragonflies, and damselflies. The great majority of bees caught are venomous workers; the few non-stinging drones (male bees) taken probably reflect their scarcity outside of the hive. A European bee-eater requires about 225 bee-sized insects to sustain it and its young every day.

Bee-eaters hunt mainly by keeping watch for flying insects from a perch. They sit alertly on a vantage point such as a treetop twig, fence, or telegraph wire, turning the head to scan on all sides, then fly out quickly to intercept a passing insect. The prey is seized adroitly in the bill, taken after a short twisting and turning pursuit; in a graceful glide the bird then returns to its perch, where it tosses the prey to grip it in the tip of the bill, and strikes it several times against the perch to left and right. A stinging insect is then held near the tip of its tail, which is rubbed against the perch with the motion of someone using an eraser. A bee's bowel fluid is squeezed out, wetting the perch, and its sting and poison sacs are torn away. Several beating and rubbing bouts alternate, after which the immobilized insect is swallowed entire.

In open country the larger, pointed-winged bee-eaters – namely the Blue-cheeked, Olive, Blue-tailed, European, Rosy, Carmine, Blue-throated and Rainbow bee-eaters – forage by "fly-catching" from a treetop or pylon cable; to varying extents they also hunt in easy, wheeling flight on high, catching insect after insect without coming to perch in between times. Blue-tailed bee-eaters in Malaysia were watched foraging over a huge

Above *A mini-flock of Rainbow bee-eaters or rainbowbirds, which live in Australia and migrate to Indonesia. This species nests in very loose colonies. Sometimes the nests are so far apart that they seem to be solitary.*

paddyfield divided into quadrants with overhead wires, which made accurate measuring of distances easy. They perched facing into the breeze and, sighting a hornet no less than 80–95m (90–100yds) away, they flew out fast, straight, and level or slightly rising on a course to pass immediately below the insect. At the last instant they reached up with the bill to seize the victim from below with deft precision.

In Africa, Carmine bee-eaters commonly hunt high in the sky. Some, netted for study immediately they had alighted to rest on bushes, had bills that were sticky with characteristically-smelling honeybee venom, so it seems that they may catch bees on high, de-venom them by nibbling them in the mandibles, and then eat them without having to come to a perch to deal with them in the usual way. Also in Africa, the beautiful White-throated bee-eater of the southern borders of the Sahara is the exception to the rule that only the large, long-winged species hunt in flight. These are small birds, quite round-winged, and they feed largely on flying ants caught high in the air. From the Sahara they migrate to winter over equatorial forests, and there they feed above the canopy, by "flycatching" from treetops and in continuous flight. A trick of theirs is to perch on the lower fronds of an oil palm, in the head of which a squirrel is systematically stripping the outer, oily fibers away to get at the kernel. The birds sally out to snatch the falling strips in midair, return to perch, and eat them just as if the strips were beetles.

Carmine bee-eaters have other feeding specializations as well. They prey largely on locusts and follow migratory swarms of the insects. They come from afar to bushfires to exploit insects put to flight by the flames, mainly locusts, grasshoppers, and mantises, catching them by wheeling around at the very edge of the flames and smoke. Carmine bee-eaters also quite often follow people and vehicles moving through the African bush, catching insects disturbed from the grass. Similarly, they also use grazing and galloping antelopes, but have gone one further evolutionary step and actually ride upon them, using them as "animate perches." The habit is commonplace in the northern tropics, where these flamboyant birds may be seen on the backs of bustards, storks, goats, or antelopes, anywhere from Senegal to Somalia. For some reason, however, the habit is not common south of the equator. Quite unexpectedly, Carmine bee-eaters, along with two or three other species in Africa and the Orient, have occasionally caught fish; they fly slowly low over the still surface of a pond or river, then make a shallow dive straight into the top few centimeters of the water, and immediately rise into the air again with a small fish in their beak. So, bee-eaters will eat fish just as their distant relatives the kingfishers will sometimes eat bees.

Above left *In tropical birds such as the Blue-cheeked bee-eater, sun-basking may take place during loafing and preening, especially early in the morning or at late evening – particularly if the nights are cold.*

Above right *A Carmine bee-eater riding on the back of a Kori bustard (Otis kori; Otidae), waiting to catch insects that are disturbed by the larger bird as it moves across the savanna.*

Below **1** *The pursuit of a bee is usually short and direct but sometimes involves twisting and turning.* **2** *The bee is caught by an upward movement of the head.* **3** *The bird glides to a perch and rubs the bee on the perch to discharge the venom and to tear away the poison sacs and sting before* **4** *swallowing it.*

Tunnel Nesters

BREEDING BIOLOGY

Again like kingfishers, bee-eaters excavate nest burrows in soil. Most species dig both in perpendicular banks and in flat ground, but Red-throated and White-fronted bee-eaters nest only in banks. Little Bee-eaters nest in tall cliffs or low banks, or commonly in a "bank" of earth at the side of a cow or buffalo hoof indentation; another favorite place is inside the large entrance of an aardvark's burrow, where the birds always drill their nest-hole in the burrow's roof. Tunnels decline in flat ground but are horizontal or inclining in cliffs, and end in a broad oval egg-chamber. Red-throated bee-eaters' tunnels have a hump separating the entrance tunnel from the egg chamber, which helps to prevent eggs from accidentally rolling out. There is no nest lining, but a blackish carpet of trodden-down, regurgitated pellets soon accumulates and can almost bury the clutch. Later, nests become fouled with feces, the debris full of scavenging beetle larvae, and a large colony has an ammoniacal stench like a seabird colony.

To excavate its nest in level ground, a bee-eater props itself on the tripod of its bill tip and wrists, then used both legs with a bicycling motion to scrabble loose earth backwards. Some burrows can be 2m (6.5ft) long. Ground nests are prone to flooding and to predation by snakes and rodents. Even nests in high cliffs are not entirely safe, and eggs and young are often taken by Monitor lizards climbing up from the foot or egg-eating snakes descending from the top.

At the end of its first year a bee-eater either breeds or, like many other tropical birds, helps a breeding pair. In most species there is little by way of courtship display, although "courtship" feeding and chasing away rival males and adjacent-nesting pairs are commonplace. White-throated bee-eaters, however, have a courtship "butterfly-flight," with raised wings, slow beats, and deep-chested appearance. In many bee-eaters, a perched bird also greets its incoming mate by raising its wings, fanning and vibrating the tail, and calling vociferously. Both sexes – and any helpers – excavate the nest, but the female does most of the incubating. Eggs are laid at one-day intervals (or up to two-day intervals in larger species), and incubation begins sporadically with the first egg and fully with the second or third egg. Hence the eggs hatch at roughly daily intervals, in the laying sequence, and the brood of young are graded in age and size, with the oldest often 2–3 times the weight of the youngest.

Both parents, and any helper(s), feed the young equally, with single insects generally larger than those that the adults themselves eat. Thereafter growth is rapid and the youngsters fledge at a weight up to 20 percent greater than the average adult weight. After fledging, they and their parents and helpers may all continue to roost in the nest-hole for a few days, but usually start roosting in vegetation some distance away. The family group – 4 in Black, about 6 in Little, 4–9 in European, or up to 12 in White-throated bee-eaters – stay together in some instances until next year's nesting. After fledging, the young accompany the adults particularly closely for some six weeks, depending on them for food.

Red-throated and White-fronted bee-eaters in Africa have some of the most complex bird societies in the world. White-throated bee-eaters in Senegal, Mali, and Nigeria have up to six helpers at the nest. Red-throated bee-eaters are densely colonial, with up to 150 birds occupying nest-holes in 1–2sq m (11–22sq ft) of cliff face. About two-thirds of nests are attended by a pair only, and pairs at the remaining third have 1–3 helpers, generally their own progeny from a previous year. White-fronted bee-eaters are similarly colonial, but have 1–5 helpers at a majority of nests, and an individual bird alternates between breeding and helping another pair in successive nestings. Certain pairs and their helpers within a colony form a clan, and a colony may comprise 3–6 clans.

Above *Little green bee-eaters occur in pairs rather than trios, suggesting that this species does not have the system of helpers that assist in tending young chicks, as happens in other bee-eater species.*

Below *Two-week-old European bee-eater chicks. At one week old, their eyes will have opened and, after having been born naked, their long, pointed, gray quills grow quickly before developing into feathers.*

The Bee-keeper's Bane

CONSERVATION AND ENVIRONMENT

No bee-eater species is greatly threatened, but some populations may be depleted if commercial bee-keeping in Africa is developed much further. Bee-eaters were known to the ancient Egyptians as pests at apiaries, and many thousands are still killed every year in Mediterranean countries. Since we now know that they consume vast amounts of hornets, bee-wolves, and other honeybee-eating insects, it might well benefit bee-keepers in the long run not to molest the birds. CHF

Right *A colony of Carmine bee-eaters can number anywhere between 100 and 1,000 individuals. The southern race* (Merops nubicus nubicoides)*, seen here, lacks the greenish-blue throat of the northern race.*

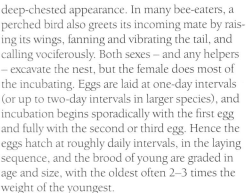

Rollers

As colorful, noisy, and beautiful birds, rollers invariably attract attention. They are perching birds of crowlike build, with rather short legs, and commonly nest in tropical gardens and houses. The order contains three families, the Coraciidae, Brachypteraciidae, and Leptosomidae.

The five genera of rollers lead somewhat different lives, yet the families are closely related to each other and – to a lesser extent – to the Cuckoo roller (Leptosomidae). They are more distantly related to kingfishers (Alcedinidae), motmots (Momotidae), todies (Todidae), and bee-eaters (Meropidae). While the birds are also similar in many ways to other hole-nesting families in both the New World and Old World tropics, including the hoopoes, hornbills, toucans, barbets, and woodpeckers, the resemblances in these cases are the result of convergent evolution.

Above *The Racket-tailed roller* (Coracias spatulatus) *is named for its long, elegant tail feathers and expanded feather tips.*

FACTFILE

Rollers

Order: Coraciiformes

Familes: Coraciidae and Brachypteraciidae

17 species in 5 genera. Genera and species include: **"true" rollers** (genus *Coracias*), **Abyssinian roller** (*Coracias abyssinica*), **European roller** (*C. garrulus*), **Indian roller** (*C. benghalensis*), **broad-billed rollers** (genus *Eurystomus*), **dollarbird** (*Eurystomus orientalis*), **Crossley's ground-roller** (*Atelornis crossleyi*), **Pitta-like ground-roller** (*A. pittoides*), **Long-tailed ground-roller** (*Uratelornis chimaera*), **Short-legged ground-roller** (*Brachypteracias leptosomus*).

Distribution Africa, Madagascar, Eurasia, Australia.

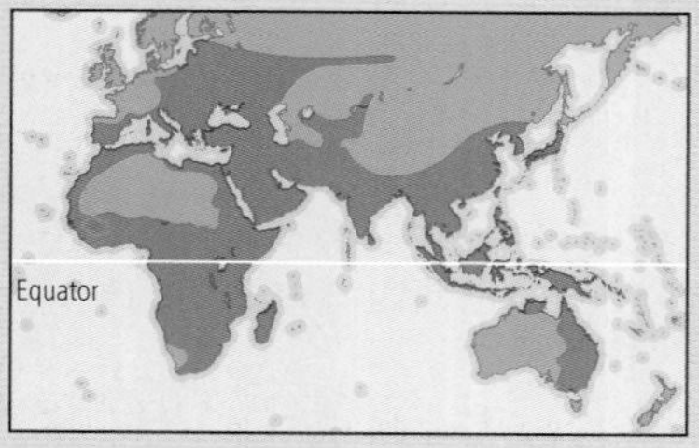

Habitat Forests, woodlands, savannas.

Size 25–45cm (10–18in).

Plumage Muted pink-browns, cinnamon, dark and light blue, or (ground-rollers) strongly variegated; sexes alike.

Voice Repeated short, gruff caws.

Nest In cavities in trees and masonry, or (in ground-rollers) in a tunnel in the ground.

Eggs 2–3 near equator, 3–6 in high latitudes; white; weight 10–17g (0.35–0.6oz). Incubation period about 18 days; nestling period 25–30 days.

Diet Mainly insects; some small vertebrates.

Conservation Status 3 ground-rollers and the Purple dollarbird (*Eurystomus azureus*) are Vulnerable.

Courting on a Roll

FORM AND FUNCTION

Rollers are so-called because of the spectacular, tumbling courtship flight of "true" rollers (*Coracias* species) and of broad-billed rollers (*Eurystomus*). The former spend much time aloft, defending their territory with raucous calls and rolling flight; but they feed mainly on the ground, dropping onto small animals from a perch. Broad-billed rollers, by contrast, feed on the wing. For most of the day a pair sits on treetops, eating little, and aggressively chasing other birds away; but in the late afternoon up to 200 gather to feed gregariously and frenziedly on winged termites. With pointed, quite long wings, large heads, short necks, and thickset bodies, they have a fast, wheeling-and-swooping flight, and can closely resemble falcons. A single bird can eat up to 800 termites (half its own weight) in the 90 minutes before dusk.

One African species is resident in rain forests; another is a migrant both within the tropical savannas and between Africa and Madagascar. The Oriental dollarbird, so-called because of the coin-sized white "windows" in its wing tips, breeds from Korea to New South Wales and winters near the equator.

All *Coracias* rollers are strongly migratory. European rollers enter Africa in September, when the closely allied and very similar Abyssinian rollers are also migrating southward within the northern tropics. Abyssinian rollers travel up to 1,000km (600mi), but European rollers go ten times as far, to arid-country wintering grounds mainly in

Left *Lilac-breasted rollers* (Coracias caudata) *can often be seen perching conspicuously atop trees, bushes, and utility poles.*

Above *Among a sea of sunflowers in Germany, the European roller customarily hunts its prey from dead branches below the crowns of trees.*

Tanzania and Namibia. In the first few days of April, tens of thousands concentrate in eastern Tanzania, and fly together in a narrow coastal corridor through Kenya and Somalia and into Arabia; thousands can be seen evenly dispersed in the sky from horizon to horizon.

Monogamous and Territorial

BREEDING BIOLOGY

All rollers appear to be monogamous, highly territorial hole-nesters. Apart from the spectacular rolling flight, the breeding biology of European and Indian rollers is not remarkable. Most African species are curiously ill-known, considering how common and eminently studiable the birds are. Madagascan ground-rollers are even less well known; they seem to be mainly active at twilight, feeding entirely on the ground, and nesting in holes in open ground (Long-tailed ground-roller), around forest tree roots (Pitta-like and Crossley's ground-rollers), or in trees (Short-legged ground-roller). There is a strong native tradition that these birds hibernate in the dry season. CHF

Cuckoo Roller

Above *The cuckoo roller differs from both rollers and ground-rollers in having very different-looking sexes. Here, a male displays his iridescent colors.*

DISTINCTIVE AND UNUSUAL, THE COUROL or Cuckoo roller lives only on Madagascar and the Comoros. In a region with many extraordinary life-forms, this bird is unique mainly for its specialized diet. It is more closely related to the "true" rollers than to the Madagascan ground-rollers.

There are three subspecies, not very different, one on Grand Comoro island, one on Anjouan island, and the third on Mohéli and Mayotte in the Comoros and throughout Madagascar.

Female cuckoo rollers with their reddish upperparts and spotted underparts look faintly like female cuckoos, which is among the reasons why the species got its common name when first described to science 200 years ago. The cuckoo roller's foot is zygodactylous, with two toes forward and two backward, as in cuckoos.

Only recently have details of cuckoo rollers' lives become quite well-known, and it is entirely coincidental that they, like cuckoos, eat hairy caterpillars, chameleons, and locusts. Feathers of the lores curl forward, then backward, in an "S" shape, not quite concealing the base of the bill. The body feathers have long aftershafts, and the whole plumage has a grayish bloom, perhaps related to the bird's two powderdown patches (downy spots filled with a talc-like powder), one on each side of the rump. The adaptive function of these patches is not known.

Cuckoo rollers generally hunt by perching motionless in the forest canopy, then, when prey is spotted, make a short, sallying flight into leaves, onto a trunk or branch, or into the air. Sometimes stomachs of the birds are lined with the fur of the hairy caterpillars they eat.

Only two nests and breeding attempts have been described in any detail. Nesting starts in September in the Comoros and November in Madagascar. Eggs are incubated for at least 20 days. Nestlings are covered in long, whitish down, and take some 30 days to fledge.

Cuckoo rollers live in pairs and are remarkably tame. On Madagascar they can live in small patches of forest and are not threatened, but on Grand Comoro there are only about 100 pairs left and conservation measures are urgently needed. CHF

FACTFILE

CUCKOO ROLLER

Order: Coraciiformes

Family: Leptosomidae

1 species: *Leptosomus discolor*

DISTRIBUTION Madagascar, Comoro Islands.

HABITAT Forests, scrub.

SIZE Length 38–50cm (15–20in).

PLUMAGE In males, iridescent, green-purple back, wings, and tail, pale gray head, white underparts; in females, mottled red-brown and greenish back, mottled brown head, red-brown tail, underparts creamy, heavily spotted with black. Bill slaty, eyes black; male with red legs, female brown.

VOICE Loud, plaintive whistle, "qui-yu" repeated.

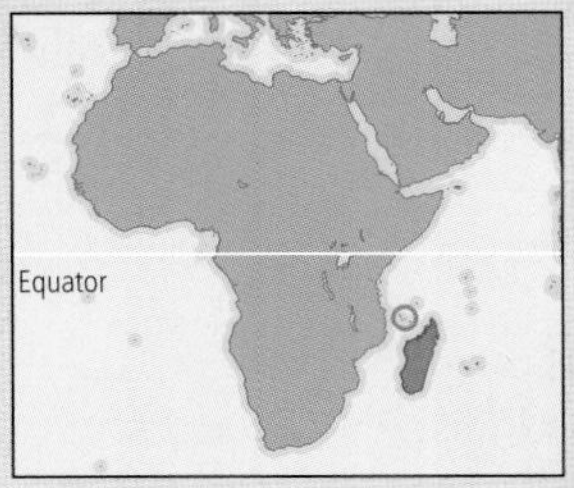

NEST Unlined tree hole 4–6m (13–20ft) above forest floor.

EGGS 2 cream-buff, rounded ovals.

DIET Chameleons, locusts, (hairy) caterpillars, small animals.

CONSERVATION STATUS The Grand Comoro subspecies is now under threat.

Hoopoe

FOR MILLENNIA, THE HOOPOE HAS HAD A special place in folklore and people's affections; as far back as Ancient Egypt and Minoan Crete, hoopoes were depicted on the walls of tombs and temples. With a vast breeding range in three continents, these birds are conspicuous and common in gardens and on cultivated land.

Hoopoes are short-legged perching and ground birds. They are usually seen singly, and forage by walking over turf and nearly bare ground under trees, probing with their long, slender, sensitive bill for grubs; they also take insects from fissured bark in trees. Their preferred food is large, hard, and often strong insects, such as cockchafers, stag beetles, mole crickets, and grasshoppers, but they also take small lizards and even centipedes, millipedes, spiders, and snails.

Standing Out from the Crowd

FORM AND FUNCTION

Hoopoes fly with irregular, butterfly-like beats of their rounded wings, and on perching often momentarily fan the crest. The flight looks weak, but they migrate great distances and can take fast evasive action in the air if need be. Even so, hundreds fall prey to Sooty falcons as they migrate through the Mediterranean. The bold pattern of broad white stripes across the black back and tail, and black wings with a large white patch in the secondaries (in southern Africa) or elsewhere with white stripes in the secondaries and a large white stripe near the wing tips, makes them conspicuous.

Southern African hoopoes are reddish, and differ from wintering Eurasian migrants in body shade and wing pattern; they were once thought a distinct species, but their voice, behavior, and biology do not differ significantly.

Adult hoopoes sometimes adopt a spread-eagled posture, squatting down, spreading the wings and tail with their tips pressed to the ground, throwing the head back, and pointing the bill, often slightly opened, straight up. This far from concealing posture was long thought to be a reaction to birds of prey. In fact, it is a sunning posture; the plumage is fluffed out and the feathers separated to maximize heat absorption. A hoopoe will often fold its wings and preen and scratch itself midway through the display.

Above *For two hundred years this posture was thought to be a defensive response by hoopoes to an aerial threat, but it is now recognized as merely the leisurely activity of sunbathing!*

FACTFILE

HOOPOE

Order: Coraciiformes

Family: Upupidae

1 species: *Upupa epops* (although the African hoopoe is sometimes considered a separate species)

DISTRIBUTION Europe, Africa, Madagascar, S Asia.

HABITAT Wooded farmlands, orchards, savannas.

SIZE Length 27–29cm (11–11.5in); weight 47–89g (1.7–2.8oz).

PLUMAGE Pink-brown, with black-tipped crest and conspicuously black-and-white-banded wings, back, and tail.

VOICE "Hoo-hoo-hoop" cry.

NEST Unlined or simply-lined cavity in tree, masonry, or ground.

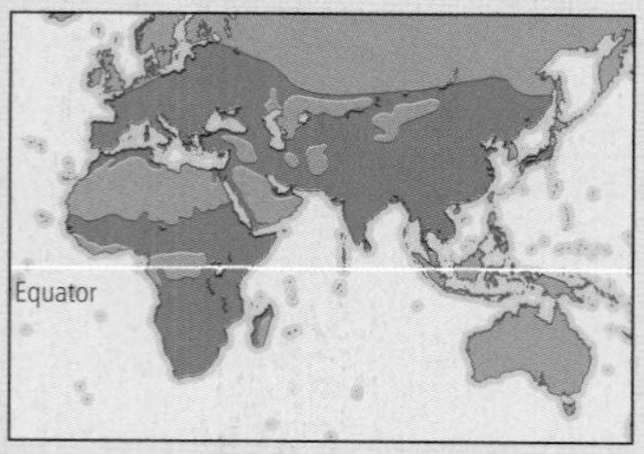

EGGS 2–5 in tropics, 7–9 in high latitudes; color very variable: yellowish, greenish, brownish; weight 4.5g (0.2oz). Incubation period 15–16 days; nestling period 28 days. Nestlings protect themselves by secreting a foul-smelling fluid.

DIET Insects, small vertebrates.

CONSERVATION STATUS The St. Helena hoopoe – a giant, flightless variety known only from bones – is extinct, apparently killed off by hunting and imported predators soon after the island of St. Helena was first discovered in 1502.

Chemical Defenses

BREEDING BIOLOGY

Nests are scantily-lined cavities in termite mounds, old woodpecker holes, rough stone walls, drainpipes, or clefts in trees; the entrance is narrow, so that the bird has to squeeze in, and the hole itself is fetid. Nests holes are found and cleared by the male. Hoopoes are not shy of people, and often nest in cavities in house walls in busy villages. Singing is mainly in the breeding season; a bird sings from a rooftop or treetop, its dovelike, far-carrying notes uttered with great persistence for minutes on end. The names "hoopoe" and *Upupa* are both onomatopoeic.

The young are sparsely covered with down. They have five methods of defense: by spraying excreta; hissing; jabbing upward with the bill; striking out with one wing; and by secreting a stinking excretion of the preen gland. During the nestling period the preen glands of the adult females, together with those of the nestlings, enlarge and produce a liquid secretion with a stench like rotting meat; the gland diminishes in size and stops secreting shortly before the young leave the nest. CHF

Wood-hoopoes

ONE OF THE FEW BIRD FAMILIES CONFINED to Africa, the wood-hoopoes are divided by beak shape into two genera. The wood-hoopoes proper are characterized by their almost straight bills, while the scimitarbills, as their name implies, have markedly curved bills.

Similar as the two genera are, genetic studies reveal that they separated some 10 million years ago, so a case can be made for placing them in separate subfamilies or, as some taxonomists insist, separate families.

Long-billed Tree-dwellers

FORM AND FUNCTION

Birds in the larger of the two genera have scarlet bills (black in a few species) that are long, strong-based, and a little decurved. In the less numerous scimitarbills – smaller birds with less bright coloration and fewer white markings – the bill is greatly downcurved, at least in the Abyssinian and Common scimitarbills (yet only somewhat decurved in the Black scimitarbill). The bill is scarlet in the Abyssinian species and black in the other two.

Although scimitarbills are rather solitary, most wood-hoopoes are gregarious, gathering in groups of 5–12. They are arboreal, finding insects in trees by probing into bark crevices with their long, slender bills. They are lively and acrobatic, often clinging crosswise to a trunk or working suspended by their strong feet underneath a stout, horizontal limb. They sometimes use both tail and wings for support when clambering up through small, twiggy branches. They fly readily, but their flight is neither strong nor sustained. Most food is soft: moths, caterpillars, pupae, termites, or insect and spider egg masses. Centipedes, millipedes, and small lizards are also eaten occasionally. Nestlings are fed mainly with caterpillars.

Helping Out with the Young

SOCIAL BEHAVIOR

Wood-hoopoes are noisy and conspicuous. Birds gathered in a tree may interrupt foraging to indulge in a mutual display or "rally." Each bird cackles vociferously – they are called *kakelaars*, or "cacklers," in Afrikaans – and rocks back and forth with wings partly opened, the tail vigorously wagging up and down. "Cackling" helps maintain the identity and cohesion of the group, which is more or less an extended family of parents, helpers, and young. In the Green wood-hoopoe, another strange ritual occurs when a bird detects a flock in its territory. Seizing a small piece of bark in its bill, it places itself at the front of its own flock and waves the "flag," thrusting it at the trespassers from a meter or two away.

A helper forgoes breeding to help defend another adult's nest and to feed its nestlings, with which it forms close bonds. Next year the helper may itself breed, its breeding success improved in turn by help given by the younger birds. CHF

*Above Two Green wood-hoopoes (*Phoeniculus purpureus*) in a social calling display. Both cackle and bow and whip their tails up and down in an exaggerated fashion.*

Left Two striking features of the Green wood-hoopoe are its red bill and its long, green-tinged tail.

WOOD-HOOPOES

Order: Coraciiformes

Family: Phoeniculidae

8 species in 2 genera. Species include: **Green wood-hoopoe** (*Phoeniculus purpureus*), **White-headed wood-hoopoe** (*P. bollei*), **Abyssinian scimitarbill** (*Rhinopomastus minor*), **Black scimitarbill** (*R. aterrimus*), **Common scimitarbill** (*R. cyanomelas*).

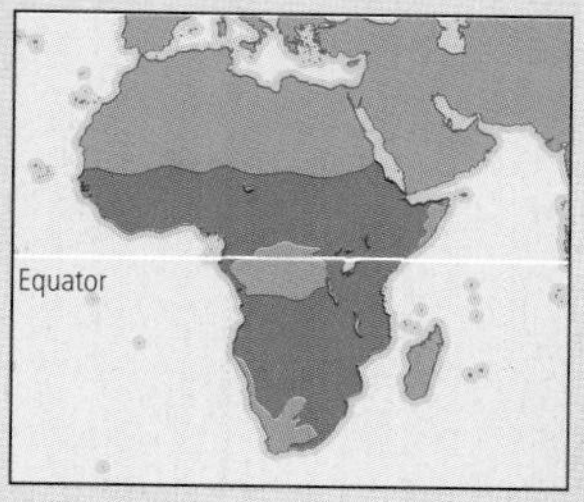

DISTRIBUTION Sub-Saharan Africa

HABITAT Forests, wooded savannas.

SIZE Length 23–46cm (9–18in).

PLUMAGE Black with strong green or violet gloss; conspicuous white marks on wing and in long tail, which is strongly graduated; some species with buff or orange-brown head; bill and legs scarlet or black, slightly or strongly decurved.

VOICE Repeated fluty notes, or cackling by flock in unison.

NEST In tree cavities, unlined.

EGGS 2–4, blue, gray, or olive, blotched with dark brown in one species. Incubation period 17–18 days; nestling period 30 days.

DIET Insects

CONSERVATION STATUS Not at risk.

Hornbills

WITH THEIR LARGE BILLS, OFTEN TOPPED by a prominent casque, their bold colors, great diversity of calls, and rushing wingbeats, hornbills are conspicuous and instantly recognizable. Their biology is no less remarkable, especially the unique breeding habit whereby the female seals herself into a hole for the majority of the nesting cycle.

Hornbills are an Old World group, unrelated to the superficially similar toucans of the New World; the parallels between the two can be ascribed to convergent evolution. Just under half of the species (24) occur in Africa south of the Sahara (excluding Madagascar), while just over half (29) are found in southern Asia. A solitary species, the New Guinea wreathed hornbill, extends east to the Solomon Islands. The larger forest species, which include all but one of the Asian species and seven of the African, are among the largest avian fruit-eaters in their habitat, and are believed to perform a vital role in the dispersal of seeds from many species of forest trees. A dozen of the African and one Asian species inhabit savanna and woodland and are mainly carnivorous; these mostly constitute the small *Tockus* species such as von der Decken's hornbill, which are primarily insectivorous, but also include two very large *Bucorvus* ground hornbills, notably the Northern ground hornbill, which are among the largest avian predators.

Hornbill Relatives

HORNBILL CLASSIFICATION

There is considerable anatomical, molecular, and behavioral evidence to suggest that the hornbills' closest relatives are the hoopoes and woodhoopoes of the families Upupidae and Phoeniculidae. Hoopoes and the small *Tockus* hornbills that they most resemble are both primarily African groups, and both have terrestrial and arboreal lifestyles, all of which evidence points to Africa as the ancestral source for all the families concerned.

The hornbills of Africa, except the two ground hornbill species, seem more closely related to each other than to any of those in the Oriental region. The smallest hornbills, in the genus *Tockus*, have diversified into 14 species, including the aberrant Long-tailed hornbill, and in several species the juveniles of both sexes most resemble the adult male, as in many other hornbills. The large African *Ceratogymna* species are unusual in that the face of young birds is colored brown, while in the larger wattled species the brown extends to the whole head and neck, is retained in the adult female, and only changes to black in the adult male. In otherwise similar Oriental forms, such as the Indian and Sri Lankan grey hornbills and *Anthracoceros* species such as the Indian pied hornbill, the young birds also resemble adult males but the adults of both sexes are very similar in the size and shape of the bill and casque. In most other Oriental hornbills in the genera *Aceros* and *Penelopides*, such as the Tarictic hornbill, the paler or browner heads are confined to juveniles and adult males, while these areas only change to black in adult females.

The two ground hornbills differ from all other species in the number of neck vertebrae, in not sealing their nest entrance, and in not showing any form of nest sanitation. They are generally considered so different and primitive as to be assigned to their own subfamily, Bucorvinae. They are unique among African hornbills in being apparently allied to Oriental species, although the link is far from obvious at first sight. One feature that they share in common with the largest Oriental forest hornbills – those grouped in the genus *Buceros* – is a preen gland covered in an especially dense tuft of feathers. The tuft serves to improve the application of the oils produced by the gland, which are used to color cosmetically the bill, casque, and white areas of plumage with red, orange, or yellow.

This one similarity might in itself be discounted as mere coincidence were it not for the fact that the ground hornbills also share a special genus of feather lice with their Oriental relatives. The other Oriental hornbills with their own genus of feather lice – the communal *Anorrhinus* species, such as the Bushy-crested hornbill – are also primitive within the family, but show most plumage similarity to various *Aceros* species.

FACTFILE

HORNBILLS

Order: Coraciiformes (suborder: Bucerotes)

Family: Bucerotidae

54 species in 9 genera

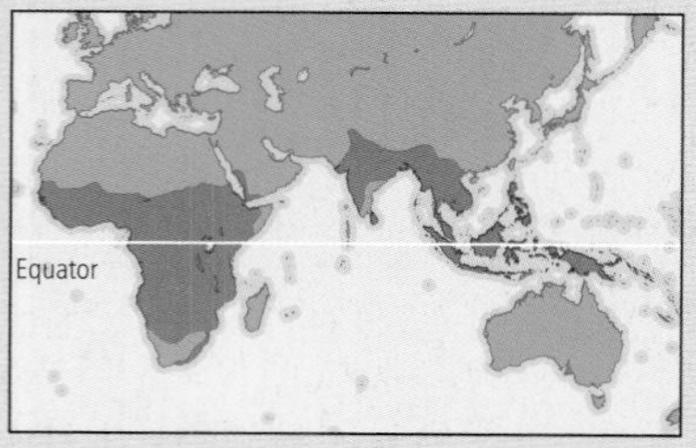

DISTRIBUTION Sub-Saharan Africa, Arabian peninsula; Pakistan, India, SE Asia and islands E to New Guinea.

HABITAT Most species in forest; one-quarter of all species, including all but one in Africa, occupy savanna.

SIZE Length 30–160cm (12–63in), including elongated central tail feathers of Helmeted hornbill; **weight** 85g–4.6kg (3oz–10.1lb); **wingspan** up to 180cm (5.9ft). Male usually 10 percent larger than female, with bill 15–20 percent longer.

PLUMAGE Mainly black and white, but gray and brown predominate in some species – there are apparently no plumage pigments other than melanin. Bill, casque, bare facial and throat skin, eyes, and/or feet often brilliantly colored in combinations of black, red, blue, and yellow. Juvenile and adult of each sex separable based on combination of plumage, facial skin, eye, bill, or casque color and structure.

VOICE Wide range, from basic clucks and whistles to soft hooting, deep booming, raucous cackling, high squealing.

NEST Natural hole in tree, cliff, or earth bank. In most species, entrance sealed to a narrow, vertical slit by female.

EGGS 1–2 in larger species, up to 8 in smaller ones; oval, white, with finely pitted shells. Incubation 25–40 days; nestling period 45–86 days, depending on body size.

DIET Insectivorous, frugivorous, or (2 species) carnivorous.

CONSERVATION STATUS 2 Philippines species – the Sulu and Visayan wrinkled hornbills – are Critically Endangered. 2 other species are Endangered and 5 Vulnerable.

See Selected Hornbill Species ▷

*▷ **Right** A Rhinoceros hornbill from Southeast Asia returns to the nest with a mouse. Like most hornbills, this species is omnivorous, varying its diet of fruit and insects with small animals caught in the foliage.*

Left *In Africa, a female Black-casqued hornbill displays the family's most remarkable feature: the huge bill with its distinctive casque. Despite its solid-looking appearance, this appendage is usually quite light, being hollow and supported by thin, bony struts.*

The Characteristic Casque

FORM AND FUNCTION

The large bill that is so characteristic of the family may explain why hornbills are the only birds with the first two neck vertebrae (axis and atlas) fused together. The bill is long and downcurved, with the tips of the mandibles meeting exactly to form a dextrous pair of forceps, and with the inner cutting edges serrated to crush and fragment the food. The casque that surmounts the bill is, in its simplest form, a narrow ridge that may reinforce the upper mandible. However, in many species the casque is elaborated into a special structure that may be cylindrical, upturned, folded, or inflated, and that sometimes even exceeds the bill itself in size.

The casque is always only slightly developed in young birds, while in the adults of most species it is larger and more elaborate in males. In all but one species, the casque is composed of a light sheath of keratin, supported internally by thin, bony supports and probably used to advertise the age, sex, and species of an individual. In the males of most African *Ceratogymna* species, such as the Black-casqued hornbill, the casque is especially large and opens to the mouth cavity; the calls are nasal in quality, suggesting that the casque may function to amplify sound. However, in the largest Asian *Buceros* species, the large, specially-formed casques, such as the bicornate and upturned casques of, respectively, the Great and Rhinoceros hornbills, may be used in fighting or to knock down fruit. The casque is most remarkable in the Helmeted hornbill, where the short, straight bill is surmounted by a solid block of keratin, termed hornbill ivory, that, together with the skull, comprises about 10 percent of the total body weight. It may function as a weighted digging tool to excavate animals from hollows and rotten wood, but it is definitely used as a battering ram by males as they crash into one another during aerial bouts of territorial defense.

Hornbill wings are broad and, in the larger species, produce a whooshing sound in flight as air rushes through the base of the flight feathers. The flow is enhanced by the lack of underwing coverts, another hornbill speciality, and in some species the noise is augmented by air whistling over the short, stiff, outermost primaries. In most species the tail is long, especially in the Long-tailed, White-crested, and Helmeted hornbills; in the latter it is further extended by a central pair of tail feathers that reach almost 1m (3.3ft) in length. The few short-tailed hornbills include several of the white-tailed *Aceros* species, such as the Bar-throated wreathed hornbill, and the two terrestrial ground hornbills.

The head and neck are often notable in color and form. Eye color often differs between species and even between sexes, as in some members of the genera *Buceros* and *Anthracoceros*, such as the Sulu hornbill. The colors of the bare skin around the eyes and on the throat can also be used to distinguish the species, sex, or age of a hornbill. In some species (*Bucorvus* ground hornbills, some *Aceros* species) the throat skin is even inflated or hangs as wattles, as in the Yellow-casqued hornbill. Hornbills are also notable for their long eyelashes, for rather stubby legs and toes with broad soles, and especially for the fusion of the bases of the three front toes.

Purveyors of Fruit

DIET

The larger forest hornbills are mainly frugivorous, and most travel widely in search of fruiting trees. The trees' patchy dispersal and irregular fruiting mean that these species are not territorial and tend to gather in large flocks when searching for food. The birds use their long bill and neck to stretch out to fruits, tossing each one back into the gullet and then using the stubby tongue to assist with swallowing. Undigested remains, such as seeds and pips, are regurgitated or defecated, often far from the parent tree, and so facilitate seed dispersal. Breeding hornbills have been observed to swallow as many as 185 small fruits at one time and then to carry them to the nest for regurgitation to the young. At one Silvery-cheeked hornbill nest, it was estimated that the male delivered 24,000 fruits in the course of 1,600 visits spanning the 120-day breeding cycle. Small items of animal food are snapped up if encountered, and in several species it appears that animal food is sought especially during breeding, probably as a source of extra protein for the growing chicks.

Selected Hornbill Species

Species include: in Africa, Northern ground hornbill (*Bucorvus abyssinicus*), Southern ground hornbill (*B. cafer*), Black-casqued hornbill (*Ceratogymna atrata*), Brown-cheeked hornbill (*C. cylindricus*), Silvery-cheeked hornbill (*C. brevis*), Trumpeter hornbill (*C. bucinator*), Yellow-casqued hornbill (*C. elata*), African pied hornbill (*Tockus fasciatus*), Bradfield's hornbill (*T. bradfieldi*), Long-tailed hornbill (*T. albocristatus*), Monteiro's hornbill (*T. monteiri*), Red-billed hornbill (*T. erythrorhynchus*), Red-billed dwarf hornbill (*T. camurus*), von der Decken's hornbill (*T. deckeni*), Crowned hornbill (*T. alboterminatus*); in Asia, Brown hornbill (*Anorrhinus tickelli*), Bushy-crested hornbill (*A. galeritus*), Tarictic hornbill (*Penelopides exarhatus*), Great hornbill (*Buceros bicornis*), Rufous hornbill (*B. hydrocorax*), Rhinoceros hornbill (*B. rhinoceros*), Helmeted hornbill (*B. vigil*), Indian grey hornbill (*Ocyceros birostris*), Sri Lankan grey hornbill (*O. griseus*), Indian pied hornbill (*Anthracoceros coronatus*), Sulu hornbill (*A. montani*), Narcondam wreathed hornbill (*Aceros narcondami*), New Guinea wreathed hornbill (*A. plicatus*), Rufous-necked hornbill (*A. nipalensis*), White-crested hornbill (*A. comatus*), Bar-throated wreathed hornbill (*A. undulatus*), Sumba wreathed hornbill (*A. everetti*), Visayan wrinkled hornbill (*A. waldeni*), Wrinkled hornbill (*A. corrugatus*),Writhed hornbill (*A. leucocephalus*).

Above *Representative species of hornbills:* ***1*** *Great hornbill, carrying fruit in its mouth;* ***2*** *Writhed hornbill, found only on the Philippine island of Mindanao and adjacent islands;* ***3*** *Red-billed dwarf hornbill, one of the Tockus genus of smaller, mainly insectivorous birds;* ***4*** *Rufous-necked hornbill, an endangered species of which less than 10,000 survive;* ***5*** *Helmeted hornbill, whose casque, unusually, is solid; the skull in this species makes up 10 percent of bodyweight;* ***6*** *Southern ground hornbill, one of two African ground-hornbill species famous for their loud, booming calls.*

Most of the smaller *Tockus* hornbills are primarily insectivorous, only taking other small animals and some fruit when available. Most are also sedentary and defend a permanent territory using elaborate displays. However, some African species that breed in open savanna during the rainy season are forced to range widely once the subsequent dry season reduces the food supply.

There are exceptions to these two main feeding strategies. Large Oriental forest species such as the White-crested and Helmeted hornbills are known to be sedentary, the former carefully searching the foliage and forest floor for animal prey and fruit, the latter possibly excavating prey from rotten wood and loose bark when not feasting on figs. Small African forest species such as the African pied and Crowned hornbills feed frequently on fruits and often gather in flocks when not breeding, even though they still include many small animals in their diet. Only the very large ground hornbills are almost entirely carnivorous, using their pickaxelike bills to subdue prey as large as hares, tortoises, snakes, and squirrels.

Below *In its Southeast Asian homeland, a Wrinkled hornbill uses its multicolored beak like food tongs to pluck fruit from a strangler fig.*

Immurement for Security

BREEDING BIOLOGY

Hornbills reach sexual maturity between the ages of 1 (*Tockus*) and 6 (*Bucorvus*, *Buceros*), depending on their body size. How long they live in the wild is unknown, but captives of the smaller species regularly exceed 20 years in captivity, while the larger species live for 50 years or more. Breeding seasons depend mainly on the birds' choice of food, with forest frugivores showing little seasonality (in accord with the year-round availability of fruit) compared with savanna insectivores, which breed mainly during the warm, wet summer.

Courtship feeding of females, mutual preening, copulation, and prospecting of nest sites are among the activities that precede breeding in larger forest species. Most species have loud calls that in the more sedentary species function to proclaim defended territories, while in the more mobile species they serve for long-distance communication. In some species the calls are accompanied by conspicuous displays, especially in small *Tockus* species in the open savanna. Territory size, in those non-fruiteaters that do not just defend an area immediately around the nest, ranges from 10ha (25 acres) for the African Red-billed hornbill to 100sq km (40sq mi) for the Southern ground hornbill.

Hornbills nest in natural cavities, usually in trees but also in rock faces and earth banks. In almost all species with the exception of the two ground hornbills, the female seals the nest entrance to form a narrow vertical slit. Initially, while building the nest from the outside, she will work with mud, but later, once she has entered the nest, uses her own droppings, often mixed with food remains. In some species the male assists by bringing lumps of mud or sticky foods, and in a few, such as *Ceratogymna*, the male forms special pellets of mud and saliva in his gullet, helps to apply these to the entrance, and regurgitates supplies to the female in the nest. The male also delivers nest-lining materials, such as dry leaves or bark flakes.

In some genera (*Ceratogymna*, *Aceros*, *Penelopides*) the male continues to feed the female and young for the rest of the nesting cycle, while in others (*Tockus*, *Buceros*, *Anthracoceros*) the female usually breaks out of the nest when the chicks are half grown and helps to feed them. In the latter case, the chicks reseal the nest unaided and only break out when ready to fly. The vertical slit gives good air circulation (through convection) for birds on the nest floor below it, while the small opening and wooden walls provide insulation. The sealed nest, and the long escape tunnel that is usually present above it, also give some protection against competitors for the site and predators.

Food is usually brought to the nest either as single items held in the bill tip (*Tockus*) or as a gulletful of fruits that are regurgitated one at a time and passed in through the slit to the female and nestlings. Only Monteiro's hornbill, which lives in

Right *In Africa, a male Red-billed hornbill delivers food to its mate, immured in a tree hole as she waits to lay a clutch of eggs. The practice of sealing up the nest during breeding is unique to the family.*

the arid Namib Desert, carries several items at a time in its bill. Food remains and debris are passed out of the nest slit, while the droppings are forcibly expelled. In most species the female simultaneously molts all her flight and tail feathers while breeding, dropping them around the time of egg laying and re-growing them again by the time that she re-emerges from the nest. The ground hornbills are exceptional to the basic hornbill pattern in that the female does not seal the nest (although she sits inside through incubation and the early nestling period, and is fed on the nest by the male and his helpers); food is delivered as a bolus of multiple items held in the bill, droppings and food remains are not expelled, and no unusual feather molt occurs.

Most hornbills are monogamous; each member of the pair shares in nest preparation and the subsequent rearing of the fledglings, with a clear division of labor between the brooding female and provisioning male. In certain species scattered through several genera, however, a form of cooperative breeding has developed in which some individuals – usually mature males and juveniles – do not breed but help a dominant pair to rear their young. These species can be recognized by the birds' habit of living in groups (of up to 25 in the Bushy-crested hornbill), and by the immature birds usually being colored very differently from the adults. Cooperative breeding has been reported for species as diverse in form and biology as the Southern ground, White-crested, Brown, Trumpeter, and Rufous hornbills. Indeed, hornbills may be the bird family with the highest proportion of cooperative breeding, practiced by possibly as many as one-third of all species.

Threatened Island Species

CONSERVATION AND ENVIRONMENT

Several hornbill species have suffered severe reductions over their entire ranges, especially in Southeast Asia and West Africa. The most endangered occur only on a single island or small archipelago, where small initial populations are easily reduced by habitat destruction. The Sulu hornbill of the Philippines is considered the most critical, but the Narcondam wreathed hornbill of India and the Sumba wreathed hornbill of Indonesia are also seriously threatened, along with other island species, especially in the Philippines. Populations of hornbills on continental mainland are generally more widespread, although there is concern about the restricted and declining Brown-cheeked and Yellow-casqued hornbills of West African rain forests, the Bradfield's hornbill of dry teak savanna in Botswana and Zimbabwe, and the Rufous-necked and Brown hornbills of Asia. AK

HORNBILLS IN HUMAN CULTURES

Hornbills are so conspicuous wherever they occur that they have been incorporated into many traditional cultures. Many Africans regard ground hornbills as sacred or taboo birds, and in many areas they thrive unmolested, even when the human population density is quite high. However, members of some ethnic groups in West Africa, such as the Hausa of Niger and Nigeria, use the stuffed heads of ground hornbills as camouflage when stalking game. Others, for example the Xhosa and Zulu peoples of South Africa, occasionally kill the birds for use in rituals to alleviate drought. Breeding individuals of these and other smaller species are also taken for the preparation of traditional medicines, and even for food when necessary, across a wide part of their range.

Hornbills are especially important in certain Asian societies. The skulls and flights feathers are used by indigenous peoples in Malaysia, Thailand, the Philippines, Indonesia, and India – as shown in the Wancho warrior's headdress (RIGHT) – and also most notably by the Dayaks of Borneo, who recognize the impressive and raucous Rhinoceros hornbill as the god of war, Singalang Burong. Elaborate images of the bird, exaggerating its recurved casque into dramatic spirals, used to be carved in wood and hoisted above longhouses, or painted on their walls. Today, this hornbill is the emblem of the Malaysian state of Sarawak, and its white tail feathers with a single black band still feature prominently in dancing cloaks and head-dresses, as do the similarly colored, much elongated central tail feathers of the Helmeted hornbill.

The solid block of "hornbill ivory" forming the front part of the casque of the Helmeted hornbill is unique. It is carved by the indigenous Kenyah and Kelabit peoples of Borneo into ear ornaments or belt toggles, and in the past was an important item of trade with Chinese visiting Brunei. The Chinese executed exquisite three-dimensional carvings on the casque and worked the ivory (which they called *hoting*) into thin sheets, onto which they fixed the golden-red pigment derived from the preen-gland oil. These sheets were cut into belt buckles that were worn by high officials of the 14th- to 17th-century Ming dynasty in China.

Conservationists eager to preserve hornbills in their natural habitats are now encouraging the use of carved effigies in place of hornbill skulls, and the substitution of wild-taken feathers for those supplied from captive birds .

Toucans

The most prominent feature of toucans is their large, often vividly colored bill. The biggest bill of any toucan is that of the male Toco toucan, which accounts for some 23cm of the bird's total length of up to 79cm (9 of 31in). Toco toucans have been illustrated so frequently that they have become an almost clichéd symbol of the warm forests of tropical America. Of all the rich bird life of the Neotropics, probably only hummingbirds are a more popular subject for artists.

Right *Representative toucan species:* ***1*** *An Emerald toucanet* (Aulacorhynchus prasinus) *calling.* ***2*** *A Black-billed mountain-toucan* (Andigena nigrirostris) *revealing a flash of yellow on its rump as it climbs the branches.* ***3*** *A Chestnut-mandibled toucan* (Ramphastos swainsonii) *tossing its head, enabling food held at the tip of the bill to be transferred to the throat.* ***4*** *A Toco toucan* (Ramphastos toco) *feeding on berries.* ***5*** *A Guianan toucanet* (Selenidera culik) *examining a possible nest cavity.* ***6*** *A Saffron toucanet* (Baillonius bailloni) *in flight.* ***7*** *A Collared aracari* (Pteroglossus torquatus) *preparing to leave its nest-hole.*

1

Toucans are closely related to barbets and arose from a common American ancestor. Some taxonomists choose to group the toucans, along with the American barbets, as a family separate from other barbets, while others place various barbets and toucans as subfamilies of a single family. Yet the fact that toucans are very uniform anatomically and genetically, and display many unique features, argues strongly in favor of according these striking birds separate family status.

A Bill of Many Parts

FORM AND FUNCTION

Toucans' bills are much lighter in weight than they appear. A thin, horny outer sheath encloses a hollow space criss-crossed by many thin, bony, supporting rods. Despite this internal strengthening, toucans' bills are fragile and sometimes break. Nevertheless, some manage to survive a long time with part of their bills conspicuously missing. Toucans have a long tongue, and strong serrations along the bills' edges, but no bristles about its base. They often have brightly colored areas of bare skin on the face and chin. Several pale-eyed species have dark marks before and behind the (black) pupil, giving the eye an appearance of being a horizontal slit.

Naturalists have speculated for centuries about the uses of the toucan's exaggerated beak. It enables these heavy, rather clumsy birds to perch inside the crown of a tree, where branches are thicker, and to reach far out to pluck berries or seeds from twigs too thin to bear their weight. Seized in the tip of the bill, food is thrown back into the throat by an upward toss of the head. This behavior explains the bill's length but not its thickness or bright coloration. The diet of toucans consists mainly of fruit but includes insects and some vertebrates; various toucans hunt actively, sometimes in pairs or groups, lizards, snakes, birds' eggs, and nestlings. Some toucans follow swarming army ants, preying upon arthropods and vertebrates flushed by the ants. When plundering bird nests the huge, vivid bill of the toucan so intimidates distressed parents that not even the boldest of them dares to attack the toucan, except when the latter flies; enraged parents may attack and even perch on the exposed back of the flying toucan, prudently withdrawing before the larger bird alights. The huge, multicolored bills of toucans

FACTFILE

TOUCANS

Order: Piciformes

Family: Ramphastidae

34 species in 6 genera. Species and genera include: **aracaris** (genus *Pteroglossus*) including **Collared** or **Spot-breasted aracari** (*P. torquatus*), **Curl-crested aracari** (*P. beauharnaesii*), **Black-billed mountain-toucan** (*Andigena nigrirostris*), **Plate-billed mountain-toucan** (*A. laminirostris*), **Emerald toucanet** (*Aulacorhynchus prasinus*), **Yellow-browed toucanet** (*A. huallagae*), **Saffron toucanet** (*Baillonius bailloni*), **Guianan toucanet** (*Selenidera culik*), **Tawny-tufted toucanet** (*S. nattereri*), **Keel-billed** or **Rainbow-billed toucan** (*Ramphastos sulfuratus*), **Toco toucan** (*R. toco*), **White-throated toucan** (*R. tucanus*), **Yellow-throated toucan** (*R. ambiguus*).

DISTRIBUTION Tropical America, from central Mexico to Bolivia and N Argentina, excluding the West Indies.

HABITAT Rain forests, woodlands, gallery forest, savannas.

SIZE Length 36–79cm (14–31in; including bill); weight 115–860g (4–31oz). Males' bills usually longer.

PLUMAGE Black with red, yellow, and white; black and green with yellow, red, and chestnut; chiefly green; or olive brown and blue, colorfully patterned with yellow, red and chestnut. Sexes similar in color except in *Selenidera* species and a few aracaris.

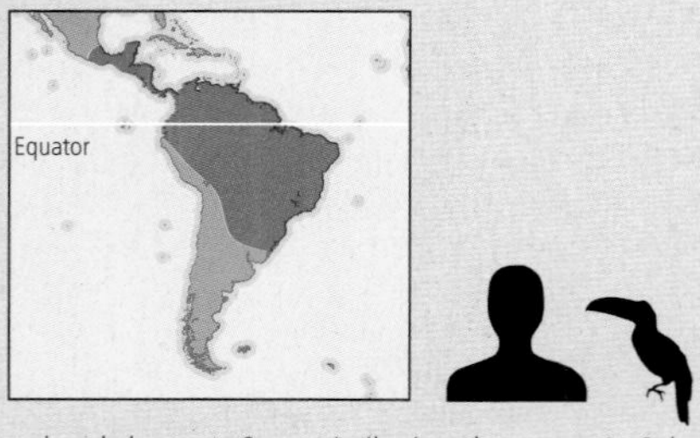

VOICE Usually unmusical, often croaks, grunts, barks, rattles, or high, sharp notes, but a few have melodic yelps or mournful calls.

NEST In natural cavities; some take over abandoned woodpecker or large barbet holes, or may evict the owners, then enlarge the hole.

EGGS 1–5, white, unmarked. Incubation period 15–18 days; nestling period 40–60 days.

DIET Chiefly fruits, supplemented by insects, other invertebrates, lizards, snakes, small birds, birds' eggs, and nestlings.

CONSERVATION STATUS The Yellow-browed toucanet, whose range is very small, is classed as Endangered.

5

2
3
4
7
6

also render them dominant over all other fruit-eating birds in the trees in which they feed. Varied bill patterns may help toucan species to recognize each other. In Central American forests, Yellow-throated and Keel-billed toucans have such similar plumages that they are only readily distinguished by their bills – and voices. The Keel-billed's beak is delicately tinted with all but one of the colors of the rainbow (its alternative name – Rainbow-billed toucan – is possibly more apt), whereas that of its relative is largely chestnut with much yellow on the upper mandible. The bills may also play a role in courtship, especially since the larger male's bill appears thinner and more scimitar-like than the shorter, broader female's bill.

Denizens of the Rain Forests

DISTRIBUTION PATTERNS

The biggest toucans – the seven species of the genus *Ramphastos* – chiefly occupy lowland rain forests, from which they make excursions into adjacent clearings with scattered trees. They are rarely seen above altitudes of 1,700m (5,600ft) above sea level. Their huge bills are well serrated, and the nostrils of adults are hidden beneath the bill base. Chiefly black or maroon-black, their calls are largely croaks and grunts, but the yelping song of the Yellow-throated toucan ("dios, te-dé, te-dé") is almost melodic when heard at a distance, as is that of the White-throated toucan ("dios-te-dé-dé"). These particular phrases are monotonously repeated.

The ten species of aracaris are smaller and more slender than *Ramphastos* toucans, with a longer, graduated tail. They too inhabit warm forests and edges, and rarely venture as high as 1,500m (4,900ft). Black or dusky green above with a crimson rump, usually with a black and chestnut head, and mainly yellow underparts, most aracaris have one or more bands of black or red below, sometimes forming a large breast spot. The coloration of their long bills comprises various mixtures of black and yellow or ivory white, or chestnut and ivory, orange, or red; their often pronounced bill serrations may be outlined in black or ivory, appearing somewhat like teeth. The Curl-crested aracari has uniquely broad, hard, and shiny feathers over the top of the head, which resemble enameled, curled shavings. Aracari calls are mainly sharp, high-pitched notes in series, or rattles that sound like a motorbike, but a few have less mechanical, wailing calls. At least some species roost in cavities throughout the year; as far as is known these are the only toucans to do so, although other toucans roost in holes when held in aviaries.

Green toucanets of the genus *Aulacorhynchus* (6 species) are small to medium with mainly green plumage. Their calls are often long, unmelodious, guttural series of croaks and barks, and dry rattles. They mostly inhabit cool mountain forests between 1,000 and 3,600m (3,300–11,800ft), but a few descend locally into warm lowlands. The Yellow-browed toucanet, of central Peru, is endangered.

Right *The Keel-billed toucan of Central America has a particularly resplendent bill. It eats mostly fruit – here, a papaw* (Carica papaya) *– but can increase its protein intake by supplementing its diet with eggs, nestlings, insects, small lizards, and tree frogs.*

Left *A Plate-billed mountain-toucan shows off its distinctive beak. Although threatened by deforestation, this endangered toucan is also under threat from the illegal international cagebird trade.*

The six species of *Selenidera* toucanets occupy rain forests at generally low altitudes (rarely to 1,500m/5,000ft) from Honduras to northeast Argentina. The birds are less social than other toucans, and their plumage is more variable; all have red undertail coverts and a yellow or gold erectile feather tuft about the ears. They and several aracaris are the only toucans in which the sexes differ markedly in color; the young can be sexed by plumage at 4 weeks of age. The reddish brown and green bill of the Tawny-tufted toucanet is marked with sky-blue and ivory. The Saffron toucanet of southeastern South America is mainly green and gold to yellow with some red. It is the only species of the genus *Baillonius*, and appears related distantly to aracaris. Sometimes a pest in orchards, it is found usually at 400–1,100m (1,300–3,600ft).

Relatively little is known about the four large species of mountain-toucans which, as their generic name *Andigena* suggests, inhabit the Andes from northwest Venezuela to Bolivia. From the subtropical zone they extend high into the altitudinal temperate zone, even reaching the tree-line at 3,650m (12,000ft). One colorful example is the Black-billed mountain-toucan, whose light blue underparts are exceptional among toucans. This toucan is black-capped, with a white throat; its back and wings are mainly brown-olive; the rump is yellow, undertail coverts crimson, and the thighs and tip of the tail are chestnut. Male or female singers lower the head and cock the tail, then raise the head and lower the tail (very like *Selenidera* species) in time to the yacking notes of the song, which is enhanced by bill-clapping sounds. The best known species, the Plate-billed mountain-toucan, with its raised ivory-yellow patch or plate on the red and black bill, is one of two in its genus that is near-threatened, as intensive logging gains pace along the western slope of the Andes. Forest clearance for cash-crop cultivation, ranching, and mining may soon threaten most toucans, as their habitat is converted to human use.

Slow Developers

SOCIAL BEHAVIOR

Toucans are variously social and non-social. Social toucans are moderately gregarious and fly in straggling flocks moving single-file, rather than in compact bands like parrots. The big *Ramphastos* species beat their wings several times, close them and drop downward, then open them in a short glide, followed by more beats to regain altitude. They rarely cross large open areas, or wide rivers, as they can only sustain long flights with difficulty. Smaller toucans beat the wings faster, aracaris appearing like long-tailed auklets, but also flying

single-file. They bathe in rainwater that collects in hollows of trunks and limbs high in trees, their preferred location. Members of pairs offer food to their mates, and, perching well apart, they gently preen one another with the tips of their long bills.

Occasionally, toucans engage in apparent play, which may be related to a dominance hierarchy that may later affect pairing. After striking their bills together, two birds may clasp each other's bills and push until one is forced backward off its perch and retreats. Another individual may then cross bills with the winner, and the victor may be challenged by yet another toucan. In another form of play, one toucan will pitch a piece of fruit, which another catches in the air, and then throws in a similar fashion to a third, who may toss it to a fourth toucan.

Left *The row of short black lines on the Lettered aracari's bill are known as the "letters." The sexes are distinguished by facial color: the male (shown here) has a black face, while the female's is chestnut-brown.*

COOPERATIVE BREEDING IN ARACARIS

Aracaris are slender, long-tailed, middle-sized toucans, which inhabit lowland forest and forest patches, sometimes amid farms. The Collared or Spot-breasted group of species sharply call "pitit" or "pseep" in often irregular series. They fly swiftly, one-by-one in small, straggling flocks. At nightfall they retire into old woodpecker holes or other cavities, tucking their head into their back feathers and pressing the tail closely over the head and back, the aracaris thus clustered like fluffy balls in the narrow chamber.

In a Panamanian forest six Collared or Spot-breasted aracaris were once observed to squeeze with difficulty through a narrow orifice in the underside of a thick horizontal branch, 30m (100ft) up in a great tree. As weeks passed the number of birds using this hole for sleeping decreased until only one remained incubating the clutch of eggs (this was probably the female or male of the breeding pair). After the eggs hatched five of the original six birds again slept in this hole, and all brought food to the nestlings, at first chiefly insects grasped in the tips of their great beaks. As the nestlings grew older, the five attendants brought increasing quantities of fruits, some of which they regurgitated.

At about 43 days the first young aracari flew from its high nursery. At nightfall its attendants led it back to sleep with them. While the fledgling tried inexpertly to enter the narrow, downward-facing doorway, a White hawk swooped down, seized the piteously crying young bird in its talons, and carried it off, followed by the calling adults.

Three of the five attendants at this nest were probably non-breeding helpers, possibly older offspring of the parents. Helpers are known in at least two aracaris, but observations of other groups in the breeding season suggest that they may also occur in at least some of the other eight species. Most toucans nest strictly in pairs, although non-breeding members of the pair's social group may occur adjacent to the breeding territory, as has been observed in the Emerald toucanet, the Plate-billed mountain-toucan, and the White-throated toucan. AFS

Toucans have uniquely modified vertebrae of the lower back and tail base, allowing the tail to lock in a position tightly applied to the head. With the head and bill tucked under the forward appressed tail, the sleeping toucan resembles a ball of fluff.

Most large *Ramphastos* species normally nest in decayed cavities in tree trunks, and if successful will use the same cavity year after year. Availability of such holes may limit the number of pairs that can breed. A favored hole is in generally sound wood with an opening just wide enough to allow the adults to squeeze through. The hole may be 17cm (6.7in) to 2m (6.5ft) deep. A suitable cavity near the base of a trunk may tempt toucans to nest closer to the ground than normal. Toco toucans may nest in a terrestrial termite mound or in an earthen bank. Smaller toucans often occupy woodpeckers' holes, sometimes evicting the owners. The large Plate-billed mountain-toucan frequently usurps the cavity of the Toucan barbet, if it is located in a tree large enough for the toucan. Some *Aulacorhynchus* species are able to carve a cavity out of very rotten wood, and *Selenidera*, *Andigena,* and *Baillonius* species regularly excavate part of the cavity they select. In fact, some excavation seems to be a major feature of breeding behavior in many toucans. The nest chamber is unlined, but the 1–5 white eggs rest upon wood chips, or upon a pebbly bed of regurgitated seeds, which grows deeper as nesting progresses.

Parents share incubation and are, for their size, often impatient sitters, rarely sitting for more than an hour. They are very shy, and they slip from the nest and fly off at the least threat, often leaving the eggs uncovered.

The nestlings hatch after about 16 days of incubation, and are blind and naked, without a trace of down. Their feet are poorly developed, but each ankle joint has a pad of large spiky projections; at first they perch as a tripod, upon the two pads and the rough-skinned, swollen abdomen. Like hatchling woodpeckers, which they closely resemble, their bills are short, with the lower mandible slightly longer than the upper. Nestlings are fed by both parents, with increasing quantities of fruits as they grow older, but they develop surprisingly slowly. By four weeks of age small toucanets are rather sparsely feathered and month-old *Ramphastos* toucans are still largely naked. Both parents brood the nestlings, with no clear pattern of one sex doing so at night. Large billfuls of waste are carried away from the nest, some species, including Emerald toucanets, keeping the nest relatively clean, whereas White-throated toucans permit decaying seeds to remain.

When they are finally fully feathered, young toucans resemble their parents, but are duller, and their bills are smaller, lacking serrations and a vertical base line, and showing little of the adult's bright colors. The bill may take a year or more to attain its adult size and features.

Small toucanets may fly from the nest when 40 days old, but large *Ramphastos* toucans may not leave until 50 days old or more, and some *Andigena* toucans do not leave until 60 days old. Some aracari fledglings are led back to the nest to sleep with their parents, but most young toucans roost amid foliage. LS/JH/HGG

Left *A pair of Many-banded aracaris* (Pteroglossus pluricinctus) *in the treetops. These are the only aracaris with a black band on their yellow breasts; they have a second red-black band across their yellow bellies.*

Below *In common with other* Ramphastos *species, the Channel-billed toucan* (R. vitellinus) *nests high in trees, mobbing intruders such as monkeys and snakes.*

Honeyguides

THE DULL PLUMAGE, FOREST AND WOODLAND habitat, and retiring disposition of honeyguides disguise a family whose behavior is among the most extraordinary and least well-known of any birds. They are named for the habit, seen in one African species, of leading people to honeybee nests.

Both from observations in the field and from experiments, it is clear that honeyguides prefer the honeycomb and bee larvae to the actual honey. Although insects form the mainstay of their diet, honeyguides can survive on pure wax for 30 days or more, relying on special enzymes in their gut to break it down. This relatively uncommon specialization is one of three exhibited by honeyguides; less unusually, they lay their eggs in other birds' nests, and – uniquely – their hatchlings are equipped with bill hooks that allow them to smash the eggs or kill the nestlings of their hosts.

Mostly Inconspicuous

FORM AND FUNCTION

Honeyguides are distantly related to barbets and woodpeckers (with which they interact, and which are major honeyguide hosts). They occur only in the Old World tropics, mostly in Africa, although two species are found in Asia, where one is a downslope migrant along the Himalayas. Their main habitat is broadleaved forest and woodland, though some *Prodotiscus* and *Indicator* species inhabit open woodlands, wooded grassland, and streamside trees in dry areas. African species number 15 among all four genera; some form several groups of closely related species. Within each such group, darker-colored species tend to dwell in evergreen forest, paler ones in woodland. The somber camouflage is relieved only by light sides to the tail; these tail patches, which are virtually alike in all African species, act as beacons to guide fledgling and immature birds to beehives. Four species depart from this drab uniformity: the Lyre-tailed honeyguide, in which the tail of both sexes (especially the male's) is curved outward from elongated undertail coverts, and the four outer pairs of feathers are short and pin-like, creating a loud tooting sound in diving flight; the Yellow-rumped honeyguide, which has yellow-orange on the head and rump; the Sunda honeyguides, the male of which has a yellow shoulder patch; and the Greater honeyguide, in which males have a black throat, white cheeks, yellow shoulder flashes, and, when breeding, a pink bill. This latter species also has the most conspicuous sexual difference and is the only species with a very different immature plumage.

Above A Greater honeyguide. In the late 1980s, a study of the nomadic Boran people of northern Kenya showed a symbiotic relationship with this species: hunting groups were three times as efficient at finding honey when guided by the bird. The honeyguides also benefited greatly, gaining entry to otherwise inaccessible nests after the humans had broken them open.

Right A Scaly-throated honeyguide, a distinctively patterned species, next to a broken honeycomb.

FACTFILE

HONEYGUIDES

Order: Piciformes

Family: Indicatoridae

17 species in 4 genera. Species include: **Greater** or **Black-throated honeyguide** (*Indicator indicator*), **Lesser honeyguide** (*I. minor*), **Scaly-throated honeyguide** (*I. variegatus*), **Malaysian** or **Sunda honeyguide** (*I. archipelagicus*), **Yellow-rumped honeyguide** (*I. xanthonotus*), **Lyre-tailed honeyguide** (*Melichneutes robustus*), **Yellow-footed** or **Eisentraut's honeyguide** (*Melignomon eisentrauti*).

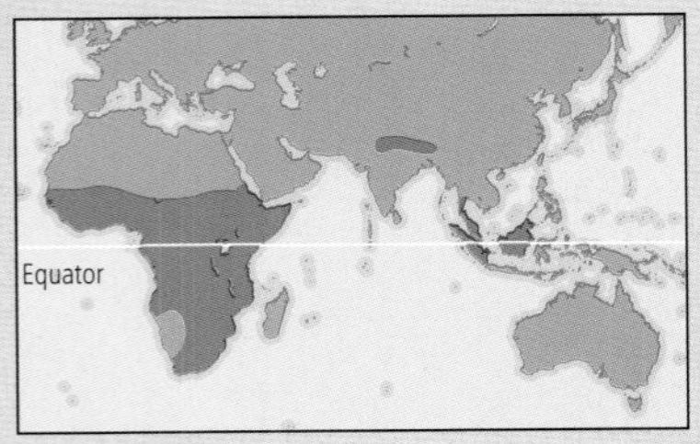

DISTRIBUTION Africa S of the Sahara, S Asia (2 species).

HABITAT Evergreen forest, montane forest, woodlands, watercourses in semi-arid areas.

SIZE Length 10–20 cm (4–8in), weight 9–62g (0.3–2.2oz).

PLUMAGE Dull to brighter yellow-olive, olive, gray or brown, paler below, outer tail with white patches (dusky or dusky white in 2 Asian species); two species have yellow lesser wing-covert patches; one has yellow-orange on head and rump; marked sexual dimorphism in only a few species.

VOICE Poorly known in some, but very diverse; males of several species utter trilling to melodic, usually repetitive songs from a high perch; also various piping, chattering, rattling, and other calls; non-vocal sounds in aerial displays.

NEST None; in all species whose breeding habits are known, eggs are laid in cavity nest or deep-cup nest of host.

EGGS Normally one per host nest, in clutches, clutch size little known; white (blue in some honeyguides of one species), thick-shelled. Incubation: 12–18 days; nestling period: 20–22 days in Prodotiscus, 30–40 days in others as far as is known.

DIET Chiefly insects, but some fruits, and all species include some form of wax, especially beeswax.

CONSERVATION STATUS Three species – the Malaysian honeyguide of SE Asia, the Yellow-rumped honeyguide of E and SE Asia, and the Dwarf honeyguide (*Indicator pumilio*) of central Africa, – are currently listed as Lower Risk: Near threatened.

Like woodpeckers and barbets, honeyguides' feet are zygodactylous. Many also have curiously raised nostrils, set on a prominent ridge, and a very thick skin, as protection against bee stings. Several species keep a close watch on human activity in woods. Though they may have a sense of smell, this visual monitoring strongly suggests that they rely on sight rather than scent to locate food.

Murderous Interlopers

SOCIAL BEHAVIOR

Honeyguides are vocally varied, with diverse songs and calls, including pre-and post- copulatory calls in several species. Circling display flights with wing sounds occur in a number of species. Males of most African honeyguides defend singing territories, attracting not just females, but also subordinate

HUMAN–HONEYGUIDE INTERACTION

Wild honeyguides can open bees' nests for themselves (since many species eat beeswax but do not guide), so it is unclear to what extent Greater honeyguides depend upon human help in obtaining honeycomb. Yet this species definitely does guide people and perhaps other large mammals, such as the ratel or Honey badger (*Mellivora capensis*), to bees' nests. Keen-eyed, most adult honeyguides know the location of every active and abandoned honeybee hive in their area. African tribal honeyhunters have long exploited this bird's special skill.

Uttering a wavering, chattering call (usually a sign of aggression), the Greater honeyguide attracts humans' attention, and then flies in short stages, stopping frequently to call and check the progress of the followers. When it nears the beehive, the bird usually falls silent; the honeyhunters then build a smoky fire to stupefy the bees, and hack open the hive with an ax or *panga* (machete). Most honeyhunters leave some honeycomb out for the honeyguides, believing that failure to do so will lead to their being guided to a dangerous animal next time, but others maintain that honeycomb spoils the bird, and leave it to find its own bits of comb.

The Greater honeyguide's bizarre behavior is especially surprising in a nest-parasite, whose opportunities for learning the guiding habit from adults are necessarily more limited than in most birds. Hand-reared Greater honeyguides eat both the larvae and the wax in the first honeycomb they are given, and wild adults prefer beeswax with larvae, although dry comb is often infested with wax-moth larvae, which honeyguides relish. The aggressive call used in guiding develops directly from begging calls. Young honeyguides do not require wax or grubs in their diet – as nestlings they are raised on the same diet (insects and fruits) that their diverse hosts feed their own young. As nestlings the honeyguides may often hear both songs and calls of various honeyguides, and when they can reach the entrance of the cavity, they can see them as well.

Most honeyguides get beeswax from abandoned hives and from those opened by ratels or people. Since African honeybees abandon hives more often than bees in temperate zones (and use cavities with larger openings), there is no shortage of live or abandoned hives accessible to honeyguides. However, Greater honeyguides in dry areas with fewer large trees tend to encounter honeybees using cavities in rocks or within smaller tree openings. Guiding would therefore be especially advantageous to such birds.

The traditional use of honeyguides for finding honey is dying out in many parts of Africa, especially near major towns and cities, as refined sugar becomes more easily available and the old lifestyles collapse. Greater honeyguides seem destined to gradually lose their unique guiding habit.

males, which will attempt to take the territory over if the primary male is absent. After displays with females the territorial male will mate in or near the territory.

Females lay their eggs (usually singly) in hosts' nests, at considerable effort and risk when the host is a social barbet. Young of well-known species hatch with temporary, membranous, hawklike hooks on the tip of the bill, with which they puncture the hosts' eggs or kill their chicks. The fledgling's insistent, loud begging calls sound like several of the host young calling together. Upon fledging, young honeyguides are fed by foster parents only for a very short time; as soon as the fledging Lesser honeyguide leaves a barbet nest, it is immediately recognized as a honeyguide and mercilessly chased from the host group's territory. It is then entirely on its own, and will follow passing honeyguides to find beeswax. Immature honeyguides are inevitably dominant to adults at beeswax sources; the distinctively plumaged immature Greater honeyguide is superdominant to all honeyguides.

Some adult honeyguides monitor hosts carefully and may even defend possible host pairs or groups against other honeyguides. Extended honeyguide–host interactions occur, even outside the breeding season, especially between honeyguides and host barbets and woodpeckers. LS/JH/AWD

Barbets

Most barbets have bristles around the gape and chin and bristles (or tufts of them) over the nostrils; in some Asian barbets the bristles are so long that they extend beyond the bill. The bearded appearance that this characteristic lends them gave rise to the common name for the family. Certain African and Asian species go by such descriptive names as "tinkerbird" and "coppersmith," from the harsh, metallic tone of their repetitive calls.

Barbets are so closely related to toucans that some taxonomists treat the various barbet groups and toucans as subfamilies of a single family (or alternatively group the American barbets with toucans in the toucan family). However, the fact remains that toucans differ in many ways from barbets, while barbets are generally similar to one another in build and behavior. Thus, since no one could mistake a toucan for a barbet, and because there are no intermediate forms, toucans are treated here as a distinct family.

Robust and Colorful

FORM AND FUNCTION

Barbets are compact, thickset birds with rather large heads and bills that are stout, conical, and sharply tipped. In larger species, the bill may be especially formidable; *Tricholaema* and *Lybius* species have notched bills that help them grip food, and in the Toucan-barbet and Prong-billed barbet the tip of the upper mandible fits into a deep cleft in the tip of the lower mandible. The tongue is brush-tipped, which helps them eat fruits, their juices, and nectar. Their legs are short and strong, the feet zygodactylic (on each foot, the second and third toes point forward, while the first and fourth point backward, as in their toucan and woodpecker relatives). The birds are more often found in the foliage than on the trunks of trees, and their short tail is used as a support in excavating nests. The large barbets appear heavy and cumbersome in their movements, but others (for example, the Red-headed barbet and the tinkerbirds) are agile, probing and gleaning much like tits. The wings are short and rounded, and unsuitable for sustained flight. Ground barbets move by inelegant hops.

The sexes are alike in a majority of barbet species, but there are some notable exceptions to this general rule. The *Eubucco* species of South

Above *Representative barbet species:* ***1*** *The Toucan-barbet* (Semnornis ramphastinus) *is found in the cloud forests of Columbia and Ecuador;* ***2*** *The central African Black-backed barbet* (Lybius minor)*;* ***3*** *The Red-headed barbet* (Eubucco bourcierii) *of South America.*

FACTFILE

BARBETS

Order: Piciformes

Family: Capitonidae

82 species in 13 genera. Species and genera include: **Black-spotted barbet** (*Capito niger*), **White-mantled barbet** (*C. hypoleucus*), **Red-headed barbet** (*Eubucco bourcierii*), **Black-backed barbet** (*Lybius minor*), **Chaplin's barbet** (*Lybius chaplini*), **ground barbets** (genus *Trachyphonus*), **D'Arnaud's barbet** (*T. darnaudii*), **Great barbet** (*Megalaima virens*), **Lineated barbet** (*M. lineata*), **Pied barbet** (*Tricholaema leucomelas*), **Prong-billed barbet** (*Semnornis frantzii*), **Toucan-barbet** (*S. ramphastinus*), **tinkerbirds** (genus *Pogoniulus*), **Yellow-fronted tinkerbird** (*P. chrysoconus*).

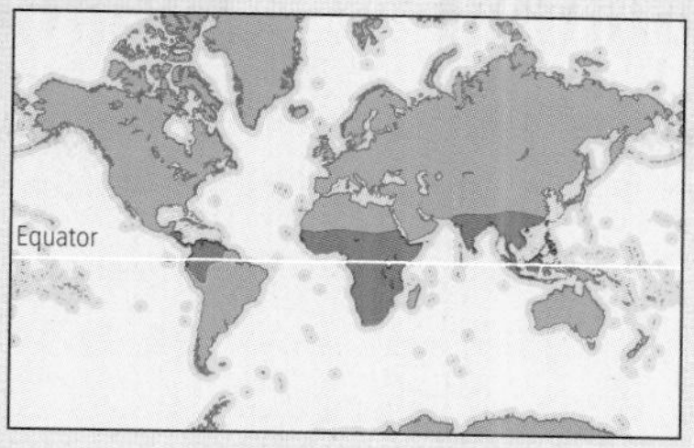

DISTRIBUTION Africa S of the Sahara, Pakistan, and Sri Lanka E to SE Asia, S China, Philippines, W Indonesia, Bali, Costa Rica to northern S America.

HABITAT Primary and secondary tropical, subtropical, and adjacent temperate forests, various woodlands, some plantations, and (in Africa) arid habitats with some trees.

SIZE Length 9–33cm (3.5–13in); weight 7–295g (0.2–10.4oz).

PLUMAGE Asian species are predominantly green with yellow, blue, red, and black markings about the head; African barbets are brown or green, or patterned white, black, red and yellow, often heavily spotted or streaked; American species are black, olive, or green with some white, red, yellow, gray, and blue markings. Well-developed differences between the sexes occur in most American but few other barbets elsewhere.

VOICE From a rapid repetition of single or series of honking, popping, or piping notes to somewhat melodic whistled and other notes; duetting is well developed.

NEST Most species excavate holes in decayed trees; others use earthen banks, burrows, or termite mounds; usually no nest lining.

EGGS 1–6, white; Incubation 12–19 days; nestling period 17–30 days in most smaller barbets, up to 42 in some larger species.

DIET Fruits, buds, flowers, nectar, insects; larger species also eat tree frogs, lizards, and small birds.

CONSERVATION STATUS The White-mantled barbet of Colombia, which has a small and very fragmented range, is threatened by habitat loss and is classed as Endangered. Nine other species are considered as Lower Risk/Near Threatened, including the Toucan-barbet and Chaplin's barbet.

America have green wings, back, and tail, and underparts of yellow streaked with green. The sexes differ from each other in the color pattern of their head, throat, and breast; the sexes are sufficiently dissimilar that the male and the female of several species were originally described as different species. The male Red-headed barbet has the whole head and throat scarlet shading to orange on the breast, and a blue collar on the nape. The female has blue on the side of the head, a gray throat and yellow-orange on the upper breast. Likewise, in the *Capito* species, the sexes differ from one another, ranging from slight to marked sexual dichromatism. Both sexes of the Black-spotted barbet have scarlet on head and throat, and black upperparts streaked with greenish yellow and creamy yellow underparts. The female differs from the male in having black spots on the throat, different markings on the inner wing, and being more heavily spotted black on the underparts. Most Asian barbets are predominantly green and differences between species lie in the head colors (brown, red, yellow, orange, sometimes blue and white) and their pattern; in all but a few species the sexes are identical in the field. Females are often larger than the males. The Great barbet, largest of all barbets, has a yellow bill, maroon-brown and bluish upperparts, a violet blue-black head, multicolored underparts (olive, brown, blue, yellow), and red undertail coverts. In contrast, there is less green in African barbets, the majority of which are patterned black, brown, yellow, red and white, and their plumage is more spotted, streaked, and barred than in Asian species. Some forest species (for instance, the species in the genus *Gymnobucco*) are a very drab brown with tufts of rictal and other bristles, and the head is more or less bare of feathers.

Right *Similar to the Blue-throated barbet* (Megalaima asiatica), *the Moustached barbet* (M. incognita) *is also known as Hume's blue-throated barbet. One distinctive feature is its greenish crown with a small red patch at the rear.*

Widespread in Africa

DISTRIBUTION PATTERNS

Any visitor to Africa is certain to hear the repetitive, often monotonous, calls of barbets, since they occur in all the continent's major vegetation zones; 7 African genera (41 species) are recognized, and within these there is a generally greater divergence in size, bill shape, and color pattern than is found in Asian and American genera. Adaptation to more arid habitats within Africa is thought to have given rise to the ground barbets and some others. In tropical Asia and America the barbets are on average larger, extremely vocal, and also mainly arboreal; 3 genera (15 species) are recognized in the Americas and 3 genera (26 species) in southern Asia (where 2 genera contain only one species each). There are some notable cases of convergence in barbets. For example, black, yellow-and-white, and brown-colored barbets occur in Africa and southern Asia; spotted barbets are found in Africa and America; and the Black-backed barbet of central Africa resembles the White-mantled barbet, which can be found in northwestern South America.

Only a few Asian barbets in China and the Himalayas migrate; other barbets may move locally as rainfall and the food availability shift.

Adept Fruit- and Insect-eaters

DIET

Most barbets feed on fruits, some of which are lost while being plucked from the tree; several species, however, have a more efficient way of feeding, using a foot to hold the fruit steady as they eat. Fruit pits or pips that are not tiny are regurgitated. Barbets thus contribute significantly to the seed dispersal of many forest trees. Petals, flower heads, and nectar are eaten by some species such as the Great barbet and the Prong-billed barbet. Most, likely all, species feed insects to newly hatched young; some regularly take insects, and probably all take emerging termites, which may be caught on the wing or gathered from the ground. Species of *Tricholaema* probe into termite earthen tunnels on trees, feeding on the termites they expose. Ants, beetles, and grasshoppers are among many insects taken, and larger species, such as the Lineated barbet and the ground barbets, will occasionally take lizards, tree frogs, and small birds. A few species (genus *Stactolaema*) use "anvils," special sites kept clean for removing the wings and legs of large insects before eating them or feeding them to the young. The D'Arnaud's barbet is highly insectivorous and feeds in low bushes and on the ground. The Red-headed barbet probes leaf litter in search of spiders and insects. Hard insect remains are regurgitated as pellets. Both fruit-and insect-eating barbets often feed with mixed foraging flocks of other species.

The majority of barbet species are extremely territorial. Barbets behave aggressively toward other birds, especially members of their own species, but also toward other hole-nesters (such as woodpeckers and starlings), honeyguides, and fruit-eating competitors.

Above *As part of its diet, the Red-fronted tinkerbird* (Pogoniulus pusillus) *eats a diverse range of berries and especially mistletoe. The pits or stones of these fruits are regurgitated with choking throat and head movements.*

Forming Duets

BREEDING BIOLOGY

The breeding behavior of barbets is varied but little known; indeed, the clutch size is not fully known for 34 species. The Prong-billed barbet from the mountains of Costa Rica lays only one clutch of eggs and has a restricted breeding season beginning in March. Many species are paired throughout the year. The breeding season may be prolonged and cover dry and wet periods in which up to three or four broods may be raised (for example, the Yellow-fronted tinkerbird), or occur in one of several rainy seasons where there are such, or may fall entirely within the main wet period, as in the case of ground barbets. Most barbets excavate a new hole yearly, but some reuse the same cavity, deepening it after each brood. Barbets roost in cavities that they excavate less deeply than a nesting cavity; if a nest is lost, such a roosting hole may be enlarged to serve as a nesting cavity. D'Arnaud's barbet, one of the ground barbets, bores a tunnel vertically downward into level ground; at a distance well above the bottom, it excavates a horizontal tunnel where the nest chamber is formed, thus avoiding flooding of the nest in the event of light or moderate rain. The lack of suitable nest-sites occasionally prompts the Pied barbet to use deserted nests of swallows or martins. In Asia large barbets may so enlarge the cavity of smaller species as to render them unusable by the latter; the intention behind this behavior may be to reduce the competition for fruits growing in the vicinity of their own nest.

Right *For a small bird, the Little green barbet or White-cheeked barbet* (Megalaima viridis) *has a surprisingly strong, loud call which involves an ascending and descending rolling "r-call."*

Left *D'Arnaud's barbets occur in pairs, family groups and small social groups. They generally move in one direction and feed loosely together, gathering at times to interact or duet, or at the threat of a predator.*

Both sexes excavate the nest cavity and share in incubation, feeding the nestlings, and nest sanitation. The young hatch blind and naked and have heel pads, as do the young of all woodpecker-like birds (only when they are feathered do the young use the feet, clinging to and climbing the cavity walls). At first the adults often swallow feces of their young; later they mix wood chips with the feces, forming small balls that are discarded away from the nest. In several African and American species extra helpers, often the young of a prior brood, may help to brood and feed the nestlings. Several ground barbets, some *Lybius* and *Stactolaema* species, and the Toucan-barbet live in social groups, the helpers assisting in nesting activities and roosting in the same cavity with the parents and young. The African *Gymnobucco* species breed colonially, with up to 150 or more pairs nesting in one large tree. Many African barbets are parasitized by honeyguides. Disruptions by honeyguides at barbet nests can cause breakage of eggs (and thus re-nesting, providing more opportunity for the honeyguide, if it failed to lay in the original clutch); indeed, a likely secondary adult of Chaplin's barbet, while the rest of its group chased a honeyguide, removed and broke, one by one, the four barbet's eggs in the nest (the probable advantage in doing this being that it might become a primary adult in laying a new clutch).

Because of the difficulty in observing barbets (many build nests near the top of very high trees), and the fact that they usually form long pair-bonds, their courtship behavior is incompletely known. The male may display aerially, and follow the female about. In social *Lybius* and *Trachyphonus* species the primary male and female duet together and prevent other members of the social group from duetting; removal of one of the primary pair results in it being replaced almost instantly by another group member as a duetting partner of the remaining male or female of the pair. In ground barbets both sexes spread their head feathers, and the tail, the male strutting around the female; they duet together, the male cocking its tail, the female often with tail lowered as they sing and turn their heads about, alert to responses by neighbors. Duetting is a common feature of barbets on all three continents. It has soft, short-distance components, and louder, long-distance components for communication within the pair or group, and with rival, neighboring groups. Duets occur throughout the year (more frequently when breeding); playback of a part of the duet involves an immediate response by the resident group, and indeed may trigger increased duetting and group interactions for several days thereafter. Duetting functions to maintain pair-bonds, to mediate within-group relations, and to maintain territories. Family parties usually roost communally, but in some *Megalaima* species the young are independent within several days and are forced out of the territory as their parents prepare to nest again.

The dependence of barbets upon suitable trees for excavating cavities, and thus for forests and woodlands that are ever diminishing, may account for the status of one species – the White-mantled barbet – as Endangered and 9 others as Near-Threatened. These forests continue to be cleared and used for livestock farming, arable cultivation, oil extraction, and mining. LS/JH/HGG

Above *Little is known about the habits and breeding of the Versicolored barbet* (Eubucco versicolor), *which is found in the forests of the eastern Andes, generally at 1,000–2,000m (3,300–6,500ft).*

Below *A pair of White-eared barbets* (Stactolaema leucotis) *will use their erect ear patches in courtship displays, which involve bowing and swinging their heads from side to side.*

Puffbirds

PUFFBIRDS ARE NAMED FOR THE LARGE heads and lax plumage evident in some species, giving them the appearance of a child's fluffy toy when they are at rest. In these species, the habit of remaining motionless and often permitting a close approach by humans led earlier observers to regard them as foolish.

Quite at variance with their undeserved reputation for dull-wittedness, these intriguing birds are actually highly accomplished "sit-and-wait" hunters, and stay still in order to conserve energy, while their keen eyes scan the surrounding vegetation for prey. As soon as a prey item is sighted, they spring into life and sally forth in energetic pursuit. The eyesight of some puffbirds is so acute that they can spot small insects at ranges of over 50m (164ft). Puffbirds are predators of the invertebrates and small amphibians and reptiles that abound in the tropical forests and open woodlands of Central and South America. In many respects, they are the Neotropical equivalents of the forest kingfishers of the Old World and Australasia. A few species supplement their diet with fruit.

Right *The Spot-backed puffbird* (Nystalus maculatus) *prefers low dry forest and the white sandy soils of the Brazilian* caatinga. *Here they build their nests, excavating in a bank or burrowing into flat ground.*

FACTFILE

PUFFBIRDS

Order: Piciformes

Family: Bucconidae

34 species in 10 genera. Species and genera include: **Russet-throated puffbird** (*Hypnelus ruficollis*), **Lanceolated monklet** (*Micromonacha lanceolata*), **White-faced puffbird** (*Hapaloptila castanea*), **Swallow-wing** (*Chelidoptera tenebrosa*); **typical puffbirds** (genera *Notharcus*, *Bucco*, *Nystalus*), including **Collared puffbird** (*Bucco capensis*); **soft-wings** (genus *Malacoptila*), including **White-whiskered puffbird** (*M. panamensis*); **nunlets** (genus *Nonnula*), including **Chestnut-headed nunlet** (*N. amaurocephala*); **nunbirds** (genus *Monasa*), including **White-fronted nunbird** (*M. morphoeus*).

DISTRIBUTION Mexico to Peru, Bolivia, and S Brazil.

HABITAT Rain forest, forest edge, dry open woodland, scrub, savanna.

SIZE Length 12–35cm (4.5–14in); weight 14–122g (0.5–4.3oz)

PLUMAGE Black, white, brown, rufous, buff, often barred, streaked or spotted (never brilliant); obvious differences between sexes only in a few species.

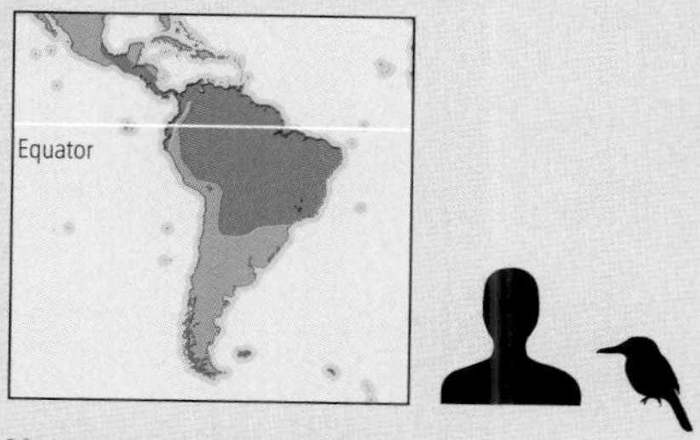

VOICE Thin and weak to loud and ringing; rarely melodious. Most species generally silent; sociable species often noisy and clamorous.

NEST Burrows in ground or cavities carved into termite mounds; occasionally abandoned nests of other birds, sometimes lined (leaves or grass).

EGGS 2 or 3, occasionally 4; unmarked, white. Incubation period unknown; nestling period 20 days (White-faced puffbird) or about 30 days (White-fronted nunbird).

DIET Chiefly invertebrates (especially insects), sometimes frogs, lizards, and small snakes; rarely fruits.

CONSERVATION STATUS 1 species, the Sooty-capped puffbird (*Bucco noanamae*) of Colombia, is classified as Lower Risk/Near Threatened.

It has been accepted for over a century that the puffbirds are the nearest living relatives of the jacamars and this relationship has been repeatedly confirmed by recent studies, most notably via comparisons of their DNA. Despite marked differences in overall form and plumage, the puffbirds and jacamars have many morphological and behavioral traits in common, especially in the structure of their feet and skeletons. Indeed, some authorities unite the two families in their own separate order. However, the relationship of these two families to other birds has been much debated. Some researchers have argued that the puffbirds and jacamars are most closely related to the woodpeckers, barbets, and toucans, whereas others have suggested a relationship with the rollers and their allies. Evidence is conflicting and further studies will be necessary to resolve this issue.

Low-key Lurkers

FORM AND FUNCTION

In contrast to the brilliant jacamars, the puffbirds are, with a few exceptions, relatively somber, with black, gray, or brown plumages predominating. These help, in part, to camouflage them while they perch on branches and wait for prey. Their "sit-and-wait" strategy might be thought to render puffbirds susceptible to attack from other predators. However, recent studies have also shown that some puffbirds are malodorous, and this may deter potential predators by rendering them unpalatable. All puffbirds have relatively short legs and feet in which the outer and inner toes project backward. These are specific adaptations to their perching lifestyle.

The largest puffbirds are the nunbirds and the White-faced puffbird. As their name suggests, the nunbirds are cloaked in svelte black or gray plumage that contrasts with their often brilliantly colored bills. The latter are moderately long and curved, well-suited to a diet of large insects, especially moths and butterflies. Nunbirds are among the most active of puffbirds and pursue their prey with fast beats of their wings followed by a glide. Prey is taken by gleaning vegetation or hovering like a flycatcher, the latter strategy aided by relatively long tails that increase lift and maneuverability. The White-faced puffbird has a grayish back and rufous underparts with a prominent white frontal band and throat. Its bill is sturdy, with a hooked tip like that of a roller. While the White-faced puffbird usually sits hunched waiting for prey, it also uses its robust bill to actively dislodge insects from decaying timber.

The seven species of *Malacoptila* puffbirds are commonly known as soft-wings on account of their loose plumage. Soft-wings are lethargic and inconspicuous but are capable of silent, darting flights on their rounded wings. Facial bristles are well-developed in some species, which help to guard the face from the movements of prey items in their death throes, especially when they are killed by being battered against a branch. The birds' robust, hooked bills enable them to capture small amphibians and lizards as well as insects. Soft-wings are amongst the few puffbirds to show marked sexual differentiation in plumage. In the White-whiskered puffbird the male is bright olive-brown above with fine cinnamon spots and with a cinnamon-streaked face and forehead. The underparts are paler and heavily streaked. In contrast, the female is much grayer.

The medium-sized puffbirds of the genera *Bucco* and *Notharcus* share a number of similarities, notably shrike-like hooked bills, and are sometimes merged together. The *Bucco* species have complex, relatively colorful rufescent plumages with white facial patterns. The Collared puffbird is unique in having a stout, bright orange beak. In contrast, the *Notharcus* puffbirds are predominantly black and white with wide, stout bills. In both genera some species have an upper bill with a forked tip into which the lower bill slots. The exact function of this feature is unknown, but it is believed to act as a "vice" to hold insects while they are being stripped of their wings and killed by the bird.

The unique Russet-throated puffbird is the sole representative of the genus *Hypnelus* and also has a forked upper bill tip. It is dull brown above with diffuse spotting, and has a broad white cheek patch and collar. The underparts are pale with a cinnamon throat and a broad black chest band. It is closely related to the five species of *Nystalus* puffbirds. These are predominantly brown, streaked birds of medium size. They sit stolidly when perch-hunting and, unlike other puffbirds, take much of their prey from the ground. Their robust, hooked bills enable them to catch insects, lizards, and even small snakes.

The smallest puffbirds are the Lanceolated monklet and the six species of nunlet. As it is small and unobtrusive, very little is known about the single species of monklet. It is also lethargic, perching for long periods in between forays after relatively large insects that it dismembers with its heavy, arched bill. Unusually for a puffbird, it is known to eat fruit. The monklet is dull brown above, while its underparts are white and heavily streaked with black. It has a pronounced white

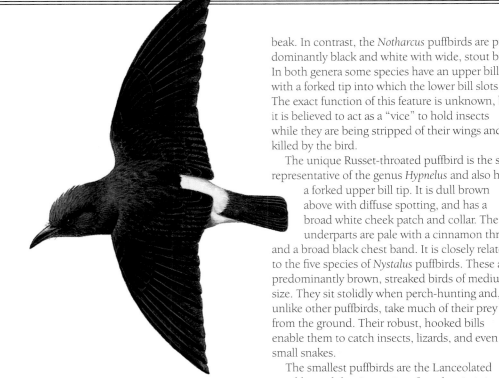

Above *When the Swallow-wing is at rest its long folded wings almost reach the tip of its short tail. It can be seen perching on tall trees in the Amazon, occasionally darting out to catch insects.*

Below *In both sexes of the Black-fronted nunbird the upper plumage, wings, and tail are black. The bill is bright orange, which is why the nunbirds are sometimes referred to as "pico de lacre" – "sealingwax bill".*

frontal band to the face. By contrast, the nunlets are much more active and glean foliage for insects with their long, slender, decurved bills. The adults of most species have conspicuously marked bills and red irises, but are otherwise rather drab. Generally, they are rufous, gray, or dull brown in plumage. When alarmed, some nunlets fan and jerk their tails from side to side. As a result of their relatively short tails and wings, their foraging flights are in the form of swooping glides, in the manner of a jay.

The most aberrant puffbird is the Swallow-wing. This small species is more aerial than the other puffbirds and has evolved some similarities to the unrelated martins and wood swallows. Swallow-wings catch insect prey in fast flight, rapidly sallying forth in sorties from prominent perches. Sometimes they glide through the air between pursuits to conserve energy. Their relatively long wings and short tails are adaptations to these fast and sailing flight strategies. They have short bills with wide gapes that enable them to swallow their aerial prey in flight. However, like other puffbirds, swallow-wings spend a lot of their time perched while watching for prey. Swallow-wings are blue-black above and also have dark throats and breasts. In contrast, their bellies are rufous. But their most intriguing plumage trait is their white back and rump. This is inconspicuous when the birds are at rest but is striking in flight, and very probably functions as a signal to advertise their presence to other swallow-wings.

Some Solitary, Some Social

BREEDING BIOLOGY

Puffbirds occur as far north as Mexico, but the highest density of species is in the tropical forests of South America, where the ranges of several species often overlap. But even when this is the case they are often separated by habitat or occur at different altitudes. Most puffbirds are confined to lowland or foothill forest, forest edges, open woodland and savanna. But some species such as the White-faced puffbird occupy humid montane forest. The Swallow-wing is catholic in its habitat preferences and often adapts to manmade environments. The Russet-throated puffbird is unusual in being frequently found in dry scrub.

Most puffbirds occur at relatively low densities, either singly or in pairs. These species tend to be silent and attract their mates with simple weak songs. But others, notably the nunbirds, are sociable and often consort in small groups. They reinforce their relationships by feeding each other and singing together in ringing choruses. These displays are usually initiated by, and focused on, a mature adult pair. The more social species frequently join mixed feeding flocks, often following army ant swarms. Troops of foraging monkeys are also sometimes followed, since they dislodge prey from vegetation. Furthermore, in some puffbirds breeding pairs are assisted at the nest by additional "helpers." The latter are very probably younger or unmated relatives, and such assistance will help ensure the survival of their family's shared genes.

Puffbirds almost invariably nest in short, descending tunnels, either excavated in the ground or in termite nests. These terminate in a nesting chamber lined with dry leaves in which 2–4 eggs are laid. In those species that have been studied, both adults take a share in nest excavation and incubation. Initially, the nestlings are blind and naked and, in some species, they have to travel to the tunnel entrance to be attended and fed by the adults. Several species conceal the tunnel entrance with vegetation, but in some instances vegetation is placed around the entrance in the form of a distinctive collar, the exact significance of which is not yet understood. In most species the young fledge within 4–5 weeks of hatching.

Cautious Optimism

CONSERVATION AND ENVIRONMENT

As many puffbirds are unobtrusive and occur at low densities they are often characterized as uncommon. However, concentrated field observations have revealed them to be more abundant than originally thought. This has recently proved to be the case with the Lanceolated monklet, which was once regarded as Near Threatened. At the time of writing, only one puffbird – the Sooty-capped puffbird – is classified by the IUCN as threatened. Habitat loss through logging and conversion to banana and oil palm plantations have caused significant enough damage to render this species as Lower Risk/Near Threatened.

Some species, such as the Chestnut-headed nunlet, certainly have very restricted ranges. Thankfully, in the case of this species, much of its primary habitat is secure and its future seems safe. Yet large-scale habitat fragmentation and deforestation remain as long-term threats to several puffbirds.

SP/AFS

Above Representative species of puffbird: **1** A Collared puffbird (Bucco capensis) *with its insect meal;* **2** *White-faced nunbirds* (Hapaloptila castanea) *raise their young along with one or two helpers. Blind, naked nestlings have to toddle up to the mouth of the nest to receive food from their parents and helpers.*

Jacamars

The solitary jacamars are the iridescent, jewel-like sprites of Neotropical forests. Although they are the closest relatives of the somber, lethargic puffbirds, jacamars by contrast display brilliant metallic colors and great vitality. They are exclusively insectivorous and many specialize in the predation of butterflies, bees, and wasps that most other birds find unpalatable. In this respect, they occupy a similar niche to the Old World bee-eaters.

Jacamars are sedentary throughout their respective ranges, which stretch from Mexico south through central America to Peru, Bolivia, and Brazil. Few species overlap in range and most favor humid, riverine forest at low altitudes. Some forms, such as the Coppery-chested jacamar, also occur in low montane cloud-forest. The latter species, together with some other *Galbula* jacamars, has a restricted range and exhibits a high degree of endemism.

Taking the Sting out of Hunting

FORM AND FUNCTION

All jacamars have elongated, pointed bills. These serve to keep the flailing wings of insect prey a safe distance from the jacamar's face, and also protect it from the venom of bees and wasps. In most species the bill is thin and straight, usually black or dull in color, but in some it can be brightly colored, as in the White-eared jacamar, with its heavy, bubblegum-pink bill. The Great jacamar has a stout, decurved bill, and this feature is correlated with this species' habit of gleaning foliage for prey rather than catching it in the open like other jacamars. Jacamars also deploy their bills in the excavation of their nesting and roosting cavities. A study of the Three-toed jacamar revealed that many birds had broken bill tips, injuries that were attributed to their use in digging. Bill fracture was highest in females, suggesting that they undertake more excavation.

Jacamars have very short legs and feet adapted mainly to perching, with both the outer and inner toe projecting backwards. The endangered Three-toed jacamar has lost the inner toe altogether. Jacamars also use their feet for digging nests or roosts.

While all *Galbula* jacamars have long, graduated tails, that of the Paradise jacamar is the longest, with projecting central feathers. These are costly ornaments for the jacamar to both grow and carry and probably function as a signal of an individuals' relative fitness to potential mates. Both males and females are iridescent blue-black, with bronze green wings and a brilliant white throat patch. Paradise jacamars hunt by diving from high perches and taking prey in twisting sorties lasting just a few seconds. Large insects are subdued by battering on a perch, and the venom of bees and wasps is removed by rubbing.

In the Rufous-tailed jacamar both sexes have dazzling metallic green upperparts but females are less rufous below. They usually perch low in shrubbery or on the forest edge, either singly or in pairs, and take prey via long sallies on their short wings. They capture large numbers of butterflies. Recent studies have shown that Rufous-tailed jacamars are able to discriminate across an impressively wide spectrum of palatable and unpalatable butterflies by memorizing their various wing patterns. So well developed is this capacity that the birds can even tell apart unpalatable

Left *Rufous-tailed jacamars are very vocal. When mated birds are together, and especially when two males are competing for a female, they call in high-pitched notes that may merge into a long, soft trill.*

Above *Representative species of jacamar:* **1** *Great jacamar* (Jacamerops aureus)*;* **2** *A Paradise jacamar* (Galbula dea) *with its prey. This species is usually found high in the forest canopy, unlike the greener members of the genus.*

FACTFILE

JACAMARS

Order: Piciformes

Family: Galbulidae

18 species in 5 genera. Species include: **White-eared jacamar** (*Galbalcyrhynchus leucotis*), **Pale-headed jacamar** (*Brachygalba goeringi*), **Three-toed jacamar** (*Jacamaralcyon tridactyla*), **Rufous-tailed jacamar** (*Galbula ruficauda*), **Coppery-chested jacamar** (*G. pastazae*), **Paradise jacamar** (*G. dea*), **Great jacamar** (*Jacamerops aureus*).

DISTRIBUTION Mexico to S Brazil, N Argentina, and Paraguay; also Trinidad and Tobago.

HABITAT Forest, thickets, savanna.

SIZE Length 12–31cm (4.5–12in).

PLUMAGE Shining, iridescent green above, mostly rufous below, otherwise dull brown or blackish and pale. Slight differences between sexes.

VOICE Often prolonged and complex song with whistles, squeals, and trills.

NEST In short burrows usually dug in roadside or streamside banks, in wooded hillsides; or in termite mounds.

EGGS 2–4, white, unmarked. Incubation period: 20–22 days; nestling period: 19–26 days (in Rufous-tailed Jacamar).

DIET Insects caught in the air, including many butterflies, beetles, and wasps.

CONSERVATION STATUS The Three-toed jacamar of Brazil is threatened by habitat loss and classed as Endangered; the Coppery-chested jacamar of NW South America is Vulnerable.

butterflies and palatable species that closely mimic the former in wing pattern.

The smallest jacamars are the four *Brachygalba* species. These are the least ornamented, being generally subdued and dark in color with short tails. Partly because of their small size they are also the most aerially agile jacamars.

Melodious Singers

SOCIAL BEHAVIOR

Although jacamars are "sit-and-wait" hunters, even when perched they are vivacious. Sitting on an exposed perch, they hold their bills angled upward and constantly turn their heads to look for prey that they then fly out to seize. Most species are extremely agile in the air, some performing acrobatic twists and loops.

The high, thin calls of jacamars are often urgent, but their songs can be melodious. Some species congregate in small groups and chorus to each other. However, when nesting they are solitary rather than colonial. Nesting burrows are dug in vertical banks or sloping ground. Some species, like the Three-toed jacamar, also excavate cavities for roosting. Occasionally jacamars breed in cavities in termite nests, but these are probably excavated by other species.

The male Rufous-tailed jacamar helps his mate to excavate and also feeds her. Nesting burrows are over 50cm (20in) long and terminate in a chamber where 2–4 white eggs are laid on the bare earth. As parents do not remove feces or food debris, the nest chamber is rapidly fouled. By day, the sexes incubate alternately, often for an hour or two at a time. The female generally tends the nest at night. The nestlings hatch with a thin coat of long white down and are fed exclusively with insects by both parents. Young Pale-headed jacamars were found to lodge in their parents' burrow for several months after fledging.

Two species of jacamar are currently classified as endangered or threatened. Most vulnerable is the Three-toed jacamar, endemic to southeastern Brazil. Although its range was relatively widespread until very recently, it is now reduced to tiny, isolated populations due to ongoing habitat destruction. Also threatened is the Coppery-chested jacamar, restricted to just a few hundred square kilometers in the eastern Andes. Its montane forest home is under severe threat, and urgent measures are required to save it. SJP

Woodpeckers

THANKS TO THEIR DISTINCTIVE CLIMBING AND *pecking habits, woodpeckers are unmistakable. Especially impressive, however – indeed unique – are their tapping and drumming communication signals, which can be heard in many of the world's woods during the breeding season.*

With their specialized pecking methods, woodpeckers are unrivaled as predators on insects that lie hidden under bark or within wood, or that live in nests with tunnels far below the surface, such as ants and termites. Woodpeckers also create dwellings for rearing their young and for daily roosting; the holes that they excavate often last for several years.

True Woodpeckers

SUBFAMILY PICULINAE

True woodpeckers are small- to medium-sized birds of powerful and stocky build. Their bill is adapted for hacking and chiseling. Their tongue, capable of extreme protrusion (extending up to 10cm/4in in the Green woodpecker) and armed with barbs at the tip, is a highly efficient catching device that enables the bird to extract insects from cracks and crevices and from the tunnels bored by insect larvae and excavated by ants and termites. Woodpecker feet are especially adapted for climbing, with two toes pointing forward and two back. The fourth toe can be bent sideways so that the crampon-shaped claws can always be positioned to best suit the curve of trunk or branch. (It is, however, pointed forward in the large ivory-bill woodpeckers of the genus *Campephilus*.) The first toe may be rather small and is absent in several species, such as the Three-toed woodpecker or the Common flameback.

Climbing movements and pecking postures are facilitated by the wedge-shaped supporting tail feathers, the shafts of which have additional strengthening. This type of tail allows the woodpecker's body to be held clear of the climbing substrate and permits a good, relaxed posture for pecking or for pauses between bouts of climbing. Special adaptations for pecking, tapping, and drumming protect internal organs, particularly the brain, against impact damage. Such protection is absolutely necessary considering the number of pecking blows executed daily (in the Black woodpecker, some 8,000–12,000).

True woodpeckers eat mainly arthropods, particularly insects and spiders, but they will also consume plant food (fruits, seeds, and berries). Nestling birds may be taken, not only from holes in trees but also from open nest-cups and from the nests of penduline tits. The Acorn woodpecker eats acorns, storing them for the winter in specially excavated holes. Sapsuckers drill holes in horizontal rows (so-called "ringing" behavior) and then lick up the exuding droplets of sap with their tongue, the tip of which is frayed and brushlike. This habit is also widespread among the pied woodpeckers of Eurasia.

Woodpeckers use crevices or branch forks to work on large food items such as large beetles, nestlings, fruits, nuts, and cones. Great spotted woodpeckers make their own so-called "anvils," into which they wedge cones in holes to peck out

FACTFILE

WOODPECKERS

Order: Piciformes

Family: Picidae

218 species in 28 genera and 3 subfamilies

DISTRIBUTION N and S America, Africa, Europe, C and S Asia, SE Asia, and Australia.

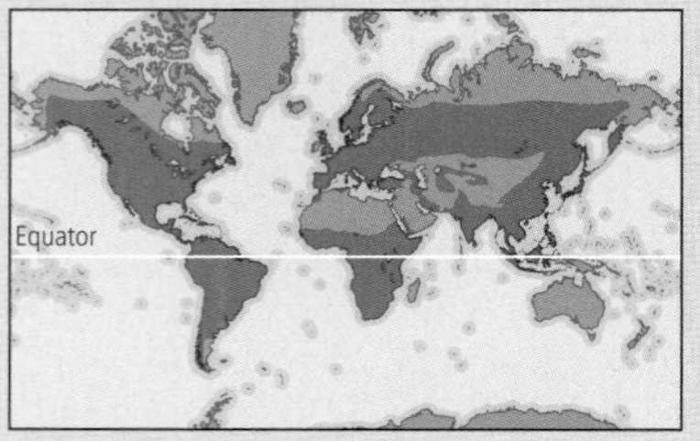

HABITAT Tropical, subtropical, and deciduous forest; orchards, parks, and grasslands.

SIZE From 8cm (3in) in the Scaled piculet to 55cm (22in) in the Imperial woodpecker; weight 8g (0.3oz) to 563g (19.9oz), in same species.

CONSERVATION STATUS 3 species – the Imperial woodpecker (*Campephilus principalis*), the Ivory-billed woodpecker (*C. principalis*), and Noguchi's woodpecker (*Sapheopipo noguchii*) – are Critically Endangered or possibly Extinct, 1 is Endangered, and 7 are Vulnerable.

See Woodpecker Subfamilies ▷

Right *Representative woodpecker species:* ***1*** *Pileated woodpecker* (Dryocopus pileatus). ***2*** *Yellow-bellied sapsucker* (Sphyrapicus varius). ***3*** *Red-headed woodpecker* (Melanerpes erythrocephalus) *at its nest-hole.* ***4*** *Great spotted woodpecker* (Dendrocopos major). ***5*** *Northern wryneck* (Jynx torquilla). ***6*** *Olive-backed three-toed woodpecker* (Dinopium rafflesii), *foraging.* ***7*** *Green woodpecker* (Picus viridis). ***8*** *Common flicker* (Colaptes auratus) *in a dance posture.* ***9*** *Three-toed woodpecker* (Picoides tridactylus).

the fat-rich seeds. Up to 5,000 cones may be found under a "primary anvil," of which there will be three or four in a territory. The ability to deal with fruits and seeds in anvils or to store them in special caches (as practiced by *Melanerpes* spp.) is a great aid to survival in areas of winter cold and consequent seasonal insect shortages.

Woodpeckers catch their prey with a great variety of different techniques, the simplest of which is the gleaning of items from leaf, branch, or trunk. Slightly more complicated is probing into bark crevices combined with the scaling of bark. Both sapsuckers and the Three-toed woodpecker obtain insects that lie hidden under bark or within wood by drilling round holes, inserting their tongues, and harpooning the item. Other "pecking" woodpeckers, along with larger species, chisel and lever off large pieces of bark and carve out deep holes in their quest for insects. A Black woodpecker may consume up to 900 bark beetle larvae or 1,000 ants at a single meal. "Ground" woodpeckers mostly peck only funnel-shaped holes in ants' nests, then extend their long, sticky tongue along tunnels and into chambers to spoon up adult ants and pupae. A Green woodpecker needs to eat about 2,000 ants daily, mostly lawn and meadow ants. When this is not possible, for example in extreme winters such as that of 1962–63 in Europe, a large part of the population will perish. Some species, for example the Yellow-tufted and Lewis's woodpeckers and related species of the genus *Melanerpes*, regularly take insects in flight.

Most woodpeckers are sedentary and may remain in the same territory for a long time. Only a few species, including the Yellow-bellied sapsucker and the Common flicker in North America, and the Rufous-bellied woodpecker (*Dendrocopos hyperythrus*) in eastern Asia, are migratory. Northern races of the Great spotted woodpecker undertake far-reaching eruptive movements at intervals of several years, when their main seed-crop diets fail. The Great spotted woodpecker penetrates into central and southern Europe in years of cone shortage; Three-toed woodpeckers invade areas of North America and Europe where the forests periodically suffer from infestations of insect pests, particularly after forest fires.

The great majority of woodpeckers are territorial, living in individual, pair, or group territories, in some cases for several years. A ringed Great spotted woodpecker showed fidelity to its 25ha (62 acre) territory for a period of 6 years; in most other species studied, most individuals remained in or close to their territory for the whole of their lives. Defending a territory helps to ensure not only breeding success but also adequate food supplies and – especially important for woodpeckers – roosting possibilities in holes affording shelter from the weather.

As a rule, woodpeckers react aggressively to intruders of their own kind. Some woodpeckers that breed in open country, such as the Andean flicker (*Colaptes rupicola*) and Campo flicker (*Colaptes campestris*), form loose colonies. Some species associate with other birds in mixed-species foraging flocks outside the breeding season. Examples are the small Lesser spotted woodpecker (*Dendrocopos minor*) in Eurasia and the Downy woodpecker (*Picoides pubescens*) in

Above *A Fernandina's flicker* (Colaptes fernandinae) *displays the upright stance and stiff, supportive tail typical of woodpeckers. This Cuban species is now endangered, with fewer than 400 pairs remaining.*

Below *Brightly-colored Asian flamebacks* (Dinopium *spp.) forage on fallen bark in a national park in Rajasthan, north-west India, in search of the insect pupae and larvae that it might contain.*

North America; both join titmice and nuthatches. This habit is not restricted to small species; the Great slaty woodpecker (*Mulleripicus pulverulentus*), for instance, is known to associate with other large woodpeckers like the White-bellied woodpecker (*Dryocopus javensis*).

Woodpecker reproductive activities generally begin with drumming, later followed by fluttering display flights and prominent calls. These signals are used by both sexes to advertise territory limits and trees with holes, to attract prospective partners to suitable nest-sites (an activity known as "nest-showing"), to stimulate the partner sexually, and to intimidate rivals. However, males are the more active sex in most cases.

A new nest-hole is not excavated every year and an old one can certainly be used for several years. Black woodpeckers may use the same hole for up to 6 years, Green woodpeckers for 10 or more. However, even these species are forced to excavate new holes when they are evicted by jackdaws or starlings. Excavation of a hole takes 10–28 days, according to species and the method employed. Both sexes participate, with the male usually taking the greater share. About 10,000 wood chips have been found under a Black woodpecker's hole. When the hole is completed, the birds chip off small pieces from the inner wall to serve as a cushion in the nest-scrape for eggs and young.

Copulation usually takes place without any special ceremony and is often associated with nest construction activities. The female assumes a precopulatory posture – crossways on a branch – and the male simply flies onto her back. Lengthy physical contact is avoided. Mutual courtship feeding, common in the related barbets and in the wryneck, has been recorded in only a few true woodpecker species, for example the Olive-backed woodpecker of Asia and the insular Guadeloupe woodpecker (*Melanerpes herminieri*) of the Antilles. The glossy white eggs are laid in the early morning, one per day until the clutch is complete. Constant guarding of the nest-hole is typical once the first egg has been laid. Such protectiveness is understandable, since many species like to breed in woodpecker holes and competition is fierce. In all species of woodpeckers the male spends the night in the nest-hole during both incubation and nestling periods. In some of the *Melanerpes* woodpeckers and in some tropical species, male and female roost there together.

During incubation and brooding, birds of a pair change over at the nest at intervals of 30 to 150 minutes. The nest-relief ceremony resembles that of nest-showing, involving calling and demonstrative tapping. The relieved bird may subsequently call and drum, but usually the incubating pair keep quiet, and are even reluctant to respond to territorial challenges. The pecking woodpeckers collect food in the bill, but the ground woodpeckers and all large species feed their young by regurgitation. Nestlings give almost ceaseless whirring or rattling food-calls. Both parents feed, and it is not unusual for the male to take the larger share; at the end of the nestling period, he may even feed the young alone.

The nestling period is 18–35 days. When they leave the nest, young woodpeckers can climb and fly. Soon afterward they follow the adults through the territory, contact being maintained by means of the same call sequences that are used by pairs for long-distance communication. Upon making close contact with the adult, the young utter loud, squeaky call notes. In some species both adults tend the young after fledging, while in others (for example, the Great spotted woodpecker and other *Dendrocopos* species, together with *Picoides* and *Campethera*, and Green and Gray-headed woodpeckers) the brood is split, each adult caring for one to three young. The family breaks up within 1–8 weeks of leaving the nest, adults increasingly using various forms of threat (ruffling of crown feathers, wing-spreading, threat calls) to drive away their offspring, which finally move off, eventually to establish their own territories.

Above *In the Amazon, a Scaly-breasted woodpecker* (Celeus grammicus) *feeds on flowers. Woodpeckers are mostly insectivorous, but a few species opportunistically take fruit and berries when available.*

However, in several, mostly tropical, species (for instance, the Ground and Great slaty woodpeckers), the young stay with the parents much longer and may be driven away only when incubation begins. Constraints on dispersal and lack of easily accessible breeding opportunities may force the young to stay within the parental territory or associate with other pairs, leading to sociality and complex breeding systems with helpers in several species. In the threatened North American Red-cockaded woodpecker (*Picoides borealis*), the social unit comprises the breeding pair and one or two additional helping adults, most frequently males. The helpers, often offspring from previous broods, take part in incubation and care for the young. The breeding pair benefits from a higher breeding success, and a helper may inherit the territory should one of the pair members die. In the

Left A familiar species on the American West Coast, the Acorn woodpecker is unique in its habit of drilling tight-fitting holes in bark in which to store acorns, providing an emergency larder for winter use.

Right A Great tit waits patiently for a Great spotted woodpecker to finish its excavation of a dead branch. When it flies off, the smaller bird will examine the holes in the hope of finding leftover food.

Acorn woodpecker, the situation is more complicated. In this species two females and three males may contribute to one brood. The young stay with the group, which may comprise as many as 15 individuals. These highly social birds use several "granary trees," each of which is drilled with up to 50,000 small holes holding single acorns; the group is extremely territorial, and defends these acorn stores vigorously against any competitors.

Piculets

SUBFAMILY PICUMNINAE

The tiny piculets of the subfamily Picumninae climb about tree branches in the manner of woodpeckers or, at times, like titmice and nuthatches. Their flight is undulating. Foraging piculets peck at bark and soft wood to get at ants, termites, and wood-boring insects. Their tail, which does not have the stiffened quills of the larger woodpeckers, shows three conspicuous white longitudinal stripes in all species of *Picumnus*. Piculets excavate a nest-hole in a tree trunk or branch, or enlarge available holes. During courtship they call and drum. The clutch consists of 2–4 eggs, and incubation takes 11–14 days. The young fledge after 21–24 days. They may stay with their parents until the eggs of the next brood hatch (as in the Olivaceous piculet). Disjunct distribution in Asia, Africa, and America indicates the piculets to be of very ancient origin, a conclusion that has been corroborated by DNA-based analyses.

Wrynecks

SUBFAMILY JYNGINAE

Wrynecks live in open woods, orchards, parks, and meadows with copses. Like woodpeckers they obtain their main food (various kinds of ants) with the help of the tongue. The name wryneck derives from their defensive behavior in the nest; when threatened by a predator, they perform snakelike twisting and swaying motions of the neck and simultaneously hiss. Filmed sequences show that such behavior is effective in intimidating small predators. A prominent feature in spring is the rather nasal *kwee* call which rises slightly in pitch and which is given by both sexes to attract a partner to prospective nest-holes. The seven or eight eggs are usually laid on the bare floor of the nest chamber (after throwing out any nest that may have already been started). Incubation takes 12–14 days, and the young spend a further 21 days in the nest, the parents feeding them with adult and pupal ants (about 8,000 individuals daily for all the nestlings); post-fledging care lasts two weeks.

From July onward, Northern wrynecks begin their migration south from breeding grounds in Europe and Asia to wintering areas in Africa and Southeast Asia. Their populations are threatened, and the species has almost completely disappeared from England in recent years. The Rufous-necked wryneck is found in southern Africa, including mountainous regions up to 3,000m (10,000ft). Brown cocktail ants (*Crematogaster* spp.) make up 80 percent of its diet.

Help and Hindrance

CONSERVATION AND ENVIRONMENT

Woodpeckers play an important role in forest ecosystems. They help to keep down the numbers of bark- and wood-boring insects, thereby contributing to the health of the tree trunk and its bark covering. Where woodpeckers have pecked, other smaller birds (tits, nuthatches, treecreepers) can forage successfully for any remaining insects and spiders, and woodpecker holes are used for nesting or roosting by many other hole-nesting insectivores. Owls, stock doves, and toucans, as well as martens and other mammals, also benefit from using woodpecker holes. Woodpeckers thus help indirectly to exert pressure on the huge populations of insects and mice or voles. Moreover, by pecking at huge amounts of dead wood, thereby making it accessible to other decomposing organisms, they play an important part in the cycle of decay and regeneration of matter.

Below Piculets, such as this White-barred piculet (Picumnus cirratus), *are wren-sized birds that tap and drum like true woodpeckers but have soft tails that are not used as props.*

Woodpeckers' activities sometimes conflict with those of humans. The birds may become a nuisance locally, for example by puncturing irrigation pipes (an unfortunate habit of the Syrian woodpecker in Israel), excavating utility poles (several species), or driving their holes into the foam insulation of modern houses (Great spotted woodpeckers). Their predilection for fruits can antagonize orchard owners (a common problem among *Melanerpes* species), and when White-backed woodpeckers delve into an apparently rich supply of their most favored feeding substrate – soft, rotten wood lying on the ground – they are sometimes liable to incur the wrath of commercial growers of Japanese shiitake mushrooms, who use this medium to cultivate their product.

Three woodpecker species are currently under a particularly acute threat of extinction. Indeed, the Ivory-billed woodpecker of southeastern Cuba was once classed as Extinct; some indications of its continued existence in the late 1990s, however, led to its reclassification as Critically Endangered, although any surviving population would be tiny. As regards the other two Critically Endangered species, there have been no confirmed sightings of the Imperial woodpecker, from the Sierra Madre range in Mexico, since 1956, while habitat loss is the principal threat to the Okinawa (or Noguchi's) woodpecker. This latter species has fallen victim to deforestation of its broadleaf forest habitat for the construction of golf courses, roads, and dams, and for commercial logging. DB/HW

Woodpecker Subfamilies

True woodpeckers
Subfamily Picinae

185 species in 24 genera. America, Africa, Eurasia. Forests; orchards, parks, grasslands, areas of cultivation with hills or earthen banks: up to 5,000m (16,400ft). Species and genera include: **Common flicker** (*Colaptes auratus*), **Imperial woodpecker** (*Campephilus imperialis*), **Great spotted woodpecker** (*Dendrocopos major*), **White-backed woodpecker** (*D. leucotos*), **Syrian woodpecker** (*D. syriacus*), **Yellow-crowned woodpecker** (*D. mahrattensis*), **Olive-backed woodpecker** (*Dinopium rafflesi*), **Common flameback** (*D. javanense*), **Pileated woodpecker** (*Dryocopus pileatus*), **Black woodpecker** (*D. martius*), **Ground woodpecker** (*Geocolaptes olivaceus*), **Acorn woodpecker** (*Melanerpes formicivorus*), **Red-bellied woodpecker** (*M. carolinus*), **Red-headed woodpecker** (*M. erythrocephalus*), **Lewis's woodpecker** (*M. lewis*), **Yellow-tufted woodpecker** (*M. cruentatus*), **Three-toed woodpecker** (*Picoides tridactylus*), **Black-backed three-toed woodpecker** (*P. arcticus*), **Gray-headed woodpecker** (*Picus canus*), **Green woodpecker** (*P. viridis*), **sapsuckers** (genus *Sphyrapicus*), including **Yellow-bellied sapsucker** (*Sphyrapicus varius*).
SIZE: Length: 16–55cm (6–22in); weight: 13–563g (0.46–19.9oz). Size and weight differences between the sexes are small, with the male most often being the larger. Small sexual differences in mensural characters are associated with sexual differences in foraging.
PLUMAGE: Upperparts usually appropriate to habitat (blackish, brownish, grayish, or greenish); head and neck mostly bright colors: red, yellow, white, or black patches and stripes; bills are black, gray, brown, or bright white. Young hatch naked and blind. Sexes differ little in plumage (sometimes almost imperceptible). These differences may affect mustache stripes, crown, and nape. In most cases, the male shows more red in these parts.
VOICE: Loud, high-pitched clicking calls; series of such clicks or squeaky notes. Most species drum, although in some, drumming is reduced to double-raps.
NEST: Excavated holes.
EGGS: 3–11, white. Incubation period 9–19 days.
DIET: Insects and their larvae, spiders, berries, fruits: acorns, seeds; sap, honey.

Piculets
Subfamily Picumninae

31 species in 3 genera. America, Africa, Eurasia. Tropical and subtropical forests, secondary forests, woods; coffee plantations; up to 3,000m (9,840ft). Species include: **Antillean piculet** (*Nesoctites micromegas*), **Scaled piculet** (*Picumnus squamulatus*), **Olivaceous piculet** (*P. olivaceus*), **Speckle-chested piculet** (*P. steindachneri*), **White-browed piculet** (*Sasia ochracea*)
SIZE: Length: 8–15cm (3–6in); weight: 8–16g (0.28–0.56oz) (28g/1oz, in the Antillean piculet).
PLUMAGE: In *Picumnus* brownish with red, orange, yellow marks on crown; three white stripes on the tail; females of the S. American species have white spots on black, white spotted crown; in the Asian species orange-yellow forecrown of male is replaced by plain brown. Sexes in *Sasia* differ only slightly, and the female of the Antillean piculet lacks the red spot on the crown.
VOICE: Sharp calls and series of calls.
NEST: Holes in rotted tree trunks and soft wood.
EGGS: 2–4, white. Incubation 11–14 days.
DIET: Insects, larvae, ants, termites, wood-boring beetles.

Wrynecks
Subfamily Jynginae

2 species of the genus *Jynx*. Africa, Eurasia. Open deciduous forests, grassy clearings, copses, gardens; in Africa up to 3,000m (10,000ft). Species: **Northern wryneck** (*Jynx torquilla*), Eurasia migrating to C Africa, SE Asia, Japan; **Rufous-necked wryneck** (*J. ruficollis*), SC and S Africa.
SIZE: Length: 16–17cm (6.3–6.7in); weight: 30–39g (1.05–1.38oz).
PLUMAGE: Chiefly brown, nightjarlike pattern of peppered and blotched markings; dark line through eye; no difference between sexes.
VOICE: Up to 18 *kwee* calls.
NEST: Natural cavities; holes excavated by woodpeckers; nest boxes; no nest material.
EGGS: 5–14, white. Incubation 12–13 days; nestling period 21 days.
DIET: Ants.

SPECIAL FEATURE

WOODLAND DRUMBEATS AND DANCES

The communication system of woodpeckers

WOODPECKERS HAVE A VARIED AND HIGHLY efficient system of communication, comprising visual and acoustic signals. They "speak" to one another by ruffling their crown feathers, flicking or spreading their wings, swaying the head or the entire body, bobbing and bowing, by giving threat and contact calls, and by tapping and drumming with the bill on tree trunks and branches. Like many other animals, woodpeckers use this "language" to express their mood and thus influence mates, competitors, and members of their social unit. Recognition of mood is important, since woodpeckers are frequently aggressive – unsurprisingly, given that most woodpeckers hold individual or pair territories containing such essential resources as roosts, foraging sites, anvils, and stores. When a male and female hold different feeding territories, they are often defended in early courtship even against the prospective mate.

In many species of woodpeckers courtship involves a great deal of aggression. Males fight each other in an attempt to secure territories rich in resources, while females compete for the best territory and male. Also, males may reject a female entering their territory. All these contests involve intense communication.

An interesting example is the ritualized threat tournament of rival Black woodpecker males. The birds threaten one another at first with *keeyak* calls, then fly to the base of a tree and attempt to drive one another upward. From time to time they thrust their bills into the air, as if on a command, and wave them about. In these movements the red crown is prominently displayed. The birds then sink into a waiting posture, only to repeat the maneuver a few minutes later. Such a tournament may last for over an hour, until one of the birds gives up. If a male and female meet, they threaten one another in similar fashion, but the male's aggression then gradually wanes. This is presumably because the smaller area of red on the female's head and her lower-intensity swaying inhibit the male's aggression. Characteristic of this threat ceremony is a quiet *ryrr* call.

Head-swaying with presentation of the head pattern is found in many species, for example the genera *Colaptes* and *Picus*, many of the sapsuckers, the Three-toed woodpecker, and the Pileated woodpecker. The behavior is especially pronounced in the Common flicker, which dances about with wings spread and tail fanned and shows off part of the head which, in the male, bears a mustache-like stripe. If such a stripe is artificially painted on a female she will be treated like a male and will provoke intense aggression.

Multiunit calls, reaching long distances, advertise the presence of a woodpecker to neighbors and mates alike. In those species in which the male and female do not forage close together (for example the Arabian woodpecker, *Dendrocopos dorae*), the mate's call is immediately answered, as if to assure it that all is well. Other loud call series are often combined with demonstrative flights among the trees or at tree-top height. These displays serve as signals to attract a partner and to advertise trees with holes when courtship peaks.

In many species, drumming is the most important general signal to announce territorial ownership. Tapping sequences advertise prospective nest sites. Most species combine vocal signals with the drumming, for which each species has its own specific pattern. In the Black woodpecker long series (43 strikes in 2.5 seconds) function as long-range signals with a great power of attraction, while quiet and shorter series are used at close range to advertise the entrance to a hole. When a female has followed a male as far as the hole or, conversely, a male has approached a female showing a nest, the active partner marks the hole entrance with long tapping sequences. Eventually the other bird is attracted nearer and gives threat calls to drive the exhibitor away, so that the nest is free for inspection. Woodpeckers sometimes advertise what prove to be unsuitable holes; in such cases the inspecting bird will leave the site, look for another tree, and attempt to lure its future mate to this new site, but success may come only after several days. In a few species, such as the Middle spotted woodpecker (*Dendrocopos medius*), drumming is less conspicuous, being virtually supplanted by a loud series of strange calls.

The basic scheme of the language of courtship – in a sense, its grammar – is common to most woodpecker species: drumming; guiding with calls and special flights; drumming, tap-drumming, tapping; hole inspection; agreement over choice

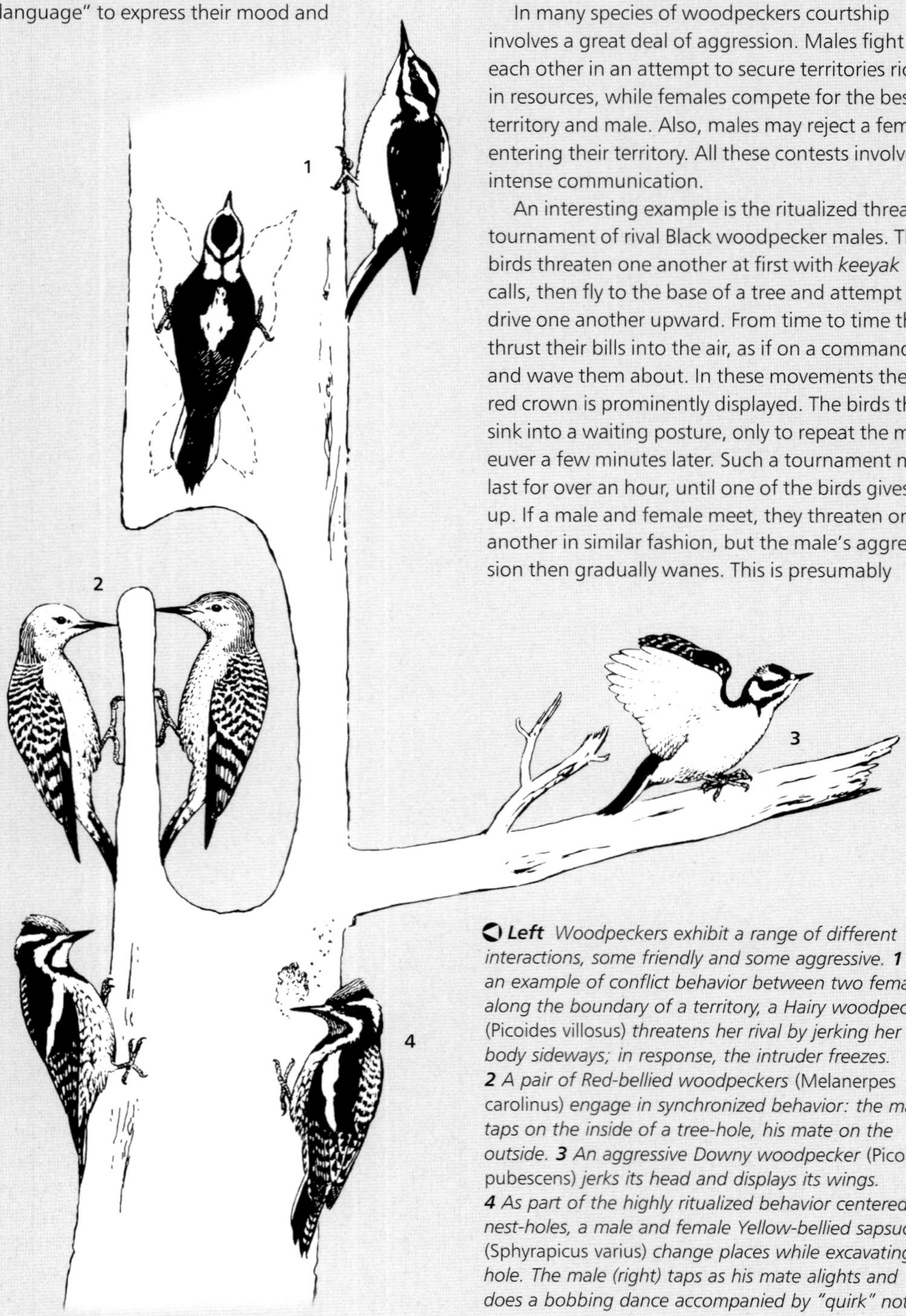

Left *Woodpeckers exhibit a range of different interactions, some friendly and some aggressive.* **1** *In an example of conflict behavior between two females along the boundary of a territory, a Hairy woodpecker* (Picoides villosus) *threatens her rival by jerking her body sideways; in response, the intruder freezes.* **2** *A pair of Red-bellied woodpeckers* (Melanerpes carolinus) *engage in synchronized behavior: the male taps on the inside of a tree-hole, his mate on the outside.* **3** *An aggressive Downy woodpecker* (Picoides pubescens) *jerks its head and displays its wings.* **4** *As part of the highly ritualized behavior centered on nest-holes, a male and female Yellow-bellied sapsucker* (Sphyrapicus varius) *change places while excavating a hole. The male (right) taps as his mate alights and does a bobbing dance accompanied by "quirk" notes.*

of hole. Excavation at the nest site is also essential, and copulations take place close to the hole.

The Red-bellied woodpecker shows a high degree of ceremony in this sequence. Male and female perform a tapping duet in precise harmony. Where a hole has only been started, they then sit close together on the trunk; if there is a completed hole one bird taps inside, the other outside. Later, there also has to be some understanding between the birds for changeovers during incubation and brooding. Nest-relief has a ceremonial character: the incoming bird gives particular calls, mostly quiet and muttering or soft and protracted. The bird in the hole confirms its readiness for a changeover by tapping on the wall of the nest-chamber and then leaves to allow its mate to take over. The same vocal signals precede copulations. In this nest-relief ceremony there is a remnant of antagonistic behavior: when, for example, the bird in the hole is reluctant to leave, its mate uses threat calls and postures to force the other's departure. Despite all this apparent hostility, woodpeckers are among the most faithful of birds. Males are almost never cuckolded and pairs may stay together for a lifetime.

If a female dies after the young have hatched, the male is able to rear the brood alone, although his initial response, in addition to feeding the young, is normally intense drumming. After a short time, however, this renewed courtship behavior wanes, and the feeding adult becomes noticeably quiet in his territory.

Above *In Colombia, two female Crimson-crested woodpeckers* (Campephilus melanoleucos) *engage in a noisy territorial dispute. Frequent vocalization and display characterize woodpecker social behavior.*

Woodpeckers that fail to acquire a mate early on (mainly males, of course) may drum and call persistently up until the end of the breeding season. Sometimes this enables them to attract a bird, pair up, and rear a family in the late spring. More social species – namely those in which the young stay with their parents for an extended period or those that form groups all year round – are typically noisier and chattier than solitary species, particularly when the group moves together or when the time for entering the roost approaches. HW/DB

New Zealand Wrens

A SMALL, OBSCURE FAMILY OF TWO EXTANT species, the New Zealand wrens have no close affinity to other passerines. Both are poor fliers – a trait possibly related to the absence of mammalian predators in New Zealand before humans arrived. With the introduction of such predators and as a result of habitat modification, at least two other species have become extinct, and the distribution of the surviving species has contracted. The male rifleman is the smallest bird in New Zealand.

Of the two species presumed extinct, the Bush wren has not been seen for decades. The Stephens Island wren from Cook Strait, which was possibly flightless, was only discovered in 1894 when a lighthouse-keeper's cat carried in 15 birds; the cat probably destroyed the whole population.

Tiny and Vulnerable

FORM AND FUNCTION

New Zealand wrens have a stocky appearance, with relatively large legs and toes, short, rounded wings, and almost no tail. The rest of their plumage is soft and downy. The bill is slender and sharp, with a slight curve upward in the rifleman.

The genus *Xenicus* is thought to be indigenous to New Zealand, with the Rock wren having evolved to occupy alpine and subalpine forest and the Bush wren, its extinct congener, adapting to lower-altitude habitats. The Rock wren occurs on or west of the mountainous divide of the South Island at altitudes between 1,200 and 2,400m (4,000–8,000ft), favoring sparsely vegetated (low shrubs and fellfields) rock outcrops, boulder screes, and moraines. Like the rifleman, it eats mainly arthropods taken from crevices under boulders and from short, tight plant swards, even when these are covered with snow (when the birds use air spaces between the snow and ground surface). Food may be cached in crevices within the birds' well-defined territories, which are advertised using a single-note call. Riflemen glean their prey from the bark of trees and the foliage of the canopy.

Helpers at the Nest

BREEDING BIOLOGY

New Zealand wrens are hole-nesters, the rifleman prefering tree cavities while the Rock wren may settle for hollow logs, rock crevices, or holes in the ground. The female rifleman lays an egg weighing about 20 percent of her own weight every two days, so a complete first clutch of five eggs, produced over nine days, may equal her entire body

Above *The rifleman lives on insects and spiders, spending much of its time foraging for them on large branches and tree trunks. It is often to be seen working its way up and around the trunks, taking a spiral route to a height of 6–9m (20–30ft). Having climbed one tree, it flies off to the foot of another.*

Left *The Rock wren is New Zealand's only truly alpine bird, and remains at high altitudes year-round. It spends much of its time on the ground and can only manage to flutter for a few dozen meters at best.*

Above *The smaller of the two extant New Zealand wrens, riflemen are also by far the most common, although they tend to avoid settled areas. Here the plain male **1** is seen above and the striped female **2** below.*

weight. The implications for the female's energy budget may have influenced rifleman behavior. For example, males feed females before and during the laying of the first clutches, providing more or less all the nutrition required for producing the eggs. The young hatch at about 1.3g (0.05oz) and require intensive parental care, most of which is provided by the male over an exceptionally long nestling period. This pattern of high male commitment to parental care is evident throughout the 60-day breeding cycle, during which two broods are produced, with the males performing most of the nest-building, the daytime incubation, and the feeding of the young.

Riflemen occasionally have one to three "helpers" to aid in feeding nestlings and fledged offspring. When this occurs in first clutches, the nestlings are about eight days old and the helpers adult birds of either sex (usually males) that may not be related to the parent birds. Their activity at the nest involves feeding the offspring, defending them against predators, and removing fecal sacs.

Two types of helpers occur at first-clutch nests: some that regularly and frequently help at a single nest, and others that assist sporadically at more than one. In one intensively studied population, most adult helpers were unpaired males that, later in the season, were seen paired with female offspring from the brood they had earlier been feeding. The increased likelihood of acquiring a mate may well be one explanation for their behavior.

Young fed by helpers do not weigh significantly more than others on leaving the nest, but they may subsequently benefit from a food supply improved in quality or quantity. The advantages may also be experienced by the parents, which thereby carry less responsibility than they otherwise would for feeding the young and defending the nest. Siblings from the first brood of the year may help with the feeding of second broods.

Nestlings reach weights considerably heavier than those of adults. If a second brood is raised, it involves about one less young, no courtship feeding occurs, the nest is less elaborate, and incubation may be delayed until the young of the first brood are independent.

Rock wrens do not display cooperative breeding, but both parents feed the young and otherwise cooperate during the breeding season, which is short – consistent with life in an alpine environment. Male Rock wrens feed females throughout the breeding season, with some reciprocation from the females.

Rock wren eggs weigh approximately 13 percent of female body weight, and clutches number about three. Only one brood is raised a year unless a relay is required due to predation. Like riflemen, Rock wrens appear monogamous, and, if an opportunity presents itself, will pair in the same season they fledge. Both species' calls are very simple, being mainly made up of single notes repeated at intervals of various lengths.

FACTFILE

NEW ZEALAND WRENS

Order: Passeriformes

Family: Xenicidae (or Acanthisittidae)

2 extant species in 2 genera: **Rock wren** (*Xenicus gilviventris*), **rifleman** (*Acanthisitta chloris*).

DISTRIBUTION New Zealand

HABITAT Forest or woodland

SIZE Length 7–10cm (3–4in); weight 5.5–20g (0.2–0.7oz).

PLUMAGE Male upperparts greenish, underparts cream, sometimes with a yellow wash on the sides; females are generally duller overall, with brownish black colors above, striped in rifleman.

VOICE Rock wren produces a whirring call of three notes and a piping sound; rifleman a repeated *zsit-zsit* sound, and a plaintive decrescendo downtrill if alarmed.

NEST Riflemen build in holes in tree trunks, including standing dead timber; Rock wrens likewise, but also in cavities in banks and other ground sites.

EGGS 2–5; white. Incubation period 19–21 days; nestling period 23–25 days.

DIET Mainly arthropods

CONSERVATION STATUS The Bush wren (*Xenicus longipes*) and the Stephens Island wren (*X. lyalli*) are both extinct, while the Rock wren is considered Lower Risk/Near Threatened.

The Threat from Stoats

CONSERVATION AND ENVIRONMENT

Since Europeans arrived in New Zealand, the rifleman has fared the best of the New Zealand wrens, even occupying habitat whose principal woody species are exotics rather than native fauna. Moreover, one of the bird's preferred natural habitats, beech forest, is still relatively abundant. Riflemen seem to be coping well with reduced or modified habitat, as well as with the threat from introduced predators.

In contrast, the Rock wren, which in the days before European settlement lived in the North Island as well as the South, is thought to have suffered a serious decline in range in recent years. The main threat comes from predation by an introduced mammal, the stoat (*Mustela erminea*). The Rock wren is now listed as Near Threatened by the IUCN; there seems little chance of protecting it in the near future, although intensive research is underway to design new control methods, involving modified traps, baits, and toxins as well as innovative biotechnological techniques. GHS

Pittas

Left *A Blue-winged pitta perches on a rock. This Southeast Asian species is migratory, and has been known to winter as far south as northwestern Australia.*

Brightly colored and stockily built, the pittas are a remarkable family of tropical forest birds centered in Southeast Asia. Their jewel-like hues, combined with their rarity, have given them a particular charisma among songbirds, similar to that of the birds of paradise.

The word "pitta" comes from the Madras area of South India, and merely signifies "bird." It was first applied to the Indian pitta in 1713.

Gorgeous Denizens of the Forest Floor

FORM AND FUNCTION

Pittas are long-legged, short-tailed birds with strong bills and feet, well adapted to life on or near the forest floor. Their bright plumage includes areas of vivid scarlet, turquoise, or metallic blue, rich and delicate greens, velvety black or porcelain white. The brightest colors are often difficult to see, since they are usually found on the underside of the body and are often hidden by the birds' habit of standing motionless with their backs toward any source of alarm, or of fleeing with rapid bounds or short flights into the vegetation.

Some pittas have patches of white on the wings or iridescent blue on the shoulders and primaries, probably to facilitate visual contact in the dim light. Larger species, such as the Rusty-naped pitta, also possess disproportionately large eyes to aid vision in the gloomy forest interior. A few are even partly nocturnal in activity, but most roost at night, perched well above ground with the bill tucked under the wing.

When calling, pittas will perch in trees up to 10m (33ft) above the ground and may throw back their heads, usually at dawn and dusk, before rainstorms or on moonlit nights, and often in chorus with neighbors. Being territorial, they respond readily to imitations of their calls, even outside the breeding season, when they are usually solitary and occupy only foraging territories. They respond to an intruder with a threat display; the Noisy pitta, for example, crouches with feathers fluffed, wings outspread, and the bill pointing upward. The Blue-rumped pitta has a similar display, but the head is bent over backwards to expose a triangular patch of white spots below the throat.

From Africa through Asia

DISTRIBUTION PATTERNS

Pittas occur widely throughout tropical Asia from sea level up to 2,500m (8,200ft), extending to Japan (the Fairy pitta), Australia, the Solomon Islands (six species), and across into tropical Africa (African pitta). All species are found in forested regions, especially the remaining tracts of lowland to mid-montane evergreen rain forest. The Asian rain forests are inhabited by the rarest and least known species (including the very distinctive Eared, Blue, and Giant pittas), among them some, such as the red-bellied Whiskered pitta and the black and sky-blue Azure-breasted pitta, that are confined to a few Philippine islands. The striking blue, yellow, and black Gurney's pitta is unique in its limited distribution, confined historically to the junction of two major faunal zones along about 500km (310 miles) of peninsular Thailand and adjacent Myanmar. It is now reduced to a single location, where only 12 pairs are known to exist, and is critically endangered.

Apart from some minor dispersion according to season and altitude, only eight pitta species (including the Indian, Blue-winged, and African pittas) are known to undertake regular migration. These pittas are nocturnal migrants, and many unusual records have resulted from their attraction to lights. The African pitta was thought to be sedentary until 50 years ago, when an ornithologist living in Tanzania found that records of birds flying into lighted houses at night over several years displayed a seasonal pattern. Further study has shown that regular migration occurs in East Africa, but records of night movements from West and Central Africa suggest

FACTFILE

PITTAS

Order: Passeriformes

Family: Pittidae

32 species, all in the genus *Pitta*. Species include: **African pitta** (*Pitta angolensis*), **Azure-breasted pitta** (*P. steerii*), **Banded pitta** (*P. guajana*), **Blue pitta** (*P. cyanea*), **Blue-banded pitta** (*P. arquata*), **Blue-headed pitta** (*P. baudii*), **Blue-rumped pitta** (*P. soror*), **Blue-winged pitta** (*P. moluccensis*), **Eared pitta** (*P. phayrei*), **Fairy pitta** (*P. nympha*), **Giant pitta** (*P. caerulea*), **Graceful pitta** (*P. venusta*), **Gurney's pitta** (*P. gurneyi*), **Hooded pitta** (*P. sordida*), **Indian pitta** (*P. brachyura*), **Noisy pitta** (*P. versicolor*), **Rusty-naped pitta** (*P. oatesi*), **Whiskered pitta** (*P. kochi*).

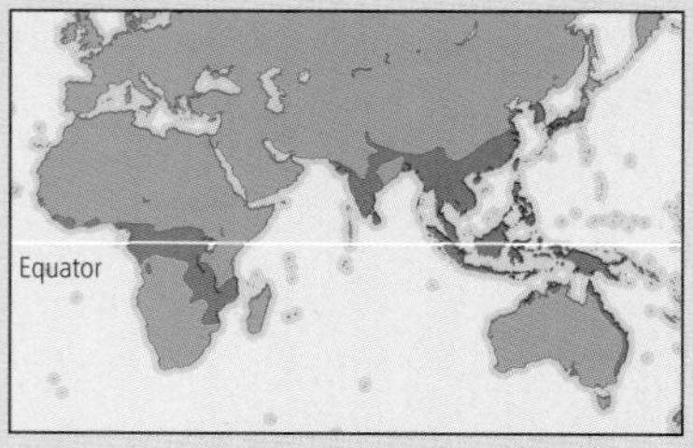

DISTRIBUTION Africa; E and S Asia to New Guinea, the Solomon Islands, and Australia.

HABITAT Evergreen and deciduous forest, bamboo, mangroves, wooded ravines, extending into secondary forest and plantations.

SIZE Length 15–29cm (5.9–11.4in); weight 42–207g (1.5–7.3oz).

PLUMAGE Very colorful, with bright blues, greens, reds, and yellows, brightest on the head and underparts; little difference between sexes, except in 4 banded species with drab females. Young birds are dull, more brownish, and mottled or spotted.

VOICE Short series of variably pitched whistles, often of two syllables; range of trilling, rolling sounds and loud barking notes when alarmed.

NEST Large, globular structure, built untidily of twigs and rootlets, often decorated with moss, lined with finer materials, with a low side entrance and a small platform in front.

EGGS 1–7, usually 2–5; vary in shape from rounded oval to spherical, some with much gloss; white or buff with fine gray or lilac streaks and reddish or purplish speckles; weight about 5–10g (0.18–0.35oz). Incubation 15–17 days; nestling period 12–21 days. Juveniles rapidly independent (within 5–24 days).

CONSERVATION STATUS Gurney's pitta is Critically Endangered. 8 other species are Vulnerable, including the Blue-headed and Graceful pittas.

Left A resident of Southeast Asian lowland forests, the Banded pitta haunts limestone cliffs in search of snails, a favored food. Often trapped for its vivid plumage, it is now listed under Appendix II of the CITES convention on the trade in endangered species.

that the picture is still incomplete, while breeding migrants arrive annually at lowland forests along the lower Zambezi River of Zimbabwe.

A survey of nocturnal migration in Malaysia has not only established the dates of movements, but also revealed that pittas are unusual in that the peaks of migration are during the new moon, not the full moon as with other song birds. Individuals of migrant species that wander beyond the normal range are regularly reported, including a small flock that on one occasion landed on a boat at sea.

Foraging in the Leaf Litter

DIET AND FEEDING

Pittas spend much of their time searching for small animal food, particularly worms, snails, and insects, among the leaf litter and humus of the forest floor. Leaves and debris are flicked over with the strong bill, or scratched aside with the feet. Occasionally prey may be located by sound, when the pitta turns its head sideways, or is flushed out, by flicking movements of the wings.

Some pittas are attracted to sites favored by snails, such as Banded pittas to limestone cliffs, and will use a rock or log as an anvil to break open the shells. A study of the food habits of a captive Hooded pitta revealed a strong preference for earthworms. This bird ate approximately its own weight in food each day.

Below Representative species of pittas: **1** Indian pitta (Pitta brachyura); **2** Gurney's pitta (P. gurneyi), listed by the IUCN as Critically Endangered.

Sharing Incubation Duties

BREEDING BIOLOGY

Pittas usually site their large, bulky nests up to 3m (10ft) above the ground, in stumps, root buttresses, fallen trees, tangled clumps of vegetation, on banks, or in rock clefts. If disturbed at the nest, the parent bird may attempt to draw the intruder away by calling. The breeding season is limited to the summer at higher latitudes, but may cover most months of the year near the equator, except for at the height of the monsoon period.

The male initiates a courtship by confronting the female in an erect posture. With wings spread, he then performs a courtship dance involving vertical movements of the body, accompanied by loud calls. If the female responds in kind, mating takes place, and the male starts to build the nest with her assistance. Both sexes share incubation and brooding duties, along with feeding of the young and the removal of fecal sacs. They may drive the young away shortly after they have fledged and lay a second clutch.

Below *A Noisy pitta* (Pitta versicolor) *supplies its insistent young with food. As in most pitta species, both parents play a part in feeding the nestlings.*

Adaptable but not Unscathed

CONSERVATION AND ENVIRONMENT

Although the Pittidae are regarded as strictly forest-dwelling birds, they are one of only a few avian families that respond dynamically to habitat alteration. Even though the birds are affected adversely by intensive logging, studies in Sabah have shown that several species will return to lightly disturbed or partially regenerated forest. Such adaptability has allowed Gurney's pitta to survive, although it is now so seriously threatened by habitat destruction in its native Thailand that its extinction is imminent unless there are new populations to be rediscovered in neighboring Myanmar, where it also used to thrive.

Several other species of pitta that have restricted mainland or island distributions are also now under threat, for the most part as a result of the extensive destruction of their habitat. In addition, pittas are under pressure from hunting, whether for the caged-bird trade or for food. Resident species as well as birds on migration are at risk, and in some areas, such as Vietnam, and for some species, such as the Fairy pitta, this additional pressure may significantly increase the birds' vulnerability. AK/MDB

Above *Like most pittas, the Red-bellied pitta* (P. erythrogaster) *spends most of its days on the forest floor, usually not far from a river or stream, hopping on sturdy legs in search of insects or worms.*

Asities

ENDEMIC TO MADAGASCAR, ASITIES APPEAR to be closely related to (and have recently been combined with) the broadbills of the family Eurylaeimidae, based on similarities of anatomy and nest structure. Male asities in breeding plumage develop fleshy wattles of a unique cellular structure around the eye; the only broadbill with a similar eye wattle occurs in the Philippines.

The birds appear in two distinct forms. The dumpy *Philepitta* species resemble broadbills or barbets; they feed mainly on fruit, but also extract nectar with their specialized tongues, and are generally quiet and sluggish. In contrast, the two species of the genus *Neodrepanis*, also known as false sunbirds, feed mainly on nectar and the insects found inside flowers with the aid of spectacularly curved bills, using an even more specialized tongue that may also enable them to feed on pollen. In marked contrast to the extreme differentiation of the two genera, the two species within each genus are so similar in size and ecology that they replace one another, rather than overlapping, in their distribution on the island.

Disparate Genera in a Single Family

FORM AND FUNCTION

The radically different bills and dissimilar plumages of the two asity genera illustrate the wide radiation that must have occurred within the broadbill family. It also suggests that intermediate forms, possibly more primitive and broadbill-like, may have existed on Madagascar in the past.

Illustrations of the parallel development of false sunbirds (on Madagascar) and true sunbirds (in Africa and Asia) include not only the development of the tubular tongue and curved bill in both groups but also a biannual molt. Indeed, the Common sunbird-asity was placed with the sunbirds for 75 years until the anatomy of its syrinx was studied in more detail. The similarities also suggest that the false sunbirds evolved on Madagascar in the virtual absence of competitors that exploited flowers.

Despite the differences between the asities, all species share an enclosed, suspended, untidy nest with a side-entrance, a feature held in common with their closest relatives, the broadbills, but also with the true sunbirds. Interestingly, DNA studies suggest that the asities' nearest relative among the broadbills is the African green broadbill.

The fifth species tentatively included in this family is, as its scientific name suggests, enigmatic in its taxonomic relationships. Studies of DNA have placed the Broad-billed sapayoa in the same region of the avian family tree as broadbills and asities, although still distinct from them. Traditionally the bird has been included with the mainly frugivorous manakins (family Pipridae) of South America, even though it is clearly not typical of the family. The lack of either broadbills or asities in the New World offers support to the manakin rather than the asity association, or else opens a whole new perspective on the origins of broadbills and asities.

The sapayoa perches for long periods in the forest understory before flying up to use its flat, broad bill to catch insects or pick fruit in flight or off foliage. It usually occurs in pairs or joins mixed-species parties, but its nest and other aspects of its biology are undescribed.

Unanswered Questions

BREEDING BIOLOGY

Asities occur alone or in pairs, except for the polygynous Velvet asity, in which the males hold adjacent small territories to attract females for copulation; in this species, only females build the nest and care for the eggs and chicks. Such breeding biology is common to several families of forest frugivores that live in the understory, including manakins. Males of Schlegel's asity may also call in close proximity, but in this species the males help at least with nest-building. The breeding biology of the sunbird-asities is virtually unknown. AK/GHS

Above The Common sunbird-asity is the better-known of the two "false sunbird" species, so called because of their striking resemblance to the "true" sunbirds of Africa and Asia. Both groups are brilliantly colored and have long, downward-curving bills used for extracting nectar.

FACTFILE

ASITIES

Order: Passeriformes

Family: Philepittidae

4 species in 2 genera, plus *Sapayoa aenigma*, of uncertain affinity: **Velvet asity** (*Philepitta castanea*), **Schlegel's asity** (*P. schlegeli*), **Common sunbird-asity** (*Neodrepanis coruscans*), **Yellow-bellied sunbird-asity** (*N. hypoxantha*), **Broad-billed sapayoa** (*Sapayoa aenigma*).

DISTRIBUTION Madagascar, except for *Sapayoa* in S America.

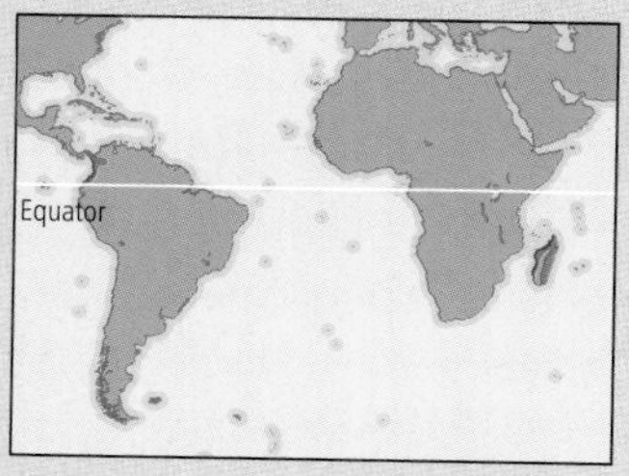

HABITAT Wet and dry forests

SIZE Length 9–16.5cm (3.5–6.5in); weight 6.2–38g (0.2–1.3oz). The sapayoa is 15cm (5.9in) long and weighs 20.8g (0.7oz).

PLUMAGE Generally olive above and yellow below; breeding males have bright green or blue wattles around the eyes. *Philepitta* species have short, broad bills; females with olive streaks below; male Velvet asity all-black in breeding plumage, with yellow edges when non-breeding. *Neodrepanis* sunbird-asities have long, tapered, decurved bills; males have metallic-blue backs. *Sapayoa* has olive plumage, with more yellow below and on brow of adult males.

VOICE Melodious whistle (*Philepitta*) or shrill trill (*Neodrepanis*). *Sapayoa* has a soft, nasal trill.

NEST An untidy ball, suspended from a twig.

EGGS 2–3 in the Velvet asity; white. Breeding undescribed for *Sapayoa*.

DIET Insects and fruit for all species

CONSERVATION STATUS The Yellow-bellied asity is regarded as Endangered, while Schlegel's asity is classified as Lower Risk/Near Threatened.

Broadbills

B*ROADBILLS ARE STOCKY BIRDS, THEIR SQUAT appearance accentuated (except in the Long-tailed broadbill) by a short, square tail. The family is named for the birds' typically wide mouths, which reach a grotesque extreme in the outsize red bill of the Dusky broadbill.*

No species of broadbill has been studied in detail, despite the uniqueness of their form and habits, the rarity of species such as the Mindanao wattled and African green broadbills, and the possibility of a broader relationship with asities. Since all are birds of the forest, a habitat that is being widely altered by human activities, they deserve more research before yet more become threatened.

Wide Bills, Sturdy Bodies

FORM AND FUNCTION

For all their bright colors, broadbills are not especially conspicuous in their forest habitat. The most prominent feature of, for example, the Black-and-red broadbill, sitting in the shade of a waterside thicket, is its almost luminous, pale-blue and yellow bill (which fades after death and cannot be appreciated in museum specimens). Several species, including those of *Smithornis*, also have on their backs one or more white, yellow, or orange patches, exposed during flight against the dark forest background. Silver-breasted broadbills have a bright chestnut rump, often fluffed out when they perch.

In most species the bill is wide, rounded along its sides and hooked, perhaps to aid the aerial capture of large arthropods by these rather slow-moving birds. This bill-form is found otherwise only in trogons and frogmouths, most of which share the broadbill's habitat and feeding behavior. The bill is most hooked in the hawklike Dusky broadbill, which has been seen snatching large grasshoppers in an upward leap from its perch. Another foraging mode involves snatching prey from foliage or bark while in flight – the manner in which a Banded broadbill captures an arboreal lizard. Most species forage at mid-levels of the forest, but the Dusky broadbill prefers the high canopy, while two other species, the African broadbill and the Asian Black-and-red broadbill, inhabit forest-edge thickets and will even drop to the ground to feed. Besides taking insects in their waterside habitat, Black-and-red broadbills sometimes also capture small aquatic organisms such as crabs and fish.

The African green broadbill and the green Asian *Calyptomena* species have bills that are still very wide at the base but have straight sides. The latter feed largely on fruit, and there is evidence that the African species also takes much fruit and buds.

The *Calyptomena* broadbills, which have been recorded eating at least 21 species of soft fig, often advertise their presence at fruiting trees by cooing rattles. Pairs will drive off members of the same species at fruit sources that are localized and defendable, but large numbers may gather at sources that are more abundant, as has been seen for Hose's broadbill. Both species live mainly in lowland forests where fruit is scattered and, like other fruit-eating birds, must wander over a sizable area if they are to find sufficient food. Their much larger relative, Whitehead's broadbill, lives exclusively above 1,200m (4,000ft) in montane forest, where fruit supplies may be less plentiful but are more stable.

Insect-eating broadbills are more sedentary, living in small flocks or joining other insectivorous birds. Banded and Black-and-yellow broadbills

FACTFILE

BROADBILLS

Order: Passeriformes

Family: Eurylaimidae

15 species in 8 genera: **African green broadbill** (*Pseudocalyptomena graueri*), **Banded broadbill** (*Eurylaimus javanicus*), **Black-and-yellow broadbill** (*E. ochromalus*), **Mindanao wattled broadbill** (*E. steerii*), **Visayan wattled broadbill** (*E. samarensis*), **Black-and-red broadbill** (*Cymbirhynchus macrorhynchus*), **African broadbill** (*Smithornis capensis*), **Grey-headed broadbill** (*S. sharpei*), **Rufous-sided broadbill** (*S. rufolateralis*), **Dusky broadbill** (*Corydon sumatranus*), **Green broadbill** (*Calyptomena viridis*), **Hose's broadbill** (*C. hosei*), **Whitehead's broadbill** (*C. whiteheadi*), **Long-tailed broadbill** (*Psarisomus dalhousiae*), **Silver-breasted broadbill** (*Serilophus lunatus*).

DISTRIBUTION S China and N India to SE Asia and sub-Saharan Africa

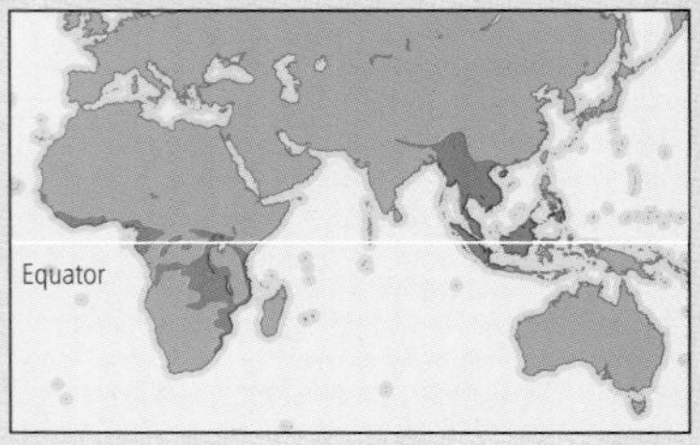

HABITAT Tropical forests and thickets

SIZE Length 11.5–27.5cm (4.5–10.8in); weight 10–163g (0.4–5.7oz).

PLUMAGE Browns, with gray to black; or else green, red, black, or silvery gray, with areas of bright color contrast, often including the bill. Little or no difference between sexes in size or coloration. Juveniles resemble adults, except that colors are duller.

VOICE Screaming whistles, explosive trills, cooing rattles, croaks.

NEST Large, untidy structure, domed and suspended, often with long tail of debris.

EGGS 2–3 for tropical species, 4–8 for Sino-Himalayan populations at higher latitudes; white or cream to pinkish, unmarked or speckled with red and purple.

DIET Chiefly either arthropods or fruit, depending on genus.

CONSERVATION STATUS 3 species – the African green broadbill and the Visayan and Mindanao wattled broadbills – are Vulnerable. The Black-and-yellow, Green, and Hose's broadbills are listed as Lower Risk/Near Threatened.

Left *With only its protruding bill revealing its presence, a Dusky broadbill peers from a nest suspended from a branch in Malaysia's Panti forest.*

Right *A Green broadbill swivels its head almost 180° to investigate noises behind its perch. The thick growth of feathers on the upper mandible is characteristic of this gregarious, fruit-eating species, which is widespread in Southeast Asian forests.*

space themselves through the forest, advertising their presence with loud, explosive trilling calls that are invariably answered by neighbors. *Smithornis* males rapidly flap their specially twisted outer primaries to produce a ripping sound that is audible up to 60m (200ft) away; they do so while making a short circuit from a low perch in which they also expose their pale back as part of a territorial display. Other broadbills are generally less noisy and give a clear, two-syllable whistle, most often when in the foraging groups that assemble after breeding. These groups are usually small, but as many as 20 Silver-breasted broadbills and up to 26 Long-tailed broadbills have been counted together. Only the Dusky broadbill is known to be permanently gregarious, in noisy parties up to 10 strong.

Birds of the Lowland Forest

DISTRIBUTION PATTERNS

Broadbills chiefly inhabit the interior of evergreen or semi-evergreen broad-leaved lowland forests, with a center of diversity in Southeast Asia. Only two species on Borneo and in Africa are exclusively mountain-dwellers, although Long-tailed and Silver-breasted broadbills are restricted to mountain foothills.

No genus is common to the African and Asian areas of broadbill distribution. The diminutive, brown and streaked African, Rufous-sided, and Gray-headed broadbills of the genus *Smithornis* look so different from their gaudy Asian relatives that they were long classified as flycatchers, the shape of the bill not withstanding. They were first recognized as broadbills in 1914, following anatomical studies, while the African green broadbill was discovered only 19 years later.

Three Asian genera are also green, with black, blue, or yellow on the head, wings, belly, or tail of the Long-tailed, Hose's and some forms of male Green broadbills. The Banded broadbill is winered with a blue bill. Other species are mainly black with areas of red, lilac, yellow and/or white, while black is replaced by rich chestnut in the Philippine broadbill.

Crude Nests in the Trees

BREEDING BIOLOGY

Broadbill nests are large, pear-shaped bags with a crudely overhung side entrance, slung by a long, woven cord of nest material from an isolated branch, creeper, or frond tip. They are roughly made of all kinds of vegetation, drawn out below into a wispy beard. Leafy creepers, lichen, moss, and liverworts are often included, and it is common for the nest chamber to be lined with fresh green leaves. The nest structure is similar to that of the four asity species (family Philepittidae) on Madagascar, and this fact, together with several anatomical structures, suggests that asities may be related to broadbills, despite their small size and quite different bill structure and feeding habits.

Usually broadbill nest-sites are well off the ground, but the Green broadbill, whose nest is broadly strapped over its support, invariably builds low, as do the *Smithornis* species. Black-and-red broadbills often use a dead stump or snag in a stream, and will sometimes take advantage of a service wire over a stream or road.

Few displays have been described for broadbills, but the male Green broadbill is known to perform nodding and gaping displays, slow, butterflylike flights, and even a strenuous aerial pirouette around the female. There are no records of helpers attending broadbill nestlings, although the Dusky broadbill flock cooperates in nest construction, and species that gather in smaller groups may do the same. In Malaysia, Black-and-yellow broadbills have been seen feeding fledglings of the Indian cuckoo (*Cuculus micropterus*) and are presumed to be its brood host. AK/DRW

Tyrant Flycatchers

Left *A Social flycatcher* (Myiozetes similis) *keeps watch from a Mexican bushtop, alert for the insects that are its main food. The species is so called because the birds often nest close to people, sometimes suspending their nests from telephone poles or wires.*

THE TYRANNIDAE, WHICH OWE THEIR NAME TO the conspicuously aggressive behavior of a few species (notably the kingbirds), are one of the largest and most remarkably diverse groups of birds in the world. The family is so large and disparate that it defies easy generalizations. Tyrant flycatchers have adapted to almost every available habitat, particularly in the forests of Amazonia and the Andes. Although some species range as far as the temperate regions of northern Canada and southern Patagonia, almost all migrate to the tropics for the winter. A "typical" flycatcher does not really exist, but within the group the small gleaning species are the most common.

Tyrant flycatchers are found throughout the Americas, but species diversity is greatest in the Neotropics, where a tenth of all bird species belong to the family. More significantly, the flycatchers represent almost 18 percent of South American land bird species: the proportion rises to 20 percent in Colombia and Venezuela and almost 26 percent in Argentina. Not only is there a high diversity of species in any given Neotropical region; at the local scale the number of species at a given site is often significantly high (sometimes over 20 percent), indicating the broad range of ecological roles to which the birds have adapted.

An Abundant New World Family

FORM AND FUNCTION

Tyrant flycatchers search for food by using modifications of a single basic procedure: they pause for varying lengths of time on a perch and then move after prey. The best known of these foraging behaviors is aerial hawking, where birds sit on an exposed perch, sally out to catch flying insects, and return to the same perch again. While aerial hawking is considered "typical" for North American flycatchers, most species (which occur in South American forests) are actually foliage or sally-gleaners: that is, they search their surroundings and either glean their food directly from them or sally out to pick it from some substrate (vegetation, branches, ground, etc .). Unlike aerial-hawkers, sally-gleaners rarely take flying prey.

Variations of sallying behavior include flying from a perch to pounce on ground prey, chasing prey on the ground, fluttering after prey from the ground or out over water, flitting along branches and taking prey by a direct strike or while hovering, or any combination of the above. Several tyrannulets like the bristle-tyrants of the genus *Phylloscartes* and the wagtail-tyrants will flick or flutter their wings and/or tail while foraging to induce their prey to move. The *Leptopogon* flycatchers perch upright and regularly lift one wing up over their back. After capturing their prey, many species will return to a prominent or exposed perch to eat it.

Many of the variations in size and body form between species are associated with maneuvers used to obtain food. Flycatchers mainly eat insects, but almost all species take some fruit, especially when not breeding. Species that forage for insects by aerial hawking tend to have short legs, long wings, and a long tail, for greater maneuverability in the air when chasing prey. Those birds that forage on the ground have strong legs for stability. Flycatchers that forage using a perch-gleaning method tend to be slight-bodied, active, warblerlike birds with long legs and tails for balancing on twigs while reaching for prey.

Bill form is also strongly associated with the manner of obtaining food. Most flycatchers have a broad, flat, triangular-shaped bill, widest at its base. This bill type is generally associated with aerial-hawking species like kingbirds and pewees.

Some bills are very broad and almost spoonlike. These tend to be associated with flycatchers that scoop prey from the undersides of leaves in an explosive upward strike, like the tody-flycatchers. A more extreme form is found in spadebills, where the bill is about as broad at the base as it is long. Species with broad bills that specialize in upward striking also tend to have prominent rictal bristles.

One of the more unusual bill forms in the Tyrannidae is the downcurved bill of the bentbills. The bend is thought to render the tip of the bill more parallel to the leaf's surface in an oblique upward strike than if it were straight, enabling the bird to catch prey with greater precision.

In contrast to the broad bills of the aerial hawkers and sally-gleaners are the thin, precise bills of the foliage- and perch-gleaners. These bills act more like tweezers and provide precision for obtaining food at close range in hidden places. Flycatchers like the elaenias that regularly eat fruit tend to have shorter, thickened bills that are broad at the base. Those species that eat large prey have relatively long, strong bills, whether they live in forests like the attilas or in open grasslands like shrike-tyrants. The largest flycatcher bill is that of the aptly named Boat-billed flycatcher.

Most flycatchers have dull green upperparts and lighter yellowish, ochraceous, or whitish underparts. Many other species are drably colored in grays or browns, especially those that live in open terrain like the shrike-tyrants and ground-tyrants. A number have bold plumage patterns with a combination of subtle and/or sometimes bright colors, like the White-cheeked tody-tyrant or White-headed marsh-tyrant. The majority of species have some subtle combination of wing

Above *A Pied water-tyrant* (Fluvicola pica) *perches by a blossom in Trinidad. The birds are usually found close to water, searching for insects on the ground or else low in bushes or among mangrove roots.*

Below *The spadebills are broad-billed birds that specialize in scooping insects from the undersides of leaves. This White-throated spadebill* (Platyrinchus mystaceus) *is from Minas Gerais state, Brazil.*

FACTFILE

TYRANT FLYCATCHERS

Order: Passeriformes

Family: Tyrannidae†

At least 436 species in 103 genera (not including the Pipridae and Cotingidae).

DISTRIBUTION Throughout the Americas and adjacent islands

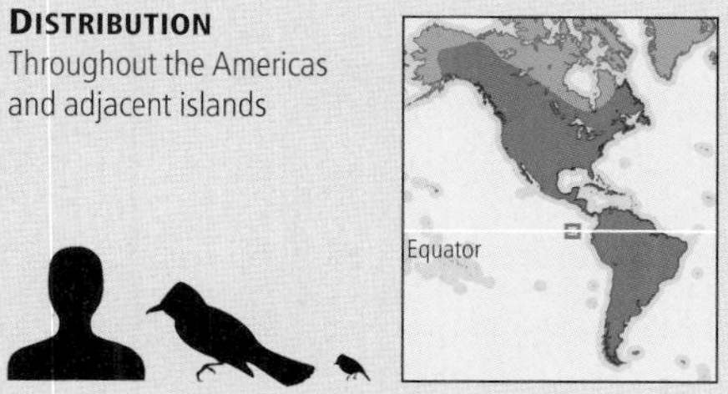

HABITAT Primarily forest or woodland; a small number of species are found strictly in grasslands.

SIZE Length 6–50cm (2.5–20in); **weight** 4.5–80g (0.2–2.8oz).

PLUMAGE Often dull, but many species are boldly patterned or colorful.

VOICE Many species very vocal; songs and calls generally simple and usually not musical, but often significant characteristics of a species.

NEST Highly varied

EGGS 2–8 (fewer in tropics); whitish, sometimes lightly to heavily mottled with reddish brown.

DIET Mainly insects, but almost all eat some fruit; occasionally small fish, lizards, snakes, tadpoles.

CONSERVATION STATUS Two Brazilian species are listed as Critically Endangered. 9 other species are currently Endangered, and 14 are Vulnerable.

† Some authors include the cotingas and manakins in this family.

See Selected Genera and Species ▷

bars, light edges of the flight feathers, eyelines, eyerings, supercilia, or tail markings.

A large number of species have a coronal patch of some type. This can be dull white or yellowish, but is often brightly colored. The most elaborate is the large transverse crest of the Royal flycatcher, which is scarlet with black and shiny blue tips. Coronal patches are frequently concealed and only revealed for display.

Only a few species are brightly colored. The male Vermilion flycatcher has a scarlet head and underparts. The Many-colored rush-tyrant, which is known locally as *Siete colores* ("Seven colors"), is a brightly-colored patchwork of green, yellow, white, black, blue, orange, and red.

The largest flycatchers are found in the high-altitude grasslands of the Andes and in the low-lands of Chile and Argentina. These birds, aptly named shrike-tyrants, scan the ground from elevated perches, searching for large insects and small lizards that they catch and tear apart with their large, hooked bills. The related ground-tyrants are mainly terrestrial birds found in open

Tools for the Job – Feather Variation in Tyrant Flycatchers

Although the majority of tyrant flycatchers are known for their dull appearance, within the group there is a broad range of subtle modifications of feather types that in some cases reaches extremes. Perhaps the most noticeable modified feathers are the tail feathers of some of the open-country species. For example, the outermost tail feathers of the Scissor-tailed flycatcher (*Tyrannus forficata*) and Streamer-tailed tyrant (*Gubernetes yetapa*) are elongated to form a deeply forked tail used in aerobatic displays. In two other open-country species, the tail feathers are the most elaborately modified and twisted of the entire family. The middle two tail feathers of the Cock-tailed tyrant (*Alectrurus tricolor*) twist perpendicularly to the back and the outer tail feathers project out to the sides, so that the tail is similar in shape to that of an airplane. During its display, the male flies slowly upward with wings fluttering and with the center part of the tail arched forward almost over its head. The outer tail feathers of the related Strange-tailed tyrant (*A. risora*) are twisted and elastic, with broadened plumes at the end. When the bird flies, the plumes are held pointed toward the ground beneath the tail. How these feathers are used in display is not yet known.

Many tyrant flycatchers have crown patches, the feathers of which are among the most varied in the group. The crowns of males are usually larger, brighter, or more elaborate than those of females. The simplest crowns are patches of feathers ranging from white to russet, orange, red, or, most commonly, yellow. Crown patches are usually concealed by duller feathers at the side and are generally only exposed for display. In one genus (*Lophotriccus*), the crown patch is a crest of elongated black feathers with gray or russet tips that are held up when the birds are agitated or displaying.

The most spectacular crown is that of the Royal flycatcher (*Onychorhynchus coronatus*), which is usually held down, giving the birds a "hammer-headed" appearance; when extended, however, it appears as a brilliant scarlet and blue fan held perpendicular to the axis of the body. The crest of the Royal flycatcher has rarely been seen extended under normal circumstances in the wild; however, when handled (by researchers using mist-nets, for example), the bird will raise its crest and slowly rock its head from side to side while opening and closing its bill rhythmically at the same time!

Perhaps the most notable feather modification in tyrant flycatchers is the tendency for species to develop peculiarly-shaped flight feathers. An example of a normal wing, that of the Blue-billed black tyrant (*Knipolegus cyanirostris*), is given for comparison in illustration **1**. In some species, however, one or more flight feathers are highly modified **2**, the Crested doradito (*Pseudocolopteryx acutipennis*); such modifications are species-specific. Although little is known about the biological significance of these modifications, field observations suggest that these feathers are used for producing sounds. In other species – for instance, the Scissor-tailed flycatcher (*Tyrannus forficata)* **3** – the ends of the outer primaries are sharply indented, improving flight for display and for hunting, as in several species of raptors. In all cases, the feather modifications are at their most extreme in males, with only some or no modification in females. DA

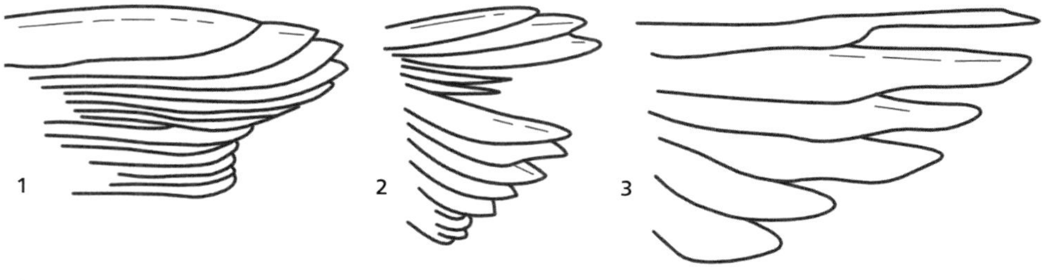

◀ *Left* *Representative species of tyrant flycatchers:* ***1*** *Royal flycatcher* (Onychorhynchus coronatus)*;* ***2*** *Vermilion flycatcher* (Pyrocephalus rubinus)*;* ***3*** *Scissor-tailed flycatcher* (Tyrannus forficata)*;* ***4*** *Great kiskadee* (Pitangus sulphuratus)*;* ***5*** *Short-tailed pygmy-tyrant* (Myiornis ecaudatus).

▶ ***Right*** *Nestlings demand food from a Willow flycatcher* (Empidonax traillii)*. Four is a typical brood size for this North American species, whose range extends across the continent from New England to the Pacific.*

grassy or rocky areas of the Andes. With slender bills and long legs, they look like pipits as they chase after insects along the ground.

Among the larger flycatchers is a group that shares a characteristic plumage pattern of yellow underparts, brown backs, and black-and-white striped crowns. The best known of these birds is the conspicuous Great kiskadee, which is widespread from North America south to central Argentina. This bold, adaptable bird eats a variety of foods, including insects, fish, tadpoles, and fruits. At the other end of the spectrum, the smallest flycatchers are the tiny tody-tyrants and pygmy-tyrants, some of which are smaller than many hummingbirds.

The majority of flycatcher species, including the tyrannulets, flatbills, and elaenias, are small- to medium-sized birds that are obscurely marked and are often confusingly similar. Many of them live high in the canopy of tropical forests and are difficult to observe. As a group, these are the hardest to identify.

There are numerous other species that do not fit into the broad categories described above. For example, the antpipits are terrestrial forest species; the Sharp-tailed grass-tyrant is found in grasslands and is the only tyrannid reported to feed to some extent on seeds; while the Streamer-tailed, Cock-tailed, and Strange-tailed tyrants are all distinguished by their spectacular tail plumes.

Most flycatchers are very vocal; however, their vocalizations consist, for the most part, of undistinguished, weakly whistled or warbled notes. Yet even these simple songs can perform the crucial function of being species-specific isolating mechanisms between very similar and closely related forms. Almost all tyrannid species have a distinctive dawn song that is given early in the morning and is not repeated at other times of day. Although some members of other families give dawn songs, they are a much more dominant feature of the Tyrannidae.

Songs are an important means of maintaining territories and are sometimes given in conjunction with courtship or territorial displays. Species such as the Vermilion flycatcher, the Streamer-tailed tyrant, and the Fork-tailed flycatcher give songs during complicated aerial displays. Many other species have equally elaborate displays. The males of most species display and/or sing regularly at the same perch, and some (for example, *Mionectes* species) do so in dispersed leks.

Vigorously Defended Territories

BREEDING BIOLOGY

Most tyrant flycatchers are monogamous, although the males of many species do not help to rear young. Many nonmigratory species remain paired all year, and dwell on permanent territories. Others, especially those that migrate, establish new pairs as the breeding season begins.

Usually the female builds the nest, incubates the eggs, and broods the young alone; this is especially the case with smaller species. The male generally stays nearby, and defends the territory and nest site. Often both sexes feed the young, but in some smaller species the females do this alone.

In most species the young reach independence within a few months of leaving the nest and are breeding on their own territories the following year. In many tropical species, the young remain with their parents until the next breeding season or even longer, sometimes helping to raise subsequent broods of siblings for a number of years. The breeding behavior of most flycatchers is, however, poorly known.

The nests of flycatchers are as varied and diverse as the rest of their behavior. The simplest shape is a cup that ranges from a loose tangle to a tightly-woven structure. Some are placed high in trees, or hidden in foliage or on the ground. The cup nest of the Many-colored rush-tyrant is a delicate inverted cone attached to one side of a reed. Several species build different types of spherical, covered nests with side entrances; others nest in holes in trees or in embankments. Some species in the hole-nesting genus *Myiarchus* are well-known for their odd habit of using sloughed reptile (especially snake) skins to line their nests.

The most elaborate tyrannid nests are those that hang from a branch or overhang. Some of these are baglike, with an entrance at the top; others have entrances below or to the side. In some species, the nest is concealed to look like moss hanging from a branch. In others, like that

Above *Found in savanna regions of central South America, the Sharp-tailed grass-tyrant* (Culicivora caudacuta) *is threatened by changing land use, as open grasslands are burned to prepare for cattle-grazing.*

Selected Tyrant Flycatcher Genera and Species

Genera and species include: **Antpipits** (genus *Corythopis*), **attilas** (genus *Attila*), including Ochraceous attila (*A. torridus*), **bentbills** (genus *Oncostoma*), **elaenias** (genera *Myiopagis*, *Elaenia*), including Noronha elaenia (*E. ridleyana*), **flatbills** (genera *Ramphotrigon*, *Tolmomyias*), **ground-tyrants** (genus *Muscisaxicola*), **kingbirds** (genus *Tyrannus*), including Eastern kingbird (*T. tyrannus*), Fork-tailed flycatcher (*T. savana*), Giant kingbird (*T. cubensis*), **pewees** (genus *Contopus*), **pygmy-tyrants** (genus *Myiornis*), **shrike-tyrants** (genus *Agriornis*), including White-tailed shrike-tyrant (*A. andicola*), **spadebills** (genus *Platyrinchus*), **tody-flycatchers** (genus *Todirostrum*), including Yellow-browed tody-flycatcher (*T. chrysocrotaphum*), **tody-tyrants** (genera *Poecilotriccus, Hemitriccus*), including White-cheeked tody-tyrant (*P. albifacies*), Fork-tailed tody-tyrant (*H. furcatus*), Kaempfer's tody-tyrant (*H. kaempferi*), **tyrannulets** (genera *Phylloscartes*, *Camptostoma*, *Mecocerculus*, *Zimmerius*), including Alagoas tyrannulet (*P. ceciliae*), Bahia tyrannulet (*P. beckeri*), Minas Gerias tyrannulet (*P. roquettei*), **wagtail-tyrants** (genus *Stigmatura*), Cock-tailed tyrant (*Alectrurus tricolor*), Strange-tailed tyrant (*A. risora*), White-headed marsh-tyrant (*Arundinicola leucocephala*), Many-colored rush-tyrant (*Tachuris rubrigastra*), Sharp-tailed grass-tyrant (*Culicivora caudacuta*), Great kiskadee (*Pitangus sulphuratus*), Boat-billed flycatcher (*Megarynchus pitangua*), Royal flycatcher (*Onychorhynchus coronatus*), Vermilion flycatcher (*Pyrocephalus rubinus*), Streamer-tailed tyrant (*Gubernetes yetapa*), Piratic flycatcher (*Legatus leucophaius*), Cocos flycatcher (*Nesotriccus ridgwayi*), Grey-breasted flycatcher (*Lathrotriccus griseipectus*).

of the Yellow-browed tody-flycatcher, it may be built next to a wasps' nest to ward off predators.

A more unusual nesting behavior is that of the Piratic flycatcher. This species occupies the active nests of other birds, often preferring the hanging bag nest of caciques (*Cacicus* spp.). It expels the original inhabitant even if it is brooding and removes the eggs.

Tyrant flycatchers generally live in territories that they defend vigorously, and those species that are migratory often also establish territories in their nonbreeding ranges. In some cases, migration is accompanied by a dramatic shift in behavior, as in the Eastern kingbird. On its breeding grounds in North America this bird is insectivorous, noisy, and aggressively territorial in its open scrubland habitat. Wintering in the Amazonian forests, however, it becomes silent and gathers in large, nomadic flocks that forage mainly for fruit.

The Risks of Restricted Ranges

CONSERVATION AND ENVIRONMENT

Tyrant flycatchers occupy almost every habitat in the Americas. In some areas, several species of flycatcher will occupy the same habitat. Competition between species is reduced by differences in species size, foraging techniques, size and type of food, foraging and nesting location, territory size, and preferred type of vegetation.

A number of species that have adapted to very particular and restricted range habitats are also the most vulnerable to population declines. Generally, all species that are declining are affected by habitat loss, but the declines are more acute for restricted-range species in areas with already limited habitat. These include species (or subspecies) found on small islands, like the Cocos flycatcher, which is restricted to the Cocos Islands off the coast of Costa Rica, and the Noronha elaenia, found only on the Fernando de Noronha archipelago off Brazil. Several of the most vulnerable flycatchers occur in southeastern Brazil. Always rare due to their extremely limited range, species such as the Minas Gerais, Alagoas, and Bahia tyrannulets and the Fork-tailed and Kaempfer's tody-tyrants are further threatened by rapid deforestation.

A more general conservation issue, as for most birds, is the loss and fragmentation of habitats over a broad area. Such developments have led to rapid population declines in species that were originally not uncommon, among them the Ochraceous attila, the Cock-tailed tyrant, and the Grey-breasted flycatcher. Finally, some species, including the White-tailed shrike-tyrant and Giant kingbird, seem to be declining for no readily apparent reason, although in their case too habitat modification and fragmented populations may well be significant factors. DA/RSR/JWF

Cotingas

THE NEOTROPICAL COTINGAS ARE AMONG THE *most diverse of all bird groupings. They exhibit the largest variation in body size of any avian family, ranging from the diminutive Kinglet calyptura at about 8cm (3in) to the crow-sized umbrellabirds. Cotingas vary in their ecology from insectivores to specialist frugivores. In plumage, they range from the drab to the spectacular, and in their breeding systems they include species that form both monogamous pairs and polygynous leks.*

Although the precise limits of the family have long been disputed, the monophyly of cotingas, with a few additions to and deletions from the traditional list of members, is supported by both anatomical and molecular evidence. Along with their close relatives the manakins and tyrant flycatchers, the cotingas are a fascinating part of the great Neotropical radiation that makes the avifauna of the American tropics the most diverse in the world.

FACTFILE

COTINGAS

Order: Passeriformes

Family: Cotingidae

94 species in 33 genera

DISTRIBUTION Mexico, C and S America.

HABITAT Forest, at all levels from tropical to temperate montane.

SIZE Length 8–50cm (3–20in); weight 6–400g (0.2–14oz).

PLUMAGE Extremely varied; males of many species brilliantly colored with reds, purples, blues, etc., and unusually modified display plumage; females usually duller, without ornamentation.

VOICE Very varied, ranging from sharp whistles and rapid trills to booming sounds and hammer- or bell-like clangs; some species make mechanical sounds with modified wing-feathers.

NEST Mainly open cup- or saucer-shaped nests, some very small and frail for the size of the bird. (Cocks-of-the-rock are exceptional in building cup nests of mud stuck on rockfaces.)

EGGS 1–3; buff or olive base, with patches of darker browns and grays. Incubation 19–28 days; nestling 21–44 days.

DIET Fruits and insects

CONSERVATION STATUS The Kinglet calyptura, restricted to one locality in Brazil, is Critically Endangered. In addition, 5 species are Endangered, and 11 are Vulnerable.

See Cotinga Subfamilies ▷

*◀ **Left** Sometimes known as the calf-bird for its lowing call, the capuchinbird owes its more familiar common name to its extraordinary coat of feathers, reminiscent of a monk's cowl and habit.*

The Four Main Groupings

FORM AND FUNCTION

The cotingas are most closely related to the manakins (Pipridae), with which they share many natural-history features. Together, the cotingas and manakins are the sister group to the diverse tyrant flycathers. The cotingas are distinguished by unique derived features of the syrinx (the avian vocal organ), among other anatomical features, and the monophyly of the group is also supported by molecular systematic studies. Some genera such as the *Rhytipterna* mourners that have traditionally been included in the cotingas are now known to be tyrant flycatchers, but others have recently been returned to the cotingas from both the tyrant flycatchers (the becards, tityras, and the *Laniocera* mourners) and the manakins (the *Schiffornis* mourners).

The cotingas are made up of four subfamilies. The first consists of the "core cotingas" (Cotinginae), including many lowland rainforest-canopy species (*Cotinga*, *Xipholena*, *Gymnoderus*, *Carpodectes*, *Lipaugus*, and others), as well as the bellbirds and fruitcrows. The second (Rupicolinae) includes the lekking cocks-of-the-rock, the red cotingas, and the mostly Andean fruiteaters. The third subfamily (Phytotominae) includes the other Andean cotingas (*Ampelion*, *Doliornis*, and *Zaratornis*), and the plantcutters, from both the Andes and temperate South America. The fourth subfamily (Tityrinae) is a diverse assemblage of genera that were formerly placed in different parts of the cotinga, manakin, and tyrant flycatcher grouping, including tityras, mourners, becards, and purpletufts.

Because of their ecological and behavioral diversity, cotingas vary extensively in body size, plumage, and morphology. In their proportions, they range from short-winged, heavily-built birds – for example the Andean fruiteaters (*Pipreola*) – through the huge, broad-winged umbrellabirds to the small, long-winged, almost swallowlike purpletufts. In color they extend from highly dimorphic species with brilliantly ornamented males to species in which both sexes are uniformly gray or brown. The brilliant colors include vivid red, yellow, pink, and purple hues created by carotenoid pigments, and blue hues that are structurally formed by interference among light waves scattered off air bubbles in the feather barbs, as in the eponymous *Cotinga* species.

Many species show elaborate, sexually dimorphic plumages and specialized feathers, such as the crests on male cocks-of-the-rock. An example of plumage modification that appears in both sexes is the long, forked tail of the highly aerial Swallow-tailed cotinga, which is produced by feathers of differing lengths. Various species have evolved fleshy wattles, feathered carbuncles, or bare skin patches that are associated with courtship display, as in the capuchinbird, the bellbirds, and the umbrellabirds. A few species, such as the red cotingas and the cocks-of-the-rock, have specialized flight feathers that are used for the production of mechanical sounds during courtship displays. Other species, including the Elegant mourner, the Scimitar-winged piha, the becards, and the tityras, have oddly-shaped primaries whose function is unknown.

Cotingas vary extensively in bill size and shape. Many frugivorous species have a wide gape to facilitate the swallowing of large fruits. The plantcutters have a conical bill with a serrate tomium (cutting edge) that they use for feeding on buds, leaves, fruits, and seeds.

The syrinx, or vocal organ, of the cotingas is also extremely variable. The family includes some of the simplest and the most complex syringes found among all birds. Many species have extremely basic syringes, with no specialized supporting elements or intrinsic muscles. The resulting vocalizations are simple as well. Other genera that have evolved loud vocalizations for their polygynous displays have highly-derived syringes with many unique features. Fruitcrows have an enlarged trachea and bronchi that create a resonance for low, booming calls. The bellbirds have a huge and highly muscular syrinx that creates some of the loudest sounds in the bird world.

The diets of cotingas are very diverse. The core cotingas, Andean cotingas, and the Rupicolinae are largely frugivorous. Smaller species eat a variety of small fruits, whereas larger species often specialize on larger fruits in the avocado family (Lauraceae). The White-cheeked cotinga specializes on High-Andean mistletoe berries. The plantcutters feed on plant buds, fruits, and seeds with their serrate, conical bills. In contrast, the tityras, mourners, becards, and sharpbill are extensively insectivorous. As in manakins, birds of paradise, bowerbirds, and others, frugivory is hypothesized to be associated with the evolution of polygynous breeding systems because it may help release males from parental care.

Above *The fruiteaters are a group of relatively short-winged, heavy-bodied cotingas from the Andes and adjoining regions that gather fruit while perched. This is a Barred fruiteater* (Pipreola arcuata).

Below *The Red-ruffed fruitcrow* (Pyroderus scutatus) *is commonly known as the Indian crow. Its red-tipped feathers were once highly prized by anglers for making flies to lure salmon.*

From the American Tropics

DISTRIBUTION PATTERNS

Cotingas are exclusively Neotropical, ranging from northern Mexico to southern South America. The Rose-throated becard is a regular but rare nesting species in the extreme southwestern United States. Most species are found in humid tropical rain forests, humid montane forests, or upper

Right *The extraordinary half-moon crest of male cocks-of-the-rock – here the Guianan species is shown – plays a central role in courtship. These polygynous birds play no further part in breeding once they have mated.*

montane forests. In contrast, plantcutters are found in shrub-dominated open country and crop lands of the Andes and southern South America.

Numerous genera contain groups of allopatric or parapatric species (respectively, occupying geographically separate, or else adjoining but distinct, ranges) that replace one another across the Neotropics. Some species are narrowly endemic to very small distributions; for example, the Yellow-billed and Turquoise cotingas are found only in western Costa Rica and Panama, while the Kinglet calyptura and the Black-and-Gold and Grey-winged cotingas are restricted to southeastern Brazilian coastal mountains.

One or Many Mates

BREEDING BIOLOGY

The breeding behaviors of the cotingas include a wide variety of those found among all birds. Andean cotingas, plantcutters, fruiteaters, tityras, and becards are apparently monogamous, with biparental care. The Purple-throated fruitcrow is highly social and lives in mixed groups of males and females. One female apparently builds the nests and does all the incubation, but all individuals cooperate to defend the territory and bring insects to the single nestling. Apparently, there is a single main pair with additional helpers of either sex. It is not known whether the helpers are young of the main pair born in previous years. The purpletufts appear in social groups and may have a similarly cooperative breeding system.

Though much remains to be learned about their breeding behavior, most cotinga species are apparently polygynous, with exclusively female parental care. Males of polygynous species often gather to perform courtship displays in leks. Some species have highly concentrated traditional leks; for example, male Guianan cocks-of-the-rock display in large groups of 10 to 20 birds or more. Each male maintains a cleared "court" on the forest floor, a system similar to that of the White-bearded manakin (*Manacus manacus*). The main courtship display is static: the male crouches in the middle of his court with his brilliant plumage spread and the head turned sideways so that the semi-circular topknot is fully displayed to the females who come into the trees above the lek. Other cotinga species have highly dispersed or solitary display areas. Males of some canopy species (for example, *Cotinga*, *Xipholena*, and *Carpodectes*) aggregate in specific treetops or adjacent trees, but these breeding systems are harder to study and are less well-known.

Cotinga display behavior can include physical elements, such as postures and stylized movements in the cocks-of-the-rock and red cotingas, or primarily vocal advertisements. Strong sexual selection in some groups, such as the bellbirds and in the Screaming piha, has led to the evolution of some of the loudest vocalizations in all birds. Bellbird vocalizations, which vary from a loud *bock* to a ringing electronic *gong-geeh*, can be heard well over 1km (0.6 miles) away. The larger fruitcrows have low, mooing calls, while the capuchinbird gives a loud whine that crescendoes like a revving chain saw or motorcycle. The Black-and-gold cotinga, which inhabits montane forests of southeastern Brazil, utters a beautiful and ethereal pure-tone whistle that slowly rises in pitch over a timespan of three seconds. Often groups of males will sing antiphonally, producing a continuous ringing whine characterized by undulating beat frequencies

Cotingas also exhibit a surprising variety of nest structures. The cryptically-colored females of many species build simple platform nests of sticks that they tend single-handed, laying a single egg.

Above *Locked together in combat, two male Andean cocks-of-the-rock* (Rupicola peruviana) *struggle for dominance by grappling with their feet. The birds compete for mates at display-grounds known as leks.*

Right *The three umbrellabirds* (Cephalopterus *spp.*) *are characterized by an exuberant quiff of head feathers and a long lappet, bright red in one species and black in the other two. This individual is an Amazonian umbrellabird* (C. ornatus).

Left *Mouth wide agape as it utters its cry, the Bare-throated bellbird* (Procnias nudicollis) *is one of four species reputed to be the world's loudest birds. It broadcasts its two-note call far and wide from the rainforest canopy.*

Cotinga Subfamilies

"Core" cotingas, Bellbirds, & Fruitcrows
Subfamily Cotinginae

36 species in 14 genera, including: bellbirds (genus *Procnias*), capuchinbird (*Perissocephalus tricolor*), umbrellabirds (genus *Cephalopterus*), Banded cotinga (*Cotinga maculata*), Turquoise cotinga (*C. ridgwayi*), Yellow-billed cotinga (*Carpodectes antoniae*), Purple-throated fruitcrow (*Querula purpurata*), Scimitar-winged piha (*Lipaugus uropygialis*), Screaming piha (*L. vociferans*).

Cocks-of-the-rock, Red cotingas, & Fruiteaters
Subfamily Rupicolinae

16 species in 4 genera, including: Guianan cock-of-the-rock (*Rupicola rupicola*), Guianan red cotinga (*Phoenicircus carnifex*), fruiteaters (genus *Pipreola*).

Plantcutters & Andean cotingas
Subfamily Phytotominae

13 species in 8 genera, including: Peruvian plantcutter (*Phytotoma raimondii*), sharpbill (*Oxyruncus cristatus*), White-cheeked cotinga (*Zaratornis stresemanni*).

Becards, Tityras, Mourners, & Purpletufts
Subfamily Tityrinae

29 species in 7 genera, including: becards (genus *Pachyramphus*) including Rose-throated becard (*P. aglaiae*), tityras (genus *Tityra*), mourners (genera *Laniisoma* and *Laniocera*) including Elegant mourner (*Laniisoma elegans*), purpletufts (genus *Iodopleura*).

Note: The position of several genera (e.g. *Tijuca*, *Phibalura*, *Calyptura*) is unknown; most classifications therefore rank them as 'incertae sedis.'

The nest is so insubstantial that the egg is often visible through the bottom. The inconspicuous nature of the nests is thought to be an adaptation to minimize the risk of predation.

Mourners, fruiteaters, and Andean cotingas build more substantial cup nests. The female cock-of-the-rock builds her nest of mud and rootlets, hardened with saliva, against a vertical rockface, perhaps as another form of protection. (As a result, cocks-of-the-rock are limited in distribution within humid forests to areas with boulders, rock faces, or cliffs.) The becards build woven, pendent nests, which they keep adding to throughout incubation, while the tityras nest in cavities. The purpletufts build a tiny cup nest on the upper surface of a branch that is reminiscent of a hummingbird's nest.

The Least-known Species

CONSERVATION AND ENVIRONMENT

Many cotinga species are poorly known, and their conservation status is not well understood. A number, however, are threatened or endangered. The Banded cotinga, which is endemic to the humid lowland forest along the Atlantic seaboard of Brazil, has become endangered through deforestation; only small, isolated pockets of suitable habitat now remain. Populations of the Peruvian plantcutter, which inhabits patchy woodlands and scrub in the coastal desert of northwestern Peru, have declined in recent decades, and the species is now threatened by cutting for firewood and agricultural development. The Grey-winged cotinga (*Tijuca condita*) is known only from the upper slopes (above 1,500m/5,000ft) of the Serra dos Orgãos in southeastern Brazil. Although its range lies within 50km (30 miles) of Rio de Janeiro, a large part of the area is a protected national park. Little is known about the health of the bird's populations, but it is officially regarded as Vulnerable, largely due to its extremely small range and the risk posed by forest fires.

Perhaps the least known cotinga is the Kinglet calyptura (*Calyptura cristata*), which was often represented in scientific collections made during the 19th century in Rio de Janeiro state. This diminutive, greenish bird, with white wing bars and a colorful, red, kingletlike crown, was not seen for more than a century, and it was feared that deforestation had driven it to extinction; however, in 1996 it was reliably spotted by observers at Teresopolis-Garrafão, just south of the Serra dos Orgãos National Park. It has not been spotted since. ROP/DWS

Manakins

Right *As its name suggests, the Wire-tailed manakin is the only family member whose tail feathers end in long filaments. These appendages are used in a tactile way, to stimulate the female during courtship.*

Left *Some ornithologists regard* Piprites *species, such as the endangered Black-capped manakin (*P. pileatus*) of southeastern Brazil, as being separate from the manakins.*

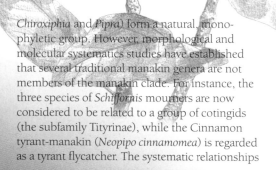

THE MANAKINS ARE A NEOTROPICAL FAMILY that are well known for their brilliant sexual dimorphism, lek breeding behavior, and elaborate courtship displays. They are most closely related to the cotingas (Cotingidae), with which they share various natural-history features.

Typical manakins are small, compact, highly active birds with short bills, large heads, broad, rounded wings, and short tails. Most species are strongly sexually dimorphic in plumage. Males display a wide variety of brilliant colors and bold patterns, and in several species have long, elaborate tail feathers, erectable throat patches, or plush crowns that curl forward over the bill. In contrast, females of all species are colored cryptic shades of green, making them some of the most challenging avian species to identify.

Colorful Males, Drab Females

FORM AND FUNCTION

Manakins are distinguished from cotingas and tyrant flycatchers by derived features of the syrinx and other morphological features, by molecular characters, and by numerous life-history traits. The majority of manakin genera (for example *Chiroxiphia* and *Pipra*) form a natural, monophyletic group. However, morphological and molecular systematics studies have established that several traditional manakin genera are not members of the manakin clade. For instance, the three species of *Schiffornis* mourners are now considered to be related to a group of cotingids (the subfamily Tityrinae), while the Cinnamon tyrant-manakin (*Neopipo cinnamomea*) is regarded as a tyrant flycatcher. The systematic relationships

Left *A female White-bearded manakin exhibits the green cryptic coloration that allows many female piprids to blend in with their forest surroundings.*

FACTFILE

MANAKINS

Order: Passeriformes

Family: Pipridae

50 species in 13 genera. Species include: **Araripe manakin** (*Antilophia bokermanni*), **Helmeted manakin** (*A. galeata*), **Blue-backed manakin** (*Chiroxiphia pareola*), **Long-tailed manakin** (*C. linearis*), **White-bearded manakin** (*Manacus manacus*), **Wied's tyrant-manakin** (*Neopelma aurifrons*), **Wire-tailed manakin** (*Pipra filicauda*).

DISTRIBUTION C and S America

HABITAT Forest, at tropical levels.

SIZE Length 9–19cm (3.5–7.5in); weight 10–25g (0.35–0.9oz).

PLUMAGE Males of most species brightly colored, black with patches of red, orange, yellow, blue, or white; females olive-green. (A few species are mainly olive-green or brown, with no difference between sexes.)

VOICE Variety of sharp whistles, trills, buzzing notes; no true songs; some species also make loud mechanical sounds with modified wing feathers.

NEST Open cups, usually in low vegetation.

EGGS Almost always 2; dull white or buff with brown markings (blackish in the Thrush-like manakin). Incubation 17–21 days; nestling period almost always 13–15 days.

DIET Small fruits; occasional insects.

CONSERVATION STATUS The Araripe manakin is listed as Critically Endangered, and Wied's tyrant-manakin is Endangered. 2 other species are Vulnerable.

seasonal altitudinal migrations, but lowland species are permanent residents. Many species of manakin have closely related allopatric or parapatric species that geographically replace them across the Neotropics.

The Switch to Fruit
DIET AND FEEDING

Manakins are extensively frugivorous. They live in the forest understory, feeding mostly on small fruits and occasional insects. Intriguingly, they pluck fruits in rapid flight sallies, apparently having evolved from a diet of insects to a derived one of fruits while retaining their primitive foraging behavior of pursuing their botanical "prey" in flight. Females have large home ranges and are not exclusively territorial. Their frequent travel among fruit sources ensures that they are commonly caught in mist-net surveys in tropical forest. Males spend a great deal of their time at traditional display sites, and during the breeding season they

Left *Seen from behind, the Blue-backed manakin reveals the mantle of sky-blue feathers to which it owes its name. In fact, the bird's red head-shield plays a more prominent part in its lekking display* (see box).

Below *A young female White-crowned manakin* (Pipra pipra). *This nominate species of the family's most speciose genus is widespread across South and southern Central America.*

of the Broad-billed sapayoa (*Sapayoa aenigma*) are enigmatic, but clearly lie outside of the manakins (see Asities).

All sexually dimorphic species of manakins exhibit delayed plumage maturation in sexually mature males, which in their first year have green, female-like plumage. Several lineages have also evolved distinctive subadult male plumages that are worn for 2–4 years until the birds acquire adult male plumage.

Anatomically, manakins are quite diverse. A few species that make mechanical wing-sounds during courtship display have modified primary and secondary feathers, wing musculature, and skeletal elements associated with sound production. Manakins are exceptional in their variation in syringeal morphology (the syrinx being the avian vocal apparatus). All genera and many species can be identified by their syrinx alone (in comparison, few oscine families could be so distinguished).

Neotropical Forest-dwellers
DISTRIBUTION PATTERNS

Exclusively Neotropical, the manakins are distributed from southern Mexico to southern Brazil, Paraguay, and northern Argentina. They are found predominantly in lowland humid tropical forests, but a few species are found in humid montane cloud forest, lowland seasonal tropical dry forests, and riparian gallery forest. As with many Neotropical frugivores, some montane species undergo

usually spend less than 10 percent of the day foraging. Manakins are semi-social at feeding sites and will sometimes join foraging flocks comprising a number of different species.

The ornithologist David Snow has hypothesized that the frugivorous manakin diet fostered the evolution of lek polygyny. Insects are often cryptic, toxic, and difficult to capture and handle. In contrast, fruit is advertised, abundant, and easy to find and process. If a species can evolve to raise its young largely on fruit, natural selection to reduce predation at the nest may select for reduced parental activity, resulting in a small clutch size and the release of males from parental care. Other examples of the association between frugivory and lek polygyny include many cotingas, the birds of paradise, the bowerbirds, and the Velvet asity of Madagascar.

Showing Off to the Females

BREEDING BIOLOGY

Manakins are well known for their lek breeding behavior. Males defend non-resource territories that vary in size from several to tens of meters in diameter. Male lek territories may be closely aggregated into concentrated leks or loosely grouped in dispersed leks. Females conduct all the parental-care duties, and males contribute only sperm to reproduction. Females choose mates from among the available males. Females prefer a few specific males, resulting in strong sexual selection by female choice. Lek behavior is primitive to the family, and the consequent persistent sexual selection has contributed to an elaborate radiation in secondary sexual characters.

Male lek displays vary tremendously among different species and genera, but the patterns in display-behavior evolution closely match the phylogenetic history of the family. In full activity, a large lek of the White-bearded manakin, one of the best-known species, is an extraordinary sight. Each male clears a small "court" on the ground, within a few meters of its neighbors, and performs on and round it an astonishing range of rapid maneuvers, accompanied by sharp calls and loud snaps of the modified wing-feathers. The females visit many leks to choose a mate, but thereafter carry out all nesting duties single-handed. Females sling their delicate cup nests between two parallel or diverging twigs of some low plant, often beside a forest stream; after an incubation period of about 19 days (unusually long for a small bird), they feed the young by regurgitation on a mixed diet of insects and fruits. Nests are not located near male leks.

Manakins of the largest genus, *Pipra*, display on higher perches, usually 3–10m (10–33ft) above ground, on which they perform rapid slides, about-faces, twists, and other maneuvers. The details vary according to species, but a swift flight to the display perch, ending in a conspicuous landing, is a feature of all of them. A unique feature of the display of the Wire-tailed manakin brings into play the elongated, wirelike filaments that project from the tips of its tailfeathers. Backing toward the female, the male raises its posterior and rapidly twists its tail from side to side, so that the filaments brush the female's chin.

The Helmeted manakin is the only manakin known to form a pair bond. Males defend territories in which females nest. Males perform indirect parental care in defending the territory, but do not attend the nest or young. This behavior, unusual in the manakin family, is apparently secondarily derived from lek polygyny. ROP/DWS

Left *The Club-winged manakin* (Machaeropterus deliciosus) *is best known for the unique mechanical sounds it produces with its modified secondary feathers.*

THE CATHERINE-WHEEL COURTSHIP OF CHIROXIPHIA MANAKINS

The courtship displays of the five *Chiroxiphia* species, including the Blue-backed and Long-tailed manakins, are among the most spectacular in the bird world. Males in this genus have a nearly unique form of cooperative lek system, displaying in pairs or trios.

The whole courtship sequence takes three distinct phases. In phase one, two males perch side by side in a tree, facing the same way and almost in contact, and utter long series of almost perfectly synchronized whistled calls. The calls are so well coordinated that they seem to come from a single bird, but in fact one bird, the dominant member of the pair, begins each note about one-twentieth of a second before the other. The function of this duet is to attract a female.

Phase two begins when a female approaches. The two males fly down to a special display perch in thick vegetation near the forest floor. Perching side by side they begin to jump up alternately, rising a few centimeters in the air and accompanying each jump with a nasal, twanging call. The female may then come to the display perch, in which case the two males turn to face her and continue their coordinated dance in a different form. The male nearer the female jumps up, facing her, and then moves back in the air, hovering, to land behind the second male, who hitches forward and in turn jumps up and moves back in hovering flight. The two males thus form a revolving Catherine wheel in front of the female. As it proceeds, the Catherine wheel dance becomes more and more rapid and the twanging calls more frenzied, until the dance is brought to a sudden end by the dominant male, who utters one or two very sharp calls, whereupon the subordinate male leaves the perch.

During phase three, the dominant male performs an aerial display around the female, criss-crossing the display perch, every now and then perching, crouching to present his red head-shield, then flying on to continue his butterfly flights. Occasionally he flies to an outlying perch and crouches there until, with a snap of his wings, he flies back toward the display perch. If the female remains on the display perch, showing that she is ready to mate, the male eventually lands beside her and mounts her.

Males in partnerships are unrelated. Male *Chiroxiphia* take three or more years to develop definitive adult plumage. Young males usually join a display partnership only after a few more years of competition, and then as a subordinate. Subordinate males later inherit display sites from dominant males. Since females make their mate choices from among many available partnerships, a very small fraction of males in the population father all the offspring in every generation. As a result, male mating success is more skewed in the Long-tailed manakin than in any other wild animal population that has been measured.

Ovenbirds

The ovenbirds take their name from the remarkable nest of the Rufous hornero, in Brazil aptly called the Mud John (João-de-Barro). The family name Furnariidae comes from furnarius, *Latin for "furnace" and the word from which the Spanish* hornero *(literally, "baker") is also derived.*

Made of mud strengthened with grasses and fibers (anything from plant matter or hair to strings and plastic), ovenbirds' nests have the rounded shape of old-fashioned baker's ovens. A narrow entrance leads to a chamber about 20cm (8in) wide that is lined with softer plant fibers (and occasionally plastics). One penalty of building this curious cave is that in some regions the ovenbird has become host to a bedbug of the family Cimiccidae, and may also host various flies that eat both feces and living nestlings' flesh.

Because of their unique nests, several legends have arisen concerning these birds. One maintains that the openings of all ovenbird nests are oriented in the same direction – a claim that can be simply disproved by a quick glance at several nests in a given area. Some people nonetheless assert that the shared orientation is indeed real, but refers to climatic conditions (prevailing winds, for example) at the time of nest building – a point that has yet to be tested, much less proven.

In Brazil, where the bird is very common, rural legend states that males seal females in the nest during incubation to ensure their fidelity. This myth may have been inspired by the fact that, during nesting, one individual of a pair is often in the nest at any given time while the other searches for food on its own. Since pairs usually forage together, it would have been easy to conclude that the other was a prisoner within the nest. However, observation of marked birds clearly shows that pairs in fact take turns in the nest.

Diverse but Drab

FORM AND FUNCTION

Ovenbirds are one of the most species-rich families of birds, and the most diverse, although they vie with the Dendrocolaptidae woodcreepers for the title of the most drab South American bird family. Their geographic distribution is exceptional in that their greatest diversity is found south of the Tropic of Capricorn; for example, the state of Rio Grande do Sul in Brazil (bordering Uruguay) has more Furnariid species than the much larger, more tropical state of Minas Gerais. They are found in a variety of habitats, some occupying niches that in Europe, Asia, or North America would belong to such groups as the tits, larks, wheatears, dippers, or nuthatches.

On the basis of ecology and behavior, three subfamilies are recognized: Furnariinae, Phylidorinae, and Synallaxinae. The subfamily Furnariinae includes the South American miners – terrestrial birds that usually inhabit open, arid lands, where they walk or run but rarely fly. They resemble drab wheatears. Earthcreepers are similar, but are even duller in color and have longer tails and lengthier, more downcurved bills; unlike the miners, they may be closely associated with water. The dipper-like cinclodes are much more strictly linked to water, and a few are even partly marine: Surf cinclodes, for example, rarely leave the water's edge, where they feed. All these species nest between rocks or in holes in the ground, either digging their own burrow or using the holes of other birds or rodents. Burrows can be up to 1.2m (4ft) long.

The true ovenbirds of the subfamily Phylidorinae are birds of savanna-type habitats, and in

Above *The Blackish cinclodes* (Cinclodes antarcticus), *also known as the Tussock bird, is one of the southernmost ovenbirds, restricted to the Falkland Islands.*

Below *Representative species of ovenbirds:* **1** *Rufous hornero* (Furnarius rufus), *with oven-shaped nest in the background;* **2** *Scale-throated earthcreeper* (Upucerthia dumetaria).

Above *Seen here in the Llanos grasslands of Venezuela, the Yellow-throated spinetail* (Certhiaxis cinnamomea) *is a small, inconspicuous ovenbird that likes wet places and is best known for its lengthy song.*

general they keep to open valleys and floodplains; many species may now, however, be expanding their geographic ranges in the wake of deforestation. The Rufous hornero, the best-known species, has adapted well to humans, and is common in farms and cities over much of southern South America. The bird's range is expanding as forests are cut down to make way for agriculture, parks, and lawns.

The subfamily Synallaxinae includes the spinetails, canasteros, and thornbirds. These are mostly rather small birds with longish, more or less graduated tails that are often forked. The tails show great variation: that of the remarkable Des Murs's wiretail has only six main feathers, the short outer pair being hidden in the tail coverts, while the middle and inner pair are very long but reduced to

FACTFILE

OVENBIRDS

Order: Passeriformes

Family: Furnariidae†

217 species in 58 genera.

DISTRIBUTION Central Mexico to S South America; Trinidad and Tobago, the Falkland and Juan Fernandez Islands.

HABITAT Virtually all encountered in the Americas, from deep forest to open arid land and from sea level to the snow line.

SIZE Length 10–26cm (4–10.3in); weight 9.5–90g (0.3–3.2oz).

PLUMAGE Generally somber browns, often rufous on the head, wings, or tail; underside streaked, spotted, or plain, often pale. A very few species are strikingly patterned, but the colors are still variations of brown.

VOICE Not noted for their musical qualities, but quite variable, generally resonant, including harsh rattles and creaks, screams, clear notes, whistling trills, etc. Some (perhaps all) species show sexual differences in songs.

NEST Among the most variable within any bird family, ranging from mud chambers to huge bundles of twigs; burrows, in which a nest is built with materials ranging from soft fibers to sticks, are also common.

EGGS 2–5, usually white, can be off-white or blue. Incubation 15–22 days.

DIET Mainly insects and other invertebrates, but may also consume small vertebrates, such as small lizards (e.g. *Anolis* and *Norops* spp.).

CONSERVATION STATUS Of 45 species listed by the IUCN, 3 – the Pernambuco spinetail, the Alagoas foliage-gleaner, and the Royal cinclodes – are Critically Endangered. In addition, 9 are classed as Endangered, 15 Vulnerable, and 18 Lower Risk/Near Threatened.

† Sometimes includes the family Dendrocolaptidae, which is treated here as a separate family (see Woodcreepers).

See Selected Ovenbird Species ▷

little more than the central shaft. Members of this subfamily mostly inhabit dense vegetation in forest edges, reed beds, scrubland, grassland, or even mangrove swamps, although a few species occur in barren areas and some others live in forests.

Nest Specialists

BREEDING BIOLOGY

The Furnariidae as a whole are famous for their unusual nests, but within the Synallaxinae nest-making is taken to extremes. The Wren-like rushbird weaves a sphere of clay-daubed grass around growing reeds. An entrance near the top is protected by a woven awning, and sometimes even by a hinged, woven trapdoor. Often a depressed clay platform on top acts as a singing perch. Other ball-like nests of grass and other materials are built near or on the ground.

The nest of the Red-faced spinetail is also globular, but is suspended from the tip of a slender hanging branch. Some species nest on a branch and enter from below onto a platform lined with feathers; others build a nest 30cm (12in) in diameter, with a side entrance giving into a small space connected by a tortuous tunnel to a larger chamber near the nest's top. There are also various forms of thorn nests featuring tunnels to the nest-chamber. *Synallaxis* spinetails may adorn their nests with owl pellets or dried carnivore feces (including those of cats or dogs). The current interest in nest predation and its evolutionary influences on nesting behaviors suggests the idea that building nests with animal feces or bones may play an important role in predator avoidance.

Cordilleran canasteros build large, exposed baskets of thorny twigs, shaped into vertical cylinders; where there are no thorny twigs, nests may be built into cactus plants. Common thornbirds construct a large, unkempt-looking mass of thorny twigs divided into two chambers. In subsequent seasons other chambers are added, to produce what appears at first sight to be a colonial nest, although in reality it is a mere accumulation of nests that may grow to be several meters long! It is not likely that more than one breeding pair of Common thornbirds ever occupy such a nest at any one time, but the surplus chambers may provide dormitories for nonbreeding members of a previous brood, or nests for other species of ovenbirds, or even for birds of other families.

The firewood-gatherer also builds a voluminous, thorny nest, and often incorporates various kinds of debris, including bones, metal, and colored rag, again perhaps to discourage predation. The lining of the neatly arched tunnel may incorporate fragments of bark or snakeskin as well as snail or crab shells. In southern Brazil the birds nest in the Araucaria pine (*Araucaria angustifolia*), whose extremely tough and pointed leaves may also serve to repel predators. Since the group is poorly known, none of the various predation-avoidance hypotheses has been scientifically tested.

Members of the subfamily Phylidorinae usually live in trees and make simpler nests. (Cachalotes are the only members to build huge, thorny nests, those of the White-throated cachalote being up to 1.5m/5ft in diameter.) Many nest in bankside tunnels that may be up to 1.8m (6ft) in depth and quite tortuous in layout. Nest-chambers may contain either well-woven nests or else small collections of loose leaves, plant fibers, or rootlets. Birds may dig their own tunnel or improve on an already existing one. Many use rock or tree fissures or holes, in which only a simple nest is built.

While standard ovenbird nests are almost always well enclosed, this general rule has its exceptions. Bay-capped wren-spinetails typically build flat, open nests of grass, lined with feathers, a few centimeters above the water in a reed bed. However, even this nest may end up enclosed, as occasionally the rim is so well developed as to leave only a small hole at the top. Some canasteros lay eggs in what is virtually a well-concealed scrape on the ground, sometimes lined with a few fragments of bone and fur from owl pellets. The very rare Rusty-backed spinetail makes a crude chamber in a tangle of drift vegetation trapped by a branch during floods, or uses the abandoned nest of another ovenbird. Tit-spinetails (*Leptasthenura* spp.) will also take over abandoned nests, but will use a variety of other sites, including holes in cacti. Clutch size in this family probably varies widely, from 2 to 5 eggs, with 4 eggs perhaps being common. If so – and data are lacking – then ovenbirds do not follow the trend common to many tropical passerines of reduced clutch size near the equator.

Comparison with another tropical bird family, the Formicariidae antbirds, suggests that of the two the Furnariidae have (possibly) lower nest predation rates, along with larger clutch sizes and relatively complex nests. This fact again suggests that the unique nest-building behaviors of the ovenbirds may serve to discourage predation.

The Threatened Minority

CONSERVATION AND ENVIRONMENT

The Furnariidae include several species that raise important conservation issues. Perhaps the rarest is the Alagoas foliage-gleaner, which was first described in 1983 and is only known from a small forest remnant in the state of Alagoas, Brazil. Another 26 species are also considered threatened or endangered, with habitat loss forming the major problem for these, as for many other, species of birds. JJR/AMH

Selected Ovenbird Species

Genera and species include: canasteros (genus *Asthenes*), including Cordilleran canastero (*A. modesta*); cinclodes (genus *Cinclodes*), including Chilean seaside cinclodes (*C. nigrofumosus*), Royal cinclodes (*C. aricomae*), Surf cinclodes (*C. taczanowskii*); earthcreepers (genera *Upucerthia*, *Eremobius*), including Scale-throated earthcreeper (*U. dumetaria*); South American miners (genera *Geobates*, *Geositta*); spinetails (genera *Synallaxis*, *Schoeniophylax*, *Hellmayrea*, *Cranioleuca*, *Certhiaxis*, *Siptornopsis*), including Creamy-crested spinetail (*Cranioleuca albicapilla*), Red-faced spinetail (*C. erythrops*), Rusty-backed spinetail (*C. vulpina*), Pernambuco spinetail (*Synallaxis infuscata*); thornbirds (genus *Phacellodomus*), including Common thornbird (*P. rufifrons*); tit-spinetails (genus *Leptasthenura*); Alagoas foliage-gleaner (*Philydor novaesi*), Bay-capped wren-spinetail (*Spartanoica maluroides*), Des Murs's wiretail (*Sylviorthorhynchus desmursii*), Eye-ringed thistletail (*Shizoeaca palpebralis*), firewood-gatherer (*Anumbius annumbi*), Great xenops (*Megaxenops parnaguae*), Rufous cachalote (*Pseudoseisura cristata*), White-throated cachalote (*P. gutturalis*), Rufous hornero (*Furnarius rufus*), Thorn-tailed rayadito (*Aphrastura spinicauda*), Wren-like rushbird (*Phleocryptes melanops*).

Below In Brazil, a Rufous hornero surveys the world from the security of its solidly-built nest. Ovenbirds take their name from these extraordinary constructions, which bear a marked resemblance to old-fashioned bakers' ovens.

Right A Bay-capped wren-spinetail shelters among long grasses on the Argentinian pampas. This monotypic species, which is thought to be closely related to the Synallaxis spinetails, is found in the marshes and bushy meadows of lowland central South America.

Woodcreepers

Formerly known as woodhewers, the woodcreepers of the family Dendrocolaptidae are tree-climbing birds of Central and South America, where they occur primarily in forested lowlands. They reach their maximum diversity in the Amazon basin, with relatively few species extending their range into adjacent mountains or even into Central America.

As their name would suggest, most woodcreepers live exclusively in trees, where they get about by using their tail as a brace while grasping the bark with their strong legs and feet. They generally work their way up trees and out onto branches in a spiraling motion, but are more adept at backing down trunks than are woodpeckers. Upon reaching the upper limit of their desired height, they fly down to the base of another tree and start upward again, much in the manner of treecreepers. Exceptionally, a few woodcreeper species spend considerable time feeding on or very close to the ground.

The affinities of woodcreepers with the ovenbird family are unclear because there are a few species that appear intermediate between the "typical" woodcreepers and the ovenbirds. Although woodcreepers are regarded by many as sufficiently distinct to warrant a separate family, this treatment is by no means unanimous. Surprisingly, older classifications that combined woodcreepers and ovenbirds in one family did so under the name Dendrocolaptidae, whereas newer combined classifications use the ovenbird family name, Furnariidae.

Above *The Red-billed scythebill* (Campylorhamphus trochilirostris) *is one of five species in its genus. All are distinguished by long, sickle-shaped bills, used to probe for insects and invertebrates in logs, treetrunks, and the rootless climbing plants known as epiphytes.*

Foragers of the Tree Trunks

Form and Function

Most woodcreepers are medium-sized, slender birds, brown or olive to rufous in color with a variety of spots, streaks, or bars on the head, back, and underparts. The tail is invariably rufous and the wings are usually this color; the tail is noticeably graduated, with the feathers having stiffened shafts that project beyond the vanes and often curve inward at the tip. Variation among the species is most obvious in plumage pattern and bill structure, which ranges from short, thick, and straight to very long, thin, and strongly decurved.

The Spot-crowned woodcreeper is a fairly typical species with a slim, slightly decurved bill of medium length, taking up 2.5cm (1in) of the bird's total length of 20cm (8in). It occurs on wooded, mountain slopes to an elevation of 3,000m (10,000ft), higher than most of its relatives. Solitarily or in pairs, the birds forage in montane forest, where they probe among epiphytes, ferns, and bromeliads, and under bark, which may be levered off with the bill; they take mainly small invertebrate prey. This is one of the first species active in the morning and the last to roost; they roost alone in a crevice or hole in a tree.

The Cocoa woodcreeper has a straight, fairly robust bill, enabling it to take quite large prey, including small lizards. These birds prefer forest edges, but also frequent clearings and open woodland, where they sometimes feed on the ground or among fallen logs. More solitary than the Spot-crowned woodcreeper, they rarely join mixed-species flocks and are rarely seen in pairs. Their tuneful whistles can be heard at all times of day.

The Plain brown woodcreeper and others in its genus are atypical in several respects, and some observers have speculated that they may represent the link between more typical woodcreepers and ovenbirds. Not only are they more subtly patterned than other woodcreepers, but their structure is more ovenbirdlike; they are also more likely to perch crosswise on branches than are more typical woodcreepers. Much of their time is spent close to the ground, often in association with swarms of army ants, taking prey flushed by the raiding ants. In the presence of larger antbirds, Plain brown woodcreepers often feed around the periphery of the swarm; in their absence, however, they can dominate mixed-species flocks. At other times they make occasional sorties from near-vertical trunks to pick a variety of small prey off trees or vegetation; they may even capture prey in mid-air. All female *Dendrocincla* woodcreepers raise their young alone without help from the male.

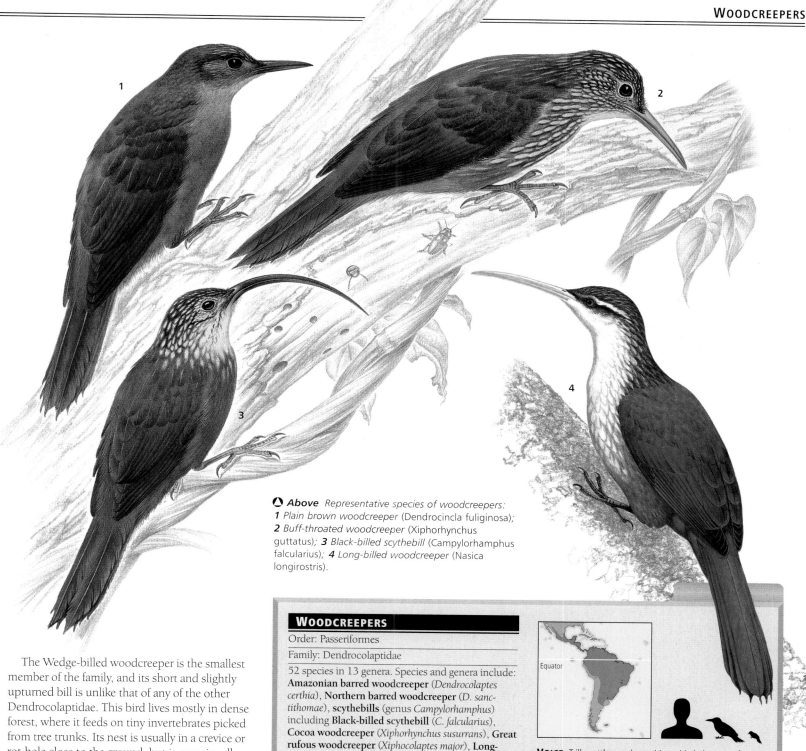

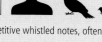

Above *Representative species of woodcreepers:* ***1*** *Plain brown woodcreeper* (Dendrocincla fuliginosa); ***2*** *Buff-throated woodcreeper* (Xiphorhynchus guttatus); ***3*** *Black-billed scythebill* (Campylorhamphus falcularius); ***4*** *Long-billed woodcreeper* (Nasica longirostris).

The Wedge-billed woodcreeper is the smallest member of the family, and its short and slightly upturned bill is unlike that of any of the other Dendrocolaptidae. This bird lives mostly in dense forest, where it feeds on tiny invertebrates picked from tree trunks. Its nest is usually in a crevice or rot-hole close to the ground, but is occasionally found up to 6m (20ft) above the forest floor. Both parents attend the nest and young. In fact, the members of a pair usually forage together, staying in contact through a distinctive *pssst* call that is often doubled. The young are fed small insects, which are carried one at a time in the adult's bill. The geographic range of the Wedge-bill is among the broadest of any woodcreeper; it occurs not only from southern Mexico south through Central America and the Amazon Basin, but also in the Atlantic forests of eastern Brazil. Only the Olivaceous woodcreeper has a larger overall range, but, as presently recognized, that species may represent several closely-related species instead of one.

WOODCREEPERS

Order: Passeriformes

Family: Dendrocolaptidae

52 species in 13 genera. Species and genera include: **Amazonian barred woodcreeper** (*Dendrocolaptes certhia*), **Northern barred woodcreeper** (*D. sanctithomae*), **scythebills** (genus *Campylorhamphus*) including **Black-billed scythebill** (*C. falcularius*), **Cocoa woodcreeper** (*Xiphorhynchus susurrans*), **Great rufous woodcreeper** (*Xiphocolaptes major*), **Long-billed woodcreeper** (*Nasica longirostris*), **Olivaceous woodcreeper** (*Sittasomus griseicapillus*), **Plain brown woodcreeper** (*Dendrocincla fuliginosa*), **Tyrannine woodcreeper** (*D. tyrannina*), **Spot-crowned woodcreeper** (*Lepidocolaptes affinis*), **Wedge-billed woodcreeper** (*Glyphorynchus spirurus*).

DISTRIBUTION N Mexico S to C Argentina.

HABITAT Forest, forest edge, open woodland; primarily in the lowlands.

SIZE Length 14–36cm (5.5–14.3in); weight 11–160g (0.4–5.6oz).

PLUMAGE Body brown or olive to rufous, often streaked or spotted on head, back, and underparts; wings and tail rufous; tail feathers with stiff, spiny shafts.

VOICE Trills, rattles, and repetitive whistled notes, often loud and sometimes musical.

NEST Tree holes or comparable hollows, sometimes behind loose bark.

EGGS 1–4 (usually 2–3); plain white. Incubation 15–21 days.

DIET Insects, other invertebrates, small vertebrates (especially lizards).

CONSERVATION STATUS Just 2 species are on the IUCN Red List: the Moustached woodcreeper (*Xiphocolaptes falcirostris*) is Vulnerable and the Greater scythebill (*Campylorhamphus pucherani*) is Lower Risk/Near Threatened.

The Northern and Amazonian barred woodcreepers were until recently considered to form a single species. Recent work using a combination of morphological and vocal characters has revealed, however, that at least two species are involved, one occurring on either side of the Andes mountains. The most obvious difference between the two is their song: in one a series of 4–6 whistles, in the other a fast trill. Careful examination of study skins has further revealed that the Northern barred woodcreeper has a black mask through the eyes, a darker throat, and a bill that is shorter and narrower than that of the Amazonian species. The latter's broader bill shape appears to correspond to a tendency to sally after prey, much like a flycatcher, rather than to feed by probing, like its Northern counterpart and most other woodcreepers. Like the Plain brown woodcreeper, the barred woodcreepers forage extensively in the forest understory over swarms of raiding army ants, sallying out to capture prey flushed by the insects.

The largest woodcreepers generally have the largest bills. The Great rufous woodcreeper, the size of a flicker (a large North American woodpecker, some 30cm long) has a massive bill, while the Long-billed woodcreeper has a long, straight bill accounting for one-fifth or more of its total length. Whereas the Great rufous and its allies use their heavy bills to dig into rotting wood, the slim bill of the Long-billed woodcreeper allows it to probe into epiphytes along horizontal branches in the canopy of flooded forests in the Amazon basin. Unlike most other woodcreepers, which are closely tied to dense forest, the Great rufous woodcreeper occurs in the comparatively open Chaco region of central South America. Both the Great rufous and Long-billed woodcreepers nest in tree cavities within 2m (6ft) of the ground.

Although not as large as either the Great rufous or Long-billed woodcreepers, scythebills have very long, decurved bills that can make up a quarter of their total length. They use their bills to probe into epiphytes, trunks, and logs, and into cracks and crevices that are beyond the reach of other woodcreepers. Although it might seem vital that such a bill should be in good condition, a Black-billed scythebill trapped in southeastern Brazil had lost the front third of its upper mandible; its weight was normal, but its plumage was in poor condition, suggesting that it was inefficient at preening and controlling the level of parasites on its body.

In many parts of their range, scythebills are closely tied to the presence of extensive patches of bamboo, to which they seem especially well adapted as their long bills are excellent for probing the hollow tubes typical of these large grasses. As with many bamboo specialists, the natural history of scythebills is poorly known, the lack of information on them probably reflecting the difficulty of working in such a localized, densely vegetated habitat. Scythebills nest in tree cavities much like other woodcreepers.

Left In Argentina, a Scimitar-billed woodcreeper (Drymornis bridgesii) *takes insect prey back to the nest to feed its young. All woodcreepers nest in cavities – usually natural holes, but sometimes abandoned woodpecker nests. This species is a bird of the Gran Chaco, the great alluvial plain extending also across areas of Bolivia, Uruguay, and Paraguay in central South America.*

Right *Perched by a treehole, a Great rufous woodcreeper displays the long, stout bill that is the most distinctive feature of the species. Unusually for woodcreepers, these birds are known sometimes to feed off the ground, while at other times they use their bills to probe into rotting wood in search of food.*

To Pair or Not to Pair

Breeding Biology

Woodcreepers are usually found alone or in pairs, but sometimes in small groups that may represent family parties, as well as in mixed-species flocks. Although they can be difficult to spot in their forested habitats, and especially to identify by sight, the woodcreepers reveal their presence by their ubiquitous songs, which are given primarily at dusk and dawn.

Although poorly known, the breeding biology of woodcreepers reveals remarkable consistency. All nest in cavities, usually sited in natural holes in trees but sometimes in old woodpecker holes. Most woodcreeper clutches comprise two to three eggs, although some species lay a single egg, and clutches may rarely contain as many as four eggs. Woodcreeper eggs are white and unmarked, like those of most other cavity-nesting birds.

Probably the most variable aspect of woodcreeper breeding biology is the degree of pair-bonding exhibited by different species. Most form extended bonds, with both sexes cooperating to raise the young. This behavior is typical not only of the larger woodcreepers, but also of the smallest species. Yet many other species do not pair up in this way, and the males of one species in particular (the Tyrannine woodcreeper of the Andes) appear to congregate in dispersed leks in which the males sing loudly from mountain ridges to attract females. Single-parent families appear to be normal only in *Dendrocincla* and *Sittasomus* species, but they also appear in other genera, although less frequently.

Typical of the pair-bonding species is the Spot-crowned woodcreeper, which lays its eggs in a concealed hole or crevice that it often enlarges. The nest cavity is lined with fragments of wood and bark, collected by both parents and augmented throughout the nesting period. Both parents attend the eggs and young. The young hatch blind and almost naked; they tend to be rather noisy. The two barred woodcreepers also remain paired for an extended period of time, and share responsibility for raising the young. Both species nest in cavities, often within 6m (20ft) of the ground.

In the Cocoa woodcreeper, however, as in the Plain brown woodcreeper, the female takes full responsibility for the nest and will often attack intruding males. She may spend up to 80 percent of daylight incubating, and will return with extra bits of bark for the nest after periods away. She also rears the young alone, bringing single food items every half an hour or so. A further contrast to the Spot-crowned woodcreeper is that the young of these single-parent families remain silent until they leave the nest; and having once departed they do not return, even to roost. CM/AMH

Antbirds

Antbirds owe their name to the habit of some species of following swarming army ants. Although the birds do not feed on the ants themselves (the levels of formic acid in the insects would likely be toxic for birds of their size), they do prey extensively on the invertebrates that are flushed from cover by the swarms.

Antbirds are divided into two very distinct families: the Thamnophilidae, or typical antbirds (204 species in 45 genera); and the Formicariidae, or ground antbirds (62 species in 7 genera). The term "ground antbird" refers to this family's favored nesting location; in fact, while foraging, most species in both families generally prefer dense undergrowth and are found on or near the ground in their rainforest habitat. The ground antbird family itself includes two distinct groups of birds, the antpittas and antthrushes, while the typical antbirds can roughly be separated into three major groups: the antshrikes, the antwrens, and the bare-faced antbirds (although these may not be natural groupings).

FACTFILE

ANTBIRDS

Order: Passeriformes

Families: Thamnophilidae, Formicariidae

266 species in 2 families and 52 genera.

DISTRIBUTION S Mexico to S South America, with thamnophilid species diversity concentrated in Amazonia, formicariid in the Andes.

HABITAT Dense undergrowth in forests; sometimes above the treeline and in the forest canopy.

SIZE Length 7.5–35cm/3–14in (typical antbirds), 10–24cm/4–9.5in (ground antbirds); **weight** 7–275g/0.2–8.8oz (typical antbirds), 20–235g/0.65–7.5oz (ground antbirds).

PLUMAGE Typical antbirds: males gray to black with varying amounts of white spots or bars, also occasional rufous feather patterns; females duller or browner; both sexes have concealed white interscapular patch. Ground antbirds: dull browns, blacks, and dark reds, but in attractive patterns.

VOICE Typical antbirds: harsh, churring alarm or contact calls. Ground antbirds: simple but loud and far-carrying.

NEST Typical antbirds: open cup in the fork of a branch. Ground antbirds: nests are found on or near the ground, on a horizontal surface or in a natural burrow.

EGGS Usually 2; white with dark spots in typical antbirds, typically blue in ground antbirds. Incubation period about 14 days; nestling period 7–14 days.

DIET Mainly insects, with some small fruit and even vertebrates taken as secondary prey.

CONSERVATION STATUS Typical antbirds: 4 species are Critically Endangered, including *Myrmotherula snowi* which is on the brink of extinction unless conservation efforts are successful; in addition, 10 species are Endangered, 11 Vulnerable, and 9 Lower Risk/Near Threatened. Ground antbirds: 6 species are Endangered, 5 are Vulnerable, and 9 are Lower Risk/Near Threatened.

See Antbird Families ▷

◐ **Left** *Great antshrikes usually live in pairs in areas of dense forest. The male, shown here, has black feathers on its back and two white wing-bars, while the female has russet coloration above. Both sexes have a head crest and the characterisric red iris.*

Birds of the Neotropical Forest Floor

FORM AND FUNCTION

The antbirds have long been thought to be closely related to the gnateaters (Conopophagidae) and tapaculos (Rhinocryptidae), and recent research in fact suggests that the two *Pittasoma* antpittas are more closely related to gnateaters than other antpittas. More distant relatives include the ovenbirds (Furnariidae) and woodcreepers (Dendrocolaptidae); together with the other four families, these constitute the "suboscine tracheophone" branch of the Order Passeriformes, an anatomical division based primarily on the number and arrangement of the muscles in the syrinx. Some modern taxonomic treatments of this group place the Furnariidae, Dendrocolaptidae, Formicariidae, Conopophagidae, and Rhinocryptidae in Parvorder Furnariida, and the typical antbirds alone in Parvorder Thamnophilida.

Of the ground antbirds of the family Formicariidae, the antpittas are long-legged birds with exceedingly short tails, giving them an appearance superficially similar to pittas (Pittidae). Antpitta species will run or hop along the forest understory, but never walk. All antpitta species have pale blue to bluish-gray legs, with the exception of Watkins' antpitta, which has pink legs. Antpittas generally do not follow army-ant swarms. The antthrushes are also long-legged birds, but have relatively long tails that are held cocked; they also usually hold their heads upright. Unlike the antpittas, antthrushes will walk in the understory instead of running or hopping.

Within the typical antbirds of the family Thamnophilidae, the antshrikes, including genera such as *Thamnophilus* and *Sakesphorus*, are largish birds with noticeably hooked bills, similar to shrikes (Laniidae). Antshrikes are generally birds of the understory to lower mid-canopy that occasionally associate with mixed-species flocks. Antwrens, including genera such as *Herpsilochmus* and *Myrmotherula*, are smaller and are typically found in the mid- to upper-canopy. Birds in this latter group have thinner bills, used for gleaning insects from leaf clusters.

The final major division of typical antbirds is the bare-faced group. Species in this grouping are considered "professional ant-followers" and spend much of their lives associated with army-ant swarms. As their lifestyle would imply, bare-faced antbirds are rarely found more that a few meters above the forest floor. Most species in this group lack feathers around the eyes and often have brightly-colored orbital skin, apparently an adaptation to keep ants from sticking to their face when foraging among swarms. Genera in this group include *Myrmeciza*, *Gymnopithys*, and *Rhegmatorhina*. Sizes in the Thamnophilidae range from the Short-billed antwren (7.5cm/3in; 7g/0.2oz) to the large Giant antshrike (35cm/14in; 275g/8.8oz).

Species within the Thamnophilidae display sexual dichromatism in plumage patterns. Males are typically gray to black with varying amounts of white spots or bars (as, for instance, in the antshrikes of the genus *Thamnophilus*), and may occasionally have rufous feather patterns, like the Rufous-capped antshrike. Females are usually duller or browner than males. In some groups it is the female plumage that is diagnostic for species identification, as is the case with antwren species in the genus *Myrmotherula*. All species of typical antbirds (and also gnateaters and the *Pittasoma* antpittas) have a concealed white interscapular patch. This is diagnostic for typical antbirds, as the ground antbirds lack this plumage character. The white patch is used in breeding displays by males and in territorial defense behavior by both sexes. Conversely, the ground antbirds (Formicariidae) are sexually monomorphic. Both males and females have relatively dull plumage colors of browns, blacks, and dark reds, but organized in attractive patterns.

Most typical antbird species are relatively unmusical, producing harsh, churring alarm or contact calls. However, a few thamnophilid species such as the Spot-backed antshrike do have attractive songs and melodious calls. Ground antbirds also have simple vocalizations, but they are usually loud and far-carrying. Given their cryptic color patterns and loud voices, these birds are more often heard than seen.

At the Core of Mixed-Species Flocks

SOCIAL BEHAVIOR

Typical antbirds form the core species for Neotropical mixed-species foraging flocks. Although many bird-watchers will be accustomed to seeing mixed flocks of birds moving through woodlands in Europe (where they might typically include tits, kinglets, nuthatches, creepers, and finches) or in North America (migrating warblers, chickadees, woodpeckers), it is in the lowland tropical forests, whose dense canopies can reach 60–80m (200–260ft), that this activity reaches a peak.

The function of such flocks is not fully understood, but the various explanations proposed fall into two main categories: firstly, that flock behavior enhances feeding efficiency by having many birds rather than single individuals searching for food, and secondly that group cohesion helps to reduce the risk of predation, with flock members giving early warning of the presence of predators to individuals that might otherwise be unaware of the danger. In either case, an individual bird would increase its survival rate by associating with a flock. Whatever the rationale, mixed flocks of insectivorous and omnivorous species work through the forest at about 0.3km/hr (0.2mph), occupying feeding levels from the ground to close to the canopy. The path followed may be very erratic and frequently crosses itself, depending on food availability.

There is usually one bird or species that maintains the cohesion and impetus of the flock. In Central America this is often a parulid warbler

Left *Representative species of antbirds:* **1** *Barred antshrike* (Thamnophilus doliatus); **2** *White-flanked antwren* (Myrmotherula axillaris); **3** *Rufous-capped antthrush* (Formicarius colma); **4** *White-plumed antbird* (Pithys albifrons).

such as the Three-striped warbler (*Basileuterus tristriatus*), but in South America it is more frequently an antbird (for instance, the Bluish-slate antshrike). As these core birds move through changing forest habitats, calling frequently, other birds are attracted to the flock for the time that it remains within their home range. As a result, there may be only one pair of each highly territorial species (perhaps together with its fledged young) in the flock at any one time.

On rare occasions mixed-species flocks will combine with flocks following army-ant swarms. In such cases, as many as 30 different antbird species may be seen in a single location. This extraordinary profusion may then be enhanced by the presence of an even greater diversity of tanagers, warblers, woodpeckers, ovenbirds, and woodcreepers, all also associating with the flock.

Birds do not always follow a flock that moves outside of their territory. Certain antbird species are more likely to join a mixed-species flock if a trespassing member of their own species is present, in which case they may do so in order to defend their home territory; participation rates also increase if the flock moves through their home range during the peak feeding and activity periods, early in the morning or late in the afternoon. Antbirds are generally less likely to associate with a mixed-species flock in the middle of the day or during inclement weather.

Diagnostic Nest Shapes

BREEDING BIOLOGY

Typical antbird nests customarily take the form of an open cup placed in the fork of a branch, although on rare occasions they have also been found in tree cavities or on the ground. This nest structure and position appear to be diagnostic to the family level, as ground antbirds are not known to place open nests in branch forks. Instead, formicarids construct fairly unstructured nests of leaves and moss that are placed on or near the ground, on a horizontal surface or occasionally in a natural burrow such as a hollow log. The enigmatic Wing-barred antbird was for many years classified as a ground antbird, but the discovery

Antbird Families

Typical Antbirds
Family Thamnophilidae

204 species in 45 genera, including: Ash-winged antwren (*Terenura spodioptila*), Bicolored antbird (*Gymnopithys leucaspis*), Bluish-slate antshrike (*Thamnomanes schistogynus*), Fringe-backed fire-eye (*Pyriglena atra*), Giant antshrike (*Batara cinerea*), Great antshrike (*Taraba major*), Restinga antwren (*Formicivora littoralis*), Rufous-capped antshrike (*Thamnophilus ruficapillus*), Short-billed antwren (*Myrmotherula obscura*), Spotted antbird (*Hylophylax naevioides*), Spot-backed antshrike (*Hypodaleus guttatus*), Ocellated antbird (*Phaenostictus mcleannani*), White-plumed antbird (*Pithys albifrons*), Wing-barred antbird (*Myrmornis torquata*).

Ground Antbirds
Family Formicariidae

62 species in 7 genera, including: antpittas (genera *Grallaria*, *Grallaricula*, *Hylopezus*, *Myrmothera*, *Pittasoma*) including Black-crowned antpitta (*Pittasoma michleri*), Stripe-headed antpitta (*Grallaria andicola*), Watkins' antpitta (*G. watkinsi*), Giant antpitta (*G. gigantea*), Hooded antpitta (*Grallaricula cucullata*); antthrushes (genera *Formicarius*, *Chamaeza*).

Right A female Slaty antwren (Myrmotherula schisticolor) watches over her nest in the Costa Rican rain forest. Like those of most typical antbird species, it is cup-shaped and set in the crook of a branch.

Left The Undulated antpitta (Grallaria squamigera), one of the larger antpitta species, inhabits the Andean foothills from western Venezuela to Bolivia, foraging on the humid forest floor.

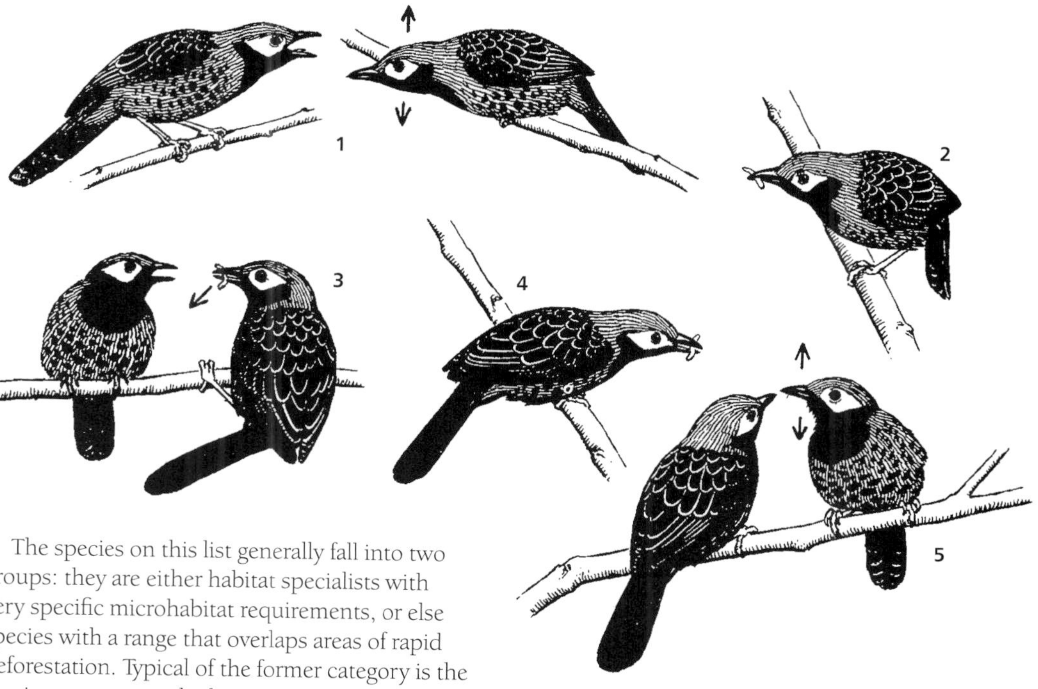

Right *Courtship feeding is an important stage in forming the monogamous pair-bond in antbirds. In the case of Ocellated antbirds, much singing is involved.* ***1*** *A female (left) screams at her mate, who bobs his head ritually.* ***2*** *Having brought food, the male sings faintly while holding the item in his mouth.* ***3*** *He then feeds the female, who takes a low posture* ***4*** *while eating the food.* ***5*** *When the session is over, the female pecks at the male's bill while he bobs his head.*

that its nest is in the form of an open cup in a forked branch provided conclusive evidence that the species in fact belongs taxonomically with the Thamnophilidae.

Typical antbirds usually lay two eggs that are white with dark spots, while ground antbird eggs are generally blue. Field research suggests that incubation is about 14 days, at which time the altricial (helpless) young hatch and are cared for by both parents for another 1–2 weeks before fledging. Young birds often stay with the parents for an extended time, and young males may even bring a female into the natal family range prior to establishing their own breeding territory.

The Retreating Forest

CONSERVATION AND ENVIRONMENT

Both typical and ground antbirds are forest birds that generally require large tracts of undisturbed habitat to maintain sustainable population sizes. Habitat destruction and loss through deforestation is the leading cause of population level declines in these families. Unfortunately, a rapidly growing list of antbird species are now either threatened or endangered.

The species on this list generally fall into two groups: they are either habitat specialists with very specific microhabitat requirements, or else species with a range that overlaps areas of rapid deforestation. Typical of the former category is the Restinga antwren, which occupies a very restricted range in a strip of dunes in Rio de Janeiro state, Brazil, where it depends on *restinga*, a beach-scrub habitat rich in bromeliads and cacti. Holiday development is now putting this environment at risk, and the bird is currently classed as Endangered. Of the species threatened by deforestation, the Fringe-backed fire-eye is now considered Critically Endangered because the region of second-growth forest to which it is restricted, located in a small area of the Brazilian state of Bahia, is rapidly disappearing. NHR/AMH

ANTBIRDS AND ARMY-ANT SWARMS

Ants of the subfamily Dorylinae, such as *Eciton burchelli*, are well-known for their frequent swarming through tropical forests, either to move their colony or in sorties for feeding. A variety of birds take advantage of their activities to feed on invertebrates and small vertebrates that are disturbed by the passage of the columns. The main bird families involved are the cuckoos, woodcreepers, typical antbirds, and tanagers.

About 50 typical antbird species – the so-called "professional" ant-followers – are regular attendants on ant swarms, but many other species will occasionally join in. Large colonies of regularly nomadic ants can attract up to 25 birds of one or two species at a time, plus scattered individuals of up to 30 other species. The professional ant-followers may obtain up to 50 percent of their food from antswarms, keeping a regular eye on the colonies to take advantage the moment the swarm starts to move.

There is often a strict hierarchy in the attendant birds, with larger species such as the Ocellated antbird holding a central "territory" just ahead of the column, while smaller species like the Bicolored antbird hold peripheral territories. Some small species, such as the White-plumed antbird, utilize the whole area by making raids from the periphery, although only at the expense of being frequently attacked. Large and potentially dominant ground antbirds, such as Black-crowned antpittas, actually occupy peripheral zones.

Territories exist vertically as well as horizontally, and many of the "non-professional" ant-followers feed in the outer or upper reaches, or even between the territories of the professionals. Many of the less regular members drop out as the antswarm moves out of their territory. In contrast, the professionals are prepared to share the resource to some extent, although individuals of a species only remain dominant over others of their own species while they are in their own territory.

Gnateaters

THE GNATEATERS ARE SMALL, SKULKING denizens of the undergrowth of South American tropical forests that occur at low densities, either singly or in pairs. Until quite recently they remained a poorly studied group, long treated as aberrant antbirds. However, modern morphological and biomolecular studies have demonstrated that they actually constitute a very distinctive family in their own right, and are among the most primitive of all songbirds.

Gnateaters are essentially ground-dwellers, and spend almost their entire lives within 2m (6.6ft) of the forest floor. Their short tails and long legs and toes are readily interpretable as adaptations to this terrestrial strategy. Thus, gnateaters share striking morphological similarities with the shortwings, which occupy a similar niche in the tropical forests of the Old World.

FACTFILE

GNATEATERS

Order: Passeriformes

Family: Conopophagidae

8 species of the genus *Conopophaga*: **Rufous gnateater** (*C. lineatus*), **Chestnut-belted gnateater** (*C. aurita*), **Hooded gnateater** (*C. roberti*), **Ash-throated gnateater** (*C. peruviana*), **Slaty gnateater** (*C. ardesiaca*), **Chestnut-crowned gnateater** (*C. castaneiceps*), **Black-cheeked gnateater** (*C. melanops*), **Black-bellied gnateater** (*C. melanogaster*).

DISTRIBUTION S America

HABITAT Primary and secondary forest undergrowth

SIZE Length 10–18cm (4–7in); weight 12.6–42.5g (0.4–1.5oz).

PLUMAGE Brown or gray above, pale, dark, or rufous below; head variously patterned (usually including a tuft of long white plumes behind the eye).

VOICE Simple, melodious songs contrasting with short, sharp alarm calls.

NEST Bowl-shaped, near ground.

EGGS 2; yellowish, with spots or smudges.

DIET Small insects

CONSERVATION STATUS Not threatened.

Dainty Ground-dwellers

FORM AND FUNCTION

With their lax, fluffy plumage, gnateaters look plump and dainty. Their upperparts are dark brown or gray, and the birds can superficially appear rather drab. However, closer observation reveals that they have distinctive plumage signals for advertising their presence amid the gloom of the forest floor. Some species have bright crown or breast feathers, while most have striking, silvery-white erectile tufts behind the eye. These tufts are more developed in males than in females, and are usually kept furled; during territorial disputes or display, however, they are dramatically erected, and appear to shine, especially in twilight.

Like many other insectivorous songbirds, gnateaters have a short, thin bill, but in some species this is also broad, enabling the birds to catch small flying insects. Their short, rounded wings are adaptations to flying in dense, crowded forest undergrowth.

The eight gnateater species are distributed across most of tropical South America, but they are highly sedentary and few species overlap. Even where the ranges of two species do come into contact, they generally occur in different habitats or at different altitudes.

Alone or in pairs, gnateaters work through the leaf litter of their territories, feeding on small insects. Occasionally, they sally forth to catch insects on trunks or in low vegetation. They do not normally join other passerines in mixed feeding flocks, and are usually very territorial. All gnateaters vigorously defend their territories with visual displays and short, sharp alarm calls.

Melodious Mating Calls

BREEDING BIOLOGY

In contrast to their harsh alarm calls, male birds attract their mates with simple, melodious, ascendant songs. Earlier accounts suggested that male Rufous gnateaters also made a mechanical noise with their wings during aerial displays at twilight. However, there have been no recent records of this behavior, and the wing feathers do not seem to be adapted to this end in the way that they are in some tyrant flycatchers. The nest is placed close to the ground and made mainly of large leaves and lined with softer plant fibers.

Both sexes incubate, and some species will feign injury to distract predators if disturbed at the nest. After fledging, the young may stay with the parents for several weeks. Although no gnateaters are threatened, recent studies reveal that they are very susceptible to habitat fragmentation. SJP/AMH

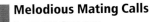

__Above__ A Rufous gnateater prepares to eat a caterpillar. The bird, which has been described as looking somewhat like a dumpy, tailless Eurasian robin, shows the erectile white feather-tufts behind the eye that are a characteristic feature of the family.

__Left__ A male Black-cheeked gnateater takes its turn to incubate eggs. Most species place their nests close to the ground; they are made mainly of large leaves and are lined with softer plant fibers.

Tapaculos

SEDENTARY, TERRESTRIAL, AND GENERALLY poorly known, the tapaculos are an exclusively Neotropical family typical of temperate South America. Most species are notoriously difficult to see, but their presence is usually given away by distinctive and often loud vocalizations.

Tapaculos range from the small, wrenlike species of the genus *Scytalopus* to the robust, quail-like *Pteroptochos* species. Four species are considered threatened (one of which was only discovered in 1997!) and two of these critically so, the main threat being loss of habitat.

Cocked Tails and Short Wings

FORM AND FUNCTION

The great English naturalist Charles Darwin commented on the birds' Spanish name when he first encountered one during his celebrated expedition on H.M.S. *Beagle* in 1834. "It is called Tapacolo, or 'Cover your posterior'," he wrote, "and well does the shameless little bird deserve its name; for it carries its tail more than erect, that is, inclined forward toward its head." Cocked tails are indeed a feature of the family, although by no means all of the 55 species currently recognized have them. In others, the tails extend horizontally from the body, ranging in size from short to long.

Other characters common to the family are short, rounded wings, stout legs, large feet, and a movable flap, or operculum, covering the nostrils. Plumage is mostly somber in hue, but patterns on some species such as the Chucao and White-throated tapaculos and the crescent chests of the genus *Melanopareia* are quite handsome.

The birds are primarily found in temperate habitats, with only one species in Amazonia. The greatest diversity of tapaculo genera is in southern South America, but the highest species diversity occurs in the northern Andes. Only *Scytalopus*, the largest genus, spans the family's entire geographical range, from the mountains of Costa Rica to islands south of Tierra del Fuego. Tapaculo taxonomy is undergoing major revision as vocalizations become better known, particularly within *Scytalopus*: 11 species were recognized in this genus in 1970, 17 in 1994, and 37 in 1997!

Agile and Secretive

SOCIAL BEHAVIOR

Most tapaculos are secretive and have legendary agility, moving with speed and yet remaining invisible. The Sandy gallito in particular is noted for its running, taking strides almost as long as its 18cm (7in) bodylength as it scurries out of the way of intruders. The birds are mostly terrestrial, scratching with their large feet among ground litter for invertebrates and seeds – the presence of certain species is often revealed by large raked areas on the forest floor. Many species favor bamboo thickets. Most forest species fly rarely, if ever, but species inhabiting semi-open country such as the Moustached turca can fly strongly, if briefly, and glide for 100m (330ft) or more across a valley.

Left *In common with many tapaculo species, the Crested gallito* (Rhinocrypta lanceolata) *has a rigidly cocked tail. Tapaculos, which are found chiefly in the cooler, humid parts of South America, spend most of their time on the ground, scratching among the leaf litter.*

FACTFILE

TAPACULOS

Order: Passeriformes

Family: Rhinocryptidae

55 species in 12 genera. Species include: **Sandy gallito** (*Teledromas fuscus*), **Moustached turca** (*Pteroptochos megapodius*), **Chucao tapaculo** (*Scelorchilus rubecula*), **White-throated tapaculo** (*S. albicollis*), **Bahia tapaculo** (*Scytalopus psychopompus*).

DISTRIBUTION S and C America

HABITAT Forest, woodland, scrub.

SIZE Length 10–27.5cm (4–11in)

PLUMAGE Colored in grays, browns, black, and white, often barred on underparts and rump; sexes alike in most species.

VOICE Usually loud and repetitive

NEST Constructed of grass, root fibers, twigs, moss etc., in burrows, tree hollows, banks, grass clumps.

EGGS 2–4; unmarked white.

DIET Insects, spiders, some vegetable matter.

CONSERVATION STATUS 2 species – Stresemann's bristlefront (*Merulaxis stresemanni*) and the Bahia tapaculo – are listed as Critically Endangered. 1 other is Endangered, and 1 Vulnerable.

All tapaculos have distinctive and often striking vocalizations. Some species, including the Chucao tapaculo and the birds of the genus *Pteroptochos*, are relatively musical; the huet-huet (*P. tarnii*) of southern Chile and Argentina takes its name from its familiar, repeated call. Others, however, including most of the *Scytalopus* species, produce monotonous repetitions of unmusical, semi-metallic, and often rather froglike notes.

The nests of most species are located on or near the ground, often in burrows or embedded in banks, although huet-huets can nest several meters up in tree hollows. Most nests are globular in shape with a small side entrance, although some cavity nests are cuplike; typically, they are built of moss, twigs, and leaves. Tapaculo eggs are relatively large and rounded, and are incubated by both parents. Both also attend the hatchlings, which are covered in sparse down. SH/AMH

Australasian Treecreepers

Australasian treecreepers fill the niche occupied by woodpeckers and the creepers of the family Certhiidae in the northern hemisphere and the Dendrocolaptidae in South America. Their modified leg muscles and extremely long toes and claws enable them to ascend tree trunks and forage on ants.

FACTFILE

Australasian Treecreepers

Order: Passeriformes

Family: Climacteridae

7 species in 2 genera: **Brown treecreeper** (*Climacteris picumnus*), **Red-browed treecreeper** (*C. erythrops*), **Rufous treecreeper** (*C. rufa*), **White-browed treecreeper** (*C. affinis*), **Black-tailed treecreeper** (*C. melanura*), **Papuan treecreeper** (*Cormobates placens*), **White-throated treecreeper** (*C. leucophaeus*).

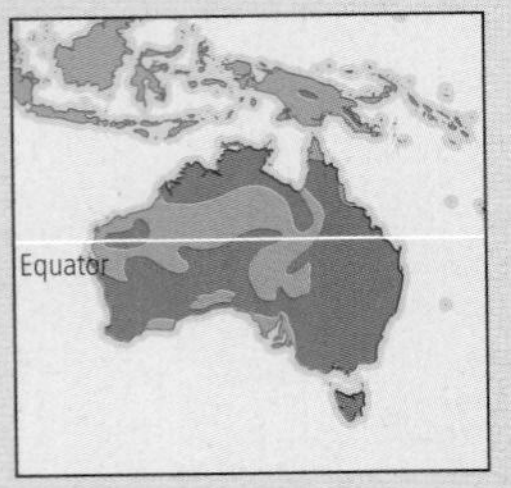

Distribution Australia and New Guinea

Habitat Forest, woodland, and tall shrubland.

Size Length 14–19cm (5.5–7.5in); weight 20–40g (0.7–1.4oz).

Plumage Upperparts brown to black, underparts buff to rufous with black and white striping or spotting; broad, pale wing-bar.

Voice Loud piping notes, trills, chatters; harsh grates, rattles.

Nest Cup in hole of tree, hollow branch, or stump.

Eggs 1–4, white to pink with varying amount of red-brown to purple-brown markings. Incubation period 14–24 days; nestling period 20–27 days.

Diet Insects (primarily ants, some beetles), spiders, nectar, rarely seeds and fungi.

Conservation Status Some species are in decline, although none are presently listed as at risk.

This small family exhibits great diversity in social organization, from the White-throated treecreeper, in which members of a pair rarely associate, to the Brown and the Rufous, in which birds live in cooperative groups but also feed at nests in up to four additional territories.

Living on Tree Trunks

Form and Function

Treecreepers are camouflaged to match bark surfaces, with brown to black upperparts and lighter underparts marked with black and white spots and stripes. Females have rufous markings on the chest or cheek. Some of the toes are partially fused to further assist in creeping. However, the birds' hind limbs function equally well on the ground, where some species spend almost half their time.

Below *A White-browed treecreeper leaves its nest in a tree-hole. The species' habit of building nests in hollow trunks or limbs is thought to make it vulnerable to predation by tree goannas.*

Only one species is found in New Guinea, and the Australian species tend to overlap little in their distributions except in the southeast, where the more generalist White-throated treecreeper co-exists with the Red-browed in eucalypt forests and the Brown in drier woodlands. The birds' diet consists primarily of ants, but beetles and spiders are also taken, and flowers of various trees may be probed for nectar.

A Complex Family Life

Breeding Biology

All species are sedentary, occupying 1.5–20ha (4–50 acre) territories announced with simple, loud, piping calls. They breed from July to February, raising 1–3 young in a nest built mainly by the female in a natural hollow. Most species engage in bill sweeping, repeatedly passing snakeskin or insect wings across bark around the nest hole, possibly as a deterrent to mammalian predators.

In all species of the genus *Climacteris*, offspring (usually males) may remain at home as nonbreeders, helping the breeding pair at subsequent nests. These cooperative groups contain up to eight birds, although in the Rufous treecreeper some nonbreeders may be unrelated immigrants. In the Brown and the Rufous, there is an additional level of social organization. Multiple cooperative groups may form a supergroup in which males (and sometimes females) feed at each other's nests and cooperate in territory defense. Both species are currently in decline due to habitat fragmentation and isolation, possibly because sociality restricts their dispersal options. VD

Lyrebirds

THE LYREBIRDS OF AUSTRALIA ARE RENOWNED for their spectacular courtship displays, singing, and vocal mimicry. The male's tail, display, and song are thought to be good examples of "extravagant" male traits, resulting from sexual selection operating through female mate choice.

The existence of the Superb lyrebird, one of the country's most spectacular birds, was first brought to the attention of the New South Wales authorities in 1797 by an ex-convict who had spent some years living in the bush. From the outset, the species' affinities were contentious, and early accounts variously labelled it a pheasant, a terrestrial bird of paradise, a domestic fowl, and a "peacock-wren"! Ironically, the name that has persisted resulted from the mistaken belief of English scientists who examined some skins sent from Australia that the male's tail was held continuously in a lyre shape during display. The only other extant lyrebird species was not discovered by Europeans until about 50 years later, when it was named in honor of Prince Albert, the consort of Queen Victoria.

We now know from DNA studies and other evidence that the lyrebirds' closest relatives are the two scrub-birds of the family Atrichornithidae. It is also probable that, like the scrub-birds, lyrebirds originated in the Australasian region, rather than being northern immigrants.

Above *Albert's lyrebird is the rarer of the two Menuridae species, being restricted to Lamington National Park, on the Queensland–New South Wales border.*

FACTFILE

LYREBIRDS

Order: Passeriformes

Family: Menuridae

2 species in the genus *Menura*: **Albert's lyrebird** (*M. alberti*) and the **Superb lyrebird** (*M. novaehollandiae*).

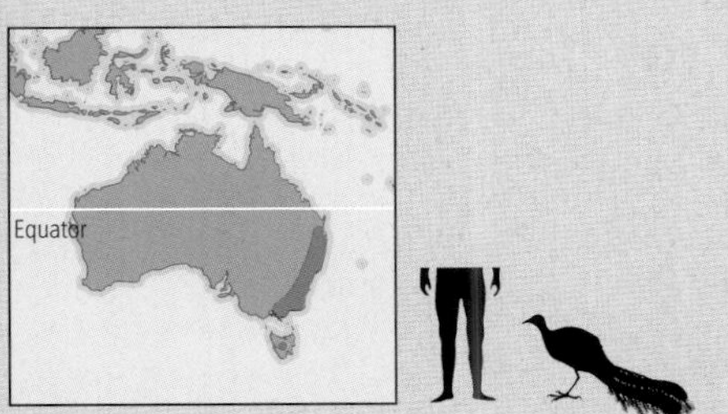

DISTRIBUTION Endemic to E Australia.

HABITAT Temperate and subtropical rain forests and wet and dry sclerophyll forests.

SIZE Length: Superb lyrebird 103cm/40in (male), 76–80cm/30–31.5in (female); Albert's lyrebird 90cm/35in. Weight: Superb lyrebird, 0.89–1.1kg/31–39oz (male), 0.72–1kg/25–35oz (female); Albert's lyrebird 0.93kg/33oz.

PLUMAGE Dark gray-brown or red-brown above, gray-brown to rufous brown below. The male's spectacular tail is very long, trainlike, and comprised of highly modified rectrices; the female's is shorter and simpler.

VOICE Loud, carrying song; much mimicry of other birds and some other sounds; also high-pitched alarm whistles, display calls.

NEST Bulky, domed chambers of sticks, bark, moss, rootlets, and fern fronds, lined with rootlets, fine plant material, and body feathers, having a side entrance. Located on the ground, earth banks, rock faces, boulders, tree buttresses, and exposed roots, logs, wire-grass clumps, and in dead and living trees to a height of about 22m (72ft).

EGGS Usually 1, oval, light gray to deep purplish brown with blackish brown or deep slate-gray spots and streaks; average weight 62g (2.2oz). Incubation period about 50 days; nestling period about 47 days; from fledging to independence, maximum 8–9 months.

DIET Mainly invertebrates in soil and rotting wood.

CONSERVATION STATUS Albert's lyrebird is listed as Vulnerable.

The Extravagant Tail

FORM AND FUNCTION

The Superb lyrebird's natural range in southeast Australia extends from southern Victoria to the extreme southeast of Queensland. The species occurs from sea level up to about 1,500m (5,000ft) in the Great Dividing Range in a variety of moist forest habitats. Some Victorian birds were introduced to two localities in Tasmania in the 1930s and 1940s, and populations have become established there. These birds have increased their ranges, whereas many mainland populations seem to have declined over the same time period.

Albert's lyrebird occupies a very restricted range, now mainly above 300m (1,000ft) in the mountains of extreme southeastern Queensland and northeastern New South Wales. The disparity in range size between the two species is intriguing; the locality in which the only known fossil lyrebird, *M. tyawanoides*, was found suggests that the genus may once have had a wider distribution.

The chicken-sized Superb lyrebird is one of the largest passerines in the world. Adults are a dark gray-brown above and dark to light gray below. The male has a spectacular, long, trainlike tail comprising two wirelike central median feathers, 12 fanlike filamentaries, and two outermost S-shaped lyrates. The filamentaries and lyrates are

Left A Superb lyrebird feeds her young. The female birds, which normally have only a single chick, show an unusual amount of parental care, continuing to nurture the fledgling for the better part of a year.

silvery white on the underside. The lyrates have a striking series of semi-transparent, rufous, moon-shaped notches, from which the name *Menura* (meaning "crescent moon") probably derives, and a black, club-shaped tip. The barbs of the filamentaries lack barbules and so have a delicate, lacy appearance. The smaller female has a shorter, simpler tail, with less dramatic notching on the lyrates. Albert's lyrebirds are slightly smaller; their plumage is more rufous-brown dorsally and on the flanks and rump, and lighter ventrally. The male's tail is shorter and less elaborate than that of the male Superb lyrebird.

Lyrebirds are mainly ground-dwelling and have long, gray legs and powerful, clawed feet that enable them to run swiftly and dig strongly. They fly only weakly, principally by gliding downhill, and their wings are short and rounded. Nonetheless, they roost high up in trees at night, ascending slowly in a series of ungainly, flapping leaps.

Topsoil Hunters

DIET

The Superb lyrebird feeds mainly on invertebrates, which it obtains by digging in the top 5–15cm (2–6in) of soil. On average, excavation sites are less than 2m (6.6ft) apart and about 0.25 to 0.5sq m (27–54sq ft) in area. Typically a site yields 25–29 prey in about 1.5 minutes of digging.

Stomach content analysis of adults has revealed a diet comprising a variety of invertebrates: earthworms, crustaceans, centipedes, millipedes, spiders, scorpions, cockroaches, beetles, flies, ants, and moths all feature prominently as adults and/or larvae or pupae. Some seeds are also consumed. Nestlings' diets resemble those of adults.

Lyrebirds' foraging behavior results in considerable movement of soil and litter and is thought to maintain bare areas of forest floor and assist in nutrient cycling and the regeneration of tree ferns.

The foraging behavior and diet of Albert's lyrebird are poorly known, but are probably broadly similar to those of its southern relative.

Territorial and Solitary

BREEDING BIOLOGY

Adult Superb lyrebirds are territorial and largely solitary. Males' territories are typically about 2.5ha (6 acres) in extent and encompass or overlap those of up to six females. Male territory defense in the breeding season mainly involves extensive loud singing, but male intruders elicit "skypointing" threat displays, prolonged chases with accompanying vocalizations, and occasionally vigorous fighting. Immature birds of both genders often form small, wide-ranging groups, within which there is much display behavior. These itinerant groups may be joined temporarily by the owners of territories that they traverse.

Superb lyrebirds can live for 20–30 years. Males take 7–8 years to attain the full adult tail plumage, but females probably commence breeding when they are 5–6 years of age. An adult male constructs many display mounds in his territory, each a slightly domed earth circle about 1.5m (5ft) in diameter. During the breeding season he spends up to 50 percent of daylight hours singing and displaying on his mounds, with performance peaking in the first three hours after dawn and in mid-afternoon.

The display features the male's elaborate tail, which is thrust up and forward over the back and head. During "invitation display," it is held closed and vibrated rapidly. In "full display," it is fully fanned so that the singing male is visible to a visiting female through a lacy curtain of feathers. The male moves rapidly from side to side and then repeatedly jumps, his wings loose against his sides. These latter displays are accompanied by vocalizations resembling twanging and galloping sounds. Copulation occurs on the mounds. Song comprises a species-typical "territorial song" and an extensive repertoire of mimicked sounds (see Vocal Mimicry in Lyrebirds). It is loud, penetrating, and delivered in long bouts.

VOCAL MIMICRY IN LYREBIRDS

Mimicry comprises 70–80 percent of the song of male lyrebirds. Both species mainly mimic the songs or calls of other native and introduced bird species. Male Superb lyrebirds mimic about 25 percent of cohabiting species, mostly very accurately; the total number can amount to 20 or so species, but they are not all necessarily mimicked by each male. Vocalizations of the Satin bowerbird (*Ptilonorynchus violaceus*) dominate mimicry by the less versatile male Albert's lyrebird, and they are broadcast more loudly than those of the other 16 or so bird species that males collectively mimic. Male lyrebirds also commonly mimic such sounds as the wingbeats, feather ruffling, bill snapping, and begging calls of other birds, as well as the calls of frogs and a few mammals.

Male Superb lyrebirds can apparently mimic several different sounds simultaneously. They also often interrupt singing to spontaneously mimic other species' calls that are being given at the time, and engage in countersinging of mimicked vocalizations with neighboring males. The lyrebird's ability to mimic man-made sounds has often been exaggerated. However, there are a few fairly persuasive records of mimicry of sounds such as human whistles and speech fragments, vehicles, cameras, and musical instruments, although some of these are by captive birds.

Mimicry is largely culturally transmitted, being learned from older males rather than directly from the model species. Thus the introduced Tasmanian birds retained mimicry of purely mainland species for more than a generation, although they did also eventually begin to mimic some of the island's endemic birds. Young male lyrebirds take some time to perfect mimicry, and adult females occasionally mimic extensively too. Not surprisingly, which species are mimicked varies geographically, while the lyrebird's own "territorial song" has learned vocal dialects too. Males at a particular locality typically share a few themes that are stable for several years and that differ from those in other, acoustically isolated areas.

Why do lyrebirds mimic? No unequivocal answer is possible yet, but a persuasive theory is that females effectively select for male vocal novelty through their mate-choice behavior. Whether novelty manifested through mimicry conveys useful information about male genetic quality or is just a good, but arbitrary, sexual attractant is an even harder question to answer.

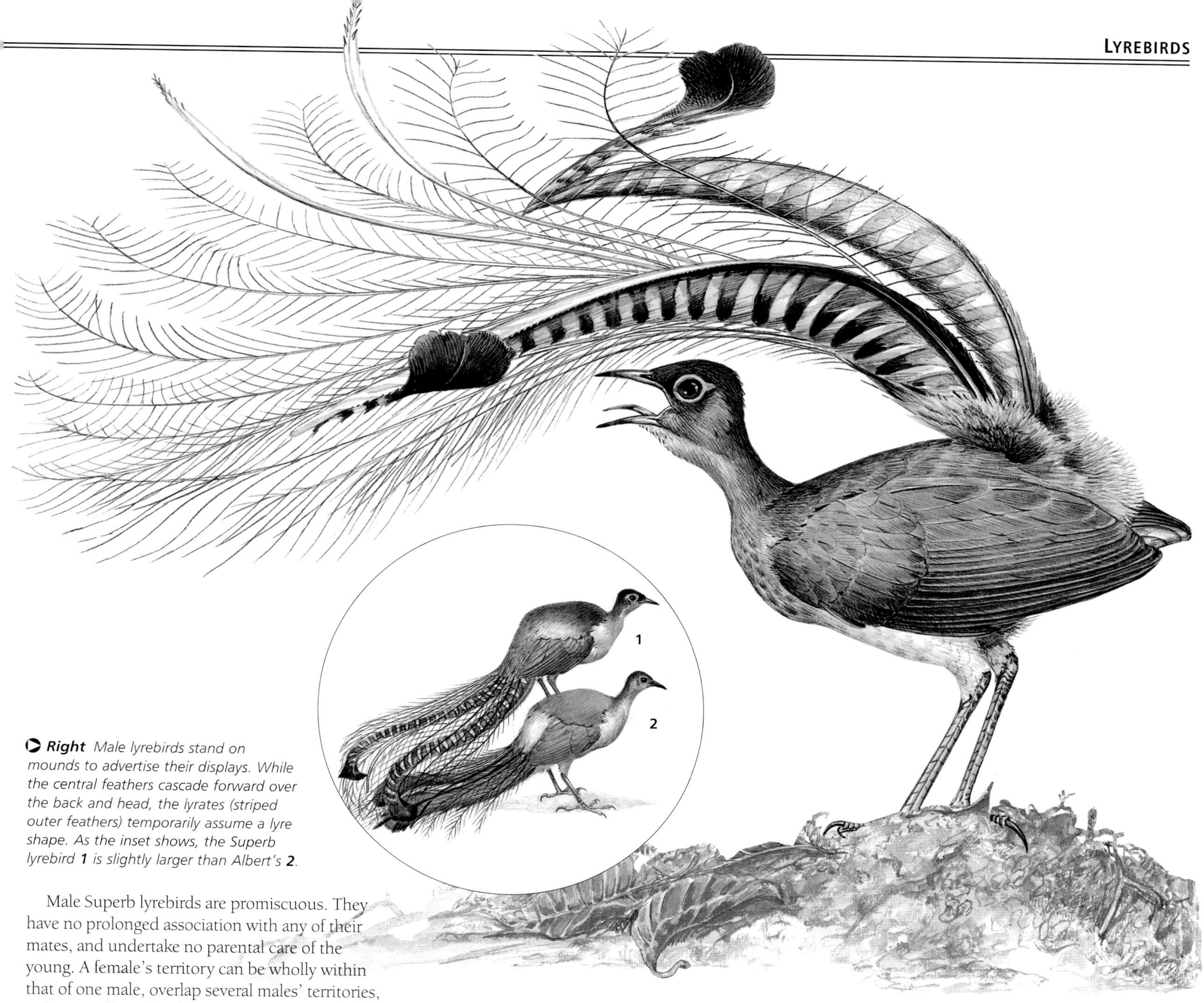

Right *Male lyrebirds stand on mounds to advertise their displays. While the central feathers cascade forward over the back and head, the lyrates (striped outer feathers) temporarily assume a lyre shape. As the inset shows, the Superb lyrebird* **1** *is slightly larger than Albert's* **2**.

Male Superb lyrebirds are promiscuous. They have no prolonged association with any of their mates, and undertake no parental care of the young. A female's territory can be wholly within that of one male, overlap several males' territories, or be spatially isolated from the territory of the male with which she copulates. The bulky nests are roofed, widely spaced, and mostly built on or fairly close to the ground.

Many features of the female's breeding regime are unusual. The clutch usually comprises one egg. Although breeding occurs in winter, incubation constancy is only 45 percent. The egg is deserted for several hours each day, during which time the embryo cools to ambient temperature. Consequently, embryonic development is very protracted, and the incubation period lasts 50 days, which is 80 percent longer than would normally be expected for a bird of this size. Nonetheless, the egg does not lose a dangerously high fraction of its water content, because the shell has a relatively low porosity and water vapor conductance. Nestling growth is also unusually slow for a passerine bird of this size, so that the nestling period is almost as long as the incubation period. Predation, mainly by exotic mammals, is a prominent cause of nesting failure, and the mother is sometimes killed too. The nestling fledges at about 63 percent of adult weight and is partly dependent on its mother for an extended period of 8–9 months after leaving the nest. She feeds it 88–138 small meals each day.

Male Albert's lyrebirds are also territorial, and display on platforms of vines and twigs on or near the ground. Their extensive singing in the winter breeding season also incorporates mimicry of other species, and full display involves tail manipulation similar to that of the Superb lyrebird and a high-stepping "dance." The male's movements cause the vines and twigs forming the platform to vibrate, sometimes shaking the foliage several meters away. A feature of the birds' song is so-called "gronking," which involves a series of short, loud notes preceded and followed by softer notes. Singing bouts typically last 30–50 minutes, but bouts of more than an hour are quite common. Nesting behavior resembles that of the Superb lyrebird, but the mating and social systems of this extremely shy species remain a mystery.

The Threat from Forest Management

CONSERVATION AND ENVIRONMENT

Both species, but particularly the Superb lyrebird, were formerly shot and trapped in large numbers for their meat and plumage. The Superb lyrebird is still abundant, but Albert's lyrebird is considered Vulnerable, with perhaps fewer than 10,000 individuals remaining. Both species have been badly affected by habitat clearing for agriculture, forestry, and settlement. Intensive forest management is currently the most serious threat to both lyrebirds, because densities are much reduced after selective logging and are lower still in pine and eucalyptus plantations. Adults and young are also preyed upon by exotic mammals.

AL

Scrub-birds

THE SCRUB-BIRD FAMILY CONTAINS ONLY TWO species, both of which are Australian, and which are separated from each other by a distance of over 3,000km (almost 2,000 miles). In their dense habitat, the most conspicuous character of scrub-birds is the loud territorial song of the male.

Scrub-birds are small, solidly-built birds that have strong, pointed bills, long, powerful legs, longish, tapered tails, short, rounded wings, and brown plumage: all ideal attributes for birds that live close to the ground in dense vegetation. Both species are fast and alert, but have very limited powers of flight.

Birds of the Bush

FORM AND FUNCTION

While scrub-birds are most closely related to another endemic Australian family, the lyrebirds, the broader affinities of these two groups within the passerines are less certain. Scrub-birds and lyre-birds share a number of unusual anatomical characteristics, including a syrinx (voice box) that is operated by three instead of four muscles and a cartilaginous furcula (wish-bone). Such peculiarities have fuelled various arguments linking these birds to such far-flung groups as the tapaculos of South America. However, recent evidence from DNA analysis suggests altogether less exotic affinities, indicating that the scrub-birds and lyrebirds may have shared a common origin with most other Australian passerines, from which they split 40–45 million years ago.

The Noisy scrub-bird is confined to a few coastal areas in the far southwestern corner of Western Australia, while the Rufous scrub-bird occurs in the ranges of northeastern New South Wales and far southeastern Queensland. Both species prefer habitat that provides very dense vegetation close to the ground, a moist micro-climate, and abundant leaf litter. Suitable habitat for the Noisy scrub-bird is confined mainly to coastal low forest and thicket, whereas that of the Rufous scrub-bird occurs mostly beneath breaks in the rainforest canopy, or in adjoining moist eucalypt forest, well-buffered from fire.

The disjunct distribution of the two current species suggests that scrub-birds were once distributed more widely, at a time when moist forests covered much of Australia during the mid-Tertiary era some 30 million years ago. Their present relict status is the result of climatic and vegetation changes that have occurred since the Miocene (24–5 million years ago), and in particular the severe climatic oscillations in the Pleistocene (1.8 million–10,000 years ago).

The diet of scrub-birds consists of a wide variety of invertebrates, with the occasional small lizard, gecko, or frog. Nestling Noisy scrub-birds are fed species from at least 18 orders of invertebrates, the most common being spiders, grasshoppers, cockroaches, and various larvae. Adults forage mainly in leaf litter, rushes, and small shrubs, where they move slowly while looking, listening, and occasionally turning over leaves with a quick flick of the head.

Defense by Singing

BREEDING BIOLOGY

Male scrub-birds occupy permanent, well-dispersed territories, within which they spend at least 80 percent of their time in a core area of 1–2ha (2.5–5 acres). There is normally one female associated with each male's territory, and she occupies a nesting area on its periphery. Males of both species defend their territories with an extremely loud and penetrating territorial song, particularly during the breeding season. The male Rufous scrub-bird is also an accomplished mimic.

Rufous scrub-birds breed in spring and early summer, while Noisy scrub-birds breed in winter. Female Noisy scrub-birds breed in their first year, males from the age of 3 onward. The birds may live until at least 9 years of age.

The female Noisy scrub-bird builds a domed nest with a small side entrance, and lines the bottom of the cavity with decayed nest material or decayed wood, which dries to form a hard,

Left *Rufous scrub-birds are the more numerous of the two species, although the total population is still only in the low thousands. Very much ground-dwellers capable only of fluttering rather than flying, these birds spend much of their time foraging through, or even under, the leaf litter in search of prey.*

Right *The Noisy scrub-bird of Western Australia has been saved from extinction by careful management aimed mainly at protecting it from bush fires, to which, as a weak flier, it is extremely vulnerable. The species was once restricted to a single, mountainside breeding site, preserved from wildfires by the nature of the terrain.*

FACTFILE

SCRUB-BIRDS

Order: Passeriformes

Family: Atrichornithidae

2 species in 1 genus, *Atrichornis*.

Distribution
Australia, in two disjunct populations.

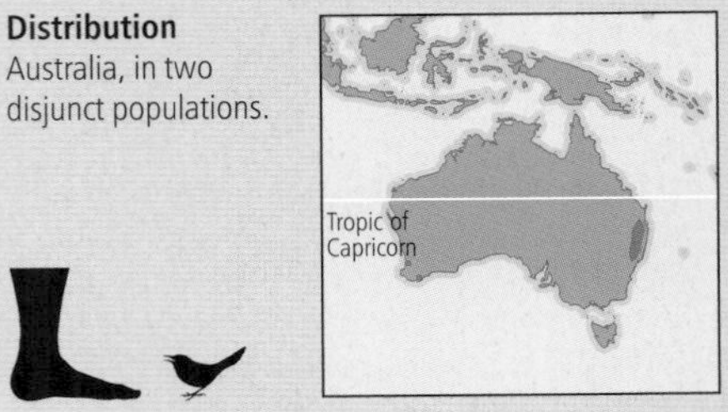

NOISY SCRUB-BIRD *Atrichornis clamosus*
The far SW corner of Western Australia. Low forest, thicket, and heath. **Size**: Length 19.5–23cm (7.7–9.1in); weight 34–52g (1.2–1.8oz); females smaller than males. **Plumage**: Upper parts brown, with fine, darker crossbars; underparts range from white on throat to rufous around the anus; male has black bar on the upper breast. **Voice**: Loud, variable song of 10–20 notes; another shorter and variable song that may incorporate mimicry of other birds; a three-note and various single-note calls. **Nest**: Domed, with side entrance. **Eggs**: 1, buff with irregular patches of brown, mainly at the larger end; 2.9 x 2cm (1.1 x 0.8in). Incubation period: 36–38 days; nestling period: 21–28 days. **Diet**: Invertebrates, with occasional small vertebrates. **Conservation status**: Vulnerable.

RUFOUS SCRUB-BIRD *Atrichornis rufescens*
NE New South Wales and far SE Queensland. Rain forest and adjacent eucalypt forest. **Size**: Length 16.5–18cm (6.5–7in): females smaller than males. **Plumage**: Upper parts rufous brown, with fine black crossbars; underparts white on throat, remainder rufous brown; male has black mottling on the throat and breast. **Voice**: Loud song of 4–20 *chip* notes; various calls; accomplished mimic of other birds. **Nest**: Domed, with side entrance. **Eggs**: 2, pink buff with blotches of brown, mainly at larger end; 2.3 x 1.8cm (0.9 x 0.7in). Incubation and nestling periods unknown. **Diet**: Invertebrates. **Conservation status**: Lower Risk/Near Threatened.

papier-mâché-like material. The nest takes up to three weeks to build, and the single egg is laid a week later. (Female Rufous scrub-birds line the nest cavity completely, and may lay two eggs.) The female incubates the egg and feeds the chick on her own. The egg hatches after 36–38 days, an exceptionally long period, but, as in the lyrebirds, it is usually left unincubated for part of the day. After the chick leaves the nest it stays with its mother, probably until after it has finished its first molt when 2–3 months old.

Back from the Brink

CONSERVATION AND ENVIRONMENT

The Noisy scrub-bird was originally found in six isolated localities between 1842 and 1889. It was then thought for over 70 years to have gone extinct, until a small population was rediscovered in 1961 at Two Peoples Bay, just east of Albany. Despite plans for the development of a new town, conservationists succeeded in having the locality set aside as a reserve for the species. Even after its rediscovery, the bird carried for many years the dubious distinction of being Australia's rarest passerine. Thanks to an intensive program of management, however, the total number of breeding territories has now been increased from less than 50 to almost 600. The conservation status of the Noisy scrub-bird has recently been downgraded from Endangered to Vulnerable.

The distribution and abundance of the Rufous scrub-bird have also declined markedly since European settlement. However, although rare, this species is not considered endangered. SF/GTS

Bowerbirds

MANY ORNITHOLOGISTS SEE BOWERBIRDS *as by far the most behaviorally complex and intriguing of all birds. Not only do the males of many species construct elaborate structures that they decorate with variable and often colorful objects – fruits, berries, fungi, tinfoil, bits of plastic – but some actually paint them with natural pigments, applied with the aid of a "paintbrush" held in the bill.*

Bowerbirds are stout, strong-footed, typically heavy-billed birds; their sizes range from that of a starling to a small crow. The family comprises three species of socially monogamous and territorial catbirds and 17 species that are known or presumed to reproduce polygynously. The males of the polygynous species are responsible for bower-building, an activity that has earned them a longstanding popular reputation for high intelligence and esthetic sensibility. Indeed, it has recently been proved that they have relatively larger brains than ecologically similar songbirds of their size and zoogeographical region, and that the bower-building species have larger brains than the non-bower-building ones.

FACTFILE

BOWERBIRDS

Order: Passeriformes

Family: Ptilonorhynchidae

20 species in 8 genera: **Adelbert** or **Fire-maned bowerbird** (*Sericulus bakeri*), **Flame bowerbird** (*S. aureus*), **Regent bowerbird** (*S. chrysocephalus*), **Archbold's bowerbird** (*Archboldia papuensis*), **Sanford's bowerbird** (*A. sanfordi*), **Fawn-breasted bowerbird** (*Chlamydera cerviniventris*), **Great bowerbird** (*C. nuchalis*), **Lauterbach's** or **Yellow-breasted bowerbird** (*C. lauterbachi*), **Spotted bowerbird** (*C. maculata*), **Western bowerbird** (*C. guttata*), **Golden bowerbird** (*Prionodura newtoniana*), **Macgregor's bowerbird** (*Amblyornis macgregoriae*), **Streaked bowerbird** (*A. subalaris*), **Vogelkop bowerbird** (*A. inornatus*), **Yellow-fronted bowerbird** (*A. flavifrons*), **Satin bowerbird** (*Ptilonorhynchus violaceus*), **Tooth-billed bowerbird** (*Scenopoeetes dentirostris*), **Green catbird** (*Ailuroedus crassirostris*), **Black-eared catbird** (*A. melanotis*), **White-eared catbird** (*A. buccoides*).

DISTRIBUTION New Guinea, Australia.

HABITAT Tropical, temperate, and montane rain forests; riverine and savanna woodland; rocky gorges; grassland; dry arid zones.

SIZE Length 21–38cm (8.5–15in); weight 70–230g (2.5–8oz); males larger than females, except in Golden and *Sericulus* bowerbirds, in which the male is smaller.

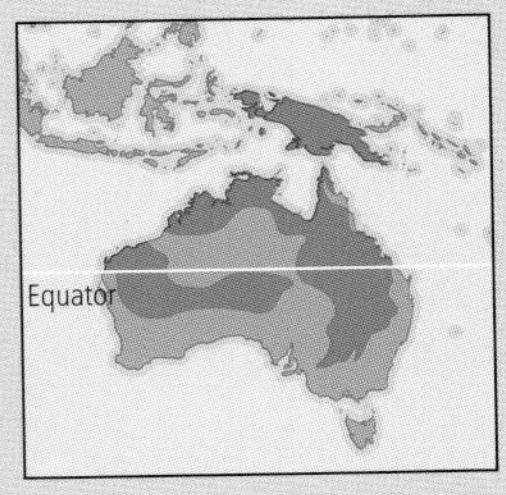

PLUMAGE 9 species with mainly brown, gray, or green cryptic coloration. Males of remaining species with yellow or orange crest or cape; or iridescent yellows, reds, or blues, with females drab brown, gray, or green, with ventral barring.

VOICE Mimicry of birds and other animals and mechanical noises, harsh churring and scolding notes, and catlike wails.

NEST Bulky, open cup- or bowl-shaped structure upon a substantial foundation of woody twigs, leaves, and tendrils built into a tree fork, vine, mistletoe tangle, or (Golden bowerbird only) in a tree crevice.

EGGS 1–2, rarely 3; plain off-white to buff or blotched and vermiculated with colored scrawing lines, predominantly at the large end. Incubation period about 21–27 days; nestling period approximately 17–30 days.

DIET Fruit, insects, other invertebrates, lizards, other birds' nestlings.

CONSERVATION STATUS The Adelbert bowerbird is Vulnerable, while Archbold's bowerbird is listed as Lower Risk/Near Threatened.

So architecturally complex and "tastefully" decorated can bowers be that early Europeans refused to believe that birds had built them; instead, they chose to believe that they were made by human mothers to entertain their children. New Guinean men admire the male birds for their industry and artistry in acquiring and displaying what they consider the birds' equivalent of the "brideprice" they themselves offer for wives. The bowers in fact have nothing to do with nesting, which is entirely the concern of females.

Master Builders

FORM AND FUNCTION

The largest bowerbird is the Great and the smallest the Golden, the latter being 30–40 percent lighter and smaller than the former. Males are typically heavier and larger in most body measurements than conspecific females, notable exceptions being the female Golden and *Sericulus* bowerbirds.

Left *In the gloom of the forest floor, the brilliant orange and yellow plumage of the male Flame bowerbird stands out strikingly. In common with other ostentatious species, it builds a modest "avenue" bower.*

Left *Bowerbirds are the supreme artisans of the bird world. The males build bowers to attract females for courting and mating; as such, they are external signs of an individual male's fitness to breed. The Great bowerbird uses a variety of natural and artificial objects to decorate his bower, augmenting the display by raising his crest (above) when a female appears.*

The family exhibits 50–60 different plumages, given that most of the 20 species have a juvenile and an adult male and female (and in some cases a subadult male) plumage. The monogamous catbirds are sexually monochromatic, but the polygynous genera are typically sexually dichromatic to varying degrees. The catbird sexes are similar, being generally green with white spotting on the breast, wings, and tail, and about the head and/or throat. Sexes of the polygynous Tooth-billed bowerbird are also identical, being olive-brown above and heavily streaked brown on dirty white below. Males of the other forest-dwelling polygynous species are brightly colored; glimmering gold, or orange, and black, as in the Flame, Adelbert, Regent, and Archbold's bowerbirds; iridescent blue-black in the well known Satin; brilliant yellow and golden-olive in the Golden; or generally brown, with a contrasting orange or yellow crest, as in most of the *Amblyornis* species. The Spotted, Great, Fawn-breasted, and Lauterbach's bowerbirds of grasslands and more arid woodlands are generally drab gray or brownish, with small, pinkish nape crests. Females of the promiscuous species wear drab, cryptically colored and marked plumage, being predominantly brown, olive, or gray, often with barring or spotting. Juvenile and immature plumages are generally similar to those of adult females.

Once adult, bowerbirds have a high average life expectancy, some individuals living for 20–30 years. Males of polygynously reproducing species take up to seven years to attain adult plumage. Bowerbirds are exceptional in having 11–14 secondaries (including the tertials), unlike typical songbirds, which have 9–10, while their enlarged lachrymal (part of the skull cranium, near the orbit) is paralleled only in the Australian lyrebirds (Menuridae).

Residents of the Rain Forests

DISTRIBUTION PATTERNS

Ten bowerbird species live only in New Guinea, eight only in Australia, while the other two are common to both. Most inhabit wet forests, up to 4,000m (13,000ft) above sea level in the case of Archbold's bowerbird (discovered as recently as 1940). Several are extremely localized, like the Adelbert (confined to the Adelbert Mountains of Papua New Guinea) and the Golden and Tooth-billed bowerbirds, which are found only in rain forests above 900m (3,000ft) on and around the Atherton Tableland in tropical north Queensland, Australia. Other species, notably New Guinea's Flame and Australia's Spotted and Great bowerbirds, have extensively continuous ranges, while those of most others are patchy and broken. Fifteen species inhabit wet tropical and montane rain forests, rainforest edges, and/or adjacent sclerophyll habitats, while the five *Chlamydera* species inhabit riverine and savanna woodland, rocky gorges, grassland, and semi-arid zones. A monogamous catbird species may occur in sympatry with one or more polygynous bowerbird species.

Fruits of the Forest

DIET

Most bowerbirds are predominantly fruit-eaters, but flowers, nectar, leaves, arthropods (mostly insects), and other animals including small vertebrates are taken. Figs are a major component of Australian catbird diets. Animal foods are important to nestlings of polygynous species, which tend to be fed one particular kind of animal (cicadas, beetles, skinks, or grasshoppers) by their mothers. Unlike birds of paradise, bowerbirds do not use their feet to hold or manipulate food or other items, and do not regurgitate food to nestlings. Catbirds store, or cache, fruits about their territories, and the males of some polygynous species do so about their bower sites.

Bills are typically the stout and powerful kind of the generalist omnivore, this family not exhibiting the specialization found in birds of paradise. Exceptions are the fine long bill of the Regent bowerbird, apparently modified for flower nectar-feeding, and the falcon-like mandibles of the Tooth-billed bowerbird, modified for leaf eating. Tooth-bills eat considerable amounts of foliage in winter, using their stout, "toothed" bill for tearing leaves and stems; complex structures on the inner mandible surfaces serve to masticate the leaves – a very unusual diet for a songbird. Some *Sericulus*, *Ptilonorhynchus*, and *Chlamydera* species form winter flocks that may invade commercial orchards. Satin bowerbird flocks will ground-feed on grasses and herbs. Most other polygynous species appear sedentary and are probably solitary in winter.

Soliciting by Building

BREEDING BIOLOGY

The courts or bowers that males (catbirds excepted) clear, build, and decorate are created to impress females and, possibly, to intimidate rival males. Males of bower-tending species are promiscuous, attracting as many females as possible to their bower by calls and/or colorful plumage. The bowers are critical to male reproductive success. Discerning females assess the frequency and intensity of male vocalizations and bower attendance, the quality and/or quantity of the bowers themselves and their decorations, and also the males' displays and plumage, before soliciting the

THE TRANSFERAL EFFECT

In bowerbirds, the gaudy display plumage of promiscuous males is said to be "transferred" to the bower displays. There are various conspicuous stages of this transfer across species: birds that have retained colorful adult male plumage build small, sparse, and dully decorated bowers, whereas the most elaborate and decorated bowers are built by duller males.

There are three basic bower types. First is the "court" of the Tooth-billed bowerbird – not a structure but a cleared area of forest floor decorated with leaves. Second are the "maypole" bowers of the Golden, Archbold's, and the *Amblyornis* bowerbirds; this latter New Guinea group were once known as "gardeners" because of the meticulous arrangements of decorations they place upon a manicured, lawnlike "mat" of moss. Maypole bowers typically consist of accumulations of woody sticks and twigs placed about the vertical trunks of sapling trees or tree ferns. Three maypole builders build complex bowers that look like small towers or thatched buildings, with a neat basal moss mat decorated with colorful flowers, fruits, beetle wing-cases (elytra), tree resin, fungi, insect castings, and other objects.

The transferal effect is clearly demonstrated by certain *Amblyornis* species and their maypole bowers, which vary from the simple stick tower of Macgregor's **1** to the complex, hutlike structure of the Vogelkop bowerbird **2**. Macgregor's, which has a large orange crest, builds the simplest bower. The Streaked has a reduced crest but a more complex bower **3**, while the Vogelkop bowerbird, the most sophisticated builder of all, completely lacks a crest.

The relationship between the relative luxuriance of the birds' crests and their bowers was first noticed in the 1960s, when the bower of the splendidly crested Yellow-fronted was still undiscovered. The transferal effect theory suggested that the bower would be simple in design. And, when they were subsequently found in remote Irian Jaya, they did indeed resemble the plain stick towers of Macgregor's bowerbird.

The third bower type are the "avenue" bowers of *Sericulus*, *Ptilonorhynchus*, and *Chlamydera* species: Those of the Flame, Satin **4**, and Spotted bowerbirds feature two parallel stick walls that form a central avenue. Here too there is clear evidence of transfer, with the crestless species building more complex and ornately decorated bowers than the crested ones.

Sexual selection through female choice has apparently caused this transference of colorful visual signals from the crest of adult males to their bowers. Females select males with superior bowers and thus enhance bower architecture, because males that make such bowers reproduce far more often than those that do not. Once such a transfer of secondary sexual characters to the bower was underway, bright plumage would have become an active disadvantage, attracting predators to the bower where the males spend much of their time.

1
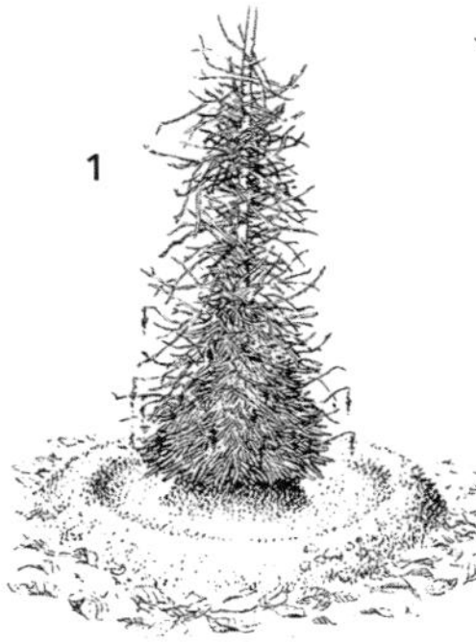

2

3
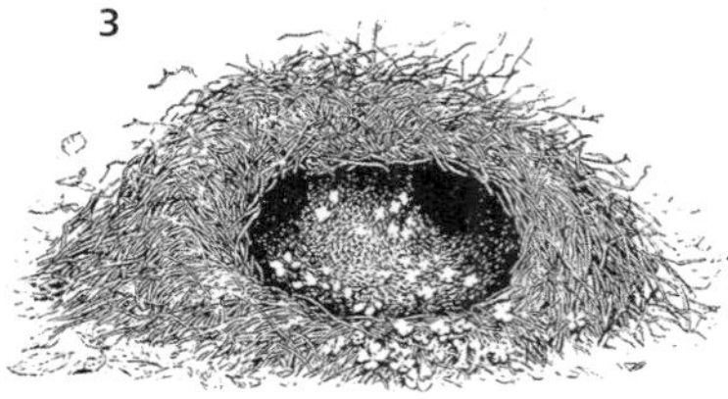

4

Above *A Green catbird with its young. Uniquely within the family, Australian catbird nests have a layer of decaying wood and/or mud placed beneath the center of the cup lining.*

Left *Regent bowerbirds feeding on grain left for them on a tree stump. The long bill of this species is specialized for eating nectar, its usual diet.*

male of their choice. It is the older males that are typically selected by females and are most likely to be successful in obtaining multiple matings. Importantly, males of the more colorful species build modest bowers, while the drabber ones construct the largest and most complex structures (see The Transferal Effect).

The bowers are located in favored spots exhibiting one or more required micro-environmental features. Until recently, promiscuous male bowerbirds were presumed to form breeding colonies or leks, their bowers being clustered in associated congregations, but this behavior has not been confirmed except for the Tooth-bill. Male Tooth-bills clear a forest floor "court" area of litter and lay green leaves on it, with the paler side uppermost; they then call almost continuously to attract females to the site. Their courts appear to be unevenly dispersed through suitable habitat to form denser aggregations that serve as leks. In contrast, studies of Regent, Satin, Fawn-breasted, Macgregor's, Archbold's, and Golden bowerbirds suggest that they do not form leks, their bowers being evenly distributed. Spotted bowerbirds may even form leks of a kind in some areas.

Bower sites are occupied for decades, and adult males exhibit longterm fidelity to them. Some Satin bowerbird sites have been used for up to 50 years. Immature males serve an "apprenticeship" of 5–6 years, visiting the bowers of other males and constructing rudimentary or "practice" bowers of their own while polishing their skills. Adequate bower-building is not innate but, to judge at least from Satin bowerbirds, is largely learned.

In those bower-building species that have been studied, females defend only their nest site, while males defend only the bower and its immediate vicinity. A single female may use a nest location over consecutive years. The nest itself typically consists of a stick foundation under a cup of dried leaves and twigs that is lined with tendrils and similar finer materials. In Macgregor's and Golden bowerbirds, the nests are set on average 2m (6.5ft) above the ground, a figure that rises to 15m (50ft) in Satin bowerbirds and Tooth-bills.

The elliptical eggs are either pale and unmarked or else colored and vermiculated. Clutch size is 1–3 eggs for both monogamous and polygynous species (with a higher average clutch size in Australia than in New Guinea). Unlike in most songbirds, eggs are laid on alternate days. Incubation periods span 21–27 days, and in both uni- and biparental species the birds spend on average 70 percent of the daylight hours incubating. Nestling periods span 17–30 days, brooding constancy averaging 31–36 percent of daylight in Australian catbirds and 16–30 percent in polygynous species. There is no evidence of two broods being successfully raised in a single season. Offspring have a long period of post-nest parental dependency.

The socially monogamous catbirds defend an all-purpose territory year-round, and pairs remain together over several seasons, if not for life. Male catbirds do not build nests, incubate the eggs, or brood nestlings, although they do feed them.

Mostly Stable

CONSERVATION AND ENVIRONMENT

While several Australian species have lost parts of previously more extensive ranges to habitat destruction or degradation, none is rare or endangered and most present populations are stable. The Action Plan for Australian Birds (2000) makes no reference to bowerbirds, except for noting that one Western bowerbird subspecies – *Chlamydera guttata carteri* – is "near threatened" because of its highly restricted range. Where New Guinean species are concerned, the principal threat to the Adelbert and Archbold's bowerbirds is habitat destruction, as well as the recent spread of domestic and feral cats and other exotic vertebrates.
CBF/DWF

Right *A male Archbold's bowerbird, with its bright yellow crest. This maypole-building species from the New Guinean highlands is at risk from logging.*

STORY

Bowerbirds – Avian Artists

① The largest member of the bowerbird family is the Great bowerbird. This Australian species, a generally drab bird (apart from its bright lilac crest), builds an elaborate "avenue"-type bower. White is a favored color; this individual has decorated its bower with snail shells and broken bottles.

② Another avenue-builder is the Satin bowerbird. This species is characterized by its strong predilection for blue objects and marked sexual dimorphism. Here, a female waits in the bower while the glossy, iridescent male (from whose plumage the species derives its common name) brings feathers. Satin bowerbirds will also paint the inside walls of the avenue with pigments made from a mixture of plant extracts and saliva.

4

❸❹ *By far the most sophisticated construction is that erected by the Vogelkop bowerbird of Irian Jaya. Shaped like thatched huts, these extraordinary structures are woven between saplings. They are carefully maintained and used season after season. The small, unobtrusively colored birds adorn their bowers with piles of flowers or a wide variety of objects, including rope, tin lids, berries and fungus.*

❺ *A Golden bowerbird working on its "twin maypole" bower. The towerlike edifice created by this species can be up to 2m (6.5ft) tall.*

❻ *The colorful male Regent bowerbird of Australia builds a relatively simple bower of interwoven twigs.*

5

6

Fairy-wrens and their Allies

A SPECIES OF FAIRY-WREN CAN BE FOUND IN nearly every kind of habitat throughout the 40 degrees of latitude covered by Australia and New Guinea, ranging from dense, tall forest in well-watered southwest Australia (the Red-winged fairy-wren) to the spinifex-covered sand dunes of the interior desert (Eyrean grasswren). Characteristically, fairy-wrens have tails longer than their bodies, and these are carried cocked jauntily most of the time.

The true fairy-wrens are diminutive birds that are endemic to Australia and New Guinea. Wherever they occur the ground cover is generally thick, and they hop rapidly through it on their long legs. Their wings are short and rounded, and they rarely make prolonged flights. Emu-wrens live deep in dense heathland: it is hard to catch more than a fleeting glimpse of them. Grasswrens are birds of the deserts, arid shrublands, and rocky plateaus, and they too are rarely seen.

Above *The Splendid fairy-wren* (Malurus splendens) *is often called Australia's most spectacular bird. This example is the Black-backed subspecies* (M. s. melanotus), *from the southeast of the continent.*

Below *Endemic to Australia, a Superb fairy-wren exhibits the long, jauntily cocked tail and vividly colored plumage typical of the true fairy-wrens of the* Malurus *genus.*

Right *Three Superb fairy-wren chicks huddle for shelter amid tussock grass in south-eastern Australia. The young fledge after about 10 days, but continue to be fed for several weeks after leaving the nest.*

Small, Jaunty, and Close to Home

FORM AND FUNCTION

Males of the genus *Malurus* are brightly colored in enameled reds, blues, blacks, and whites, while the females and immatures are usually brown. In the emu-wrens also there are clear differences between the sexes, but in the other three genera the sexes are hard to tell apart in the field.

Most of the fairy-wrens depend largely on insects for food, but the grasswrens with their sturdier bills eat large quantities of seeds as well. Foraging is largely done as the birds bound along on the ground, although some species search through shrubs and even in the canopies of tall trees. Because their wings are small and rather inefficient, fairy-wrens only rarely fly sorties after airborne insects.

Throughout a wide variety of environments most species spend their lifetime within a prescribed area, which they defend from trespass by members of their species outside the immediate family; within it they find all their requirements for feeding, breeding, and shelter. Such "territories" cover from 1–3ha (2.5–7.4 acres), and tend to persist from year to year with little change.

Living in Extended Families

BREEDING BIOLOGY

Despite their small size the birds may live for a surprisingly long time (more than 10 years) compared with their counterparts in the northern hemisphere. A long life and residential status allows them to maintain family bonds beyond the usual period during which offspring depend on their parents for food and protection; most fairy-

FACTFILE

FAIRY-WRENS AND ALLIES

Order: Passeriformes

Family: Maluridae

27 species in 5 genera. Species and genera include: **true fairy-wrens** (genus *Malurus*), **grass-wrens** (genus *Amytornis*), **emu-wrens** (genus *Stipiturus*), **Purple-crowned fairy-wren** (*Malurus coronatus*), **Red-winged fairy-wren** (*M. elegans*), **Superb fairy-wren** (*M. cyaneus*); **Black grasswren** (*Amytornis housei*), **Carpentarian grasswren** (*A. dorotheae*), **Eyrean grasswren** (*A. goyderi*), **Gray grasswren** (*A. barbatus*), **Kalkadoon grasswren** (*A. ballarae*), **Short-tailed grasswren** (*A.merrotsyi*), **White-throated grasswren** (*A. woodwardi*), **Mallee emu-wren** (*Stipiturus mallee*).

DISTRIBUTION New Guinea and Australia

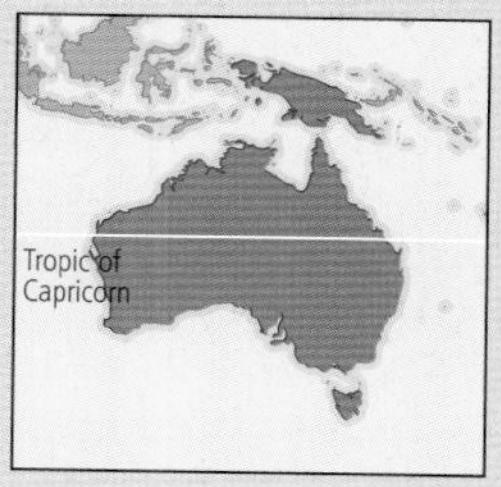

HABITAT From margins of rain forest to desert steppes, salt pans, coastal swamp, heathland, spinifex tussocks, desert sandplains.

SIZE Length 14–22cm (5.5–9in); weight 7–37g (0.25–1.3oz).

PLUMAGE Varies from bright blue in some males to plain brown.

VOICE Brief contact calls, churrs, and sustained reels of song.

NEST Domed with side entrance in *Malurus* and *Stipiturus*; half-domed or truncated spheres in *Amytornis*.

EGGS 2–4, whitish with red-brown speckling. Incubation period 12–15 days; nestling period 10–12 days.

DIET Insects and seeds

CONSERVATION STATUS The Mallee emu-wren and the White-throated grass-wren are both Vulnerable.

wrens tend to live in groups with more than two adults capable of breeding, yet only one female lays the eggs. Although the senior male in a group may fertilize some of his partner's eggs, a surprisingly large percentage are fathered by other males, often from outside the social group. Careful experimentation and radiotelemetry has shown that it is the female who chooses her partner, and that she seeks out her mate before sunrise during her fertile period.

In most species of fairy-wren the female builds the nest, lays and incubates the eggs, and broods the young when they first hatch. Her partner may call her off the nest and escort her while she forages hurriedly before returning to brood. The young fledge after 10 days or so in the nest, and at first are unable to fly, although they will scuttle fast over the ground and hide. If a predator approaches a nest, all group members perform a frenzied "rodent run" display: instead of hopping, their usual mode of progression, they run, keeping close to the ground, trailing their tails, and squeaking – hence the name of the performance. This behavior is surprisingly effective in distracting a variety of predators, from snakes to humans.

The young are fed for several weeks after they have left the nest. Fairy-wrens nest several times during a season, and the nonbreeding members of the group may take over the raising of an early brood to allow the breeding female to start a second (or third) nest. Later in the season, young hatched earlier will attend their nestling-siblings and help to raise them, so that the habit of "helping" appears early in life.

A Future in Parks and Gardens?

CONSERVATION AND ENVIRONMENT

Most of the fairy-wrens and their allies have managed to survive the environmental impact of European settlement, along with its accompanying exotic pests, the feral cat and the fox. While only two entire species are currently threatened (the White-throated grasswren and the Mallee emu-wren), four others – the Variegated fairy-wren, White-winged fairy-wren, Gray grasswren, and Southern emu-wren – have individual taxa that are considered Vulnerable.

One of the attractions of the family is that some species have thrived in mushrooming suburbia. The Superb fairy-wren in particular is common in the parks and gardens of six Australian state capitals and brightens the lives of those who live there.

In the last 50 years three new species of grasswren have been recognized: the Gray, Kalkadoon, and Short-tailed. In addition, one has been rediscovered after a gap of 85 years (the Eyrean grasswren), and three others (the Carpentarian, White-throated, and Black) have become much better known. IR/ER

Below *If threatened by a predator, a fairy-wren will abandon its normal standing posture **1** and run forward, tail down and squeaking **2–4**. As a distraction, this "rodent-run" display is surprisingly effective.*

Honeyeaters and Australian Chats

Honeyeaters are the most successful passerine family in the Australasian region, with over 70 species in Australia alone. They have diversified to occupy a wide variety of habitats, from mangroves and rain forests to subalpine forests and semi-arid woodlands.

These birds are often the most numerous species present in an area, and there may be more than 10 different species of honeyeater in a single hectare (2.5 acres). Exploiting the resources of nectar-producing plants and carbohydrate-rich invertebrate exudates, they exhibit patterns of social organization that range from simple monogamous pairs to some of the most complex societies so far described for birds anywhere in the world.

On the Tip of the Tongue

FORM AND FUNCTION

All honeyeaters have a long, protrusible tongue with a brushlike tip that they use to extract nectar from flowers. They are important pollinators of flowers, and many are likely to have coevolved with certain species of plants.

In general they are slender-bodied, streamlined birds with long, pointed wings and undulating flight, but species vary greatly in size and habits. Bold and vigorous, they have strong legs and sharp claws that enable them to clamber agilely around flowers and foliage as they feed, sometimes upside-down. Many have long, down-curved, sharply-pointed bills, with variations in shape associated with differences in diet. Most honeyeaters are drab but a few are brightly colored, resembling their counterparts, the sunbirds of Africa and Asia and the hummingbirds of America. Almost all species have bare patches of colored skin that range from modest gape-stripes, eyerings, or eyepatches to large, brightly-colored, bare facial patches. In the vast majority of species the sexes are monochromatic in plumage, but males are usually larger than females. Among the few strikingly sexually dichromatic species, the plumage of juveniles may resemble adult females, as in the case of the Pied honeyeater, or be sexually dichromatic upon fledging, like that of the Crescent honeyeater.

The Australian chats also possess a brushlike tip to the tongue but are almost entirely insectivorous. In contrast to the majority of honeyeaters, chats are terrestrial foragers and are all sexually dichromatic in plumage. Males in three species have strikingly bright red or orange plumage. Juveniles' plumage resembles that of adult females.

Birds of the Southwest Pacific

DISTRIBUTION PATTERNS

Honeyeaters are endemic to the southwest Pacific, centered in Australia, New Guinea, Indonesia, New Zealand, and Hawaii, whereas the chats are endemic to Australia.

Honeyeater diversification has been so great that 14 of the 40 genera contain only a single species, and 10 genera only two species each. A few genera have many species: *Lichenostomus* has 20 and *Meliphaga* 13. *Myzomela* is the most widespread genus, occurring from Sulawesi in the west to Micronesia in the north and Fiji in the east. The friarbirds of the genus *Philemon* and species in the genus *Lichmera* both occur in Australia, but have undergone major radiations in Wallacea.

Molecular analyses have confirmed that the Australian chats (formerly Ephthianuridae) and Macgregor's bird of paradise (*Macgregoria pulchra*) are actually honeyeaters, whereas the Sugarbirds (*Promerops* spp.) are not. Moreover, two species previously regarded as honeyeaters – the Bonin Island honeyeater (*Apalopteron familiare*) and the Golden honeyeater (*Cleptornis marchei*) of Saipan Island – are actually White-eyes (Zosteropidae).

The Nectar Hierarchy

DIET

All honeyeaters will probably consume nectar if it is available. Similarly, the vast majority of honeyeaters will feed on invertebrates. Species differ greatly in their dependence upon either nectar or insects, although all take insects for the essential nutrients not present in nectar, and some are almost entirely insectivorous. Insects and spiders are obtained by a variety of methods, including leaf gleaning, probing under bark, and sallying in midair. Many honeyeaters also get a significant proportion of their carbohydrates from sources other than nectar, including lerp (the sugary protective covering over the nymphs of some psyllid bugs), manna (a sugary exudate from eucalypt foliage and branches), and honeydew (excretions of psyllid and coccid bugs). In rain forests, fruit can also be a major food source for honeyeaters

FACTFILE

HONEYEATERS & AUSTRALIAN CHATS

Order: Passeriformes

Family: Meliphagidae (including subfamily Epthianuridae)

182 species in 42 genera. There are 177 species of honeyeaters in 40 genera, including the **Blue-faced honeyeater** (*Entomyzon cyanotis*), **Bell miner** (*Manorina melanophrys*), **stitchbird** (*Notiomystis cincta*), **Western spinebill** (*Acanthorhynchus superciliosus*), and the **wattlebirds** of the genus *Anthochaera*. The Australian chats of the subfamily Ephthianuridae comprise 5 species in 2 genera: the **Crimson chat** (*Epthianura tricolor*), **Orange chat** (*E. aurifrons*), **White-fronted chat** (*E. albifrons*), **Yellow chat** (*E. crocea*), and **Gibber chat** (*Ashbyia lovensis*).

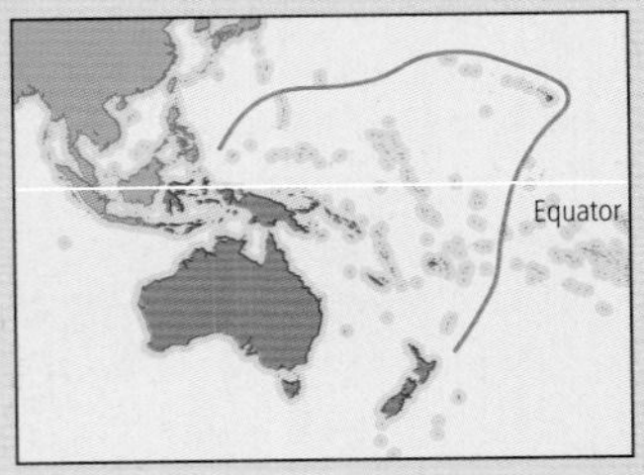

DISTRIBUTION Endemic to the SW Pacific, centered in Australia, New Guinea, Indonesia, New Zealand, and Hawaii. The Australian chats are limited to Australia.

HABITAT Honeyeaters are found in all habitats except open, arid country and grasslands; Australian chats in open scrubland, dry woodland, and desert, and around water margins.

SIZE Length: honeyeaters 8–48cm (3–19in), Australian chats 10–13cm (4–5in); weight: honeyeaters 6.5–150g (0.2–5oz), Australian chats 10–11g (0.4oz).

PLUMAGE Most honeyeater species are dull green, gray, or brown, some with black, white, or yellow markings. Australian chats are red, yellow, or black and white; males are more brightly marked than females.

VOICE Small honeyeater species are often musical, larger ones raucous. Australian chats utter metallic, twanging contact calls, aggressive chattering calls, and high-pitched whistles.

NEST Cup-shaped; those of Australian chats are set in bushes close to the ground or on the ground.

EGGS Honeyeaters lay 1–5 eggs (average 2); white, pinkish, or buff with reddish-brown spots. Incubation period 12–17 days; nestling period: 10–30 days. Australian chats usually lay 3–4 eggs, white or pinkish white with reddish brown spots.

DIET For honeyeaters, invertebrates, nectar and other sweet secretions, and sometimes fruit; for Australian chats, insects taken on the ground.

CONSERVATION STATUS 6 species are currently classed as Vulnerable, and 4 as Endangered – the mao (*Gymnomyza samoensis*), Crow honeyeater (*G. aubryana*), Regent honeyeater (*Xanthomyza phrygia*), and Black-eared miner (*Manorina melanotis*). In addition, 3 of 5 known Hawaiian species are certainly extinct, and the other 2 – the Kauai oo (*Moho braccatus*) and Bishop's oo (*M. bishopi*) – are thought to have disappeared in the course of the 1990s.

(for instance, for *Meliphaga* in New Guinea). The Painted honeyeater is almost entirely fruit-eating – it is nomadic in its search for mistletoe berries.

Ecologists often divide honeyeaters into long-billed and short-billed forms. Small to medium-sized honeyeaters with rather short, straight bills (for example, *Melithreptus* and *Manorina* species) are chiefly insectivorous. By contrast, long-billed honeyeaters eat more nectar than insects. Species like the spinebills use their long, curved bills to feed from tubular flowers. This group also includes medium-sized honeyeaters such as the *Phylidonyris* species, which are generalized nectar-eaters visiting a wide range of flowers, and the wattlebirds, which prefer eucalypts and banksias.

How can several species of long-billed honeyeaters live in the same area when they are competing for a limited nectar supply? The answer seems to lie in a balance between small, efficient honeyeaters and large, aggressive ones. The larger

HONEYEATERS AS POLLINATORS

Many Australian flowers rely on honeyeaters for pollination and reward them with nectar. Some plants such as eucalypts have generalized flowers that are also visited by insects, but others, especially in the families Proteaceae and Epacridaceae, have flowers specifically adapted to birds. Long, narrow, tubular corollas, sometimes defended by hairs, deter insects but not birds, and are often yellow or red in coloration. Bird-pollinated flowers are often clumped, and inflorescences of banksias (*Banksia* spp.) 20–40cm (8–16in) long may contain as many as 5,000 or more flowers.

Pollen deposited on the forehead, face, chin, and beak of a honeyeater, often visible as a yellow patch, is transferred to the stigmas of the next flowers visited. Honeyeaters typically carry thousands of pollen grains from several different plants at a time, and their relationships with Australian flowers are much less specific than those between hummingbirds and flowers in tropical America. However, a process of coevolution appears to have produced some elaborate mutual adaptations, such as the long, curved beak that Western spinebills use to probe the tubular flowers of kangaroo-paws and jug-flowers.

Some plants produce only a few flowers at a time, albeit over prolonged periods, which forces honeyeaters to move between plants and promotes crossfertilization, a genetic advantage. Indeed, many bird-pollinated plants are probably incapable of self-pollination. The greater mobility of birds makes them more effective at outcrossing than insects, and they are less affected by adverse weather. The unpredictable climate and flowering patterns in much of Australia result in many honeyeaters being mobile species that suddenly appear in, and then disappear from, a region as they opportunistically follow the nectar flow. RDW

Right *As various plants flower, honeyeaters migrate to exploit their nectar. Here, a male Scarlet honeyeater* (Myzomela sanguinolenta), *perches on a Red bottlebrush plant* (Callistemon citrinus).

species, such as wattlebirds, aggressively exclude other honeyeaters from dense clumps of flowers where nectar levels are highest, but cannot defend all the flowers over a wider area. This limitation allows the smaller honeyeaters, which can still feed profitably on the poorer nectar sources, to coexist with them. Thus a hierarchy of aggression based on size maintains a diversity of birds even where nectar abundance varies greatly. Extreme levels of interspecific territoriality are exhibited by the predominantly insectivorous honeyeaters, including the miners and *Manorina* species, which achieve almost a total monopoly of the sites their colonies occupy through cooperative defense of their territories from almost all avian intruders.

Given their diversity and that of the food types and habitats they exploit, it is not surprising that honeyeaters exhibit a very broad range of movement patterns. Movements tend to be associated with the flowering patterns of the major food plants; some of these flower regularly, but many do not, being dependent upon unpredictable rainfall. A few species are able to exploit regular flowering patterns and exhibit extensive annual migrations. For example, in southeastern Australia, the Yellow-faced honeyeater and the White-naped honeyeater regularly migrate northward each austral autumn and return in the spring. Although some individuals remain in the south for the winter, it is quite likely that resident and migratory populations intermingle during the austral summer. Some honeyeaters and chats of the more arid regions, such as Black honeyeaters and Crimson chats, may undergo large-scale, less predictable movements as they track rainfall and the flushes of nectar and insects that follow. Whether there are longterm patterns to these movements remains to be determined. Many honeyeaters probably confine their movements to a local region up to 100km (62 miles) in radius, as they track local flowering events, while many others like the Bell miner are year-round residents of home ranges less than 1ha (2.5 acres) in area.

Above *An Eastern spinebill* (Acanthorhynchus tenuirostris) *with its young. This common Australian species breeds between October and January.*

Right *Representative species of honeyeaters and chats:* **1** *Yellow wattlebird* (Anthochaera paradoxa)*;* **2** *Regent honeyeater* (Xanthomyza phrygia)*;* **3** *Blue-faced honeyeater* (Entomyzon cyanotis)*;* **4** *Western spinebill* (Acanthorhynchus superciliosus)*;* **5** *Crimson chat* (Epthianura tricolor).

Corroborees in the Colonies

BREEDING BIOLOGY

Most honeyeaters and chats appear to be socially monogamous, and the handful of species in which parentage has been examined genetically indicate that genetic monogamy is most likely to be the rule. The polygynous stitchbird of New Zealand seems to be an exception. In addition, the mating system of another sexually dichromatic honeyeater, the Crescent honeyeater, has also been found to have high levels of extra-pair fertilizations.

Honeyeaters will sometimes defend both foraging and breeding territories from conspecifics and other species. Foraging territories can be confined to just a part of a flowering tree that will only be defended for the short period that the plant yields nectar. Pairs of some species only establish and defend territories when breeding, otherwise foraging in loose aggregations or, like the Yellow-faced honeyeater, in single or multispecies flocks; others, like the Bell miner, defend their territories year-round. The territories of pairs may be dispersed, as with the White-eared honeyeater, loosely associated in dispersed neighborhoods like those of the Helmeted honeyeater, or tightly packed in communally defended colonies, as with the Bell miner.

The honeyeaters and chats have long breeding seasons, and many are capable of raising more than one brood in a season. Most build cup-shaped nests. The only known exceptions are the two *Ramsayornis* honeyeaters, which build nests with domes, and the stitchbird and at least one species of *Moho*, which nest in tree cavities. The typical clutch size is two (range 1–5). In most species the female incubates alone, but in some both sexes incubate and both parents feed the young. Parents in the genus *Manorina* are regularly assisted by helpers in caring for young, whereas helpers are only occasionally present in several other genera, for instance *Melithreptus*, *Entomyzon*, and *Lichenostomus*. Honeyeaters have a typical passerine molt at the end of the breeding season.

Honeyeaters that breed in colonies, like miners, or loose neighborhoods (for example, the New Holland honeyeater) sometimes indulge in elaborate communal displays called "corroborees." Ten or more birds will gather in a tight group, fluttering their half-spread wings and calling repeatedly at one another or at focal individuals. These displays seem to be triggered by either a predator or new individuals entering a neighborhood.

COMPLEX COMMUNITIES OF MINERS

All four species of miner exhibit an extremely complex form of cooperative breeding in which as many as 30 different individuals may assist a single breeding pair in raising a single brood of young. Miners live in colonies of from 6 to about 250 birds. Colonies are usually made up of between one and three coteries. Breeding males within a coterie tend to be close relatives (for instance brothers), but will be more distantly related or not related at all to males in adjoining coteries. Male offspring rarely disperse from the coterie in which they were raised. In contrast, females disperse upon reaching sexual maturity at 8–9 months of age to gain breeding positions in neighboring colonies. Consequently, the female breeders within a coterie are not usually close relatives of one another or of the breeding males.

Each coterie typically has several breeding pairs with abutting home ranges. Helpers are predominantly males, and they confine their aid to breeding pairs within their coterie. The contingent of helpers attending a nest can include fledglings from one brood feeding their siblings in the next as well as uncles, aunts, grandparents, and even neighboring breeding males who may simultaneously be raising their own young in a nearby nest. Most cooperatively breeding birds confine their aid to a single breeding pair, but miners may deliver food to up to three nests belonging to different breeding pairs within the coterie on the same day!

On occasion, birds from several coteries will unite to mob predators or expel other birds from the colony's territory. Through communal defense, miners often achieve an almost complete monopoly of the habitat their territory occupies.

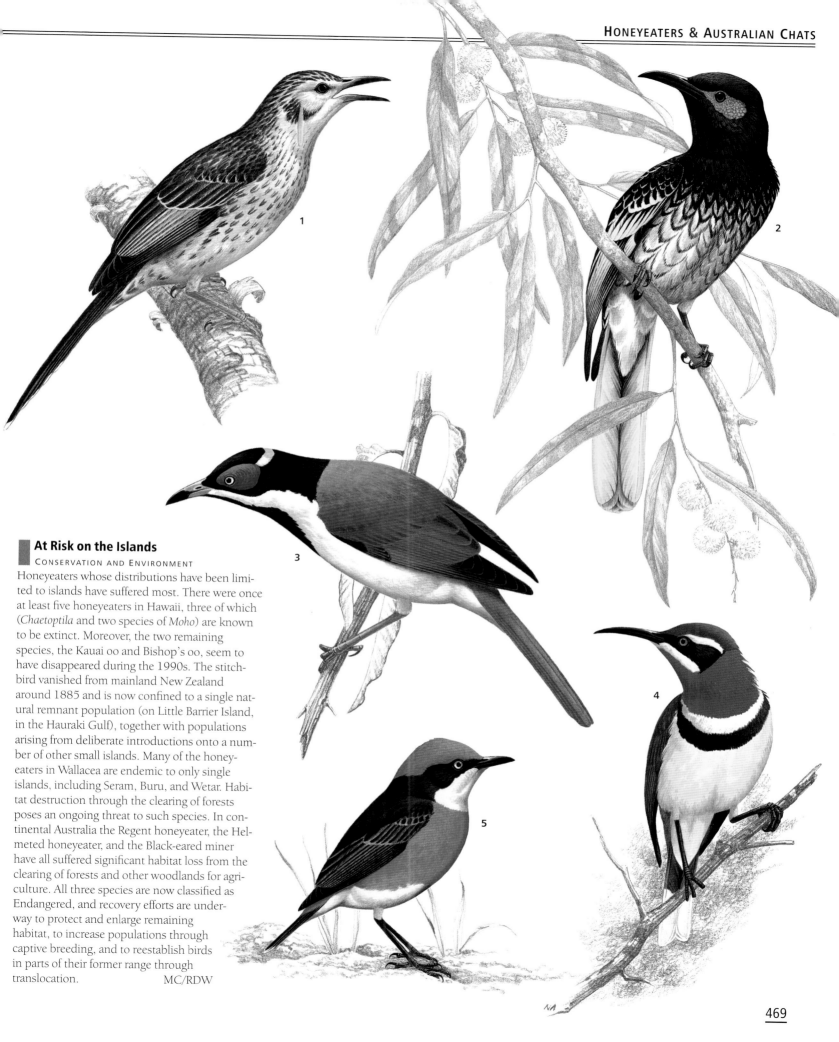

At Risk on the Islands

CONSERVATION AND ENVIRONMENT

Honeyeaters whose distributions have been limited to islands have suffered most. There were once at least five honeyeaters in Hawaii, three of which (*Chaetoptila* and two species of *Moho*) are known to be extinct. Moreover, the two remaining species, the Kauai oo and Bishop's oo, seem to have disappeared during the 1990s. The stitchbird vanished from mainland New Zealand around 1885 and is now confined to a single natural remnant population (on Little Barrier Island, in the Hauraki Gulf), together with populations arising from deliberate introductions onto a number of other small islands. Many of the honeyeaters in Wallacea are endemic to only single islands, including Seram, Buru, and Wetar. Habitat destruction through the clearing of forests poses an ongoing threat to such species. In continental Australia the Regent honeyeater, the Helmeted honeyeater, and the Black-eared miner have all suffered significant habitat loss from the clearing of forests and other woodlands for agriculture. All three species are now classified as Endangered, and recovery efforts are underway to protect and enlarge remaining habitat, to increase populations through captive breeding, and to reestablish birds in parts of their former range through translocation. MC/RDW

Australian Warblers

THE AUSTRALIAN WARBLERS OF THE FAMILY Acanthizidae (or Pardalotidae) are also sometimes referred to as the pardalotids, the lack of a commonly agreed name reflecting the diversity of form, habitat, and behavior of the 67 living species. Usually subdued in color, they are small to medium-sized birds, ranging from the tiny weebill, weighing just 5g (0.2oz), to the Rufous bristlebird at 80g (3oz).

Left Some pardalotes, also known as diamondbirds, dig nests 0.5m (1.5ft) or more into the ground. Here a Spotted pardalote (Pardalotus punctatus) stands by its burrow.

Right The Grey warbler (Gerygone igata) is a New Zealand forest species best known for its melodious song, often heard in early spring. The sweet, high-pitched trill won the bird its Maori name of riroriro.

Australian warblers are common in habitats ranging from tropical rain forest to desert, from coast to alps, and from ground to canopy. The family is notable for the variety of its social organization and vocalizations; cooperative breeding is common, and many species have mellifluous or striking songs, sometimes including mimicry or duetting.

Sweet-voiced Insect-hunters

Form and Function

Morphological variation reflects foraging methods and diet, leading to convergence in form with ecologically similar birds. Most species have straight, slender bills, suitable for a diet dominated by insects and other small arthropods. Those gleaning insects in the canopy are small, warblerlike birds, while species feeding primarily on the ground tend to be larger, and look similar to wrens or thrushes. In contrast, the brightly-colored pardalotes are reminiscent of flowerpeckers and have short, stubby bills, which they use to pick tiny insects and their sugary exudates off the surface of eucalypt leaves. Weebills have similar bills, probably for a similar function. Whitefaces have strong, titlike bills, reflecting the importance of seeds in their diet. The resemblance of various Acanthizidae to warblers, wrens, thrushes, tits, and flowerpeckers confused early taxonomists, but is now known to be the result of evolutionary convergence, not ancestry.

Most species are restricted to Australia and are resident year-round, but New Guinea has all three species of mouse-warblers (*Crateroscelis*) as well as about half of the scrubwrens and gerygones. *Gerygone* is the only genus with a wider distribution, with three species in Southeast Asia, two in New Zealand, and one on South Pacific islands. Many species are sedentary, although pardalotes can be nomadic and the White-throated gerygone is an almost complete migrant in the south of its range.

Evolutionary relationships among species in the family are still debated, but recent work suggests 16 genera in four major groups: pardalotes; bristlebirds; scrubwrens (*Sericornis*) and their allies; and a final grouping made up of the thornbills (*Acanthiza*), the gerygones (*Gerygone*), and the weebill. These three genera contain some two-thirds of the family's species. Some authorities place the four species of pardalote in their own family, Pardalotidae, with the other birds in a separate family, the Acanthizidae.

A Long Breeding Cycle

Breeding Biology

Australian warblers typically have a long breeding cycle. Most nest in all-purpose territories, although Forty-spotted pardalotes form loose colonies. Eggs are laid at two-day intervals, even in tiny species, and the clutch is usually two or three, although some species can lay four eggs. Incubation periods are long for the size of the birds; the White-browed scrubwren, for example, which weighs just 13g (0.5oz), incubates for 17–21 days. Nestling periods are more typical of birds of their size, but the young are usually looked after for six to eight weeks after leaving the nest, and never for less than three weeks. Many species are multibrooded and have breeding seasons of three to five months or more. Individuals of even tiny species can be long-lived; for example, a Brown thornbill weighing just 7g (0.25oz) was recaptured in the wild 17 years after marking!

Australian warblers build covered nests with a side entrance, which vary among species in ingenious ways. Nests built on or near the ground can be notoriously well-hidden. White-browed

Left A Yellow-rumped thornbill perches by its nest. The birds sometimes build a second, dummy nest on top of the real one as a decoy to fool egg thieves such as the currawongs (Strepera spp.).

scrubwrens can build nests in hollows under leaf litter, and Speckled warblers make a scrape in the ground so that the base of the nest is below ground level. Pardalotes build their nests in tree hollows or in chambers at the end of tunnels that they excavate in loose soil. Yellow-throated scrubwrens are unique in their genus for building suspended nests, up to 1m (3.3ft) long, often set over streams and resembling flood debris. Gerygones build purselike suspended nests with long tails and a hooded entrance, sometimes close to wasps' nests. The Rock warbler's nest is often suspended from the roof of a cave. Entrance to the Banded whiteface's nest chamber is through a funnel up to 20cm (8in) long and 3cm (1.2in) in diameter; its function is unknown. Yellow-rumped thornbills build the most curious nest. It is domed with a concealed side entrance and a false "cup nest" built conspicuously on top, possibly to fool predators or brood parasites.

Despite their domed nests, Australian warblers are frequent victims of cuckoo parasitism. Fan-tailed cuckoos (*Cacomantis flabelliformis*) often parasitize scrubwrens and thornbills, and all of the bronze-cuckoos commonly lay in the nests of the smaller species. Rufous bronze-cuckoos (*Chrysococcyx ruficollis*) lay specifically in suspended nests, such as those of gerygones, and Black-eared cuckoos (*C. osculans*) parasitize a variety of Australian warblers, particularly those species that lay uniform brown eggs similar to the cuckoo's own.

Many Australian warblers have sweet or complex songs, which can include mimicry or duetting. The simple, bell-like notes of pardalotes ring from eucalypt forests and woodlands throughout Australia. Gerygones (the name itself means "born of sound") have simple but delightful songs, often in a descending melody. Several species are gifted mimics, including the Yellow-throated scrubwren, the Speckled warbler, and the Chestnut-rumped hylacola, which weaves the songs of many species into a canary-like song. Pilotbirds and bristlebirds have clear, far-carrying songs that often include duetting. Not all Australian warblers delight the human ear, however; some have mechanical or buzzy voices.

Cooperative breeding and pair-breeding are both common, even within genera. In many species, some sons remain on the natal territory and become "helpers" in its defense, and in raising subsequent young. Such cooperative breeding is the norm in scrubwrens and thornbills, and also occurs in pardalotes, gerygones, and probably other genera; it may be the ancestral breeding condition in this family. Speckled warblers are perhaps unique in that males always disperse from their natal territory, but may then become resident on the territory of an unrelated breeding pair. In the non-breeding season, some species form flocks composed of adjacent breeding groups.

Subspecies in Danger

CONSERVATION AND ENVIRONMENT

One species of Australian warbler, the Lord Howe gerygone, is known to have become extinct in historical times. Once abundant in the forests of Lord Howe Island off Australia's eastern coast, it fell victim to the introduction of rats to the island following a shipwreck in 1918.

Of extant species, the Forty-spotted pardalote, the Biak gerygone, and the Eastern bristlebird all have extremely restricted ranges, and are considered endangered as a result. Among subspecies, the western population of the Rufous bristlebird is extinct, the King Island populations of the scrubtit and Brown thornbill are Critically Endangered, and the Mount Lofty population of the Chestnut-rumped hylocola is Endangered. These birds too have very restricted ranges, exacerbated by habitat degradation. RM

FACTFILE

AUSTRALIAN WARBLERS

Order: Passeriformes

Family: Acanthizidae (Pardalotidae)

67 species in 16 genera. Species include: **Banded whiteface** (*Aphelocephala nigricincta*), **Brown thornbill** (*Acanthiza pusilla*), **Yellow-rumped thornbill** (*A. chrysorrhoa*), **Chestnut-rumped hylacola** (*Hylacola pyrrhopygia*), **Eastern bristlebird** (*Dasyornis brachypterus*), **Rufous bristlebird** (*D. broadbenti*), **Forty-spotted pardalote** (*Pardalotus quadragintus*), **pilotbird** (*Pycnoptilus floccosus*), **Biak gerygone** (*Gerygone hypoxantha*), **Lord Howe gerygone** (*G. insularis*), **White-throated gerygone** (*G. olivacea*), **Rock warbler** (*Origma solitaria*), **scrubtit** (*Acanthornis magnus*), **Speckled warbler** (*Chthonicola sagittata*), **weebill** (*Smicrornis brevirostris*), **White-browed scrubwren** (*Sericornis frontalis*), **Yellow-throated scrubwren** (*S. citreogularis*).

DISTRIBUTION Australia and New Guinea region, with *Gerygone* also in SE Asia, New Zealand, and SW Pacific islands.

HABITAT All terrestrial habitats, from tropical rain forest to alps and arid scrub.

SIZE Length 9–27 cm (3.5–11in); weight 5–80g (0.2–3oz).

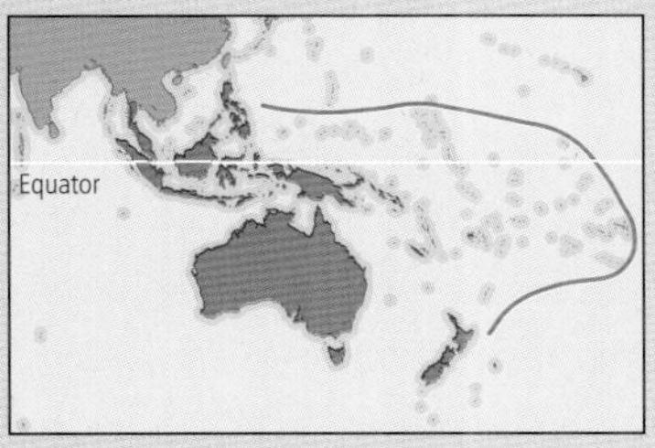

PLUMAGE *Pardalotus* gray and brown, spotted or streaked white with bright yellow or red patches; others brown or gray, with pale green, yellow, or rufous washes; some with strong facial patterns or yellow rumps.

VOICE Some species buzzy or mechanical, but many mellifluous; can include mimicry or duetting.

NEST Domed nests, some with concealed entrance; suspended, placed in or under thick vegetation, in tree hollows, or at end of excavated tunnel.

EGGS 2 or 3, occasionally 4; many speckled, some white or uniform brown. Incubation period 11–22 days; nestling period 10–25 days; 3–8 or more weeks of post-fledging care.

DIET Mostly arthropods, some seeds.

CONSERVATION STATUS 1 species, the Lord Howe gerygone, was last seen in 1936 and has now been declared extinct due to rat predation on nests. 3 species, including the Eastern bristlebird, are Endangered, and 4 are Vulnerable.

Australo-Papuan Robins

Above *Exhibiting the red breast that earned the Eopsaltridae their common name, a Flame robin* (Petroica phoenicea) *holds a captured insect.*

THE EOPSALTRIDAE FAMILY (ALSO SOMETIMES referred to as the Petroicidae) consists of 44 species in 13 genera, the majority of them found in Australia and New Guinea. Of the rest, three species, including the New Zealand and Chatham Islands robins (Petroica australis *and* P. traversi), *are endemic to New Zealand, while a single species, the Yellow-bellied robin* (Eopsaltria flaviventris), *is endemic to New Caledonia.*

Although called "robins," the members of this family are in no way related to the true robins (Turdidae) of the USA or Europe. They are, however, considered convergently similar, with some species (notably *Monachella* and *Microeca*) resembling Old World flycatchers, others (*Drymodes* and *Amalocichla*) recalling the true thrushes, and the rest resembling Old World robins.

Perch-and-Pounce Insectivores

FORM AND FUNCTION

Throughout their range, the Eopsaltridae have evolved to occupy a wide range of niches, from the dry, open, savanna regions of Australia (for example, the jacky-winter) and the banks of fast-flowing streams (the Torrent robin) to the montane forests and alpine shrubberies of New Guinea (the Alpine robin). Throughout these habitats, most species occupy the lower stratum of the vegetation where they forage. A large number of species are perch-and-pounce insectivores, often remaining motionless on horizontal branches, rocks, or logs before pouncing onto the ground or trunks of trees for arthropod prey. *Microeca* species also hawk for flying insects from perches.

The plumage of the Eopsaltridae varies, with the majority of species (for example, those in the genera *Microeca*, *Eopsaltria*, *Poecilodryas*, and *Tregellasia*) being olive-green to olive-brown with yellow. Both *Petroica* and *Eugerygone* comprise red-breasted robins, although, interestingly, the species in these two genera are the only members of the Eopsaltridae to exhibit notable sexual dichromatism. A number of species have black and white plumage, while one species, the Blue-grey robin (*Peneothello cyanus*), exhibits the coloration its name indicates. Brown mottled plumage on a pale breast is a diagnostic characteristic of the juveniles of the Eopsaltridae.

Below *A jacky-winter nesting in a forked branch. The derivation of the bird's name is disputed, but "jacky" is probably a rendering of its repeated call.*

Right *A Yellow robin* (Eopsaltria australis) *feeding its young. The birds typically lay two or three eggs in a clutch. Incubation lasts about 15 days, and the young then take 10–14 days to fledge. Both parents bring food for the nestlings.*

FACTFILE

AUSTRALO-PAPUAN ROBINS

Order: Passeriformes

Family: Eopsaltridae

44 species in 13 genera. Species include: **jacky-winter** (*Microeca fascinans*), **Lemon-bellied flyrobin** (*M. flavigaster*), **Olive flyrobin** (*M. flavovirescens*), **Alpine robin** (*Petroica bivittata*), **Torrent robin** (*Monachella muelleriana*).

DISTRIBUTION Indonesia, New Guinea, Australia, New Zealand, and New Caledonia, with *Petroica multicolor* also found in Fiji, Samoa, Solomon Islands, and Vanuatu.

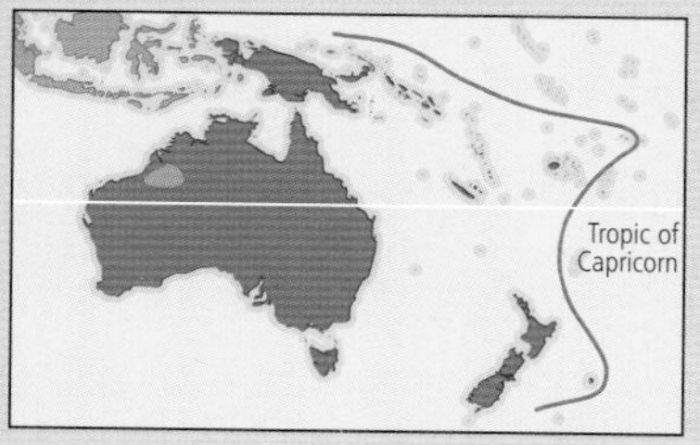

HABITAT Mangroves, coastal scrubs, open woodland to lowland forest, rain forest to montane forest and alpine shrubbery.

SIZE Length 10–23cm (4–9 in).

PLUMAGE Most species olive green to olive brown with yellow; pink to red; brown; black and white; 1 species (*Peneothello cyanus*) overall slaty blue-gray. Plumage alike between sexes, except in *Petroica* and *Eugerygone*. Mottled brown on pale breast is characteristic of juveniles.

VOICE Often high-pitched, short, and staccato to drawn-out whistles or trilling, usually descending or ascending in pitch; scolding chatter.

NEST Small, shallow cup placed low in the vegetation, usually 1–5m (3.3–16.5ft) above ground level, but occasionally up to 20m (65ft). *Drymodes* and *Amalocichla* species nest on the ground in shallow, bowl-shaped depressions. Majority breed between July and January, although northern representatives range between June and March; some (e.g. *Microeca griseceps*) noted breeding in May.

EGGS Usually 2 (occasionally 3), buff-white to light olive to greenish-blue to grayish-white, all with brown, reddish, gray-brown, or blue-gray spots. Larger markings along middle of egg, with smaller, more numerous spots at larger end.

DIET Mainly arthropods, occasionally grass seeds and small reptiles.

CONSERVATION STATUS Near to extinction as recently as 1981, the Chatham Islands robin is now listed as Endangered.

Drymodes and *Amalocichla* species form shallow, raised nests on the ground, lined with fine twigs and rootlets. The remainder of the Eopsaltridae construct small, cup-shaped nests (those of the Lemon-bellied flyrobin are the smallest in Australia), usually low in the vegetation – typically between 1 and 5m (3.3–16.5ft) up, although occasionally up to 20m (65ft) – in vertical forks of small trees or vines. The outsides of the nests are camouflaged with strips of bark, lichen, or moss, held together with spider web threads. The inside is usually lined with small dead leaves, feathers, soft rootlets, fibrous stems, or else mammal hair, as in the case of the Olive flyrobin.

Saving the Black Robin

CONSERVATION AND ENVIRONMENT

Many species of the Eopsaltridae, such as the New Guinea robins, have highly contracted distributions and are thus considered rare. As a result of clearing and modification of woodlands throughout southern Australia, two Hooded robin (*Melanodryas cucullata*) subspecies are classed as vulnerable or near threatened, with five species being of conservation concern due to their declining distribution and abundance.

At one time the Eopsaltridae included one of the world's rarest birds, the Chatham Islands robin. The population on Chatham Island itself, southeast of New Zealand, was wiped out, and by 1976 only seven individuals remained on nearby Little Mangere Island. In 1979, all the surviving birds were transported to Mangere Island, where the population declined further to only five individuals, four males and one female, in 1981. Intensive management through cross-fostering and nest protection subsequently saw numbers increase, and by 1999 the population had reached 259. Accidental introduction of predators to their sanctuaries remains an ever-present danger; moreover, the limited gene-pool of the birds means that the whole population is at risk should disease strike. There are plans to move the growing population to the larger Pitt Island. JAC

Above *Thanks to predation by introduced rats and cats, the Chatham Islands or Black robin* (Petroica traversi) *came close to extinction, reaching a nadir of just five birds. However, careful conservation management has enabled a viable population to develop from the last breeding pair, "Old Blue" and "Old Yellow."*

Logrunners and their Allies

SHY BIRDS OF THE UNDERBRUSH, THE logrunners and their allies are a group of insectivorous ground-dwellers confined to the Australasian region. The logrunners properly so called comprise two species in the genus Orthonyx: *the logrunner itself* (O. temminckii), *and its sister species, the chowchilla* (O. spaldingii).

Relationships within the group are uncertain; genetic evidence now suggests that the *Orthonyx* species may warrant a family of their own, with the whipbirds, wedgebills, quail-thrushes, and jewel-babblers placed in a separate family, the Cinclosomatidae. Another four species that have sometimes been included in the Orthonychidae – the Malaysian rail-babbler, the ifrit, and the two melampittas – have even more uncertain affinities, with their ground-dwelling habits apparently the main reason for their past placement in this group; here they are not included in the family.

Leaf-Litter Hunters

FORM AND FUNCTION

Logrunners occur in rain forests on the east coast of Australia and in New Guinea, while whipbirds and wedgebills are distributed across Australia in habitats ranging from rain forests to deserts. Quail-thrushes are found in Australia in dry habitats from woodland to desert, but in rain forest in New Guinea, where the jewel-babblers also occur. Most species are probably sedentary, although some of the desert-dwelling species may be partially nomadic. All are relatively weak fliers, relying on their cryptic habits and coloration to avoid detection. As a result, very little is known about most of them, the chowchilla being the only species to have been studied in any detail.

All species have strong legs and bills for scratching and digging in leaf litter and soil, while searching for their mainly invertebrate prey. Logrunners have an unusual pelvis that allows sideways movement of their legs, and tails in which the shafts of the feathers are modified at the tip into stiff spines. They make good use of both these features, resting on one leg and their strong, spine-tipped tails while scratching away leaf litter with the other leg before pecking at exposed food items. Small vertebrates such as skinks and frogs are eaten by some species, while the desert species may also eat some seeds.

More Often Heard than Seen

BREEDING BIOLOGY

Most species are probably territorial and monogamous, although cooperative breeding has been recorded in some. Chowchillas live in groups of two to six birds year-round, and although they do not appear to breed cooperatively, they do cooperate in foraging and territory defense. Most species have loud and distinctive territorial songs; some also duet. Given their cryptic nature, they are more readily found and identified by their songs than by visual observation.

Left *The quail-thrushes are thrushlike in size and plumage but, as ground-dwelling birds living off seeds and insects, resemble quails in their behavior. This is the Cinnamon quail-thrush* (Cinclosoma cinnamomeum).

FACTFILE

LOGRUNNERS

Order: Passeriformes

Family: Orthonychidae

16 species in 5 genera; an additional 4 species in 3 genera are sometimes placed in the family. Species and genera include: **logrunner** (*Orthonyx temminckii*), **chowchilla** (*O. spaldingii*), **whipbirds** and **wedgebills** (genera *Psophodes*, *Androphobus*) including **Western whipbird** (*P. nigrogularis*), **quail-thrushes** (genus *Cinclosoma*), **jewel-babblers** (genus *Ptilorrhoa*).

DISTRIBUTION SE Asia, New Guinea, Australia.

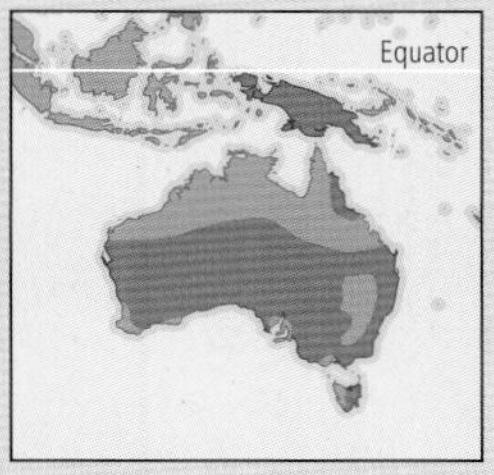

HABITAT Desert scrub, woodland, rain forest.

SIZE Length 17–31cm (6–12in); weight 45–210g (1.6–7.5oz).

PLUMAGE Most species are brown, black and white, or olive, some with blue; differences in appearance between sexes vary from slight to marked.

VOICE Calls range from buzzing to whistling and bell-like sounds.

NEST Cup-shaped or domed, placed in dense shrubs or on ground.

EGGS 1–3; white or pale blue. Incubation 17–25 days; nestling period 12–29 days.

DIET Predominantly invertebrates; some small vertebrates and seeds.

CONSERVATION STATUS The Western whipbird is currently listed as Lower Risk/Near Threatened.

Most species lay two to three eggs in cuplike nests set in shrubs or on the ground; the breeding season is from June to December, or after rain for the desert-living species. Logrunners, however, build bulky, dome-shaped nests on or close to the ground, and only lay one or two eggs. Little is known of the behavior or breeding habits of the New Guinean species, and their incubation and fledging periods are also unknown.

Above A logrunner returns to her nest after a successful foraging trip. The robinlike red breast indicates that this is a female bird; the males of both Orthonyx species are white-chested.

Below About 25cm (10in) long, the chowchilla or Northern logrunner is much larger than the southern species. Both birds take their alternative name of spinetails from the spiny shafts that protrude from the tip of the tail; these are used for support.

Populations under Threat

CONSERVATION AND ENVIRONMENT

The small ranges and specific habitat requirements of many species make them vulnerable to habitat degradation. Overgrazing and habitat loss – specifically, the clearing of mallee scrub to make way for agriculture – are threatening the Western whipbird, which is restricted to four isolated populations in south-west and southern Australia; of these, two are now listed as Vulnerable and one as Endangered. One population of the Spotted quail-thrush (*Cinclosoma punctatum*) has also been nominated for listing. Most other species appear secure, although little is currently known of the conservation status of the New Guinean species. AJ

Australo-Papuan Babblers

Above *A White-browed babbler holds a caterpillar in its bill. The birds often form noisy groups as they forage on the ground for insects, seeds, and spiders.*

FACTFILE

AUSTRALO-PAPUAN BABBLERS

Order: Passeriformes

Family: Pomatostomidae

5 species of 1 genus, *Pomatostomus*: **Rufous babbler** (*P. isidori*, New Guinea), **Grey-crowned babbler** (*P. temporalis*), **Hall's babbler** (*P. halli*), **Chestnut-crowned babbler** (*P. ruficeps*), **White-browed babbler** (*P. superciliosus*).

DISTRIBUTION Throughout Australia and lowland New Guinea.

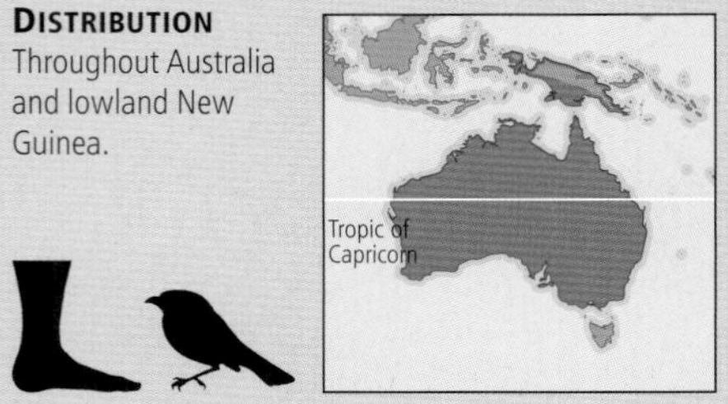

HABITAT In Australia, eucalypt forest, acacia shrubland, mulga, mallee/sclerophyll scrub; in New Guinea, eucalypt savanna and lowland rain forest up to 500m (1,650ft).

SIZE Length 28–36cm (11–14in).

PLUMAGE Generally brownish, with buff belly fading into whitish breast; crowns brown, gray, or chestnut; dull orange all over (*P. isidori*); slight size differences between sexes.

VOICE Bubbly warbles, low growls, and chattering, usually by several members of a group simultaneously, particularly in smaller-bodied species; loud duetting in *P. temporalis*; sharp, quick, high-pitched warning call in most species when approached or captured.

NEST Domed nest with entrance at the front, made of sticks; pendulous nest made of dead leafy masses and sticks (*P. isidori*).

EGGS 2–4, pale with dark, curvy streaks.

DIET Chiefly insects and larvae, gleaned from the ground, fallen logs, or lower tree trunks.

CONSERVATION STATUS Not threatened.

Babblers are conspicuous elements of the Australian and New Guinea avifaunas. Like many Asian babblers of the family Timaliidae, they are generally quite vocal and, in the case of the Grey-crowned babbler, sometimes duet. However, biochemical studies have shown that they are unrelated to the Timaliidae, and instead are part of the large radiation of corvoid passerines that now dominate the Australian perching-bird fauna. For this reason they are sometimes called "pseudo-babblers."

All species have decurved bills and an elongated profile. Their wings are short and wide, adapted for brief forays between trees and the ground, where they routinely forage, and between fallen logs. All species have relatively long tails, frequently with white tips.

The Rufous babbler is endemic to New Guinea. In Australia, Grey-crowned babblers are the most widespread species, ranging all the way from inland Victoria up the east coast to Cape York, west across the Top End to the Kimberley, and south to central Western Australia. However, they are considered near-threatened in the extreme southern part of their range. Grey-crowns also occur in the open eucalypt forest of the Fly River region of New Guinea. By contrast, Hall's babbler is relatively restricted, being found in a band of country dominated by mulga (acacia scrub) stretching from northwestern New South Wales to central Queensland.

Chestnut-crowned babblers are found in a large, crescent-shaped range in western New South Wales, eastern South Australia, and southern Queensland. White-browed babblers are found throughout southern Australia. Whereas Chestnut-crowns inhabit open country with only low shrubs, White-broweds occur from forest edges and thick, isolated scrub patches as far north as the Tanami Desert in the Northern Territory. All four Australian species occur sympatrically in some areas of southwest Queensland.

Cooperative and Communal

SOCIAL BEHAVIOR

The Grey-crowned babbler is the best-studied Pomatostomidae species in terms of behavior. It exhibits ontogenetic shifts in eye color, with young birds having dark brown irises (also found in the adults of all other Australian species), subadults dark yellow, and adults cream irises. These changes are thought to signal status within a group. Chestnut-crowned and White-browed babblers are smaller than Grey-crowned and Rufous babblers, but larger than Hall's babbler, which was discovered only in the late 1960s.

All of the *Pomatostomus* species are cooperative breeders; young birds from a previous year often remain with older relatives, even when they are capable of reproducing themselves, to help raise the young of the year. The result is that babblers are most often seen in large social groups of 5–15 birds. They frequently build multiple nests in an area and change their roosting site on successive evenings. In the evening, an individual bird, often a subadult, will stand watch outside the communal nest where the rest of the family are roosting.

The Rufous babbler of New Guinea builds pendulous nests that often hang about 1m (3–4ft) below a branch in the lower canopy. The birds are vocal leaders of the "brown and black" mixed-species foraging flocks that move through the rainforest mid-canopy. SE

Whistlers

WITH THEIR LARGE, ROUNDED HEADS, *many pachycephalids were traditionally known as thickheads, hardly a complimentary name for some of Australasia's finest songsters. In some species the sexes are of similar, relatively drab coloring, but in most the males are more brightly colored, with a yellow or rufous breast and bold black and white markings around the head.*

Most genera are best represented in Australia and New Guinea, but the large *Pachycephala* whistlers are also found from India through to many southwestern Pacific islands. This genus does not occur in New Zealand, where the family is represented by other species, such as the whitehead.

Clear-voiced Songbirds

FORM AND FUNCTION

All pachycephalids have strong feet and a relatively large hooked bill, which is used for catching prey; the most formidable bill belongs to the Crested shrike-tit, which uses it to prize off bark, exposing insects sheltering below. Most species are sedentary, but some, such as the Golden whistler, migrate from breeding grounds in southeastern Australia to winter in more tropical areas. The Crested shrike-tit and Crested bellbird, the sittellas, and the piopio are of questionable lineage, and have sometimes been placed in other families.

Most species glean insects from foliage or pounce on invertebrates from perches. Sittellas pick insects from the bark of trees as they travel headfirst down the trunks. Some species, particularly the pitohuis, also eat fruit. Shrike-thrushes generally take larger prey, even up to the size of frogs and small skinks.

Breeding occurs from July to January, although the Grey whistler has been recorded breeding year-round in the tropical part of its range. Pair territories are defined by song, although some species (for example the whitehead) also breed in groups. During the breeding season Rufous whistlers can be prompted to sing by loud noises such as thunderclaps or rifle shots. Usually only the female builds the nest, an open cup in the fork of a tree or shrub. The Crested bellbird places paralyzed hairy caterpillars around the rim of the nest cup, perhaps as a defensive measure.

The piopio may be extinct, having last been seen in 1963; introduced predators are blamed for its demise. Of the other species, only the Red-lored whistler is seriously threatened, due to clearing of its mallee habitat in southeastern Australia. Reserves have recently been established in mallee areas to help stop the bird's decline. PMcD/HAF

Right The Golden whistler is a familiar bird of Australian woodlands, also sometimes seen in parks and orchards. It is prized for its strong, musical song.

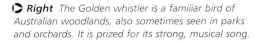

Below A Grey shrike-thrush (Colluricincla harmonica) barely in its tightly-woven nest. Nests are often constructed in the same site year after year.

FACTFILE

WHISTLERS

Order: Passeriformes

Family: Pachycephalidae

62 species in 14 genera. Species and genera include: **Crested bellbird** (*Oreoica gutturalis*), **Crested shrike-tit** (*Falcunculus frontatus*), **Golden whistler** (*Pachycephala pectoralis*), **Grey whistler** (*P. simplex*), **Red-lored whistler** (*P. rufogularis*), **Rufous whistler** (*P. rufiventris*), **Tongan whistler** (*P. jacquinoti*), **piopio** (*Turnagra capensis*), **pitohuis** (genus *Pitohui*) including **White-bellied pitohui** (*P. incertus*), **shrike-thrushes** (genus *Colluricincla*), **sittellas** (genus *Daphoenositta*), **whistlers** (genera *Rhagologus*, *Pachycare*, *Hylocitrea*, *Coracornis*, *Aleadryas*, *Pachycephala*), **whitehead** (*Mohoua albicilla*).

DISTRIBUTION Australasia and SE Asia

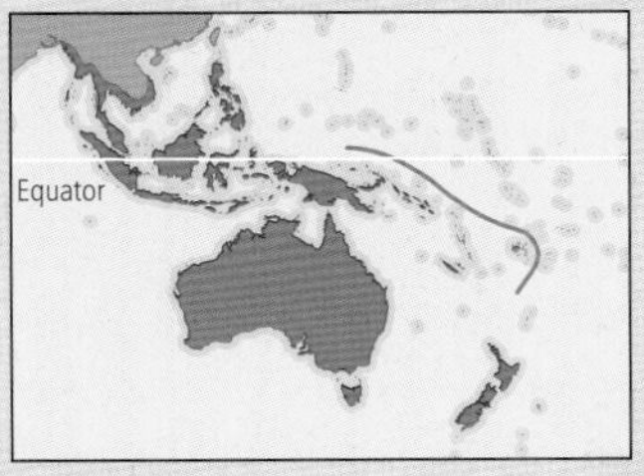

HABITAT All habitats from rain forest to arid scrub.

SIZE Length 11–26cm (4–10in); weight: 12–80g (0.5–3oz).

PLUMAGE Gray or brown, but adult males of many species brightly colored.

VOICE Distinctive, melodious, whistling or bell-like songs; harsher contact calls.

NEST Carefully woven cups of grasses, lichen, twigs, and bark chips, held together by spider's webs, usually placed in fork of a tree or shrub.

EGGS 2–4, white to buff, usually with a ring of brown to gray blotches at the larger end. Incubation period 14–18 days; nestling period 13–17 days.

DIET Mainly insects, small vertebrates, bird eggs and nestlings, fruit.

CONSERVATION STATUS The piopio is thought to be extinct. No other species are endangered, although the Red-lored and Tongan whistlers and the White-bellied pitohui are listed as Lower Risk/Near Threatened.

Vireos

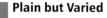

A *SMALL FAMILY OF NINE-PRIMARIED BIRDS exclusively restricted to the New World, vireos are most closely related to the corvoid assemblage (crows). The family can be divided into three groups: the true vireos (genus* Vireo*), the greenlets (genus* Hylophilus*), and the pepper-shrikes and shrike-vireos (genera* Cyclarhis *and* Vireolanius*).*

The vireos reside mostly in the northern parts of South America, Central America or Mexico, and most are non-migratory. Twelve of the 13 species breeding in the United States migrate southward during the fall. Only the true vireos are represented in North America, while all three groups are represented in Latin America. The distance of migration varies from as little as 160km (100 miles) in Gray vireos to more than 4,800km (3,000 miles) in Red-eyed vireos.

Plain but Varied

FORM AND FUNCTION

The family is notable for their generally dull plumages and stout, fairly heavy to quite heavy beaks; those of pepper-shrikes, shrike-vireos, the Blue Mountain vireo, and some races of the Black-whiskered vireo are almost massive. The beak of most true vireos is characterized by a tiny hook on the cutting edge at the tip of the upper mandible – a feature found in pepper-shrikes and shrike-vireos as well, but lacking in greenlets.

Greenlets are more uniform in color – greenish on the back with varying amounts of yellow buff and white on the face and underparts – than vireos, which are similarly colored, but also include species that are brown (Cozumel vireo) or gray (Grey vireo) above. True vireos differ further from greenlets by the presence of whitish or yellowish eye-stripes or eye-rings. The *Vireo* genus can be divided on the basis of plumage morphology into those with wing-bars and eye-rings (the Blue-headed vireo) and those with plain wings and eye-stripes (the Red-eyed vireo). However, the recently discovered Chocó vireo exhibits characteristics of both groups (wing-bars and eye-stripe). Wing-bars are lacking in greenlets, pepper-shrikes and shrike-vireos, but the pepper-shrikes have a distinctive, reddish stripe above the eyes and shrike-vireos have strongly patterned facial and crown markings of yellow, bluish-green, or chestnut.

Insectivores that Stay Close to Home

DIET

Vireos rely on a diet of insects and arthropods, and incessantly search in dense foliage in various layers for their food. All vireos appear to eat some fruit in varying degrees. Arthropods are taken mostly from leaves and twigs by Red-eyed and White-eyed vireos. Out of the whole family, only the Grey vireo takes prey from the ground and then only in about 5 percent of its foraging bouts. Greenlets generally take arthropods in scrub, but can also forage high in the canopy. Vireos also fly-catch, taking insects on the wing. Pepper-shrikes and shrike-vireos glean prey from forest canopy leaves and twigs. Other than on migration, vireos are not noted for long distance flights; the Grey vireo, however, may fly several hundred meters at a time within its territory – up to 8ha (20 acres) – in the desert scrub and canyon country it favors.

Above *Representative species of vireos:* **1** *Red-eyed vireo* (Vireo olivaceus)*;* **2** *the Yellow-throated vireo* (V. flavifrons) *of eastern North America.*

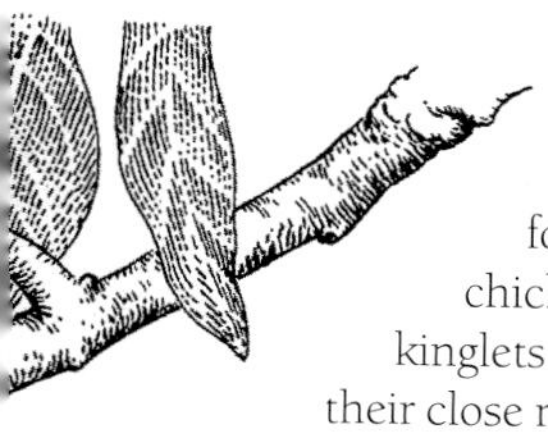

The temperate zone resident, Hutton's vireo, forms winter flocks with chickadees, nuthatches and kinglets. Red-eyed vireos and their close relatives form small flocks in winter, moving from one fruiting tree to another. The remaining migrant species keep winter territories that are defended against other individuals of the same species.

Tireless Singers

BREEDING BIOLOGY

Breeding biology is known primarily for species that breed in the United States and Canada, while the behavior and nesting of tropical species, especially the greenlets, are very poorly known. Generally, tropical and subtropical vireos and their allies are territorial all the year round, gathering in pairs or family groups. Solitary and White-eyed vireos and their close relatives sing throughout the year. In summer, Red-eyed and Bell's vireos are among the most persistent singers of all New World songbirds. Although as a rule members of the vireos are not known for their elaborate or beautiful voices, the Chocó Vireo has a relatively elaborate and attractive voice for the genus. Individual peppershrikes also often have pleasant warbled, albeit repetitive, song types within their repertoires. Such song stands in sharp contrast to the police-whistle trill of the Blue Mountain vireo and the monotonous chatter song of several Caribbean and circum-Caribbean scrub-dwelling species.

Left *A typical vireo nest is baglike and hangs from the crotch of a thin branch. Two is a typical brood size in the tropics, but clutches in the temperate zone may be twice as large.*

Right *Solitary vireos forage mostly in treetops, as do Red-eyed and Yellow-throated vireos; but White-eyed vireos and other closely related scrub-dwelling species forage low in vegetation.*

Female song has only been documented in the Grey vireo in which it is a regular feature of nest changeover by incubating or brooding adults. Males of most vireos sing when on the nest, probably as a reminder to the female of the nest location and as a stimulus for her to return to it once her hunger is satisfied.

In tropical and temperate regions in the northern hemisphere nesting begins between late April and mid-May and in all except the Red-eyed, Black-whiskered, Yucatan and Philadelphia vireos nests are built by both sexes. In the aforenamed species a singing male accompanies his female as she builds, but does not actually participate in construction. Nest building requires 4–5 days in Bell's vireos and most other temperate species and up to 25 days in the Chestnut-sided shrike-vireo.

The vireos all build a pendent nest, which is attached to branches and hangs below. The nest has an outer layer of coarse strips of bark and leaves, or in some species moss, bound together by spider silk and decorated with whitish spider egg cases, and an inner layer of fine grass stems carefully coiled around the bowl of the nest. In species in which both sexes build, males are capable of building rough bag nests by themselves, but the lining is done by the female.

In temperate species eggs are laid within a day of nest completion. Males of all species except the Red-eyed vireo and closely related forms sit on eggs at intervals during the day when the female is not incubating. When hatching occurs the male Red-eyed vireo and his Black-whiskered and Yucatan vireo relatives finally participate in the care of the young and share in feeding arthropods to the nestlings. Upon fledging individual young appear to be fed exclusively by one parent or the other for up to 20 days after leaving the nest.

Fighting Parasitic Perils

CONSERVATION AND ENVIRONMENT

In North America, some vireos are highly susceptible to nest parasitism by the Brown-headed cowbird, and their breeding success is accordingly reduced. Bell's vireo, a heavily parasitized species, often buries cowbird eggs laid in its nest by adding additional nesting material to the interior of its nest thereby effectively walling in eggs of the social parasite. Solitary vireos have been observed tossing cowbird eggs from the nest, although cowbird young may also be raised by this species.

Of the 32 species of vireos, five are globally threatened with extinction: the Chocó vireo, San Andrés vireo, and Black-capped vireo are all threatened by habitat loss and are restricted to very small ranges, with population estimates of less than 10,000 birds. The Blue Mountain vireo and Noronha vireo are island endemics that are considered near-threatened. PS/JCB

FACTFILE

Vireos

Order: Passeriformes

Family: Vireonidae

52 species in 4 genera. Species include: **Black-billed pepper-shrike** (*Cyclarhis nigrirostris*), **Rufous-browed pepper-shrike** (*Cyclarhis gujanensis*), **Bell's vireo** (*Vireo bellii*), **Black-capped vireo** (*V. atricapillus*), **Black-whiskered vireo** (*V. altiloquus*), **Blue mountain vireo** (*V. osburni*), **Cozumel vireo** (*V. bairdi*), **Grey vireo** (*V. vicinior*), **Hutton's vireo** (*V. huttoni*), **Jamaica vireo** (*V. modestus*), **Philadelphia vireo** (*V. philadelphicus*), **Red-eyed vireo** (*V. olivaceus*), **Solitary vireo** (*V. solitarius*), **Warbling vireo** (*V. gilvus*), **White-eyed vireo** (*V. griseus*), **Chocó vireo** (*V. masteri*), **Yucatán vireo** (*V. magister*), **San Andrés vireo** (*V. caribaeus*), **Scrub greenlet** (*Hylophilus flavipes*), **Grey-headed greenlet** (*H. decurtatus*), **Yellow-browed shrike-vireo** (*Vireolanius eximius*), **Chestnut-sided shrike-vireo** (*V. melitophrys*).

DISTRIBUTION N, C, and S America, West Indies (Vireoninae only)

HABITAT Scrub, woodlands, forests

SIZE Length Vireoninae (true vireos and greenlets): 10.2–15.3cm (4–6in); Cyclarhinae (pepper-shrikes) and Vireolaniinae (shrike-vireos): 12.7–16.5cm (5–6.5in); weight respectively, 9–22g (0.3–0.7oz); 20–39g (0.6–1.3oz); 22–36g (0.7–1.2oz): males and females similar in size.

PLUMAGE Chiefly green above, but some species gray or brown on the back; yellow or white on belly. Sexes similar but male Black-capped vireo has a black crown (female gray), and male Chestnut-sided shrike-vireo much wider and brighter barring on throat, breast, and "face" than females.

VOICE Rarely musical, repetitive song of the same or different whistled or "burry" notes (a gravelly roll to certain syllables); up to 15 different calls in some species.

NEST Baglike, suspended by rim from fork

EGGS 2 in tropical species; 4–5 in northern species. Whitish with brown spots at the broad end; incubation period 11–13 days; nestling period 11–13 days.

DIET Arthropods and some fruit in summer and winter.

CONSERVATION STATUS The San Andrés vireo, endemic to the small Caribbean island of this name, is Critically Endangered. Due to its extremely limited range in the West Andes of Colombia, the Chocó vireo is classified as Endangered. The Black-capped vireo is Vulnerable.

Crows

M*EMBERS OF THE FAMILY CORVIDAE ARE familiar to most people who live where they occur because the birds are usually large, noisy, and obvious. Some, such as the House crow, are specialists at coexisting with people and have been living in towns for centuries. Others, including the American and the Torresian crows, have only recently moved into urban areas, but are now living there in large numbers. Blue jays are one of the most common birds at feeders throughout North America, and are spreading their range westward, taking advantage of human hospitality.*

Because of their size, color, apparent intelligence, and scavenging ways, crows figure prominently in folklore. In Europe, ravens and crows have often been seen as birds of ill omen, probably because of their role as scavengers on battlefields. Common ravens have a more positive role in native North American tradition, where they are seen as creators and folk heroes.

Today people still have strongly divided opinions on the subject of crows. Many see them as evil, noisy nest predators that need to be controlled; others, however, admire them for their humanlike intelligence and sociality.

FACTFILE

CROWS

Order: Passeriformes

Family: Corvidae

118 species in 24 genera

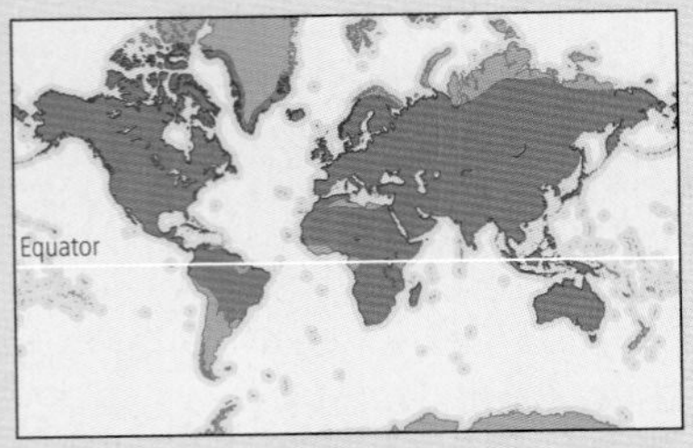

DISTRIBUTION Worldwide, except for the high Arctic, Antarctic, southern S America, New Zealand, and most oceanic islands.

HABITAT Varied, including forests, farmland, grasslands, desert, steppes, and tundra.

SIZE Length 19–70cm (7–28in), including the long tails of some magpie species; **weight** 40–1,500g (2–50oz).

PLUMAGE Often all black, or black marked with white or gray; many with bold markings on wings or tail; many jays brightly marked with blue, chestnut, buff, or green. Sexes usually similar in coloration.

VOICE Varied range of harsh or more musical calls; some species capable of mimicry.

NEST Bowl-shaped structures of twigs with lining of fine materials, placed in tree; some domed nests or nest in holes.

EGGS Usually 2–8; whitish, buff, cream, light blue, or light green, often marked with dark spots or blotches. Incubation 16–22 days; nestling period 18–45 days.

DIET Varied, including fruit, seeds, nuts, insects, terrestrial invertebrates, small vertebrates, often eggs of other birds or carrion; many, perhaps most, species store food.

CONSERVATION STATUS The Hawaiian crow is Critically Endangered; 4 species are Endangered and 8 Vulnerable.

See Selected Crow Species ▷

Below In normal years, the Spotted nutcracker largely keeps to its breeding grounds in the forests of northern Europe, wherever there are hazel trees or Arolla pines. On rare occasions, however, it irrupts into more southerly and westerly regions, including Britain.

Large, Robust, and Clever

FORM AND FUNCTION

The crow family contains the largest of all passerines, the ravens, as well as a wide variety of smaller jays, magpies, and other species. Some are regarded as the most intelligent of all birds. Many of the species are woodland dwellers; indeed, most of the jays and magpies of Asia and South America are almost confined to forests. However, most of the familiar species of Europe and North America prefer more open habitats, and none of the African or Australian representatives are forest birds.

The most widespread and familiar group comprises the typical crows and ravens of the genus *Corvus*. These are large birds with tails of short or medium length and plumage that is all black, black and white, black and gray, or entirely sooty-brown. In Europe the genus is represented by the Common raven, the Carrion and Hooded crows, the rook, and the Eurasian jackdaw; in southern Asia by the House and Jungle crows among others; and in Africa by species including the Pied crow and White-necked raven. In North America and again in Australia there are a number of all-black crows that resemble each other quite closely in structure and appearance but differ in their voices. Thus the American, Fish, Sinaloa, and Tamaulipas crows are more readily separable by voice than appearance, and in Australia the Torresian crow, Little crow, Australian raven and others are difficult to identify except by their calls. The genus *Corvus* has been more successful than

◐ ***Left*** *Representative crow species:* ***1*** *Rook* (Corvus frugilegus), *a Eurasian species famous for large breeding colonies, or "rookeries."* ***2*** *Common raven* (C. corax), *the biggest member of the family, now no longer common in populous areas as a result of persecution by shooting and poisoning.* ***3*** *Eurasian jay* (Garrulus glandarius), *found in temperate woodlands from Britain to Japan.* ***4*** *Blue jay* (Cyanocitta cristata) *holding an acorn, its favorite food.* ***5*** *Common or Black-billed magpie* (Pica pica), *carrying a berry.*

others of the family in colonizing remote islands, resulting in the development of species with local distributions in the West Indies, Indonesia, the southwest Pacific, and Hawaii.

The Red-billed and Alpine choughs resemble *Corvus* in their glossy, all-black plumage, but have relatively slender, downcurved bills, colored red or yellow. They were previously thought to be closely related to *Corvus*, but recent genetic evidence places choughs apart from the rest of the family. They are mainly mountain birds, extending their range to elevations of nearly 9,000m (27,000ft) in the Himalayas, but also occurring near rocky sea-cliffs in some regions.

Two species of nutcrackers inhabit Eurasia and North America respectively. The Eurasian Spotted nutcracker is mainly chestnut with white streaks, whereas the American Clark's nutcracker is mostly gray. Both feed largely on seeds or nuts, relying on hidden supplies during the winter.

A number of long-tailed crows are given the name "magpie," although they are unlikely to be closely related. These include not only the familiar piebald magpies of Europe, Asia, and North America, but also a number of more brightly-colored species from southern Asia such as the Green and Blue magpies. They all have short, strong bills and very long, graduated tails, which in the Asian magpies are decorated with white and black spots. The dividing line between the Asian magpies and jays relies mainly on the length of the tail, but the real relationships of these species remain to be worked out. The treepies of southeast Asia all have relatively short but strongly curved upper beaks, and long tails whose central tail feathers have rounded tips that are slightly to extremely expanded. The Ratchet-tailed treepie has flared, pointed tips to all its graduated tail feathers, resulting in an unusual, jagged appearance.

The American jays (not including the Gray jay, a representative of the Old World group) are distinct from the rest of the family. Many of the species are of rather small size, some of them no bigger than large thrushes. The Brown jay, however, is as large as a small crow, and the two species of magpie-jays have extremely long, ornate tails, much like the Asian magpies. Most of the American jays have blue in their plumage, although a few are brown or bluish-gray.

Among several atypical groups placed in the family, the ground-jays of central Asia are unusual in being predominantly ground-dwelling. They inhabit dry semi-desert and steppe regions and usually run from danger rather than taking to the air. (The much smaller Hume's ground-jay, once thought to be related to the ground-jays, is now assigned to the family Paridae – see True Tits.)

Left *On the Indonesian island of Bali, a Racket-tailed treepie shows off the long, spatula-shaped tail that makes it one of the most distinctive corvids. The birds inhabit forest edges across much of Southeast Asia.*

Allowing for this and a very few other exceptions, however, the crow family is fairly well defined. The combination of large to very large size, robust build, strong legs, and a strong bill, along with external nostrils covered by bristlelike feathers, serves to distinguish most crows from other songbirds, although certain starlings, drongos, and birds-of-paradise share some of these features. The nasal bristles usually are quite pronounced, and in the Racket-tailed treepie they are extremely dense, short, and velvetlike. Only the Pinyon jay has no nasal bristles at any time in its life. The rook and the Gray crow have covered nostrils as young birds, but gradually lose them and show bare skin on the face as they age.

The adaptability and versatility of crows show most clearly in their diets and feeding behavior. Most species take both animal and plant foods, with large insects and small nuts being especially important. Many are quick to exploit new and artificial food sources. The manipulation of food is made easier by the robust, generalized bill that is widespread in the family, and in most species also by use of the feet to hold food while it is dismembered. Many species use only the lower mandible to peck objects held between the feet, and the American jays have developed a special, bony buttress on the lower jaw to make this motion even more efficient. Many species have been recorded "dunking" or washing food, perhaps to counter stickiness or to soften hard items. Food-hiding is also prevalent in the family (see Squirreling Crows). The New Caledonian crow constructs tools to assist in food gathering, modifying twigs and leaves to probe into tree holes and cavities in search of insect larvae. New Caledonian crows construct different types of tools in different localities, and some are unique in the animal world in fashioning hooks to extract prey.

It has often been suggested that crows can survive on almost any food, but the poor physical condition of many captive specimens strongly implies that their nutritional requirements are similar to those of most other birds. In fact, being omnivorous is no guarantee of food availability, and many crow species cannot find enough food to feed all the chicks they hatch.

The longevity of crows has probably been overestimated by casual observers because of the birds' tendency to persist from generation to generation in suitable territories. Thus we have the old folk saying that "A crow lives three times as long as a man, and a raven lives three times as long as a crow." In reality, the maximum age recorded in captivity for a Common raven is 29 years, and that bird died of senile decay, suggesting that wild birds do not often live as long.

Left *The Green magpie is one of several long-tailed, brightly colored Asian species that are probably more closely related to the jays than to the more familiar magpies of the USA and Europe.*

Recoveries of ringed birds of several crow species show that one-third to one-half of young birds may die in their first year, and that few adults live to be older than 10 years. Still, such survival rates are quite high for birds, and some of the larger crows would thus appear by passerine standards to be long-lived.

Several studies of marked birds have shown that individuals of most species do not start to breed until they are at least 2 years old, and some American crows may not breed until they are 6 or 7. Carrion crows and Eurasian magpies, however, may be paired and holding territories during the second year of life. This deferment of sexual maturity may reflect the scarcity of breeding opportunities, or it may allow the young birds to gain additional experience before attempting to breed.

Cooperative Breeders

BREEDING BIOLOGY

A majority of members of the crow family defend exclusive breeding territories in which they nest. As examples, the Common raven, Eurasian jay, and Western scrub-jay all defend territories from which both birds of the pair threaten intruders. A few species nest colonially, notably the Eurasian jackdaw, which has rather loosely spaced colonies nesting in holes, and the rook, which nests in denser colonies in the tops of trees. The colonial nesters are gregarious throughout the year, and many of the species that hold breeding territories flock outside the breeding season, some of them occupying large communal roosts. Others, such as the Florida scrub-jay, defend their territories vigorously throughout the year. American crows hold territories year-round, but gather in large foraging groups and roosts at certain times of year, defending the territory during the day but joining others to sleep off-territory at night. Studies of marked birds of several species have shown that the same territories are occupied year after year and that the pair-bond often lasts for life. Some Florida scrub-jays may never leave their natal territory throughout their lives, breeding in the very spot where they were hatched.

Several different crow species have breeding seasons that are timed to take advantage of peak food supplies for the nestlings. Thus, rooks lay in March in England to take advantage of an April peak in earthworm abundance, whereas the Eurasian jay lays in late April or May to benefit

Right *A Siberian jay resting with its young. These birds are sedentary, rarely venturing far from their breeding grounds in Nordic and Russian conifer forests.*

from the late May and early June peak in the numbers of defoliating caterpillars on trees.

A number of species are cooperative breeders, with more than two birds attending a nest and feeding the young. Most frequently, the extra birds are offspring of the breeding pair that have stayed at home for one or more years. Such a lifestyle is especially prevalent in the American jays. Among the *Corvus* crows, cooperative breeding is known only in certain populations of American and Carrion crows. American crow families can include 15 individuals, all offspring of a single breeding pair, remaining at home for as many as six or more years. Several pairs of Mexican jays may have active nests concurrently in a single group territory, and individual jays may feed young in several nests within their flock's territory. Once the chicks fledge, all adults in the group feed the young from all nests. Studies examining the DNA of these groups indicate that young in one nest may, in fact, be the offspring of several different sets of parents. In contrast, the closely related Florida scrub-jay is a more simple cooperative breeder, in that all the young in a nest are the offspring of the single breeding pair.

The female alone incubates in most species (both sexes in nutcrackers), and the female is usually fed on the nest by the male and any helpers. Because incubation often starts before the last egg of the clutch is laid, the nestlings can hatch over a period of several days and so may differ in size. When food is short, the smallest of the brood often dies. In some *Corvus* species, the smallest nestlings may be discarded just after hatching, decreasing competition among the rest for limited food supplies. Both parents feed the young on food that is mostly carried to the nest concealed in the birds' throats. The fledglings of most, if not all, species are fed by their parents for some weeks after they leave the nest, and in at least some species they may remain in the parents' territory for many months after they become independent. In cooperatively breeding species they may stay for years, or even leave for weeks or months and then return.

Above *A Carrion crow, the most common Eurasian species, lives up to its name by feeding two almost fully-grown juveniles scraps from a dead rabbit. Adult birds are usually solitary, prompting the adage "One rook is a crow; a flock of crows are rooks."*

Below *The Florida scrub-jay, an isolated relative of the common Scrub jay of the American west, is now classed as Vulnerable by the IUCN as housing estates and citrus groves have cut into its Sunbelt habitat. By the year 2000, only about 10,000 birds remained.*

A Mixed Picture

Conservation and Environment

Several crow species are agricultural pests. The rook makes severe inroads into cereal sowings in winter and early spring; during the Second World War, rooks in Britain were reckoned to cause damage amounting to £3 million (US$4.8 million) a year. American crows have long been stereotyped as despoilers of corn (maize) crops, and they can have a significant impact on small plots in some areas. As with rooks, however, their foraging is not solely destructive; they may also benefit farmers by eating the larvae of insect pests.

While a few species such as Blue jays and Fish and House crows are expanding their ranges, a number of species with restricted ranges and habitats are threatened. Especially at risk are some of the island-dwelling *Corvus* species. The Hawaiian crow is Critically Endangered; extensive habitat alteration, together with introduced avian diseases, have reduced the wild population to only a few individuals on the island of Hawaii. A captive breeding program has not been effective; most of the young birds released from captivity soon died. Perhaps the species evolved to require prolonged contact with parents and other adults, allowing young crows to learn survival skills over the course of a few years, and without this social learning the introduced young are vulnerable to Hawaiian hawks and other predators. KJM/DH

Selected Crow Species

Species include: chough or Red-billed chough (*Pyrrhocorax pyrrhocorax*), Alpine or Yellow-billed chough (*P. graculus*), American crow (*Corvus brachyrhynchos*), Australian raven (*C. coronoides*), Carrion crow (*C. corone corone*), Common raven (*C. corax*), Eurasian jackdaw (*C. monedula*), Fish crow (*C. ossifragus*), Grey crow (*C. tristis*), Hawaiian crow (*C. hawaiiensis*), Hooded crow (*C. corone cornix*), House crow (*C. splendens*), Jungle crow (*C. macrorhynchos*), Little crow (*C. bennetti*), Marianas crow (*C. kubaryi*), New Caledonian crow (*C. moneduloides*), Pied crow (*C. albus*), rook (*C. frugilegus*), Sinaloa crow (*C. sinaloae*), Tamaulipas crow (*C. imparatus*), Torresian crow (*C. orru*), White-necked raven (*C. albicollis*), Biddulph's or Xinjiang ground-jay (*Podoces biddulphi*), Blue jay (*Cyanocitta cristata*), Steller's jay (*C. stelleri*), Blue or Red-billed blue magpie (*Urocissa erythrorhyncha*), Brown jay (*Cyanocorax morio*), Clark's nutcracker (*Nucifraga columbiana*), Spotted nutcracker (*N. caryocatactes*), Eurasian jay (*Garrulus glandarius*), Florida scrub-jay (*Aphelocoma coerulescens*), Western scrub-jay (*A. californica*), Mexican jay (*A. ultramarina*), Grey or Canada jay (*Perisoreus canadensis*), Siberian jay (*P. infaustus*), Pinyon jay (*Gymnorhinus cyanocephalus*), Turquoise jay (*Cyanolyca turcosa*), Green magpie (*Cissa chinensis*), Common or Black-billed magpie (*Pica pica*), Racket-tailed treepie (*Crypsirina temia*), Ratchet-tailed treepie (*Temnurus temnurus*).

SQUIRRELING CROWS

How crows cache food

MOST MEMBERS OF THE CROW FAMILY THAT have been studied in detail in the wild have been seen to hide food. They typically use small holes in the ground under debris or vegetation, but may also use sites above ground on trees or buildings. The birds usually make a deliberate effort to cover the hidden food by raking loose material on top of it, or by walking a short distance to find a stone, leaf, or other object to place on top. Young crows often hide stones, sticks, or other inedible items in a manner identical to that used for food, perhaps as a way of developing their skills. In captivity, many ravens, magpies, and jays appear to have a compulsion to conceal food and other objects.

The extent to which food is deliberately stored for later use varies widely among species. Some crows, such as the nutcrackers and the Pinyon jay, are highly specialized for hiding and recovering food, mostly tree seeds and nuts. Food hidden in this way may be extremely important in allowing species to exist in areas of harsh winters where little food is available. The Grey jay inhabits the coniferous forests of northern Canada and the United States, where winter supplies are scarce and the thick and extensive snow cover precludes the possibility of hiding food in the ground for later retrieval. Instead, the Grey jay uses the excretions from its unusually large salivary glands to glue food items such as spruce seeds onto the foliage of trees. Every season, these caching crows may hide tens of thousands of seeds, each usually in a different location.

Where no seasons of major shortage exist, the hiding and later recovery of food is on a small scale, as for example with magpies and jackdaws in England. Still, locally abundant foods are cached. Several instances have been reported of very hungry ravens and crows taking the trouble, when suddenly faced with an abundance of food, to hide a large quantity before beginning to eat. Such behavior makes sense when the birds face competition from other species. The available food is harvested quickly, hidden from the reach of competitors, and then eaten at leisure. Preferred items may be eaten first and the less preferred cached for later consumption.

Crows use memory to find their cached food at a later date. Laboratory experiments have demonstrated that the species that depend most on cached food have the best memories for location. Not only can crows remember thousands of different locations, but they can also recall which item was placed in a given spot and when it was hidden. Perishable items and preferred foods are recovered first. The area of the brain responsible for spatial memory (the hippocampus) is larger in those species that do the most food-hiding. Crows apparently remember each location not by its appearance, which can change over time, but rather by its proximity to local landmarks. In addition, the position of the cache from the landmark relative to the sun is important. Crows appear to have a sun-related compass that they use for remembering caches, indicating that birds use the direction from the sun in local as well as in migratory orientation.

The crows' memory for the whereabouts of hidden food is impressive, but by no means infallible. Some of the hidden nuts and seeds are not recovered, and these may germinate, having been both dispersed and "planted" by the nutcrackers or jays. The movement of tree seeds by crows may thus be important for the distribution of many tree species.

Social crows such as the rook may have difficulty in hiding food where it will not be discovered by other members of the flock. Other members of a group may watch food being hidden and then try to pilfer it, and this hazard of social life may influence the crows' learning abilities (see Calculating Crows and Judicious Jays). Clark's nutcrackers are highly adapted for hiding food, but live alone or in small family groups, while Mexican jays are nonspecialized cachers that live in highly structured social groups. Laboratory studies showed that, although the nutcrackers were better at remembering their own cache locations, the Mexican jays were better at remembering where they had seen other birds hide food.

KJM/DH

Left *A Common magpie prepares to cache an acorn against hard times. The birds eat preferred items first, storing the less favored for later consumption.*

Calculating Crows and Judicious Jays

① *Various corvid species use their intelligence to good effect in securing food. The New Caledonian crow, which lives on the islands of Grande Terre and Maré in the South Pacific, has evolved sophisticated techniques for obtaining prey items. Remarkably, these birds have learned to use tools to flush out insects and grubs, their staple diet, from the crevices of trees. For this purpose, they not only employ simple twigs found on the forest floor, but also fashion tools from diverse materials, such as bamboo and the midribs and edges of leaves. Common implements include a variety of hooks made by different methods (for example, the barbed tool above – a forest vine stripped of all but its final thorn) and a tapered, stepped probe torn with the bill to a standard pattern from the fibrous leaf of the Pandanus plant. Tool manufacture is extremely rare in the animal kingdom.*

② *Carrion crows in Sendai City, Japan, have discovered an ingenious method of smashing open walnuts, which will not yield to their beaks. Perched on traffic lights, the crows wait for the signal to turn red and the vehicles to stop before placing their nuts on the road. When the lights turn green, the birds retire to safety while the cars crack open their meal. At the next signal change, the birds hastily pick up the kernels. Another example of crow problem-solving by exploiting human activity is provided in the anecdotal evidence of ice-fishermen in northern Europe: Hooded crows learned to recognize the signal flag indicating that a baited hook hung on a line through an ice hole had been taken by a fish. Flying down and pulling up the lines with their beaks, while preventing them from slipping back with their feet, the crows deftly secured their illicit catch.*

❸ *At the University of Cambridge, laboratory studies with Western scrub-jays have tested the hypothesis (formulated from observations of this species in the wild) that individuals who pilfer food cached by other birds are more likely to move their own food to another site should they themselves be observed in the act of hiding it. In abstract terms, the thieves demonstrated an ability to use their memories of past experiences to plan for the future – a capacity the research team calls "mental time-travel." This finding suggests that the "theory of mind" (the ability to project one's own experiences onto another individual's intentions and beliefs) may not be an exclusively human preserve. It appears that the scrub-jay's competitive social world, where memorizing one's own actions and predicting those of others come at a premium, has fostered the development of sophisticated thought processes.*

4

5

❹❺ *To investigate whether crows employ cognitive skills – the complex mental processes of adaptive reasoning – when using and making tools, a research team from Oxford University devised a series of experiments for two captive New Caledonian crows, Betty and Abel. In a test to ascertain if they understood basic physical laws and the function of tools, the birds were required to choose between a straight and a hooked stick to retrieve a bucket of food from the bottom of a vertical tube. Both of them selected and successfully used the appropriate tool (left). Moreover, when supplied with only straight wires, Betty repeatedly bent them to create her own hooks (right). These experiments suggest that the birds are not simply following a learned sequence of actions, but rath are capable of adjusting their behavior to the demands of specific tasks. Biologists hope that fu ther studies on social learning among this specie and on the anatomy of their brains will reveal mo about how they acquire their extraordinary aptitu*

Birds of Paradise

Birds of paradise – or at least the magnificently plumed adult males, such as the Black sicklebill with its sabre-like central tail feathers 1m (3.3ft) long – are considered by many to be the most ornate and beautiful of all birds. They are typically highly animated and vocal, crow- or starling-like, strong-footed birds, with species that range from sexually and cryptically monochromatic to dramatically dichromatic. The family includes birds with both monogamous and polygynous mating systems.

Birds of paradise are so named because the first specimens of the adult males to reach the Western world consisted of empty, legless skins prepared by Papuans for the plume trade (an important business in and around New Guinea for thousands of years). This led naturalists of the 16th century to conclude that, lacking a stomach and feet, these beautiful creatures must float in the airs of "paradise," falling to earth only upon death.

The birds have served as a focus of myth, ceremony, personal adornment, and dance for the peoples of New Guinea and neighboring islands for tens of thousands of years. No other avian family exhibits the diversity of feather structure and color found in the Paradisaeidae, particularly males of the polygynous species. Their spectacular and ornate plumages, which serve to impress females choosing a partner, represent an extreme expression of the processes of sexual selection.

Magnificent Males

Form and Function

Monogamous species are sexually monochromatic, and the polygynous ones typically sexually dichromatic from a slight to an extreme degree. The five uniformly blue-black manucodes (*Manucodia* spp.), the similarly dully plumaged paradise-crow, and the generally black Macgregor's bird of paradise show no sexual differences in plumage and are thought to be monogamous. The two sexually monochromatic species of paradigalla were also presumed to be monogamous, but one is now known to breed polygynously. The other more colorful and sexually dichromatic species appear to be polygynous.

Male plumages of polygynous species (in which males mate with more than one female during a breeding season) are extremely varied, from black with brilliant areas of metallic iridescence to brilliant combinations of yellows, reds, blues, and browns, with rich pastel areas of specialized display plumes or weird head or tail "wires" of modified feathering. Each genus of the polygynously reproducing species has a basic male plumage structure peculiar to itself that is manipulated in certain ways during courtship displays in order to

Below *The King bird of paradise is the smallest species of the family, while the Curl-crested manucode is the largest. The King bird of paradise builds its cup nest within a tree crevice.*

FACTFILE

Birds of Paradise

Order: Passeriformes

Family: Paradisaeidae

42 species in 17 genera. Species include: **Blue bird of paradise** (*Paradisaea rudolphi*), **Raggiana bird of paradise** (*P. raggiana*), **Black sicklebill** (*Epimachus fastuosus*), **Brown sicklebill** (*E. mayeri*), **King bird of paradise** (*Cicinnurus regius*), **King of Saxony bird of paradise** (*Pteridophora alberti*), **Lawes' parotia** (*Parotia lawesii*), **Wahnes' parotia** (*P. wahnesi*), **Long-tailed paradigalla** (*Paradigalla carunculata*), **Magnificent bird of paradise** (*Diphyllodes magnificus*), **Magnificent riflebird** (*Ptiloris magnificus*), **Paradise riflebird** (*P. paradiseus*), **Victoria's riflebird** (*P. victoriae*), **paradise-crow** (*Lycocorax pyrrhopterus*), **Ribbon-tailed astrapia** (*Astrapia mayeri*), **Standardwing bird of paradise** (*Semioptera wallacei*), **Macgregor's bird of paradise** (*Macgregoria pulchra*), **Yellow-breasted bird of paradise** (*Loboparadisea sericea*), **Curl-crested manucode** (*Manucodia comrii*), **Trumpet manucode** (*M. keraudrenii*).

Distribution N Moluccan islands of Indonesia, New Guinea; E and NE Australia.

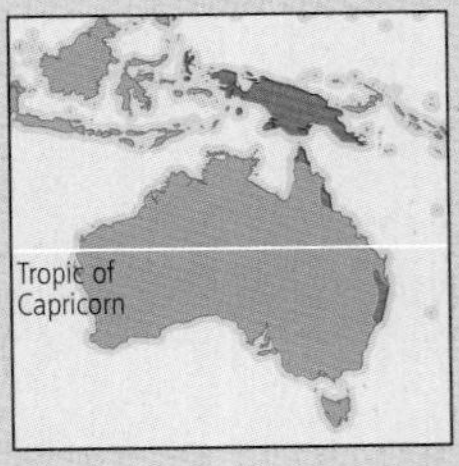

Habitat Tropical and montane to subalpine forests, savanna woodland, mangroves.

Size Length 15–110cm (6–44in); **weight** 0.11–1 lb (50–450g). Males typically larger than females except in the Yellow-breasted bird of paradise.

Plumage Most males are highly colorful with iridescing and ornately elaborated feathers; females and immature males are contrastingly cryptically plumaged and typically barred ventrally. Some socially monogamous species are black or iridescent blue-black all over.

Voice Varied, including crow-like notes, loud gunfire- or bell-like sounds.

Nest Typically bulky open cup- or bowl-shaped structures of stems of epiphytic orchids and/or ferns, vines, and leaves, placed in tree branch forks or among vines.

Eggs 1–2, rarely 3; pale, often of a pinkish base, colorfully spotted, blotched and, typically, smudged by elongate brushstroke-like markings that are most dense about their larger end. Incubation period 14–27 days; nestling period 14–30 days.

Diet Largely fruit-eaters, some more insectivorous; leaves, buds, flowers, arthropods, and small vertebrates.

Conservation Status 4 species – the Blue and Macgregor's birds of paradise, Wahnes' parotia, and the Black sicklebill – are Vulnerable; 8 including the Long-tailed paradigalla, are Lower Risk/Near Threatened.

best present it to potential mates. Several species exhibit colorful areas of pigmented bare skin (head, wattles, legs, feet) that, because they tend to be much brighter in males than in females, may relate to courtship while also acting as species-specific social signals. In adult males of several genera, some outer primaries are modified in shape either slightly or highly, probably for the production of sound in flight. Females of known monogamous species have identifiable vocalizations, whereas among polygynous species all loudly-broadcast cries are male advertisement calls, while the females are virtually mute. Manucodes have an elongated, coiled trachea that is displaced to sit subcutaneously above the pectoral muscles; this produces low, far-carrying, tremulous call-notes that are unique within the group.

Birds of paradise were long considered most closely related to bowerbirds, and some ornithologists joined the two groups in a single family, Paradisaeidae. As a result of exponentially increasing biological and molecular appreciation, however, it is now clear that the birds are most closely related

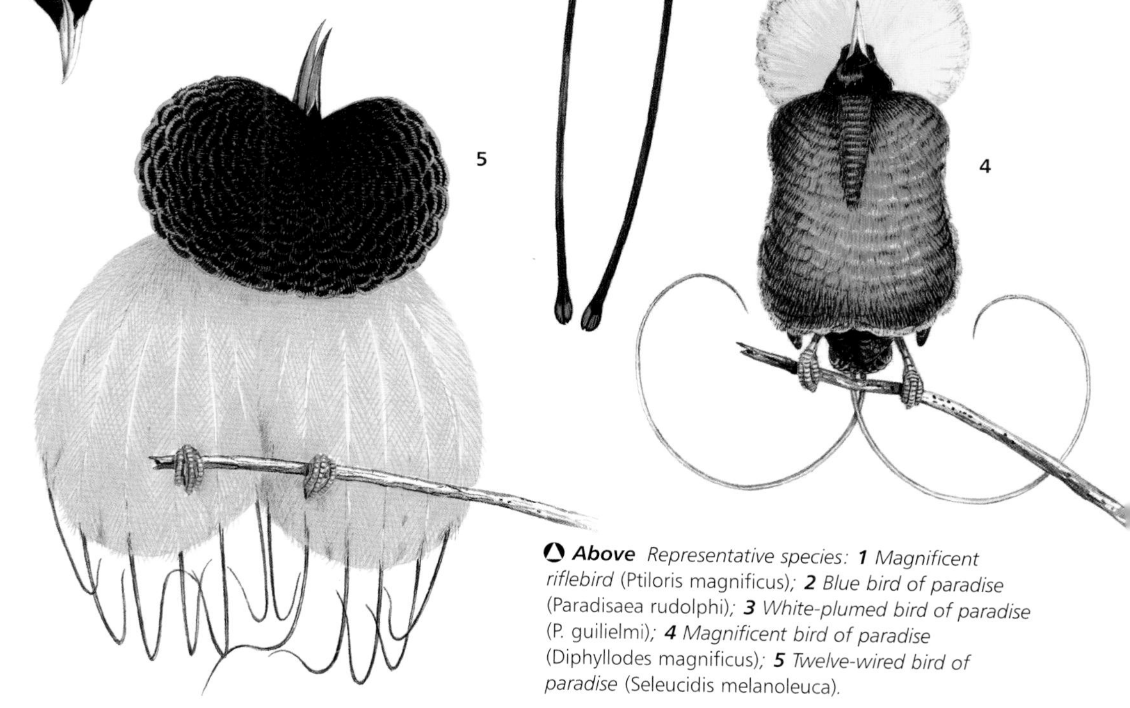

Above *Representative species:* ***1*** *Magnificent riflebird* (Ptiloris magnificus)*;* ***2*** *Blue bird of paradise* (Paradisaea rudolphi)*;* ***3*** *White-plumed bird of paradise* (P. guilielmi)*;* ***4*** *Magnificent bird of paradise* (Diphyllodes magnificus)*;* ***5*** *Twelve-wired bird of paradise* (Seleucidis melanoleuca).

to crows and other crow-like birds (the higher songbirds or passerines), while bowerbirds are relatively distant relations.

The Crested, Loria's, and Yellow-breasted birds of paradise form a group of three "wide-gaped" species that build domed nests and have juvenile and female plumages quite unlike those of the other species. These three are also unusual in apparently being exclusively fruit-eating at all ages (their wide gapes are an adaptation to this diet), and in having relatively weak legs and feet. They were long thought to constitute a distinctive subfamily (the Cnemophilinae) of the Paradisaeidae, but as a result of recent molecular studies they are now considered a separate, unrelated family of more primitive songbirds. Macgregor's bird of paradise is also now doubtfully a member of the family; molecular research indicates that it probably belongs in the Australasian honeyeater family (Meliphagidae) instead.

An examination of the DNA of a single, old specimen of a Lesser melampitta (*Melampitta lugubris*), a pitta-like bird of the New Guinea highlands, concluded that it belonged to the Paradisaedae. However, a subsequent study of its nesting biology and external morphology strongly indicates that it has little in common with the family.

Out of New Guinea

DISTRIBUTION PATTERNS

Most bird of paradise species are confined to New Guinea and adjacent islands, where the family doubtless originated, but the paradise-crow and Standardwing are confined to the northern Moluccan Islands of Indonesia, and the Paradise and Victoria's riflebirds to limited parts of eastern Australia. The extensive New Guinea ranges of both the Magnificent riflebird and the Trumpet manucode also just reach the wet forests of the northeasternmost tip of Australia.

Some New Guinea species have wide lowland distributions, but most have restricted and/or patchy ranges, at discrete altitudinal zones in the mountains. A few are confined to offshore islands. Most species are wet tropical and montane to subalpine forest birds, although a few occur in subalpine woodlands, lowland savanna, or mangroves.

A Diversity of Diets

DIET

Birds of paradise are omnivorous, encompassing a diversity of diets that is unsurprisingly wide in view of their range of body sizes and bill morphologies. Bills vary from short, stout, and crow-like through to finer, starling-like beaks, to extremely long, decurved, and sickle-shaped tools used to probe under moss and bark and between otherwise inaccessible bases of tree fronds for arthropods, insect larvae, and other prey. While most species are predominantly fruit-eaters that also take a variety of arthropods, small vertebrates, leaves, and buds, the sicklebills and riflebirds are highly specialized insectivores that eat relatively little fruit. In the latter, longer-billed species, the bill of adult females is for the most part larger than that of adult males. This fact is noteworthy, as many birds of paradise have to cope with limited resources during nonbreeding months. At such times sexual differences in bill size may significantly reduce intersexual competition for limited arthropod populations, because each sex would take differing species or sizes of prey accordingly.

Typical birds of paradise use their feet to hold and manipulate food items (something bowerbirds do not do). Parents regurgitate food to their offspring (also unlike bowerbirds). They initially feed nestlings on arthropods, but then switch to mostly fruit or a mix of fruit and arthropods.

Above *The Twelve-wired bird of paradise inhabits the mangrove swamps of the western part of the island of New Guinea and is partial to the sago palm that grows there.*

Right *The habits of Wilson's bird of paradise* (Cicinnurus respublica) *are not well known but it can easily be identified by the bare blue skin on its crown.*

The Ultimate Displays

BREEDING BIOLOGY

The various bird of paradise genera display a range of reproductive behavior, from pair-bonded monogamy to polygyny. Monogamous species exhibit a bond that appears to be year-round, the pair defending an all-purpose breeding territory and attending the nest and nestlings. The majority of species, however, have polygynous mating systems, with promiscuous males and exclusively female nest attendance. Females build nests, incubate, and raise offspring alone and unaided. Displays of promiscuous males range from solitary and non-territorial to communal, or lekking, mating systems, with a range of intermediate manifestations. Males may occupy a single courting site – either a terrestrial court or a tree perch or perches – or else an all-purpose territory that includes one or more such sites.

The displays of promiscuous males are undertaken to impress females or, as in the *Paradisaea* species that congregate on leks (communal breeding grounds), to establish a male dominance hierarchy. Like many males, the six-wired parotias (*Parotia* spp.) display solitarily at courts or perches that are used by individuals down the generations year after year. Younger males wait the opportunity to occupy such traditional display sites, during which time they must, like young male bowerbirds, spend years (perhaps as many as seven) in immature, female-like plumage. With adult males mating with many females, there are few males in the breeding population relative to the combined numbers of immature males and females. Pressure by immature males as well as from rival adults ensures that only the fittest males are able to maintain a place in the breeding community and

Right *The polygamous male Raggiana bird of paradise begins his most elaborate display once he has attracted a female to his lek. He leans forward, tail up and head down, extending and clasping his wings.*

that only the most vigorous immatures come to inherit display sites. Interestingly, captive young male birds of paradise have bred at a relatively early age in the absence of adult-plumaged males, suggesting seasonal hormonal activity may be restricted by the presence of dominant adult males. No such inhibition affects females, which are capable of breeding when 2–3 years old.

While a breeding system in which few males fertilize many females has brought about evolutionarily rapid divergence in the appearance of the various male Paradisaeidae species, they nonetheless remain genetically close. Thus, species markedly different in male appearance are not genetically isolated, and as a result hybrids occur where two species meet. In all, 13 intergeneric and seven intrageneric hybrid crosses are documented, occurring where species ranges and favored habitats overlap.

Not only do species within a genus hybridize, but species from different genera, in which the males may be of utterly different appearance, also hybridize. Adult male offspring of such hybridizations inherit some plumage traits of both parents which are discernible. Many such hybrids, mostly known from only one or two individuals, were originally described erroneously as new species.

It seems that the predominance of tropical forest fruits in the diet of some birds is important to the development of polygyny, and that the quality of fruit and/or its dispersion in space and time within the forest may dictate the kind of breeding system and the way in which males disperse and display. A seasonal hyperabundance of fruits permits promiscuous males to spend inordinate amounts of time at display sites, and enables females to nest and tend to their young unaided, while their spatial/temporal dispersion makes an all-purpose territory economically undefendable.

Birds of paradise typically build open cup- or bowl-shaped nests on tree branches, some species favoring densely-foliaged small crowns of isolated saplings within a small forest gap – a situation that may reduce predation upon the eggs, nestlings, or brooding adults by tree-climbing predators. Nests consist of the stems of epiphytic orchids and/or ferns, vines, and leaves.

Eggs are typically elliptical and pinkish to buff, with long, broad, brushstroke-like markings of browns, grays, lavender, or purplish-gray; a clutch consists of one or two, rarely three. They appear to be laid on successive days (unlike in bowerbirds, which lay on alternate days). Incubation periods span 14–27 days and nestling periods 14–30 days, the longer durations being those of higher-altitude species. There is no evidence of two broods being successfully raised in one season.

Vulnerable but Not Endangered

CONSERVATION AND ENVIRONMENT

No bird of paradise species is currently considered endangered, although four species (Macgregor's and the Blue birds of paradise, the Black sicklebill, and Wahnes' parotia) are listed as vulnerable, and eight species as near-threatened. These birds may still be secure in the large areas of their ranges which are inaccessible and largely uninhabited. However, some species populations, including several in West Papua (western New Guinea) – formerly Irian Jaya – are still unsurveyed and await objective assessment, so it is possible that several distributionally restricted species may be threatened by habitat destruction.

The most vulnerable species of all is perhaps the most striking, the Blue bird of paradise, because mid-montane forests vital to its survival are being reduced by encroaching agriculture and because of the demand for its beautiful feathers. The bird may be further threatened by potential competition with the Raggiana bird of paradise, which abuts the lower altitudinal limit of the Blue bird and is more adaptable. CBF/DWF

Wood Swallows

WOOD SWALLOWS ARE A DISTINCT GROUP of small birds with robust wings that enable them to stay aloft for hours, scooping up insects in their broad bills. Besides hawking high in the sky, wood swallows have been seen feeding among the blossoms of trees, and since they have brushlike tips to their tongues, it is thought that they gather nectar as well as insects.

The birds' triangular wing silhouette closely resembles that of the Common starling, earning them the German name of *Schwalbenstare* ("swallow-starlings"). The title is apt, since each year several wood swallows are shot in mistake for the "pest" exotic starlings (*Sturnus vulgarus*) that are currently trying to colonize Western Australia.

Long-distance Flyers

FORM AND FUNCTION

Wood swallows are magnificent flyers, but hop clumsily on the ground with their short legs. Some of the birds remain as residents all the year round, but others are regular migrants, returning annually to the same place to breed.

The truly nomadic species, the White-browed and the Masked wood swallows, form mixed flocks that annually travel thousands of kilometers between breeding attempts, rarely breeding in the same place two years running even if there appears to be plenty of food. In Australia, the proportion of these mixed flocks varies from east (where White-browed wood swallows make up 75 percent of the total) to west (where they account for just 5 percent), with an approximately equal mix around the middle of the continent. Despite their very similar ecology and behavior, hybridization between the two species is rare.

Strong Family Ties

BREEDING BIOLOGY

Both members of a breeding pair build the nest, incubate the eggs, and feed the young for at least a month. In the Dusky, Little, Black-faced, and White-breasted wood swallows, groups of more than two birds have been known to attend the nest, sharing in all duties. Although the birds' open-cup nests might appear to be very exposed, wood swallows defend them aggressively, and usually one group member remains nearby on watch for predators.

Wood swallows in temperate regions breed in spring, those in the tropics during the wet season. Species of the arid inland, such as the Black-faced wood swallow, may breed at any time, responding very rapidly to heavy falls of rain. They may nest in loose colonies, but rarely within 3m (10ft) of one another. Family parties remain together long after the breeding season and are very sociable, frequently preening each other. Several families often huddle together for roosting even when the night temperature remains above 30°C (86°F). In cold weather, wood swallows may even cluster during the day, and as many as 200 have been seen to gather, like a swarm of bees, on a tree trunk.

A male may courtship-feed his female; copulation is preceded by a characteristic, very beautiful display in which both birds flutter their partly-open wings and rotate their half-spread tails until they achieve coordination, at which point mating takes place. IR

Below Although unrelated to the true swallows, artamids like this Masked wood swallow are fine flyers that capture most of their insect prey on the wing.

FACTFILE

WOOD SWALLOWS

Order: Passeriformes

Family: Artamidae

11 species of the genus *Artamus*: **Ashy wood swallow** (*A. fuscus*), **Bismarck wood swallow** (*A. insignis*), **Black-faced wood swallow** (*A. cinereus*), **Dusky wood swallow** (*A. cyanopterus*), **Fiji wood swallow** (*A. mentalis*), **Great wood swallow** (*A. maximus*), **Little wood swallow** (*A. minor*), **Masked wood swallow** (*A. personatus*), **Ivory-backed wood swallow** (*A. monachus*), **White-breasted wood swallow** (*A. leucorhynchus*), **White-browed wood swallow** (*A. superciliosus*).

DISTRIBUTION India, SE Asia, Indonesia, Melanesia, New Guinea, Australia, and Fiji.

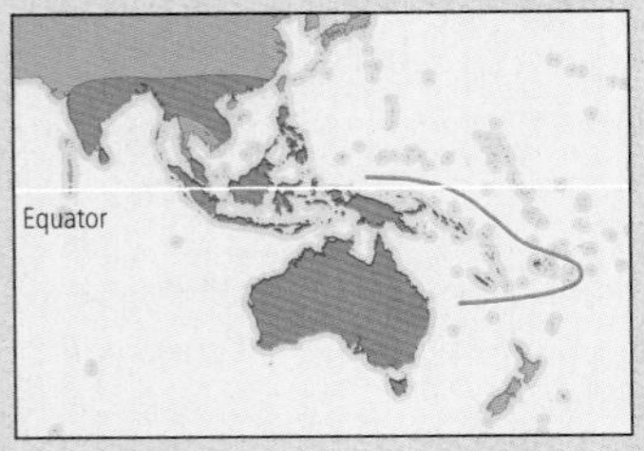

HABITAT Woodland, shrubland, grassland.

SIZE Length 12–23cm (5–9in); weight 14–73g (0.5–2.5oz).

PLUMAGE Gray and brown, piebald, multi-colored gray, white, reddish, and black; juveniles mottled or speckled.

VOICE: Brisk "preet preet."

NEST Open cup, flimsily built of sticks.

EGGS 2–4, creamy white, spotted red-brown. Incubation period 12–16 days; nestling period 16–20 days.

DIET Insects

CONSERVATION STATUS No wood swallow species are threatened.

Magpie-larks and Australian Mudnesters

Left *The nest of the Magpie-lark shows how the species acquired its alternative name of "mudlark." Here, two large chicks prepare to leave the nest.*

Intense social and territorial behavior patterns bind these four endemic Australian and New Guinean species. Although all four were linked in the past solely on the basis of their mud nests, it is now known that the two magpie-larks comprise their own family (Grallinidae), leaving the White-winged chough (no relation to European choughs) and the apostlebird as the Corcoracidae.

Both the Magpie- and the Torrent-lark are black and white, with slight differences between the sexes. Typically for passerines, the feathering in both species is smooth and glossy. In contrast, the plumage of both Australian mudnesters is soft and fluffy (as in many babblers). These birds spend most of their time foraging on the ground, and consequently have well-developed legs.

Problems of Adaptation

DISTRIBUTION PATTERNS

While the Torrent-lark is limited to upland New Guinea, where it lives close to fast-flowing streams, the Magpie-lark is common over a much wider range, and has adapted well to partially cleared farmland and urban areas. The White-winged chough and apostlebird are more susceptible to fragmentation of their open woodland habitat. Many get hit by vehicles, and their slow rate of reproduction means they are disappearing near major roads and highways. The mudnesters are popular with humans due to their amusing social interactions, although their habit of digging into, and tossing, soil and leaf litter makes them somewhat less endearing to gardeners.

The two Australian mudnester species have overlapping ranges on the eastern Australian mainland. White-winged choughs range from inland semi-arid regions to the coast, whereas apostlebirds are restricted to the western side of the Great Dividing Range. Their invertebrate and seed diets also overlap, but White-winged choughs prefer to probe the soil and leaf litter for beetle larvae and worms with their long, curved bills, whereas apostlebirds use their shorter, chunkier, "finchlike" bills mostly for eating seeds

Pair-dwellers and Family Groups

BREEDING BIOLOGY

Magpie-larks are pair-dwelling and have their own highly sophisticated territorial behavior. Throughout their wide range pairs can be seen advertising their territories with coordinated antiphonal duets, often given with a synchronized wing display from a conspicuous tree or telegraph pole. Magpie-larks defend their all-purpose territories throughout the year, and may attack other black and white birds, and even their own reflections!

The Australian mudnesters are very rarely found in simple pairs, but rather in groups of up to 20 members. Usually this is the result of young from previous years staying with the group instead of dispersing, but new groups also form when a coalition of related birds supporting a dominant male joins another supporting a dominant female. All members of the group help to build the nest, to incubate and brood the nestlings, and to feed the young, both in the nest and for many months after they fledge. Normally only one female lays in a nest, although occasionally two may do so, resulting in a very large clutch.

In White-winged choughs, sexual maturity is not reached until 3–4 years of age, as young birds take a long time to master the foraging skills necessary for breeding. Even established breeders must have at least two helpers if they are to successfully fledge young; groups have even been known to kidnap the fledglings of others and raise them to be helpers (or slaves!). It is only on the death of the female breeder that group members will eventually disperse. RH

FACTFILE

MAGPIE-LARKS AND MUDNESTERS

Order: Passeriformes

Families: Grallinidae, Corcoracidae

4 species in 3 genera

Distribution
Australia, Timor, Lord Howe Island, New Guinea.

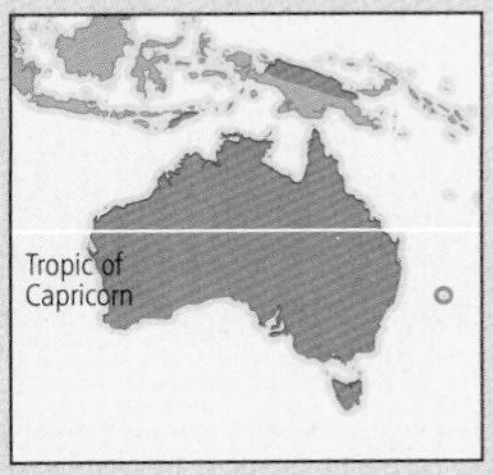

MAGPIE-LARKS Family Grallinidae
2 species of the genus *Grallina*: **Magpie-lark** (*Grallina cyanoleuca*) and **Torrent-lark** (*G. bruijni*). Australia, Timor, Lord Howe Island, New Guinea. Open woodland and forest. **Length** 20–30cm (8–12in); **weight** 40g/1.4oz (*G. bruijni*); 90g/3.2oz (*G. cyanoleuca*). **Plumage**: gray, or black and white. **Voice**: harsh buzzing and piping calls; male and female perform synchronized duet. **Eggs**: 3–5, white to pink in color, blotched with red, brown, or gray; incubation period 17–18 days; nestling period 19–23 days. **Diet**: insects, other invertebrates including snails, some seeds.

AUSTRALIAN MUDNESTERS Family Corcoracidae
2 species in 2 genera: **White-winged chough** (*Corcorax melanorhamphos*) and **apostlebird** (*Struthidea cinerea*). E Australia. Woodland and grassland. **Length** 33–47cm (13–19in); **weight** 110g/4oz (*S. cinerea*); 365g/13oz (*C. melanorhamphos*). **Plumage**: soft and fluffy; sooty black in chough, gray and black in apostlebird. **Voice**: contact call piping whistles, alarm call a harsh screech. **Eggs**: 2–5, creamy white with brown, gray, or black blotches; incubation period 18–19 days; nestling period 18–29 days. **Diet**: seeds and insects.

Below *A White-winged chough nest starts out as a saddle of mud plastered over a branch **1**. Group members then raise it to form first a platform **2**, then a saucer **3**, before it takes final shape as a bowl **4**.*

Butcherbirds and their Allies

CRACTICIDS – THE BUTCHERBIRDS AND THEIR allies – are endemic to Australia, Papua New Guinea, and Borneo. All are insectivorous. Three widespread genera – the Australasian magpie, the various Cracticus *butcherbirds, and the* Strepera *currawongs – are basically robust, black-and-white birds with loud, tuneful songs.*

These three genera are probably related to the wood swallows (treated here as a separate family), and to the monotypic Bornean bristlehead and two species of *Peltops*, the Lowland peltops and Mountain peltops from Papua New Guinea. The bristlehead is a small, stumpy, black-and-red bird that utters harsh, crowlike calls, while the plumage of the two peltops is basically black, with small areas of white and red, and they generally feed by catching their prey in flight.

Butcherbirds, Magpies, Currawongs

FORM AND FUNCTION

The three widespread genera differ from one another in their methods of prey capture and social behavior. Butcherbirds fly between perches and only drop to the ground to catch an occasional prey item. They also impale their prey on twigs or hooks – the habit from which they get their name. Their long, massive, blue-black bills, with hooked tips, enable them to capture and dismember insects, frogs, and small birds. They live in pairs or family groups. The Pied butcherbird lives in groups of up to ten birds and breeds cooperatively, with more than one adult female laying in the group nest and all members of the group helping to rear the young. All species have beautiful, piping calls that are performed by the pair or group. Grey butcherbirds in particular are splendid mimics and will often imitate many other bird species, as well as horses and dogs.

The Australasian magpie is the best known bird of the family, probably because it prefers open grassland for feeding and nesting and has increased in densities around human habitations. Also, it has a beautiful, carolling song. It is larger than the butcherbirds and almost always feeds on the ground, locating prey on the surface with its keen eyesight and beneath the surface by listening for invertebrate movements within the soil. The species is fiercely territorial; a small proportion of the birds are even known to attack humans during the breeding season.

Mating systems vary widely. Across much of Australia, the magpies live in simple pairs. In some southern populations, however, territorial groups can contain as many as 20 birds, and up to five females can build nests and lay eggs within a territory. Genetic studies have revealed that, despite the territorial nature of the species, many females mate with males from outside their own group. Nevertheless, both the males and other members of the group help to feed the young.

Currawongs are larger again, and although they spend some time on the ground, they are adept at foraging in the forest, where they are known to steal eggs and nestlings from other birds' nests, even those of Australasian magpies. Unlike the other two genera, they tend not to live in year-round territories, but roam about instead, depending on food availability. JH/IR

FACTFILE

BUTCHERBIRDS AND THEIR ALLIES

Order: Passeriformes

Family: Cracticidae

14 species in 5 genera. Species include: **Tagula butcherbird** (*Cracticus louisiadensis*), **Australasian magpie** (*Gymnorhina tibicen*), **Bornean bristlehead** (*Pityriasis gymnocephala*), **Lowland peltops** (*Peltops blainvillii*), **Mountain peltops** (*P. montanus*), **Pied currawong** (*Strepera graculina*).

DISTRIBUTION Australia, New Guinea, Borneo; 1 species introduced to New Zealand and Fiji.

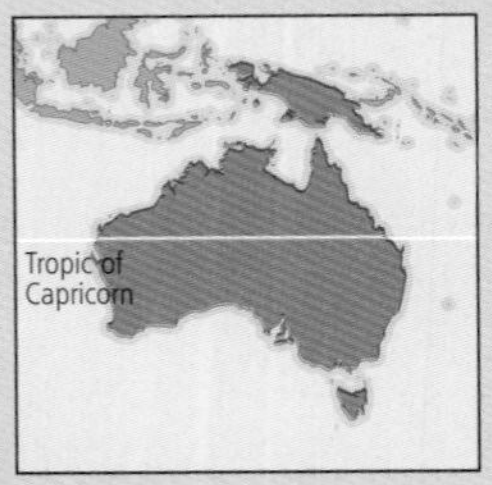

HABITAT Woodland, shrub, and grassland.

PLUMAGE Mostly gray, white, and black; 2 genera have red.

SIZE Length 18–53cm (7–21in); weight 60–140g (2–5oz).

VOICE Loud, varied carolling.

NEST Built of sticks, lined with fiber, and set in a tree fork. Australian magpies sometimes also include substantial amounts of wire.

EGGS 3–5; white, blue, or green, blotched and streaked brown. Incubation period 20 days; nestling period 28 days.

DIET Omnivorous, but mainly insects.

CONSERVATION STATUS No species is currently endangered, although the Bornean bristlehead is listed as Lower Risk/Near Threatened, and the little-known Tagula butcherbird as Data Deficient.

Right *Australasian magpies share the colors of their Western namesakes but are not at all closely related. Sturdy birds with fine voices, they frequent parks and gardens across Australia, and have been introduced to New Zealand.*

New Zealand Wattlebirds

THERE ARE RELATIVELY FEW SPECIES OF BIRDS restricted to New Zealand. Many of them have found it difficult to cope with the loss of habitat due to massive clearing for forestry and agriculture, and the efficient, introduced predators and avian competitors that accompanied European settlement. The New Zealand wattlebirds typify this conflict; all have declined in abundance over the past 100 years and one, the huia, has almost certainly become extinct; it was last seen alive in 1907.

The three species that make up this family were all forest-dwellers that spent a proportion of their time foraging at ground level: this behavior and their readily accessible nests are thought to have made them very susceptible to predation by cats, possums, and rats. However, it would seem that collecting by both Maoris and the early European settlers significantly hastened the demise of the huia, which, unfortunately, was sought for ornamentation by both cultures.

Forest Foragers

FORM AND FUNCTION

The family gains its name from the conspicuous hanging face wattles that adorn each species. These are orange-colored except in the North Island race of the kokako, which has blue wattles. The huia was one of the very few birds in which there was a pronounced difference in bill shape between the sexes.

The kokako and saddleback both eat a wide variety of fruits, berries, and insects gathered at all levels of the forest. Their legs are well developed, and their wings, while not large, are quite adequate for short flights. Breeding is usually in the spring and early summer. The female builds the nest, and she alone incubates the eggs. The male feeds the female on the nest and escorts her when she leaves it to forage. Both parents feed the nestlings. The young stay with their parents for several months after leaving the nest, but there is no evidence of longer-term associations of a type that might lead to cooperative breeding groups.

Left *Saddlebacks are famed songsters, renowned for their innovative melodic lines. Groups develop recognizable "dialects," distinct from their neighbors' calls.*

The Fight to Save the Kokako

CONSERVATION AND ENVIRONMENT

Both living wattlebird species have been the subject of much concern to ornithologists in New Zealand, and the Department of Conservation has been active in preventing either from following the huia into extinction. The establishment of the saddleback on several predator-free islands forms one of the few major success stories in the management of endangered species.

It is hoped that efforts along similar lines with the kokako will be as successful. The birds once occurred throughout the lowland forests of the North and South Islands, but are now restricted to 15 scattered populations on the North Island. These were in decline; they had poor nesting success and a sex imbalance, with a preponderance of males thought to be due to predation of incubating females. Recently, management has targeted potential predators – introduced rats and possums – for poisoning, resulting in more pairs breeding and general population increases. It is now planned to translocate kokako to new sites where predators are similarly managed. IR

Above *A kokako raises its head to utter its organ-like song, revealing the disk-shaped blue wattles. The South Island subspecies, which had orange wattles, has not been seen since 1967.*

FACTFILE

NEW ZEALAND WATTLEBIRDS

Order: Passeriformes

Family: Callaeidae

3 species in 3 genera (one probably extinct): **huia** (*Heteralocha acutirostris*), **kokako** (*Callaeas cinerea*), **saddleback** (*Philesturnus carunculatus*).

HABITAT Dense forest

SIZE Length 22–50cm (9–20in); weight 77–240g (3–8.5oz).

VOICE A variety of whistles, clicks, mews, and pipes.

PLUMAGE Gray, black, and white in kokako; black and reddish in saddleback.

NEST Shallow, open, loosely built with twigs and leaves, lined with moss and fern-scale; placed in a hollow or ledge up to 10m (33ft) above ground.

DISTRIBUTION Endemic to New Zealand's North Island and to a dozen or so islands off North Island and Stewart Island.

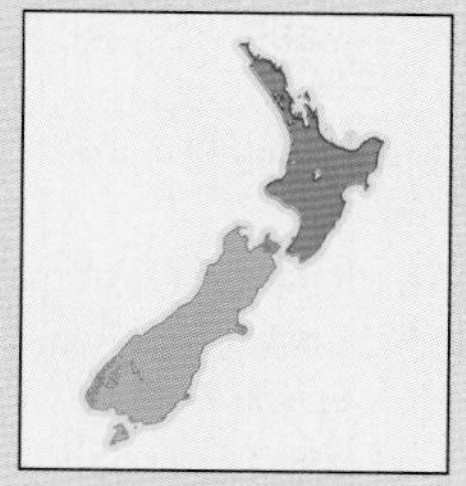

EGGS 2–4; whitish with purplish-brown blotches and spots. Incubation period 18–25 days; nestling period 27–28 days.

DIET Fruit and insects

CONSERVATION STATUS The huia was last recorded in 1907 and is now believed to be Extinct by the IUCN. The kokako is Endangered, while the saddleback is Lower Risk/Near Threatened.

Old World Orioles

Left *Heavily dusted with pollen from feeding, an African black-headed oriole perches on an aloe branch. This bird displays the striking yellow plumage that is a feature of the males of many oriole species. All the Oriolidae are forest-dwellers that live, feed, and breed high in the tree canopy.*

FACTFILE

OLD WORLD ORIOLES

Order: Passeriformes

Family: Oriolidae

28 species in 2 genera. Species include: **African black-headed oriole** (*Oriolus larvatus*), **African golden oriole** (*O. auratus*), **Black oriole** (*O. hosii*), **Golden oriole** (*O. oriolus*), **Green-headed oriole** (*O. chlorocephalus*), **Isabella oriole** (*O. isabellae*), **Maroon oriole** (*O. traillii*), **São Tomé oriole** (*O. crassirostris*), **Silver oriole** (*O. mellianus*), **Green figbird** (*Sphecotheres viridis*), **Wetar figbird** (*S. hypoleucus*).

DISTRIBUTION Africa, Asia, the Philippines, Malaysia, New Guinea, and Australia; 1 species present in Europe. Most species present in the E quarter of the family's range.

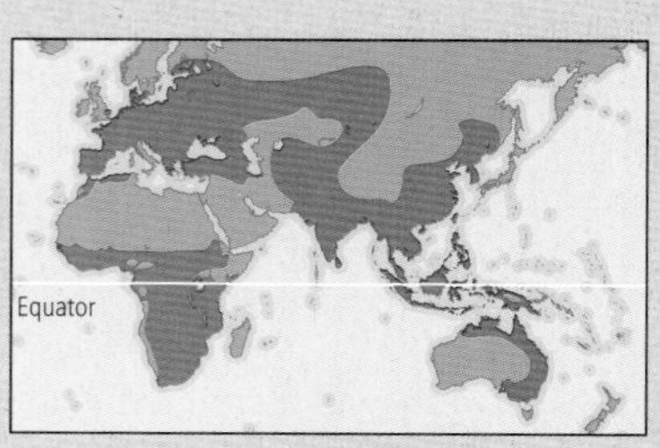

HABITAT Woodlands and forest

SIZE Length 20–30cm (8–12in); weight 50–135g (1.8–4.8oz).

PLUMAGE Predominantly yellow, or yellow and black, occasionally crimson and black; female orioles, with few exceptions, are less brightly colored than males, and in a number of species are also streaked. The figbirds are duller olive green, gray, and yellow, with bare red skin around the eyes in the male.

VOICE Orioles have clear, liquid calls and a growling or bleating call; some orioles are capable mimics. Figbirds have peculiar chattering calls.

NEST Open, cup-shaped nests high in trees.

EGGS 2–4 in orioles, usually 3 in figbirds; apple to dull olive-green, with red, reddish-purple, purplish-brown, and brown markings.

DIET Fruit and insects, including large hairy caterpillars.

CONSERVATION STATUS The Isabella oriole is Endangered and the São Tomé and Silver orioles are Vulnerable. 3 species, including the Black oriole, are considered Lower Risk/Near Threatened.

ORIOLES FORM A GROUP OF SPECTACULARLY striking birds, the males of which are boldly marked with large splashes of yellow or red and black. The name "oriole" is thought to be derived from the Latin aureolus, *meaning "golden." Despite their vivid colors, the birds are suprisingly difficult to spot, since they tend to inhabit the forest or woodland canopy. However, their far-carrying, flutey songs and calls often alert birdwatchers to their presence before they catch the first glimpse of gold or red.*

The related figbirds of Australasia, with their green and gray coloration, are far more cryptic, but make themselves more obvious through their flocking habits. The Oriolidae are not closely related to the New World orioles, which belong to an entirely different family, the Icteridae.

Birds of the Forest Canopy

FORM AND FUNCTION

All orioles are remarkably similar in shape and size. The greatest species diversity has developed on the islands of Indonesia and New Guinea, and the widest range of plumage colors may be seen there. In contrast to the African orioles, in which the plumage is yellow and black (or, in one species only, yellow and olive green), the plumages of orioles in Australasia range from completely black with chestnut undertail coverts in the Black oriole, through the crimson and black of the Maroon oriole, to the dull yellowish-green of the Australian orioles. The majority of species are sedentary, but others range widely in search of fruits, and a few are truly migratory. The Golden oriole migrates from Europe to nonbreeding grounds in Africa in winter, and central Asian species winter in India.

All the species occur in woodland or forest, where they are restricted to feeding in trees, although both the Golden and the African black-headed orioles will feed on the ground on fallen fruits or insects in the grass layer. Orioles are among the few birds that consume quantities of hairy caterpillars. Large insects are killed – and hairy caterpillars skinned – by beating them vigorously against a branch.

Most orioles are solitary, or found in pairs or family parties. In Africa the African golden, African black-headed, and Green-headed orioles occasionally join mixed-species foraging flocks, and, when they do so, move slowly through the forest or woodland with the other birds. When foraging alone, orioles often fly long distances, as much as 1–2km (0.6–1.2 miles), from fruiting tree to tree or to other food sources. The fruit-eating habits of orioles can bring them into conflict with humans when they come to feed in orchards for cherries, figs, or loquats.

The figbirds are duller in plumage than the orioles, and are heavier built and more sluggish. In the orioles the bill is slightly decurved, while the figbirds have short, stout bills, hooked at the tip. Figbirds are more sociable than orioles, and are usually encountered in small, noisy flocks of up to

30 birds. They feed on temporarily plentiful supplies of fruit that appear unpredictably on different trees in the forest. They can damage commercial fruit crops, such as mulberries and figs, bringing them into conflict with people.

On a number of Indonesian islands, species pairs of orioles and friarbirds (*Philemon* spp.) show such strikingly similar plumage that it is hard to tell them apart. These birds have similar ecologies, and it appears that the mimicry allows the smaller orioles to escape aggression from the larger friarbirds when feeding on the fruit of the same tree.

Secrets Still to be Revealed

BREEDING BIOLOGY

Relatively little is known about the breeding behavior of many species, because they are quite secretive and live in the upper forest canopy; in fact, the nests and eggs of several still remain to be discovered. One of the best studied species is the European Golden oriole, which holds large territories and seems to be essentially monogamous, although it can have up to four male helpers at nests.

Oriole nests are deep, neatly-woven baskets of fine material including grass and beard lichens. They are slung below a branch, near its end, somewhat like a hammock. The lining is of softer, finer material. Nests, particularly those built from beard lichens, often have material trailing down that serves to camouflage them. In the African orioles, nest sites are more often inside the tree, and are seldom on the outer edge of the canopy. In the northern oriole species, both sexes build the nest and share the incubation and rearing duties. In the few tropical species that have been studied, incubation is largely by the female, who is provisioned by the male.

Figbird nests are shallower and flimsier than oriole nests, and are placed in the canopy of trees, in forks at the ends of slender branches. Figbirds construct their nests of twigs and grass, and do not weave the materials together as elaborately as the orioles do.

Under Threat

CONSERVATION AND ENVIRONMENT

Three species of oriole are considered globally threatened. The Isabella oriole is only found on the island of Luzon in the Philippines, where forest destruction is seriously fragmenting the population. The São Tomé oriole is limited to the small island of that name off the coast of West Africa, where its forest habitat is threatened by clearance for cocoa plantations. The Silver oriole breeds in small patches of evergreen broadleaved forest in southern China, which are under pressure from logging interests. HC/WRJD

Above *The* Sphecotheres *figbirds are widespread in Australasia; for example, the Green figbird occurs all along Australia's eastern and northern coasts, and is especially common on Cape York Peninsula. Figbirds are stockier and more gregarious than other orioles.*

Below *Watched by its mate, a male Golden oriole feeds its young. This is the only Old World oriole species to be found in Europe, although its range also extends to northern Africa and Asia.*

Cuckoo-shrikes

CUCKOO-SHRIKES, SOMETIMES KNOWN AS caterpillar-birds, are not related to either shrikes or cuckoos, although the majority have shrike-like bills and resemble cuckoos in size and plumage colors or patterns. They are a family of two distinct groups: the cuckoo-shrikes (8 genera, 72 species), which are generally drab in color and range from sparrow- to pigeon-sized; and the brightly colored minivets of the genus Pericrocotus *(13 species), which are much more active, gregarious, and wagtail-like in size and shape.*

Orioles (Orioloidae) appear to be the closest relatives of the Campephagidae, and sometimes the two families are combined. Woodshrike and flycatcher-shrike genera have traditionally been included in the family, and are so treated here, although recent evidence suggests that they might be more closely related to the African bush-shrikes of the families Malaconotidae and Vangidae.

Above *Representative cuckoo-shrike species:* ***1*** *Large cuckoo-shrike* (Coracina novaehollandiae); ***2*** *Red-shouldered cuckoo-shrike* (Campephaga phoenicea).

Neither Shrikes nor Cuckoos

FORM AND FUNCTION

Two genera of cuckoo-shrikes occur in Africa; one (*Campephaga*, 6 spp.) is endemic, while the other (*Coracina*, with a total of 47 species, but only 5 in Africa) also occurs from East Pakistan through Southeast Asia to New Guinea and Australia. The monotypic Ground cuckoo-shrike is endemic to Australia, the Black-breasted triller or fruithunter to Borneo, and the Golden cuckoo-shrike to New Guinea. The remaining species are distributed throughout the Indian subcontinent, Southeast Asia, Malaysia, Indonesia, and Australasia, northward to eastern China and Russia, as well as to several oceanic islands.

Cuckoo-shrikes have long, pointed wings, moderately long tails (either graduated or rounded), and well-developed rictal bristles that in many species cover the nostrils. Many species, including *Campephaga*, *Coracina*, and the *Pericrocotus* minivets, have spinelike shafts to erectile feathers on the rump and lower back. These are not normally visible, but are raised in defensive display and are easily shed. The newly hatched young of some genera such as

Right *A Varied triller* (Lalage leucomela) *broods a large nestling on its nest in northern Australia. The* Lalage *trillers are a family of relatively small cuckoo-shrikes, typically 15–20cm (6–8in) long, that take their name from the loud whistling notes uttered by the males when courting. In most species, both sexes share nest duties.*

Campephaga, *Hemipus*, and *Tephrodornis* are covered with white or gray down and, later, finely barred feathers that blend in perfectly with the nest and its environment. The sexes are usually separable, sometimes strikingly so, as for example in the African *Campephaga* species, but the fledglings are similar to adult females.

Trillers are small birds, often pied in color. The White-winged triller is unique in the family in that the male molts from a black and white breeding dress into a non-breeding dress that is similar to the female's plumage, with brown above and white lightly streaked with brown below.

In contrast, the minivets are dainty and strikingly colored, and have narrow wings, a prominent wing bar, and a long, strongly graduated tail. There is a marked difference between the sexes, while juveniles are similarly colored and patterned to adult females but not so prominently. The Scarlet minivet is bright red, heightened by black areas on the head, throat, back, most of the wings, and central tail feathers. The female is just as striking, with yellow replacing the red, and the black extending to the chin, throat, and forehead. The least striking of the group, but still handsome, is the Ashy minivet; in this species the male has a gray back and rump, a black and white tail, a prominent black nape and crown, and white on the forehead and underparts.

Foraging in Groups

DIET AND FEEDING

Most of the family are gregarious at times. Minivets, in particular, are usually found in noisy, colorful parties of 20 or more birds as they move through the treetops in search of insects. Most species form part of the mixed feeding flocks characteristic of the forests and open woodlands of Africa, India, and Southeast Asia. Flycatcher-shrikes do the same, but they and some trillers also catch many insects on the wing, using their relatively short bill and wide gape. Wood-shrikes are more cumbersome and slower-moving when feeding; they catch some insects in flight, but will also feed on the ground when necessary.

The larger *Coracina* cuckoo-shrikes also form loose feeding parties, eating some fruit as well as insects as they pass through the

forest canopy. Most trillers also eat fruit and insects, but the Black-breasted triller, a montane forest species, is thought to eat only fruit. Some trillers feed mainly on the ground, and the Ground cuckoo-shrike forages entirely terrestrially, walking about in small groups with a diet almost entirely of small animals.

The Ashy minivet is the only long-distant migrant of the family; flocks of up to 150 birds may form on the breeding grounds in China and Russia prior to departure for overwintering in Southeast Asia. The rest of the family is mainly sedentary or nomadic, but Australian species move north–south over large distances in response to rainfall, while some Indian species undergo altitudinal migrations.

FACTFILE

CUCKOO-SHRIKES

Order: Passeriformes

Family: Campephagidae

85 species in 9 genera. Species and genera include: **minivets** (genus *Pericrocotus*) including **Ashy minivet** (*P. divaricatus*), **Long-tailed minivet** (*P. ethologus*), **Scarlet minivet** (*P. flammeus*), **Small minivet** (*P. cinnamomeus*), **Black-breasted triller** or **fruithunter** (*Chlamydochaera jefferyi*), **Black cuckoo-shrike** (*Campephaga flava*), **flycatcher-shrikes** (genus *Hemipus*), **Ground cuckoo-shrike** (*Pteropodocys maxima*), **Large cuckoo-shrike** (*Coracina novaehollandiae*), **Long-billed cicada-bird** (*C. tenuirostris*), **Golden cuckoo-shrike** (*Campochaera sloetii*), **trillers** (genus *Lalage*) including **White-winged triller** (*L. sueurii*), **wood shrikes** (genus *Tephrodornis*).

DISTRIBUTION Sub-Saharan Africa and Madagascar, Pakistan through SE Asia, S China, Russia, to Japan, Philippines, Indonesia, Australia, and some other Pacific and Indian Ocean islands.

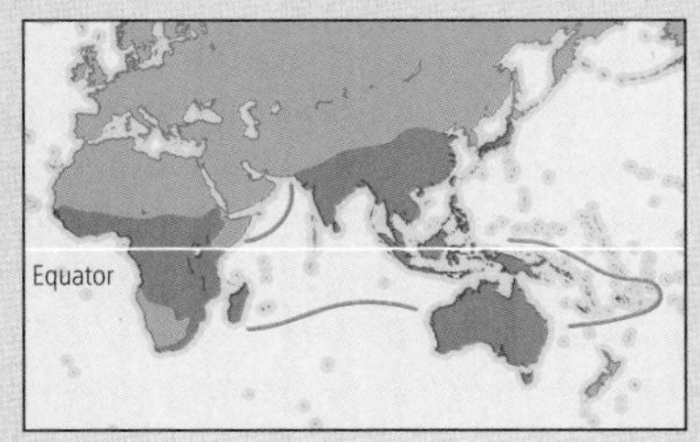

HABITAT Dense primary and secondary forest; a few species in forest edge, deciduous woodland, or coastal scrub.

SIZE Length 12–34cm (4.7–13in); weight 20–111g (0.7–3.9oz).

PLUMAGE Most species are some shade of gray, often with areas of black or white; females often paler than males or barred below; in some African species, males are mainly black, females yellow; Asian minivets are brightly colored, males mainly red and black, females yellow and black.

VOICE Calls range from loud, high-pitched, musical whistles to harsh shrikelike notes, often elaborated into songs; trillers trill, while cicadabirds (*Coracina* spp.) sound like their namesakes.

NEST A neat cup high in the fork of a tree or bonded to the top of a horizontal branch, built of twigs, rootlets, and spiderweb; often flimsy, but well concealed with moss and lichen. The nests of some Australian cicada-birds are occasionally grouped together.

EGGS 2–5; white or pale green, blotched with brown, purple, or gray. Incubation, where known, 14 days (small *Lalage* triller species) to 20–23 days (*Coracina* and *Campephaga* cuckoo-shrikes); nestling period 12 and 20–25 days respectively for the same two groups.

DIET Mainly arthropods, especially caterpillars; some species take fruit, others a few lizards and frogs.

CONSERVATION STATUS The Réunion cuckoo-shrike (*Coracina newtoni*) is Endangered, and 3 other species, including the White-winged cuckoo-shrike (*C. ostenta*), are Vulnerable. In addition, 9 species are listed as Lower Risk/Near Threatened.

Courtship Flights and Wing-flicking

BREEDING BIOLOGY

The courtship and breeding behaviors of the family have been little studied. The male Scarlet minivet pursues the female into the air, seizes her tail in his bill, and they then spiral down together, only breaking apart just before returning to their perches. Such flights above the canopy seem to be a feature of all minivets, but may also be territorial in function, as a similar spiraling descent has been recorded for a male White-winged triller when defending its territory. The male Black cuckoo-shrike performs a mothlike, fluttering flight, with tail fanned and depressed, during its courtship display. In the courtship display of some larger cuckoo-shrikes, the male lifts each wing alternately every few seconds, calling vigorously and repeating the sequence at intervals. However, females also flick their wings, and wing-flicking seems characteristic of the group as a whole, usually as a comfort movement immediately after landing but also at times of excited interaction.

In some species, such as the White-winged triller, the Larger cuckoo-shrike, and the flycatcher shrikes, both sexes take part in nest building, incubation, and feeding the young. In others, such as the Scarlet and Flame-colored minivets, males feed the nestlings and help a little in nest building, whereas in a few (notably the Black cuckoo-shrike) the male only helps feed the nestlings, although he may accompany the female while she builds the nest and guards the nest area during incubation.

Above *A Large cuckoo-shrike brings food to its hungry young. This Australian species typically lays three eggs in a nest of twigs set 6–12m (20–40ft) up in the fork of a tree.*

Some species are single-brooded, but others, like the Small minivet, have two broods in rapid succession, and some, including the Large cuckoo-shrike in India, have two breeding seasons a year (in February–April and August–October). Helpers have been recorded at nests of the Small minivet and the Ground cuckoo-shrike, and are suspected in other communal species. The White-winged triller defends a large territory in coastal areas of Australia, but in the interior pairs may breed in close proximity to each other. AK/LGG

Fantail Flycatchers

FANTAILS OFTEN FLY STRAIGHT AT HUMAN observers and hover a meter or two from them. Consequently they are regarded with great affection, to the extent that one species is called the Friendly fantail. In truth they are probably more interested in the flies buzzing around us than in us.

The fantails are related to the monarch flycatchers (Monarchidae) and the drongos (Dicruridae), with which they are sometimes combined. Many are common and popular, and some adapt well to human disturbance. Some species or subspecies on islands have small populations – for example, the Grey fantail is extinct on Lord Howe Island and Vulnerable on Norfolk Island.

Hyperactive Insectivores

FORM AND FUNCTION

All the fantails are remarkably similar in body form. Their most striking character is a long tail, which can be spread into an impressive fan and waved from side to side. The body is small, the legs delicate, and the bill short but broad, with conspicuous bristles. The willie-wagtail is black and white; other species range from all black through brown to rufous and gray. Many species have conspicuous white eyebrows, throat patches, or tail spots.

Almost half the fantails occur in New Guinea, with three or four species coexisting in the rain forest, foraging at different levels. Five species occur on the Asian mainland, from India through Southeast Asia to southern China. Australia has five species, with the willie-wagtail occupying much drier and more open habitats than the other fantails; the birds have successfully invaded the Indonesian and Pacific islands as far east as Fiji and Samoa. Rufous fantails occur from Sumba in the west and Saipan in the north through New Guinea to southeastern Australia, where they are summer migrants.

Flies, beetles, and other insects are captured in acrobatic sallies from the understory or canopy. Willie-wagtails often perch on sheep's backs, and frequently take insects from the ground. The fanned tail and hyperactivity of fantails probably help to flush insects. They often follow mixed flocks of insectivorous birds, probably capturing insects flushed by flock members.

Above *A Grey fantail exhibits the splayed tailfeathers and drooped-wing stance typical of these small, active birds. This species is familiar in Australia and New Zealand, often flying close to people to snap up any insects they may disturb.*

Delicate but Durable Nests

BREEDING BIOLOGY

The breeding biology of willie-wagtails and Gray fantails is well-known, but little is known about the breeding of the species that live in New Guinea. Fantails build nests of bark and moss, lined with hair, wool, or thistledown, usually on a low, thin branch. They breed from August to February in Australasia, and from January to August in Asia, and lay several clutches.

Three or four eggs, colored whitish to pink, are typical in temperate regions, with two being the normal number in the tropics. Eggs and nestlings are often taken by large birds or introduced mammals. The birds' nests are sometimes parasitized by cuckoos. HAF

FACTFILE

FANTAIL FLYCATCHERS

Order: Passeriformes

Family: Rhipiduridae

42 species in the genus *Rhipidura*, including: **Friendly fantail** (*R. albolimbata*), **Grey fantail** (*R. fuliginosa*), **Rufous fantail** (*R. rufifrons*), **willie-wagtail** (*R. leucophrys*).

DISTRIBUTION India, S China, SE Asia, New Guinea, Australia, New Zealand, Pacific islands.

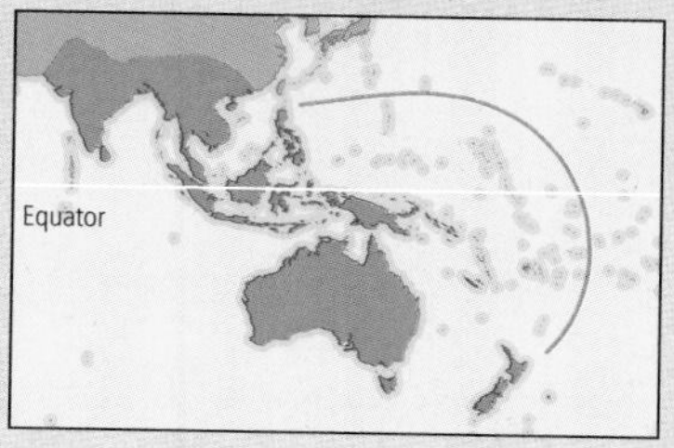

HABITAT Woodland and forest

SIZE Length 13–21 cm (5–8in); weight 7–25g (0.25–0.9oz).

PLUMAGE Black, gray, white, yellow, or rufous; sexes similar in most species.

VOICE Chattering or squeaky calls and songs

NEST Cup- or goblet-shaped

EGGS 2–5; white, pink, or cream, with darker markings. Incubation period 12–16 days; nestling period 11–17 days.

DIET Insects

CONSERVATION STATUS The Malaita fantail (*Rhipidura malaitae*) and the Manus fantail (*R. semirubra*) are Vulnerable. 5 other species are classed as Lower Risk/Near Threatened.

Right *In India, a White-browed fantail flycatcher* (Rhipidura aureola) *incubates a clutch of eggs. Flycatcher nests are neat, solidly-constructed cups, bound together so tightly with cobwebs that they can withstand relatively harsh weather conditions.*

Drongos

Left *The Crested drongo is restricted to Madagascar and the adjacent Comoro Islands. It can imitate not only the calls of other birds, but also those of lemurs!*

DRONGOS ARE A DISTINCTIVE FAMILY OF medium-sized black birds that often perch in the open and offer a distinctive silhouette with their stout, hooked bill, long rictal bristles, short legs, and long, often distinctively-shaped tail. All are placed in the genus Dicrurus, *except for the Papuan mountain drongo of New Guinea with its square tail of 12 (not 10) feathers and exceptionally long rictal bristles.*

Drongos appear to be most closely related to orioles, crows, bulbuls, birds of paradise, flycatchers, and shrikes. The word "drongo" itself comes from a Madagascan tribe's indigenous name for the endemic species *Dicrurus forficatus*.

Black, Long-tailed, and Conspicuous

FORM AND FUNCTION

Drongos usually occur in pairs or, after breeding, in families accompanied by gray-barred juveniles with undeveloped tails and crests. They often sally forth from a perch, with spectacular swoops, curves, and twists, in pursuit of some passing insect. They are characterized by their long, forked, often lyre-shaped tail, the exact shape and structure of which varies from almost square to deeply forked. In a few species, the outer feathers are elongated, curled, or denuded of barbs to end in spoon-shaped rackets, as in the Greater and Lesser racket-tailed drongos of India and Southeast Asia. Other Asiatic species also have conspicuous crests, but the seven species of drongo that occur in Africa and the adjacent islands of the Indian Ocean are more conventional in appearance; their most extreme development is a deeply forked tail or brown wings.

The Fork-tailed drongo of sub-Saharan Africa and the Black drongo or "King crow" of the Indo-Malaysian region are so similar in appearance and biology that they are sometimes considered to be a single species. Both are abundant and familiar birds in a range of habitats, from gardens and fields to open woodland. Both accompany grazing animals in order to snatch up insects disturbed by their movements, sometimes even riding on their backs. This association is most prevalent during the dry season, when food is in short supply, and it is then that fruit and nectar are also most frequent in the diet. The Black drongo is even considered important to agriculture for the quantities of insect pests that it consumes around livestock, especially during winter when it congregates and roosts in flocks.

Both species are bold and pugnacious and will pirate food from other birds and small mammals, retrieving the prey in their bill from mid-air or the ground, and then flying back to a perch, where they hold the food underfoot and tear it to pieces with the bill like a small hawk. Both species are also excellent mimics, and the Fork-tailed Drongo will even utter the alarm call of other animals to scare them off and steal their food. Both species display their agile flight to best advantage at a grass fire, hunting the escaping insects, or at an emergence of winged termites, flying vertically upward to catch the ascending alates.

Other drongo species are all similar in their biology, where known, although some larger species, such as the Crow-billed drongo and the Spangled drongo, add more small vertebrates, such as lizards, birds, and bats, to their diet. Many other species also use animals as beaters, including

Above *The Fork-tailed drongo is an African species known for its bold foraging behavior. It has even been observed to drive meerkats away from lizards and beetles they have excavated by mimicking their alarm call.*

Above *The sturdy bill and long, forked tail of this Spangled drongo (the only species that occurs in Australia) are also characteristic of the family as a whole. The tail gives the birds extra maneuverability as they chase insect prey on the wing.*

monkeys, squirrels, and mixed-species bird parties, particularly in forest habitats. All species drink water when available, scooping it up in their bill or maybe sipping it as they plunge-bathe in pools, but only the Crow-billed drongo has been reported to wet its plumage on rain-soaked bark and twigs.

Each drongo species has a characteristic distribution of gloss on the feathers, from over the entire plumage to isolated spots on each feather, as in the Spangled drongo. Several Asian species such as the Hair-crested, Ribbon-tailed, and racket-tailed drongos have elaborate crests and tail feathers; these delicate structures may have an attraction for the opposite sex. The Hair-crested drongo also feeds more on nectar than any other species, however, so the crest may have an additional role in pollination. The development of these structures varies geographically within species, as does the extent of gray, white, or gloss in the plumage, for example in the Ashy or White-bellied drongos, suggesting that there are complex local selection pressures in operation.

From Africa to Australia

DISTRIBUTION PATTERNS

The Black drongo ranges in seven subspecies from southeastern Iran through India to China, Java, and Bali. However, the Hair-crested drongo has the most extensive range and migratory populations of any drongo, with some 30 recognized races extending east from the northern Himalayas to China, south through the whole of Indo-Malaysia to Indonesia, and east to the Solomons and Australia. The Fork-tailed drongo of Africa is more conservative, with only 4–5 subspecies, but it is replaced by the very similar Shining drongo in the lowland forests of West Africa, by the Square-tailed drongo along east African forests, by the Crested drongo on Madagascar, and by the small island populations of the browner Comoro drongo, the deeply fork-tailed Mayotte drongo, and the Aldabra drongo. These species on small islands, also including the Príncipe, Andaman, and Sumatran drongos, are among the rarest and most endangered members of the family.

Fierce Defenders of the Nest

BREEDING BIOLOGY

Black and Fork-tailed drongos sing complex songs, including mimicked phrases that, in the Fork-tailed drongo, are incorporated by a pair into complex duets that may form part of their courtship melody. Later, both sexes take part in nest construction, incubation, and care of the young. A nesting pair of drongos will attack with great ferocity any crow or raptor that crosses its territory, often while mimicking the call of the intruder. Despite these defenses, the Fork-tailed drongo is the principal host to the African cuckoo, while the Black drongo is parasitized by the drongo-cuckoo and Common koel. AK/PRC

Below The Black drongo of India and Southeast Asia – also commonly known as the King crow – is famously aggressive in defense of its nest, fearlessly confronting hawks, eagles, and other predators much larger than itself.

FACTFILE

DRONGOS

Order: Passeriformes

Family: Dicruridae

24 species in 2 genera. Species include: **Papuan mountain** or **Pygmy drongo** (*Chaetorhynchus papuensis*), **Aldabra drongo** (*Dicrurus aldabranus*), **Andaman drongo** (*D. andamanensis*), **Ashy drongo** (*D. leucophaeus*), **Black drongo** (*D. macrocercus*), **Crested drongo** (*D. forficatus*), **Crow-billed drongo** (*D. annectens*), **Fork-tailed drongo** (*D. adsimilis*), **Comoro drongo** (*D. fuscipennis*), **Greater racket-tailed drongo** (*D. paradiseus*), **Hair-crested drongo** (*D. hottentottus*), **Lesser racket-tailed drongo** (*D. remifer*), **Mayotte drongo** (*D. waldenii*), **Shining drongo** (*D. atripennis*), **Spangled drongo** (*D. bracteatus*), **Square-tailed drongo** (*D. ludwigi*), **Príncipe** or **Velvet-mantled drongo** (*D. modestus*), **Ribbon-tailed drongo** (*D. megarhynchus*), **Sumatran drongo** (*D. sumatranus*), **White-bellied drongo** (*D. caerulescens*).

DISTRIBUTION Sub-Saharan Africa, India, SE Asia, Philippines, Malaysia, Indonesia, to Solomon Islands and Australia.

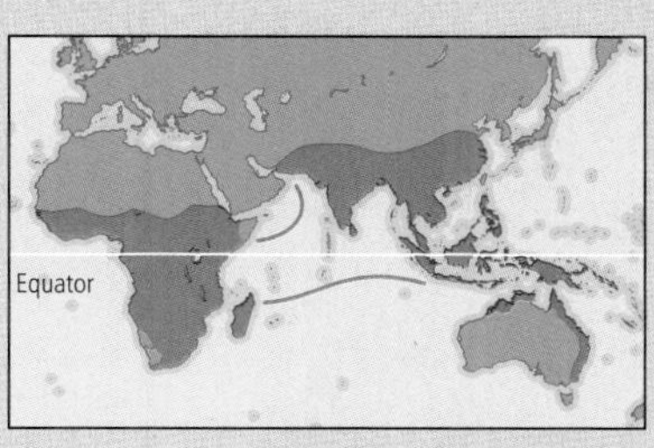

HABITAT Wooded habitats with perches, from dry savanna to rain forest, including cultivation, secondary growth, coastal scrub, and mangroves, up to 3,300m (10,800ft).

SIZE Length 18–38cm (7–15in); up to 72cm (28in) if tail of long-tailed species is included.

PLUMAGE Adults black, with greenish or purplish gloss; iris deep red, with some gray or white areas in two species; juveniles brown-eyed in all; tail forked, or square in two species; outer feathers elongated or racket-tipped in three species; others have elaborate crest feathers. Sexes alike; females slightly smaller.

VOICE Jumble of harsh, metallic notes, musical calls, and whistles; several species are accomplished mimics.

NEST Frail cradle, often hung in horizontal fork of branch, bound with spider web and concealed with lichens.

EGGS 2–5, white or pink, plain or marked with brown, red, and gray. Incubation 15–17 days; nestling period 16–21 days.

DIET Predominantly insects; some nectar, fruit, and small vertebrates.

CONSERVATION STATUS 2 species – the Grand Comoro and Mayotte drongos – are Endangered; 4 others are listed as Lower Risk/Near Threatened.

Monarch Flycatchers

Monarch flycatchers occupy forest, open woodland, and forest edges, and are usually seen singly or in pairs, although some species live in small groups. They attract attention by their energetic aerobatics in pursuit of insect prey. With characteristically fanned tails and quivering wings, they are constantly on the move. Sometimes noisy, they have distinctive harsh, grating calls.

The paradise-flycatchers are the most spectacular species, with tail feathers more than twice the length of their body that are thought to be sexually selected. A range of species is found from New Guinea, through Australasia, and across tropical Asia and southern Africa.

The Most Striking Flycatchers

FORM AND FUNCTION

Monarchs are small- to medium-sized birds that typically have flat, broad beaks, small feet, and steep foreheads with a slight crest. Several species have brightly colored flaps or patches of bare skin around their eyes. The plumage can be strikingly colored, with various combinations of black, gray, white, rufous, blue, or yellow. In some species the sexes are indistinguishable, while in others there are plumage differences between, and sometimes even within, the sexes. For example, male Madagascar paradise flycatchers appear similar to the orange-rufous females when immature, but molt into the adult plumage of either the black and white or the rufous morph over the first 3–5 years of life. Their long tails may have a role in sexual selection. Males intrude on neighboring territories when females are fertile, perhaps seeking extra-pair copulations, and the intruders generally have tails that are longer than average, which may be preferred by the females. Longer tails may, however, be costly, causing decreased agility and thus reducing foraging efficiency.

Many species forage in mixed-species groups. They usually feed by gleaning or hovering and snatching, and have well-developed bristles around their bills (rictal bristles) that help in catching their insect prey.

Different species specialize at different levels in the canopy, subcanopy, or understory. The sexes of the Frilled monarch differ strikingly in behavior; the rufous females sally after insects in the subcanopy, while the black and white males glean from trunks, branches, and vines, and have claws that are more curved than the females'. The Restless flycatcher hovers above the ground making a strange grinding noise that has given rise to its other name, the scissors-grinder.

Of Australo-Papuan origin, the Monarchidae are now found in southern and central Africa, India, and Asia to Japan, south to Australia and east to the Hawaiian Islands. Papua New Guinea

FACTFILE

MONARCH FLYCATCHERS

Order: Passeriformes

Family: Monarchidae

98 species in 18 genera. Species and genera include: **Spot-winged monarch** (*Monarcha guttula*), **Restless flycatcher** or **scissors-grinder** (*Myiagra inquieta*), **Shining flycatcher** (*M. alecto*), **Frilled monarch** (*Arses telescophthalmus*), **African blue monarch** (*Elminia longicauda*), **paradise-flycatchers** (genus *Terpsiphone*) including **Madagascar paradise-flycatcher** (*T. mutata*), **Caerulean paradise-flycatcher** (*Eutrichomyias rowleyi*).

DISTRIBUTION Africa, tropical and eastern Asia, Australasia.

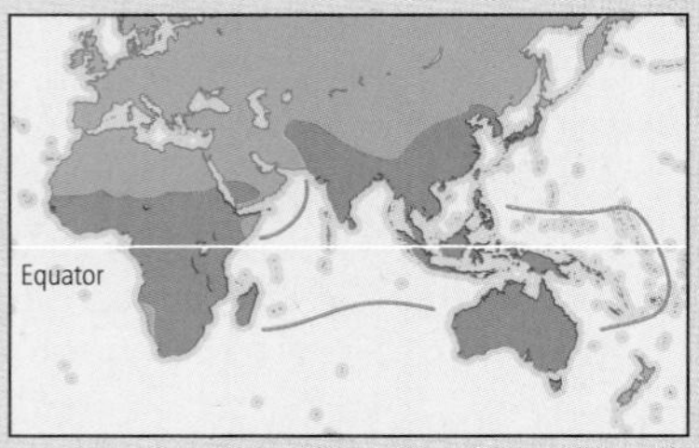

HABITAT Forest and woodland

SIZE Length 12–30cm (5–12in), including tail; **weight** 5–40g (0.2–1.4oz).

PLUMAGE Often metallic black or gray, or chestnut with white underparts; sexes differ in appearance in some species.

VOICE Harsh or whistling calls

NEST Cup-shaped, often decorated with lichen, moss, bark, or spiders' webs.

EGGS 1–5; white, with red-brown spots or blotches.

DIET Most species eat insects; a few also take fruit.

CONSERVATION STATUS In total, 36 species are listed by the IUCN. 2 species – the Maupiti monarch (*Pomarea pomarea*) and the Guam flycatcher (*Myiagra freycineti*) – are Extinct. 5 species, including the Tahiti monarch (*Pomarea nigra*) and the Seychelles paradise-flycatcher (*Terpsiphone corvina*), are Critically Endangered; 6, including the Biak monarch (*Monarcha brehmii*), are Endangered; and 8 are Vulnerable. The others are classed as Lower Risk/Near Threatened or Data Deficient.

Left *In Sri Lanka, a Black-naped monarch* (Hypothymis azurea) *incubates its eggs. This handsome species is widespread and quite common from India as far east as southern China.*

Right *Unencumbered by his long tail, a male Asian paradise-flycatcher* (Terpsiphone paradisi) *takes a turn on a cocoon-adorned nest in India's Western Ghats.*

is the center of diversity, with 28 species in five genera, while Africa has 16 species in four genera. Some members of the group appear superficially similar to Old World flycatchers, but DNA studies have shown that they are unrelated. The two species in the Southeast Asian genus *Philentoma* are included in Monarchidae here, although some observers think they may be more closely related to the Vangidae.

Displays for Defense and Courtship

SOCIAL BEHAVIOR

A few species migrate, but many are sedentary and territorial, some defending their territories throughout the year. Vocalizations, and sometimes visual displays, are important in territorial defence. Often described as noisy birds, the monarch flycatchers have a variety of harsh, buzzy, churring notes, as well as sweet warbling or whistling notes. During aggressive interactions, some species, among them the Spot-winged monarch, adopt a posture with the tail raised and fanned, wings drooping and held out from the body, and bill pointing upward. Aroused birds may also raise their crests.

Most monarch flycatchers are monogamous, although some, like the African blue monarch, have polygamous and communal breeding systems. Courtship displays include bowing and tail-fanning. Male Shining flycatchers strut and bow to females, raising and fanning their tails, raising their erectile crown feathers, calling, and exposing their bright red mouths.

In many species both sexes contribute to the construction of the nest, the incubation of the eggs, and the feeding of the young. The nest of the Frilled monarch is a delicate basket suspended between vines. Other species use upright or horizontal forks to place their neat cups of fiber, rootlets, fine grass, pieces of bark or leaf, lichen, and spiders' web. Eggs are laid at intervals of a day, and the clutch is incubated for 12–18 days. Chicks leave the nest after 10–18 days. New Guinean and Australian species nest in spring and summer (September to December).

At Risk on the Islands

CONSERVATION AND ENVIRONMENT

There are 19 threatened species in this family, many of which are endemic to small islands and therefore are naturally vulnerable to extinction by chance events such as cyclones. This vulnerability is exacerbated by human activities including deforestation, development, and settlement. Severe loss of habitat, as well as competition and predation by introduced species, is causing already small populations to go into rapid decline. The beautiful Caerulean paradise-flycatcher, until recently presumed extinct, is one of five of these threatened species facing an extremely high risk of extinction in the wild that have been listed as Critically Endangered. MH/HAF

Leafbirds

THE LEAFBIRDS ARE BRILLIANTLY-COLORED forest birds of Southeast Asia. Golden-fronted leafbirds inhabit deciduous monsoon forest, but all the other species live in evergreen forest, and so are mainly restricted in the heavily-logged areas west of Myanmar. All species are confined to trees, and feed in the forest mostly at canopy level.

There used to be three genera in the family, but the ioras are now treated separately (see opposite). Leafbirds share characteristics with the bulbuls (Pycnonotidae), drongos (Dicruridae), and caterpillar birds (Campephagidae), and DNA studies seem to confirm the relationships. Fairy bluebirds, in particular, need further taxonomic study; the two species may actually be conspecific. Their combination of brilliant blue and black is, nevertheless, repeated in the throat pattern of most male leafbirds (*Chloropsis* spp.).

Colorful Forest-dwellers

FORM AND FUNCTION

Both leafbirds and fairy bluebirds have short, thick tarsi with small toes, and shed body feathers profusely when handled, as do bulbuls. This trait may have escape value by confusing predators, particularly snakes. The fairy bluebirds are substantially bigger than the leafbirds, and adults have red eyes not shared by the *Chloropsis* species. As their name implies, the leafbirds are green, with or without blue on the wing-coverts and tail and blue, yellow, and/or orange on the head and underparts.

FACTFILE

LEAFBIRDS

Order: Passeriformes

Family: Irenidae

10 species in 2 genera. Species include: **Asian fairy bluebird** (*Irena puella*), **Philippine fairy bluebird** (*I. cyanogaster*), **Blue-masked leafbird** (*Chloropsis venusta*), **Blue-winged leafbird** (*C. cochinchinensis*), **Golden-fronted leafbird** (*C. aurifrons*), **Orange-bellied leafbird** (*C. hardwickii*), **Philippine leafbird** (*C. flavipennis*).

DISTRIBUTION From Pakistan through India and SE Asia to the Philippines.

HABITAT Evergreen forest to dry scrub.

SIZE Length 12–27cm (4.8–11in); weight about 10–90g (0.4–3.2oz); differences between sexes slight to moderate.

PLUMAGE Blue and black, or mostly green; juveniles resemble females.

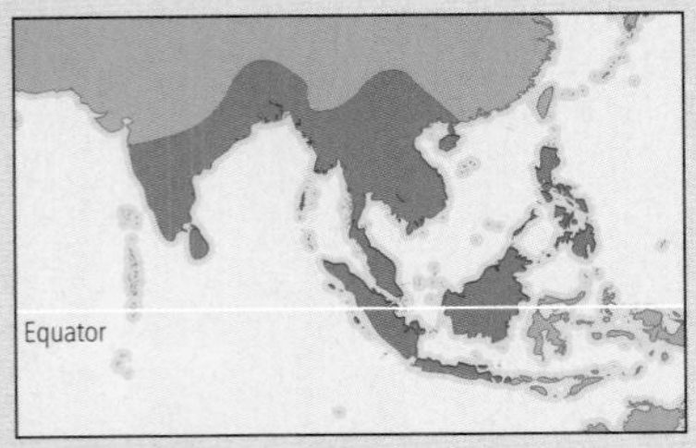

VOICE Melodious songs include mimicry of other species and calls with whistles and chattering.

NEST Open cup in tree or bush.

EGGS 2–3 (occasionally 4); pinkish, speckled and lined with red and purple; in Asian fairy bluebird, greenish white to stone, streaked brown, gray, and purple.

DIET Insects, fruit, nectar.

CONSERVATION STATUS The Philippine leafbird is considered Vulnerable. 2 other leafbirds are classed as Lower Risk/Near Threatened.

Some Irenidae species occasionally join mixed-species foraging flocks, but others are distinctly possessive of the fruiting and flowering trees and shrubs where they feed. Fairy bluebirds are very much fruit-eaters, roaming the forest in small flocks generally numbering up to six or eight birds and advertising their presence with loud, liquid (but percussive) whistles. Leafbirds take both insects and fruit (in the latter case by sticking their bill in and sucking out the contents). They also take nectar, and may help to pollinate some trees.

Fairy bluebird songs are inadequately recorded, but leafbirds, especially the Orange-bellied and Golden-fronted, are fine songsters, and several are notorious mimics.

The few leafbird nests that have been described often incorporate cobwebs. Some are suspended from thin horizontal shoots, but others hang by the rim from twin vertical twigs. Asian fairy bluebirds form a cup of rootlets, moss, and liverworts on a platform of twigs set in a sapling well into the forest. Only the female builds the nest and incubates the eggs, but both sexes feed the young, which are at first covered in thick, fluffy, gray-brown down.

Three leafbirds – the Orange-bellied, a form of the Blue-winged, and the Blue-masked – are mountain-dwellers, but they are found no higher than 2,500m (8,200ft). The Philippine leafbird is endangered, with no recent sightings on several of the islands where it was previously recorded. CJM/DRW

Left *A Golden-fronted leafbird in the Himalayan foothills. Leafbirds' skillful mimicry can confuse bird-watchers, since they are able to reproduce perfectly the calls of other birds, and even of small mammals.*

Below *The range of the beautiful Asian fairy bluebird (here, a female) extends from eastern India to the western Philippines, but the species is now extinct on Sri Lanka.*

Ioras

FORMERLY, THE IORAS WERE REGARDED AS a third genus of the leafbird family (Irenidae), but they differ considerably and seem not to form a natural assemblage with the other two genera. Ioras are smaller than the leafbirds, with proportionately longer bills and more slender legs. They are basically green or green and yellow, sometimes with marked plumage differences between the sexes.

Ioras cover a wide range of habitats, from dry acacia scrub (Marshall's iora) through forest edge and cultivated areas to closed canopy forests (Great and Green ioras); Common ioras are also regularly seen in gardens. They search through foliage for insects, and the Green iora is a regular core member of mixed-species foraging flocks in forest areas. It is unclear, however, whether the individual pairs move far from their own territories, and they may not be big wanderers.

Arboreal Insect-eaters

Form and Function

Great and Common ioras vary in the extent that males develop a black dorsal breeding plumage. Green ioras have yellow "spectacles" around the eyes, more conspicuous in males than in females.

The birds are classic foliage gleaners, searching the thinnest outer leafy twigs of trees and bushes in a very agile way. Common ioras are made conspicuous by their loud, varied calls, which start with a long, sweet whistle, speedily modulating to a shorter, lower sound – *weeeeeee-tu*. The males of both Common and Great ioras, and possibly other species, perform elaborate displays that end in a parachutelike fall; the birds have been described as looking like green balls of fluff as they descend.

Above *A female Common iora. Among the more widespread species such as this, there are considerable racial differences between different populations.*

Ioras build compact cup nests that are felted to the branches of a tree or shrub with cobwebs and that may take as little as five days to build. Incubation is by both sexes, with the female taking the night-time stint; the young hatch after about 14 days. Common ioras may separate their two fledglings, the parents tending one each. This species is the only known nest host of the Banded bay cuckoo (*Cacomantis sonneratii*), a widespread species across India and Southeast Asia.

The Green iora is not yet threatened, but its main, primary growth habitat is dwindling fast; its ability to live in secondary growth may be the key to its survival. CJM/DRW

FACTFILE

IORAS

Order: Passeriformes

Family: Aegithinidae

4 species of the genus *Aegithina*: **Common iora** (*A. tiphia*), **Great iora** (*A. lafresnayei*), **Green iora** (*A. viridissima*), **Marshall's** or **White-tailed iora** (*A. nigrolutea*).

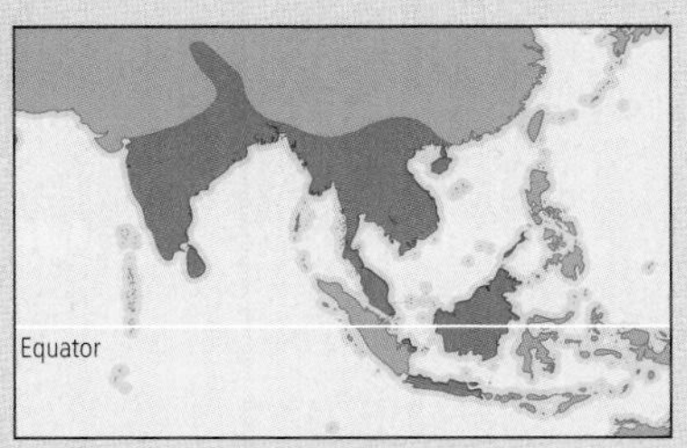

DISTRIBUTION Pakistan, India, SE Asia, and the Philippines.

HABITAT In trees and shrubs of evergreen forest and forest-edge, mangroves to dry scrub and farmland.

SIZE Length 12–17cm (4.8–7in); weight about 10–20g (0.4–0.8oz); slight to moderate differences between sexes.

PLUMAGE Mostly green and black, and most ioras have white double wing-bars; juveniles resemble females.

VOICE Strident whistles and melodious songs, including mimicry.

NEST Open cup in tree or bush

EGGS 2–3 (4 recorded rarely); speckled pinkish and lined red and purple.

DIET Insects (all stages) and other invertebrates.

CONSERVATION STATUS The Green iora is classified as Lower Risk/Near Threatened.

Shrikes

Bold predators that kill with hooked beaks, shrikes mostly prey on insects, but will also take frogs, lizards, rodents, and other birds, some as large as themselves. They are noted for their habit of impaling their prey on thorns, or sometimes on barbed wire, creating larders that can be revisited later when live food becomes scarce.

The taxonomy of the shrikes is currently in flux. Once regarded as a single family, they are now usually assigned, as here, to three separate ones: the Laniidae ("true" shrikes, or simply "shrikes"), including the genus *Eurocephalus,* which was formerly placed with the helmet-shrikes; the Malaconotidae (bush-shrikes); and the Prionopidae (helmet-shrikes). One species, the Bornean bristlehead, which used to be assigned to the shrikes in a subfamily of its own, the Pityriasinae, is now generally agreed to have nothing to do with the group; in this volume it is included with the butcherbirds of the family Cracticidae.

Pouncing Predators

FORM AND FUNCTION

All shrikes have a powerful, raptorlike, more or less sharply hooked bill that is used for killing prey; species like the Northern grey shrike, which catch small vertebrates, kill them by striking them on the back of the head. The birds' legs and feet are strong, and their claws are sharp for holding prey. The tail is long in many species, particularly in the two *Corvinella* shrikes; in the magpie-shrike, it can reach a length of 30cm (12in). That shrike inhabits central–eastern and southern Africa, and is mainly black, with white on the wings and flanks; although markedly different in color, it is generally thought to be closely related to the Yellow-billed shrike because of its social behavior and distribution. The latter, a bird of central Africa, is brown profusely streaked with black above and buff below; its bill is yellow, and there is a chestnut patch on each wing.

In Europe, the most common species is the rather small Red-backed shrike. Remarkably, the sexes are markedly different in this species; the male shows a bright chestnut back, a gray head and rump, pinkish underparts, and a black and white tail, whereas the female's ground color varies between red-brown and gray-brown; the plumage is also heavily vermiculated below, a feature, incidentally, shared with most young shrikes of the genus *Lanius*. Many *Lanius* species, such as the Northern grey shrike and the Common fiscal, are a mixture of black, gray, and white; the sexes are alike, or almost so.

Shrikes characteristically search the ground from a vantage point and pounce on their prey. They may also, however, catch insects on the wing. Many of them store food by impaling their prey on thorns or barbed wire or hanging it from the fork of a branch. Impaling or wedging prey is useful, as it enables species like the Northern grey shrike to dismember small vertebrates that it would be unable (unlike raptors) to hold down with its talons; the practice also provides a food supply for times of bad weather when insects are less active and more difficult to locate.

Radiating Out of Africa

DISTRIBUTION PATTERNS

Two genera are confined to sub-Saharan Africa: *Eurocephalus* and *Corvinella* (both 2 spp.). The main genus, *Lanius*, has nine species restricted to Africa and six others that have populations on that continent or that visit it in winter; however, it also contains species that have spread throughout the northern temperate and even the Arctic regions. Thus, the Northern grey shrike is found throughout Europe and Russia; it also breeds in the northern parts of North America where, further south, it is replaced by the rather similar Loggerhead shrike, whose range extends as far south as Mexico.

In drier parts of southern Eurasia, the Middle East, and North Africa, the Northern grey shrike is replaced by various races of the Southern grey shrike, now generally regarded as a distinct species. In eastern and central China, it is replaced by the Chinese grey shrike, of which one large race, *L. sphenocercus giganteus*, is found up to 5,000m (16,500ft) in Tibet. The Long-tailed shrike also has a large range, from Turkmenistan through Asia and on to New Guinea; in contrast, the Strong-billed shrike is endemic to

Left *A Southern grey shrike dismembers a mouse to feed its young. Shrikes are well-known for their habit of impaling prey – both insects and small mammals – in "larders" for later use. The Loggerhead shrike of the southern USA and Mexico is even reported to make noxious lubber grasshoppers* (Romalea *spp.*) *edible by regularly storing them in this way for a day or so, until the poison in their bodies has degraded.*

Right *The Long-tailed shrike* (Lanius schach), *which is divided into about nine races, has a vast breeding area in Asia, extending from Turkmenistan in the west to the Pacific coast of China and New Guinea in the east. It inhabits lightly wooded and cultivated areas as well as scrubland.*

FACTFILE

SHRIKES

Order: Passeriformes

Family: Laniidae

30 species in 3 genera. Species include: **Northern grey shrike** (*Lanius excubitor*), **Southern grey shrike** (*L. meridionalis*), **Loggerhead shrike** (*L. ludovicianus*), **Chinese grey shrike** (*L. sphenocercus*), **Lesser grey shrike** (*L. minor*), **Red-backed shrike** (*L. collurio*), **Woodchat shrike** (*L. senator*), **Masked shrike** (*L. nubicus*), **Long-tailed shrike** (*L. schach*), **Mountain shrike** (*L. validirostris*), **Souza's shrike** (*L. souzae*), **Common fiscal** (*L. collaris*), **Grey-backed fiscal** (*L. excubitoroides*), **Newton's fiscal** (*L. newtoni*), **magpie-shrike** (*Corvinella melanoleuca*), **Yellow-billed shrike** (*C. corvina*), **White-rumped shrike** (*Eurocephalus rueppelli*), **White-crowned shrike** (*E. anguitimens*).

DISTRIBUTION Most widespread in Africa, but the range of the genus *Lanius* extends to Europe, Russia, India, Asia, the Philippines, Japan, Borneo, New Guinea, and N America.

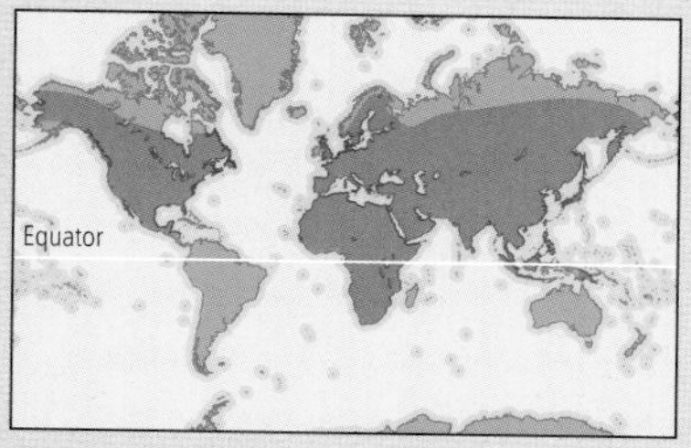

HABITAT In Africa, savanna areas, farmland, and open woodland. Outside Africa, semi-open habitats, orchards, meadows, and hedges, open pine and oak forests.

SIZE Length 15–30cm (5.9–11.8in); weight 20–100g (0.7–3.5oz)

PLUMAGE Often mixtures of black, white, and gray, but brighter colors also occur; sexes similar or almost so except in a very few species.

VOICE Often more or less melodious warbles, sometimes containing a lot of mimicry; also harsh, discordant calls.

NEST In trees or shrubs

EGGS Usually 4–7; wide range of ground color, with streaks or blotches of brown, purple-brown etc.. Incubation 12–15 days, or more in some species; nestling period 12–20 days, according to species and meteorological conditions.

DIET Chiefly insects and other invertebrates; some species regularly take small vertebrates.

CONSERVATION STATUS 1 species – Newton's fiscal – is Critically Endangered. The only other listed species is the Mountain shrike (*Lanius validirostris*), which is Lower Risk/Near Threatened.

the Philippines, and the threatened Newton's fiscal to São Tomé in the Gulf of Guinea.

In Africa, shrikes are largely confined to savanna areas and to farmland. They may also occur in open woodland, like Souza's shrike, a species that is almost endemic to miombo (*Brachystegia* spp.) woodland. Outside Africa, shrikes can be met with in various types of semi-open habitats that are rich in insects and dotted with perches; most species have adapted well to low-intensity farming areas, where they appreciate the presence of orchards, meadows, and hedges. At least one species, the Masked shrike, which breeds in the Middle East, regularly occurs in open pine and oak forests.

After breeding in northern latitudes, all populations of the mainly insectivorous Woodchat, Lesser grey, and Red-backed shrikes migrate to Africa. Interestingly, the western populations of the last two species exhibit a loop migration; in autumn, they fly southeast, mainly toward Greece and its islands and then on to Egypt and to southern Africa; in spring, on their way back, they pass further east, via the Arabian peninsula, Israel, Syria, and Turkey. Other shrikes, such as the Long-tailed shrike in Nepal, may undertake altitudinal movements; they can also be partial migrants like the Northern grey shrike, which catches small vertebrates in winter. The various races of the Common fiscal, living in tropical climes, also find enough food all the year round and are, mainly if not exclusively, resident. On migration, *Lanius* species are territorial; some, like the Red-backed shrike, also defend territories in their winter quarters. The males tend to return to breeding grounds before the females.

Elaborate Displays

BREEDING BIOLOGY

The majority of the *Lanius* shrikes breed in pairs, but at least one African species, the Grey-backed fiscal, is a cooperative breeder; only one pair in a group nests, assisted by a varying number of helpers. The same behavior occurs in the two *Eurocephalus* species (the White-rumped and White-crowned shrikes), and in *Corvinella*; thus, in southern Ghana, the Yellow-billed shrike lives in groups averaging 12 birds throughout the year, and various individuals defend the territory and feed the breeding female, nestlings, and fledglings.

In Africa, the breeding season is often linked to the beginning of the rains and so to the abundance of insects, and two or three successive broods are normal. Breeding in northern latitudes is confined to the short summer period (May–July); one brood is the norm, but replacement clutches are frequent.

Above *Seen here in the Moremi Nature Reserve in Botswana's Okavango Delta region, the magpie-shrike obviously takes its name from its unusual black and white coloration and its long, magpielike tail.*

The courtship display in *Lanius* is accompanied by much wing-shivering, tail-spreading, movements of the head, and other activities; thus the male Woodchat shrike nods his head rapidly up and down, ruffles his head feathers, bends his legs, and flutters his wings while singing to the female. This behavior may lead both partners to join in a duet. Courtship feeding of the female by the male has been observed in most *Lanius* species and in *Corvinella*.

Both sexes help build the nest and feed the nestlings. As a rule the female alone incubates. *Corvinella* shrikes build substantial, loose nests; *Lanius* species also have rather bulky nests, made of twigs and lined with fibers, tendrils, and grass or other materials. Most nests look rather untidy, but those of the Masked shrike at least are neat, compact cups. The nests are either placed in trees or bushes. The nests of *Eurocephalus* are also neatly built, generally plastered with spiders' webs, and usually fixed onto the horizontal fork of a slender branch.

Victims of Intensified Agriculture

CONSERVATION AND ENVIRONMENT

The only Critically Endangered species is Newton's fiscal, which is confined to an African island and, remarkably for a *Lanius* shrike, only occurs in primary lowland and mid-altitude forest. In North America and Europe, many shrike populations are suffering badly from the intensification of agriculture, and are declining. NL/LGG

Right *A pair of Red-backed shrikes keep watch over their brood. Birds of this species are unusual in choosing to site their nests in thickets and brambles; most shrikes place them high up in trees.*

Helmet-shrikes

THE MOST OBVIOUSLY DISTINCTIVE FEATURES of the helmet-shrikes grouped in the single genus Prionops *are the prominent crests from which they take their name. All eight species are restricted to sub-Saharan Africa, where they live chiefly on insects, notably beetles, caterpillars, grasshoppers, and mantises; the White helmet-shrike, however, also occasionally takes small vertebrates, particularly geckos.*

Many tropical bird species are gregarious in both the breeding and non-breeding seasons. Such sociability is one of the most important field characteristics of the helmet-shrikes, which are generally found in parties of up to 12 or more birds.

Above *One of the most widespread* Prionops *species, the White helmet-shrike is found across much of Africa, from the Sahara's southern rim to Botswana and the Transvaal.*

Gregarious Insect-eaters of Africa

FORM AND FUNCTION

The bill of helmet-shrikes is strong, sharply hooked at the tip, and either black or red. The tail is long and rounded, and the feet are strong; they have scales (scutellations) on both the side and front of the tarsus. The White helmet-shrike has a black back, gray head, white crest, and white underparts. Most red-billed species are mainly gray-brown with a black head and breast, the exception being the Chestnut-bellied helmet-shrike, which has a black back, throat, and tail, a white head and breast, and a chestnut belly. The Yellow-crested helmet-shrike is wholly black apart from its crest.

Helmet-shrikes are mainly sedentary; only local or altitudinal movements are known. Although the savanna species have a wide distribution, the ranges of the different species usually do not overlap, and, when they do, they are ecologically separate. Thus the White helmet-shrike has several subspecies in its wide range of latitude (15°N–25° S), but is replaced in parts of Kenya by the Grey-crested helmet-shrike, and in the highlands of the eastern Congo and southwestern Angola by the Yellow-crested helmet-shrike. In areas where it occurs together with both Retz's and the Chestnut-fronted helmet-shrikes, the White searches low down on or near the ground, while Retz's is more arboreal, usually searching for insects high up in the canopy. Since it is smaller than the other two, the Chestnut-fronted helmet-shrike is thought to feed on different prey.

All helmet-shrikes are presumed to be cooperative breeders, a dominant pair being assisted by a varying number of helpers. Breeding occurs primarily in the dry season, but extends into the wet season, at least for some species. The nests are compact cups generally fixed onto a horizontal branch and often decorated with spiders' webs. The clutch comprises 2–5 eggs, often 4.

At least two species are threatened: the Yellow-crested helmet-shrike of Congo and Uganda, and the Gabela helmet-shrike, which is only found in a small area of central-western Angola. Neither bird has been intensively studied, but both are understood to have small populations in fragmented ranges that are increasingly at risk from deforestation.

NL/LGG

FACTFILE

HELMET-SHRIKES

Order: Passeriformes

Family: Prionopidae

8 species in the genus *Prionops*: **Chestnut-bellied helmet-shrike** (*P. caniceps*), **Chestnut-fronted helmet-shrike** (*P. scopifrons*), **Gabela** or **Angola helmet-shrike** (*P. gabela*), **Gabon helmet-shrike** (*P. rufiventris*), **Grey-crested helmet-shrike** (*P. poliolophus*), **Retz's helmet-shrike** (*P. retzii*), **White helmet-shrike** (*P. plumatus*), **Yellow-crested helmet-shrike** (*P. alberti*).

HABITAT Wooded savanna; dense lowland forest.

SIZE Length 19–25cm (7.5–10in); weight 33–52g (1.2–1.8oz).

PLUMAGE Boldly marked black or brown or white; prominent crests and colored wattles in most species; sexes similar or slightly different.

VOICE Whistling notes; rasping, nasal calls; bill-snapping.

DISTRIBUTION Sub-Saharan Africa

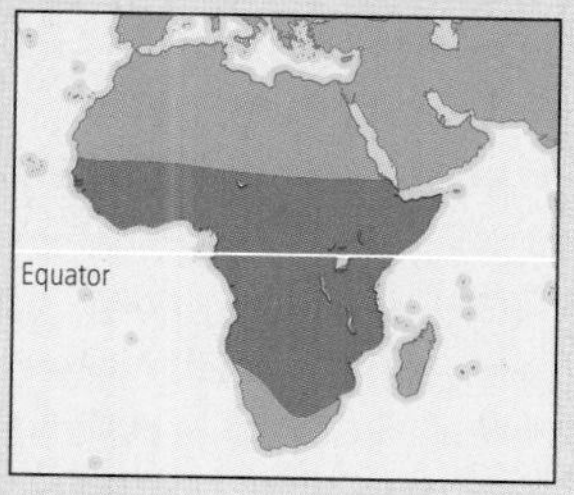

NEST In trees or shrubs; generally on a horizontal branch.

EGGS 3–5; wide range of ground color, streaked or blotched brown. Incubation 12–15 days, or more in some species; nestling period 12–20 days, according to species and meteorological conditions.

DIET Chiefly insects and other invertebrates; some species regularly take small vertebrates.

CONSERVATION STATUS The Gabela helmet-shrike is Endangered. The Yellow-crested helmet-shrike is Vulnerable.

Bush-shrikes

Left *A Crimson-breasted gonolek* (Laniarius atrococcineus) *weaves a neat nest from strips of bark. Nests are normally placed in a tree-fork from 2–7m (6.5–23ft) up, and may also incorporate lichen, tendrils, grass, and cobwebs.*

GIVEN THEIR HABITATS, BUSH-SHRIKES ARE globally much less conspicuous than shrikes, but their songs and calls betray their presence – for example the far-carrying whistles of the Malaconotus *species or the extraordinary duets of the gonoleks. Like the shrikes, all bush-shrikes have a sharply hooked and notched bill. Most of the birds are extraordinarily beautiful.*

The taxonomy of bush-shrikes remains controversial. A recent landmark work included within the family up to 84 species in 18 genera, including the helmet-shrikes of *Prionops*, which are here treated separately.

Above *A Rosy-patched bush-shrike* (Rhodophoneus cruentus) *raises its head in song, revealing the crimson-splashed breast that gives it its name.*

FACTFILE

BUSH-SHRIKES

Order: Passeriformes

Family: Malaconotidae

46 species in 7 genera. Species include: **Grey-headed bush-shrike** (*Malaconotus blanchoti*), **Sulphur-breasted bush-shrike** (*M. sulfureopectus*), **Fiery-breasted bush-shrike** (*M. cruentus*), **Green-breasted bush-shrike** (*M. gladiator*), **Yellow-crowned gonolek** (*Laniarius barbarus*), **Tropical boubou** (*L. aethiopicus*), **Sooty boubou** (*L. leucorhynchus*), **Bulo burti boubou** (*L. liberatus*), **Black-crowned tchagra** (*Tchagra senegala*), **Northern puffback** (*Dryoscopus gambensis*), **Sabine's puffback** (*D. sabini*), **brubru** (*Nilaus afer*), **Mt. Kupé** or **Serle's bush-shrike** (*Telophorus kupeensis*).

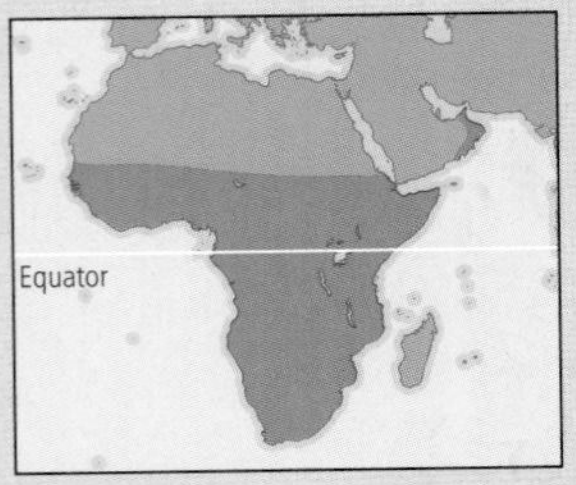

DISTRIBUTION Sub-Saharan Africa; isolated races of the Black-crowned tchagra found in N Africa and in SE of Arabian peninsula.

HABITAT Dense tropical forest, lowland or montane, or open deciduous woodland; savanna areas.

SIZE Length 12–25cm (4.7–10in); weight 20–100g (0.7–3.5oz).

PLUMAGE Many are brightly colored crimson, yellow, green; sexes often similar or almost so, sometimes different.

VOICE Very variable; sometimes mournful whistles or melodious songs; many species duet, some mimic.

NEST In trees or shrubs

EGGS 2–3; wide range of ground color, streaked brown or purple-brown. Incubation 12–15 days; nestling 12–20 days.

DIET Chiefly insects and other invertebrates; some species regularly take small vertebrates.

CONSERVATION STATUS The Bulo burti boubou is Critically Endangered. 4 species, including the Mt. Kupé bush-shrike, are Endangered. The Green-breasted bush-shrike is Vulnerable.

Flashes of Color in the Forest

FORM AND FUNCTION

Bush-shrikes are notable for their brilliantly-colored plumage. The widespread Grey-headed bush-shrike, for instance, has a green back and yellow underparts; it is rather similar to the mainly West African Fiery-breasted bush-shrike, but the latter has different color morphs; there are birds with scarlet, yellow, or orange underparts. The Yellow-crowned gonolek, which is restricted to western Africa, is crimson below and black above, apart from a golden crown and undertail-coverts. Another member of the genus *Laniarius*, the Tropical boubou, is mainly black above and white below, whereas the Sooty boubou bears its name well and is uniformly black. The sexes are alike in the first two species, but in the Sooty boubou the female is somewhat duller, without gloss.

Bush-shrikes are mainly insectivorous. They are not, however, sit-and-wait predators like shrikes; instead, they skulk in dense habitat, where they generally behave like oversized warblers and feed by gleaning the vegetation inside branches and foliage. According to species, prey may be taken at all forest levels; some, like the tchagras, hop on the ground and capture most of their prey items there. The large birds of the genus *Malaconotus* are equipped with a very powerful bill and are able to catch small vertebrates; like shrikes of the genus *Lanius*, they may also store their prey.

Sub-Saharan Dazzlers

DISTRIBUTION PATTERNS

Almost all bush-shrikes are endemic to Africa south of the Sahara; the only exception is the Black-crowned tchagra, also widespread in sub-Saharan Africa but with isolated races respectively in North Africa and in the southeast of the Arabian peninsula. Most genera have species that either inhabit dense tropical forest, both lowland and montane, or else open deciduous woodland. Thus the Northern and Sabine's puffbacks occur throughout West and Central Africa, the former in savanna woodland, the latter rather in lowland forests. However, some species, such as the brubru, are confined to savanna areas. Bush-shrikes are mostly sedentary, but local movements, including altitudinal ones, have been proved or suspected in some species.

The African genus *Malaconotus* is remarkable in that pairs of species, one large and one small, live in the same habitat and are color replicas of each other. Thus the Sulphur-breasted bush-shrike, sometimes placed in the genus *Teleophorus*, is a small edition of the Grey-headed bush-shrike, both occurring in the savanna of West Africa. The function of such duplication is uncertain, particularly as the species involved are ecologically separate.

Right *A Sulphur-breasted bush-shrike attends its almost fully-grown nestlings. The birds share their territory with a similarly-colored but much larger species, the Grey-headed bush-shrike.*

Below *Seen here in South Africa's Rustenburg Nature Reserve, the Southern boubou* (Laniarius ferrugineus) *is known for its duetting songs.*

Shrinking Numbers

BREEDING BIOLOGY AND CONSERVATION

As far as known, all bush-shrikes appear to be normally monogamous and territorial. Their breeding season is favored by the rains. The courtship display of the majority of species has not been described, but it is well-known in a few, for instance in puffbacks, in which males puff out their rump feathers and look like puffballs, and in tchagras, which have a typical display flight (wings extended, tail fanned out) accompanied by melodious phrases. Nests are often neatly-formed cups; they generally receive only 2 or 3 eggs. The breeding ecology of many species remains virtually unknown.

The most threatened bush-shrike is probably the Bulo burti boubou, known only from one individual caught in 1988 in Somalia. Other species are also very rare, for instance the Mt. Kupé bush-shrike, which is confined to a few tiny areas in western Cameroon, and the Green-breasted bush-shrike, also occurring in Cameroon and having a small population in eastern Nigeria. These two species, as well as many others, are threatened by deforestation. NL/LGG

Vangas, Wattle-eyes, and Batises

Left *The Sickle-billed vanga uses its extraordinarily long curved bill to probe for food in cracks and crevices, rather in the manner of the African wood-hoopoes (Phoeniculidae).*

VANGAS ARE A GOOD EXAMPLE OF WHAT happens when a unique stock of birds becomes established on a large, isolated island such as Madagascar, and diverges to fill several quite different ecological niches. As a result, the birds differ markedly in size and color, and even more in the shape of the bill, although all share the same skull shape and structure of the bony palate.

The vangas, wattle-eyes, and batises have several features in common with the African helmet-shrikes of the family Prionopidae, and some recent authors consider them to be part of that family. Members of the genus *Batis* are most similar, as a uniform group of small, flycatcherlike shrikes; the pied males usually have a broad black breast-band, the grayer females a brown one. Wattle-eyes are similar to batises, some with pied males and females, others with browner females, and a few with yellow bellies, but all are distinguished by the fleshy red, blue, or mauve wattle around the eyes.

From Madagascar to the Mainland

FORM AND FUNCTION

Vangas are found from wet evergreen forests to semi-desert scrub. Wattle-eyes and batises occupy wooded habitats, from evergreen forest to drier savanna. The size and shape of the bill reflect the size of prey taken, the location, and the mode of capture. Batises and wattle-eyes, with short, broad bills, snap up small flying insects on the wing, mainly in the air, less often from foliage. The larger vangas have a hooked, shrikelike bill, culminating in the massive bright blue beak of the Helmet vanga, which is surmounted by a large casque. This bird searches the foliage for large insects, and also captures small vertebrates such as chameleons and amphibians.

Wattle-eyes and batises are usually found in pairs or small family parties, but most vangas feed and move in the nonbreeding season in loose flocks of 4–12 individuals, ranging up to 25 or more in the Sickle-billed vanga. Most also join mixed-species flocks, which may include other vangas or bush-shrikes. In contrast, the Hook-billed and Lafresnaye's vangas are usually solitary.

In those species studied, the male courtship-feeds and later provisions the female, while both sexes help in nest construction, incubation, and feeding the young. Most nest in pairs, but an extra helper has been seen to build at a nest of a Chabert vanga, and Sickle-billed vangas also live in groups. The nests of the majority are neat cups made from small leaves, roots, fibers, and bark, all bound to the supporting branches with spiders' web and often covered in lichen. In marked contrast, the Sickle-billed vanga builds a crowlike nest of sticks. AK/LGG

Left *A Black-throated wattle-eye exhibits a striking, salmon-colored wattle, the fleshy outgrowths above the eyes from which these African birds take their name. This handsome species was formerly known as the Wattle-eyed flycatcher.*

FACTFILE

VANGAS, WATTLE-EYES, & BATISES

Order: Passeriformes

Family: Vangidae

41 species in 14 genera (vangas 15 species in 11 genera; wattle-eyes 10 species in 2 genera; batises 16 species in 1 genus). Species include: **Chabert vanga** (*Leptopterus chabert*), **Helmet vanga** (*Euryceros prevostii*), **Hook-billed vanga** (*Vanga curvirostris*), **Lafresnaye's vanga** (*Xenopirostris xenopirostris*), **Blue vanga** (*Cyanolanius madagascarinus*), **Sickle-billed vanga** (*Falculea palliata*), **Black-throated wattle-eye** (*Platysteira peltata*), **Ituri batis** (*Batis ituriensis*).

DISTRIBUTION Wattle-eyes and batises are confined to sub-Saharan Africa; vangas are endemic to Madagascar (with 1 subspecies of Blue vanga that extends to Moheli in the Comoros archipelago).

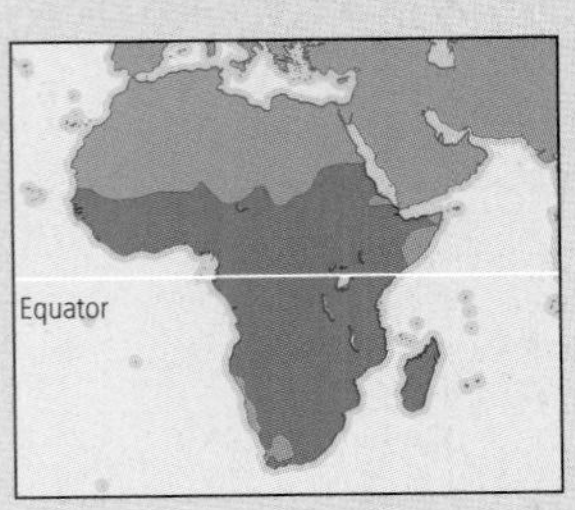

HABITAT Evergreen and deciduous forest, bushy scrub.

SIZE Length 8–30cm (3–12in); weight 5–97g (0.2–3.4oz).

PLUMAGE Many species are black above and white below, some with additional areas of chestnut and gray; 3 species of vanga predominantly blue; sexes differ in all batises and wattle-eyes and most vangas.

VOICE Varies between species; mainly whistles, short or drawn out, given singly or repeated, also various churring notes; the calls of the Sickle-billed vanga has been likened to the cry of a child.

NEST Known nests are cuplike and built with twigs.

EGGS 2–4; various pale ground colors (white, cream, pink, blue-green), profusely spotted with red, brown, gray, or lilac. Incubation 16–18 days; nestling period 13–18 days.

DIET Arthropods and some small vertebrates, earthworms and millipedes.

CONSERVATION STATUS Van Dam's vanga (*Xenopirostris damii*) and the Banded wattle-eye (*Platysteira laticincta*) are Endangered; 3 other species are Vulnerable.

Rockjumpers and Rockfowl

ROCKJUMPERS AND ROCKFOWL ARE AS FAR apart in appearance and behavior as any two groups of birds could be. Rockjumpers are small, warblerlike birds with short, cocked tails that occur on mountainous outcrops and build solitary grass nests on the ground. In contrast, rockfowl are up to twice the size, have brightly colored bald heads, follow army ants in the forest, and build mud nests colonially on rocks. Surprisingly, these two groups of rock-frequenting specialists are now thought to be each other's closest relatives. What is not in doubt is that they are unusual forms with uncertain affinities.

Rockfowl have in the past been classified with the crows, starlings, flycatchers, babblers, or warblers, but DNA analysis now links them most closely to the rockjumpers, themselves previously placed among warblers, babblers, or thrushes. In Sibley and Monroe's classification, both groups are listed under "Parvorder incertae sedis" in an expanded Corvidae. They are probably isolated remnants of archaic avian orders, with no close relatives.

Above *A Grey-necked rockfowl on its mud nest. Rockfowl habitually nest and roost in caves.*

An Unlikely Pairing

FORM AND FUNCTION

Small and warblerlike, rockjumpers have a relatively short tail, usually held cocked or half-cocked as the birds run or hop over rocks. Rockfowl are bizarre-looking birds with conspicuous black parietal patches and a large black bill. They hop silently through the forest on long, silver-gray legs, rarely using their moderately-sized wings for flight but often balancing on vines with the aid of a tail that is 2–3 times the length of the body.

Rockjumpers are confined to mountainous areas of southwest and southern Africa; the rockjumper is endemic to South Africa, the Damara rockjumper to Namibia and coastal Angola, while the Orange-breasted rockjumper occurs from the Eastern Cape in South Africa to Lesotho in the east. The White-necked rockfowl is found from Guinea to Ghana in the Upper Guinea Forest of West Africa, while the Grey-necked rockfowl is endemic to the Lower Guinea forest of Nigeria, Cameroon, Gabon, and Bioko. Both groups inhabit hillside terrain, but while rockjumpers frequent rocky outcrops, finding insects on the ground, rockfowl forage for invertebrates in leaf litter.

Rockjumpers are solitary nesters, while rockfowl nest colonially, with from 1 to as many as 40 nests on a single rock face. Both groups are monogamous, but rockfowl sometimes engage in communal displays at roosting sites, chasing and "bowing" to each other.

Rockfowl lay one or two eggs in the wet season. Nestlings fledge at about 70 percent of adult size. Egg and nestling loss to predators such as raptors and snakes is further increased by the birds' habit of destroying each others' nests in order to gain scarce nesting sites. Rockjumpers lay two to four eggs; as in the case of rockfowl, nestling care is by both parents, and the young leave the nest at a relatively early age.

Rockjumpers as a group are not threatened, although the eastern race has a restricted range. Rockfowl are considered Vulnerable by the IUCN; White-necked rockfowl populations are small and isolated, but the Grey-necked rockfowl is more common. HST

FACTFILE

ROCKJUMPERS AND ROCKFOWL

Order: Passeriformes

Family: Picathartidae

4 species in 2 genera

Distribution
SW and S Africa (*Chaetops*); W to C Africa (*Picathartes*).

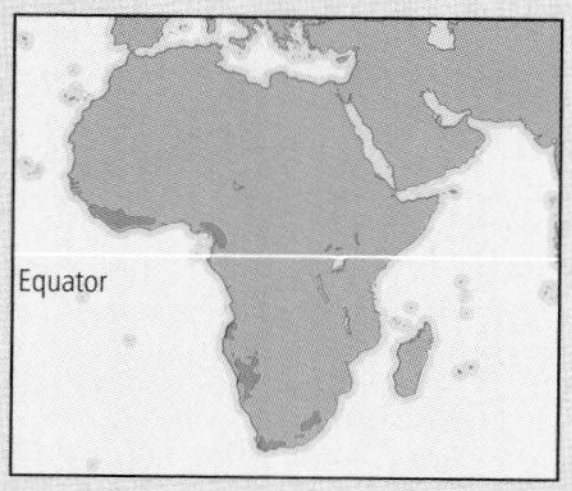

ROCKJUMPERS Genus *Chaetops*
2 species: **Damara rockjumper** (*Chaetops pycnopygius*), **rockjumper** (*Chaetops frenatus*), with two races, the **Cape rockjumper** (*C. f. frenatus*) and the **Orange-breasted rockjumper** (*C. f. aurantius*). Thornveld and *fynbos*.

Length 20–27cm (8–11in); **weight** 24–30g (0.8–1oz). **Plumage:** streaked brownish above and dark rufous below; rufous rump, white-tipped dark tail, white eyebrows and malar stripe. Sexes slightly dissimilar. **Voice:** melodious warble or tuneless varied song. Various alarm, anxiety, and contact calls. **Nest:** untidy bowl of grass on the ground or in low vegetation. **Eggs:** 2–4 pale pink or white eggs spotted with brown. **Diet:** insects; occasionally lizards.

Note: Western (*C. f. frenatus*) and eastern (*C. f. aurantius*) populations of *Chaetops frenatus* have generally been treated as separate species but this account follows Fry et al., treating them as conspecifics. Similarly, *C. pycnopygius* is regarded as congeneric to *C. frenatus*.

ROCKFOWL Genus *Picathartes*
2 species: **White-necked rockfowl** (*Picathartes gymnocephalus*) and **Grey-necked rockfowl** (*P. oreas*). Lowland rain forest. **Length** 38–41cm (15–16in); **weight** 200–250g (7–9oz). **Plumage:** mainly blackish above, white or lemon yellow below; bare head, yellow or crimson and powder blue with black parietal patches. Sexes similar. **Voice:** mostly silent; soft clucks or continuous "chirr"; raucous alarm call. **Nest:** cup-shaped mud structures set on cliffs, rockfaces, or cave roofs. **Eggs:** 1–2 (26 x 38mm/ 1 x 1.5in), white marked with brown blotches; incubation period 20–25 days; nestling period 25–26 days. **Diet:** forest floor invertebrates; frogs and lizards. **Conservation status:** Both rockfowl species are currently listed as Vulnerable.

Palmchat

PALMCHATS ARE GREGARIOUS, CONSPICUOUS, and extremely common birds found only on the island of Hispaniola in the West Indies. Large, clumsily-built communal nests, attended by noisy gangs of birds, are a ubiquitous feature of the Hispaniolan lowlands. Genetic studies have shown that the palmchat is most closely related to the waxwings and silky flycatchers, and some authors include the bird with them in the family Bombycillidae.

Like its closest relatives, the palmchat is gregarious, arboreal, and feeds mainly on fruit, also catching insects on the wing. The palmchat differs from waxwings in having rougher plumage, a heavier bill, and larger feet.

Palmchats are found in a wide range of open habitats from sea level to altitudes of 2,000m (6,500ft), although they are most abundant at lower elevations. The palmchat is probably the most common landbird on Hispaniola, and is the national bird of the Dominican Republic.

The most striking aspect of palmchat social behavior is the birds' unusual nesting habit. Bulky stick nests, often over 1m (3.3ft) in diameter and height, are built communally by groups of palmchats. Within this communal structure, each pair has its own separate nest chamber, crudely lined with dried grass and strips of palm leaves.

Each pair apparently lives independently of the others, although occasionally more than one female may lay eggs in the same chamber (whether this is due to parasitism or cooperation is unknown). Exceptionally large nests may be several meters high and wide and be occupied by over 50 pairs, although a more typical group size is 4–10 pairs, and some nests are built by one pair only. In the lowlands the usual nest-site is in the frond bases of a Royal palm, but conifers are used at higher altitudes.

The nest is used for roosting outside the breeding season (breeding takes place from February to August), and may remain occupied for several years. Only a handful of bird species worldwide build multichambered communal nests, the best-known being the Sociable weaver (*Philetairus socius*) of southern Africa. HT/MA

Right *Palmchats are a common sight in parks and gardens in Haiti and the Dominican Republic, the two nations that share the island of Hispaniola.*

FACTFILE

PALMCHAT

Order: Passeriformes

Family: Dulidae

1 species: *Dulus dominicus*

DISTRIBUTION Hispaniola (W Indies).

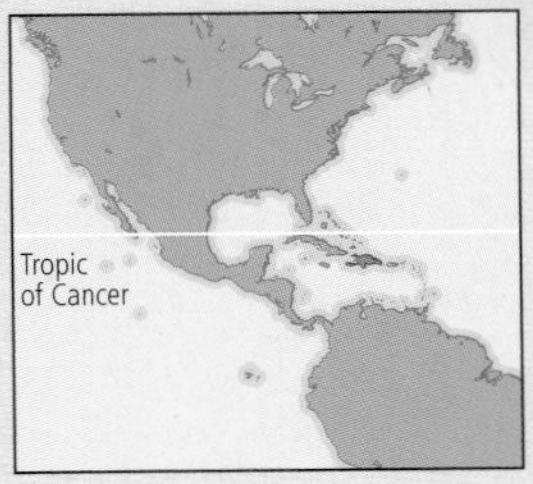

HABITAT Open woodland, farms, suburban areas.

SIZE Length 20cm (8in).

PLUMAGE Coarser than waxwing's; upperparts olive-brown, underparts buffy white, boldly streaked with brown.

VOICE A variety of short, harsh notes and a distinctive alarm call (a musical whistle that drops in pitch, often given in chorus by a whole group); but no true song.

NEST Large, communal nest of twigs, placed high (5–25m/15–80ft) up in a palm or other tree.

EGGS 2–7; quite variable in color, typically off-white with purple-brown markings. Incubation 15 days; nestling period 32 days.

DIET Berries, flowers, insects.

CONSERVATION STATUS Not threatened.

Grey Hypocolius

THE TAXONOMIC POSITION OF THE GREY hypocolius has long been a puzzle, and, pending conclusive evidence from molecular analysis, it remains controversial to this day. Most modern authorities have tended to treat this monotypic genus as a desert relative of the waxwings, placing it in a subfamily of the Bombycillidae.

Recently, however, the trend has been to place the species in a family of its own, the Hypocoliidae, which some suggest may be more closely related to the bulbuls (*Pycnonotidae*) than the waxwings. The uncertain affinities of this enigmatic species are reflected in its common names; for example, it is sometimes referred to as the Bulbul shrike in English, while its German name is *Seidenwürger* ("Waxwing shrike").

The unique appearance of the species also underscores its uncertain systematic position. It is about 23cm (9in) long, somewhat shrikelike in form but slimmer, with a longer tail, a smaller head, and a berry-eating bill that resembles a bulbul's, without the pronounced hook of a shrike's beak. The wings are short and sometimes held in a drooped position, as for example in the White-throated robin (*Irania gutturalis*). Both sexes are essentially gray, but the male is generally brighter and has a blacker tip to the tail, a black face mask that joins in a small crest on the nape, and black primaries with white tips that are striking in flight but obscured when perched.

Breeding is largely confined to the Tigris and Euphrates river valleys in Iraq and to the hot, coastal lowlands of southern Iran, where loose colonies occur locally in patches of bushes and trees, tamarisk scrub, palm groves, irrigated gardens, and similar terrain. Nests are lined with soft leaves and fibers, and sometimes with hair. The birds eat mainly fruit (such as dates and figs) and berries (especially mulberries, *Lycium* spp.), which they often take while concealed in vegetation; since they are usually silent, they can easily be overlooked. Some insects are also taken, particularly for feeding to the young.

The species migrates a short distance south to winter on both the Persian Gulf coast (in Bahrain, Qatar, and the United Arab Emirates) and the Red Sea coast (Hejaz) of the Arabian peninsula. In these locations, the birds gather in small, sometimes noisy, nomadic parties, typically of 3–20 birds, but sometimes up to 80 strong. The groups move about in search of food supplies, which may be in different areas in different years, depending on the distribution of rainfall. Where food is plentiful numbers build up, and spectacular communal roosts of up to 500 birds have been reported from Bahrain and the Hejaz. Irregular migrations to northwestern India and to northeastern Africa have also been reported.

The Maquabah roosting site in northern Bahrain, a semi-arid area of date-palm scrub, is particularly favored by the Grey hypocolius, but this region is under increasing pressure from human encroachment. LC

Below *Having migrated south to warmer climes to see out the winter months, a male Grey hypocolius perches on a branch on the island of Bahrain in the Persian Gulf. Female birds lack the black face mask, and also the black tip of the tail.*

FACTFILE

GREY HYPOCOLIUS

Order: Passeriformes

Family: Hypocoliidae

1 species: *Hypocolius ampelinus*

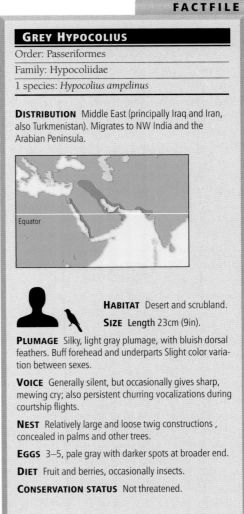

DISTRIBUTION Middle East (principally Iraq and Iran, also Turkmenistan). Migrates to NW India and the Arabian Peninsula.

HABITAT Desert and scrubland.

SIZE Length 23cm (9in).

PLUMAGE Silky, light gray plumage, with bluish dorsal feathers. Buff forehead and underparts Slight color variation between sexes.

VOICE Generally silent, but occasionally gives sharp, mewing cry; also persistent churring vocalizations during courtship flights.

NEST Relatively large and loose twig constructions , concealed in palms and other trees.

EGGS 3–5, pale gray with darker spots at broader end.

DIET Fruit and berries, occasionally insects.

CONSERVATION STATUS Not threatened.

Waxwings and Silky Flycatchers

THE WAXWINGS AND SILKY FLYCATCHERS RELY mainly on fruit, but also eat insects and flowers. Their life-history traits are strongly influenced by their fruit diets. Most species are non-territorial and gregarious even during breeding, behavior associated with feeding on patchy, abundant fruits. Song is absent or reduced.

The Bombycillidae locate their breeding sites in areas with abundant fruit and breed late in the season, to coincide with the ripening of summer fruits. The birds digest only fruit pulp, leaving the seeds unharmed, making them prime seed dispersers. Insects are caught by fly-catching from high, exposed perches. Most of the Bombycillidae are gregarious. The family is closely related to the palmchat of Hispaniola, and perhaps also to the Grey hypocolius of the Tigris–Euphrates Valley. These species also are frugivorous and gregarious, and lack song.

Waxwings

TRIBE BOMBYCILLINI

Waxwings have short, stout bills and legs, medium-length, square tails, and pointed wings. The sexes are similar in appearance. Flight is typically strong and undulating. The name "waxwing" refers to the waxlike red "droplets" at the tips of the secondary feathers of Bohemian and Cedar waxwings (rarely on the tail). Japanese waxwings do not have the waxlike nubs at the tips of their secondaries; instead, the webs at the tips of these wing feathers are pigmented with red.

Right *Viewed in profile, a Cedar waxwing clearly shows the waxlike red tips to the secondary flight feathers – actually prolongations of the shafts – from which the birds take their name. Waxwings are highly social, and can sometimes be seen passing berries down a line until one bird finally eats the food.*

Below *A Bohemian waxwing* (Bombycilla garrulus) *in flight.*

The Cedar waxwing of North America nests in open deciduous and mixed coniferous–deciduous forests. Bohemian waxwings nest in open, boreal spruce, pine, and birch forests of Europe, Asia, and North America. Japanese waxwings nest in the coniferous forests of eastern Asia. Waxwings range widely from fall to spring, sometimes in very large flocks of hundreds or even thousands of birds, feeding on fruit in virtually any habitat. Irregularly, large numbers of waxwings invade regions outside of their normal winter range, presumably driven by fruit shortages. These irruptive appearances were at one time considered a bad omen in parts of Europe, earning the waxwing the name "pest-bird."

Fruits are typically plucked from perches (or occasionally snatched while hovering) and swallowed whole. Waxwings sometimes fly to the ground to feed, but more often to drink. During spring months, as the previous season's fruit supplies dwindle, the birds feed on buds, flowing sap, and flowers. In the warmer months insects are eaten, captured by fly-catching and gleaning bark and foliage. Waxwings often nest near water,

Right *The Long-tailed silky flycatcher is a central American species, confined to the highlands of Costa Rica and Panama. The bird spends much of its time on high branches, from which it darts out on sudden sallies to catch passing insects. Like all the Bombycillidae, it also eats many fruits and berries.*

feeding on abundant aquatic insects like mayflies, caddisflies, damselflies, stoneflies, and even such agile prey as dragonflies. The birds turn to fruits as soon as they ripen in early summer, and eat mainly fruit throughout the year, with particular species consumed during each season. Early-summer fruits include strawberries, mulberries, and serviceberries (*Amelanchier* spp.); summer-to-fall fruits include raspberries, blackberries, cherries, and honeysuckles; fall-to-winter fruits are grapes, crabapples, and the berries of hawthorn, some dogwoods, viburnums, rowan or mountain ash, juniper, and mistletoe. Waxwings can become intoxicated from eating fermenting berries, but Bohemian waxwings have a relatively high capacity to metabolize ethanol and so can tolerate alcohol better than many birds, indicating an evolved response to this dietary hazard.

Waxwings are monogamous. Pairing may occur as early as the winter. A courtship ritual takes place in which a male and female pass an object, usually a fruit but sometimes something inedible, back and forth several times. The display ends when one bird eats the fruit; it is not known whether it is always one sex or the other that terminates the sequence. Copulation may follow. This behavior appears to be derived from the courtship feeding of females by males.

Pairs do not defend territories, and nests may be aggregated, associated with feeding on locally abundant and ephemeral fruits. The lack of territoriality probably accounts for the absence of true song in the tribe. Aggressive encounters near the nest are probably mate-guarding. Nest materials are gathered by both mates, but the female does most of the construction. Nests are typically built away from the main trunk of a tree, on a horizontal limb or crotch. The height is variable, from 1–17m (3–55ft) above ground, depending on the structure of the nesting tree. The nest is loose and bulky, made of twigs, grass, and lichens, and lined with fine grasses, mosses, and pine needles. The exterior is sometimes adorned with dangling grasses, flower heads, or lichens and mosses to create a cryptic effect, resembling jumbles of moss and twigs. The females do nearly all the incubation, and are fed on the nest by their mates. Both parents feed the young. Most waxwings are single-brooded, but Cedar waxwings can have two broods in a year.

The red and yellow pigments of waxwing plumage are carotenoids, chemicals derived solely from dietary sources. In Bohemian and Cedar waxwings the size and number of red tips increase with age, as does the brightness and width of the tail band. Cedar waxwings mate assortatively (non-randomly, on the basis of shared characters) by the development of the red feather tips, and older mates fledge more young, indicating that diet-derived plumage characters are adaptive signals in mate selection.

In North America the establishment of Eurasian honeysuckle shrubs has caused some Cedar waxwings to grow orange, rather than the normal yellow, tail bands. The honeysuckle fruits contain an unusual red carotenoid pigment that is deposited with yellow carotenoids, producing orange coloration when waxwings eat honeysuckle fruits during feather molt.

Silky Flycatchers

TRIBE PTILOGONATINI

Silky flycatchers have prominent crests and long tails. Phainopeplas are sexually dichromatic: the male is black with red eyes and white wing-patches, while the female is olive-gray. Sexual dimorphism in coloration is less pronounced in the other species. The Black-and-yellow silky flycatcher lacks a crest, has a thrushlike body form and posture, and occurs singly or in pairs in dense growth near the ground.

The phainopepla occurs in the arid scrublands of southwestern North America. The Gray silky flycatcher is found in upland forests in Mexico, while the Long-tailed and Black-and-yellow species inhabit similar terrain in Costa Rica and western Panama.

FACTFILE

WAXWINGS

Order: Passeriformes

Family: Bombycillidae

7 species in 4 genera: **waxwing** or **Bohemian waxwing** (*Bombycilla garrulus*), **Cedar waxwing** (*B. cedrorum*), **Japanese waxwing** (*B. japonica*), **phainopepla** (*Phainopepla nitens*), **Black-and-yellow silky flycatcher** (*Phainoptila melanoxantha*), **Grey silky flycatcher** (*Ptilogonys cinereus*), **Long-tailed silky flycatcher** (*P. caudatus*).

DISTRIBUTION Europe, Asia, N and C America.

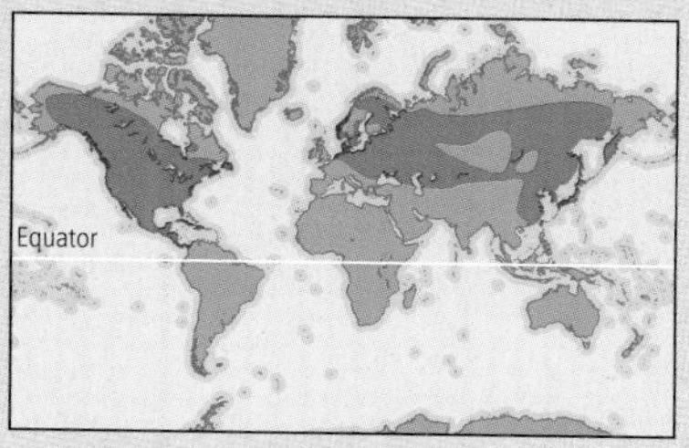

HABITAT Woodlands

SIZE Length 18–24cm (7–9.5in).

PLUMAGE Soft and silky; chiefly brown, gray, and black with some red and yellow; crested.

VOICE Song absent or inconspicuous. Varied calls; trills and whistles. Calls can be noisy, especially when the birds are flying and feeding.

NEST In trees; cup built mainly from twigs, lined with fine fibers.

EGGS 2–7; gray, blue, or whitish green. Incubation 12–16 days; nestling period 16–25 days.

DIET Fruits, insects, flowers.

CONSERVATION STATUS No waxwing is currently endangered, but the Japanese waxwing is listed as Lower Risk/Near Threatened.

The phainopepla is monogamous. The open-cup nest, held together by spider silk, is built by the male and used to attract females. The female may help to line the nest, and both sexes incubate.

Phainopeplas are unusual in breeding twice a year in two distinct habitats. Their breeding biology varies dramatically between the two, illustrating the influence of food dispersion on life-history traits. From February to April they breed in the Sonoran Desert of Arizona, where they establish territories harboring mistletoes, which are common in the area, producing an abundant supply of fruits. As mistletoe supplies decline and temperatures rise, the birds move to the oak and sycamore scrublands of Arizona and California, where they breed from May to July, feeding on fruits that are locally abundant and ephemeral; here they are not territorial but loosely colonial. It is not known whether the same birds breed sequentially in these habitats, but the coincidence between population movements and breeding activity would suggest this is the case. MW/MA

Dippers

Dippers are unique among passerines in being adapted to forage actively underwater. Their morphology allows them to swim and wade seemingly unimpeded along the beds of fast-flowing streams, into which they plunge in search of aquatic invertebrates. Despite these stream-dipping habits, their name is derived from a display, given from the mid-stream perches from which their foraging bouts are initiated, in which they bob the entire body up and down and repeatedly flash their white eyelids by blinking.

The rare Rufous-throated dipper, which occurs only in a restricted South American range, has lost both the underwater foraging behavior and the dipping display that characterize other members of the family. Nevertheless, a first encounter with any of the dippers is likely to be either the sight of a small, plump, dark bird disappearing down a swift-flowing watercourse or else the sound of a high-pitched, metallic call audible over the babbling waters.

Freshwater Divers

Form and Function

All of the dippers are dark, rotund birds with stumpy tails, thrushlike bills, strong legs, and feet with well-developed claws. The five species are morphologically very similar except in the distribution of white or colored patches in their plumage. Like the auks of the oceans they are streamlined yet bulky, due to the extra-thick body plumage that aids waterproofing and insulation and the highly developed chest musculature that allows the short wings to be flapped underwater. A large preen gland has developed, allowing the feathers to be heavily waterproofed; when the birds emerge from the water, droplets simply roll off their plumage without wetting the feathers or skin. Dippers can survive winter temperatures down to –45°C (–49°F) if the river is not completely frozen; they can even feed under ice. They walk into shallow water when foraging, and may swim or dive; certain species can remain underwater for 30 seconds, although most dives are much shorter. Prey items, which are taken from among rocks, stones, and weeds, are mostly aquatic insects and mollusks, although fish can comprise a significant proportion of the biomass.

Although dippers were long thought to be most closely related to wrens, molecular evidence has now shown that their nearest relatives are actually the thrushes and flycatchers. Widely distributed relative to their number, the five dipper species span five continents, the White-breasted Dipper just reaching northwest Africa. The only place where two species are sympatric is in eastern Central Asia, where the White-breasted occurs at higher altitudes than the Brown dipper. All five occupy the same sort of habitat.

The divergent plumages of the dipper taxa are important in displays and suggest that sexual selection following geographical expansion, rather than niche divergence, has been important in their diversification. Most populations are sedentary, although in several species altitudinal movements are common between summer and winter, while some White-breasted dipper populations move more than 1,000km (625 miles).

Nests by the Riverside

Breeding Biology

Dippers are highly territorial in summer and winter, using a neck-stretching display and chasing to maintain territorial boundaries. Territory size is mainly determined by the extent of stream bed available for feeding. The birds breed in early spring when food is most abundant, and are usually monogamous. Both sexes construct the nest (in 14–21 days), although females do most of the building. Nests are placed among tree roots, on small cliffs, under bridges, or in walls – sometimes even behind waterfalls, where the adult must enter and exit through the falling water. The bulky nests are often inconspicuous, either because they are built in crevices or because the mossy exterior closely resembles the surroundings.

The nestling period is relatively long, but nesting success is usually high, at 70 percent. If disturbed, nestlings may explode from the nest after 14 days (when full body weight is attained); remarkably, they can swim and dive expertly before they can fly. Dippers are unusual in using the same nest (relined) for second broods, and

*◀ **Left** In its usual habitat close to fast-flowing water, a White-breasted dipper pauses with food for its young in its bill. The birds take dragonflies, stoneflies, mayflies, and other insects from the river-bank, but also hunt out their larvae, along with tiny mollusks and crustaceans, underwater. Other aquatic prey include newts, tadpoles, and fish fry.*

*▶ **Right** In Colorado, an American dipper dives for food. Adaptations that equip the birds for diving include highly developed nictitating membranes (third eyelids) and movable flaps over the nostrils, both serving to keep out water, as well as a preen gland for waterproofing the plumage that is ten times the size of that of any other passerine.*

Above *Three American dipper chicks peer out of a riverside nest. Both parents feed nestlings for a relatively long fledging period of 22–23 days, after which the young emerge ready to seek out their own food.*

they frequently re-use it for up to four successive years. The birds become very secretive during molt, which is rapid, and the American dipper can become flightless. Post-fledging mortality is high (over 80 percent in the first six months), but thereafter annual mortality is between 25 and 35 percent. In suitable breeding areas the habitat is continuously occupied, and a nonbreeding surplus of birds either does not exist or is small and frequents areas unsuitable for breeding. It is therefore unusual for an adult that dies in the breeding season to be replaced.

A Need for Clean Water

CONSERVATION AND ENVIRONMENT

Dippers are vulnerable to the deterioration of water quality through various forms of pollution – acidification from industry, silting due to mining and agriculture, or eutrophication due to agricultural runoff. In addition, the diversion of water from rivers and streams can destroy habitat.

These problems may be particularly prevalent in the mountains of Asia and South America, where four dipper species occur. The Rufous-throated Dipper is especially vulnerable due to its small population size within a largely unprotected and restricted range. JV/CH/DRL

FACTFILE

DIPPERS

Order: Passeriformes

Family: Cinclidae

5 species of the genus *Cinclus*: **American dipper** (*C. mexicanus*), **Brown dipper** (*C. pallasii*), **dipper** or **White-breasted dipper** (*C. cinclus*), **Rufous-throated dipper** (*C. schultzi*), **White-capped dipper** (*C. leucocephalus*).

DISTRIBUTION Western N and S America, Europe, N Africa, Asia.

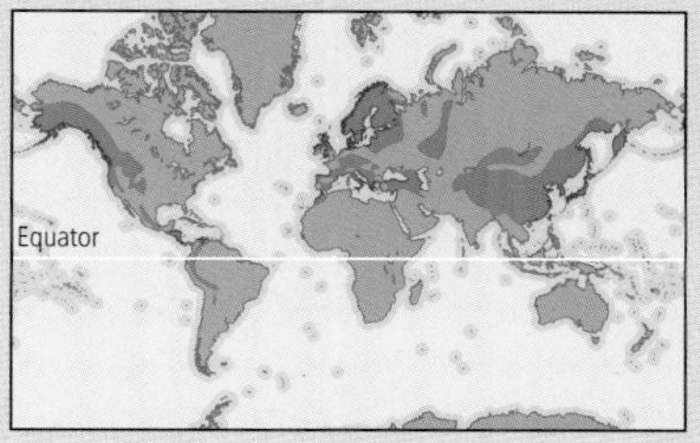

HABITAT Clear running water, usually among hills and mountains.

SIZE Length 15–17.5cm (6–7in); weight 60–80g (2.1–2.8oz).

PLUMAGE Chiefly black, brown, or gray, sometimes with a white bib, back, or cap; no difference between sexes.

VOICE All species produce a harsh *zit* call and a rich warbling song.

NEST Large domed structures, about 20 by 20 by 15cm (8 x 8 x 6in) deep with an opening 6cm (2.5in) in diameter, usually over running water; constructed of moss and lined with grass and dead leaves.

EGGS 4–6, usually 5; white. Incubation period 16–17 days; nestling period 17–25 days, usually 22–23 if undisturbed.

DIET Chiefly aquatic larvae, especially stoneflies, mayflies, caddisflies; also crustaceans, mollusks; occasionally small fish, tadpoles.

CONSERVATION STATUS The Rufous-throated dipper, restricted to a small and declining population on the Bolivia–Argentina border, is Vulnerable.

Thrushes

THE THRUSHES ARE A LARGE AND WIDESPREAD group, with representatives on every continent. However, they share few characters in common that clearly separate them from such related groups as the babblers, warblers, and flycatchers. There is no easy definition of a thrush, but most species show the following features: a short, slender bill; 10 primary feathers, of which the outer one is much reduced in length; generally 12 tail feathers (although, exceptionally, some have 10 or 14); a tarsus that is "booted" (not divided into separate scales on the leading edge); strong and well-developed feet; juvenile plumage that is spotted, except in the whistling-thrushes; and cup-shaped nests. In addition, thrushes typically forage on the ground for animal food, supplemented with fruit taken from trees and shrubs.

Distinctive features of certain true thrushes in the genera *Turdus*, *Catharus*, and *Zoothera*, along with the two nightingales and two of the solitaires, are their highly developed and complex songs, marked by the ability to simultaneously produce different notes. This trait sets these birds apart not only from other apparently similar species such as babblers but also from other thrushes in the same genera that are comparatively poor singers.

FACTFILE

THRUSHES

Order: Passeriformes

Family: Turdidae

304 species in 49 genera

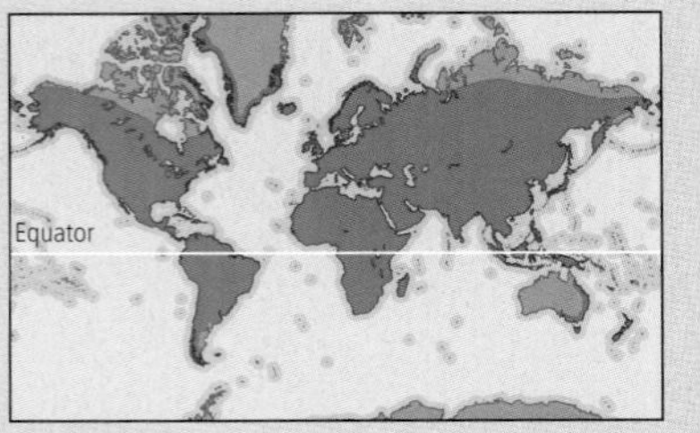

DISTRIBUTION Worldwide including many oceanic islands; absent from New Zealand (2 species introduced from Europe).

HABITAT Wide range of wooded habitats from tropical, temperate, and montane forests to urban gardens, parks, open heaths, and desert edges.

SIZE Length 11–33cm (4.3–13in); weight 8–220g (0.28–7.7oz).

PLUMAGE Predominantly shades of gray, brown, and white, but many (esp. males) with bright patches of all colors except yellow; several species all black or (in the whistling-thrushes) mainly dark blue with patches of iridescent blue on the head and wing coverts; females similar to males but generally lacking bright colors, with contrasting patterning reduced or absent. Chats generally browner and duller, but several species show both blue and red, particularly the redstarts, which have large areas of red in the tail; wheatears have camouflaged plumage as a defense in open terrain; forktails are largely black and white. Cochoas have highly colorful plumage, from deep green to blue and purple. Several species, especially the blackbird, produce aberrant leucistic (showing pigment loss) or entirely albinistic individuals.

VOICE Many species very musical, producing some of the most beautiful of all bird songs; alarm calls usually sharp and staccato, churring, or thin and high-pitched. 1 species, Lawrence's thrush, is a particularly adept mimic.

NEST Open cups in various concealed sites, e.g. trees, shrubs, or in holes in rocks or in the ground, or more rarely on open ground; usually built solely by the female.

EGGS 2–6, rarely 7 (occasionally more in hole-nesters); whitish, blue, greenish, or buff, unmarked or with brown or black spots or more diffuse markings. Incubation 12–15 days; nestling period usually 11–18 days; both parents feed and care for nestlings.

DIET Invertebrates of many kinds, especially insects and their larvae, also earthworms; most species also take fruit.

CONSERVATION STATUS 7 species are listed as Critically Endangered; 7 more species are Endangered, and 23 are Vulnerable.

▷ See Selected Thrush Species

Right *The Common nightingale has long been famed for its song, heard to best effect on early-summer evenings. In fact, the birds also sing by day, but in the daylight hours their song is often drowned out by that of other species.*

Widespread and Diverse

FORM AND FUNCTION

It seems certain that the thrushes had their evolutionary origins in East Asia, and it is there that the family is present in greatest variety. The whistling-thrushes, which include the largest members of the family, live along fast-flowing streams in the Himalayas and other mountain ranges of eastern Asia. They have strongly hooked bills and forage for animal food among rocks at the water's edge. The forktails – slender birds with long tails – and the water-redstarts are also specialists in foraging along the banks of mountain torrents.

The grandala, a long-winged, short-legged bird with blue plumage, is so unlike a typical thrush that its inclusion in the family is at first sight surprising. Grandalas are highly aerial, social birds that live at very high altitudes above the timberline in the Himalayas and associated mountains, where they feed in flocks on bare mountain slopes. Almost equally unthrushlike are the three species of cochoas. These are wide-billed birds of the southeast Asian tropical forest, with plumage patterned with green, blue, and violet. Although little known, they are probably ecological equivalents of the cotingas of tropical America.

With some 60 species, the *Turdus* or true thrushes constitute by far the largest genus in the family. They occupy a central place in that they appear to be unspecialized and to show the basic type from which the various more specialized groups have radiated. Moreover, they are familiar to almost everyone, since in every continent except Australia there is at least one common garden species. No other genus of land birds is so widespread. Eurasia is especially rich in species, the Song thrush, Mistle thrush, and blackbird being among the best known of European birds.

Elsewhere in the world, their place on garden lawns and playing fields is taken by, for example, the Olive thrush in southern Africa, the Rufous-bellied thrush in Brazil, the Clay-colored thrush in Central America, and the American robin in North America. By contrast, the Island thrush comprises at least 50 subspecies occurring on islands between Indonesia and the western Pacific; each one differs from the race closest to it, while often showing closer similarity with other species on more distant islands. The Island thrush has undoubtedly spread from a common ancestor but, with the exception of one or two subspecies on New Guinea, has avoided colonization of any areas within the larger continents.

Among the best-known features of the true thrushes are their songs, which characteristically are composed of a succession of

Left *Representative thrush species:* ***1*** *Eurasian robin* (Erithacus rubecula), *familiar across the Palearctic region.* ***2*** *Song thrush* (Turdus philomelos), *taking a snail.* ***3*** *White-browed robin chat* (Cossypha heuglini), *one of 15 African robin chat species.* ***4*** *Fieldfare* (Turdus pilaris), *another widespread Palearctic bird.* ***5*** *American robin* (Turdus migratorius).

short, richly warbled or fluty phrases. A few species, such as the Common nightingale, the Hermit thrush, and several of the solitaires, are particularly known for the clarity of their rich, melodious songs, which include exquisitely modulated whistles, buzzes, and rising, flutelike trills; listening to them, it is often easy to imagine that many birds are singing at one time, when in fact the singer may be just a single individual. Other species are remarkable mimics. In Europe, blackbirds and song thrushes have been known to imitate ringing telephones, while Lawrence's thrush from South America has been recorded mimicking the calls and songs of over 170 different species.

The birds' substantial cup nests are usually made of dry grass or root fibers, in some cases strengthened by a layer of mud that is often mixed with decaying leaves; they are usually finished with an inner lining of finer grasses or similar material. In some species, however, almost anything that is in the vicinity of the nest will be incorporated, including paper, string, rags, wool, and animal hair. Most nest sites are usually well-concealed, either on the ground (as in many of the African chats) or low down in vegetation, while others are placed higher up in trees. Some of the rock thrushes, the wheatears, and the cliff-chats are hole nesters and will frequently share whatever holes are available, either on the faces of sheer cliffs or else in flatter terrain, with rabbits and rock hyraxes; in the open deserts of Central Asia, Isabelline and Desert wheatears often make use of the holes of the gerbil-like pikas. The large and frequently reused nests of the whistling-thrushes, which are added to each year and accumulate to quite a size, are often placed on a ledge near, or sometimes even behind, waterfalls, which provide an effective defense against predators.

Above *A brood of Northern wheatear* (Oenanthe oenanthe) *chicks wait hungrily to be fed. Six is a typical clutch size for these birds, whose fur-lined cup nests are usually concealed in holes or crevices.*

Below *A Varied thrush perches on a cedar branch. This strikingly colored species is endemic to western North America, and breeds from Alaska to California, where many of the birds also go to winter.*

Northern species are long-distance migrants, but tropical species are generally nonmigratory, although they often undertake considerable dispersal movements after breeding. Some species from middle latitudes are partial migrants, with some individuals moving south or else to lower altitudes in winter, while others remain within their breeding area providing they can find sufficient food. Several species are particularly noteworthy long-distance travelers; some Grey-cheeked thrushes breed in eastern Siberia and winter in the Amazon basin in South America, while two of the wheatears (the Isabelline and Desert) and the Rufous-tailed rock thrush make annual journeys from Central Asia and eastern Siberia to sub-Saharan Africa – a distance of almost 9,000km (5,500 miles). Even this stretch may be exceeded by some Northern wheatears that breed at the extreme east of their range in northwestern Alaska and return via the Bering Sea, eastern and central Asia, and Arabia to winter in East Africa.

The *Zoothera* ground-thrushes are probably of earlier descent than the *Turdus* thrushes, but are nonetheless closely related to them. They have the same general build, although their bills and legs are generally stouter; they are further distinguished by a striking, usually black and white underwing pattern. Shy and elusive birds, they include some species that are very poorly known, sometimes only from one or two specimens from remote or unexplored islands. They live near the ground in forests, and are mainly confined to restricted ranges in parts of Asia and Africa, although several species inhabit very small ranges in Indonesia, the Philippines, and on other islands in the southwest Pacific.

The principal exceptions are the six species in the White's thrush complex, a group of closely-related species that have clearly separated from each other or a common ancestor only relatively recently and that were long regarded as taxonomically forming a single species. Their breeding

Right *The beautiful Eastern bluebird* (Sialia sialis) *used to be a familiar sight in the USA east of the Rockies, but their numbers have declined over the past 25 years, perhaps as a result of competition from House sparrows and starlings.*

range extends from Siberia, northern China, and the Himalayas to Vietnam, southern India, and Sri Lanka, and from New Guinea to western Australia. Other "outposted" species are the Varied thrush of western North America and the Aztec thrush of Mexico. It is not easy to trace the route by which these birds, together with the 35 *Turdus* species that now occupy differing areas of South America, arrived at their present distribution from their origins in the Old World, except that at some stage, given the obvious problems of traversing the Pacific, their ancestors must have crossed a now-disappeared land-bridge to the New World.

The large group of robins, robin-chats, and related species (including the nightingale) are small thrushes of woodland and tropical forest, although many also occur on open hillsides and heaths with scattered trees. They are mainly ground feeders, with proportionately longer legs than the true thrushes. It has recently been proposed from molecular analysis that the group's origins may be closer to the Old World flycatchers than to the true thrushes, and some taxonomists have separated them into their own tribe, the Saxicolini, within the Old World flycatcher family (Muscicapidae). This treatment has not gained general acceptance, however, since there are plausible ecological and behavioral grounds for considering them to have equally close affinities to the thrushes.

While many of these species are principally insect-eaters, they take their prey by foraging in vegetation or else by dropping from a prominent perch onto ground-dwelling insects, in the manner of the rock-thrushes. A few, such as the alethes of tropical Africa, are habitual followers of army ants, feeding on insects flushed by the ants. This group of small thrushes have their headquarters in Asia and Africa and are entirely unrepresented in the New World.

Protective Pairs

SOCIAL BEHAVIOR

Most thrushes that have been studied are similar in their social systems. Monogamous pairs defend nesting territories in the breeding season; in resident species, pairs may remain together all year. In the nonbreeding season some species, particularly the European and North American thrushes, tend to be highly social, feeding in flocks and roosting communally, especially in cold weather. Although generally considered placid and peaceful in their nature, the birds can become aggressive in defense of territory or recently-fledged young, and several species have loud warning cries that alert other species to the presence of predators.

The larger thrushes may defend their nests pugnaciously, and one among them, the fieldfare, is apparently unique in the way in which it does so.

Above *Whinchats* (Saxicola rubetra) *inhabit meadows and heaths. Their nests are usually well hidden in grass tussocks, but are still vulnerable to predation.*

Above right *The Seychelles magpie-robin* (Copsychus sechellarum) *is considered Critically Endangered, with less than 50 birds surviving. A recovery program begun in 1990 is seeking to reintroduce the bird to other islands in the Seychelles group.*

Left *The Common redstart* (Phoenicurus phoenicurus), *a widespread Eurasian species, is only distantly related to the American redstart, a New World warbler.*

Selected Thrush Species

Species and genera include: **alethes** (genus *Alethe*), **cochoas** (genus *Cochoa*), **forktails** (genus *Enicurus*), **grandala** (*Grandala coelicolor*), **Common nightingale** (*Luscinia megarhynchos*), **nightingale-thrushes** (genus *Catharus*) including **Hermit thrush** (*C. guttatus*), **Grey-cheeked thrush** (*C. minimus*) and **Wood thrush** (*C. mustelinus*), **robin-chats** (genus *Cossypha*), **robin** or **Eurasian robin** (*Erithacus rubecula*), **rock thrushes** (genus *Monticola*) including **Rufous-tailed rock thrush** (*M. saxatilis*), **solitaires** (genera *Entomodestes*, *Myadestes*) including **kama'o** (*M. myadestinus*) and **oloma'o** (*M. lanaiensis*), **water-redstarts** (genus *Rhyacornis*), **wheatears** (genus *Oenanthe*) including **Desert wheatear** (*O. deserti*) and **Isabelline wheatear** (*O. isabellina*), **whistling-thrushes** (genus *Myophoneus*). Included in the **true thrushes** (genus *Turdus*) are the **American robin** (*T. migratorius*), **blackbird** (*T. merula*), **Clay-colored thrush** (*T. grayi*), **fieldfare** (*T. pilaris*), **Island thrush** (*T. poliocephalus*), **Lawrence's thrush** (*T. lawrencii*), **Mistle thrush** (*T. viscivorus*), **Olive thrush** (*T. olivaceus*), **Rufous-bellied thrush** (*T. rufiventris*), and **Song thrush** (*T. philomelos*); the **ground thrushes** (genus *Zoothera*) include the **Aztec thrush** (*Z. pinicola*), **Varied thrush** (*Z. naevia*), and **White's thrush** (*Z. dauma*).

Fieldfares nest semi-colonially (an unusual habit in the family) and attack predators that approach their nests by bombarding them with their feces. In exceptional cases, hawks have been known to become so plastered with feces in this way that they have been unable to fly and eventually have succumbed to starvation. Some of the chats, particularly those such as stonechats that inhabit open country, and some of the wheatears, which spend long periods on high vantage points guarding their territory, also boldly mob and chase off potential predators or else feign injury to divert them from the nest or young.

Isolated Species at Risk

CONSERVATION AND ENVIRONMENT

As the thrushes, robins, and robin-chats are principally woodland- or forest-dwelling birds, many have declined in number or are currently under threat from habitat destruction and fragmentation caused by the increased pressure on forests in Asia and Africa. In addition, uncontrolled hunting is affecting several species. Perhaps of greater concern, however, is the introduction of non-native species into new areas, particularly islands, where the diseases and parasites that accompany them can have drastic consequences. On Hawaii, two of the endemic solitaires, the kama'o and the oloma'o, may now be extinct, neither having been sighted for over 20 years; both are thought to have fallen victim to diseases carried by mosquitoes brought in by feral pigs.

Concerns have also been expressed about the fate of more common species such as the Song thrush in Europe, where recent declines in the population have been attributed to the widespread use of insecticides, as well as to uncontrolled hunting in parts of the bird's range where it is still considered a delicacy and is widely trapped or shot. In North America, recent decreases in the Wood thrush population have followed on the spread of the Brown-headed cowbird (*Molothrus ater*), a brood parasite that has had a severe impact on breeding success in parts of the thrush's range. The taking of some solitaires for the trade in caged songbirds has aggravated the problems of habitat destruction, and is the most likely reason for the extinction of one race of the Cuban solitaire (*Myadestes elisabeth*). PC/DWS

Above *A group of Mistle thrushes feed on berries. Although the birds rely on fruit for a substantial part of their diet, owing their common name to their fondness for mistletoe berries, they are equally dependent on insects and worms.*

Old World Flycatchers

SMALL WOODLAND OR FOREST BIRDS, THE Old World flycatchers can often be recognized from their way of capturing flying insects. They use a sit-and-wait strategy, sallying out from a low perch to capture prey in mid-air. In this way Spotted flycatchers may capture a prey on average every 18 seconds. In colder weather, with few flying insects, they have to hover among the foliage in the tree canopy, which is much more demanding of energy.

Flycatchers occur over most of Europe, Africa, Asia, Australia, and the Pacific Islands, from coastal shrubs to high-altitude forests up to 4,000m (13,100ft). In Europe they are popular garden birds, and some species (those of the genus *Ficedula*, for example) can easily be attracted to nest boxes. The majority of species, however, are found in Southeast Asia and New Guinea. In tropical areas some flycatcher species, like the Mariqua flycatcher, may form small flocks outside the breeding season, or, like the Ashy alseonax, may join mixed-species flocks.

Above *A Pied flycatcher sallies out after prey. The birds can outmaneuver almost any winged insect, dealing with each appropriately on returning to their perch; wasps and bees may be rubbed against a branch to make them discharge their stings.*

Patient Insect Eaters

Form and Function

Flycatchers vary in appearance from very colorful to almost uniformly brown or gray. In many species the sexes differ considerably in plumage but not in size, while in others, most often the duller-colored ones, the sexes are very similar. Typical flycatchers have relatively broad, flat bills, with modified feathers called "rictal bristles" around the nostrils to help in catching flying insects. The legs and feet are often weak, possibly because the birds' feeding technique demands no more than sitting and waiting. Not all flycatchers feed solely on insects; many also take berries and fruits. One African species, the White-eyed slaty flycatcher, even takes nestlings of smaller birds.

Most tropical species are resident, but some perform seasonal movements, and high-altitude species migrate to lower altitudes after the breeding season. Species in Europe and Asia migrate to winter in Africa, India, or Southeast Asia. Spotted flycatchers from Britain winter in South Africa, where, uniquely among all birds, they molt their primary feathers from the outside in. Pied flycatchers from as far east as Moscow fly west to put on migratory fat in northern Portugal in the autumn. There they defend territories for up to three weeks while gaining 70 percent extra weight to enable them to fly directly across the Sahara.

Left *A Fiscal flycatcher* (Sigelus silens) *in South Africa's Addo Elefant National Park. This species owes its common name to its resemblance to the Common fiscal* (Lanius collaris), *an African shrike.*

Right *The Spotted flycatcher, a summer migrant to Europe from Africa, has a distinctive upright stance. It frequently catches butterflies as they feed on buddleia.*

Deceivers Ever

BREEDING BIOLOGY

The majority of flycatchers build small cup nests set in the forks of tree branches; exceptions to this general rule are the Pied and Collared flycatchers of Eurasia and some African species in the genus *Muscicapa*, which nest in tree-holes.

Most species are thought to be monogamous. The Pied and Collared flycatchers, however, are sometimes polygamous, having a remarkable mating system in which some males defend two or more territories in succession, to each of which they try to attract a different female. A male Pied flycatcher, after arriving on the breeding grounds in spring, will set up a territory around a nest-hole and try to acquire a female. If successful, he may occupy a second territory a few hundred meters away, and attempt to attract another, secondary female; there are even cases on record of a bird managing to obtain a third. The distance between a male's first and second territories averages 200m (650ft), but spans up to 3.5km (2.2 miles) have been known, with many territories of other males lying in between.

By maintaining two territories, males can hide from arriving females the fact that they are already mated. When trying to attract a secondary female, they behave exactly as if unmated. They often desert the secondary females after egg-laying, devoting most of their efforts to helping the primary female to feed the young. Left single-handed, the secondary female may see some of her young starve. Thus the deceptive males may increase the number of their own offspring, but only at a considerable cost to the females. In some studies, approximately 15 percent of the males succeeded in attracting more than one female, while many more tried to do so but failed. Even so, the deceivers remained in a minority.

In most bird species, the males guard their females before they lay eggs in order to prevent other males from copulating with them. However, bigamous Pied flycatcher males cannot guard their primary mate while they are visiting their second territory, and as a result they run the risk of other males inseminating the primary female and siring some of her young, leading to shared fatherhood within broods. Yet even monogamous males are not safe from this danger, since they often have to absent themselves from the nest to chase intruding males from the territory; observations suggest that there is a considerable risk of extra-pair copulations if the male travels more than 10m (33ft) from the female. In all broods, the attendant male is the true father of only about 75 percent of the young. Even though polygynous males are more at risk, they produce more young by having a secondary female. Bigamy is thus an adaptive feature in the species.

In contrast, experiments with Collared flycatchers (close relatives of the Pied) have shown that male birds may visit other birds' nests not in order to breed there immediately, but rather to assess which part of a breeding colony is most productive for future use. The experimenters augmented the number of nestlings in one area of a colony and decreased the numbers in another. In the following year, a majority of the male visitors chose to breed in the area that had exhibited the bigger broods. CJM/ALu

Above *The Mugimaki flycatcher* (Ficedula mugimaki) *is a far-eastern forest species whose name means "sowing wheat" in Japanese.*

FACTFILE

OLD WORLD FLYCATCHERS

Order: Passeriformes

Family: Muscicapidae

115 species in 17 genera. Species include: **Ashy alseonax** (*Muscicapa caerulescens*), **Dusky alseonax** (*M. adusta*), **Spotted flycatcher** (*M.striata*), **Blue and-white flycatcher** (*Cyanoptila cyanomelana*), **Collared flycatcher** (*Ficedula albicollis*), **Pied flycatcher** (*F. hypoleuca*), **Mariqua flycatcher** (*Bradornis maniquensis*), **White-eyed slaty flycatcher** (*Dioptrornis fischeri*), **Rueck's blue flycatcher** (*Cyornis ruckii*).

DISTRIBUTION Europe, Asia, Africa, Australia, Pacific islands.

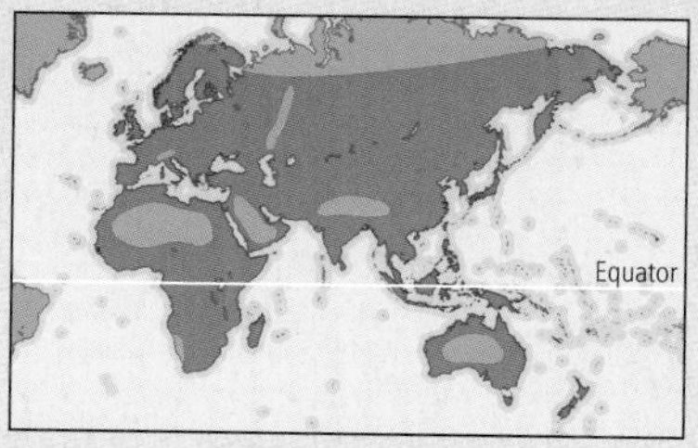

HABITAT Chiefly woodlands, forests, shrubs.

SIZE Length 10–21cm (4–8in)

PLUMAGE Varies considerably; some species plain gray or brown, other black and white or bright blue, yellow, or red; little difference between sexes in dull species, marked differences in brightly colored species.

VOICE Wide range of notes; songs vary from the simple and monotonous to the complex.

NEST Most species build cup nests on tree branches; a few are hole nesters. although they do not excavate their own holes.

EGGS Usually 2–6, range 1–8; whitish, greenish, or buff, most often with spots; in hole-nesting species, eggs are bluish without spots. Incubation period 12–14 days; nestling period 11–16 days.

DIET Mainly insects

CONSERVATION STATUS Rueck's blue flycatcher, known only from 2 specimens collected in Sumatra in 1917–1918, is listed as Critically Endangered. 3 other species are Endangered and 12 Vulnerable.

Starlings and Mynas

WHILE SHARING MANY FEATURES, THE starling family is remarkable in comprising species that range from among the world's most common birds to some of the rarest and most endangered. Throughout the Old World range of the family, many species interact closely with the human race. Some are valued as pets, especially those like the Hill myna with a capacity for mimicry, including of human speech. Some are taken by humans for food, while others are serious agricultural pests.

Other starlings are valued as agents of pest control. Throughout history, one of the greatest scourges of human crops has been the locust, and the taste some species such as Rose-colored and Wattled starlings and the Common myna show for this pest brought them to human attention many centuries ago. The Wattled starling has also had medical significance: its ability to reabsorb its fleshy wattles has been studied in cancer research, while its capacity to regrow feathers has been investigated by optimistic seekers of cures for human baldness!

Active and Adaptable

FORM AND FUNCTION

Starlings are small to medium-sized birds that make their presence felt near human habitations by their ceaseless activity, loud calling, and squabbling. In general appearance they are rather variable, since forest-dwelling forms, like the Hill myna and the African glossy starlings, tend to have broad, rounded wings, whereas those species that live in drier, more open habitats, such as the Common and Wattled starlings, have longer and more pointed wings. The legs and feet are fairly large and strong, and the birds tend to walk rather than hop. In the oxpeckers, two African species in the genus *Buphagus*, the toes are also long and sharp, to enable the birds to cling to the pelts of large mammals.

The bill is rather stout, and usually straight and reasonably long. Such a bill allows starlings to be catholic in their choice of food, and most eat both invertebrates and fruit. Some are even more omnivorous, including nectar and seeds in their diets. The tongue of the Brahminy starling bears a brushlike tip that is used for collecting pollen and nectar, while the brushlike crests of some of the mynas are important in pollination. The bill of the oxpeckers is used in a scissorlike fashion to remove ticks from the pelts of wild and domestic animals. Many starling species have brightly colored, usually yellow, eyes.

Some of the Southeast Asian starlings have areas of bare skin on their heads, especially around the eye; these areas are yellow in the White-headed starling and Common myna, blue in the Bali myna, and red in the Helmeted myna and coleto. The amount of naked skin reaches its maximum in the coleto, where head feathering is restricted to a narrow strip of bristles running down the center of the crown. The Hill myna and the Wattled starling develop fleshy wattles on the head; in the latter species the wattles appear, and head feathering is lost, mainly on birds coming into breeding condition, but the wattles are subsequently reabsorbed and feathers grow anew. The Rose-colored and Brahminy starlings have long feathers on the head that can be raised into a crest, and the Sulawesi myna has a stiff crest that is permanently erect.

Most starlings are gregarious, breeding in colonies, feeding in flocks, and roosting communally at night. Several species may roost together, and they may also roost among other birds. Roosts are usually located in trees, but the Common starling has recently adopted a habit of roosting in

Below A Red-billed oxpecker eats insects from an impala in the Kruger National Park, South Africa. Large mammals mostly tolerate the birds, which rid them of parasites such as ticks, even though they also take considerable amounts of blood from wounds.

Left The Hill myna of India and Southeast Asia is generally considered the bird world's finest mimic, producing sounds that range from whistles and hoarse chuckles to accurate imitations of human speech.

cities in flocks that can contain over a million birds. Such roosts, and the precision formation flying of the birds as they assemble, can present city dwellers with one of the most amazing spectacles of the bird world. On the other hand, the roosting birds deposit copious quantities of droppings on the roads and pavements below, presenting hazards for citizens.

Nomads and Homebodies

DISTRIBUTION PATTERNS

Most species of starling are resident, being restricted to small islands or island groups, or else inhabiting forests where their range of movements is restricted to that required to provide them with an adequate supply of fruiting trees. Other starlings exhibit everything from local to long-distance migration, and some are nomadic.

The Amethyst and Blue-eared glossy starlings undertake local migrations in Africa, while the Brahminy starling makes similar movements in India. The Grey starling migrates from its breeding areas in eastern Russia, northern China, and Japan to winter in southern China and the Philippines. Northern European and Asiatic populations of the Common starling migrate to milder climates for the winter, those from Siberia heading south toward the northern shores of the Indian Ocean while Scandinavian birds migrate southwest toward the Atlantic seaboard.

FACTFILE

STARLINGS AND MYNAS

Order: Passeriformes

Family: Sturnidae

114 species in 29 genera. Species include: **Common starling** (*Sturnus vulgaris*), **Grey starling** (*S. cineraceus*), **White-headed starling** (*S. erythropygius*), **coleto** (*Sarcops calvus*), **Pohnpei starling** (*Aplonis pelzelni*), **Shining starling** (*A. metallica*), **Santo mountain starling** (*A. santovestris*), **Blue-eared glossy starling** (*Lamprotornis chalybaeus*), **Superb starling** (*L. superbus*), **Brahminy starling** (*Temenuchus pagodarum*), **Slender-billed starling** (*Onychognathus tenuirostris*), **Rose-colored starling** (*Pastor roseus*), **Bank myna** (*Acridotheres ginginianus*), **Common myna** (*A. tristis*), **Hill myna** (*Gracula religiosa*), **Bali myna** (*Leucopsar rothschildi*), **Helmeted myna** (*Basilornis galeatus*), **Sulawesi myna** (*B. celebensis*), **Amethyst starling** (*Cinnyricinclus leucogaster*), **Wattled starling** (*Creatophora cinerea*), **Grosbeak myna** (*Scissirostrum dubium*), **Red-billed oxpecker** (*Buphagus erythrorhynchus*).

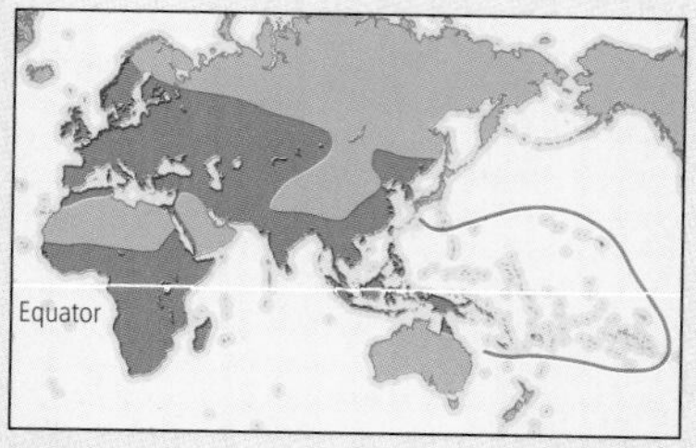

DISTRIBUTION Africa, Europe, Asia, Oceania (just into Australasia); introduced to N America, New Zealand, southern Australia, and many tropical islands.

HABITAT Forest, savanna, steppes, temperate grassland.

SIZE Length 16–45cm (6–18in); weight 45–170g (1.5–6oz).

PLUMAGE Chiefly dark, but usually with iridescent sheens of green, purple, and blue; some with brilliant orange and yellow, some with dull gray, some with bare skin or fleshy wattles. Males and females usually similar, but males sometimes brighter.

VOICE Wide range of whistles, squawks, and rattles. Some mimic other animal sounds, including human speech.

NEST Most in holes in which a bulky nest of dried grass is built; some build domed or pendulous nests; many colonial or loosely so.

EGGS Usually 1–6, pale blue with brownish spots, but some genera without spots. Incubation period 11–18 days; nestling period 18–30 days.

DIET Most eat fruit and insects, some also seeds, nectar, and pollen; oxpeckers specialize on insects parasitic on large mammals.

CONSERVATION STATUS The Pohnpei mountain starling and the Bali myna are both Critically Endangered; 2 other species are Endangered, and 5 Vulnerable.

Below The Common starling – in the inset, an adult bird feeds a large fledgling – often breeds singly, but may also form large colonies, typically in city centers. In such circumstances, the sky can be darkened by huge flocks of birds, like this one sweeping over an abandoned pier.

Left *Tristram's starling* (Onychognathus tristramii) *is a desert-dwelling species of the western Palearctic. This individual is foraging in an arid region of Israel.*

Some starlings are nomadic, particularly the Wattled starling, which settles to breed where locusts abound but moves on when the insects disappear. The breeding sites of the Rose-colored starling are also determined by the abundance of insects, especially locusts, and an area that has a large colony of birds one year may be deserted the next. After the breeding season, all Rose-colored starlings vacate their central European breeding area and migrate to India for the winter.

Diversifying from Fruit

DIET

Originally starlings were probably frugivorous, as many of what are regarded as the more primitive forms are to this day. These forest-dwelling species, like the Hill myna, feed mainly on the fruits of trees and tend to live in pairs, although larger groups may congregate when trees such as *Ficus* figs carry abundant ripe fruit.

As the family has evolved, however, their diet has diversified to include not only nectar and seeds but also invertebrates, especially insects. This development has been accompanied by the adoption of more terrestrial habits by the birds, and, in the case of species more dependent on insects, by summer migrations to areas where insects are more abundant. For example Common starlings that winter in southern Asia, southern Europe, and north Africa migrate north to breed in areas where they capitalize on the spring abundance of such soil-dwelling insects as crane-fly larvae. To exploit this food source, the birds have developed a highly specialized feeding technique, inserting the closed bill into the soil surface or grass roots, and then opening it forcefully to create a hole in which they seek their prey. The generalization of the diet has also led some starlings to form a commensal relationship with humans, exploiting such crops as cereals and fruits, and in the process causing much agricultural damage.

Refuges in Holes

BREEDING BIOLOGY

Most starlings breed in holes, in which they build a bulky nest. Apertures in trees and cliffs are most often used, although niches in buildings and other artificial structures may also serve. The Slender-billed starling nests in holes behind waterfalls, and several species use those made by other birds such as barbets and woodpeckers. Some starlings bore their own holes, among them the Bank myna in river banks and the Grosbeak myna, which sets holes about 30cm (12in) apart in the trunks of dead trees. There are, however, a few exceptions to the hole-nesting norm: the Superb starling builds a domed nest in bushes, while the Shining starling creates pendulous nests, weaver fashion, that hang in compact groups from the outer branches of tall trees.

Pairs of forest starlings, such as the Hill myna, tend to breed in comparative isolation from one another, but other species exhibit varying states of colonial behavior. In species that nest in naturally occurring cavities, the density of nesting pairs is limited by the availability of nest sites. Thus Common starlings aggregate in loose colonies in which pairs nest between 1 and 50m (3.3–165ft) apart, but the birds within such colonies are highly synchronized in their breeding activities, indicating that there is considerable social interaction between colony members. The greatest degree of coloniality of all is shown in species such as the

Right *Male Wattled starlings famously shed their head feathers in the breeding season to reveal a vivid yellow patch beneath; at the same time, they grow long black wattles that are reabsorbed when the season is over. Here a male (left) feeds with two females.*

Above *The Bali myna first became known to science in 1912, when its population numbered in the hundreds. Since then, collection for the exotic bird trade and destruction of its habitat for plantations have seen the species plummet to just 32 individuals.*

Left *The metallic sheens common in the Sturnidae are nowhere more in evidence than in African* Lamprotornis *species like this Blue-eared glossy starling.*

Grosbeak myna and the Shining starling that make their own nests.

In many of the species that have been studied, both sexes participate in care of the eggs but the male usually plays the lesser role. In no species are males known to feed the female on the nest, but nestlings are fed by both parents, although the role of the male can be variable. In Common starlings and some other species, males may be polygynous, having two, and rarely up to five, mates simultaneously. Cooperative breeding, where three or more fully grown birds may feed a brood of chicks in one nest, has been demonstrated in some African starlings.

Glut and Dearth

CONSERVATION AND ENVIRONMENT

The Common starling causes extensive damage in Eurasia and North America by eating grapes, olives, cherries, germinating cereals, and cattle food, while in northern Europe and central Asia and in New Zealand it is held to be useful on account of its destruction of insects. It is one of the most successful birds, with a world population running into hundreds of millions. The most dramatic example of its success comes from its introduction to North America. Despite the failure of several earlier attempts, about 60 individuals were released in New York's Central Park in 1890. A first nest was recorded that year under the eaves of the American Museum of Natural History. A further 40 birds were released the following year, and from then on the species never looked back. Within a century, the Common starling had become one of the continent's most numerous birds.

Other starling species have fared less well, and during the last 400 years four species, two from Indian Ocean islands and two from islands in the Pacific, are known or thought to have become extinct. The Bali myna survives in a small population in a forest nature reserve, but despite a multifaceted conservation plan including the release of captive-bred birds, its future looks bleak. It is subject to intense pressure from bird collectors for the pet trade, and even released captive birds have subsequently fallen victim to the trappers. It may well soon become extinct in the wild, leaving only a handful of the world's zoos to maintain living birds. Other island species, like the Pohnpei and Santo mountain starlings, are also considered to be under threat because of their small populations and damage to their limited habitats.

Even in the case of the most numerous starlings, however, there is no room for complacency. In the last quarter of the 20th century the Common starling underwent a remarkable decline of over 50 percent in Britain and parts of northern Europe, probably as a result of various facets of agricultural intensification.

CJF

Mockingbirds

THE MOCKINGBIRD FAMILY'S NAME IS DERIVED *from the ability of several members, especially the Northern mockingbird, to copy noises made by other animals. Although other birds are the main source for their mimicry, mockingbirds have also been recorded imitating frogs, pianos, and even human voices. Their songs are powerful and far-carrying.*

The mockingbirds (also known as mimic-thrushes) are a fairly distinct group of New World birds that occur over much of North, South, and Central America except for the northern parts of Canada; only the Patagonian mockingbird occurs in the southern third of South America. They are mostly thrush-sized, and are thought to be closely related to thrushes and wrens.

New World Mimics

FORM AND FUNCTION

Mockingbirds tend to have longer tails than thrushes and wrens, and also have longish beaks that are often strongly downcurved. Many are marked rather like a "standard" thrush, brown above and paler below, with heavy streaking, but several are darker and more uniformly gray. The brightest is probably the Blue mockingbird, which is a bright, grayish-blue all over except for a black mask. The Grey catbird is one of the smaller members of the family and is somewhat aberrantly colored: it is a uniform gray all over (darker above than below), with a black cap and bright chestnut undertail feathers. Many mockingbirds cock and fan their long tail in a conspicuous manner, especially as part of their vigorous displays.

Above *Named for its markedly downcurved beak, the Curve-billed thrasher is a bird of the dry lands of Mexico and the southwestern USA.*

Below *On the Galápagos Islands, a Hood mockingbird feeds off the chick of a ground-nesting Masked booby* (Sula dactylatra). *The species is unusual in its habitual taste for blood.*

Most mockingbirds spend most of their time hopping through the undergrowth on their long, powerful legs. They take a wide variety of foods. Through much of the year they eat ground-living arthropods, but in season they also take many fruits and berries. The Galápagos mockingbird also feeds on small crabs along the shoreline and frequently takes carrion. Also in the Galápagos, on Española (Hood Island), the Hood mockingbird has acquired a reputation for pecking open the unattended eggs of a wide variety of seabird species and for stealing eggs of the Galápagos dove and of both the Land and Marine iguanas; in addition, this bird regularly drinks blood, mostly from wounds on iguanas, sealions, and nestling seabirds. Mockingbirds take most of their prey from the ground using their powerful beak, which serves either as a probe or (as in the case of the eggs) for breaking into potential food items.

Found on the Ground

DISTRIBUTION PATTERNS

Mockingbirds have successfully colonized many Caribbean islands and the Galápagos archipelago, and have been introduced to Hawaii and Bermuda. Many birds in the northern part of their range move south for the winter; for example, most Grey catbirds and Brown thrashers leave Canada, and a large majority of the Sage thrashers that breed in the USA spend the winter in Mexico. Some Northern mockingbirds, however, winter in Canada.

The main habitat of the family is scrub or forest understory, including high-altitude grasslands above the treeline; many species inhabit dry, near-desert regions. All use low vegetation as cover, and most forage on the ground. The main exceptions are the two trembler species in the Lesser Antilles, which live in rain forest, and the Black-capped mockingthrush, an aberrant species now sometimes placed with the wrens.

Untidy Nest-builders

BREEDING BIOLOGY

Resident species spend most of the year in their territory, which they defend strongly against other members of the species. Usually they live alone or in pairs, but some species live in groups of 40

individuals, several of which may help in raising the young. While the exact relationships of the birds in such groups have yet to be established, the extra helpers are known in some cases to be young from an earlier brood of the same pair. In the Galápagos mockingbird, incestuous relationships between family members have been recorded.

As far as is known, all species build rather bulky, untidy nests of twigs in dense vegetation. In most cases the nest is either on the ground or within about 2m (6.6ft) of it, although sometimes pairs may build at heights of 15m (50ft) or more. Two to five (rarely six) eggs are laid, and these hatch in 12–13 days; the chicks are raised to the point of leaving the nest in about the same length of time. Breeding commences in the spring or, in some arid areas such as the Galápagos, shortly after the start of the rainy season. The breeding season can be prolonged, with two or even three broods being raised. Pairs often remain together in successive seasons, although Grey catbirds have been shown to be more likely to separate and/or leave the territory if they fail to raise young. This behavior is thought to be an adaptation against predators; since most nests that fail do so because they are raided by predators, moving after a nest is lost might result in the parent birds being able to find a safer refuge.

The Brown trembler of Dominica and other islands of the Lesser Antilles is an aberrant species easily recognized by its habit of trembling its wings (probably a social signal to others of its species, since it is most frequently performed in groups). It also spends much of its time in rainforest trees, where it forages while clinging to the trunk on its rather short legs. It may have taken over the woodcreeper niche of hunting for insect prey on tree-trunks, as woodcreepers are absent from the islands where it lives. PC/CMP

Above *In Ohio, an immature Northern mockingbird plucks a berry. The species is celebrated for its beautiful singing voice as well as its ingenious mimicry.*

Below *The Black-capped mockingthrush inhabits thick, often marshy vegetation around pools and rivers in the Neotropical lowlands.*

FACTFILE

MOCKINGBIRDS

Order: Passeriformes

Family: Mimidae

36 species in 12 genera. Species include: **Black-capped mockingthrush** (*Donacobius atricapillus*), **Grey catbird** (*Dumetella carolinensis*), **Blue mockingbird** (*Melanotis caerulescens*), **Brown thrasher** (*Toxostoma rufum*), **Cozumel thrasher** (*T. guttatum*), **Curve-billed thrasher** (*T. curvirostre*), **Sage thrasher** (*T. montanus*), **Brown trembler** (*Cinclocerthia ruficauda*), **Galápagos mockingbird** (*Nesomimus trifasciatus*), **Hood mockingbird** (*N. macdonaldi*), **mockingbird** or **Northern mockingbird** (*Mimus polyglottos*), **Patagonian mockingbird** (*M. patagonicus*), **Socorro mockingbird** (*Mimodes graysoni*).

DISTRIBUTION New World, from S Canada to Tierra del Fuego.

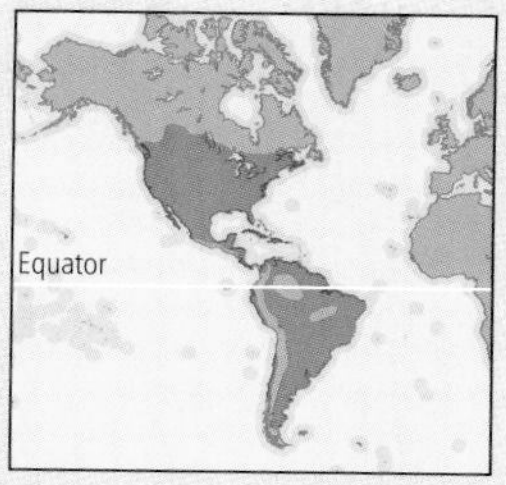

HABITAT Scrub (often arid), keeping mainly close to the ground.

SIZE Length 20–30cm (8–12in), including a fairly long tail in most species.

PLUMAGE Most are primarily brown or gray above, usually with pale or white underparts, often heavily streaked or spotted. A few are brighter, either rich brown (Black-capped mockingthrush) or bright gray-blue (Blue mockingbird). Many have strikingly colored eyes: red, yellow, or white. Sexes are similar.

VOICE Powerful, complex songs, which in some species include many noises copied from other birds or animals.

NEST Largish, untidy cup-nest of grass and twigs, usually fairly close to (or on) the ground, but sometimes high up in a tree.

EGGS Usually 2–5; color varies from pale and whitish to dark greenish-blue, often heavily streaked or spotted with darker markings. Incubation 12–13 days; fledging period usually 12–13 days, but may be longer in some tropical species.

DIET Fruits, berries, seeds, arthropods.

CONSERVATION STATUS The Socorro mockingbird and the Cozumel thrasher are both Critically Endangered; 2 other species are Endangered, and 1 is Vulnerable.

Nuthatches

NUTHATCHES ARE AMONG THE VERY FEW birds that are able not only to climb up trees but also to run down them, headfirst. In addition, they are among the liveliest birds of many north-temperate forests. The word "nuthatch" is derived from the Eurasian species' fondness for hazelnuts.

Most nuthatches forage on insects and spiders, either gleaning them from the surface of trunks, branches, and twigs, or else using their bills to probe into crevices and under bark flakes. Four of the 24 species are currently considered threatened and two are near-threatened, mainly because they occur in very small ranges and/or their habitat is subjected to large-scale deforestation, particularly in Southeast Asia.

Agile Foragers

FORM AND FUNCTION

The true nuthatches belonging to the main genus, *Sitta*, are so similar in form and habits that they are easily recognized. They have, in the main, gray-blue upperparts (although some tropical species range to bright blue or purple), with longish bills, short necks, and a short tail. Their climbing method differs from that of woodpeckers and treecreepers in that they do not use their tail for support. When climbing, the nuthatch does not position its feet parallel to one another; rather, one is placed high, from which to hang, and the other low, for support.

Nuthatches living in temperate-zone forests take tree seeds from the fall onward and store them one by one in bark crevices or under moss for later consumption. The rock nuthatches and the wallcreeper forage in the same manner as the tree-dwelling species, but on rocks.

The wallcreeper is similar to the true nuthatches in size and color but has a curved bill similar to the treecreepers', and has bright red patches on its wings that it flashes during climbing. Although the Australian sittellas of the genus *Daphoenositta* share the same basic morphology and behavior and were formerly incorporated in the nuthatch family, they are now considered to be unrelated (see Thickheads and Whistlers).

Across the Northern Hemisphere

DISTRIBUTION PATTERNS

The Eurasian nuthatch has the most extensive range of all the species that make up the family, stretching from North Africa and Spain across to Japan. The White-breasted nuthatch breeds in most of North America, and the Chestnut-bellied nuthatch, a close relative of the Eurasian species, throughout the Indian subcontinent and Southeast Asia. The wallcreeper is found from Spain across Eurasia to the eastern Himalayas, but has a restricted habitat in high mountains.

Other species have smaller distributions, sometimes limited to small mountain ranges, as in the Corsican nuthatch, the Algerian nuthatch (only discovered in 1975), and the White-browed nuthatch, which is only found around Mount Victoria in Myanmar; both the latter two species are classed as Endangered by the IUCN. Nineteen species occur in Asia, Europe has five (four of them shared with Asia), North America four, and two species (one shared with Europe) are found in the Atlas mountains of North Africa. The true nuthatches prefer mature woods, both broadleaved and coniferous, from high altitudes down to sea level, whereas the rock nuthatches and the wallcreeper live in open, rocky areas and on mountain slopes.

No nuthatch is a regular migrant, but a few species that breed in boreal forests undertake invasive movements in some years. The Red-breasted nuthatch migrates south from the woods of Canada in varying numbers, sometimes as far as Texas. The Siberian subspecies of the Eurasian nuthatch moves west as far as Finland and occasionally on into Sweden.

*Above Typical nuthatch postures: **1** Corsican nuthatch feeding; the right foot takes the bird's weight. **2** Courtship feeding in Corsican nuthatches. **3** Eurasian nuthatch in a threat display, and **4** in a defensive pose.*

FACTFILE

NUTHATCHES

Order: Passeriformes

Family: Sittidae

25 species in 2 genera. Species and genera include: **Algerian nuthatch** (*Sitta ledanti*), **Brown-headed nuthatch** (*S. pusilla*), **Chestnut-bellied nuthatch** (*S. castanea*), **Corsican nuthatch** (*S. whiteheadi*), **Eastern rock nuthatch** (*S. tephronota*), **Eurasian nuthatch** (*S. europaea*), **Pygmy nuthatch** (*S. pygmaea*), **Red-breasted nuthatch** (*S. canadensis*), **Western rock nuthatch** (*S. neumayer*), **White-breasted nuthatch** (*S. carolinensis*), **White-browed nuthatch** (*S. victoriae*), **White-cheeked nuthatch** (*S. leucopsis*), **wallcreeper** (*Tichodroma muraria*).

DISTRIBUTION Asia, Europe, N America, N Africa.

HABITAT Woodlands, parks, rocks.

SIZE Length 9.5–20cm (4–8in); weight 10–60g (0.35–2.1oz).

PLUMAGE Upperparts blue-gray (bright blue or purple in some SE Asian species); many species have black eye-stripes; underparts grayish white to brown. Little difference between sexes, and no distinct breeding plumage except in the wallcreeper, where breeding males have a black throat.

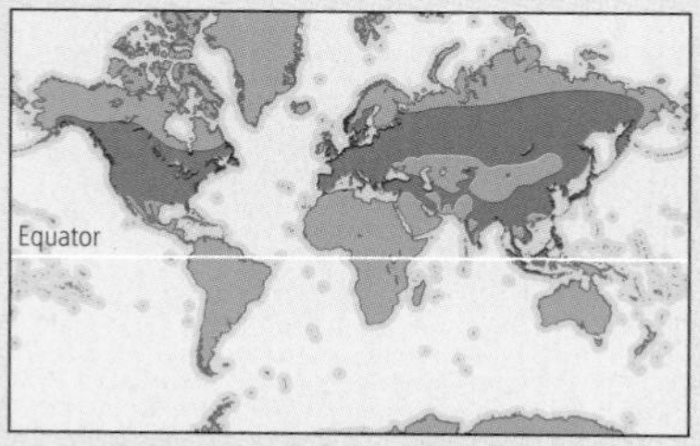

VOICE Repeated piping phrases, chattering calls.

NEST Holes in trees or rocks, often with mud or other materials added. The 2 rock nuthatch species build nests out of mud.

EGGS 4–10; white with reddish spots; weight 1–2.5g (0.04–0.09oz). Incubation period 14–18 days; nestling period 20–25 days.

DIET Insects and spiders, also seeds in autumn and winter.

CONSERVATION STATUS 2 species – the Algerian and White-browed nuthatches – are listed as Endangered. In addition, the Giant and Beautiful nuthatches, both of which have fragmented populations in the China–Myanmar–Thailand border region, are considered Vulnerable.

Cavity Nesters

BREEDING BIOLOGY

Nuthatches are usually seen in pairs or small groups occupying fairly large home ranges. The Eurasian nuthatch is strictly monogamous, as is probably the case with many other species, and pairs defend their territories throughout the year. Wallcreepers also defend individual territories in their winter quarters, which are often at lower altitudes than their breeding areas. The North American Pygmy and Brown-headed nuthatches live in small groups including young birds that help the breeding pair at the nest, and roost communally in winter. Roosting aggregations of Pygmy nuthatches may include over 100 birds in large cavities. In winter, nuthatches often associate with mixed feeding flocks of tits, woodpeckers, and other forest birds.

All species nest in tree or rock cavities, but there is a bewildering variety in the details of nest construction behavior and the materials used, mainly directed at keeping out nest competitors or predators. Some of the smaller species excavate their own nest-chamber in rotten wood. The White-breasted and White-cheeked nuthatches rub noxious insects as a repellent around the entrance hole, while the Red-breasted uses smeared resin as a protection. Rock nuthatches make a nest-chamber by closing up a rock niche with a hemispherical mud wall and entering through a specially constructed tube. The Eurasian nuthatch and its relatives use mud to reduce the size of the cavity entrance, leaving just enough space to accommodate the width of the bird's body. In dry periods rock nuthatches (and also other species in Asia) use animal dung as a building material, but they also mix in insects, berries, and feathers.

Above *The White-breasted nuthatch (Sitta carolinensis) is widespread throughout North America. Unlike other tree-climbers such as woodpeckers and tree-creepers, this agile bird can descend trunks headfirst just as quickly as it ascends them.*

Inside the cavity, a wide variety of materials are employed. Some species bring in bark flakes, seed wings, or other woody material arranged in a loose mass. Others construct a more typical passerine nest lining it with moss, grass, feathers, and hair.

Most pairs produce a single brood during the breeding season, which is incubated by the female alone. In the course of incubation, the male brings food for the female. When the young hatch, they are fed by both parents on insects and spiders. The young can fly and climb well by the time they fledge, at which point they need just one or two weeks to become independent. EM/HL

Holarctic Treecreepers

TREECREEPERS ARE SMALL, MOSTLY BROWN birds that are usually seen climbing steadily up the trunk of a tree, often in a spiral path; they will hide behind the trunk if alarmed. When they have reached the desired height, they glide down to the base of another tree to repeat the process. Holarctic treecreepers climb with their feet parallel to each other and moving simultaneously, whereas the Australasian treecreepers of the family Climacteridae always have one foot in front of the other, progressing by bringing the lower foot up to the level of the upper before the latter is moved higher.

Treecreepers have long toes with claws deeply curved for climbing, and a long, slightly down-curved bill for probing into crevices and under flakes of bark in search of insects. The five Holarctic species belonging to the genus *Certhia* all have pointed tail feathers with stiffened shafts, which the birds use as a prop when climbing. The tail feathers are molted quickly, except for the central pair, which are retained to provide support and are only molted once the new surrounding feathers are sufficiently well grown to perform this function. The adaptation of stiff tail feathers is also present in the unrelated woodpeckers and woodcreepers.

Agile Climbers

FORM AND FUNCTION

The five *Certhia* species are all very similar in appearance and habits, and they are mainly solitary. While scurrying up a tree searching for food, they look like agile brown mice. The circumpolar Common treecreeper, known as the Brown creeper in America (where the bird is sometimes treated as a separate, sixth species, *Certhia americana*), to some extent overlaps the range of all four of the other species. In Britain, where it is the only breeding treecreeper, this species mainly inhabits open deciduous woodlands, and is often found in gardens. In continental Europe, however, this habitat is occupied by the Short-toed treecreeper (which in some languages is even known as the "Garden treecreeper"), and the Common treecreeper is mainly confined to coniferous forests at higher altitudes.

Four species occur in the Himalayas up to the timberline (about 3,500m/11,500ft); all move down to the warmer foothills and plains in winter. The Himalayan treecreeper apparently favors conifers and avoids oak forest, where it is replaced by the Brown-throated treecreeper. Stoliczka's treecreeper has the most restricted distribution of all, and is not at all well known. However, specimens have come from oak, mixed deciduous, and conifer forests up to the tree line.

Detailed measurements of the bills and claws of hundreds of museum specimens have shown that there are subtle but consistent differences between the Common and Short-toed species. The figures varied between different populations,

FACTFILE

HOLARCTIC TREECREEPERS

Order: Passeriformes

Family: Certhiidae

6 species in 2 genera: **Common treecreeper** (*Certhia familiaris*), **Brown-throated treecreeper** (*C. discolor*), **Himalayan treecreeper** (*C. himalayana*), **Short-toed treecreeper** (*C. brachydactyla*), **Stoliczka's treecreeper** (*C. nipalensis*), **Spotted creeper** (*Salpornis spilonotus*).

DISTRIBUTION Eurasia to S and SE Asia; Africa; N America.

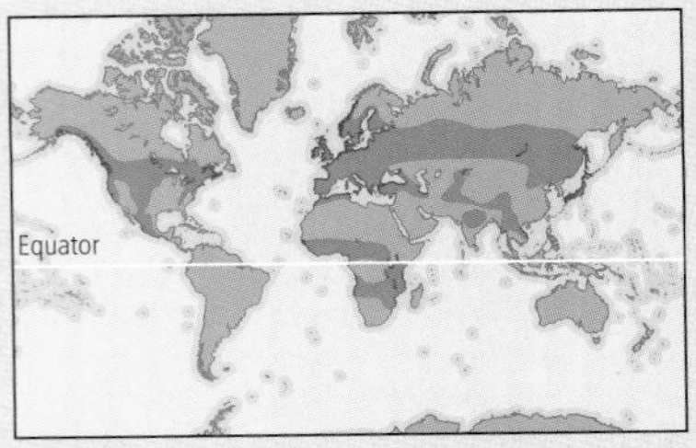

HABITAT Forest, woodland.

SIZE Length 12–15cm (5–6in); weight 7–16g (0.25–0.6oz).

PLUMAGE Upperparts brown streaked, underparts paler in *Certhia*; blackish spotted white in *Salpornis*.

VOICE Thin, high-pitched whistles and songs.

NEST Cup on a loose platform of twigs, usually wedged against the trunk of a tree behind loose bark (*Certhia*); neat cup, decorated on the outside, attached to a horizontal branch with cobwebs (*Salpornis*).

EGGS 3-9 (usually 5 or 6), white with red-brown dots (*Certhia*); 2 or 3, pale with black and lilac markings (*Salpornis*). Incubation period 14–15 days; nestling period 15–16 days.

DIET Insects and spiders

CONSERVATION STATUS Not threatened.

Left *A Short-toed treecreeper demonstrates the way in which the birds move up trees, with their feet roughly in line, as they probe for soft-bodied insects under the bark.*

Right *A Common treecreeper carrying a cranefly back to its nest in a tree cavity.*

clamped to the trunk of a tree and provided with two entrances are readily occupied. One entrance is used by the bird on entering the box, and the other for leaving. There are generally 5 or 6 eggs in a clutch, and they are pale with red speckling and dotting. Incubation, by the female, lasts for a fortnight; fledging takes a little longer. Second broods are not general, but neither are they rare, and they may overlap with the first, in which case the male takes over provisioning the young from the first clutch while the female finishes the second nest; she may even start to lay the new set of eggs before the first brood has fledged. There are a few records of serial polygyny, with the male mating with a second female once the first is settled on the first clutch of eggs. There are even records of two females sitting side by side in the same nest on separate clutches. CJM

Left *A Common treecreeper feeds its young. A typical clutch size for this species is six, although ranges from three to nine have been recorded. The birds build untidy nests, often behind loose bark, that have at their base a neat cup of bark and feathers.*

and there are some races of each species that are very similar in measurements. Studies with live birds, and with other species like the Great tit (*Parus major*), have in fact shown that the bill and claw can vary between seasons, and are adapted by the birds to suit the prevailing feeding conditions. A bird seen wiping its beak on a branch may not be cleaning it but, instead, honing it down to become thinner and shorter. It is quite possible that there are good reasons for treecreepers to have longer or shorter bills and claws at different times of the year to enable more efficient foraging. One surprising result of the study was to show that some races show reverse sexual dimorphism, with the shorter-winged females having longer bills and claws.

Snug Roosting Cavities

BREEDING BIOLOGY

The roosting habits of the Common treecreeper were investigated in detail in Britain following the introduction of soft-barked American redwood trees (*Sequoiadendron* spp.) in 1853. As soon as the trees were big enough, the birds began to excavate small, oval roosting cavities behind peeling sections of bark, just as the Brown creepers of North America do. Such holes may be occupied singly or by a whole family; in either case, the birds snuggle tightly into the hollow and conserve heat by using their wings and upper parts as vertical blankets. Many roosting hollows may be found on the same tree. It seems that newly-fledged broods often roost together, and in very cold winter weather more than a dozen birds have been recorded in a single huddle!

Although the *Certhia* treecreepers hardly ever breed in conventional nest boxes, special ones

THE ABERRANT SPOTTED CREEPER

Although the Spotted creeper is included in the Holarctic treecreepers, it lacks the modified tail of the main genus *Certhia* and differs markedly from all other treecreepers in its nest. This is built on a horizontal branch, usually in a fork (*below*), and is beautifully camouflaged; the outside is decorated with spiders' eggbags, lichen, and caterpillar frass (excrement). Its uniqueness suggests that the bird may have closer relatives in other, nonclimbing families.

The species has an interesting, disjointed distribution, with populations across much of central India and various parts of sub-Saharan Africa. The birds are not very much bigger than the *Certhia* treecreepers, but are chunkier in build and weigh twice as much. They have very strong feet, and their strategy in foraging on trees is much like that of nuthatches; the tail is held away from the trunk, and the bird sometimes hangs below the branch. The eggs are laid in clutches, generally of two (in India) or three (Africa). The colors are also different on the two continents: in India they are greenish or gray with darker brown spots and paler blotches, but in Africa they are blue or greenish with gray or lavender blotches and darker brown markings.

Philippine Rhabdornises

IN THE PAST THE MEMBERS OF THIS ENDEMIC Philippine family were commonly known as Philippine treecreepers because of their superficial resemblance to other creepers. However they clearly lack the morphological features that characterize the Certhia *treecreepers, notably stiffened tail feathers, elongated toes, and adaptive toe pads. Coincidentally, they rarely, if ever, "creep"!*

Although they share similarities with some babblers of the family Timaliidae, particularly the genus *Ptilocichla*, the Rhabdornithidae have been retained as a distinct family. Three species are currently recognized, but the taxonomy is controversial. Some observers consider the Grand rhabdornis to be a subspecies of the Stripe-breasted rhabdornis. Others suspect that there may be as many as five species: the three currently recognized, plus the Stripe-headed population from Samar south to Mindanao as a possible fourth, and the Negros and Panay Stripe-breasted population as the fifth. The family is currently under review both morphologically and molecularly.

Little-known Forest Foragers

FORM AND FUNCTION

All rhabdornises tend to forage in the middle and upper canopy of forests, at the forest edge, and in trees in clearings, traveling in small flocks of usually less than 20 individuals. Often they join mixed-species flocks that include Philippine bulbuls (*Hypsipetes philippinus*), Elegant tits (*Parus elegans*), fantails (*Rhipidura* spp.), and Velvet-fronted nuthatches (*Sitta frontalis*). They move through the trees, hopping and jumping from branch to branch but rarely creeping, gleaning insects from bark or under leaves or bark, or plucking small fruits and seeds from the branches. The birds seem to be opportunistic feeders, as evidenced by their varied diet. One dead specimen of Grand rhabdornis was found to have eaten nothing but small fruits; however, observers have noticed a Stripe-breasted rhabdornis catch and eat a small tree frog. Once, in eastern Mindanao, over 100 Stripe-breasted rhabdornises gathered to flycatch what appeared to be termites or some closely-related flying insect.

Both Stripe-headed and Stripe-breasted rhabdornises have been recorded roosting in large numbers in the canopies of trees in clearings or at the forest edge. One Stripe-breasted rhabdornis roost in Davao Oriental Province in Mindanao started to form at dusk, with individuals and small groups coming in from all directions until it was too dark for observers to see the birds. Several hundred individuals entered the roost.

Typically, two species of rhabdornises occur on most of the islands within their restricted range. When this happens, the Stripe-headed rhabdornis usually occupies the lowlands to the height of 1,200m (4,000ft), while either the Stripe-breasted or Grand rhabdornis may occupy elevations above 1,000m (3,300ft). On occasion two species overlap in altitude and can be seen foraging together in the same trees. On the small island of Biliran, north of Leyte, only the Stripe-breasted rhabdornis has been recorded. There it has been found at elevations below 500m (1,650ft). RSK

Above *A Stripe-breasted rhabdornis catches an insect. The birds only rarely work their way up tree trunks in the manner of the Holarctic treecreepers; more often they perch crosswise on branches, as here, or hop among them, gleaning insects or fruit from leaves and twigs.*

FACTFILE

PHILIPPINE RHABDORNISES

Order: Passeriformes

Family: Rhabdornithidae

3 species of the genus *Rhabdornis*: **Grand rhabdornis** (*R. grandis*), **Stripe-breasted rhabdornis** (*R. inornatus*), **Stripe-headed rhabdornis** (*R. mystacalis*).

DISTRIBUTION The Philippines, on Luzon, Masbate, Samar, Leyte, Bohol, Panay, Negros, Mindanao, and others.

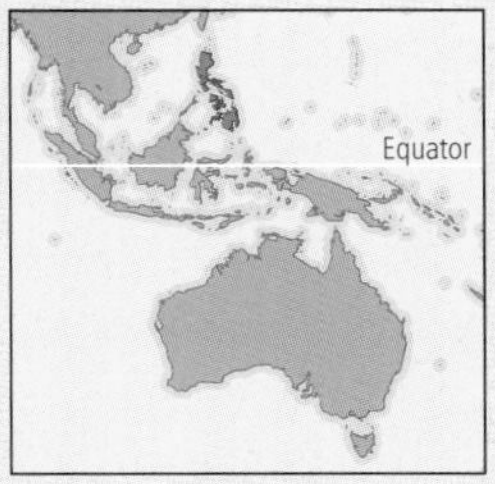

HABITAT Forest, second growth, and clearings adjacent to forests.

SIZE Length 14–17cm (5.5–7in).

PLUMAGE Upperparts generally brown, some with white shaft streaks; underparts whitish with brown streaks on flanks or across entire breast; distinctive dark brown mask, from bill through eye to nape.

VOICE Poorly known, but call of all species seems to consist of four rather nondescript notes or syllables, *tsee* or *zip*, often with the third note a louder, more distinctive *wick* or *zeeet*.

NEST Poorly known, but recorded in a few instances in holes in trees.

EGGS Unknown. Birds with enlarged gonads recorded in March, April, and May.

DIET Insects, small fruits, and seeds. On one occasion a rhabdornis was seen to eat a small tree frog.

CONSERVATION STATUS Not threatened.

Wrens

WRENS ARE RETICENT BIRDS; THEY ARE rarely brightly colored, and can easily be overlooked. However, the dull plumage belies a group of birds that emphasize their presence through full-bodied and rich songs, and that practice a huge range of social behaviors.

All wrens build elaborate, roofed nests, which they use not only to house eggs and nestlings but also as communal roosts and as aids to male courtship. The Latin name for the whole group, *Troglodytidae,* is derived from this habit; it means "cavedwellers," in reference to the habit of building enclosed nests, and probably also to the birds' generally skulking and secretive behavior.

In several species the males are prodigious nest-builders, and are also fine, energetic singers. Field studies of the polygynous wrens of Europe and North America suggest that the form and extent of both these activities may be extreme in this group, perhaps a result of strong sexual selection through female mate choice or male–male competition.

In contrast, many of the Central American species are thought to be monogamous, while those in the Cactus-wren group (genus *Campylorhynchus*) live in family parties and have evolved a cooperative breeding system; the independent young help their parents to raise further broods. The wrens are thus a family of great social diversity, even allowing for the fact that little is known about the habits of many of the tropical species.

Small Foragers in the Scrub

FORM AND FUNCTION

Members of the family are often relatively small – the largest species, the Giant wren of southern Mexico, is only about the size of a thrush. Most species are around 10cm (4in) long, weighing about 12g (0.4oz). The only temperate birds that are lighter in weight are some warblers, goldcrests, kinglets, and hummingbirds. The birds' small size

FACTFILE

WRENS

Order: Passeriformes

Family: Troglodytidae

83 species in 14 genera. Species include: **Apolinar's marsh-wren** (*Cistothorus apolinari*), **Long-billed marsh-wren** (*C. palustris*), **Cactus wren** (*Campylorhynchus brunneicapillus*), **Giant wren** (*C. chiapensis*), **Common**, **European**, or **Winter wren** (*Troglodytes troglodytes*), **Cobb's wren** (*T. cobbi*), **House wren** (*T. aedon*), **Flutist wren** (*Microcerculus ustulatus*), **Musician** or **Song wren** (*Cyphorhinus aradus*), **Zapata wren** (*Ferminia cerverai*), **Niceforo's wren** (*Thryothorus nicefori*).

DISTRIBUTION N, C, S America; 1 species – the Common wren – in Eurasia and just into N Africa.

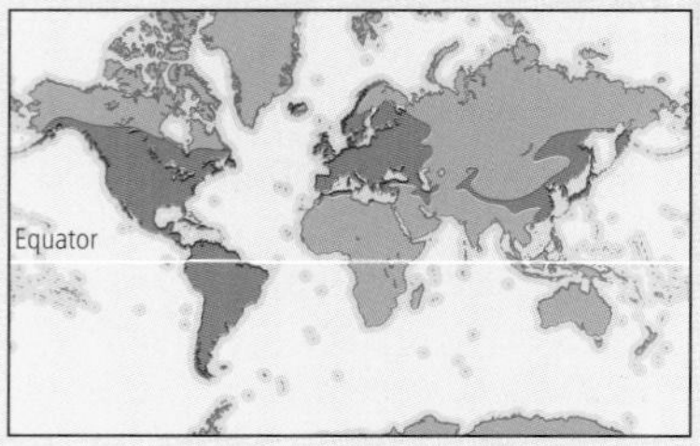

HABITAT Dense, low undergrowth in forest or by watercourses; rocky and semi-desert localities.

SIZE Length 7.5–12.5cm (3–5in); weight 8–15g (0.3–0.5oz). The largest is the Giant wren, 20–22cm (8–9in) long; females are usually slightly smaller than males.

PLUMAGE Brown, cinnamon, or rufous, with dark barring above; paler and sometimes spotted below; no striking differences between sexes.

VOICE Varies from long single whistles to songs containing hundreds of notes and melodious and intricate alternating duets.

NEST Suspended in vegetation, holes, or under overhangs; always roofed, with a side entrance and sometimes with an access tunnel; typically 8–12cm (3–5in) high, 6–10cm (2.5–4in) wide; Cactus wrens, however, build structures up to 60cm by 45cm (24x18in).

EGGS Maximum of 10 in N temperate species; 2–4 in the tropics; white with red or reddish flecking; 1.3 by 1.8cm (0.5 x 0.75in) to 1.8 by 2.4cm (0.75 x 1in). Incubation 12–20 days; nestling period 12–18 days.

DIET Exclusively invertebrates, mainly insects, spiders, etc; also butterfly and moth larvae and adults.

CONSERVATION STATUS Niceforo's wren, found only on 1 known site in Colombia, is Critically Endangered; 2 other species are Endangered, and 3 are Vulnerable.

*◀ **Left** Found across the northern hemisphere, the small, busy Common or Winter wren is one of the world's best-loved birds, celebrated in legend and folklore.*

facilitates their tendency to inhabit dense undergrowth and scrub.

Wrens generally have the blunt, round wings typical of birds living in dense, crowded habitats. This type of wing morphology gives them good maneuverability but relatively poor straight-flight performance. They usually have rather dull plumage, in which browns, black, and white predominate. Several species, notably in the *Thryorchilus* and *Henicorhina* genera, put this limited palette to good effect, and almost all species are extremely attractive when seen close to or in the hand. Once again, the dull plumage is characteristic of birds that inhabit dense habitats where visual communication cues are of little value.

Wrens live in a variety of habitats, including boreal and sub-arctic scrub, coniferous and deciduous forest, reed beds, deserts, rocks and cliffs, lowland tropical rain forest, and montane forests. Within these broad habitats they typically occupy the denser parts in the undergrowth and scrub.

The dietary habits of wrens are relatively straightforward; all species are insectivorous, which obviously makes them vulnerable in areas where insect food is scarce. Presumably in response to the threat of scarcity, migratory behavior has evolved in several species – notably in the Winter and House wrens and the Long-billed marsh-wren, the three species that live at the highest latitudes.

Mostly in America

DISTRIBUTION PATTERNS

The greatest diversity of wrens can be found in Central America and the northwestern states of South America, a pattern that suggests that their origins were in the southern parts of North America or the northern parts of South America. The majority of species are found in the mountainous areas on the west of the continent, with species diversity dropping off rapidly in the lowlands and further north and south. The birds' apparently poor flying ability, and the small size of most species, have not prevented them from invading some islands; for instance Cuba has an endemic species in the Zapata wren and the Falkland group has the endemic Cobb's wren, while distinct subspecies of the Winter wren are found on Taiwan and on St. Kilda, off northwest Scotland.

The Winter wren is the sole representative of the family in the Old World. It is thought to have migrated west from Alaska to Siberia, and its range now stretches westward all around the globe from the eastern USA as far as Iceland. Recent DNA analysis has suggested that the species may be sufficiently distinct to justify its own genus.

Another species with a transcontinental distribution is the House wren, which occurs from the eastern USA to Patagonia. There has been some controversy about the bird's phylogeny. Originally it was described as three separate species, *Troglodytes aedon*, *T. brunneicollis*, and *T. musculus*, found in North, Central, and South America respectively. However, ecological studies in areas where two of the species apparently overlapped suggested that there was in fact little if any difference between them, with the result that all three were lumped together as *T. aedon*, a single transcontinental grouping with the widest natural latitudinal range of any bird species. Recent biochemical analyses, however, have now suggested that the three species should be kept separate. Other wrens, and especially the numerous forms endemic to Central America, occupy much narrower ecological niches and are often very restricted in their distribution.

Above *A Carolina wren* (Thryothorus ludovicianus) *comes to terms with the winter cold. The birds, which are most easily distinguished by their white eyebrows and "tea-kettle, tea-kettle" call, often suffer severely in harsh winters; extended spells of snow cover can severely reduce their numbers.*

Tuneful and Territorial

BREEDING BIOLOGY

Wrens fall into two distinct groups on the basis of behavior. The majority are small, cryptically colored, secretive, and rather solitary inhabitants of dense forest understory. They flutter and climb among tangled vegetation in search of the tiny

Below The Common wren's courtship sequence begins when a male, sitting on an exposed perch **1** or foraging on the ground, sees a female in his territory. **2** He reacts by flying straight toward her. **3** She often flies off and a chase ensues, the male pursuing in a twisting, rapid flight. **4** These chases often end in a "pounce," in which the male tries, usually without success, to make physical contact or to mate with the female. **5** After this episode the male sings his soft, abbreviated courtship songs, and then attempts to lead the female to one of his nests. **6** When she is within a few meters of the nest, the male may motion her to enter by repeatedly inserting and withdrawing his head. **7** The female may then go inside, the male singing meanwhile from a nearby perch. **8** After a period of at least 10 seconds and sometimes several minutes, she will emerge, and some time later, **9**, will bring material for lining the nest. Copulation takes place outside the nest during or after this elaborate ritual.

insects and other animals that make up their diet. In a minority are the much bigger Cactus wren and its allies, living in the more open semi-desert habitats of Central America. Although they have a diet similar to that of their smaller relatives, they move much more boldly, perhaps because they often gather in small family flocks that may afford some protection from predators.

All wrens studied in detail seem to be territorial, at least during the breeding season. The role of song in the defense of space is uncertain in most species, but as a family the wrens are renowned songsters. Several of the monogamous forest-dwelling species live in pairs all year round and some, including the Song wren, produce melodic and beautifully coordinated alternating duets. The singing prowess of some wren species is emphasized in their names, as for instance in the Flutist wren and the two nightingale-wrens.

The song of the males in polygynous species, such as the Common wren and Long-billed marsh-wren, may serve both to defend territory and to attract mates, up to five mates in a season being not uncommon. Neighboring males spend a large proportion of their time each morning answering each other across well-defined territory boundaries. When a female enters a territory she is courted vigorously by the occupant, who sings and leads her round the nests he has already built.

Common and House wrens may have three or four nests ready simultaneously for use both in these displays and by females making breeding attempts, which they commence promptly after courtship and the collection of lining materials for the nest cup – the extent of their contribution to nest construction.

Males of these species may build from 6 to 12 nests in the course of a three-month breeding season, but their efforts are paltry beside those of male Long-billed marsh-wrens, which construct as many as 25–35 nests over a similar period. The nests are built in clusters, some even being semi-detached, and appear to have a primarily ceremonial role. Males sing vigorously from these collections of nests, and lead any female that appears to several of them in succession. Subsequently the male builds yet another nest, in which the female attempts to breed, usually away from his conspicuous courtship center. Predation rates from nests are very high, often affecting up to 80 percent of all breeding attempts.

Observations suggest that females in these species have a free choice of where to breed, subject to the availability of usable nests. One could therefore predict that any trait in males that enhances their ability to attract females would be subject to intense sexual selection. It is no surprise, then, to find that males of polygynous species build more nests, have more elaborate songs and courtship displays, and spend more time singing in the breeding season than do those of monogamous species. These adaptations may be seen as the results of an evolutionary "arms race," driven over the millennia by a combination of male self-advertisement or salesmanship and female sales resistance. Because they spend so much time trying to obtain mates, male Common wrens and Long-billed marsh-wrens never incubate and only help feed their nestlings at the end of the season when, presumably, the chances of achieving further productive matings are negligible.

In complete contrast, the Cactus wren is monogamous and uses a cooperative system. Parents produce up to four broods a year, later ones being fed in the nest by both the parents and their independent young from earlier broods. All members of these family groups assist the breeding male in territorial defense against other families, but, rather paradoxically, all but recently fledged juveniles sleep alone in one of the many large nests dotted about in the cacti on their territory.

Left *The Rufous-breasted wren* (Thryothorus rutilus) *is a species from Central and northern South America. Like all wrens, it is highly vocal, song being an effective means for these small, cryptic birds of maintaining territorial boundaries and keeping in touch with their mates.*

Right *Familiar birds of the American desert, Cactus wrens build their nests amid the spines of cactus plants, which provide an effective defense against predators both terrestrial and aerial. Measuring 18–21cm (7–8in) from head to tail, the birds are the size of starlings and rank among the largest of the Troglodytidae. A typical clutch size is 4 or 5 eggs.*

The Threat from Rats and Cats

CONSERVATION AND ENVIRONMENT

There are no recorded extinctions of wren species, but some races have been lost. One race of the Rock wren, *Salpinctes obsoletus exsul*, became spectacularly extinct when its home – the island of San Benedicto, one of the Revillagigedo group off western Mexico – erupted in 1952. The other extinction events, and the majority of other conservation concerns, are anthroprogenic (brought about by human agency).

Several races are endangered, threatened mainly by habitat loss; examples include the races of Southern house wren on Guadeloupe and St. Lucia. Of possibly greater concern are two Colombian species, Apolinar's marsh-wren and Niceforo's wren. The former is found only in lakeside habitat in a well-populated area of central Colombia, much of which is under pressure from development. Niceforo's wren is extremely rare, and is very vulnerable to extinction; first seen in 1945, it was not observed again until 1989.

On the Falkland Islands, Cobb's wren has undergone a significant range contraction in recent years, due to the encroachment of feral cats and rats, but is still apparently abundant on some islands in the group. The Cuban Zapata wren, whose total population probably numbers only 100 individuals, requires active conservation measures to be taken to ensure its longterm survival. Island populations of wrens can in fact naturally undergo massive population fluctuations; small populations of Winter wrens are particularly prone to large variations, probably due to winter weather conditions. These populations continue to persist, however, as the high rate of production during the summer usually means that populations only remain low for a season. MRE/PJG

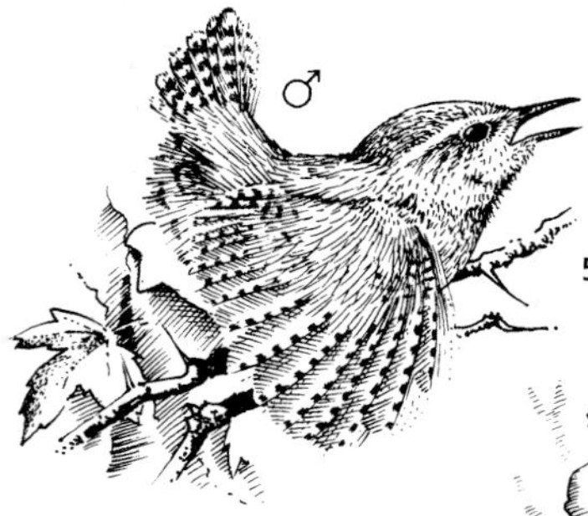
5

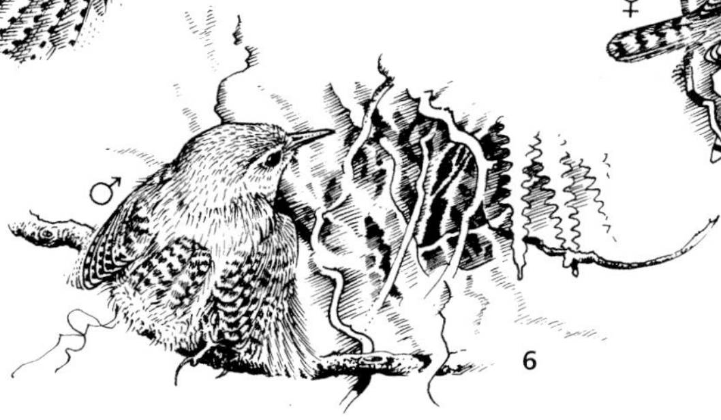
6

7

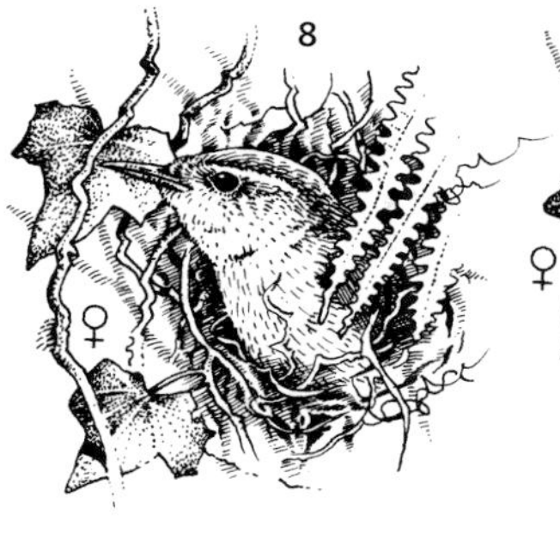
8

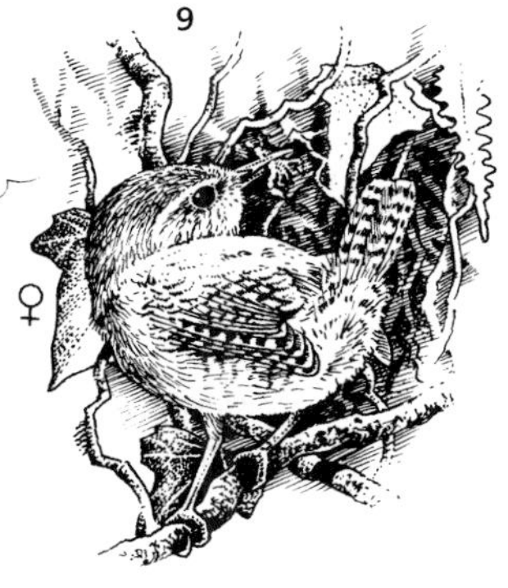
9

Gnatcatchers

The family Poliotilinidae brings together 11 species of gnatcatchers, all grouped in the genus Polioptila, *with three gnatwrens in two separate genera. The gnatcatchers are tiny, elegant, long-tailed gray birds with paler underparts; gnatwrens are, as their name suggests, rather wrenlike in appearance, often cocking their long tails in the manner of their namesakes.*

For a short time the verdin (*Auriparus flaviceps*) was placed with the gnatcatchers, following preliminary DNA analyses. Recent studies, however, have confirmed that its true relations lie rather with the penduline tits, which it resembles in size, shape, and habits.

Dainty Forest Foragers

FORM AND FUNCTION

The gnatcatchers mainly differ from one another in the males' head patterns – Tropical gnatcatchers, for instance, have black and masked gray heads with a broad black band over the bill, through the eyes, and onto the ear coverts, while the Blue-gray has a gray head with an eyestripe that is only slightly darker. The group includes species found in dry scrub, secondary woodland, and humid forest, where the birds are very active and conspicuous high up in the cover of shrubs and bushes; the exception is the Masked gnatcatcher, which is normally found near the ground.

Swishing and cocking their long tails for balance, gnatcatchers glean insects from the outer foliage. Frequently excitable, they sometimes form small flocks, or join mixed flocks of other species. Their songs and calls are simple, high, whistles, often given with the songster standing erect with its bill pointing skyward. Some species produce nasal calls, and even sounds that have been described as kittenlike mews.

The birds' range stretches from the northern USA through to Argentina. In Mexico, near the center of their distribution, six main forms are encountered. The Blue-grey gnatcatcher inhabits hilly woodland and savanna, although not down to sea level. The White-browed, which favors similar habitats, is generally considered a subspecies of the Tropical gnatcatcher, found across much of central America and northern South America. The White-lored is named for its striking white lores – the spaces on birds' heads between the eyes and the bill – and occurs on the Pacific coast down to sea level. The Black-tailed gnatcatcher, which favors dry habitats like the coastal slopes and sagebrush, is a distinct species with an almost wholly black tail, whereas the Black-capped gnatcatcher, which is found throughout the country, is considered by some authorities to be merely a race of the Tropical or White-browed species. The relationships of the Yucatán gnatcatcher, which is found in dry parts of the northern Yucatán peninsula, are unclear. It is thought to be a subspecies of either the Tropical, White-browed, or even the White-lored gnatcatcher.

Below The Long-billed gnatwren uses its elongated beak to pick out insects from scrubby undergrowth in the forests of Central and South America.

FACTFILE

GNATCATCHERS

Order: Passeriformes

Family: Poliotilinidae

14 species in 3 genera. Species include: **Blue-grey gnatcatcher** (*Polioptila caerulea*), **Masked gnatcatcher** (*P. dumicola*), **Tropical gnatcatcher** (*P. plumbea*), **Creamy-bellied gnatcatcher** (*P. lactea*), **Collared gnatwren** (*Microbates collaris*), **Tawny-faced gnatwren** (*M. cinereiventris*), **Long-billed gnatwren** (*Ramphocaenus melanurus*).

DISTRIBUTION Tropical and subtropical N, C, and S America, including parts of the Caribbean.

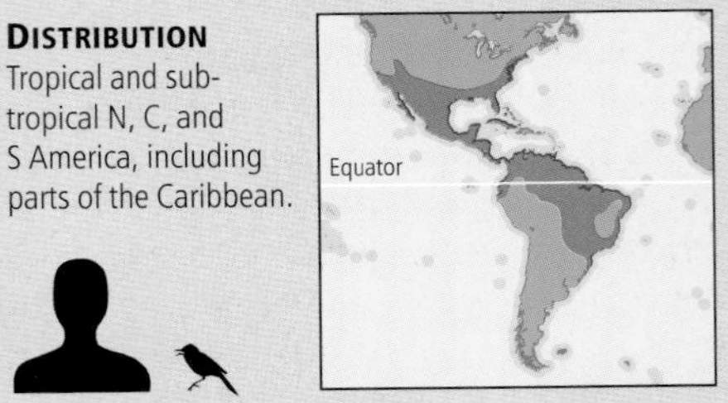

HABITAT In forest, woodland, and scrub, including quite dry areas.

SIZE Length 9–12cm (3.7–4.8in); **weight** about 6–10g (0.25–0.4oz); slight difference between sexes in some species.

PLUMAGE Gray and black (gnatcatchers); brown with pale underparts (gnatwren).

VOICE Simple whistles and calls, chips, and rather nasal soft notes.

NEST Substantial pile of dead leaves with deep cup, or neat saucer on branch.

EGGS 2–6; pale ground (blue. olive, or stone) with a few darker dots and speckles.

DIET Insects and other arthropods.

CONSERVATION STATUS The Creamy-bellied gnatcatcher is listed as Lower Risk/Near Threatened.

Gnatwrens are forest birds and are very active in the undergrowth. The Long-billed gnatwren may be seen in clearings, in secondary growth, or at the forest edge. These busy little birds can be difficult to spot in the tangled vegetation, and the best way of finding them is often to listen for their clear, rising, musical trill. They may join mixed-species flocks, but seem not to follow them outside their own territory.

In contrast, the two *Microbates* species are birds of primary-growth forest, where they live low down rather than in the canopy. They too may join mixed-species flocks, and sometimes even attend antswarms. Their calls are often nasal and scolding, although the song may be a series of pleasant whistles.

Neat Nests in the Trees

BREEDING BIOLOGY

Male and female gnatcatchers cooperate to build very neat little nests that are attached to a branch, often quite high up, and are excellently camouflaged with lichens and other material attached by spiders' webs. They lay up to five small, white eggs with chestnut markings; incubation (by both parents) takes 13 days, and the chicks, which are born naked, fledge after a fortnight. The birds often reuse material from a first nest when building a new nest for the second clutch.

Gnatwren nests are set close to the ground and consist of a neat cup placed on top of a substantial, rough pile of leaves and other vegetation. The cup of the Long-billed gnatwren is so deep and compact that the incubating bird has to hold its head upright, and the long, protruding bill looks like a vertical twig. Two eggs are laid, with a yellowish ground color and red speckling. Incubation takes 17 days and involves both parents. The young, which hatch naked like gnatcatchers, leave the nest after 12 days or so.

The Creamy-bellied gnatcatcher lives in a severely threatened habitat – the lowland Atlantic forest of the eastern South American seaboard – and is now rare. CJM

Above *One of two* Microbates *species, the Tawny-faced gnatwren is a lively resident of the lower levels of Central and South American rain forests.*

Right *A Masked gnatcatcher rests on a branch on the Argentinian pampas. One author has remarked that the gnatcatchers "strongly suggest tiny mockingbirds."*

Penduline Tits

THE 13 SPECIES OF PENDULINE TITS ARE ALL very small, and have finer, more needle-pointed, conical bills than the other tits. Penduline tits get their name from their hanging nests. They live in small parties for most of the year, many of the species in rather open, scrubby woodland. The Eurasian penduline tit, however, lives in marshes, in small trees such as willows and tamarisks, and spends much of its time hunting for food among the reeds.

The birds forage in the tops of vegetation (in arid areas, often in quite low trees or scrub), taking a diet mainly of insects, although the Eurasian species takes many small seeds at certain times of year. They have relatively powerful legs and feet, and frequently hang upside down in their search for prey. They often hold prey under a foot while preparing to eat it, and some may even use their feet to pull small twigs within range of the beak.

Small Birds with Hanging Nests

FORM AND FUNCTION

The Eurasian penduline tit has by far the greatest range of the birds in the family, stretching from southern Europe to eastern China. In Europe its range is extending slowly west and north. It varies markedly in plumage, and some people have suggested that it would best be considered as four separate species. The African *Anthoscopus* species make up a genus of tiny birds, often the smallest in their area, measuring only 8–10cm (3–4in) and weighing 7–9g (0.3oz). They are widespread, mainly in open, dryish *Acacia* woodland or very dry thorn scrub; only the Forest penduline tit occurs in rain forest. The verdin, the only North American representative of the family, occurs in dry, bushy country in the southwestern United States and the northern half of Mexico. The tit-hylia, a very small bird that inhabits the forests of west and central Africa, lives in small flocks and builds a domed nest similar in form to those of the penduline tits. Apart from this, little is known of the behavior of this species, and its exact relationship to the family are uncertain.

All the tropical species seem to be resident, although birds may disappear from certain places at certain times of year, suggesting that there is some local movement. In the case of the Eurasian species, those birds that live in the northern parts of the range migrate south in order to spend the winter in a milder climate; for example, some of those in northeastern China fly to Japan. The Fire-capped tit is also a migrant. It lives in montane forests of the Himalayas and China in the summer, but in the winter some birds drop down to lower altitudes nearby, while others migrate southward into warmer woodlands in central India and elsewhere. Uniquely for this family, but like members of the true tits, the Fire-capped tit nests in lined holes in trees.

Purselike Structures with False Entries

BREEDING BIOLOGY

The Eurasian penduline tit and all *Anthoscopus* species build purselike nests of a strong, feltlike construction; these last for some time and are often conspicuous after the breeding season in areas where the trees lose their leaves. Indeed, old nests are occasionally used as purses by certain African tribes.

The entrance to *Anthoscopus* nests is through a tunnel near the top of the nest; below this is an indentation that appears to be the entrance, but which is in fact blind. The parents close the real entry while they are inside the nest and again when they leave, making it hard for would-be predators to find their way in. This trick seems to work not only against potential intruders such as snakes, but also foils marauding ants.

The verdin builds a domed nest that, unlike those of the other species, is built of thorny twigs, up to 2,000 of which are woven into a structure that seems rather bulky for a bird of its size. The entrance is quite well-concealed, and the thorns keep many predators at bay. The nest is so impregnable that the bird makes no attempt to hide it but places it conspicuously near the end of a

FACTFILE

PENDULINE TITS

Order: Passeriformes

Family: Remizidae

13 species in 5 genera: **Fire-capped tit** (*Cephalopyrus flammiceps*), **Chinese penduline tit** (*Remiz consobrinus*), **Eurasian penduline tit** (*R. pendulinus*), **White-crowned penduline tit** (*R. coronatus*), **verdin** (*Auriparus flaviceps*), **African penduline tit** (*Anthoscopus caroli*), **Buff-bellied penduline tit** (*A. sylviella*), **Cape penduline tit** (*A. minutus*), **Forest penduline tit** (*A. flavifrons*), **Mouse-colored penduline tit** (*A. musculus*), **Sennar penduline tit** (*A. punctifrons*), **Yellow penduline tit** (*A. parvulus*), **tit-hylia** (*Pholidornis rushiae*).

DISTRIBUTION N and C America, Africa, Eurasia.

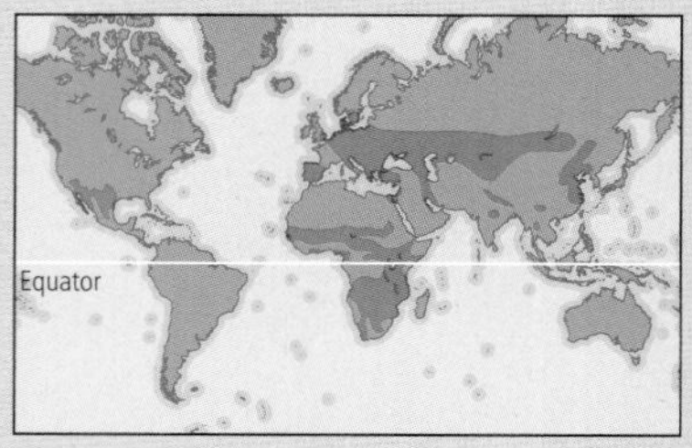

HABITAT Open country, in trees and bushes; reedbeds.

SIZE Length 10–11cm (4in).

PLUMAGE Mostly pale grays, white, and yellows, but striking black mask and rich chestnut "saddle" in adult Penduline tit; a few species with bright yellow or red.

VOICE Fairly quiet *ti-ti-ti*; thin whistles.

NEST Purselike; verdin uses prickly twigs.

EGGS White, with or without red spots, except verdin (bluish-green). Incubation period 13–14 days; nestling period about 18 days.

DIET Chiefly insects; some species also take small seeds.

CONSERVATION STATUS Not threatened.

Above right *A Eurasian penduline tit outside its nest. Males of the family seem to do most of the nest-building while females incubate and rear the young.*

Right *An entrance leading nowhere* ***1*** *is built into the nest of the Cape penduline tit. The blind chamber to which it leads is thought to mislead predators.* ***2*** *The true entrance lies above.*

Left *The tiny desert-dwelling verdin is rarely seen drinking but is thought to gain essential moisture from insects, seeds and berries.*

1

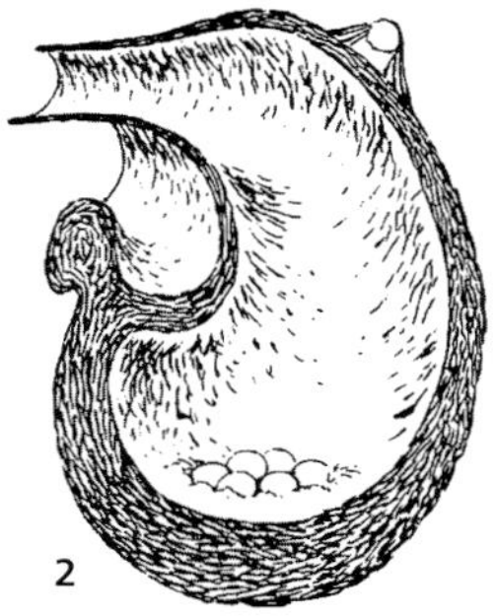

2

branch of a shrub or low tree. Unlike all other members of the family, the verdin has bluish eggs, spotted with red. It also does not have helpers at the nest. Temperature often drops quite low in the deserts where the verdin lives, and this species overcomes some of the problems of surviving the cold nights by roosting in its nest; in fact, some nests seem to be built outside the breeding season specifically for roosting.

Cape penduline tits start a second brood after fledging the first. The young from both broods continue to roost in the nest after fledging, along with the parents; in fact, as many as 18 birds may be found there, which is more than the pair could have raised, indicating that at least some are not from the family. If the nest survives this treatment, the birds may roost in it for at least four months after the end of breeding. Daytime foraging parties seldom exceed eight birds, so the roosting group must split up during the day. CMP

True Tits

Left *A lone Black-capped chickadee. In the winter chickadees form the nucleus of mixed flocks of woodpeckers, nuthatches, creepers, and kinglets.*

Tits are small, active woodland and scrub dwellers. Most are gregarious and vocal. The North American and European species include some of the most popular of all birds, frequently visiting bird-feeders in winter and nesting in boxes in summer. They rarely cause any damage, but provide hours of interest and enjoyment to home-bound observers.

The word "tit" derives from "titmouse," in Britain the name for all the "true" tits of the family Paridae but in North America for only one group of Parus species (another group are called the chickadees). Other, unrelated bird species are also called tits, but only members of the three families Paridae, Aegithalidae, and Remizidae are thought to be closely related; they form a group probably closely allied to the nuthatches and creepers. Fifty of the 53 species are currently put in the genus Parus, but there is a move to break up this rather unwieldy group into about ten different genera.

Nimble Insect-hunters

FORM AND FUNCTION

In form and general appearance, most of the true tits are fairly uniform and easily recognized all over the world. Many have pale or white cheeks contrasting with black or dark caps; a number are crested. They have short, sturdy bills and short legs. All spend most of their time in trees and bushes, although they will also forage on the ground. They are extremely nimble and readily hang upside down on small twigs. Most are year-round residents.

Most tits are primarily insect-eaters. Many also take seeds and berries, particularly species in colder climates, where seeds are the main item of the winter diet. An abundance of an alternative food source is the reason why tits are so common in gardens and at bird feeders during the winter months. Some tits store food, primarily seeds, but sometimes also insects; such items are usually put behind cracks in the bark of trees, but may also be buried under moss. The cache may not be used for some time, or the bird may store food and collect it within several hours.

In the warmer breeding season, all species feed insects to their young. A pair of Blue tits may feed caterpillars to their nestlings at the rate of one a minute while the young are growing most rapidly, and bring well over 10,000 such items while they are in the nest. Tits have been thought – although the evidence is not convincing – to be important in the control of forest pests, and large numbers of nesting-boxes have been put up for this reason.

Tits are very versatile and quick to learn from one another. In 1929 some tits in Southampton, England, were observed to remove the tops from milk bottles and drink the cream. This habit spread very rapidly throughout England by tits copying the skills from one another.

Little is known about the rather drab, greenish Yellow-browed tit, which lacks the clear distinctive

Below *Representative tit species:* ***1*** *Rufous-bellied tit* (Parus rufiventris)*;* ***2*** *Yellow-cheeked tit* (P. spilonotus)*;* ***3*** *Blue tit* (P. caeruleus)*;* ***4*** *Azure tit* (P. cyanus)*;* ***5*** *Enlarged head of Bridled titmouse (P. wollweberi).*

FACTFILE

TRUE TITS

Order: Passeriformes

Family: Paridae

53 species in 4 genera. Species include: **Black-capped chickadee** (*Parus atricapillus*), **Black tit** (*P. leucomelas*), **Blue tit** (*P. caeruleus*), **Bridled titmouse** (*P. wollweberi*), **Coal tit** (*P. ater*), **Crested tit** (*P. cristatus*), **Great tit** (*P. major*), **Marsh tit** (*P. palustris*), **Oak titmouse** (*P. inornatus*), **Juniper titmouse** (*P. ridgwayi*), **Siberian tit** (*P. cinctus*), **Tufted titmouse** (*P. bicolor*), **White-naped tit** (*P. nuchalis*), **Willow tit** (*P. montanus*), **Yellow-browed tit** (*Sylviparus modestus*), **Sultan tit** (*Melanochlora sultanea*), **Hume's ground-jay** (*Pseudopodoces humilis*).

DISTRIBUTION Europe, Asia, Africa, N America into Mexico

HABITAT Chiefly woodland and forests

SIZE Length 11.5–14cm (4.5–5.5in); weight 6–20g (0.2–0.7oz), except for Sultan tit, which is 22cm (9in) long and weighs about 40g (1.4oz).

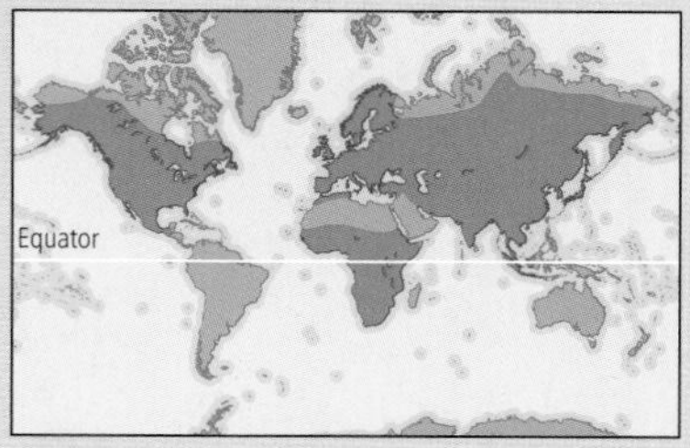

PLUMAGE Chiefly brown, white, gray, and black, some with yellow; 3 species have bright blue. Only slight differences between sexes – some females duller than males.

VOICE Wide range of single notes, chattering calls, and very varied complex songs, many whistled.

NEST In holes, excavated in soft wood by some species.

EGGS Usually 4–12, whitish with reddish-brown spots. Incubation period 13–14 days; nestling period 17–20 days.

DIET Chiefly insects, but also seeds and berries; some species store food for later retrieval.

CONSERVATION STATUS White-naped tit is Vulnerable

patterning of most species; it is put in a separate genus, *Sylviparus*. It lives in high-altitude forests above about 2,000m (6,500ft). It was not until its nest was found in a hole in a rhododendron tree in 1969 that its breeding habits were known to resemble those of other tits.

Two other species may belong to this family. The Sultan tit from Southeast Asia is an enormous bird for a tit; about 22cm (8.7in) long and weighing close to 40g (1.4oz), it is almost twice the size of the next largest species. It is predominantly a glossy blue-black (the female is a little duller), with a bright yellow crown, an erectile crest, and a yellow belly. It lives in rich forests and is not well known.

More extraordinary still is Hume's ground-jay. This bird occurs above the treeline on and around the Tibetan plateau. It is a drab brown bird with a medium-length decurved bill. It nests down rodent burrows or in holes in banks. It looks nothing like a tit, but recent, independent studies of its morphology and its DNA have concluded that this is the family to which it belongs.

Mostly North of the Equator

DISTRIBUTION PATTERNS

The true tits are by far the largest and most widespread of the three families, occurring from sea level to high mountains wherever there are trees: apart from treeless areas and offshore islands, only South America, Madagascar, Australia, and the Antarctic are without true tits. Eleven species occur in North America (one of which also occurs in the Old World), 13 in Africa south of the Sahara, and the remainder are primarily Eurasian.

Many species have extensive ranges. The Great tit, Coal tit, and Willow tit breed from the British Isles across Asia to Japan. The Marsh tit also breeds at both ends of this range, but has a gap of some 2,000km (1,250 miles) in its range in Central Asia. The Siberian tit ranges from Scandinavia across Asia into Alaska and Canada. The Willow tit of Europe and Asia is very similar to the North American Black-capped chickadee; probably in prehistoric times a single species encircled the northern hemisphere, only later diverging into two species.

Prolific Hole-nesters

BREEDING BIOLOGY

True tits are monogamous in temperate areas, the male defending a territory against all comers. These territories are usually established in winter and early spring, and may break down by the time that the parents are busy raising young, although in some species there is a brief resurgence of territorial behavior in the fall after the molt. Other species maintain their territories throughout the year. In Scandinavia the Willow tit may winter in groups of up to six in one territory; mortality in winter can be high, and some territories do not have a pair by spring.

THE TUFTED TITMOUSE

Some of the 10 New World species of true tits are well known to the casual birdwatcher. The Black-capped chickadee and the Tufted titmouse (shown here) are common sights at bird feeders, and the latter is particularly familiar since, unlike the chickadee, it will also readily use nesting boxes.

Although never reaching densities as high as those of the Blue or Great tits in the Old World, the Tufted titmouse is very common throughout much of the eastern USA. It used to be scarcer in the northern states, probably because, as a resident species, it found the winters too severe. In recent years, however, it has gradually become more common, especially in urban areas. It first appeared in Ontario in 1914, and since then has steadily increased and established itself along the southern edge of the Canadian province. It seems almost certain that the widespread provision of food at bird-feeders is the key to this species being able to survive in such cold areas.

The Tufted titmouse is the only crested member of the family found in the eastern half of the United States; in Texas and Mexico these birds have a striking black crest and are thought by some to be a separate species, the Black-crested titmouse. Its counterpart in the west, the Plain titmouse, lacks the rich orange-brown flank and the black above the beak. Recent studies have shown this "species" in fact to be two separate ones, the Oak titmouse, largely confined to California, and the Juniper titmouse, which lives further inland. A third crested species, the Bridled titmouse, is found in the states bordering northwest Mexico. The crested tits are the ones refered to as "titmice" in North America, as opposed to the chickadees that include all the other members of the family.

In other species the birds may join up in flocks for much of the year, roving over large areas of woodland. Parties of mixed species of tits, often together with other small woodland birds, are a common feature of woodlands in Europe, Asia, and North America. The behavior of tropical and African species is less well known. However, in the African Black tit, territories are occupied by three or four birds during the breeding season, and all help to raise the brood. The "extra" birds are usually males that have been raised in the same territory the previous year.

Some species, such as the Siberian tit, remain on their breeding grounds throughout the year despite very low winter temperatures (as low as –45°C/–49°F overnight). A few temperate species may migrate over long distances, especially when the seed crops on which they are dependent in winter fail. Great tits from northern Russia have been known to winter as far afield as Portugal.

As far as is known, all *Parus* species are hole-nesters. A few nest in nesting-boxes in gardens; these species are well known and have been studied extensively (see The Great Tit). The majority probably search for a pre-existing hole that is suitable, and do not seem to enlarge it in any way. However some, including the Crested tit, Willow tit, and Black-capped chickadee, excavate their own nest-chamber in a soft piece of dead timber. This habit seems so fixed that they will excavate a new chamber even if the previous year's chamber is standing unused in the same tree. These species

Left *Perched on a decaying tree stump this Crested tit is sporting its distinctive pointed black and light gray speckled cap. Crested tits give a high, rapid series of cries as a warning against enemies.*

will normally not use nesting boxes, although if the boxes are filled with wood chippings they may then "excavate" them and find them acceptable! When suitable tree sites are in short supply, holes in the ground may be used.

Most species line their nests with moss, some adding hair or feathers; the female does the work, although the male may accompany her on trips to collect material. The eggs are laid at daily intervals. Clutches tend to be large, 4–5 in tropical species and more in temperate areas; as with other hole-nesting species, large clutches are thought to be related to the safety from predators provided by hole-nesting, enabling large numbers of young to be raised. The average clutch of Blue tits in oak woodlands is about 11 eggs (exceptionally, birds may lay as many as 18 or 19); these are the largest known clutches of any passerine bird.

In some species clutch size has been shown to vary with a number of factors: first-year birds lay smaller clutches than older, more experienced ones; clutches are smaller in the poorer habitats of gardens than in woodland, and are also smaller later in the season when caterpillars are scarcer, and when breeding density is high. Most species have a single brood, but some raise two broods in favorable seasons. Incubation is by the female alone. After leaving the nest, the large brood is cared for by both parents, for a week or so in temperate species and probably for much longer in some tropical species. CMP

Above *An adult Coal tit brings a caterpillar meal back to its nest. The white patches on the nape distinguish this species from similar tits, such as the Marsh tit. The Coal tit prefers conifer wood habitats and nests in holes in banks and tree stumps.*

Below *Siberian tits are especially attracted to bird-feeders containing suet or unshelled sunflower seeds. They roost in cavities in trees or even in mouseholes in the snow, and go slightly torpid during the night, regaining their normal temperature at dawn.*

THE GREAT TIT

The world's most studied bird?

THE GREAT TIT IS PERHAPS THE MOST STUDIED wild bird in the world. The first person to realize its usefulness was H. Wolda in the Netherlands, who started to keep careful records of breeding numbers in 1912. Many of his results were published, together with new data, by H. N. Kluijver in 1951. This classic work inspired many other studies.

The Great tit is common over most of Europe and Asia from Ireland to Japan; except in the very coldest areas the birds are usually resident. In the wild it nests in holes in trees, but it readily accepts nest-boxes; indeed, it often seems to prefer them to natural sites, and so all the birds within an area may nest in boxes. It is these characteristics that have made the Great tit (and to a lesser extent the Blue tit) such a convenient bird to study.

Great tits eat a wide variety of food. Although primarily insect-eaters, they readily turn to seeds and nuts in winter when insects become scarce. The Great tit has a powerful beak with which it can hammer open seeds as large as hazelnuts, something that most of the smaller tits cannot do. Outside the breeding season, the birds often join small foraging parties containing both other species of tits and other birds, each bird keeping an eye on where the others are feeding and what they are taking. As soon as one bird finds a new source of food, the others will change their foraging technique to include the new item in their searches.

Great tits tend to settle in broadleaved deciduous woodland at densities of around one pair per ha (2.5 acres) and in coniferous woodland at about one pair per 2–5ha (5–12.5 acres). Long-term studies show that although the numbers of breeding pairs are relatively stable, there is a tendency to "see-saw" up and down from year to year. The most important factor is the presence or absence of beech mast. The beech tree tends to produce a rich crop of seeds at intervals of two years (or more), and breeding numbers increase after a winter when the crop has been available, and decrease when there is no crop. Birds in northern areas may emigrate ("irrupt") southward in years without a crop, but are more likely to remain on the breeding grounds in years when there is a good crop. Birds also come to bird-feeders in gardens less often when beech mast is abundant, which leads some people to conclude that tits are very scarce in years when the reverse is in fact true.

Great tits stake out their territories in late winter and early spring, taking up the best areas first so that latecomers may have to settle in marginal areas. Some birds are even excluded from obtaining a territory at all. This fact can be demonstrated by removing established territory-holders; the empty territories are usually filled by newcomers within 24 hours. The replacement pairs take up much the same areas as the original occupants, except that some occupants of adjacent territories expand their domain into the territory temporarily left vacant. If a tape-recording of Great tit song is played in the vacant territory, it apparently "fools" would-be immigrants into thinking that the area is still occupied, as reoccupation by new owners is delayed. Territorial behavior breaks down when the birds have young, when they are too busy collecting food to be able to defend their territory.

In Central Europe Great tits start to build nests in early to mid-April. The male brings food to the laying female. Each egg weighs about 10 percent of the female's bodyweight, and she may lay 10 eggs, so she needs a plentiful supply of food in addition to her normal requirements to form her own bodyweight in eggs over a period of just 10–14 days. She also needs to lay as early as possible, since young born early in the season stand a better chance of surviving than those from later nests. However, the female's requirement for food is such that she cannot always lay at the time that would be best from the chicks' point of view; instead, she has to wait until food is sufficiently abundant for her. Evidence for this is provided by the fact that, where food is put out for them just prior to laying, Great tits lay earlier than in adjacent areas where the birds are not fed. Great tits breeding in gardens also tend to lay eggs earlier than those in woodland, probably again because the food put out for them by humans enables them to do so.

In many areas of Europe, Great tits are now breeding earlier than they did in the 1970s. Global warming has resulted in the trees coming into leaf earlier than they used to, so caterpillars are present sooner, and the tits breed earlier so as to have the best chance of getting food for their young.

After 13–14 days of incubation, the eggs hatch. Now the parents have to work exceedingly hard to feed their large and hungry brood. In the first days, the female may need to brood the young to keep them warm, but once they are 4–5 days old both parents spend almost all the daylight hours bringing food to the nest. Caterpillars collected from the trees are the main item. If caterpillars are sufficiently plentiful, the parents may between them bring them at a rate of one a minute; at the height of the nestling period, 1,000 feeding visits

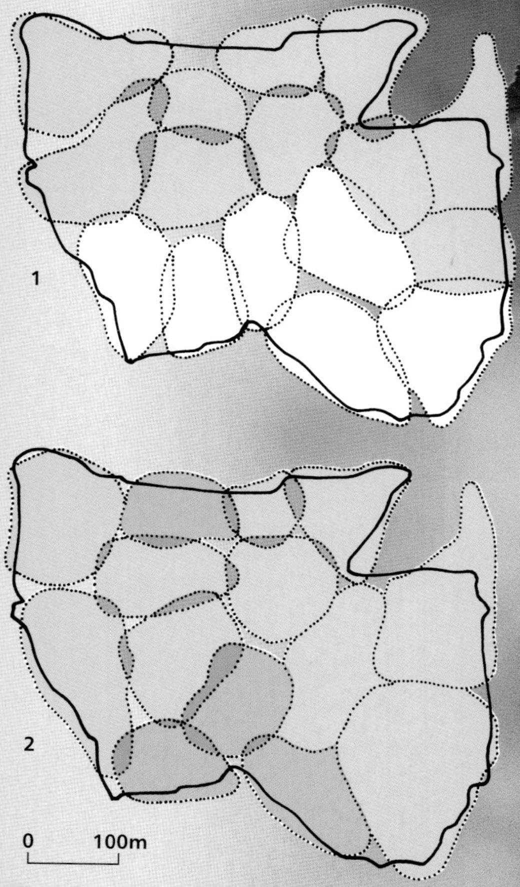

Above Rapid territorial expansion in the Great tit which tends to breed and winter in woods, urban parks, gardens, orchards and hedgerows. These maps show the distribution of Great tit territories in a small wood with an area of 18ha (45 acres). **1** Six pairs of birds were removed from their territories. **2** Within three days four new pairs had taken up residence (amber), while other residents of the wood had expanded their territories where their neighbors had been removed.

Right During severe winters when there is a shortage of its natural diet of seeds and nuts, the Great tit turns to the convenient bird-feeders that are set up by humans in their gardens.

Left In deciduous woodland, the breeding Great tit often relies heavily on one species of insect to feed its young and itself. In oak woods in spring, moth caterpillars are an abundant source of sustenance.

may be made in one 16-hour day. Even so, the young continue to beg for more.

In a large brood the young leave the nest lighter in weight than in smaller broods, showing that they have not received sufficient food to bring them to full weight. This has important consequences, as the heavier young have a better chance of surviving to the next breeding season. The higher survival rate comes about not merely due to better nutrition in the nest, but also because the heaviest chicks (and the earliest fledged) become dominant in the feeding flocks and so have the best opportunities for displacing the weaker ones in disputes over food. In many species of tit only one brood is raised each year, but the Great tit may raise two if feeding conditions are good when the first brood leaves the nest.

Few of the huge number of young produced survive. About 1 in 10 of the eggs laid, and about 1 in 6 of the newly-fledged young, survive to become breeding adults. Roughly 50 percent of adult birds survive the winter to breed the following summer. Thus, on average, about one bird per pair survives to breed, and one egg per brood survives to become a breeding adult. Of 1,000 adult birds entering their first winter in Central Europe, perhaps one will live to reach the age of 10. CMP

Long-tailed Tits

ALL SEVEN SPECIES OF LONG-TAILED TITS ARE *highly social, living in flocks of 6–12 birds for much of the year. They are very small, their tail making up perhaps half their length. Five of the species live in the Himalayan regions and in China and are not well-known. The other two, the Long-tailed tit and the bushtit, have been quite well studied.*

The Pygmy tit (*Psaltria exilis*), which lives in Java, is a tiny bird in its own genus that has been put in this family by some authors. Apart from the fact that it lives in flocks and builds nests similar in form to those of the long-tailed tits, little is known of the behavior of this species, and its relationships to the rest of the family are uncertain.

Short Bodies, Long Tails

FORM AND FUNCTION

The Long-tailed tit occurs across the whole of the Palearctic region from Ireland to Japan; it has been divided into some 19 subspecies, and shows quite varied plumage across its range. The northern subspecies that ranges from northern Europe to Japan is mostly much pinker, especially on the wings, than the others and has a striking, completely white head; other races have a broad, dark line above the eye that breaks up the white head pattern. Similarly, the bushtit, which is found in western areas of North America from southwestern British Columbia southward into central America as far as Guatemala, has a number of races. The populations that occur from Texas

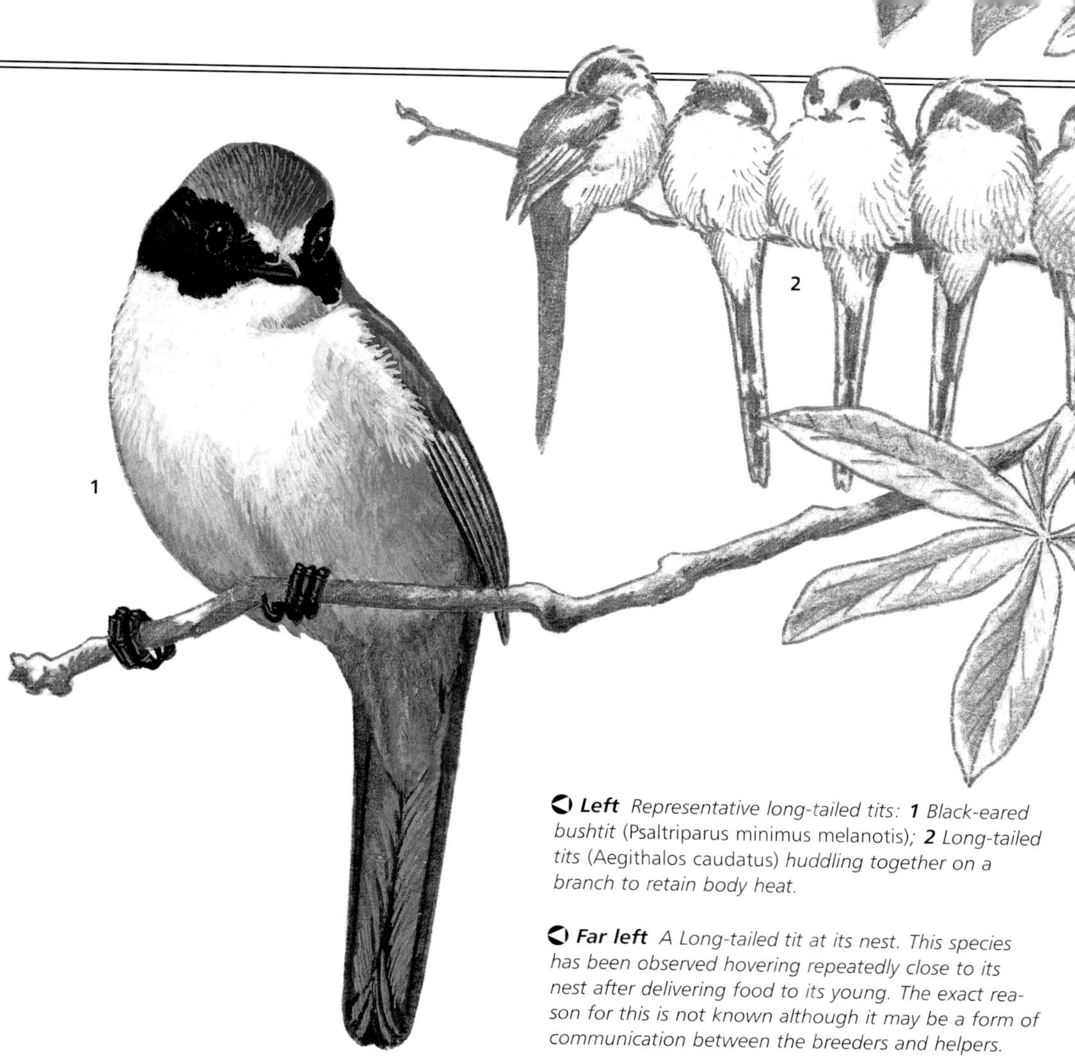

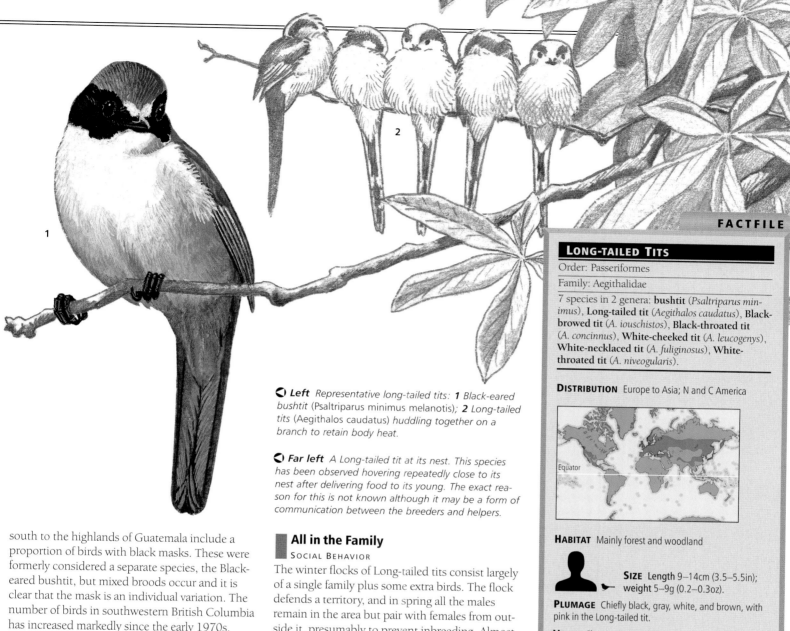

Left *Representative long-tailed tits:* ***1*** *Black-eared bushtit* (Psaltriparus minimus melanotis)*;* ***2*** *Long-tailed tits* (Aegithalos caudatus) *huddling together on a branch to retain body heat.*

Far left *A Long-tailed tit at its nest. This species has been observed hovering repeatedly close to its nest after delivering food to its young. The exact reason for this is not known although it may be a form of communication between the breeders and helpers.*

FACTFILE

LONG-TAILED TITS

Order: Passeriformes

Family: Aegithalidae

7 species in 2 genera: **bushtit** (*Psaltriparus minimus*), **Long-tailed tit** (*Aegithalos caudatus*), **Black-browed tit** (*A. iouschistos*), **Black-throated tit** (*A. concinnus*), **White-cheeked tit** (*A. leucogenys*), **White-necklaced tit** (*A. fuliginosus*), **White-throated tit** (*A. niveogularis*).

DISTRIBUTION Europe to Asia; N and C America

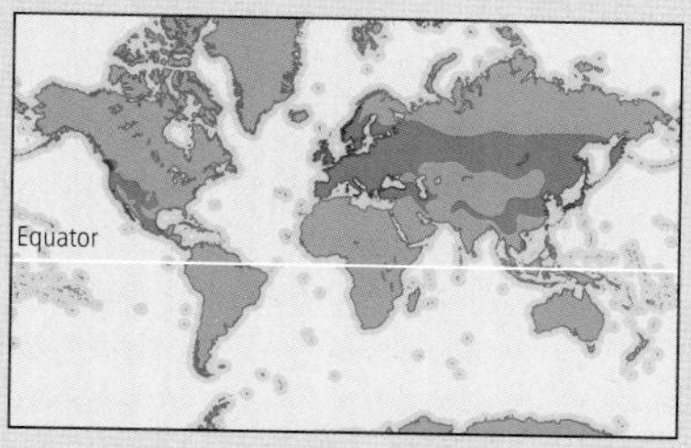

HABITAT Mainly forest and woodland

SIZE Length 9–14cm (3.5–5.5in); weight 5–9g (0.2–0.3oz).

PLUMAGE Chiefly black, gray, white, and brown, with pink in the Long-tailed tit.

VOICE Churring contact calls and subdued songs.

NEST Purselike structure of moss, feathers, and lichens.

EGGS Usually 6–10, white, speckled with red spots in many species. Incubation period 13–14 days: nestling period 16–17 days.

DIET Mainly insects

CONSERVATION STATUS Not threatened

south to the highlands of Guatemala include a proportion of birds with black masks. These were formerly considered a separate species, the Black-eared bushtit, but mixed broods occur and it is clear that the mask is an individual variation. The number of birds in southwestern British Columbia has increased markedly since the early 1970s.

Long-tailed tits weigh only some 7–9g (0.3oz), but the bushtit is even smaller, only some 4.5–6g (0.2oz). Since they maintain a body temperature of just over 40°C (104°F), they find it very difficult to store sufficient energy to survive long winter nights, especially when it is very cold. At 20°C (68°F), bushtits eat 80 percent of their own bodyweight in insects daily, and when it is colder they need even more if they are to survive. As a result, many Long-tailed tits and bushtits die in very cold weather; they find it just too difficult to gather enough food to keep their tiny bodies warm.

The birds have one trick that helps them to survive; on cold nights the flocks roost and huddle together, so reducing heat loss. A single bird, roosting on its own, uses about 25 percent more energy during the night than one of a pair huddled together, and almost certainly a single bird could not survive a cold night on its own.

Left *Bushtits usually flock in small bands and keep contact through a series of constant light "tsip" and "pit" notes. Their alarm call is a high trill.*

All in the Family

SOCIAL BEHAVIOR

The winter flocks of Long-tailed tits consist largely of a single family plus some extra birds. The flock defends a territory, and in spring all the males remain in the area but pair with females from outside it, presumably to prevent inbreeding. Almost all pairs attempt to raise their young unaided. Should they fail, however, as may happen if the nest is raided by a predator, it may be too late to be worth trying to lay a replacement clutch. In that case, the birds go and help at another nest, usually one belonging to a relative. In the case of males such a nest is easy to find, because the males at adjacent nests are likely to be their brothers. For females, however, it usually means going back to their winter territory, where their relatives are to be found. In this way the helpers all assist at nests where the parent birds share their genes.

The Long-tailed tit builds an elaborate, purselike nest of feathers and moss – more than 2,000 feathers have been counted in a single nest. The beautifully constructed structure, which is bound together with spider's web and camouflaged with a covering of lichen, takes many days to complete; it may be 18cm (7in) or so deep in Long-tailed tits and up to 30cm (12in) or more in the bushtit.

The two genera construct their nests slightly differently. In the Long-tailed tit, the birds find a suitable branch or grouping of twigs that can serve as the foundation for the nest. At first the structure is the conventional cup shape, but the parents gradually build up the sides until, reaching up as far as they can, they close in the top to complete a dome. In contrast, bushtits' nests are initially less well-supported at the bottom, and gradually, by trampling the base progressively lower while adding more material to the sides, the birds end up with a hanging nest. In both species, both members of the pair start to roost in the nest at night once it has a dome. Probably only the female incubates, since it is often possible to see pairs of birds in which one has a bent tail from sitting in the tiny nest.

CMP

Swallows

ALMOST EVERYBODY IS FOND OF SWALLOWS, thanks to their supreme powers of flight, their attractive plumage, their symbolic role as harbingers of the arrival of summer, the fact that they eat insects, and their penchant for nesting in close proximity to humans.

The most recent taxonomic review of the Hirundinidae family recognizes 89 species in 14 genera. However, an accurate assessment is made difficult by the morphological constraints imposed by an aerial lifestyle, and some taxonomists recognize more genera. In a few cases species limits are also debated, and some subspecies, such as the African and Asian forms of the Wire-tailed swallow, may be valid species in their own right. The terms "swallow" and "martin" are not used consistently with respect to taxonomy, although in general martins have shorter tails.

Spectacular Fliers

FORM AND FUNCTION

Swallows are easily recognized by their slender bodies combined with long, narrow, pointed wings and forked tails, the outer feathers of which are often elongated to form streamers. These features, which help them to pursue a highly specialized lifestyle feeding on airborne invertebrates, are shared with other, unrelated species with a similar way of life such as the swifts.

The slender body reduces the amount of drag (wind resistance) experienced. The wingshape has a high aspect ratio, which means that a large amount of lift is generated with little drag. Such aerodynamic efficiency comes at the cost of reduced maneuverability, in comparison with short and broad wings, but this weakness is partly compensated for by the forked tail, which increases maneuverability.

Some species have tail streamers that increase the amount of lift; the streamers act like airplane flaps, maintaining a smooth flow of air over the wings and so delaying the point at which the bird would stall without increasing the drag. Most species have short tarsi and small, weak feet that are adapted to perching rather than walking, although species that excavate their own burrows or nest on cliff faces have strong claws.

The *Pseudochelidon* river martins are exceptions to this generalized morphology; they more closely resemble other passerines, and may be an ancestral link between other passerines and swallows. The legs and feet of river martins are larger, and the associated musculature is less reduced in terms of size, number of muscles, and complexity. Their bills are stouter and thicker than the wide, flattened bills of other swallows, and the syrinx has much less complete bronchial rings. The rough-winged swallows of the genera *Stelgidopteryx* and *Psalidoprocne* have a series of barbules on the outer edge of the outer primary; these produce a hooklike thickening, the function of which is unknown.

Great Migrators

DISTRIBUTION PATTERNS

Swallows occur on all continents except Antarctica, and are also absent from northern polar regions and a few remote oceanic islands. Africa, where the family probably evolved, has the largest number of endemic breeding species (29), followed by South and Central America with 21. A few species breed in more than one continent; for example, the Sand martin breeds in Europe, Asia, North America, and locally in north Africa.

FACTFILE

SWALLOWS

Order: Passeriformes

Family: Hirundinidae

89 species in 14 genera. Species and genera include: **Barn swallow** (*Hirundo rustica*), **Blue swallow** (*H. atrocaerulea*), **Cave swallow** (*H. fulva*), **Cliff swallow** (*H. pyrrhonota*), **Ethiopian swallow** (*H. aethiopica*), **Grey-rumped swallow** (*H. griseopyga*), **Red Sea swallow** (*H. perdita*), **Rufous-chested swallow** (*H. semirufa*), **White-tailed swallow** (*H. megaensis*), **Wire-tailed swallow** (*H. smithii*), **Sand martin** (*Riparia riparia*), **Bahama swallow** (*Tachycineta cyaneoviridis*), **Golden swallow** (*T. euchrysea*), **Mangrove swallow** (*T. albilinea*), **Tree swallow** (*T. bicolor*), **White-thighed swallow** (*Neochelidon tibialis*), **river martins** (genus *Pseudochelidon*) including **White-eyed river martin** (*P. sirintarae*), **New World martins** (genus *Progne*) including **Purple martin** (*P. subis*), **rough-winged swallows** (genera *Stelgidopteryx* and *Psalidoprocne*).

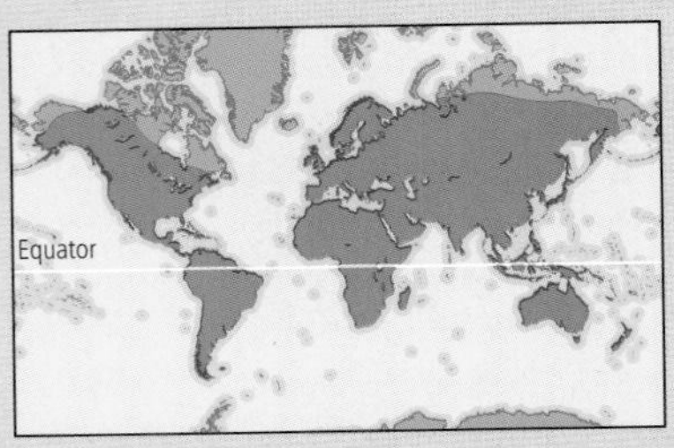

DISTRIBUTION Worldwide except the Arctic, Antarctic, and some remote islands.

HABITAT All open areas including bodies of water, mountains, deserts, and above forest canopies.

SIZE Length typically 15cm (6in); weight 20g (0.7oz); however measurements range from 10 to 24cm (4–9.5in) and from 10 to 60g (0.3–2.1oz).

PLUMAGE Upperparts are typically metallic blue-black, green-black, or brown; some species have a contrasting rump color. Underparts are normally paler (often white, buff, or chestnut). Sexual differences are usually slight, but males are sometimes brighter and have longer tails than females. Juveniles are often duller and have shorter tails than adults.

VOICE A simple, rapid twittering or buzzing song, usually consisting of a longer and more varied sequence of the birds' call notes.

NEST Mud nests (open or enclosed), or a simple cup made of vegetation. Mud nests are usually attached to buildings, cliff faces, or placed in caves. Natural holes (e.g. tree cavities) and burrows (often self-excavated) are also frequently used.

EGGS 4–5 in most temperate species (up to 8 in some) and 2–3 in most tropical species; usually white, sometimes spotted with red, brown, or gray. Incubation averages 14–16 days (range 11–20), but may be extended in bad weather; nestling period 16–24 days, but 24–28 days in the larger New World martins. When food is scarce the nestling phase is extended, especially in those species whose nestlings exhibit torpor in bad weather.

DIET Almost exclusively airborne invertebrates.

CONSERVATION STATUS The White-eyed river martin, known only from one area of central Thailand, is listed as Critically Endangered, but has not been seen for over 20 years and may be extinct. 4 other species are Vulnerable.

Above *Representative swallow species:* ***1*** *Bank swallow or Sand martin* (Riparia riparia), *a colonial nester.* ***2*** *Southern rough-winged swallow* (Stelgidopteryx ruficollis), *from central and South America.* ***3*** *Blue swallow* (Hirundo atrocaerulea), *listed as Vulnerable.* ***4*** *Tree swallow* (Tachycineta bicolor).

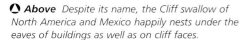

Above *Despite its name, the Cliff swallow of North America and Mexico happily nests under the eaves of buildings as well as on cliff faces.*

In temperate regions the birds' food supply is severely reduced during the winter months, so many species are migratory. Unlike most passerine migrants, hirundines travel by day and at low altitude; they also regularly feed en route, and thus have lower levels of fat reserves than is typical for migrants of their size. African breeders often migrate with the rains, but swallow movements are poorly understood, and other species like the Grey-rumped swallow appear to be opportunistic nomads with no set migration route.

Many species have expanded their distributions in recent years, as the birds' increasing use of buildings for nest-sites has introduced them to localities where they were previously unknown. For example, the Ethiopian swallow has expanded south into Kenya and Tanzania, and the Cave swallow has spread from Mexico into the southern USA. Environmental change has also caused distribution patterns to shift. British populations of Barn swallows winter in South Africa, where their range has expanded to the west in response to increased rainfall in this region.

Hunting Insects on the Wing

DIET

All swallows feed almost exclusively on airborne invertebrates, mainly insects. Plant material has been found in the diet of a few species, but only rarely. Only the Tree swallow consumes plant material, mainly berries, at all regularly, and this only at times when insects are in short supply.

Swallows are not opportunistic feeders, aimlessly flying around with their mouths open consuming the aerial plankton; instead, they actively hunt specific items. Sympatric species tend to specialize on different size classes of invertebrates. Individual birds of a particular species will often select the largest prey items available. Migratory species usually switch their diet between wintering and breeding areas; for example, the Barn swallow increases the proportion of ants in its diet

when in Africa. A species' preferred foraging height may also change between wintering and breeding areas. These changes are thought to be the result of competition with resident species on the wintering grounds.

Above *Two almost-fledged Barn swallows look out from a cup of mud and straw lined with feathers; as the birds' name suggests, the nests are usually placed on buildings. These sparrow-sized birds are the USA's most widespread swallow species, and the only one to have a deeply-forked tail.*

Monogamous for the Most Part

BREEDING BIOLOGY

Social monogamy is the rule in swallows, with both sexes exhibiting parental care, although nest-building and incubation are often largely performed by the female. However, males are often promiscuous, actively seeking copulations with females other than their mates. In some species like the Purple martin, polygyny may be exhibited, with the same male being paired to two females.

While nest-sites are defended against conspecifics, only a few species such as the Mangrove swallow defend exclusive feeding territories; this behavior is probably a consequence of the ephemeral nature of the birds' food supply. The lack of defended territories therefore allows colonies to form. These can be very large in some species (up to 4,000 pairs in the Cliff swallow); other species typically nest in much smaller, loose aggregations or sometimes singularly (for example, the Barn swallow), while others like the Blue swallow are obligate solitary nesters.

The reasons for interspecific variation in colony size are poorly understood, but intraspecific studies reveal that colonial nesting has both advantages and disadvantages. Colonies may act as "information centers," in which individuals benefit from other birds' knowledge of the location of high-quality feeding grounds. Such a situation exists in Cliff swallow colonies but not in all colonial hirundines (for example Barn swallows). All colonial nesting species probably benefit from increased rates of predator detection, a reduced probability of predation (probability theory implies that in a large group the chance of any one individual being taken is reduced), and more effective mobbing of predators. Within a colony, high-quality individuals may benefit from increased extra-pair copulations, resulting in more offspring, while very low-quality individuals may have their best chance of reproducing by gaining "sneaky copulations."

The disadvantages of colonial nesting include increased rates of parasite infection and disease (due to the high population density), an increased probability of being cuckolded, the need to invest more time in nest defense, and, when there is no information-center effect, a decreased per-capita food supply. Some studies have found that, even in species such as Barn swallows that exhibit colonial nesting, no net benefit accrues from the trait. In these cases, colonies may simply form because young birds choose to nest where there is evidence of successful breeding and because adults are site-faithful.

The site-fidelity of adults provides the opportunity for nest reuse, which is common in species that build mud-nests. Species that nest in burrows very rarely reuse them, even for second broods, possibly because the probability of collapse is too great. A pair may also decide not to reuse a nest if the levels of parasite infection within it are high; this factor can cause formerly large colonies to suddenly be deserted, a frequent phenomenon in Cliff swallow colonies.

In colonial species there is intense competition between males for mates, and this fact has led to a number of sexually selected traits. A well-studied example is the Barn swallow; females prefer long-tailed males, and this sexual selection pressure has extended tail length beyond the optimum for aerodynamic properties. There is nonetheless a logic to the preference, as long tails are a signal of male quality, exhibiting a positive correlation with survival probability. Although testosterone (the steroid hormone responsible for tail growth) is at high levels an immunosuppressant, a long tail nevertheless acts as an indicator of the efficiency of a bird's immune system, because the ability to grow one demonstrates unequivocally that the bird has sufficient steroids both to resist infection and to support the extra tail growth.

Females also select males with symmetrical tails. Aerodynamic models of swallow flight show that symmetry improves flight performance – a fact that may explain why adult survival rates are positively correlated to tail symmetry. For example, male Barn swallows with very symmetrical tails are less likely to be caught by sparrowhawks (*Accipiter nisus*) than those with asymmetrical tails. The degree of symmetry may also be an honest indicator of the amount of parasite infection. Barn swallows have a series of white spots on the outer tail feathers that are sexually selected and act as an indicator of individual quality.

A male's nest-building ability is also sexually selected. Females that select males based on their nest-building ability also gain high levels of paternal investment in the offspring. Females paired to high-quality males, as assessed by tail length and symmetry, actually suffer from a reduced level of paternal care, but gain other benefits instead, most of them arising from the offspring inheriting "good" genes. The offspring will benefit by inheriting a high resistance to parasites as well as inheriting their father's sexiness and associated high reproductive rate (this is known as the "sexy-son hypothesis"). There is also a reduced probability of the partner dying during the breeding season.

Species at Risk

CONSERVATION AND ENVIRONMENT

Six swallow species are threatened with extinction. The Bahama swallow has a small and fragmented range that has been reduced by logging. Introduced predators are probably reducing breeding success, while House sparrows and starlings now compete for nest sites. Renewed logging and housing development are expected to result in rapid declines in the near future. The Golden swallow occurs in Haiti, Jamaica, and the Dominican Republic. Its range and population has declined considerably since the 19th century, probably as a result of loss of its breeding habitats (humid montane and pine forests).

The Blue swallow is a migratory species threatened by destruction and degradation of its grassland habitat in both its breeding and wintering areas. The main threats in its case are afforestation, intensive grazing, grass burning, and invasive non-native trees and bracken. The White-tailed swallow has a very restricted range in southern Ethiopia. Little is known about its ecology, but it is thought to be at risk from development and the clearance of *Acacia* thorn scrub. The White-eyed river martin and the Red Sea swallow are both so poorly known that their conservation status remains in doubt (see Mystery Swallows). KE/AKT

Above *A male Violet-green swallow* (Tachycineta thalassina) *brings food back to the nest. These birds, which breed from Alaska south to the Mexican border, often nest in woodpecker holes, although they also use nest-boxes and the eaves of houses.*

Below *A pair of Lesser striped swallows* (Hirundo abyssinica) *perch on branches in the Pilanesburg National Park, South Africa.These birds are widely distributed across much of the continent from the Cape up to Ethiopia.*

MYSTERY SWALLOWS

In May 1984 the decomposing remains of a swallow were found at a lighthouse on the Red Sea coast of Sudan. The bird turned out to be new to science, and remains the only individual of its species that has been seen. Its Latin name, *Hirundo perdita*, means "lost swallow," reflecting the fact that the individual probably wandered from its usual range (the species is thought to breed in the hills of Sudan or Ethiopia). Similar birds have since been observed in Ethiopia on five occasions, but slight differences in plumage suggest that these may represent yet another species awaiting formal scientific description.

The White-eyed river martin was discovered in a large roost of swallows at a reservoir in central Thailand in January 1968. People in the region catch birds for sale in local markets, either as food or for Buddhists to release (a meritorious action believed to secure good karma). In the first few years after the bird's discovery, over 100 individuals were caught, but since then only a few birds have been seen at the lake and in markets, and no sightings have been reliably reported since 1980. The bird's breeding grounds are probably along northern Thai rivers (or possibly in China), but an expedition in 1969 failed to locate them. The population level must be very small, and the species may be on the verge of extinction.

Bulbuls

BULBULS ARE A WELL-DEFINED GROUP OF small birds that vary only from sparrow- to thrush-sized. The long, oval nostrils have a thin sheet of bone at the back that is usually pierced by one or two small nerve canals, and is their most definitive feature. More obvious characteristics are the fine, hairlike feathers that protrude from the nape (sufficiently long in some species to form a distinct crest), and the long, fluffy feathers on the back.

Bulbuls are mainly forest-dwelling birds, but a number of species prefer the forest edge, and some have adapted to drier woodlands, to savanna, and to cultivated areas. All exist in the Old World, and they are often the commonest, or at least the most frequently heard, birds in the Afrotropical (including the Malagasy) and the Oriental regions. The name "bulbul" is Arabian in origin and is probably imitative of the bird's calls, but is also used as a common name for the family in Iraq and India.

Active, Alert, and Frequently Heard

FORM AND FUNCTION

In addition to the distinctive bone and feather features already mentioned, bulbuls can also be recognized by their short wings, more curved from front to back than in most other birds and each with 10 primary feathers, of which the outermost is much shorter than the rest. The tail of 12 feathers is medium in length and may be square or slightly rounded, forked or graduated.

Bulbuls have a short neck, and in most species the bill is relatively slender, slightly downcurved, weakly notched or hooked, and with very well-developed bristles around the gape. Two seed- and insect-eating species of Asia, the finch-billed bulbuls in the genus *Spizixos*, are notable exceptions that also have a forward-pointing crest. The legs of bulbuls are short, and the rather weak feet that support the slender body are suited mainly to an arboreal rather than a terrestrial existence, even though several forest species feed mainly while hopping on the ground or clinging to bark and lichen.

The plumage of bulbuls is generally dull green, yellow, olive, brown, or gray, and the bill and legs are brown or black. The upperparts are usually plainer and darker than the underparts, and often the tail is more rufous than the back. Only in a few species are the underparts, vent, face, eye rings, or iris more brightly colored, in red, white, or yellow. In some species there are black, white, or yellow markings on the face or underparts, usually in the form of patches, streaks, or bars, and in others the black crown feathers are elongated to form a dense (and, in a few species, a long) pointed crest. Most attractive are the *Hypsipetes* bulbuls of Southeast Asia and various Indian Ocean islands; some are black and gray with white heads, or else all dark like the Black bulbul, and almost all have a red bill and legs.

Bulbuls have been allied to the cuckoo-shrikes of the family Campephagidae through their voluminous rump feathers, as well as to drongos (Dicruridae), frugivorous orioles (Oriolidae), starlings (Sturnidae), and, most recently and emphatically though DNA analysis, to African warblers (Cisticolidae); by anatomy, they are linked to the very similar nectar-eating *Chloropsis* leafbirds of the family Irenidae. It is not clear whether bulbuls evolved in Africa (which currently has 66 species in 13 genera) or in Asia (with 63 species in 10). Even the ten Malagasy species share one genus with each of these continental regions, although this and various other evidence generally supports an Asian origin.

Only two genera occur in both Asia and Africa. The 40 species in the genus *Pycnonotus* are the most widespread and best-known bulbuls in both continents; Asia is home to 34 species, among

Left *In South Africa, a Black-fronted bulbul* (Pycnonotus nigricans) *visits an aloe flower. The birds forage on fruit and nectar, eating the equivalent of a quarter of their bodyweight a day.*

them the common Black-crested bulbul, while Africa has six, including the Common bulbul. Many species occupy forest edge and open woodlands, along with the two closely allied finch-billed bulbuls of Asia; a few species extend to savanna and even into semi-desert scrub. These are the bulbuls of dark crests, red faces, red, yellow, or white vents, and a reputation for finding and mobbing potential predators such as snakes and owls. Their cheerful song, flexible behavior, and confiding nature has made several of them popular cage birds, a practice that has led to their introduction into other parts of the world, either intentionally or accidentally. Two species have adapted particularly well: the Red-whiskered bulbul has established itself in Florida, the south of the Malayan peninsula, southeastern Australia, the Comoros, Mauritius, Réunion, Singapore, and the Nicobar and Hawaiian islands, while the Red-vented bulbul has been successfully introduced to several Pacific islands.

The attractive bulbuls of *Hypsipetes* – the other genus shared by Africa and Asia – occur in Africa only on the Malagasy islands adjacent to the continent, where they include the largest of all bulbuls, the Seychelles bulbul. The distribution and diversity of species in both these genera once again supports an Asian origin for the family.

The forest bulbuls of Africa occur in three main genera: the greenbuls and brownbuls of *Chlorocichla* (6 species), *Andropadus* (16 species), and *Phyllastrephus* (24 species); within each genus,

Above *The Red-whiskered bulbul's attractive appearance has made it a popular cage bird, and has led to its introduction from its Southeast Asian homeland to far-flung parts of the world, including Florida.*

individual species are often so similar in appearance that they are hard to tell apart, although their calls are usually distinctive. They include some of the smallest bulbuls, in particular the Tiny greenbul, as well as some of the commonest birds in the forest, including the Yellow-whiskered and Little greenbuls in Gabon, and they all succeed in subdividing the forest by details of foraging location and habitat. As many as 22 species may co-occur in the rain forests of west and central Africa, feeding at different elevations or catching food in different ways, such as searching under bark, turning over leaves on the forest floor, plucking off fruit, or snatching insects. Similar species from less speciose, even monotypic, genera also occupy these forests. These include the Swamp greenbul, the Spotted and Golden greenbuls, three species of *Bleda* bristlebill, five species of bearded *Criniger* greenbuls, and, in particular, two species of *Baeopogon* greenbuls that mimic honeyguides (Indicatoridae) with their white outer tail feathers.

At least four small genera of Afrotropical birds are included with bulbuls, even though they do not have the bony nasal ridge and may not belong in the family at all. The Black-collared bulbul of the west-central African savannas might be the most primitive bulbul. Its calls, general behavior,

FACTFILE

BULBULS

Order: Passeriformes

Family: Pycnonotidae

140 species in 23 genera. Species include: **Black-collared bulbul** (*Neolestes torquatus*), **Black bulbul** (*H. madagascariensis*), **Seychelles bulbul** (*H. crassirostris*), **Brown-eared bulbul** (*Ixos amaurotis*), **Common** or **Black-eyed bulbul** (*Pycnonotus barbatus*), **Red-vented bulbul** (*P. cafer*), **Red-whiskered bulbul** (*P. jocosus*), **Black-fronted** or **Red-eyed bulbul** (*P. nigricans*), **Black-crested bulbul** (*P. melanicterus*), **Yellow-browed oxylabes** (*Crossleyia xanthophrys*), **dapple-throat** (*Arcanator orostruthus*), **Spotted greenbul** (*Ixonotus guttatus*), **Golden greenbul** (*Calyptocichla serina*), **Swamp greenbul** (*Thescelocichla leucopleura*), **leaf-love** (*Phyllastrephus scandens*), **Tiny greenbul** (*P. debilis*), **Terrestrial brownbul** (*P. terrestris*), **Yellow-whiskered greenbul** (*Andropadus latirostris*), **Little greenbul** (*A. virens*), **Yellow-bellied greenbul** (*Chlorocichla flaviventris*), **Yellow-spotted nicator** (*Nicator chloris*), **malia** (*Malia grata*), **Hook-billed bulbul** (*Setornis criniger*), **Ashy bulbul** (*Hemixos flavala*),**Chestnut bulbul** (*H. castanonotus*), **Hairy-backed bulbul** (*Tricholestes criniger*).

DISTRIBUTION From Africa E to India, SE Asia, China, Japan, Malaysia, the Philippines, and Indonesia; introduced elsewhere, especially to islands.

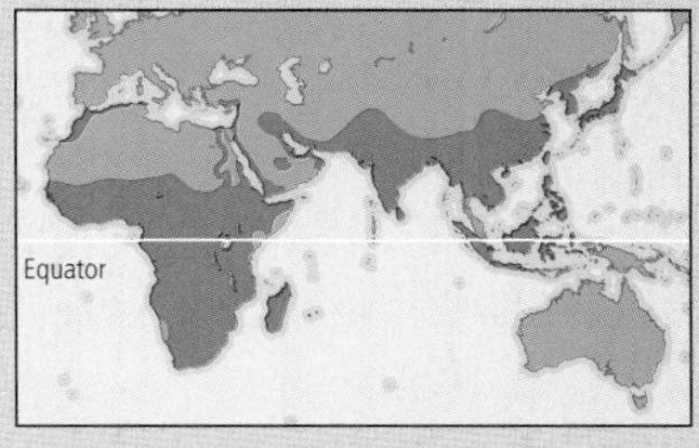

HABITAT Mainly forest, woodlands, and thickets, a few in scrub; several species adapted to rural cultivation and suburban areas.

SIZE Length 13–28cm (5–11in); weight 12–80g (0.4–2.8oz). Sexes similar in size, except some greenbul females much smaller than males.

PLUMAGE Dull brown, gray, or green, rarely black. A few with black and/or white on head, several with bright patches of red, white, or yellow on the undertail coverts or, less often, on the head. Sexes similar in appearance, but females and juveniles sometimes duller.

VOICE Wide range of single and double notes, whistles, and husky chattering; many species are noisy and best identified by call.

NEST A neat, shallow cup in the fork of a tree or bush, usually substantial, built of twigs, leaves, fungi, spiders' webs, and other materials; lined with fine roots or grass.

EGGS 1–5, pink or white, usually attractively blotched with various shades of purple, brown, or red. Incubation 11–14 days; nestling period about 11–18 days.

DIET Fruit, berries, buds, nectar, pollen, and arthropods; even birds' eggs and beeswax. Some species mainly frugivorous, others more insectivorous.

CONSERVATION STATUS 1 species – the Liberian greenbul (*Phyllastrephus leucolepis*) – is Critically Endangered; 2 are Endangered, and 10 are Vulnerable.

Above *Restricted to the Seychelles archipelago, the Seychelles bulbul is the largest of all the bulbuls, about the size of a mockingbird or a large blackbird.*

and DNA suggest that it is more bulbul-like than the bush-shrikes (Malaconotidae) that it otherwise resembles. Three other shrike-like species, the most widespread of which is the Yellow-spotted nicator, have unusually explosive calls and build a flimsy nest, often secured in a tree fork by a *Marasmius* fungus. The Yellow-browed oxylabes of Madagascar may have its real affinites with Asian babblers, while the rare dapple-throat from Tanzania may be more robin- or babbler-like. In similar vein, the malia from Sulawesi probably belongs best with the Asian babblers that it so resembles, and five Madagascan species currently included in *Phyllastrephus* probably deserve their own genus *Bernieria*, and may also be babblers.

There is not so great a diversity of bulbul genera in Asian evergreen forests as in Africa, since several *Pynconotus* species occupy both forest and more open habitats. The Asian forest species are often as dull-colored and confusing to identify as African species, in particular the puff-throated *Alophoixus* (7 species, sometimes united with the African *Criniger*), the buff-vented *Iole* (4 species), and the streaky *Ixos* (7 species). Not so many species co-occur in each forest type or area, but there are once again a few special genera that swell the bulbul numbers, such as the Hook-billed and Hairy-backed bulbuls and the Ashy and Chestnut bulbuls of the genus *Hemixos*. If the eight species of brilliant green and blue leafbirds (but not the ioras or fairy-bluebirds) are indeed just nectivorous bulbuls, as anatomy suggests, then the ranks of Asian bulbuls would be swelled even further.

Most open-country bulbuls are active, alert, noisy, gregarious birds, full of character and movement. Some forest bulbuls attract attention by special foraging movements, such as flicking a wing or tail, but many are cryptic as they move through the foliage and undergrowth. Bulbuls often feed in flocks with other species, usually as followers rather than leaders, but they are almost always the first to give warning of a potential predator, and they often attract the attention of naturalists to rarer birds such as owls. Those

Above *Representative bulbul species:* **1** *Crested finchbill* (Spizixos canifrons), *from mountainous regions of southern Asia.* **2** *Black bulbul* (Hypsipetes madagascariensis). **3** *Common or Black-eyed bulbul* (Pycnonotus barbatus), *the most widespread African species.*

species that feed on or near the ground often follow columns of army ants for the insects they disturb, while some of the more arboreal species follow monkeys and squirrels.

As one would expect of birds with such short, rounded wings, bulbuls are feeble flyers, poor at colonizing islands and not very migratory. A few species migrate from one altitude to another, especially in Asia where they occur from sea level to 3,000m (10,000ft) up in the Himalayas. Only one species, the Brown-eared bulbul that occurs in Japan and northeastern China, appears to be a true migrant, and even then it is only its more northerly populations that migrate, wintering as far south as Korea and moving by day, often in large flocks of up to a thousand birds.

Leks, Pairs, and Cooperative Groups

SOCIAL BEHAVIOR

The nesting of bulbuls is relatively straightforward, and their nuptial displays make the most of their distinctive features. The male Red-vented bulbul, whose mating display is typical of the many *Pycnonotus* species, depresses and spreads its tail laterally to show off the bright crimson undertail coverts while fluttering its spread wings up and down above its head. In the frugivorous Yellow-whiskered greenbul of Africa, the normally solitary males gather to sing in social leks, which females visit for mating but thereafter ignore as they rear the young alone. Several other bulbuls are known to breed in cooperative groups, among them the Swamp, Yellow-breasted, and Spotted greenbuls and the Terrestrial brownbul, and this behavior may well be more widespread given the number of species that normally live in cohesive social groups.

Most species that are not social live as territorial pairs, at least for the duration of the breeding season. Both sexes usually take part in building the nest, which is located in the fork of a tree or a bush, often in an open, poorly concealed location. The nest is a simple but substantial cup, loosely built of plant materials through which rainwater can easily drain away, and sometimes bound together by fungi. Most nests are situated relatively low down, at a height of 0.5–9m (5–30ft), although nests of the Black and Seychelles bulbuls have been recorded at over 15m (50ft). Many nests are found by predators, including snakes, cats, crows, and lizards, and a number of bulbul species are parasitized by different species of cuckoos. Both parents take turns to incubate the eggs, which in many species are extremely beautiful, often having unusually thick shells for birds of their size. Both parents feed the chicks once they hatch, and often more than one brood is reared in a year. Despite the birds' superficially dull appearance, further study of the feeding and breeding biology of bulbuls is expected to reveal considerable diversity, as well as many new adaptations. AK/CW

Above *The Yellow-bellied greenbul belongs to the genus* Chlorocichla *– fruit-eating, forest-dwelling African bulbuls. Most are shy and secretive birds, the exception being the Joyful greenbul* (C. laetissima), *which is often seen in noisy flocks.*

Above *Seen here in Singapore, where it is a familiar visitor to parks and gardens, the Yellow-vented bulbul* (Pycnonotus goiavier) *is a common and widespread Southeast Asian species.*

White-eyes

WHITE-EYES – SMALL, GREENISH BIRDS WITH white eye-rings – have short, pointed bills and brush-tipped tongues with which to collect nectar. They also hunt insects and spiders by gleaning foliage, probing into small crevices, and hawking.

The birds forage in gardens and at forest edges, and flock around bird tables. They appear in orchards, and eat fruits as well as aphids. They have versatile feeding habits and exploit a variety of resources to survive, even managing to breed on small wooded islands where most other passerines fail to establish themselves.

From Japan to the Cape of Good Hope

DISTRIBUTION PATTERNS

White-eyes are found in parts of Africa, Asia, New Guinea, Australia, New Zealand, and on various West and South Pacific islands. Some continental white-eyes migrate regularly in winter to lower latitudes, although part of the population often remains resident in the colder region. As flocking white-eyes are easy to trap or mist-net, they have become the most popular species among the bird-banders of southern Africa, Australia, and New Zealand, providing information on molt, weight change, movements, longevity, and the survival of local populations. In Australia, 285,345 silvereyes were banded between 1953 and 1997, of which 35,889 (12.6 percent)were later recovered.

White-eyes also disperse in flocks to remote islands. In 1832, silvereyes from Tasmania colonized New Zealand across 2,000km (1,250 miles) of sea. The Maori, who had names for all the native birds in New Zealand, named the species *tauhou*, meaning "stranger."

Isolated populations on oceanic islands and high mountains – for example, the Black-capped speirops on the mountains of São Tomé in the Gulf of Guinea – have differentiated into new forms by growing in size and/or by losing certain pigments from their plumage. Such has been the case with the Grey white-eye in the East Caroline Islands, while the Mountain blackeye in the mountains of north Borneo has even dispensed with the family's distinctive white eye-ring.

As such differentiations take place over a relatively short geological timescale, successive invasions of original stock have led to the present coexistence of two or three species on some islands. Yet the similarities between some distant species, resulting from convergence, are so remarkable that only molecular studies can establish their true affinities.

Partners for Life

BREEDING BIOLOGY

Most white-eyes pair for life and breed in small territories. The two sexes are alike, but call-notes have sexual differences that appear to facilitate mate selection. Members of a pair often perch together and preen each other (allopreening). Males sing a complex song at dawn throughout the breeding season.

The same ritualized form of aggression that the birds exhibit in territorial defense is used in winter fighting over food. Aggressive birds flutter their wings at their opponents; equally matched birds may take to the air to fight, following a bout of mutual display. Sometimes they will supplant a feeding bird, or attack and chase an approaching bird with beak clatter or challenge calls. Within a flock a social hierarchy is maintained, and the dominant bird establishes the right of access to concentrated food sources such as ripe figs by giving signals of intended attack.

Endangered on the Islands

CONSERVATION AND ENVIRONMENT

Some island species are in danger of extinction, as their population size is small and their available habitats are being destroyed through local development. Predation by introduced animals also constitutes a big threat to vulnerable species. For example, the Robust white-eye, which was limited to Norfolk Island, became extinct within 10 years of the arrival of the Black rat in 1918, and on the same island today the largest member of the genus, the White-chested white-eye, is itself at risk. Similar threats underlie the Endangered status of the Mauritius olive white-eye. On Saipan, once-abundant Golden white-eyes have become Endangered as a result of the recent introduction of the Brown tree snake (*Boiga irregularis*). Although white-eyes are considered pests by orchard-keepers, they also consume large quantities of pest insects wherever they occur. On the debit side, in Australia they are known to spread environmental weeds, such as bridal creeper and lantana.

White-eyes seem to have developed immunity against arthropod-borne diseases. Both a pox virus and avian malaria hematozoa have been isolated from silvereyes in New Zealand. On Hawaiian islands where native honeycreepers of the Drepanidinae subfamily disappeared from mosquito-infested lowlands, Japanese white-eyes that had been introduced in the 1920s and 1930s from mainland Japan spread without apparent distress from malaria. JK

Right A Chestnut-flanked white-eye (Zosterops erythropleurus) *exhibits the distinctive eye-ring that gives the birds their name.*

FACTFILE

WHITE-EYES

Order: Passeriformes

Family: Zosteropidae

88 species in 12 genera. Species include: **Black-capped speirops** (*Speirops lugubris*), **Grey white-eye** (*Zosterops cinereus*), **Japanese white-eye** (*Z. japonicus*), **silvereye** (*Z. lateralis*), **Mauritius olive white-eye** (*Z. chloronothos*), **Golden white-eye** (*Cleptornis marchei*), **Mountain blackeye** (*Chlorocharis emiliae*).

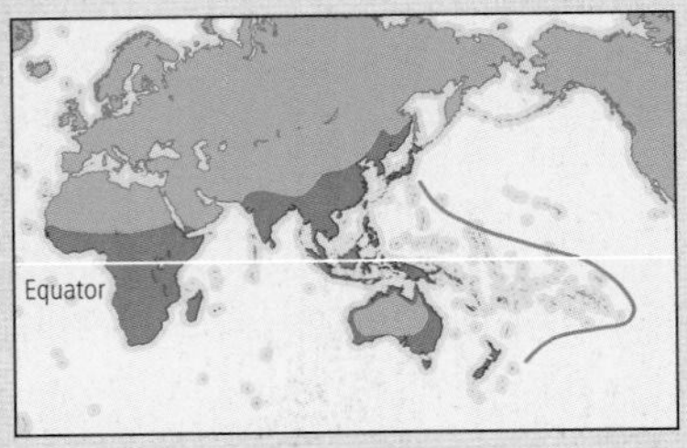

DISTRIBUTION Africa, Asia, New Guinea, Australia, Oceania (introduced to Hawaii and Tahiti).

HABITAT Woodland, forest, gardens.

SIZE Length 10–14cm (4–5.5in); weight 8–31g (0.3–1.1oz). In some species the females are smaller than the males.

PLUMAGE Greenish with yellow, gray, white, and brown parts; most species have a conspicuous white ring round the eye. In some species the males are more brightly-colored than the females.

VOICE Males produce a rich, warbling song at dawn; a high-pitched plaintive note is produced for keeping contact across long distances. Other distinct notes are used in alarm or distress and for courtship; wing flutter and beak clatter are used in aggression.

NEST Cup-shaped, slung in a tree fork under cover.

EGGS 2–4, whitish or pale blue without spots (4 species with spotted eggs); size 1 x 1.4cm–1.5 x 2cm (0.4 x 0.6in–0.6 x 0.8in). Incubation period 10–12 days; nestling period 11–13 days; fledgling period 2 weeks.

DIET Insects, spiders, nectar, berries; fruits in winter.

CONSERVATION STATUS 6 species, including the White-chested white-eye (*Zosterops albogularis*), are currently listed as Critically Endangered, 5 as Endangered, and 10 as Vulnerable. The Robust white-eye (*Z. strenuus*), which lived on Lord Howe Island off Australia, became extinct in the 1920s after the Black rat was introduced to the island.

Old World Warblers

THE OLD WORLD WARBLERS ARE ONE OF THOSE groups that really test the birdwatcher. Many of the birds form groups of very similar-looking species that are puzzling to beginner and experienced birdwatcher alike, but can be spectacularly different in their calls. They are generally small birds that tend to remain hidden in dense vegetation and only emerge briefly to provide glimpses before disappearing again in their search for their favored insect prey. This large family also, however, contains numerous distinctive and brightly colored species; these mostly occur in the tropics.

In general, warblers are small birds with fine, narrowly pointed bills. Their feet are strong and well-suited for perching. Some, such as the Dartford warbler, have long tails that counterbalance the body as the birds thread their way through dense foliage, inspecting leaves and twigs in their tireless search for insects.

The Major Groupings

FORM AND FUNCTION

There are a number of main groups of warblers. The reed and bush warblers include the large genus *Acrocephalus*, with 32 species. These are often found in marshes, reedbeds, and swamps, and are usually uniformly brown, robustly-built birds with large feet and bills that clamber about among the reeds and produce readily identifiable harsh, chattering songs. The whirring songs of *Locustella* species sound like insects, and have earned them the name of "grasshopper warblers." Then there are the scrub and woodland warblers, including the 7 species of *Hippolais* and the 24 species of *Sylvia*, the latter being unusual among warblers in having different plumages between the sexes. The leaf warblers include the second largest genus, *Phylloscopus*, with 46 species of small, greenish, short-billed, very similar-looking birds that tend to inhabit tree canopies and feed by gleaning insects from the undersides of leaves.

There is a wide variety of African tree warblers, including those in the genera *Apalis*, *Eremomela*, *Camaroptera*, and *Sylvietta*. The latter are curious-looking birds that forage up and down tree-trunks and branches, probing into crevices with the stout bills that earned them the name of "nuthatch-warblers" in the past. Then there are a number of more aberrant warblers, such as the almost tailless *Tesia* ground-warblers from the forests of Southeast Asia, and the emu-tails of Madagascar and fernbird of New Zealand, both of which have tail feathers in which the barbs are not joined together, giving them a spiky appearance.

The largest genus, *Cisticola*, includes some 45 species and forms part of the grass warbler grouping. The cisticolas are characteristic of grassy habitats in Africa and are a very confusing group of streaky warblers that are best differentiated by their songs. Indeed, these differences are often reflected in their common names, for example Whistling, Chattering, Trilling, Bubbling, Rattling, Wailing, Churring, and Tinkling cisticolas, to name just a few! Cisticolas are unusual among warblers in having longer tails during the breeding season, a double annual molt, and strong size dimorphism between the sexes, with the males being larger than the females. The prinias make up another major group of grass warblers; they are noisy, dull-colored birds with long, graduated tails.

Another major grouping is the tailorbirds of India and Southeast Asia. There are 15 species of tailorbird, which derive their name from their habit of using vegetable fibers or spiders' webs to sew large leaves together into a cone in which they build their nest. They puncture the margins of the leaves with their sharp beak to form the holes through which they thread their "twine." Tailorbirds have relatively long bills that curve downward, and carry their tails (which are longer in males than females) cocked up in a characteristic manner.

Not Strictly of the Old World

DISTRIBUTION PATTERNS

The great majority of species are Eurasian or African. Some are extremely common and widespread. For example, the Willow warbler is the most numerous migrant breeding bird in the UK, and is found at high densities over a large geographical range from northern Europe to eastern Russia. However, no fewer than 33 of the 64 genera contain only one species. Some of these have for many years been considered to be of doubtful affinity, but the use of Sibley and Ahlquist's classification, based on DNA, has helped to clear up

FACTFILE

OLD WORLD WARBLERS

Order: Passeriformes

Family: Sylviidae

389 species in 64 genera.

DISTRIBUTION Chiefly Europe, Asia, Africa; small numbers in the New World.

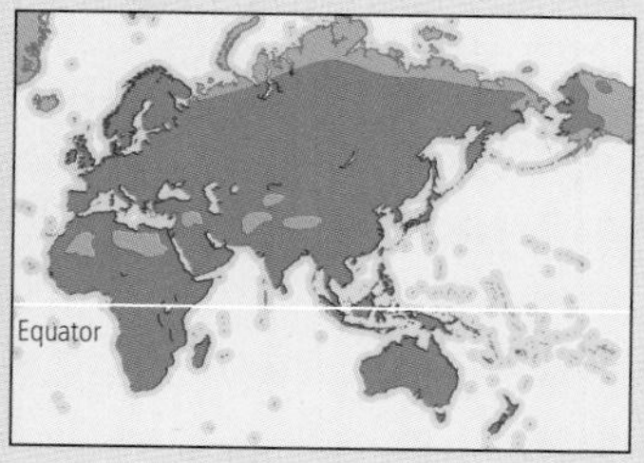

HABITAT All types of vegetation; grassland to forest.

SIZE Length 9–16cm (3.5–6.3in) in most species; weight 5–20g (0.2–0.7oz). There are, however, several large exceptions, e.g. the grassbird, with maximum length 23cm (9in), weight about 30g (1oz).

PLUMAGE Chiefly brown, dull green, or yellow, often streaked darker; some tropical species (e.g. White-winged apalis) are brightly colored. In most species the sexes are similar; exceptions include the blackcap and some tailorbirds.

VOICE Calls varied, often harsh in the larger species; songs of some species are simple and stereotyped, but in others complex, varied, and melodious.

NEST Elaborate, carefully-woven cup-shaped or spherical structures placed low in dense vegetation; tailorbirds and some other species stitch leaves together with cobwebs to form a cone in which the nest is sited.

EGGS Usually 2–7; pale ground colors with dark spots or blotches. Incubation period 12–14 days; nestling period 11–15 days, although young may leave the nest earlier (from 8 days), unable to fly, and are then attended by their parents for some days afterwards.

DIET Predominantly insects; some species also eat fruit; many take nectar occasionally.

CONSERVATION STATUS 3 species – the Taita apalis, the Long-billed tailorbird, and the millerbird – are currently listed as Critically Endangered; 10 are Endangered, and 29 are Vulnerable. In addition, the Aldabra warbler (*Nesillas aldabrana*) from the Seychelles, which was only discovered in 1967, is now considered extinct.

See Selected Old World Warbler species ▷

◀ ***Left*** *Male Dartford warblers are grayer above and darker pink below than females. Dartford warblers usually hold territories of between 2–6 ha, depending on habitat quality and their nests are located in dense gorse or deep heather.*

▶ ***Right*** *The grassbird is sometimes known as the lollipop bird because of its most prominent feature – a long, pointed chestnut and brown tail, which is usually held bunched together like a stick. Grassbirds enjoy sunning themselves on perches among tall grasses and flowers.*

some of these issues. Even so, the past taxonomic confusion surrounding warblers is highlighted by some of the birds' common names: there is an Oriole warbler, for example, as well as wren-warblers and tit-warblers.

To compound the confusion, there are also the New World warblers of the family Parulidae, which are not closely related, having nine primary feathers where the Old World warblers have 10. However, some species of Old World Warblers do occur in the New World! The Arctic warbler, for example, has extended its breeding range from Siberia into western Alaska, although even these birds return to the Old World to winter in southern Asia. Mainland Australia has eight resident species, including the distinctive spinifex-bird, and New Zealand only one, the fernbird. The archipelagos of the Pacific Ocean and Indian Ocean contain a variety of unique warblers, often with tiny populations.

Warblers' dependence on insect prey is the main reason why most species that breed at high latitudes are strongly migratory. Most north Eurasian warblers winter in Africa or tropical Asia, some performing prodigious journeys. For example, Willow warblers nesting in Siberia travel up to 12,000km (7,500 miles) twice a year, to and from sub-Saharan Africa. Before they fly, these long-distance migrants accumulate substantial reserves of fuel in the form of fat deposits; it is not unusual for them to double their bodyweight. Climate warming has allowed some migrant species to return to their breeding grounds earlier, and some may now be breeding 1–2 weeks earlier than in previous decades. However, it is possible that climate change might adversely affect long-distance migrants if desert barriers such as the Sahara grow wider or if the timing of the availability of food supplies used en route changes, so that the birds' movements no longer synchronize with them.

Above *Representative species of Old World warblers: **1** A Blackcap* (Sylvia atricapilla) *in shrubs; **2** A Sedge warbler* (Acrocephalus schoenobaenus) *in reed swamp; **3** Reed warbler* (Acrocephalus scirpaceus)*; **4** Paddyfield warbler* (Acrocephalus agricola) *from Southeast Asia; **5** Red-faced crombec* (Sylvietta whytii).

Below *Young Wood warblers* (Phylloscopus sibilatrix) *are fed by both parents, who forage high in the foliage but nest on the ground. Adults breed in broadleaved woodland with sparse undergrowth and their second broods tend to be smaller than their first.*

Those few warblers that remain in cold climates in winter are sometimes badly affected by food shortages in harsh weather. Dartford warblers in Britain, for example, often suffer large population decreases during severe winters. However, there is good evidence that these birds have been able to take advantage of recent climate warming, and numbers have increased rapidly in recent years.

The blackcap has been the subject of intense study with respect to the genetics of migration, and these studies have revealed many important facts that are likely to be applicable to other warblers and other migrant birds more generally. The use of careful experimental breeding programs has shown that the direction of breeding is subject to genetic control, as is migratory distance. Thus, when birds that show a predisposition to migrate in a southeasterly direction are mated with birds that tend to migrate southwesterly, the offspring show a tendency to migrate directly southward. Furthermore, when blackcaps from the Canary Islands, which are essentially non-migratory, are mated with migratory birds from Germany, their offspring tend to show an intermediate strength of migratory tendency.

Blackcaps from Germany provide good evidence for the rapid natural selection of a novel migratory behavior. Increasing numbers now migrate, not south to Iberia, as was usually the case in the past, but in a westerly direction to winter in the British Isles, where they take advantage of an increasingly warm climate and of increased provisioning of food for birds on garden bird-

tables. These birds may well survive better, and may then arrive back early on their breeding grounds in Germany, because they have less far to migrate than those that migrate to Iberia. Thus they are more likely to mate with others with the same genetic propensity, thereby increasing the prevalence of this behavior in the population.

Singing for a Partner

SONG AND SOCIAL BEHAVIOR

Warblers are typically monogamous, although instances of polygamy are known for a number of species, including the Sedge warbler. Voice is often important for mate-attraction and mate-selection, in addition to being a primary means of advertising territory boundaries. Male Sedge warblers may cease to sing after pairing, and individual Reed warblers with elaborate song repertoires tend to succeed in attracting females sooner than less accomplished singers. Some species, such as the Icterine warbler, extend their repertoires by mimicking other bird species, for reasons that are not fully understood. The Marsh warbler has gone further: its song consists entirely of imitations of

Right *An Olivaceous warbler* (Hippolais pallida). *This species often flicks its tail downward in a nervous manner and when alarmed, calls out in a series of quiet, hard tongue-clicking notes.*

other species. Each Marsh warbler mimics on average 80 species, over half of them African birds heard in the warbler's winter quarters.

Voice also plays an important part in distinguishing species. For example, chiffchaffs and Willow warblers look almost identical, but have unmistakably distinct songs. Songs are generally delivered from perches, but warblers of low vegetation often use song flights as a means of broadcasting their songs over long distances; examples of this behavior can be seen in scrub warblers such as the whitethroat and many of the cisticolas, including the Cloud-scraper.

All warblers seem to be competing for the same basic food: small insects. In practice there is often a high degree of spatial separation that minimizes competition between species and between individuals. Species that occur at high densities, notably the temperate-zone ones, are characteristically territorial. Typically the males defend territories against members of their own species, and sometimes also against members of closely-related ones. For example, blackcaps and Garden warblers defend territories against each other as well as against other members of their own species. This behavior may help ensure that a pair has enough food for themselves and their broods, as well as serving perhaps to space out their very similar nests to make it less profitable for predators to specialize in searching for them.

Isolated Species at Risk

CONSERVATION AND ENVIRONMENT

More than 10 percent of Old World warblers are considered to be globally threatened. There are three main types of threat. First, there are 19 species that occur on small, isolated, oceanic islands, where their tiny populations make them highly vulnerable to extinction. An example is the Seychelles warbler, which has evolved cooperative breeding behavior (by which birds other than the breeding pair help to raise the young in a nest) because of the over-crowded conditions and the lack of available breeding space.

The second group comprises 11 species that are cut off in a different way. They inhabit isolated mountain habitats that are separated from other mountains by large areas of unsuitable lowland. An example is the Namulu apalis, found only on Mount Namulu in northern Mozambique.

Finally, there are 12 species that occur in localized patches of habitat that are threatened by destruction due to man's activities. For example, the Grey-crowned prinia from the region around Nepal is suffering from the conversion of its shrubby grassland and forest habitat to agriculture; and the Aquatic warbler in Europe is losing habitat due to extensive drainage of marshes for conversion to farmland. HC/EFJG

Selected Old World Warbler Species

Species and genera include: Aquatic warbler (*Acrocephalus paludicola*), Marsh warbler (*A. palustris*), millerbird (*A. familiaris*), Reed warbler (*A. scirpaceus*), Sedge warbler (*A. schoenobaenus*), Arctic warbler (*Phylloscopus borealis*), chiffchaff (*P. collybita*), Willow warbler (*P. trochilus*), scrub warblers (18 species belonging to the genera *Bradypterus*, *Sylvia*), blackcap (*Sylvia atricapilla*), Dartford warbler (*S. undata*), Garden warbler (*S. borin*), whitethroat (*S. communis*), Brown emu-tail (*Dromaeocercus brunneus*), grass warblers (*Cisticola*) including Cloud-scraper cisticola (*C. dambo*), fernbird (*Bowdleria punctata*), grassbird (*Sphenoeacus afer*), Grey-crowned prinia (*Prinia cinereocapilla*), Tawny-flanked prinia (*P. subflava*), Icterine warbler (*Hippolais icterina*), Namulu apalis (*Apalis lynesi*), Taita apalis (*A. fuscigularis*), White-winged apalis (*A. chariessa*), Oriole warbler (*Hypergerus atriceps*), Seychelles warbler (*Bebrornis sechellensis*), spinifexbird (*Eremiornis carteri*), tailorbirds (genus *Orthotomus*) including Long-billed tailorbird (*O. moreaui*), tit-warblers (genus *Leptopoecile*), wren-warblers (genus *Calamonastes*).

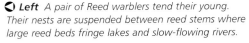

Left *A pair of Reed warblers tend their young. Their nests are suspended between reed stems where large reed beds fringe lakes and slow-flowing rivers.*

Firecrests and Kinglets

BIRDWATCHERS ARE COMMONLY ALERTED TO the presence of firecrests and kinglets by their thin calls and tinkling songs. It is much less easy to see them as they flit restlessly through the canopy, searching incessantly for their minuscule invertebrate prey. Their coloration provides effective camouflage, and their brightly-colored crowns are only visible when they display them.

The Regulidae form a neat group of miniature birds that, surprisingly given their small size, inhabit some of the colder parts of the northern hemisphere. Generally found in the coniferous forests of Europe, Asia, and North America, they seem to go against the trend for animals to grow larger as they move north. However, the birds appear to have adopted a niche and evolved a range of adaptations allowing them to thrive.

Small Birds of Cold Climates

FORM AND FUNCTION

The northernmost population of kinglets are migratory, and their abundance in the north is revealed by the large numbers that are recorded or caught for ringing at migration hotspots. Some, however, remain resident at high latitudes. In Scandinavia it has been calculated that goldcrests can survive 18 hours of darkness at –25°C (–13°F) by burning off fat equivalent to 20 percent of their bodyweight while huddling with other goldcrests in the depths of a tree's foliage.

Although they have evolved numerous adaptations to living in cold climates, prolonged periods of severe weather can have serious impacts on population size. Thus, longterm population monitoring of goldcrests in the United Kingdom show that their abundance is highly variable. They are very productive, however, and although a clutch of eggs can represent 1–1.5 times the female's weight, she often starts to lay a second clutch in another nest, built earlier by the male, when the first brood is only half-grown. This behavior presumably allows the birds to fit in two broods during the relatively short breeding season that they face at higher latitudes.

Keeping Snug in the Nest

BREEDING BIOLOGY

The birds' nesting habits also appear to be adapted to minimizing heat loss, for the parents as well as for the eggs and young. Their compact nests are well-insulated cups of moss and lichen and are lined by a warm layer of feathers. Some of these feathers form a loose umbrella over the nest cup, which may help to keep cold air out. The diameter of the nest is relatively small, again presumably to minimize heat loss from the top, so the eggs (up to 12 in some species) lie in layers, and the incubating female is obliged to push her hot, well-vascularized legs deep into the pile to provide warmth to those eggs not in contact with her brood patch. The nestlings are also adapted to minimize heat loss by burrowing to the lower parts of the nest after feeding, pushing the nestlings that need to be fed to the higher, but more exposed, part. HC

Below *The aptly named Golden-crowned kinglet generally associates in groups and feeds in company along with the titmice, nuthatches, and Brown creepers.*

FACTFILE

FIRECRESTS AND KINGLETS

Order: Passeriformes

Family: Regulidae

6 species of the genus *Regulus*: **Canary Islands kinglet** (*R. teneriffae*), **firecrest** (*R. ignicapillus*), **flamecrest** (*R. goodfellowi*), **Golden-crowned kinglet** (*R. satrapa*), **goldcrest** (*R. regulus*), **Ruby-crowned kinglet** (*R. calendula*).

DISTRIBUTION Europe to Asia, including Japan and Taiwan; also in N America down to Mexico.

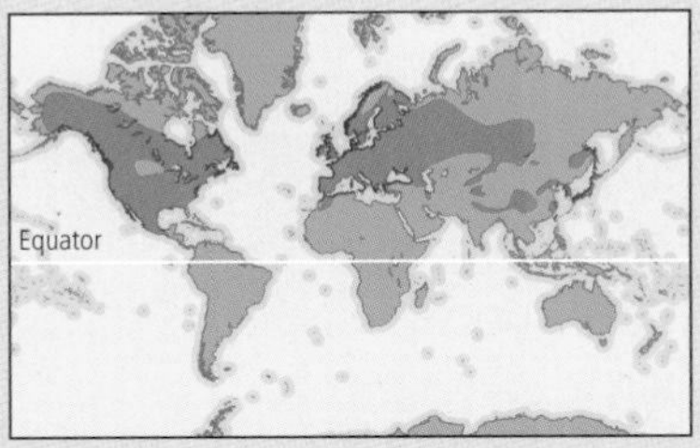

HABITAT Generally coniferous forest, but also in deciduous woodland.

SIZE Length 9–10cm (3.5–3.9in); weight 4–8g (0.15–0.3oz).

PLUMAGE Mostly olive green above and paler buff below; wings with marked pale bars; some with bold facial stripes. Males have characteristic red, orange, or yellow central crown that can be erected in display. Females and younger birds may have less marked or no bright colors on crown.

VOICE Generally high-pitched and thin (though melodious) songs and calls.

NEST Very neat, small, spherical cups of moss and lichens, held together by cobwebs and lined with feathers and hair; usually high in trees, suspended from twigs near the end of a conifer branch.

EGGS 7–12; white to pale buff ground colors, some with marked fine spotting in a darker brown. Incubation 15–17 days; nestling period 19–24 days.

DIET Small insects, caterpillars, and spiders; occasionally small seeds.

CONSERVATION STATUS Not threatened.

Babblers and Laughingthrushes

SOME 285 DIFFERENT SPECIES ARE LINKED *in the family Timaliidae, a diverse collection of mostly Old World forest birds that includes such subgroups as the African mountain babblers, the Southeast Asian tit-babblers, the Madagascan jerys, and the primarily Himalayan barwings, parrotbills, babaxes, sibias, yuhinas, and fulvettas. However, the laughingthrushes of the subfamily Garrulacinae form the family's main subdivision. A single monotypic species, the wrentit, is included here as the family's only New World representative.*

The birds are mostly insect-eaters (although some also take fruit), and they have soft, fluffy plumage. They also tend to be gregarious and noisy – hence the family's common name. For the rest, they are more notable for their differences than their similarities, ranging in size from tiny, wrenlike birds to others as big as thrushes, and in plumage from relatively dull, cryptic species of the underbrush like the Jungle babbler to the brightly-colored mesia and fulvettas.

Babblers

FAMILY TIMALIIDAE

Babblers are a varied group of small to medium-sized songbirds that form an important constituent of the bird populations of tropical Asia. Behavior and feeding ecology within the family are very diverse; it includes active leaf-gleaners flitting in the forest canopy like warblers and long-legged genera hopping on the ground like thrushes. The "average" babbler falls between a warbler and a thrush in size, with short, rounded wings and a longish tail. The birds forage in bushes, in low vegetation, and on the ground, but normally are not found far from some sort of cover. All are essentially sedentary, although a few mountain species may shift to lower altitudes in winter.

In Africa the family is represented by 10 genera, only one of which (*Turdoides*) is found also in Asia. An endemic genus of three species occurs in Madagascar. In North Africa, the Middle East, and Iran, the Fulvous and Arabian babblers and their allies live in sparsely vegetated wadis (dried-up river beds) in open desert. In the Negev Desert of

Below *The seven sibias, including this Long-tailed sibia* (Heterophasia picaoides), *are magpie-sized inhabitants of Asian forests.*

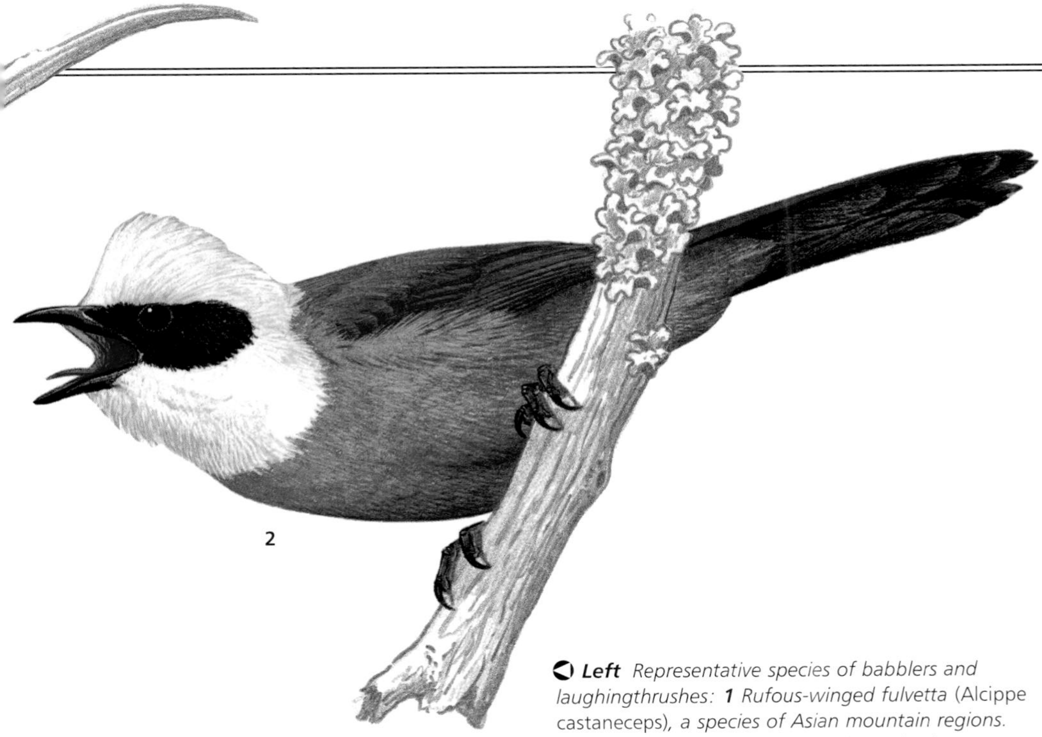

Left *Representative species of babblers and laughingthrushes:* ***1*** *Rufous-winged fulvetta* (Alcippe castaneceps), *a species of Asian mountain regions.* ***2*** *White-crested laughingthrush* (Garrulax leucolophus), *from India and Southeast Asia.* ***3*** *Southern pied babbler* (Turdoides bicolor), *from Africa.*

Israel the Arabian babbler is the commonest resident bird, occurring wherever there are a few acacia bushes to provide cover and nest-sites. In contrast, the Iraq babbler inhabits the extensive swamps of the Tigris–Euphrates delta, and similar species inhabit reed beds and tall grass habitats throughout southern Asia.

Just a single species reaches the New World, and it is not absolutely certain that it belongs in this family; the wrentit is found only in the Pacific coastal strip from California to Oregon. Like all babblers, these tiny, wrenlike birds, with stubby wings and long, narrow tails, are sedentary and territorial. Family flocks persist long after breeding, moving together through scrub and chapparal.

In the middle altitudes of the Himalayas, at about 1,500–3,000m (5,000–10,000ft), babblers are the dominant passerines, with 50 species breeding in Nepal alone out of a resident passerine community of about 320 species. Here they are represented by a great diversity of types, from tiny wren babblers skulking among rotting logs on the forest floor to the large scimitar babblers, probing with their hoopoelike bills, and the Black-headed sibia, drinking the sap oozing from holes in the trunks of oak trees, after the fashion of sapsuckers. Other species scratch among leaf litter on the forest floor (Puff-throated babbler). The tiny White-browed fulvetta is among the commonest birds in subalpine scrub in the western Himalayas of India. Geographically close, but ecologically different, the mockingbird-like babaxes inhabit buckthorn scrub on the edge of the Tibetan plateau, in arid, high-altitude desert.

In the evergreen rain forests of Southeast Asia and the temperate forests of the Himalayas, babblers are important members of the mixed feeding flocks of small, insect-eating birds that are characteristic of those environments. Species involved include minlas, yuhinas, tree babblers, tit-babblers, and shrike-babblers, which mix freely with warblers, tits, minivets, treecreepers and woodpeckers to form loosely organized parties, sometimes numbering several hundred birds, moving steadily through the forest throughout the day and only breaking up in the evening to roost in separate, single-species groups.

In India the Common and Jungle babblers are familiar birds of the garden and roadside, moving in noisy bands from tree to tree or hopping about energetically on the ground. Parties of 5–15 birds consisting of extended families are the rule, with the whole group collaborating to incubate eggs laid by the dominant female (and presumably fertilized by the dominant male). All members of the

FACTFILE

BABBLERS AND LAUGHINGTHRUSHES

Order: Passeriformes

Family: Timaliidae

285 species in 52 genera

Distribution Asia, Africa, Madagascar; 1 species in western N America.

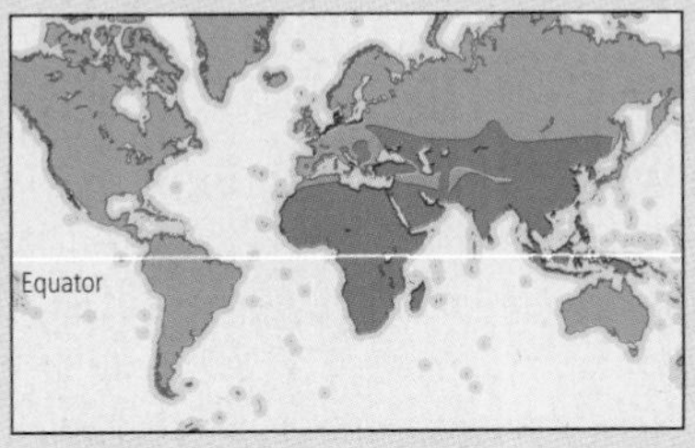

BABBLERS Family Timaliidae

231 species in 50 genera. Species and genera include: **Arabian brown** or **Arabian babbler** (*Turdoides squamiceps*), **Common babbler** (*T. caudatus*), **Fulvous babbler** (*T. fulvus*), **Iraq babbler** (*T. altirostris*), **Jungle babbler** (*T. striatus*), **babaxes** (genus *Babax*), **barwings** (genus *Actinodura*), **Bearded parrotbill** or **Bearded tit** or **reedling** (*Panurus biarmicus*), **Black-headed sibia** (*Heterophasia capistrata*), **fulvettas** (genus *Alcippe*), **minlas** (genus *Minla*), **Puff-throated babbler** (*Pellorneum ruficeps*), **scimitar babblers** (genera *Pomatorhinus*, *Xiphirhynchus*), **shrike-babblers** (genera *Pteruthius*, *Gampsorhynchus*), **tit-babblers** (genus *Macronus*), **tree babblers** (genus *Stachyris*), **White-browed fulvetta** (*Alcippe vinipectus*), **wren babblers** (genera *Kenopia*, *Napothera*, *Pnoepyga*, *Ptilocichla*, *Rimator*, *Spelaeornis*, *Sphenocichla*), **wrentit** (*Chamaea fasciata*), **yuhinas** (genus *Yuhina*). Asia, Africa, Madagascar; 1 species in western N America. Desert scrub to swamp, tropical forest to alpine dwarf shrubs. **Size**: Length 10–35cm (4–14in); weight 5–150g (0.2–5.3oz). **Plumage**: Many species are cryptic browns and grays; species living in dense forest often include bright yellows, reds, and blues; no differences between sexes. **Voice**: Great variety of calls; some species have antiphonal duets. **Nest**: Usually above ground in bushes or trees; open or domed. **Eggs**: 2–7; variety of colors, but often white or blue; unmarked. Incubation period 14–15 days; nestling period 13–16 days. **Diet**: Mainly insects and other invertebrates, also fruit and nectar. **Conservation status**: 4 species are Endangered and 12 Vulnerable.

LAUGHINGTHRUSHES Subfamily Garrulacinae

54 species in 2 genera (*Garrulax* and *Liocichla*). Species include: **Striated laughingthrush** (*Garrulax striatus*), **White-crested laughingthrush** (*G. leucolophus*), **White-throated laughingthrush** (*G. albogularis*). S and E Asia, in tropical evergreen and semi-evergreen forest from sea level to 4,000m (13,200ft). **Size**: Length 15–35cm (6–14in); weight 40–150g (1.6–5.3oz). **Plumage**: Includes many bright colors, especially on the wings; head sometimes crested. **Voice**: Wide range of whistling and chattering sounds, as well as the eponymous "laughing." **Nest**: Open, well-made cup in small fork or against trunk of tree. **Eggs**: 2–5, deep blue. **Diet**: Insects and fruit; larger species may take small lizards and frogs. **Conservation status**: The Collared and Rufous-breasted laughingthrushes are Endangered; 5 other species are Vulnerable.

Above *Known in Britain as the Bearded tit, the Bearded parrotbill is not easily confused with the true tits of the family Paridae. An active, gregarious bird of the reedbeds, it owes its name to the black markings extending down from between its bill and eyes.*

family assist in feeding the nestlings, which fledge after about 14 days.

These babblers commonly host brood parasites, especially the Pied cuckoo (*Oxylophus jacobinus*); as many as three cuckoo eggs may be laid in a single nest. Unlike European cuckoos, the hatchling Pied cuckoo does not evict its nest mates, so that several young babblers may be reared alongside one or more cuckoos. The cuckoo eggs are identical in color to the bright blue eggs of the babblers, distinguished only by a smoother texture and slightly more spherical shape.

Groups of Jungle babblers defend collective territories; encounters between neighboring groups are very noisy, with most of the birds on each side calling excitedly. Fighting sometimes breaks out in these skirmishes, and antagonists can be seen rolling on the ground with claws locked together. Similar behavior is seen in mock fights that take place within groups, usually among the younger birds. Members of a group roost side by side at night on a horizontal branch, taking a long time to settle down because of constant switching of position. The eventual order leaves the senior members closest to the trunk, the immatures at the far end, and any juveniles in the middle.

The distinctive parrotbills are found mainly in bamboo thickets or other tall-grass habitats (reeds, elephant grass) in northeastern India, northern Myanmar, and southern China, from the plains up to the subalpine zone of the Himalayas. Like many babblers, they are found in flocks of 5–25 throughout the year but little is known of their biology. The best-known is the Bearded parrotbill

(also known as the Bearded tit or reedling), a small, long-tailed bird with short wings that breeds in large reedbeds from eastern England across central Europe and Asia. Unlike other babblers, the sexes differ in plumage, only the male showing the eponymous black "beard" (really more of a mustache). Although generally rather sedentary and poor fliers, these birds periodically "irrupt," with flocks suddenly appearing outside their normal range. Possibly this is the mechanism by which the species disperses among the "islands" of its chosen habitat, often separated by vast areas of desert.

Laughingthrushes

Subfamily Garrulacinae

Laughingthrushes are robust, thrushlike birds of the forest floor and understory, with very strong legs and short, rounded wings. All are sedentary, although a few descend seasonally from higher altitudes. Most species are brightly colored, with patches of red or yellow on the wings or head, or

Cooperative Breeding in Babblers

One of the most striking characteristics of babbler behavior is the birds' tendency to occur in small, compact groups. These are not temporary groupings but permanent coalitions, usually forming an extended family, that remain together throughout the birds' lives and that cooperate in defense of a common territory.

What has brought about this widespread cooperation? Being poor fliers, babblers tend to remain close to where they were born; considering the large number of species involved, they are remarkable in being wholly nonmigratory. A few of the high-altitude species move small distances between their summer and winter altitude zones, but usually no more than a few kilometers at most. This sedentary behavior makes it easier for groups to defend a year-round territory.

Because babbler groups defend territories year-round, there is no annual competition for space, and it is more difficult for new territories to become established. A young bird that remains with its parents until its abilities are fully developed, perhaps after several years, will stand a better chance of reproducing eventually than one that disperses immediately. In this situation, it is easy for cooperative behavior, such as communal rearing of the nestlings, to evolve.

In order to form a new territory in the face of strong competition, it may be necessary for several birds to collaborate. This is best achieved by groups of siblings; as they share many of the same genes, help given to other group members may benefit the survival of their own.

In the Jungle babbler, for example, groups of sibling males or females sometimes split off from their parental group and roam about in marginal areas. If they encounter a similar coalition of the opposite sex, originating from another group, then they may combine to try to establish a new territory. Alternatively, single females may try to join unrelated groups and sometimes succeed, despite opposition from the dominant female. Sibling and parent–offspring mating have not been recorded and breeding between unrelated individuals seems to be the rule, so some interchange between groups is essential.

The importance of kin associations in setting up new territories provides another incentive for birds to remain in their parental group until they can form a coalition of sufficient strength to attempt to defend their own ground. It also gives some incentive for nonbreeders to assist in rearing younger siblings, as these may eventually help them in establishing a new territory.

Above *Well-known for its whistling song, the Silver-eared mesia* (Leiothrix argentauris) *is a hardy, handsome inhabitant of forest and scrub. It can be found from Pakistan east through Indochina, moving around in small flocks in search of insects and berries.*

Left *The shy, furtive scimitar babblers of India and Southeast Asia use their long, curved bills to search for insects in the leaf litter, rather in the manner of American threshers. Here a Rusty-cheeked scimitar babbler* (Pomatorhinus erythrogenys) *looks up in alarm.*

Below *Like most of the Asian* Garrulax *species, the Red-tailed laughingthrush* (Garrulax milnei) *spends much time on the forest floor, often betraying its presence only by its loud, cackling call.*

white heads or breasts. They are highly social and are rarely found on their own, although the heavy-bodied, fruit-eating Striated laughingthrush is often seen in pairs only. Their behavior is similar to that of the larger babblers, but White-throated laughingthrushes may occur in single-species flocks of up to 100 birds. Being in the midst of one of these flocks is a remarkable experience; a wide variety of ringing calls emerge from bushes and trees all around, often with little sign of the birds themselves, which are surprisingly cryptic.

The spectacular White-crested laughingthrush performs a remarkable communal display in which several members take part. The birds prance together on the forest floor, flapping their wings, with their white crests raised like helmets, and utter a series of hoarse, laughing calls that gradually mount to a crescendo. At dawn and dusk these calls are normally answered by neighboring groups, producing a chorus of choruses that echo among the densely forested hills where the birds live. AJG

Larks

A FAMILIAR FEATURE OF BIRD COMMUNITIES of most of the open areas of the Old World, larks are particularly varied and plentiful in the arid parts of Africa, where a dozen or more species can sometimes be present in a single region. Many species have evolved elaborate songs, and these are often given in flight.

Although many lark species are associated with very arid desert or semi-desert areas, the birds are by no means confined to hot climates. The Horned or Shore lark breeds on open arctic tundras and high on mountain ranges, as well as through much of North America. Recently a national survey found the skylark to be Britain's most widespread breeding bird.

Old World Songsters

FORM AND FUNCTION

Most larks are basically streaked, brown birds. Some have dark markings and white patches on their plumage, often on the wings or tail, which in most cases are only readily visible when the birds are in flight. One of the most colorful species, the Bifasciated or Hoopoe lark, has black and white markings on the wings and pinkish buff plumage on the body, and looks very like the bird after which it is named. Clearly, the larks' normal plumage is cryptic, serving to conceal the birds when they are on the ground and, particularly, when they are incubating.

Several species have races whose color is closely related to that of the substrate where they live. The Desert lark carries this tendency to extremes, and sandy-colored birds living in sandy deserts are often found close to dark ones that have evolved to match the dark background of old lava flows. The Horned lark has black patterns on the breast and face and a rakish black, horseshoe-shaped mark round the front of the head, ending in small, backswept horns of feathers. Most species have fairly strong bills, although one, the Thick-billed lark, has a monstrous beak similar in size to that of a hawfinch; others, like the Bifasciated lark, have rather long, downcurved bills. The thick bills are an adaptation for cracking the hard coat of seeds, and the curved beaks for digging in earth to uncover food. In fact many species – not just those with curved bills – have been seen to dig in the ground when feeding, searching for insects or, more often, for seeds.

In common with many other predominantly ground-dwelling birds, larks generally have fairly long legs with long hind claws that give them additional stability when standing. Although some species fly at the slightest sign of danger, even far away, many prefer to escape by walking or running. These species often make masterly use of the contours of the ground, and of any nearby vegetation, for concealment during their retreat. Some habitually crouch when threatened, relying on camouflage for defense. Many species have no need of trees or bushes within their territory, but others regularly perch on posts, bushes, or trees. Many species, including the varied *Mirafra* bush-lark genus, use open scrubland.

Left *The larks that inhabit the southern African veld divide into two groups: those that mainly eat insects and those that are primarily seed-eaters. Insect-eaters like* **1** *the Spike-heeled lark* (Chersomanes albofasciata) *and* **2** *Fawn-colored* (Mirafra africanoides) *larks have longish bills; seed-eating* **3** *Stark's* (Calandrella starki) *and* **4** *Pink-billed* (C. conirostris) *larks have short, stubby ones.*

Left *The Spike-heeled lark of southern Africa uses its long bill to dig into soft soil and sand in search of insects.*

Right *In Kenya's Masai Mara National Park, a Rufous-naped lark* (Mirafra africana) *gives song. Unlike many larks, this one has a repetitive, single-phrase call.*

FACTFILE

LARKS

Order: Passeriformes

Family: Alaudidae

About 80 species in 13–15 genera. Species and genera include: **Bifasciated** or **Hoopoe lark** (*Alaemon alaudipes*), **bush-larks** (genus *Mirafra*) including **Singing bush-lark** or **Australasian lark** (*Mirafra javanica*), **Archer's lark** (*Heteromirafra archeri*), **Calandra lark** (*Melanocorypha calandra*), **Black lark** (*M. yeltoniensis*), **Crested lark** (*Galerida cristata*), **Fischer's sparrow-lark** (*Eremopterix leucopareia*), **Horned** or **Shore lark** (*Eremophila alpestris*), **Desert lark** (*Ammomanes deserti*), **Red-capped lark** (*Calandrella cinerea*), **Lesser short-toed lark** (*C. rufescens*), **Rudd's lark** (*Heteromirafra ruddi*), **Eurasian skylark** (*Alauda arvensis*), **Raso lark** (*A. razae*), **Thick-billed lark** (*Ramphocoris clotbey*), **woodlark** (*Lullula arborea*).

DISTRIBUTION Europe, Asia, Africa, America, Australia.

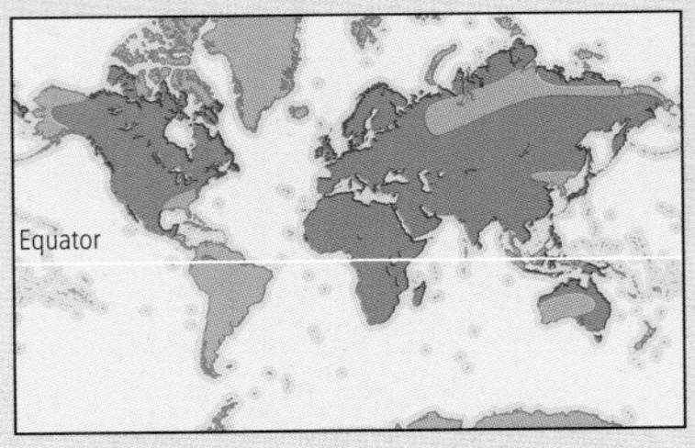

HABITAT Open country

SIZE Length 12–24cm (5–9in); weight 15–75g (0.5–2.6oz).

PLUMAGE Most are brown and streaked, some with black and white markings; the Black lark is completely black.

VOICE Melodious songs, some short but others prolonged warblings, often given in flight.

NEST Most build cups of dead grass on the ground; some build more complex, partly-domed structures.

EGGS 2–6; speckled in most species. Incubation 11–16 days.

DIET Seeds, insects.

CONSERVATION STATUS 2 species – Rudd's and the Raso lark – are Critically Endangered; 2 others are Endangered and 4 Vulnerable.

Mostly Seeds

DIET

Although most adult larks are predominantly seed-eaters, all take some invertebrate food, particularly when feeding their nestlings; the animal protein seems to be essential for the growing young. In many habitats – for instance, in desert conditions – the availability of seeds may be severely limited, and so the birds may be few and far between. However, where there has been a particularly productive set of seeds – immediately after sporadic rains in normally arid areas, for example, or where crops have successfully been cultivated – flocks of dozens or even hundreds of birds may settle. These will often be of a single species, but mixed-species flocks are not uncommon, since the conditions that suit one lark will generally suit others also.

Above *Display flights are a prominent feature of the behavior of the male Bifasciated lark of North Africa and the Middle East. In them, the bird soars upward on a spiral path and then glides back down.*

Below *A southern African species, the Large-billed lark (*Galerida magnirostris*) is a solitary bird that forages by walking about on the ground, pecking up seeds, which it crushes with its stout beak.*

Advertising through Song

BREEDING BIOLOGY

Most species are highly territorial when breeding; males defend their territories against rivals and advertise for mates by singing in flight. Many species, including the skylark, the woodlark, and the Bifasciated and Calandra larks, have pleasing songs; the latter regularly sings from the ground, and in the past was often kept as a songbird in the Mediterranean region. The birds' recognition and warning callnotes are also pleasant to the human ear, and are more elaborate than those found in some other groups, such as the sparrows.

In many areas the larks' breeding season is strongly related to rainfall. Breeding starts very quickly after the rains to ensure that the young can be hatched as weed seed stocks reach their peak. In such circumstances only a single brood may be raised, and the birds may then move, or simply wait until the next rains. However, temperate species often raise two or even three broods in a single season.

Almost all species nest on the ground, sometimes in the open but usually at least partly concealed in vegetation. A few species, mostly from the hottest desert areas, build nests just off the ground in bushes, where the circulation of the air may serve to cool the nest slightly. Sometimes incubating birds spend long periods standing over their nests in the middle of the day to shade the eggs from the sun.

Clutch sizes are often low in very hot and dry areas; for example, Fischer's sparrow-lark, which nests in equatorial East Africa, has a clutch of only two eggs. However, temperate-breeding species

Below *The gaping mouths of two hungry Horned lark chicks could hardly be missed by any adult bird. The species is a widespread one, familiar across much of the USA, and 3–5 is the normal clutch size.*

Left Representative species of larks: **1** Black lark (Melanocorypha yeltoniensis), *a central Eurasian species.* **2** *Eurasian skylark* (Alauda arvensis), *renowned as a songster.* **3** *Horned lark* (Eremophila alpestris). **4** *Fischer's sparrow-lark* (Eremopterix leucopareia), *from Africa.* **5** *Australasian lark* (Mirafra javanica).

such as the skylark, the woodlark, and European-breeding Crested lark regularly lay four, five, or even six eggs. A few of the desert-dwelling lark species have been described as building a buttress of stones below the lower edge of their nests, which are generally built on slopes. It has been suggested that the purpose may be to allow the nest to drain quickly in the event of flash floods, but it would seem more likely that the stones simply serve as a windbreak.

The young always receive some insect food in their first days, but many species revert to a vegetable diet well before the chicks move away from the nest, while still flightless within a week or two of hatching. This abrupt change to vegetable food may cause the fledglings to have rather poor-quality feathers. Certainly, all lark species so far studied undergo a complete postjuvenile molt, whereas most other passerines only molt their body plumage at this stage, retaining the wing and tail feathers grown in the nest until after their first breeding season. The advantage of molting immediately may be to save parental effort; thanks to it, parent birds are able to raise their young without having to provide the extra food needed for really good-quality feathers. Instead, the youngsters may provide these for themselves at a slower rate immediately after they have started on their independent lives.

The Dangers of Drought and Cats

CONSERVATION AND ENVIRONMENT

Several larks are listed as threatened. The most critical one is the Raso lark, which is confined to the tiny island of Raso in the Cape Verde group. Here it nests on an area of volcanic plain just a few kilometers in extent. The population reached 250 birds by 1990, but the bird can only breed after rains and may not have had the opportunity to do so for some time; certainly, only 92 birds were found in the course of a census of the island in 1998. The species is protected, but the census found evidence of cats on the island, so its prospects may not be bright. Males and females of this species show a huge difference in bill size, possibly evolved to enable the birds to exploit a wider ecological niche within their territory.

All the other rare larks live on the African mainland, where they have similarly restricted ranges. One, Archer's lark from inland northwest Somalia, has not been recorded since 1955. It is very secretive, however, and so is by no means necessarily extinct!

CJM

SPECIAL FEATURE

THE SKYLARK'S DECLINE

The threat to a countryside icon

THE SKYLARK (*ALAUDA ARVENSIS*) IS AN ICON of the British countryside. In appearance a rather drab little bird, its song is inspiring, varied, and sung for much of the year. In the past it was common over much of Britain, so that, at dawn, large numbers of birds were often within earshot at once, shattering the silence with a sudden, joyous chorus known as "larkrise." As a result, the skylark secured a special place in music, folklore, and literature, as epitomized in Shelley's celebrated tribute:

Hail to thee, blithe Spirit!
Bird thou never wert,
That from Heaven or near it,
Pourest thy full heart
In profuse strains of unpremeditated art.

It was therefore alarming when statistics collected in the late 20th century revealed that skylark populations on lowland farmland had tumbled by 60 percent over 25 years. The birds are still common and widespread, so their decline had not been obvious to the casual birdwatcher.

The intensification of modern agricultural technology seems to be the cause of these population changes, and a great deal of research effort has been expended in finding out the mechanisms of decline and how to reverse them. National surveys have been undertaken by the British Trust for Ornithology annually for 40 years to track the population indices of common birds. Each year the government publishes an official index of bird populations, calculated from these and other sources. It has officially acknowledged the problem, and is committed to reversing the declines through direct policy changes and agri-environment schemes associated with the Common Agricultural Policy (CAP) of the European Union.

The problems affecting larks are not restricted to birds living in areas where crops are being maximized, as breeding birds from wild places also come to farmland for the winter. As a result of modern agricultural developments, they find there very little surplus productivity left to support natural ecosystems. In fact the birds' demands are small. Just over 5kg (11lb) of wheat is enough to feed a skylark for a whole year – an amount that represents well under 0.1 percent of the wheat now produced from a single hectare (2.5 acres) of good farmland, and that is currently worth all of 50 US cents, 35 pence, or half a Euro.

Only a very small part of the problem is due to poisoning as a result of the misuse of chemicals or direct ingestion of properly-used materials. The ecological implications of modern farming for birds are far wider and more all-embracing. The following developments are relevant to the skylark:

- Herbicide use has reduced the number of seeds in arable land from an average of 2,000 per square meter (185/sq ft) 60 years ago to 200 (18.5) at the present moment. Also, new varieties of cereal and modern harvesting machinery have greatly reduced the spillage of grain. At current levels there are large areas of arable fields that cannot support skylarks in winter.
- The change from spring to autumn sowing has meant that most of the arable ground is covered by growing crops in winter, at a time when much arable farmland used to be left as stubble that afforded the birds excellent feeding.
- The use of modern fertilizers and of new varieties of crop have reduced or eliminated the need for fallow years, when the ground was rested, or traditional rotations, when different crops were introduced that were often skylark-friendly.
- Autumn-sown crops are too well-established, and the cover too thick, for the birds to be able to breed in fields in the spring.
- Use of pesticides to eradicate invertebrate pests has greatly reduced the numbers of insects in arable land. These are an essential part of the skylark's ecology, since the chicks are fed on insects and their larvae for the first week after hatching.

All of these changes have been brought about by developments in modern agriculture, and farmers who wish to maximize their returns from their lands find it difficult to buck the trends. However, modern agriculture is not just a matter of market forces; in Europe it is highly regulated through the CAP. Up to the present, CAP subsidies have simply encouraged farmers to increase production by supporting food prices. This policy has, however, been extended to include schemes such as "set-aside," under which areas of farmland were removed from production and farmers were given subsidies not to produce food that would inevitably be surplus to requirements. This program has in fact been quite beneficial for skylarks, but unfortunately the need for set-aside has been greatly reduced in recent years.

Now the CAP is being forced to change, as international agreements on trade outlaw the use of subsidies that upset the balance of world trade. In future, Europe's farmers will receive payments that aim to protect the environment; they will be rewarded for not spraying field edges, for example, and for planting farm woodlands, for allowing hedges to grow for longer before trimming, and even for putting areas of farmland into longterm set-aside. These schemes will certainly help wildlife in general; but as skylarks breed in the middle of fields, they will benefit from only some of the new measures.

CJM

Right *The sight of a skylark singing high in the sky is becoming increasingly rare in Britain. The birds often start to sing from a fence or post, and then fly upward, maintaining a constant melodic outpouring for 10–15 minutes without a pause. The songs serve to mark the birds' territories.*

Below *Eurasian skylarks are ground-nesters, which makes them particularly vulnerable to changes in land use. Modern agricultural methods not only deprive them of suitable nest sites in the spring breeding season, but also drastically reduce the amount of food available to the birds.*

Flowerpeckers

Left *The Crested berrypecker of New Guinea is a thrush-sized bird that differs in several respects from the typical flowerpeckers, having a simplified tongue structure and no serrations on its bill. In addition, it builds a cup-shaped nest and lays only a single egg.*

Right *In Australia, a Mistletoe bird surveys the world from its neat hanging nest. The nests are built by the female birds from vegetable fibers, roots, and grasses, bound together with cobwebs. Inside, they are lined with silky down.*

Many of the smaller birds commonly seen in the Asian and Australian biogeographic regions belong to the Dicaeidae. Besides the flowerpeckers properly so called, the family is here taken to include the berrypeckers and longbills of New Guinea.

Flowerpeckers

GENERA DICAEUM, PRIONOCHILUS

Flowerpeckers are small, dumpy, very active birds associated with mistletoes, berry-bearing shrubs, trees, and vines. They resemble the closely-related sunbirds, but they are smaller and most of them have shorter and less curved bills, with serrated edges. Most species are sexually dimorphic, with adult males having brightly-colored patches (usually red). Like sunbirds they also visit flowers – hence the name. Thirty-seven species belong to the main genus *Dicaeum*, distributed in southern Asia from India through southern China and southeast Asia to Taiwan and the Philippines in the north and to New Guinea, the Bismarck Archipelago, the Solomon Islands, and Australia in the south. Three or four of the Philippines species are now considered at risk.

The birds frequent high trees and feed on berries, insects, and spiders, which they often snatch while hovering. They also feed on nectar, and their short tongue is divided into two grooved tips, considered to be an adaptation to nectar feeding rather like that of sunbirds. The stomach is also specialized in fruit-eating species, with the duodenum connecting to the pro-ventriculus at the end of the esophagus to permit the direct passage of ingested fruit. The gizzard section (stomach) is a blind sac in which insects and spiders are digested. The birds are very active while feeding, and give metallic or high-pitched call notes similar to those of sunbirds.

Many flowerpeckers are notable as seed dispersers. In Australia, the Mistletoe bird swallows the mistletoe berry whole, bypassing the stomach, and the seed is excreted within half an hour of ingestion. The seed adheres to the tree on which it is dropped and germinates there. The Mistletoe bird occurs wherever mistletoes (mostly Loranthaceae spp.) grow, be it in the arid center of the continent or the rain forest of the tropical coast, but is absent from Tasmania, where mistletoes are not found. In the Thick-billed flowerpecker of Southeast Asia, the sticky seed is separated from the flesh before ingestion and deposited on the branch. Six other species of thick-billed flowerpeckers, each with 10 primary wing feathers, are grouped in the genus *Prionochilus*.

Flowerpeckers generally breed in pairs, and remain in pairs or small flocks outside the breeding season. Most species living in or on the edges of rain forests are sedentary, but Mistletoe birds are nomadic in some parts of Australia and may congregate in large numbers on fruiting shrubs.

FACTFILE

FLOWERPECKERS

Order: Passeriformes

Family: Dicaeidae

55 species in 7 genera. Species include: **Yellow-breasted flowerpecker** (*Prionochilus maculatus*), **Mistletoe bird** (*Dicaeum hirundinaceum*), **Thick-billed flowerpecker** (*D. agile*), **Green-crowned longbill** (*Toxorhamphus novaeguineae*), **Crested berrypecker** (*Paramythia montium*).

DISTRIBUTION S Asia, New Guinea, Australia.

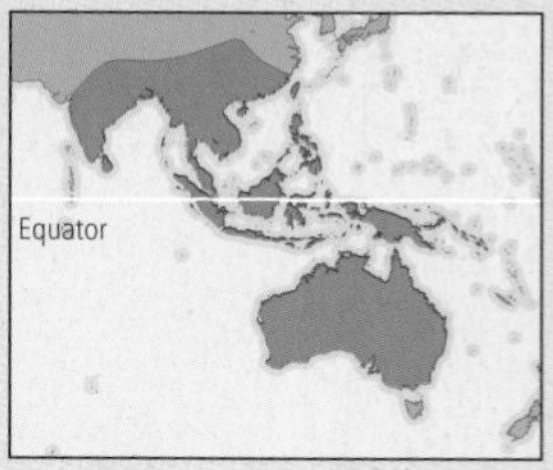

HABITAT Woodland, forest.

SIZE Length 7.5–15cm (3–6in); **weight** 5–20g (0.2–0.7oz). Crested berrypecker is 21cm (8in) long and weighs 42g (1.5oz).

PLUMAGE Upperparts dark and glossy, underparts light. In species with dull plumage, no difference between sexes; in others, males have patches of bright colors.

VOICE Faint metallic notes and high-pitched twittering; some species produce a series of rapid, oscillating notes.

NEST Open and cup-shaped (berrypeckers) or pendent with a side entrance (flowerpeckers).

EGGS 1–3, white with or without brownish blotches; size 1.5 by 1cm to 3 by 2.1cm (0.6 x 0.4in to 1.2 x 0.8in). Incubation period 12 days; nestling period about 15 days.

DIET Berries, nectar, insects, spiders.

CONSERVATION STATUS One species – the Cebu flowerpecker (*Dicaeum quadricolor*) – was thought to have become extinct early in the 20th century, but was rediscovered at three sites in the 1990s. 2 other *Dicaeum* species are listed as Vulnerable, and 4 are classed as Lower Risk/Near Threatened, as is the Scarlet-breasted flowerpecker (*Prionochilus thoracicus*).

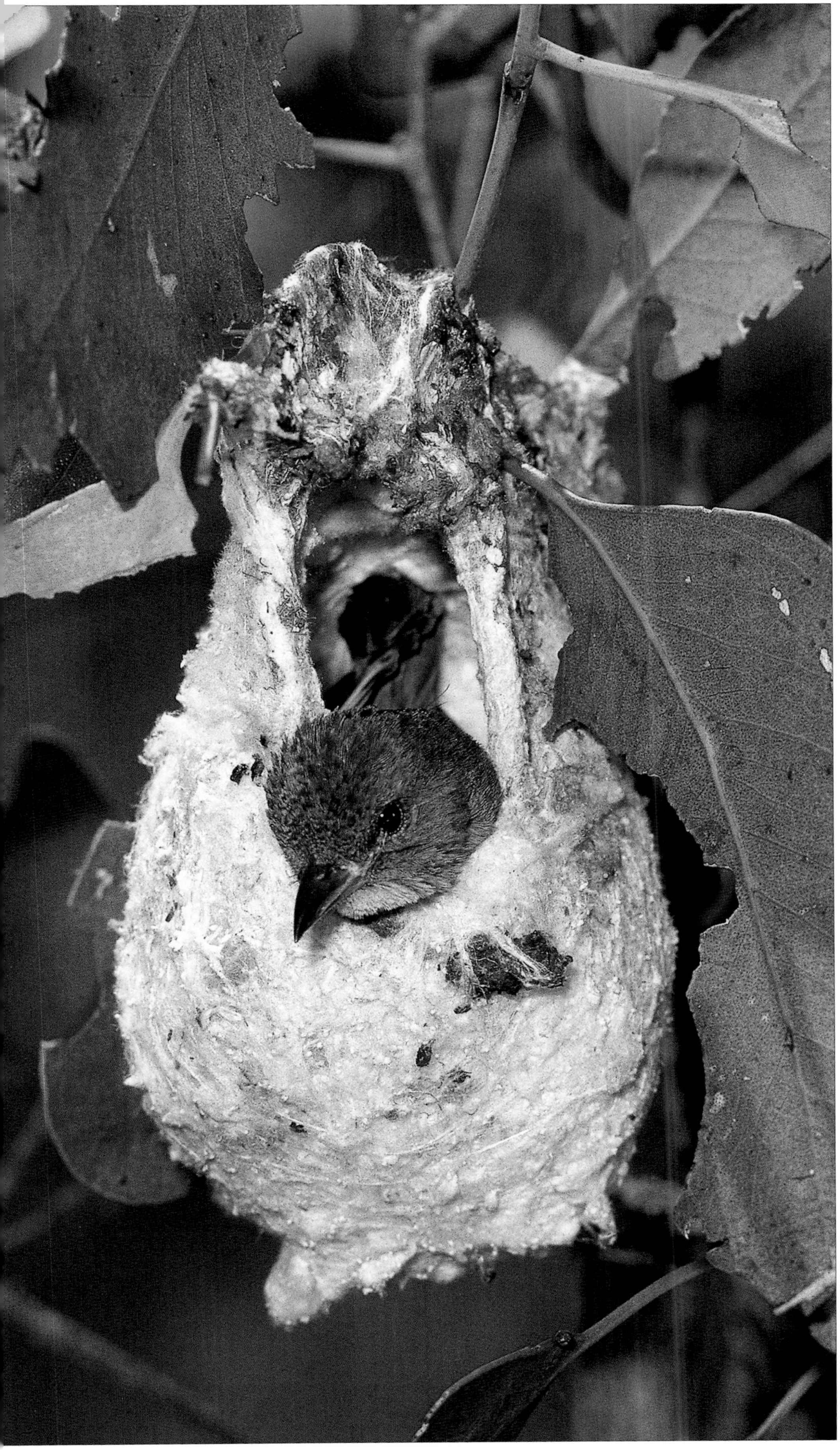

Mistletoe birds breed from October to March (the southern summer) when mistletoes fruit. Their nests are similar to those of sunbirds, and are typically made of plant fibers and down, woven together with cobwebs and suspended from an outer branch; they are pear-shaped, with a side entrance near the top. The birds lay a clutch of two or three generally whitish eggs, although in a few species like the Plain flowerpecker (*Dicaeum concolor*) they are speckled. Both parents are known to attend the nest.

Berrypeckers

GENERA MELANOCHARIS, OREOCHARIS, PARAMYTHIA

The berrypeckers are confined to the New Guinea region. Their tongue is not as specialized as in flowerpeckers. Most are sexually dimorphic, but in two species no brightly-colored male is known. The female Fan-tailed berrypecker (*Melanocharis versteri*) is darker and larger than the male, unusually for passerine birds. The Crested berrypecker is much larger than the others, and is sexually monomorphic. It is the only crested member of the family, and is sometimes placed in a separate family of painted berrypeckers (Paramythiidae) together with the Tit berrypecker (*Oreocharis arfaki*), so called because the male resembles a tit.

In the eastern highlands an altitudinal sequence exists: the lowland Black berrypecker (*M. nigra*) is replaced first by the mid-montane Lemon-breasted berrypecker (*M. longicauda*) and then by the high-montane Fan-tailed berrypecker, which is found up to the timberline. In the Black berrypecker, a slight increase in size with altitude has been reported.

Calls are high-pitched and faint, but the song is a rapid succession of twittering notes. The diet includes fruit, insects, and spiders. The birds build a neat, cup-shaped nest made of fern scales and plant fibers, woven with animal silk and decorated with lichen; it is usually anchored on a fork of an outer branch. The breeding season usually extends through the dry season, but for some species breeding has not yet been described.

Longbills

GENERA TOXORHAMPHUS, OEDISTOMA

Longbills are small, long-billed berrypeckers that physically resemble short-tailed honeyeaters; they were in fact treated as members of the honeyeaters' Meliphagidae family until DNA studies demonstrated their affinity with berrypeckers. They live in lowland and low-montane forests of New Guinea and its satellite islands. They are primarily foliage-gleaning insectivores, but also visit flowers in the canopy. They build berrypecker-type nests, mainly in the dry season. Of the four known species, the Pygmy longbill (*Oedistoma pygmaeum*) is the smallest, weighing only 5g (0.2oz) and measuring 7.3cm (3in) in length, with a relatively short bill of 13mm (0.5in). The bird's nest and eggs are undescribed. JK

Sunbirds, Spiderhunters, and Sugarbirds

Left *The spiderhunters of the genus* Arachnothera *are, by sunbird standards, relatively dull birds that use their long bills to pick spiders from their webs as well as to sip nectar.*

In all the tropics there are small, brightly colored birds with long bills that drink nectar from flowers and often pollinate them in the process. In the New World these are the hummingbirds (Trochilidae), but in Africa and Asia they are the sunbirds, an unrelated family.

Most sunbirds are quite common within their range and do not appear to be in any danger. Five African species are considered to be at risk, however, largely because of their limited habitat range; these include the Giant sunbird, which is the largest true sunbird and is endemic to the island of São Tomé. The Elegant sunbird, which is thought to survive only on the island of Sangihe, north of Sulawesi in Indonesia, is also now listed as Endangered.

Old World Nectar-feeders

FORM AND FUNCTION

Male sunbirds are typically dark and iridescent, colored blue, purple, green, or almost black on the head and upperparts and with bright red, yellow, orange, or white underparts. Some have prominent, usually red or yellow tufts on the flanks at the sides of a colored breast band, and in a few species the males have elongated central tail feathers, a feature especially of some of the *Aethopyga* species of India and surrounding countries. The females are usually duller, being mainly olive-green or brown on the upperparts and yellow, dirty white, or greenish, often with darker spots or streaks, on the underparts. In some species, however, the sexes are alike, although only in those in which the male is not so brightly colored. Some males have a nonbreeding or eclipse plumage that more closely resembles that of females. Among the most spectacular species are the Golden-winged, Tacazze, and Scarlet-tufted malachite sunbirds of the East African mountains, the Superb, Splendid, and Johanna's sunbirds of West Africa, and several Himalayan species. Duller ones include the olive and mouse-colored sunbirds of African forests. In general, forest species are less brightly colored than their open-country relatives. There is considerable variation over the group, however, and there are several subgroups whose members are very similar to one another.

The 10 species of spiderhunters are much duller and lack the iridescence of the sunbirds. They are predominantly greenish on the back and yellow or white on the underparts, and have long, strongly downcurved bills. The two sexes are similar, except that females may be slightly smaller and lack any pectoral tufts. The two sugarbirds are mainly streaked brownish on the back, and have a streaked chestnut breast and yellow undertail coverts. They also have long, graduated tails (those of the males are longer than the females'), and they are markedly larger than the other birds in the family.

All members of the family have beaks and tongues adapted to feeding on nectar. Even quite subtle differences in beak size can affect the birds' ability to reach the nectaries of different flowers, and there is evidence in a few species of slight differences between the sexes in bill length, suggesting that they may divide the available flower resources between them. The nectar is drawn up by capillary action; the tongue, which can protrude from the tip of the

FACTFILE

SUNBIRDS AND ALLIES

Order: Passeriformes

Family: Nectariniidae

132 species in 17 genera. Species and genera include: **Splendid sunbird** (*Cinnyris coccinigaster*), **Johanna's sunbird** (*C. johannae*), **Olive-backed sunbird** (*C. jugularis*), **Palestine sunbird** (*C. oseus*), **Superb sunbird** (*C. superba*), **Western olive sunbird** (*Cyanomitra obscura*), **Eastern olive sunbird** (*C. olivacea*), **Mouse-colored sunbird** (*C. veroxii*), **Golden-winged sunbird** (*Drepanorhynchus reichenowi*), **Giant sunbird** (*Dreptes thomensis*), **Pygmy sunbird** (*Hedydipna platura*), **Scarlet-tufted malachite sunbird** (*Nectarinia johnstoni*), **Tacazze sunbird** (*N. tacazze*), **spiderhunters** (genus *Arachnothera*), **Cape sugarbird** (*Promerops cafer*), **Gurney's sugarbird** (*P. gurneyi*).

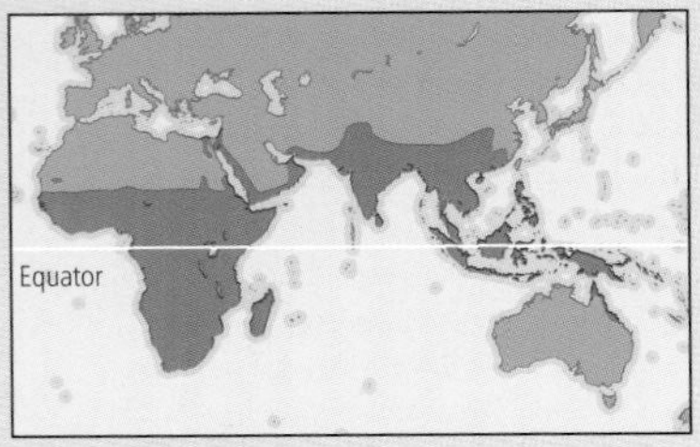

DISTRIBUTION Old World tropics from Africa to NE Australia, including Himalayas.

HABITAT Lowland and montane forest, second growth, savannas, gardens, thornscrub, moorlands, rhododendron forest.

SIZE Length 9–30cm (3.5–12in), including the tail, which accounts for about a third of the length; **weight** 5–25g (0.2–0.9oz). Sugarbirds are 25–44cm (10–17in) long, and weigh 30–45g (1–1.6oz).

PLUMAGE Sunbird males are bright iridescent blue, green, or almost black, often with bright red, yellow, or orange underparts; females are usually duller – olive green, gray, or brown, with tinges of yellow below and some streaks or spots; the colors of some males are highlighted by yellow or red display tufts at the bend of the wings and by long central tail feathers. Male spiderhunters lack metallic colors, as do some forest sunbirds of Africa. Sugarbirds are streaked brownish, with a chestnut breast, paler underparts, and yellow undertail.

VOICE Sharp, metallic songs – loud, high-pitched, fast, and tinkling.

NEST Purse-shaped in sunbirds; embedded or suspended; side entrance is often covered with a porchlike projection; often decorated or held together by spider webs. Nests of spiderhunters and sugarbirds are cup-shaped.

EGGS 2, sometimes 3, whitish or bluish white with dark spots, blotches, or streaks. Incubation period 13–15 days (17 days in sugarbirds); nestling period 14–19 days.

DIET Flower nectar and arthropods, especially spiders; rarely fruit.

CONSERVATION STATUS 2 sunbirds – the Amani sunbird (*Anthreptes pallidigaster*) of East Africa and the Elegant sunbird (*Aethopyga duyvenbodi*) of Indonesia are currently listed as Endangered; 4 other African species are Vulnerable.

bill and is split near the tip, curls laterally to form a tube to aid this process. The detailed structure of the tongue has taxonomic significance in the designation of genera. All species also have strong feet and sharp claws for perching on and near flowers. Most species have a fast, high-pitched, rather tinkly song, and some rather metallic calls.

Widespread in the Warmer Regions

DISTRIBUTION PATTERNS

The sunbirds are now usually divided into 15 genera, although previous authors have listed as few as four, absorbing most species into either the short-billed *Anthreptes* or, especially, the longer-billed *Nectarinia*. Two-thirds of the species are confined to Africa south of the Sahara, but the birds range up the Nile to the Middle East (in the case of the Palestine sunbird), and through Madagascar and nearby islands (where six species live) on to southern Arabia and all of India including the Himalayas. They are also found throughout the Southeast Asian mainland and islands, with one species, the Olive-backed sunbird, reaching the extreme northeast of Australia. The 10 species of spiderhunters, a closely related genus, occur in Southeast Asia, with one extending to western India. The two sugarbird species, the Cape and Gurney's, are confined to southern Africa; they are now included in the Nectariniidae, although at various times in the past they have been thought to be starlings (Sturnidae) or honeyeaters (Meliphagidae), or even assigned to a separate family of their own.

The Nectariniidae occupy almost all habitats from forests to semi-arid thorn scrub, and are even found above the treeline on mountains. They rarely occur in desert areas, however, except where there is a temporary abundance of flowers. In forests they inhabit both the canopy and undergrowth. The sugarbirds are very closely associated with *Protea* vegetation (known as *fynbos* in the Cape region of South Africa), and they will congregate in areas where these are flowering, especially on mountainsides.

The great majority of sunbirds are sedentary, although many make nomadic movements in response to the availability of flowers. The Pygmy sunbird is one of the few to exhibit long-distance seasonal migration, in its case from the arid south of the Sudan to the Uelle district of the Democratic Republic of Congo, where it breeds, timing its arrival to benefit from the flowering of local trees and bushes. Some species show seasonal altitudinal shifts; in Kenya, ringed Golden-winged sunbirds have been recovered at distances of 65km (40 miles) and 101km (63 miles) that included a considerable altitude shift as well.

Right *In a Cape Town botanic garden, a Southern double-collared sunbird* (Cinnyris chalybea) *perches on a flowering plant. This gorgeously colored species is relatively common in the Cape area of South Africa.*

A Diet of Nectar and Spiders

FEEDING BEHAVIOR

Nearly all Nectariniidae species eat both nectar and small arthropods, especially spiders. A few also eat fruits. The arthropods may well be taken from flowers, but the birds also take them from leaves, or even, like flycatchers, in the air. Fine serrations on the edges of the birds' delicate bills help them to capture and hold insects.

It is nectar, however, that is clearly the most important factor determining the birds' shape, habitat, and behavior, and most species are fairly catholic in their choice. The long-billed species are particularly prevalent at large red or orange flowers, and some of these, such as *Erythrina*, *Spathodea*, *Symphonia*, and others, have become especially adapted to have sunbirds as their principal pollinator. Few, if any, plants, however, are exclusively dependent on a particular sunbird species for pollination.

When a bird visits a flower, pollen sticks to its beak, feathers, or tongue, and so can be carried to the next flower. To assist the process, sunbirds, unlike most other birds, have opercula (flaps) to cover their nostrils, keeping pollen out. Sunbirds rarely hover in front of a flower in the manner of the Neotropical hummingbirds, preferring to feed from a perch, either on the flower itself or a nearby twig or stem.

Short-billed species, which tend to be more insectivorous, cannot always reach the nectaries of the flowers they visit by entering them from the front. So, like certain hummingbirds, they sometimes try to cheat the flower by piercing it through the corolla in order to steal its nectar. These holes may then be used by other birds and insects for the same purpose.

Many mistletoe flowers, which also depend on sunbirds for pollination, literally explode when a sunbird visits them. A new flower houses spring-like filaments and anthers bearing pollen. When a sunbird pokes its long bill into one of the slits on the side of the flower, the trap is sprung and the flower bursts open to spray a cloud of fresh pollen onto the forehead of the visiting bird. Studies in Kenya of sunbirds feeding at one *Loranthus* mistletoe revealed that, whereas young sunbirds exploded flower after flower in their own faces and became covered with pollen, adults often ducked quickly after they tripped the trigger.

Some nectar sources are defended vigorously by individual sunbirds; for example Golden-winged sunbirds have been known to defend patches of *Leonotis nepetifolia* in Kenya against both conspecifics and several other species. Although many sunbirds aggregate in large numbers at suitable flowers, they rarely form cohesive flocks. Some, especially among the insect-eating *Anthreptes*, participate in parties of mixed species.

Hanging, Purse-shaped Nests

BREEDING BIOLOGY

The breeding season is related to rainfall and to peaks in the abundance of insects and nectar. Some species may breed at almost any time of the year, however, and pairs may re-nest up to five times in succession. Most species defend territories with vigorous singing and displays; indeed, breeding male sunbirds have a reputation for being extremely pugnacious, and subordinate species in East Africa may breed successfully only when large, dominant species do not usurp their nectar supplies. The sugarbirds, however, often nest quite close together in loose colonies. Male Cape sugarbirds have a conspicuous display flight involving wing-clapping, in which the tail is held over the back.

Right *The Cape sugarbird is endemic to the* fynbos *botanic region around the Cape of Good Hope, South Africa. It specializes in taking nectar from* Protea *plants, visiting some 300 flowerheads each day.*

Below *Hungry nestlings demand food from a Yellow-breasted sunbird* (Nectarinia jugularis). *Sunbirds build untidy-looking hanging nests, often held together by spiders' webs.*

The nests of all true sunbirds are enclosed. Typically they are purse-shaped and suspended from a twig or leaf of a tree; they have, however, also been seen attached to an electric-light flex above a bar. Certain nests are located in bushes. Some species seem to nest close to wasp or bee nests, and there is some anecdotal evidence that the nest entrance is often pointed away from prevailing winds. The nests usually look ragged, presumably to help disguise their real nature, and they may be extensively decorated with spiders' webs and other debris. They have a side entrance that may have a porch projecting above the entrance hole, and there is often more material hanging below the actual nest cup. Many (perhaps all) spiderhunters and sugarbirds build cup nests; the latter place them in the middle of dense *Protea* bushes.

The nests are built mainly or exclusively by the female, who also does nearly all the incubation of the eggs, usually two in number. Incubation lasts from 11 to 15 days (17 days in sugarbirds). The fledging period is 2–3 weeks, and the young may be dependent for a further three weeks or so after that period, during which time the female may start the next clutch. Males of many species help feed the young, although the female plays the major part, and in some spiderhunters they also do a little of the incubation. Males are always nearby, however, and help defend the nest and its contents, as well as the flowers that are their mate's energy supply. In contrast to hummingbirds, which are often promiscuous, sunbirds are thought to be monogamous, although there are suggestions that a few may be cooperative breeders, or even polygynous, as in the case of the Giant sunbird. PL/FBG

Sparrows and Snowfinches

THE SPARROWS HAVE A WELL-DEVELOPED tendency to associate with humans, and are probably the best-known of all the world's birds, since most cities in Europe, Asia, and much of North America support large populations of either House or Tree Sparrows. In total, at least eight species regularly nest in, or in close proximity to, inhabited buildings.

The House sparrow is the most persistent of all, and is rarely found breeding away from humans; some birds, in fact, may never come into contact with a natural habitat. House sparrows have even been known to nest 600m (2,000ft) down a coal mine in Yorkshire, England; the species is only absent from the equatorial rain forests, deserts, and tundra.

The Most Familiar Birds

FORM AND FUNCTION

Because of similarities to the weavers in feeding behavior, courtship, and nest construction, the birds in this group were once regarded as a subfamily of the large family Ploceidae, which also included weavers, whydahs, and indigobirds, but they are now separated as a closely allied family.

The Rock sparrows or petronias are streaked gray and brown with a yellow patch on the throat, and usually occur in more arid areas than the true sparrows. However, the African species, including the Yellow-throated sparrow, which also extends to India, are more arboreal. The Rock and Pale rock sparrows better match their names, since they typically nest in holes in rocks or walls. They are fairly sociable birds, nesting in loose colonies of up to 100 pairs, the former often in company with House sparrows. In parts of its range the Pale rock sparrow occasionally becomes a pest when large numbers gather to feed on ripening millet.

Snowfinches live almost entirely at high altitudes at or above the snowline. In parts of Europe the White-winged snowfinch is often seen feeding around ski slopes, where it readily takes food scraps left by skiers. Several other species, including the Black-winged, Red-necked, and Plain-backed snowfinches, are much more elusive and rarely come into contact with humans.

From Cities to the Snowline

DISTRIBUTION PATTERNS

Some city-dwelling House sparrows may spend their entire lives within 1.5km (1 mile) of their birth place, while others may make longer journeys in their first year but then settle in a preferred location. The Tree sparrow fills the House sparrow's niche in the eastern parts of its range, where the House sparrow itself is largely absent; in parts of southern Asia where the two species overlap, however, the Tree sparrow occurs in towns and villages and the House sparrow occupies the surrounding countryside. In treeless regions of Mongolia, the Tree sparrow becomes the truly montane bird that its Latin name (*Passer montanus*) would suggest, as a ground-loving bird demonstrating affinities with the snowfinches.

◐ **Left** *A White-winged snowfinch perched on a rock high in the French Alps. These birds are commonly found around mountain hostels, and are a familiar sight to climbers.*

The Desert sparrow, in contrast, is a bird of the true desert, occurring in oases and wadis, occasionally near human settlements. It lives mainly on the small seeds of desert plants, and builds its nest in the crown of tall palm trees. Five species of gray-headed sparrows, in which the sexes are identical, are widely distributed throughout Africa and are very similar to one another, differing only subtly in plumage tones, calls, and bill size.

Apart from the White-winged snowfinch, which extends its range into western Europe, the snowfinches are birds of the extensive plateaulands of central and western China, and are among the highest-nesting of all birds, since they occur at altitudes ranging from 1,800 to 4,600m (5,900–15,000ft). In winter all snowfinches become gregarious, forming flocks that may descend to lower altitudes

FACTFILE

SPARROWS AND SNOWFINCHES

Order: Passeriformes

Family: Passeridae

36 species in 4 genera. Sparrow species include: **Desert sparrow** (*Passer simplex*), **House sparrow** (*P. domesticus*), **Tree sparrow** (*P. montanus*), **Rock sparrow** (*Petronia petronia*), **Yellow-throated rock sparrow** (*P. superciliaris*). Snowfinches (genus *Montifringilla*) include: **Black-winged snowfinch** (*M. adamsi*), **Plain-backed snowfinch** (*M. blanfordi*), **Red-necked snowfinch** (*M. ruficollis*), **White-winged snowfinch** (*M. nivalis*).

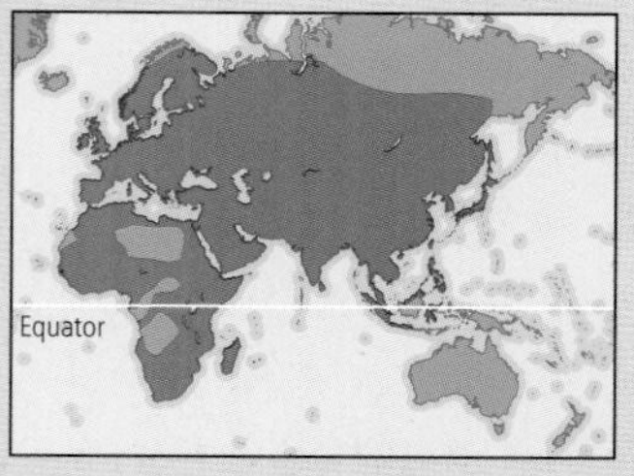

DISTRIBUTION Africa, Europe, Asia; introduced to the Americas, Australasia, and many islands.

HABITAT Almost anywhere in open forest, thornscrub, heaths, parks, gardens, edges of agricultural land, riverine vegetation; 1 species is restricted to desert oases; at least 2 species occur widely in villages and towns. Snowfinches are birds of the alpine zone.

SIZE Length 11.5–18cm (4.5–7in).

PLUMAGE Mainly brown and gray, but sometimes with black or (in 3 species) creamy to bright yellow. Snowfinches are slightly plumper in shape and generally pale brown with varying amounts of white.

VOICE Loud chirpings, twitterings, and some simple, trilled songs.

NEST Grassy nests constructed in a dome- or ball-like shape; also holes in trees, rocks, or buildings; many species gregarious, breeding in colonies in trees or bushes. Several snowfinch species make their nests in holes in the ground or animal burrows.

EGGS 3–7; length 1.8–2.2cm (0.6–0.8in); whitish, creamy, or pinkish, suffused with mauve-brown, grayish, or lilac markings.

DIET Predominantly seeds, vegetable matter, and some insects. True sparrows (*Passer*) are largely seed-eaters, with a marked preference for cereals, but they also take rice and the seeds and buds of a wide variety of plants; House sparrows, especially those living in cities and suburbs, can exist mainly on a diet of bread and household scraps.

CONSERVATION STATUS Not threatened.

in severe weather. Unless food becomes scarce, however, they rarely leave the mountains altogether, instead spending their time roaming the alpine meadows and boulderfields in search of wind-blown seeds.

Opportunistic Nest-builders

BREEDING BIOLOGY

Most sparrows build a roughly-constructed, ball-shaped nest. Many species are gregarious and breed in colonies, although the nests themselves are not communal like those of buffalo-weavers. Usually the nests are placed in trees; more rarely, they occupy holes in cliffs, from which sparrows have been known to evict both Sand (*Riparia riparia*) and House (*Delichon urbica*) martins.

Like those of sparrows, snowfinches' nests are large, bulky structures, made mostly of dead grasses and moss. White-winged snowfinches generally place their nests in rocky crevices, but have also been known to nest in buildings. The more elusive species, such as the Black-winged and Red-necked, are often dependent on nest holes made by pikas (*Ochotona* spp.), which they share with family parties of the lagomorphs. PC/PRC

Left *Far from the crowded haunts of its urban cousins, the Cape sparrow* (Passer melanurus) *inhabits arid and semi-arid savanna regions of southern Africa. It forages on the ground for insects and seeds.*

Below *House sparrows are gregarious by nature. Native to Eurasia and North Africa, they were first introduced to North America in 1852, when 100 birds were released in Brooklyn, New York City.*

Weavers

WEAVERS ARE NAMED FOR THEIR UNIQUE *nest-building behavior, in which the breeding male constructs a nest that is among the most intricate produced by any animal. Some nests are extraordinary communal structures, whose construction involves a high degree of social co-operation.*

Above *The Rufous-tailed weaver* (Histurgops ruficauda) *inhabits a limited area of northern Tanzania. Its dome-shaped nest of bark strips and thin twigs, once abandoned, may be reused by Fischer's lovebird.*

Below *The male Grosbeak weaver is identified by its white patches. It uses thin fibers shredded from grass to build its nest and tucks the edges of the side entrance to form a border which will not fray easily.*

Most weavers build a nest of a design that is peculiar to the species. Nests differ mainly in size, in the materials used, in the workmanship of the weaving, and in the length of the entrance tube.

The Bird World's Expert Builders

FORM AND FUNCTION

Nest materials include fine grass (in the Lesser masked weaver), coarse grass (Holub's golden-weaver), lichens (Olive-headed weaver), or leaf petioles (Red-headed weaver). The entrance tube is up to 0.5m (1.6ft) long in the Spectacled weaver, an insectivorous species that prefers to nest solitarily and in which the sexes are similarly colored. Polygamous species, such as the Cape weaver, in which the male adopts a colorful breeding dress, prefer to nest either in small groups or, like the Village weaver, in large colonies of several hundred nests.

In most species of the largest and most widespread genus (*Ploceus*, 63 species), thin strips are torn from the edges of grass blades or palm fronds and then woven into a stiff, kidney-shaped container with a bottom entrance, using the conical bill as a shuttle. The nest is suspended from the tip of a branch or frond, and once the first few strips have been knotted to the support, the male hangs below, using his legs as braces, and weaves a vertical ring. He then perches on the ring and weaves the walls of the nest chamber on one side and the nest entrance on the other.

The breeding males of many species appear so similar to one another that they are only separable by details of the markings or coloration of the feet, bill, and eyes. The females generally lack bright colors and appear dull and sparrowlike in shades of brown, olive, gray, and cream, making it difficult even to separate the females, juveniles, and nonbreeding males of the same species. In the monomorphic and monogamous species, the pair select a nest site and then build the nest and raise the young together. In the sexually dimorphic and polygamous species, the dull female visits males as they display at their nests, pops in and out, and then selects one in which to lay her eggs. She then completes the chosen nest by adding the lining, usually of soft seedheads and feathers, and by finishing off the entrance tube. She mates with the owner but then lays and incubates the eggs all alone, and rears the chicks. The male, meanwhile, starts to build a new nest in the hope of extending his polygamy to more females. Often he tears down a rejected nest to free up the site, builds a new one, and once again displays below it with fluttering wings and urgent calls.

The craftsmanship of a nest is not obviously related to the size or shape of the weaver's bill. Neatly woven nests are built by the fine-billed and insectivorous malimbes or Spectacled weaver, but also by the stout bills of the Little, Compact, and Grosbeak weavers. The latter two species show an earlier stage in nest evolution, where the nest is woven upright between reeds for support and has a side entrance, even though the weaving and

Above *Representative species of weaver:* ***1*** *Golden palm weaver* (Ploceus bojeri), *male displaying at nest;* ***2*** *Golden bishop* (Euplectes afer) *male courtship flight;* ***3*** *White-headed buffalo weaver* (Dinemellia dinemelli)*;* ***4*** *Social weaver* (Philetairus socius) *with nests in background;* ***5*** *Mauritius fody* (Foudia rubra).

strips used are still extremely fine. Such upright, side-opening nests are also characteristic of non-arboreal weavers in the second-largest genus, *Euplectes*, which includes the bishops, of reeds, weeds, and sedges, and the widows of open heath- and grasslands. Fodies (genus *Foudia*), a special group of bright red and yellow finches that is confined to Madagascar and adjacent islands in the Indian Ocean, several of which are endangered, also weave a side-opening nest.

The grass-nesting widows show a further variation in nest construction, where the male builds only a skeletal shell, well concealed among the tufts, while the female forms the main structure later when she adds the lining materials. This leaves the elaborately dressed males free to spend long periods in aerial patrol and exaggerated display over their nesting territories, each attempting to attract as many females as possible, as exemplified by Long-tailed or Jackson's widows.

Each female lays eggs of a particular color and markings, but while some species have plain eggs

FACTFILE

WEAVERS

Order: Passeriformes

Family: Ploceidae

118 species in 17 genera.

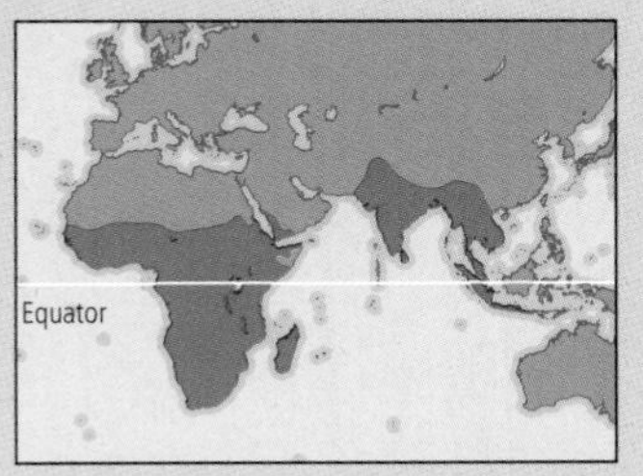

DISTRIBUTION Mainly sub-Saharan Africa, including Madagascar, with 1 species extending to Arabia; 5 species in India and SE Asia, extending E to China and S to Indonesia.

HABITAT Evergreen forest to semi-desert, with most species in open woodland, savanna, or thorn scrub.

SIZE Length 11–26cm (4–10in), a few up to 65cm (26in), as in breeding male of Long-tailed widow; weight 9–80g (0.3–3oz).

PLUMAGE Mainly yellow, brown, black, and white; some species with extensive red. Males of most species have bright, sometimes elaborate breeding dress, including changes to bill and eye color; females, juveniles, and non-breeding males often dull and sparrowlike.

VOICE Breeding males noisy, with variety of harsh and soft chattering, churring, and swizzling calls. Females and juveniles mainly have sharp contact and loud begging notes.

NEST Various structures of different degrees of complexity, from simple domes of grass stems, through bundles of thorny sticks or grass stems, to carefully-woven suspended chambers.

EGGS 2–7 (usually 3–5); ground color of white, pink, blue, gray, brown, or green, unspotted in some species, plain or variously marked in others, but generally consistent for individual female. Incubation 9–17 days, nestling period 11–24 days. Parasitic weaver lays 1–2 eggs to match those of small grass-warbler hosts.

DIET Mainly grass seeds and insects, but some fruit and buds; many species largely granivorous, a few largely insectivorous, but all feed insects to chicks. Most drink daily.

CONSERVATION STATUS 1 species – the Mauritius fody (*Foudia rubra*) – is Critically Endangered; 6 more are Endangered, and 7 are Vulnerable.

See Weaver Subfamiliies ▷

of a single color, others show considerable variation both between and within species. This diversity of appearance is thought to counter the extensive within-species parasitism that is reported for some species, making it more difficult for parasitic females to match their eggs to those of a host. Only the odd little Parasitic weaver, with its stout black bill, has become a true parasite, cuckolding the nests of various species of grass-warbler (Cisticolidae) and closely matching their eggs in color. The variation in color of weaver eggs as well as the diversity in the length of the nest's entrance tube may also be a defense against the parasitic Diederik cuckoo. This is a specialist parasite of weavers and widows that also lays eggs of different colors and has several times been found tangled in the tightly-woven tubes.

A variety of weavers reside in more arid areas where green, flexible grass is at a premium, and so each builds a nest that is thatched rather than woven. Males of the largest of all weavers, the starling-sized Red-billed buffalo-weaver, compile a great bundle of thorny twigs at the end of a branch, within which each attracts females to form neat nesting chambers lined with dry grass and green leaves. These males are unique among birds in having a false penis, used to stimulate the females to copulation and themselves to orgasm. The smallest of all weavers, the bearded Scaly weaver, builds a simple thatched dome with a side entrance; the nests are often joined in small clusters. It uses grass stems and panicles in a construction very similar to the nest of the waxbills (Estrildidae), which the weavers so resemble in size and shape. This aberrant little weaver is also unusual among granivores in being able to create water internally by metabolism of dry seed, so freeing itself from having to drink at regular intervals in its arid environment.

The various sparrow-weavers, including the unusually large Rufous-tailed weaver endemic to Tanzania, live in groups and build shaggy, tubular nests of stiff grass stems that they drape over branch tips, closing one end when they want to lay eggs but otherwise roosting individually in each of the open tubes. Most species live in small territorial groups of 5–10 birds, each with its own cluster of nests, but in the two small social weavers, such as the Grey-headed social-weaver, several groups may combine to cluster 50 or more nests in one tree. Only one pair of the White-browed sparrow-weaver builds a breeding nest within their cluster of roosting nests, but the Sociable weaver uses stiff grass stems to form a massive, haystacklike nest, on the underside of which are formed separate chambers to house as many as 100 breeding pairs. All of these thatched nests are used secondarily by various other birds for breeding and roosting, especially the huge nest of the Sociable weaver, which is home to the African pygmy falcon and is also co-opted by owls, geese, eagles, finches, and chats.

Above *Village weavers are large, noisy birds and are the most widely distributed weaver in Africa. They are often found close to human settlements.*

Right *The Red-billed quelea lives in such massive flocks, some numbering millions, that it is nicknamed the "locust bird." It nests in colonies that cover up to 100ha (250 acres) and contain more individuals than any other bird colony in the world.*

Weaver Subfamilies

True weavers
Subfamily Ploceinae

105 species in 10 genera. Species include: Black-headed weaver (*Ploceus melanocephalus*), Vieillot's black weaver (*P. nigerrimus*), Forest weaver (*P. bicolor*), Lesser masked weaver (*P. intermedius*), Holub's golden-weaver (*P. xanthops*), Olive-headed weaver (*P. olivaceiceps*), Spectacled weaver (*P. ocularis*), Cape weaver (*P. capensis*), Village weaver (*P. cucullatus*), Little weaver (*P. luteolus*), Baya weaver (*P. philippinus*), Streaked weaver (*P. manyar*), Giant weaver (*P. grandis*), Strange weaver (*P. alienus*), Clarke's weaver (*P. golandi*), Black-chinned weaver (*P. nigrimentum*), Bates's weaver (*P. batesi*), Rüppell's weaver (*P. galbula*), Yellow weaver (*P. megarhynchus*), Preuss's weaver (*P. preussi*), Chestnut weaver (*P. rubiginosus*), Heuglin's masked weaver (*P. heuglini*), Long-tailed widow (*Euplectes progne*), Yellow-crowned bishop (*E. afer*), Jackson's widow (*E. jacksoni*), Parasitic weaver (*Anomalospiza imberbis*), Red-billed quelea (*Quelea quelea*), Bob-tailed weaver (*Brachycope anomala*), Crested malimbe (*Malimbus malimbicus*), Red-headed weaver (*Anaplectes rubriceps*), Grosbeak weaver (*Amblyospiza albifrons*), Compact weaver (*Pachyphantes superciliosus*).

Buffalo-weavers
Subfamily Bubalornithinae

3 species in 2 genera: Red-billed buffalo-weaver (*Bubalornis niger*), White-billed buffalo-weaver (*B. albirostris*), White-headed buffalo-weaver (*Dinemellia dinemelli*).

Sparrow-weavers
Subfamily Plocepasserinae

8 species in 4 genera: Rufous-tailed weaver (*Histurgops ruficauda*), White-browed sparrow-weaver (*Plocepasser mahali*), Chestnut-crowned sparrow-weaver (*P. superciliosus*), Chestnut-backed sparrow-weaver (*P. rufoscapulatus*), Donaldson sparrow-weaver (*P. donaldsoni*), Grey-headed social-weaver (*Pseudonigrita arnaudi*), Black-capped social-weaver (*P. cabanisi*),Sociable weaver (*Philetairus socius*).

Bearded weavers
Subfamily Sporopipinae

2 species of the genus *Sporopipes*: Scaly weaver (*S. squamifrons*) and Speckle-fronted weaver (*S. frontalis*).

Predominantly African

DISTRIBUTION PATTERNS

Some weaver species are widespread, such as the Village weaver and Yellow-crowned bishop in Africa or the Baya and Streaked weavers in Asia. Others have more restricted ranges, including the fody species on the islands of the Indian Ocean, several of which are restricted to a single island, or the Giant weaver on São Tomé. More weaver species occur in Kenya and Uganda than in the rest of Africa combined, but several are very local and rare. Some are found only at restricted localities in montane forest along the Rift Valley, such as the Strange weaver, or in patches of special Kenyan habitat, such as Clarke's weaver in the Sokoke Forest and Jackson's widow on the high grasslands. Other species, such as the Black-chinned weaver of west Africa or Bates's weaver of Cameroon, are rare for reasons that are unclear. One species, Rüppell's weaver, has spread to western Arabia. The Asian weavers form a separate group, most of them on the Indian subcontinent, but some are rare, such as the Yellow weaver.

Most weavers are resident, especially the territorial insectivores of the forests, which include the most specialized of all weavers, the Preuss's and Olive-headed weavers that creep about on the bark of trees like nuthatches. Others move locally as the availability of food and water varies with the savanna seasons. Only the species of very arid areas are regularly nomadic, such as the Chestnut weaver of southwest and northeast Africa or Heuglin's masked weaver of sub-Saharan Nigeria. True migration is only reported for populations of Baya weavers that move down from the high altitudes of the Himalayas in winter.

The exception is the highly mobile Red-billed quelea which wanders across huge areas of African savanna in huge flocks, tracking the intertropical rain fronts that produce the crops of fresh grass seeds and small insects required for breeding. They settle briefly to nest and are highly synchronized, ready to move on only six weeks after the first egg is laid. This is in part due to the briefest incubation period for any bird of just 9–10 days. The offspring are ready to breed before they are a year old, breeding males in different color morphs depending on the blondness of the head and the extent of the black facial mask, and breeding females with the bill color yellow instead of red.

Controlling the Locust Bird

CONSERVATION AND ENVIRONMENT

Flocks of seed-eating birds have been agricultural pests across Africa since the time of the Egyptian pharaohs, especially for subsistence farmers growing limited areas of millet and rice. Large-scale efforts to control Red-billed quelea began in western Africa in the 1940s, and in 1955 the first conference on the quelea problem was organized in Senegal. Extensive research has continued ever since, along with control measures including firebombs and poisonous spray. These have met with limited success due to the mobility of the target species and its high reproductive rate. The Red-billed quelea is a source, though irregular, of food for many people, and it does attract and support a wide range of natural predators in the wilder parts of the continent. There are two other species of quelea, and a closely related Bob-tailed weaver, but these are no more pestilential than any other weaver species. Only the Red-billed quelea has become superefficient at exploiting the unpredictable supplies of seeds and insects so typical of Africa, even though similar supplies also support the majority of other weaver species. AK/PRC

STORY

The Industrious Weaver

❶ *The Southern masked weaver* (Ploceus velatus) *is the most widespread African weaver species, its range extending from southwestern Tanzania to the far south of the continent. To prepare a nest site, the male first strips a branch of all its leaves – to make it harder for snakes like the boomslang to approach the nest unnoticed – and then begins construction by weaving grasses into a simple ring.*

1

2

3

❷❸ *This highly adaptable species nests in a wide variety of indigenous and exotic trees, including acacias, planes, palms, cedars, and mesquites. It is even recorded as suspending its nests from barbed-wire fences. Prodigious builders, males work swiftly with grasses and reeds to form the initial ring into a doughnut shape, which is then further expanded into a substantial spherical or oval ball. Once the construction is complete, after about five days, the male hangs beneath it and sings to advertise his availability. As many as five nests may have to be built before the male successfully attracts a mate to take up residence.*

❹ *Siting the entrance at the base of the nest helps deter predators; protection is sometimes enhanced by a vertical entrance tunnel some 8–12cm long. In addition to suffering egg and chick theft, the Southern masked weaver is the main target host of the parasitic Diederik cuckoo* (Chrysococcyx caprius).

❺ *A gregarious species, the Masked weaver's large breeding colonies can comprise several trees, each containing dozens of nests. They are often near water, both for sustenance and for ready access to fresh grass-leaves. Here, both weavers and foam-nesting frogs* (Chiromantis sp.) *have occupied a tree in the middle of an ephemeral rain-pool*

5

4

Wagtails and Pipits

Because of their wide distribution, conspicuous behavioral displays, and association with open, often agricultural landscapes, wagtails and pipits are some of the best-recognized small passerines. The contrasting black-and-white or yellow-and-black coloration of the birds is made more noticeable by the conspicuous wagging of their long tail while flashing white outer tail feathers, and by the habit of using elevated perches for observation and flycatching.

All of the family are known for their very aggressive territorial flight displays and, especially, for their courtship song flights – a fluttering ascent from the ground or a perch, followed by a slow, parachuting descent accompanied by song. The cryptically colored pipits hold a record among birds in their prolonged flight displays, in which males ascend to more than 100m (330ft) above the ground and fly into the wind while singing loudly for up to three hours. Wagtails and longclaws routinely follow large grazing mammals or agricultural machinery during foraging on wintering and breeding grounds. The propensity of migrating wagtails and some species of pipits to form huge aggregations at migratory stopovers and massive communal roosts (of up to 70,000 birds in some wagtails) has facilitated banding and made these species some of the most studied migrants of the Old World.

Slender-bodied Ground-birds

Form and Function

Most wagtails and pipits are small, slender birds with characteristically long tails and legs. Asian White wagtails and longclaws are somewhat larger and bulkier. In most species the bill is slim and long, but in longclaws it is rather more robust. All species have long toes and often very elongate hind claws, especially in longclaws where the hind claw is thought to aid in walking over grass. The sexes are generally similar in size, but males often have slightly longer wings. Wagtails, many pipits, and longclaws all wag their tails when walking, especially at the end of runs and when disturbed by intruders.

In wagtails, the plumage is gray, black-and-white, black-and-yellow, or gray-and-yellow. Males in most species have brighter or more colorful plumage than females. In pipits, the plumage is brown above, usually streaked, while the underparts are mostly gray or whitish with streaks on the breast and sides. Both sexes are similarly plumed, with the exception of three species in which the males have either reddish coloration on their face and throat or bright yellow underparts. In the Golden pipit, the sexes are distinct; the underparts are bright black and yellow in males but only yellowish in the females. In longclaws, males and females both have cryptic dorsal plumage, but the underparts are brightly colored and variable among populations, being yellow, orange, or red, with a black pectoral band in some species.

Below *Pipits such as this Red-throated pipit* (Anthus cervinus) *forage on marshes or on drying seaweeds on coasts, as well as in drier, open areas.*

Wintering South

Distribution Patterns

Wagtails are mostly confined to the Palearctic, but three species crossed to the New World when the Bering land bridge supported tundra–steppe vegetation, and are now regular breeders in the arctic coastal uplands and islands of western North America. The eight species of longclaws and the Golden pipit are largely confined to savanna grasslands in Africa. Pipits occur on all continents, excluding Antarctica.

Most Palearctic wagtails and pipits winter in the

Left *Representative species of wagtails and pipits:* **1** *Yellow-throated longclaw* (Macronyx croceus); **2** *Yellow wagtail* (Motacilla flava) *holding an insect. This is the black-headed form;* **3** *Richard's pipit* (Anthus novaeseelandiae).

tropics of Africa, the Middle East, southeastern Asia, the Indian subcontinent, or Australia. Pipits in North America winter in Southeast Asia, the southern United States, or Central America, while North American wagtails share their wintering grounds in southeast Asia with Asian wagtails.

Most wagtails and pipits are middle- to long-range migrants, while some south Asian, Australian, and African species and populations are resident. Even in resident populations, pipits often undertake elevational and short-distance seasonal movements to avoid unfavorable weather conditions. Interestingly, all North American wagtail species have retained their ancestral migratory routes, which developed before the most recent submergence of the Bering land bridge, and migrate to their wintering ground in southeast Asia along the Asian coastline. Golden pipits and longclaws are resident or short-distance migrants.

The Yellow and White wagtail migrate during the day, in small, loose flocks, often undertaking long, nonstop flights of up to 70 hours' duration over desert and water. Prior to such crossings, the birds may form huge aggregations and forage at stopover locations. Other wagtails migrate singly or in small groups, and many are highly territorial during migration. Most pipits are diurnal migrants and migrate singly or in small flocks of 5–8 birds, although a few species occasionally form flocks several thousand strong.

Most wagtails and pipits winter on irrigated and grazed grounds in open grasslands, in fields of sugar cane or rice, or along the banks of streams, lakes, or coastlines. Wagtails, and especially the White and Yellow wagtail, routinely associate with zebras, antelopes, and grazing stocks on wintering grounds in Africa and Australia. White wagtails primarily winter in or near human settlements. In pipits, species that are closely associated with forested areas for breeding tend to winter in open oak forests, coffee plantations with large trees, mango groves, or in wooded areas along roads.

Most Yellow and White wagtail species and some pipits maintain winter territories along creeks and rivers. Territory size directly depends on food abundance, being smaller when supplies are plentiful. Wagtails and pipits frequently form large communal roosts on wintering grounds. In wagtails wintering in southeast Asia and Japan, these roosts can include several hundreds and even thousands of birds, and are often located on building roofs, in tall trees in city parks, along streets, or in industrial parks, as well as in reed beds and sugar cane fields. Communal roosts of wintering pipits are smaller (up to 20–30 birds) and are on the ground or in tall grass.

FACTFILE

WAGTAILS AND PIPITS

Order: Passeriformes

Family: Motacillidae

65–70 species in 5 genera. Species and genera include: **wagtails** (genus *Motacilla*), **Black-backed wagtail** (*M. lugens*), **Cape wagtail** (*M. capensis*), **Japanese wagtail** (*M. grandis*), **White wagtail** (*M. alba*), **Yellow wagtail** (*M. flava*), **Forest wagtail** (*Dendronanthus indicus*), **pipits** (genus *Anthus*), **American pipit** (*A. rubescens*), **Rock** or **Water pipit** (*A. spinoletta*), **Golden pipit** (*Tmetothylacus tenellus*), **longclaws** (genus *Macronyx*).

DISTRIBUTION Worldwide. Wagtails mostly in the Old World; pipits on all continents; longclaws and the Golden pipit in Africa.

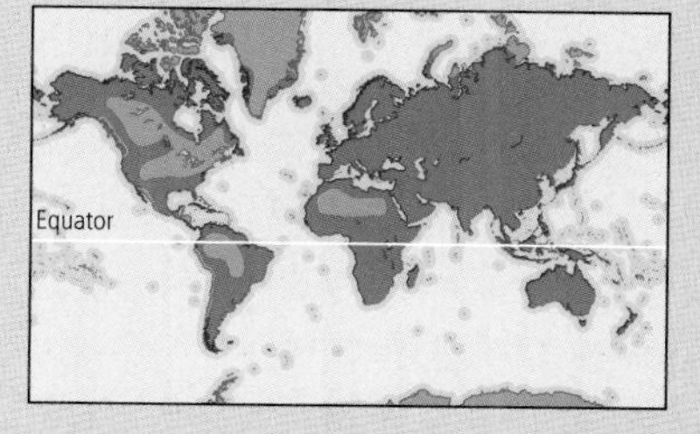

HABITAT Tundra, grassland, steppe, open woodlands.

SIZE Length 12.5–22cm (5–9in); weight 12–50g (0.4–1.8oz).

PLUMAGE Wagtails are gray, black and white, black and yellow, or gray and yellow, with pronounced sexual and seasonal variation. Pipits are mostly brown, often heavily streaked, paler below; some species with red or yellow colorations in males, otherwise no pronounced sexual or seasonal variation. Longclaws and the Golden pipit are mostly gray and brown above with bright yellow, yellowish, or reddish underparts (often with a contrasting dark pectoral band).

VOICE Sharp callnotes; simple and repetitive song; prolonged courtship song flights.

NEST An open cup built of dry leaves and grass stems, lined with fine grasses, leaves, hair, and feathers; built by female in most species. May be on the ground under overhanging vegetation; in cavities or crevices of cliffs; or in burrows or buildings. Incubation by both sexes in most wagtails, by female in most pipits; both sexes feed nestlings.

EGGS 2–7, white, gray, or brown, typically with brown or black speckles. Incubation 11–16 days; nestling period 11–17 days.

DIET Almost entirely arthropods; some grasshoppers, small mollusks and earthworms, infrequently some seeds.

CONSERVATION STATUS Two Kenyan species – the Sokoke pipit (*Anthus sokokensis*) and Sharpe's longclaw (*Macronyx sharpei*) – are Endangered as a result of loss of habitat; 3 other pipits are listed as Vulnerable.

Pickers and Flycatchers

DIET AND FEEDING

Wagtails and pipits mostly feed on terrestrial and aquatic invertebrates, especially arthropods. Diets vary among the species: beetles, grasshoppers, and crickets comprise most of the diet in longclaws and are frequently consumed by pipits, whereas various small mollusks, earthworms, and flies and their larvae are common in the wagtail diet. Wintering birds sometime feed on termites, and may consume some seeds and berries.

The birds' preferred method of foraging is picking and short-pursuit picking while walking or running through vegetation, or, less frequently, flycatching from perches (which is particularly rare in some pipits and longclaws). Yellow and White wagtails, and some longclaws, are regular commensal followers of grazing mammals; typically, a small group of birds will position themselves near an animal's head or feet and will move with the animal, constantly changing position while picking flashed insects off vegetation. Wagtails often flycatch while perching on the backs of grazing animals, or pick insects directly from an animal. The foraging success of wagtails associated with grazers is often twice as high as that of birds which forage alone.

Spectacular Song Flights

BREEDING BIOLOGY

Throughout their range, wagtails breed mostly in open, shrubby areas, in moist grassland or in tundra landscapes, especially along creeks, roadsides, lakesides, in human settlements, or along coasts, particularly in association with seabird colonies.

Across their vast geographical range, most pipits breed in open landscapes with low vegetation – lowland steppe, shrubby tundra, or along rivers and coasts. Some pipit species prefer a mixture of tree and grassland vegetation, forest margins, clearings, and burnt patches. A few pipits breed in taiga, relatively dense coniferous forests, and coastal evergreen forest, but even then mostly along creeks and in small clearings. Golden pipits breed in arid and semi-arid grasslands with bushes and small trees, while longclaws are common in wet meadows, densely covered grasslands, and often in high-altitude grasslands up to 3,400m (11,000ft).

Wagtails, pipits, and longclaws are strongly territorial during the breeding season, and use the breeding territory for feeding and nesting. Usually, both sexes patrol and defend the boundaries of breeding territories, undertaking territorial defense flights and frequently calling from song perches on the territory perimeter; they spend up to 3 hours a day performing these displays.

Although most birds nest in isolated pairs, in many species breeding territories may sometimes be clustered together. In such nesting neighborhoods, pairs may nest as close as 20m (65ft) to one another.

Some migratory pipits and wagtails, especially in the northern parts of their range, pair during migration or shortly before their arrival at the breeding ground. For example, pairing in high-elevation Water and American pipits may occur during prolonged stopovers at lower elevations early in the spring. In some wagtails, pairs are known to form on wintering grounds. In resident populations of Cape and Japanese wagtails, pairs often occupy a year-round territory and can remain together for several years.

Wagtails and pipits are mostly socially monogamous, but copulations outside of the social pair are observed in some populations. About 4 percent of Japanese wagtail males breed with two females. In some pipit populations, 6–7 percent (sometimes up to 20 percent) of males are paired with two females.

In wagtails, most courtship songs are given during display flights or from song perches. In courtship displays, wagtails may fly 30m (100ft)

Right *African Orange-throated longclaws* (Macronyx capensis) *are similar to the unrelated North American meadowlarks in their cryptic upperparts and bright underparts crossed with dark breast bands.*

Left *This Grey wagtail* (Motacilla cinera) *is on its way to deliver an insect meal to its brood. Unlike other wagtails the Grey wagtail regularly perches in trees, especially when it is disturbed during feeding.*

Below *White wagtail feeding its young. Wagtail nests are abundantly lined with hair from domestic or wild mammals, and sometimes with large feathers. Forest wagtails build very accurate small nests camouflaged on a branch with small pieces of lichen.*

up (although 8–12m/25–40ft is more typical), and then descend to the ground fluttering their wings and uttering the courtship song. The songs themselves are multiple, high-pitched repetitions of territorial and contact calls. Wagtails are often heard singing while foraging, and seem to sing especially vigorously when predators are in sight. In White and Black-backed wagtails, most courtship displays take place on the ground – the male spreads and twists his tail and opens both wings while approaching the female in circles.

In most pipits, the song is a continuously repeated series of syllables given while the male ascends up to 100m (330ft) and then flies, repeatedly, into the wind. The flights can last for as long as three hours in the Sprague's pipit (*Anthus spragueii*), although in most species a range from a few minutes to a half of an hour is normal. During the breeding period, birds may spend up to two hours a day singing. Occasionally, the song is delivered from the ground or a perch, while in the Golden pipit a song display is often delivered from the top of a bush. In longclaws, courtship displays are mostly delivered in flights lasting up to 45 minutes, or, more rarely, from perches.

Wagtails and pipits usually build nests on the ground, often in a slight hollow, which the females may excavate themselves. White and Black-backed wagtails are known to nest in small mammal burrows, beaver huts, or cavities and crevices of rock cliffs or buildings, sometimes up to 50m (165ft) above the ground. One exception,the Forest wagtail, nests on large trees, typically near the trunks 4–5m (13–16ft) above the ground. Most wagtail and pipit species build a relatively open nest, protected on one or two sides by a tussock, rock, or overhanging vegetation. Ground-nesting longclaws occasionally nest on top of high tussocks, especially in areas with a high risk of flooding.

In wagtails and pipits, the female usually does most of the nest-building, but in some wagtails the male contributes significantly. The male often brings the nest material, and closely guards the nest and the female during this period. Pairs may start building in several places within a territory before choosing the final nest site.

Wagtails and pipits lay 3–7 eggs; larger species have smaller clutches. Longclaws usually lay 2–3 eggs. The ground color of the shells varies from almost white to dark olive with light brown spotting, with the speckles sometimes concentrated to solid color at the larger end. In northern and high-elevation parts of their range, wagtails and pipits typically raise only one brood a season, whereas most species elsewhere have 2–3 broods a year.

In most wagtails both parents incubate; usually the female incubates at night and dusk, while either sex incubates in the middle of the day. In pipits and longclaws the female does most of the incubation, although both sexes may incubate in some pipits. Males usually feed their incubating females, either at or near the nest. Incubation lasts 11–15 days in most species. When approaching their nests, wagtails and pipits usually fly to the vicinity and then run or walk the last 10–20m (30–65ft).

Shortly after eggs hatch, the female either eats or carries away the shell fragments. In all species the nestlings at hatching are altricial (dependent) and nidicolous (they remain in the nest). They develop rather fast – in most species the eyes are open and contour feathers start to open on the fifth day after hatching. By 11 days after hatching, young wagtail and pipits can walk around the nest and can fly up to 30cm (12in). Nestlings usually leave the nest when 12–14 days old, but some pipits can leave the nests prematurely – as early as 9 days old if disturbed.

Both parents participate equally in providing food for the nestlings. In addition, the female warms the young on cold mornings and evenings up to the seventh day after hatching. Some wagtails and pipits deliver food as often as 8–12 times an hour, totaling as many as 300 deliveries a day. Both parents feed fledglings for 2–3 weeks after their departure from the nest.

A Mixed Picture

CONSERVATION AND ENVIRONMENT

Human activity has affected different species in different ways, and has also had contrasting effects in the wintering and breeding grounds. A growing number of human settlements, roads, bridges, and other constructions along the coastlines and in northern regions have actually increased the number of suitable breeding habitats for wagtails. Moreover, logging has provided abundant nest sites, including stumps, piles of trunks, and logging waste. In northern Europe, White wagtails recently colonized new habitat along forest creeks and reed marshes following the reintroduction of beavers and muskrats.

At the same time, disturbance and overharvesting by humans on the birds' wintering grounds seems to be a major cause of decreasing population numbers in some wagtail species, especially the Yellow wagtail. The birds are often killed en masse on their wintering grounds in Africa, where they share roosts with granivorous species that are controlled by avicides and explosives. In Southeast Asia, communal roosts of wintering wagtails are frequently disturbed and are overharvested by commercial trappers for food, leading to population declines in North American and northeast Asian wagtails. In some pipits and longclaws, the loss of native grasslands to agriculture, and especially to cattle grazing, burning, and haying, has reduced suitable breeding habitat and led to significant population decline. AVB/BW

Accentors

ACCENTORS DO NOT HAVE SPECTACULAR plumage, they behave unobtrusively, and most species live at high altitude and are poorly known. However, drab appearances can be deceptive, and the two species that have been investigated in any detail, the dunnock and the Alpine accentor, have both proved to have extraordinary reproductive lives.

Accentors are small, plainly-colored birds, mostly confined to mountainous regions. They are sparrowlike in appearance, but with a more slender and pointed bill. The sexes are similar in plumage, but males are slightly larger and a little brighter. Until recently they were thought to be most closely related to the thrushes, but recent biological studies suggest that their nearest relatives are in fact the wagtails and pipits (Motacillidae), sparrows (Passeridae), and finches (Fringillidae).

Small Birds in High Places

DISTRIBUTION PATTERNS

The accentors have the rare distinction of being almost exclusively restricted to the Palearctic region (incorporating Europe, Africa north of the Sahara, and Asia north of the Himalayas). The Alpine accentor has a wide but patchy distribution across this region, occurring from Western Europe and North Africa to Japan, but it occurs only at high altitudes of 1,500–5,000m (5,000–16,500ft). This distribution has resulted in the evolution of many races or subspecies: nine have so far been described. Other species, such as the Yemeni accentor (*P. fagani*) and Japanese accentor (*P. rubida*), have very restricted ranges, being confined to the mountainous regions of the countries from which they take their names.

All species of accentor except the dunnock breed in mountains. The Himalayan accentor can be found 5,000m (16,500ft) above sea level; the Robin accentor also resides at high altitudes, although it prefers to live in dwarf rhododendron and other scrub, or in willows and sedge in damp meadows. The Mongolian accentor breeds in the high country of Mongolia and north-central China. The Maroon-backed accentor is found from Nepal to western China, and shows a preference for damp areas deep in coniferous forest. The dunnock is the most cosmopolitan of the accentors in its distribution, being found in a variety of habitats from subarctic regions of Scandinavia and Russia to the Mediterranean.

Some species are migratory, such as the more northerly populations of dunnocks and the Siberian accentor, which breeds in boreal and subarctic zones and winters in Korea and eastern China. Other species show only altitudinal movements, descending to lower levels in winter. Accentors tend to live in remote, difficult terrain, so most species are poorly known.

FACTFILE

Accentors

Order: Passeriformes

Family: Prunellidae

13 species of the genus *Prunella*. Species include: **Alpine accentor** (*P. collaris*); **Mongolian accentor** (*P. koslowi*); **dunnock** (*P. modularis*); **Himalayan accentor** (*P. himalayana*); **Maroon-backed accentor** (*P. immaculata*); **Robin accentor** (*P. rubeculoides*); **Siberian accentor** (*P. montanella*).

DISTRIBUTION Europe, Africa N of the Sahara, Asia except S peninsulas.

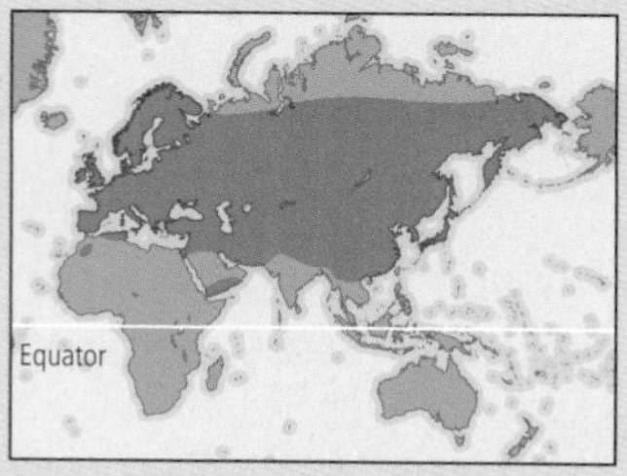

HABITAT Mountainous regions (except dunnock).

SIZE Length 13–18cm (5–7in); weight 15–40g (0.5–1.4oz).

PLUMAGE Upperparts rufous or brownish gray, streaked or striped in most species: underparts grayish, usually with rufous markings. Sexes are similar.

VOICE Little known; the dunnock has a high-pitched "tseep" call and a complex song structure; in Alpine accentors, females as well as males produce complex songs.

NEST Open cup of plant fragments and feathers, on the ground or in low scrubs or in a rock crevice.

EGGS 3–6; color ranges from light bluish green to blue; unmarked. Incubation 11–15 days; nestling period 12–14 days.

DIET Mainly insects in summer; seeds and berries in winter.

CONSERVATION STATUS The Yemeni accentor (*Prunella fagani*) is listed as Lower Risk/Near Threatened.

Left *The Maroon-backed accentor lives in the Himalayas, where it is found at altitudes of 2,900–4,600m (9,500–15,000ft). Even though the birds live far from people, they are not shy; mountain-climbers have reported accentors coming close in search of crumbs from meals.*

Right *Easily the best-known of the accentors, the dunnock is a common garden visitor across much of western Europe. In Britain it was long known, misleadingly, as the "Hedge sparrow," although its only real link with the sparrow family is its coloration.*

Above *The Robin accentor takes its common name from its robinlike red breast. Another Himalayan species, it needs thick plumage to keep it warm at altitudes up to 5,300m (17,400ft).*

A Complex Battle of the Sexes

BREEDING BIOLOGY

Dunnocks have an exceptionally variable social organization. Some breed monogamously in pairs, but other males defend the territories of two females (polygyny); alternatively, some females have two males sharing defense of their territory (polyandry), and sometimes two or more males share in the defense of two or more females (polygynandry). This variation depends on the ability of males to monopolize females, which in turn depends on the size of female territories.

There is intense conflict among males and females over their preferred mating system. Males do best in a polygynous situation, which permits them to father offspring by two or more females, and they do worst in polyandry, in which they generally have to share the paternity of a single brood. In contrast, females do best in polyandry because, provided they copulate with both males, both will help to raise the offspring and more chicks are likely to survive; they do worst in polygyny, because they have to share the male's care with another female. Monogamy is a compromise in which neither sex produces the maximum number of young.

These conflicts of interest within and between sexes generate some fascinating behavior, including the dunnock's unique precopulatory display. The female stands with her wings drooped and quivering, and her tail raised and vibrating rapidly from side to side. In response to this display the male hops from side to side behind the female and pecks at her cloaca. The display lasts a minute or so, during which time the male pecks the cloaca about 30 times. The female's pink and distended cloaca can be seen making pumping movements, until she eventually ejects a drop of sperm from previous copulations. At this point, the male jumps towards the female and the pair make cloacal contact for a fraction of a second. This performance is repeated once or twice an hour throughout the female's 10-day mating period. The likely purpose of the display is for males to increase the chance that their sperm, rather than that of a rival male, will reach the site where females store sperm, thereby enhancing their chance of gaining paternity.

The Alpine accentor also has a bizarre mating system; two to four males share a large range with a similar number of females. Within these polygynandrous groups, the dominant male attempts to monopolize matings with fertile females, but females seek copulations with every male in order to maximize the amount of male care that will be bestowed on their brood. There is intense competition among males over access to females, and also competition among females to attract males.

These conflicts have two remarkable consequences. First, Alpine accentors have an even higher copulation rate than dunnocks, and to cope with the demand the birds' testes constitute as much as 8 percent of male body weight. Second, females compete with each other to attract males by producing complex songs during their fertile period. BJH/MEB

Waxbills and Whydahs

WAXBILLS ARE SMALL BIRDS, NOTABLE FOR their diversity of plumage and, in many species, a bright, sealing-wax-colored bill. Many species are brightly colored or handsomely marked and this, together with the ease with which most can be maintained in captivity, has made them familiar cagebirds across the world.

The birds belong to an Old World family that occurs naturally in three areas. Most species are found in the Afrotropical region of sub-Saharan Africa, including Madagascar (19 genera and 78 species, plus 16 parasitic viduines – whydahs – in one genus), and in Australasia, including Wallacea and several Pacific islands (10 genera and 43 species). Fewer species occur in the intervening Indo-Malayan region (4 genera and 19 species). A number of domesticated species have also escaped in or been introduced to regions outside their normal range where, as successful colonists, they have become even more widely known.

Spritely and Brightly Colored

FORM AND FUNCTION

Some species of waxbills have somber colors of gray, brown, black, and white, but often make up for this with attractive markings, as illustrated by the Double-barred finch, Pictorella munia, and Bronze munia. Others are richly colored, including the cordon bleus and firefinches from Africa and the Green and Red avadavats from Asia. The most colorful of all are the beautiful grassfinches of Australasia, of which the Gouldian finch, with its combination of green, yellow, blue, turquoise, purple, and white, has the most exotic and gaudy appearance. This species also shows interesting variations in head coloration: about 75 percent of wild Gouldian finches have black heads, most of the remainder are red-headed, but about one in every thousand is yellow-headed. Several other waxbill species also have color morphs or geographical variants in plumage color and markings.

The true waxbills of Africa (genus *Estrilda*, 16 species), such as the Common and Swee waxbills are among the smallest species in the family, along with the 5 species of the genus *Uraeginthus*, including the Red-cheeked cordon bleu, and the 10 *Lagonosticta* species, among them the Red-billed firefinch. Most have a pointed bill and wedge-shaped tail, including even those few, such as the Common grenadier, with elongated tails,

Below *An assortment of different estrildid species gathered at a watering place – (top to bottom) two Black-cheeked waxbills* (Estrilda erythronotos)*, a male Green-winged pytilia* (Pytilia melba)*, and a Jameson's firefinch* (Lagonosticta rhodopareia).

◐ ***Left*** *There are two naturally-occurring color morphs of the Gouldian finch – some 25 percent are red-headed (far left), while the majority are black-headed. This Australian species is hugely popular in aviculture, and many different color variants have been produced by selective breeding.*

and all are light, agile, and easily able to perch on grass stems as they feed. A special ground-living form of waxbill (genus *Ortygospiza*, 3 species) occurs on the treeless grasslands, including the African quailfinch, which has longer legs and a shorter tail than most waxbills but still builds a typical nest, even if it is set under a grass tuft and on the ground. Three other African waxbills are unusual in that they breed in abandoned nests of weavers and other waxbills rather than building their own. These include the large Cut-throat and Red-headed finches (genus *Amadina*) of the bushy savannas, as well as the smallest of all waxbills, the Zebra waxbill of marshy habitats.

The Zebra waxbill, which is widespread in Africa and adjacent Arabia, has much in common with the quail-finches but may have its closest affinities with the Red and Green avadavats, two munialike species from Asia. Of these, the Red avadavat is unique within the family in that the male molts into a brighter plumage when breeding. The only group linking waxbills across their entire range are the speciose but dully-colored munias, mannikins, and silverbills, most of them in the genus *Lonchura*. These birds span across the continents of Africa (6 species), Asia (13 species), and Australasia (17 species), with, for good measure, an extra, monotypic Madagascar munia at one end and the Pictorella munia of northern Australia at the other.

Australia and adjacent islands, including New Guinea and Pacific islands as far east as Fiji, are especially well known for their colorful grassfinches. Many have long, pointed tails, but it is the short-tailed Zebra finch, one of the least brightly colored, that is best known (see The Mouse of the Avian World). Australasia has no fine-billed insectivorous finches like those of Africa, but some species, such as the Blue-faced parrotfinch and the Red-eared firetail, occur in forest or forest edge and exist secretively as pairs. The rest of the Australian finches live in open grasslands and savanna, close to water and often in large flocks. Only the Painted firetail and Zebra finch extend into the arid semi-deserts of the interior, where isolated waterholes or artificial boreholes supply their liquid requirements. These and several other species of dry-country Australian finches are unusual in drinking by "double-scooping," a special technique that conveys water directly into the crop rather than as repeated sips, as is the case of most other birds. This behavior allows the birds to exploit minute water sources, including dew-drops, and to drink as quickly as possible to avoid predators. No waxbill has yet extended its abilities to create internal, metabolic water from dry seeds, as has the waxbill-like but unrelated Scaly weaver (*Sporopipes squamifrons*; Ploceidae) of Africa.

The Red-browed firetail is said to be the Australian species most similar to the true waxbills of Africa, but whether it is a close relative remains debatable. It is close to the twinspot-like Crimson and Star finches and to other firefinch-like firetails such as the Painted and Mountain firetails. Allied to these birds, and also to the munias, are the unusual Plum-capped finch and the lovely forest-dwelling parrotfinches (genus *Erythrura*, 10 species), of which the Tawny-breasted and Pin-tailed parrotfinches extend to the bamboo thickets of Asia. Another ally to this group of birds might be their open-country cousin, the unique and gaudy Gouldian finch.

The true grassfinches of Australia, such as the Long-tailed and Masked finches, are similar to parrotfinches but less gaudy and most similar in behavior to the short-tailed Zebra and Double-

FACTFILE

WAXBILLS AND WHYDAHS

Order: Passeriformes

Family: Estrildidae

156 species in 30 genera (including parasitic subfamily Viduinae, 1 genus and 16 species).

DISTRIBUTION Sub-Saharan Africa and adjacent Arabia, Asia, Australasia, and several Pacific islands; viduines only in sub-Saharan Africa.

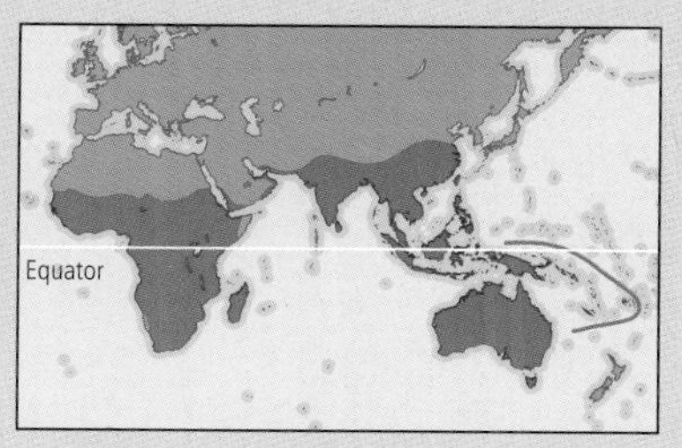

HABITAT Tropical forests, reedbeds, savanna, thorn scrub, open grassland, and semidesert.

SIZE Length 9–17cm (3.5–6.7in); weight 5.2–30g (0.2–1.1oz); some male viduines up to 41cm (16in) long due to elongated central tail feathers.

PLUMAGE A great variety of colors, patterns, and markings, often bright and attractive. Males more colorful than females in some species, similar in others. Male viduines (and one estrildine) have separate breeding plumage. Male viduines mainly black or steely-blue, some with white or yellow and a few with very long tail feathers; females, juveniles, and non-breeding males dull, brown, and sparrowlike.

VOICE Range of sharp, high-pitched calls for contact and flock synchronization, but no territorial song; males utter soft chirping and warbling songs when alone and during courtship. Male song of viduines mimics that of its host species.

NEST An untidy, enclosed dome with a side entrance, usually built of grass seed-heads and well-lined; from ground level to tree canopy, depending on species, most low down; a few species nest inside a hole or an old nest of another bird; most solitary, a few colonial.

EGGS Usually 3–8, white. Incubation 11–18 days, nestling period 16–25 days. Both parents share in all stages of nesting cycle. Chicks have distinctive, contorted begging posture and characteristic mouth markings. Viduines parasitic on nests of other waxbills; eggs also white, and chicks mimic mouth markings of host offspring.

DIET Mostly fresh and dry grass seeds. All species also feed on insects, particularly when rearing young, and a few species are mainly insectivorous.

CONSERVATION STATUS 2 species – the Gouldian finch and Pink-billed parrotfinch – are Endangered, and 8 are Vulnerable.

See Selected Waxbill and Whydah Species ▷

barred finches. A distinct form of Zebra finch also occurs in eastern Indonesia, close to the largest of all waxbills, the attractive and well-known Java sparrow, and the little-known Timor sparrow found on Timor and Roti north of Australia.

Ground-feeding Seed-eaters

DIET AND FEEDING

Waxbills feed mainly on grass seeds (Graminaceae). They occur alongside a number of other seed-eating families of birds, and have always been classified adjacent to or even within the diverse family of African and Asian weavers (Ploceidae). Dry seeds offer a diet rich in carbohydrates and some fat, but poor in protein and water. This means that most seed-eating birds have to drink at regular intervals and to consume some insects for protein, especially when producing eggs and rearing chicks. Dry grass seeds are abundant on the ground for much of the year, although they become in short supply when rain falls and they germinate. As they grow and a new crop of seeds ripens on the plant, food becomes especially abundant, nutritious, and accessible to the smaller waxbills that can land directly on a grass stem to feed off the seed head or slip between the stems to the ground. Agricultural cereals are just domesticated grasses, and so a few waxbill species have increased in numbers and become agricultural pests, especially on rice in Asia. Some species have been hunted for food or traded as cagebirds and, while some species remain unaffected, others are now rare or threatened.

Grasses that seed annually produce the most seeds and are abundant in the drier habitats of savanna and steppe, which is where most waxbills occur outside of cultivated areas, especially in Africa and Australia. Grasses are scarce inside forests, so the few forest-dwelling waxbills are either insectivorous or live close to clearings and along the forest edge. Indeed, the African antpeckers (genus *Parmoptila*, 2 species) look so like insect-eating warblers with their finely pointed bills that they were only classified as waxbills once it was realized that their ball-shaped nests and chicks with conspicuous mouth spots showed them to be estrildines. Woodhouse's antpecker feeds mainly on ants, using its brushlike tongue to sweep up workers and maybe also to add some nectar and pollen to its diet. In the same way,

Above *A pair of Blue-breasted cordon bleus* (Uraeginthus angolensis) *foraging. They often breed close to the nests of aggressive wasps for protection and to warn dangerous ants not to use the same bush.*

negrofinches (genus *Nigrita*, 4 species), such as the Gray-headed negrofinch, have a fine, insect-eating bill that they also use to take some fruits from among the foliage of the African forests. The African olivebacks (genus *Nesocharis*, 3 species) feed mainly on insects taken from the forest canopy, but one species with a thick bill, the White-collared oliveback, eats only the seeds from flowers of a single species of daisy (Compositae).

At the other extreme, also from Africa, are forest species that feed on seeds much larger than those of grasses, which they take with their especially large and powerful bills from legumes and other plants. Such species are typified by the aptly named seedcrackers (genus *Pyrenestes*, 3 species), especially the Black-bellied seedcracker, and bluebills (genus *Spermophaga*, 2 species), notably the Red-headed bluebill. The same trend is shown to a lesser extent by the related crimson-wings (genus *Cryptospiza*, 4 species) and twin-spots (6

Above *Originally from Asia, the Red avadavat was introduced to Europe as a cagebird. In Tuscany, Italy it is thriving in dense colonies among the reedy wetlands after only appearing in the wild in the late 1980s.*

Selected Waxbill and Whydah Species

Species include: Zebra waxbill (*Amandava subflava*), Red avadavat (*A. amandava*), Cut-throat finch (*Amadina fasciata*), Red-headed finch (*A. erythrocephala*), Gouldian finch (*Chloebia gouldiae*), Painted firetail (*Emblema pictum*), Blue-faced parrotfinch (*Erythrura trichroa*), Tawny-breasted parrotfinch (*E. hyperythra*), Pin-tailed parrotfinch (*E. prasina*), Pink-billed parrotfinch (*E. kleinschmidti*), Common waxbill (*Estrilda astrild*), Swee waxbill (*E. melanotis*), Pictorella munia (*Heteromunia pectoralis*), Pink-throated twinspot (*Hypargos margaritatus*), Red-billed firefinch (*Lagonosticta senegala*), Madagascar munia (*Lemuresthes nana*), Bronze munia (*Lonchura cucullata*), Bengalese finch (domesticated) or White-rumped munia (*L. striata*), Red-browed firetail (*Neochmia temporalis*), Crimson finch (*N. phaeton*), Star finch (*N. ruficauda*), Plum-headed finch (*N. modesta*), White-collared olive-back (*Nesocharis ansorgei*), Grey-headed Negrofinch (*Nigritia canicapilla*), Mountain firetail (*Oreostruthus fuliginosus*), African quailfinch (*Ortygospiza atricollis*), Java sparrow (*Padda oryzivora*), Timor sparrow (*P. fuscata*), Woodhouse's antpecker (*Parmoptila woodhousei*), Long-tailed finch (*Poephila acuticauda*), Masked finch (*P. personata*), Black-bellied seedcracker (*Pyrenestes ostrinus*), Green-winged pytilia (*Pytilia melba*), Red-headed bluebill (*Spermophaga ruficapilla*), Red-eared firetail (*Stagonopleura oculata*), Zebra finch (*Taeniopygia guttata*), Double-barred finch (*T. bichenovii*), Red-cheeked cordon bleu (*Uraeginthus bengalus*), Common Grenadier (*U. granatina*), Pin-tailed whydah (*Vidua macroura*), Village indigobird (*V. chalybeata*), Eastern paradise whydah (*V. paradisaea*).

Above *Representative species of waxbills and whydahs:* ***1*** *The White-backed munia* (Lonchura striata)*;* ***2*** *the Pin-tailed whydah* (Vidua macroura).

species in the genera *Mandigoa*, *Clytospiza*, *Euschistospiza*, and *Hypargos*), which take their name from their star-spotted breast feathers. The seedcrackers are also famous for their considerable variation in bill size, even within local populations of the same species. Bill size is related to the size of seeds chosen as food, but the ratio of small to large bills varies over time as the availability of different-sized seeds changes, and as large- and small-billed birds interbreed. All of these, and many other waxbills, also dig for food, both seed and insect, by flicking the bill sideways, although the bluebills more often hammer directly into the substrate. Their repertoire includes breaking into the runways of termites, a source of food especially favored by the Green-backed pytilia and it relatives (genus *Pytilia*, 5 species)

Most waxbills track any annual changes in food availability by forming large flocks during the nonbreeding season, sometimes of several species together. Individuals within flocks tend to synchronize their behavior, feed together, take off simultaneously, show coordinated flight movements, and all preen or bathe at the same time. Their coordination appears to be organized principally by calls in forest-inhabiting species, where visibility is poor, and by sight in species that inhabit open country. The pair bond is strong in most species, and members of pairs usually keep together even during the nonbreeding season. Pairs, family groups, and flocks also socialize, by clumping together when perched or by preening one another, and this tactile behavior probably serves to maintain and strengthen the bonds between them.

Parasitic Whydahs and Willing Waxbills

BREEDING BIOLOGY

Breeding usually occurs well after the first rains, once fresh grass seeds and small insects are available to nourish the nestlings and when seed heads can be used in nest construction. Waxbills are well-known for their complex courtship behavior patterns, which differ between the genera. The domed nest is usually built by both members of the pair and is characteristic of the family. It may be constructed for roosting or nesting and, during breeding, may include a flimsy cock's nest added on top for the male to sleep in. Both sexes also incubate the eggs and later feed the chicks in a unique posture – the chick's head is turned upside down to receive the adult's bill and the regurgitate from within its crop. The nestlings of each species have distinctive markings on the palate and tongue, revealed whenever they gape for food and probably developed to help parents locate the nestling's mouths in the semi-darkness of the nest and stimulate them to feed the chicks. The parents often eat their chicks' droppings at the beginning of the nesting cycle, but later carry them away or just leave them at the entrance.

Each species of viduine indigobird or whydah parasitizes a particular species of waxbill, by laying one or more identically-sized white egg in the waxbill's nest; the only exception is the widespread Pin-tailed whydah, which may cuckold several different host species. Viduine chicks mimic the mouth markings of their host's chicks with extraordinary accuracy and are reared alongside them, entirely by the host parents. Male viduines learn their host's song, and later sing it for use in their own courtship. All appear to be polygamous, attracting as many females as possible by their conspicuous plumage, as in the case of the Village indigobird, or by their spectacular tail feathers and aerial displays, as for the Eastern paradise whydah. Viduine species do not match exactly the diversity and relationships of their waxbill hosts, suggesting that this intrafamilial form of parasitism arose relatively recently in waxbill evolution, possibly as an extension of the within-species parasitism that is know to occur commonly in several waxbill species.

The Pink-billed parrotfinch and the Gouldian finch are both Endangered. Endemic to the island of Viti Levu, the Pink-billed parrotfinch is protected under Fijian law. Its forest habitat has been significantly reduced by around 50 percent while logging and clearance for agriculture still occurs. The Gouldian finch faces similar dangers although its main threat is from cattle-grazing which prevents grass from seeding. AK/SME

Above *As with other birds in their genus, the Black-crowned waxbills* (Estrilda nonnula) *of central Africa use their feet to clamp to a perch when feeding or gathering nest material. Estrildine finches typically nest on the ground, where their constructions are hidden in tall grass or at the base of a bush.*

Right *Aptly named the Beautiful firetail finch* (Emblema bella)*, this species has piercing hazel eyes with a light blue periophthalmic ring (the circle surrounding each eye). Shy and solitary birds, the female resembles the male but lacks the black patch in the center of the belly.*

The Mouse of the Avian World

The relative ease with which the 12g (0.4oz) Zebra finch can be kept and bred in captivity helps to explain why it is used so widely for experimental work in many parts of the world, particularly in laboratories in Europe and the United States. It is often placed in its own genus *Taeniopygia*, or united with the closely-related Double-barred finch and the typical Australian grassfinches in the genus *Poephila*. The Zebra finch is a specialist on grass seeds, rarely including insects in its diet, even when in the wild and when raising chicks, which means that it is easy to feed. An individual only requires about 3g (0.1oz) of seed and 4ml (0.14fl oz) of water per day for maintenance, although half-ripe seeds and more supplies are necessary when the birds are breeding or living under extreme temperatures. Pairs nest readily in colonies – the incubation and nestlings periods are only 11–15 and 17–18 days respectively – and sexual maturity is achieved by males just 70, and by females 100 days later.

Zebra finches have been the avian model for many studies, in the same way that the white laboratory mouse has provided a mammalian model. The finches have been especially important in studies on the ontogeny and role of communication, both vocal and visual, including the use of ultraviolet sensitivity. Their ability to withstand climatic extremes has made them important in studies of physiology, especially concerning the role of heat, food, and water. Their monogamous behavior and strong pair bond have involved them in evolutionary studies of mate choice and sexual selection, especially the roles of coloration and markings. Their tolerance of cross-fostering to the equally domesticated Bengalese finch has extended these studies to the influence of parents in mate choice, communication development, and species recognition. Young Zebra finches reared in this way become sexually imprinted on their fosterparents and, as a consequence, show abnormal mate choice when they become adult. They prefer to court and pair with members of the foster rather than their own species, and the effects of parental imprinting are remarkably stable. Similar experiments have shown that the male Zebra finch learns features of its courtship song during early development by listening to the songs of its father. Birds reared by natural fathers imitate their songs, while males foster-reared by Bengalese finches sing like members of that species.

Above *The Pin-tailed parrotfinch – the red-bellied male is shown here – is becoming increasingly uncommon due to persecution resulting from its fondness for rice in the paddyfields of Southeast Asia.*

Chaffinches

THIS IS A SMALL FAMILY, CONTAINING ONLY three species. All chaffinches are similar in shape, live in woodland, and take a varied diet of seeds and insects. In many ways chaffinches are the prototype seedeaters, representing a link between the insectivorous songbirds and the more specialized cardueline finches.

FACTFILE

CHAFFINCHES

Order: Passeriformes

Family: Fringillidae

3 species of the genus *Fringilla*: **chaffinch** (*F. coelebs*), **brambling** (*F. montifringilla*), **Blue** or **Canary Islands chaffinch** (*F. teydea*).

DISTRIBUTION Eurasia, Canary Islands; introduced to South Africa and New Zealand.

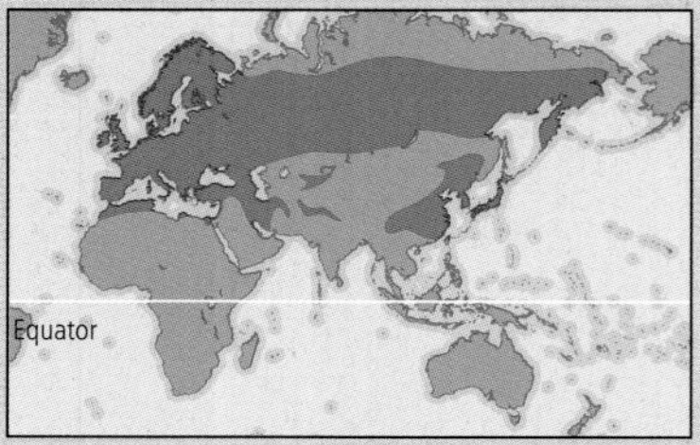

HABITAT Woodland and forest

SIZE Length about 15cm (6in); weight 26–30g (0.9–1oz).

PLUMAGE Males very colorful: in chaffinch, blue head, greenish back, and pink breast; in brambling, black and buffish back and orange underside; Canary Islands chaffinch mainly bluish. Females generally duller, and in chaffinch mainly pale green. All species have conspicuous shoulder patches, wing and tail markings.

VOICE Chaffinch has "spink, spink" call, and loud, musical song, lasting 2–3 seconds and consisting of a succession of "chip" notes followed by a flourish. Brambling has a harsh "tswark" note, and a softer "tchuck," mostly used on the wing, and a long-drawn-out "dwee" note, which constitutes the song.

NEST Mainly of grass, moss, and other vegetation, usually in a tree or bush.

EGGS 3–5, dark greenish-blue with purple-brown streaks and spots that have a paler rim. Incubation period 12–14 days; nestling period 11–17 days.

DIET Seeds; and the young feed on insects, especially caterpillars.

CONSERVATION STATUS The Blue chaffinch is listed by the IUCN as Lower Risk/Conservation Dependent, having been protected in its Canary Islands home since 1980.

Above *On its territory an unmated male chaffinch sings more than 3,300 times in a 12-hour day! Once he finds a mate, he sings less often.*

Woodland Residents

FORM AND FUNCTION

The breeding range of the chaffinch extends through Europe into Mediterranean Africa and western Siberia, almost to the tree limit. Throughout much of its range it is one of the commonest species in deciduous woodlands. At high latitudes, the chaffinch is replaced by the brambling, which is most numerous near the tree limit in birch or open conifer wood. Unlike the chaffinch, its range extends across Asia. Both the chaffinch and the brambling are migratory. Indeed, for most the brambling is a winter visitor, almost completely vacating its breeding haunts and migrating, by night, into southern Europe and Asia. Conversely, only some populations of the chaffinch migrate; those in the south are largely sedentary. Females tend to migrate further than males, leading the botanist Linnaeus to name the species *coelebs* (Latin for "bachelor"), since most of the birds wintering in his native Sweden were males.

Adult birds feed mainly on grain, seeds of the smaller weeds, and beechmast, supplemented by a wide range of invertebrates. Numbers in any one area may vary markedly between winters, depending on the amount of beechmast available; in good years a single brambling roost may number millions of birds, but in poor years few if any may be present. In Britain, wintering British chaffinches tend to forage singly or in small groups near their summer territory, but immigrants, mostly Scandinavian, roam over large areas in big flocks.

Brightening Up for Spring

BREEDING BIOLOGY

In late winter the feather tips of the nonbreeding plumage are worn away, revealing the brighter breeding dress. Breeding generally occurs between late April and June, and both chaffinches and bramblings can raise two broods, though in more northern areas one is usual. Both species are strongly territorial, with territory size varying between 0.1 and 4 hectares (0.25–10 acres), depending on habitat and food resources.

Fringillid nests are neatly constructed and well camouflaged; the cup is normally lined with soft hair or feathers. The young are dependent for about 5 weeks and are fed mostly caterpillars by both parents unless the female has a second brood, in which case the male will take over parental duties.

The Blue chaffinch, which is restricted to high-altitude pine forests in the Canary Islands, is similar in most aspects to its congeners. A full census has not been conducted yet, but there are an estimated 900–1,370 pairs on Tenerife and 185–260 individuals on Gran Canaria. Although it is not currently at high risk, fragmentation of its habitat and increased water use, particularly on Gran Canaria (which has an endemic race), remain potential threats. RR

Finches

CELEBRATING THE ENGLISH SPRING, WILLIAM *Wordsworth wrote: "Thou, Linnet! In thy green array, Presiding spirit here today, Dost lead the revels of the May." Like the poet, few people can be unfamiliar with finches – if not his Green linnet (better known today as the siskin), then maybe the goldfinch of North America, the canaries of Africa, or the rosefinches of Asia. With the exception of a few species, the Carduelidae are tree-dwelling forest birds, more so than their allies, the Fringillidae.*

Finches exhibit a range of beak sizes, each species adapted to foraging on the seeds of particular plant species, from thistledown for the goldfinch to the cherrystones favored by the hawfinch, whose beak can exert as much power as a person can with a pair of pliers. Many species frequently come to bird tables, taking advantage of the seeds put out for them. Certain species are also renowned for their song; early settlers took them far afield to remind them of home.

Old World Origins

Form and Function

Together with the fringillid finches and the Hawaiian honeycreepers to which they are closely allied, the carduelids are distinguished by the presence of nine (not ten) primaries, 12 tail feathers, and certain skull features. They occur throughout Europe, Africa (though not in Madagascar), and Asia, and some groups have colonized the Americas, predominantly in the north; a group of 12 remarkably similar siskin species in Central and South America represent something of an outpost and may be derived from a single, windblown flock. Particular species have also been introduced, with varying degrees of success, in various parts of the world.

Relationships between the three finch families (Fringillidae, Carduelidae, and Drepanididae) and the other recently evolved passerine families, particularly the buntings (Emberizidae) and the tanagers (Thraupidae), have been the subject of much debate. This confusion is, at least in part,

FACTFILE

FINCHES

Order: Passeriformes

Family: Carduelidae

About 136 species in 19 genera. Species include: **American goldfinch** (*Carduelis tristis*), **European goldfinch** (*C. carduelis*), **greenfinch** (*C. chloris*), **linnet** (*C. cannabina*), **redpoll** (*C. flammea*), **siskin** (*C. spinus*), **Red siskin** (*C. cucullata*), **Common** or **Red crossbill** (*Loxia curvirostra*), **Parrot crossbill** (*L. pytopsittacus*), **Scottish crossbill** (*L. scotica*), **Crimson-winged finch** (*Rhodopechys sanguinea*), **hawfinch** (*Coccothraustes coccothraustes*), **bullfinch** (*Pyrrhula pyrrhula*), **House finch** (*Carpodacus mexicanus*), **Island canary** (*Serinus canaria*), **Pine grosbeak** (*Pinicola enucleator*), **São Tomé grosbeak** (*Neospiza concolor*).

DISTRIBUTION N and S America, Eurasia (including Indonesia and Philippines), Africa (except Madagascar); introduced to New Zealand and Australia.

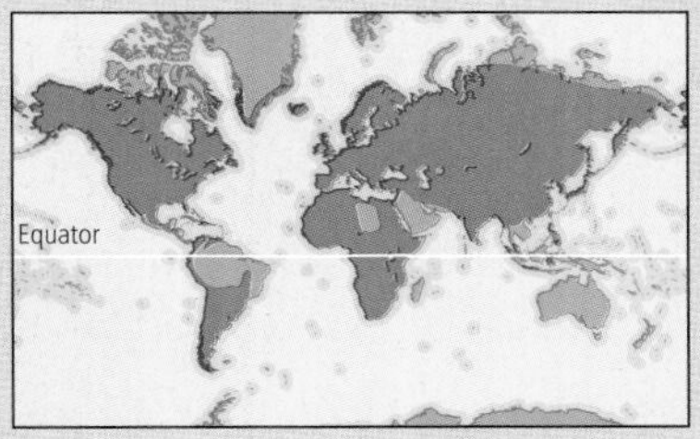

HABITAT Woodland and forest

SIZE Length 11–19cm (4–7.5in); weight up to 100g (3.5oz).

PLUMAGE Varied in color, but typically green, yellow, or red, with prominent wing and tail markings; many species streaked, especially in juvenile plumage.

VOICE Very varied, but most have pleasant, musical songs of pure notes; a few, such as the bullfinch, have rather coarse, creaky songs.

NEST Built mainly of grass, moss, and other vegetation, usually in a tree or a bush.

EGGS 3–5, whitish with brown spots. Incubation period 12–14 days; nestling period 11–17 days.

DIET Seeds; young fed on seeds and insects or seeds alone.

CONSERVATION STATUS The little-known São Tomé grosbeak is Critically Endangered, and 5 other species – the Red siskin of Venezuela and Colombia, the Hispaniolan crossbill, the Azores bullfinch, and two species from the Horn of Africa – the Ankober serin and the Warsangli linnet – are Endangered. 3 other species are Vulnerable.

Left *This Bullfinch is drinking water from a pond probably not too far from its nest; bullfinches rarely move more than a few miles from their home territory.*

attributable to the recent origin of these groups and their relatively rapid speciation; as a result, differences between them tend to be much smaller than between the nonpasserine families. Consequently, different authors treat groups as subfamilies within the same family or, alternatively, as separate families within a superfamily.

Traditional taxonomies (namely, those based on morphology rather than on molecular analysis) considered the fringillids and carduelids as the closest relatives, but recent DNA evidence indicates that in fact the carduelids are actually closer to the drepanididines, both apparently having derived from a fringillid-like ancestor. As a group, the three families are probably most closely related to the emberizid buntings, with which they are sometimes joined in a single family. The finches almost certainly arose and radiated in the Old World, whereas the emberizids probably have a New World origin, only later spreading into the Old World.

Crossbills are distributed throughout the northern hemisphere; however, the taxonomic status of one particular species excites some interest. The Scottish crossbill has a bill that is intermediate in size between the smaller Red crossbill and the larger Parrot crossbill, seemingly adapted to feeding particularly on cones of the Scots pine (*Pinus silvestris*). It is sometimes considered as a form of the Red (or occasionally Parrot) crossbill. If it is in fact a true species, it is the only bird endemic to the British Isles, and one of very few found only in Europe. The specific status of the endangered Hispaniolan crossbill, which occurs in Haiti and the Dominican Republic several thousand kilometers away from the usual crossbill range, is similarly uncertain.

Left *As well as sunflowers, this breeding male American goldfinch also feeds on thistles and dandelions. Between 11–13cm (4.5–5in) in length, it is able to balance easily on seed-heads.*

Right *Representative species of finches:* ***1*** *Pine grosbeak* (Pinicola enucleator) *in flight. When food is scarce, for example if seed crops fail, this species may show irruptive movements southward;* ***2*** *The Common or Red crossbill* (Loxia curvirostra) *is an erratic and nomadic species.*

Above *A European goldfinch has devised a way to get the berry at the end of the string: it is lifting the string with its bill while holding the loop with its feet.*

Seedeaters Par Excellence

DIET

The cardueline finches feed almost entirely on seeds; some species, such as the linnet, even raise their young entirely on seeds. The largely northern *Carduelis* species tend to be fairly agile and have relatively fine bills for picking seeds out of seedpods and plant heads (typically from the family Asteraceae – for example, thistles and dandelions). The canaries and seedeaters of the other large genus, *Serinus*, mostly dwell in the relatively arid landscapes of Africa and have heavier bills, more often feeding on the ground on fallen seeds. The remaining species are a fairly diverse group including the hawfinch, which can crack open cherrystones with its massive bill, the crossbills, which are uniquely adapted for extracting seeds from pine cones, and the rosefinches, which inhabit the mountain ranges of Asia.

The carduelines' seed-dominated diet contrasts with that of the fringillid finches, which are much more catholic, taking a range of invertebrates in addition to seeds. Consequently, they show some differences in skull structure and the associated musculature, and have much more specialized feeding habits and adaptations (see The Beak of tthe Finch). They also tend to be less territorial; most species, with the notable exception of the bullfinches, nest in loose colonies and forage in flocks, even during the breeding season.

Many species migrate to separate wintering quarters. Those in Europe and North America often migrate south during the winter as seeds become unavailable, although the extent to which this occurs varies between species and even between populations of the same species. The rosefinches of central Asia, which may be found at altitudes of 4,000–5,000m (13,000–16,500ft) during the breeding season, typically move to lower elevations during the winter, although some species are remarkably hardy and remain at higher altitudes. The African seedeaters, on the other hand, tend to be either mostly sedentary or else rove in nomadic flocks searching for seed-rich areas. The northern tree-feeding species – siskins, redpolls, grosbeaks, and crossbills – also tend to be nomadic, following a fluctuating food supply.

In years when the seed crops fail, particularly following periods when the crop has been good, birds may irrupt, moving out of their normal

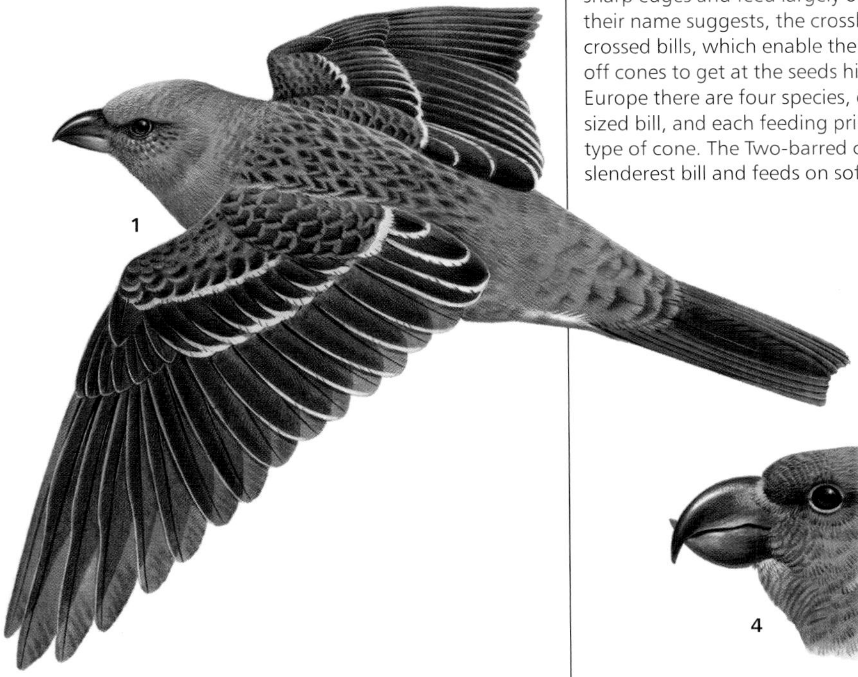

THE BEAK OF THE FINCH

The beak of a finch is modified internally for husking seeds. Each seed is wedged in a groove on the side of the palate and crushed by raising the lower jaw. The husk is peeled off with the aid of the tongue and then discarded, while the kernel is swallowed. The more specialized cardueline finches have developed musculature that allows much more lateral movement of the bill, making the husking process more efficient.

Ancestral finches probably had general-purpose bills similar to that found in the chaffinch today. From this basic shape a number of modifications have evolved. Goldfinches and siskins have long, tweezerlike bills for probing into seedheads and small cones. The goldfinch is the only species able to eat the seeds of teasel, which lie at the bottom of long, spiked tubes. They are also very agile, enabling them to take seeds remaining on the plant. Siskins also have tweezerlike bills and feed largely from seeds in small cones, such as alder. The African canaries and Asian rosefinches have stouter bills and take a much wider range of seeds and berries; they are also less agile, and generally feed on the ground or in bushes.

Bullfinches and grosbeaks have rounded bills with sharp edges and feed largely on buds and berries. As their name suggests, the crossbills have unique crossed bills, which enable them to prize the scales off cones to get at the seeds hidden underneath. In Europe there are four species, each with a different-sized bill, and each feeding primarily on a different type of cone. The Two-barred crossbill has the slenderest bill and feeds on soft larch cones; the medium-billed Red and Scottish crossbills differ slightly in their bill shape, with the former feeding on spruce seeds and the latter on those of Scots pine; the large, heavy-billed Parrot crossbill also feeds on pine seeds.

Among the finches, the hawfinch has the largest bill, enabling it to crack open very hard-shelled seeds. To deal with these, the hawfinch has two finely serrated knobs on either side of the midline on both the upper and lower mandibles. These spread the strain of cracking hard seeds evenly over all muscles. With these tools hawfinches, which weigh just 55g (2oz), can generate the loads of up to 45kg (100lb) required to crush olive stones.

Below *Heads and bills of the finches:* ***1*** *Hawfinch* (Coccothraustes coccothraustes); ***2*** *Siskin* (Carduelis spinus); ***3*** *European goldfinch* (C. carduelis); ***4*** *Parrot crossbill* (Loxia pytopsittacus); ***5*** *Two-barred crossbill* (L. leucoptera).

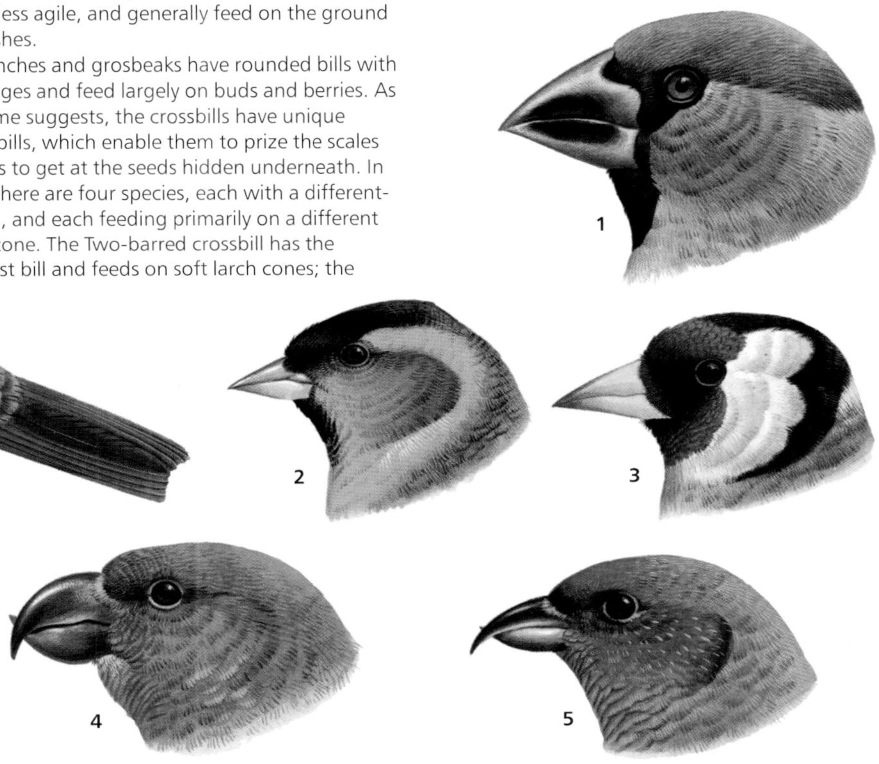

ranges in enormous numbers in search of food. As long ago as 1251, the chronicler Matthew Paris wrote about strange birds (crossbills) that had invaded England in great numbers and devastated the apple crop (there being few conifers to feed on). Such irruptions occur throughout the birds' range, but are best documented in Europe, where they occur at irregular intervals up to 17 years apart. Reports of irruptive crossbills ringed in Switzerland have come from as far afield as Spain later in the same winter (the birds having continued in their search for food), or, in subsequent years, from birds breeding in Siberia, 4,000km (2,500 miles) to the northeast.

Pairing Up

BREEDING BIOLOGY

Unlike the chaffinch or brambling, in which the male first establishes a territory and then attracts a mate, carduelines will pair up first, then find a nest site and establish a small territory round it. Pair-formation may take some time, with a male bird repeatedly approaching a female, often in a crouching posture, dropping or raising his wings and/or spreading his tail feathers, depending on the species. Many species also make display flights – some, such as the American goldfinch and linnet, fly over the nest, others, like the redpoll, ranging more widely. In the *Carduelis* and *Serinus* species, the display is often rather slow and butterflylike, but in the trumpeter finches and Crimson-winged finch the bird circles high with an undulating flight, alternating fluttering ascents with gliding descents.

Because they eat a wide variety of seeds, many finches have extended breeding seasons, and individual pairs may raise two or three broods (unlike the fringillids, which feed their young mainly on caterpillars and are single- or double-brooded). The goldfinches of both Europe and North America prefer to feed on thistles and breed later in the season, the American goldfinch being one of the latest nesters on that continent. The crossbills, however, may nest in any month; depending largely on the erratic cone crops of conifers, this may be during the spring in pine forests, in late summer in larch forests, or even in winter in spruce forests, while if a mix of trees is available, breeding may occur almost continuously. Breeding in winter has been recorded near Moscow in air temperatures of –19°C (–2.2°F); the incubating female maintained the nest at a cosy 38°C (100.4°F). Many of the African species have similarly extended breeding seasons, again to exploit seed crops that arise after the rainy periods.

The nest is usually a cup made of mosses, grass, animal hair, or feathers, and is typically built by the female, although male bullfinches and grosbeaks help in the early stages. In general, the female alone is responsible for incubating the eggs and brooding the young, for a period of around 3–4 weeks. During this time the male, who also feeds the young when they hatch, feeds her. As the hatchlings get older, the female will join in feeding them. Parents will often forage some distance from the nest in search of suitable seeds or insects for their young, returning at infrequent intervals (perhaps every 20–60 minutes). Although most finches carry food back to nestlings in their gullet, the grosbeaks, trumpeter finches, and bullfinches develop special sacs in the floor of their mouths, on either side of the tongue, to hold food. These, when full, extend back as far as the neck, giving a noticeably bulging throat. They seem to be formed anew each spring. Although some of the larger species can live in excess of 15 years, most birds probably only live for 2–3 years.

Below Both hawfinch parents feed their young after hatching. The difference between the male and female can be seen clearly: the male (on the left) is more brightly colored than the female.

Songsters and Visitors

CONSERVATION AND ENVIRONMENT

Finches have a long history of association with humans. The Island canary, common in wooded parts of the Azores, the Canary Islands, and Madeira, was first brought to mainland Europe in the 16th century and was rapidly domesticated to become the ancestor of all cage canaries. Although the plumage of the wild canary (a rather dull streaky green) differs between the sexes, those of cage birds seldom do. A wide range of forms have been carefully bred from chance mutations, differing widely in shape, color, and song (which is also more elaborate than in the birds' wild cousin). For example, the "Norwich" is a bulbous bird with rather fluffy feathers, whereas the "Yorkshire" is a much slimmer type with a sleek plumage. A number of other finches, particularly the goldfinches, have been popular as cage birds for both their bright plumage and delightful song; it is no coincidence that the name for a flock of these alluring birds is a "charm." They are crossbred with canaries (producing hybrids, or "mules") in the search for new color forms and better songsters.

The birds have also been put to other uses. They are more sensitive than humans to carbon monoxide (which we cannot smell), and were used by miners and by tunneling soldiers in the First World War to warn of the buildup of this gas. Canaries were particularly popular because their bright yellow plumage made them easy to see in the gloom. More recently, canaries' singing behavior has been studied with a view to improving the treatment of neurological disorders. The birds sing only in the spring and summer, and the part of the midbrain (the song control nucleus) that controls the voicebox or syrinx is much larger in males, particularly those with extended song repertoires, than in females. It is also larger in spring than in winter, and this seasonal change in size is caused

species, such as the siskin, have learned to join in. The Pine grosbeak is extremely tame, allowing approaches almost to within touching distance.

Only nine species of finches are considered to be at risk, most of them because they have very restricted ranges and small populations, with the result that although they may not be immediately threatened, they are vulnerable to habitat loss and degradation. The São Tomé grosbeak, for example, was only known from three 19th-century specimens until it was rediscovered in 1991; today its population is thought to number less than 50 birds. Although much of the trade in wild finches has ceased, two species are still threatened by illegal trapping and trade in captive individuals: the Red siskin, which once roamed in semi-nomadic flocks over much of northern South America but is now confined to parts of Venezuela and Colombia, and the Yellow-faced siskin of Brazil. RR

__Above__ The House finch is indigenous to the western USA but in the early 1940s several caged birds were illegally released in New York City. Within the space of 60 years these House finches have expanded across the entire eastern half of the States.

__Right__ During long winter nights at high latitudes, redpolls seek shelter by sleeping in snow tunnels. They are such lively birds that even when they are resting some flock members are constantly active.

both by the growth of individual cells and by the largescale birth and death of neurons, something that had previously not been thought to occur. If this process can be understood, it may be possible to apply the knowledge in the treatment of neurological wasting diseases, such as Alzheimer's.

Some species are pests of various crops. In Britain, the bullfinch has been a major pest of orchards, taking the flowerbuds of many fruit trees, particularly those of pears and plums. In earlier times the bird had a price on its head – in the 16th century, a penny per bird. In the 1960s and 1970s the problem was particularly acute, as an increasing population expanded out of the birds' traditional woodland habitats. Following a precipitous decline in numbers in the late 1970s and 1980s, however, damage is much less of a problem, and the species is now even seen as a target for conservation measures. In America the House finch is also a major pest of fruit crops, particularly in California, and in Africa some species can become locally abundant in particular areas and do serious damage to crops.

Many temperate species are likely to have extended their ranges as humans have cleared the forests for agriculture, creating a mix of open and wooded habitats. The serin, for example, has increased its range within Europe dramatically in the last 200 years. More recently, these birds have learned to exploit the seed-feeders many people put out during winter; even typically woodland

Hawaiian Honeycreepers

THE HAWAIIAN HONEYCREEPERS ARE A remarkable group of birds that evolved in the Hawaiian Islands from a single ancestral finch that colonized this most isolated of archipelagos around 5 million years ago. The rapid evolution of many highly varied species resulted in the most spectacular avian example of adaptive radiation, one that far exceeds the range of variation among Darwin's finches of the Galápagos (see Tanagers and Tanager Finches) and spans nearly the entire range of passerine – and some nonpasserine – adaptation.

The divergence between species appears greatest in the feeding apparatus, but includes plumage, vocalizations, and ecology. The disparate species nevertheless exhibit several underlying threads of similarity, among them details of skull and tongue and a distinctive, musty odor, and recent DNA studies show them all to be very closely related.

Evolutionary Opportunists

FORM AND FUNCTION

Some drepanidines remain finchlike in bill and tongue structure. The Laysan and Nihoa finches have highly varied diets that include seabird eggs, insects, and many kinds of vegetable matter, but others are more conventional seed eaters, specializing on particular trees: the palila on mamane (*Sophora chrysophylla*); the Kona grosbeak on naio (*Myoporum sandwicense*); and the Greater and Lesser koa finches on koa (*Acacia koa*). The finchlike bill is lengthened and modified for fruit-eating in the hook-billed 'o'u and the (extinct) tanager-like 'ula-'ai-hawane, for gleaning land snails and other invertebrates from tree bark in the po'o-uli, and for purposes unknown in the Lana'i hookbill, known from only a single specimen with a large gap between the opposite-curved mandibles.

Among early divergences from the finchlike bill are the Maui 'alauahio and the kakawahie, warblerlike birds with thin, straight bills; and the 'akikiki and Hawaii creeper, which pick bark like nuthatches using a slightly curved bill. The remaining drepanidines, with a wide variety of bill shapes, share a unique tongue structure, tubular with a brushlike tip, that probably first evolved for nectarivory (the basis of the name "honeycreeper"). The nectar-adapted tongue proved equally adept for capturing insects, and was the apparent key to the greatest burst of adaptation.

Songs of drepanidine finches are canarylike and highly complex, while those of the insectivores tend to be simple trills or warbles. The latter also have quiet songs that are quite complex and may include mimicry. Those of the nectarivores stand apart and resemble the songs of some honeyeaters (Meliphagidae), with haunting overtones, reedy and bell-like notes, rusty-hinge squeaks, and mechanical sounds.

Many Hawaiian honeycreepers are primarily yellow-green, but the finchlike ones often have yellow or red heads, and the nectarivores exhibit brilliant crimson, vermilion, yellow, and black patterns. A few species show brown or gray tones. Juvenile plumages often include pale wing-bars, retained in the adults of a few species. Even the brighter colors are usually cryptic. Males are usually brighter than females, some strikingly so, but among the nectarivores, creepers (*Oreomystis*), and a few others, the two sexes are identical.

Drepanidines closely resemble the related cardueline finches in breeding biology, with movable, female-centered territories early on, and small defended areas once the nest is built. Nests are mostly open cups hidden in terminal leaf clumps of tree branches, but some are in cavities. Eggs are whitish, variably blotched and scrawled with brown, rusty or purple around the larger end. Females do most of the nest construction, incubation, and brooding, but males bring materials and food. Breeding seasons are unusually long, stretching from November through August, with a peak in the spring. Juveniles may follow and be fed by parents for up to a year. Nest helpers are typical only of the Maui 'alauahio, but occur occasionally in a few other species.

The Ravages of Introduced Diseases

CONSERVATION AND ENVIRONMENT

Over half of historically known Hawaiian honeycreepers are extinct, and most of the remainder are endangered, some critically so (only three po'o-uli survived in 2001). The most significant agent of extinction, and the primary limiting factor today, is the presence of avian diseases (malaria and pox), spread by introduced mosquitoes from introduced birds. 'I'iwi exhibit nearly 100 percent mortality to malaria. On the main islands, Hawaiian honeycreepers are now restricted to the cooler uplands, where mosquitoes are less abundant. However, adaptation by the insects to cooler regions, coupled with probable global warming, poses a threat to even these last sanctuaries.

Research into mosquito control in remote areas holds out some hope, and a few species, such as the Oahu 'amakihi, have become resistant to disease and are repopulating lowland areas. Habitat loss was also a major factor in both aboriginal and post-contact times, while in the 20th century feral pigs became a serious problem by destroying undergrowth and creating pools in which mosquitoes bred. HDP

FACTFILE

HAWAIIAN HONEYCREEPERS

Order: Passeriformes

Family: Drepanididae

About 22 surviving species in 13 genera. Species include: **Laysan finch** (*Telespiza cantans*), **Nihoa finch** (*T. ultima*), **palila** (*Loxioides bailleui*), **Greater koa finch** (*Rhodacanthis palmeri*), **Lesser koa finch** (*R. flaviceps*), **'o'u** (*Psittirostra psittacea*), **akepa** (*Loxops coccineus*), **po'o-uli** (*Melamprosops phaeosoma*), **Lana'i hookbill** (*Dysmorodrepanis munroi*)†, **Maui 'alauahio** (*Paroreomyza montana*), **kakawahie** or **Molokai creeper** (*P. flammea*), **'akikiki** (*Oreomystis bairdi*), **Hawaii creeper** (*O. mana*), **'i'iwi** (*Vestiaria coccinea*), **Oahu 'amakihi** (*Hemignathus flavus*).

HABITAT Wet and dry forests on the larger islands, open habitats on Laysan and Nihoa.

SIZE Length 10–20cm (4–8in); weight 10–45g (0.4–1.6oz).

PLUMAGE Mainly yellow, olive green, black, red, and orange, with accents of white, brown, and gray. Males brighter than females in most (but not all) species; some with distinctive juvenile plumages.

VOICE Complex, canarylike songs, simple warbles or trills, or variable songs with reedy and mechanical notes and squeaks.

DISTRIBUTION Hawaiian islands, from Laysan E to Hawaii.

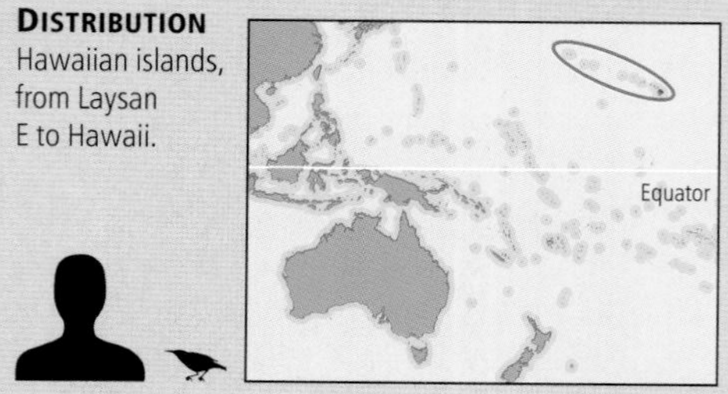

NEST Simple open cups, usually in terminal leaf clumps, sometimes in cavities.

EGGS 1–5, usually 2, pale with earth-tone blotches and scrawls concentrated at larger end. Incubation period 14–18 days; nestling period 15–27 days, highly variable among species.

DIET Mainly nectar and invertebrates (including insects, arachnids, and snails), but also fruit and fruit juices, tree sap, seeds and seed pods, carrion, and seabird eggs.

CONSERVATION STATUS Thirteen species have gone extinct in historic times; 6 are Critically Endangered, 2 of which – the nukupu'u (*Hemignathus lucidus*) and the O'ahu 'alauahio (*Paroreomyza maculata*) – may also be extinct. Five further species are Endangered, and 7 Vulnerable.

† Extinct species

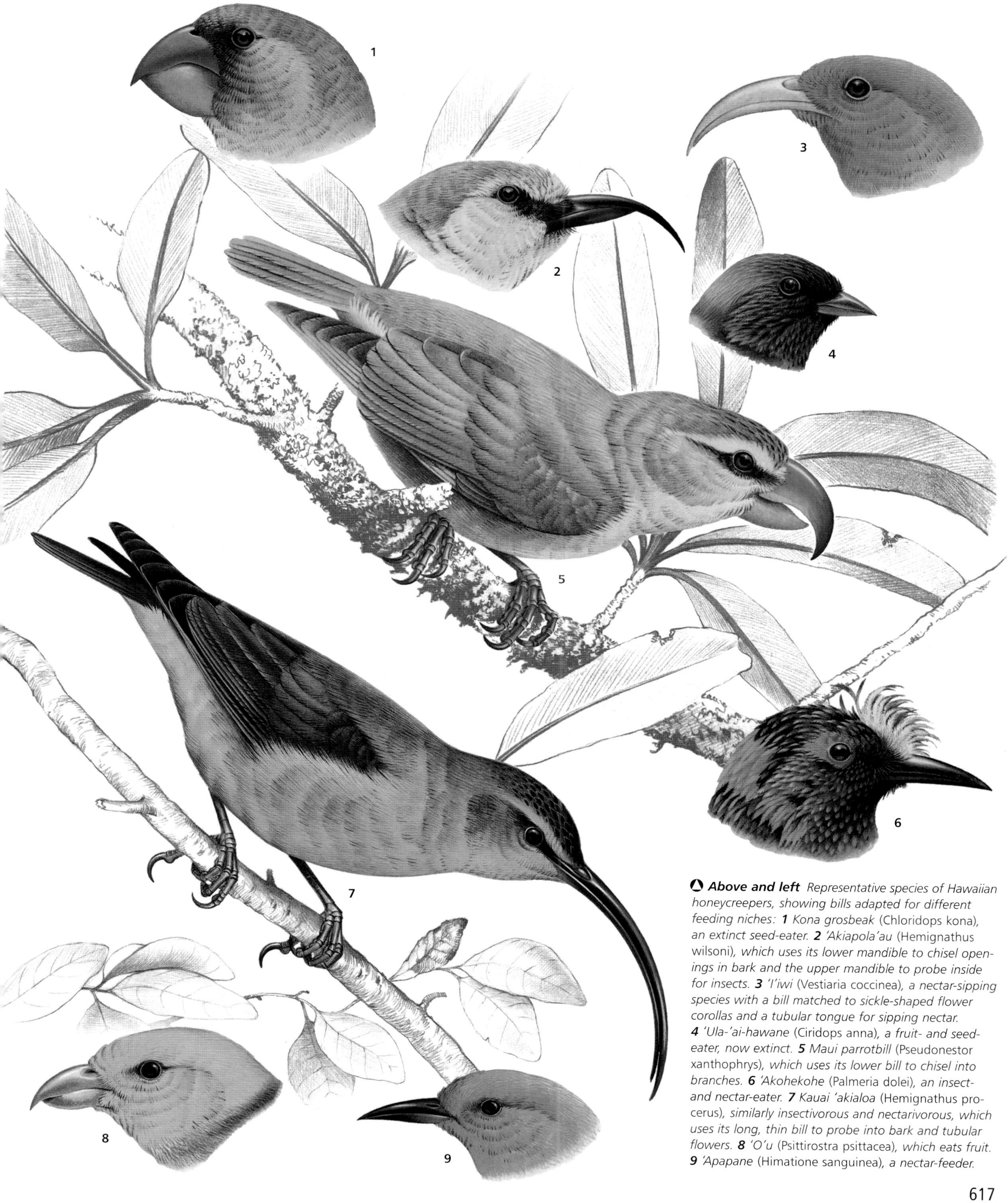

Above and left *Representative species of Hawaiian honeycreepers, showing bills adapted for different feeding niches:* **1** *Kona grosbeak* (Chloridops kona), *an extinct seed-eater.* **2** *'Akiapola'au* (Hemignathus wilsoni), *which uses its lower mandible to chisel openings in bark and the upper mandible to probe inside for insects.* **3** *'I'iwi* (Vestiaria coccinea), *a nectar-sipping species with a bill matched to sickle-shaped flower corollas and a tubular tongue for sipping nectar.* **4** *'Ula-'ai-hawane* (Ciridops anna), *a fruit- and seed-eater, now extinct.* **5** *Maui parrotbill* (Pseudonestor xanthophrys), *which uses its lower bill to chisel into branches.* **6** *'Akohekohe* (Palmeria dolei), *an insect- and nectar-eater.* **7** *Kauai 'akialoa* (Hemignathus procerus), *similarly insectivorous and nectarivorous, which uses its long, thin bill to probe into bark and tubular flowers.* **8** *'O'u* (Psittirostra psittacea), *which eats fruit.* **9** *'Apapane* (Himatione sanguinea), *a nectar-feeder.*

Buntings and New World Sparrows

THE TERM "BUNTING" IS DERIVED FROM AN old English word "buntyle," the original meaning of which is somewhat obscure. Whatever its significance, the name was given to several seed- and insect-eating, ground-feeding birds that were resident in western Europe.

The name was later carried by early settlers from Britain to other parts of the world and applied there to some not particularly closely related birds; in North America, for example, it served to designate some members of the cardinal grosbeaks (family Cardinalidae) such as the Indigo bunting. Confusingly, most true buntings of the New World are called sparrows, although they bear little relation to the familiar English birds, such as the House sparrow (*Passer domesticus*), also known by that name.

Stout Bills for Seed-eating

FORM AND FUNCTION

The true buntings almost certainly evolved in the New World. About 85 percent of the world's species are found in the Americas, where they occupy a diverse and broad range of habitats. Among the 60 or so American species found north of Mexico, for example, there are birds that inhabit arctic tundra, boreal forest, prairies and meadows, deserts, alpine meadows, salt and freshwater marshes, and deciduous and coniferous woods. The ancestor of the Old World buntings probably crossed the Bering Sea into Asia, with the genus *Emberiza* evolving in temperate Asia, where it is best represented, and spreading westward into Europe and Africa. Interestingly, there are only a couple of buntings, the Crested and the Slaty, breeding in tropical Asia, and the group has failed to penetrate or persist in the East Indies–Australasia region, although the Cirl bunting and the yellowhammer have been introduced with some success into New Zealand.

Buntings are characterized by a stout, conical bill adapted for crushing and taking the husks off seeds. The upper and lower parts of the bill can be moved sideways in some species. Juncos, for example, are particularly adept at manipulating, cracking, and discarding husks with their bills. Sparrows and buntings usually forage on the ground. Some, such as the towhees, scratch for food with a very distinctive "double scratch," in which the bird remains stationary while scratching backward simultaneously with both feet.

Buntings and New World sparrows show considerable diversity in plumage and voice; somberly plumaged species such as the Corn bunting, which is a dull grayish-brown with heavy streaks, contrast with the more brightly-plumed birds, such as the yellowhammer, with its bright yellow underparts and streaked yellow head, or the Lapland longspur, with its black, chestnut and white head markings. Except in those species in which the male has a bright display plumage, the sexes are alike in coloration.

The Courtship Song

BREEDING BIOLOGY

In temperate and arctic regions, most buntings are monogamous. In a majority of the species studied, including, for example, the Reed bunting, few males attract more than one female. Extra-pair paternity is common, however, and up to about 70 percent of broods contain extra-pair offspring. Reed buntings generally nest in marshes, often in high density, which makes mate-guarding hard. A small proportion of males are bigamous.

The Eurasian Corn bunting is often polygamous, with some males within a population

Left In recent years the Yellowhammer has declined rapidly in Britain. Chemical fertilizers and pesticides used in crop growing have contributed to this, but new intensive grazing practices have now left fields barren with no cover for nests or seeds for food.

Above *The North American Rufous-sided towhee* (Pipilo erythrophthalmus) *shows a marked variation in voice between races in the east and the west.*

FACTFILE

BUNTINGS & NEW WORLD SPARROWS

Order: Passeriformes

Family: Emberizidae

291 species in 72 genera.

DISTRIBUTION Practically worldwide; absent from extreme SE Asia and Australasia (introduced to New Zealand). Almost cosmopolitan in the New World.

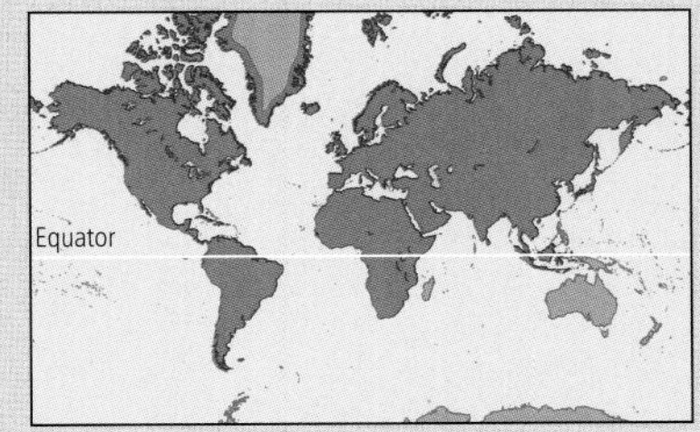

HABITAT Open woodlands, grasslands, arctic tundra and alpine meadows, desert regions; in Eurasia, primarily in open country, hedgerows, parkland, "edge" habitats.

SIZE Length 10–24cm (4–9.5in); weight 7–52g (0.25–1.9oz).

PLUMAGE Ranges from dull brown and gray to black, white, orange, bright blue-green, yellow, or red; several groups have sharply patterned plumages, particularly on the head.

VOICE Alarm calls usually loud and easily localized, anxiety calls frequently ventriloquial; songs short and simple to long and melodious, containing whistles, chatters, and trills.

NEST Woven, cup-shaped nests, usually well-concealed on ground or in low bush.

EGGS Usually 2–7; base color off-white, light brown, or light blue, usually with brownish, reddish, or blackish marks. Incubation period: 10–14 days; nestling period: 10–15 days.

DIET Primarily grains, invertebrates, fruit, and buds; adults eat seeds and berries; nestlings are fed almost exclusively on arthropods.

CONSERVATION STATUS The Entre Ríos seedeater (*Sporophila zelichi*), Hooded seedeater (*S. melanops*), and Tumaco seedeater (*S. insulata*) are Critically Endangered. Worthen's sparrow (*Spizella wortheni*) and the Zapata sparrow are both Endangered.

See Selected Bunting Species ▷

reportedly attracting up to three females at a time and others none at all. It is generally supposed that this mating system evolves when there are large differences in the quality of territory among males, so that a female is better off pairing with an already mated male in a good territory than with a bachelor male in a poor one.

The Lark bunting of North America is also occasionally polygamous, probably especially where the birds occur in high densities. A male with an apparently high-quality territory is more likely to attract a second mate than one with a lower-quality territory. Lark buntings nest in dry, open fields with sparse vegetation, so a well-shaded site in which to build the nest is important to a female, and a territory that contains such a site may be selected even if there is another female already nesting there. Occasionally, more than one male Lark bunting will help at a nest.

Smith's longspurs, which breed in subarctic tundra bordering on the northern edge of the tree-line, form no pair bonds. When females are ready to lay their eggs, they mate repeatedly (up to 350 times in a week) with two or three males. The males, in turn, mate with more than one female. Studies confirm that, within broods, more than one male fathers the young, and two or more males may assist the female in raising them.

The reproductive behavior of the Saltmarsh sharp-tailed sparrow of the east coast of North America appears to be similar to that of Smith's longspur. Males do not defend a territory, although they may guard a female by chasing away other males; no pair bond is formed, and males do not assist in feeding the young.

Most sparrows and buntings are territorial. In migratory species, the male characteristically arrives on the territory before the female and defends it against other males. Often, the male reoccupies the territory he held the year before. Less commonly, females return to the same territory in which they nested the previous year. Most breeding activities – courting, pairing, nesting, and raising young – occur within the territory.

Collecting food for young may or may not occur within territory boundaries. For example, American tree sparrows defend large territories, usually of more than 1ha (2.5 acres), within which food is collected, whereas Clay-colored sparrows defend small territories of less than 1,000 sq m (11,000 sq ft) and forage exclusively outside the territory, often on communal feeding grounds.

The territories of the Eurasian yellowhammer tend to be linear, extending along a hedgerow of the edge of a wood; they are about 60m (200ft) long, and the defended area may extend 10–15m (33–50ft) into the adjacent field. The yellowhammer's territory is important for pair-formation and nesting, but foraging largely takes place in neutral ground beyond the territorial borders.

Even within species, the size of a territory varies both geographically and in different habitats. For example, sparrows like the American tree sparrow defend a territory on which they feed. In Ohio, where the birds live in shrubby thickets in residential areas, in abandoned fields, and other brushy habitats, breeding territories average about 1ha (2.5 acres), whereas in the San Francisco Bay area, where they breed in saltmarshes along creeks that are flooded at high tide, their territories are about 0.4ha (1 acre) in extent. In Ohio, many individuals do not migrate south in winter, and these birds defend winter territories that are six to ten times larger than the breeding territories. Once the breeding season is over, territorial boundaries of migratory individuals break down and adults and young gather together in flocks.

Outside of the breeding season, most migratory buntings and American sparrows form flocks that are essentially feeding aggregations. These flocks often contain individuals of several different species. Wintering flocks of sparrows are com-

Selected Bunting Species

Species and genera include: American tree sparrow (*Spizella arborea*), Clay-colored sparrow (*S. pallida*), Black-headed bunting (*Emberiza melanocephala*), Cirl bunting (*E. cirlus*), Corn bunting (*E. calandra*), Red-headed bunting (*E. bruniceps*), Reed bunting (*E. schoeniclus*), Rock bunting (*E. cia*), Rustic bunting (*E. rustica*), Yellow bunting (*E. sulphurata*), yellowhammer (*E. citrinella*), Black-throated finch (*Melanodera melanodera*), Chestnut-collared longspur (*Calcarius ornatus*), Lapland longspur (*C. lapponicus*), Smith's longspur (*C. pictus*), Crested bunting (*Melophus lathami*), Gough Island bunting (*Rowettia goughensis*), Grasshopper sparrow (*Ammodramus savannarum*), Henslow's sparrow (*A. henslowii*), Saltmarsh sharp-tailed sparrow (*A. caudacutus*), Nelson's sharp-tailed sparrow (*A. nelsoni*), Seaside sparrow (*A. maritimus*), Lark bunting (*Calamospiza melanocorys*), Savannah sparrow (*Passerculus sandwichensis*), Slaty bunting (*Latoucheornis siemsseni*), Snow bunting (*Plectrophenax nivalis*), towhees (genus *Pipilo*), Zapata sparrow (*Torreornis inexpectata*).

Left *Representative species of buntings and New World sparrows:* ***1*** *Black-headed bunting* (Emberiza melanocephala)*;* ***2*** *Corn bunting* (E. calandra) *chirping its rattling "bunch-of-keys" song;* ***3*** *White-throated sparrow* (Zonotrichia albicollis).

monly rather small and loose, but in some species, especially the longspurs, Lark buntings, and Snow buntings, flocks of from a few dozen to thousands of individuals can occur. Longspurs characteristically flush from the shortgrass habitat in which they winter as a group and circle the area, often to return to near the place from which they flushed. When on the ground, even in sparse habitats, wintering longspurs are difficult to see.

Courtship in buntings usually involves a male advertising his presence by singing, often from a conspicuous perch. When a female approaches, the male may chase her. During these courtship chases the male may buffet the female, and both birds may tumble to the ground. Song flights occur in open-country species. Males of Lapland longspurs and Snow buntings in the Arctic, and Chestnut-collared longspurs and Lark buntings on the North American prairies, for example, typically rise a few meters above the ground and then slowly circle back to earth, holding their slowly-beating wings at an angle above the body as they utter their song. The flight song of Nelson's sharp-tailed sparrow of the American prairies is particularly spectacular. The male rises as much as 20m (65ft) up while uttering "tic" notes. He then sings as he glides forward and downward, sings for a second time, then drops into the vegetation perhaps 100m (330ft) from where he first took off.

The songs of many buntings are among the more melodious of all bird songs; that of the Bachman's sparrow, for example, consists of a long, sweet note followed by a clear trill that carries a long distance. Other species, however, are less impressive. The song of the Corn bunting reminds one of jingling keys, while that of the Grasshopper sparrow is a high, thin, insectlike "chip chip scheeeeeee, tzick tzick." Henslow's sparrow has one of the least impressive songs of all, an unobtrusive, short, insectlike "tsi-lick."

Nests are usually placed on the ground or low in a bush, and tend to be neat, compact cups built of dried vegetation (typically grass and weeds) and lined with hair, mosses, fine vegetation fibers,

Below *Snow buntings feed their young on a diet of insects and arachnids. Their search for food sometimes leads them to urban areas and they have even been found at beach parking lots looking for weed seeds.*

wool, and/or feathers. Females are usually solely responsible for building the nest, incubating the eggs, and brooding the young; males, however, contribute substantially to feeding the young, at both the nestling and the fledgling stages.

In parts of their range where the seasons is long enough to permit more than one brood to be produced, male Savannah sparrows attend the fledged young of the first brood while their mate incubates a second set of eggs. When those eggs hatch, the male helps the female feed those young as well. On occasion, the male may be mated to two different females at the same time. Both within and among species, clutch sizes tend to be largest at high latitudes.

Populations in Retreat

Conservation and Environment

Population sizes of several species of buntings and New World sparrows are declining, generally as a consequence of habitat destruction associated with agricultural practices and urbanization. The Corn bunting, for example, has declined in many areas of Europe, probably due to changes in land use, and many of the North American grassland sparrows are similarly declining because of habitat loss. The Zapata sparrow is endemic to Cuba, where there are three different local populations. Because of their small ranges, all three are susceptible to natural threats as well as human activities, and the species is under threat.

Many of the South American sparrows have very small ranges, some known from only a few sites. Because of their limited ranges and habitat destruction, most of these species are endangered, and a few may even have become extinct. For example, the Pale-headed brush-finch traditionally inhabited oases in arid, inter-Andean valleys in southern Ecuador. It has always been extremely local in distribution, and today most areas in the region with sufficient water are extensively farmed, resulting in a near-total removal of natural vegetation. There are no recent records of the species, which is most probably extinct.

The "Cape Sable" Seaside sparrow, although generally recognized as a subspecies, is distinctive in many ways, and perhaps would better be treated as a separate species. Its range is limited to a few freshwater marshes in southern Florida (other populations of Seaside sparrows breed only in coastal saltmarshes) and it could easily be driven to extinction by a natural disaster such as a hurricane or wildfire. Fire and habitat destruction were largely responsible for the extinction of another well-marked Seaside sparrow population, that of the "Dusky" Seaside sparrow.

Above *Breeding male Lark buntings are black with large white wing patches. This one is at his nest on the ground surrounded by flowering prickly pear cacti.*

Within historic times, the clearing of immense tracts of hardwood forests in eastern North America for agricultural purposes created habitat suitable for many species, including the Bachman's, Lark, and Vesper sparrows, that previously had not been found there, at least commonly. More recently, many of these man-made eastern grasslands have been swallowed up by urbanization or have reverted to second-growth woodland, and the ranges of the species have contracted accordingly. Today, the Lark sparrow is a rare straggler in the east, Bachman's sparrow's range has contracted to the southeast, where it is primarily found in grassland in open, mature pine woods, and the Vesper sparrow has seriously declined in numbers in most eastern states and provinces. Henslow's sparrow, found in dense grasslands, has also declined in numbers in the same region, and has been extirpated from many areas, but seems to be increasing in parts of the Midwest where tallgrass prairies are being protected. On the other hand, populations of sparrows that breed at high latitudes, north of areas of intense human activity, are generally healthy. JDR/RWK

Cardinal Grosbeaks

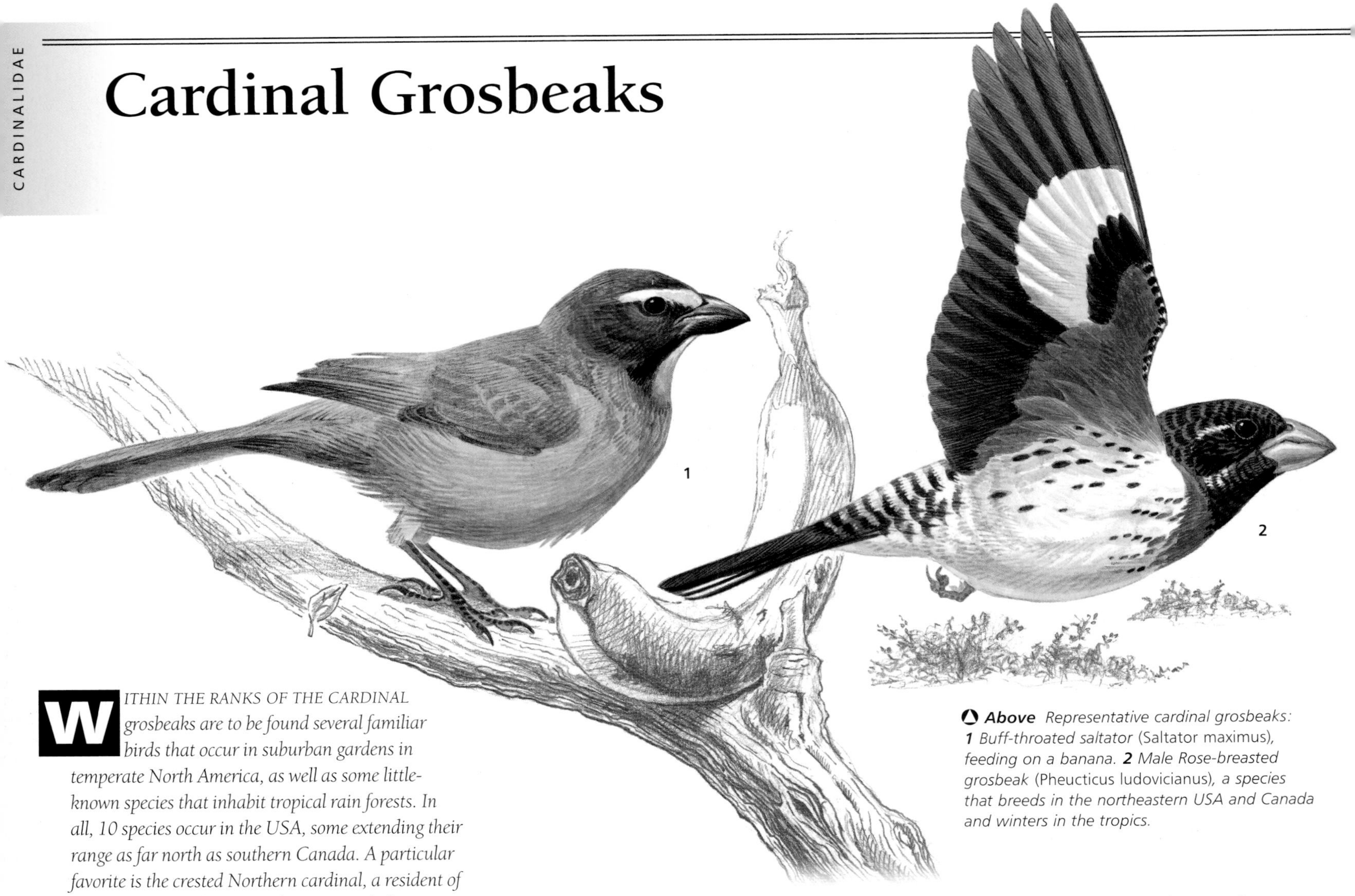

Above *Representative cardinal grosbeaks:* ***1*** *Buff-throated saltator* (Saltator maximus), *feeding on a banana.* ***2*** *Male Rose-breasted grosbeak* (Pheucticus ludovicianus), *a species that breeds in the northeastern USA and Canada and winters in the tropics.*

WITHIN THE RANKS OF THE CARDINAL grosbeaks are to be found several familiar birds that occur in suburban gardens in temperate North America, as well as some little-known species that inhabit tropical rain forests. In all, 10 species occur in the USA, some extending their range as far north as southern Canada. A particular favorite is the crested Northern cardinal, a resident of the eastern USA that sports his warm red plumage amid winter's snow. His mate is much duller.

Thanks largely to people who provide seeds in winter, the cardinal has extended its breeding range over the last century from the Ohio Valley to north of the Great Lakes in southern Canada. In the southwestern USA and much of Mexico, it coexists with the equally high-crested and thick-billed pyrrhuloxia, more gray than red. A third species with a rosy red crest, the Vermilion cardinal, lives in the desert scrub along the Caribbean coast of northeastern Colombia and Venezuela. The males also have a rose-colored tail and wings; the females share the rosy crest, but otherwise are mostly sandy-brown.

Tuneful and Brightly Colored

FORM AND FUNCTION

The six or seven species of buntings of the genus *Passerina*, which live chiefly in the USA and Mexico, are among the most colorful of songbirds. One of the most elegant, the Painted bunting, has a blue head, a yellow-green mantle, a red rump and underparts, and dark wings and tail. The almost solid-blue male Indigo bunting, which nests in bushy places through much of the eastern half of the USA, wears a brownish dress much like the female's in its winter home in southern Mexico and Central America. In the American west, the Lazuli bunting replaces the Indigo bunting. The male has a turquoise-blue head and back, with a cinnamon-colored breast, white belly, and two white wing-bars. The female is much like the female Indigo bunting in color, but has two faint wing-bars. The ranges of the Lazuli and Indigo buntings overlap on the Great Plains of the US Midwest, and the two species hybridize.

In southwestern Mexico the Orange-breasted bunting is locally common. The male has a green cap with a turquoise-blue head and back, and golden-orange throat and belly. The female is lime green above with yellow underparts.

The Rose-breasted grosbeak is another migratory species. After nesting in open woodlands and similar habitats in the northeastern USA and southern Canada, the birds fly off to spend the colder months as far away as Venezuela and Peru. In their winter plumage, the males retain enough red on their breasts to distinguish them from the browner-colored females.

The closely-related Yellow grosbeak is a resident of Mexico; most populations are not migratory. The males are bright yellow, with black wings that are spotted with white; the females are duller yellow, with streaks on the crown and back, and olive rather than black wings. They are often found in pairs or small groups, typically feeding rather high in fruiting trees.

The dickcissel, which sings its name in open fields chiefly in the Mississippi Valley region and which winters in vast numbers as far south as Venezuela and Trinidad, sometimes causes heavy losses where rice is grown. Dickcissels are at present declining in numbers, perhaps in part because they are persecuted in South America as agricultural pests.

Among the nonmigratory tropical members of this subfamily are the Blue-black grosbeak and his brown mate, both of which sing beautifully in the rain forests and bushy clearings of southern Mexico and Central America. They eat corn (maize), but are not gregarious, and consequently do only slight damage to crops. The closely-related Ultramarine grosbeak is found in semi-arid scrub in northern South America, as well as in dense thickets and woodland edge farther to the southeast. Like Blue-black grosbeaks, the males are dark blue, the females dark brown above and paler brown below. The smaller Glaucous-blue grosbeak of southern Brazil and Uruguay is similar in color. All of these "blue" grosbeaks are shy and retiring birds, but they will sometimes sing from a conspicuous perch on the top of a bush.

FACTFILE

CARDINALS

Order: Passeriformes

Family: Cardinalidae

43 species in 12 genera. Species include: **Black-headed grosbeak** (*Pheucticus melanocephalus*), **Rose-breasted grosbeak** (*P. ludovicianus*), **Yellow grosbeak** (*P. chrysopeplus*), **Blue-black grosbeak** (*Cyanocompsa cyanoides*), **Ultramarine grosbeak** (*C. brissonii*), **Buff-throated saltator** (*Saltator maximus*), **Golden-billed saltator** (*S. aurantiirostris*), **Green-winged saltator** (*S. similis*), **Crimson-collared grosbeak** (*Rhodothraupis celaeno*), **dickcissel** (*Spiza americana*), **Glaucous-blue grosbeak** (*Cyanoloxia glaucocaerulea*), **Indigo bunting** (*Passerina cyanea*), **Lazuli bunting** (*P. amoena*), **Orange-breasted bunting** (*P. leclancherii*), **Painted bunting** (*P. ciris*), **Rose-bellied bunting** (*P. rositae*), **Northern cardinal** (*Cardinalis cardinalis*), **pyrrhuloxia** (*C. sinuata*), **Vermilion cardinal** (*C. phoeniceus*), **Slate-colored grosbeak** (*Pitylus grossus*), **Yellow-green grosbeak** (*Caryothraustes canadensis*), **Yellow-shouldered grosbeak** (*C. humeralis*).

DISTRIBUTION C Canada to C Argentina

HABITAT Temperate-zone woodlands, tropical rain forests, thickets, arid scrub, plantations, gardens, fields.

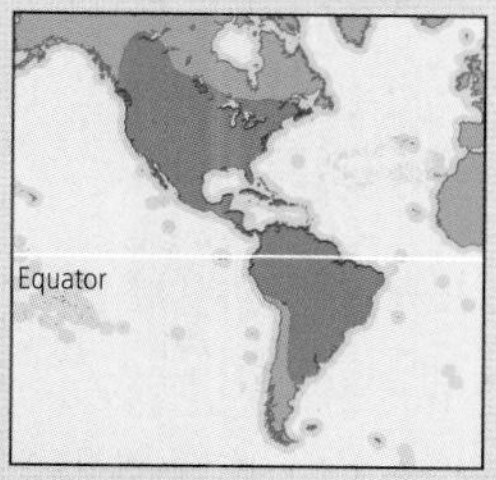

SIZE Length 11.5–22cm (4.5–8.5in).

PLUMAGE Brilliant and varied, or olive, gray, blue-black. Males and females alike in some species, very different in others.

VOICE Many are superb and persistent songsters.

NEST Massive or loosely-built open cups in trees and shrubs, rarely on ground.

EGGS 2–5, white, greenish, bluish, or blue, unmarked or speckled or scrawled. Incubation period 11–14 days; nestling period 9–15 days.

DIET Seeds and grains, fruits, flowerbuds, blossoms, insects.

CONSERVATION STATUS Not threatened.

The Slate-colored grosbeak is more closely confined to mid and upper levels of rain forests, where its bluish-gray plumage, almost uniformly dark, contrasts with a heavy, bright-red bill. The female is similar in color, but slightly paler and a more olivaceous gray. The birds are usually found in pairs, and often join mixed-species flocks. The male Crimson-collared grosbeak of northwestern Mexico has a dull, pinkish-red collar and breast, with a black hood and dark wings. The female is similar in color, but dull olive-green where the male is pinkish-red. A few of the birds wander north to the lower Rio Grande Valley in winter.

Most cardinal grosbeaks consume soft fruits, flower buds, and insects as well as weed seeds and grains. The pyrrhuloxia of Mexico and the US Southwest uses its strong, parrotlike beak to crush the beans of the spiny mesquite brush (*Prosopis* spp.) common in the region. The 15 or so saltator species, like tanagers, show a preference for fruits. Most saltators, which inhabit semi-open and scrub country through much of tropical America, are rather large and fairly plain grosbeaks, often with white eyebrows; unusually for grosbeaks, both genders are similar in color. The widespread Buff-throated saltator is a frequent attendant at feeders where bananas are offered. When eating, these birds and some of their relatives rest a fruit precariously on a horizontal branch while they bite off pieces.

Fighters and Helpers

SOCIAL BEHAVIOR

The social habits of cardinal grosbeaks vary greatly. Solitary and pugnacious in the breeding season, the lovely male Painted buntings may occasionally wound and even kill their adversaries. At the other extreme Yellow-green grosbeaks, which travel at all seasons through rain forests and shady clearings in loose flocks, display no territorial exclusiveness. Parents feeding nestlings are joined by one or more helpers.

The female usually builds the cup-shaped nest, but male cardinals and Blue-black grosbeaks share the task. Although in most species only the female incubates, male Rose-breasted and Black-headed grosbeaks take turns on the eggs, often singing while they sit. Male cardinals, Buff-throated saltators, and Blue-black grosbeaks bring food to their incubating partners. In most species the father helps feed the young, but male Painted buntings are unreliable attendants, and the polygamous male dickcissel neglects his offspring.
JDR/AFS

Right *Northern cardinals (here, a male in its striking plumage) are so-called "commensals of civilization" – species that thrive with the spread of human activity. Garden habitats, and the provision of feeders in winter, have greatly benefited Northern cardinals.*

New World Warblers

Remarkably diverse, New World warblers are a group of small, insectivorous birds known for their colorful plumages and behavioral variation. These warblers occur in most forested habitats from northern Canada south to Argentina. Many North American species are familiar to birders and ornithologists alike, but much remains to be learned about the distribution, ecology, and behavior of the resident tropical species.

Ornithologists currently recognize 116 species, placed in their own family, the Parulidae. The group is also commonly referred to as the "wood-warblers." The New World warblers are closely related to a number of other families of songbirds, including the Emberizidae (sparrows and buntings), Thraupidae (tanagers and New World honeycreepers), and Icteridae (orioles and blackbirds). The evolutionary boundaries between the families are controversial, and recent genetic studies have shown that a few species long classified as parulid warblers (including the Olive warbler, Yellow-breasted chat, and wrenthrush) are actually more closely related to these other families than they are to the Parulidae. Despite their many similarities in appearance and behavior as well as their group names, the New World warblers are not closely related to the Old World warblers (family Sylviidae).

Similar Shapes, Different Plumage

FORM AND FUNCTION

The pattern of morphological variation seen in the New World warblers presents an interesting contrast to that of many other bird groups. On the one hand, the warblers are highly diverse in plumage color and pattern. This fact is especially true of the species that breed in North America, which have a palette of plumage colors that includes blue, red, orange, yellow, and green; tropical resident warblers tend to be more cryptically colored in yellow and brown. On the other hand, the gross morphology of most warblers is fairly similar: species tend to differ from one another in overall size, but vary only slightly in traits like body proportions or bill shape.

Above *Representative species of New World Warblers:* ***1*** *Male Common yellowthroat* (Geothlypis trichas)*;* ***2*** *A male American redstart* (Setophaga ruticilla) *displaying its black and orange plumage; females are grey and yellow;* ***3*** *Kirtland's warbler* (Dendroica kirtlandii) *is currently classified as Vulnerable by the IUCN.*

Left *The Yellow warbler is the most widely distributed* Dendroica *species and is the only Parulid warbler that breeds in the Galápagos Islands.*

American redstarts have an unusual development pattern called "delayed plumage maturation": male redstarts retain a drab, female-like plumage through their first summer as adult breeders, and then molt into a much more dramatic black and orange pattern during their second fall.

The genus *Dendroica* is the best-known. Most of the 27 *Dendroica* species breed in North America and migrate south to overwinter, but a handful are year-round residents of islands in the West Indies. The *Dendroica* wood-warblers are famous as an example of behavioral adaptation, in that multiple species may coexist by foraging in different parts of the same trees. This group is also well known among birdwatchers as the "confusing fall warblers" that seem always to stay high in the tallest trees, making it hard to get a clear look at their subtle field marks.

The second largest group is the genus *Basileuterus*, which contains 24 nonmigratory species that occur from northern Mexico to South America. The highest diversity is found in the Andes, where up to four species may occupy different elevational zones on a single mountain slope. The remaining 24 Parulidae genera all contain fewer than 10 species each; 14 contain just a single species.

It was long thought that the New World warblers diversified during the repeated cycles of glaciation of the late Pleistocene period (roughly 1.5 million to 15,000 years ago). Recent studies based on DNA variation have, however, shown that the Parulidae started

The sexes of most tropical warblers have similar plumages, but most migratory warblers show some degree of sexual dimorphism. In some such species, the female is only slightly duller than the male, but in others the two sexes have markedly different plumage colors and patterns. In a few species such as the Chestnut-sided and Blackpoll warblers, there are also striking seasonal differences between summer and winter plumage. Young birds in their first year often have a somewhat duller plumage than older birds, and individuals hatched during the summer are often the most challenging to identify during fall migration.

FACTFILE

NEW WORLD WARBLERS

Order: Passeriformes

Family: Parulidae

116 species in 26 genera. Species include: **American redstart** (*Setophaga ruticilla*), **Bachman's warbler** (*Vermivora bachmanii*), **Black-and-white warbler** (*Mniotilta varia*), **Blackpoll warbler** (*Dendroica striata*), **Chestnut-sided warbler** (*D. pensylvanica*), **Golden-cheeked warbler** (*D. chrysoparia*), **Kirtland's warbler** (*D. kirtlandii*), **Yellow warbler** (*D. petechia*), **Yellow-rumped warbler** (*D. coronata*), **Buff-rumped warbler** (*Basileuterus fulvicauda*), **Grey-headed warbler** (*B. grisieceps*), **Louisiana waterthrush** (*Seiurus motacilla*), **ovenbird** (*S. aurocapillus*), **Northern parula** (*Parula americana*), **Prothonotary warbler** (*Protonotaria citrea*), **Semper's warbler** (*Leucopeza semperi*), **Tennessee warbler** (*Vermivora peregrina*).

HABITAT Forests and brushlands

SIZE Length 10–18cm (4–7in); weight: 7–25g (0.2–0.9oz); both sexes similar.

PLUMAGE Highly variable; some species very bright, others drab. Sexes similar in some species, highly dimorphic in others.

DISTRIBUTION N and S America, West Indies

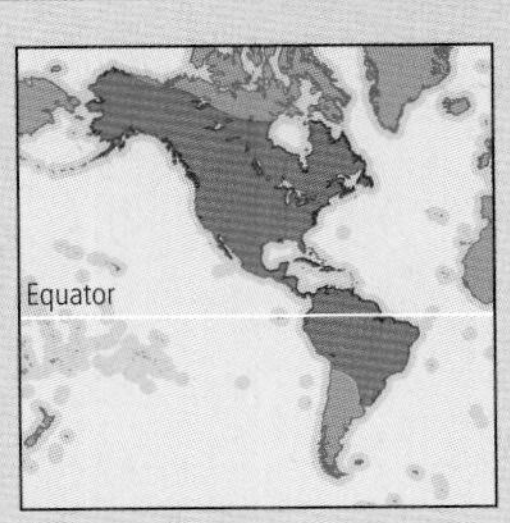

VOICE Distinct musical songs, often more than one per species; a wide variety of call notes.

NEST Well-built, in tree or shrub or on ground.

EGGS 2–8 (usually 3–5); white to green, usually with brown spots or splashes. Incubation period 10–14 days; nestling period 8–12 days.

DIET Invertebrates, especially insects; some fruit and nectar.

CONSERVATION STATUS 3 species are currently listed as Critically Endangered: Belding's yellowthroat (*Geothlypis beldingi*), and the Bachman's and Semper's warblers, both of which have not been sighted for several decades and could well be already extinct. In addition, 5 species are Endangered and 7 Vulnerable.

to diversify much earlier, and many species probably originated more than 5 million years ago – a fact especially true for the *Dendroica* warblers, many of which arose during an early period of rapid speciation.

Traveling South for the Winter

DISTRIBUTION PATTERNS

Most forested habitats in the New World support at least one species of warbler. Almost all of the species that breed in North America are migratory, but those that breed in tropical or subtropical regions are not. The highest diversities occur in northeastern North America, where it is possible to find more than six species with overlapping breeding territories and to spot over 20 species on a good day during migration. In winter, very high concentrations of warblers occur in places such as western Mexico and the islands of the Greater Antilles such as Cuba and Jamaica, where overwintering migrants may temporarily outnumber local resident birds of all species.

Northern-breeding warblers show a high diversity of migratory patterns, but species that breed in western North America tend to overwinter in Mexico and northern Central America, while species that breed in eastern North America head for the Caribbean and southern Central America. Variation in migratory behavior is pronounced even within some species. For example, in winter a mangrove forest in Costa Rica is likely to contain both Yellow warblers from the local nonmigratory population and migratory individuals from breeding sites as far away as northern Canada. Perhaps the most dramatic migration is that of the Blackpoll warbler: in fall, the birds move out over the Atlantic Ocean off the northeastern coast of the United States and fly nonstop, pushed by the prevailing winds, to the northeastern coast of South America.

Insectivores that Sometimes Take Fruit

DIET

Most New World warblers are almost exclusively insectivorous, but a few consume substantial amounts of fruit or flower nectar. Caterpillars are a preferred food of most wood-warblers, particularly during the breeding season. In the boreal forests of North America, some wood-warblers become particularly abundant during outbreaks of the Spruce budworm caterpillar.

Fruit-eating species include the Yellow-rumped warbler. The eastern population of this species was previously known as the "Myrtle warbler," because in winter the birds commonly rely on the fruits of the wax myrtle tree and other waxy fruits such as those of the poison ivy vine. Yellow-rumped warblers have unusual digestive specializations that allow them to break down and assimilate the energy-rich wax in these fruits. Other species such as the Northern parula, Tennessee warbler, and Chestnut-sided warbler visit flowers to feed on nectar in all seasons, but most commonly in winter.

Warblers capture insects using a variety of search behaviors. The majority move rapidly through the foliage of bushes or trees and glean insects off leaves or stems. Others, such as the ovenbird, walk about the forest floor and pick insects out of the leaf litter or from low vegetation. The Black-and-white warbler resembles a nuthatch as it creeps along tree trunks or branches searching for insects hidden in bark crevices. The migratory Louisiana waterthrush and tropical resident Buff-rumped warbler share a predilection for foraging on the ground at the margin of fast-moving streams. Species such as the American redstart and the various tropical whitestarts in the genus *Myioborus* specialize on flying insects, and the bright, highly contrasting plumages of these species may be an adaptation for flushing this type of prey.

Right *In Kentucky warblers* (Oporornis formosus), *only the female broods but both parents tend to the nestlings. If disturbed, these young birds may leave the nest when only 7 days old.*

Below *Totally unrelated to the Neotropical family of ovenbirds – the Furnariidae – the parulid ovenbird can be heard from quite a distance singing its distinctive and ringing song, "Teac-cher, Tea-cher, Tea-cher."*

Multiple Mates

BREEDING BIOLOGY

New World warblers are intensely territorial during the breeding season, and nonmigratory species defend year-round territories. Warblers tend to breed in single male–female pairs, but instances of polygamy are known from a number of species. In most species studied to date, females do most of the nest construction and incubation, but both parents feed the nestlings and fledglings. Despite this apparent pair bond, genetic studies of the nestlings from single nests have shown that females may often mate with several males within a single breeding season. Males sing most actively during breeding. Many species have two distinct song types, which probably function respectively in mate attraction and territorial defense against other males.

Warblers construct a variety of nests. Most species build open cup nests that are variously placed on the ground, in low bushes, or in trees. Prothonotary and Lucy's warblers are unusual in using tree cavities as nest sites, and Prothonotary warblers will readily accept manmade nest boxes. The ovenbirds' tightly woven, covered nests are placed on the ground and resemble old-fashioned baker's ovens. Tropical warblers usually lay three eggs, while most temperate-zone breeders lay four.

Migratory species have a variety of social systems during the non-breeding season. For example, Yellow-rumped warblers often aggregate in large flocks that move about the landscape, while other species join mixed-species flocks. Many species, including the American redstart, defend individual territories at their overwintering sites and frequently return to the same winter territory.

The Effect of Habitat Changes

CONSERVATION AND ENVIRONMENT

A number of New World warbler species are threatened or endangered by the human modification of their specialized habitats. These include the Kirtland's warbler, which breeds only in young jack-pine habitats in Michigan; the Golden-cheeked warbler, whose breeding grounds are limited to oak–juniper woodlands in central Texas; the Grey-headed warbler, found only in a small montane region of northeastern Venezuela that is now largely deforested; and Belding's yellowthroat, which is restricted to marshes in Baja California. About a dozen other warblers are similarly at risk because they have very restricted ranges or habitat requirements. Two species have not been sighted for several decades and are probably extinct: Semper's warbler, which was endemic to the island of St. Lucia in the West Indies, and Bachman's warbler, which bred in bamboo-dominated swamps in the southeastern United States and wintered in Cuba.

Continuing deforestation places many populations of tropical resident warblers at increasing risk. There is also widespread concern about the recent population declines seen in many more common migratory species, which are likely to result from broad-scale habitat changes in both their breeding and overwintering areas. IJL/DHM

Tanagers and Tanager Finches

Left *The vividly colored Summer tanager specializes in catching bees and wasps, and is considered a pest by beekeepers.*

THE SIGHT OF A BRILLIANTLY-COLORED FLOCK of tanagers foraging in the treetops is one of the greatest pleasures of a birding trip to the New World tropics. Bright hues are what most people imagine when they think of these birds, for literally every color of the rainbow is represented within the group. Even so, many species are not very colorful at all and are in fact quite drab in plumage. Bill size and shape also vary, from large for seed-cracking to thin and decurved for feeding on nectar.

In view of the birds' diversity, ornithologists have noted for decades that no single trait can be used to describe the group. But it is precisely their disparity that makes them so interesting from an evolutionary standpoint. All the variation has evolved over the last 25 million years; and while it parallels that seen in adaptive radiations of island birds, in tanagers it has occurred on a continental scale.

Medium-sized and Multicolored

FORM AND FUNCTION

Tanagers belong to the group of nine-primaried songbirds, so named because they have only nine well-developed primary feathers, with the tenth being much reduced. For the most part they are medium-sized; species in the genus *Tangara* are typical, weighing about 20g (0.7oz). However, there is considerable variation; some species of flower-piercers, honeycreepers, and dacnises weigh around 10g (0.35oz), while many tanagers weigh 30g (1oz) or more, and the White-capped tanager exceeds 100g (3.5oz).

Plumage color can be quite spectacular. The aptly-named Paradise tanager has a black back and wings with a contrasting green head, blue belly, turquoise throat, and scarlet rump. Even so, some showy species can actually be quite cryptic when encountered in their dark forest or tropical canopy habitat. Others have much drabber plumage, with brown, gray, dull yellow, and dull green colors predominant. In roughly half of all species females are less colorful than males, but in many others both sexes are alike, with females as well as males having strikingly brilliant plumage. In species where the males are more colorful, the male typically retains colorful plumage all the year round.

Bill size and shape are quite variable and reflect dietary specializations. Some species have evolved adaptations to feed on either fruit, nectar, insects, or seeds, but most are omnivorous and will take a combination of these food items. As a group, the tanagers' diet consists of substantially more fruit than most other birds, and some species do not digest the seeds that they swallow. Thus, tanagers are probably by far the most important disseminators of tropical American trees and shrubs.

While some species include only a little nectar in their diet, several species have evolved unique specializations to exploit this rich energy source. Some species, such as honeycreepers, have long and slightly downcurved bills. Flower-piercers have a hook on the end of their upper mandible, which they use to hold a tubular flower while the lower mandible pierces the corolla. Thus, they take nectar from the flower without pollinating it, and so are often called nectar-robbers. Along with bill adaptations associated with nectar-feeding, many of these species also have unusual tongues. Some are tubular in shape, some are lined with fringes, and others are bifurcated at the tip.

Similarly, insect-eaters have evolved specialized behaviors and bill characteristics that help them obtain their prey. For example, the Swallow tanager resembles other tanagers, but differs in its broad, flat bill and pointed, swallowlike wings. Like other tanagers, the birds eat fruit, but, like swallows, they also catch many insects in flight, assisted by the wide gape of their bill.

Among birds that capture insects on a substrate, alternative searching adaptations enable the different species to co-exist in the same location.

FACTFILE

TANAGERS AND TANAGER FINCHES

Order: Passeriformes

Family: Thraupidae

DISTRIBUTION Western hemisphere from Canada to S tip of South America, including Antilles, also Gough, Inaccessible, and Nightingale Islands in the S Atlantic; mostly tropical.

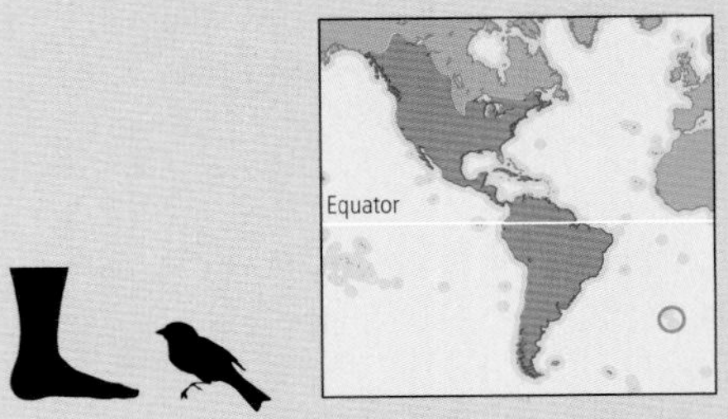

HABITAT Forests, scrub, thickets, plantations, parks, gardens, grasslands; lowlands to high mountains.

SIZE Length 9–28cm (3.5–11in); **weight** 8.5–114g (0.3–3.7oz).

PLUMAGE Exceedingly varied; many bright colors to gray, olive, black, and white. Sexes alike or very different.

VOICE On the whole, poorly developed; some species songless, a few persistent and pleasing songsters.

NEST Usually well-made open cups in trees and shrubs; several species build covered nest with side entrance.

EGGS Usually 2, but up to 4–5 in euphonias and the few species that breed in northern temperate latitudes; blue, blue-gray, gray, or white, spotted, blotched, and scrawled with lilac, brown, or black. Incubation period 12–18 days; nestling period 11–24 days.

DIET Typically insects, fruit, seeds, and nectar.

CONSERVATION STATUS Among 45 threatened species, the Cone-billed and Cherry-throated tanagers are listed as Critically Endangered, as are 3 seedeater species – *Sporophila insulata*, *S. melanops*, and *S. zelichi*.

See Selected Tanager Species ▷

For example, some species forage in large clusters of moss and some on the undersides of branches, while others prefer leaf surfaces. A few species, such as the Rosy thrush-tanager, will take insects from the ground. Others, such as the Grey-headed tanager, will accompany the mixed flocks of small birds that follow army ants to capture insects that the ants drive up from the ground litter.

Seeds make up an important part of the diet of many species, particularly those tanager-finches that were traditionally classified in the Emberizidae. Seedeaters in the genus *Sporophila* use their

Left *A Green-headed tanager* (Tangara seledon). *These social birds can often be seen in small flocks high in the tree canopy.*

Below *Flecked emerald-green, the Spotted tanager* (Tangara punctata) *is a South American species that is found in deep rain forest across the continent, from Suriname to Peru.*

thick bills to feed on grass seeds while clinging to stems. Some seed-eating tanager-finches such as the Large-billed seed-finch have particularly massive bills. The Darwin's finches, perhaps one of the best known examples of an adaptive radiation, include many thick-billed seed-eaters (see box).

Across the Americas

DISTRIBUTION PATTERNS

Tanagers occur in South, Central, and North America as well as the islands of the Caribbean. In addition, three species are endemic to islands in the South Atlantic. However, most species are found in South America, and the Andes have a particularly high concentration.

Wholly tropical tanagers are nonmigratory, but may wander within their range with the changing seasons. Many species are altitudinal migrants; for example, the Swallow tanager travels from the warm lowlands, where the birds live when not breeding, into the mountains of northern South America to nest at heights ranging from 800 to 1,800m (2,600–5,900ft).

Several species migrate to temperate latitudes to breed. These include both species that breed in North America, such as the Western, Scarlet, and Summer tanagers, and others that breed in southern South America, including the Blue-and-yellow tanager and several species of seedeaters.

Predominantly Social

BREEDING BIOLOGY

Little is known of the breeding biology of most tanager species. In the temperate zones, tanager breeding occurs in the spring and early summer, when temperatures are warmer and insect abundance is at its highest. Tropical species time their breeding to coincide with the changes between wet and dry seasons. In some of the nectivorous species, breeding is associated with the timing of flowering. In all cases, reproduction takes places during the season when food resources are at their most abundant.

The little Blue-black grassquit of Central and South America has a distinctive courtship display, using its wings to jump 30–45cm (12–18in), then dropping to the same perch; often it gives a buzzy *tzee-ep* call during this display.

North American species, such as the Western tanager, defend breeding territories by singing. They are also thought to defend territories on their wintering grounds. Other territorial species include some of the flower-piercers, which defend their territories not only from conspecifics but also from other species. In general, however, the extent of territoriality in tanagers is not well known.

Monogamy appears to be the rule for most tanagers, although polygyny has been observed in some species such as the Blue-grey and Scarlet-rumped tanagers. Most tanagers build open, cup-shaped nests, which are normally placed in a tree or shrub, although a few species are ground-nesters. The Swallow tanager is unique in that it

Darwin's Finches

Darwin's finches provide perhaps the best-known example of an adaptive radiation in birds. The 14 species all have very similar plumage, but differ greatly in their bill adaptations. Found only on Cocos Island and the Galápagos, they have evolved their great variety of feeding specializations within a short timeframe of less than 5 million years.

The Warbler finch, so named for its narrow bill, gleans insects and other arthropods off surfaces. The thick bills of the ground finches are used to crack seeds of various sizes. Some species of ground finches also use their bills to pluck ticks from tortoises and iguanas, while one species, the Sharp-beaked, pecks at the developing feathers of seabirds and drinks the blood that results. Woodpecker finches engage in tool-use, employing twigs or cactus spines to pry insects from bark and decaying wood, and in consequence are able to exploit the food niche normally occupied by woodpeckers. The tree finches eat insects as well as seeds, but the Vegetarian finch consumes only leaves, fruits, and buds.

The Darwin's finches may have originated from a Caribbean species, as most of their close relatives are found in the Caribbean and many are Caribbean endemics. Interestingly, these relatives also show a great diversity of bill types; for example, the bananaquit has a thin, decurved bill and feeds on nectar, whereas the bullfinches have large, seed-crushing bills. All of these birds build covered, domed-shaped nests similar to those of the Darwin's finches, and this concordance, together with DNA evidence, provides the evidence for their close relationship.

uses natural or artificial cavities in cliffs, earthen banks, or masonry. This species places its shallow cup nest inside long tunnels up to 2m (6.5ft) deep. In place of open-cup nests, many species, including the 29 species of euphonias and chlorophonias, make globular, dome-shaped nests with side entrances. Covered nests are also built by the Darwin's finches and their close relatives (see box).

In most species, the female alone builds the nest; however, there are many species in which both sexes participate. Usually two to three eggs are laid, but in euphonias and chlorophonias the average number is four or five. The eggs are laid early in the morning, usually on consecutive days. In the vast majority of species they are incubated by the female alone. One exception is the Rosy thrush-tanager, in which both males and females have been seen incubating the eggs.

The incubation period varies with the form and situation of the nest. In low, open, thick-walled nests, such as those of the Scarlet-rumped and Crimson-backed tanagers, it lasts for 12 days, but the period stretches to 13 or 14 in the smaller, mossy nests of Silver-throated tanagers and other *Tangara* species, which are usually higher and less conspicuous. In the covered nests with a side entrance that euphonias hide in crannies, incubation is prolonged to 15 to 18 days.

Nearly always, both males and females help feed the young, but only the female broods. Insects are the main food items supplied. Flower-piercers, euphonias, and chlorophonias feed their nestlings by regurgitation rather than directly from the bill. Cooperative breeding is a feature of some species; helpers are assumed to be young from a previous brood. Thus, for example, a young Golden-hooded tanager in immature plumage will sometimes help its parents to feed a later brood. In this and some other species in the genus *Tangara*, three or four adults may attend one or two nestlings.

The nestling period varies in the same way as the incubation period. Low, open nests that are more susceptible to predation have shorter periods of 11 or 12 days, while in nests that are higher off the ground, nestlings remain for 14 or 15 days. In the covered nests of euphonias and cholorphonias, the young stay in the nest for 19 to 24 days.

Tanagers are relatively social birds. When not breeding, they often occur in small groups of less than ten. Larger single-species flocks are unusual,

Above *Adaptive radiation in Darwin's finches:* ***1*** *Woodpecker finch* (Camarhynchus pallidus)*: insects from trunks and branches.* ***2*** *Sharp-beaked ground finch* (Geospiza difficilis)*: arboreal seed-eater.* ***3*** *Vegetarian finch* (C. crassirostris)*: buds and leaves.* ***4*** *Common cactus-finch* (G. scandens)*: arboreal seed-eater.* ***5*** *Large tree finch* (C. psittacula)*: arboreal insect-eater.* ***6*** *Large ground finch* (G. magnirostris)*: ground-feeding seed-eater.* ***7*** *Large cactus-finch* (G. conirostris)*: seed-eater.* ***8*** *Warbler finch* (Certhidea olivacea)*: small insects from trees.* ***9*** *Small ground finch* (G. fuliginosa)*: ground-feeding seed-eater.*

however, and a few species, including some flower-piercers, are solitary. In the tropics, most tanagers forage in mixed-species flocks, often constituting an important core element of these assemblages.

Tanagers communicate through songs and call notes as well as behavioral displays. These displays, whether for courtship or territorial defense, often emphasize aspects of their showy plumage. For example, male Red-legged honeycreepers will lift their wings, exposing the yellow undersides, while Silver-beaked tanagers raise their bills vertically, showing the bluish-white expanded portion of their lower bill. Swallow tanagers engage in mass displays, all simultaneously "curtseying," or bowing deeply down and up, while facing one another or perched close together.

Running Out of Space

CONSERVATION AND ENVIRONMENT

Many tanagers are threatened with extinction, mostly from habitat destruction. In addition, a few species, especially of the seedeaters in the genus *Sporophila*, are trapped extensively for the cage-bird trade. Many species, including the Darwin's finches, the St. Lucia black finch, and the Gough finch, are island endemics and so are susceptible to introduced predators and competitors as well as being at risk because of their restricted distribution. Many of the continental species, such as the Black-and-gold, Cherry-throated, and Seven-colored tanagers, are also known from only a few localities.
KJB/AFS

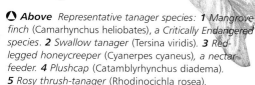

Above *Representative tanager species:* ***1*** *Mangrove finch* (Camarhynchus heliobates), *a Critically Endangered species.* ***2*** *Swallow tanager* (Tersina viridis). ***3*** *Red-legged honeycreeper* (Cyanerpes cyaneus), *a nectar-feeder.* ***4*** *Plushcap* (Catamblyrhynchus diadema). ***5*** *Rosy thrush-tanager* (Rhodinocichla rosea).

Selected Tanager Species

413 species in 104 genera. Species and genera include: bananaquit (*Coereba flaveola*), Black-and-gold tanager (*Bangsia melanochlamys*), Blue-and-yellow tanager (*Thraupis bonariensis*), Blue-grey tanager (*T. episcopus*), Blue-black grassquit (*Volatinia jacarina*), bullfinches (genera *Loxigilla, Melopyrrha*), Cherry-throated tanager (*Nemosia rourei*), chlorophonias (genus *Chlorophonia*), Cone-billed tanager (*Conothraupis mesoleuca*), Crimson-backed tanager (*Ramphocelus dimidiatus*), Scarlet-rumped tanager (*R. passerinii*), Silver-beaked tanager (*R. carbo*), dacnises (genus *Dacnis*), Darwin's finches (genera *Geospiza*, *Camarhynchus*, *Certhidea*, *Pinaroloxias*), Sharp-beaked ground finch (*Geospiza difficilis*), Large tree finch (*Camarhynchus psittacula*), Mangrove finch (*C. heliobates*), Vegetarian finch (*C. crassirostris*), Woodpecker finch (*C. pallidus*), Warbler finch (*Certhidea olivacea*), euphonias (genus *Euphonia*), flower-piercers (genus *Diglossa*), Golden-hooded tanager (*Tangara larvata*), Paradise tanager (*T. chilensis*), Seven-colored tanager (*T fastuosa*), Silver-throated tanager (*T. icterocephala*), Gough finch (*Rowettia goughensis*), Grey-headed tanager (*Eucometis penicillata*), honeycreepers (genus *Cyanerpes*), Red-legged honeycreeper (*C. cyaneus*), Large-billed seed-finch (*Oryzoborus crassirostris*), Rosy thrush-tanager (*Rhodinocichla rosea*), Scarlet tanager (*Piranga olivacea*), Summer tanager (*P. rubra*), Western tanager (*P. ludoviciana*), seedeaters (genus *Sporophila*), St. Lucia black finch (*Melanospiza richardsoni*), Swallow tanager (*Tersina viridis*), White-capped tanager (*Sericossypha albocristata*).

Icterids

Icterids are a diverse family of perching birds exclusive to the New World that include meadowlarks, bobolinks, American orioles, troupials, caciques, oropendolas, cowbirds, American blackbirds, and grackles. They are one of the most successful groups of New World birds in terms of species diversity and population numbers, with over 100 species occurring throughout the Americas from Alaska in the north to Cape Horn in the south. Most species are tropical, with centers of species richness in Mexico (24 species) and Colombia (27 species). However, high species densities also occur in temperate grasslands and marshes.

Some icterids are among the commonest birds in the world, with populations in the tens of millions, and social species congregate in vast flocks during the non-breeding season. The most abundant is the Red-winged blackbird, whose populations approach 200 million individuals. The icterids exhibit a remarkable range of foraging adaptations, behavioral flexibility, and innovative capabilities, the latter sure signs of their intelligence. This adaptability has enabled them to exploit most terrestrial environments, from arctic muskeg to hot deserts, tropical forests, grasslands, wetlands, and oceanic islands. The same resourcefulness has also allowed many icterids to conquer artificial environments; Common grackles and Brewer's blackbirds are now as ubiquitous in shopping malls and urban parks as they are in their natural habitats.

Below *After the breeding season, Red-winged blackbirds amass with other icterids in flocks reaching hundreds of thousands or even millions. These large flocks are now regarded as a health hazard.*

Left *A male Long-tailed meadowlark displaying its conspicuous red front. Although the bird's melodic song is music to human ears, in reality many of its songs are used as primary weapons to defend territory.*

Right *Yellow orioles* (Icterus nigrogularis) *sometimes have the luxury of instant meals when they feed from the nectar feeders that people set up in gardens.*

Familiar and Flexible

Form and Function

Icterids range in size from the crow-sized male Montezuma oropendola down to the most diminutive of the New World orioles, some of which are as small as sparrows. Many icterids, notably the grackles, blackbirds, baywings, and cowbirds, are predominantly black or brown in plumage, whereas others such as the New World orioles and troupials are often fabulously clothed in brilliant orange, red, and golden hues. In sedentary and monogamous icterids the sexes are generally alike, but striking sexual differences in size and plumage characterize many of the migratory and polygynous species. Dimorphism in size reaches extreme expression in the Montezuma oropendola, in which the males are more than twice the size of the females in their harems.

Behavioral flexibility in icterid foraging strategies is also coupled with functional flexibility thanks to their "gaping." This adaptation results from a special configuration of the skull that allows icterids to open their beaks powerfully rather than passively, and thus force open holes in fruit, bark, soil, or other substrates to expose food. Gaping has enabled icterids to exploit almost every food niche occupied by other perching birds. The practice is especially developed in some oropendolas and caciques, which, aided by their bladelike mandibles, employ gaping to cut their way through fruit skins to expose the flesh, a favored food. Many orioles and troupials also exploit fruits in similar fashion.

Most icterids are relatively long-billed, a correlate of their reliance on insect prey when breeding. However, the bobolink and some cowbirds have robust, finchlike bills for cracking the seeds they take for much of the year. In contrast, specialized species such as the Jamaican blackbird and Solitary cacique use gaping to pry open bark, dead wood, and bromeliads to expose invertebrates. The meadowlarks are also powerful gapers, cap-

turing invertebrates by forcing roots and soil apart with their bills. Some marsh-nesting blackbirds also employ gaping to split open sedge stems to expose small prey.

All icterids have relatively short, deep wings that can be rapidly flapped for lift. Some temperate icterids are accomplished long-distance migrants; bobolinks migrate 20,000km (12,500 miles) a year from their North American breeding grounds to winter on South American pampas. Some grackles have elongate, wedge-shaped tails with pronounced keels. These have evolved partly for sexual display, but also for improving lift in slow flight. Many arboreal caciques and orioles have relatively long tails that act as balancing aids. In contrast, meadowlarks, which spend much of their time foraging on the ground, have relatively short tails.

DNA analyses are revolutionizing our understanding of icterid relationships. They have already demonstrated that many similarities previously thought to indicate close relationships between species have actually evolved independently in response to similar selective pressures.

FACTFILE

ICTERIDS

Order: Passeriformes

Family: Icteridae

102 species in 26 genera.

DISTRIBUTION N and S America

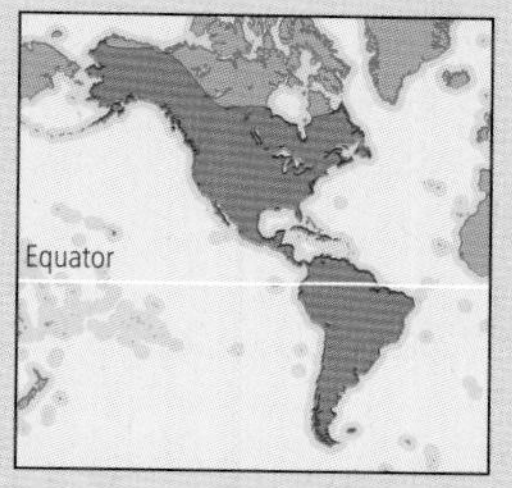

HABITAT Grasslands, savannas, marshes, woodlands, and forests.

SIZE Length 15–54cm (6–21in); weight 16–528g (0.6–19oz).

PLUMAGE Chiefly black, with bold patches of yellow, orange, or red; brown common among both sexes of grassland species and females of many others. Differences between males and females pronounced in temperate, migratory species, and among polygynous species at all latitudes.

VOICE A wide range of single notes and chattering calls. Songs range from simple and harsh to long, complex, and musical.

NEST Trees, shrubs, on ground, and in emergent aquatic vegetation; occasionally on cliffs.

EGGS Variable in background color and amount of spotting; weight 2.1–14.2g (0.07–0.5oz). Incubation period 12–15 days; nestling period 9–35 days.

DIET Invertebrates, seeds, fruit, nectar, and small vertebrates.

CONSERVATION STATUS Forbes's blackbird, the Colombian mountain grackle, and the Montserrat oriole are Critically Endangered; 4 other species are Endangered, and 4 are Vulnerable.

See Selected Icterid Species ▷

A classic example of this convergent evolution is the extreme similarity in morphology and behavior between the North American meadowlarks and the distantly-related longclaw pipits of Africa. Both groups have independently evolved the similarities as adaptations to grassland ecosystems. Likewise, the similarities shared by the New and Old World orioles are also due to convergence rather than relationship.

DNA studies indicate extensive convergence between the icterids. For example, they demonstrate that many of the American blackbirds are not close relatives, being in fact part of a much larger group that includes the grackles and cowbirds. Indeed, the Yellow-headed blackbird is actually the nearest relative of the meadowlarks and the bobolink, and is only distantly related to the other "blackbirds." These analyses also show that the baywing, long thought to be a primitive cowbird, is not a cowbird at all but rather the closest relative of the aberrant Bolivian blackbird.

A Marked Division of the Sexes

BREEDING BIOLOGY

Mating systems in icterids range from monogamous pair bonding as in most orioles through to the harem polygyny of some oropendolas. Males usually perform elaborate displays combining ritual movements and vocalizations. The songs of the chopi and Melodious blackbird are particularly beautiful. In almost all icterids the female is wholly responsible for nest building, and only females incubate eggs. Neither do males feed their incubating mates, although they usually assist in provisioning the young. Baywings and Bolivian blackbirds reduce breeding costs by employing additional "helpers" to tend the young birds. These assistants may be relatives of the parents, and their involvement helps replicate their shared family genes.

The troupials and baywings have avoided some breeding costs altogether by abandoning nest building and utilizing old nests of other species; alternatively, they forcibly pirate nests by evicting the owners. The cowbirds minimize breeding costs even further by laying their eggs in the nests of other species and relying on these hosts to rear their young. In the Screaming and Giant cowbirds, this brood parasitism is largely focused on other icterids, notably baywings, caciques, and oropendolas. In contrast, the Brown-headed and Shiny cowbirds employ "shotgun" parasitism and lay their eggs in the nests of hundreds of other species. Cowbird adaptations for brood parasitism include mimicry of the eggs and young of host species, reduced incubation times for their own eggs, and removal of host eggs.

Many icterids nest in colonies, and those of the Tricolored blackbird may contain 100,000 individuals. Most colonial breeders are marsh or savanna nesters, but the tropical oropendolas and

Above *Icterids obtain much of their food by inserting the closed bill into some potential food source, and then forcibly opening it. They "gape" into rotting wood, flowers, curled leaves, clumps of grass, and soil. In all cases, gaping exposes food, usually arthropods, not available to a bird gleaning prey from the surface.*

Below *Representative icterid species:* ***1*** *Male Northern oriole* (Icterus galbula)*;* ***2*** *Common grackle* (Quiscalus quiscula).

Selected Icterid Species

Species include: Baudó oropendola (*Gymnostinops cassini*), Montezuma oropendola (*G. montezuma*), Selva cacique (*Cacicus koepckeae*), Solitary cacique (*C. solitarius*), troupials (*Icterus* spp.), Martinique oriole (*I. bonana*), Montserrat oriole (*I. oberi*), St. Lucia oriole (*I. laudabilis*), Jamaican blackbird (*Nesopsar nigerrimus*), Yellow-headed blackbird (*Xanthocephalus xanthocephalus*), Saffron-cowled blackbird (*Xanthopsar flavus*), Red-winged blackbird (*Agelaius phoeniceus*), Tricolored blackbird (*A. tricolor*), Yellow-shouldered blackbird (*A. xanthomus*), Pampas meadowlark (*Sturnella defillippi*), Red-bellied grackle (*Hypopyrrhus pyrohypogaster*), Forbes's blackbird (*Curaeus forbesi*), chopi (*Gnorimopsar chopi*), Bolivian blackbird (*Oreopsar bolivianus*), Colombian mountain grackle (*Macroagelaius subalaris*), Melodious blackbird (*Dives dives*), Common grackle (*Quiscalus quiscula*), Slender-billed grackle (*Q. palustris*), Brewer's blackbird (*Euphagus cyanocephalus*), baywing (*Molothrus badius*), Screaming cowbird (*M. rufoaxillaris*), Shiny cowbird (*M. bonariensis*), Brown-headed cowbird (*M. ater*), Giant cowbird (*Scaphidura oryzivora*), bobolink (*Dolichonyx oryzivorus*).

Left *As brood parasites, young cowbirds grow quickly at the expense of the host bird's young. Here, a group of young Shiny cowbirds are being fed by a female Rufous-collared sparrow* (Zonotrichia capensis).

caciques nest in mixed colonies in forest clearings. Colonial nesting is defensively advantageous for many species, as members will join together to vigorously attack nest predators and cowbird parasites. Colonies may also function as "information centers," whereby the movements of successful foragers can be monitored and followed by other individuals to locate productive food sources.

Most icterids build intricate nests in trees. The nests of orioles, oropendolas, and many caciques are especially elaborate, consisting of pendulous hanging baskets. Many species protect their nests from predators by building them over water or in the vicinity of wasps' nests.

Restricted-Range Risks

CONSERVATION AND ENVIRONMENT

Eleven icterid species are currently listed as threatened, of which three are critically endangered. This number represents over 10 percent of all recognized species. Tropical icterids are sedentary, and often have restricted ranges in forest environments or small populations on islands, rendering them particularly susceptible to a diversity of threats. The Slender-billed grackle is already extinct due to marsh drainage in the early 20th century. More perplexingly, some icterids are threatened by the parasitism of their close relative, the Shiny cowbird.

The critically endangered Montserrat oriole provides a classic example of the threats to restricted-range species. This oriole is confined to the old-growth forest of the small island of Montserrat. Consequently, it has always had a small population, but this was massively reduced by volcanic eruptions in the 1990s that destroyed over two-thirds of the birds' remaining habitat. Tragically, the surviving orioles have been further reduced by the increasing frequency of hurricanes. This combination of natural cataclysms has brought the species to the very brink of extinction, and captive breeding may represent its only hope. Four other West Indian icterids – the Martinique and St. Lucia orioles and the Yellow-shouldered and Jamaican blackbirds – are similarly threatened by habitat destruction and brood parasitism by the Shiny cowbirds. Cowbird control measures and the establishment of protected areas represent the best hope for these species.

Forest clearance is the primary threat to the restricted-range icterids of tropical South America. The Endangered Red-bellied grackle and the Critically Endangered Colombian mountain grackle are now confined to tiny remnants of montane forest in the Colombian Andes. In contrast, the threatened Selva cacique and the Endangered Baudó oropendola are denizens of the ravaged lowland forest tracts of Peru and Colombia. The Critically Endangered Forbes's blackbird is now reduced to a handful of isolated populations in eastern Brazil, and has been brought to the edge of extinction by habitat destruction and cowbird parasitism. Extensive habitat protection is the only long-term measure that will ensure the survival of these species in the wild. In contrast, the Pampas meadowlark and Saffron-cowled blackbird had wide ranges in the once-extensive grasslands of South America, but these ecosystems have been dramatically reduced by intensive cattle rearing, and both species are now also threatened. SP/GHO

Right *An Eastern meadowlark* (Sturnella magna) *stretches on its perching post from where it delivers its song. Along with its Western counterpart, it is one of the best-known birds of American farmlands.*

Glossary

Adaptation features of an animal that adjust it to its environment. NATURAL SELECTION favors the survival of individuals whose adaptations adjust them better to their surroundings than other individuals with less successful adaptations.

Adaptive radiation where a group of closely related animals (e.g. members of a family) have evolved differences from each other so that they occupy different NICHES and have reduced competition between each other.

Adult a fully developed and mature individual, capable of breeding but not necessarily doing so until social and/or ecological conditions allow.

Air sac thin walled structure connected to the lungs of birds and involved in respiration; extensions of these can occur in hollow bones.

Albino a form in which all dark pigments are missing, leaving the animal white, usually with red eyes.

Allopatry condition in which populations of different species are geographically separated (cf PARAPATRY, SYMPATRY).

Alpine living in mountainous areas, usually above 1,500m (5,000ft).

Altricial refers to young that stay in the nest until they are more or less full grown (as opposed to PRECOCIAL). See also NIDICOLOUS.

Aquatic associated with water.

Arboreal associated with or living in trees.

Avian pertaining to birds.

Beak see BILL.

Bill the two MANDIBLES with which birds gather their food and preen their feathers. Synonymous with beak.

Blubber fat, usually that lying just beneath the skin.

Bolus a ball (of food).

Boreal zone the area of land lying just below the north polar region and mainly covered in coniferous forest.

Broadleaved woodland woodland mainly comprising angiosperm trees (both deciduous and evergreen), such as oaks, beeches, and hazels; characteristic of many temperate areas of Europe and North America.

Brood group of young raised simultaneously by a pair (or several) birds.

Brood parasite a bird that has its eggs hatched and reared by another species (e.g. certain cuckoo species and cowbirds).

Cache a hidden store of food; also (verb) to hide food for future use.

Call short sounds made by birds to indicate danger, threaten intruders, or keep a group of birds together. See also SONG.

Canopy a fairly continuous layer in forests produced by the intermingling of branches of trees; may be fully continuous (closed) or broken by gaps (open). The crowns of some trees project above the canopy layer and are known as emergents.

Carpal the outer joint of the wing, equivalent to the human wrist.

Casque bony extension of the upper MANDIBLE.

Cecum diverticulation or sac of the hindgut.

Clade a set of species derived from a single common ancestor.

Class a taxonomic level. All birds belong to the class Aves. The main levels of a taxonomic hierarchy (in descending order) are: Phylum, Class, Order, Family, Genus, Species.

Climax (of vegetation) the final stage in a plant "succession," in which the community of plants reaches a stable equilibrium in its environment.

Cloaca terminal part of the gut into which the reproductive and urinary ducts open. There is one opening to the outside of the body, the cloacal aperture, instead of separate anus and urinogenital openings.

Clutch the eggs laid in one breeding attempt.

Colonial living together in a COLONY.

Colony a group of animals gathered together for breeding.

Comb a fleshy protuberance on the top of a bird's head.

Communal breeder species in which more than the two birds of a pair help in raising the young. See COOPERATIVE BREEDING.

Congener a member of the same genus.

Coniferous forest forest comprising largely evergreen conifers (firs, pines, spruces etc.), typically in climates either too dry or too cold to support DECIDUOUS FOREST. Most frequent in northern latitudes or in mountain ranges.

Conspecific a member of the same species.

Contact call CALLS given by males in competition.

Contour feathers visible external covering of feathers, including flight feathers of tail and wings.

Convergent evolution the independent acquisition of similar characters in evolution, as opposed to the possession of similarities by virtue of descent from a common ancestor.

Cooperative breeding a breeding system in which parents of young are assisted in the care of young by other adult or subadult birds.

Coverts the smaller feathers that cover the wings and overlie the base of the large FLIGHT FEATHERS (both wings and tail).

Covey a collective name for groups of birds, usually gamebirds.

Crèche a gathering of young birds. especially in penguins and flamingos; sometimes used as a verb.

Crest long feathers on the top of the heads of birds.

Crop a thin-walled extension of the foregut used to store food; often used to carry food to the nest.

Crustaceans invertebrate group that includes shrimps, crabs and many other small marine animals.

Cryptic camouflaged and difficult to see.

Deciduous forest temperate and tropical forest with moderate rainfall and marked seasons. Typically trees shed leaves during either cold or dry periods.

Desert areas of low rainfall, typically with sparse scrub or grassland vegetation or lacking vegetation altogether.

Dichromatism the condition in which members of a species show one of two distinct color patterns.

Dimorphic literally "two forms." Usually used as "sexually dimorphic " (i.e. the two sexes differ in color or size).

Dimorphism the existence of two distinct forms within a species. Sexual dimorphism is the existence of marked morphological differences between males and females.

Disjunct distribution geographical distribution of a species that is marked by gaps. Commonly brought about by fragmentation of suitable habitat, especially as a result of human intervention.

Dispersal the movements of animals. often as they reach maturity, away from their previous HOME RANGE. Distinct from dispersion, that is the pattern in which things (perhaps animals, food supplies, nest-sites) are distributed or scattered.

Display any relatively conspicuous, stereotyped pattern of behavior that conveys specific information to others, usually to members of the same species; often associated with courtship but also in other activities, e.g. "threat display."

Display ground the place where a male (or males) tries to attract females.

DNA deoxyribonucleic acid; the key substance of chromosomes – important for inheritance.

Dominance hierarchy a "pecking order"; in most groups of birds, in any pair of birds each knows which is superior and a ranking of superiors therefore follows.

Double-brooded (also triple or multiple brooded) birds that breed twice or more each year, subsequent nests following earlier successful ones, excluding those when the first or all earlier nests fail, in which case the term "replacement nests" applies.

Echolocation the ability to find one's way by emitting sounds and gauging the position of objects by timing the returning echo (e.g. in swifts).

Erectile of an object, e.g. a crest, that can be raised.

Facultative optional. See also OBLIGATE.

Family either a group of closely related species, e.g. penguins, or a pair of birds and their offspring. See CLASS.

Feces excrement from the digestive system passed out through the CLOACA.

Fledge strictly to grow feathers. Now usually used to refer to the moment of flying at the end of the nesting period when young birds are more or less completely feathered. Hence fledging period, the time from hatching to fledging, and fledgling, a recently fledged young bird.

Flight feathers the large feathers of the wing, which can be divided into PRIMARY FEATHERS and SECONDARY FEATHERS.

Fossil any remains, impression, cast, or trace of an animal or plant of a past geological period, preserved in rock.

Fossorial burrowing.

Frontal shield a fleshy area covering the forehead.

Frugivore eating mainly fruits.

Gallery forest a thin belt of woodland along a riverbank in otherwise more open country.

Generalist an animal whose lifestyle does not involve highly specialized strategems (cf SPECIALIST), for example, feeding on a variety of foods which may require different foraging techniques.

Genus the taxonomic grouping of species. See CLASS.

Gizzard the muscular forepart of the stomach. Often an important area for the grinding up of food, in many species with the help of grit.

Glycogen a polysaccharide that is the principal form in which carbohydrate is stored in the tissues of animals.

Gondwanaland the supercontinent incorporating the modern continents of the southern hemisphere – Africa, S. America, Australia, Antarctica, as well as (during the Paleozoic and Mesozoic) India, Madagascar, and New Zealand.

Gregarious the tendency to congregate into groups.

Guano bird excreta. In certain dry areas the guano of colonial sea birds may accumulate to such an extent that it is economic to gather it for fertilizer.

Gular pouch (or gular sac) an extension of the fleshy area of the lower jaw and throat, e.g. in pelicans.

Habitat the type of country in which an animal lives.

Hallux the first toe. Usually this is small and points backward, opposing the three forward-facing toes.

Harem a group of females living in the territory of, or consorting with, a single male.

Hatchling a young bird recently emerged from the egg.

Helper an individual, generally without young of its own, which contributes to the survival of the offspring of others by behaving parentally towards them. See COOPERATIVE BREEDING.

Herbivore an animal which eats vegetable material.

Holarctic realm a region of the world including North America, Greenland, Europe and Asia apart from the Southwest, Southeast, and India.

Homeothermic warm-blooded; ability to keep body temperature constant.

Home range an area in which an animal normally lives (generally excluding rare excursions or migrations), irrespective of whether or not the area is defended from other animals.

Honeydew the sugary fluid, derived from plant sap, excreted by aphids, leafhoppers and treehoppers.

Hormone a substance secreted within the body that is carried by the blood to other parts of the body where it evokes a specific response, such as the growth of a particular type of cell.

Hybrid the offspring of a mating between birds of different species.

Hypothermy a condition in which internal body temperature falls below normal.

Incubation the act of incubating the egg or eggs, ie keeping them warm so that development is possible. Hence incubation period, the time taken for eggs to develop from the start of incubation to hatching.

Insectivore an animal that feeds on insects.

Introduced of a species that has been brought from lands where it occurs naturally to lands where it has not previously occurred. Some introductions are natural but some are made on purpose for biological control, farming, or other economic reasons.

Irruption sudden or irregular spread of birds from their normal range. Usually a consequence of a food shortage.

Karyotype the characteristic number and shape of the chromosomes of a cell, individual or species.

Keratin a tough, fibrous protein from which feathers are formed (also reptile scale, human hair, fingernails etc.).

Krill small shrimplike marine CRUSTACEANS, which are an important food for certain species of seabirds.

Lamellae comblike structures that can be used for filtering organisms out of water.

Lanceolate (of feathers) referring to lance-like (pointed) shape.

Lek a display ground where two or more male birds gather to attract females. See DISPLAY.

Life history the history of an individual organism, from the fertilization on the egg to its death.

Littoral referring to the shoreline.

Mallee scrub small scrubby eucalyptus that covers large areas of semi-arid country in Australia.

Mandible one of the jaws of a bird which make up the BILL (upper or lower).

Melanin a dark or black PIGMENT.

Metabolic rate the rate at which the chemical processes of the body occur.

Migration usually the behavior in which birds fly (migrate) from one part of the world to another at different times of year. There is also local migration and altitudinal migration where birds move, e.g. on a mountainside, from one height to another.

Molt the replacement of old feathers by new ones.

Monoculture a habitat dominated by a single species of plant, often referring to forestry plantations.

Monogamous taking only a single mate (at a time).

Monotypic the sole member of its genus, family, order etc.

Montane pertaining to mountainous country.

Montane forest forest occurring at middle altitudes on the slopes of mountains, below the alpine zone but above the lowland forest.

Morph a form, usually used to describe a color form when more than one exist.

Morphology the study of the shape and form of animals.

Natural selection the process whereby individuals with the most appropriate ADAPTATIONS are more successful than other individuals, and hence survive to produce more offspring and so increase the population.

Neotropics the tropical part of the New World; includes South America, Central America, part of Mexico and the West Indies.

Nestling a young bird in the nest, hence nestling period, the time from hatching to flying (see FLEDGE).

Niche specific parts of a habitat occupied by a species, defined in terms of all aspects of its lifestyle (e.g. food, competitors, predators and other resource requirements).

Nidicolous young birds that stay in the nest until they can fly. See ALTRICIAL.

Nidifugous of young birds that leave the nest soon after hatching. See PRECOCIAL.

Nomadic wandering (as opposed to having fixed residential areas).

Obligate required, binding. See also FACULTATIVE.

Oligotrophic of a freshwater lake with low nutrient levels; such lakes are usually deep and have poor vegetation.

Omnivore an animal that eats a wide variety of foods.

Opportunistic an animal that varies its diet in relation to what is most freely available. See GENERALIST, SPECIALIST.

Order a level of taxonomic ranking. See CLASS.

Organochlorine pesticides a group of chemicals used mainly as insecticides, some of which have proved highly toxic to birds: includes DDT, aldrin, dieldrin.

Pair bond the faithfulness of a mated pair to each other.

Palearctic a zoogeographical area roughly comprising Europe and Asia (but excluding the Indian subcontinent and Southeast Asia).

Pampas (sing. pampa) grassy plains of South America.

Parapatry a condition in which the geographical ranges of two or more different species adjoin one another, but are separate. (cf ALLOPATRY, SYMPATRY)

Parasitize in the ornithological sense, usually to lay eggs in the nests of another species and leave the foster parents to raise the young. See BROOD PARASITE.

Passerine strictly "sparrow-like" but normally used as a shortened form of Passeriformes, the largest ORDER of birds. (See Notes on Classification.)

Pecten a structure lying on the retina of the eye.

Pigment a substance that gives color to eggs and feathers.

Pod a group of individuals, especially juvenile pelicans, with a temporary cohesive group structure.

Polyandry a mating system in which a female mates with several males.

Polygamy a mating system in which a bird of one sex takes several mates.

Polymorphic where a species occurs in two or more) different forms (usually relating to color). See MORPH, DIMORPHIC.

Polygyny a mating system in which a male mates with several females.

Population a more or less separate (discrete) group of animals of the same species.

Prairie North American STEPPE grassland between 30°N and 55°N.

Precocial young birds that leave the nest after hatching. See ALTRICIAL.

Predator a bird that hunts and eats other vertebrates, hence "anti-predator behavior" describes the evasive actions of the prey.

Preen gland a gland situated above the base of the tail. The bird wipes its bill across this while preening the feathers, so distributing the waxy product of the preen gland over the feathers. The exact function of this is not known; some groups of birds do not possess preen glands.

Primary feather one of the large feathers of the outer wing.

Primary forest forest that has remained undisturbed for a long time and has reached a mature (CLIMAX) condition; primary rain forest may take centuries to become established. See also SECONDARY FOREST.

Promiscuous referring to species where the sexes come together for mating only and do not form lasting pair bonds.

Pyriform pear-shaped.

Quartering act of flying back and forth over an area, searching it thoroughly.

Race a subsection of a species which is distinguishable from the rest of that species. Usually equivalent to SUBSPECIES.

Radiation see ADAPTIVE RADIATION.

Rain forest tropical and subtropical forest with abundant and year-round rainfall. Typically species are rich and diverse.

Range (geographical) area over which an organism is distributed.

Raptor a bird of prey, usually one belonging to the order Falconiformes.

Ratites members of four orders of flightless birds (ostrich, rheas, emu and cassowaries, kiwis) that lack a keel on the breastbone.

Relict population a local group of a species that has been isolated from the rest for a long time.

Resident an animal that stays in one area all the year round.

Rictal bristles stiff, modified feathers that surround the gape of several insectivorous birds (e.g. nightjars). Their function is much debated, though they are thought to act like the sensitive mammals' whiskers, aiding prey capture.

Roosting sleeping.

Sahara–Sahelian zone the area of North Africa comprising the Sahara Desert and the arid Sahel zone to its south.

Savanna a term loosely used to describe open grasslands with scattered trees and bushes, usually in warm areas.

Scrape a nest without any nesting material where a shallow depression has been formed to hold the eggs.

Scrub a vegetation dominated by shrubs – woody plants usually with more than one stem. Naturally occurs most often on the arid side of forest or grassland types, but often artificially created by humans as a result of forest destruction.

Secondary feather one of the large flight feathers on the inner wing.

Secondary forest an area of rain forest that has regenerated after being felled. Usually of poorer quality and lower diversity than PRIMARY FOREST and containing trees of a more uniform size.

Sedentary nonmigrating. See RESIDENT.

Sequential molt where feathers (usually the wing feathers) are molted in order, as opposed to all at once.

Sexual selection an evolutionary mechanism whereby females select for mating only males with certain characteristics, or vice versa.

Siblicide the killing by young birds of siblings (e.g. in some raptor species).

Sibling group a group containing brothers and sisters.

Sibling species closely related species, thought to have only recently separated.

Single-brooded birds which only make one nesting attempt each year, although they may have a replacement clutch if the first is lost. See DOUBLE-BROODED.

Song a series of sounds (vocalization), often composed of several or many phrases constructed of repeated elements, normally used by a male to claim a territory and attract a mate.

Specialist a bird whose lifestyle involves highly specialized strategems, e.g. feeding with one technique on a particular food.

Species a population, or series of populations, which interbreed freely, but not with those of other species. See CLASS.

Speculum a distinctively colored group of flight feathers (e.g. on the wing of a duck).

Spur the sharp projection on the leg of some game birds; often more developed in males and used in fighting. Also found on the carpal joint of some other birds.

Staging ground/place area where birds may pause to feed during migration.

Steppe open grassy plains, with few trees or bushes, of the central temperate zone of Eurasia or North America (PRAIRIES), characterized by low and sporadic rainfall and a wide annual temperature variation. In cold steppe temperatures drop well below freezing point in winter, with rainfall concentrated in the summer or evenly distributed throughout the year, while in hot steppe, winter temperatures are higher and rainfall concentrated in winter months.

Stooping dropping rapidly (usually used of a bird of prey in pursuit of prey).

Strutting ground an area where male birds may display.

Subadult no longer juvenile but not yet fully adult.

Sublittoral the sea shore below the lowtide mark.

Suborder a subdivision of an order. See CLASS.

Subspecies a subdivision of a species. Usually not distinguishable unless the specimen is in the hand; often called RACE. See also CLASS.

Subtropics the area just outside the tropics (i.e. at higher latitudes).

Sympatry a condition in which the geographical ranges of two or more different species overlap. (cf ALLOPATRY, PARAPATRY)

Syrinx the vocal organ of a bird.

Taiga the belt of forests (coniferous) lying below (at lower latitudes to) the TUNDRA.

Tarsus that part of the leg of a bird which is just above the foot. Strictly the tarsometatarsus, bones formed from the lower leg and upper foot.

Taxon (pl. taxa) a group of organisms of any taxonomic rank.

Temperate zone an area of climatic zones in mid latitude, warmer than the northerly areas but cooler than the subtropical areas.

Terrestrial living on land.

Territorial defending an area, in birds usually referring to a bird or birds that exclude others of the same species from their living area and in which they will usually nest.

Territory area that an animal or animals consider their own and defend against intruders.

Thermal an area of (warm) air which rises by convection.

Thermoregulation the regulation and maintenance of a constant internal body temperature.

Torpor a temporary physiological state, akin to short-term hibernation, in which the body temperature drops and the METABOLIC RATE is reduced. Torpor is an ADAPTATION for reducing energy expenditure in periods of extreme cold or food shortage.

Totipalmate feet feet in which three webs connect all four toes. (Most birds have only two webs between the three forward pointing toes, with the hind claws free.)

Tribe a term sometimes used to group certain species and/or genera within a family. See CLASS.

Tropics strictly, an area lying between 22.5°N and 22.5°S. Often because of local geography, birds' habitats do not match this area precisely.

Tundra the area of high latitude roughly demarcated by its being too cold for trees to grow.

Upwelling an area in the sea when, because of local topography, water from deep down in the sea is pushed to the surface. Usually upwellings are associated with rich feeding conditions for birds.

Vermiculation (on feathers) fine markings.

Wallace's line an imaginary line passing between the Philippines and the Moluccas in the north and between Sulawesi and Borneo, and Lombok and Bali, in the south. First defined by the naturalist Alfred Russel Wallace (1823–1913), it separates the Oriental and Australian zoogeographical regions. The zone of mixing between the regions is known as "Wallacea."

Wattle a fleshy protuberance, usually near the base of the BILL.

Wetlands fresh- or salt-water marshes.

Wing front limb of a bird transformed into an organ for flight.

Wing formula statement of relative lengths of wing feathers, especially of primary feathers. Used as a defining characteristic for many species.

Wing loading the body mass of a bird divided by the total surface area of its wings.

Wing spur a sharp projection at or near the bend of the wing. See SPUR.

Wintering ground the area where a migrant spends the nonbreeding season.

Yolk sac a sac that hangs from the ventral surface of the vertebrate embryo. In birds (and reptiles, but not most mammals), the yolk sac contains yolk as nourishment for the embryo.

Zygodactylous having two toes directed forwards and two backwards.

Bibliography

The following list of titles indicates key reference works used in the preparation of this volume and those recommended for further reading. The list is divided into two sections: general and regional books about birds and books dealing with particular families or groups.

General and Regional

Ali, S. and Ripley, S.D. (1983) *A Pictorial Guide to the Birds of the Indian Subcontinent*, Bombay Natural History Society/Oxford University Press, Delhi.

Ali, S. and Ripley, S.D. (1978–1999, 2nd edn) *Handbook of the Birds of India and Pakistan*, Vols 1–10. Oxford University Press, Delhi.

Berthold, P. (2001, 2nd edn) *Bird Migration*, Oxford University Press, Oxford, UK.

BirdLife International (2000) *Threatened Birds of The World*, Lynx Edicions, Barcelona, Spain.

Blakers, M., Davies, S.J.J.F. and Reilly, P.N. (1984) *The Atlas of Australian Birds*, Melbourne University Press, Melbourne, Australia.

Bock, W.J. and Farrand, J. (1980) *The Number of Species and Genera of Recent Birds: a Contribution to Comparative Systematics*, American Museum of Natural History, New York.

Bond, J. (1979, 5th edn) *Birds of the West Indies: a Guide to the Species of Birds that Inhabit the Greater Antilles, Lesser Antilles and Bahama Islands*, Collins, London, UK.

Borrow, N. and Demey, R. (2001) *Birds of Western Africa*, Christopher Helm, London, UK.

Brown, L.H., Urban, E.K. and Newman, K. (1982–2003) *The Birds of Africa, vols I–VII*, Academic Press, London, UK.

Campbell, B. and Lack, E. (1985) *A New Dictionary of Birds*, T. & A.D. Poyser, London, UK.

Clements, J. (1981) *Birds of the World: a Checklist*, Croom Helm, London, UK.

Cramp, S. (ed) et al (1978–1994) *Handbook of the Birds of Europe, the Middle East and North Africa: the Birds of the Western Palearctic, vols I–IX*, Oxford University Press, Oxford, UK.

del Hoyo, J., Elliott, A. and Sargatal, J. (eds) (1992–2002) *Handbook of Birds of the World, vols I–VII*, Lynx Edicions, Barcelona, Spain.

Dementiev, G.P. et al (1966) *Birds of the Soviet Union, vols I–VI*, Jerusalem.

Dunning, J.S. (1982) *South American Land Birds: a Photographic Aid to Identification*, Harrowood, Pennsylvania.

Eastwood, E. (1967) *Radar Ornithology*, Methuen, London.

Ehrlich, P. and A. (1982) *Extinction*, Gollancz, London, UK.

Elkins, N. (1983) *Weather and Bird Behavior*, T. & A.D. Poyser, London, UK.

Farner, D.S., King, J.R. and Parkes, K.C. (1971–83) *Avian Biology, vols I–VII*, Academic Press, New York and London.

Finlay, J.C. (1984) *A Bird Finding Guide to Canada*, Hurtig, Edmonton, Alberta, Canada.

Flint, V.E., Boehme, R.L., Kostin, Y.V. and Kuznetzov, A.A. (1984) *A Field Guide to Birds of the USSR*, Princeton University Press, NJ.

Fry, C.H. – see Brown, L.H.

Fuller, E. (2000) *Extinct Birds*. Oxford University Press, UK.

Gallagher, M. and Woodcock, M.W. (1980) *The Birds of Oman*, Quartet, London, UK.

Gibbons, D.W., Reid, J.B. and Chapman, R.A. (1993) *The New Atlas of Breeding Birds in Britain and Ireland: 1988–1991*, T. & A.D. Poyser, London, UK.

Gill, F.B. *Ornithology* (1994, 2nd edn), W.H. Freeman and Company, New York.

Glenister, A.G. (1971) *The Birds of the Malay Peninsula, Singapore and Penang*, Oxford University Press, Kuala Lumpur, Malaysia.

Godfrey, W.E. (1966) *The Birds of Canada*, National Museum of Canada, Ottawa.

Grimmet, R., Inskipp, C. and Inskipp, T. (1998) *Birds of the Indian Subcontinent*, A. & C. Black, London, UK.

Gruson, E.S. (1976) *A Checklist of the Birds of the World*, Collins, London.

Halliday, T. (1978) *Vanishing Birds: their Natural History and Conservation*, Sidgwick and Jackson, London, UK.

Harris, M. (1982) *A Field Guide to the Birds of Galapagos, revised edn*, Collins, London, UK.

Harrison, C.J.O. (1975) *A Field Guide to the Nests, Eggs and Birds, with north Africa and Middle East*, Collins, London, UK.

Harrison, C.J.O. (1978) *A Field Guide to the Nests, Eggs and Nestlings of North American Birds*, Collins, London, UK.

Harrison, C.J.O. (1982) *An Atlas of the Birds of the Western Palaearctic*, Collins, London.

Harrison, P. (1983) *Seabirds – an Identification Guide*, Croom Helm, London, UK.

Heather, B. and Robertson, H. (1997) *The Field Guide to the Birds of New Zealand*, Oxford University Press, UK.

Higgins, P.J. – see Marchant, S.

Hilty, S.L. and Brown, W.L. (1986) *A Guide to the Birds of Colombia*, Princeton University Press, NJ.

Howard, R. and Moore, A. (1991, 2nd edn) *A Complete Checklist of the Birds of the World*, Oxford University Press, Oxford, UK.

Howell, S.N.G. Webb, S. (1995) *A Guide to the Birds of Mexico and Northern Central America*, Oxford University Press, Oxford, UK.

Irby Davis, L. (1972) *A Field Guide to the Birds of Mexico and Central America*, Texas University Press, Austin.

Kaufman, K. (2000) *The Birds of North America*, Houghton Mifflin, New York.

Keith, S. – see Brown, L.H.

King, A.S. and McLelland, J. (1975) *Outlines of Avian Anatomy*, Baillière Tindall, London, UK.

King, B., Woodcock, M. and Dickinson, E.C. (1975) *A Field Guide to the Birds of South-East Asia*, Collins, London, UK.

Krebs, J.R. and Davies, N.B. (1989, 2nd edn) *An Introduction to Behavioral Ecology*, Blackwell Scientific Publications, Oxford, UK.

Lack, D. (1968) *Ecological Adaptations for Breeding in Birds*, Methuen, London, UK.

Leahy, C. (1982) *The Bird Watcher's Companion: an Encyclopedic Handbook of North American Birdlife*, Hale, London, UK.

Maclean, G.L. (1993, 6th edn) *Roberts' Birds of South Africa*, John Voeckler Bird Book Fund, Cape Town, South Africa.

Marchant, S. et al (1990–2003) *Handbook of Australian, New Zealand and Antarctic Birds. Vols 1–6*, Oxford University Press, Melbourne, Australia.

Mayr, E. and Diamond, J. (2001) *The Birds of Northern Melanesia*. Oxford University Press, UK.

Meinertzhagen, R. (1954) *Birds of Arabia*, Oliver and Boyd, Edinburgh, Scotland.

Monroe, B.L. and Sibley, C.G. (1997) *World Checklist of Birds. English Names and Systematics*, Yale University Press, New Haven, CT.

Moreau, R.E. (1972) *The Palaearctic – African Bird Migration Systems*, Academic Press, London, UK.

National Geographic Society (1999, 3rd edn) *Field Guide to the Birds of North America*, NGS, Washington, DC.

Newman, K. (1983) *The Birds of Southern Africa*, Macmillan, Johannesburg, South Africa.

Nightingale, T. and Hill, M. (1993) *Birds of Bahrain*, Immel Publishing Ltd., London, UK.

O'Connor, R.J. (1984) *The Growth and Development of Birds*, Wiley, New York.

Penny, M. (1974) *The Birds of the Seychelles and the Outlying Islands*, Collins, London, UK.

Perrins, C.M. and Birkhead, T.R. (1983) *Avian Ecology*, Blackie, London, UK.

Peters, J.L. et al. (1931–87) *Checklist of Birds of the World*, Museum of Comparative Zoology, Cambridge, Massachusetts.

Peterson, R.T. (1980) *A Field Guide to the Birds East of the Rockies*, Houghton Mifflin, Boston.

Peterson, P.T., Mountford, G. and Hollom, P.A.D. (1983, 4th edn) *A Field Guide to the Birds of Britain and Europe*, Collins, London, UK.

Pizzey, G. (1980) *A Field Guide to the Birds of Australia*, Collins, Sydney, Australia.

Poole, A. and Gill, F. (eds) (1992–2003) *The Birds of North America*, The Birds of North America Inc., Philadelphia, PA.

Ridgely, R.S. and Tudor, G. (1989, 1994) *The Birds of South America, vols I and II*, University of Texas Press, Austin.

Ridgely, R.S. (1976) *A Guide to the Birds of Panamá*, Princeton University Press, NJ.

Schauensee, R.M. de (1982) *A Guide to the Birds of South America*, Academy of Natural Sciences of Philadelphia, PA.

Schauensee, R.M. de and Phelps, W.H. (1978) *A Guide to the Birds of Venezuela*, Princeton University Press, Princeton, NJ.

Schauensee, R.M. de (1984) *The Birds of China Including the Island of Taiwan*, Oxford University Press, Oxford, Smithsonian Institution Press, Washington DC.

Serle, W., Morel, G.J. and Hartwig, W. (1977) *A Field Guide to the Birds of West Africa*, Collins, London, UK.

Sibley, C.G. and Monroe, B.L. (1990) *Distribution and Taxonomy of Birds of the World*, Yale University Press, New Haven, CT.

Sibley, C.G. and Ahlquist, J.E. (1990) *Phylogeny and Classification of Birds: a study in molecular evolution*, Yale University Press, New Haven, CT.

Sibley, D.A. (2000) *National Audubon Society The Sibley Guide to Birds*, Alfred Knopf, New York.

Sick, H. (1993) *Birds in Brazil*, Princeton Press, Princeton, NJ.

Sinclair, I. and Langrand, O. (1998) *Birds of the Indian Ocean Islands*, Struik, Cape Town, South Africa.

Skutch, A.F. (1975) *Parent Birds and their Young*, University of Texas Press, Austin, Texas.

Slater, P. (1971, 1975) *A Field Guide to Australian Birds, vol I*, Oliver and Boyd, Edinburgh; *vol II*, Scottish Academic Press, Edinburgh.

Snow, D.W. and Perrins, C.M. (1998) *The Birds of the Western Palearctic. Concise Edition. Vols 1 and 2*, Oxford University Press, Oxford, UK.

Statz, D.F., Fitzpatrick, J.W., Parker III, T.A. and Moskovits, D.K. (eds) (1996) *Neotropical Birds: Ecology and Conservation*, University of Chicago Press, Chicago.

Stevenson, T. and Fanshawe, J. (2002) *Field Guide to the Birds of East Africa*. T. & A.D. Poyser, London, UK.

Steyn, P. (1982) *Birds of Prey of Southern Africa: their Identification and Life Histories*, David Philip, Cape Town, South Africa.

Stiles, F.G. and Skutch, A.F. (1989) *A Guide to the Birds of Costa Rica*, Cornell University Press, Ithaca.

Urban, E.K. – see Brown, L.H.

Svensson, L. and Grant, P.J. (1999) *Collins Bird Guide*, HarperCollins, London, UK.

Watson, G.E. (1975) *Birds of the Antarctic and Sub-Antarctic*, American Geophysical Union, Washington, DC.

Wernham, C.V. et al (2002) *The Migration Atlas*, T. & A.D. Poyser, London, UK.

Wild Bird Society of Japan (1982) *A Field Guide to the Birds of Japan*, Wild Bird Society of Japan, Tokyo.

Williams, J.G. and Arlott, N. (1980) *A Field Guide to the Birds of East Africa*, Collins, London, UK.

Wilson, E. (1967) *Birds of the Antarctic*, Blandford Press, Poole, UK.

Zimmerman, D.A., Turner, D.A. and Pearson, D.J. (1996) *Birds of Kenya and Northern Tanzania* Princeton University Press, NJ.

Families or Groups

Brown, L. and Amadon, D. (1968) *Eagles, Hawks and Falcons of the World, 2 vols*, Country Life Books, Feltham, Middlesex, UK.

Clement, P., Harris, A. and Davies, J. (1993) *Finches and Sparrows. An Identification Guide*, Christopher Helm, London, UK.

Davies, S.J.J.F. (2002) *Bird Families of the World. Ratites and Tinamous*, Oxford University Press, Oxford.

Delacour, J. (1977, 2nd edn) *The Pheasants of the World*, Spur Publications, Hindhead, UK.

Delacour J. and Amadon, D. (1973) *Curassows and Related Birds*, American Museum of Natural History, New York.

Ferguson-Lees, J. and Christie, D.A. (2001) *Raptors of the World*, Christopher Helm, London, UK.

Forshaw, J.M. (1978, 2nd edn) *Parrots of the World*, David and Charles, Newton Abbot, UK.

Forshaw, J.M. and Cooper, W.T. (1977) *The Birds of Paradise and Bower Birds*, Collins, Sydney and London, UK.

Frith, C.B. and Beehler, B.M. (1998) *Bird Families of the World. The Birds of Paradise*, Oxford University Press, Oxford, UK.

Fry, C.H. (1984) *The Bee-eaters*, T. & A.D. Poyser, London, UK.

Gaston, A.J. (1998) *Bird Families of the World. The Auks*, Oxford University Press, Oxford, UK.

Gibbs, D., Barnes, E. and Cox J. (2001) *Pigeons and Doves*, Pica Press, London, UK.

Goodwin, D. (1976) *Crows of the World*, British Museum (Natural History), London, UK.

Goodwin, D. (1982) *Estrildid Finches of the World*, British Museum (Natural History), London, UK.

Goodwin, D. (1983, 3rd edn) *Pigeons and Doves of the World*, British Museum (Natural History), London, UK.

Grant, P.R. (1999) *Ecology and Evolution of Darwin's Finches*, Princeton University Press, NJ.

Greenwalt, C.H. (1960) *Hummingbirds*, American Museum of Natural History, New York.

Hancock, J. and Kushlan, J. (1984) *The Herons Handbook*, Croom Helm, London, UK.

Harris, T. and Franklin, K. (2000) *Shrikes and Bush-shrikes*, Christopher Helm, London, UK.

Hayman, P., Marchant, J. and Prater, A. (1985) *Shorebirds: an Identification Guide to the Waders of the World*, Croom Helm, London, UK.

Holyoak, D.T. (2001) *Bird Families of the World. Nightjars and their allies*, Oxford University Press, Oxford.

Isler, M.L. and Isler, P.R. (1999) *The Tanagers*, Smithsonian Institution Press, Washington, DC.

Johnsgard, P.A. (1983) *Cranes of the World*, Croom Helm, London, UK.

Johnsgard, P.A. (1978) *Ducks, Geese and Swans of the World*, University of Nebraska Press, Lincoln, Nebraska.

Johnsgard, P.A. (1983) *The Grouse of the World*, University of Nebraska Press, Lincoln, Nebraska.

Johnsgard, P.A. (1981) *The Plovers, Sandpipers and Snipes of the World*, University of Nebraska Press, Lincoln, Nebraska.

Juniper, T. and Parr, M. (1998) *Parrots*, Pica Press, London, UK.

Kear, J. and Duplaix-Hall, N. (eds) (1979) *Flamingos*, T. & A.D. Poyser, London, UK.

Kemp, A.C. (1995) *Bird Families of the World. Hornbills: Bucerotiformes*, Oxford University Press, Oxford.

Lack, D. (1956) *Swifts in a Tower*, Methuen, London, UK.

Lambert, F. and Woodcock, M. (1996) *Pittas, Broadbills & Asities*, Pica Press, London, UK.

Matthysen, E. (1998) *The Nuthatches*, T. & A. D. Poyser, London, UK.

Mikkola, H. (1983) *Owls of Europe*. T. & A.D. Poyser, London, UK.

Newton, I. (1972) *Finches*, Collins, London, UK.

Newton, I. (1979) *Population Ecology of Raptors*, T. & A.D. Poyser, London, UK.

Nørgaard-Oleson, E. (1973) *The Tanagers*, Skibby Books, Denmark.

Perrins, C.M. (1979) *British Tits*, Collins, London, UK.

Ripley, S.D. (1977) *Rails of the World*, M.F. Feheley, Toronto, Canada.

Rowley, I. and Russell, E. 1997. *Bird Families of the World. Fairy-Wrens and Grasswrens*, Oxford University Press, Oxford, UK.

Short, L.L. and Horne, J.F.M. (2001) *Toucans, Barbets and Honeyguides*, Oxford University Press, Oxford, UK.

Short, L.L. (1982) *Woodpeckers of the World*, Delaware Museum of Natural History, Greenville, Delaware.

Simpson, G.G. (1976) *Penguins*, Yale University Press, New Haven, Connecticut.

Skutch, A. (1989) *Life of the Tanager*, Cornell University Press, Ithaca.

Snow, D.W. (1982) *The Cotingas*, British Museum (Natural History), London, UK.

Soothill, E. and R. (1982) *Wading Birds of the World*, Blandford Press, Poole, UK.

Williams, T.D. (1995) *Bird Families of the World. The Penguins*, Oxford University Press, Oxford, UK.

Winkler, H., Christie, D.A. and Nurney, D. (1995) *Woodpeckers*. Pica Press, London, UK.

Wyllie, I. (1981) *The Cuckoo*, Batsford, London, UK.

Zann, R.A. (1996) *The Zebra Finch. A Synthesis of Field and Laboratory Studies*, Oxford University Press, Oxford, UK.

Index

Page numbers in *italic* type refer to picture captions. Page numbers in **bold** indicate extensive coverage or feature treatment of a subject.

A

Aburria 197
  aburri 196, 197
Acanthisitta chloris 416–17, *416*, *417*
Acanthisittidae **416–17**
Acanthorhynchus
  superciliosus 466, 467, *468*
  tenuirostris *468*
accentor (*Prunella*) **602–03**
Accipiter 166
  brachyurus 172
  butleri 172
  gentilis 162–63, *167*, 170, 172, 173
  gundlachii 172
  haplochrous 172
  imitator 172
  luteoschistaceus 172
  melanochlamys *168*
  nisus 160, 163, *168*, 172, 561
  superciliosus 162, 172
  tachiro 170, 172
Accipitridae **162–74**
Aceros 384, 386, 388
  comatus 386, 388, 389
  corrugatus 386, 388
  everetti 386, 389
  leucocephalus 386, *387*
  narcondami 386, 389
  nipalensis 386, *387*
  plicatus 384, 386
  undulatus 386
  waldeni 386
Acridotheres
  ginginianus 531, 532
  tristis 530, 531
Acrocephalus 568
  agricola *570*
  familiaris 569, 572
  paludicola 572
  palustris 31, 571–72
  schoenobaenus 316, *570*, 571, 572
  scirpaceus 316, *317*, *570*, 571, 572, *572*
Acryllium vulturinum 191, *191*
Actenoides hombroni 367, 370
Actinodura 575
Actophilornis
  africana 244, 245
  albinucha 245
adaptive radiation 514, 616, 629, **630**, *630*
Aechmophorus
  clarkii 62, 63, 65
  occidentalis 62, *62*, 63, *64*, 65
Aegithalidae 550, **556–57**
Aegithalos
  caudatus **556–57**, 557, *557*
  concinnus 557
  fuliginosus 557
  iouschistos 557
  leucogenys 557
  niveogularis 557
Aegithina
  lafresnayei 507
  nigrolutea 507
  tiphia 507, *507*
  viridissima 507
Aegithinidae **507**
Aegolius 324
  acadicus 330, *330*, 331
  funereus 324, *324*, 331
Aegotheles
  albertisi 343, *343*
  archboldi 343
  bennettii 343
  crinifrons 343
  cristatus 343
  insignis 343
  savesi 343
  tatei 343
  wallacii 343
Aegothelidae **343**
Aegypius monachus *164*, 172
Aepyornis maximus *20*, **21**
Aepypodius 192
Aeronautes saxatilis 346, *347*
Aethia
  cristatella 276, 277
  pusilla 276, 277, 278
  pygmaea 276, 277, *283*
Aethopyga duyvenbodi 586
Afropavo congensis 178, 180
Agamia agami 98, 100, 104
Agamiinae 104
Agapornis
  fischeri *298*, 301
  roseicollis 304, 305
Agelaius
  phoeniceus 632, 634
  tricolor 634
  xanthomus 634, 635
Agelastes
  meleagrides 190, 191
  niger 190, 191
Aglaeactis 356
Aglaiocercus 356, 361
Agriornis 424, 425, 428
  andicola 427, 428
Agyrtria versicolor 356
Ailuroedus
  buccoides 458
  crassirostris 458, *461*
  melanotis 458
Aix
  galericulata *138*, 139, 140, 143
  sponsa *138*, 140, *144*, 332
Ajaia ajaja 112, 113, 114, 115, *115*, 116, *118*
'Akiapola'au *see Hemignathus wilsoni*
akepa *see Loxops coccineus*
'akikiki *see Oreomystis bairdi*
'Akohekohe *see Palmeria dolei*
Alaemom alaudipes 578, 580, *580*
Alauda
  arvensis 578, 580, 581, *581*, **582–83**
  razae 578, 581
Alaudidae **578–83**
albatross 19, 21, 26, **66–69**, 268
  Amsterdam *see Diomedea amsterdamensis*
  Atlantic yellow-nosed *see Thalassarche chlororhynchos*
  Black-browed *see Thalassarche melanophris*
  Black-footed *see Phoebastria nigripes*
  Chatham *see Thalassarche eremita*
  Grey-headed *see Thalassarche chrysostoma*
  Laysan *see Phoebastria immutabilis*
  Light-mantled sooty *see Phoebetria palpebrata*
  Northern Buller's *see Diomedea bulleri*
  Royal *see Diomedea epomorphora*
  Short-tailed *see Phoebastria albatrus*
  Sooty *see Phoebetria fusca*
  Steller's *see Phoebastria albatrus*
  Wandering *see Diomedea exulans*
  Waved *see Phoebastria irrorata*
albinistic plumage 522
Alcae **276–81**
Alca torda *261*, 276, 277, *277*, 278, 279, 282, 283, *283*
Alcedininae 369
Alcedo 366
  atthis 366, 367, 369, *369*, 370, 371, *371*
  bougainvillei 367, 371
  cristata *366*
  meninting 367, 370
  monachus 367, 370
  princeps 367, 370
alcid 268
Alcidae **276–81**
Alcippe 575
  castaneceps *575*
  vinipectus 575
Aleadryas 477
Alectoris
  chukar 177, *179*, 180
  rufa *179*, 180, *180*
Alectroenas 289
Alectrurus
  risora 426, 427, 428
  tricolor 426, 427, 428
Alectura lathami 192, *192*, *194*
Aledinidae **366–71**
Alethe 526
Alisterus scapularis 305
Alle alle 260, 276, 277, *277*, 278, 282
allopreening 208, 566
Alophoixus 564
Alopochen aegyptiacus 137, *138*, 140
alseonax (*Muscicapa*) 528–29
alula *151*
Amadina
  erythrocephala 605, 607
  fasciata 605, 607
Amalocichla 472, 473
Amandava
  amandava 604, 605, 607
  subflava 607
Amaurornis flavirostris 207, 209, 210
Amazilia 359
  castaneiventris *361*
  tzacatl 356
Amazona
  auropalliata 296, *302*
  brasiliensis 305
  guildingii 298
  imperialis 305
  ventralis 305, 306
  vittata 305, 306
Amblyornis 459, 460
  flavifrons 458
  inornatus 458, 460, *463*
  macgregoriae 458, 460, 461
  subalaris 458, 460
Amblyospiza albifrons 592, *592*, 594
Ammodramus
  caudacutus 619, 620
  henslowii 620, 621
  maritimus 620, 621
  nelsoni 620
  savannarum 620
Ammomanes deserti 578
Ampelion 429
Amytornis 465
  ballarae 465
  barbatus 465
  dorotheae 465
  goyderi 464, 465
  housei 465
  merrotsyi 465
  woodwardi 465
Anaplectes rubriceps 592, 594
Anarhynchus frontalis 225, 227, 228
Anas
  acuta *138*, 139, 140
  americana *138*
  aucklandica 139, 140
  carolinensis 139, 140
  cyanoptera 140
  eatoni 140
  flavirostris 140, 143
  laysanensis 137, *138*, 139, 140
  nesiotis 140
  penelope 139, 140
  platyrhynchos *138*, 140
Anastomus
  lamelligerus 107, *108*
  oscitans 107, 109, *110*
Anatidae **132–45**, 210
Anatini 137, 139, 140
Andigena 393, 395
  laminirostris 390, *392*, 393, 395
  nigrirostris 390, *390*, 393
Androdon 352, 353
  aequatorialis 356
Andropadus 563
  latirostris 563, 565
  virens 563
Androphobus 475
Anhima cornuta 130–31, *130*
Anhimidae **130–31**
Anhinga
  anhinga 96–97, *96*
  melanogaster 96–97, *96*
Anhingidae *28*, **96–97**
ani, Groove-billed *see Crotophaga sulcirostris*
Anodorhynchus
  glaucus 305, 306
  hyacinthinus *298*
Anomalospiza imberbis 594
Anorrhinus 384
  galeritus 386, 389
  tickelli 386, 389
Anous 264, 266
  stolidus 264
  tenuirostris 264, *264*, 269
Anser
  albifrons 140, 142, 144
  anser *135*, 271
  brachyrhyncus *135*
  caerulescens 140, 144, *144*
  canagicus *135*
  indicus 26, *135*
Anseranas semipalmata 132, *135*, 137, 140
Anseranatini 132, 140
Anseriformes 19, **132–45**
Anserini 134, 140
antbird **446–49**
  and army-ant swarms 448, **449**
  bare-faced 446, 447
  Bicolored *see Gymnopithys leucaspis*
  ground 446–47, 448
  Ocellated *see Phaenostictus mcleannani*
  Spotted *see Hylophylax naevioides*;
  typical 446, 447, 448
  White-plumed *see Pithys albifrons*
  Wing-barred *see Myrmornis torquata*
Anthochaera 466
  paradoxa *468*
Anthoscopus 548
  caroli 548
  flavifrons 548
  minutus 548, *548*, 549
  musculus 548
  parvulus 548
  punctifrons 548
  sylviella 548
Anthracoceros 384, 386, 388
  coronatus 384, 386
  montani 386, 389
Anthracothorax 356
  mango *352*
Anthreptes 587, 588
  pallidigaster 586
Anthropoides 198, 199
  paradisea 198, *198*
  virgo 27, *201*
Anthus
  cervinus *598*
  novaeseelandiae *599*
  rubescens 599, 600
  sokokensis 599
  spinoletta 599, 600
  spragueii 601
Antilophia
  bokermanni 434
  galeata 434, 437
antpecker, Woodhouse's *see Parmoptila woodhousei*
antpipit 427, 428
antpitta (*Grallaria*) **446–49**, 447
antshrike 446, 447
  Barred *see Thamnophilus doliatus*
  Bluish-slate *see Thamnomanes schistogynus*
  Giant *see Batara cinerea*
  Great *see Taraba major*
  Rufous-capped *see Thamnophilus ruficapillus*
  Spot-backed *see Hypodaleus guttatus*
antthrush 446, 447, 448
  Rufous-caped *see Formicarius colma*
antwren (*Formicivora*, *Myrmotherula*, *Terenura*) 446, 447
Anumbius annumbi 440
anvils
  barbets 401
  woodpeckers 408, 410, 414
Apalis 568
  chariessa 572
  fuscigularis 569, 572
  lynesi 572
Apaloderma 362
  narina 362
  vittatum 362
Apalopteron familiare 466
'apapane *see Himatione sanguinea*
Apatornis *20*
Aphelocoma
  californica 484, **487**
  coerulescens 483, 484, *484*
  ultramarina 484
Aphrastura spinicauda 440
Aplonis
  metallica 531, 532, 533
  pelzelni 531, 533
  santovestris 531, 533
Apodidae **346–50**
Apodiformes 19, **346–61**
apostlebird *see Struthidae cinerea*
Aptenodytes
  forsteri 31, 50, *50*, 51, 52, 53, 54–55, **58–59**, *58*
  fosteri 292
  patagonicus 31, *50*, 51, 52, 53, *53*, 54, 55, 56, 59
apteria 184
Apterygidae **46–47**
Apterygiformes 19, **46–47**
Apteryx
  australis *29*, 46–47, *46*
  haasti 46–47, *46*
  owenii 46–47
Apus 346
  acuticauda 346
  apus 346–47, *346*, 348, *348*
  melba 346, 347, *348*
Aquila 163, 166
  adalberti *164*, 172
  audax 163, 172, 173
  chrysaetoc 163, 169, 170, 172, 173, *173*
  clanga 172
  gurneyi 172
  heliaca 172
  rapax *167*, 172
  verreauxii 162, 163, *164*, *167*, 169, 172, 173
Ara
  ambigua *306*
  ararauna 304, 305
  macao *296*
  militaris *304*
aracari (*Pteroglossus*) **390–95**
Arachnothera 586, *586*
Aramidae **212**
Aramides calopterus *208*
Aramidopsis plateni 207, 208, 211
Aramus guarauna 212, *212*
Aratinga pertinax 299–300, 305
Arborophila rufipectus 180, 181
Arcanator orostruthus 563, 564
Archaeopteryx 18–19, *20*, 22
Archboldia
  papuensis 458, 459, 460, 461, *461*
  sanfordi 458
Archilochus colubris 354, 356, *356*
Ardea
  alba 98–99, 104, *104*
  cinerea 99, *101*, 104, 271
  cocoi 99, 104
  goliath 98, 99, *99*
  herodias 98, 99, *101*, *102*, 104
  humbloti 104
  ibis 98, 101, *102*, 103, 104, *104*
  insignis 99, 103, 104, 155
Ardeidae **98–105**
Ardeinae 99, 104
Ardeola
  grayii 104
  ralloides 104
Ardeotis 215
  arabs 214, 215
  australis 214, *214*, 215, 216, *216*, 217
  kori 214, *214*, 215, *216*
  nigriceps 214, 215, *216*, 217
Arenaria interpres 234, *234*, 235, 236
Argentavis magnificens *20*, 21, 146

argus
  Crested *see Rheinardia ocellata*
  Great *see Argusianus argus*
Argusianus argus 180
army-ant swarms 448, **449**, 526
Arses telescophthalmus 504, 505
Artamidae **492**
Artamus
  cinereus 492
  cyanopterus 492
  fuscus 492
  insignis 492
  leucorhynchus 492
  maximus 492
  mentalis 492
  minor 492
  monachus 492
  personatus 492, *492*
  superciliosus 492
Arundinicola leucocephala 424, 428
Ashbyia lovensis 466
Asio 324
  flammeus 329, 331, *334*
  otus 326, 329, 331
asity (*Philepitta*) **421**
Aspatha gularis 372
Asthenes 440
  modesta 440
astrapia, Ribbon-tailed *see Astrapia mayeri*
Astrapia mayeri 488
Asturina plagiata 172
Atelornis
  crossleyi 380
  pittoides 380
Athene 324
  blewitti 331
  noctua 324, 327, 331
Atrichornis
  clamosus 456–57, *456*
  rufescens 456–57, *456*
Atrichornithidae **456–57**
Attagis
  gayi 254, *254*
  malouinus 254, 255
attila 428
  Ochraceous *see Atilla torridus*
Attila 428
  torridus 427, 428
Augastes 356
auk 26, 31, 76, **276–81**
  conservation status 283
  developmental patterns **282**
  Great *see Pinguinis impennis*
  Little *see Alle alle*
auklet **276–83**
  Cassin's *see Ptychorampus aleuticus*
  Crested *see Aethia cristatella*
  Least *see Aethia pusilla*
  Parakeet *see Cyclorhynchus psittacula*
  Rhinoceros *see Cerorhinca monocerata*
  Whiskered *see Aethia pygmaea*
Aulacorhynchus 392, 395
  huallagae 390, 392
  prasinus 390, *390*, 395
Auriparus flaviceps 546, 548, *548*, 549
avadavat
  Green 604, 605
  Red *see Amandava amandava*
Aves 19
Aviceda
  cuculoides 162, 172
  subcristata *162*, 172
avocet (*Recurvirostra*) 29, 30, **242–43**
Avocettula recurvirostris *356*
awlbill, Fiery-tailed *see Avocettula recurvirostris*
Aythya
  fuligula 140, *141*
  valisineria 140
Aythyini 139, 140

B

Babax 574, 575
babbler **574–77**
  Arabian *see Turdoides squamiceps*
  Australo-Papuan **476**
  Brown *see Turdoides plebejus*
  Chestnut-crowned *see Pomatostomus ruficeps*
  Common *see Turdoides caudatus*
  cooperative breeding **576**
  Fulvous *see Turdoides fulvus*
  Grey-crowned *see Pomatostomus temporalis*
  Hall's *see Pomatostomus halli*
  Iraq *see Turdoides altirostris*
  Jungle *see Turdoides striatus*
  mountain 574
  pseudo **476–77**
  Puff-throated *see Pellorneum ruficeps*
  Rufous *see Pomatostomus isidori*
  Rusty-cheeked scimitar *see Pomatorhinus erythrogenys*
  scimitar 575, *577*
  Southern pied *see Turdoides bicolor*
  tree 575
  White-browed *see Pomatostomus superciliosus*
  wren 575
Baeopogon 563
Baillonius bailloni 390, *390*, 392–93, 395
Balaeniceps rex 29, 122–23, *122*, *123*
balance 24
baldpate *see Anas americana*
Balearica 198
  pavonina 198, *201*
  regulorum *198*, 201
bananaquit *see Coereba flaveola*
Bangsia melanochlamys 631
barbet **398–402**
  D'Arnaud's *see Trachyphonus darnaudii*
  anvils 401
  Black-backed *see Lybius minor*
  Black-spotted *see Capito niger*
  Blue-throated *see Megalaima asiatica*
  Chaplin's *see Lybius chaplini*
  cooperative breeding 402
  Double-toothed *29*
  dueting 402
  Great *see Megalaima virens*
  ground *see Trachyphonus*
  and honeyguide 396, 397, 402
  Lineated *see Megalaima lineata*
  Little green *see Megalaima viridis*
  Moustached *see Megalaima incognita*
  Pied *see Tricholaema leucomelas*
  Prong-billed *see Semnornis frantzii*
  Red-headed *see Eubucco boucierii*
  tongue 398
  Toucan-barbet *see Semnornis ramphastinus*
  Versicolored *see Eubucco versicolor*
  White-cheeked *see Megalaima viridis*
  White-mantled *see Capito hypoleucus*
barbthroat *see Threnetes*
Bartramia longicauda 235, 237
barwing 574, 575
Basileuterus 625–26
  fulvicauda 625, 626
  grisieceps 625, 627
  tristriatus 448
Basilornis
  celebensis 530, 531
  galeatus 530, 531
Bassanus see *Morus*
Batara cinerea 447, 448
bateleur *see Terathopius ecaudatus*
batis **514**
  Ituri *see Batis ituriensis*
Batis 514
  ituriensis 514
Batrachostominae 342
Batrachostomus 342
  auritus 342
  septimus *342*
  stellatus 342
baywing *see Molothrus badius*
baza
  African *see Aviceda cuculoides*
  Crested *see Aviceda subcristata*
Bebrornis sechellensis 572
becard 429–33
  Rose-throated *see Pachyramphus aglaiae*
bee-eater (*Merops*, *Nyctyornis*, *Meropogon*) 31, **374–79**
  nests 378
  sunbasking 374, *377*
bellbird 30, 429–33
  Bare-throated *see Procnias nudicollis*
  Crested *see Oreoica gutturalis*
bentbill 428
Bernieria 564
berrypecker (*Melanocharis*, *Paramythia*, *Melanocharis*) 584, 585
  Tit *see Oreocharis arfaki*
bill 22, *22*, **28–29**, 29, 30
  adaptation by birds 540
  filter-feeders 126
  filtering lamellae 70
  finches **613**
  forked 404
  frontal shield **210**
  gaping adaptation 632–33, *634*
  Hawaiian honeycreeper 616, *617*
  hornbills 384, 386, 389
  nectarvores 421, *421*, 586–87, 616, *617*
  parrots 299, 302
  sawtooth *141*
  sexual dimorphism 495
  tactile foraging **118–19**
  tomium 430
  toucans 390, 392
bill hook 396, 397
bill sweeping 452
biocides 153, **160–61**, 170, 177
bird of paradise 31, 430, 437, **488–91**
  Blue *see Paradisaea rudolphi*
  Crested 490
  King *see Cicinnurus regius*
  King of Saxony *see Pteridophora alberti*
  Loria's 490
  Macgregor's *see Macgregoria pulchra*
  Magnificent *see Diphyllodes magnificus*
  Raggiana *see Paradisaea raggiana*
  Standardwing *see Semioptera wallacei*
  Twelve-wired *see Seleucides ignotus*
  White-plumed *see Paradisaea guilielmi*
  Wilson's *see Cicinnurus republica*
  Yellow-breasted *see Loboparadisea sericea*
bishop (*Euplectes*) 593–95
bittern (*Botaurus*, *Ixobrychus*) **98–105**
  Sun *see Eurypyga helias*
Biziura lobata 132, 134, 140
blackbird *see Turdus merula*
blackbird
  American 632
  Bolivian *see Oreopsar bolivianus*
  Brewer's *see Euphagus cyanocephalus*
  Forbes's *see Curaeus forbesi*
  Jamaican *see Nesopsar nigrimus*
  marsh-nesting 633
  Melodious *see Dives dives*
  New World 31
  Red-winged *see Agelaius phoeniceus*
  Saffron-cowled *see Xanthopsar flavus*
  Tricolored *see Agelaius tricolor*
  Yellow-headed *see Xanthocephalus xanthocephalus*
  Yellow-shouldered *see Agelaius xanthomus*
blackcap *see Sylvia atricapilla*
blackeye, Mountain *see Chlorocharis emiliae*
black-hawk
  Great *see Buteogallus urubitinga*
Bleda 563
bluebill, Red-headed *see Spermophaga ruficapilla*
bluebird
  Eastern *see Sialis sialis*
  fairy *see* fairy bluebird
bluebonnet *see Northiella haematogaster*
bobolink *see Dolichonyx oryzivorus*
bobwhite (*Colinus*) 176, *179*, 180
body plan **22–23**
Boissonneaua 356
Bolborhynchus aymara 300, 305
Bombycilla
  cedrorum 518, *518*, 519
  garrulus 518, *518*, 519
  japonica 518, 519
Bombycillidae 516, 517, **518–19**
Bombycillini 518–19
Bonasa 184
  bonasia 185, 186
  sewerzowi 185, 186, 187
  umbellus 185, 186, 187
bones 21, 22, **23**, 25
bonxie *see Catharacta skua*
booby (*Sula*, *Papasula*) **82–87**, 94, 268
booming bowl *311*
Bostrychia 112
  carunculata 113, 115, 116
  hagedash 115, *115*, 117
  olivacea 114, 115
  rara 114, 115, 117
bosun bird *see* tropicbird
Botaurinae 104
Botaurus
  lentiginosus 104
  pinnatus *102*, 104
  poiciloptilus 104
  stellaris *99*, 100, 102, 104
boubou *see Laniarius*
Bowdleria punctata 570, 572
bowerbird 430, 437, **458–63**
  Adelbert *see Sericulus bakeri*
  Archbold's *see Archboldia papuensis*
  brain size 458
  Fawn-breasted *see Chlamydera cerviniventris*
  Fire-maned *see Sericulus bakeri*
  Flame *see Sericulus aureus*
  Golden *see Prionodura newtoniana*
  Great *see Chlamydera nuchalis*
  Lauterbach's *see Chlamydera lauterbachi*
  Macgregor's *see Amblyornis macgregoriae*
  Regent *see Sericulus chrysocephalus*
  Sanford's *see Archboldia sanfordi*
  Satin *see Ptilonorhynchus violaceus*
  Spotted *see Chlamydera maculata*
  Streaked *see Amblyornis subalaris*
  Tooth-billed *see Scenopoeetes dentirostris*
  Vogelkop *see Amblyornis inornatus*
  Western *see Chlamydera guttata*
  Yellow-breasted *see Chlamydera lauterbachi*
  Yellow-fronted *see Amblyornis flavifrons*
Brachycope anomala 594, 595
Brachygalba 407
  goeringi 407
Brachypteracias leptosomus 380
Brachypteraciidae **380**
Brachyramphus
  brevirostris 276, 277, 279, 283
  marmoratus 276, 277, 279, 283
  perdix 277, 279, 283
Bradornis maniquensis 528, 529
Bradypterus 572
brain size
  bowerbirds 458
  crows 485
brambling *see Fringilla montifringilla*
brant *see Branta bernicla*
Branta
  bernicla 132, 140
  canadensis *135*
  leucopsis 140, *144*, 145
  ruficollis *135*
  sandvicensis *135*, 140, 143
brilliant *see Heliodoxa*
  Green-crowned *see Heliodoxa jacula*
bristlebill 563
bristlefront, Stresemann's *see Merulaxis stresemanni*
bristlehead, Bornean *see Pityriasis gymnocephala*
bristles
  barbets 398
  puffbirds 404
  rictal 336, *336*, 342, 424, 498, 502, 504, 528
bristle-tyrant *see Phylloscartes*
broadbill 421, **422–23**
  African *see Smithornis capensis*
  African green *see Pseudocalyptomena graueri*
  Banded *see Eurylaimus javanicus*
  Black-and-red *see Cymbirhynchus macrorhynchus*
  Black-and-yellow *see Eurylaimus ochromalus*
  Dusky *see Corydon sumatranus*
  Green *see Calyptomena viridis*
  Grey-headed *see Smithornis sharpei*
  Hose's *see Calyptomena hosei*
  Long-tailed *see Psarisomus dalhousiae*
  Mindanao wattled *see Eurylaimus steerii*
  Philippine 423
  Rufous-sided *see Smithornis rufolateralis*
  Silver-breasted *see Serilophus lunatus*
  Visayan wattled *see Eurylaimus samarensis*
  Whitehead's *see Calyptomena whiteheadi*
bronzewing, Australian *see Phaps*
brood parasitism 312, 313–15, *313*, **316–17**, 396–97, 402, 503, 507, 527, 576, 634, *635*
  cowbird 479
  Parasitic weaver 594
  whydah 608
  within-species 594
brood reduction 55, *85*
  Cain and Abel syndrome 163
brownbul, Terrestrial *see Phyllastrephus terrestris*
brubru *see Nilaus afer*
brush-finch, Pale-headed 621
brush-turkey 192, 194
  Australian *see Alectura lathami*
Bubalornis
  albirostris 594
  niger 594
Bubalornithinae 594
Bubo 324
  bubo *322*, 324, 326, 329, 330, 331
  sumatranus *324*
  virginianus 328, 331
Bucco 403, 404
  capensis 403, 404, *405*
  noanamae 403, 405
Bucconidae **403–05**
Bucephala
  albeola 140, 143
  clangula 140, *141*, 143
Buceros 384, 386, 388
  bicornis 386, *387*
  hydrocorax 386, 389
  rhinoceros *385*, 386, 389
  vigil 386, *387*, 388, 389
Bucerotes **384–89**
Bucerotidae **384–89**
Bucorvinae 384
Bucorvus 384, 386, 388, 389
  abyssinicus 384, 386
  cafer 384, 386, *387*, 388, 389
budgerigar *see Melopsittacus undulatus*
buffalo-weaver (*Bubalornis*, *Dinemellia*) 155, *593*, 594
bufflehead *see Bucephala albeola*
Bugeranus
  carunculatus 198, 199, 200, 201, *201*
  leucogeranus 198, 199, 200, 201, *201*
bulbul **562–65**
  Ashy *see Hemixos flavala*
  Black *see Hypsipetes leucocephalus*
  Black-collared *see Neolestes torquatus*
  Black-crested *see Pycnonotus melanicterus*
  Black-fronted *see Pycnonotus nigricans*
  Brown-eared *see Ixos amaurotis*
  Chestnut *see Hemixos castanonotus*
  Garden *see Pycnonotus barbatus*
  Hairy-backed *see Tricholestes criniger*
  Hook-billed *see Setornis criniger*
  Liberian *see Phyllastrephus leucolepis*
  Madagascar *see Hypsipetes madagascariensis*
  Philippine *see Hypsipetes philippinus*
  Red-vented *see Pycnonotus cafer*
  Red-whiskered *see Pycnonotus jocosus*
  Seychelles *see Hypsipetes crassirostris*
  Yellow-vented *see Pycnonotus goiavier*
bullfinch *see Pyrrhulla pyrrhula*
  Azores 611
bunting 611, **618–21**
  Black-headed *see Emberiza melanocephala*
  Cirl *see Emberiza cirlus*
  Corn *see Emberiza calandra*
  Crested *see Melophus lathami*
  Gough Island *see Rowettia goughensis*
  Indigo *see Passerina cyanea*
  Lark *see Calamospiza melanocorys*
  Lazuli *see Passerina amoena*
  Orange-breasted *see Passerina leclancherii*
  Painted *see Passerina ciris*
  Red-headed *see Emberiza bruniceps*
  Reed *see Emberiza schoeniclus*
  Rock *see Emberiza cia*
  Rose-bellied *see Passerina rositae*
  Rustic *see Emberiza rustica*
  Slaty *see Latoucheornis siemsseni*
  Snow *see Plectrophenax nivalis*
  Yellow *see Emberiza sulphurata*
buoyancy 26, 240, 279
Buphagus 530
  erythrorhynchus 530, *530*, 531
Burhinidae 250, **251**
Burhinus
  bistriatus 251
  capensis 251
  grallarius 251
  oedicnemus 251, *251*
  senegalensis 251
  superciliaris 251
  vermiculatus 251
Busarellus nigricollis 172
bush-lark (*Mirafra*) 578
bush-quail, Manipur *see Perdicula manipurensis*
bush-shrike 498, **512–13**
  Fiery-breasted *see Malaconotus cruentus*
  Green-breasted *see Malaconotus gladiator*
  Grey-headed *see Malaconotus blanchoti*

Mt Kupé *see Telophorus kupeensis*
Rosy-patched *see Rhodophoneus cruentus*
Sulphur-breasted *see Malaconotus sulfureopectus*
bushtit *see Psaltiparus minimus*, Black-eared *see Psaltiparus minimus melanotis*
bustard **214–17**
Arabian *see Ardeotis arabs*
Australian *see Ardeotis australis*
Black-bellied *see Eupodotis melanogaster*
Blue *see Eupodotis caerulescens*
Buff-crested 214
courtship display **216**, 216
Denham's *see Neotis denhami*
Great *see Otis tarda*
Great Indian *see Ardeotis nigriceps*
Kori *see Ardeotis kori*; *Otis kori*
Little *see Tetrax tetrax*
Little brown *see Eupodotis humilis*
Nubian *see Neotis nuba*
Red-crested *see Eupodotis ruficrista*
Savile's 214, 215
White-bellied *see Eupodotis senegalensis*
Butastur rufipennis 163, 167, 172
butcherbird **494**
Grey 494
Tagula *see Cracticus louisiadensis*
Buteo 166
augur 163, 172
buteo 163, 166, 170, 172
galaagoensis 172
jamaicensis 163, *170*, 172
lagopus 170, 172
magnirostris 163, 172
regalis 172
ridgwayi 163, 172
rufinus *164*
solitarius 172
swainsoni *164*, 166, 172
Buteogallus urubitinga 172
Butorides 99
striatus 101, 104
virescens 101, 103, 104
buttonquail (*Turnix, Ortyxelos*) *216*, **218**
buzzard 162, 163, 172
Augur *see Buteo augur*
Black-breasted *see Hamirostra melanosternon*
Black-collared 163
Common *see Buteo buteo*
Grasshopper *see Butastur rufipennis*
Lizard *see Kaupifalco monogrammicus*
Long-legged *see Buteo rufinus*
Rough-legged *see Buteo lagopus*
Steppe *see Buteo buteo*
true 163
buzzard-eagle, Black-chested *see Geranoaetus melanoleucus*

C

Cacatua moluccensis 300
Cacatua galerita 300, 305
Cacatuinae 305
cachalote (*Pseudoseisura*) 440
Cacicus 428
koepckeae 634, 635
solitarius 632, 634
cacique (*Cacicus*) 428, 632, 633, 635
Cacomantis sonneratii 507
cactus-finch (*Geospiza*) 630–31
cahow *see Pterodroma cahow*
Cain and Abel syndrome 163
Cairina moschata 140
Calamonastes 572
Calamospiza melanocorys 619, 620, *621*
Calandrella
cinerea 578
rufescens 578
starki *578*
Calcarius
lapponicus 618, 620
ornatus 620
pictus 619, 620
calfbird *see Perissocephalus tricolor*
Calidris
acuminata 235, 236
alba 232, *232*, 234, 235, 236
alpina 232, *232*, 234, 235, 236
canutus 232, *232*, 234, 235
ferruginea *234*, 235, 236
mauri 26, 235
melanotos 235, 236
minuta 232, 234, 235, *236*
tenuirostris 230
Callaeas cinerea 495, *495*
Callaeidae **495**
Callipepla gambelii *176*
Calliphlox amethystina 354
Callocephalon fimbriatum 305
Caloenas nicobarica 291
Calonectris diomedea 71, *71*
Calypte anna *352*, 356
Calyptocichla serina 563
Calyptomena 422
hosei 422, 423
viridis 422, 423, *423*
whiteheadi 422
Calyptorhynchus funereus *300*
calyptura, Kinglet *see Calyptura cristata*
Calyptura 433
cristata 429, 432, 433
Camarhynchus 631
crassirostris 630, *630*, 631
heliobates 619, 631, *631*
pallidus 630, *630*, 631
psittacula *630*, 631
Camaroptera 568
camouflage 25
eggs and nests 31
parrots 299
puffbirds 404
sandgrouse 284
white winter coat 26, *184*
Campephaga 498, 499
flava 499, 500
phoenicea *498*
Campephagidae **498–500**
Campephilus
imperialis 408, 413
melanoleucos *415*
principalis 408
Campethera 411
Campochaera sloetii 498, 499
Camptorhynchus labradorius 140, 143
Camptostoma 428
Campylopterus 356
Campylorhamphus 443, 444
falcularius 443, *443*, 444
pucherani 443
trochilirostris *442*
Campylorynchus
brunneicapillus 542, 544, 545, *545*
chiapensis 542
canary 613, 614–15
Island *see Serinus canaria*
canastero 439, 440
Cordilleran *see Asthenes modesta*
canvasback *see Aythya valisineria*
capercaillie (*Tetrao*) 184–87
Capito 399
hypoleucus 399, 402
niger 399
Capitonidae **398–402**
Capito niger *157*
Caprimulgidae **336–41**
Caprimulgiformes 19, **336–45**
Caprimulgus 337
affinis 337
anthonyi 337, 339
carolinensis 337, 338, 339
centralasicus 337, 339
europaeus 336, 337, *338*, 339, 340, 341, *341*
longirostris 337
noctitherus 337, 339
parvulus 337, 339
pectoralis 337, 338, 339, 341, *341*
prigoginei 337, 339
rufigena *337*
rufus 337
saturatus 337, 339
vociferus 337, **339**, 341
capuchinbird *see Perissocephalus tricolor*
caracara (*Daptrius, Polyborus, Phalcoboenus, Milvago*) 40, **154–55**, 159
cardinal grosbeak (*Cardinalis*) **622–23**
Cardinalidae **622–23**
Cardinalis
cardinalis 622, 623, *623*
phoeniceus 622, 623
sinuata 622, 623
Carduelidae **611–15**
Carduelis 613
cannabina 611, 614
carduelis 611, *612*, *613*, 614
chloris 611
cucullata 611
flammea 611, 613, 614, *615*
spinus 611, 613, *613*
tristis 611, *612*, 614
Cariama cristata 222, *222*
Cariamidae **222**
carib *see Eulampis*
carinates 37
Carpococcyx viridis 313, 315
Carpodacus mexicanus 611, 615, *615*
Carpodectes 429, 432
antoniae 432, 433
Caryothraustes
canadensis 623
humeralis 623
casque 384, 386, 389
cassowary (*Casuarius*)19, 25, **44–45**
Casuariidae **44–45**
Casuariiformes 19, **41–43**, **44–45**
Casuarius
bennetti 44–45
casuarius 44–45, *44*
unappendiculatus 44–45, *44*
Catamblyrhynchus diadema *631*
catbird (*Ailuroedus, Dumetella*) 458, 459
caterpillar-bird *see* cuckoo-shrike
Catharacta
antarctica 270, 272, *272*
chilensis 270, 272
maccormicki 270, *271*
skua 270, *271*, 272, 273
Cathartes
aura *146*, 147
burrovianus 147
melambrotus 147
Cathartidae **146–47**
Catharus 522, 526
guttatus 524, 526
minimus 524, 526
mustelinus 526, 527
Catreus wallichii 178, 180
Caudipteryx zoui 18, *20*
cave-dwelling species 348, *348*, 350
cecum 216
Celeus grammicus *411*
Centrocercus 184, 186
minimus 185, 186, 187
urophasianus 185, 186, *186*
Centropus
phasianinus *314*, 315
steerii 313, 315
Cephalopterus 433
ornatus *432*
Cephalopyrus flammiceps 548
Cepphus
carbo 277
columba 276, 277, 278, 279, 282, 283
grylle 276, 277, *277*, 278, 279, 282, 283
Ceratogymna 384, 386, 388
atrata 386, *386*
brevis 386
bucinator 386, 389
cylindricus 386, 389
elata 386, 389
Cercibis oxycerca 115
Cereopsini 132, 134, 140
Cereopsis novaehollandiae *132*, 134, 140
Cerorhinca monocerata 276, 277
Certhia 538, 540
americana 538
brachydactyla 538, *538*
discolor 538
familiaris 528, 538, 540, *540*
himalayana 538
nipalensis 538
Certhiaxis 440
cinnamomea *439*
Certhidea 631
olivacea 630, *630*, 631
Certhiidae 452, **538–40**
Cerylinae 369
Ceyx 366
erithaca 367, 371
fallax 367, 370
lecontei 366, 367
lepidus 367
pusillus 367, *371*
rudis 366, 367, 369, *369*, 370, 371
chachalaca 196–97
Plain *see Ortalis vetula*
Chaetocercus 356
Chaetops
frenatus 515
frenatus aurantius 515
frenatus frenatus 515
pycnopygius 515
Chaetoptila 469
Chaetorhynchus papuensis 502, 503
Chaetura pelagica 346, 347, 349, 350
chaffinch (*Fringilla*) 610
Chalcopsitta atra 305
Chalybura 356
Chamaea fasciata 574, 575
Chamaepetes 197
goudotii 196, 197
Chamaeza 448
chanting-goshawk (*Melierax*) 163, *168*
Chapin, W.L. 178
Charadriidae 253
Charadriiformes 19, 225, **232–81**
Charadriinae **225–29**
Charadrius
alexandrinus 227
alexandrinus nivosus 231
australis 226, 227, 228
bicinctus 227, 228, 229
dubius 226, 227
peroni 227
ruficapillus 226, 227, 229
santahelenae 227, 229
semipalmatus 225, 227
vociferus *227*
Charmosyna
papua 300, 305
wilhelminae 305
chat 522
Australian **466–69**
cliff 524
Crimson *see Epthianura tricolor*
Gibber *see Ashbyia lovensis*
Orange *see Epthianura aurifrons*
robin *see Cossypha*
White-browed robin *see Cossypha heuglini*
White-fronted *see Epthianura albifrons*
Yellow *see Epthianura crocea*
Yellow-breasted 624
Chauna
chavaria 130–31, *130*
torquata 130–31, *130*
Chelidoptera tenebrosa 403, *404*, 405
Chersomanes albofasciata *578*
chickadee 447, 479, 550
Black-capped *see Parus atricapillus*
chiffchaff *see Phylloscopus collybita*
Childonias 264
hybrida 264
leucoptera 264
nigra 264, *264*, 268
chimango 40
Chionididae **256–57**
Chionis
alba 256–57, *256*
minor 256–57, *257*
Chiroxiphia 434, **437**
linearis 434, 437
pareola 434, *436*, 437
Chlamydera 459, 460
cerviniventris 458, 459, 461
guttata 458, 461
lauterbachi 458, 459
maculata 458, 459, 461
nuchalis 458, 459, *462*
Chlamydochaera jefferyi 498, 499
Chlamydotis undulata *214*, 215, 216, *216*, 217
Chloebia gouldiae 604, 605, *605*, 607, 608
Chloephaga rubidiceps 140, 143
Chloridops kona 616, *617*
Chloroceryle
aenea 367, 369
amazona 366, 369
americana 367, 369
inda 367, 369
Chlorocharis emiliae 566
Chlorocichla 563
flaviventris 563, *565*
laetissima *565*
Chlorophonia 630, 631
Chloropsis 506
aurifrons 506, *507*
cochinchinensis 506, 507
flavipennis 506, 507
hardwickii 506, 507
venusta 506, 507
Chlorostilbon 356
Chondrohierax
uncinatus 172
wilsoni 163
chopi *see Gnorimopsar chopi*
Chordeiles minor 336, 337, *337*, *339*
Chordeilinae 336, 337
chough (*Pyrrhocorax, Corcorax*) 484, 493
chowchilla *see Orthonyx spaldingii*
Chrysococcyx klaas 312, 313
Chrysolampis mosquitus *356*
Chrysolophus
amherstiae 178, *179*, 180, *180*
pictus 178, *179*, 180, *180*, 181
Chuck-will's widow *see Caprimulgus carolinensis*
chukar *see Alectoris chukar*
Chunga burmeisteri 222
cicadabird 499
Long-billed *see Coracina tenuirostris*
Cicinnurus
regius 488, *488*
republica *490*
Ciconia
abdimii 107, 109, *110*
boyciana 107
ciconia *32*, 106, 107, *108*, 109, 110, *110*
episcopus 107, *109*
maguari 107, 109
nigra 106, *106*, 107, 109, 110, 111
stormi 107
Ciconiidae **106–11**
Ciconiiformes 19, **99–129**, **146–47**
Ciconiini 107
Cinclidae **520–21**
Cinclocerthia ruficauda 534, 535
cinclode (*Cinclodes*) 26, 440
Cinclodes 440
antarcticus *438*
aricomae 439, 440
nigrofumosus 440
taczanowskii 440
Cinclosoma 475
cinnamomeum *474*
punctatum 475
Cinclosomatidae 474
Cinclus
cinclus 520, *520*, 521
leucocephalus 521
mexicanus 520, 521, *521*
pallasii 520, 521
schultzi 520, 521
Cinnyricinclus leucogaster 531
Cinnyris
chalybea 587
coccinigaster 586
johannae 586
jugularis 586, 587
oseus 586, 587
superba 586
Circaetus
fasciolatus 172
gallicus 165, 172
gallicus pectoralis *168*
Circus
aeruginosus *162*, 166, 167, *168*, 172
assimilis 166, 172
cyaneus 167, 170, 172
macrosceles 172
macrourus 172
maillardi 172
maurus 172
melanoleucus *168*
pygargus 166, 167, 172
Ciridops anna 616, *617*
Cissa chinensis 483, 484
cisticola 568
Cloud-scraper *see Cisticola dambo*
Cisticola 568, 572
dambo 572
Cisticolidae 594
Cistothrus
apolinari 542, 545
palustris 542, 543, 544, 545
Cittura cyanotis 367, 370
Cladorhynchus leucocephalus 242, 243
Clamator glandarius 313, 314
Clangula hyemalis 140
clay-lick *303*
Cleptornis marchei 466, 566
cliff-chat 524
Climacteridae **452**, 538
Climacteris
affinis 452, *452*
erythrops 452
melanura 452
picumnus 452
rufa 452
climate change 231, 616
cloaca 22, *27*
Clytoceyx rex 366, 367, *371*
Clytospiza 607
Cnemophilinae 490
Coccothraustes coccothraustes *29*, 611, 613, *613*, *614*
Coccyzus
americanus *314*
erythropthalmus 313
Cochleariinae 104
Cochlearius cochlearius 98, 100, *100*, 104
cochoa 522, 523, 526
Cochoa 526
cockatiel *see Nymphicus hollandicus*
cockatoo **296–307**
Gang-gang *see Callocephalon fimbriatum*

Palm *see Probosciger aterrimus*
Rose-breasted *see Eolophus roseicapillus*
Salmon-crested *see Cacatua galerita*
Sulfur-crested *see Cacatua galerita*
Yellow-tailed *see Calyptorhynchus funereus*
cock-of-the-rock **429–33**
Andean *see Rupicola peruviana*
Guianan *see Rupicola rupicola*
leks 429, 432, *432*
Cocorax melanorhamphos 493, *493*
Coeligena 356
Coenocorypha aucklandica 234, 235
Coereba flaveola 630, 631
cognitive skills **486–87**
Colaptes 414
auratus *409*, 410, 413, 414
campestris 410
fernandinae *410*
rupicola 410
coleto *see Sarcops calvus*
Colibri 352, 356, 359, *359*
Coliidae **364–65**
Coliiformes 19, **364–65**
Colinus virginianus 176, 177, 179, *179*, 180
Colius
castanotus 364, 365
colius 364
leucocephalus 364, 365
striatus 364, *364*, 365, *365*
Collocalia 346, 348–49
bartschi 346
elaphra *348*
esculenta affinis *33*
fuciphaga 346, *348*
spodiopygia *348*
unicolor *348*
Colluricincla 477
harmonica *477*
colonies, mixed-species 268
coloration 23, 25
albinistic 522
color morphs 65, 90, 595, 604, *605*
cryptic *see* camouflage
and diet *118*, 127
feather structure 25, 430
green 320
iridescence 25, 320, 352, *352*
leucistic 522
pigmentation 25, 320, 430
ultraviolet 30
winter coat 26, *184*
see also plumage
Columba
livia 288, 289, 291, *291*, **294–95**
mayeri 289, 293, *293*
oenas 289, 291, 292
palumbus *293*
Columbidae **288–95**
Columbiformes 19, **288–95**
Columbina passerina 289, 292
comet, Red-tailed *see Sappho sparganura*
communal feeding 83, 90, 101
communal roosts 296, 483, 492, 517, 598, 599, 601
communication 29, 30
parrots **300**
woodpeckers 408, **414–15**
see also display
condor 21
Andean *see Vultur gryphus*
Californian *see Gymnogyps californianus*
Confuciusornis 18
Conopophaga
ardesiaca 450
aurita 450
castaneiceps 450
lineatus 450, *450*
melanogaster 450
melanops 450, *450*
peruviana 450
roberti 450
Conopophagidae 446, **450**
Conothraupis mesoleuca 628, 631
Contopus 424, 428
Conuropsis carolinensis 300, 302, 305–06
cooperative behavior
feeding 90
honeyguides **397**
hunting 80
cooperative breeding 31, 427, 500, 510, 511, 533, 589, 616, 634
aracaris **395**
Australasian treecreepers 452
babblers **576**
barbets 402
bee-eaters 378
bulbuls 565
Cactus wren 545
crow family 483, 484
honeyeaters 468
hornbills 389
long-tailed tit 557
miners **468**
mudnesters 493
Pied butcherbird 494
puffbirds 405
Purple-throated fruitcrow 432
raptors 156
rifleman 417
tanagers 630
wood-hoopoes 383
coot (*Fulica*) 25, 26, 206, **207–11**, 209, 210
coppersmith 398
Copsychus sechellarum *526*
coquette *see Lophornis*
Coracias 380
abyssinica 380
benghalensis 380
caudata *380*
garrulus 380, *381*
spatulatus *380*
Coraciidae **380**
Coraciiformes 19, **366–81**, **384–89**
Coracina 498, 499
newtoni 499
novaehollandiae *498*, 499
ostenta 499
tenuirostris 499
Coracopsis nigra 305
Coracornis 477
Coragyps atratus 147, *147*, 174, 175
Corcoracidae **493**
cordon bleu (*Uraeginthus*) 604
Corividae 515
Cormobates
leucophaeus 452
placens 452
cormorant (*Phalacrocorax*) 26, **90–93**, 114, 260, 268
corncrake *see Crex crex*
coronet *see Boissonneaua*
corroborees 468
Corvidae **480–87**
Corvinella
corvina 508, 509, 510
melanoleuca 508, 509, 510, *510*
Corvus 480, 482, 484
albicollis 480, 484
albus 480, 484
bennetti 484
brachyrhynchos 480, 484
corax 480, *481*, 483, 484
corone cornix 480, 484, 486
corone corone 480, 483, 484, *484*, 486
coronoides 484
frugilegus 158, 480, *481*, 483, 484
hawaiiensis 480, 484
imparatus 484
kubaryi 484
macrorhynchos 480, 484
monedula 480, 483, 484, 485
moneduloides 483, 484, **486–87**
orru 480, 484
ossifragus 484
sinaloae 484
splendens 480, 484
tristis 483, 484
Corydon sumatranus 422, *422*, 423
Corythaeola cristata 321
Corythaixoides leucogaster *320*, 321
Corythopis 428
Coscoroba coscoroba 134, 140
Cossypha 525, 526, 527
heuglini *523*
cotinga 425, **429–33**, 434, 437
Andean 429, 430, 433
Banded *see Cotinga maculata*
Black-and-gold *see Tijuca atra*
core 429, 430, 433
Grey-winged *see Tijuca condita*
Guianan red *see Phoenicircus carnifex*
polygyny 430, 432
red 429, 430, 433
Swallow-tailed *see Phibalura flavirostris*
Turquoise *see Cotinga ridgwayi*
White-cheeked *see Zaratornis stresemanni*
Yellow-billed *see Carpodectes antoniae*
Cotingidae **429–33**
Cotinginae 429, 433
Cotinga 429, 430, 432
maculata 433
ridgwayi 432, 433
Coturnix 177
adansonii 177, 180, 181
chinensis 177, 180
coromandelica 177, 180
coturnix 177, *179*, 180
delegorguei 177, 180
japonica 177, 180
novaezealandiae 180, 181
coua, Crested *see Coua cristata*
Coua cristata *312*
coucal (*Centropus*) 312
courser **252–53**
Bronze-winged *see Rhinoptilus chalcopterus*
Cream-colored *see Cursorius cursor*
Double-banded *see Smutsornis africanus*
Indian *see Cursorius coromandelicus*
Jerdon's *see Rhinoptilus bitorquatus*
Temminck's *see Cursorius temminckii*
Three-banded *see Rhinoptilus cinctus*
Two-banded *see Smutsornis africanus*
courtship feeding *449*
covey 179–80, *180*
cowbird (*Molothrus, Scaphidura*) 632, 634–35
Cracidae **196–97**
Cracticidae **494**
Cracticus louisiadensis 494
crake 206, 207, 209
African 210
Black *see Amaurornis flavirostris*
Red-necked 208
Spotted *see Porzana porzana*
Striped 210
crane (*Balearica, Grus, Anthropoides, Bugeranus*) 19, **198–205**
conservation status 198, 200–201, **202–03**
display 199, **204–05**
migration 200, 201
Cranioleuca 440
albicapilla 440
erythrops 440
vulpina 440
Crax 197
blumenbachii 196, 197
rubra 196, *196*, *197*
Creagrus furcatus 259, 260, 262, 263
Creatophora cinerea 530, 531, *532*
creeper 447, 616
Brown *see Certhia familiaris*
Hawaii *see Oreomystis mana*
Molokai *see Paroreomyza flammea*
Spotted *see Salpornis spilonotus*
Crex crex 207, 211, *211*
crimson-wing 606
Criniger 563, 564
Crocodile bird *see Pluvianus aegyptius*
crombec, Red-faced *see Sylvietta whytii*
crop 22, *22*, 27
milk **292**
painted snipe 247
crossbill (*Loxia*) 612, 613, 614
Crossleyia xanthophtys 563, 564
Crossoptilon mantchuricum 176, 178, 180
Crotophaga sulcirostris 313, *314*
crow (*Corvus*) 287, **480–87**
Crypsirina temia *482*, 483, 484
Cryptospiza 606
Crypturellus
parvirostris 49
saltuarius 49
tataupa 49
cuckoo 19, **312–17**, 449
African 503
Banded bay *see Cacomantis sonneratii*
Black-billed *see Coccyzus erythropthalmus*
brood parasitism 312, 313–15, *313*, **316–17**
Channel-billed *see Scythrops novaehollandiae*
defensive excretion 315
Diederik 594
drongo-cuckoo 503
European *see Cuculus canorus*
Great spotted *see Clamator glandarius*
ground 212
Guira *see Guira guira*
Indian *see Cuculus micropterus*
Klaas's *see Chrysococcyx klaas*
Pied *see Oxylophus jacobinus*
Red-chested *see Cuculus solitarius*
Sumatran ground *see Carpococcyx viridis*
Yellow-billed *see Coccyzus americanus*
Cuckoo roller *see Leptosomus discolor*
cuckoo-shrike **498–500**
Black *see Campephaga flava*
Black-faced *see Coracina novaehollandiae*
Golden *see Campochaera sloetii*
Ground *see Pteropodocys maxima*
Large *see Coracina novaehollandiae*
Red-shouldered *see Campephaga phoenicea*
Réunion *see Coracina newtoni*
White-winged *see Coracina ostenta*
Cuculidae **312–17**
Cuculiformes 19, **312–21**
Cuculinae 312
Cuculus
canorus 312, 313, *314*, **316–17**
micropterus 423
solitarius *313*
Culicivora caudacuta 427, 428, *428*
Curaeus forbesi 633, 634, 635
curassow **196–97**
Alagoas *see Mitu mitu*
Great *see Crax rubra*
Helmeted *see Pauxi pauxi*
Horned *see Pauxi unicornis*
Nocturnal *see Nothocrax urumutum*
Razor-billed 197
Red-billed *see Crax blumenbachii*
curlew (*Numenius*) 30, **232–37**
Stone *see* stone curlew
currawong, Pied *see Strepera graculina*
Cursorius
coromandelicus 253
cursor *252*, 253
temminckii 253
Cyanerpes 631
cyaneus 631, *631*
Cyanochen cyanopterus 137, 140
Cyanocitta
cristata 480, *481*, 484
stelleri 484
Cyanocompsa
brissonii 622, 623
cyanoides 622, 623
Cyanocorax morio 482, 484
Cyanolanius madagascarinus 514
Cyanoliseus patagonus 304, 305
Cyanoloxia glaucocaerulea 622, 623
Cyanolyca turcosa 484
Cyanomitra
obscura 586
olivacea 586
veroxii 586
Cyanopsitta spixii 304–05
Cyanoptila cyanomelana 529
Cyanoramphus unicolor 300, 305
Cyanthus latirostris 356
Cyclarhinae 479
Cyclarhis 478
gujanensis 479
nigriostris 479
Cyclopsitta diophthalma 305
Cyclorhynchus psittacula 276, 277, *277*, 278
Cygnini 134, 140
Cygnus
atratus 140
buccinator 140
columbianus 140
cygnus *135*, *136*, 140
melanocorypha *135*, 140
olor *135*, 139, 140
Cymbirhynchus macrorhynchus 422, 423
Cyornis ruckii 529
Cyphorhinus aradus 30, 31, 542, 544
Cypseloides niger 346, 350
Cypsiurus parvus 346, *348*, 349
Cyrtonyx ocellatus 178, 180

D

Daceloninae 369
Dacelo novaeguineae *29*, 31, 367, 370, 371, *371*
Dacnis 628, 631
Daphoenositta 477, 536
dapple-throat *see Arcanator orostruthus*
Daptrius 159
americanus 155, 159
ater 154–55, 159
darter **96–97**
Darwin, Charles 38, 165, 182, 451
dawn songs 427
DDT 153, 159, 160, 170, *171*
defense strategies
alarm calls *29*
bill sweeping 452
bittern 100
Cape penduline tit nests *548*
communal defense 468, 527
distraction displays *227*, 252, *252*, 273, *275*, 338–39
dive-bombing 272–73
erectile plumage 498
fairy-wrens *465*
feigned injury 527
fieldfares 526–27
freezing 344, *344*
hummingbirds 359
magic eye 344
malodorous chemicals 382, 404
mobbing 312, 322, *324*, 527, 560
nests behind waterfalls 524
nests near wasp's nests 635
nests over water 635
petrels 71
singing 456–57
threat displays 328, *329*
wrynecks 412
Delichon urbica 591
Dendragapus 184
canadensis 185, 186, *186*
obscurus 185, 186
Dendrocincla 444
fulginosa 442, 443, *443*, 444
tyrannina 443, 444
Dendrocolaptes
certhia 443, 444
sanctithomae 443, 444
Dendrocolaptidae 438, 439, **442–45**, 446, 452
Dendrocopos 411
dorae 414
hyperythrus 410
leucotus 413
mahrattensis 413
major 408, *409*, 410, 411, *412*, 413
medius 414
minor 410–11
syriacus 413
Dendrocygna
bicolor 133, 140
viduata *138*
Dendrocygnini *138*, 140
Dendroica 625–26
chrysoparia 625, 627
coronata 625, 626, 627
kirtlandii *624*, 625, 627
pennsylvanica 625, 626
petechia 625, *625*, 626
striata 625, 626
Dendronanthus indicus 599, 601
desert habitats
Great roadrunner **315**
sandgrouse 284–87
Diatryma steini *20*
Dicaeidae **584–85**
Dicaeum 584
agile 584
concolor 585
hirundinaceum 584, *584*, 585
dickcissel *see Spiza americana*
Dicruridae 501, **502–03**
Dicrurus 502–03
adsimilis 502, *502*, 503
aldabranus 503
andamanensis 503
annectens 502, 503
atripennis 503
bracteatus 502, *502*, 503
caerulescens 503
forficatus 502, *502*, 503
fuscipennis 503
hottentottus 503
leucopheaus 503
ludwigi 503
macrocercus 502, 503, *503*
megarhynchus 503
modestus 503
paradiseus 503
remifer 503
sumatranus 503
waldenii 503
Didunculus strigirostris 288, 289, 290, 291
dieldrin 159, 170
digestive system **22**, 27
bustards 216
cecum 216
cellulose 27, 184
flowerpeckers 584
gizzard 139, 140, 216
grain-eaters 27
grit 216
herbivores 27, 318
hoatzin 318

owl pellets **326**, 327
waterfowl 139, 140, 144
Diglossa 631
dikkop (*Burhinus*) 251
dimorphic plumage 270
Dinemellia dinemelli 155, *593*, 594
Dinopium 410
javanense 408, 413
rafflesi 409, 411, 413
dinosaurs 18
Diomedea 66
amsterdamensis 67
bulleri 66
epomorphora 67, 68
exulans 21, 67, *67*, 68, *68*, 69
Diomedeidae **66–69**
Dioptrornis fischeri 528, 529
Diornis giteus 20
Diphyllodes magnificus 488, *489*
dipper (*Cinclus*) 26, **520–21**
Discosura 352, 356
display 25, 31, 430, *430*, 437
birds of paradise 490–91, *491*
booming bowls *311*
bowerbirds **458–63**
broken wing 227, 252, *252*, 273, 275
bustards *30*, **216**, 216
Chiroxiphia manakins **437**
cocks-of-the-rock 430, *430*, 432, *432*
Common mure *283*
corroborees 468
courtship song flights 598, 600–601
cranes 199, **204–05**
distraction displays *227*, 252, *252*, 273, 275, 338–39
drumming *232*, 237
dunnocks 603
erectile plumage 498
frigatebirds 94, *95*
grebes 62, *64*, 65
gulls 262, *263*
larks 580, *580*
lorikeets 308, *308*
nightjars *339*
ostriches 36, *37*
oystercatchers 249
painted snipe 247
peacock **182–83**
redirected aggression 85
rollers 380
sexual selection 453
skimmers *274*, 275
storks 109, **110–11**
terns *269*
territorial flight 598, 600
territorial singing 629
threat displays 328, *329*, 454
transferal effect **460**
woodpeckers *409*, **414–15**
see also lekking
dive-bombing 272–73
diver (*Gavia*) 26, **60–61**
Dives dives 634
diving **26**
Atlantic gannet *85*
auks 276, **279**, *279*
Brown pelican 80, *80*
osprey 150, **152–53**
penguins **50–51**, 52
uncinate processes 22
waterfowl 132, 139
diving petrel (*Pelecanoides*) **76–77**
Dodo 25, **291**
Dolichonyx oryzivorus 632, 633, 634
Doliornis 429
dollarbird (*Erystomus*) 380
Donacobius atricapillus 534, 535, *535*
doradito, Crested *see Pseudocolopteryx acutipennis*
Dorilinae 449
Doryfera 352, 356
Doryferinae 352
dotterel
Australian *see Peltohyas australis*
Black-fronted *see Elseyornis melanops*
Eurasian *see Eudromias morinellus*
Inland *see Charadrius australis; Peltohyas australis*
Red-kneed *see Erythrogonys cinctus*
Tawny-throated *see Oreopholus ruficollis*
double-jointed leg joints 166
dove
American mourning *see Zenaida macroura*
Bar-shouldered *see Geopelia humeralis*
Eared *see Zenaida auriculata*
Eurasian collared *see Streptopelia decaocto*
European turtle *see Streptopelia turtur*
Galápagos 534
ground 288
Magnificent fruit *see Ptilinopus magnificus*
migration 290
Namaqua *see Oena capensis*
Orange *see Ptilinopus victor*
Rock *see Columba livia*
Scaly-breasted *see Columbina passerina*
Socorro *see Zenaida graysoni*
Stock *see Columba oenas*
dovekie *see Alle alle*
dowitcher (*Limnodromus*) **234–35**
Drepanididae 611, 612, **616–17**
Drepanorhynchus reichenowi 586, 587, 588
Dreptes thomensis 586, 589
Dromadidae **250**
Dromaeocercus brunneus 572
Dromaiidae **41–43**
Dromaius novaehollandiae **41–43**
Dromas ardeola **250**, *250*
drongo (*Dicrurus, Chaetorhynchus*) 501, **502–03**
drongo-cuckoo 503
drumming display *232*, 237
Drymodes 472, 473
Drymornis bridgesii 444
Dryocopus
javensis 411
martius 143, 408, 410, 411, 413, 414
pileatus 409, 413, 414
Dryoscopus
gambensis 512, 513
sabini 512, 513
Dryotriorchis spectabilis 165, 172
duck 26, 27, **132–45**, 260, 268
American wood 143
Andean *see Oxyura ferruginea*
Black-headed *see Heteronetta atricapilla*
Blue *see Hymenolaimus malacorhynchus*
Bronze-winged *see Speculanas specularis*
comb 136, 140
dabbling 137, *138*, 139, 140, 142
eclipse plumage 26
Falkland steamer *see Tachyeres brachypterus*
Flightless steamer *see Tachyeres pteneres*
Freckled *see Stictonetta naevosa*
Fulvous whistling duck *see Dendrocygna bicolor*
Hartlaub's *see Pteronetta hartlaubii*
Labrador *see Camptorhynchus labradorius*
Laysan *see Anas laysanensis*
Long-tailed *see Clangula hyemalis*
Mandarin *see Aix galericulata*
molting 23
Muscovy *see Cairina moschata*
Musk *see Biziura lobata*
Pink-eared *see Malacorhynchus membranaceus*
Pink-headed *see Rhodonessa caryophyllacea*
Ruddy *see Oxyura jamaicensis*
Salvadori's *see Salvadorina waigiuensis*
sea *see* seaduck
shelduck *see* shelduck
steamer 132, 134, 136, 140, *142*
surface-feeding 137, *138*, 139, 140
Torrent *see Merganetta armata*
tree 133
Tufted *see Aythya fuligula*
whistling 132–33, *138*, 140
White-backed *see Thalassornis leuconotus*
White-faced whistling *see Dendrocygna viduata*
White-headed *see Oxyura leucocephala*
Wood *see Aix sponsa*
Ducula 289
goliath 291
dueting 299, 476, 493, 503, 512, *513*
barbets 402
Dulidae **516**
Dulus dominicus 516, *516*
Dumetella carolinensis 534, 535
dunlin *see Calidris alpina*
dunnock *see Prunella modularis*
Dysmorodrepanis phaeosoma 616

E

eagle 27, **162–74**
Aquiline 163, 172
Bald *see Haliaeetus leucocephalus*
Black *see Ictinaetus malayensis*
Black-and-Chestnut *see Oroaetus isidori*
Booted *see Hieraaetus pennatus*
Crested *see Morphnus guianensis*
Crowned *see Harphyaliaetus coronatus*
feet *164*
fish 150, 165, 172
Golden *see Aquila chrysaetoc*
Greater spotted *see Aquila clanga*
Great Philippine *see Pithecophaga jefferyi*
Gurney's *see Aquila gurneyi*
Harpy *see Harpia harpyia*
Imperial *see Aquila heliaca*
Long-crested *see Lophaetus occipitalis*
Martial *see Polemaetus bellicosus*
New Guinea *see Harpyopsis novaeguineae*
snake and serpent 165–66, 172
Spanish Imperial *see Aquila adalberti*
Tawny *see Aquila rapax*
true 163, 172
Verreaux's *see Aquila verreauxii*
Wedge-tailed *see Aquila audax*
White-tailed *see Haliaeetus albicilla*
ears 29, *29*
owls **334–35**
earthcreeper 440
Scale-throated *see Upucerthia dumetaria*
echolocation 29, 338, 344, *348*
Eciton burchelli 449
Eclectus roratus 298, 299, 305
Ectopistes migratorius 293
eggs 27, 29, 31
male incubation 39–40, 42, 46, 48, 69
mimicry in brood parasitism 313–15, **316–17**
ostriches 36
roc **21**
egret (*Ardea, Egretta*) **98–105**
Egretta
ardesiaca 101, *101*, 104
eulophotes 98–99, 103, 104
garzetta 98–99, 102, 104
rufescens 98, 99, 101, 104
thula 99
vinaceigula 99, 104
eider (*Somateria*) 140, 143
Eippiorhynchus
asiaticus 106, 107, 109, 111
senegalensis 106, 107, 109, *110*, 111
elaenia 427, 428
Elaenia 428
ridleyana 427, 428
Elanoides 169
forficatus 168, 172
Elanus
axillaris 167, *170*, 172
caeruleus 165, 172
scriptus 167, 172
Electron carinatum 372
elephant bird *see Aepyornis maximus*
Elminia longicauda 504, 505
El Niño Southern Oscillation 56, 87, 262
Elseyornis melanops 225, 227
Emberiza 618
bruniceps 620
calandra 618–19, 620, *620*, 621
cia 620
cirlus 618, 620
citrinella 618, *618*, 619, 620
melanocephala 620, *620*
rustica 620
schoeniclus 618, 620
sulphurata 619, 620
Emberizidae 611, **618–21**, 628
Emblema
bella 608
pictum 605, 607
emerald *see Chlorostilbon*
Versicolored *see Agyrtria versicolor*
emigration 554
Empidonax traillii 427
emu 19, 25, **41–43**
emu-tail (*Dromaeocercus*) 568
emu-wren (*Stipiturus*) 465
endemism 406
Enicognathus ferrugineus 305
Enicurus 526
Ensifera 353
ensifera 29, 352, *355*, 356, *356*
Entomodestes 526
Entomyzon 468
cyanotis 466, 468
Eolophus roseicapillus 300, *300*, *303*, 305
Eopsaltria 472
australis 473
flaviventris 472
Eopsaltridae **472–73**
Ephthianuridae 466
Epimachus
fastuosus 488, 491
mayeri 488
Epthianridae **466–69**
Epthianura
albifrons 466
aurifrons 466
crocea 466
tricolor 466, 468, *468*
Eremiornis carteri 570, 572
Eremophila alpestris 578, *580*, *581*
Eremopterix leucopareia 578, 580, *581*
Eriocnemis 356, 359
Erithacus rubecula 523
Erystomus 380
azureus 380
orientalis 380
Erythrogonys cinctus 227, 228, 229
Erythrotriorchis
buergersi 172
radiatus 172
Erythrura 605
hyperythra 605, 607
kleinschmidti 605, 607, 608
prasina 605, 607, *608*
trichroa 605, 607
Esacus
magnirostris 251
recurvirostris 251
Estrilda 604
astrild 604, 607
erythronotos 604
melanotis 604, 607
nonnula 608
Estrildidae **604–09**
Eubucco 398–99
boucierii 398, *398*, 399, 401
versicolor 402
Eucometis penicillata 628, 631
Eudocimus
albus 112, 115
ruber 112, 115, *115*, *118*
Eudromias morinellus 225, 227, 229
Eudynamys 313
scolopacea 313, 314, *314*
Eudyptes
chrysocome 51, 53, *53*, *55*
chrysolophus 50, 51, 53, 54, 55
pachyrhynchus 51, 55
robustus 51, 54, 55
schlegli 51, 54
sclateri 51, 55, 57
Eudyptula minor 51, 52, 53, *53*, 54, 55, *55*
Eugenes fulgens 360
Eugerygone 473
Eulampis 352, 356
Eulipoa 192
Euphagus cyanocephalus 632, 634
Eupherusa eximina 360
Euphonia 630, 631
Euplectes 593
afer 593, 594, 595
jacksoni 593, 594, 595
progne 593, 594
Eupodotis 214, 216
caerulescens 215, 217
humilis 215, 217
melanogaster 214, 215, *216*
ruficrista 214, *214*, 215, *216*
senegalensis 214, *214*, 215
Euptilotis neoxenus 362
Eurocephalus
anguitimens 508, 509, 510
rueppelli 508, 509, 510
Eurostopodus 337
diabolicus 336
Euryceros prevostii 514
Eurylaeimidae 421, **422–23**
Eurylaimus
javanicus 422, 423
ochromalus 422, 423
samarensis 422
steerii 422
Eurynorhynchus pygmeus 232, *232*, 234, 235
Eurypyga helias 221, *221*
Eurypygidae **221**
Euschistospiza 607
Eutoxeres 353, 356
aquila 29, *356*
Eutrichomyias rowleyi 504, 505
Eutriorchis astur 165, 170, 172
evolution 18–20, *20*, 182
adaptive radiation 514, 616, 629, **630**, *630*
coevolution 466
convergent 38, 166, 380, 384, 401, 634
flight 18–19, 20
flightless species 18–19, 20, 25
natural selection 18, 166
plumage 18
polygynous species 430
reptilian ancestry of birds 318
responses to dietary hazards 519
sexual selection 453, 460, 504
size constraints 20–22
species multipication 369
extinct species *20*
Aldabra warbler 469
Arabian ostrich 34, 35, 37
Canary Island black oystercatcher 249
Dodo **291**
emus 41, 43
flightless birds **25**
Glaucous macaw 306
Great auk 276, 277, 283
grebes 63
Guadalupe caracara 155, 159
Hawaiian honeycreeper 616
herons 99
honeycreeper 326
honeyeaters and oos 466
huia 495
hummingbirds 361
ibises 112
Jamaican poorwill 339
Laughing owl 322, 331
lyrebird 453
megapodes 192, 195
monarch flycatchers 504
New World vultures 146, 147
New Zealand quail 181
New Zealand wrens 416, 417
Pale-headed brush finch 621
Passenger pigeon 293
pigeons 289
piopio 477
rails 206, 207, 211
Robust white-eye 566
Tuamotu kingfisher 371
waterfowl 133, 136, 139, 143
White-eyed river martin 558
woodpeckers 408
wrens 545
eye-blazing 303
eyes 22, *23*, 29
all-round vision 234
binocular vision 29, 334, 369
color vision 29, 30
kingfishers 367, 369
magic 344
nictating membrane *520*
night vision **334–35**, **340–41**
nocturnal species 30
orbital ring 90
ostriches 35, *35*
owls 322, **334–35**
periophthalmic ring *608*
peripheral vision *38*
puffbirds 403
tapetum lucidum 340, *341*
ultraviolet sensitivity 30, 608
underwater vision 90
vertical pupils 274

F

fairy bluebird (*Irena*) **506–07**, 564
fairy-wren (*Malurus*) **464–65**
Falcipennis falcipennis 184, 185, 186, 187
Falco 155, 159
alopex 156, 159
amurensis 156, 158, 159
araea 156, 159
ardosiaceus 156, 158, 159
berigora 158, 159
biarmicus 158, 159, 160, 287
cenchroides 156, 159
cherrug 158, 159
chicquera 158, 159
columbarius 158, 159
concolor 156, 159, 382
cuvierii 159
deiroleucus 158, 159

dickinsoni 156, 159
eleonorae 156, 158, 159
fasciinucha 158, 159
femoralis *157*, 158, 159
hypoleucus 159
jugger 158, 159
longipennis 159
mexicanus 158, 159
moluccensis 156, 159
naumanni 158, 159
newtoni 156, 159
novaeseelandiae 158, 159
pelegrinoides 158, 159
peregrinus 154, 156, *157*, 158, 159, *159*, 160, *160*
punctatus 156, *157*, 159, 160
rufigularis 158, 159
rupicoloides 156, 159
rusticolus *157*, 158, 159
severus 159
sparverius 154, 156, *157*, 159
subbuteo 156, 159
subniger 158, 159
tinnunculus 154, *154*, 156, *157*, 159
vespertinus 156, *157*, 158, 159
zoniventris 156, 159
falcon 19, **154–61**, 166
Aplomado *see Falco femoralis*
Barbary *see Falco pelegrinoides*
Bat *see Falco rufigularis*
Black *see Falco subniger*
Brown *see Falco berigora*
desert 158
Eastern red-footed *see Falco amurensis*
Eleonora's *Falco eleonorae*
forest *see* forest-falcon
Grey *see Falco hypoleucus*
Laggar *see Falco jugger*
Lanner *see Falco biarmicus*
Laughing *see Herpetotheres cachinnans*
New Zealand *see Falco novaeseelandiae*
Peregrine *see Falco peregrinus*
Prairie *see Falco mexicanus*
Pygmy *see Polihierax semitorquatus*
Red-necked *see Falco chicquera*
Saker *see Falco cherrug*
Sooty *see Falco concolor*
Taita *see Falco fasciinucha*
true 154, 155–56, 158, 159
Western red-footed *see Falco vespertinus*
White-rumped *see Polihierax insignis*
falconet (*Microhierax, Spiziapteryx*) 154, 155–56, 158, 159
Falconidae **154–61**, 166
Falconiformes 19, **146–75**
Falconinae 154, 155–56, 158, 159
Falculea palliata 514, *514*
fantail (*Rhipidura*) **501**
fantail flycatcher, Whitebrowed *see Rhipidura areola*
fat, storage 347
feathers *see* plumage
feet
adaptations for swimming **26**
evolution 18
mousebirds 364
parrots 298, 299–300, 302
scutellations 511
vultures and eagles *164*
webbed 26, 276
woodpeckers 408
zygodactyly 298, 299–300, 302, 312, 320, 381, 396, 398
Ferminia cerverai 542, 543, 545
fernbird *see Bowdleria punctata*
Ficedula 528
albicollis 529
hypoleuca 528, *528*, 529
mugimaki *529*
fieldfare *see Turdus pilaris*
figbird (*Sphecotheres*) **496–97**
fig-parrot
Double-eyed *see Cyclopsitta diophthalma*
Large *see Psittaculirostris desmarestii*
filoplume 23, *23*
filter-feeders 70, 126
finch 27, 447, **611–15**
Bengalese *see Lonchura striata*
bill **613**
Black-throated *see Melanodera melanodera*
Crimson *see Neochmia phaeton*
Crimson-winged *see Rhodopechys sanguinea*
Cut-throat *see Amadina fasciata*
Darwin's 629, **630**, 631
Double-barred *see Taeniopygia bichenovii*
Gough *see Rowettia goughensis*
Gouldian *see Chloebia gouldiae*
Greater koa *see Rhodacanthis palmeri*
ground 630, *630*
House *see Carpodacus mexicanus*
Large ground *see Geospiza magnirostris*
Large tree *see Camarhynchus psittacula*
Laysan *see Telespiza cantans*
Lesser koa *see Rhodacanthis flaviceps*
Long-tailed *see Poephila acuticauda*
Mangrove *see Camarhynchus heliobates*
Masked *see Poephila personata*
Nihoa *see Telespiza ultima*
Plum-capped 605
Plum-headed *see Neochmia modesta*
Red-headed *see Amadina erythrocephala*
St. Lucia black *see Melanospiza richardsoni*
Sharp-beaked ground *see Geospiza difficilis*
Small ground *see Geospiza fuliginosa*
Star *see Neochmia ruficauda*
tanager **628–31**
trumpeter 614
Vegetarian *see Camarhynchus crassirostris*
Warbler *see Certhidea olivacea*
Woodpecker *see Camarhynchus pallidus*
Zebra *see Taeniopygia guttata*
finchbill 563
Crested *see Spizixos canifrons*
finfoot 26, **223**
African *see Podica senegalensis*
Masked *see Heliopais personata*
firecrest *see Regulus ignicapillus*
fire-eye, Fringe-backed *see Pyriglena atra*
firefinch (*Lagonosticta*) 604
firetail 605
Beautiful *see Emblema bella*
Mountain *see Oreostruthus fuliginosus*
Painted *see Emblema pictum*
Red-browed *see Neochmia temporalis*
Red-eared *see Stagnopleura oculata*
firewood-gatherer *see Anumbius annumbi*
fiscal
Common *see Lanius collaris*
Grey-backed *see Lanius excubitoroides*
Newton's *see Lanius newtoni*
fish-eagle (*Haliaeetus*) 165, *167*, 172
Fisher, Ronald 182
fishing-eagle (*Ichthyophaga*) 165
flameback
Common *see Dinopium javanense*
Greater *29*
flamecrest *see Regulus goodfellowi*
flamingo 19, **124–29**
Andean *see Phoenicoparrus andinus*
Caribbean *see Phoenicopterus ruber ruber*
Chilean *see Phoenicopterus chilensis*
Greater *see Phoenicopterus ruber ruber*
James' *see Phoenicoparrus jamesi*
Lesser *see Phoeniconaias minor*
milk **127**
Puna *see Phoenicoparrus jamesi*
flatbill 427, 428
flicker (*Colaptes*) 32, 409–14
flight
evolution 18–19, 20, 25
hummingbirds 352, *352*, 354–55, *355*
morphological adaptation **22–23**, *22*, *24*
size and weight constraints 21, 22, 27, 29
swifts 347
take-off 21, **24**
see also wings
flightless species 18, 21, **25**, **37**, 343
cassowaries **44–45**
convergent evolution 38
Dodo **291**
emus **41–43**
evolution 18–19, 20
Falkland steamer duck 140, *142*
Galápagos cormorant 90
Great auk 276, 277
grebe 62
kagu 220, *220*
kakapo 300, **310–11**
kiwis 37, **46–47**
ostriches **34–37**
penguins **50–59**
rails 209, 211
rheas **38–40**
Stephens Island wren 416
flocks, mixed species 410–11, 447–48, 449, 479, 496, 498, 501, 504, 528, 537, 541, 547, *550*, 552, 631
florican
Bengal *see Houbaropsis bengalensis*
Lesser *see Sypheotides indica*
Florisuga 356
flowerpecker **584–85**
Plain *see Dicaeum concolor*
Thick-billed *see Dicaeum agile*
Yellow-breasted *see Prionochilus maculatus*
flower-piercer 628, 630, 631
flufftail
Buff-spotted 208
White-winged *see Sarothrura ayresi*
Fluvicola pica *425*
flycatcher 27
Blue and white *see Cyanoptila cyanomelana*
Boat-billed *see Megarynchus pitangua*
Cocos *see Nesotriccus ridgwayi*
collared *see Ficedula albicollis*
fantail **501**
Fiscal *see Sigelus silens*
Fork-tailed *see Tyrannus savana*
Grey-breasted *see Lathrotriccus griseipectus*
Mariqua *see Bradornis mariquensis*
monarch *see* monarch flycatcher
Mugimaki *see Ficedula mugimaki*
Old World 526, **528–29**
paradise *see* paradise-flycatcher
Pied *see Ficedula hypoleuca*
Piratic *see Legatus leucophaius*
Restless *see Myiagra inquieta*
Royal *see Onychorhynchus coronatus*
Rueck's blue *see Cyornis ruckii*
Scissor-tailed *see Tyrannus forficata*
Shining *see Myiagra alecto*
silky *see* silky flycatcher
Social *see Myiozetes similis*
Spotted *see Muscicapa striata*
tyrant **424–28**, 430
Vermilion *see Pyrocephalus rubinis*
Wattle-eye *see Platysteira peltata*
White-winged slaty *see Dioptrornis fischeri*
Willow *see Empidonax traillii*
flycatcher-shrike *see Hemipus*
flyrobin (*Microeca*) 472–73
fody 593
Mauritius *see Foudia rubra*
foliage-gleaner, Alagoas *see Philydor novaesi*
food caches 410, 412, *412*, 483, **485**, 508, *508*
forest-falcon (*Micrastur*) 154–55, 159
forest-rail
Chestnut 209
Forbes' 208
forktail 522, 523, 526
Formicariidae **446–49**
Formicarius 448
colma *447*
Formicivora littoralis 448, 449
fossilized birds 18–19, *20*, 21, 299
fostering 608
Foudia 593
rubra 593, *593*
francolin (*Francolinus*) 177–78, 179, *179*, 180
Francolinus
afer 180
leucoscepus *179*
ochropectus 176
Fratercula
arctica *271*, 276, 277, *277*, 279, *279*, 282
cirrhata 276, 277, 283
corniculata 276, 277
Fregata
andrewsi 94, 95
aquila 94, 95
ariel 94, 95
magnificens 95, *95*
minor 94, 95
Fregatidae **94–95**
friarbird 466, 497
frigatebird (*Fregata*) **94–95**
Fringilla
coelebs 610, *610*, 614
montifringilla 610, 614
teydea 610
Fringillidae 611, 612, 613
frogmouth (*Batrachostomus, Podargus*) **342**
frontal shield 206, **210**
frugivory and polygyny 432
fruitcrow 429–33
Purple-throated *see Querula purpurata*
Red-ruffed *see Pyroderus scutatus*
fruiteater 429–33, *430*
Barred *see Pipreola arcuata*
fruithunter *see Chlamydochaera jefferyi*
Fulica
americana 207, *208*, *210*
atra 207, 208, 210, *210*, 319
cornuta 206, 207, 209, *210*
cristata *208*
gigantea 206, 207
fulmar (*Fulmarus*) 66, 70
Fulmarus
glacialis 70, 71, *71*, *72*
glacialoides 70, 71
fulvetta 574, 575
Rufous-winged *see Alcippe castaneceps*
White-browed *see Alcippe vinipectus*
Furnariida 446
Furnariidae 155, **438–41**, 442, 446, 626
Furnarius rufus 438, *438*, 439, 440, *440*

G

galah *see Eolophus roseicapillus*
Galbalcyrhynchus leucotis 406, 407
Galbula 406
dea 406, 407, *407*
pastazae 406, 407
ruficauda 406, *406*, 407
Galbulidae **406–07**
gale bird *see Phalaropus*
Galerida
cristata 578
magnirostris *580*
Gallicrex cinerea 207
Galliformes 19, **176–97**, 318
Gallinago
gallinago 230, *232*, 235, 236, 237
hardwickii 230
Gallinula
chloropus 207, *208*, 209, 210, *210*, 211
pacifica 207
gallinule 206, 207, 209, 210
Common *see Gallinula chloropus*
Purple *see Porphyrula martinica*
Gallirallus
australis 207, 208, 209
lafresnayanus 207
gallito
Crested *see Rhinocrypta lanceolata*
Sandy *see Teledromas fuscus*
Gallus gallus *179*, 180
Gama, Vasco da 50
game birds 19, 176–97
Gampsonyx 169
swainsonii 162, 172
Gampsorhynchus 575
gannet (*Morus*) **82–87**, 266
gaping 632–33, *634*
Garrodia nereis 75
Garrulacinae **575–77**
Garrulax
albogularis 575, 577
leucolophus 575, *575*, 577
milnei *577*
striatus 575, 577
Garrulus glandarius *482*, 483, 484
Gavia
adamsii 61
arctica 60, 61, *61*
immer 60, 61, *61*
pacifica 60, 61
stellata 60, *60*, 61, *61*
Gaviidae **60–61**
Gaviiformes 19, **60–61**
Geobates 440
Geococcyx californianus 312, *314*, **315**
Geocolaptes olivaceus 411, 413
Geopelia humeralis *289*
Geophaps plumifera *288*
Geopsittacus occidentalis *298*, 301
Geositta 440
Geospiza 631
conirostris *630*
difficilis 630, *630*, 631
fuliginosa *630*
magnirostris *630*
scandens *630*
Geothlypis
beldingi 625, 627
trichas *624*
Geranoaetus melanoleucus 172
Geranospiza caerulescens 165, 172
Geronticus 112
calvus 112, 114, 115
eremita 112, 113, *113*, 114, 115, 116, 117
ginae 413
gizzard 22, *22*, 27, 139, 140, 188, 284, 290, 308
Glareola
cinerea 253
lactea 253
maldivarum 253
nordmanni 253
nuchalis 253
ocularis 253
pratincola 252, 253
Glareolidae 250, **252–53**
Glaucidium 324, 326
brasilianum *334*
minutissimum 324, 331
passerinum 324, 330, 331
Glaucis 353
hirsuta 356
gleaning 512
global warming *see* climate change
Glossopsitta porphyrocephala 303, 305
Glyphorynchus spirurus 442, 443
gnatcatcher (*Polioptila*) **546–47**
gnateater (*Conopophaga*) 446, 447, **450**
gnatwren (*Microbates, Ramphocaenus*) **546–47**
Gnorimopsar chopi 634
goatsucker *see Caprimulgus europaeus*
go-away bird, White-bellied *see Corythaixoides leucogaster*
godwit (*Limosa*) 30
bills **234–35**
goldcrest *see Regulus regulus*
goldeneye, Common *see Bucephala clangula*
goldenthroat *see Polytmus*
goldfinch (*Carduelis*) 613
Goldmania violiceps 356
gonolek (*Laniarius*) **512–13**
goose 27, **132–45**, 319
African comb *see Sarkidiornis melanotos*
Bar-headed *see Anser indicus*
Barnacle *see Branta leucopsis*
Blue-winged *see Cyanochen cyanopterus*
Brent *see Branta bernicla*
Canada *see Branta canadensis*
Cape Barren *see Cereopsis novaehollandiae*
Egyptian *see Alopochen aegyptiacus*
Emperor *see Anser canagicus*
Graylag *see Anser anser*
Hawaiian *see Branta sandvicensis*
Knob-billed *see Sarkidiornis melanotos*
Magpie *see Anseranas semipalmata*
migration **144–45**
nests 140, *145*
Orinoco 143
Pink-footed *see Anser brachyrhyncus*
pygmy 139
Red-breasted *see Branta ruficollis*
Ruddy-headed *see Chloephaga rubidiceps*
sheldgoose *see* sheldgoose
Snow *see Anser caerulescens*
Spur-winged *see Plectropterus gambensis*
true 134, 140
White-fronted *see Anser albifrons*
Gorsachius 99
goisagi 103, 104
magnificus 99, 103, 104
goshawk 172
African *see Accipiter tachiro*
Chestnut-shouldered *see Erythrotriorchis buergersi*
Doria's *see Megatriorchis doriae*
Northern *see Accipiter gentilis*
Red *see Erythrotriorchis radiatus*
White-bellied *see Accipiter haplochrous*
Goura 288, 289, 290, 291
victoria *290*
grackle 632, 633
Colombian mountain *see Macroagelaius subalaris*
Common *see Quiscalus quiscula*
Red-bellied *see Hypopyrrhus pyrohypogaster*
Slender-billed *see Quiscalus palustris*
Gracula religiosa 530, 531, *531*, 532
Grallaria 448
andicola 448
gigantea 448
squamigera *448*
watkinsi 447, 448

Grallaricula 448
  cucullata 448
Grallina
  bruijni 493
  cyanoleuca 493, *493*
Grallinidae **493**
Grandala coelicolor 523, 526
grassbird *see Sphenoeacus afer*
grassfinch 604, 605–06
grasshopper bird *see Ciconia ciconia*
grassquit, Blue-black *see Volatinia jacarina*
grass-tyrant, Sharp-tailed *see Culicivora caudacuta*
grass-warbler 594
grasswren (*Amytornis*) **464–65**
grebe 19, 26, *26*, **62–65**
  Alaotra *see Tachybaptus rufolavatus*
  Atitlán *see Podilymbus gigas*
  Black-necked *see Podiceps nigricollis*
  Clark's *see Aechmophorus clarkii*
  Colombian *see Podiceps andinus*
  Great crested *see Podiceps cristatus*
  Hoary-headed *see Poliocephalus poliocephalus*
  Hooded *see Podiceps gallardoi*
  Horned *see Podiceps auritus*
  Junín flightless *see Podiceps taczanowskii*
  Least *see Tachybaptus dominicus*
  Little *see Tachybaptus ruficollis*
  Madagascar *see Tachybaptus pelzelnii*
  New Zealand *see Poliocephalus rufipectus*
  Pied-billed *see Podilymbus podiceps*
  Red-necked *see Podiceps grisegena*
  Silvery *see Podiceps occipitalis*
  Slavonian *see Podiceps auritus*
  Western *see Aechmophorus occidentalis*
greenbul
  Golden *see Calyptocichla serina*
  Joyful *see Chlorocichla laetissima*
  Little *see Andropadus virens*
  Spotted *see Ixonotus guttatus*
  Swamp *see Thescelocichla leucopleura*
  Tiny *see Phyllastrephus debilis*
  Yellow-bellied *see Chlorocichla flaviventris*
  Yellow-whiskered *see Andropadus latirostris*
greenfinch *see Carduelis chloris*
greenlet (*Hylophilus*) 478
grenadier, Common *see Uraeginthus granatina*
griffon (*Gyps*) 169, 170, 174–75
grosbeak 613, 614
  Black-headed *see Pheucticus melanocephalus*
  Blue-black *see Cyanocompsa cyanoides*
  cardinal **622–23**
  Crimson-collared *see Rhodothraupis celaeno*
  Glaucous-blue *see Cyanoloxia glaucocaerulea*
  Kona *see Chloridops kona*
  Pine *see Pinicola enucleator*
  Rose-breasted *see Pheucticus ludovicianus*
  São Tomé *see Neospiza concolor*
  Slate-colored *see Pitylus grossus*
  Ultramarine *see Cyanocompsa brissonii*
  Yellow *see Pheucticus chrysopeplus*
  Yellow-green *see Caryothraustes canadensis*
ground-jay
  Biddulph's *see Podoces biddulphi*
  Hume's *see Pseudopodoces humilis*
  Xinjiang *see Podoces biddulphi*
ground-tyrant 428
grouse 27, **184–87**
  Black *see Tetrao tetrix*
  Blue *see Dendragapus obscurus*
  Caucasian black *see Tetrao mlokosiewiczi*
  Chinese *see Bonasa sewerzowi*
  diet 318
  Gunnison sage *see Centrocercus minimus*
  Hazel *see Bonasa bonasia*
  leks 184, *185*, 186–87, *186*
  Red 184, 187
  Ruffed *see Bonasa umbellus*
  Sage *see Centrocercus urophasianus*
  Sharp-tailed *see Tympanuchus phasianellus*
  Siberian *see Falcipennis falcipennis*
  Spruce *see Dendragapus canadensis*
Gruidae **198–205**
Gruiformes 19, **198–223**, 255
Grus 199
  americana 198, 199, 200, 201, *201*, **202–03**
  antigone 198, 199, 200, *201*
  canadensis 198, 199, *200*, *201*
  grus *31*, 198, *198*, 199, 201
  japonensis 198, 199, 200, **204–05**
  nigricollis 198, 200
  vipio 198, 199, 200
guan **196–97**
  Black-fronted piping *see Pipile jacutinga*
  Chestnut-bellied *see Penelope ochrogaster*
  Crested *see Penelope purpurascens*
  Highland *see Penelopina nigra*
  Horned *see Oreophasis derbianus*
  Sickle-billed *see Chamaepetes goudotii*
  Trinidad piping *see Pipile pipile*
  Wattled *see Aburria aburri*
  White-winged *see Penelope albipennis*
guanay *see Phalacrocorax bougainvillii*
Guaruba guarouba 304, 305
Gubernetes yetapa 426, 427, 428
guillemot (*Cepphus, Uria*) 22, **276–83**
guineafowl 163, **190–91**
  Black *see Agelastes niger*
  Crested *see Guttera pucherani*
  Helmeted *see Numida meleagris*
  Plumed *see Guttera plumifera*
  Vulturine *see Acryllium vulturinum*
  White-breasted *see Agelastes meleagrides*
Guira guira 312, 313
Gulf Stream 227
gull 19, 26, **258–63**, 268, 282
  Arctic latitudes 263
  Audouin's *see Larus audouinii*
  Black-billed *see Larus bulleri*
  Black-headed *see Larus ridibundus*
  Bonaparte's *see Larus philadelphia*
  Chinese black-headed *see Larus saundersi*
  Common *see Larus canus*
  dark-hooded group 258–59
  Franklin's *see Larus pipixcan*
  Glaucous *see Larus hyperboreus*
  Grey *see Larus modestus*
  Grey-headed *see Larus cirrocephalus*
  Great black-backed *see Larus marinus*
  Great black-headed *see Larus ichthyaetus*
  Hartlaub's *see Larus novaehollandiae*
  Heermann's *see Larus heermanni*
  Herring *see Larus argentatus*
  Iceland *see Larus glaucoides*
  inland species 258, 259
  Ivory *see Pagophila eburnea*
  Laughing *see Larus atricilla*
  Lava *see Larus fuliginosus*
  Lesser black-backed *see Larus fuscus*
  Little *see Larus minutus*
  masked group 258–59
  Mediterranean *see Larus melanocephalus*
  Mew *see Larus canus*
  migration 259
  Olrog's *see Larus atlanticus*
  pelagic species 259
  Relict *see Larus relictus*
  Ring-billed *see Larus delawarensis*
  Ross's *see Rhodostethia rosea; Rodostethia*
  Sabine's *see Xema sabini*
  Saunders's *see Larus saundersi*
  Silver *see Larus novaehollandiae*
  Slender-billed *see Larus genei*
  Sooty *see Larus hemprichii*
  Southern black-backed *see Larus dominicans*
  Swallow-tailed *see Creagrus furcatus*
  temperate latitudes 263
  tropical latitudes 263
  Western *see Larus occidentalis*
  White-eyed *see Larus leucophthalmus*
  white-headed group 258
Guttera
  cristata *29*
  plumifera 190, 191
  pucherani 190, 191, *191*
Gygis alba *18*, 264, *264*, 268, *268*, 333
Gymnobucco 399, 402
Gymnoderus 429
Gymnogyps californianus 146–47, *147*
Gymnomyza
  aubryana 466
  samoensis 466
Gymnopithys 447
  leucaspis 448, 449
Gymnorhina tibicen *312*, 494, *494*
Gymnorhinus cyanocephalus 483, 484, 485
Gymnostinops
  Cassini 634, 635
  montezuma 632, 634
Gypaetus barbatus *164*, 165, 172, 173, 175
Gypohierax angolensis *164*, 165, 172
Gyps 169
  africanus 172
  bengalensis 163, *164*, 172
  coprotheres 169, 172, 174
  fulvus *167*, 172, 173
  himalayensis 172
  indicus 163, 172
  rueppellii 172, 174, *174*
gyrfalcon *see Falco rusticolus*

H

Habroptila wallacii 207
hacking 159
Haematopodidae **248–49**
Haematopus
  ater 248, 249
  bachmani 248, *248*, 249
  chathamensis 249
  finschi 249
  fuliginosus 248, 249
  leucopodus 249
  longirostris 249
  moquini 248, 249
  ostralegus *248*, 249
  ostralegus osculans 248
  palliatus 249
  unicolor *248*, 249
hairy caterpillars 496
Halcyon 369, 370, 371
  chloris 367, 369, 370
  coromanda 367, 370, *371*
  leucocephala 367
  malimbica 366, 369
  pileata 367
  saurophaga 367, 369
  senegaloides 367, 369
  senegalensis 367, 369
Haliaeetus
  albicilla 160, 169, 170, 172, 173
  leucocephalus 160, *164*, 165, 169, 170, *171*, 172, 173
  leucogaster 172
  leucoryphus 172
  pelagicus 162, 172, *172*
  sanfordi 172
  vocifer 172
  vociferoides 165, 170, 172
Haliastur indus 165, 172
hallux 124
Halobaena caerulea 70, 71
Halocyptena microsoma 75
Hamirostra melanosternon 163, 172
hammerhead *see Scopus umbretta*
hammerkop *see Scopus umbretta*
hanging-parrot 296, 304
  Blue-crowned *see Loriculus galgulus*
Hapalopsittaca 302
  fuertesi 302, 305
Hapaloptila castanea 403, 404, 405, *405*
harem 39, *39*
Harpactes 362, *363*
  kasumba 362
  oreskios 362
Harphyhaliaetus coronatus 172
Harpia harpyja 163, 172
Harpyopsis novaeguineae 172
harrier (*Circus*) 166, 167, 169, 172
harrier-hawk (*Polyboroides*) 172
hawfinch *see Coccothraustes coccothraustes*
hawk 19, **162–74**
  Bat *see Macheiramphus alcinus*
  Black-collared *see Busarellus nigricollis*
  buteonine 163
  Cooper's 160
  Crane *see Geranospiza caerulescens*
  Ferruginous *see Buteo regalis*
  Fish *see Pandion haliaetus*
  Galápagos *see Buteo galaagoensis*
  Grey *see Asturina plagiata*
  Grey-backed *see Leucopternis occidentalis*
  Gundlach's *see Accipiter gundlachii*
  Harris's *see Parabuteo unicinctus*
  Hawaiian *see Buteo solitarius*
  Long-tailed *see Urotriorchis macrourus*
  Mantled *see Leucopternis polionota*
  Plumbeous *see Leucopternis plumbea*
  Red-tailed *see Buteo jamaicensis*
  Ridgway's *see Buteo ridgwayi*
  Roadside *see Buteo magnirostris*
  Sharp-shinned 160
  Swainson's *see Buteo swainsoni*
  White-necked *see Leucopternis lacernulata*
hawk-eagle (*Spizastur, Stephanoaetus, Spizaetus*) 163
hawking 337–38, 340, 342, 343, 424, 492
hearing 29, *29*, 30, 31
  owls **334–35**
Hedydipna platura 586, 587
Heliangelus 356
Heliodoxa 356
  jacula *355*
Heliomaster 356
Heliopais personata 223, *223*
Heliornis fulica 223
Heliornithidae **223**
Hellmayrea 440
helmetcrest, Bearded *see Oxypogon guerinii*
helmet-shrike (*Prionops*) **511**, 512, 514
Hemignathus
  flavus 616
  lucidus 616
  procerus *617*
  wilsoni *617*
hemipode 218
Hemiprocne
  comata 351, *351*
  coronata 351, *351*
  longipennis 351
  mystacea 351
Hemiprocnidae **351**
Hemipus 498, 499, 500
Hemitriccus 428
  furcatus 428
  kaempferi 427, 428
Hemixos
  castanonotus 563, 564
  flavala 563, 564
Henicopernis
  infuscatus 172
  longicauda 172
Henicorhina 543
Herbst's corpuscles 234
hermit *32*, **352–53**, **356**
  Green *see Phaethornis guy*
  Hairy *see Glaucis hirsuta*
  leks 353
  Reddish *see Phaethornis ruber*
  Saw-billed *see Ramphodon naevius*
  Scale-throated *see Phaethornis eurynome*
Herodotus 253
heron 19, **98–105**, 114, 160
  Agami *see Agamia agami*
  Bare-throated night *see Tigrisoma mexicanum*
  Black *see Egretta ardesiaca*
  Black-crowned night *see Nycticorax nycticorax*
  Boat-billed *see Cochlearius cochlearius*
  Capped *see Pilherodius pileatus*
  Cocoi *see Ardea cocoi*
  conservation status 103, 160
  Fasciated tiger *see Tigrisoma fasciatum*
  Goliath *see Ardea goliath*
  Great blue *see Ardea herodias*
  Green *see Butorides virescens*
  Grey *see Ardea cinerea*
  Indian pond *see Ardeola grayii*
  Japanese night *see Gorsachius goisagi*
  long-necked 98
  Madagascar *see Ardea humbloti*
  Malagasy pond *see Ardeola idea*
  night 99, 103, 104
  Rufous night *see Nycticorax caledonicus*
  Squacco *see Ardeola ralloides*
  Striated *see Butorides striatus*
  tiger 100, 102, 104
  typical 99, 102, 103, 104
  Whistling *see Syrigma sibilatrix*
  White-bellied *see Ardea insignis*
  White-crested tiger *see Tigriornis leucolophus*
  White-eared night *see Gorsachius magnificus*
  Yellow-crowned night *see Nyctinassa violacea*
  Zigzag *see Zebrilus undulatus*
Herpetotheres cachinnans 154, 155, *157*, 159
Herpsilochmus 447
Hesperornis 19, *20*
Heteralocha acutirostris 495
Heteromirafra
  archeri 578, 581
  ruddi 578
Heteromunia pectoralis 604, 605, 607
Heteronetta atricapilla 140, 142
Heterophasia
  capistrata 575
  picaoides *574*
hibernation 339, 347
Hieraaetus pennatus 172
hill-partridge, Sichuan *see Arborophila rufipectus*
hillstar, Andean *see Oreotrochilus estella*
Himantopus
  himantopus 242, *242*, 243
  novaezelandiae 243
Himatione sanguinea *617*
Hippolais 568
  icterina 572
  pallida *571*
Hirundapus caudacutus 346, 347
Hirundinidae **558–61**
Hirundo
  abyssinica *561*
  aethiopica 558, 559
  atrocaerulea 558, *559*, 560, 561
  fulva 558, 559
  griseopyga 558, 559
  megaensis 558, 561
  perdita 558, **561**, 561
  pyrrhonota *33*, 558, 559, 560
  rustica 558, 559, 560, *560*, 561
  semirufa 558
  smithii 558
Histurgops ruficauda 594
hoatzin *see Opisthocomus hoatzin*
hobby (*Falco*) 156
hoechko *see* poorwill
homosexuality 210
Honey badger *see Mellivora capensis*
honey-buzzard (*Henicopernis, Pernis*) 163, 165, 172
honeycreeper 628, 631
  Hawaiian 326, **616–17**
  Red-legged *see Cyanerpes cyaneus*
honeydew 466
honeyeater **466–69**, 490
  Black 468
  Blue-faced *see Entomyzon cyanotis*
  Bonin Island *see Apalopteron familiare*
  cooperative breeding 468
  Crescent 466, 468
  Crow *see Gymnomyza aubryana*
  Golden *see Cleptornis marchei*
  Helmeted 468, 469
  New Holland 468
  Painted 467
  Pied 466
  as pollinators 466, **467**
  Regent *see Xanthomyza phrygia*
  Scarlet *see Myzomela sanguinolenta*
  white-eared 468
  White-naped 468
  Yellow-faced 468
honeyguide **396–97**
  Black-throated *see Indicator indicator*
  brood parasitism 396–97, 402
  Dwarf *see Indicator pumilio*
  Greater *see Indicator indicator*
  hatchling bill hook 396, 397
  human-honeyguide interaction 396, **397**
  Lesser *see Indicator minor*
  Lyre-tailed *see Melichneutes robustus*
  Malaysian *see Indicator archipelagicus*
  Scaly-throated *see Indicator variegatus*
  Sunda *see Indicator archipelagicus*
  Yellow-footed *see Melignomon eisentrauti*
  Yellow-rumped *see Indicator xanthonotus*
hookbill, Lana`i *see Dysmorodrepanis phaeosoma*
hoopoe *see Upupa epops*
  African 382
  migration 382
  wood *see* wood-hoopoe
Hoploxypterus 225
hornbill 31, **384–89**
  African pied *see Tockus fasciatus*
  Bar-throated wreathed *see Aceros undulatus*
  bill and casque 384, 386, 389
  Black-casqued *see Ceratogymna atrata*
  Bradfield's *see Tockus bradfieldi*
  Brown *see Anorrhinus tickelli*
  Brown-cheeked *see Ceratogymna cylindricus*
  Bushy-crested *see Anorrhinus galeritus*

Crowned *see Tockus alboterminatus*
Giant 29
Great *see Buceros bicornis*
ground *see Bucorvus*
Helmeted *see Buceros vigil*
in human cultures **389**
Indian grey *see Ocyceros birostris*
Indian pied *see Anthracoceros coronatus*
ivory 386, 389
Long-tailed *see Tockus albocristatus*
Monteiro's *see Tockus monteiri*
Narcondam wreathed *see Aceros narcondami*
nests 388
New Guinea wreathed *see Aceros plicatus*
Northern ground *see Bucorvus abyssinicus*
Red-billed *see Tockus erythrorhynchus*
Red-billed dwarf *see Tockus camurus*
Rhinoceros *see Buceros rhinoceros*
Rufous *see Buceros hydrocorax*
Rufous-necked *see Aceros nipalensis*
Silvery-cheeked *see Ceratogymna brevis*
Southern ground *see Bucorvus cafer*
Sri Lankan grey *see Ocyceros griseus*
Sulu *see Anthracoceros montani*
Sumba wreathed *see Aceros everetti*
Tarictic *see Penelopides exarhatus*
Trumpeter *see Ceratogymna bucinator*
Visayan wrinkled *see Aceros waldeni*
von der Decken's *see Tockus deckeni*
White-crested *see Aceros comatus*
Wrinkled *see Aceros corrugatus*
Writhed *see Aceros leucocephalus*
Yellow-casqued *see Ceratogymna elata*
hornero, Rufous *see Furnarius rufus*
Horwich, Robert 203
houbara *see Chlamydotis undulata*
Houbaropsis bengalensis 214, 215, 217
huia *see Heteralocha acutirostris*
human cultures
honeyguides 396, **397**
hornbills **389**
Humboldt, Alexander von 345
hummingbird 19, 25, **352–61**, 586
Allen's *352*
Anna's *see Calypte anna*
Antillean crested *see Orthorhynchus cristatus*
Bee *see Mellisuga helenae*
Blue-chested *see Polyerata amabilis*
Blue-throated *see Lampornis clemenciae*
breathing rate 355
brilliants *see Heliodoxa*
Broad-billed *see Cyanthus latirostris*
caribs *see Eulampis*
Chestnut-bellied *see Amazilia castaneiventris*
coquettes *see Lophornis*
coronets *see Boissonneaua*
emeralds *see* emerald
Fiery-throated *see Panterpe insignis*
flight 352, *352*, 354–55, *355*
Giant *see Patagona gigas*
goldenthroats *see Polytmus*
incas *see Coeligena*
Indigo-capped *see Saucerottia cyanifrons*
jacobins *see Florisuga*
lancebills *see Doryfera*
leks 352
Magnificent *see Eugenes fulgens*
mangos *see Anthracothorax*
metaltails *see Metallura*
migration 353, 356
nests 360, *360*, 361, *361*
plumage 352, *352*, 353, 359, *359*
plumeleteers *see Chalybura*
pufflegs *see Eriocnemis*
Ruby-throated *see Archilochus colubris*
Ruby-topaz *see Chrysolampis mosquitus*
Rufous *see Selaphorus rufus*
Rufous-tailed *see Amazilia tzacatl*
sabrewings *see Campylopterus*
sapphires *see Hylocharis*
social behavior 359
starthroats *see Heliomaster*
streamertails *see Trochilus*
Stripe-tailed *see Eupherusa eximina*
sunangels *see Heliangelus*
sunbeams *see Aglaeactis*
Sword-billed *see Ensifera ensifera*
sylphs *see Aglaiocercus*
thermoregulation 355–56, 359
thorntails *see Discosura*
Tooth-billed *see Androdon aequatorialis*
torpidity 20, 347, 356
trainbearers *see Lesbia*
Vervain *see Mellisuga minima*
Violet-capped *see Goldmania violiceps*
violet-ears *see Colibri*
Violet-headed *see Klais guimeti*
woodnymphs *see Thalurania*
woodstars *see Chaetocercus*
Huxley, Sir Julian 65
Hydrobatidae **74–75**
Hydrophasianus chirurgus 244, 245
Hylocharis 356
Hylocitrea 477
Hylomanes momotula 372, *372*
Hylopezus 448
Hylophilus 478
decurtatus 479
flavipes 479
Hylophylax naevioides 448
Hymenolaimus malacorhynchus 134, 136, 140
Hypargos 607
margaritatus 607
Hypergerus atriceps 570, 572
Hypnelus ruficollis 403, 404, 405
Hypocoliidae **517**
hypocolius, Grey *see Hypocolius ampelinus*
Hypocolius ampelinus 517, *517*
Hypodaleus guttatus 447, 448
Hypopyrrhus pyrohypogaster 634, 635
Hypothymis azurea 504
Hypsipetes 562
crassirostris 563, *564*, 565
leucocephalus 562, 563
madagascariensis 563, *564*, 565
philippinus 541

I

Ibidorhyncha struthersii 242, 243
ibis 30, 103, **112–19**
American white 113, 114, 116
Andean *see Theristicus branickii*
Australian *see Threskiornis molucca*
Bald *see Geronticus calvus*
Black *see Pseudibis papillosa*
Black-faced *see Theristicus melanopis*
Black-headed *see Threskiornis melanocephalus*
Buff-necked *see Theristicus caudatus*
Dwarf olive 113
Giant *see Pseudibis gigantea*
Glossy *see Plegadis falcinellus*
Green *see Mesembrinibus cayennensis*
Hadada *see Bostrychia hagedash*
Indian black 114
Madagascar crested *see Lophotibis cristata*
Northern bald *see Geronticus eremita*
Olive *see Bostrychia olivacea*
Oriental crested *see Nipponia nippon*
Plumbeous *see Theristicus caerulescens*
Puna *see Plegadis ridgwayi*
Sacred *see Threskiornis aethiopicus*
Scarlet *see Eudocimus ruber*
Sharp-tailed *see Cercibis oxycerca*
Spot-breasted *see Bostrychia rara*
Straw-necked *see Threskiornis spinicollis*
tactile foraging 112, 114, **118–19**
Waldrapp *see Geronticus eremita*
Wattled *see Bostrychia carunculata*
Whispering *see Phimosus infuscatus*
White *see Eudocimus albus*
White-faced *see Plegadis chihi*
White-shouldered *see Pseudibis davisoni*
ibisbill *see Ibidorhyncha struthersii*
Ichthyophaga
humilis 162, 172
icthyaetus 165, 172
Ichthyornis 20
Icteridae 210, 496, **632–35**
Icterus 634
bonana 634, 635
galbula 634
laudabilis 634, 635
nigroguaris 632
oberi 633, 634, 635
Ictinaetus malayensis 163, 172
Ictinia 169
mississippensis 172
plumbea 172
'i'iwi *see Vestiaria coccinea*
imprinting **202–03**, 608
inca *see Coeligena*
incest 210
Indicator 396
archipelagicus 396
indicator 396, *396*, 397
minor 396
pumilio 396
variegatus 396, *396*
xanthonotus 396
Indicatoridae **396–97**
indigobird, Village *see Vidua chalybeata*
interbreeding 132, **261**
Iodopleura 433
Iole 564
iora (*Aegithina*) 506, **507**, 564
Irania gutturalis 517
Irediparra gallinacea 245, *245*
Irena
cyanogaster 506
puella 506, 507, *507*
Irenidae **506–07**
irruption 410, 554
isolation rearing 203
Ixobrychus
exilis 100, *102*, 104
minutus 99, 104
Ixonotus guttatus 563, 565
Ixos 564
amaurotis 563, 565

J

Jabiru mycteria 106, 107, *109*, 111
jacamar 403, 404, **406–07**
Coppery-chested *see Galbula pastazae*
Great *see Jacamerops aureus*
Pale-headed *see Brachygalba goeringi*
Paradise *see Galbula dea*
Rufous-tailed *see Galbula ruficauda*
Three-toed *see Jacamaralcyon tridactyla*
White-eared *see Galbalcyrhynchus leucotis*
Jacamaralcyon tridactyla 406, 407
Jacamerops aureus 406, 407, *407*
jacana **244–45**
African *see Actophilornis africana*
Bronze-winged *see Metopidius indicus*
Comb-crested *see Irediparra gallinacea*
Lesser *see Microparra capensis*
Madagascar *see Actophilornis albinucha*
Northern *see Jacana spinosa*
Pheasant-tailed *see Hydrophasianus chirurgus*
Wattled *see Jacana jacana*
Jacana
jacana 244, 245
spinosa 244, 245
Jacanidae **244–45**
jackdaw, Eurasian *see Corvus monedula*
jacky-winter *see Microeca fascinans*
jacobin *see Florisuga*
jaeger (*Stercorarius*) **270–73**
jay 480, 482
Blue *see Cyanocitta cristata*
Brown *see Cyanocorax morio*
Canada *see Perisoreus canadensis*
Eurasian *see Garrulus glandarius*
Grey *see Perisoreus canadensis*
Mexican 484
Pinyon *see Gymnorhinus cyanocephalus*
Siberian *see Perisoreus infaustus*
Steller's *see Cyanocitta stelleri*
Turquoise *see Cyanolyca turcosa*
Jenner, Edward 314
jerdoni 226
jery 574
jewel-babbler **474–75**
junco 618
junglefowl 178, 180
Red *see Gallus gallus*
Jynginae 412, 413
Jynx
ruficollis 412, 413
torquilla 409, 412, 413

K

kagu *see Rhynochetos jubatus*
kaka *see Nestor meridionalis*
Norfolk Island *see Nestor productus*
kakapo 25, 296, 300, 304, **310–11**
kakawahie *see Paroreomyza flammea*
kakelaar *see* wood-hoopoe
kama'o *see Myadestes myadestinus*
Kauai 'akialoa *see Hemignathus procerus*
Kaupifalco monogrammicus 172
kea *see Nestor notabilis*
Kenopia 575
kestrel (*Falco*) 156, 287
Ketupa 324
zeylonensis 324, 331
killdeer *see Charadrius vociferus*
kingbird 428
Eastern *see Tyrannus tyrannus*
Giant *see Tyrannus cubensis*
kingfisher 19, 27, 143, **366–71**
African dwarf *see Ceyx lecontei*
African mangrove *see Ceyx pusillus*
Amazon *see Chloroceryle amazona*
American pygmy *see Chloroceryle aenea*
Beach *see Halcyon saurophaga*
Belted *see Megaceryle alcyon*
Black-capped *see Halcyon pileata*
Blue-breasted *see Halcyon malimbica*
Blue-capped *see Actenoides hombroni*
Blue-eared *see Alcedo meninting*
Collared *see Halcyon chloris*
Common *see Alcedo atthis*
Common paradise *see Tanysiptera galatea*
Crested *see Megaceryle lugubris*
dive 367, 369, *369*
Earthworm-eating *see Clytoceyx rex*
eyesight 367, 369
Giant *see Megaceryle maxima*
Grey-headed *see Halcyon leucocephala*
Green *see Chloroceryle americana*
Green-and-rufous *see Chloroceryle inda*
Green-backed *see Alcedo monachus*
Lilac-cheeked *see Cittura cyanotis*
Malachite *see Alcedo cristata*
Mangrove *see Halcyon senagaloides*
Marquesas *see Todirhamphus godeffroyi*
Mustached *see Alcedo bougainvillei*
Oriental dwarf *see Ceyx erithaca*
Pied *see Ceyx rudis*
Red-backed *see Todirhamphus pyrrhopygia*
Ringed *see Megaceryle torquata*
Ruddy *see Halcyon coromanda*
Sacred *see Todirhamphus sanctus*
Scaly-breasted *see Alcedo princeps*
Shovel-billed *see Clytoceyx rex*
Stork-billed *see Pelargopsis capensis*
Sulawesi dwarf *see Ceyx fallax*
Tuamotu *see Todirhamphus gambieri*
Variable dwarf *see Ceyx lepidus*
woodland *see Halcyon*
kinglet (*Regulus*) 447, 479, **573**
king-parrot 299
Australian *see Alisterus scapularis*
kiskadee, Great *see Pitangus sulphuratus*
kite 163, 165, 166, 167, 169, 172
Black *see Milvus migrans*
Black-shouldered *see Elanus axillaris*
Black-winged *see Elanus caeruleus*
Brahminy *see Haliastur indus*
Cuban *see Chondrohierax wilsoni*
Hook-billed *see Chondrohieraxuncinatus*
Letter-winged *see Elanus scriptus*
Mississippi *see Ictinia mississippensis*
Pearl *see Gampsonyx swainsonii*
Plumbeous *see Ictinia plumbea*
Red *see Milvus milvus*
Slender-billed *see Rostrhamus hamatus*
Snail *see Rostrhamus sociabilis*
Square-tailed *see Lophoictinia isura*
Swallow-tailed *see Elanoides forficatus*
White-collared *see Leptodon forbsei*
kittiwake *see Rissa tridactyla*
Black-legged *see Rissa tridactyla*
Red-legged *see Rissa brevirostris*
kiwi (*Apteryx*) 19, 29, 30, 37, **46–47**
Klais guimeti 356
kleptoparasitism 80, 228, 260, 271–72, *271*, 273
frigatebird **94**
Kluijver, H.N. 554
Knipolegus cyanirostris 426
knot
Great *see Calidris tenuirostris*
Red *see Calidris canutus*
koel
Asian *see Eudynamys scolopacea*
Common 503
kokako *see Callaeas cinerea*
kookaburra 366
Laughing *see Dacelo novaeguineae*
korhaan *216*

L

Lagonosticta 604
rhodopareia 604
senegala 604, 607
Lagopus 184
lagopus 184, 185, *185*, 186, *186*, 187
leucurus 184, *184*, 185
mutus 184, 185, 186, *186*
Lalage 499
leucomela 498
sueurii 498, 499, 500
lammergeier *see Gypaetus barbatus*
Lampornis clemenciae 356
Lamprotornis
chalybaeus 531, *533*
superbus 531, 532
lancebill *see Doryfera*
land reclamation 230–31
Laniarius (boubou) 512
aethiopicus 512
atrococcineus 512
barbarus 512
ferrugineus 513
leucorhynchus 512
liberatus 512, 513
Laniidae **508–10**
Laniisoma 433
elegans 430, 433
Laniocera 429, 433
Lanius 508–09, 512
collaris 508, 509, 510
collurio 508, 509, 510, *510*
excubitor 508, *508*, 509, 510
excubitoroides 509, 510
ludovicianus 508, 509
meridionalis 508, 509
minor 509, 510
newtoni 509, 510
nubicus 509, 510
schach 508, *509*, 510
senator 509, 510
souzae 509, 510
sphenocercus 508, 509
validirostris 509
lappets 180, *181*
lapwing (*Vanellus*) **224–29**, 225, *228*, 237, 260
Laridae **258–63**
lark **578–83**
Archer's *see Heteromirafra archeri*
Bifasciated *see Alaemom alaudipes*
Black *see Melanocorypha yeltoniensis*
Calandra *see Melanocorypha calandra*
conservation status 581, **582**
Crested *see Galerida cristata*
Desert *see Ammomanes deserti*
Fawn-colored *see Mirafra africanoides*
Hoopoe *see Alaemom alaudipes*
Horned *see Eremophila alpestris*
Large-billed *see Galerida magnirostris*
Lesser short-toed *see Calandrella rufescens*
Raso *see Alauda razae*
Red-capped *see Calandrella cinerea*
Rudd's *see Heteromirafra ruddi*
Rufous-naped *see Mirafra africana*
Shore *see Eremophila alpestris*
Spike-heeled *see Chersomanes albofasciata*
Stark's *see Calandrella starki*
Thick-billed *see Ramphocoris clotbey*
Larosterna inca 264, *264*, 268
Larus
argentatus 258, 260, 261, 262, *262*, 263, 269
atlanticus 262, 263
atricilla 263
audounii 263
bulleri 262, 263
canus 260, 261, 263
cirrocephalus 263
delawarensis 260, 263
dominicans 263
fulginosus 258, 261, 262, 263
fuscus 258, 260, 261, 262, 263
genei 263
glaucoides 263
heermanni 262
hemprichii 258, 263
hyperboreus 258, 260, 263
ichthyaetus 259, 263
leucophthalmus 263
marinus 258, 259, 260, *261*, 263
melanocephalus 263
minutus 258, 259, 260, *261*, 262, 263
modestus 261, 263
novaehollandiae 260, 263
occidentalis 263
philadelphia 259, 263
pipixcan 258, 259, 260, 263
relictus 259, 262, 263
ridibundus 242, 259, 260, 262, 263

saundersi 262, 263
Lathamus discolor 300, 305
Lathrotriccus griseipectus 427, 428
Latoucheornis siemsseni 618, 620
laughingthrush (*Garrulax*) **574–77**
lead poisoning **139**
leafbird (*Chloropsis*) **506–07**, 564
leaf-love *see Phyllastrephus scandens*
Legatus leucophaius 428
legs 23, 24
Leiothrix argentauris *577*
Leipoa ocellata 192, 194, *194*
lek-breeding species 31, 232, **238–39**, 437
  birds of paradise 490–91, *491*
  bowerbirds 461
  bustards **216**, 216
  cocks-of-the-rock 429, 430, 432, *432*
  grouse 184, *185*, 186–87, *186*
  hermits 353
  hummingbirds 352
  kakapo 304, *311*
  manakins **437**, 437
  Mionectes 426
  Tyrannine woodcreeper 444
  Yellow-whiskered greenbul 565
Lemuresthes nana 605, 607
Lepidocolaptes affinis 442, 443, 444
Leptasthenura 440
Leptodon forbsei 170, 172
Leptopoecile 572
Leptopogon 424
Leptopterus chabert 514
Leptoptilos
  crumeniferus 21, 106, 107, *108*, 109, 110, *110*
  dubius 106, 107, 109, 110
  javanicus 106, 107, 110
Leptosomatidae 380, **381**
Leptosomus discolor 380, **381**
Lesbia 356, 359
leucistic plumage 522
Leucopeza semperi 625, 627
Leucopsar rothschildi 530, 531, 533, *533*
Leucopternis 163
  lacernulata 172
  occidentalis 172
  plumbea 172
  polionota 172
Lewinia mirificus *208*
Lichenostomus 466, 468
Lichmera 466
lilytrotter **244–45**
  African *see Actophilornis africana*
Limicola falcinellus *234*, 235
Limnodromus
  griseus 232, 235
  scolopaceus 235
Limosa
  haemastica 231
  lapponica 232, 235, 236
  limosa 234, 235
limpkin *see Aramus guarauna*
Linnaeus, Carolus 610
linnet *see Carduelis cannabina*
  Green *see Carduelis spinus*
  Warsangli 611
Lipaugus 429
  uropygialis 430, 433
  vociferans 432, 433
Loboparadisea sericea 488, 490
locust bird *see Quelea quelea*
Locustella 568
Loddigesia mirabilis 356, *361*
logrunner *see Orthonyx temminckii*,
  Northern *see Orthonyx spaldingii*
Lonchura 605
  cucullata 604, 607
  striata 607, *607*, 608
longbill 584, 585
  Green-crowned *see Toxorhamphus novaeguineae*
  Pygmy *see Oedistoma pygmaeum*
longclaw (*Macronyx*) **598–601**, 634
longspur (*Calcarius*) 618–20
loon (*Gavia*) 19, **60–61**
Lophaetus occipitalis 172
Lophodytes cucullatus 140, 143
Lophoictinia isura 163, 172
Lophophorus impejanus 180
Lophornis 352, 356
Lophornithinae 352
Lophostrix cristata *330*
Lophotibis cristata 114, 115
Lophura
  bulweri *181*
  leucomelanos 178, 180
Loriculus galgulus 298, 305
lorikeet **296–307**
  diet **308–09**
  Papuan *see Charmosyna papua*
  Purple-crowned *see Glossopsitta porphyrocephala*
  Pygmy *see Charmosyna wilhelminae*
  Rainbow *see Trichoglossus haematodus*
  Scaly-breasted *see Trichoglossus chlorolepidotus*
  Stephen's *see Vini stepheni*
  Varied *see Psitteuteles versicolor*
Lorinae 305
lory **296–307**
  Black *see Chalcopsitta atra*
  Black-capped *see Lorius lory*
  diet **308–09**
lovebird 302, 304
  Fischer's *see Agapornis fischeri*
  Rosy-faced *see Agapornis roseicollis*
  Yellow-collared *29*
Loxia
  curvirostra 29, 611, 612, *612*, 613
  leucoptera 613, *613*
  pytopsittacus 611, 612, 613, *613*
  scotica 611, 612, 613
Loxigilla 631
Loxioides bailleui 616
Loxops coccineus 616
Lullula arborea 578, 580, 581
lunar cycle, nightjars and **340–41**
lungs **22**, 26, *27*
Lurocalis semitorquatus 337, 338
Luscinia megarhynchos *523*, 524, 526
Lybius 398, 402
  chaplini 399, 402
  minor *398*, 399
Lycocorax pyrrhopterus 488, 490
lyrebird (*Menura*) 31, **453–55**

M

macaw (*Ara, Anodorhynchus, Cyanopsitta*) 296, 300, 302
Macgregoria pulchra 466, 488, 490, 491
Machaeropterus deliciosus *437*
Macheiramphus alcinus 163, 165, 172
Machrochires 346
Macroagelaius subalaris 633, 634, 635
Macrodipteryx
  longipennis 336, 337, 339, *339*, *341*
  vexillarius *29*, 336, 337, 339
Macronectes
  giganteus 70, 71, *71*
  halli *72*
Macronus 575
Macronyx 598
  capensis *600*
  croceus *599*
  sharpei 599
Macropsalis forcipata 336, 337
Magellan, Ferdinand 50
magpie (*Pica, Cissa, Urocissa*) **480–85**
  Australasian *see Gymnorhina tibicen*
magpie-jay 482
magpie-lark *see Grallina cyanoleuca*
magpie-robin
  Seychelles *see Copsychus sechellarum*
magpie-shrike *see Corvinella melanoleuca*
Malaconotidae 498, **512–13**
Malaconotus 512–13
  blanchoti 512, 513
  cruentus 512
  gladiator 512, 513
  sulfureopectus 512, 513, *513*
Malacoptila 403, 404
  panamensis 403
Malacorhynchini 137, 140
Malacorhynchus membranaceus 137, 140
maleo 192
Malia grata 563, 564
malimbe, Crested *see Malimbus malimbicus*
Malimbus malimbicus 594
mallard, Northern *see Anas platyrhynchos*
malleefowl *see Leipoa ocellata*
Maluridae **464–65**
Malurus 464–65
  coronatus 465
  cyaneus *464*, 465
  elegans 464, 465
  splendens *464*
Manacus manacus 432, 434, *434*, 437
manakin 421, 425, 429, 430, **434–37**, 434
  Araripe *see Antilophia bokermanni*
  Black-capped *see Piprites pileatus*
  Blue-backed *see Chiroxiphia pareola*
  Club-winged *see Machaeropterus deliciosus*
  Helmeted *see Antilophia galeata*
  Long-tailed *see Chiroxiphia linearis*
  syringeal morphology 436
  White-bearded *see Manacus manacus*
  White-crowned *see Pipra pira*
  Wire-tailed *see Pipra filicauda*
Mandigoa 607
mangos (*Anthracothorax*) *352*, 356
mannikin 605
man-of-war *see* frigatebird
Manorina 467, 468
  melanophrys 466, 468
  melanotis 466, 469
manucode 488, 489
Manucodia
  comrii 488, *488*
  keraudrenii 488, 490
mao *see Gymnomyz samoensis*
Marmaronetta angustirostris *138*
marsh-harrier, Western *see Circus aeruginosus*
marsh-tyrant, White-headed *see Arundinicola leucocephala*
marsh-wren 542
martin **558–61**
  House *see Delichon urbica*
  New World 558
  Purple *see Progne subis*
  river 558
  sand *see Riparia riparia*
  White-eyed river *see Pseudochelidon sirintarae*
Marvellous spatuletail *see Loddigesia mirabilis*
Maui 'alauahio *see Paroreomyza montana*
meadowlark (*Sturnella*) 632–33, *632*, 634
Mecocerculus 428
Megaceryle
  alcyon *366*, 367, 369, 370
  lugubris 367, 369
  maxima 367, 369
  torquata 367, 369, 370
Megadyptes antipodes 51, 52, 53, *53*, 54, 55, 56–57
Megaegotheles novaezealandiae *343*
Megalaima 402
  asiatica *399*
  incognita *399*
  lineata 399, 401
  virens 399, 401
  viridis *401*
megapode (*Megapodius*) **192–95**
Megapodiidae **192–95**
Megapodius 192
  eremita 192, 194
  laperouse 192, 195
  pritchardii 192, 195
  reinwardt 192, *192*
Megarynchus pitangua 424, 428
Megatriorchis doriae 172
Megaxenops parnaguae 440
Melamprosops phaeosoma 616
Melanerpes 410, 411, 413
  carolinus 413, *414*, 415
  cruentatus 410, 413
  erythrocephalus *409*, 413
  formicivorus 408, 412, *412*, 413
  herminieri 411
  lewis 410, 413
Melanitta nigra 140
Melanocharis 585
  nigra 585
  versteri 585
Melanochlora sultanea 550, 551
Melanocorypha
  calandra 578, 580
  yeltoniensis 578, *581*
Melanodera melanodera 620
Melanodryas cucullata 473
Melanopareia 451
Melanoperdix nigra 178, 179, 180
Melanospiza richardsoni 631
Melanotis caerulescens 534, 535
Meleagrididae **188–89**
Meleagris
  gallopavo 188–89, *189*
  ocellata 188–89, *188*
Melichneutes robustus 396
Melierax
  canorus 163, 172
  poliopterus *168*, 172
Melignomon eisentrauti 396
Meliphaga 466, 467
Meliphagidae **466–69**, 490, 587
Melithreptus 467, 468
Mellisuga
  helenae 353
  minima 20
Melophus lathami 618, 620
Melopsittacus undulatus 296, 300, 301, *304*, 305
Melopyrrha 631
Menura
  alberti 453–55, *453*, *455*
  novaehollandiae 453–55, *454*, *455*
  tyawanoides 453
Menuridae **453–55**
Merganetta armata 134, 136, 140, 143
Merganettini 134, 136, 140
merganser (*Mergus, Lophodytes*) 140
Mergellus albellus 140, *141*, 143
Mergini 139, 140
Mergus
  australis 133, 139, 140, 143
  merganser 140, *141*
  octosetaceus 133, 139, 140
  serrator 132, 140, *141*
  squamatus 140, 143
merlin *see Falco columbarius*
Meropidae **374–79**
Meropogon forsteni *374*, 375
Merops
  albicollis 375, 377, 378
  apiaster 374, *374*, 375, 376, 378, *378*
  breweri *375*
  bullocki 375, 378
  bullockoides 375, 378
  gularis 375, *375*, 378
  hirundineus 374, *375*
  malimbicus 374, 375, 376
  muelleri *375*
  nubicus 374, 375, 376, 377, *377*, *378*
  nubicus nubicoides *378*
  orientalis 375, *378*
  ornatus 374, 375, 376, *376*
  persicus 375, 376
  philippinus 375, 376–77
  pusillus *374*, 375, 378
  superciliosus 375, 376, *377*
Merulaxis stresemanni 451
Mesembrinibus cayennensis 114, 115, 117
mesia 574
  Silver-eared *see Leiothrix argentauris*
mesite (*Mesitornis, Monias*) **219**
Mesitornis
  unicolor 219, *219*
  variegata 219, *219*
Mesitornithidae **219**
Metallura 356, 359, 361
metaltail *see Metallura*
Metopidius indicus 244, 245
Micrastur 159
  buckleyi 159
  gilvicollis 159
  mirandollie 159
  plumbeus 159
  ruficollis *157*, 159
  semitorquatus 159
Micrathene whitneyi *32*, 324, *324*, 331
Microbates
  cinereiventris 546, 547, *547*
  collaris 546, 547
Microcerculus ustulatus 542, 544
Microchera albacoronata 356, *356*
Microeca 472
  fascinans 472, *472*, 473
  flavigaster 473
  flavovirescens 473
  griseceps 473
Microhierax 155, 159
  caerulescens 156, 159
  erythrogenys 155, 159
  fringillarius 155, 159
  latifrons 159
  melanoleucus 159
Micromonacha lanceolata 403, 404, 405
Microparra capensis 244, 245
Micropsitta pusio 304, 305
Micropsittinae 305
migration
  avocets 242
  conservation and migratory species 627
  cranes 200, 201
  crows 485
  doves 290
  emus 41, **43**
  falcons 156, 160
  finches 613
  hawks 166
  honeyeaters 468
  hoopoes 382
  hummingbirds 353, 356
  ibis 117
  loop 510
  monarch flycatchers 505
  navigation **236**, 485
  nocturnal 418–19
  osprey 150
  owls 329
  parrots 300
  penguins 53
  phalaropes 241
  plovers and lapwings 225, 226–28, 229, **236**
  rail 209
  respiratory system 26
  rollers 380–81
  Ruddy turnstone *236*
  sandgrouse 286
  sandpipers and snipes 232, 234, **236**, 237
  shrikes 510
  site fidelity 236
  skuas *271*, 272
  starlings 531–32
  storks 106, 107
  swallows 559, 561
  swifts 347, 348
  terns 266
  tyrant flycatchers 424, 428
  waders 230, **236**, 241
  wagtails and pipits 598–99, 600
  warblers 568, 570, 626, 627
  waterfowl 139, 142, **144–45**
  white-eyes 566
milk
  flamingos **127**
  pigeons **292**
millerbird *see Acrocephalus familiaris*
Milvago 159
  chimachima 159
  chimango 159, *159*
Milvus
  migrans 165, 167, 172
  milvus 167, *168*, 172, 173
mimic-thrush *see* mockingbird
Mimidae **534–35**
Mimodes graysoni 535
Mimus
  patagonicus 534, 535
  polyglottos 535, *535*
miner
  Bell *see Manorina melanophrys*
  Black-eared *see Manorina melanotis*
  cooperative breeding **468**
  South American 440
minivet (*Pericrocotus*) **498–500**, 499
Minla 575
Mionectes 427
Mirafra 578
  africana *578*
  africanoides *578*
  javanica 578, *581*
mistletoe bird *see Dicaeum hirundinaceum*
Mitu mitu 196, 197
Mniotilta varia 625, 626
mobbing 312, 322, *324*, 560
mockingbird (*Melanotis, Nesomimus, Mimus, Mimodes*) **534–35**
mockingthrush, Black-capped *see Donacobius atricapillus*
Moho 468, 469
  bishopi 466, 469
  braccatus 466, 469
Mohoua albicilla 477
mollymauk *see Thalassarche*
Molothrus
  ater 479, 527, 634
  badius 632, 634
  bonariensis 634, 635, *635*
  rufoaxillaris 634
molting 23, 25, 276–77
  breeding plumage 225
Momotidae **372**
Momotus momota 372, *372*
Monachella 472
  muelleriana 472, 473
monal, Himalayan *see Lophophorus impejanus*
Monarcha
  brehmii 504
  guttula 504, 505
monarch flycatcher 501, **504–05**
  African blue *see Elminia longicauda*
  Black *see Monarcha brehmii*
  Black-naped *see Hypothymis azurea*
  Frilled *see Arses telescophthalmus*
  Maupiti *see Pomarea pomarea*
  Spot-winged *see Monarcha guttula*
  Tahiti *see Pomarea nigra*
Monarchidae 501, **504–05**
Monasa 403
  morphoeus 403
Monias benschi 219, *219*

monklet **403–05**
Lanceolated *see Micromonacha lanceolata*
monogamy 31
Monticola 524, 526
saxatilis 524, 526
Montifringilla 591
adamsi 590, 591
blanfordi 590, 591
nivalis 590, *590*, 591
ruficollis 590, 591
moorhen (*Gallinula*) 206, 207, 209, 314
Morphnus guianensis 172
morphs 65, 90, 595, 604, *605*
moruk *see Casuarius bennetti*
Morus
bassanus 83, *85*, 86, *86*
capensis 82, 83, 86, 87
serrator 83
Motacilla
alba 598, 599, 600, 601, *601*
capensis *313*, 599, 600
cinera *601*
flava 29, 599, *599*, 600, 601
grandis 599, 600
lugens 599, 601
Motacillidae **598–601**
motmot **372**
Blue-crowned *see Momotus momota*
Blue-throated *see Aspatha gularis*
Keel-billed *see Electron Carinatum*
Tody *see Hylomanes momotula*
mourner (*Laniisoma, Tityra*) 429–33
mousebird (*Colius, Urocolius*) 19, **364–65**
mudlark *see Grallina cyanoleuca*
Mulleripicus pulverulentus 411
munia 605
Bronze *see Lonchura cucullata*
Madagascar *see Lemuresthes nana*
Pictorella *see Heteromunia pectoralis*
White-rumped *see Lonchura striata*
murre **276–83**
Common *see Uria aalge*
Thick-billed *see Uria lomvia*
murrelet (*Synthliboramphus, Brachyramphus*)) **276–83**
Muscicapa 529
adusta 529
caerulescens 528, 529
striata 33, 528, *528*, 529
Muscicapidae 526, **528–29**
Muscisaxicola 425, 428
musculature 22, 24
Musophaga violacea 321, *321*
Musophagidae 210, **320–21**
Myadestes 526
elisabeth 527
lamaiensis 526, 527
myadestinus 526, 527
Mycteria 110
americana 107, *108*, 109
cinereus 107, 110
ibis 107, 110, *110*
leucocephalus *106*, 107, 110, *110*
Mycterini 107
Myiagra
alecto 504, 505
inquieta 504
Myiarchus 417
Myioborus 626
Myiopagis 428
Myiopsitta monachus 143, 155, 301, 304, 305
Myiornis 428
ecaudatus *427*
Myiozetes similis *424*
myna 31, **530–33**
Bali *see Leucopsar rothschildi*
Bank *see Acridotheres ginginianus*
Common *see Acridotheres tristis*
Grosbeak *see Scissirostrum dubium*
Helmeted *see Basilornis galeatus*
Hill *see Gracula religiosa*
Sulawesi *see Basilornis celebensis*
Myophoneus 522, 523, 524, 526
Myrmeciza 447
Myrmornis torquata 448
Myrmothera 448
Myrmotherula 447
axillaris *447*
obscura 447, 448
schisticolor *448*
snowi 446
Myzomela 466
sanguinolenta *467*

N

Napothera 575
Nascia longirostris 443, *443*, 444
navigation **236**, 288, **294–95**, 485
Necrosyrtes monachus 172
Nectarinia 587
johnstoni 586
jugularis *588*
tacazze 586
Nectariniidae **586–89**
nectar-robber *see* flower-piercer
nectarvores 584, 586–87, 616, *617*
bills 421, *421*, 460
flower-piercers 628
honeyeaters 466–68
hummingbirds **352–61**
lories and lorikeets 298, **308–09**
tongue 298, 308, *308*, 398, 466, 492, 566, 584, 586–87, 616
white-eyes 566
negrofinch
Grey-headed *see Nigritia canicapilla*
Nemosia rourei 628, 631
Neochelidon tibialis 558
Neochmia
modesta 607
phaeton 605, 607
ruficauda 605, 607
temporalis 605, 607
Neodrepanis
coruscans 421, *421*
hypoxantha 421
Neognathae 19
Neolestes torquatus 563–64
Black *see Hypsipetes leucocephalus*
Neopelma aurifrons 434
Neophema
chrysostoma 300, 305
petrophila 304, 305
Neophron percnopterus 35, 165, 172, 173, 175
Neopipo cinnamonea 434
Neospiza concolor 611, 615
Neotis
denhami 215, 216
nuba 215, 217
Nesillas aldabrana 569
Nesocharis 606
ansorgei 606, 607
Nesoctites micromegas 413
Nesofregetta fuliginosa 75
Nesomimus
macdonaldi 534, *534*, 535
trifasciatus 534, 535
Nesopsar nigrimus 632, 634, 635
Nesotriccus ridgwayi 427, 428
Nestor
meridionalis 305
notabilis *298*, 300–301, 302, 304, 305, *306*
productus 305
Nestorinae 305
nests 31, **32–33**
burrows 71–72, 279, 304
Cape penduline tit *548*
cavity-nesting species 444
down-lined 140, *145*
falcons 155, 158–59
flamingos 124, 127, *128*
floating 245, *245*, 266, 268
ground-nesting species 580, 582, *582*
Gygis alba 268, *268*
hammerhead *121*
magpie-larks and mudnesters 493, *493*
mammal burrows 143
in mammal burrows 524, 601
megapodes 193–95, *194*
multichambered communal 516, 593, 594
ovenbirds 438, 440, *440*
palmchat 516
pirating 634
rock crevices or ledges 31, 143, 257, 261, 279, 291
scrapes 31, 35, 48–49, 71, 215, 225, 243, 249, 251, *251*, 253, *253*, 254, *270*
sealed 388, *389*
tailorbirds 568
in termite mounds 304, 382
treecreepers 540
tree holes 31
tunnels and burrows 250, 328, *329*, 373, 378, 401, 405
waterfowl *142*, 143, *145*
weavers 516, 592, 594, **596–97**
woodpeckers 31, 143, 408, 411, 412
nest-showing 411
Netta erythrophthalma 140
Nettapus coromandelianus 140
Nicator chloris 563, 564
nicator, Yellow-spotted *see Nicator chloris*
niche, ecological, adaptive radiation to fill 514, **630**, *630*
nictitating membrane *520*
nidicolous young *31*, 282
nidifugous young *31*
nighthawk
Common *see Chordeiles minor*
Nacunda *see Podager nacunda*
Short-tailed *see Lurocalis semitorquatus*
nightingale 31, 522, 526
Common *see Luscinia megarhynchos*
nightingale-thrush *see Catharus*
nightjar 19, 25, **336–41**
Band-winged *see Caprimulgus longirostris*
Brown *see Veles binotatus*
European *see Caprimulgus europaeus*
Fiery-necked *see Caprimulgus pectoralis*
Itombwe *see Caprimulgus prigoginei*
Little *see Caprimulgus parvulus*
Long-tailed *336*
Long-trained *see Macropsalis forcipata*
lunar cycle **340–41**
Nechisar 339
nocturnal lifestyle 336, **340–41**
owlet *see* owlet-nightjar
Pennant-winged *see Macrodipteryx vexillarius*
Puerto Rican *see Caprimulgus noctitherus*
Rufous *see Caprimulgus rufus*
Rufous-cheeked *see Caprimulgus rufigena*
Satanic eared *see Eurostopodus diabolicus*
Savanna *see Caprimulgus affinis*
Scrub *see Caprimulgus anthonyi*
Sooty *see Caprimulgus saturatus*
Standard-winged *see Macrodipteryx longipennis*
torpidity 347
Vaurie's *see Caprimulgus centralasicus*
vocalizations **339**
Nigritia 606
canicapilla 606, 607
Nilaus afer 512
Ninox 324
boobook 324, 330, 331
connivens *324*
natalis 331
scutulata 324, 331
Nipponia nippon 112, 113, 114, 115
nocturnal species
frogmouths 342
kakapo 304
night heron 99, 104
nightjars 336–41
oilbird 345
owls 322, 324
potoos 344
senses 30
noddy **264–69**
Black or Lesser *see Anous tenuirostris*
Blue-gray *see Procelsterna cerulea*
Brown *see Anous stolidus*
Grey 269
nomadic species 177, 286, 290, 300, 329, 467, 492, 531, 532, 584, 613–14
emu 41, 42, **43**
non-native species, introduction 25, 46, 56, 76, 211, 331, 417, 473, 566, 618
Nonnula amaurocephala 403, 405
Northiella haematogaster 305
Notharcus 403, 404
Nothocercus bonapartei 49
Nothocrax urumutum 196, 197
Nothura maculata 49
Notiomystis cincta 466, 468, 469
Nucifraga
caryocatactes *480*, 482, 484, 485
columbiana 482, 484, 485
nukupu'u *see Hemignathus lucidus*
Numenius
americanus 232, *234*, 235, 236
arquata *232*, 235
borealis 232, 235, 237
tahitiensis 234, 235
Numida meleagris **190**, 190, 191
Numididae **190–91**
nunbird **403–05**
Black-fronted *404*
White-faced *see Hapaloptila castanea*
White-fronted *see Monasa morphoeus*
nunlet **403–05**
Chestnut-headed *see Nonnula amaurocephala*
nutcracker (*Nucifraga*) 480–85
nuthatch (*Sitta*) 411, 447, 479, **536–37**
Nyctea scandiaca 322, 327, 328, *328*, 331, 334
Nyctibiidae **344**
Nyctibius
aethereus 344
bracteatus 344
grandis 344, *344*
griseus 344, *344*
jamaicensis 344
leucopterus 344
maculosus 344
Nycticorax 99
caledonicus 99, 104
nycticorax 99, 101, *102*, 104
Nycticryphes semicollaris 246–47, *246*
Nyctinassa violacea 98, 104
Nyctiperdix 285, 287
Nyctyornis
amictus 375
athertoni *364*, 375
Nymphicinae 305
Nymphicus hollandicus 305
Nystalus 403, 404
maculatus *403*

O

O'ahu 'alauahio *see Paroreomyza maculata*
O'ahu 'amakihi *see Hemignathus flavus*
Oceanites oceanicus 74, *74*, 75
Oceanodroma
castro 75
hornbyi 74, 75
leucorhoa 74, 75, *75*
markhami 75
tethys 75
Ochotona 591
Ocreatus underwoodii 356
Ocyceros
birostris 384, 386
griseus 384, 386
Odontophoridae 180
Odontophorus strophium 176
Oedistoma 585
pygmaeum 585
Oena capensis 288, 289, 290
Oenanthe 522, 524, 526, 527
deserti 524, 526
isabellina 524, 526
oenanthe 524, *524*
Oeophasis derbianus 196, *197*
Ognorhynchus icterotis 305, 306
oilbird *see Steatornis caripensis*
oliveback (*Nesocharis*) 606
oloma'o *see Myadestes lamaiensis*
Oncostoma 424, 428
Onychognathus
tenuirostris 531, 532
tristramii 532
Onychorhynchus coronatus 425, 426, *427*, 428
oo (*Moho*) 466–69
opercula 254, 451, 588
Ophrysia superciliosa 176, 180
Opisthocomidae **318–19**
Opisthocomus hoatzin 19, **318–19**
Oporonis formosus *626*
Oreocharis 585
arfaki 585
Oreoica gutturalis 477
Oreomystis 616
bairdi 616
mana 616
Oreopholus ruficollis 225, 227
Oreopsar bolivianus 634
Oreortyx picta *179*
Oreostruthus fuliginosus 605, 607
Oreotrochilus estella 356
organochlorines 160
oriole (*Oriolus, Icterus*)
New World **632–35**
Old World **496–97**
Oriolidae **496–97**, 498
Oriolus
auratus 496
chlorocephalus 496
crassirostris 496, 497
hosii 496
isabellae 496, 497
larvatus 496, *496*
mellianus 496, 497
oriolus 496, 497, *497*
traillii 496
ornithophilous plants 354
Oroaetus isidori 172
oropendola (*Gymnostinops, Psarocolius*) *33*, 210, 632, 634–35
Ortalis 196
vetula 196
Orthonychidae **474–75**
Orthonyx
spaldingii 474–75, *475*
temminckii 474–75, *475*
Orthorhynchus cristatus 356
Orthotomus 572
moreaui 569, 572
Ortygospiza 605
atricollis 605, 607
Ortyxelos meiffrenii 218
Oryzoborus crassirostris 629, 631
osprey *see Pandion haliaetus*
Osteodontornis orri 21
ostrich (*Struthio*) 34–37
Otididae **214–17**
Otis
kori *377*
tarda 21, *30*, 214, *214*, 215, 216, *216*, 217
Otus 324
asio 331
ireneae 331, *331*
kennicottii 331
leucotis *324*
scops 329, 331
siaoensis 331
'o'u *see Psittirostra psittacea*
ovenbird 155, **438–41**
Neotropical 442, 446, 448
parulid *see Seiurus aurocapillus*
owl 19, 27, 31, 166, **322–35**, 412
African bay *see Phodilus prigoginei*
Barking *see Ninox connivens*
Barn *see Tyto alba*
barn 331, **332–33**
Barred *see Strix varia*
bay **332–33**
Boreal *see Aegolius funereus*
Brown fish *see Ketupa zeylonensis*
Brown hawk *see Ninox scutulata*
Burrowing *see Speotyto cunicularia*
Christmas Island hawk *see Ninox natalis*
Common bay *see Phodilus badius*
Common scops *see Otus scops*
conservation status 322, 331, 333
Crested *see Lophostrix cristata*
diet 322, **326**, 326–28, 332–33, 334
distribution 324
diurnal species 330, 334
eagle *see Bubo bubo*
ears and hearing 332, **334–35**
"ear tufts" 322
Eastern screech-owl *see Otus asio*
Elf *see Micrathene whitneyi*
eyes and vision 322, **334–35**
Ferruginous pygmy *see Glaucidium brasilianum*
fishing 324, 327, 335
Great grey *see Strix nebulosa*
Great horned *see Bubo virginianus*
hawk 324, 330, 331
Itombwe *see Phodilus prigoginei*
Laughing *see Sceloglaux albifacies*
Least pygmy *see Glaucidium minutissimum*
Little *see Athene noctua*
Long-eared *see Asio otus*
Madagascar grass (Madagascar red) *see Tyto soumagnei*
Malaysian eagle *see Bubo sumatranus*
Manus masked *see Tyto manusi*
Minahasa masked *see Tyto inexspectata*
nocturnal species 322, 324, 330, 334
Northern hawk *see Surnia ulula*
pellets 27, **326**, 327
Pel's fishing *see Scotopelia peli*
perch-hunting 326
pygmy *see Glaucidium passerinum*
Saw-whet *see Aegolius acadicus*
scops 324, 331
screech 324, 330
Short-eared *see Asio flammeus*
Siau scops *see Otus siaoensis*
Snowy *see Nyctea scandiaca*
Sokoke scops *see Otus ireneae*
Southern boobook *see Ninox boobook*
Spectacled *see Pulsatrix perspicillata*
Spotted *see Strix occidentalis*
Spotted wood *see Strix seloputo*

Sula barn *see Tyto nigrobrunnea*
Tawny *see Strix aluco*
Tengmalm's *see Aegolius funereus*
vision 29, 30
vocalization 330, 333
Western screech-owl *see Otus kennicottii*
White-faced scops *see Otus leucotis*
owlet, Forest *see Athene blewitti*
owlet-nightjar (*Aegotheles*) **343**
oxpecker 530
Red-billed *see Buphagus erythrorhynchus*
oxylabes, Yellow-browed *see Crossleyia xanthophtys*
Oxylophus jacobinus 576
Oxypogon guerinii 356
Oxyruncus cristatus 22, 430, 433
Oxyura
ferruginea 140
jamaicensis 138, 140
leucocephala 140
Oxyurini 134, 140
oystercatcher (*Haematopus*) 228, *231*, **248–49**

P

Pachycare 477
Pachycephala 477
jacquinoti 477
pectoralis 477, *477*
rufiventris 477
rufogularis 477
simplex 477
Pachycephalidae **477**
Pachyphantes superciliosus 592, 594
Pachyramphus 433
aglaiae 430, 433
Padda
fuscata 607
oryzivora 607
Pagodroma nivea 70, 71
Pagophila eburnea 259, *259*, 260, 261, *261*, 263
painted snipe 232, 244, **246–47**
Greater *see Rostratula benghalensis*
South American *see Nycticryphes semicollaris*
Palaeognathae 19, 48
palila *see Loxioides bailleui*
palmchat *see Dulus dominicus*
Palmeria dolei 617
Pandion haliaetus 150–51, *150*, *151*, **152–53**, 160
Pandionidae **150–53**
Panterpe insignis 356
Panurus biarmicus 575, 576, *576*
Panyptila 346
cayennensis 348
Papasula abbotti 82, 83, 85, 86, 87
papilla 25
Parabuteo unicinctus 166, 172
Paradigalla carunculata 488
Paradisaea 490
guilielmi 489
raggiana 488, 491, *491*
rudolphi 488, *489*, 491
Paradisaeidae **488–91**
paradise-crow *see Lycocorax pyrrhopterus*
paradise-flycatcher (*Eutrichomyias, Terpsiphone*) **504–05**
parakeet
Antipodes *see Cyanoramphus unicolor*
Austral 300, 305
Blue-throated *see Pyrrhura cruentata*
Brown-throated *see Aratinga pertinax*
Burrowing *see Cyanoliseus patagonus*
Carolina *see Conuropsis carolinensis*
Derbyan *see Psittacula derbiana*
Golden *see Guaruba guarouba*
Grey-hooded *see Bolborhynchus aymara*
Mauritius *see Psittacula echo*
Monk *see Myiopsitta monachus*
Plum-headed *see Psittacula cyanocephala*
Rose-ringed *see Psittacula krameri*
Slaty-headed *see Psittacula himalayana*
Paramythia 585
montium 584, *584*, 585
Paramythiidae 585
paraseptal processes 76
parasites 65, 82, 438, 560
Paridae **550–55**, 550
Paris, Matthew 614
Parmoptila woodhousei 607
Paroreomyza
flammea 616
maculata 616
montana 616
parotia (*Parotia*) 490
Parotia
lawesii 488
wahnesi 488, 491
parrot 19, 31, **296–307**
African grey *see Psittacus erithacus*
Black *see Coracopsis nigra*
Blue-headed *see Pionus menstruus*
Blue-winged *see Neophema chrysostoma*
coloration and plumage 30, 296, 299, 300, *304*
communication **300**
Eclectus *see Eclectus roratus*
Fuerte's *see Hapalopsittaca fuertesi*
Golden-shouldered *see Psephotus chrysopterygius*
Great-billed *see Tanygnathus megalorynchos*
Ground *see Pezoporus wallicus*
Hispaniolan *see Amazona ventralis*
Imperial *see Amazona imperialis*
kakapo **310–11**
Night *see Geopsittacus occidentalis*
Owl *see* kakapo
Puerto Rican *see Amazona vittata*
Red-capped *see Purpureicepahlus spurius*
Red-rumped *see Psephotus haematonotus*
Red-tailed *see Amazona brasiliensis*
Rock *see Neophema petrophila*
St. Vincent *see Amazona guildingii*
Swift *see Lathamus discolor*
vocalization 296, 299, **300**, 305
Yellow-eared *see Ognorhynchus icterotis*
Yellow-fronted *see Poicephalus flavifrons*
Yellow-naped amazon *see Amazona auropalliata*
parrotbill 574, 576
Bearded *see Panurus biarmicus*
Maui *see Pseudonestor xanthophrys*
parrotfinch (*Erythrura*) 605
partridge **176–81**
Grey *see Perdix perdix*
Red-legged *see Alectoris rufa*
Stone *see Ptilopachus petrosus*
Udzungwa forest *see Xenoperdix udzungwensis*
Parula americana 625, 626
Parulidae 570, **624–27**
Parus 552
ater 550, 551, *553*
atricapillus 550, *550*, 552
bicolor 550, **552**
caeruleus 550, *550*, 553
cinctus 550, 552, *553*
cristatus 550, *552*
elegans 541
inornatus 550, 552
leucomelas 550, 552
major 412, 540, 550, 551, 552, **554–55**
montanus 550, 552
nuchalis 550
palustris 550, 551–52
ridgwayi 550, 552
rufiventris 550
spilonotus 550
wollweberi 550, *550*, 552
Passer
domesticus 314, 590, 591, *591*
melanurus 591
montanus 590, 591
simplex 590, 591
Passerculus sandwichensis 620, 621
Passeridae **590–91**
Passeriformes *19*, **416–635**
Passerina 622
amoena 622, 623
ciris 622, 623
cyanea 618, 622, 623
leclancherii 622, 623
rositae 623
passerine *324*
Pastor roseus 530, 531
Patagona gigas 353, 356, *356*, *360*
Pauxi
pauxi 196, 197
unicornis 196, 197
Pavo
cristatus 25, 31, 176, 177, *179*, 180, **182–83**
muticus 176, 180
peacock *see Pavo cristatus*
peacock-pheasant, Malaysian *see Polyplectron malacense*
peafowl (*Afropavo, Pavo*) 178
pecten *334*
Pedionomidae **255**
Pedionomus torquatus 232, 247, 255, *255*
peep *236*
pelagic species 259, 264
Pelagodroma marina 74, *74*, 75
Pelargopsis capensis 367, 371
Pelecanidae **78–81**
Pelecaniformes 19, **78–97**
Pelecanoides
garnotti 76, 77
georgicus 76, *76*, 77, *77*
magellani 76, 77
urinatrix 76, *76*, 77, *77*
Pelecanoididae **76–77**
Pelecanus
conspicillatus 78–79, *78*, 80
crispus 78, 79, 80, 81
erythrorhynchos 78–79, 80
occidentalis 78–79, 80, *80*
onocrotalus 29, 78, *78*, 79, 80
philippensis 78, 79, 80, 81
rufescens 78, 79, 81, *81*
pelican (*Pelecanus*) 19, 21, 26, **78–81**, 160, 260
Pellorneum ruficeps 575
Peltohyas australis 252, *252*, 253
Peltops
blainvillii 494
montanus 494
Penelope 197
albipennis 196, 197
ochrogaster 196, *196*
purpurascens 197
Penelopides 384, 388
exarhatus 384, 386
Penelopina nigra 196, 197
Peneothello cyanus 472, 473
penguin 19, 25, 26, **50–59**, *256*, 257, 271, *272*
Adélie *see Pygoscelis adeliae*
burrows 143
Chinstrap *see Pygoscelis antarctica*
diving and swimming **50–51**, 52
Emperor *see Aptenodytes forsteri*
Erect crested *see Eudyptes sclateri*
Fiordland *see Eudyptes pachyrhynchus*
Galápagos *see Spheniscus mendiculus*
Gentoo *see Pygoscelis papua*
Humboldt *see Spheniscus humboldti*
Jackass *see Spheniscus demersus*
King *see Aptenodytes patagonicus*
Little *see Eudyptula minor*
Macaroni *see Eudyptes chrysolophus*
Magellanic *see Spheniscus magellanicus*
migration 53
Rockhopper *see Eudyptes chrysocome*
Royal *see Eudyptes schlegli*
Snares crested *see Eudyptes robustus*
Yellow-eyed *see Megadyptes antipodes*
penis, false 594
Pepperberg, Irene 300
pepper-shrike 478, 479
Black-billed *see Cyclarhis nigriostris*
Rufous-browed *see Cyclarhis gujanensis*
perch-and-pounce insectivores 472
perch-gleaning 424
Perdicinae 180
Perdicula manipurensis 176, 180
Perdix perdix 29, 177, *179*, 180, 181
'perdiz cordillerana' *see Attagis gayi*
Pericrocotus 498, 499
cinnamomeus 499, 500
divaricatus 498, 499
flammeus 498, 499, 500
thologus 499
periophthalmic ring *608*
Perisoreus
canadensis 482, 484
infaustus 483, 484
Perissocephalus tricolor 429, 430, 433
Pernis apivorus 163, *168*, 172
pesticides 153, 159, **160–61**, 170, *171*, 177, *618*
petrel 19, 30, 31, **70–73**, 271
Beck's *see Pseudobulweria becki*
Bermuda *see Pterodroma cahow*
Blue *see Halobaena caerulea*
burrows 257
Chatham *see Pterodroma axillaris*
diving *see* diving petrel
gadfly 70, 72
Hall's *see Macronectes halli*
MacGillivray's *see Pseudobulweria macgillivrayi*
Magenta *see Pterodroma magentae*
Northern giant *see Macronectes halli*
Snow *see Pagodroma nivea*
Southern giant *see Macronectes giganteus*
storm *see* storm petrel
Petroica 473
australis 472
bivittata 472, 473
phoenicea 472
traversi 472, 473, *473*
Petroicidae 472
Petronia 590–91
petronia 590, 591
superciliaris 590, 591
pewee 428
Pezoporus wallicus 296, 301, 304, 305
Phacellodomus 440
rufifrons 440
Phaenostictus
mcleannani 448, 449, *449*
Phaethon
aethereus 88, 89, *89*
lepturus 88, 89, *89*
rubricauda 88, 89
Phaethontidae **88–89**
Phaethornis 32, 359
eurynome 356
guy 356
ruber 353, *356*
Phaethornithinae 352, **356**
Phaetusa simplex 264
Phainopepla nitens 519
Phainoptila melanoxantha 519
Phalacrocoracidae **90–93**
Phalacrocorax
africanus 90
aristotelis 91
atriceps verrucosus 93
auritus 90, 91, *91*
bougainvillii 82, 91, 92
capensis 91
carbo 90, 91, *91*, *92*, *93*
harrisi 90, 91, 92
olivaceus 91
onslowi 91
pelagicus 91
penicillatus 91
perspicillatus 91
punctatus 90, 91
pygmaeus 90, 91, *92*
urile 91
Phalaenoptilus nuttallii 337, 339, 340–41
phalarope (*Phalaropus*) 26, **240–41**
Phalaropus
fulicarius 240–41, *241*
lobatus 240–41, *240*
tricolor 240–41, *241*
Phalcoboenus
albogularis 159
australis 154, 159
carunculatus 159
megalopterus 159
Phaps 289
Pharomachrus mocinno 362, 363, *363*
Phasianidae **176–83**
Phasianinae 180
Phasianus colchicus 176, *179*, 180, 181
pheasant **176–83**
Argus 31
Brown-eared *see Crossoptilon mantchuricum*
Bulwer's *see Lophura bulweri*
Cheer *see Catreus wallichii*
Common *see Phasianus colchicus*
Golden *see Chrysolophus pictus*
Kalij *see Lophura leucomelanos*
Lady Amherst's *see Chrysolophus amherstiae*
Ring-necked *see Phasianus colchicus*
Pheucticus
chrysopeplus 622, 623
ludovicianus 622, *622*, 623
melanocephalus 623
Phibalura 433
flavirostris 430
Philemon 466, 497
Philentoma 505
Philepitta
castanea 421, 427
schlegeli 421
Philepittidae **421**
Philesturnus carunculatus 495, *495*
Philetairus socius 155, 516, *593*, 594
Philomachus pugnax 235, 236, **238–39**
philopatric species 267
Philydor novaesi 439, 440
Phimosus infuscatus 115
Phleocryptes melanops 440
Phodilus
badius 332, *332*, 333
prigoginei 332, 333
Phoebastria 66
albatrus 67, 69
immutabilis 67, 68, 69
irrorata 66–67
nigripes 67, 68
Phoebetria 66
fusca 67, 68, 69
palpebrata 67, *67*, 68
Phoenicircus carnifex 433
Phoeniconaias minor 124–29
Phoenicoparrus
andinus 124, *124*, 126, *128*, 129
jamesi 124, *124*, 126, 129
Phoenicopteridae **124–29**
Phoenicopteriformes 124
Phoenicopterus
chilensis 124, *124*, 126, 128, 129
ruber 29, 124, *124*, 126, *126*, 127, 128, *128*, 129, 292
Phoeniculidae **383**
Phoeniculus
bollei 383
purpureus 383, *383*
Phoenicurus phoenicurus 527
Pholidornis rushiae 548
Phylidonyris 467
Phylidorinae 440
Phyllastrephus 563, 564
debilis 563
leucolepis 563
scandens 563
terrestris 563, 565
Phylloscartes 424, 428
beckeri 427, 428
ceciliae 427, 428
roquettei 427, 428
Phylloscopus 568
borealis 570, 572
collybita 572
sibilatrix 338, *570*, **624–27**
trochilus 568, 570, 572
Phytotoma raimondii 433
Phytotominae 429, 433
Pica pica 481, 484, *485*
Picathartes
gymnocephalus 515
oreas 515, *515*
Picathartidae **515**
Picidae **408–15**
Piciformes 19, **390–415**
Picinae 413
pico de lacre *404*
Picoides 411
arcticus 413
borealis 411
pubescens 410–11, *414*
tridactylus 408, *409*, 410, 413, 414
villosus 414
piculet (*Nesoctites, Picumnus, Sasia*) **408–15**
Picumninae 412, 413
Picumnus 412
cirratus 412
olivaceus 412, 413
squamulatus 408, 413
steindachneri 413
Picumnuscirratus 412
Picus
canus 411, 413
viridis 408, *409*, 410, 411, 413
pigeon 19, 27, 31, *216*, **288–95**, 355
African green *see Treron calva*
breathing rate 355
crowned *see Goura*
Feral *see Columba livia*
fruit-eating 288, 289, 290
Giant *see Ducula goliath*
imperial *see Ducula*
milk **292**
navigation and homing 288, **294–95**
Nicobar *see Caloenas nicobarica*
Notu *see Ducula goliath*
Passenger *see Ectopistes migratorius*
Pink *see Columba mayeri*
Purple-crowned *see Ptilinopus superbus*
Racing *see Columba livia*
Spinifex *see Geophaps plumifera*
Tooth-billed *see Didunculus strigirostris*
Victorian crowned *see Goura victoria*
Wood *see Columba palumbus*
piha
Scimitar-winged *see Lipaugus uropygialis*
Screaming *see Lipaugus vociferans*
Pilherodius pileatus 102, 104
Pinaroloxias 631

Pinguinis impennis 25, 26, 276, 277, 283
Pinicola enucleator 611, *612*, 615
pintail
  Kerguelen *see Anas eatoni*
  Northern *see Anas acuta*
Pionus menstruus 303
piopio *see Turnagra capensis*
Pipile 197
  jacutinga 196, 197
  pipile 196, 197, *197*
Pipilo 620
  erythrohthalmus 619
pipit (*Anthus, Tmetothylacus*) **598–601**
Pipra 434, 437
  filicauda 434, *434*, 437
  pira 436
Pipreola 430, 433
  arcuata 430
Pipridae 421, 429, **434–37**
Piprites pileatus 434
piracy *see* kleptoparasitism
Piranga
  ludoviciana 629, 631
  olivacea 19, 629, 631
  rubra 628, 629, 631
Pitangus sulphuratus 427, *427*, 428
Pithecophaga jefferyi 163, 170, 172, 173
Pithys albifrons 447, 448, 449
pitohui 477
  White-bellied *see Pitohui incertus*
Pitohui 477
  incertus 477
pitta (*Pitta*) **418–20**
Pitta
  angolensis 418
  arquata 418
  baudii 418
  brachyura 418, *419*
  caerulea 418
  cyanea 418
  erythrogaster 420
  guajana 418, 419, *419*
  gurneyi 418, *419*, 420
  iris 420
  kochi 418
  moluccensis 418
  nympha 418, 420
  oatesi 418
  phayrei 418
  sordida 418, 419
  soror 418
  steerii 418
  venusta 418
  versicolor 418
Pittasoma 447, 448
  michleri 448
Pittidae **418–20**
Pitylus grossus 623
Pityriasinae 508
Pityriasis gymnocephala 494, 508
Plains wanderer *see Pedionomus torquatus*
plantain-eater *see* turaco
plantcutter 429–33
  Peruvian *see Phytotoma raimondii*
Platalea
  alba 114, 115
  flavipes 114, 115
  leucorodia 29, 112, *112*, 115, *115*
  minor 113, 114, 115
  regia 112, 115
Plataleinae 115
Platycercus
  elegans 296, *298*, 305
  venustus 304
Platyrinchus 424, *425*, 428
  mystaceus 425
Platysteira
  lacticincta 514
  peltata 514, *514*
play 394
Plectrophenax nivalis 620, *620*
Plectropterini 136, 140
Plectropterus gambensis 136, 140
Plegadis 112
  chihi 115, *115*
  falcinellus 112, 114, 115, *115*
  ridgwayi 112, 114, 115, *116*
Ploceidae 590, **592–97**
Ploceinae 594
Plocepasser
  donaldsoni 594
  mahali 594
  rufoscapulatus 594
  superciliosus 594
Plocepasserinae 594
Ploceus 592
  alienus 594, 595
  batesi 594, 595
  bicolor 594
  bojeri 593
  capensis 592, *592*, 594
  cucullatus 594, *594*, 595
  galbula 594, 595
  golandi 594, 595
  grandis 594, 595
  heuglini 594, 595
  intermedius 592, 594
  luteolus 592, 594
  manyar 594, 595
  megarhynchus 594, 595
  melanocephalus 594
  nigerrimus 594
  nigrimentum 594, 595
  ocularis 592, 594
  olivaceiceps 592, 594, 595
  philippinus 594, 595
  preussi 594, 595
  rubiginosus 594, 595
  velatus **596–97**
  xanthops 592, 594
plover **224–29**
  American golden *see Pluvialis domestica*
  Black-bellied *see Pluvialis squatarola*
  Blacksmith *see Vanellus armatus*
  Brown-chested 227
  Crab *see Dromas ardeola*
  Double-banded *see Charadrius bicinctus*
  Egyptian *see Pluvianus aegyptius*
  Eurasian golden *see Pluvialis apricaria*
  Grey *see Pluvialis squatarola*
  Grey-headed 227
  Hooded *see Thinornis rubricollis*
  Kentish *see Charadrius alexandrinus*
  Little ringed *see Charadrius dubius*
  Malasian *see Charadrius peroni*
  migration 225, 226–28, 229, **236**
  Pied 225
  Quail *see Ortyxelos meiffrenii*
  Red-capped *see Charadrius ruficapillus*
  St. Helena *see Charadrius santahelenae*
  Semipalmated *see Charadrius semipalmatus*
  Shore *see Thinornis novaeseelandiae*
  Snowy *see Charadrius alexandrinus nivosus*
  Sociable 227
  Spur-winged 227, *228*
  Wattled *see Vanellus senegallus*
  White-tailed 227
plovercrest *see Stephanoxis lalandi*
plumage **22–23**, 24–26
  breeding 225
  color morphs 65, 90, 595, 604, *605*
  contour feathers 23, *23*, 25
  dimorphic 270
  down *23*, 25, 140, *145*, 284
  eclipse 26, 586
  erectile 498
  evolution 18
  filoplumes 23, *23*
  fluorescent *304*
  goose down 140, *145*
  hummingbirds 352, *352*, 353, 359, *359*
  iridescence 25, 320, 352, *352*
  mimicry by different species 497, *563*
  molting 23, 25, 276–77
  sandgrouse water-carrying mechanism **287**
  sound production using 430, 436, *437*
  thermoregulation 18, 23
  tyrant flycatcher **426**
  waterbirds 240
  waterproofing 23, 64, 82, 88, 214, *520*
  winter coat 26, *184*
  see also coloration
plumeleteer *see Chalybura*
plunge-diving **266**
  kingfishers 367, 369, *369*
plushcap *see Catamblyrhynchus diadema*
Pluvialis
  apricaria 227, *227*, 228
  domestica 24
  squatarola 226, 227, 228, 229
Pluvianus aegyptius 252, *252*, 253
Pnoepyga 575
pochard 132, 139, 140, 142
  Madagascar 133
  South American *see Netta erythrophthalma*
Podager nacunda 338
Podargidae **342**
Podargus 342
  ocellatus 342
  papuensis 342
  strigoides 342, *342*
Podica senegalensis 223, *223*
Podiceps
  andinus 63
  auritus 62, 63, *63*
  cristatus 29, 62, 63, 65
  gallardoi 63, 65
  grisegena 63, *63*, 64
  nigricollis 63, 65
  occipitalis 63, 64, 65
  taczanowskii 63, 64, 65
Podicipedidae **62–65**
Podicipediformes 19, **62–65**
Podilymbus
  gigas 63
  podiceps 62, 63
Podoces biddulphi 484
Poecilotriccus 428
  albifacies 424, 428
Poephila 608
  acuticauda 605, 607
  personata 605, 607
Pogoniulus 398, 399
  chrysoconus 399, 401
  pusillis 401
Poicephalus flavifrons 300, 305
Polemaetus bellicosus 170, 172, 173
Polihierax 155, 159
  insignis 159
  semitorquatus 155, *157*, 159, 594
Poliocephalus
  podiocephalus 63, 65
  rufipectus 63, 65
Polioptila
  caerulea 546
  dumicola 546, *547*
  lactea 546, 547
  plumbea 546
Poliotilinidae **546–47**
pollinators 530, 586, 588
  coevolution 466
  honeyeaters 466, **467**
  hummingbirds 352, 354
  ornithophilous plants 354
pollution 160, 170, 173
Polo, Marco 21
polyandry 166, 210, 603, 627
  Brown skua 272
  buttonquail 218
  Eurasian dotterel 229
  Greater painted snipe 247
  jacanas 245
  phalaropes 241
Polyborinae 154–55, 159
Polyboroides
  radiatus 166, *168*, 172
  typus 165, 172
Polyborus 159
  lutosus 155, 159
  plancus 154, *157*, 159
polychlorinated biphenyls (PCBs) 160, 170
Polyerata 359
  amabilis 356
polygamy *see* polyandry; polygynandry; polygyny
polygynandry 603
polygyny 31, 166, 178, 180, 194, 210, 304, 421, 430, 488, 489, 490, 529, 542, 544–45, 589, 603, 630, 634
  bowerbirds 458, 459, 461
  cotingas 430
  and diet 430
  Eurasian bittern 102
  grouse 184, *185*, 186–87, *186*
  lekking *see* lek-breeding species
  Long-trained nightjar 336
  weavers 592
Polyplectron malacense 180
Polytmus 356
Pomarea
  nigra 504
  pomarea 504
Pomatorhinus 575
  erythrogenys 577
Pomatostomidae **476**
Pomatostomus 476
  halli 476
  isidori 476
  ruficeps 476
  superciliosus 476, *476*
  temporalis 476
poorwill
  Common *see Phalaenoptilus nuttallii*
  hibernation 339, 347
  Jamaican *see Siphonorhis americana*
po'o-uli *see Melamprosops phaeosoma*
Porphyrio
  mantelli 206, 207, *208*, 209
  porphyrio 207, 209, 210
Porphyrula martinica 206, 207, 208, 209, 211
Porzana porzana 208
potoo (*Nyctibius*) **344**
powderdown patches 99, 381
prairie chicken (*Tympanuchus*) 27, 176
pratincole (*Glareola, Stiltia*) 251, **252–53**
preen gland 82, 214, *520*
preening, mutual *304*
Prinia
  cinereocapilla 572
  subflava 572
prion 70, 71, 72
  Broad-billed 70
Prionochilus 584
  maculatus 584
Prionodura newtoniana 458, 459, 460, 461, *463*
Prionopidae 514
Prionops 512
  alberti 511
  caniceps 511
  gabela 511
  plumatus 511, *511*
  poliolophus 511
  retzii 511
  rufiventris 511
  scopifrons 511
Priotelus temnurus 362
Probosciger aterrimus 300, 305
Procellariidae **70–73**
Procellariiformes 19, **66–77**
Procelsterna cerulea 264
Procnias 433
  nudicollis 432
Prodotiscus 396
Progne 558
  subis 558, 560
prolactin 127, 292
Promerops 466
  cafer 586, 587, 588, *588*
  gurneyi 586, 587
Prosobonia cancellata 232, 235
Protarchaeopteryx robusta 18
Protonotaria citrea 625, 627
proventriculus 22, *22*
Prunella
  collaris 602, 603
  fagani 602
  himalayana 602
  immculata 602, *602*
  koslowi 602
  modularis 602, *602*, 603
  montanella 602
  rubeculoides 602, *603*
  rubida 602
Prunellidae **602–03**
Psalidoprocne 558
Psaltiparus
  minimus 557, *557*
  minimus melanotis 557, *557*
Psaltria exilis 556
Psarisomus dalhousiae 422, 423
Psarocolius decumanus 33
Psephotus
  chrysopterygius 304, 305, *306*
  haematonotus 296, 305
Pseudibis
  davisoni 112, 113, 115
  gigantia 112, 113, 115, 117
  papillosa 112, 114, 115, 116, 117
pseudo-babbler **476–77**
Pseudobulweria
  becki 71, 72
  macgillivrayi 71, 72
Pseudocalyptomena graueri 421, 422, 423
Pseudochelidon 558
  sirintarae 558, **561**, 561
Pseudocolopteryx acutipennis 426
Pseudonestor xanthophrys 617
Pseudonigrita arnaudi 594
Pseudopodoces humilis 482, 550, 551
Pseudoseisura
  cristata 440
  gutturalis 440
Psittacidae **296–307**
Psittaciformes 19, **296–307**
Psittacinae 305
Psittacula
  cyanocephala 305
  derbiana 300, 305
  echo 305, 306
  himalayana 300, 305
  krameri 301, 305
Psittaculirostris desmarestii 305
Psittacus erithacus 300, 305
Psitteuteles versicolor 303, 305
Psittirostra psittacea 616, *617*
Psophia
  crepitans 213, *213*
  leucoptera 213
  viridis 213, *213*
Psophiidae **213**
Psophodes 475
  nigrogularis 475
ptarmigan (*Lagopus*) 26, 158, 184, 185, 186
Pteridophora alberti 488
Pterocles
  alchata 285, 286, 287
  bicinctus 285
  burchelli 285
  coronatus 285, 287
  decoratus 285
  exustus 285, 286
  gutturalis 285, 286, *286*
  indicus 285, *285*
  lichtensteinii 285
  namaqua 284, 285, *285*, 286, 287, *287*
  orientalis 285, 286, 287
  personatus 285
  quadricinctus 285
  senegallus 285, 287
Pteroclididae **284–87**
Pteroclidiformes 19, **284–87**
Pterocnemia pennata 38–40, *39*, *40*
Pterodroma
  axillaris 71
  cahow 71, *71*, 72
  magentae 71, 72
Pteroglossus 390
  beauharnaesii 390, 392
  pluricinctus 395
  torquatus 390, *390*, 395
Pteronetta hartlaubii 140
Pterophanes 353
Pteropodocys maxima 498, 499, 500
Pteroptochos 451
Pteroptochus megapodius 451
Pteruthius 575
Ptilinopus
  magnificus 290
  superbus 289
  victor 288, 289
Ptilocichla 541, 575
Ptilognatini 519
Ptilogonys
  caudatus 519, *519*
  cinereus 519
Ptilonorhynchidae **458–63**
Ptilonorhynchus 460
  violaceus 458, 459, 460, 461, *462*
Ptilopachus petrosus 177, 180
Ptiloris
  magnificus 488, 489, 490
  paradiseus 488
  victoriae 488, 490
Ptilorrhoa 474–75
Ptychorampus aleuticus 276, 277, 278
puffback
  Northern *see Dryoscopus gambensis*
  Sabine's *see Dryoscopus sabini*
puffbird **403–05**
  Collared *see Bucco capensis*
  Russet-throated *see Hypnelus ruficollis*
  Sooty-capped *see Bucco noanamae*
  Spot-backed *see Nystalus maculatus*
  Swallow-wing *see Chelidoptera tenebrosa*
  White-faced *see Hapaloptila castanea*
  White-whiskered *see Malacoptila panamensis*
puffin (*Fratercula*) 260, **276–81**
Puffinus
  auricularis 71
  gravis 71, *71*, 72
  griseus 71, 72, 302
  heinrothi 71, 72
  lherminieri 71
  puffinus 70, *71*
  tenuirostris 70, 71, 72
puffleg *see Eriocnemis*
pukeko *see Porphyrio porphyrio*
Pulchrapollia gracilis 299
Pulsatrix perspicillata 324
purpletuft 429–33
Purpureicepahlus spurius 298
Pycnonotidae **562–65**
Pycnonotus 562, 564, 565
  barbatus 563, *564*
  cafer 563, 565

goiavier 565
jocosus 563, *563*
melanicterus 563
nigricans 562
pygmy-parrot 295, 300, 302, 305
Buff-faced *see Micropsitta pusio*
pygmy-tyrant 426, 428
Short-tailed *see Myiornis ecaudatus*
Pygoscelis
adeliae 50, 51, 52, 53, *53*, 54, 55
antarctica 50, 51, 52, 53, 54, 55, 56, *256*
papua 50, 51, 52, 53, 54, 55, *55*
Pyrenestes 606
ostrinus 606, 607
Pyriglena atra 448, 449
Pyrocephalus rubinis 425, 427, *427*, 428
Pyroderus scutatus 430
Pyrrhocorax
graculus 482, 484
pyrrhocorax 482, 484
Pyrrhula pyrrhula 611, *611*, 613, 614, 615, 630, 631
pyrrhuloxia *see Cardinalis sinuata*
Pyrrhura 302
cruentata 302, 305
pytilia, Green-winged *see Pytilia melba*
Pytiliga 607
melba 604, 607

Q

quail **176–81**
Blue *see Coturnix adansonii*
Blue-breasted *see Coturnix chinensis*
Bobwhite 177, 179, 180
bustard *see* buttonquail
Common *see Coturnix coturnix*
Gambel's *see Callipepla gambelii*
Harlequin *see Coturnix delegorguei*
Himalayan *see Ophrysia superciliosa*
Japanese *see Coturnix japonica*
Mountain *see Oreortyx picta*
New World 177, 178–79, 180
New Zealand *see Coturnix novaezealandiae*
Ocellated *see Cyrtonyx ocellatus*
Old World 177, 180
Rain *see Coturnix coromandelica*
quailfinch 605
African *see Ortygospiza atricollis*
quail-thrush (*Cinclosoma*) **474–75**
quelea, Red-billed *see Quelea quelea*
Quelea quelea 365, 594, *594*, 595
Querula purpurata 432, 433
quetzal 363
Resplendent *see Pharomachrus mocinno*
Quiscalus
palustris 634, 635
quiscula 632, 634, *634*

R

rachis 25
racquet-tail
Booted *see Ocreatus underwoodii*
rail 19, 25, 26, **207–11**
Brown-banded *see Lewinia mirificus*
Californian clapper *see Rallus longirostris obsoletus*
flightless species 209, 211
Guam *see Rallus owstoni*
Invisible *see Habroptila wallacii*
King 209
New Caledonian *see Gallirallus lafresnayanus*
Nkulengu 208
Snoring *see Aramidopsis plateni*
Virginia *see Rallus limicola*
Water *see Rallus aquaticus*
White-throated 209
Yellow 210
rainbowbird *see Merops ornatus*
Rallidae **207–11**
Rallus
aquaticus 206, *206*, 207, *208*
limicola 206, 207, *208*
longirostris obsoletus 211
owstoni 207, 209, 211
Ramphastidae **390–95**
Ramphastos 392, 393, 395
ambiguus 390, 392
sulfuratus 390, 392, *392*
swainsonii 390
toco 29, 390, *390*, 395
tucanus 390, 392, 395
vitellinus 395
Ramphocaenus melanurus 546, *546*, 547
Ramphocoris clotbey 578
Ramphodon 353
naevius 356
Ramphomicron microrhynchum 352
Ramphotrigon 428
Ramsayornis 468
ratites 19, *20*, **37**, 48
raven (*Corvus*) 480
rayadito, Thorn-tailed *see Aphrastura spinicauda*
razorbill *see Alca torda*
Recurvirostra
americana 243
andina 243
avosetta 242, 243, *243*
novaehollandiae 243
Recurvirostridae **242–43**
redpoll *see Carduelis flammea*
redshank 236
Common *see Tringa totanus*
redstart 522
American *see Setophaga ruticilla*
Comon *see Phoenicurus phoenicurus*
water-redstart *see Rhyacornis*
reedling *see Panurus biarmicus*
Regulidae **573**
Regulus
calendula 573
goodfellowi 29, 573
ignicapillus **573**, *573*
regulus 573
satrapa 573, *573*
teneriffae 573
regurgitation 22, 27, 75, 90, 102–03, 262, 266, 320, 345, 386, 395, 401, 608, 630
Remiz
consobrinus 548
coronatus 548
pendulinus 548, *548*
Remizidae **548–49**, 550
reproductive behavior 31
reproductive system 27, 29
reptiles 18
respiratory system **22**, 26
Rhabdornis
grandis 541
inornatus 541, *541*
mystacalis 541
Rhabdornithidae **541**
Rhagologus 477
Rhamphocelus
carbo 631
dimidiatus 630, 631
passerinii 630, 631
rhea (*Rhea, Pterocnemia*) 19, 25, **38–40**
Rhegmatorhina 447
Rheidae **38–40**
Rheiformes 19, **38–40**
Rheinardia ocellata 180
Rhinocrypta lanceolata 451
Rhinocryptidae 446, **451**
Rhinopomastus
aterrimus 383
cyanomelas 383
minor 383
Rhinoptilus
bitorquatus 253
chalcopterus 253
cinctus 253
Rhipidura 541
albolimbata 501
areola 501
fulginosa 501, *501*
leucophrys 501
malaitae 501
rufifrons 501
semirubra 501
Rhipiduridae **501**
Rhodacanthis
flaviceps 616
palmeri 616
Rhodinocichla rosea 628, 630, 631
Rhodonessa caryophyllacea 140, 143
Rhodopechys sanguinea 611, 614
Rhodophoneus cruentus 512
Rhodostethia rosea 261, 263
Rhodothraupis celaeno 623
Rhyacornis 523, 526
Rhynchotus rufescens 48, 49
Rhynochetidae **220**
Rhynochetos jubatus 220, *220*
Rhytipterna 429
rictal bristles *see* bristles
riflebird (*Ptiloris*) 490
rifleman *see Acanthisitta chloris*
Rimator 575
Riparia riparia 31, 558, *559*, 591
Rissa
brevirostris 259
tridactyla 258, 259, *259*, 261, *261*, *262*, 263, 270, 271
roadrunner, Great *see Geococcyx californianus*
robin 526, 527
Alpine *see Petroica bivittata*
American *see Turdus migratorius*
Australo-Papuan **472–73**
Blue-grey *see Peneothello cyanus*
Chatham Islands *see Petroica traversi*
Eurasian *see Erithacus rubecula*
Flame *see Petroica phoenicea*
Hooded *see Melanodryas cucullata*
New Guinea 473
New Zealand *see Petroica australis*
Torrent *see Monachella muelleriana*
White-throated *see Irania gutturalis*
Yellow *see Eopaltria australis*
robin-chat *see Cossypha*
roc **21**
rockfowl (*Picathartes*) **515**
rockjumper (*Chaetops*) **515**
Rodostethia 259
roller 19, **380–81**
Abyssinian *see Coracias abyssinica*
broad-billed *see Erystomus*
Crossley's ground-roller *see Atelornis crossleyi*
Cuckoo *Leptosomus discolor*
European *see Coracias garrulus*
Indian *see Coracias benghalensis*
Lilac-breasted *see Coracias caudata*
Long-tailed ground-roller *see Uratelornis chimaera*
Pitta-like ground-roller *see Atelornis pittoides*
Racket-tailed *see Coracias spatulatus*
Short-legged ground-roller *see Brachypteracias leptosomus*
true *see Coracias*
Rollulus roulroul 177, 180
rook *see Corvus frugilegus*
roosting
communal 296, 483, 492, 517, 598, 599, 601
starling 530–31
thermoregulation 557
rosefinch 613
rosella 302
Crimson *see Platycercus elegans*
Northern *see Platycercus venustus*
Rostratula benghalensis 246–47, *247*
Rostratulidae 244, **246–47**
Rostrhamus 169
hamatus 165, 172
sociabilis 165, *168*, 169, 172
Rowettia goughensis 619, 620, 631
ruff *see Philomachus pugnax*
Rupicola
peruviana 432
rupicola 430, 432, 433
Rupicolinae 429, 430, 433
rushbird, Wren-like *see Phleocryptes melanops*
rush-tyrant, Many-colored *see Tachuris rubrigastra*
Rynchopidae **274–75**
Rynchops
albicollis 29, 274–75
flavirostris 274–75
niger 274–75, *274*

S

sabrewing *see Campylopterus*
saddleback *see Philesturnus carunculatus*
Sagittariidae **148–49**
Sagittarius serpentarius 148–49, *148*, *149*
Sakesphorus 447
sallying 337–38, 340–41, 343, 403, 424, 528
Salpinctes obsoletus exsul 545
Salpornis spilonotus 538, **540**, *540*
Saltator
aurantiirostris 623
maximus 622, 623
similis 623
Salvadorina waigiuesis 137, 140
sandbathing 286
sanderling *see Calidris alba*
sandgrouse (*Pterocles, Syrrhaptes*) 19, *216*, 286
sandpiper 227, 228, **232–37**
bills **234–35**
Broad-billed *see Limicola falcinellus*
Common *see Tringa hypoleucos*
courtship and display 232, *232*, 234, 236–37, **238–39**
Curlew *see Calidris ferruginea*
Green *see Tringa ochropus*
migration 232, 234, **236**, 237
Pectoral *see Calidris melanotos*
ruff *see Philomachus pugnax*
Sharp-tailed *see Calidris acuminata*
Solitary *see Tringa solitaria*
Spoonbilled *see Eurynorhynchus pygmeus*
Tuamotu *see Prosobonia cancellata*
Upland *see Bartramia longicauda*
Western *see Calidris mauri*
Wood *see Tringa glareola*
sand-plover 225
Sapayoa aenigma 421, 434
Sapheopipo noguchii 408, 413
Sappho sparganura 356, *356*
sapsucker **408–15**
Yellow-bellied *see Sphyrapicus varius*
Sarcogyps calvus 172
Sarcops calvus 530, 531
Sarcoramphus papa 147, *147*
Sarkidiornis melanotos 136, 140
Sarothrura ayresi 207, 211
Sasia ochracea 413
Saucerottia cyanifrons 356
sawbill 141, 143
Saxicola rubetra 527
Saxicolini 526
Scaphidura oryzivora 634
scaup 132, 139, 140
Sceloglaux albifacies 322, 331
Scelorchilus
albicollis 451
rubecula 451
Scenopoeetes dentirostris 458, 459, 460, 461
Schiffornis 429, 434
Schoeniophylax 440
Schoutedenapus schoutendeni 346
scimitarbill (*Rhinopomastus*) **383**
Scissirostrum dubium 531, 532, 533
scissorbill *see Rynchops*
scissors-grinder *see Myiagra inquieta*
Scolopacinae **232–37**
Scolopax rusticola 235
Scopidae **120–21**
Scopus umbretta 120–21, *120*, *121*
scoter 132
Common *see Melanitta nigra*
Scotopelia 324
peli 324, *324*, 331
screamer (*Chauna, Anhima*) **130–31**
scrub 572
scrub-bird (*Atrichornis*) **456–57**
scrub-jay (*Aphelocoma*) 484–87
scutellations 511
Scytalopus 451
psychopompus 451
scythebill (*Campylorhamphus*) 443–44
Scythrops novaehollandiae 312, 313, 314
seaduck 132, 139–40, 141, *141*
sea-eagle (*Haliaeetus*) 172
seal 257
secretarybird *see Sagittarius serpentarius*
seedcracker 606, 607
Black-bellied *see Pyrenestes ostrinus*
seedeater 613, 628, 629, 631
seed-eating species
finch 613–14
Scaly weaver 594
seed dispersal 290, 320, 386, 401, 485, 584
seed-finch, Large-billed *see Oryzoborus crassirostris*
seed snipe (*Attagis, Thinocorus*) 232, **254–55**
Seiurus
aurocapillus 625, 626, 627
motacilla 625, 626
Selaphorus 359
rufus 356, *359*
Selenidera 392, 393, 395
Culik 390, *390*
nattereri 390, 392
Seleucides ignotus 489, 490
Semioptera wallacei 488, 490
Semnornis
frantzii 398, 399, 401
ramphastinus 395, 398, *398*, 399, 402
senses 29–31
Sericossypha albocristata 628, 631
Sericulus 458, 460
aureus 458, *458*, 459
bakeri 458, 459, 461
chrysocephalus 458, 459, 460, 461, *461*, *463*
seriema (*Cariama, Chunga*) **222**
Serilophus lunatus 422, 423
serin 615
Ankober 611
Serinus 613
canaria 611, 614
serpent-eagle (*Spilornis, Dryotriorchis, Eutriorchis*) *162*, 165, 172
Setophaga ruticilla 624, 625, 627
Setornis criniger 563
sexual role-reversal 218
sexual selection 453, 460, 504
sexy-son hypothesis 561
shag (*Phalacrocorax*) **90–93**
sharpbill *see Oxyruncus cristatus*
shearwater (*Puffinus, Calonectris*) 19, 26, **70–73**
sheathbill (*Chionis*) **256–57**
sheldgoose 132, 136–37, 140, 142, 143
shelduck (*Tadorna*) 133, 136–37, *138*, 139, 140, 141, 143, 271
Shizoeaca palpebralis 440
shoebill *see Balaeniceps rex*
shorebirds 225–83
shrike **508–10**
Bay-backed *see Lanius vittatus*
Bulbul *see Hypocolius ampelinus*
bush *see* bush-shrike
Chinese grey *see Lanius sphenocercus*
helmet *see* helmet-shrike
Lesser grey *see Lanius minor*
Loggerhead *see Lanius ludovicianus*
Long-tailed *see Lanius schach*
Masked *see Lanius nubicus*
Mountain *see Lanius validirostris*
Northern grey *see Lanius excubitor*
Red-backed *see Lanius collurio*
Southern grey *see Lanius meridionalis*
Souza's *see Lanius souzae*
Strong-billed 508, 510
true 508
White-crowned *see Eurocephalus anguitimens*
White-rumped *see Eurocephalus rueppelli*
Woodchat *see Lanius senator*
Yellow-billed *see Corvinella corvina*
shrike-babbler 575
shrike-thrush 477
Grey *see Colluricincla harmonica*
shrike-tit, Crested 477
shrike-tyrant 428
White-tailed *see Agriornis andicola*
shrike-vireo (*Vireolanius*) 478–79
Sialis sialis 524
sibia (*Heterophasia*) 574
sicklebill (*Epimachus, Eutoxeres*) 353, 356, 490
Sigelus silens 528
silky flycatcher (*Phainoptila, Ptilogonys*) **518–19**
silverbill 605
silvereye *see Zosterops lateralis*
Sinosauropteryx 18
Siphonorhis americana 337, 339
Siptornopsis 440
siskin (*Carduelis*) 615
site fidelity 236
Sitta
canadensis 536, 537
carolinensis 536, 537, *537*
castanea 536
europaea 536, *536*, 537
frontalis 541
ledanti 536
leucopsis 536, 537
neumayer 536
pusilla 536, 537
pygmaea 536, 537
tephronota 536
victoriae 536
whiteheadi 536, *536*
Sittasomus 444
griseicapillus 443
sittella 477, 536
Sittidae **536–37**
skeleton 21, 22, **23**
skimmer (*Rynchops*) 258, 268, **274–75**
skua (*Catharacta, Stercorarius*) 258, **270–73**
skull 22, 23
skylark
Eurasian or European *see Alauda arvensis*
skypointing 454
smell, sense of 29, 30, 46
smew *see Mergellus albellus*

Smithornis 422–23
  capensis 422, 423
  rufolateralis 422, 423
  sharpei 422, 423
Smutsornis africanus 253, *253*
snakebird *see* darter
snake-eagle (*Circaetus*) 165, *168*, 172
snipe 30, **232–37**
  bills **234–35**
  Common *see Gallinago gallinago*
  Greater painted *see Rostratula benghalensis*
  Japanese *see Gallinago hardwickii*
  migration 232, 234, **236**, 237
  Painted 232, 244, **246–47**
  Seed *see* seed snipe
  South American painted *see Nycticryphes semicollaris*
  Subantarctic *see Coenocorypha aucklandica*
Snowcap *see Microchera albacoronata*
snowcock 177, 178, 179, 180
  Himalayan *see Tetraogallus tibetanus*
Snow, David 437
snowfinch (*Montifringilla*) **590–91**
social cooperation
  palmchat 516
  weaver 592
social-weaver, Grey-headed *see Pseudonigrita arnaudi*
soft-wing *see Malacoptila*
solitaire 522, 524, 526, 527
  Cuban *see Myadestes elisabeth*
Somateria
  mollissima 140, 141, *141*
  spectabilis 141
Somaterini 139
spadebill 428
  White-throated *see Platyrinchus mystaceus*
sparrow 27, **590–91**
  American tree *see Spizella arborea*
  Bachman's 620, 621
  Cape *see Passer melanurus*
  Clay-colored *see Spizella pallida*
  Desert *see Passer simplex*
  Grasshopper *see Ammodramus savannarum*
  Grey-headed 590
  Henslow's *see Ammodramus henslowii*
  House *see Passer domesticus*
  Java *see Padda oryzivora*
  Lark 621
  Nelson's sharp-tailed *see Ammodramus nelsoni*
  New World **618–21**
  Pale rock 590
  Rock *see Petronia petronia*
  Rufous-collared *see Zonotrichia capensis*
  Saltmarsh sharp-tailed *see Ammodramus caudacutus*
  Savannah *see Passerculus sandwichensis*
  Seaside *see Ammodramus maritimus*
  Timor *see Padda fuscata*
  Tree *see Passer montanus*
  Vesper 621
  White-throated *see Zonotrichia albicollis*
  Worthen's 619
  Yellow-throated rock *see Petronia superciliaris*
  Zapata *see Torreornis inexpectata*
sparrowhawk (*Accipiter*) *28*, 172
sparrow-lark, Fischer's *see Eremopterix leucopareia*
sparrow-weaver (*Plocepasser*) 594
Spartanoica maluroides 440, *440*
Speculanas specularis 140
speirops, Black-capped *see Speirops lugubris*
Speirops lugubris 566
Spelaeornis 575
Speotyto cunicularia 324, 327, 328, 329, *329*, 331
Spermophaga 606
  ruficapilla 606, 607
Sphecotheres
  hypoleucus 496
  viridis 496, 497
Spheniscidae **50–59**
Sphenisciformes 19, **50–59**
Spheniscus
  demersus 50, 51, 52, 53, *53*, 54, 55, 57
  humboldti 51, 57
  magellanicus 50, 51
  mendiculus 51, 52, 54, 56
Sphenocichla 575
Sphenoeacus afer 569, 572
Sphyrapicus varius 409, 410, 413, *414*
spiderhunter **586–89**
Spilornis
  cheela 162, 172
  elgini 172
  kinabaluensis 172
  minimus 172
spinebill (*Acanthorhynchus*) 467
spinetail 439, 440
  Creamy-crested *see Cranioleuca albicapilla*
  Pernambuco *see Synallaxis infuscata*
  Red-faced *see Cranioleuca erythrops*
  Rusty-backed *see Cranioleuca vulpina*
  Yellow-throated *see Certhiaxis cinnamomea*
spinifex-bird *see Eremiornis carteri*
spinning 240, *240*
Spiza americana 622, 623
Spizaetus
  bartelsi 172
  nanus 172
  ornatus 164
  philippinensis 172
Spizastur melanoleucus 172
Spizella
  arborea 619, 620
  pallida 619, 620
Spiziapteryx circumcinctus 155, 156, 159
Spizixos 562
  canifrons 564
spoonbill (*Ajaia, Platalea*) 30, 103, **112–19**
  tactile foraging 112, 114, **118–19**
Sporophila 619, 628–29, 631
  insulata 628
  melanops 628
  zelichi 628
Sporopipes 594
  frontalis 594
  squamifrons 594, 605
Sporopipinae 594
spurfowl, Yellow-necked *see Francolinus leucoscepus*
Stachyris 575
Stactolaema 401, 402
  leucotis 402
Stagnopleura oculata 605, 607
starling 260, 314, 355, **530–33**
  Amethyst *see Cinnyricinclus leucogaster*
  Blue-eared glossy *see Lamprotornis chalybaeus*
  Brahminy *see Temenuchus pagodarum*
  Common *see Sturnus vulgaris*
  glossy 530, 531
  Grey *see Sturnus cineraceus*
  migration 531–32
  Pohnpei *see Aplonis pelzelni*
  roosts 530–31
  Rose-colored *see Pastor roseus*
  Santo mountain *see Aplonis santovestris*
  Shining *see Aplonis metallica*
  Slender-billed *see Onychognathus tenuirostris*
  Superb *see Lamprotornis superbus*
  Tristram's *see Onychognathus tristramii*
  Wattled *see Creatophora cinerea*
  White-headed *see Sturnus erythropygius*
starthroat *see Heliomaster*
Steatornis caripensis 345, *345*
Steatornithidae **345**
Stelgidopteryx 558
  ruficollis 559
Stephanoaetus coronatus 162, 172
Stephanoxis lalandi 356
Stercorariidae **270–73**
Stercorarius
  longicaudus 270, 271
  parasiticus 270, 271, *271*, 272
  pomarinus 270, 271, *271*, 272
Sterna
  albifrons 264, 269
  aleutica 264
  anaethetus 264, 267, 269
  balaenarum 264, 268, 269
  bernsteini 264, 269
  caspia 264, *264*, 266
  dougallii 264, 266, 269
  eurygnatha 264, 269
  fuscata 264, *264*, 267, 268, *268*, 269
  hirundo 264, 266, *266*, 268, *269*
  lorata 264, 269
  maxima 264, 269
  nilotica 264, 267
  paradisaea 241, 264, *264*, 266, 267, 267, *269*
  sandvicensis 264, **266**, *269*
Sternidae **264–69**
Stictonetta naevosa 134, 140
Stictonettini 134, 140
stifftail 132, 134, 140, 141, 142
Stigmatura 424, 428
stilt (*Cladorhynchus, Himantopus*) **242–43**
Stiltia isabella 253
stinker *see* petrel
stint **234**
  Little *see Calidris minuta*
Stipiturus 465
  mallee 465
stitchbird *see Notiomystis cincta*
stock dove 412
stonechat 527
stone curlew (*Burhinus, Esacus*) 250, **251**
stork 19, 103, **106–11**, 114
  African open-bill *see Anastomus lamelligerus*
  American wood *see Mycteria americana*
  Asian open-bill *see Anastomus oscitans*
  Black *see Ciconia nigra*
  Black-necked *see Eippiorhynchus asiaticus*
  courtship 109, **110–11**
  Greated adjutant *see Leptoptilos dubius*
  Hammerhead *see* hammerhead
  Jabiru *see Jabiru mycteria*
  Lesser adjutant *see Leptoptilos javanicus*
  Maguari *see Ciconia maguari*
  Marabou *see Leptoptilos crumeniferus*
  Milky *see Mycteria cinereus*
  Oriental white *see Ciconia boyciana*
  Painted *see Mycteria leucocephalus*
  Saddle-bill *see Eippiorhynchus senegalensis*
  Storm's *see Ciconia stormi*
  typical 106, 107, 111
  Whale-headed *see Balaeniceps rex*
  White *see Ciconia ciconia*
  White-bellied *see Ciconia abdimii*
  wood 107, 110, 111
  Woolly-necked *see Ciconia episcopus*
  Yellow-billed *see Mycteria ibis*
storm petrel 29, **74–75**, 272
  Grey-backed *see Garrodia nereis*
  Hornby's *see Oceanodroma hornbyi*
  Leach's *see Oceanodroma leucorhoa*
  Least *see Halocyptena microsoma*
  Madeiran *see Oceanodroma castro*
  Markham's *see Oceanodroma markhami*
  Polynesian *see Nesofregetta fuliginosa*
  Ringed *see Oceanodroma hornbyi*
  Swinhoe's 75
  Wedge-rumped *see Oceanodroma tethys*
  White-faced *see Pelagodroma marina*
  Wilson's *see Oceanites oceanicus*
streamertail *see Trochilus*
streamlining 25
Strepera graculina 312, 494
Streptopelia 290
  decaocto 288, 289, 293
  turtur 289, 290, *290*
Strigidae 166, **322–31**
Strigiformes 19, **322–35**
Strix 324
  aluco 324, 326, 327, 328, 331, 334
  nebulosa 324, 327, *327*, 328, 331
  occidentalis 324, 331
  seloputo 324
  varia 324, 331
Struthidae cinerea 493
Struthio camelus 19, 25, 29, **34–37**, 35
Struthionidae **34–37**
Struthioniformes 19, **34–37**
Sturnella
  defillippi 634, 635
  magna 635
Sturnidae **530–33**, 587
Sturnus
  cineraceus 531
  erythropygius 530, 531
  vulgaris 530, 531, *531*, 532, 533
sugarbird (*Promerops*) **586–89**
Sula
  dactylatra 82, 83, *85*, 86
  leucogaster 82, 83, 86, *86*, *87*
  nebouxii 82, 83, 85, 86, *87*
  sula 82, 83, 87, *87*
  variegata 82, *82*, 83, 86, 87
Sula dactylatra 534
Sulidae **82–87**
sunangel *see Heliangelus*
sunbasking 65, 208, 364, 374, *377*, 382, *382*
sunbeam *see Aglaeactis*
sunbird 421, **586–89**
  Amani *see Anthreptes pallidigaster*
  Eastern olive *see Cyanomitra olivacea*
  Elegant *see Aethopyga duyvenbodi*
  false 421, *421*
  Giant *see Dreptes thomensis*
  Golden-winged *see Drepanorhynchus reichenowi*
  Johanna's *see Cinnyris johannae*
  Mouse-colored *see Cyanomitra veroxii*
  Olive-backed *see Cinnyris jugularis*
  Palestinian *see Cinnyris oseus*
  Pygmy *see Hedydipna platura*
  Scarlet-tufted malachite *see Nectarinia johnstoni*
  Southern double-collared *see Cinnyris chalybea*
  Splendid *see Cinnyris coccinigaster*
  Superb *see Cinnyris superba*
  Tacazze *see Nectarinia tacazze*
  Western olive *see Cyanomitra obscura*
  Yellow-breasted *see Nectarinia jugularis*
sunbird-asity (*Neodrepanis*) 421
sungrebe *see Heliornis fulica*
Surnia ulula 324, *327*, 331
swallow 27, 31, **558–61**
  Bahama *see Tachycineta cyaneoviridis*
  Bank *see Riparia riparia*
  Barn *see Hirundo rustica*
  Blue *see Hirundo atrocaerulea*
  Cave *see Hirundo fulva*
  Cliff *see Hirundo pyrrhonota*
  Ethiopian *see Hirundo aethiopica*
  Golden *see Tachycineta euchrysea*
  Grey-rumped *see Hirundo griseopyga*
  Lesser striped *see Hirundo abyssinica*
  Mangrove *see Tachycineta albilinea*
  Red Sea *see Hirundo perdita*
  rough-winged 558
  Rufous-chested *see Hirundo semirufa*
  Southern rough-winged *see Stelgidopteryx ruficollis*
  Tree *see Tachycineta bicolor*
  Violet-green *see Tachycineta thalassina*
  White-tailed *see Hirundo megaensis*
  White-thighed *see Neochelidon tibialis*
  Wire-tailed *see Hirundo smithii*
  wood *see* wood swallow
swamphen, Purple *see Porphyrio porphyrio*
swan (*Cygnus, Coscoroba*) 21, **132–45**
swift 19, 31, **346–50**, 558
  Alpine *see Apus melba*
  Black *see Cypseloides niger*
  Chimney *see Chaetura pelagica*
  cold weather survival **347**
  Common *see Apus apus*
  Dark-rumped *see Apus acuticauda*
  Fork-tailed palm *see Tachornis squamata*
  Lesser swallow-tailed *see Panyptila cayennensis*
  Palm *see Cypsiurus parvus*
  Schouteden's *see Schoutedenapus schoutendeni*
  swallow-tailed 346, 349
  treeswift *see* treeswift
  White-throated *see Aeronautes saxatilis*
  White-throated spine-tailed *see Hirundapus caudacutus*
swiftlet (*Collocalia*) 346, 348–49
swimming
  adaptations for **26**
  auk 276, **279**, *279*
  buoyancy 26, 240, 279
  cormorant 90, *91*
  diving petrel 76
  loon or diver 60, *60*
  penguin **50–51**, 52
  waterfowl 140–42
sylph *see Aglaiocercus*
Sylvia 568, 572
  atricapilla 570, *570*, 572–73
  borin 572
  communis 572
  undata 568, 569, 570, 572
Sylvietta 568
  whytii 570
Sylviidae **568–72**, 624
Sylviorthorhynchus desmursii 439, 440
Sylviparus 551
  modestus 550
symbiosis 352
Synallaxis 439, 440
  infuscata 439, 440
Synthliboramphus
  antiquus 277, 282
  craveri 276, 277, 282, 283
  hypoleucus 276, 277, 282, 283
  wumizusume 277, *277*, 282, 283
Sypheotides indica 214, *214*, 215, 217
Syrigma sibilatrix 102, 103, 104
syrinx 429, 430, 436, 456
Syrrhaptes
  paradoxus 285, *285*, 286
  tibetanus 285

T

Tachornis 349
  squamata 348
Tachuris rubrigastra 425, 427, 428
Tachybaptus
  dominicus 63
  pelzelnii 63
  ruficollis 62, 63, 65
  rufolavatus 63, 65
Tachycineta
  albilinea 558, 560
  bicolor 558, 559, *559*
  cyaneoviridis 558
  euchrysea 558, 561
  thalassina 561
Tachyeres
  brachypterus 134, 136, 140, *142*, 143
  pteneres 134, 136, 140
tactile foraging 112, 114, **118–19**
Tadorna
  cana 140, 143
  ferruginea 138
  tadorna 138, 140, 143
  variegata 140, 143
Tadornini 136–37, *138*, 140
Taeniopygia
  bichenovii 604, 607, 608
  guttata 605, 607, **608**
tail 22, *23*
  parrots 300
  toucans 394
  treecreepers 538
  woodcreepers 442
  woodpeckers 408
tailorbird 31, 568, 572
  Long-billed *see Orthotomus moreaui*
tail streamers 558
takahe *see Porphyrio mantelli*
Talegalla 192
tanager 448, 449, 611, **628–31**
  Black-and-gold *see Bangsia melanochlamys*
  Blue-and-yellow *see Thraupis bonariensis*
  Blue-gray *see Thraupis episcopus*
  Cherry-throated *see Nemosia rourei*
  Cone-billed *see Conothraupis mesoleuca*
  Crimson-backed *see Rhamphocelus dimidiatus*
  Golden-hooded *see Tanagara larvata*
  Grey-headed *see Eucometis penicillata*
  Green-headed *see Tanagara seledon*
  Paradise *see Tanagara chilensis*
  Rosy thrush-tanager *see Rhodinocichla rosea*
  Scarlet *see Piranga olivacea*
  Scarlet-rumped *see Rhamphocelus passerinii*
  Seven-colored *see Tangara fastuoa*
  Silver-beaked *see Rhamphocelus carbo*
  Silver-throated *see Tangara icterocephala*
  Spotted *see Tanagara punctata*
  Summer *see Piranga rubra*
  Swallow *see Tersina viridis*
  Western *see Piranga ludoviciana*
  White-capped *see Sericossypha albocristata*
Tangara 628, 630
  chilensis 628, 631
  fastuoa 631
  icterocephala 630, 631
  larvata 630, 631
  punctata 629
  seledon 629
Tanygnathus megalorynchos 299, 305
Tanysiptera galatea 367, 370
tapaculo (*Scelorchilus, Scytalopus*) 446, **451**
tapetum lucidum 340, *341*
Taraba major 446, 448
tarsometatarsus (tarsus) *23*, 24
Tasmanian native hen 210
taste, sense of 30

Tauraco
bannermani 321
hartlaubi *320*
macrorhynchus 321
ruspolii 321
tchagra 512–13
Black-crowned *see Tchagra senegala*
Tchagra senegala 512, 513
teal
Auckland Island *see Anas aucklandica*
Campbell Island *see Anas nesiotis*
Cinnamon *see Anas cyanoptera*
Cotton *see Nettapus coromandelianus*
Green-winged *see Anas carolinensis*
Marbled *see Marmaronetta angustirostris*
Speckled *see Anas flavirostris*
Teledromas fuscus 451
Telespiza
cantans 616
ultima 616
Telophorus 513
kupeensis 512
Temenuchus pagodarum 530, 531
Temnotrogon roseigaster 362
Temnurus temnurus 482, 484
Tephrodornis 498, 499
Teraornis *20, 21*
Terathopius ecaudatus *162*, 165–66, 172
Terenura spodioptila 448
tern 26, 94, 258, 260, **264–69**, 270, 271, 282
Aleutian *see Sterna aleutica*
Arctic *see Sterna paradisaea*
Black *see Chlidonias nigra*
black-capped group 264, 266
Bridled *see Sterna anaethetus*
Caspian *see Sterna caspia*
Cayenne *see Sterna eurygnatha*
Chinese crested *see Sterna bernsteini*
Common *see Sterna hirundo*
crested group 268
Damara *see Sterna balaenarum*
Fairy *see Gygis alba*
Forster's 268
Grey-backed 267
Gull-billed *see Sterna nilotica*
Inca *see Larosterna inca*
Kerguelen 269
Large-billed *see Phaetusa simplex*
Least *see Sterna albifrons*
Little *see Sterna albifrons*
marsh 264, 266, 267, 268
pelagic species 264
Peruvian *see Sterna lorata*
River 269
Roseate *see Sterna dougallii*
Royal *see Sterna maxima*
Sandwich *see Sterna sandvicensis*
Saunders's 269
Sooty *see Sterna fuscata*
Whiskered *see Chlidonias hybrida*
White *see Gygis alba*
White-winged black *see Chlidonias leucoptera*
Terpsiphone 504
corvina 504
mutata 504
paradisi *504*
territorial behavior 31
territorial expansion **554–55**
Tersina viridis 628, 629, 630, 631, *631*
Tesia 568
testosterone 210
Tetrao 184
mlokosiewiczi 185, 186, 187
parvirostris 185, 186
tetrix 184, 185, *185*, 186
urogallus 185, 186, *186*
urogallus cantabricus 187
urogalloides 186
Tetraogallus tibetanus 180
Tetraonidae **184–87**
Tetrax tetrax 214, 215, 216, 217
Thalassarche 66
chlororhynchos 67, 69
chrysostoma 67, 69, *69*
eremita 67
melanophris 66, 67, *67*, 69
Thalassornis leuconotus 133–34, 140
Thalassornithini 133–34, 140
Thalurania 356
Thamnomanes schistogynus 448
Thamnophilida 446
Thamnophilidae **446–49**
Thamnophilus 447
doliatus 447
ruficapillus 447, 448
Theristicus
branickii 115
caerulescens 115
caudatus 112, 114, 115, 116, 117
melanopis 115
thermoregulation *101*, 146, 573
communal roosting 492, 557
Great roadrunner 315
hummingbirds 355–56, *359*
penguins 52, **58–59**
plumage 18, 23
poorwill 339
size constraints 20
snow-burrows 184, *615*
sunbasking 364, 374, *377*, 382, *382*
swifts 347
Thescelocichla leucopleura 563, 565
thickhead 477
thick-knee (*Burhinus*) 251
Thinocoridae **254**
Thinocorus
orbignyianus 254–55
rumicivorus 254–55, *254*
Thinornis
novaeseelandiae 227, 229
rubricollis 230
thistletail, Eye-ringed *see Shizoeaca palpebralis*
thornbill, Purple-backed *see Ramphomicron microrhynchum*
thornbird 439, 440
Common *see Phacellodomus rufifrons*
thorntail *see Discosura*
thrasher (*Toxostoma*) 534–35
Thraupidae 611, **628–31**
Thraupis
bonariensis 629, 631
episcopus 630, 631
Threnetes 353, 356
Threskiornis
aethiopicus 112, 114, 115, *115*, 116, *116*, 117
melanocephalus 115, *116*
molucca 115
spinicollis 115
Threskiornithidae **112–19**
thrush **522–27**
Aztec *see Zoothera pinicola*
Clay-colored *see Turdus grayi*
Grey-cheeked *see Catharus minimus*
ground *see Zoothera*
Hermit *see Catharus guttatus*
Island *see Turdus poliocephalus*
Lawrence's *see Turdus lawrencii*
Mistle *see Turdus viscivorus*
Olive *see Turdus olivaceus*
rock *see Monticola*
Rufous-bellied *see Turdus rufiventris*
Rufous-tailed rock *see Monticola saxatilis*
Song *see Turdus philomelos*
true 522, 523, 526
Varied *see Zoothera naevia*
whistling *see Myophoneus*
White's *see Zoothera dauma*
Wood *see Catharus mustelinus*
thrush-tanager, Rosy *see Rhodinocichla rosea*
Thryorchilus 543
Thryothorus
ludovicianus *543*
nicefori 542, 545
rutilus *545*
Tichodroma muraria 536, 537
Tigriornis leucolophus 104
Tigrisoma
fasciatum 104
mexicanum *102*
Tigrisomatinae 104
Tijuca
atra 432, 433
condita 432, 433
Timaliidae **574–77**
Tinamidae **48–49**
Tinamiformes 19, **48–49**
tinamou **48–49**
Great *see Tinamus major*
Highland *see Nothocercus bonapartei*
Magdalena *see Crypturellus saltuarius*
Red-winged *see Rhynchotus rufescens*
Small-billed *see Crypturellus parvirostris*
Tataupa *see Crypturellus tataupa*
tinamous 19
Tinamus major 48, 49
tinkerbird (*Pogoniulus*) 398–401
tit 447
African penduline *see Anthoscopus caroli*
Bearded *see Panurus biarmicus*
Black *see Parus leucomelas*
Black-browed *see Aegithalos iouschistos*
Black-throated *see Aegithalos concinnus*
Blue *see Parus caeruleus*
Buff-bellied penduline *see Anthoscopus sylviella*
bushtit *see Psaltiparus minimus*
Cape penduline *see Anthoscopus minutus*
Chinese penduline *see Remiz consobrinus*
Coal *see Parus ater*
Crested *see Parus cristatus*
Elegant *see Parus elegans*
Eurasian penduline *see Remiz pendulinus*
Fire-capped *see Cephalopyrus flammiceps*
Forest penduline *see Anthoscopus flavifrons*
Great *see Parus major*
long-tailed *see Aegithalos caudatus*
Marsh *see Parus palustris*
Mouse-colored penduline *see Anthoscopus musculus*
penduline **548–49**
Pygmy *see Psaltria exilis*
Rufous-bellied *see Parus rufiventris*
Sennar penduline *see Anthoscopus punctifrons*
Siberian *see Parus cinctus*
Sultan *see Melanochlora sultanea*
true **550–55**
White-cheeked *see Aegithalos leucogenys*
White-crowned penduline *see Remiz coronatus*
White-naped *see Parus nuchalis*
White-necklaced *see Aegithalos fuliginosus*
White-throated *see Aegithalos niveogularis*
Willow *see Parus montanus*
Yellow-browed *see Sylviparus modestus*
Yellow-cheeked *see Parus spilonotus*
Yellow penduline *see Anthoscopus parvulus*
tit-babbler 574, 575
tit-hylia *see Pholidornis rushiae*
titmouse (*Parus*) 411, 550, 552
tit-spinetail 440
tit-warbler 570, 572
tityra 429–33
Tityra 430, 433
Tityrinae 429, 433
Tmetothylacus tenellus 598, 599, 600, 601
Tockus 384, 388
albocristatus 384, 386
alboterminatus 386, 388
bradfieldi 386, 389
camurus 386, *387*
deckeni 384, 386
erythrorhynchus 388, *389*
fasciatus 386, 388
monteiri 386, 388–89
Todidae **373**
Todirhamphus 369
gambieri 367, 371
godeffroyi 367
pyrrhopygia *370*
sanctus 367, 369, 370
Todirostrum 424, 428
chrysocrotaphum 428
Todus
angustirostris 373
mexicanus 373
multicolor 373, *373*
subulatus 373
todus 373, *373*
tody (*Todus*) **373**
tody-flycatcher 428
Yellow-browed *see Todirostrum chrysocrotaphum*
tody-tyrant (*Hemitriccus, Poecilotriccus*) 427, 428
tongue 30
nectarvores *298*, 308, *308*, 398, 466, 492, 566, 584, 586–87, 616
woodpeckers 408
wryneck 412
tools, use of 101, 165, 175, 483, **486–87**
Topaza pella 356
topaz, Crimson *see Topaza pella*
Torgos tracheliotus 162, 172, 174, *175*
torpidity 20, 347, 356, 558
torrent-lark *see Grallina bruijni*
Torreornis inexpectata 619, 620, 621
toucan (*Andigena, Ramphastos*) 19, 384, **390–95**, 398, 412
toucanet
Emerald *see Aulacorhynchus prasinus*
Guianan *see Selenidera Culik*
Saffron *see Baillonius bailloni*
Tawny-tufted *see Selenidera nattereri*
Yellow-browed *see Aulacorhynchus huallagae*
touch, sense of 30
filoplumes 23, *23*
towhee 618, 620
Rufous-sided *see Pipilo erythrothalmus*
Toxorhamphus 585
novaeguineae 584
Toxostoma
curvirostre *534*, 535
guttatum 535
montanus 534, 535
rufum 534, 535
Trachyphonus 398, 399, 401, 402
darnaudii 399, 401, *401*
tragopan (*Tragopan*) 177, 178, 180
Tragopan
blythii *176*
melanocephalus 180
temminckii *181*
trainbearer *see Lesbia*
treecreeper
Australasian **452**, 538
Black-tailed *see Climacteris melanura*
Brown *see Climacteris picumnus*
Brown-throated *see Certhia discolor*
Common *see Certhia familiaris*
Garden *see Certhia brachydactyla*
Himalayan *see Certhia himalayana*
holarctic **538–40**
Papuan *see Cormobates placens*
Philippine *see* rhabdornis
Red-browed *see Climacteris picumnus*
Rufous *see Climacteris rufa*
Short-toed *see Certhia brachydactyla*
Spotted *see Salpornis spilonotus*
Stoliczka's *see Certhia nipalensis*
tail 22
White-browed *see Climacteris affinis*
White-throated *see Cormobates leucophaeus*
treepie
Racket-tailed *see Crypsirina temia*
Ratchet-tailed *see Temnurus temnurus*
treeswift (*Hemiprocne*) **351**
trembler, Brown *see Cinclocerthia ruficauda*
Treron 289
calva *288*
Trichoglossus
chlorolepidotus 305, 308, *308*
haematodus 298, 308, *308*
Tricholaema 398, 401
leucomelas 399, 401
Tricholestes criniger 563, 564
Trigonoceps occipitalis 172
triller (*Chlamydochaera, Lalage*) **498–500**
Tringa
glareola 235, 236
hypoleucos *232*, 234, *234*, 235
ochropus 234, 235, 236
solitaria 234, 235
totanus *232*, 235
Tringinae **232–37**
Trochili **352–61**
Trochilidae **352–61**, 586
Trochilinae **356**
Trochilus 356
Troglodytes
aedon 542, 543, 545
cobbi 542, 543, 545
troglodytes 542, *542*, 543, 544, *544*, 545
Troglodytidae **542–45**
trogon 19, **362–63**
Bar-tailed *see Apaloderma vittatum*
Collared *see Trogon collaris*
Cuban *see Priotelus temnurus*
Eared *see Euptilotis neoxenus*
Elegant *see Trogon collaris*
Hispaniolan *see Temnotrogon roseigaster*
Narina *see Apaloderma narina*
Orange-breasted *see Harpactes oreskios*
Red-headed *363*
Red-naped *see Harpactes kasumba*
Trogon collaris 362, *362*, 363
Trogonidae **362–63**
Trogoniformes 19, **362–63**
tropical latitudes 263
tropicbird (*Phaethon*) **88–89**
troupial 632, 634
trumpeter (*Psophia*) **213**
tubenoses 19
turaco 19, 210, **320–21**
Bannerman's *see Tauraco bannermani*
Great blue *see Corythaeola cristata*
Guinea *see Tauraco macrorhynchus*
Hartlaub's *see Tauraco hartlaubi*
Prince Ruspoli's *see Tauraco ruspolii*
Violet *see Musophaga violacea*
turca, Moustached *see Pteroptochus megapodius*
Turdidae **522–27**
Turdoides 574
altirostris 575
bicolor *575*
caudatus 575
fulvus 574, 575
plebejus 314
squamiceps 574, 575
striatus 574, 575, 576
Turdus 522–27
grayi 523, 526
lawrencii 522, 524, 526
merula 522, 523, 524, 526
migratorius 523, *523*, 526
olivaceus 523, 526
philomelos 523, *523*, 524, 526, 527
pilaris 523, 526–27
poliocephalus 523, 526
rufiventris 523, 526
viscivorus 523, 526, *527*
turkey (*Meleagris*) **188–89**
Turnagra capensis 477
Turnicidae **218**
Turnix
everetti 218
melanogaster 218, *218*
olivii 218
sylvatica 218
tanki 218
turnstone *226*, *234*
Ruddy *see Arenaria interpres*
twinspot 606–07
Pink-throated *see Hypargos margaritatus*
Tympanuchus 184, 186
cupido 184, 185, 186, *186*
pallidicinctus 184, 185, 186
phasianellus 185, *186*
Tyrannidae 352, **424–28**
tyrannulet (*Phylloscartes*) 427, 428
Tyrannus 424
cubensis 427, 428
forficata 426, *426*, 427
savana 427, 428
tyrannus 427, 428
tyrant
Blue-billed black *see Knipolegus cyanirostris*
Cock-tailed *see Alectrurus tricolor*
dawn songs 427
Strange-tailed *see Alectrurus risora*
Streamer-tailed *see Gubernetes yetapa*
tyrant flycatcher 352, **424–27**, 429
feather variation **426**
migration 424, 428
tyrant-manakin
Cinnamon *see Neopipo cinnamomea*
Wied's *see Neopelma aurifrons*
tystie *see Cepphus grylle*
Tyto
alba 326, 332–33, *332*, 334
inexspectata 332, 333
manusi 333
nigrobrunnea 332, 333
soumagnei 332, 333
Tytonidae 166, **332–33**

U

'Ula-'ai-hawane *see Ciridops anna*
umbrellabird 429–33
Amazonian *see Cephalopterus ornatus*
uncinate processes 22, *23*
Upucerthia 440
dumetaria *438*, 440
Upupa epops **382**
Upupidae **382**
Upupiformes **382–83**
Uraeginthus 604
angolensis *606*
bengalus 604, 607
granatina 604, 607
Uratelornis chimaera 380

urban habitats
nightjars 337
swifts 346, *350*
Uria
aalge 276, 277, 278, 279, *279*, 283, 283
lomvia 276, 277, 279, 283
Urocissa erythrorhyncha 484
Urocolius
indicus 364, 365, *365*
macrourus 364
uropygial gland 214
Urotriorchis macrourus 168, 172

V

Vanellinae **225–29**
Vanellus 225
armatus 225, 228
cayanus 227
chilensis 227
crassirostris 226, 227, 229
leucurus 227, 228
macropterus 225, 229
miles 225, 227
senegallus 227
superciliosus 227
tricolor 227
vanellus 225, 226, 227, 229
vanga **514**
Blue *see Cyanolanius madagascarinus*
Chabert *see Leptopterus chabert*
Helmet *see Euryceros prevostii*
Hook-billed *see Vanga curvirostris*
Lafresneye's *see Xenopirostris xenopirostris*
Sickle-billed *see Falculea palliata*
Van Dam's *see Xenopirostris damii*
Vanga curvirostris 514
Vangidae 498, 505, **514**
Veles binotatus 337, 338
vent 22
ventriculus *see* gizzard
verdin *see Auriparus flaviceps*
Vermivora bachmanii 625, 627
Vestiaria coccinea 616, *617*
Vidua
chalybeata 607, 608
macroura 607, *607*, 608
paradisaea 607, 608
Viduinae 605
viduine 604, 608
Vini 308
stepheni 301, 305, 308
violet-ear *see Colibri*
vireo (*Vireo*) **478–79**
Vireo 478
altiloquus 478, 479
atricapillus 479
bairdi 478, 479
bellii 479
caribaeus 479
flavigrons 478
gilvus 479
griseus 478, 479
huttoni 479
magister 479
masteri 478, 479
modestus 479
olivaceus 478, *478*, 479
osburni 478, 479
philadelphicus 479
solitarius 479, *479*
vicinior 478, 479
Vireolaniinae 479
Vireolanius 478
eximius 479
melitophrys 479
Vireonidae **478–79**
Vireoninae 479
vision 29–30, 31
see also eyes
visorbearer *see Augastes*
vocalization 29, 30, 31, 196
dialects *495*
dueting 299, 402, 476, 493, 503, 512, *513*
gronking 455
loudest 430, 432, *432*
mimicry 31, 296, **300**, 305, **454**, 456, 480, 494, 502, *502*, 506
territorial defense 456
Volatinia jacarina 629, 631
vulture 19, 21
Bearded *see Gypaetus barbatus*
Black *see Coragyps atratus*
Cinereous *see Aegypius monachus*
Egyptian *see Neophron percnopterus*
feet *164*
Greater yellow-headed *see Cathartes melambrotus*
Griffon *see* griffon
Himalayan *see Gyps himalayensis*
Hooded *see Necrosyrtes monachus*
King *see Sarcoramphus papa*
Lappet-faced *see Torgos tracheliotus*
Lesser yellow-headed *see Cathartes burrovianus*
Long-billed *see Gyps indicus*
New World 30, **146–47**, 154
Old World 146, **162–75**
Palm-nut *see Gypohierax angolensis*
Red-headed *see Sarcogyps calvus*
tool use 175
Turkey *see Cathartes aura*
White-backed *see Gyps africanus*
White-headed *see Trigonoceps occipitalis*
White-rumped *see Gyps bengalensis*
Vultur gryphus 146–47, *147*

W

waders **230–43**
conservation **230–31**
foot-trembling 228
migration **236**
spinning 240, *240*
wagtail (*Dendronanthus, Motacilla*) **598–601**
wagtail-tyrant 428
wallcreeper *see Tichodroma muraria*
warbler 447, 448, 625
African tree 568
Aldabra *see Nesillas aldabrana*
Arctic *see Acrocephalus paludicola; Phylloscopus borealis*
Bachman's *see Vermivora bachmanii*
Black-and-white *see Mniotilta varia*
Blackpoll *see Dendroica striata*
Buff-rumped *see Basileuterus fulvicauda*
Chestnut-sided *see Dendroica pensylvanica*
Dartford *see Sylvia undata*
Garden *see Sylvia borin*
Golden-cheeked *see Dendroica chrysoparia*
grass 572
grasshopper 568
Grey-headed *see Basileuterus griseiceps*
ground 568
Icterine *see Hippolais icterina*
Kentucky *see Oporonis formosus*
Kirtland's *see Dendroica kirtlandii*
leaf 568
Lucy's 627
Marsh *see Acrocephalus palustris*
migration 568, 570, 626, 627
Myrtle 626
New World 570, **624–27**
nuthatch 568
Old World **568–72**, 624
Olivaceous *see Hippolais pallida*
Olive 624
Oriole *see Hypergerus atriceps*
Paddyfield *see Acrocephalus agricola*
parulid 447–48
Prothonotary *see Protonotaria citrea*
reed *see Acrocephalus scirpaceus*
scrub 568
sedge *see Acrocephalus schoenobaenus*
Semper's *see Leucopeza semperi*
Seychelles *see Bebrornis sechellensis*
Tennessee 626
Three-striped *see Basileuterus tristriatus*
Willow *see Phylloscopus trochilus*
Wood *see Phylloscopus sibilatrix*
woodland 568
Yellow *see Dendroica petechia*
Yellow-rumped *see Dendroica coronata*
'watch-run-peck' foragers 228–29, *228*
water
drinking methods *216*, 605
excretion 27
metabolism from dry seed 594
watercock 211
Asian *see Gallicrex cinerea*
waterfowl 19, 130–45, **132–45**, 210
conservation status 132, 143
digestive system 139, 140, 144
extinct species 133, 136, 139, 143
hybrids 132
lead poisoning **139**
migration 139, 142, **144–45**
molting and flightlessness 134
nests *142*, 143, *145*
water-redstart *see Rhyacornis*
waterthrush, Louisiana *see Seiurus motacilla*
water-tyrant, Pied *see Fluvicola pica*
wattlebird 468
New Zealand **495**
Yellow *see Anthochaera paradoxa*
wattle-eye (*Platysteira*) **514**
wattles *181*, *185*, 188, 190, *196*, 197, 430
eye 421, 514
hornbill 384, 386
New Zealand wattlebird 495, *495*
Wattled starling 530, 531, *532*
waxbill (*Amandava, Estrilda*) **604–09**
waxwing (*Bombycilla*) **518–19**
weaver 31, **592–97**
Bates's *see Ploceus batesi*
Baya *see Ploceus philippinus*
bearded 594
Black-chinned *see Ploceus nigrimentum*
Black-headed *see Ploceus melanocephalus*
Bob-tailed *see Brachycope anomala*
Cape *see Ploceus capensis*
Chestnut *see Ploceus rubiginosus*
Clarke's *see Ploceus golandi*
Compact *see Pachyphantes superciliosus*
Forest *see Ploceus bicolor*
Giant *see Ploceus grandis*
Golden palm *see Ploceus bojeri*
Grosbeak *see Amblyospiza albifrons*
Heuglin's masked *see Ploceus heuglini*
Holub's golden-weaver *see Ploceus xanthops*
Lesser masked *see Ploceus intermedius*
Little *see Ploceus luteolus*
nests 516, 592, 594, **596–97**
Olive-headed *see Ploceus olivaceiceps*
Parasitic *see Anomalospiza imberbis*
Preuss's *see Ploceus preussi*
Red-headed *see Anaplectes rubriceps*
Rufous-tailed *see Histurgops ruficauda*
Rüppell's *see Ploceus galbula*
Scaly *see Sporopipes squamifrons*
Sociable *see Philetairus socius*
Social *see Philetairus socius*
Southern masked *see Ploceus velatus*
Speckle-fronted *see Sporopipes frontalis*
Spectacled *see Ploceus ocularis*
Strange *see Ploceus alienus*
Streaked *see Ploceus manyar*
true 594
Vieillot's black *see Ploceus nigerrimus*
Village *see Ploceus cucullatus*
White-headed buffalo *see Dinemellia dinemelli*
Yellow *see Ploceus megarhynchus*
wedgebill **474–75**
weka *see Gallirallus australis*
wetlands 230–31, 232–47
wheatear (*Oenanthe*) 526
whimbrel 228
whinchat *see Saxicola rubetra*
whipbird **474–75**
Western *see Psophodes nigrogularis*
Whip-poor-will *see Caprimulgus vociferus*
whistler **477**
Golden *see Pachycephala pectoralis*
Grey *see Pachycephala simplex*
Red-lored *see Pachycephala rufogularis*
Rufous *see Pachycephala rufiventris*
Tongan *see Pachycephala jacquinoti*
white-eye (*Cleptornis, Zosterops*) 466, **566–67**
whitehead *see Mohoua albicilla*
whitestart 626
whitethroat *see Sylvia communis*
whydah (*Vidua*) **604–09**
Eastern paradise *see paradisaea*
Pin-tailed *see Vidua macroura*
widow 593
Jackson's *see Euplectes jacksoni*
Long-tailed *see Euplectes progne*
wigeon
American *see Anas americana*
Eurasian *see Anas penelope*
willie-wagtail *see Rhipidura leucophrys*
wing claws **318**, 321
wings 22, **23**, 24
spurs *130*
see also flight
wiretail, Des Murs's *see Sylviorthorhynchus desmursii*
wishbone *23*
Wolda, H. 554
Wompoo *see Ptilinopus magnificus*
woodcock 29, **232–37**
bills **234–35**
Eurasian *see Scolopax rusticola*
vision 234
woodcreeper **442–45**, 446, 448, 449
Amazon barred *see Dendrocolaptes certhia*
Buff-throated *see Xiphorhynchus guttatus*
Cocoa *see Xiphorhynchus susurrans*
Great rufous *see Xiphorhynchus major*
Long-billed *see Nascia longirostris*
Moustached *see Xiphocolaptes falcirostris*
Northern barred *see Dendrocolaptes sanctithomae*
Olivaceous *see Sittasomus griseicapillus*
Plain brown *see Dendrocincla fulginosa*
Scimitar-billed *see Drymornis bridgesii*
Spot-crowned *see Lepidocolaptes affinis*
tail 22
Tyrannine *see Dendrocincla tyrannina*
Wedge-billed *see Glyphorynchus spirurus*
woodhewer *see* woodcreeper
wood-hoopoe (*Phoeniculus*) **383**
woodlark *see Lullula arborea*
woodnymph *see Thalurania*
wood-partridge
Black *see Melanoperdix nigra*
Crested *see Rollulus roulroul*
woodpecker 19, **408–15**, 447, 448
Acorn *see Melanerpes formicivorus*
anvils 408, 410, 414
Arabian *see Dendrocopos dorae*
Black *see Dryocopus martius*
Black-backed three-toed *see Picoides arcticus*
communication 408, **414–15**
Crimson-crested *see Campephilus melanoleucos*
Downy *see Picoides pubescens*
feet 408
food caches 410, 412, *412*
Grey-headed *see Picus canus*
Great slaty *see Mulleripicus pulverulentus*
Great spotted *see Dendrocopos major*
Green *see Picus viridis*
Ground *see Geocolaptes olivaceus*
Guadeloupe *see Melanerpes herminieri*
Hairy *see Picoides villosus*
and honeyguide 396, 397
Imperial *see Campephilus imperialis*
Ivory-billed *see Campephilus principalis*
Lesser spotted *see Dendrocopos minor*
Lewis's *see Melanerpes lewis*
Middle spotted *see Dendrocopos medius*
nests 31, 143, 408, 410, 412
Noguchi's *see Sapheopipo noguchii*
Okinawa *see Sapheopipo noguchii*
Olive-backed *see Dinopium rafflesi*
pied 408
Pileated *see Dryocopus pileatus*
Red-bellied *see Melanerpes carolinus*
Red-cockaded *see Picoides borealis*
Red-headed *see Melanerpes erythrocephalus*
Rufous-bellied *see Dendocopos hyperythrus*
Scaly-breasted *see Celeus grammicus*
Syrian *see Dendrocopos syriacus*
tail 22, 408
Three-toed *see Picoides tridactylus*
tongue 408
White-backed *see Dendrocopos leucotus*
White-bellied *see Dryocopus javensis*
Yellow-crowned *see Dendrocopos mahrattensis*
Yellow-tufted *see Melanerpes cruentatus*
wood-quail, Gorgeted *see Odontophorus strophium*
wood-rail, Red-winged *see Aramides calopterus*
woodshrike *see Tephrodornis*
woodstar *see Chaetocercus*
Amethyst *see Calliphlox amethystina*
wood swallow (*Artamus*) **492**
wren 417, **542–45**
Apolinar's marsh-wren *see Cistothrus apolinari*
Bush *see Xenicus longipes*
Cactus *see Campylorynchus brunneicapillus*
Carolina *see Thryothorus ludovicianus*
Cobb's *see Troglodytes cobbi*
Common or European *see Troglodytes troglodytes*
Flutist *see Microcerculus ustulatus*
Giant *see Campylorynchus chiapensis*
House *see Troglodytes aedon*
Long-billed marsh-wren *see Cistothrus palustris*
Musician *see Cyphorhinus aradus*
New Zealand **416–17**
Niceforo's *see Thryothorus nicefori*
Rock *see Salpinctes obsoletus exsul; Xenicus gilviventris*
Rufous-breasted *see Thryothorus rutilus*
Song *see Cyphorhinus aradus*
Stephens Island *see Xenicus lyalli*
Winter *see Troglodytes troglodytes*
Zapata *see Ferminia cerverai*
wren-spinetail, Bay-capped *see Spartanoica maluroides*
wrenthrush 624
wrentit *see Chamaea fasciata*
wren-warbler 50, 572
wrybill *see Anarhynchus frontalis*
wryneck (*Jynx*) **408–15**

X

Xanthocephalus xanthocephalus 634
Xanthomyza phrygia 466, *468*, 469
Xanthopsar flavus 634, 635
Xema sabini 259, *261*, 263
Xenicidae **416–17**
Xenicus
gilviventris 416–17, *416*
longipes 416, 417
lyalli 416, 417
Xenoperdix udzungwensis 178, 180
Xenopirostris
damii 514
xenopirostris 514
xenops, Great *see Megaxenops parnaguae*
Xiphirhynchus 575
Xiphocolaptes falcirostris 443
Xipholena 429, 432
Xiphorhynchus
guttatus *443*
major 443, 444, *444*
susurrans 442, 443, 444

Y

yellowhammer *see Emberiza citrinella*
yellowthroat (*Geothlypis*) *624*,625, 627
Yuhina 574, 575

Z

Zaratornis 429
stresemanni 430, 433
Zebrilus undulatus 104
Zenaida
auriculata 289, 293
graysoni 289
macroura 289, 290, *290*, 293
Zimmerius 428
Zonotrichia
albicollis 620
capensis 635
Zoothera 522, 524, 526
dauma 524–25, 526
naevia 526
pinocola 526
Zosteropidae 466, **566–67**
Zosterops
albogularis 566
chloronothos 566
cinereus 566
erythropleurus 566
japonicus 566
lateralis 566
strenuus 566
zygodactyly 298, 299–300, 302, 312, 320, 381, 396, 398